KU-334-585

INTERNATIONAL SOCIETY FOR ROCK MECHANICS

SOCIETE INTERNATIONALE DE MECANIQUE DES ROCHES

INTERNATIONALE GESELLSCHAFT FÜR FELSMECHANIK

International Congress on Rock Mechanics

Congrès International de Mécanique des Roches

Internationaler Kongress der Felsmechanik

PROCEEDINGS / COMPTES-RENDUS / BERICHTE

VOLUME / TOME / BAND   1

Editors / Editeurs / Herausgeber
G.HERGET & S.VONGPAISAL

Montréal / Canada / 1987

 # 178479 17

Proceedings
Sixth International Congress on Rock Mechanics

Comptes-rendus
Sixième Congrès International de Mécanique des Roches

Berichte
Sechster Internationaler Kongress der Felsmechanik

Montréal / Canada / 1987

A.A.BALKEMA / ROTTERDAM / BOSTON / 1987

and   Canadian Rock Mechanics Association / CIM / CGS
et   L'Association canadienne de mécanique des roches
und   Kanadische Vereinigung für Felsmechanik

D
6 24·15/3
INT

For the complete set of three volumes, ISBN 90 6191 711 5
For volume 1, ISBN 90 6191 712 3
For volume 2, ISBN 90 6191 713 1
For volume 3, ISBN 90 6191 714 X
© 1987 by the authors concerned
Published by A.A.Balkema, P.O.Box 1675, Rotterdam, Netherlands
Distributed in USA and Canada by: A.A.Balkema Publishers, P.O.Box 230, Accord, MA 02018

Foreword

At the International Symposium on Weak Rock which took place in Tokyo, Japan, in 1981, the Council of the International Society for Rock Mechanics accepted the invitation of the Canadian Rock Mechanics Group – with support from the Canadian Institute of Mining and Metallurgy and the Canadian Geotechnical Society – to host the 6th International Congress for Rock Mechanics in Montreal, Canada. The final date for the Congress was set for August 30th – September 3rd, 1987.

Based on a number of inputs from national and international organizations, especially the ISRM Advisory Board, four general Themes were chosen for the Congress, which would enable a comprehensive exchange of knowledge and experience related to recent advances in rock mechanics. The four Themes are:

Fluid flow and waste isolation in rock masses
Rock foundations and slopes
Rock blasting and excavation
Underground openings in overstressed rock

The first two volumes of the proceedings contain all publications concerning the four Themes which reached the office of the Organizing Committee in time. All papers which were submitted have been examined and selected by the respective national groups which have, in accordance with the statutes of ISRM, the responsibility for the content of the papers from their countries. Some of the papers were deficient in regard to the translation requirements. This was remedied wherever possible by the Proceedings Committee.

For each Theme an internationally recognized speaker was asked to present a lecture of broad significance. These introductory lectures, including the lecture on Rock Engineering in Canada, will be reproduced in volume three of the Congress proceedings. This volume will also contain the discussion contributions, the official presentations for the Congress (opening speeches), one page summaries of poster sessions, and individual papers which unfortunately arrived too late to be printed in volumes one and two.

The proceedings of the Sixth International Congress for Rock Mechanics are of particular significance because they appear on the twenty-fifth anniversary since formation of the International Society for Rock Mechanics. These proceedings represent, therefore, a very important milestone in the development of the science of rock mechanics.

The compilation presented here would not have been possible without the generous help of many individuals and organizations, especially the Canada Centre for Mineral and Energy Technology (CANMET), that provided assistance in typing, translation and editing.

Branko Ladanyi
Honorary Chairman

Gerhard Herget
Congress Chairman

Norbert Morgenstern
Technical Program

Préface

Lors du symposium international sur les roches faibles qui avait lieu à Tokyo au Japon, en 1981, le comité de la Société internationale de mécanique des roches acceptait l'invitation du Groupe national canadien – appuyé par l'Institut canadien des mines et de la métallurgie et la Société canadienne de géotechnique – de tenir le 6e Congrès international de mécanique des roches à Montréal au Canada. Il fut décidé que le congrès aurait lieu du 30 août au 3 septembre 1987.

A partir de suggestions d'organisations nationales et internationales ainsi que du comité consultatif de la SIMR, on a choisi pour ce congrès quatre thèmes généraux qui devraient favoriser un vaste échange de connaissances professionnelles et de récents progrès dans le domaine de la mécanique des roches. Ces quatres thèmes sont les suivants:

Ecoulement de fluides et enfouissement de déchets dans les massifs rocheux
Fondations et talus rocheux
Sautage et excavation
Souterrains en massifs rocheux sous grandes contraintes

Les deux premiers tomes des comptes rendus contiennent toutes les publications se rapportant aux quatre thèmes du congrès qui sont parvenues à temps au comité organisateur. Toutes ces publications ont été examinées et sélectionnées par les groupes nationaux respectifs lesquels, selon les statuts de la SIMR, sont responsables du contenu de tous les travaux provenant de leur pays. Les articles qui ne rencontraient pas les exigences de traduction furent corrigés par le sous-comité des comptes rendus.

Un spécialiste de réputation internationale fut choisi pour présenter une conférence d'intérêt général pertinant à chacun des thèmes. Ces conférences ainsi que celle portant sur la mécanique des roches au Canada seront publiées dans le troisième tome des comptes rendus du congrès. Ce tome comprendra aussi les notes des discussions soumises au comité organisateur du congrès, les présentations officielles du congrès (discours d'ouverture, etc), des sommaires d'une page des séances d'affichage ainsi que toutes les communications qui seront arrivées trop tard pour être publiées dans les tomes 1 et 2.

Les comptes rendus du Sixième Congrès international de Montréal revêtent un intérêt tout particulier à cause de leur coïncidence avec le vingt-cinquième anniversaire de fondation de la Société internationale de mécanique des roches. Ainsi, ils devraient marquer une étape importante dans le développement de la science de la mécanique des roches.

La compilation présentée ici n'aurait pas été possible sans l'aide généreuse de nombreux individus et organisations, spécialement le Centre canadien de la technologie des minéraux et de l'énergie (CANMET), qui a assisté à la dactylographie, la traduction et l'édition.

Branko Ladanyi
Président honoraire

Gerhard Herget
Président du congrès

Norbert Morgenstern
Programme technique

Vorwort

Während des Internationalen Symposiums über Weichgesteine in Tokyo, Japan, 1981, entschied die Versammlung der Internationalen Gesellschaft für Felsmechanik (IGFM), die Einladung der Kanadischen Felsmechanik Gruppe – unterstützt vom Kanadischen Institut für Bergbau und Metallurgie und der Kanadischen Gesellschaft für Geotechnik – anzunehmen, und den Sechsten Internationalen Kongress der Felsmechanik in Montreal, Kanada zu organisieren. Als Datum für den Kongress wurde der 30. August – 3. September, 1987 gewählt.

Auf Grund von Besprechungen mit nationalen und internationalen Organisationen, insbesondere dem Beirat der IGFM, wurden vier allgemeine Themen ausgewählt, die einen umfassenden Austausch von Fachwissen und neuen Errungenschaften in der Felsmechanik ermöglichen. Diese vier Themen sind:

Flüssigkeitsbewegung und Abfallisolierung im Fels
Felsgründungen und Böschungen
Sprengen und Ausbruch
Untertägige Hohlräume im überbeanspruchten Gebirge

Die ersten zwei Bände enthalten alle Veröffentlichungen der vier Themen, die beim Organisationskomitee rechtzeitig eintrafen. Alle Beiträge sind von den entsprechenden Nationalgruppen überprüft und ausgewählt worden, da sie die Verantwortung für die Veröffentlichungen ihrer Länder haben. Einige der eingereichten Arbeiten hatten Mängel in Bezug auf Übersetzungen der Titel und der Zusammenfassungen, die soweit wie möglich behoben wurden.

Für jedes Thema wurde ein international anerkannter Sprecher gebeten, einen Vortrag von allgemeiner Bedeutung zu halten. Diese Einführungsvorträge, einschliesslich des Vortrages über Felsbau in Kanada, werden im dritten Band der Kongressbeiträge erscheinen. Dieser Band enthält auch die Diskussionsbeiträge, die Beiträge zur Eröffnungssitzung, Zusammenfassungen (eine Seite) für die Postersitzungen und Veröffentlichungen, die zu spät eintrafen, um im ersten oder zweiten Band zu erscheinen.

Die Sitzungsberichte des sechsten internationalen Kongresses für Felsmechanik sind von besonderer Bedeutung, weil sie zum 25. Jahrestag des Bestehens der IGFM erscheinen. Diese Sitzungsberichte stellen daher einen wesentlichen Markstein in der Entwicklung der Felsmechanik dar.

Die Zusammenstellung der Sitzungsberichte wäre nicht möglich gewesen ohne die grosszügige Hilfe von vielen Personen und Organisationen, besonders dem Kanadischen Zentrum für Mineral und Energie Technologie (CANMET), das Schreibkräfte und Übersetzer zur Verfügung stellte.

Branko Ladanyi
Ehrenvorsitzender

Gerhard Herget
General-Vorsitzender

Norbert Morgenstern
Technisches Program

Organization of the Sixth ISRM Congress

ORGANIZING COMMITTEE

Dr G.Herget, Congress Chairman
Dr B.Ladanyi, Honorary Chairman/Fundraising
Dr D.E.Gill, Congress Co-Chairman/Local Functions
Dr N.R.Morgenstern, Technical Program
Dr W.F.Bawden, Exhibits
Dr J.Bourbonnais, Hospitality
Mr T.Carmichael, Finance
Dr J.A.Franklin, Publicity
Mr L.Geller, Translations
Dr A.T.Jakubick, Technical Tours
Mrs J.Robertson, Accompanying Persons Program
Mrs C.Michael, Registration, CIM

ADVISORY COMMITTEE

Prof. E.T.Brown, President ISRM
Mr N.F.Grossman, Secretary General ISRM
Vice-Presidents
Dr H.Wagner (Africa)
Prof. Tan Tjong Kie (Asia)
Dr W.M.Bamford (Australasia)
Dr S.A.G.Bjurström (Europe)
Mr A.A.Bello Maldonado (North America)
Dr F.H.Tinoco (South America)

CARMA EXECUTIVE

Dr J.E.Udd, Chairman
Dr J.Curran, Vice-Chairman
Mr T.Carmichael, Secretary Treasurer

ISRM Congress 1987					
Sunday, August 30	Monday, August 31	Tuesday, September 1	Wednesday, September 2	Thursday, September 3	
Pre-Congress Workshop: Young *(Canada):* Monitoring and interpretation techniques for mining induced seismicity (extra charge) (8:30 - 18:00)	**OFFICIAL OPENING** Hoek *(Canada):* Rock Engineering in Canada	THEME 2: **Rock Foundations and Slopes** Chairman: Tinoco *(Venezuela)* Speaker: Panet *(France)* Reinforcement of rock foundations and slopes by active or passive anchors 4 Presentations	THEME 3: **Rock Blasting and Excavation** Chairman: Tan Tjong-Kie *(China)* Speaker: McKenzie *(Australia)* Blasting in hard rock: Techniques for diagnosis and modelling for damage and fragmentation 4 Presentations	THEME 4: **U/G Openings in Overstressed Rock** Chairman: Kidybinsky *(Poland)* Speaker: Wagner *(S. Africa)* Design and support of underground excavations in highly stressed rock 4 Presentations	08:30 10:00
	Coffee				10:30
	THEME 1: **Fluid Flow and Waste Isolation in Rock Masses** Chairman: Bjurström *(Sweden)* Speaker: Doe *(USA)* Design of borehole testing programs for waste disposal sites in crystalline rock 5 Presentations	9 Presentations	9 Presentations	9 Presentations	12:00
	Lunch Break and Poster Sessions				13:30
	9 Presentations	Panel & Floor Discussions Moderator: Kovari *(Switzerland)* Panelists: MacMahon *(Australia)* Ribacchi *(Italy)* Yoshinaka *(Japan)*	Panel & Floor Discussions Moderator: Lindqvist *(Sweden)* Panelists: Favreau *(Canada)* Gehring *(Australia)* da Gama *(Brazil)*	6 Presentations	14:30
		MODERATOR SUMMARY	**MODERATOR SUMMARY**	Coffee	15:00
	Coffee	**WORKSHOPS**		Panel & Floor Discussions Moderator: Whittaker *(UK)* Panelists: Sharma *(India)* Kaiser *(Canada)* Maury *(France)*	
	Panel & Floor Discussions Moderator: Barton *(Norway)* Panelists: Langer *(FRG)* Hudson *(UK)* Rissler *(FRG)*	Swelling Rock Constitutive Laws for Salt Rock Numerical Methods as a Practical Tool	Failure Mechanisms Around Underground Workings Rock Cuttability and Drillability Rock Testing and Testing Standards		16:00
				MODERATOR SUMMARY	16:30
	MODERATOR SUMMARY			**CLOSING ADDRESS:** Wittke *(FRG)*	17:00
	Poster Sessions				18:00
					19:00
Reception by Congress Chairman	**Reception ISRM** President			Reception	19:30
				Dinner	20:30
	Ballet Jazz			Silver-Jubilee Addresses	21:30
				Les Sortilèges	22:00

Organisation du Sixième Congrès de la SIMR

COMITE D'ORGANISATION

Dr G.Herget, Président du Congrès
Dr B.Ladanyi, Président Honoraire
Dr D.E.Gill, Coprésident du Congrès
Dr N.R.Morgenstern, Programme technique
Dr W.F.Bawden, Exposition
Dr J.Bourbonnais, Acceuil
M. T.Carmichael, Finances
Dr J.A.Franklin, Publicité
M. L.Geller, Traduction
Dr A.T.Jakubick, Visites techniques
Mme J.Robertson, Programme pour les personnes
   accompagnant les participants
Mme C.Michael, Inscription, CIM

COMITE CONSEIL

Prof. E.T.Brown, Président, SIMR
M. N.F.Grossman, Secrétaire Général SIMR
Vice-présidents
Dr H.Wagner (Afrique)
Prof. Tan Tjong Kie (Asie)
Dr W.M.Bamford (Australasie)
Dr S.A.G.Bjurström (Europe)
M. A.A.Bello Maldonado (Amérique du Nord)
Dr F.H.Tinoco (Amérique du Sud)

EXECUTIF DU CARMA

Dr J.E.Udd, Président
Dr J.Curran, Vice-Président
M. T.Carmichael, Secrétaire Général

Congrès SIMR 1987					
Dimanche 30 août	**Lundi 31 août**	**Mardi 1 septembre**	**Mercredi 2 septembre**	**Jeudi 3 septembre**	
Atelier pré-congrès Young *(Canada)*: Séismicité induite dans les mines : techniques de surveillance et d'interprétation (coût additionnel) (8:30 - 18:00)	**OUVERTURE OFFICIELLE** Hoek *(Canada)*: Ingénierie des roches au Canada	THÈME 2: **Fondations et talus rocheux** Président: Tinoco *(Vénézuela)* Coordonnateur: Panet *(France)* Renforcement des fondations et des talus à l'aide d'ancrages actifs et passifs 4 présentations	THÈME 3: **Sautage et excavation** Président: Tan Tjong-Kie *(Chine)* Coordonnateur: McKenzie *(Australie)* Sautage en roches dures : techniques de diagnostique et de modélisation des nuisances et de la fragmentation 4 présentations	THÈME 4: **Souterrains en massifs rocheux sous grandes contraintes** Président: Kidybinsky *(Pologne)* Coordonnateur: Wagner *(Afrique du Sud)* Conception et soutènement des souterrains dans les massifs rocheux sous grandes contraintes 4 présentations	08:30 10:00
	Pause Café				10:30
	THÈME 1: **Ecoulement de fluides et enfouissement de déchets dans les massifs rocheux** Président: Bjurström *(Suède)* Coordonnateur: Doe *(EU)* Conception de programmes d'essais en forage pour les sites d'enfouissement des déchets dans les roches cristallines 5 présentations	9 présentations	9 présentations	9 présentations	12:00
	Pause déjeuner et sessions d'affichage				13:30
	9 présentations	Discussions/table ronde/plénière Modérateur: Kovari *(Suisse)* table ronde: MacMahon *(Australie)* Ribacchi *(Italie)* Yoshinaka *(Japon)*	Discussions/table ronde/plénière Modérateur: Lindqvist *(Suède)* table ronde: Favreau *(Canada)* Gehring *(Australie)* da Gama *(Brésil)*	6 présentations	14:30
		Résumé du modérateur	**Résumé du modérateur**	**Pause café**	15:00
	Pause café	**Ateliers**		Discussions/table ronde/plénière Modérateur: Whittaker *(GB)* table ronde: Sharma *(Indes)* Kaiser *(Canada)* Maury *(France)*	
	Discussions/table ronde/plénière Modérateur: Barton *(Norvège)* table ronde: Langer *(RFA)* Hudson *(GB)* Rissler *(RFA)*	Essais, analyse et conception en matière de roche gonflante Lois de comportement pour la modélisation du sel gemme Les méthodes numériques, un outil commode	Les mécanismes de rupture autour des ouvrages souterrains Essai de la taillabilité et de la forabilité des roches Essai des roches et normes d'essai	**Résumé du modérateur**	16:00 16:30
	Résumé du modérateur			**Cérémonie de clôture:** Wittke *(RFA)*	17:00
	Sessions d'affichage				18:00
					19:00
Réception du président du congrès	**Réception du président de la SIMR**			Réception	19:30
				Dîner	20:30
	Ballet Jazz			Présentations du jubilé d'argent	21:30
				Les Sortilèges	22:00

Organisation des Sechsten Kongresses der IGFM

ORGANISATIONSKOMITEE

Dr G.Herget, Kongressvorsitzender
Dr B.Ladanyi, Ehrenvorsitzender
Dr D.E.Gill, 2. Kongressvorsitzender
Dr N.R.Morgenstern, Wissenschaftliches Programm
Dr W.F.Bawden, Ausstellungen
Dr J.Bourbonnais, Gastgeberfunktionen
T.Carmichael, Finanzen
Dr J.A.Franklin, Öffentlichkeitsarbeit
L.Geller, Übersetzungen
Dr A.T.Jakubick, Fachexkursionen
Frau J.Robertson, Programm für Begleitpersonen
Frau C.Michael, Anmeldung, CIM

BERATUNGSKOMITEE

Prof. E.T.Brown, IGFM-Präsident
N.F.Grossman, IGFM-Generalsekretär
Vizepräsidenten:
Dr H.Wagner (Afrika)
Prof. Tan Tjong Kie (Asien)
Dr W.M.Bamford (Australien/Ozeanien)
Dr S.A.G.Bjurström (Europa)
Prof. A.A.Bello Maldonado (Nordamerika)
Dr F.H.Tinoco (Südamerika)

CARMA GESCHÄFTSFÜHRUNG

Dr J.E.Udd, Vorsitzender
Dr J.Curran, 2. Vorsitzender
T.Carmichael, Generalsekretär

Internationaler Felsmechanik Kongress 1987

Sonntag, 30. August	Montag, 31. August	Dienstag, 1. September	Mittwoch, 2. September	Donnerstag, 3. September	
Pre-Kongress Arbeitskreise Young *(Kanada)*: Technik der Messungen und Analyse für die Bestimmung der durch Bergbau verursachten Seismizität (extra Gebühr) (8:30 - 18:00)	**Begrüssung und Eröffnung** Hoek *(Kanada)*: Felsbau in Kanada	THEMA 2: **Felsgründungen und Böschungen** Vorsitzender: Tinoco *(Venezuela)* Sprecher: Panet *(Frankreich)* Felsverbesserungen für Gründungen und Böschungen mit vorgespannten und nicht vorgespannten Ankern 4 Vorträge	THEMA 3: **Sprengen und Ausbruch** Vorsitzender: Tan Tjong-Kie *(China)* Sprecher: McKenzie *(Australien)* Sprengen im Hartgestein: Diagnose und Modelle für Sprengschäden und Haufwerk 4 Vorträge	THEMA 4: **Untertägige Hohlräume im überbeanspruchten Gebirge** Vorsitzender: Kidybinsky *(Polen)* Sprecher: Wagner *(S. Afrika)* Planung und Ausbau von Untertagehohlräumen im Fels mit hohen Druckspannungen 4 Vorträge	08:30
					10:00
			Pause		10:30
	THEMA 1: **Flüssigkeitsbewegung und Abfallisolierung im Fels** Vorsitzender: Bjurström *(Schweden)* Sprecher: Doe *(USA)* Entwurf von Bohrlochuntersuchungsprogrammen für Abfallagerung im kristallinen Gestein 5 Vorträge	9 Vorträge	9 Vorträge	9 Vorträge	
					12:00
		Mittagspause und Postersitzung			13:30
	9 Vorträge	Allgemeine Diskussion Diskussionsleiter. Kovari *(Schweiz)* Podiumsdiskussion: MacMahon *(Australien)* Ribacchi *(Italien)* Yoshinaka *(Japan)*	Allgemeine Diskussion Diskussionsleiter: Lindqvist *(Schweden)* Podiumsdiskussion: Favreau *(Kanada)* Gehring *(Australien)* da Gama *(Brasilien)*	6 Vorträge	
					14:30
		Zusammenfassung des Diskussionsleiters	**Zusammenfassung des Diskussionsleiters**	**Pause**	
					15:00
	Pause	**Arbeitskreise**		Allgemeine Diskussion Diskussionsleiter: Whittaker *(England)* Podiumsdiskussion: Sharma *(Indien)* Kaiser *(Kanada)* Maury *(Frankreich)*	
	Allgemeine Diskussion Diskussionsleiter: Barton *(Norwegen)* Podiumsdiskussion: Langer *(West-Deutschland)* Hudson *(England)* Rissler *(West-Deutschland)*	Quellendes Gestein Materialgesetze für Salzgesteine Numerische Methoden als praktisches Hilfsmittel	Bruchvorgänge um untertägige Hohlräume Gesteinschneid- und Bohrfähigkeit Gesteinsprüfung und Prüfnormen		
					16:00
				Zusammenfassung des Diskussionsleiters	
	Zusammenfassung des Diskussionsleiters				16:30
				Abschlussvortrag: Wittke *(W-Deutschland)*	
					17:00
	Postersitzung				18:00
					19:00
Empfang des Kongress Vorsitzenden	**Empfang des Presidenten der IGFM**			Empfang	19:30
				Festessen	20:30
	Jazz Ballett			Ansprachen zum 25. Jahrestag	21:30
				Les Sortilèges	22:00

Contents / Contenu / Inhalt
Volume 1 / Tome 1 / Band 1

1 Fluid flow and waste isolation in rock masses
Ecoulement de fluides et enfouissement de déchets dans
les massifs rocheux
Flüssigkeitsbewegung und Abfallisolierung im Fels

2 Rock foundations and slopes
Fondations et talus rocheux
Felsgründungen und Böschungen

XXIII

3 Rock blasting and excavation
Sautage et excavation
Sprengen und Ausbruch

Contents / Contenu / Inhalt
Volume 2 / Tome 2 / Band 2

**4 Underground openings in overstressed rock
Souterrains en massifs rocheux sous grandes
contraintes
Untertägige Hohlräume im überbeanspruchten Gebirge**

Author index
Index des auteurs
Authoren Verzeichnis

1

Fluid flow and waste isolation in rock masses
Ecoulement de fluides et enfouissement de déchets dans les massifs rocheux
Flüssigkeitsbewegung und Abfallisolierung im Fels

Behaviour of the intra-thrust zone along water conductor system of Yamuna Hydel Scheme, stage II, part II

Comportement de l'intrazone de chevauchement le long du tunnel d'amenée d'eau de l'aménagement Yamuna Hydel, stade II, partie II
Felsverhalten der Nahan und Krol Verschiebungszone entlang des Wasserzuleitungssystems des Yamuna Hydel Projektes, Stufe II, Teil II

P.P.AGRAWAL, Engineer-in-Chief (Retd), Irrigation Department, Ultar Pradesh, India
S.C.JAIN, Executive Engineer, Irrigation Department, Ultar Pradesh, India

ABSTRACT: Yamuna Hydel Scheme Stage II in district Dehradun of Uttar Pradesh, India was the maiden attempt of executing underground works of complicated nature in the young Himalayan rocks, generally of poor to very poor quality. This scheme has been implemented in two parts. The head race tunnel of Part II of this scheme traverses through over-stressed rocks of Nahan, Subathu and Mandhali formations separated by Nahan and Krol thrusts (intra-thrust zone) and three tear faults. To monitor the rock behaviour in intra-thrust zone, an intermediate adit (inspection gallery) and an observation gallery parallel to the main tunnel was constructed in this zone. Different types of instruments were installed in these galleries for the purpose. Analysis of the observed data from the various instruments have revealed satisfactory behaviour of this zone. Further observations are continuing to keep a close watch on the behaviour of this zone.

RESUME: Le projet de Yamuna Hydel Scheme Stage II dans le district de Dehradun de l'Etat de Uttar Pradesh en Inde, est la première entreprise des travaux souterrains d'une nature très complèxe dans le terrain rècent d'Himalaya, engènèral, de qualitè faible à très faible. On a exècutè ce projet dans deux phases. Le tunnel à l'eau à pression de partie II de ce projet se traverse par les roches sur-chargées de Nahan, Subathu et des plateformes de Mandhali séparées par des poussées de Nahan et Krol et des trois défauts lactymères. Afin de suivre la nature de la roche dans la zone intra-poussée, on a construit une galerie intermédiàire et une galerie d'observation parallele au tunnel principal. De différentes sortes des instruments étaient établis dans ces galeries pour cet objet. L'analyse des périodes observées auprès de divers instruments a indiqué une nature satisfaissante de cette zone. Encore des observations sont èn train de surveiller la nature de cette zone.

ZUSAMMENFASSUNG: Der Yamuna Hydel Plan Stufe II im Bezirk Dehradun, Bundesland Uttar Pradesh, Indien war der erste Versuch komplizierte, untergrundliche Arbeit in den jungen, himalaiischen Gesteinen zu erledigen; diese Gesteine sind normalerweise von arm bis sehr armer Qualität. Dieser Plan wurde in zwei Teilen verwirklicht. Der Hauptrenntunnel von Teil II dieses Plans durchquert übergedrückte Gesteine von den Nahan, Subathu und Mandhali Formationen, die durch die Nahan und Krol Drücke (innere Druckzone) und drei Rissfehler getrennt sind. Ein Zwischenbereich (Untersuchungsgalerie) und eine Beobachtungsgalerie, die dem Haupttunnel parallel laufen, wurden in der inneren Druckzone gebaut, um das Gesteinenverhalten in dieser Zone zu steuern. Um diesen Zweck zu erfüllen, wurden verschiedene Sorten von Instrumenten in diesen Galerien installiert. Eine Analyse der beobachteten Daten der verschiedenen Instrumenten zeigt zufriedenstellendes Verhalten in dieser zone. Weitere Beobachtungen werden gemacht um das verhalten hier eng zu untersuchen.

1 INTRODUCTION

Yamuna Hydel Scheme Stage-II in district Dehradun of Uttar Pradesh, India is a run-of-the-river hydro-electric scheme envisaging utilization of a total drop of 186 m in river Tons, a major tributary of river Yamuna which flows along a circuitous route forming S-curves, which have been cut across to make available the drop for generation of power by constructing a long water conveyance system of 7.5 m diameter tunnels. A study of the geological features along the alignment of the proposed tunnel had indicated adverse geological features in some length of the alignment which, it was apprehended, could result in delay of the completion of the work if the entire scheme was implemented in one stage. Hence the scheme had been decided to be implemented in two parts. Part I of the scheme utilizes a drop of 124 m by constructing a 60 m high concrete diversion dam at Ichari across river Tons, 6 Km long 7.5 m diameter lined tunnel and an underground power house at Chibro with an installed capacity of 240 M.W. The outlet discharge from the draft tubes of Chibro Power House is led through another 6 Km long tunnel system upto Khodri Power House to utilize the balance drop of 62 m to generate 120 M.W. power at a surface power station. The paper mainly deals with part II of the scheme, the details of which are given in fig. 1.

2 GEOLOGY ALONG HEAD RACE TUNNEL OF YAMUNA HYDEL SCHEME STAGE-II PART II

The head race tunnel of the scheme passes through the rocks of Nahan, Subathu and Mandhali formations separated by Nahan and Krol thrusts (intra-thrust zone) respectively as shown in fig. 2. Three tear faults F_1-F_1, F_2-F_2 and F_3-F_3 were also encountered in the tunnel alignment. The Mandhali rocks are exposed at the northern end of the tunnel while the Nahan rocks are exposed at the Southern end of the tunnel. The Mandhali rocks comprise boulder slates, quartzitic slates and quartzites, the Subathu rocks comprise purple coloured shales, silt stone and quartzite with occasional gypsum and the Nahan rocks comprise greenish grey coloured sand stone, purple coloured clay stone and silt stone. The Krol and Nahan thrusts, which extend in the Himalayan region, are considered to be still active and prone to tectonic movements.

3 CONSTRUCTION AND LINING OF HEAD RACE TUNNEL IN INTRA-THRUST ZONE

The tunnelling was initially started from the two ends. It was anticipated that rock in the intra-thrust zone shall be crushed, sheared and highly brecciated red shale and Subhatu clays. Therefore,

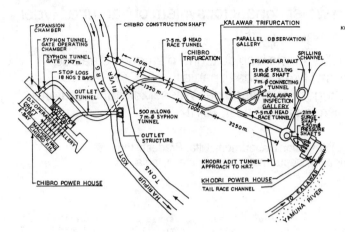

Figure 1. Layout plan of Yamuna Hydel Scheme Stage II Part II.

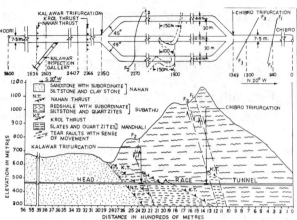

Figure 2. Geological plan and section along head race tunnel showing intra-thrust zone.

to enable a study of behaviour of rocks in the intra-thrust zone and to collect data required for the design of supports and lining for the main tunnel in this reach and to provide possible access for future inspections, an intermediate adit (inspection gallery) about 3 m diameter and sloping down from ground level at EL 543 m to EL 467 m at the tunnel grade was constructed at Kalawar, an intermediate point about midway in the length of the tunnel (see fig. 3).

After completing the excavation of inspection gallery, the excavation of head race tunnel was started at this third heading also on both sides of its junction with the tunnel to get an idea of the behaviour of rock in intra-thrust zone. Due to adverse geological conditions, it became impossible to construct a single 7.5 m diameter tunnel only in a length of about 1 Km and finally three smaller tunnels of 4.8 m diameter each instead of single tunnel of 7.5 m diameter were constructed. For this purpose, trifurcate junctions had to be constructed at either end of this zone. The location, design and construction of trifurcate junctions which are unique in the world, posed serious problems both from the design and construction angles.

As some movements were expected along the thrust planes, it was decided to provide flexible lining in the head race tunnel in the intra-thrust zone. The flexible lining probably first of its kind in the world, comprised concrete shells of 1.5 m length near the two thrust planes flanked by two shells of 3 m length on either side and shells of 6 m length in the rest of the intra-thrust zone. These were separated by joints filled with plastic concrete of suitable design to serve as flexible joints. 25 to 35 mm thick guniting with chicken wire mesh was done in the entire length over the concrete shells and the flexible joints. Plastic concrete consisted of pulverized clay in addition to normal constituents. The above provisions of flexible lining in the intra-thrust zone were made to allow for free movement between the various concrete shells consequent to any anticipated movement of rock along thrust planes in this zone. During the last about 3 years of running of the tunnel this has worked satisfactorily.

4 MONITORING OF THE BEHAVIOUR OF INTRA-THRUST ZONE DURING CONSTRUCTION

Monitoring of the behaviour of intra-thrust zone was done during the progress of construction and subsequent paras.

During construction, monitoring was done by installation of instruments in the inspection gallery and in the head race tunnel. Besides, an observation

gallery of 3 m diameter parallel to head race tunnel and at a distance of 45 m from the main tunnel was also constructed at tunnel grade. This observation gallery leads to a triangular vault constructed across Nahan thrust at tunnel grade (see fig. 3).

Lining of the galleries was designed safe for full anticipated hydrostatic pressure of 0.5 N/mm². Extensive grouting of the rock mass between the main tunnel and the galleries was done in three stages at pressures varying from 0.2 N/mm² to 1.0 N/mm².

Instruments were installed in the observation gallery and triangular vault also to study the nature and extent of creep etc. in this zone to provide necessary guidance in the design of tunnels passing through similar thrusts elsewhere in the Himalayan region. A grid system is being established in the Himalayan region using this vault as one of the reference points and the observed data shall be used to make a long term prediction regarding behaviour of the intra-thrust zone.

4.1 Instrumentation in the inspection gallery

During excavation of inspection gallery, it was observed that after opening of the cavity, the rock started squeezing inside the cavity due to which steel supports comprising 150 mm x 150mm steel I-sections underwent displacement, got twisted and tilted down the grade gradually in one and half months after installation. A series of tests were therefore conducted in this gallery during its construction. Flat Jack Tests were carried out for determining stress field in black clay and its modulus of deformation. Cancellation pressure of 1.08 N/mm² in vertical direction and modulus of deformation of 0.06 x 10⁴ N/mm² was observed.

Some closure bolts were also installed in this gallery. The location was so chosen that bottom half of cross section was in Nahan formations and top half in black clay. It was observed by measuring the distance between bolts from time to time that squeezing took place only in black clay and the rate of squeezing decreased with time and the excavated section became fairly stable after 6 to 12 months period.

For measurements of creep across Nahan thrust, three tiltmeter bases were installed in the cross drift constructed in the left wall of inspection gallery. One base was located on black clay and two across Nahan thrust on sand stone (see fig. 4). Measurements were taken between bases 1 and 2 and between bases 1 and 3. The results are given in figure 4. On the basis of these observations, the rate of vertical component of creep movement across Nahan thrust was found to be about 5 mm per year.

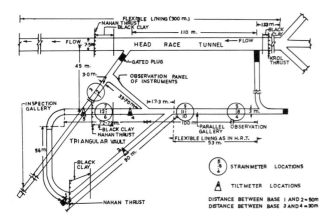

Figure 3. Layout plan of inspection and observation galleries in intra-thrust zone.

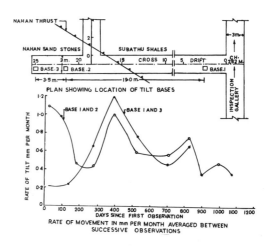

Figure 4. Tilt observations in cross drift in the left wall of inspection gallery.

4.2 Instrumentation in the head race tunnel

During excavation of the head race tunnel in the intra-thrust zone, heavy over-breaks of the order of 10 m with rock flows were encountered. Some of the supporting steel ribs of 150 mm x 150 mm I-section placed at 500 mm c/c got deformed. It was observed that black clay strata encountered could not withstand the external water pressure contained in the rock mass. At one stage, this resulted in severe collapse of the tunnel accompanied with flooding of the tunnel and a portion of the inclined inspection gallery. Several tests were carried out in the tunnel to study the behaviour of rock in this zone. Tiwag's Radial Press Tests were carried out in red shale to find out the modulus of deformation of rock, which was assessed as 0.1×10^4 N/mm^2. Eight number closure bolts were installed in red shale and a few more in black clay zone at various locations in the main tunnel. The maximum closure observed in red shale during 5 months was of the order of 95 mm along the vertical axis and about 45 mm along the horizontal axis and in the black clay zone, of the order of 110 mm and 71 mm along vertical and horizontal axis respectively during this period. Bore hole extensometers were also installed to measure bore hole extensions. Maximum bore hole extensions of 30.58 mm was observed at crown and 20.52 mm on left and 17.06 mm on right sides respectively in red shale. In black clay, the maximum observed readings were 97.08 mm at crown and 43.85 mm and 22.54 mm on right and left sides respectively.

4.3 Seismological observations

A seismological observatory was also set up on the surface on the Nahan sand stone close to Krol and Nahan thrusts at Kalawar. One vertical and two horizontal electromagnetic seismographs each having a magnification of one lakh were installed. Observations made in these seismographs over a period of about one year, however, did not reveal any tremors which could be correlated with any crustal movement in this zone.

5 MONITORING OF THE BEHAVIOUR OF INTRA-THRUST ZONE AFTER COMMISSIONING OF THE WATER CONDUCTOR SYSTEM

A system of instruments has been installed to monitor the behaviour of the intra-thrust zone along water conductor system since its commissioning in late 1983. These instruments have been installed in the inspection and observation galleries which are accessible even after commissioning. Crustal movement

study is also proposed to be carried out in this zone.

5.1 Instrumentation in the inspection and observation galleries

Twelve resistance type strainmeters were installed in the inspection and parallel observation galleries to monitor the behaviour of intra-thrust zone by observing the stresses being developed in the lining of these galleries. Three strainmeters were installed on one steel support in the inspection gallery and nine strainmeters were installed on three steel supports in the parallel observation gallery. Locations of the instruments are shown in figure 3. Four strainmeters (Nos. 4,7,8 and 9) out of twelve installed originally, went out of order during observations. The results of the instruments which are still in order are shown in figure 5. The maximum compressive and tensile stresses that have developed so far in the lining are 3.3 N/mm^2 and 1.5 N/mm^2 respectively, which are well within the safe permissible limits. This study has thus induced a confidence in the safety of the structures in the intra-thrust zone during actual operational conditions.

5.2 Instrumentation in the triangular vault

To study the deformation characteristics of Krol Nahan intra-thrust zone during and after filling of the head race tunnel, ground deformation studies employing water tube tilt meters and silica-tube strain meter were undertaken in the triangular vault. In addition, one high gain high frequency mobile seismograph was also installed in the environ of this zone with a view to record local seismic activity if any, associated with the filling of the tunnel.

Tilt observations employing water tube tilt meters are being taken between pair of bases 1 and 2 and 3 and 4 and are presented in figure 6. No conspicuous visual trends have been noticed in tilt signals so far.

Horizontal strain measurements employing silica tube strain meter have also been taken during filling of the tunnel. No signs of any movements were noticed.

5.3 Earthquake recording

Short term micro earthquake recording carried out in the environ of intra-thrust zone at Kalawar did not indicate any activity associated with the filling of the tunnel. Devices for recording the tremors with

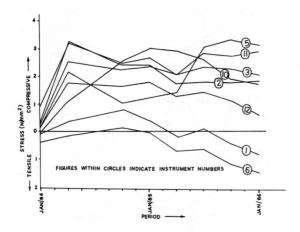

Figure 5. Observed stresses in the lining of the inspection and observation galleries in intra-thrust zone (see fig. 3 for location of instruments).

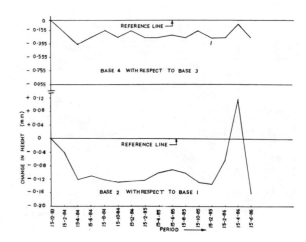

Figure 6. Tilt observations in the triangular vault (see figure 3 for locations of tilt meters).

their epicantres in intra-thrust zone at Kalawar are being arranged for long term monitoring of the behaviour of intra-thrust zone.

5.4 Crustal movement studies

Rock movement (crustal movement) studies in the intra-thrust zone are also proposed to be done. A net work of the closely spaced geodetic survey points across Krol and Nahan thrusts at Kalawar has been established to monitor the behaviour of intra-thrust zone at the surface.

5.5 Visual observations

Visual observations are also being made in the entire length of the galleries and triangular vault. No cracks, abnormal leakage/seepage has been observed so far indicative of any movement along the thrust planes in the intra-thrust zone.

6 CONCLUSION

Monitoring of the various instruments and analysis of the data received so far has not indicated any visual movement along the Krol & Nahan thrust planes in the intra-thrust zone along water conductor system of Yamuna Hydel Scheme Stage-II part II either during the filling operations of the tunnel or since commissioning in November 1983. Further observations are continuing and will continue to keep a close watch on the behaviour of the intra-thrust zone. It is also proposed to set up a grid system of similar instruments on similar other hydroelectric projects which are under construction in the Himalayan region to monitor and analyse the behaviour of this intra-thrust zone in the region as a whole

ACKNOWLEDGEMENTS

The above studies have been done at the project site in close collaborations with School of Earthquake Engineering, University of Roorkee, Roorkee, U.P., Geological Survey of India, Central Mining Research Station, Dhanbad and Survey of India, Organisations, to whom we express our grateful thanks for all the help rendered.

Hydraulic conductivity in basaltic discontinuities in the foundation of Taquaruçu Dam – Brazil

Conductivité hydraulique dans les discontinuités des roches de fondation du barrage de Taquaruçu – Brésil
Hydraulische Konduktivität in Klüften des basaltischen Untergrundes des Taquaruçu Dammes – Brasilien

A.A.AZEVEDO, Geologist, Researcher at the Instituto de Pesquisas Tecnológicas do Estado de São Paulo (IPT), Brazil

D.CORRÊA Fo, M.Sc. Geologist, Researcher at the Instituto de Pesquisas Technológicas do Estado de São Paulo (IPT), Brazil

E.F.QUADROS, M.Sc. Engineer, Researcher at the Instituto de Pesquisas Technológicas do Estado de São Paulo (IPT), Brazil

P.T.DA CRUZ, D.Sc. Engineer, Consulting at the Instituto de Pesquisas Tecnológicas do Estado de São Paulo (IPT), Brazil

ABSTRACT: Artesian pressure was observed in the contact between two basaltic flows in the foundation of the Taquaruçu dam-Brazil, at depths of 4 to 5 meters. A series of inflow and outflow water tests were performed with small heads and compared with conventional water loss tests. Differences in the flow regimens were quite evident showing the importance of working at low heads to identify limit head for laminar flow.

RESUME: Dans la discontinuité A/B à la profondeur de 4 a 5 m, dans les roches basaltiques des fondations du barrage de Taquaruçu (São Paulo, Brèsil), on a observé d'écoulement artesiène. Plusieurs éssays ont eté realizés dans cette discontinuité. Ces essays on eté du type d'injection et de drainage. Les resultats ont eté mise en rapport aux essays d'eau conventionales (type Lugeon). Les differences ont montrée l'importance d'utilization des basses pressions des essays pour identifier le limite d'écoulement laminaire.

ZUSAMMENFASSUNG: "Artesianism" ist im kontakt A/B (4 bis 5m Tief) der basaltiches unterguind des Taquaruçu Dammes abgeschlossen. Injektioneversuche und drainageversuche wurden gemachen und die ergebnissen mit lugeonversuche vergleicht. Es sind große unterschieden in den strom festgeschtellt, was verlaugt die benutzung von versuchen unter kleinen drucken.

1. INTRODUCTION

The Taquaruçu dam of Companhia Energética do Estado de São Paulo (CESP), under construction, is situated at Paranapanema river, West region of São Paulo-Brazil. The dam is 30 m high and has a crest length of 2 000 m (Figure 1).

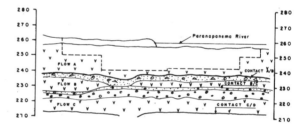

Figure 2. Geological cross – section at the dam axis, showing the excavation for the concrete structure, and the main geological contacts between basaltic flows.

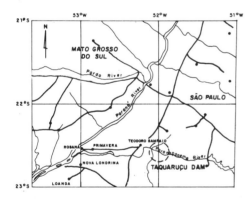

Figure 1. Location of Taquaruçu dam-São Paulo, Brasil

The flow through the rock foundation at the concrete structures is mainly affected by the contact of basaltic flows A and B. This contact is continuous with a large oppening (milimetric to centimetric, with an average value of 0,50 cm) practically without any fill material and constitutes a preferencial horizontal flow feature.

The contact is 4 to 5 meters below the surface within the foundation level of the Spillway and Intake Structures of the dam. During the excavations

the initial flow into the area was about 12 000 l/min, dropping to 6 000 l/min after stabilization. This very high discharge has caused some problems for the execution of concrete structures.

During the excavations and at different drainage conditions the A/B contact was tested with special attention and details. Multiple stage tests both of inflow (infiltration) and outflow (drainage), associated with conventional water loss tests were performed.

The multiple stage with low heads made possible an accurate analysis of the flow characteristics at the basaltic feature. Plots of flow (Q) versus effective pressure or head (H_o) and plots of Q/H_o allowed a much better interpretation of the flow characteristics and of the transfer of laminar to transitional and turbulent flow. A good estimative of both critical head (H_{cr}) and corresponding critical flow (Q_{cr}) that characterizes this change of flow could be done.

These techniques have shown a great advantage in the prediction of flows into the excavation area that is normally under estimated when test results of conventional water loss tests area used.

2. FIELD TESTS

Three types of tests were run in the contact A/B. Conventional water loss tests, using however a larger number of pressure; inflow tests or infiltration (injection of water with very low pressures) and outflow tests or drainage also with low pressures.

The conventional water loss tests followed known techniques similar to Lugeon test (ABGE, 1975). Infiltration and drainage tests were performed by increasing the pressure test by means of longer casing (above the top of the rock) to test the artesianism flow or by reducing the pressure test using shorter tubes (below the artesian head). Figure 3 shows a sketch for this kind of test.

The discharge Q, for each head was carefully recorded. The duration of each test was about 10 minutes for each pressure used. The test was repeated three times or more until the flow became aproximatelly constant.

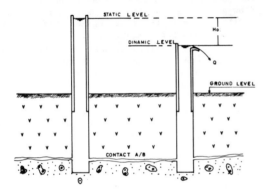

Figure 3. Set-up for outflow or drainage tests. Inflow or infiltration tests were run with a similar set-up by raising the water level above the artesian level.

Four drainage and nine infiltration tests were executed in differents points of the contact. Conventional water loss tests were also run in the exploratory phase of the project.

Multiple stages tests were performed according to Rissler (1978) and also reported by Cruz and Quadros (1983) in conection with Nova Avanhandava dam in Brazil.

3. TEST RESULTS

Test results were ploted in two ways; flow Q (1/min.) versus effective head H_o (also expressed as effective pressure in kg/cm²) and the ratio Q/H_o versus H_o.

Five representative test results are shown in Figures 4,5,6,7 and 8, where both diagrams were superimposed. As it can be seen in each test, 6 to 8 pressure stages were used in the rabge of a few centimeters up to a maximum of 10 meters in the inflow or infiltration tests.

Whenever test results follow a straight line through the origin in the Q versus H_o or a horizontal line in the Q/H versus H_o diagram the flow regimen should be laminar; when this line gets curved we have either a transitional or turbulent flow regimen.

In all figures but specially in Figure 7, it becomes clear that for very low pressures the flow regimen changes to transitional or turbulent and that test results of conventional water loss tests stays out of laminar regimen range.

The point in which the regimen changes corresponds to what has been called the critical head (H_{cr}) and the corresponding flow is Q_{cr} (critical flow).

Estimated values of both critical head (H_{cr}) and flow (Q_{cr}) are shown in Table 1.

From Table 1 one can observe that the critical head H_o varies within the range of 20 to 50 cm (0,02 to 0,05 kg/cm²) and the critical flow varies within the range of 45 to 190 1/min.

The value of 2,1 1/min. obtained in test nº 2 (borehole nº 352 C-III - 240,0 to 237,0 m) corresponds to a test run on the fractured zone just above the contact A/B.

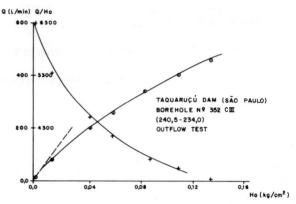

Figure 4. Outflow test results

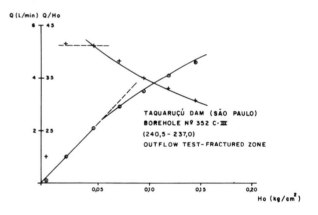

Figure 5. Outflow test results

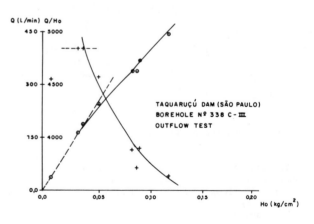

Figure 6. Outflow test results

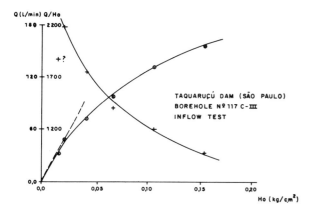

Figure 7. Inflow test results

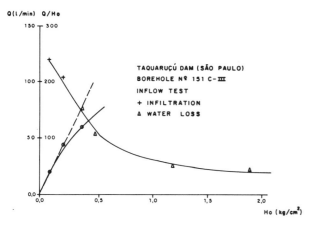

Figure 8. Inflow test results

Table 1. Estimative values of critical head (H_{cr}) and critical flow (Q_{cr}) from tests results.

BOREHOLE Nº	CRITICAL PRESSURE, H_{cr} (kg/cm²)	CRITICAL FLOW, Q_{cr} (l/min)	NOTES
352 C-III (240,6 A 234,0 M)	0,002	12,6	OUTFLOW TEST FIGURE 4
352 C-III (240,5 A 237,0 M)	0,051	2,1	FRACTURED ZONE ABOVE CONTACT FIGURE 5
356 C-III	0,039	189	OUTFLOW TEST FIGURE 6
356 C-III	0,040	56	OUTFLOW TEST
316 C-III	0,016	63	INFLOW TEST
304 C-III	0,017	92	INFLOW TEST
117 C-III	0,022	46	INFLOW TEST FIGURE 7
151 C-III	0,06	20	INFLOW TEST (INFILTRATION AND WATER LOSS TEST) FIGURE 8

4. ANALYSIS OF TEST PROCEDURE AND RESULTS

The test procedure showed to be appropriate to the analysis of the flow regimen and the results obtained were satisfactory.

In open fractures, it became evident that small effective heads are necessary to properlly define the flow regimen or the limits of laminar flow.

The heads must be measured to less than 1 cm,

because a 10% difference in head introduces significative variation in the Q/H_o ratio.

Even in tests run with good accuracy the critical head (H_{cr}) could not be defined preciselly due to mechanical restrictions of the devices used in the performed tests. It is very important that plots of Q versus H_o or Q/H_o versus H_o be done during the tests to introduce intermediary heads when necessary to define the flow regimen.

Both critical flows and heads vary in a relative wide range, probably due to local differences in the oppening of the contact.

Predictions of flow into excavations in large areas will require a large number of tests and other investigations like impression packers in boreholes to give an idea of the persistency or continuity of the geological feature, associated with the natural (field) gradient that will result in the area.

5. CONCLUSIONS

Test results showed the convenience of using the artesian pressure as a "head" for outflow or drainage test. Long time duration tests may be necessary in some cases.

The purpose of this paper was solely to call the attention on the importance of using low heads to analyse the configuration of flow in very permeable geological features.

ACKLOWLEDGEMENTS

Authors wants to express the acknowledgments for the Instituto de Pesquisas Tecnológicas do Estado de São Paulo - IPT and Companhia Energética de São Paulo CESP for the opportunity of presenting this paper.

REFERENCES

ASSOCIAÇÃO BRASILEIRA DE GEOLOGIA DE ENGENHARIA (ABGE). 1975. Ensaio de perda d'água sob pressão; diretrizes. São Paulo. 16 p. (ABGE, Boletim 02).

Corrêa Fº, D. and Quadros, E.F. 1986. Metodologias para determinação do comportamento hidrogeotécnico dos maciços rochosos. In: II Symposium Sul Americano de Mecânica de Rochas, Porto Alegre (RS), Brasil, 1986.

Corrêa Fº, D. 1986. Ensaios de perda d'água sob pressão. São Paulo. (Dissertação de Mestrado - Universidade de São Paulo).

Cruz, P.T. and Quadros, E.F. 1983. Analysis of water losses in basaltic rock joints. In: 5th International Congress on Rock Mechanics, Nelbourne, Austrália, 1983.

INSTITUTO DE PESQUISAS TECNOLÓGICAS DO ESTADO DE SÃO PAULO (IPT). 1981. Síntese do conhecimento atual das características hidrogeotécnicas da área de implantação da barragem de Taquaruçu. São Paulo. Relatório IPT nº 14720. São Paulo, 1981.

Quadros, E.F. 1982. Determinação das características de fluxo de água em fraturas de rochas. São Paulo. (Dissertação de Mestrado, Escola Politécnica da Universidade de São Paulo).

Rissler, P.C. 1978. Determination of the water permeability of jointed rock. Aachen, Institute for Foundation Engineering Soil Mechanics, Rock Mechanics and Water Ways Construction.

Stress relaxation behaviour of rock salt: Comparison of in situ measurements and laboratory test results

Relaxation des contraintes dans le sel gemme: Comparaison des résultats obtenus par mesures in situ et en laboratoire

Relaxationsverhalten von Steinsalz: Vergleiche von Laboruntersuchungen und in-situ-Messungen

K.BALTHASAR, Institute of Soil and Rock Mechanics, University of Karlsruhe, FRG
M.HAUPT, Institute of Soil and Rock Mechanics, University of Karlsruhe, FRG
CH.LEMPP, Institute of Soil and Rock Mechanics, University of Karlsruhe, FRG
O.NATAU, Institute of Soil and Rock Mechanics, University of Karlsruhe, FRG

ABSTRACT: In a rock salt mine stress measurements with prestressed hard inclusion cells were performed during which remarkable stress relaxation effects were observed. Consequently a special uniaxial relaxation test apparatus was developed to investigate the phenomenon of stress relaxation in laboratory. The in situ and the laboratory test results are compared.

RESUME: Des mesures des contraintes selon la méthode de l'inclusion rigide par des jauges precontraintes ont été réalisés dans une mine de sel gemme. On a mesuré des relaxations de contraintes voyantes. Pour l'exploration profonde de ce phénomène, on a développé un appareil d'essai pour la relaxation uniaxiale. Les résultats des essais in situ on été comparés avec les résultats du laboratoire.

ZUSAMMENFASSUNG: In einem Steinsalzbergwerk wurden Spannungsmessungen nach der Methode des vorgespannten harten Einschlusses durchgeführt. Dabei wurde eine auffällige Spannungsrelaxation gemessen. Zur weiteren Untersuchung dieses Phänomens wurde ein spezieller einaxialer Relaxationsprüfstand entwickelt. Die Ergebnisse der Labor- und der In-situ-Untersuchungen werden verglichen.

1. INTRODUCTION

In situ stress measurements according to the method of hard inclusion which were carried out in a rock salt mine in northwestern Germany by the way have detected the phenomenon of stress relaxation in the rock salt masses. During the evaluation of the measurement data at last the necessity arose to explore the stress relaxation behaviour on the basis of the stress relaxation phenomena found in situ. Now the measurement results as well from in situ as from laboratory tests should be compared with respect to distinct differences of the stress conditions and measurement methods.

Theoretically the following boundary conditions are present during the measurements in situ and in the laboratory, respectively (fig. 1). Differences mainly are based on the fact that in laboratory uniaxial tests were performed whereas in situ a three-dimensional state of stress appears.

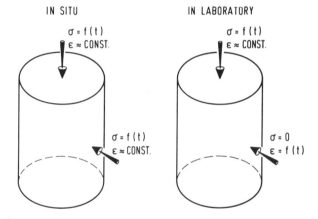

Fig. 1: Stress relaxation measurements: Stress - strain relations under in-situ and laboratory conditions

2. BASIS OF COMPARISON FOR LABORATORY AND IN SITU

Due to the different stress conditions a direct comparison between uniaxial relaxation tests and the relaxation behaviour in situ becomes problematic. However, a comparison appears to be justified because there can be observed some conformities: The axial stress release found in the uniaxial relaxation test is accompanied by a creep strain in lateral direction (radial extension). I. e. radial strain overlays the relaxation process once the axial strain is kept constant. Hence, purely uniaxial stress conditions do not obtain even in the laboratory specimen.

Moreover, stress measurement with hydraulic hard inclusion pressure cells - a technique which is particularly well suited for viscoplastical rock salt masses - allows to measure each normal stress component separately without having to postulate strains in the rock mass. Thus, with regard to the measurement techniques relaxation processes in rock salt masses can be recorded correctly in situ. This is achieved by installing oriented hydraulic pressure cells in the rock masses. These gauges record several normal stress components of the stress tensor. With

the hard inclusion technique stress relaxation can be observed for any given spatial orientation. The prevailing multi-axial stress condition and its changes are studied individually. Both the stress relaxation recorded in laboratory tests and in situ can be accompanied by creep strains. Normally, both processes take place simultaneously. Measurement technique and test conditions, therefore, are designed to "extract" and record separately the stress relaxation process.

3. IN SITU MEASUREMENT METHOD

Stress measurements in situ by means of hydraulic hard inclusion gauges work particularly well in viscoplastical rock salt masses because the pressure cells can be embedded completely and free of pressure transmission losses. Fundamental developments for the embedding technique of hard inclusion gauges in rock salt are due to WÖHLBIER and NATAU (1966): In a first step, the hydraulic pressure cells which are oriented in a predetermined position in the borehole are brought into contact with the surrounding rock masses by filling the entire borehole with rock-salt-like mortar leaving no hollow space. The mortar in its hardened state shows mechanical properties comparable to those of rock salt. The pressure cells are now enclosed by viscous rock salt.

In a second step, prestressing of the pressure cells embedded in the rock masses is obtained by injecting synthetic resin (two-component blend) into the refilled borehole. Prestressing the cells is done at pressure higher than the actual rock pressure. The pressure can be raised to the level of the fracturing pressure of the rock masses. The synthetic resin injection creates a stress field on the level of the frac pressure (comparable to the instantaneous shut-in pressure during hydraulic fracturing). Once the injection is completed, the pressure changes monotonously to the actual pressure level of the rock mass. Simultaneously, a direction-dependent differentiation of the stress components takes place.

An application for stress measurement in a permian salt dome in northwestern Germany with this type of prestressed pressure cells is discussed in NATAU et al. (1985). Also included are further details of this measuring technique (installation procedures in boreholes, prestressing techniques). NATAU et al. (1985) use the recorded normal stress components which are independent from one another to compute the entire stress tensor in rock salt deposits. The final in situ stress measurement, however, is executed only when all pressure cells record practically constant pressures, i. e. when the pressure of the rock masses has become stabilized. Before this condition is reached, the following observations can be made: The rock salt masses stressed hydraulically with injections beyond their actual stress level relax once the injection is completed. The eventual stress release in its temporal extension can be recorded at each pressure cell for the different spatial directions individually. In the example quoted above (NATAU et al., 1985), the measurement of the various stress components is done in boreholes which are free of interferences from mining cavities in undisturbed rock salt masses. Consequently, it can be assumed that the stress release recorded at the pressure cells occurs without detectable strains in the surrounding salt. This stress relaxation is recorded for several independent spatial orientations depending on the gauge configuration selected. In this case an orthogonal system of pressure directions measured in three different boreholes have been selected as examples to demonstrate in situ stress developments after prestressing of the rock mass (fig. 2).

4. RESULTS OF MEASUREMENTS IN SITU

All pressure cells record a decrease in pressure when the resin injection is completed. This decrease follows a natural order: The stress rate decreases steadily. In a semilogarithmical scale (logarithmical time axis) an almost linear change in stress results until finally a stable pressure level is reached and the relaxation process completed (fig. 2). Relaxation time (the time necessary to reach the stationary condition) and stress differences between maximum injection pressure and actual rock pressure form the basic data which in case of need can be depicted in a standardized form as a stress change/time diagramm (NATAU et al., 1986).

Pressure cells of equal orientation situated in boreholes with orthogonal orientation show slight differences in their relaxation behaviour while their final pressure levels coincide well. A comparison of measurements from the three boreholes with pressure cells of different orientation shows fairly uniform ascending lines for all directions recorded. Only the relaxation time at different sites respectively orientations differs in some cases (fig. 2).

The resulting relaxation curves can be described mathematically as hyperbolical sinus functions (NATAU et al., 1986).

5. STRESS RELAXATION TEST SYSTEM

The significant conditions at the construction of a stress relaxation test system were low costs and an uncomplicated design, guaranteeing not much service. On the other side, there were also special requirements on the accuracy of the device. The result is a good compromise between these two poles: Our test system is able to stabilize the specimen's length within a limit of ± 0.5 μm.

The basic part of the test device is a four-column load frame with high stiffness (fig. 3). The loading of the specimen is done with a screw jack, which is driven by an electric motor with reduction gear. To measure the specimen's load, the screw jack has a built-in electronic load cell.

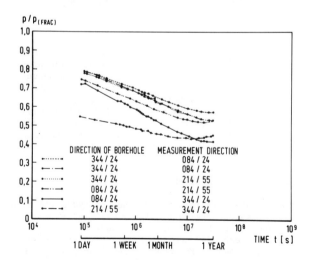

Fig. 2: Results of in-situ stress measurements

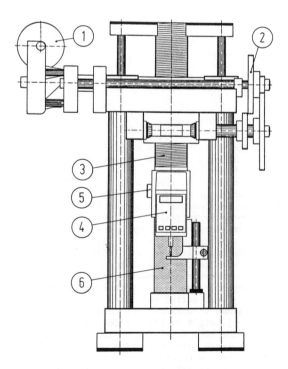

Fig. 3: Relaxation device: (1) Electric motor, (2) reduction gear, (3) screw jack, (4) digital strain gauge, (5) load cell, (6) specimen.

Because it is not possible to construct a load frame with unlimited stiffness, even if using heavy designs, it was essential to regulate the specimen's deformation. Therefore, a microprocessor-supported closed loop control system for the electric driven screw jack was developed which gets its information about the actual specimen's length from a digital strain gauge with a resolution of 1 μm.

The maximum regulation speed of this device depends on the engine output as well as the stiffnesses of the load frame, the screw jack and the specimen. Since there appear very high stress rates in the relaxation tests immediately after stopping the loading, the device needs some seconds for reaching the displacement value to be kept constant. In this moment an exact regulation is not ensured. Later on the displacement regulations amounting to 0.5 μm become apparent by a small break of the stress curve which can be neglected.

The relaxation behaviour is extremly influenced by the ambient temperature. Even temperature differences of less than 1° K evidently took effect on the stress. For this reason the whole test device was installed within a temperature controlled box keeping constant the temperature with an accuracy of ± 0.1° K. The registration of the test results are accomplished by an x-t-recorder.

6. RESULTS OF LABORATORY MEASUREMENTS

All in-situ tests showed a continuously decreasing stress. Immediately after reaching and keeping constant the strain a fast decrease of stress could be observed. In most cases a stress decrease of more than 30 percent of the initial stress occurred within a relaxation time of 1 hour. Some times even a stress decrease of more than 50 percent was reached within this time. But due to the decreasing stress rate this process slackened rapidly. Hence mostly the stress came not up to about 25 percent of the initial stress within the test duration amounting to 100 days maximal.

As the in-situ measurements the laboratory tests yield an almost linear curve in a semilogarithmical scale, too. Some of the test results are represented in fig. 4. This depiction illustrates another observation: So far no dependency of the relaxation behaviour on the strain kept constantly during relaxation phases could be found. This fact eases the comparison between in situ and laboratory results because the in situ state of strain is not known exactly.

In contrast to the in-situ measurements never an unambiguously constant stress level was reached in laboratory tests within the test duration. This is the result of the different stress conditions described above and leads to the assumption that a possibly existing residual stress in the rock salt samples must be very low or even zero. By this way it is proved that the asymptotically stabilized state of stress measured in situ really is the present state of stress and not a possibly existing residual state of stress caused by prestressing.

REFERENCES

Natau, O., Lempp, Ch., Balthasar, K., 1984. Monitoring System for Stress Measurement. Proc. Int. Congr. IAEG on Management of Hazardous Chemical Waste Sites. Winston-Salem, N.C.

Natau, O., Lempp, Ch., Borm, G., 1986. Stress Relaxation monitoring by prestressed Hard Inclusions. Proc. Int. Symp. on Rock Stress and Rock Stress Measurements, 509–514, Stockholm.

Wöhlbier, H., Natau, O., 1966. Die Entwicklung einer neuartigen Einbautechnik für Bohrlochdruckgeber. Proc. 1st Int. Congr. ISRM, 4.4: 25–30, Lisbon.

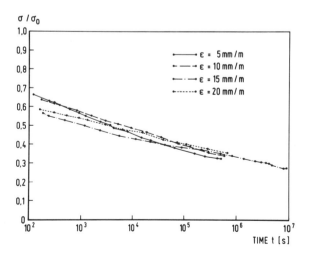

Fig.4: Stress relaxation in uniaxial tests

Caractérisation hydrodynamique à différentes échelles d'un massif granitique fracturé

Hydrodynamic characterization of a fractured granite body at various scales
Charakterisierung der Hydrodynamik für unterschiedliche Grösseneinheiten in einem zerklüfteten Granitgebirge

A. BARBREAU, Commissariat à l'Energie atomique, Institut de protection et de sûreté nucléaire, Fontenay-aux-Roses, France
M.C.CACAS, Ecole des Mines de Paris, Fontainebleau, France
E.DURAND, Bureau de Recherches géologiques et minières, Orléans, France
B.FEUGA, Bureau de Recherches géologiques et minières, Orléans, France

ABSTRACT : With a view to studying the influence of the "scale effect" on the determination of hydraulic characteristics in fractured rocks, detailed investigations have been carried out in an experimental section of an adit in granitic rocks in the Fanay-Augères mine, France. Ten 50 m long boreholes were drilled in this adit, where permeability measurements on various scales ranging from an individual fracture to a cubic hectometer and punctual head measurements were carried out, as well as the recording of the adit's discharge rate. A detailed study of fracturing was realized on site. The data collected enabled models of fractured areas to be established, whose hydraulic characteristics were calibrated with the results of the permeability measurements. On the basis of these models a permeability tensor of "equivalent continuous medium" type could be determined which enabled the flow (head and discharge rate) around the adit to be reproduced satisfactorily by calculation. The validity, shown in this case, of the equivalent continuous medium concept is due to the strong fracture network connectivity, which is situated far above the percolation threshold.

RESUME : En vue d'étudier l'influence de l' "effet d'échelle" sur la détermination des caractéristiques hydrauliques d'un milieu rocheux fracturé, des investigations approfondies ont été réalisées dans un tronçon expérimental de galerie dans un granite dans la mine de Fanay-Augères (France). Dans cette galerie ont été forés dix sondages de 50 m de longueur unitaire, qui ont fait l'objet de mesures de perméabilité à différentes échelles, depuis celle de la fracture individuelle jusqu'à celle de l'hectomètre cube, et de mesures de charge ponctuelles, cependant qu'étaient enregistrés les débits drainés par la galerie. Une étude approfondie de la fracturation a été réalisée sur le site. Les données recueillies ont permis d'établir des modèles de champs de fractures, dont les caractéristiques hydrauliques ont été calées sur les résultats des mesures de perméabilité. Sur la base de ces modèles a pu être déterminé un tenseur de perméabilité de type "milieu continu équivalent" qui a permis de reproduire de manière satisfaisante, par le calcul, les écoulements autour de la galerie (charges et débits drainés). La validité, démontrée dans ce cas, du concept du milieu continu équivalent est due à la forte connectivité du réseau de fractures, située très au-dessus du seuil de percolation.

ZUSAMMENFASSUNG : Zur Erforschung der Einflussnahme des "Grössenfaktors" auf die Bestimmung der geohydraulischen Parameter eines zerklüfteten Festgesteinsaquifers wurden in dem Bergwerk von Fanay-Augères (Frankreich) an einer Versuchsstrecke eines Stollens im Granit eingehende Untersuchungen durchgeführt. In diesem Stollen wurden 10 Bohrungen von je 50 m Tiefe niedergebracht, an denen Durchlässigkeitsmessungen gefahren wurden, die so abgestuft waren, dass sie von der Einzelkluft bis zur Grösseneinheit von 10^6 Kubikmetern reichten. Ausserdem wurden punktuelle Druckhöhenmessungen vorgenommen und gleichzeitiges Messen der über den Stollen abgeleiteten Abflussraten. An diesem Versuchsfeld wurde auch eine eingehende Untersuchung der Kluftverhältnisse durchgeführt. An Hand der erzielten Daten konnten Modelle der Kluftzone aufgestellt werden, deren geohydraulische Parameter an den Ergebnissen der Durchlässigkeitsversuche geeicht wurden. Auf Grund dieser Modelle konnte ein Durchlässigkeitstensor vom Typ eines "äquivalenten Kontinuums" bestimmt werden, der die rechnerische Wiedergabe der Fliessvorgänge (Druckhöhen und abgeleitete Abflussraten) im Umkreis des Stollens zufriedenstellend ermöglichte. Die an diesem Beispiel bewiesene Gültigkeit des Konzeptes eines äquivalenten Kontinuums beruht auf der starken "Kreuzungsfähigkeit" des Klüftesystems, welche weitaus über der "percolation threshold" liegt.

INTRODUCTION

Les projets de stockage ou d'enfouissement de déchets radioactifs dans des roches massives mais fracturées telles que les granites ont suscité de nombreuses recherches sur l'hydraulique des milieux rocheux fracturés. Le problème majeur auquel sont confrontés les projets de stockage dans de tels milieux est en effet celui du risque de contamination de la biosphère par des radionucléides relâchés par le dépôt et qui gagneraient le jour entraînés par les circulations d'eau dans le réseau de fractures. Les études de sûreté visant à quantifier ce risque nécessitent une modélisation de ces circulations, ce qui suppose une connaissance aussi exacte que possible des caractéristiques hydrodynamiques du milieu.

Or, si les techniques de détermination de ces caractéristiques par essais in situ peuvent être considérées comme bien maîtrisées pour ce qui est des milieux considérés comme continus classiquement étudiés par les hydrogéologues (alluvions, grès perméables, etc.) ou les ingénieurs de génie civil (sols constitutifs de barrages en terre), il n'en va pas de même pour les milieux rocheux fracturés, et ce d'autant moins que ces milieux sont moins perméables. En effet, leur faible perméabilité limite le volume de terrain affecté par les essais, ce qui entraîne que ceux-ci sont en général fortement influencés par des hétérogénéités ou des discontinuités locales et que l'on ne sait pas au juste quelle signification attacher à leur résultat. En d'autres termes, ces essais sont affectés par un effet d'échelle qu'il est bien difficile d'appréhender sans investigations approfondies.

C'est dans le but de tenter de préciser, sur un cas réel, l'influence de cet effet d'échelle sur la détermination des caractéristiques hydrodynamiques d'un milieu rocheux fracturé qu'un programme d'investigations structurales et hydrogéologiques a été lancé dans la mine de Fanay-Augères, située dans le massif granitique de Saint-Sylvestre (Haute-Vienne, France) (fig. 1).

Figure 1 - Localisation de la mine de Fanay-Augères

L'idée de base de ce programme était de déterminer la relation débit-gradient de charge (et sa variabilité) de l'échelle de la fracture à celle de l'hectomètre cube et de voir dans quelle mesure il était possible de passer d'une échelle à l'autre à partir de données portant sur la fracturation d'une part, sur les caractéristiques hydrauliques des fractures d'autre part.

Une quantité considérable de données portant sur la fracturation et l'hydrogéologie (perméabilités, charges hydrauliques, débits) de la zone étudiée a été accumulée. Ces données, confrontées aux données de même nature provenant d'autres sites, constituent un apport à la connaissance des milieux granitiques. Elles constituent en outre une base solide pour le développement de nouvelles méthodes d'approche des écoulements en milieu fracturé, évoquées à la fin de cet article.

PRESENTATION DU SITE EXPERIMENTAL

La mine de Fanay-Augères, appartenant à la COGEMA, exploite des failles minéralisées en uranium qui recoupent le granite de Saint-Sylvestre, d'âge namurien (Carbonifère moyen). Ce granite est un leucogranite calco-alcalin, à grains grossiers ou moyens, parfois fins et à structure relativement équante. Il est affecté par une fracturation assez dense, apparue pour l'essentiel au Tardi-hercynien mais ayant pu rejouer lors des épisodes tectoniques ultérieurs.

La mine comporte un réseau de plusieurs dizaines de kilomètres de galeries qui se situent pour la plupart dans le granite sain et offrent des conditions d'observation de celui-ci tout à fait exceptionnelles.

C'est dans l'une d'elles, située au niveau 320 de la mine, à une profondeur de 175 à 200 m sous le flanc d'une colline, qu'a été mis en place le dispositif d'étude.

La section de cette galerie, de direction N 10°E, est de l'ordre de 10 m².

Les figures 2 et 3 précisent la position du tronçon de 100 m de long retenu pour les expériences par rapport aux travaux miniers existants et aux principaux accidents cassants affectant le massif.

DESCRIPTION DU DISPOSITIF D'ETUDE MIS EN PLACE ET DES TRAVAUX REALISES

Le choix du tronçon expérimental a été précédé :
- d'une étude des grands accidents, réalisée à partir des données disponibles (sondages miniers et galeries de mine) ; la zone retenue, qui n'est recoupée par aucun accident majeur, est encadrée par deux failles :

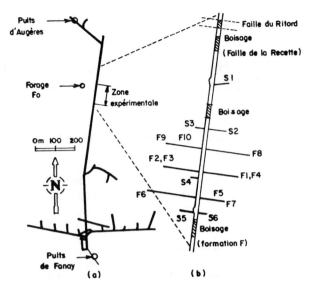

Figure 2 - Extrait du plan du niveau 320 et localisation de la zone expérimentale (a). Implantation des sondages (b)

l'une, au sud, subverticale et de direction N 115°E (dénommée "formation F") et l'autre, au nord, de direction N 165-175E et de pendage 60°E (faille de la Recette). Cette faille passe à une dizaine de mètres de l'extrémité nord du tronçon expérimental.
- et d'un suivi des charges hydrauliques dans le terrain pendant une durée d'un an à l'aide de sondages peu profonds, S1 à S6, réalisés à partir de la galerie. Ce suivi avait pour but de vérifier la saturation du milieu avant le début des expériences.

Une fois sélectionné le tronçon de galerie expérimental, celui-ci a fait l'objet, sur un de ses parements (surface de 100 m x 2 m) d'un relevé détaillé de la fracturation.

L'équipement proprement dit de ce tronçon comportait :
- l'installation de dispositifs de captage et de mesure en continu des débits drainés par la galerie à l'entrée et à la sortie de la zone expérimentale, ainsi que dans deux sections intermédiaires. Ceci permettait de connaître l'évolution dans le temps des débits drainés par les trois portions du tronçon expérimental délimités par les dispositifs de mesure, ainsi que du débit total drainé par ce tronçon.
- la réalisation de dix sondages radiaux de 50 m de longueur, F1 à F 10, répartis suivant trois profils (fig. 4).

Ces sondages, forés en carottage continu avec orientation des carottes, ont fait l'objet de 251 mesures de perméabilité par injection d'eau entre obturateurs, ou entre un obturateur et le fond du forage. Ces essais ont été interprétés en régime permanent.

Les sondages ont ensuite été équipés pour la mesure des charges hydrauliques dans le massif : dans chacun d'entre eux ont été isolées sept chambres de mesure de cinq mètres de longueur, séparées les unes des autres par des obturateurs gonflables de 2 m de long.

Enfin, un forage vertical, F0, profond de 223 m, réalisé à partir du jour à 75 m à l'ouest de l'extrémité nord de la zone expérimentale, a été lui aussi équipé en piézomètre multiple.

L'ensemble de tous les capteurs était relié à une centrale d'acquisition de données installée dans la galerie.

La campagne de mesures a duré de 1983 à 1986.

Les résultats recueillis sont présentés dans les paragraphes qui suivent.

LA FRACTURATION DE LA ZONE EXPERIMENTALE

Au total, 1014 fractures de trace supérieure à 0,20 m ont été relevées sur le parement de la galerie, et

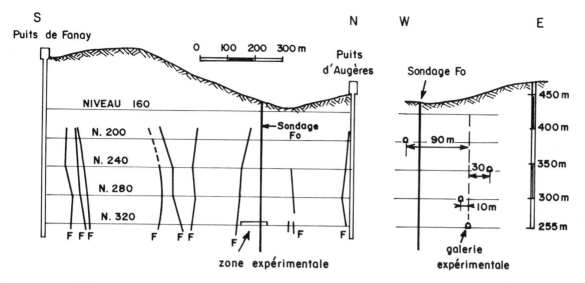

Figure 3 - Localisation de la zone expérimentale sur des coupes verticales N-S et E-W (F : grande faille)

4023 sur les carottes orientées des sondages F1 à F10 (longueur cumulée : 501 m)

Pour chaque fracture ont été notés, quand c'était possible, les paramètres suivants :
- localisation spatiale de la fracture (coordonnées du centre de sa trace sur le parement de la galerie ou distance par rapport à la tête du sondage)
- direction et pendage
- type de fracture (diaclase, faille, filon, etc.)
- longueur observable de la fracture
- nombre d'extrémités visibles
- continuité
- épaisseur (distance entre les épontes, qu'il y ait ou non remplissage)
- ouverture libre
- type de remplissage
- venue d'eau
- décompression
- type de terminaison
- morphologie à l'échelle métrique
- rugosité (morphologie à échelle centimétrique)

Certains de ces paramètres ne sont pas indépendants les uns des autres, ce qui a permis d'effectuer des tests de cohérence permettant de vérifier la bonne qualité des données recueillies.

Ces données ont fait l'objet des traitements suivants :
- report des pôles des plans de fractures sur diagrammes de Schmidt

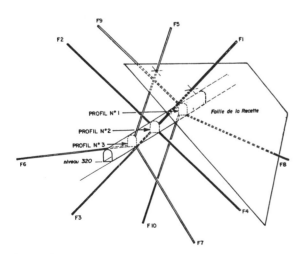

Figure 4 - Vue en perspective du tronçon de galerie expérimentale de Fanay-Augères et des trois profils de sondages radiaux

- délimitation des différentes familles de fractures
- étude, par famille, de la distribution d'un certain nombre de paramètres, parmi lesquels les écartements entre fractures voisines.

L'analyse du diagramme de Schmidt correspondant aux levés en galerie (fig. 5a) fait apparaître quatre familles principales de fractures, bien connues par ailleurs des géologues, et dont les plans moyens ont les orientations suivantes :

1 N 80° à 130°E, vertical
2 N 160° à 30°E, vertical (subparallèle à la galerie)
3 N 170°E - 50°W
4 N 150°E - 30°E

Les mesures sur carottes de sondages (fig 5b) font apparaître les mêmes familles. La comparaison des deux diagrammes fait toutefois clairement ressortir l'effet de biais dû à l'orientation du plan ou de la ligne de mesure par rapport aux directions des familles de fractures.

C'est ainsi que les fractures de la famille 1, subperpendiculaires à la galerie, sont très nombreuses sur le diagramme des relevés effectués dans celle-ci et peu nombreuses sur celui des mesures faites sur carottes. Inversement, la famille 2, subparallèle à la galerie, est mal représentée sur le premier diagramme et très bien sur le second.

L'histogramme des longueurs de traces de fractures, toutes familles confondues (fig. 6) montre que plus de 50% de ces longueurs sont inférieures au mètre.

A noter que les indications concernant les longueurs de trace, combinées à celles relatives aux extrémités, visibles ou non, de ces traces, ont permis à Massoud (1987) de reconstituer les distributions des longueurs réelles des fractures.

Par ailleurs, plus de 60 % des fractures sont jointives (épaisseur non déterminable à l'oeil nu), cependant que les fractures à ouverture libre visible représentent seulement 2 % de l'ensemble.

Les relevés sur carottes ont permis d'évaluer des densités de fracturation, toutes familles confondues. Ces densités varient significativement en fonction du faciès du granite : dans le faciès fin, la densité moyenne est de 9,7 m^{-1} (valeurs extrêmes : 8,1 à 13,3 m^{-1}) alors que dans le faciès grossier elle n'est que de 6,0 m^{-1} (3,0 à 11,4 m^{-1}).

Sur le parement, plusieurs paramètres peuvent être utilisés pour caractériser la densité de fracturation: par exemple, elle peut être évaluée à partir du nombre de fractures recoupées par une ligne horizontale, ce qui donne une valeur de 3,2 m^{-1} ; elle peut également être prise égale au rapport de la longueur cumulée de toutes les fractures sur la surface de relevé; dans ce cas, on obtient 5,7 m^{-1}.

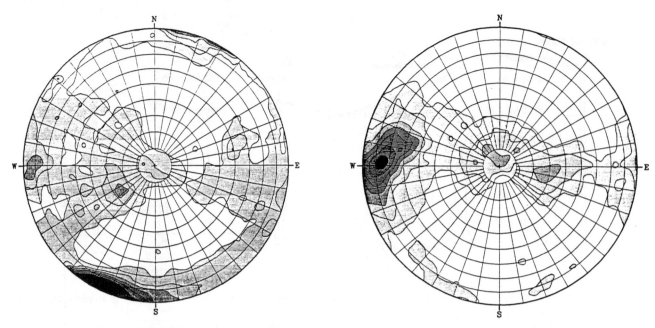

Figure 5 - Représentation sur diagramme de Schmidt (hémisphère inférieur) des densités de pôles de plans de fractures relevés (a) en galerie, (b) sur les carottes de sondages F1 à F10

Ces valeurs, plus faibles que celles obtenues sur carottes, s'expliquent sans doute en partie par le fait qu'on a ignoré, dans le relevé en galerie, les fractures de petite taille qui, par contre, ont été prises en compte sur les carottes. Mais elles traduisent peut-être aussi la variabilité spatiale de la fracturation.

L'ensemble des données de fracturation recueillies a servi de base à des recherches portant sur la structure et la modélisation de la fracturation (Long et Billaux, 1987 ; Massoud, 1987) dont certaines applications sont évoquées à la fin de cet article.

LES PERMEABILITES DE FRACTURES

La répartition et les résultats sommaires des injections d'essai entre obturateurs effectuées dans les sondages sont indiqués dans le tableau ci-contre.

Ces résultats mettent en évidence , comme on pouvait s'y attendre, une grande hétérogénéité de la perméabilité.

La comparaison des profils de perméabilité avec ceux de densité de fracturation (fig. 7) montre qu'il n'y a pas de relation apparente entre l'une et l'autre : une forte densité de fracturation ne suffit

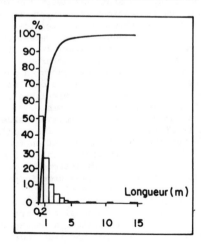

Figure 6 - Distribution des longueurs de traces de fractures mesurées en galerie.

Longueur de la chambre	Nombre d'essais	Perméabilité minimale (m/s)	Perméabilité maximale (m/s)
Totalité du forage	10	3.10^{-8}	1.10^{-6}
10 m	60	$2.10^{-10}*$	8.10^{-6}
2 m ou 2,5 m	181	$6.10^{-10}*$	5.10^{-5}

* seuil de sensibilité de la méthode

pas à créer la perméabilité "en grand" ; encore faut-il pour cela qu'un certain nombre de fractures soient ouvertes et interconnectées, le degré d'interconnexion ne dépendant pas seulement de la densité des fractures, mais aussi de leur taille.

Par contre, les accidents de quelque importance ("zones très fracturées" sur les carottes) sont clairement responsables des perméabilités élevées.

Des perméabilités de type "milieu continu équivalent" déduites des injections, appuyées par la connaissance des profils de fracturation des sondages, ont été tirées, moyennant quelques hypothèses simplificatrices, des distributions d'ouvertures hydrauliques des différentes familles de fractures (Long et Billaux, 1987).

Les caractéristiques de ces distributions sont les suivantes :

A	B	C	D	E
1	1098	369	3,4	4,1
2	436	157	2,8	2,9
3	102	26	3,1	3,6
4	185	68	2,4	2,2
5*	599	167	3,6	4,1

(* la famille 5 est une famille subhorizontale très peu représentée en galerie)

A = numéro de famille
B = nombre total de fractures
C = nombre de fractures non conductrices
D = ouverture hydraulique équivalente moyenne des fractures conductrices (10^{-5} m)
E = écart-type (10^{-5}m)

LES INDICATIONS TIREES DES MESURES DE DEBIT

Les débits stabilisés produits par les six petits sondages de reconnaissance préliminaires S1 et S6 ont servi de base à un calcul approché des perméabilités ; les valeurs trouvées sont comprises entre 10^{-7} et 10^{-8} m/s.

Les débits drainés par les trois portions du tronçon expérimental de galerie sont restés peu différents les uns des autres, le rapport du plus élevé sur le plus faible ne dépassant jamais 1,6. En outre, ils ont suivi des évolutions parfaitement parallèles pendant toute la durée des mesures. Ces évolutions sont caractérisées par une diminution des débits au cours du temps.

C'est ainsi que le débit global initial, très stable autour de 5,75 l/min (9,6.10^{-5} m^3/s) pendant une année, a diminué progressivement pour atteindre 1,35 l/min seulement après trois ans. Cette diminution s'explique par le drainage des terrains lors de la phase de creusement des sondages F1 à F10, mais aussi par des travaux miniers entrepris à proximité de la galerie expérimentale, à un niveau inférieur ; elle est à mettre en parallèle avec la diminution de charge hydraulique observée pendant la même période.

A noter que les débits mesurés en galerie semblent totalement insensibles aux variations de la pluviométrie.

La réalisation des forages a également été mise à profit pour faire des mesures de débit.

C'est ainsi que les dix forages F1 à F10, avant leur équipement, fournissaient un débit cumulé de 50 l/min (8,3.10^{-4} m^3/s).

Plus intéressantes sont les mesures faites le long de ces sondages. La figure 7 donne un exemple du profil des débits suivant un des forages (le F3). On constate que les venues d'eau y sont très localisées, observation qui est valable pour tous les forages.

Ces venues d'eau sont clairement en relation avec certaines "zones très fracturées", sièges de perméabilités importantes, identifiées par ailleurs.

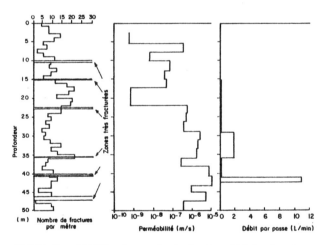

Figure 7 - Profils comparés de la densité de fracturation, de la perméabilité et du débit produit pour le sondage F3

LES RENSEIGNEMENTS FOURNIS PAR LES MESURES DE CHARGE HYDRAULIQUE

Les valeurs de charges hydrauliques mesurées dans les piézomètres S1 à S6 lors de l'étude préliminaire avaient été interprétées comme témoignant de la saturation du massif, bien que ces valeurs fussent plus faibles que celles déduites d'un premier calcul, de type milieu continu équivalent, supposant le milieu parfaitement saturé.

Les mesures réalisées à des distances de la galerie nettement plus importantes, dans les sondages F1 à F10, allaient permettre de constater que cette saturation n'était que partielle.

La figure 8 illustre les mesures effectuées dans le profil central du tronçon expérimental et leur interprétation en termes de réseau d'écoulement.

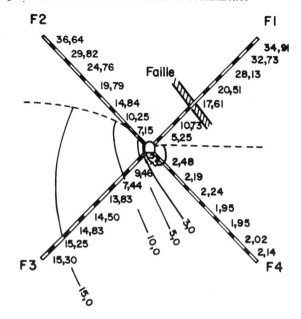

Figure 8 - Réseau des lignes équipotentielles déduit des mesures de charge en sondages. (charges comptées à partir du radier de la galerie).
La ligne en pointillés représente la limite entre la zone à la pression atmosphérique, située au-dessus, et non totalement saturée, et la zone totalement saturée, située en dessous.

Dans toutes les chambres des sondages remontants, les charges mesurées sont égales à la cote, ce qui signifie que la pression y est égale à la pression atmosphérique (prise comme zéro des pressions).

Une telle situation se rencontre en zone non saturée (aux tensions capillaires près), comme le montre l'exemple théorique de la figure 9a, mais la distribution des charges en milieu totalement saturé entre le toit d'une galerie et la surface libre peut en être également très proche (fig. 9b).

Dans le cas de Fanay, toutes les chambres de mesure contenant de l'eau, certaines contenant toutefois également de l'air, on est vraisemblablement dans la situation illustrée par la figure 9a. Ceci s'expliquerait par la désaturation du massif au-dessus de la galerie lors de la réalisation des sondages (du fait du drainage provoqué par ceux-ci), suivie d'une resaturation incomplète des fractures les plus grosses après fermeture des sondages.

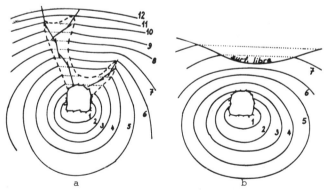

Figure 9 - Illustration théorique de deux cas possibles de présence d'une zone à pression (quasi-) égale à la pression atmosphérique au-dessus d'une galerie. (Isopièzes gradués en mètres à partir du radier. Les lignes d'égale cote, en pointillés, se confondent avec les isopièzes dans la zone non saturée).

A contrario, dans les sondages descendants, F3 et F4, les charges sont supérieures à la cote, et la pression supérieure à la pression atmosphérique, ce qui témoigne d'une saturation totale du milieu sous la galerie.

Sur les trois profils, on observe une dissymétrie marquée de l'écoulement, ce qui traduit peut être une anisotropie des perméabilités en grand ou un drainage dont l'origine n'est pas élucidée.

On remarque également une importante irrégularité dans la distribution des charges, ce qui n'est pas pour surprendre, s'agissant d'un milieu fracturé.

La prise en compte des mesures effectuées dans le sondage F0, réalisé depuis le jour, amène à proposer un schéma d'écoulement tel que celui représenté sur la figure 10.

Ce schéma suppose l'existence d'une vaste zone à pression atmosphérique, ce qui pourrait s'expliquer par un dénoyage partiel du terrain par la présence des galeries (anciennes) des niveaux 200, 240, 280 et 320 (galerie expérimentale).

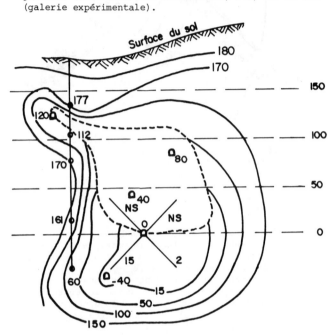

Figure 10 - Tentative d'interprétation des mesures de charge hydraulique disponibles dans le massif.

La ligne pointillée marque la limite entre la zone partiellement désaturée (à la pression atmosphérique) et la zone totalement saturée.

La charge élevée mesurée dans la chambre du forage F0 la plus proche de la surface amène à supposer que la partie supérieure des terrains peut être totalement saturée, ce qui revient à admettre que la perméabilité de ces terrains (hormis certainement la frange la plus superficielle), est plus faible que celle de la zone des galeries. Ces considérations doivent être tempérées par le fait qu'un réseau de lignes équipotentielles tel que celui proposé sur la figure 10 s'inspire de ce que l'on peut être amené à tracer dans un milieu continu. En réalité, dans un milieu fracturé, les choses sont beaucoup plus complexes et ne peuvent être représentées simplement à une telle échelle.

En ce qui concerne l'évolution des charges au cours du temps, deux types de comportements bien distincts sont mis en évidence : pour les points situés dans la zone à la pression atmosphérique, on n'observe évidemment aucune variation ; pour les autres, on observe une diminution due très probablement au drainage des terrains par le creusement d'une nouvelle galerie au niveau 360, ce creusement n'ayant toutefois pas encore eu le temps de provoquer une désaturation des terrains.

ELEMENTS DE MODELISATION

Le but recherché est de déterminer numériquement les propriétés hydrauliques d'un massif de granite d'échelle hectométrique, à partir d'observations à l'échelle métrique : caractéristiques de la géométrie de la fracturation et mesures de perméabilités locales par essais à l'eau.

Le traitement statistique des longueurs de traces observées et des directions des plans de fractures permet d'engendrer des réseaux tridimensionnels de disques interconnectés (représentation simplifiée de fractures), de géométrie statistiquement compatible avec celle du réseau réel.

Le modèle du milieu fracturé ainsi obtenu est utilisé pour simuler les mesures de perméabilité réalisées entre obturateur. Les calculs d'écoulement dans les réseaux tridimensionnels engendrés sont effectués en admettant que l'eau circule non pas dans la totalité des plans de fracture, mais dans des chenaux inclus dans ceux-ci, suivant l'hypothèse de la "chenalisation" ("channeling"). La conductivité hydraulique de ces chenaux est calée afin de reproduire les résultats des mesures in-situ de perméabilité. Par exemple, on cherche à caler une loi log-normale de conductivités hydrauliques, permettant de reconstituer la loi log-normale des débits d'injection obtenue d'après les mesures.

La phase suivante de l'étude constitue une validation du modèle. Les conductivités calées sont injectées dans une simulation tridimensionnelle d'un réseau de fractures de plus grande échelle, afin d'évaluer un tenseur anisotrope de perméabilité du milieu continu et homogène équivalent. Ce tenseur est ensuite introduit dans un calcul par éléments finis de l'écoulement établi dans un très grand domaine comprenant la galerie et les forages. Les conditions aux limites en pression proviennent des mesures piézométriques effectuées sur le site.

La validité du modèle sera prouvée si le débit de drainage mesuré dans la galerie correspond au débit de drainage calculé par la procédure décrite ci-dessus.

Cette modélisation, actuellement en cours, est pratiquement achevée. Les premiers résultats obtenus permettent une validation avec un facteur de l'ordre de deux entre les débits calculés et mesurés.

La figure 11 donne une comparaison des densités de fractures observées et simulées statistiquement, sur diagrammes de Schmidt.

La figure 12 montre un exemple de section bidimensionnelle d'un réseau tri-dimensionnel simulé.

La figure 13 donne un schéma de l'allure du tenseur anisotrope de perméabilité sur quelques simulations du réseau tridimensionnel, en coupe perpendiculaire à l'axe de la galerie.

La figure 14 donne une image de la distribution des potentiels calculés par éléments finis sur une section perpendiculaire à la galerie.

CONCLUSION

Le problème de l'"effet d'échelle" semble en milieu fracturé avoir reçu, par les mesures de terrain réalisées et la modélisation effectuée, un début de réponse conceptuellement satisfaisante. En effet, les mesures à différentes échelles s'intègrent dans une suite de modèles adaptés à leur interprétation, et fournissant des valeurs successives de paramètres de plus en plus globaux, adaptés au modèle de l'échelle supérieure, et aboutissant finalement au milieu continu équivalent.

Cette approche a été rendue possible à Fanay Augères car le milieu fracturé étudié est très fortement connecté, bien au-delà du seuil de percolation. De plus, grâce au très grand nombre de mesures réalisées (en particulier les essais d'injection d'eau en sondage), il a été possible de caractériser les propriétés du milieu statistiquement (distribution log-normale des débits d'injection par exemple), puis de rechercher la moyenne statistique du comportement hydraulique, qui est ensuite assimilée, par hypothèse d'ergodicité, à la moyenne spatiale à grande échelle du phénomène étudié.

Signalons que les essais à Fanay-Augères se sont poursuivis par des traçages, qui étant en cours d'interprétation, ne seront pas abordés dans cet article.

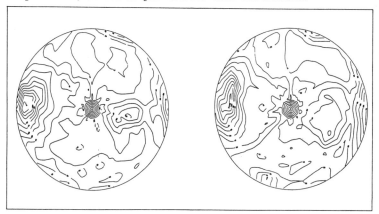

Figure 11 - Comparaison des courbes "isodensité" des diagrammes de Schmidt d'un réseau de fractures simulé dans le plan des forages (à droite) et du réseau observé (à gauche).

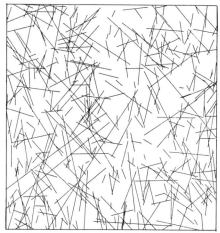

Figure 12 - Exemple de coupe d'un réseau de fractures simulé.

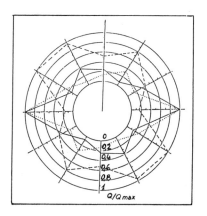

Figure 13 - Exemple de perméabilités directionnelles pour trois simulations.

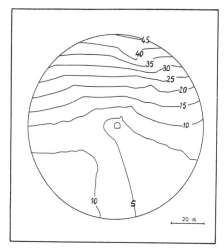

Figure 14 - Piézométrie calculée dans le milieu (courbes isopièzes graduées en mètres.

BIBLIOGRAPHIE

Barbreau A, Derlich S., Côme B., Peaudecerf P., Durand E., Marsily G. de. 1985. Experiments performed on granite in the underground research laboratory at Fanay-Augères, France. Int. Symp. on coupled processes affecting the performance of a nuclear waste repository, Lawrence Berkeley Laboratory, Sep. 18-20, 1985. A paraître dans un ouvrage édité par C.F. Tsang, Academic Press, New-York, 1987.

Barbreau A., Marsily G. de, Peaudecerf P., Durand E. 1985. The evolution of large scale permeability in deep granite. An. nuclear Soc. International meeting on nuclear waste disposal, Pasco, Wash. Sep. 24-26, 1985.

Beucher H., Marsily G. de. 1984. Approche statistique de la détermination des perméabilités d'un massif fracturé. Rapport Ecole des Mines de Paris LHM/RD/84/16.

Blanchin R., Chilès J.P. 1984. Etude statistique et géostatistique de la fracturation de la mine de Fanay-Augères. Rapport BRGM 84 RDM 074 IM.

Cacas M.C. 1986. Etude de l'effet d'échelle en milieu fissuré. Rapport Ecole des Mines de Paris. LHM/RD/86/80.

Long J.C.S., Billaux D. 1987. From field data to fracture network modelling. An example incorporating spatial structure. A paraître dans Water Resources Res.

Marsily G. de,1985. Flow and transport in fractured rocks : Connectivity and scale effect. IAH International Symposium on the Hydrogeology of low permeability rocks. Tucson, USA. Jan 7-12, 1985.

Massoud H. 1987. Modélisation de la petite fracturation par des méthodes géostatistiques. Thèse Ecole des Mines de Paris.

Rouleau A., Gale J.E. 1985. Fracture size, interconnectivity and groundwater flow : a site-specific stochastic analysis. Symposium de Montvillargenne (Ecole des Mines de Paris). Juin 1985.

Wilke S., Guyon E., Marsily G. de. 1984. Water penetration through fractured rock : test of a tridimensional percolation description. Mathematical geology, 17, 1, p. 17-27.

REMERCIEMENTS

Les auteurs tiennent à remercier la COGEMA pour l'aide qu'elle leur a apportée au cours de cette étude. Celle-ci a été réalisée avec le concours financier de la Commission des Communautés Européennes.

Permeability determinations for a discontinuous, crystalline rock mass
Déterminations de la perméabilité d'un massif rocheux cristallin discontinu
Permeabilitätsbestimmungen an klüftigen, kristallinen Gesteinskörpern

E.T.BROWN, Imperial College, London, UK
P.I.BOODT, Imperial College, London, UK

ABSTRACT: A versatile, modular borehole pressure test system was developed and used in a series of constant head injection and pressure drop tests in a granite containing three orthogonal sets of widely spaced discontinuities. Small scale tests on individual or very small numbers of discontinuities gave similar results to those obtained in laboratory tests. The results of large scale tests in test sections up to 28 m long showed that, at this scale, continuum concepts may not be used to study flows through the rock mass.

RESUME: Un système polyvalent de pressurisation des sections d'un trou de sande a été mis au point et utilisé lors d'une serie de tests dans un massif rocheux granitique contenant trois familles de discontinuités ortho-gonales largement espacées. Lors des tests, la pression d'injection ainsi que la diminution de pression sub-séquente étaient mesurées. Les tests à petite échelle sur une ou un petit nombre de discontinuités donnent des résultats simulaires à ceux obtenus au laboratoire. Les résultats des tests à grande échelle sur des sections mesurant jusqu'a 28 m de long montrent que le concept de milieu continu ne peut par être utilisé pour étudier les écoulements dans le massif rocheux.

ZUSAMMENFASSUNG: Ein vielseitig verwendbares modular aufgebautes Druckmessystem für Bohrloecher wurde entwickelt, sowie in einer Reihe von Druckverlustmessungen und von Injektionen bei konstantem Druck erprobt. Die Tests erfolgten an Granit mit 3 weitraeumigen, aufeinander Senkrecht Stehenden Kluftsystemen. Kleinmassstaebliche Versuche an einzelnen oder einer geringen Zahl von Klueften ergaben den Laborversuchen vergleichbare Ergebnisse. Die Resultate von grossmassstaeblichen Versuchen mit Messtrecken von bis zu 28 m zeigten, dass in dieser Groessenordnung Kontinuumbedingungen nicht angewendet werden koennen, um die Durchstroemung des Gesteinskoerpers zu untersuchen.

INTRODUCTION

Interest in the transport properties of discontinu-ous, crystalline rock masses has developed considera-bly in recent years, largely because of the import-ance of these properties in studies of the geologi-cal isolation of radio-active waste. This concern has generated laboratory studies of the thermo-hydro-mechanical behaviour of single discontinuities, the development of numerical models for simulating flow in single discontinuities and in networks of fract-ures, and field studies of the in situ hydraulic characteristics of individual discontinuities and of discontinuous rock masses. Studies of the latter type are expensive and fraught with practical diffi-culty. As a result, few sets of reliable data are available in the open literature; this paper seeks to add to their number.

The work described was carried out as part of a research programme on site assessment methodology for radioactive waste repository design, coordinated by the Building Research Establishment (BRE) for the United Kingdom Department of the Environment (Hudson, 1983). It involved pressure testing of a hard, joint-ed, crystalline rock mass in which fluid flow was dominated by natural fractures. A fundamental question exists as to the extent to which continuum theories may be applied to such discontinua. Can flow through these rock masses be described adequate-ly by an equivalent continuum permeability tensor? If so, what representative elementary volume (Bear, 1972) is required for continuum theory to apply?

THE TEST SITE

The tests were carried out at an experimental site in the Carnmenellis granite near Troon, Cornwall, in south-west England. This underground facility which

has been used for a range of other purposes as part of the BRE research programme (e.g. Cooling, Tunbridge and Hudson, 1984), consists of a series of tunnels roughly at right angles to each other(Fig.1)

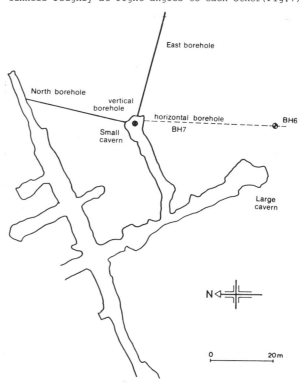

Fig. 1. Arrangement of excavations and boreholes at the test site.

entered from the base of a disused quarry. The tests were carried out in three orthogonal boreholes drilled from a small chamber 7 m long, 6 m wide and 3 m high at the location shown in Fig. 1. Cover to ground surface was 34 m. Monitoring of water levels in boreholes showed the initial groundwater level to be approximately 10 m below ground surface.

Two major sets of sub-vertical discontinuities with strike concentrations at 103° and 194° are apparent in the quarry and underground excavations. They have spacings in the order of one metre and may have clean, closely matching walls or may be lined with variable thicknesses of secondary minerals. A set of sub-horizontal fractures have spacings of less than one metre immediately below ground surface. This spacing increases to in the order of 5 - 10 m at depths of greater than 100 m. Rare mineralised veins containing sulphides and cassiterite, and a number of porphyry dykes, strike at 050° parallel with a regional trend. Discontinuities with dips of 45 - 70° and variable dip directions are much less frequent than the steeply dipping discontinuities (Bourke et al, 1981).

Discontinuity surveys using horizontal scanlines were carried out at the test site in this and other parts of the overall BRE programme. This, together with the results of surveys carried out in the quarry, allowed determination of the central trends of the three dominant discontinuity sets present. Three 100 mm diameter and 30 m long test boreholes called the north, east and vertical boreholes, were drilled from the experimental chamber orthogonal to these three discontinuity sets (Fig. 1). For practical reasons, the north and east holes were drilled rising at 5° above the horizontal.

The locations and orientations of all discontinuities intersecting the three test boreholes were then established using the borehole impression packer probe (B.I.P.P.) developed by Hinds (1974) and described further by Brown, Harper and Hinds (1979). Some 33 discontinuities were recorded in the sub-horizontal east borehole, 16 in the sub-horizontal north borehole, and seven in the vertical borehole which intersects the more widely spaced set of sub-horizontal discontinuities.

TEST METHOD SELECTION

The test methods had to be suitable for the physical constraints imposed by the site and for the predicted hydraulic characteristics of the rock mass. It was desirable to be able to perform both small scale tests on single discontinuities and large scale tests influencing large numbers of discontinuities. Because the hydraulic characteristics of low permeability, discontinuous, crystalline rock masses may be expected to be anisotropic, it was concluded that a three dimensional method of testing involving multiple boreholes would be required. Attention was restricted to the hydraulic conductivities or permeabilities of the discontinuities and of the rock mass; their storativities were not considered.

The test methods available for applications such as this have been reviewed by Wilson et al (1979) among others. The anticipated permeabilities and water flow rates were too low to suggest that pumping tests would be practicable. Tracer tests were also found to be impracticable for the available test site. As a general principle, steady state methods were to be preferred to transient methods such as slug or pulse tests because of the greater uncertainties involved in the interpretation of the results of transient tests. In an attempt to eliminate the considerable influence of non-linear flow on the results of borehole tests on rock discontinuities (Elsworth and Doe, 1986), heads and flow velocities were to be restricted to those corresponding to laminar flow. Consideration of these various factors led to the choice of the constant head injection test (CHIT) as the primary test type. Because it may be carried out conveniently at the conclusion of a constant head injection test, the pressure drop test (PDT) was also used to provide additional data for comparison with the CHIT results.

The requirements for three-dimensional permeability testing of discontinuous rock masses have been discussed by Snow (1966), Louis and Maini (1970) and, more recently, by Boodt, Maini and Brown (1982). It is generally argued that test holes should be oriented perpendicular to the mean planes of the main discontinuity sets present so that the hydraulic characteristics of each set may be determined independently. In a three hole pressure test, the borehole axes should coincide with the principal permeability directions (assuming that this concept is applicable). If, as in the present case, the three main discontinuity sets are approximately orthogonal, each test hole can be oriented parallel to the intersection of two discontinuity sets and almost orthogonal to the third set.

TEST EQUIPMENT AND PROCEDURES

Downhole equipment

No commercially available downhole equipment was found which was suitable for use in the proposed tests. Consequently, new equipment was designed and constructed. Criteria used in designing the equipment included the requirement that it must be suitable for use in a range of tests and that it must be capable of being set up quickly in the possible test configurations from restricted underground sites. The system must be suitable for use in the test configuration proposed by Sharp (1970) in which flow from two separate sections, one on either side of the test section, is used to ensure that the flow from the test section is radial; this concept is illustrated in Fig. 2. It was also desirable to

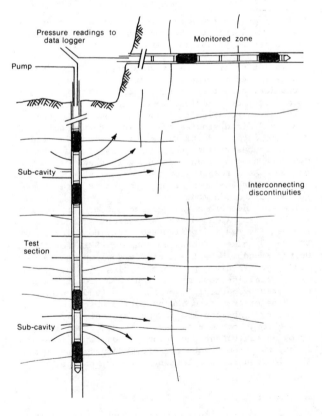

Fig. 2. Typical test arrangement

be able to measure the fluid pressure and flow rate
in the test section down the hole. In the event, no
no flowmeter could be identified that had the required resolution and was small enough to be installed in
the test units, and so flows were monitored at the
borehole collars.

A modular design was chosen because of its flexi-
bility and ease of use. The system consisted of
three types of unit - monitoring or distribution
units, packer units and spacer units. The monitoring or distribution units contained perforations
through which water was distributed. They were
designed so that the nominal minimum test section
length was 0.5 m. Piezoresistive pressure transducers capable of measuring absolute pressures in the
range 0 to 1 MPa (± 0.0002 MPa) were mounted in each
unit.

The pneumatic packer units were 2.0 m long, producing a sealed length of 1.25 m when inflated in a
100 mm diameter borehole. To prevent the packer
material (ductube) from tearing because of excessive
shortening, the ends of the inflatable sections were
mounted on sliders which were free to move axially as
the packers were inflated. Spacer units in lengths
of 0.5 m and 1.5 m were used to increase the test
section length and to carry water and air to the test
section and the packers.

Each unit had an individual electrical lead with
waterproof connectors which could be used to connect
downhole transducers with their power supplies and
the data logger. Couplings were designed which
could connect two separate water supplies (one for
the test section and one for the "linearizing" sections), an air supply for the packers and the electrical leads without letting water into or out of the
system.

Surface equipment

A schematic diagram of the "surface" equipment is
shown in Fig. 3. The electrically driven centri-
fugal pump delivered 22.25 l/min against a head of
1.0 MPa. The flow rate was measured by one of three
flowmeters - a magnetic flowmeter accurate to 1.0%
for flows in the range 0.07 to 7.0 l/min, a turbine
flowmeter with a range of 0.02 to 1.3 l/min and an
accuracy of ± 1.0% of the full scale deflection, and
a rotameter for flows from 0.001 to 0.+25 l/min
accurate to 0.0003 l/min.

The pressures were controlled by two pressure regu-
lators, one for the test section and one for the
linearizing sections on either side of the test section. Provision was also made for measuring water
pressures at the surface. Preliminary tests showed
that the steady state flow rates achieved in the constant head injection tests rarely exceeded 0.5 l/min
and were often considerably less. Calculations
given in detail by Boodt and Brown (1985) showed that
the head losses in the supply pipework at this flow
rate were less than 1.5% of the lowest excess pressure used in the tests. This is within the range of
accuracy of the pressure transducers indicating that
it was not unreasonable to measure the pressures at
the "surface" (i.e. in the underground test chamber).

Test procedures

In the more numerous small scale tests, individual
discontinuities intersecting the test hole were isolated between the two pneumatic packers. In a few
cases, the discontinuities occurred in closely
spaced groups of two or more which prevented them
from being tested individually. In the large scale
tests carried out in the east borehole and in the
vertical borehole, multiple discontinuities intersected the test section. Initially, a single packer
and a distribution unit were positioned 5 m from the

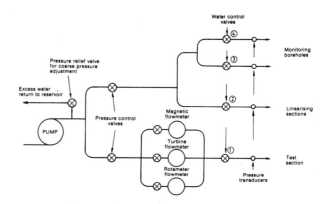

Fig. 3   Schematic diagram of surface equipment.

bottom of the hole and a sequence of pressure tests
carried out. The packer and distribution unit were
then moved out by another 5 m and the test sequence
repeated. This procedure was repeated until the
packer reached the top of the borehole.

When the test equipment had been positioned in the
hole, the water control valves (Fig. 3) were opened
to flood the test hole. Once the hole was flooded,
the pump was turned off and the packers were inflated with the water control valves being kept open to
prevent build up of pressure in the test section.
The monitoring holes were flooded using a similar
procedure, but with a single packer located at the
top of the hole.

To carry out a constant head injection test, the
desired test pressure was set using the pressure control valve connected to the test section via valve 1.
In order to obtain low but measurable flows in the
laminar range and to minimise the influence of discontinuity deformation, excess test pressures were
limited to the range 0.15 - 0.50 MPa following preliminary testing. With valve 1 open, the flow was
directed through the magnetic flowmeter and then, at
lower flow rates, through the turbine and rotameter
flowmeters. When the flow rate became steady, the
test was complete. To carry out a pressure drop
test, valve 1 was closed and the decline in borehole
pressure monitored.

Throughout the test programme, CHIT's and PDT's
were carried out in a particular sequence. For a
given test configuration, three CHIT's were carried
out at increasing pressures with a PDT being carried
out after the third CHIT. This was followed by two
or more CHIT's with a PDT at the final pressure. When
the conclusion of the test sequence coincided with
the end of a day's work, the final PDT was left to
run overnight.

TEST RESULTS

Small-scale tests

Boodt and Brown (1985) give the complete set of recorded flow rate-time curves for the CHIT's and the
$\ln(P/P_o)$ - time curves for the PDT's. The curves
for the tests at a depth of 5.25 m in the east borehole are shown in Figs. 4 and 5. For the CHIT's, a
graph of steady state flow rate ($\dot{Q}$) against excess
pressure (P_o) was plotted and the gradient determined. This value was used to calculate the equivalent

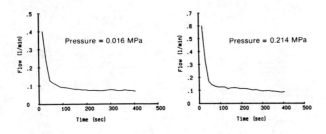

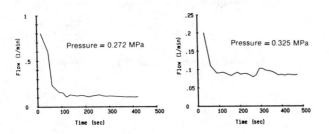

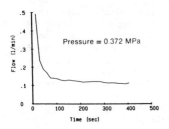

Fig. 4. Flow rate-time curves for constant head injection tests at a depth of 5.25 m, east borehole.

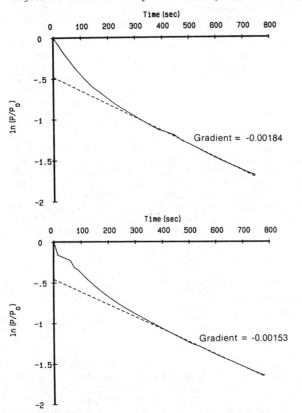

Fig. 5. Ln(P/P$_o$) - time curves for pressure drop tests at a depth of 5.25 m, east borehole. (P$_o$ is the initial excess pressure in the borehole at time t$_o$; P is the pressure at time t$_1$).

parallel plate aperture of the discontinuity, e, from

$$e = \left[\frac{6\nu Q}{\pi g n H_o} \ln\left(\frac{R}{r_o}\right)\right]^{\frac{1}{3}} \qquad (1)$$

where ν = kinematic viscosity of water,
   R = radius of influence of the test arbitrarily assumed to be 30 m,
   r_o = radius of the borehole (51 mm),
   n = number of discontinuities intersecting the test section,
and H_o = steady state excess pressure head in the test section.

The hydraulic conductivity of the discontinuity was then calculated as

$$K_j = \frac{e^2 g}{12\nu} \qquad (2)$$

For the PDT's, the equivalent parallel plate apertures were calculated from the equation derived by Maini (1971),

$$e = \left[\frac{6\nu r_o^2}{gn} \ln\left(\frac{R}{r_o}\right)\frac{\ln(H_o/H_1)}{(t_1-t_o)}\right]^{\frac{1}{3}} \qquad (3)$$

where H_o = excess pressure head at the centre of the test section at the initiation of the PDT at time t_o,
and H_1 = excess pressure head at the centre of the test section at time t_1.

Values of $[\ln(H_o/H_1)]/(t_1/t_o)$ for each test were obtained from the straight line sections of semi-log curves such as that shown in Fig. 5. The discontinuity hydraulic conductivities were again calculated from equation (2).

Equations (1)-(3) were derived assuming that
(i) the test section is vertical and is intersected by horizontal discontinuities;
(ii) the discontinuities may be represented by constant apertures between smooth, parallel plates;
(iii) radial, laminar flow occurs within each discontinuity.

Clearly, these assumptions do not always apply in field pressure tests of the type described here. The influence of variations in borehole and discontinuity orientations have been analysed by Rissler(1978) and by Bourke et al (1981). Both studies showed that for a vertical borehole intersecting inclined discontinuities, flows deviated from those in the comparable horizontal discontinuity case by more than 5% only when the angle of dip exceeded 45°. Rissler (1978) also obtained numerical solutions for cases in which the borehole is perpendicular to an inclined discontinuity. He found that for discontinuity dips of up to 45°, flows deviated negligibly from those for horizontal discontinuities; for discontinuity dips of 85°, flows were up to 20% less than those for horizontal discontinuities.

These results indicated that no orientation corrections were required for many of the small-scale tests. Corrections could be required for the tests involving sub-vertical discontinuities intersected by the east and north boreholes. However, because no means existed of determining such corrections other than by numerical simulation in each case, no orientation corrections were applied in calculating the values of equivalent parallel plate aperture, e, and discontinuity hydraulic conductivity, K$_j$, listed in Tables 1, 2 and 3. The ranges of these two parameters shown in Tables 1 - 3 are similar to the ranges of 20 - 1000 x 10^{6} m and 0.5 - 10 x 10^{-3} m/s measured in laboratory tests on single natural discontinuities in the same rock (Boodt and Brown, 1985; Elliott et al, 1985). The laboratory tests showed both parameters to be highly pressure dependent.

Table 1.   Small scale test results, North borehole.

Test depth (m)	No. of discontinuities	Aperture, e(x10^{-6}m) CHIT	PDT 1	PDT 2	Conductivity, K$_j$(x10^{-6}m/s) CHIT	PDT 1	PDT 2	Discontinuity orientatons (dip dirn./dip)
3.5	2	10.1	79.0	82.3	64.5	3920	4260	092/21, 095/22
8.5	2	15.2	81.4	87.0	144	4170	4750	099/20, 101/15
10.1	1	30.1	116	110 77.7*	568	8410	7610 3790*	180/85
22.8	1	42.9	55.0	72.0 61.0*	1160	1900	3260 2340*	208/36
26.4	1	24.0	-	54.2	361	-	1840	124/54

*Long term test

Table 2.   Small scale test results, East borehole.

Test depth (m)	No. of discontinuities	Aperture, e(x10^{-6}m) CHIT	PDT 1	PDT 2	Conductivity, K$_j$(x10^{-6}m/s) CHIT	PDT 1	PDT 2	Discontinuity orientations (dip dirn./dip)
2.40	2	-	70.1	68.5	-	3090	2950	185/76, 194/84
6.15	5	20.7	165	153	270	17100	14700	114/89, 111/89 226/60, 149/80 152/78
7.40	1	-	43.9	50.6	-	1210	1610	175/69
14.10	2	26.9	52.6	60.1	454	1740	2270	177/34, 149/73
16.00	2	6.72	25.8	50.8	28.4	419	1620	207/76, 170/90
19.50	1	16.1	73.2	-	170	3370	-	240/35
20.40	1	13.8	119	125.5	119	8860	9900	261/82
22.30	1	10.1	74.6	88.2	64.0	3500	4890	257/62
23.70	1	8.6	41.8	51.2 30.1*	46.7	1100	1670 570*	198/76

* Long term test

Table 3.   Small scale test results, vertical borehole.

Test depth (m)	No. of discontinuities	Aperture, e(x10^{-6}m) CHIT	PDT 1	PDT 2	Conductivity, K$_j$(x10^{-6}m/s) CHIT	PDT 1	PDT 2	Discontinuity orientations (dip dirn./dip)
4.75	1	74.7	108	136	3510	7300	11500	262/82
5.25	2	35.5	221	216	500	19400	18500	351/82, 284/88
6.00	3	38.5	225	205	768	26300	21800	284/88, 333/89 322/72
7.85	6	-	-	69.3	-	-	3020	286/90, 298/75 300/50, 299/68 284/74, 291/72
10.20	2	-	-	125	-	-	9760	320/88, 258/88
11.50	6	5.95	39.4	38.7	22.2	975	944	271/88, 263/59 267/65, 274/76 273/88, 259/90
12.20	3	8.88	158	161	49.5	15600	16300	273/88, 259/90 253/78
14.35	2	8.55	43.0	46.0	46.0	1160	1330	293/48, 282/73
18.15	2	40.1	111	115	1010	7680	8360	292/62, 259/70
19.30	1	-	275	265	-	47500	44200	330/82
25.20	2	78.8	172	121 24.5*	3900	18700	9180 379*	288/77, 299/75
26.80	2	-	57.7	54.6	-	2070	1870	291/85, 356/31
28.40	1	-	74.5	95.7 80.4*	-	3490	5750 4060*	253/71

* Long term test

Large scale tests

Figure 6 shows the flow rate-time curves recorded in the large scale CHIT's carried out over the 2 - 30m test interval in the east subhorizontal borehole. For each CHIT, the slope of a constant flow rate - excess pressure head plot was determined and used to calculate the equivalent porous medium coefficient of permeability from the equation

$$K_e = \frac{Q}{H_o} \cdot \frac{\ln(R/r_o)}{2\pi l} \qquad (4)$$

where l is the test section length.

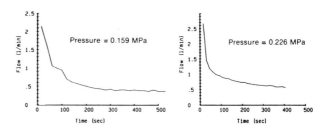

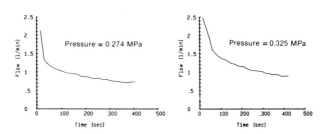

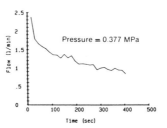

Fig. 6.   Flow rate - time curves for large scale constant head injection tests in the 2 - 30 m test section, east borehole.

Figure 7 shows the $\ln(P/P_o)$ - time curves recorded in the two large scale PDT's carried out over the 2 - 30 m test interval in the east borehole.   The gradients of the straight line portions of curves such as these were used to calculate equivalent porous medium permeabilities from equation (2) using values of e given by the equation derived by Maini (1971),

$$e = \left[\frac{6\nu r_o^2}{gl}\ln\left(\frac{R}{r_o}\right)\frac{\ln(H_o/H_1)}{(t_1-t_o)}\right]^{\frac{1}{3}} \qquad (5)$$

Tables 4 and 5 list the values of K$_e$ calculated from the results of the large scale tests carried out in the east and vertical boreholes respectively.

27

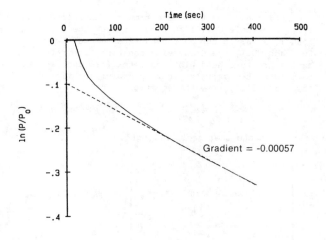

Gradient = -0.00057

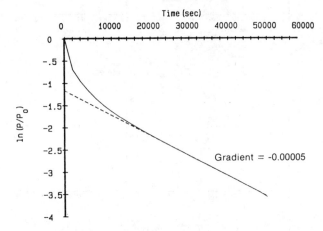

Gradient = -0.00005

Fig. 7. Ln(P/P_O) - time curves for large scale pressure drop tests in the 2 - 30 m test east borehole.

Table 4. Equivalent porous medium coefficient of permeability (K_e) for large scale tests in the East borehole.

Test interval (m)	K_e (x 10^{-9} m/s)		
	CHIT	PDT 1	PDT 2
25 - 30	47.8	2930	1500
20 - 30	26.3	1020	708
15 - 30	22.4	79.6	528
10 - 30	17.2	-	-
5 - 30	74.4	152	106
2 - 30	20.9	170	13.5

Table 5. Equivalent porous medium coefficients of permeability (K_e) for large scale tests in the vertical borehole.

Test interval (m)	K_e (x10^{-9} m/s)		
	CHIT	PDT 1	PDT 2
25 - 30	2.42	59.5	-
20 - 30	1.19	719	43.8
15 - 30	0.21	776	573
10 - 30	0.28	374	273
5 - 30	6.41	173	1150
2 - 30	0.22	386	-
5 - 30 (repeat)	0.37	370	322

Figures 6 and 7 illustrate an aspect of the testing procedure which is important in interpreting and reconciling the test results. Figure 6 shows that the testing times of 6 - 9 minutes used in CHIT's were sometimes insufficient to enable steady state conditions to be reached. Although more marked in this case, a similar feature was exhibited by a few of the small scale CHIT's. The test times were certainly insufficient to permit interpretations of discontinuity geometry to be made using the approach proposed by Doe and Osnes (1985).

Figure 7 shows the results of two PDT's carried out on different time scales. The gradients of the linear portions of the two ln(P/P_O) - time curves differ by an order of magnitude. As shown in Table 4, the equivalent porous medium permeability calculated from the second or longer term PDT for the 2 - 30 m test interval corresponds more closely with that calculated from the CHIT results. Tables 1,2 and 3 show that the small scale PDT's gave values of equivalent parallel plate aperture and hydraulic conductivity that were always greater than those given by the CHIT's. This discrepancy could result from the short test times used in the PDT's.

The calculations of the values of the equivalent porous medium coefficient of permeability listed in Tables 4 and 5 are based on a number of assumptions which may not always be fulfilled. Figures 8 and 9 show average values of this parameter for the large scale CHIT's carried out in the east and vertical boreholes plotted against test section length. For each test section, the steady state values of $Q/P_O/l$ for each test pressure were calculated and the five values averaged. Exceptions were made for the 25 - 30 m and 20 - 30 m test sections in the vertical borehole because only a single test value was obtained in each case. The results of the repeated rather than the initial series of tests for the 5 - 30 m section in the vertical borehole are plotted in Fig. 9.

Figures 8 and 9 show that the 30 m borehole length used in these tests was inadequate to define the representative elemental volume (REV) for this rock mass. The REV is the minimum volume of rock at which an increase in volume will not produce a significant change in the rock mass property under investigation. The relative variation in the $Q/P_O/l$ values for the vertical borehole is more marked than that for the east borehole in which there is a suggestion that there may be little variation in $Q/P_O/l$ for $l > 20$ m. This could be interpreted as a reflection of the fact that the east borehole intersected a total of 33 discontinuities

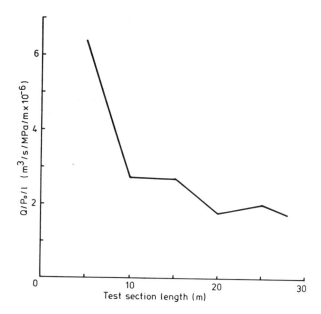

Fig. 8. Average flow rate per unit excess pressure per unit length against test section length for large scale constant head injection tests in the east borehole.

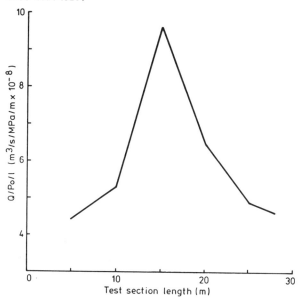

Fig. 9. Average flow rate per unit excess pressure per unit length against test section length for large scale constant head injection tests in the vertical borehole.

whereas the vertical borehole intersected only seven. However, Long and Witherspoon (1985) have shown that the problem cannot be analysed in such simple terms; the fracture length and degree of fracture network interconnection also exert important influences on the transport properties of a rock mass and on whether or not it may be considered to act like a porous medium.

Assuming that the flows in the individual discontinuities intersected by a given borehole were mutually independent, Boodt and Brown (1985) simulated the large scale tests in all three test holes from the results of the relevant small scale tests. These calculations confirmed the essential conclusion drawn from the results of the large scale tests, namely that the 30 m long test holes were too short to define the REV for this rock mass. However, the equivalent porous medium permeabilities per unit length of borehole determined in the simulations from

the results of the small scale CHIT's were approximately an order of magnitude greater than those recorded in the large scale CHIT's. Similarly, the equivalent porous medium permeabilities per unit length of borehole measured in the small scale CHIT's were about an order of magnitude greater than those measured in the large scale tests. These results suggest that the assumption of independent flow in all discontinuities may be invalid.

By assuming further that the discontinuity intersections recorded in each of the boreholes would be repeated cyclically if the borehole lengths were increased, Boodt and Brown (1985) extended their simulations of large scale pressure tests to section lengths of up to 120 m. The results suggest that in crystalline rock masses containing widely spaced discontinuities, the test section lengths required to define REV's may become impractically large.

Long (1983) concluded that in order for it to be valid to represent flow through a fractured rock mass in terms of an equivalent continuum permeability, (i) there must be no significant change in the equivalent continuum permeability in response to a small increase or decrease in the rock volume considered (i.e. the REV must be reached); and (ii) a single equivalent symmetric continuum permeability tensor must exist which is invariant with changes in the direction of the applied hydraulic gradient.

Despite their deficiencies, the results reported herein indicate that neither of these conditions are met in the Carnmenellis granite at the Troon test site for test lengths of up to 30 m. The significant differences in the results obtained in the large scale tests in the vertical and horizontal boreholes demonstrate that condition (ii) is not satisfied. Accordingly, it is concluded that, on the scale of the tests, the continuum concept of a permeability tensor may not be used in calculating flows through this rock mass. For flow calculations in the near field of an underground excavation, for example, numerical simulations using discontinuum concepts (e.g. Long, 1983; Samaniego and Priest, 1984) are to be preferred.

CONCLUSIONS

1. A versatile, modular borehole pressure test system was developed for use from restricted underground locations and used in a series of small and large scale tests in a jointed granite.
2. The ranges of values of equivalent parallel plate aperture and discontinuity hydraulic conductivity measured in the small scale tests on single or very small numbers of discontinuities were similar to those measured in laboratory tests.
3. Pressure drop tests consistently gave higher values of discontinuity aperture and hydraulic conductivity than constant head injection tests. This could be because many of the pressure drop tests were carried out over insufficiently long periods of time.
4. The magnitudes of, and fluctuations in, flow rate per unit excess pressure per unit test section length generally decreased as the test section length was increased in the large scale tests. However, the maximum borehole length of 30 m was too short to permit the representative elementary volume to be determined for the rock mass concerned.
5. In rock masses such as that studied, discontinuum rather than continuum methods of analysis are required to calculate flows in the near fields of underground excavations.

ACKNOWLEDGEMENTS

The work described herein was funded by a research grant from the Science and Engineering Research Council and a research contract from the Department of the Environment. The authors wish to acknowledge

the support of the Radioactive Waste Professional
Division of the DOE and especially that of the BRE's
Nominated Officers, Dr. J.A. Hudson and Miss C.M.
Cooling, and to thank Mr. J.W. Dennis of Imperial
College who manufactured the test equipment and help-
ed carry out the field tests.

REFERENCES

Bear, J. 1972. Dynamics of fluids in porous media
New York: American Elsevier.
Boodt, P.I. & E.T. Brown (1985). Some rock mass
assessment procedures for discontinuous crystalline
rock. Report No. DOE/RW/85.133, Dept. of the
Environment.
Boodt, P.I., T. Maini & E.T. Brown (1982). Three
dimensional water pressure testing of fractured
rock. Proc. 1st Int. Mine Water Congr., Budapest,
Sect. A, p. 165 - 183.
Bourke, P.J., A. Bromley, J. Rae & K. Sincock (1981).
A multi-packer technique for investigating resis-
tance to flow through fractured rock. Proc. NEA
Workshop on Siting Radioactive Waste Repositories
in Geological Formations, Paris, p. 173 - 189.
Brown, E.T., T.R. Harper & D.V. Hinds (1979). Dis-
continuity measurement using the borehole impress-
ion probe - a case study. Proc. 4th Congr., Int.
Soc. Rock Mech. 2 : 57 - 62. Rotterdam: Balkema.
Cooling, C.M., L.W. Tunbridge & J.A. Hudson (1984).
Some studies of rock mass structure and in situ
stress. In E.T. Brown & J.A. Hudson (eds),
Design and performance of underground excavations,
p. 199 - 206. London: British Geotechnical
Society.
Doe, T.W. & J.D. Osnes (1985). Interpretation of
fracture geometry from well test. In O. Stephansson (ed.), Fundamentals of rock joints, p. 281-292
Lulea: Centek Publishers.
Elliott, G.M., E.T. Brown, P.I. Boodt & J.A. Hudson
(1985). Hydromechanical behaviour of joints in the
Carnmenellis granite, S.W. England. In O. Stephansson (ed.), Fundamentals of rock joints, p. 249 -
258. Lulea: Centek Publishers.
Elsworth, D. & T.W. Doe (1986). Application of non-
linear flow laws in determining rock fissure
geometry from single borehole pumping tests. Int.
J. Rock Mech. Min. Sci. & Geomech. Abstr. 23: 245-254.
Hinds, D.V. (1974). A method of taking an impress-
ion of a borehole wall. Rock Mech. Report N28,
Imperial College, London.
Hudson, J.A. (1983). UK rock mechanics research for
radioactive waste disposal. Proc. 5th Congr.,
Int. Soc. Rock Mech. 2 : E161 - 165. Rotterdam:
Balkema.
Long, J.C.S. (1983). Investigation of equivalent
porous medium permeability in networks of continuous fractures. Ph.D. thesis, Univ. of California,
Berkeley.
Long, J. & P. Witherspoon (1985). The relationship
of the degree of interconnection to permeability in
fracture networks. J.Geophys. Res. 90: 3087 -
3098.
Louis, C. & T. Maini (1970). Determination of in
situ hydraulic parameters in jointed rock. Proc.
2nd Congr., Int. Soc. Rock Mech., Paper 1 - 32.
Maini, Y.N.T. (1971). In situ hydraulic parameters
in jointed rock: their measurements and interpretation. Ph.D. thesis, Univ. of London.
Rissler, P. (1978). Determination of water permeability of jointed rock. Publns of the Institute
for Foundation Engineering, Soil Mechanics, Rock
Mechanics & Waterways Construction, RWTH Aaachen,
Vol. 5.
Samaniego, J.A. & S.D. Priest (1984). The prediction of water flows through discontinuity, networks
into underground excavations. In E.T. Brown &
J.A. Hudson (eds.), Design and performance of
underground excavations, p. 157 - 164. London:
British Geotechnical Society.

Sharp, J.C. (1970). Fluid flow through fissured
media. PhD thesis, Univ. of London.
Snow, D.T. (1966). Three-hole pressure test for
anisotropic foundation permeability. Rock Mech.
Engrg Geol. 4: 298 - 316.
Wilson, C.R., T.W. Doe, J.C.S. Long & P.A. Witherspoon (1979). Permeability characterization of
nearly impermeable rock masses for nuclear waste
repository siting. Proc. Workshop on Low-Flow,
Low-Permeability Measurements in Largely Impermeable Rocks, p. 13 - 27. Paris: OECD.

Estimation of minimum mechanical property values and sample volume dependence for brittle rock

Estimation de valeurs minimales de propriétés mécaniques et effet du volume de l'éprouvette pour des roches fragiles

Die Bestimmung von Kleinstwerten mechanischer Eigenschaften und der Einfluss von Probengrösse im spröden Fels

B.M.BUTCHER, Sandia National Laboratories Albuquerque, N.Mex, USA
R.H.PRICE, Sandia National Laboratories Albuquerque, N.Mex., USA

ABSTRACT: Minimum possible strength and Young's modulus values, and the influence of sample size on the unconfined strength, were determined for geological units of a volcanic rock with widely varying functional porosity. Least values of 36.6 MPa for unconfined strength and 12.1 GPa for Young's modulus, for 0.2 functional porosity, were determined by extreme value statistical analyses. Agreement of the unconfined strength data, σ, with the Weibull relationship $(\sigma - \sigma_{min})V^{0.58} = b$ was observed when the least value of the measured strength was corrected to the common functional porosity for each data set: V is the sample volume, and b is a constant. In applying this equation, our prediction of the variation of unconfined strength with sample size was found to be slightly less than the strengths of different diameter samples, and was therefore conservative.

RÉSUMÉ: On a établi les valeurs minimales de résistance possible et de module de Young ainsi que l'influence de la taille de l'échantillon sur la résistance à la compression sans contrainte latérale pour des unités géologiques d'une roche volcanique dont la porosité fonctionnelle varie considérablement. Suite à une analyse statistique de veleurs limites, on a obtenu des valeurs minimales de 36,6 MPa pour la résistance à compression sans contrainte latérale et de 12,1 GPa pour le module de Young pour une porosité fonctionnelle de 0,2. Une correspondance des données de résistance à la compression sans contrainte latérale, σ, à la relation de Weibull $(\sigma - \sigma_{min})V^{0.58} = b$, fut observée lorsque la valeur minimale de la résistance mesurée fut corrigée à la porosité fonctionnelle commune pour chaque ensemble de données: V est le volume de l'échantillon et b est une constante. En appliquant cette équation, notre prédiction de la variation de la résistance à la compression sans contrainte latérale en fonction de la taille de l'échantillon s'avéra être légèrement inférieure aux résistances d'échantillons de diamètre différent et était donc conservatrice.

ZUSAMMENFASSUNG: Die kleinstmöglichen Druckfestigkeits und Elastizitötsmodul-Werte, sowie der Einfluß der Probengröße auf die Druckfestigkeit bei unbehinderter Seitenausdehnung wurden an geologischen Proben eines vulkanischen Felsgesteins mit stark variierender funktioneller Porosität bestimmt. Eine statistische Grenswert-Analyse ergab Mindestwerte von 36,6 MPa für die genannte Drukfestigkeit und 12,1 GPa für den Elastiztätsmodul, bei einer funktionellen Porosität von 0,2. Es wurde eine Übereinstimmung zwischen den Daten für die Druckfestigkeit bei unbehinderter Seitenausdehnung, σ, mit der Weibull Gleichung $(\sigma - \sigma_{min})V^{0.58} = b$ beobachtet, als der kleinste gemessene Wert der Druckfestigkeit jeder Daten-gruppe in die allgemeine funktionelle Porosität abgeändert wurde: V ist das Volumen der Probe, b ist eine Konstante. Bei Anwendung dieser Gleichung zeigte sich, daß die Abhängigkeit der Druckfestigkeit bei unbehinderter Seitenausdehnung von der Größe dieser Proben laut unserer Voraussage etwas geringer ausfiel als die Druckfestigkeit von Proben mit unterschiedlichem Durchmesser, unsere Voraussage war somit konservativ bemessen.

INTRODUCTION

The decrease in laboratory-defined strength of intact rock observed when sample volumes are increased remains a very challenging problem in rock mechanics. Understanding the sample size effect is important for the design of underground structures because the designer must extrapolate strength information from the laboratory to the much greater volumes of intact rock surrounding such structures. Without such information it is difficult to predict when cracks will start to grow and eventually cause the structure to be unstable. (The rock is considered to have failed when a crack starts to grow in it.) When joints are present, our analysis applies to the intact rock between them.

This paper summarizes the results of a study to determine statistical estimates of the influence of sample volume on mechanical properties. Results are based on the study described by Yegulalp and Wane (1968), derived from the statistical theory of extremes (Gumbel 1958). The data for our analysis of unconfined compressive strength and Young's modulus were obtained in connection with evaluation of possible underground sites for the storage of nuclear waste (Nimick et al 1985; Olsson 1981; Price et al 1982a-d, 1984). The functional porosity (porosity plus montmorillonite volume fraction) of the sample sets ranged from 0.1 to 0.4 volume fraction.

Yegulalp and Wane's paper (1968) is based on the fact that when information about mechanical properties is derived, either in the laboratory or *in situ*, the usual outcome is data scatter in the test results. The standard methodology for examining this variation involves construction of a statistical frequency distribution for the results and its application to find the variability in terms of standard deviations and other well established statistical measures. Often times a lower bound to strength data, as determined by the mean +/- a multiple of the standard deviation, is used in stress analysis to provide an estimate of structural performance. Another observation by Yegulalp and Wane is that most testing programs involve ''a pathetically small sample of the geologic universe of interest and not enough testing is carried out to determine the exact form of the distribution of test results''. Usually a normal distribution is assumed to estimate probabilities.

Extreme value statistics deal with the analysis of the extremes of a distribution and the forecasting of further extremes. For example, suppose that a number of concrete samples are taken daily during operation of a concrete batching plant, and, for quality control, several samples from each batch are tested after the appropriate curing time to determine their unconfined strength. Each batch would have a data set associated with it from a certain number of samples. In addition, each data set would have a lowest strength value. Extreme value analysis of the collection of lowest values from all the data sets then could be used to obtain the probability of obtaining a future strength less than or equal to a specified lowest value of strength. Also, the number of tests that would be required, on the average, to find any other prescribed strength could be estimated. An extension of these concepts to sample volume dependence is suggested by posing the question of how large the volume of rock would have to be, on the average, instead of how many samples of a given size, would be required to encounter this value of strength.

The analysis in our paper is a three step process. Extreme value statistics are first applied to estimate lowest possible bounds for unconfined strength and Young's modulus. Instead of statistically treating the data as a single set, the distribution of the lower bounds of a number of separate sets of data is defined. This distribution is then used to predict a minimum value for the property under consideration. (For some stress calculations a maximum possible value of Young's modulus is also needed, and can be estimated from extreme value statistics). The minimum values then are compared with results from an analysis of all the data as a single set, assuming a normal distribution. Finally, Weibull statistical theory is applied to obtain a prediction for sample-size dependence and compared with unconfined compression data for different diameter samples.

Another consequence of the theory is that the strength corresponding to a given probability of failure, for example one chance in a million, can be estimated. This type of information is relevant in stress analyses related to probabilistic studies of potential adverse consequences of the storage of nuclear waste.

DATA DESCRIPTION

The data sets (Nimick et al 1986; Olsson 1981; Price et al 1982a-d, 1984) were taken from test results for various geological units under examination for an underground nuclear waste repository in volcanic rock. The geological units vary in functional porosity from 0.1 to 0.4. This variation, at first, appears not to satisfy the condition for statistical analysis that all samples represent the same material (and have the same functional porosity). Price and Bauer (1985) have found, however, that results for different functional porosities can

be related, a procedure that is possible because of the dominant influence of functional porosity on mechanical properties. They find that the best fit empirical relationship between unconfined compressive strength, C_0 (MPa), and functional porosity ϕ is:

$$C_0 = 4.04\phi^{-1.85}, \qquad (1)$$

and that Young's modulus E (GPa) depends on functional porosity according to:

$$E = 85.5\exp(-6.96\phi). \qquad (2)$$

Equations 1 and 2 were used to adjust the data in each set to a common porosity basis, $\phi = 0.2$, and the minimum value for each set was determined. For example, given the unconfined compressive strength C_{01} at functional porosity ϕ_1, the strength C_{02} for an assumed ϕ_2 can be estimated from:

$$C_{02} = C_{01}\left(\frac{\phi_2}{\phi_1}\right)^{-1.85} \qquad (3)$$

In applying this adjustment, each data set was considered independently. Comparison on a common basis then is possible. Later in the report, when the application of Weibull statistics to the data is explored, the data points in each set will be corrected to the average functional porosity typical of its particular data set. While the number of data points in each set (typically 10 to 20) is probably insufficient for a precise statistical analysis, a meaningful trend exists.

EXTREME VALUE STATISTICS

In application of extreme value statistics to the mechanical data, the asymptotic theory, the second part of the theory of extremes (Gumbel 1958), will be used. This part differs from the exact theory in that it requires no knowledge of the initial distribution and is still valid if a few neighboring observations are dependent. Three possible distributions are obtained (Gumbel 1958) for the asymptotic probability $\pi(y)$ that the smallest value is equal to or larger than y: the ExponentiaL, Cauchy, and Limited distributions. The Limited distribution for least values is referred to as a Weibull distribution and has the form:

$$\pi_3(y) = \exp\left\{-\left(\frac{y-\epsilon}{\theta-\epsilon}\right)^k\right\}, \qquad (4)$$

with ϵ, θ, and k constants such that $\epsilon < \theta$, $y > \epsilon$ and $k > 0$.

These distributions are plotted by taking natural logarithms of each side of the equation twice and defining the reduced variable z:

$$z = -\ln(-\ln(\pi(y))). \qquad (5)$$

The Exponential distribution plots as a straight line in y-z coordinates, the Cauchy distribution is concave downward, and the Limited distribution is concave upward, as shown in Figure 1. The quantity $(\theta - \epsilon)$ in the limited distribution is a scaling factor, and ϵ in $(y - \epsilon)$ moves the curve down in regard to the y coordinate, biasing it without major change in shape. The data will determine which distribution is appropriate.

DATA ANALYSIS

Values for minimum unconfined compression strength and Young's modulus from each data set, adjusted to $\phi = 0.2$, were ranked in order of decreasing value, in order to see which

32

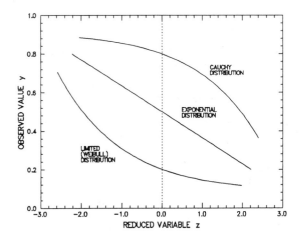

Figure 1. Permissable extreme value probability distributions.

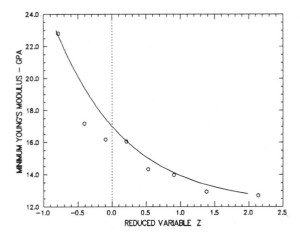

Figure 2b. Young's modulus data can be also be represented by a Weibull distribution function.

asymptotic distribution best represents the data (Figure 2a and 2b). The minimums were not quite equivalent because the number of data points varied from set to set. Next, the values corresponding to y, the mechanical property values, were plotted against z, where z is computed from:

$$z = -\ln(-\ln(\pi(y))) = -\ln(-\ln(\frac{j}{N+1})), \qquad (6)$$

j is the rank and N is the number of values in the lowest value set. Figures 2a and 2b show that both the strengths and Young's moduli define curves that are concave upward (cf Figure 1), and therefore are represented by Equation (4). Approximate lowest possible values of ϵ are 36.6 MPa for the unconfined strength ($k = 1.66$ and $\theta = 54.1$ MPa) and 12.1 GPa for Young's modulus ($k = 1.02$ and $\theta = 17.0$ GPa), and are interpreted as the estimated lowest values ever to be encountered in an infinite number of tests. The constants were obtained by adjusting their values until the theoretically predicted curves, the solid curves in Figures 2a and 2b, approximated the data.

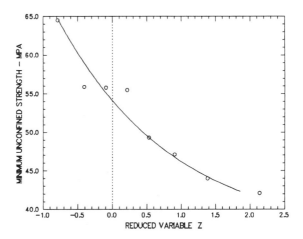

Figure 2a. The unconfined strength data are best represented by a Weibull distribution function.

The following examples show how extreme value data might be used (Yegulalp and Wane 1968). First, we might be interested in the probability of observing an unconfined strength of 40 MPa or greater in a test. According to Equation (4), a probability of 0.936 would be expected and therefore the chances of observing a lesser strength would be 0.064, or 64 times out of 1000. Second, an estimate of how many samples, on the average, would have to be tested to encounter an unconfined strength of 40 MPa or less might be of interest. This prediction is made by computing the value of the return period, $T(y) = \frac{1}{1 - \pi_3(y)} \cong 16$. To obtain this value of compressive strength, at least 16 samples, on the average, would have to be tested. This quantity might also be interpreted, on much more tenuous grounds, as the volume a sample would have to have, on the average, to exhibit this magnitude of unconfined strength. The test specimens for all of the test data were 25.4 mm in diameter and 50.8 mm long, with an average volume of 2.57×10^4 mm^3, or approximately 2.6×10^{-5} m^3. Thus a sample volume 16 times the original sample volume, or 4.2×10^{-4} m^3, would have an average unconfined strength of 40 MPa. This estimate is, however, dependent on the assumption that strength is controlled by weakest links in the form of flaws. Even if the larger volume did contain such a flaw, it is not clear that instability would arise at exactly 40 MPa, because of factors such as the orientation and location of the flaw in relation to the ends of the test specimen.

If the data are all considered to be part of a large data set and assumed to be part of a normally distributed data universe, then mean ($\bar{x}$) and standard deviation (s) rules can be used to estimate probability. For the 101 Young's moduli, a mean of 21.4 GPa and a standard deviation of 5.7 GPa result. The 103 ultimate strength values yield a mean, $\bar{C}_0$, of 80.5 MPa and a standard deviation, s, of 25.8 MPa. As an example, we might decide that a failure probability of 0.05 is suitable for design. The value of strength identified with this probability, according to normal distribution statistics, is therefore:

$$C_0 = \bar{C}_0 - 1.645s = 38.1$$

Similarly, a probability of failure of 0.005 corresponds to a strength of 14.0 MPa. The examples show that definition of material

constants for stress analysis in this manner is
very difficult because a probability for failure
must be specified. In fact, the minimum value
of strength for a given application can be made
as small as desired simply by reducing the
probability of failure. Further, there is no
real basis for selecting how small the failure
probability should be for a given application.
On the other hand, extreme value statistics
provide guidance for design in the sense that an
asymtotic value of the strength is defined, in
this case 36.6 MPa, below which failure is very
improbable. Design analysis is considerably
simplified if the statistical theory of extremes
can be shown to apply for a given rock, because
only one material constant value, rather than a
range of values, need be specified.

WEIBULL STATISTICS

The mechanical properties of intact brittle rock
are controlled in most cases by the distribution
of microcracks, porosity, or other types of
flaws. The Weibull distribution of tensile
strength, based on this type of weakest link
criterion, postulates that the probability of
failure at stress σ is dependent on the sample
volume V:

$$P(\sigma) = 1 - \exp\left[\left(\frac{\sigma - \sigma_{min}}{\sigma_0}\right)^n V\right], \qquad (7)$$

where σ_{min}, σ_0 and n are constants. Assuming
that the stress distributions in the samples are
uniform, the failure stress - volume
relationship for different size samples is given
by:

$$(\sigma - \sigma_{min})V^{\frac{1}{n}} = b, \qquad (9)$$

where b is a constant.

In general, models developed for tensile
failure are based on the weakest link concept,
and are not generally applicable to compressive
loading (Costin 1985). The special case of
uniaxial compression is a possible exception.
To see if Weibull theory has any relationship to
unconfined strength, the test results are ranked
in order, this time from the lowest failure
stress to the highest. The order, j, of the N
ranked values of failure stress is then used to
compute the probability:

$$P(\sigma_j) = \frac{j}{(N+1)}, \qquad (10)$$

with the reduced variable z defined as

$$z = \ln(-\ln(1 - P(\sigma_j))). \qquad (11)$$

A plot of $\ln(\sigma_j)$ versus z should show a linear
dependency with slope -1/n according to Weibull
Theory.

In applying Weibull theory to the unconfined
compression data, the proceedure of correcting
all strength values to a common functional
porosity of $\phi = 0.2$ appears inappropriate, in the
sense of the less extrapolation the better.
Instead, the values of each set were simply
corrected to the average functional porosity of
the particular data set they represented, to
place them on an equivalent basis for
examination. The least possible value of the
strength determined from least value analysis,
at $\phi = 0.2$, was also extrapolated to the value
expected for the average functional porosity of
each individual data set. The alternate case
where σ_{min} was taken as zero for all porosities
was also examined.

In general, the results in Figures 3-10 show
that each data set approximated a linear
relationship between the logarithm of the

unconfined strength and the reduced variable z.
An exception is the plot for Topopah Spring data
set #1 (Figure 7), which appears to show no
correlation. On the other hand, one of the best
fits with the Weibull theory was obtained for
Topopah Spring data set #2, and this data should
be nearly identical in features to set #1. A
similar comparison is apparent for Bullfrog data
sets #1 and #2. No systematic trend is evident
to discredit this correlation, even though the
number of data points differ from set to set,
and minor discrepancies are apparent at the end
points of the Weibull plots (Figures 3 to 10).
We conclude, therefore, that Weibull statistics
appear to apply to all of these data sets in a
similar fashion. The average value of the
exponent, n, is approximately 1.7 for the
nonzero values of σ_{min} given in the figures, so
that the sample volume dependence is:

$$(\sigma - \sigma_{min})V^{0.58} = b. \qquad (12)$$

For example, using Topopah Spring data set #2
with $\sigma_{min} = 82.4$ MPa , an average value of the
unconfined strength C_0 of 182 MPa, and a sample
volume of 2.6×10^{-5} m^3, the average values of the
unconfined strength, C_0, varies with sample
volume according to:

$$(C_0 - 82.4)V^{0.58} = 0.218. \qquad (13)$$

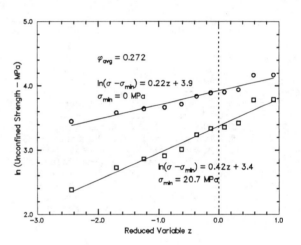

Figure 3. Weibull analyses of Bullfrog #1 data
(Olsson 1981).

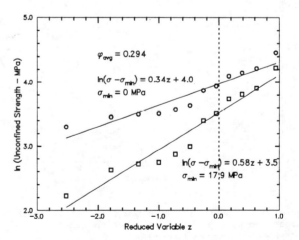

Figure 4. Weibull analyses of Bullfrog #2 data
(Price et al 1982a).

Another way of showing this dependence is to predict that doubling the sample volume would reduce the average value of the unconfined strength C_0 by 18%.,

$$C_0 = 82.4 + (182 - 82.4)(1/2)^{0.58} = 149 MPa.$$

The sample volume change defined by Weibull theory is different from the sample volume change suggested by least value theory because the Weibull volume equation is not defined for the lowest possible value of strength ($\sigma = \sigma_{min}$). Nevertheless, both descriptions are physically consistant because they predict that very large volume changes are associated with relatively minor differences in strength values near the lower limit of strength: as the stress difference term in Equation 12 becomes small, the volume term becomes very large, so that their product remains constant. In other words, the dependence of volume on stress in the vicinity of the lowest strength threshhold is very strong, as observed in the earlier section on least value analysis. The practical interpretation of this observation is that the probability of failure of a large volume of rock is very likely when the stress in an engineering structure slightly exceeds the lowest value threshold. Therefore, a single minimum value of permissible stress, from extreme value analysis, should be acceptable for stress analysis.

COMPARISON WITH EXPERIMENT

The data for testing the sample size dependence of unconfined strength (Price 1986) was acquired independently of the data for the statistical analysis. Rock from an outcropping of Topopah Spring tuff was quarried for samples with 5 sizes ranging from 0.025 to 0.229 m in diameter.

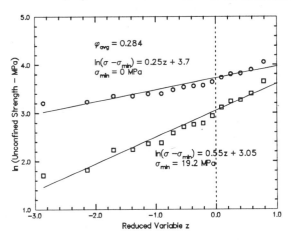

Figure 5. Weibull analyses of Calaco Hills data (Price and Jones 1982c).

The functional porosity of these samples was approximately 0.12.

The experimentally determined average values of the unconfined strength from each different diameter test series are compared in Figure 11 with curves calculated from Equation 12. Theoretical curves for two functional porosities, 0.12 and 0.14 show that the predictions are very sensitive to small changes in functional porosity within this range. Figure 11 also shows that the observed strengths are slightly greater than predictions of the Weibull equation. Our method for estimating volume dependence is therefore conservative, in the sense that the observed average values are not any lower than what we would estimate, i.e. parameters such as factor of safety values

computed from the experimental data, would actually be greater than values from these statistics. The use of minimum strength values in Weibull theory appears to be a useful procedure, therefore, with the caution that data has been very limited and a number of critical assumptions have been made.

ACKNOWLDGEMENTS
The authors are indepted to Dan Shelton for his assistance in regard to the statistical methods used in this paper.

This work was performed at Sandia National Laboratories, supported by the U. S. Department of Energy (DOE) under Contract # DE-AC04-76-DP00789.

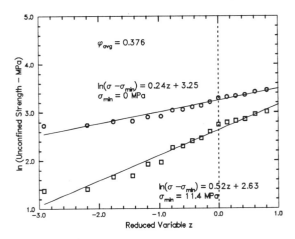

Figure 6. Weibull analyses of Tram data (Price and Nimick 1982b).

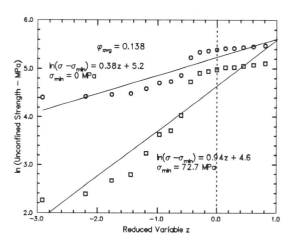

Figure 7. Weibull analyses of Topopah Spring #1 data (Price et al 1984).

REFERENCES

Costin, L. S. 1985. Time-Dependent Deformation and Failure. in Rock Fracture Mechanics, B. K. Atkinson (ed.). London: Academic Press.

Gumbel, E. J. 1958. Statistics of Extremes. New York: Columbia University Press.

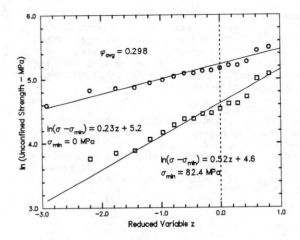

Figure 8. Weibull analyses of Topopah Spring #2 data (Nimick et al 1985).

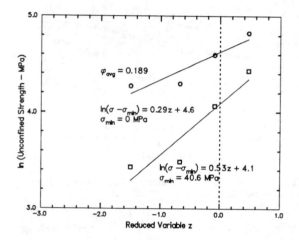

Figure 9. Weibull analyses of Topopah Spring #3 data (Price et al 1982d).

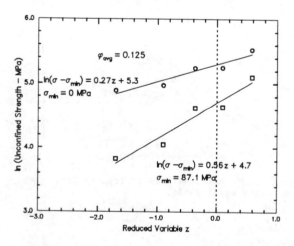

Figure 10. Weibull analyses of Topopah Spring #4 data (Price et al 1985).

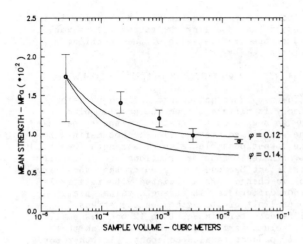

Figure 11. Our predicted dependence of the unconfined strength of tuff on sample volume is conservative because it gives a lower value of strength than observed (Price 1986).

Nimick, F. B., R. H. Price, R. G. Van Buskirk, and J.R. Goodell 1985. Uniaxial and Triaxial Compression Test Series on Topopah Spring Tuff from USW G-4, Yucca Mountain, Nevada. Sandia National Laboratories Report SAND85-1101. Albuquerque: Sandia National Laboratories.

Olsson, W. A. July 1 1981. Unpublished data, Sandia National Laboratories, Albuquerque, NM.

Price, R. H., A. K. Jones, and K. G. Nimick 1982a. Uniaxial Compression Test Series on Bullfrog Tuff. Sandia National Laboratories Report SAND82-0481. Albuquerque: Sandia National Laboratories.

Price, R. H., and K. G. Nimick 1982b. Uniaxial Compression Test Series on Tram Tuff. Sandia National Laboratories Report SAND82-1055. Albuquerque: Sandia National Laboratories.

Price, R. H., and A. K. Jones 1982c. Uniaxial and Triaxial Compression Test Series on Calico Hills Tuff. Sandia National Laboratories Report SAND82-1314. Albuquerque: Sandia National Laboratories.

Price, R. H., K. G. Nimick, and J. A. Zirzow 1982d. Uniaxial and Triaxial Compression Test Series on Topopah Spring Tuff, Sandia National Laboratories Report SAND82-1723. Albuquerque: Sandia National Laboratories.

Price, R. H., S. J. Spence, and A. K. Jones 1984. Uniaxial Compression Test Series on Topopah Spring Tuff from USW GU-3, Yucca Mountain, Southern Nevada. Sandia National Laboratories Report SAND83-1646. Albuquerque: Sandia National Laboratories.

Price, R. H., and S. J. Bauer S. J. 1985. Analysis of the Elastic and Strength Properties of Yucca Mountain Tuff, Nevada. in Proc. of the 26th U.S. Symp. on Rock Mech., E. Ashworth (ed.), p. 89-96. Rotterdam: Balkema.

Price, R. H. 1986. Effects of Sample Size on the Mechanical Behavior of Topopah Spring Tuff. Sandia National Laboratories Report SAND 85-0709. Albuquerque: Sandia National Laboratories.

Yegulalp, T. M., and M. T. Wane 1968. Application of Extreme Value Statistics to Test Data. Transactions of the Society of Mining Engineers, AIME 241: 372-376.

Geotomography for rock mass characterization and prediction of seepage problems for two main road tunnels under the city of Oslo

Utilisation de la géotomographie pour la caractérisation du rocher et la prédiction de problèmes d'écoulement pour deux tunnels routiers majeurs sous la ville d'Oslo

Seismische Tomographie zur Felscharakterisierung und Vorhersage von Sickerproblemen für zwei Autobahntunnel unter der Stadt von Oslo

TORE LASSE BY, Norwegian Geotechnical Institute, Taasen, Oslo

ABSTRACT: Seismic tomography is a rather new method for preinvestigations in rock and soil. The method was used to detect faults and major fracture zones at a tunnel project leading the E-18/E-6 highways under the city of Oslo. These preinvestigations were carried out in 8 sections by means of 55 m long boreholes from the surface.

RESUME: La tomographie séismique a été utilisée pour la détection de failles et zones majeures de fracture dans un projet de tunnel menant les autoroutes E-18 et E-6 saus la ville d'Oslo. Ces avant-reconnaissances ont été faites sur 8 sections au moyen de 55 trous de forage menés de la surface.

ZUSAMMENFASSUNG: Seismische Tomographie ist eine Voruntersuchungsmethode zu Bestimmung der Qualität des Baugrundes. Bei der Planung der 6-Spurigen Stadt-Autobahn Tunneln E6/E18 durch den sehr wechselhaften Baugrund des Oslo'er Zentrum's wurde, um Defekte und Bruchzonen im Fels zu lokalizieren, seismische Tomographie benutzt. Die Grundlagendaten wurden durch Aufnahme von 8 seismischen Profielen mit bis zu 55 Metern tiefen Bohrlöchern ermittelt.

1. INTRODUCTION

The traffic situation in the center of Oslo is approaching untenable conditions. Endless queuing represents great costs for trade, industry and the society in general. Air pollution from exhaust threatens to be a serious health problem for people with daily business down town. Several concepts have been put foreward during the last years to give Oslo a quit new main traffic system. All of them intend to remove the highways E-18/E-6 from the heart of Oslo by means of different tunnelsystems. Political authorities have chosen the so called "Rockline" concept in hard competition with other alternatives.

1.1 The Rockline concept

The E-18/E-6 Rockline is planned as two 2300 m long three-lane highway tunnels each 110 m^2, separated by a 20 m pillar. Only 1500 m of the Rockline are actually going to be built in rock. 400 m are established as concrete culverts in soil and 400 m as sinktunnel on the seafloor, where crossing the harbour basin. The total costs of the rock part are estimated to be 110 mill US $ in 1986.

To avoid serious fault zones and deep grooves crossing the trasé, it will be necesarry to go down to 45-50 m below the surface. This bottom level is reached under the square in front of the City Hall of Oslo. The geology in this area is extreamly mixed up, Løset (1986). In the fault zone, however, there are clayey alterated alun shist. Figure 1 gives the length section of the tunnel axis.

Towards east, the tunnels will mainly be driven in Precambrian gneisses. The western parts will go in Cambro-Silurian sedimentary rocks. The boundary between these two rock-formations forms a marked gouge crossing the tunnel alignment at its lower point. This is a serious north-south permian fault where the western part is lowered compared to the eastern side.

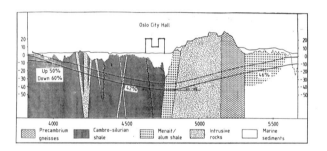

Figure 1. Length section of the tunnel axis. The deep gourge between the main east and west rock formation coerces the tunnels down to 50 m below surface.

1.2 Seepage control and groundwater lowering

To avoid groundwater lowering caused by waterleakage into the tunnels followed by ground subsidence and damage in built-up areas, comprehensive efforts will be put forewards.

Water infiltration from the surface will act as a temporary reinforcement preceding extensive preinjection and concrete lining of the tunnels. As description of the crosshole method and results from seismic tomography is the main object in this paper, the prementioned methods are not further discussed.

In the area for the tunnels passing the main fault zone beneath the City Hall square, the ground will be stabilized by ground freezing from the tunnels, Berggren (1986). In this conection it is seriously important to know the exact location and extension of the zone. Further, the depth and position of the tunnel axes are given by detailed examination of overburden, the rock surface and the rock mass quality in the ground freezing area. For estimation of seepage problems during the freezing operation, waterbearing zones are detected by means of seismic tomography.

2. CROSSHOLE SEISMICS AND SEISMIC TOMOGRAPHY

The crosshole method including seismic tomography is a rather new approach in the field of in-situ ground investigations, Pihl (1985), Cosma (1984). By means of acoustic scanning of the actual area, detailed information enhancement is achieved.

2.1 Description of the method

Stress wave propagation in general depends upon the material properties of the medium in which the waves are transmitted. A simple example is the music from a record player. When the sound propagates from one room to another, what you finally hear will probably only be the lower frequencies, the bass. All the high frequencies are attenuated because of the properties of the partition between your ear and the sound source.

In the sence of a seismic site investigation, we are taking benefit of this and other material properties causing a substantial change in several acoustic parameters. The main parameters in high-resolution seismic measurements are travel times, amplitudes and frequency content of both P-and S-waves.

What we by a common name denote as "crosshole seismic measurements", are in general a seismic transmitter/receiver system consisting of one or several combinations of wave paths based on instruments located in boreholes, tunnels and/or at the surface.

In the Rockline project both transmitter and receivers were located in boreholes. Nominal horizontal spacing between transmitter and receiver-boreholes were 50 m.

2.2 Device description and instrument specifications

Several types of signal sources as well as signal receivers are avaiable. By stating demands to measurement progress, resolution, accuracy, reliability and project costs, one has to choose a device combination that fits these requirements best for the individual projects.

NGI has through several years good experience with seismic detonators as signal sources in combination with wide-range waterborne-sound transducers for making sound measurements over a large frequency range. For this purpose, the Brüel & Kjær hydrophone 8101 has proved to be an excellent choice.

The B&K 8101 hydrophone covers the frequency range 1 Hz to 125 kHz with a receiving sensitivity of - 184 dB re 1V/μPa or 630 μV/Pa. With the 1 gram of explosive seismic source, this combination treats the range between transmitters and receivers up to a 100 m spacing in rock. Maximum borehole depth is in principle unlimited. However maximum operating static pressure for B&K 8101 is 400 m ocean depth.

The presupposition for using hydrophones for signal detection is of cause that there is water in the boreholes to achieve a good acoustic coupling between transducer and rock. Other solutions based on geophones or accelerometers with electrical or hydraulic borehole clamping are available for dry or upwards holes.

Figure 2 shows a principle sketch of the crosshole seismic performance at the Rockline project. From one signal source, acoustic signals were transmitted to receivers in three other boreholes and then

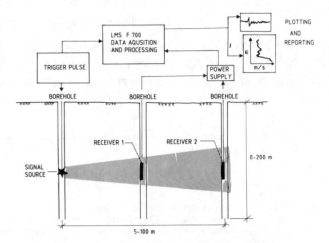

Figure 2. Principle sketch of the cross hole seismic layout for the Rockline. From one borehole acoustic signals initiated by seismic detonators were transmitted to receivers in other boreholes. Data acquisition and processing were executed by LMS F 700 Fourier Systems.

transferred to a data acquisition system at the surface.

Data were recorded at a digital storage oscilloscope with 15 bit dynamic range. As an option, where more than 4 channels were recorded simultaneously, NGI's powerful LMS F700 Fourier Systems took over data acquisition and further analyses.

The LMS-system can be extended to 1024 channels. The maximum effective global sampling frequency is 320 kHz based on a 12 bit ADC.

2.3 New concepts in signalexitation

The seismic detonator gives rise to a broad-band high-energy pulse well fitted for most crosshole seismic investigations, By (1986).

Like the hydrophones, however, measurements can only proceed on the assumption that there is a good acoustic coupling between the source and th rock. This can only be obtained through a borehole liquid, locally applied or as a total borehole filling.

NGI is looking into new concepts based on sources with controlled frequencies, phase and energy. This exitator should be applicable both for dry and wet holes independent of the inclination. Range and resolution will vary depending upon the searched features in the rock mass.

To gain high resolution to detect minor features like cavities, boreholes, internal waterways etc. the frequency of the seismic pulse should be adjusted to the actual size of these irregularities.

Explosives as well as mechanical based transmitters are rather limited what concerns frequency-bands exceeding 5-10 kHz. So far, borehole mounted piezosources seem to represent an important improvement in signal exitation.

2.4 Detailed ground assessment by means of seismic tomography

Traditional crosshole seismics give a clear indication of major changes in geological structures. However, uncertain location of features as well as problems including appropriate presentation of the results is a serious drag. Therefore computerized geophysical tomography is applied to produce a clear image of internal ground structures, based on the sound velocity distribution in the rock mass. NGI has adapted the Vibrovision PC program from

Geoseismo OY which enables the user to accomplish a detailed tomographic scanning of the interborehole area. Further details of the exploitation of seismic tomography to this project, including additional explanations of the in-situ lay-out are given in the next chapters.

3. FIELD MEASUREMENTS

The two tunnel axes are crossing under the City Hall Square at an altitude of ÷ 45 m, which also almost corresponds to the vertical depth below surface.

To investigate thickness of overburden, main faults and other quality reducing zones and features within the area, seismic tomography was performed in 5 boreholes constructing 8 profiles or sections, figure 3.

The signal source was moved in 5 meters steps in one borehole while hydrophones were kept in fixed positions in 3 other holes. Then the hydrophones were raised 5 meters, and the signal exitation procedure was repeated. The complete raypath design, leading to a detailed scanning of the interborehole area, is illustrated in figure 4.

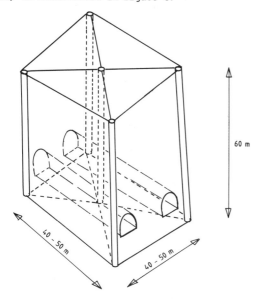

Figure 3. Cross hole seismics were performed in 8 sections by means of 5 vertical boreholes from the surface to a depth of 60 m. The tunnel floor will be at level - 46.8 m below surface. Detected weakness zones could be followed between the seismic sections in the 3-D model.

The upper 10-20 meters of the boreholes were lined through the overburden. For this purpose, a continous polyethylen tube were installed. These plastic pipes were run half a meter into the rock and the connection was grouted to ensure gravel and rock fragments from falling into the boreholes.

The polyethylen tubes were entered through the steel casings forming the holes in the overburden part. The steel casings were redrawn and the soil then collapsed around the tubes. As all activies took place under the groundwater level, this method secured a good coupling between the transmission medium and the source/receiver device.

3.1 Data acquisition

The set of data from each section or profile consists of a velocity matrix based on continous borehole coordinates and exact travel times. Borehole deviation measurements and very precise registration of pulse exiation time as well as arrival times, leads to a theoretical P-wave velocity matrix within

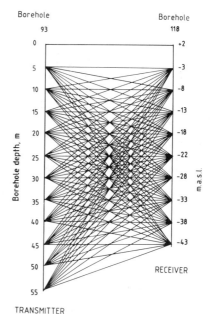

Figure 4. Idealized interborehole scanning. Sections from the Rockline. Raypaths are represented by straight lines, which according to figure 5 is obviously wrong.

a 2-3% margin. This is under the presupposition of straight rays travelling the shortest distance between source and receiver. We know, however, that within layered media, several arrivals representing direct and reflected as well as refracted P- and S-waves occure and mix up the signaltrace. A singel first arrival P-wave peak can be hopeless to pick out. By means of advanced ray-tracing routines it is, however, possible to increase the accuracy of the velocity matrix representing the actual ground structures. Figure 5 illustrates the alternative pathways.

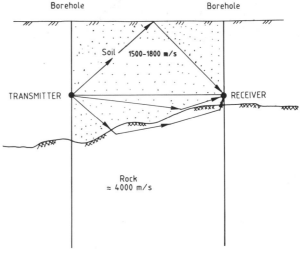

Figure 5. In bedded ground the raybending will be extensive. The wavelets are reflected from bedding planes as well as from the surface and discontinuities intersecting the area.

4. RESULTS

After the data acquisition process including travel time registrations and borehole deviation measurements had taken place, the final velocity matrix for each profil was treated by the "Vibrovision" PC program leading to the tomographic plots of the interborehole area, By (1986 b).

Figure 6 represents one of these plots. Dark sha-
dings represent low P-wave velocities. The overburden is clearly defined by the black hatching in
the upper part.

Faults or major fracture zones within the more
competent rock mass can be closer examined by evaluating signal attentuation and the global frequency
spectra of stress waves propagating through sound
rock compared to the actual formation. By means of
such additional information, a more detailed characterization of intermittent features can be given.

By introducing the parameter "rise velocity" composed from first arrival rise times and first arri-
vel amplitudes, the infilling material of quality
reducing zones can be given a discription. Infilling materials composed from water and clayey materials leads to lower rise velocities. More rigid
materials as quartz, calcite etc. result in high
velocities, Sancar (1986).

What concerns the Rockline project, at this
stage no further interpretation of the data was
achieved. The seismic tomography had given unique
and reliable information about the rock surface, dip
and strike of the main fault and of fracture zones
intersecting the investigated sections.

The extent of the ground freezing operation could
now be planed in detail as water leakage zones of
probable importance had been located.

```
Nmb. of rays       :  126
Nmb. of iterations :   50
RMS error          :   .754708
Mean velocity      :  3446.
Cell size X,Y      :  2.00  2.00
```

2500 2813 3125 3438 3750 4063 4375 4688 5000

P-WAVE VELOCITY

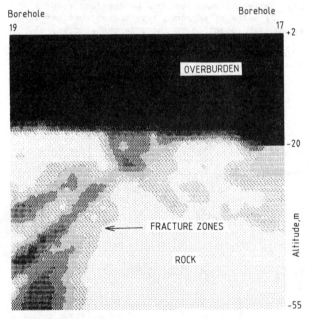

Figure 6. Tomographic plot of section between
boreholes 17 and 19. P-wave velocity limits
(i.e. the gray scale limits) are defined by
assigning values of minimum and maximum velocities, here set to 2500 ^{m}/s and 5000 ^{m}/s. In
this specific section, the rock face is rather
horizontal at an altitude of ÷ 20 meter. The
more competent rock beneath is intersected by
several fracture zones. A more descriptive
assessment of these zones can be given by interpreting the attenuation and frequency-information.
Strike and dip of the fracture zones in a 3-D picture are found by means of a combination of several tomographic plots or sections.

ACKNOWLEDGEMENT

This version of the project scope and the obtained
results is given by permission of the Road Authorities of the City of Oslo. The paper preparation is
partly supported by the Royal Norwegian Council for
Scientific and Industrial research. (NTNF).

REFERENCES

Berggren, A.-L., Rohde, J.K.G. and Aas, G. 1986:
Ground freezing as rock support - Design study
for a highway tunnel in Oslo (In Norwegian).
Fjellsprengningsteknikk - Bergmekanikk - Geotek-
nikk, pp 38.1-38.14. Trondheim: Tapir.

By, T.L. 1986:
Signalexitation for crosshole seismics by means of
seismic detonators (In Norwegian). Internal
Report 54103-7, Norwegian Geotechnical Institute,
28 pp.

By, T.L. 1986 b:
Crosshole seismics for investigation of fault zone
under the Oslo City Hall Square (In Norwegian)
Client Report 85614-10, Norwegian Geotechnical
Institute, 72 pp.

Cosma, C. 1984:
Crosshole seismic method. The Finnish Association
of Mining and Metallurgical Engineers, Serie A
nr. 73.

Løset, F. 1986:
Engineering geological description of fault zone
under the Oslo City Hall Square (In Norwegian)
Client Report 86036-2, Norwegian Geotechnical
Institute.

Pihl, J., Gustavsson, M., Ivansson, S. and Morén, P.
1985:
Tomographic analysis of crosshole seismic measurements. Proc. of the Symposium on In situ experiments in granite associated with the disposal of
radioactive waste, pp 146-159, International
Stripa Project, Stockholm.

Sancar, B. 1986:
A geological description of the Storglomvass dam
site area, interpretation of karstification of the
area, and detection of possible endokarst forms by
means of cross-hole seismics. M.Sc. thesis,
Dep. of geology, Univ. of Oslo, 165 pp.

Finite-element simulation of groundwater flow and heat and radionuclide transport in a plutonic rock mass

Simulation par la méthode des éléments finis de l'écoulement de l'eau souterraine et du transfert de la chaleur et des radionucléides dans un massif rocheux plutonique
Simulation mittels Elementenmethode von Grundwasserbewegung und Wärme- und Radionukleidentransport in einem plutonischen Fels

T.CHAN, Atomic Energy of Canada Ltd, Whiteshell Nuclear Research Establishment, Pinawa, Manitoba
N.W.SCHEIER, Atomic Energy of Canada Ltd, Whiteshell Nuclear Research Establishment, Pinawa, Manitoba

ABSTRACT: A three-dimensional finite-element code, MOTIF, has been developed to model the coupled processes of groundwater flow, heat transfer and solute transport in the vicinity of a hypothetical nuclear fuel waste vault in plutonic rock. A conceptual model of such a system has been constructed using geological, geophysical and hydrogeological data from the Whiteshell Research Area in southeastern Manitoba, Canada. Results of two-dimensional simulations of the system using MOTIF indicate the importance of the distance of the radionuclide source from a major fracture zone and the possible influence of fracture zone grouting on radionuclide travel times.

RÉSUMÉ: On a mis au point un programme de calcul à éléments finis en trois dimensions, MOTIF, pour modéliser les processus associés de l'écoulement d'eau souterraine, de transfert de chaleur et de migration de solutés dans le voisinage d'une enceinte hypothétique de stockage permanent de déchets de combustible nucléaire située dans la roche plutonique. On a construit un modèle conceptuel de ce système à l'aide de données géologiques, géophysiques et hydrogéologiques provenant de la Zone de Recherches de Whiteshell dans l'est du Manitoba, au Canada. Les résultats des simulations en deux dimensions du système à l'aide de MOTIF indiquent l'importance de la distance de la source de radionuclides à partir d'une zone de fractures principale ainsi que l'influence possible du jointoyage (ou scellement) au mortier sur la durée de migration des radionuclides.

ZUSAMMENFASSUNG: Ein drei-dimensionales Finite Element Computer Program "MOTIF" wurde entwickelt um die verbundenen Prozesse des Wasser-, Warme und Lösstuff Transportes in der Nähe eines Nuklearen Endlagers in plutonischem Gestein zu modellieren. Das Konzept des Modelles ist auf geologischen, geophysikalischen und hydrogeologischen Daten basiert die in dem Whiteshell Forschungsgebiet im Osten Manitobas, Kanda, gesammelt wurden.
Simulationen eines zwei-dimensionales modelles des Endlagers mit "MOTIF" zeigen den Einfluβ der Nähe von Kluftzonen und deren Verdichtung an die Migrationszeit freigelussener Radionuklide.

1 INTRODUCTION

The Canadian Nuclear Fuel Waste Management Program is currently assessing the concept of disposal of used nuclear fuel waste deep within plutonic rock. The most likely path by which radionuclides might return to the biosphere is by dissolution and transport in flowing groundwater. A three-dimensional finite-element code, MOTIF, has been developed by Atomic Energy of Canada Limited to solve the coupled equations of groundwater flow, conductive and convective heat transport, and radionuclide transport (Guvanasen 1984, Chan et al. 1985). The numerical accuracy of the code has been verified by comparison with analytical and independent numerical solutions (Chan et al. 1985). Its ability to correctly model groundwater flow in a fractured rock mass is being validated by comparison with field measurements of drawdown associated with the sinking of the access shaft for the Underground Research Laboratory (URL) in the Lac du Bonnet pluton, Manitoba, Canada (Davison & Guvanasen 1985, Davison 1986).

In this paper, we present a brief description of the mathematical and numerical formulations for the MOTIF code along with some results of two-dimensional simulations of groundwater flow, heat transport and convective contaminant transport in the vicinity of a hypothetical nuclear fuel waste vault. In particular parameter sensitivity analyses are reported, which illustrate the effect of proximity of the radionuclide source to a fracture zone and the influence of fracture-zone grouting on contaminant transport.

2 GOVERNING EQUATIONS

The governing equations for groundwater flow, heat transport and solute transport can be derived from the conservation laws of mass, momentum and energy (Bear 1972).

Assuming the generalized Darcy's law is valid, the conservation of fluid mass in a nonisothermal, partially saturated porous medium can be expressed, in indicial notation, as

$$\rho \rho_o g \left\{ n \, c_\phi + S_w \left[n \alpha_f + (1-n) \alpha_s \right] \right\} \frac{\partial h}{\partial t}$$

$$+ S_w n \left[\frac{\partial \rho}{\partial C} \frac{\partial C}{\partial t} + \frac{\partial \rho}{\partial T} \frac{\partial T}{\partial t} \right] \tag{1}$$

$$- \frac{\partial}{\partial x_i} \left[\rho k_r \frac{k_{ij} \rho_o g}{\mu} \frac{\partial}{\partial x_i} \left(h + \frac{\Delta \rho}{\rho_o} x_3 \right) \right] - I = 0$$

where ρ = density, ρ_o = reference density, g = gravitational acceleration, n = porosity, $c_\phi = \partial S_w / \partial p$ = moisture capacity, p = pore pressure, S_w = degree of saturation, C = solute concentration, α_f = fluid compressibility, α_s = bulk solid compressibility, t = time, T = temperature, k_r = relative permeability, x_3 = Cartesian coordinates with x_3 vertical, k_{ij} = permeability tensor, μ = dynamic viscosity, $\Delta \rho = \rho - \rho_o$, $h = p/(\rho_o g) + x_3$ = reference piezometric head, and I = specific production rate. The density and viscosity can, in general, depend on T, p and C.

The conservation of mass for a radioactively decaying solute leads to the transport equation:

$$\frac{\partial}{\partial t}\{[S_w n + (1-n)a_1] C\} + \frac{\partial}{\partial x_i}(n S_w v_i C)$$

$$-\frac{\partial}{\partial x_i}(S_w D_{ij} n \frac{\partial C}{\partial x_j}) - I_c \qquad (2)$$

$$+ [n S_w + (1-n)a_1] \lambda C = 0$$

where a_1 = sorption constant, v_i = i-th component of interstitial velocity, D_{ij} = tensor of hydrodynamic dispersion coefficients $= a_{ijkl} v_k v_l/v + D_d T_{ij}^*$, $v = (v_i v_i)^{\frac{1}{2}}$, a_{ijkl} = dispersivity tensor, D_d = molecular diffusion coefficient, T_{ij}^* tortuosity tensor, λ = radioactive decay constant, and I_c = specific solute mass production rate.

For an isotropic medium, the tensor of hydrodynamic dispersion coefficients becomes

$$D_{ij} = a_T v \delta_{ij} + (a_L - a_T) \frac{v_i v_j}{v} + D_d T^* \delta_{ij} \qquad (3)$$

where a_L = longitudinal dispersivity, a_T = transverse dispersivity, T = scalar tortuosity and δ_{ij} = kronecker delta.

Conservation of energy leads to

$$\frac{\partial}{\partial t}[(\rho c)_e T] + \frac{\partial}{\partial x_i}(S_w n\rho c_f v_i T) - \frac{\partial}{\partial x_i}[E_{ij}\frac{\partial T}{\partial x_j}]$$

$$- I_H = 0 \qquad (4)$$

where $(\rho c)_e = S_w n\rho c_f + (1-n)\rho_s c_s$, c_f = specific heat capacity of fluid, ρ_s = density of solid skeleton, c_s = specific heat capacity of solid skeleton, $E_{ij} = S_w n\rho c_f a_{ijkl} v_k v_l/v + \lambda_{ij}^T$ where λ_{ij}^T = effective bulk thermal conductivity tensor, and I_H = specific heat production rate. Equations (1), (2) and (4) above are coupled through the convective terms and through the variation of fluid density and viscosity with temperature, pressure and concentration.

3 NUMERICAL METHODS

Using the Galerkin method of weighted residuals in conjunction with finite-element spatial discretization, Equations (1), (2) and (4) above are reduced to three coupled systems of ordinary differential equations. Application of finite-difference temporal discretization further reduces these into three coupled systems of algebraic equations, which can be solved by standard algorithms such as those based on Gaussian elimination. A time-stepping factor has been incorporated to allow the choice of implicit, Crank-Nicholson or explicit finite difference, or any scheme in between.

The three sets of equations for flow, solute transport and heat transport are solved sequentially. Solution of the flow equation provides the spatial distribution and time evolution of hydraulic head or, equivalently, pressure and velocity. The velocity is fed into the convection terms in the solute transport and heat transport equations. The pressure, p, temperature, T, and concentration, C, solutions are substituted into the constitutive equations to update the values of fluid density, ρ, and viscosity, μ, before advancing to the next time step. A Picard iteration loop is incorporated within the time-marching cycle to handle nonlinear problems.

Three types of isoparametric elements are available in MOTIF: a hexahedron, a 2-D quadrilateral and a 1-D line element. These elements are all defined in a 3-D space, thus the hexahedron element can be used to represent porous media in a 3-D model while the quadrilateral element can be used either to represent porous media in a 2-D model or planar

fractures or fracture zones in a 3-D model. Similarly the line element can be used to represent porous media in a 1-D model, or planar fractures or fracture zones in a 2-D model, or narrow channels and pipes in a 3-D model. A combination of these can be employed in a single model.

4 CONCEPTUAL HYDROGEOLOGICAL MODEL

A conceptual hydrogeological model has been constructed based on field data from the Whiteshell Research Area. This research area has been selected for the modelling study because of the large amount of relevant field data which is currently available to develop a representative and physically-based conceptual model. Although there is no intention of locating an actual disposal vault in the research area, the modelling study considers various scenarios involving a hypothetical disposal vault located at a depth of 500 m.

The Whiteshell Research Area covers about 750 km^2 in southeastern Manitoba (Figure 1) including the sites of the Whiteshell Nuclear Research Establishment (WNRE) and the Underground Research Laboratory (URL). A major portion of the area is part of the Lac du Bonnet Batholith, a large granitic pluton.

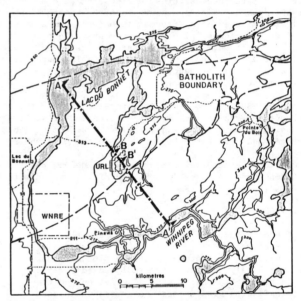

Figure 1. Topographic map of the Whiteshell Research Area (contours in metres above sea level).

The Winnipeg River system is assumed to provide a stable hydrogeological boundary nearly surrounding the area. In general, the topography slopes from southeast to northwest. The water table closely follows the topography.

Subsurface investigations in boreholes have been limited primarily to the WNRE and URL sites. Features and properties controlling groundwater flow were determined from borehole logs and hydrogeological testing and monitoring (Davison, 1984). Figure 2 shows a NW-SE oriented cross section of the geology at the URL site. Three low-dipping fracture zones appear to be the predominant controls of groundwater flow.

The area that must be modelled to analyze the impact of the hypothetical disposal vault is larger than that for which detailed data is currently available. Therefore the conceptual model of the region has been developed from detailed studies at the WNRE and URL sites and from geological features interpreted from airphotography and regional geophysical surveys (Scheier & Whitaker 1986).

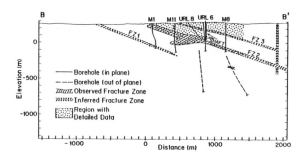

Figure 2. Section BB' (see Figure 1) showing observed and inferred geology near the URL.

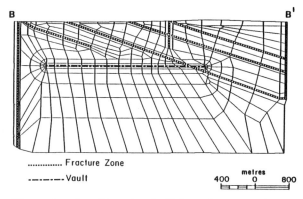

Figure 3. Two-dimensional finite-element mesh – vault region.

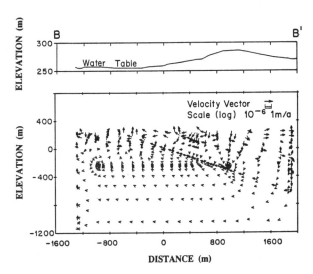

Figure 4. Water table profile and predicted steady state velocity distribution near the vault.

The model consists of a low-permeability granitic rock mass traversed by a number of more permeable low-dipping and vertical fracture zones. The low-dipping fracture zones are assumed to intersect the vertical zones. This is to permit the analysis of a regionally continuous fracture zone system, a conservative assumption with regards to radionuclide transport. The fracture zones are considered to be porous media and uniform in thickness. The rock mass is divided into five layers of porous media with permeability and porosity decreasing with depth. The permeability of the fracture zones is two to four orders of magnitude higher than that of the rock mass.

5 FLOW SIMULATION

Coupled fluid flow and heat transport in the conceptual model are being simulated using the MOTIF code. Several two-dimensional and three-dimensional finite-element representations of the conceptual model have been constructed. Details can be found in Chan et al. (1986).

This paper discusses only the 2-D model corresponding to cross section AA' in Figure 1. This 27 km x 4 km section is assumed to be approximately along a flow line, i.e., there is little flow perpendicular to the plane. All major fracture zones and stratigraphy of the conceptual model are included.

The top boundary of the model has prescribed head values equal to water table elevations. The bottom boundary is assumed to be impermeable and the sides are assumed to be no-flow boundaries. All boundaries have prescribed temperatures based on geothermal measurements at the URL.

A hypothetical used-fuel vault, 1.9 km x 1.9 km, is located at a depth of 500 m (Figure 3). It is purposely located to intersect a fracture zone in order to assess the effect of fracture proximity.

To check numerical convergence of predictions, two meshes and three time-stepping sequences of differing refinement have been investigated. Figure 3 illustrates the vault region of the mesh selected for subsequent analyses. Small elements are used near the vault and fracture zones with a progressive increase in size further away. This is to ensure accurate calculation of groundwater velocities along likely radionuclide pathways. The mesh consists of a total of 914 elements and 985 nodes.

Prior to simulating waste emplacement in the vault, a steady-state condition under the combined influence of topography and natural geothermal gradient has been modelled. Figure 4 shows the predicted velocity vector pattern near the vault. The pattern is semi-circular with recharge in the higher area in the southeast, lateral movement above the lower permeability layers at depth and discharge up through the left side of the vault to the low-lying area in the northwest. The local surface

topography (as reflected in the water table) drives the flow. Flow is up the low-dipping fracture zone intersecting the vault. For this simulation, the velocities near the surface and in fracture zones near the vault are of the order of 1 m/a. In the rock mass near the vault, velocities are typically four orders of magnitude lower.

Transient simulations with decaying heat output corresponding to 10-year-old used CANDU (Canadian Deuterium Uranium Reactor) fuel in the vault have been conducted for a period of one million years using geometrically increasing time steps. In general, the perturbation in both temperature and velocity lasts about 100,000 years. The temperature rise extends about 2000 m from the vault, with convective heat transport being negligible. The peak temperature is 70°C at the vault centroid after 60 years. The peak temperature rapidly decreases and is delayed at points further away. The velocity peak slightly lags that in temperature. At 10,000 years, the overall perturbation in velocity field is near its maximum. The flow pattern is now nearly vertical through almost the entire vault due to the buoyancy effect of the waste heat. In the rock mass at the vault centroid, the peak velocity is approximately 20 times the initial value. This increase drops off rapidly with distance. In fracture zones near the vault, the velocities increase to about twice the steady-state values. Details of the velocity patterns for the transient simulation have been reported previously (Chan et al. 1986).

Sensitivity analyses have been performed to estimate the effect of grouting the fracture zone where it intersects the vault. Grouting has been represented by lowering the permeabilities and porosities of the two elements at the intersection. The permeability has been lowered to 10^{-15} m^2. This value is equal to 1% of the original longitudinal permeability, and 2% of the original transverse permeability of the fracture zone. The porosity has been halved, to a value of 0.05.

Figure 5 illustrates the impact of this grouting on the steady-state velocity predictions. Groundwater flow through the intersection is reduced by a factor of four. Most of the fluid bypasses the obstruction in a semi-circular pattern on the left. The downward flow on the right is also slightly reduced and directed horizontally to the left. Velocities in the fracture zone above the obstruction average about 75% of those for the ungrouted case. The impact of grouting on the transient velocity field is similar to that for the steady state.

It must be noted, however, that the impact for the 3-D case may be different. Unlike for the 2-D case, flow can remain in the relatively permeable fracture zone while bypassing the grouted regions around the excavated tunnels. Thus the perturbation in flow in the rock mass beside the fracture may be less and flow through the grouted region may also be lower.

travel times in this simulation vary from 990 years to 26 million years, the fastest path being up the low dipping fracture zone that intersects the vault. Predictions for convective transport of particles released in the vicinity of the fracture intersection are summarized in Table 1 and illustrated in Figure 6. The paths C and C* (not illustrated) originate at the intersection and follow path R* to the surface.

For a particle released to the right of the fracture zone, paths R and R* (Figure 6a), over 98% of the travel time is spent traversing the short segment through layer 3 to reach the fracture zone. The alteration in the flow field due to grouting causes a significant increase in the length of this segment and results in a fivefold increase in total travel time.

For a particle released in the fracture zone (paths C and C*) grouting increases the travel time by only about 12%, although the average velocities along the path are reduced by about 25% at any particular time. This is because there is a significant increase in velocity with time due to heating from the vault.

Table 1. Impact of grouting the fracture zone/vault intersection on predicted convective transport of particles released from the vault.

Path Segment	No Grouting	Grouting
	Path R	Path R*
Layers 3 & 2		
travel time (a)	6.2 x 10^4	3.1 x 10^5
% of total	98%	99%
Fracture & Layer 1		
travel time (a)	9.6 x 10^2	1.7 x 10^3
% of total	2%	1%
Total		
travel time (a)	6.3 x 10^4	3.1 x 10^5
distance (m)	1405	1634
average velocity (m/a)	2.2 x 10^{-2}	5.2 x 10^{-3}
	Path C	Path C*
Fracture & Layer 1		
travel time (a)	9.9 x 10^2	1.1 x 10^3
distance (m)	1451	1463
average velocity (m/a)	1.5 x 10^0	1.3 x 10^0
	Path L	Path L*
Layers 3 & 2		
travel time (a)	2.0 x 10^7	4.1 x 10^6
% of total	100%	100%
Fracture & Layer 1		
travel time (a)	1.0 x 10^3	2.7 x 10^3
% of total	0%	0%
Total		
travel time (a)	2.0 x 10^7	4.1 x 10^6
distance (m)	1536	1469
average velocity (m/a)	7.6 x 10^{-5}	3.6 x 10^{-4}

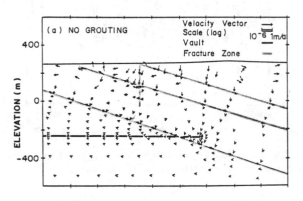

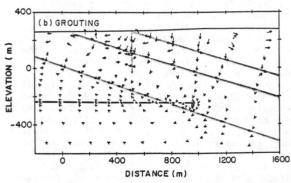

Figure 5. Impact of grouting the fracture zone/vault intersection on the predicted velocity distribution.

6 TRANSPORT SIMULATION

As an initial step in transport modelling, particle tracking codes are being used to calculate radionuclide pathlines and travel times based only on groundwater velocity distributions predicted using MOTIF. Using the previously described transient 2-D flow analysis results, the calculated path lengths for inert contaminant particles from the vault to the surface do not vary a great deal, being two to three times the depth of the vault. However, the

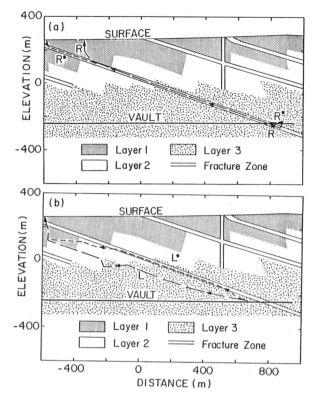

Figure 6. Impact of grouting the fracture zone/ vault intersection on predicted paths for convective transport of particles released from the vault: Paths R and L - no grouting; Paths R* and L* - with grouting.

For a particle released to the left of the fracture zone, paths L and L* (Figure 6b), grouting results in a fivefold decrease in total travel time. For the ungrouted case (path L) a major portion of the path is through the low permeability rock mass. However, the alteration in the flow field due to grouting (Figure 5) results in the particle being swept into the fracture zone and quickly carried to the surface layer.

These results indicate that a wide range of radio-nuclide transport scenarios may have to be consi-dered. One extreme involves rapid convection in major fracture zones with transient thermal effects being important. In the other extreme convective transport through the bulk rock matrix may be negli-gible compared to diffusion, dispersion and retarda-tion mechanisms. The relative importance is very sensitive to the proximity of major fracture zones to the vault. Table 2 shows the rapid increase in convective travel time with respect to distance from the fracture zone for particles released from the vault just to the right of the fracture zone inter-section (Figure 5a). Therefore, a simple yet effec-tive safety measure is to place the waste containers at a certain distance away from the fracture zone.

To ensure adequate analysis of diffusion and dis-persion a detailed two-dimensional, finite-element, convective-dispersive transport model for the vault region is now being analyzed using MOTIF. Prelimi-nary results indicate a rapid convection of contami-nants up the low-dipping fracture zone and then significant dispersion in the surface layers. Dispersion and diffusion significantly accelerate the first arrival of contaminants along the slow paths but have only minor influence on transport along the faster paths.

Table 2. Impact of source proximity to fracture.

Distance to Fracture (m)	Convective Travel Time To Fracture (a)
2.5	930
3.4	4,000
6.7	11,000
14.2	30,000
33.0	89,000

7 CONCLUSIONS

Two-dimensional finite-element simulations have been performed to study groundwater flow, conductive and convective heat transfer and convective solute transport in the vicinity of a hypothetical nuclear fuel waste vault in plutonic rock. The results indicate that the proximity of the radionuclide source to a major fracture zone may strongly deter-mine the travel time to the surface and the role of diffusion in the radionuclide transport process. Grouting of a major fracture zone intersecting the vault may significantly increase or possibly decrease the travel time. However, the impact for the 3-D case may be different because, unlike for the 2-D case, flow can remain in the fracture zone while bypassing the grouted regions around the exca-vated tunnels.

REFERENCES

Bear, J. 1972. Dynamics of fluids in porous media. New York: Elsevier.

Chan, T., V. Guvanasen & J.A.K. Reid. 1985. Numer-ical modeling of coupled fluid, heat and solute transport in deformable fractured rock. Presented at the International Symposium on Coupled Proces-ses Affecting the Performance of a Nuclear Waste Repository, 1985 September 18-20, Berkeley, California. Proceedings in press. San Francisco: Academic Press.

Chan, T., N.W. Scheier and J.A.K. Reid. 1986. Finite-element thermohydrogeological modeling for Canadian nuclear fuel waste management. In Pro-ceedings of the Second International Conference on Radioactive Waste Management, p. 653-660. Toronto: Canadian Nuclear Society.

Davison, C.C. 1984. Hydrogeological characteriza-tion of the site of Canada's Underground Research Laboratory. In Proceedings of International Association of Hydrogeologists Symposium on Groundwater Resource Utilization and Contaminant Hydrogeology, pp. 310-335. Montreal.

Davison, C.C. 1986. URL drawdown experiment and comparison with model predictions. In Proceedings of 20th Information Meeting of the Nuclear Fuel Waste Management Program. Atomic Energy of Canada Limited Technical Record TR-375*.

Davison, C.C. & V. Guvanasen. 1985. Hydrogeologi-cal characterization, modeling and monitoring of the site of Canada's Underground Research Labora-tory. Presented at International Association of Hydrogeologists 17th International Congress on Hydrogeology of Rocks of Low Permeability, January 7-12, 1985, Tucson.

Guvanasen, V. 1984. Development of a finite-element hydrogeological code and its application to geoscience research. In Proceedings of 17th Information Meeting of the Nuclear Fuel Waste Management Program. Atomic Energy of Canada Limited Technical Record TR-299*.

Scheier, N.W. & S.H. Whitaker 1986. Development of a geosphere model for use in assessing the performance of a nuclear fuel waste disposal vault. In Proceedings of the Joint Annual Meeting of the Geological Association of Canada, Mineralogical Association of Canada and the Canadian Geophysical Union. Ottawa.

---

* Unrestricted, unpublished report available from SDDO, Atomic Energy of Canada Limited Research Company, Chalk River, Ontario K0J 1J0

Risques de formation de dômes après stockage de déchets nucléaires dans une couche de sel

Saltdoming hazard related to HLW disposal in salt layer
Salzdombildungsrisiko in Folge von HLW Einlagerung in einem Salzlager

L.CHARO, Laboratoire de Mécanique des Solides, Ecole Polytechnique, Palaiseau, France
A.ZÉLIKSON, Laboratoire de Mécanique des Solides, Ecole Polytechnique, Palaiseau, France
P.BÉREST, Laboratoire de Mécanique des Solides, Ecole Polytechnique, Palaiseau, France

ABSTRACT : Formation of salt domes has been studied in the scope of safety analysis of a radioactive waste disposal in rock salt. Closed form solutions (limit loads theory), numerical computations (F.E.M.) and laboratory experiments (centrifuge) have been used.

RESUME : Le phénomène de formation de dômes de sel a été étudié dans le cadre de l'analyse de sécurité d'un stockage de déchets nucléaires dans une couche de sel. L'étude comprend principalement des approches analytique (charges limites), numérique (éléments finis) et expérimentale (modèles centrifugés).

ZUSAMMENFASSUNG : Hier wurde, innerhalb der Sicherheitsanalyse einer Einlagerung von radioactive Abfälle in einem Salzlager, das Erscheinung von Salzdombildung studiert. Diese Untersuchung besteht hauptsächlich in analytische Betrachtungen (Höchstlastheorie), numerische Rechnen (Finite-Element) und Modellversuche (Zentrifuge).

1 INTRODUCTION

Les dômes de sel constituent une figure caractéristique de la tectonique salifère : la faible densité du sel, et sa plasticité peuvent rendre instables les formations salifères de grande épaisseur et provoquer leur mouvement vers la surface du sol. De telles formations étant envisagées pour la réalisation d'un enfouissement de déchets nucléaires de haute activité, les conditions d'apparition d'un dôme doivent faire l'objet d'une étude particulièrement attentive. Cette étude doit considérer les caractéristiques actuelles du gisement, mais aussi les possibilités d'évolution à court et à long terme après la mise en place du stockage. En effet, les déchets de haute activité ont une très longue vie active (10^5 à 10^6 ans) et constituent initialement ($\sim 10^2$ ans) une puissante source de chaleur.

2 INTERPRETATION MECANIQUE

Supposons que dans la couche de sel représentée sur la figure 1 existe un état de contrainte isotrope qui résulte du poids des terrains surincombants considérés comme des liquides :

$$\sigma_{ij} = P\,\delta_{ij} \qquad \text{avec} \qquad P = \sum_k g\,\rho_k\,h_k \,,$$

où g est l'accélération de pesanteur, ρ_k est la densité moyenne du matériau k et h_k est l'épaisseur de la couche k.

On remarque que les mesures de contraintes in-situ dans le sel, bien que peu nombreuses, autorisent à estimer que l'état de contrainte naturel est souvent peu différent de celui supposé ici.

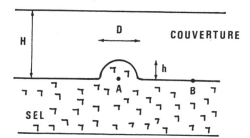

Figure 1 : Irrégularité initiale.

On peut ainsi estimer les contraintes aux points A et B de la figure 1 :

$$P_B = g\,\rho_C\,H$$

$$P_A = g\,\rho_C\,(H - h) + g\,\rho_S\,h$$

d'où un gradient de pression horizontal entre ces points donné par $P_B - P_A = g\,(\rho_C - \rho_S)\,h$.

Quand la densité du sel est inférieure à celle de la couverture ($\rho_S < \rho_C$), alors $P_B > P_A$: si l'écart des densités et les dimensions de l'irrégularité sont suffisamment importants et si les matériaux présentent une faible résistance à l'écoulement, il se produit dans la couche de sel un mouvement horizontal qui alimente l'irrégularité et permet son développement.

La densité du sel peut être considérée à peu près constante et égale à environ 2,15. Par contre, la densité des terrains sédimentaires augmente en général rapidement avec la profondeur. On peut estimer que le sel devient plus léger que sa couverture sédimentaire (inversion des densités) à une profondeur de l'ordre de 500 à 700m.

Cette interprétation simple permet de dégager les principaux paramètres intervenant dans le phénomène : écart des densités, dimensions de l'irrégularité de l'interface sel-couverture, épaisseur de la couverture, résistance des matériaux, que l'on caractérisera ici par leurs seuils de plasticité (S_S et S_C). La viscosité des matériaux n'intervient pas dans la possibilité d'occurrence du phénomène, mais joue un rôle fondamental sur la vitesse du mouvement, quand celui-ci peut avoir lieu.

3 MODELE LIQUIDE

Dans le cas de deux couches de liquides visqueux, l'instabilité due à l'inversion des densités se traduit par la formation sur l'interface d'ondulations dont l'amplitude, initialement faible, va croître exponentiellement dans le temps, jusqu'à ce qu'un équilibre stable soit établi, c'est-à-dire jusqu'à ce que le liquide moins dense soit totalement passé en position haute.

De nombreux auteurs (Biot, Odé, 1963 ; Danes, 1964 ; Ramberg, 1968 ; Selig, 1965 ; Taylor, 1950) ont abordé le problème des dômes en analysant cette instabilité à l'aide de la dynamique des fluides classiques. Ils obtiennent ainsi une relation entre le taux de croissance de l'amplitude (ω) des ondulations et leur longueur d'onde ($\lambda = 2\pi/k$). On met en évidence l'existence d'une certaine longueur d'onde dominante, qui détermine l'évolution du système puisque les vitesses de croissance des perturbations augmentent exponentiellement avec le temps ($v = f(e^{\omega t})$).

Cette approche est intéressante, car elle explique en partie le phénomène des dômes de sel : la pesanteur est l'élément moteur du mouvement. Cependant, elle s'avère insuffisante dans le cadre de notre problème, car elle ne permet pas d'expliquer l'absence de mouvement qu'on rencontre souvent dans la nature, même lorsque les densités sont inversées. Nous devons donc prendre en compte, pour les matériaux, une loi de comportement plus réaliste, et en particulier leur attribuer un certain seuil de plasticité qui doit être dépassé pour que l'écoulement ait lieu.

4 CALCULS DE CHARGE LIMITE

L'application de la théorie des charges limites permet d'estimer l'ordre de grandeur que doivent avoir les principaux paramètres pour qu'une irrégularité de l'interface puisse, ou ne puisse pas, se développer.

On a supposé que le sel se comporte comme un fluide (seuil plastique nul) : la couche de sel est remplacée par les contraintes qu'elle exerce sur la couverture, à laquelle on attribue une cohésion C et un critère de plasticité de Tresca.

4.1 Méthode cinématique :

Deux types de mécanismes de déformation ont été étudiés (figure 2).

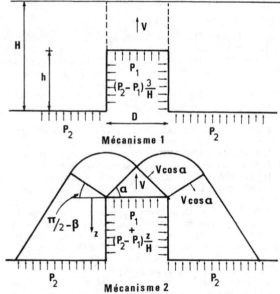

Figure 2 : Mécanismes cinématiques.

. Mécanisme 1 : Une certaine zone de la couverture, poussée par l'irrégularité, tend à s'écouler vers la surface libre.
. Mécanisme 2 : L'écoulement est dirigé globalement vers la couche de sel.

Le calcul montre qu'on est sûr que l'irrégularité se développe si :

. pour le Mécanisme 1 : $g\,\Delta\rho\,h > 2\,\dfrac{(H-h)}{D}\,C$

. pour le Mécanisme 2 :

$$g\,\Delta\rho\,h > \left(\frac{tg\,\alpha - 2\,\alpha + \frac{3\pi}{2} + 1 + 2\sqrt{2}\,\frac{h}{D}\,\cos\alpha}{1 + \frac{1}{2}\,\frac{h}{D}}\right)\,C$$

(avec $\beta = \pi/4$).

Le paramètre de chargement est, dans les deux cas, le terme $g\,\Delta\rho\,h$: pour chaque géométrie, la plus petite des deux expressions précédentes fournit un majorant de la charge limite. Pour des profondeurs de gisement relativement faibles, en particulier pour celles qu'on peut attendre dans un site de stockage, le mécanisme 1 apparaît plus favorable à la montée du dôme.

4.2 Méthode statique :

Après division de la couverture en 14 zones homogènes, (figure 3), on a trouvé pour chacune d'elles un champ de contraintes statiquement admissible. Les expressions obtenues étant relativement compliquées, on ne les exploitera pas ici.

Le calcul montre qu'on est sûr que l'irrégularité initiale ne se développe pas si :

$$g\,\Delta\rho\,h \leqslant 2\sqrt{2}\ C\ .$$

Cette expression fournit un minorant de la charge limite.

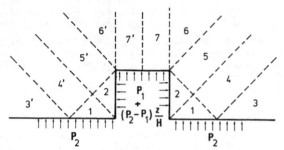

Figure 3 : Méthode statique.

4.3 Exemple :

Soit : H = 700m , h = 100m, D = 450m , $\Delta\rho$ = 0,55 .

En appliquant les résultats précédents, on obtient

$$g\,\Delta\rho\,h \leqslant 2,67\ C\quad.$$

La cohésion minimale de la couverture nécessaire pour empêcher le développement de l'irrégularité est donc de l'ordre de 0,2 MPa. C'est une valeur très faible, surtout si on considère que les valeurs considérées pour h et $\Delta\rho$ sont très importantes par rapport à ce que l'on peut espérer trouver in-situ.

5 ETUDE NUMERIQUE

L'étude plus complète du phénomène des dômes de sel nécessite une étude numérique, qui a été menée à l'aide de la méthode des éléments finis (programme ASTREA). Le seul chargement imposé est le poids propre des matériaux, conjointement avec l'existence sur l'interface sel-couverture d'une irrégularité initiale. Le problème est traité en symétrie de révolution.

On a fait l'hypothèse que les matériaux ont un comportement élasto-viscoplastique (Bingham) dont la loi peut s'exprimer de la façon suivante : le tenseur vitesse de déformation est la somme d'un terme élastique et d'un terme viscoplastique :

$$\underline{\dot{\varepsilon}} = \underline{\dot{\varepsilon}}^e + \underline{\dot{\varepsilon}}^{vp}\quad \text{avec}\quad \underline{\dot{\varepsilon}}^{vp} = \frac{<F>}{\eta}\frac{\partial Q}{\partial\underline{\sigma}}\ ,$$

où F et Q sont respectivement un critère et un potentiel plastique de von Misès (<F> = 0 si le critère n'est pas atteint) et η est une constante de viscosité. Le tenseur vitesse de contrainte est donné par $\underline{\dot{\sigma}} = [\,D\,]\,\underline{\dot{\varepsilon}}^e$ où [D] est la matrice d'élasticité.

Un grand nombre de calculs a été effectué, afin d'analyser l'influence des divers paramètres. On présente dans la suite les principaux résultats.

5.1 Vitesse de montée du dôme :

J. Mandel (cité dans Cormeau, 1976) a démontré que pour un corps élastoviscoplastique soumis à un chargement extérieur constant, le champ de vitesse de déformation tend, à long terme, soit vers zéro, soit vers une valeur constante non nulle. On retrouve ceci avec les résultats numériques, comme le montre la figure 4 qui met en évidence le rôle du seuil plastique de la couverture sur l'existence d'une vitesse asymptotique non nulle. On observe que cette dernière est très vite atteinte (à l'échelle du problème).

5.2 Evolution du dôme :

L'application successive de cette méthodologie de calcul a permis de déterminer l'évolution d'une irrégularité initiale (figure 5). On a supposé que la surface du sol est érodée à mesure qu'elle se soulève. On constate la formation d'un synclinal périphérique (affaissement de la couverture aux alentours du dôme) phénomène souvent observé dans la nature. Cet amincissement progressif de la couche de sel peut devenir à plus long terme suffisamment important pour empêcher l'alimentation et donc la croissance du dôme. On constate aussi un affaissement relatif du centre du sommet du dôme : on verra que ce phénomène est aussi observé expérimentalement.

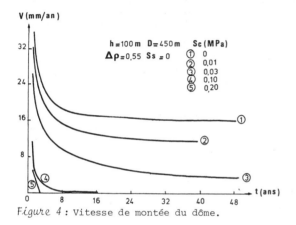

Figure 4 : Vitesse de montée du dôme.

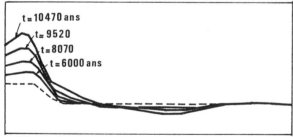

Figure 5 : Evolution de l'irrégularité.

5.3 Influence des divers paramètres :

La vitesse de montée du dôme (V) augmente linéairement avec sa hauteur (h) et son diamètre (D), avec l'écart de densité ($\Delta\rho$) et avec l'épaisseur de la couche de sel (E) (figure 6).

Pour chacun de ces paramètres, une valeur minimale est nécessaire pour que le dôme puisse se développer.

Si on néglige l'effet de la compaction sur la densité, la vitesse de croissance du dôme diminue d'abord rapidement avec l'épaisseur de la couverture, puis devient constante à partir d'une certaine profondeur (figure 7). Ceci correspond au passage du mécanisme 1 au mécanisme 2 du paragraphe 4.1. Cependant, en corrigeant les vitesses en fonction de la densité correspondant à chaque profondeur, on constate que l'épaisseur de la couverture n'a pas d'influence sur la vitesse de montée du dôme (droite V_c de la figure 7).

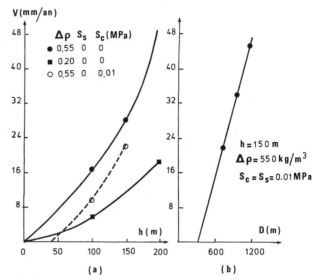

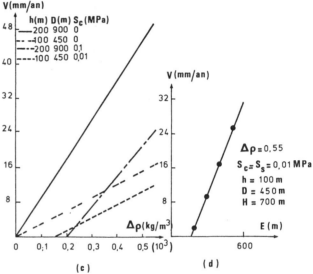

Figure 6 : Vitesse verticale du dôme fonction de la hauteur ; du diamètre de l'irrégularité ; de l'écart de densité et de l'épaisseur du sel.

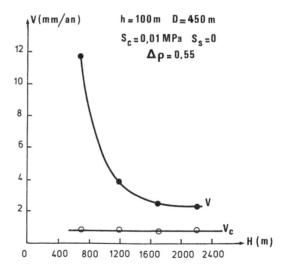

Figure 7 : Vitesse verticale du dôme, fonction de la profondeur de l'interface.

49

5.4 Influence d'une couche rigide dans la couverture :

Les calculs ont montré qu'une couche rigide dans la couverture est capable d'empêcher le développement d'un dôme qui sous une couverture homogène aurait une vitesse de croissance de 28mm/an. Cependant, la poussée du dôme induit dans le banc rigide des contraintes de traction importante : 14MPa aux abords de l'axe de symétrie. Le banc est donc susceptible de se rompre et de ne plus constituer ultérieurement une barrière pour la croissance du dôme.

5.5 Présence d'une source de chaleur :

Les calculs ont montré que si l'augmentation de température n'affecte qu'une zone limitée de la couche de sel, la vitesse de montée du sel n'augmente que très légèrement. En effet, le mouvement est contenu par la zone du massif dont les caractéristiques n'ont pas été modifiées.

6 ETUDE EXPERIMENTALE

Le phénomène de formation de dômes a été étudié expérimentalement à l'aide d'essais sur modèles réduits centrifugés. La force centrifuge subie par le modèle simule et amplifie la pesanteur, élément moteur du mouvement, ce qui permet d'obtenir des temps d'expérimentation raisonnables.

La centrifugeuse utilisée, celle du CESTA*, installée à Bordeaux, est une des plus puissantes existant aujourd'hui : ses caractéristiques principales sont :
- distance entre l'axe de rotation et le modèle : 10m. Ceci permet d'obtenir un champ d'accélération uniforme ;
- accélération maximale : 100 g ;
- poids maximum du modèle : 1000 kg. Ceci permet d'utiliser des modèles de dimensions suffisamment grandes pour que :
  . les effets de paroi ne jouent qu'un rôle mineur,
  . on puisse construire des structures particulières sans difficultés majeures,
  . l'observation des détails (fissuration, plis, fractures) soit aisée.

Le coût des essais sur centrifugeuse étant très élevé, les matériaux à utiliser ont été choisis et testés à l'aide d'essais sous presse. Trois matériaux de base se sont avérés bien adaptés aux besoins : argile grise de Provins, argile blanche, gélatine photo-élastique.

Les couleurs différentes des argiles permettent de suivre visuellement l'évolution de l'interface. Leurs viscosité et cohésion sont facilement modifiables en variant la quantité d'eau du mélange. De même, leur densité peut être augmentée en ajoutant au mélange du corindon, ou diminuée en ajoutant des micro-billes de verre. La gélatine photo-élastique, placée en haut ou en bas du modèle, permet de visualiser l'état de contrainte dans la "couverture" ou dans la "couche de sel".

Les résultats de ces essais se présentent sous forme de photos et films vidéo. On se limitera ici à citer quelques résultats qualitatifs obtenus :
- il a été prouvé que sous certaines conditions, la force de pesanteur constitue un chargement suffisant pour provoquer la remontée du matériau plus léger ;
- les essais réalisés sur un modèle dont l'interface "sel-couverture" était horizontal n'ont donné, après plusieurs heures de centrifugation à 100g, aucun signe de mouvement. La mise en place, sur le même modèle, d'une irrégularité sur l'interface, a suffi pour provoquer, à environ 30g, la remontée du matériau moins lourd ;
- la montée du dôme s'accompagne de la formation du synclinal périphérique et de l'affaissement relatif du centre du sommet du dôme, phénomènes également observés au moyen des calculs numériques ;
- un essai réalisé sur un modèle comprenant une couche plus résistante au sein de la couverture, en contact avec le sommet de l'irrégularité initiale, a permis d'observer la rupture de cette couche et le développement de l'irrégularité. On rappelle qu'à l'aide des calculs numériques on a constaté, pour le même cas, l'absence de mouvement initial, mais l'apparition de contraintes importantes dans la couche rigide capable de rompre celle-ci et donc de permettre ultérieurement le développement de l'irrégularité initiale.

7 CONCLUSIONS

On considère que l'ensemble des résultats obtenus est assez favorable à la sécurité d'un stockage de déchets dans une couche de sel, du point de vue du risque de formation de dômes. En effet, il existe en France des gisements de sel réunissant les conditions nécessaires pour que ce risque soit très faible et même inexistant. Ceci concerne principalement les conditions relatives à l'écart des densités et aux dimensions des irrégularités citées au paragraphe 5.

A ce sujet, il faut remarquer que l'un des sites français, susceptible d'être choisi comme lieu de stockage, a fait l'objet d'une étude détaillée des densités des terrains. Les résultats obtenus sont très favorables, l'écart des densités étant très faible : 2,18 pour le sel, 2,19 pour la couverture.

Il faut aussi s'assurer que ces conditions favorables ne seront pas modifiées après la mise en place des déchets. De ce point de vue, il convient que le site de stockage soit choisi dans une zone géologiquement et tectoniquement stable, et que la conception du stockage soit telle que les modifications apportées par la présence des déchets soient peu significatives. En particulier, on rappelle que la modification des propriétés du sel par l'augmentation de la température n'affecte pas les risques de façon sensible, à condition qu'elle ne concerne qu'une zone limitée de la couche de sel.

Remerciements :

Cette étude a été partiellement financée par la Commission des Communautés Européennes, l'Agence Nationale de Gestion des Déchets Radioactifs et l'Association ARMINES.

REFERENCES :

Biot, M.A., Ode, H. 1963, "Theory of gravity instability with variable overburden and compaction", Geophysics, vol. XXX, n° 2, pp. 213-227.

Danes, Z.F. 1964, "Mathematical formulation of salt-dome dynamics", Geophysics, vol. XXIX, n° 3, pp. 414-424.

Ramberg, H. 1968, "Instability of layered systems in the field of gravity", Phys. Earth. Planet. Interiors 1, pp. 427-447.

Selig, F. 1965, "A theoretical prediction of salt-dome patterns", Geophysics, vol. XXX, n° 4, pp. 633-643.

Taylor, G. 1950, "The instability of liquid surface when accelerated in a direction perpendicular to their planes", Proc. Roy. Soc. London, vol. 201, pp. 192-196.

Cormeau, J.C. 1976, "Viscoplasticity and Plasticity in the finite element method", Thesis for the degree of Doctor of Philosophy, University of Wales.

Goguel, J. 1965, "Traité de Tectonique", Masson et Cie.

* CESTA : Centre d'Expérimentation Scientifique et Technique d'Aquitaine (C.E.A.).

A comprehensive analysis of the influential factors and mechanism of water-inrush from floor of coal seam

Analyse détaillée des effets et des mécanismes des venues d'eau soudaines du plancher d'exploitations houillères

Eine umfassende Analyse der Einflüsse und Mechanismen der Wasserzuführung aus der Sohle von Kohlflözen

CHEN GANG, Beijing Mining Institute, Central Coal Mining Research Institute, People's Republic of China
LUI TIAN-QUAN, Beijing Mining Institute, Central Coal Mining Research Institute, People's Republic of China

ABSTRACT: In this paper, the effects of water-inrush from floor stratum caused by coal mining have been deeply studied, and the mechanism and influential factors of water-inrush have been analysed comprehensively by the finite element method and in-situ measurements. It is shown that the results calculated by F.E.M. agree well with those obtained in situ. The calculation has been done with the CBKD elasto-plastic finite element programme developed by the authors.

RESUME: Cet article va présenter une recherche approfondie sur l'influence d'une exploitation sur le dégagement instantané d'eau dans le mur d'une couche de charbon, par la méthode de calcul des éléments fini et les mesures pratiquées dans les mines. On va voir aussi l'analyse du mécanisme de dégagement instantané d'eau et les facteurs d'influence sur le dégagement. Les résultats de calcul et de mesure sont essentiellement égales. D'ailleurs touts les calculs sont réalisés par notre programme élas-plastique d'élément fini (CBKP).

ZUSAMMENFASSUNG: In diesem Aufsatz werden druch Finite Elemente Berechnung und betriebliche Beobachtung der Einfluß des Abbaus auf die Wasserzuführung erforscht, und auch das Prinzip der Wasserzuführung und der Einflußfaktor analysiert. Die Berechnungsergebnisse sind mit den betrieblichen Meßdaten grundsätzlich übereinstimmt. Außerdem werden die ganz Berechnungen durch das eigene aufgestellte CBKP elastisch-plastisch Finit-Elemente Programm durchgeführt.

INTRODUCTION

In this paper, the actual information collected from nine coal fields in North China has been used. It includes hundreds of inrushing cases (see ref.1). The observational data of water-inrush are taken into consideration during the processes of numerical analysis and the results obtained agree well with the data. An example selected is face 2701 in coal mine No.2 of Fengfeng Mining Administration where the geological conditions are typical in North China. They geological conditions are as follows:

The length along face advancing direction is 350m, the length along face varies from 70 to 120m, the inclination ranges from 10° to 20°, the mining depth is 132 to 167m and the mining height is 1.5m. The mining area is surrounded by unmined coal seams and is free from large tectonics. The immediate roof consists of limestone 0–1.2m in thickness.

the main roof consists of mudstone and sandy-mudstone with thicknesses of 2.5m and 1.7m respectively. The floor strata are composed of fine-grained sandstone and mudstone 1–2m and 4m in thickness respectively. The seam and the karst strata are 52m apart. The water pressure in the karst is 1.23 MPa. The distance between the coal seam and the so called Large Green limestone is about 17m. The thickness of the Large Green limestone is 6m. It has been known as a highly water-bearing stratum. In the mining process, metal props were used to support the roof. The spaces between adjacent props are 0.8m along the strike and 0.6m along the face line respectively.

A special observation tunnel was driven within the Large Green coal seam which is located 24m below the Small Green coal seam. It includes a drift 80m in length and an ascending working which is 70m long as shown in figs.1 and 2. In each of the upper, middle and lower parts of the ascending working, these is a set of inclined boreholes to be used for strata movement observation, geophysical measurements and water pressure testing. All of these holes were drilled upward. Eight surveying ways were adopted:

1. The loading test for the props.
2. The stress distribution measurement of rock mass with oil pressure cells.
3. The strata movement surveying.
4. The tunnel movement surveying.
5. The fault movement surveying.
6. The water injecting test.
7. The borehole ultrasonic measurement.
8. The radio profile measurement.

The results of comprehensive analyses of the data in situ are as follows:

1st zone: it ranges from 0 to 32m ahead of the coal face, called the ground pressure

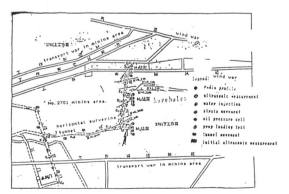

Fig.1 2701 mining area test figure comprehensively

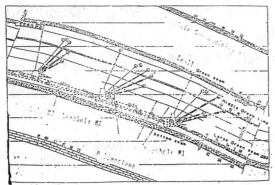

Fig.2 A cross-sectional view of 2701 mining area

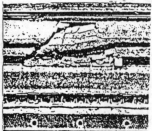

Fig.3 The similar material model

Fig.4 Design of the stress ring trancducer

increasing area. The effects of the ground pressure in this area are as follows: with the face advancing, the roof pressure in-creases gradully. It may amount to several times of the original stress. Finally, it decreases to zero. The roof is compressed and subsides. The seam and the floor strata are also compressed. The deformations are about 1 to 14mm/m, mostly 2 to 8mm/m. The influenced depth is 15m below the seam being mined.

2nd zone: it ranges from 0 to 32m behind the face and is called the ground pressure decreasing area. In this area, because the roof loading is transferred to surroundings, the ground pressure decreasing area is for-med. The roof subsides and sometimes sus-pends. Then, the immediate roof will fall down and the floor will expand. The tensile deformations are about 2 to 14mm/m and the stratum structure is destroyed.

3rd zone: it is located at 32 to 65m be-hind the coal face and is called pressure-increasing area. In this area, the main roof which has already collapsed will rest on the floor through the caved immediate roof, and there will be forces acting among them. So the roof pressure increases, but the value is not larger than that in the 1st zone. The separated strata is closed again. There will be 2 to 10mm/m residual deformations.

4th zone: it ranges from 65m behind the coal face to the end of the mining section, called pressure-recovery area. In this area, the strata located above the gob has been stabilized. There are still 1 to 5mm/m resi-dual deformations.

The calculative model, selected parameters and boundary conditions are determined on the basis of the above comprehensive ana-lysis.

In the simulating model test, four kinds of models formed by soft and hard model ma-terials, with and without filling were desig-ned. The materials used in the test are mix-tures of sand, cement, gypsum etc.. The si-mulated rock strength of the models was de-termined by material strength tests. The resistance displacement transducer was used for surveying displacement (see fig.3). All surveying data were automaticlly collected with the 7V07 monitor made in Japan. The sur-veying results of models were given in re-ference 2. They are close to those obtained by finite element method.

1. CALCULATING MODEL

Three-dimensional problem was simplified as two-dimensional problem because the face is long enough. The dimensions of model are

100m along the strike and 49.5m in height. The bottom of the model was assumed as being fixed in both x and y directions, whilst the left and right borders were assumed as being free in Y direction. The overburden strata were taken as a uniform force acting on the top of model. The calculating model was di-vided into five sections based on the obser-vation data in situ. The five basic parame-tres for F.E.M. are listed in table 1. The model was divided into 400 elements with 441 nodes altogether. The equal parameter element with four joints was adopted. Model materials were considered as perfectly elasto-plastic ones and the Drucker-Prager criterion was used. The loading factors were 0.4, 0.3, 0.2 and 0.1 in turn.

Table 1. Mechanical parameters

Rock type	E (10^9Pa)	μ	ρ $(10^3kg/m)$	c (Pa)	ϕ
sandy shale	23.52	0.3	2.5	68.6	25
mudstone	14.7	0.35	2.3	39.2	22
limestone	49.0	0.24	2.5	137.2	30
fine sandstone	41.2	0.28	2.4	98.0	26
sandy mudstone	21.6	0.32	2.4	58.8	23
coal seam	7.84	0.4	2.0	29.4	20

2. ANALYSIS OF THE CALCULATING RESULTS

The main factors which cause water-inrush from floor may be geological structure, water pressure, mining depth, the characters of floor and ground pressure activities in-duced by mining. An important point is that whether the impervious strata can prevent water from inrushing or not after mining under certain geological conditions. But in the case of sudden water-inrush, the damaged depth in floor must be emphatically discussed.

2.1 Simulating calculation for NO.2701 mining section in Fengfeng by F.E.M.

According to the calculations, the subsidence

curves distribute in radial forms in the roof strata. And in the floor strata, compressive deformation will be induced in front of the coal face within 35m, the maximum value occurs at 10m in front of the coal face. There are expansion and consolidation zone respecbively in the floor behind the coal face. The expansion zone occupies 32m long behind the face line with a maximum value of -33mm, which locates at 14m apart from the face line, see figs.5 and 6.

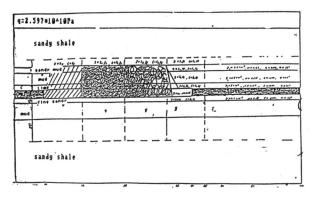

Fig.5 Calculative model in F.E.M.

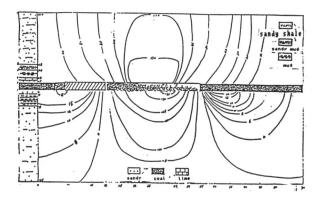

Fig.6 Egual subsidence curve in 2701 mining area

It can be seen that the vertical stress concentration factor is 2.2 during the face moving in front of the coal face. The stress concentration factor is less than 1.0 at the location 25m behind the face line. But the factor is less than 1.8 when the distance behind the face is larger than 30m. It is usually estimated in theory that the maximum stress concentration factor is 3.5 before the first falling of the main roof. From the point of view of ground pressure, it is most probable that water-inrush from floor occurs before the first falling of the main roof, and the most possible point of water-inrush is located at the inflection point between expansive and compressive zone of floor. The location of the point is about 5m behind coal face based on the calculation. This has been proved by the observation data in situ. The maximum fissure ratio at the point is 5‰.

2.2 The effect of mining depth on the floor

In the calculation of F.E.M., the change of mining depth can be easily simulated by changing the uniform load acting on the upper border of the model. Seven models with mining depths of 100m, 200m, 300m, 500m, 700m, 1000m, and 1500m were calculated under otherwise identical conditions. The result is shown in fig.8.

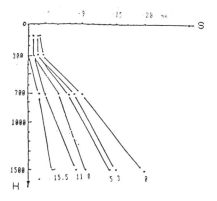

Fig.8 Floor maximum expansion curve

It can be seen that when the depth changes within 300m, the expansion on the floor changes little. However, when the mining depth exceeds 300m, the maximum value of expansion on the floor will be directly proportional to the depth. Regression equation is:

$$S_{max} = -0.413-0.0128H$$

in which: S_{max} -- maximum expanssion value of the floor stratum; H -- mining depth. The correlation coefficient is 0.99. If we take the expansion value of the floor as an equivalent criterion for water-inrush from the floor, then, when the mining depth is over 300m, water-inrush would become with increasing mining depth.

2.3 The effects of water pressure and floor charaters

For the water-inrush from floor, geologic structure will be one of the most important influencing factors. According to statistics, 80% to 90% of water-inrush from floor are related to fault. From engineering stand point, it is more meaningful to investigate the action of underground water flow through concealed fissures than to study permeation flow. In this paper, the action of water from floor was simplified as one set of uniformly distributed forces acting on the surface of impervious stratum. According to the principle of orthogonal design, the simulated calculation was done, see table 2. It is shown

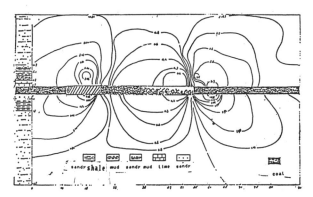

Fig.7 Vertical stress concentration factor curve in 2701 mining area

that when the floor stratum is soft comparatively, the pressure will have greater effect on the floor, but when the floor is hard and with good structure, it will give small effect to the floor. It can be seen that the modulus of elasticity of the floor has a hyperbolic relation with the expansion as shown in fig 9.

Table 2 Orthogonal design

	E (10^9Pa)	θ (10^9Pa)	E.θ	Smax (mm)
1	1,20.6	1,0	1	-2.7
2	2,2.9	1,0	2	-14.3
3	1,20.6	2,1666	2	-47.0
4	2,2.9	2,1666	1	-177.0
I	-49.7	-17.0	-179.7	
II	-191.3	-224.0	-61.3	
R	-141.6	-207.0	+118.4	

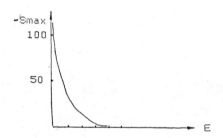

Fig.9   Relation between the maximum expansion and the floor character

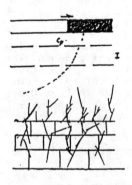 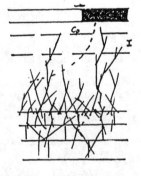

Fig.10 Geologic structure in delayed water-inrush

Fig.11 Geologic structure in sudden water-inrush

3. DISCUSSION ON THE MECHANISM OF WATER-INRUSH

According to the stress field obtained by finite element method, the zone with stress concentration factor larger than 1.5 is taken as compressive failure zone that with stress concentration factor less than 0.5 is taken as tensile failure zone. By this way, it is easy to explain the mechanism of sudden water-inrush and delayed water-inrush shown in fig.10. The difference between them is that whether concealed fissures have reached to the failure zone or not. For the type of sudden water-inrush, the concealed fissures extend into the fractured zone. In this case, underground water will inrush from the floor when the face advances just beyond that point. For the delayed water-inrush type shown in fig.10, a large mined out area is still needed to form a large pressure decreasing zone since the concealed fissure has not reached to the fractured zone, before the acurrence of water-inrush under the combined action of water pressure and ground pressure.

The empirical formula that has been used for water-inrush prediction in China is:

$$T_s = P/(M - C_p)$$

In which, T_s -- water-inrush coefficient; P -- water pressure; M -- thickness of protecting stratum; C_p -- fracture depth caused by mining.

Based on the elasto-plastic F.E.M. analysis we recommend the modified formula:

$$T_s = P/(M + 1.482 - 0.021H)$$

In this formula, the effect of mining depth is included.

4. CONCLUSIONS

Based on the simulating model test, F.E.M. analysis and the observation data in situ, the following conclusions are obained:

1. The water-inrush from floor is more likely to occur when the main roof falls down for the first time there after that.The most possible point of water-inrush is located at 5m behind coal face.

2. When the mining depth is over 300m, water-inrush will be more influenced by the increasing mining depth.

3. When the floor stratum of coal seam is hard, water pressure will give less effect on water-inrush.

4. The floor has been divided into compressive zone and tensile fracture zone with values of vertical stress concentration factor larger than 1.5 and less than 0.5 respectively. This will benefit the explanation of the mechanism of different types of water-inrush.

5. A modified formula developed for predicting water-inrush is recommended which is suitable for medium hard rocks.

REFERENCES

Wang Zhenan. 1980. The Summary Report for Surveying Mining Pressure at No.2701 Small Green Mining Section In Fengfeng No.2 Mine.

Chen Gang. 1985. The Influential Factors Analysis for Mining.

Benchmarking rock-mechanics computer codes: The Community Project COSA

Evaluation de codes de calculs géomécaniques: Le projet COSA de la Communauté européenne

Ein Vergleich für felsmechanische Computerprogramme: Das COSA Projekt der Europäischen Gemeinschaft

B.CÔME, Commission of the European Communities, Brussels, Belgium

ABSTRACT: Twelve research teams in the European Community participated in a benchmark exercise for rock-mechanics computer codes about disposal of radioactive waste in salt. Two cases were considered: firstly, a complex hypothetical problem; secondly, a laboratory model test. This work was carried out within the Commission of the European Communities' R & D programme on "Management and Storage of Radioactive Waste".

RESUME: Douze équipes de la Communauté Européenne ont participé à un exercice d'intercomparaison de codes de calcul géomécaniques concernant l'évacuation de déchets radioactifs dans le sel. Deux cas ont été examinés: d'abord, un problème théorique complexe; ensuite, un essai de laboratoire sur maquette. Ce travail a été réalisé dans le cadre du programme de R & D de la Commission des Communautés européennes sur "la gestion et le stockage des déchets radioactifs".

ZUSAMMENFASSUNG: Zwölf Forschungsgruppen der Europäischen Gemeinschaft haben an einem Benchmark für felsmechanische Computerprogramme teilgenommen, das die Endlagerung radioaktiver Abfälle im Salz betraf. Zwei Probleme sind betrachtet worden: erstens, ein theoretisches Problem; zweitens, ein Modellversuch im Labor. Diese Arbeit wurde im Rahmen des F + E Programms der Kommission der Europäischen Gemeinschaften über "Bewirtschaftung und Lagerung radioaktiver Abfälle" durchgeführt.

1 INTRODUCTION

Among the various studies relating to geological disposal of radioactive waste, the calculation of mechanical stresses in the host formation is of great importance. This investigation is necessary because of the two following requirements: first, the design and construction of stable underground repositories for the operational phase, i.e. the deposition of waste; then, the prediction of the large-scale mechanical behaviour of the host rock after repository closure, mainly under the influence of heat emission from high-level waste.

Owing to the complexity of the geometrical arrangements considered, and of the rheological behaviour of geological media, these calculations are carried out by computer codes, generally using the finite element method.

For about ten years, the Commission of the European Communities has been participating in the development of such calculation tools, in the framework of its R & D programme about "Management and Storage of Radioactive Waste". A next useful step was then to perform, at Community level, an exercise which would allow to assess the capabilities of these tools. In a first step, the exercise should focus on the visco-plastic calculation of stresses in salt, as this material was the subject of the largest body of geomechanical research, both theoretical and experimental. At the end of 1984, the Commission launched the Community Project COSA ("Comparison of Computer COdes for SAlt"), which brought together the main European teams having expertise in this field. The project was completed by the middle of 1986 (1). This exercise was the first one of this kind at Community level; it was given the following quantitative objectives: assessing the numerical accuracy of the codes involved, and verifying their capabilities to adequately replicate real world's situations. More qualitatively, it was hoped that increased exchanges would take place between the various teams, and that the methodology used for the exercise would be tested so that it could be improved for further steps.

Similar projects were already performed outside the Community, particulary in the USA in the framework of the "WIPP Benchmark" managed by the SANDIA laboratories (2). For COSA, one tried to avoid duplications with these predecessors; however, profit was taken from this experience.

2 IMPLEMENTATION OF THE EXERCISE

Ten European computer teams participated in the COSA project; the British engineering firm ATKINS R & D, Epsom, acted as scientific secretary and impartial co-ordinator on behalf of the Commission. These teams are:

In Belgium: The company FORAKY, together with the Centre d'Etude de l'Energie Nucléaire (CEN/SCK) and the Laboratoire de Génie Civil of the University at Louvain-la-Neuve (LGC).

In Germany: The Rheinisch-Westfälische Technische Hochschule (RWTH), Aachen; The Gesellschaft für Strahlen- und Umweltforschung (GSF), Braunschweig; The Kernforschungszentrum Karlsruhe (KfK), Karlsruhe.

In Denmark: The Engineering Department of the RISØ Laboratory (RISØ), Roskilde.

In France: The Laboratoire de Mécanique des Solides of the Ecole Polytechnique (LMS), Palaiseau; The Centre de Mécanique des Roches of the Ecole des Mines (EMP), Fontainebleau; The Laboratoire d'Analyse Mécanique des Structures, Département des Etudes Mécaniques et Thermiques, of the CEA-Saclay (CEA-DEMT).

In Italy: The Istituto Sperimentale Modelli e Strutture (ISMES), Bergamo.

In the Netherlands: The Energieonderzoek Centrum Nederlands (ECN), Petten.

The codes participating in the exercise are listed in Table 1; those between brackets were specifically used for thermal calculations.

Following discussions between the co-ordinator and the participating teams, the exercise was set up in two benchmarks.

Benchmark 1 aimed at testing the numerical capabilities of the computer codes used, by solving a relatively complex hypothetical problem, designed

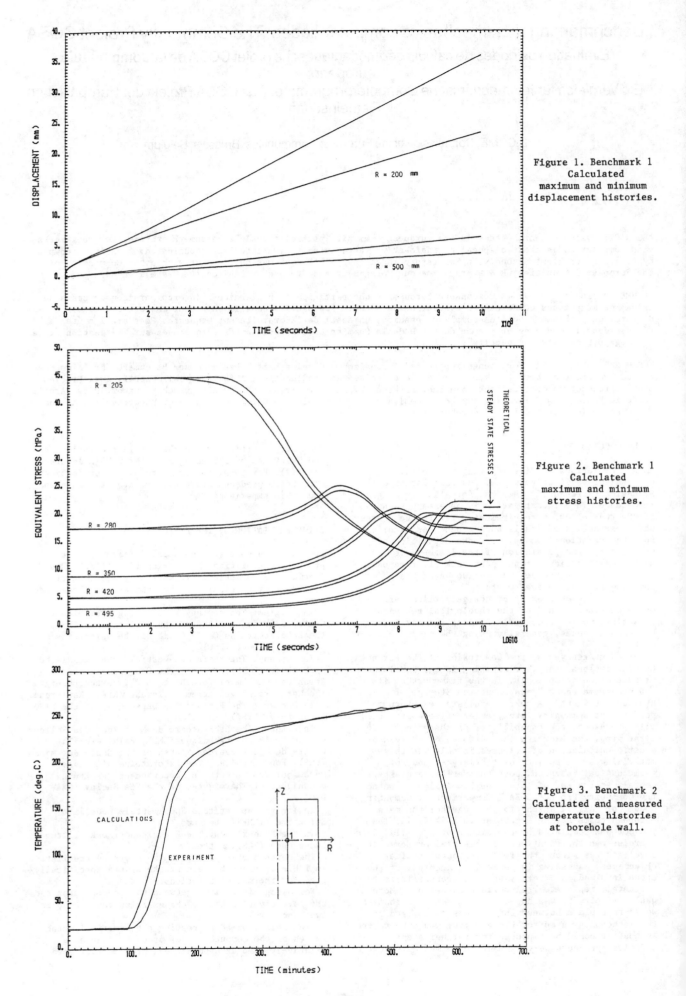

Figure 1. Benchmark 1 Calculated maximum and minimum displacement histories.

Figure 2. Benchmark 1 Calculated maximum and minimum stress histories.

Figure 3. Benchmark 2 Calculated and measured temperature histories at borehole wall.

Table 1. Participating teams and codes.

Firms	Codes
FORAKY+CEN/SCK + LGC	CREEP
RWTH	MAUS (FAST-BEST)
GSF	ADINA/ANSALT (ADINAT/ANTEMP)
KfK	ADINA (ASYTE)
RISØ	ADINA
LMS	ASTREA (ASTHER)
EMP	CYSIPHE/VIPLEF (CHEF)
CEA-DEMT	INCA (DELFINE)
ISMES	GAMBLE
ECN	GOLIA/MARC/ANSYS

in such a way that all codes could solve it, and that semi-analytical solutions would exist to allow at least partial verification.

For Benchmark 2, the objective was to test the codes's capabilities to replicate real world's situations. For this purpose, it may prove useful to back-calculate in situ and large-scale experiments carried out in salt. In order to avoid duplications with other similar work performed at national level in some Member-States, it was found more appropriate to select scaled laboratory experiments. This procedure was also suggested in the USA (3) and was thought to have the advantage of allowing better definition of the boundary and loading conditions, because all these factors are under the experimentalist's control. Similarly, a greater number of measurement can be performed at a reduced cost, which facilitates the comparison of calculated and measured results.

In addition to the purely technical aspects, it was initially felt useful to assess the cost effectiveness of the codes and of the computers on which they are run. For this purpose, each team had to run a small programme, the cost of which was taken as a basis for cost assessment and comparison for the abovementioned calculations. It was the "Computational Cost Parameter" programme, already used for the WIPP Benchmark by SANDIA (2). However, this objective became a relatively unimportant part of the project, and will not be reported here.

It is noteworthy that a representative of ATKINS's staff attended the benchmark 1 calculations together with the specialists in each team, so that the "user-friendliness" of codes and computers could be assessed. Benchmark 2 calculations were much longer, and were carried out by the various teams according to their own work schedule.

3 BENCHMARK 1

As explained earlier, a hypothetical problem was considered here in order to test the "mathematical" and "numerical analysis" aspects of the participating codes. The input data were therefore not to be considered as real life's parameters. They were selected in order to avoid undue numerical difficulties. A one-dimensional case with spherical symmetry was considered.

3.1 Description

A thick spherical shell made of homogeneous elasto-visco-plastic material was loaded by a steady-state temperature distribution and constant pressures

applied on the inner and outer boundaries. Except the creep law, the material's properties were taken as constant and temperature independent. It was proposed to calculate the evolution of the equivalent stress at specified locations, as well as the displacements of the inner and outer boundaries, as functions of time, over 10^{10} seconds, i.e. about 300 years. Each team was free to select the finite element mesh density, as well as the type of elements.

Input data were given to each team at the time of calculation by ATKINS's representatives:
- radii of spherical shell: internal: 200 mm; external: 500 mm
- Temperature distribution: stationary; inner temperature 500°K, outer temperature 400°K
- Pressure applied at the inner wall: 30 MPa (constant); at the outer wall: 0
- Material properties: Young's modulus 10^4 MPa; Poisson's ratio 0.25; coefficient of thermal expansion: 0; creep law: $\dot{\varepsilon}_{cc} = A \cdot \sigma_{eq}^n \cdot \exp(-B/T)$
with $A = 3.10^{-6}$ s^{-1}; $n = 5.5$; $B = 12500$ J; $T =$ absolute temperature in °K

3.2 Results

Figures 1 and 2 respectively depict the maximum and minimum calculated evolutions of wall displacements, and those of the equivalent stress at the specified locations. There is a reasonably good agreement between codes and also with the theoretical asymptotic values for stresses obtained from analytical formulae. It was therefore possible to conclude that the codes showed a good numerical behaviour.

4 BENCHMARK 2

This constituted the core of the exercise; it was chosen to compare results of calculations not only between themselves, but also with measurements from a reduced-scale laboratory test. The test itself - called RTA - was performed at the Technical University Delft, Mining Department; it was the fourth of a test series which began under a Community research contract (4). It must be noted that, because of internal reasons outside the exercise, the RISØ laboratory withdrew from the project after completion of Benchmark 1, and therefore did not participate in Benchmark 2.

4.1 Description of the test

A salt cube, 304 mm edge, with a central axial hole, 66 mm dia, was compressed in three directions by the stiff platens of a press; then, a 35 mm dia heater was inserted in the hole and induced a temperature rise. The heat output was known, as well as the temperature histories at selected locations (via thermocouples), and the displacements of the borehole wall and press platens. The test duration was 13.5 hrs. The salt was sampled from the "Na$_2$" seam at the Asse mine (FRG) and had been the subject of intensive rheological and thermal studies for several years. These works were carried out mainly at the GSF, and at the Bundesanstalt für Geowissenschaften und Rohstoffe in Hannover (FRG).

Given the loading and heating histories, and the properties of the salt, it was asked to calculate (i) the temperature evolutions at specified locations; (ii) the internal displacement of the borehole wall and the displacements of specified points in the salt; (iii) the evolution of stresses at specified locations. Temperatures and displacements were of course not disclosed to the participants before the calculations.

It could be noticed immediately that, like in many real world's situations, this problem was a typical three-dimensional one. An ideal model should therefo-

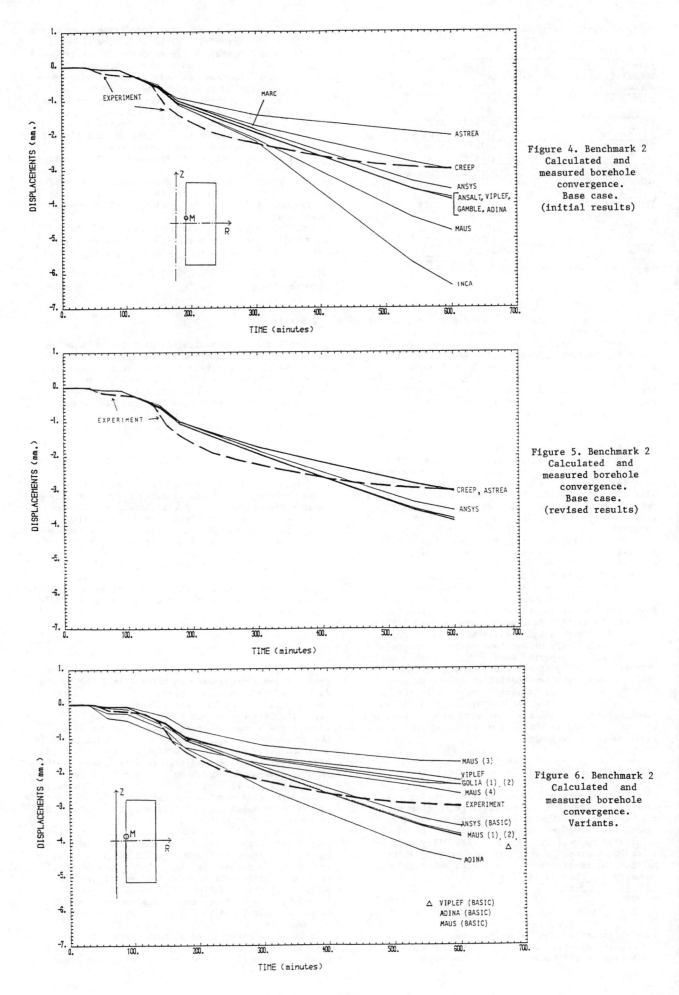

Figure 4. Benchmark 2
Calculated and
measured borehole
convergence.
Base case.
(initial results)

Figure 5. Benchmark 2
Calculated and
measured borehole
convergence.
Base case.
(revised results)

Figure 6. Benchmark 2
Calculated and
measured borehole
convergence.
Variants.

re have been calculated with three- dimensional codes. However, at least presently, this type of calculation necessitates long and expensive runs on very large computers, and the results are not always satisfactory. Therefore, the group of participants came to the idea of a two-dimensional "base case", with acceptable simplifications, which would be common to all teams; additionally, each team would be free to consider variants. This remark mainly applies to the stress and displacement calculations.

The properties of the Na_2 salt were described as follows (with temperature dependence when relevant):
- Specific gravity: 2.187 Kg/m^3
- Thermal conductivity: $\lambda = 5.734 - 1.838.10^{-2}\theta + 2.86.10^{-5}\theta^2 - 1.51.10^{-8}\theta^3$ (with λ in W/m/C°, θ in °C)
- Heat capacity Cp = $1.8705.10^6 + 3.8772.10^2\theta$ (J/m^3/°C)
- Young's modulus E = 24 GPa
- Poisson's ratio ν = 0.27
- Coefficient of thermal expansion α = $4.2.10^{-5}$ °C^{-1}

Other mechanical properties, mainly creep functions, are discussed in a later paragraph. As regards the steel platens, the thermal and mechanical (temperature - independent) properties were:
Specific gravity 7,830 Kg/m^3; Thermal conductivity 60 W/m/°C; Heat capacity 3,758.10^6 J/m^3/°C; Young's modulus 210 GPa; Poisson's ratio 0.3.

4.2 Thermal analysis

Before calculating stresses and displacements, a first step of the exercise consisted in determing the temperature distribution in the salt block. For this purpose, the cube was simplified into a cylinder of diameter 304 mm; therefore, the problem could be considered as axissymmetric. Furthermore, it was considered that the mid-plane perpendicular to the hole axis was a plane of symmetry. Steel platens were introduced respectively as equivalent disks and concentric thick ring with suitable heat capacities and volumes.

For the sake of simplification, temperatures measured at the outer surface of the steel platens were taken as prescribed boundary conditions. The heat flux generated by the heater was supposed to be applied to the total hole wall, as follows:

Time (min)	0	90	150	540	600	810
Heat flux (W/cm^2)	0	0	1.61	1.61	0	0

Betweeen these dates, the flux varied linearly with time. The maximum flux corresponded to a power of 1 kW in the heater.

Figure 3 compares the predicted and measured temperature histories at the borehole wall. All codes gave quite comparable results; the agreement between codes and measurement was excellent at the borehole wall, less satisfactory however at other locations. This was due to the numerous approximations made to achieve reasonable modelling.

4.3 Mechanical analysis - Base case

For the base case, a half portion of the cube was calculated and simplified into a cylinder. Only the upper steel platen was considered; the lateral boundary was directly submitted to a prescribed pressure history.

The applied forces were as follows:

Time (min)	0	30	60	750	780	810
Applied forces (kN)	250	250	2500	2500	250	250

Loads varied linearly between these dates.

It was agreed that there should be no friction between steel and salt. In addition to this "external" loading came the thermally-induced stresses.

The most sensitive problem was related to the material rheology. Only primary (transient) creep was considered. This was described by:

$$\dot{\varepsilon}_{cr} = m \cdot B \cdot exp(-mt) \cdot \sigma_{eq}^n \cdot exp(-Qt/RT)$$

with m = 0.35 day^{-1}; B = 0.21; Q_t = 44.8x10^{3}J
Due to the limitation of some codes, a simplification consisted in setting mt at 0, so that a "pseudo-steady" creep law was obtained.

A first set of results (displacements and stresses) was obtained for the base case; figure 4 depicts the comparison of measurement and predictions for the convergence of the borehole wall. Although the order of magnitude of the predictions could be considered as correct, due to the various approximations made, the discrepancies between individual predictions could not be easily explained. It became apparent that some of these discrepancies were the result of input data errors at the time of calculations, and that one was due to a coding reason in one of the codes. Subsequent corrected runs were therefore performed, with the consequence of a much better agreement between predictions, as shown on figure 5.

4.4 Mechanical analysis- Variants

Some kind of "sensitivity analysis" was performed after the "base case" by exploring some variants concerning the salt rheology and the idealisation of the experiment. Figure 6 puts together the various convergence curves obtained with these variants; for comparison, the "base case" answer is also given for VIPLEF, ADINA, MAUS.

Four variants dealt with salt rheology
- a LEMAITRE strain hardening creep model was fitted on the results of two sample creep tests by EMP on the same salt as the test block.
The new creep equation used in VIPLEF was (5):

$$\dot{\varepsilon}_{cr} = \frac{\alpha \left[(\sigma/K)^\beta \cdot exp(-A/T) \right]^{1/\alpha}}{(\varepsilon_{cr})^{(1-\alpha)/\alpha}}$$

where α = 0.463, β = 3.57, K = 0.166, A = 2540
The corresponding curve is labelled "VIPLEF" on figure 6.
- a value of 7 GPa for Young's modulus was introduced in a revised ADINA run instead of the 24 GPa for the base case (curve "ADINA").
- another secondary creep law, derived from in situ tests in the ASSE salt, was considered for the MAUS code. The corresponding equation was:

$$\dot{\varepsilon}_{cr} = 4.85 \cdot 10^{-6} \cdot \sigma_{eq}^5 \cdot exp(-6880/T)$$

Curve "MAUS (3)" give the corresponding convergence prediction.
- Finally, a primary creep equation with strain hardening was used for the MAUS code (curve "MAUS (4)"). This creep equation is:

$$\dot{\varepsilon}_{cr} = 4.85 \cdot 10^{-6} \langle \sigma_q - \sigma_o \rangle^{14} \cdot exp(-6880/T)$$

where $\langle \sigma - \sigma_o \rangle = \begin{cases} \sigma - \sigma_o & \text{if} \geqslant 0 \\ 0 & \text{if} < 0 \end{cases}$

and $\sigma_o = K \left[\int_{t_o}^{t} \dot{\varepsilon}_{eff} \, dt \right]^{0.36}$ with K = 100 MPa
Also, a number of variants were performed about the idealisation of the test case.
- curve "GOLIA (1)" was obtained assuming that there was no sliding between the steel platens and the salt, whereas "GOLIA(2)" corresponds to a pressure directly applied onto the salt (no steel platen)
- variant "MAUS(1)" was calculated with 64 time steps instead of 133 for the base case, whereas "MAUS (2)" took only 47 time steps
- a plane strain analysis in the mid-plane of the cube was performed using ADINA, with no significant difference compared to the base case (no curve given).

To sum up, it could be seen that predictions of the
borehole convergence were not particularly sensitive
to the various boundary conditions tested, nor to
time discretisation. On the other hand however,
considerable modelling improvement was obtained - for
this short-term experiment - by specifying strain-
hardening creep equations for salt. Nevertheless, the
approximation made for the base case appears to have
been reasonably good.

5 CONCLUSIONS AND PERSPECTIVES

Project COSA actually gave a "snapshot" of the
current combined expertise of Community organisations
in the modelling of thermo-mechanical salt behaviour.
Whereas simplified, hypothetical problems with
constant loading conditions seem to be easily solved
by all codes, complex non-linear analyses with
time-dependent temperature and loading histories
would need to be assessed independently by several
teams in order to be confident about the results.
   Not surprisingly, creep equations for salt were
identified as being the most influential parameters
on calculations; these rheological aspects still need
to be investigated if reliable long-term predictions
of in situ behaviour of salt repositories are to be
made, as already pointed out at various occasions
(6). It is therefore intended to continue this
project with a further phase of benchmarking, progressing towards the simulation of a large scale, long
duration problem akin to repository design. This will
constitute COSA II project, now being defined.

6 ACKNOWLEDGEMENTS

It is a pleasure for the author of this paper to
gratefully acknowledge the very valuable contribution
of all firms and organisations involved in the COSA
project.

REFERENCES

(1) Lowe, M.J.S. & Knowles, N.C. 1986. The Community
    Project COSA: comparison of geo-mechanical computer
    codes for salt. CEC Report no. EUR 10760. Luxem-
    bourg (in press).
(2) Morgan, H.S.; Krieg,R.D. & Matalucci, R.V. 1981.
    Computer analysis of nine structural codes used in
    the Second WIPP Benchmark problem. SANDIA Report
    SAND81-1389, Sandia Laboratories. USA.
(3) Office of Nuclear Waste Isolation. 1984. Perfor-
    mance assessment plans and methods for the salt
    repository project. Technical Report BMI/ONWI-545.
(4) Roest, J.P.A. & Gramberg, J.. 1985. Acoustic
    crosshole measurements of cataclastic thermo-me-
    chanical behaviour of salt. Proceedings of a CEC-
    NEA workshop on "Design and instrumentation of in
    situ experiments in underground laboratories".
    A.A. Balkema Publishers.
(5) Tijani, S.M.; Vouille, G. & Hugout, B. 1983. Le
    sel gemme en tant que liquide visqueux. Proceedings
    of 5th ISRM Symposium, Melbourne. A.A. Balkema
    Publishers.
(6) Rolnik, H. 1984. Dimensionner aujourd'hui un
    stockage de déchets de haute activité dans le sel
    gemme: quelle rhéologie?. PhD Thesis. Ecole Poly-
    technique. Paris.

Identification et caractérisation hydraulique des fractures recoupées par un forage
Identification and hydraulic characterization of fractures in boreholes
Identifizierung und hydraulische Kennzeichnung von Trennflächen in einem Bohrloch

F.H.CORNET, Institut de Physique du Globe de Paris, Université Paris VI, France
J.JOLIVET, Institut de Physique du Globe de Paris, Université Paris VI, France
J.MOSNIER, Laboratoire de Géophysique Appliquée (CNRS), Orléans, France

ABSTRACT: Various logs have been run in boreholes drilled in granite in order to identify and obtain a hydraulic characterization of the fractures intersected by these boreholes.

A new electrical log has been developed to identify the depth, dip and strike of fractures. Thermal and hydrochemical logs have outlined those fractures in which a natural flow occured.

Results from Darcilog (related to P wave attenuation) and from logs of attenuation of S waves and Stoneley waves (obtained with Elf Aquitaine's EVA tool) have been compared to those of a spinner log and of a thermal log run during injections in these boreholes. It is concluded that whilst the logs of attenuation of S waves and Stoneley waves do outline those fractures which are hydraulically conductive near the borehole, they cannot identify systematically which fractures are the most conductive.

RESUME: Differentes diagraphies ont été mises en oeuvre pour identifier et obtenir une caractérisation des propriétés hydrauliques des fractures recoupées par des forages réalisés dans un massif granitique.

Une nouvelle diagraphie électrique a été mise au point pour déterminer la profondeur, le pendage et l'azimut des fractures. Des diagraphies thermiques et hydrochimiques ont permis de mettre en évidence les fractures où intervenaient un écoulement naturel.

Les résultats d'un Darcilog (supposé assimilable à une diagraphie d'atténuation des ondes P) et de diagraphies d'atténuation des ondes S et des ondes de Stoneley (obtenus avec la sonde EVA de Elf Aquitaine) ont été comparés à ceux obtenus lors d'une débitmétrie et d'une thermométrie réalisées durant l'injection d'eau dans ces forages. Il a été observé que si les diagraphies d'atténuation des ondes S et des ondes de Stoneley permettent bien de mettre en évidence les fractures conductrices au niveau du forage, elles ne permettent pas de déterminer systématiquement lesquelles de ces fractures sont les plus conductrices.

1. Introduction

Un programme de recherche sur la possibilité d'exploiter la chaleur des roches profondes peu perméables a été entrepris sur le site granitique du Mayet de Montagne, à 25 km au sud-est de Vichy dans le Massif Central français.

Il s'agit d'étudier la possibilité de créer entre deux forages profonds de 800 m et distants l'un de l'autre de 100m, un échangeur thermique ayant une aire efficace de l'ordre de 200 000m^2 accomodant un débit de circulation d'eau de l'ordre de 60 m^3/h avec un taux de perte de fluide de circulation inférieur ou égal à 10 % pour une impédance hydraulique de l'ordre de 1 GPa/m^3/sec.

Le principe retenu pour la création de cet échangeur est d'utiliser la fracturation naturelle du massif soit telle qu'elle est, soit après stimulation hydraulique de certaines fractures préexistantes. Il importe donc d'obtenir tout d'abord une connaissance aussi précise que possible de cette fracturation naturelle puis de mettre en évidence les fractures susceptibles de participer efficacement à l'échangeur thermique recherché.

Nous présentons ici les études par diagraphies en forage qui ont été réalisées dans ce but. Cette reconnaissance a été complétée par une détermination du champ de contrainte régional par la méthode H.T.P.F. (hydraulic tests on preexisting fractures) décrite par Cornet et Valette (1984) et Cornet (1986). Cette dernière ne sera pas rediscutée ici, les résultats ayant déjà été présentés (Cornet, 1986).

Les deux forages (INAG 3-8 profond de 780m et INAG 3-9 profond de 840m, de diamètre variant entre 165mm et 156mm selon la profondeur) ont été réalisés au marteau fond de trou avec un débit d'air comprimé de l'ordre de 50m^3/min pour une pression de 70 bars. Ils ne sont pas tubés. Les cuttings ont été récoltés tous les mètres d'avancement ce qui a permis d'obtenir une première connaissance de la qualité des terrains. L'analyse détaillée de ces "cuttings" (Couturié et al., 1984) montre que ces forages sont entièrement réalisés dans le granite du mayet de Montagne mais qu'ils ont recoupé un certain nombre de zones bréchiffiées ou simplement faillées.

Cette technique de forage a été choisie de préférence au carottage continu du fait de son faible coût de mise en oeuvre (1 520 000 FF pour les deux forages) d'une part, d'autre part parce qu'il ne nous a pas paru possible d'extraire les informations que nous recherchions de l'étude des carottes.

Un ensemble de huit diagraphies a été réalisé dans le premier forage (INAG 3-8) à savoir :
- une diagraphie thermique,
- une diagraphie de la géochimie des eaux du forage
- trois diagraphies électriques : latérolog de Schlumberger, SHDT (ou pendagimétrie) de Schlumberger, diagraphie Mosnier (Mosnier,1981),
- une diagraphie du spectre du rayonnement γ naturel,
- deux diagraphies acoustiques (B.H.C. de Schlumberger et sonde EVA d'Elf Aquitaine).

Les deux premières diagraphies devaient permettre de mettre en évidence les zones sujettes à des circulations naturelles d'eau(avant forage). Les trois diagraphies électriques avaient pour objet la détermination des cotes où le forage recoupe des fractures ainsi que l'identification du pendage et de l'azimut de ces fractures. La diagraphie spectrale du rayonnement γ naturel devait faire apparaître les variations de composition en uranium, thorium et potassium.

Ces dernières étaient supposées mettre en évidence les zones de concentration ou de lixiviation de l'uranium associées à des circulations anciennes de fluide dans les fractures. Cette idée avait été suggérée par l'analyse des carottes prélevées dans un forage de 150m de profondeur réalisé sur ce site. Malheureusement le manque de résolution des appareils utilisés n'a pas permis d'atteindre cet objectif.

Les deux diagraphies soniques devaient permettre d'obtenir une estimation relative de la conductivité hydraulique des fractures.

Les résultats obtenus sur ce premier forage ont montré que les diagraphies thermiques et géochimiques permettaient bien de mettre en évidence les zones de circulation naturelle d'eau et fournissaient de plus une information sur l'interconnection hydraulique de certaines d'entre elles.

Des trois diagraphies électriques, seule la diagraphie Mosnier s'est révélée être efficace pour déterminer le pendage et l'azimut des fractures.

Pour le deuxième forage (INAG 3-9) n'ont été réalisées que les diagraphies suivantes :
- une diagraphie thermique,
- deux diagraphies électriques (pendagémétrie avec la sonde SHDT de Schlumberger et diagraphie Mosnier),
- deux diagraphies acoustiques (outil BHC de Schlumberger et sonde EVA de Elf-Aquitaine).

Puis des diagraphies de débitmétrie ont été réalisées dans les deux forages en injectant de l'eau à débit constant en tête de puits. Ces résultats ont été complétés en effectuant une débitmétrie et une thermométrie dans le forage 3-8 alors qu'une injection à débit constant intervenait dans le forage 3-9. Ces résultats ont été confrontés aux résultats des diagraphies acoustiques.

Nous présentons ci-après les résultats des diagraphies thermiques et géochimiques réalisées après que le massif soit retourné à son état stationnaire du point de vue des circulations naturelles d'eau. Nous décrivons ensuite le principe de la diagraphie électrique Mosnier et présentons quelques résultats. Puis nous comparons les résultats des mesures de conductivité hydraulique de fractures déduites des diagraphies acoustiques à ceux de débitmétrie et de thermométrie obtenues lors d'injection d'eau.

2. DIAGRAPHIES GEOCHIMIQUE ET THERMOMETRIQUES SUR LE FORAGE 3-8

Ces deux ensembles de diagraphies ont été réalisées après que les forages aient été laissés au repos durant plus d'un mois après la foration et avant toute autre intervention. Le massif était ainsi supposé être retourné à un état stationnaire du point de vue des écoulements naturels d'eau.

Pour la diagraphie géochimique, des échantillons d'eau de 300 cm^3 ont été prélevés tous les 20m en commençant par les cotes les plus superficielles de façon à ne pas apporter de perturbation pour les prélèvements les plus profonds. les mesures de PH et de l'alcalinité (Hco_3^-) ont été effectués sur le terrain alors que les dosages concernant S_iO_2, cl^-, NO_3^-, SO_4^{--}, Ca^{++}, Mg^{++}, Na^+, k^+, Li, et M_n

ont été réalisés au laboratoire (Bidaux,1986). Il apparait que l'eau est de faciès dominant bicarbonaté calcique, légèrement acide (PH variant de 6.52 à 6.93). Les variations de composition au long du forage sont en général faibles et irrégulières (Figure 1). On relève cependant d'une part un niveau réducteur marqué à 480 m lié apparemment à une circulation d'eau dans le massif et à la présence de sulfures mise en évidence dans les cuttings, d'autre part des teneurs anormalement faibles en Na^+ et k^+ à 580 m qui pourraient

être dues à la présence de granite altéré (suggéré par l'analyse des cuttings). De légères anomalies de concentration des divers constituants testés ont en outre été observés entre 520m et 540m, 600 et 620m, 660 et 680m.

Une première diagraphie thermique a été réalisée jusqu'à 500m de profondeur, 45 jours après la fin de la foration. La sonde utilisée était constituée d'une thermistance unique ayant une constante de temps inférieure à la seconde, une sensibilité de l'ordre du millième de degré et une fidélité meilleure que le centième de degré. Cette diagraphie a été utilisée pour établir des diagraphies de gradient thermique calculées sur des bases de 2, 5 et 10m.

Le gradient thermique est défini comme la valeur moyenne obtenue pour une longueur de référence (2,5 et 10m) et la diagraphie correspond aux variations avec la profondeur (incrément de 0.25cm) de cette valeur moyenne. La diagraphie obtenue pour une base de 5m est présentée sur la figure 2a.

Pour ces diagraphies de gradient thermique, le bruit est de l'ordre de 4^0/1000m ; on observe en outre des variations brutales pouvant atteindre 60^0/1000m dans les deux cents premiers mètres. On relève plus particulièrement, en dessous de 200m de profondeur, des "accidents" entre 223 et 236m, 363 et 384m, 409 et 415m et à 471m où le gradient atteint 56^0/1000m.

Cette diagraphie a été répétée un an plus tard après qu'aient été injectés au moyen d'un double obturateur gonflable centré à 440m de profondeur, $110m^3$ d'eau à un débit moyen de 1.3 m^3/h. Sur cette nouvelle diagraphie du gradient thermique (Figure 2b) on remarque que l'accident majeur intervient à 440m, que les anomalies observées entre 200m et 440m, si elles sont toujours présentes, sont très atténuées mais que l'anomalie à 471m reste très marquée. On relève plus particulièrement un réchauffement de la roche, par rapport à la valeur moyenne attendue, entre 550m et 471m.

De ces trois diagraphies nous avons tiré les conclusions suivantes :
- La zone réductrice observée à 480m (pas d'échantillonnage de 20m) sur la diagraphie géochimique est associée à une perturbation thermique notoire (réchauffement) du massif à 471m. Elle correspond donc à une zone d'écoulement ascendant naturel.
- L'anomalie thermique à 440m observée sur la deuxième diagraphie thermique, coincide exactement avec la cote où a eu lieu l'essai de stimulation hydraulique. Ce dernier a donc établi une connection hydraulique efficace entre le forage et la zone d'écoulement ascendant naturel.
- Le fait que les anomalies thermiques observées entre 440m et 200m soient nettement moins marquées sur la deuxième diagraphie que sur la première suggèrent que ces anomalies étaient associées à l'écoulement naturel ascendant observé à 471m : le drainage par le forage, grâce à la stimulation hydraulique à 440m, a fortement diminué le gradient hydraulique dans ce courant ascendant au dessus de cette cote. En outre, ces anomalies ne sont observées que sur une faible longueur de forage (5 à 10m) ce qui suggère qu'elles correspondent à des écoulements induits par la présence du forage et non à des écoulements préexistants avant la foration, contrairement à l'anomalie à 471m.

Enfin il est intéressant d'observer qu'en dessous de 550m, aucune anomalie du gradient thermique n'est observée ce qui permet de conclure que les anomalies de composition chimique des eaux du forage, ou bien ne sont pas dues à des circulations d'eau mais à une réaction avec l'encaissant, ou bien sont associées à des circulations suffisamment lentes pour qu'aucune perturbation thermique ne soit décelable.

Cette dernière observation suggèrerait donc que la vitesse d'écoulement naturel dans les fractures diminue avec la profondeur.

3. DIAGRAPHIES ELECTRIQUES

Les diagraphies électriques classiques utilisées pour mettre en évidence des fractures dans un milieu homogène comme le granite sont le latérolog et la pendagemétrie. Le latérolog mesure les variations de résistivité électrique pour des "tubes" de roche coaxiaux au forage et dont l'épaisseur est une fonction de la distance entre les électrodes. La pendagemétrie correspond à des mesures de résistivité électrique dans quatre azimuts différents au moyen de patins disposés aux extrémités de deux diamètres perpendiculaires entre eux et frottant contre la paroie. Par corrélation entre les résultats obtenus pour chacun des patins, il est possible de calculer le pendage de la structure conductrice recoupée par le forage en supposant que celle-ci soit plane. Cette diagraphie est couplée à celle de la détermination de la géométrie du forage (azimut et pendage de l'axe du forage, variations de diamètre dans deux directions perpendiculaires entre elles).

Ces diagraphies sont classiques et ne seront pas discutées plus avant. Notons simplement que les résultats obtenus avec ces deux techniques sont tout à fait comparables à ceux de la diagraphie électrique Mosnier, décrite ci-après, quant à la détermination des profondeurs des fractures recoupées par le forage mais que la diagraphie Mosnier s'est révélée beaucoup plus précise pour la détermination des pendages et azimuts des fractures.

Le principe de la diagraphie électrique Mosnier (Mosnier, 1981) utilisée pour cartographier les fractures est comparable à celui de la pendagemétrie mais la sonde Mosnier permet une "cartographie" de la totalité de la paroi du forage contrairement à la pendagémetrie.

Ce principe consiste à mesurer la différence de conductivité électrique entre la matrice rocheuse (résistante) et le matériau qui remplit la fracture (plus conducteur du fait de sa teneur élevée en eau). La méthode pratique consiste à faire une carte de l'impédance électrique de la paroi du forage en la mesurant à la fois en fonction de l'azimut et de la profondeur.

A partir de cette idée, une sonde de 26 électrodes (2 de garde EGS et EGI, et 24 de mesures EM) a été construite (voir figure 3). Le courant est injecté dans le terrain par la sonde de 26 électrodes, l'électrode de recueil étant l'armature métallique externe du câble de diagraphie. Pour obliger les lignes de courant issues des électrodes de mesure à pénétrer dans le terrain, au lieu de se refermer immédiatement sur le câble, une certaine longueur de ce dernier, juste au dessus de la sonde, est isolée par une gaine. Les deux électrodes de garde, disposées de part et d'autre des 24 électrodes de mesure, permettent d'assurer que les lignes de courant au niveau des électrodes de mesures sont parfaitement normales à la paroi du forage. Les 24 électrodes de mesures permettent de déterminer l'intensité du courant qui pénètre dans la formation dans tous les azimuts (une électrode de mesure tous les 15°), l'orientation de la sonde étant repérée par rapport au Nord magnétique.

Un exemple de résultat obtenu pour le forage 3-8 entre les cotes 472m et 470m est présenté sur la Figure 4. On a représenté pour des incréments de profondeur de 12cm, les amplitudes de l'intensité du courant mesuré pour les 24 électrodes sur des diagrammes polaires.

On remarque plus particulièrement l'apparition d'un conducteur dans la direction N 70°E à 471.9m. Ce conducteur se dédouble ensuite pour des cotes décroissantes puis réapparait solitaire à 470,7m dans la direction N 250°E et disparait pour les cotes supérieures. Ces résultats permettent d'identifier une fracture dans le direction N 160°E ∓ 7° et de pendage 85° (fracture sub-verticale).

Elle correspond à la cote où était apparue l'anomalie thermique significative décrite ci-dessus.

Un deuxième type de présentation des résultats consiste à associer à chaque groupe de 24 valeurs de l'intensité i(z), 24 carrés (de 1.5mm de côté), dont la noirceur dépend de l'intensité émise par l'électrode correspondante. Une nouvelle ligne est tracée pour la cote z + Δz, ou Δz est l'incrément de profondeur entre deux scrutations azimutales et ainsi de suite. On obtient ainsi une cartographie électrique de la paroi du forage. Un exemple de résultat est présenté sur la Figure 5 où apparait nettement la fracture à 471m. Cette cartographie du forage est obtenue en temps réel sur le terrain.

Cette diagraphie a été réalisée sur la hauteur totale des deux forages 3-8 et 3-9 et il serait fastidieux d'énumérer ici l'ensemble des fractures mises en évidence. Seule une approche statistique de ces résultats peut permettre d'obtenir une représentation de la fracturation naturelle ainsi observée. Cet aspect ne sera pas développé ici. Soulignons cependant que seules les fractures conductrices d'électricité sont identifiées avec cette méthode, et que toutes celles qui sont recimentées ne peuvent pas être identifiées de la sorte.

4. CARACTERISATION DE LA CONDUCTIVITE HYDRAULIQUE DES FRACTURES A PARTIR DES DIAGRAPHIES ACOUSTIQUES

L'atténuation des ondes mécaniques se propageant dans une roche saturée de fluide, dépend de façon significative des mouvements de fluide dans l'espace poreux (Biot,1956 ; Morlier et Sarda,1971; Mavko et Nur, 1979). Aussi a-t-il été proposé d'utiliser des mesures de l'atténuation des ondes P (Lebreton et al., 1977 ; Cheung, 1984) ou de l'atténuation des ondes de Stoneley (Cheng and Toksöz, 1981 ; Paillet, 1980 ; Mathieu and Toksöz, 1984) pour obtenir une caractérisation soit de la perméabilité d'un milieu poreux, soit de la conductivité hydraulique des fractures.

Nous présentons une comparaison des résultats obtenus à partir de diagraphies acoustiques (Darcilog établi à partir des premières arrivées des ondes P, diagraphies des variations d'atténuation des ondes S et des ondes de Stoneley obtenues à partir des logs acoustiques réalisés avec la sonde EVA multiémetteurs-multirécepteurs de Elf Aquitaine, Arditty et Arens, 1985) et des résultats de débitmétrie et de thermométrie obtenus lors d'injections à débit constant dans les forages INAG 3-8 et INAG 3-9. Nous n'avons pas utilisé pour cette comparaison les valeurs de conductivité hydraulique déduites des diagraphies d'atténuation mais avons préféré nous en tenir aux valeurs brutes Le Darcilog est établi en calculant pour chaque cote où un signal est reçu l'indice I_c défini comme l'inverse du rapport entre l'amplitude de la première arche du signal reçu sur la somme des amplitudes des deux arches suivantes ($I_c = A_2+A_3/A_1$ où A_i est l'amplitude de la ième arche du signal pour i = 1, 2, 3). Il a été proposé (Ducomte, 1984) que plus la conductivité hydraulique est élevée plus la valeur de I_c devrait l'être. La diagraphie de l'indice I_c a été calculée à partir de la diagraphie acoustique réalisée avec l'outil BHC de Schlumberger.

Les atténuations des ondes S et des ondes de Stoneley, sont calculées à partir de rapports d'amplitude d'ondes pointées sur des signaux acoustiques provenant d'un émetteur commun et de deux récepteurs proches ou d'un récepteur commun et de deux émetteurs proches (Mathieu et Arditty, 1986).

Afin d'identifier les fractures les plus conductrices du point de vue hydraulique dans le forage INAG 3-8 (non tubé) un obturateur gonflable

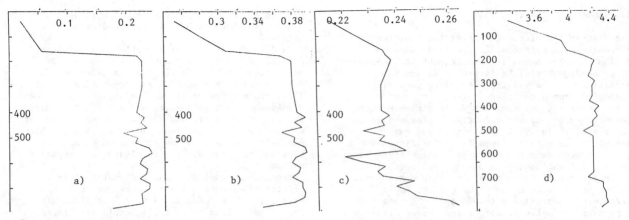

Figure 1. Résultat de l'analyse chimique des eaux du forage INAG 3-8 pour un pas d'échantillonnage de 20m. L'axe vertical représente la profondeur en mètres, l'axe horizontal la concentration de l'élément correspondant en milligrammes par litre. a) NO_3^- , b) cl^-, c) Na^+, d) HCO_3^-.

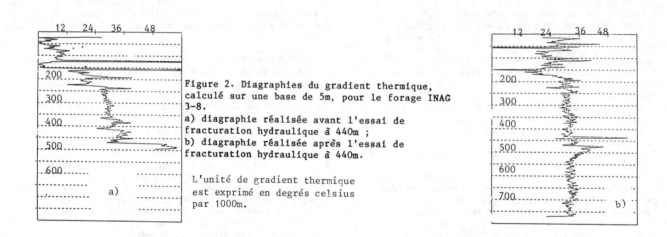

Figure 2. Diagraphies du gradient thermique, calculé sur une base de 5m, pour le forage INAG 3-8.
a) diagraphie réalisée avant l'essai de fracturation hydraulique à 440m ;
b) diagraphie réalisée après l'essai de fracturation hydraulique à 440m.

L'unité de gradient thermique est exprimé en degrés celsius par 1000m.

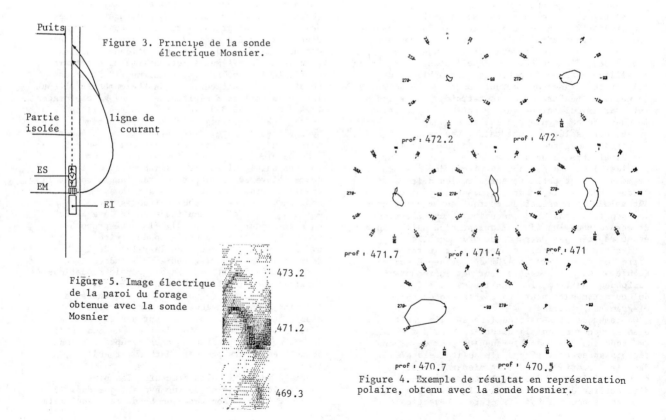

Figure 3. Principe de la sonde électrique Mosnier.

Figure 5. Image électrique de la paroi du forage obtenue avec la sonde Mosnier

Figure 4. Exemple de résultat en représentation polaire, obtenu avec la sonde Mosnier.

64

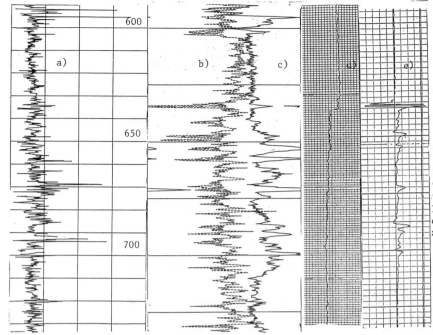

Figure 6. Comparaison des résultats obtenus entre 600m et 725m dans le forage INAG 3-8, pour a) l'indice I_c, b) l'atténuation des ondes S, c) l'atténuation des ondes de Stoneley (échelle inversée par rapport à b), d) la débitmétrie lors de l'injection d'eau dans INAG 3-8, e) le gradient thermique lors de l'injection dans INAG 3-9. Les cotes ont été corrigées des décalages de zéro.

a été descendu au bout d'un train de tige à 251 m de profondeur et une injection à débit constant de 28 m³/h a été entreprise en-dessous de cette cote. Les variations du débit d'écoulement dans le forage (débitmétrie) ont été mesurées à l'aide d'un micro-moulinet.

Enfin, une injection à débit constant (60 m³/h) a été entreprise dans le forage INAG 3-9 (situé à 100m de INAG 3-8) entre 605m et le fond du forage (840m). Après 1 h 50 de pompage le forage INAG 3-8 s'est mis à produire de l'eau en surface et une thermométrie y a alors été effectuée afin d'identifier les fractures productrices d'eau. Ces fractures sont identifiées par les variations de gradient thermique qu'elles engendrent. En effet, l'eau produite par la formation s'écoule dans le forage à la température de la roche à cette profondeur puis voit sa température décroître au fur et à mesure qu'elle remonte dans le forage. Plus le débit de production est élevé, plus l'influence sur le gradient thermique local est marqué.

La figure 6 présente pour le forage INAG 3-8 entre les profondeurs 600 et 720m, les variations de l'indice I_c, les variations d'atténuation des ondes S et des ondes de Stoneley, les variations de vitesse de rotation du micromoulinet lors de l'essai d'injection dans INAG 3-8, les variations de gradient thermique lors de l'injection dans INAG 3-9.

Il apparaît tant sur la débitmétrie que sur la thermométrie que la fracture la plus conductrice dans cet intervalle se trouve à 638m. Elle correspond à celle où l'atténuation des ondes S est la plus marquée ainsi qu'à une atténuation des ondes de Stoneley qui n'est cependant pas parmi les plus importantes. Elle n'apparaît pas sur le log d'indice I_c mais ceci semble être dû à un problème instrumental à cette cote. La thermométrie fait clairement ressortir des venues d'eau à 648 m, 658m, 672,5m entre 683 et 685 m et à 700m. Ces cotes bien que visibles sur la débitmétrie n'y sont pas très marquées. Les cotes 648m et 658m sont bien individualisées sur la diagraphie des ondes S mais les venues d'eau à 672.5m et 683m sont moins clairement marquées. les cotes 658m, 672m 50 et 700m ressortent très nettement sur la diagraphie d'atténuation des ondes de Stoneley, la cote 683m y

étant aussi visible. Par contre il existe des zones de forte atténuation des ondes de Stoneley qui ne se manifestent que très modestement sur la thermométrie et la débitmétrie. Aucune de ces cotes, sauf une, n'est visible sur la diagraphie de l'indice I_c. En revanche les cotes où de fortes anomalies de l'indice I_c sont observées ne correspondent pas à des zones hydrauliqueemnt conductrices.

Les résultats obtenus entre 400 et 500m de profondeur permettent d'identifier tant sur la débitmétrie que sur la thermométrie les cotes 440m et 471m déjà mentionnées (440m site d'une fracturation hydraulique, 471m site d'une fracture où intervient une circulation naturelle d'eau). Si la cote 440m marque très nettement sur la diagraphie d'atténuation des ondes S, la cote 471m, bien que visible n'y apparaît pas comme particulièrement marquée. Soulignons que la cote 440m était à peine visible sur la diagraphie équivalente réalisée avant fracturation. Pour l'atténuation des ondes de Stoneley, la cote 440m n'apparaît pas tandis que la cote 471m n'y apparaît que modestement. Sur la diagraphie de l'indice I_c réalisée avant l'essai de fracturation hydraulique, la cote 471m n'apparaît pas.

Ainsi les trois cotes les plus significatives du point de vue hydraulique pour le forage 3-8 (440m, 471m et 638m) correspondent pour deux d'entre elles aux cotes où l'atténuation des ondes S est la plus forte mais la troisième (471m) n'apparaît pas comme une valeur particulièrement forte.

L'atténuation des ondes de Stoneley, normale à la cote 440m, à peine supérieure à la moyenne aux cotes 471 et 638m n'aurait pas permis d'individualiser ces trois cotes vis à vis d'autres zones fracturées associées à de fortes atténuations, mais de conductivité hydraulique faible, voire quasiment nulle. La seule étude de l'indice I_c n'aurait pas non plus permis d'identifier les zones les plus conductrices.

Si l'indice I_c est représentatif de l'atténuation des ondes de compression, ce dernier paramètre n'est pas sensible qu'à la conductivité hydraulique, il dépend aussi d'autres facteurs tels que les variations de lithologie. Il n'est donc pas surprenant que de forts indices I_c aient été observés en l'absence de fracture.

Alors que ce raisonnement pourrait aussi s'appliquer à l'atténuation des ondes S, tel ne semble pas avoir été le cas pour le site du Mayet de Montagne. Cette observation confirme des résultats comparables obtenus par Elf Aquitaine dans d'autres types de formation (Mathieu et Arditty, 1986).

On relève d'autre part que tous les indices de conductivité hydraulique sauf un (440m) coïncident avec une atténuation plus ou moins forte des ondes de Stoneley sans que l'on puisse relever une certaine proportionnalité entre ces deux paramètres. L'absence d'atténuation à 440m peut s'expliquer par le fait qu'il s'agit de fractures très fines. En effet on peut montrer théoriquement (Mathieu et Toksöz, 1984) qu'une fracture d'épaisseur inférieure à 100 microns n'induit pas d'atténuation sensible des ondes de Stoneley. En outre, le manque de corrélation entre forte atténuation des ondes de Stoneley et fort indice de conductivité hydraulique peut s'expliquer par la faible profondeur de pénétration de l'onde de Stoneley. Cette dernière, inférieure à quelques dizaines de centimètres ne permet que de donner une image de la conductivité à proximité du puits.

5. DISCUSSION ET CONCLUSION

Lorsque l'on compare les résultats obtenus avec les diagraphies thermiques initiales, c'est-à-dire effectuées après avoir laissé le forage au repos, et ceux de la débimétrie, on remarque que les anomalies thermiques observées entre 300m et 415m, qui avaient été attribuées à des venues d'eau dans le forage, n'ont pas été observées lors de la débitmétrie. Par contre elles coincident toutes sauf une (384m) avec une atténuation notoire des ondes de Stoneley et parfois avec une atténuation des ondes S.

De même les intervalles de profondeur à l'intérieur desquels des anomalies ont été détectées dans la géochimie des eaux du forage comportent tous sauf un (520-540m) des zones où l'on observe une forte atténuation des ondes de Stoneley.

En outre toutes les zones où de fortes atténuation des ondes de Stoneley ont été observées correspondent à des zones reconnues comme brechiffiées ou très fortement hydrothermalisée lors de l'analyse des cuttings.

Ainsi il semble bien que le forage ait recoupé un certain nombre de zones hydrauliquement conductrices au niveau du forage mais leur rôle hydraulique lors de la mise sous pression du forage est resté négligeable devant l'importance de deux ou trois fractures majeures.

En fait cette notion de conductivité hydraulique fait intervenir l'interconnection des fractures entre elles et leur orientation relative vis à vis du champ de contrainte régional de sorte que cette propriété ne peut être étudiée à partir des seules observations des propriétés des fractures au niveau de leur intersection avec le forage.

Par contre il semble bien que l'étude combinée de l'atténuation des ondes S et des ondes de Stoneley permette de mettre en évidence toutes les fractures potentiellement conductrices même si elles ne permettent pas de faire apparaître à elles seules une hiérarchie dans le rôle hydraulique de ces fractures au niveau du massif.

Pour le problème posé par le développement d'un échangeur thermique en roche chaude sèche cette information sur les fractures potentiellement conductrices pourrait permettre de définir les fractures préexistantes qui ne participent pas dans leur configuration naturelle à l'échangeur thermique mais qui serait susceptible d'améliorer le rendement de cet échangeur après une stimulation appropriée.

REMERCIEMENTS

Ce travail a été financé par l'Agence Française pour la Maîtrise de l'Energie et le Centre National de la Recherche Scientifique (PIRSEM et INSU) ainsi que par la Mission Scientifique et Technique (Département Génie Civil). Nous tenon à remercier la Société Elf Aquitaine et plus particulièrement P. Arditty et F.Mathieu pour les résultats obtenus avec la sonde EVA.

REFERENCES

Arditty, P. et G. Arens 1985. EVA-pour une interprétation lithogogique et pétrophysique des réservoirs. Séminaire sismique de puits, Monaco, Arditty Editor (Elf Aquitaine, Paris La Défense, France).

Bidaux, P. 1986. Contribution à l'étude des circulations profondes en milieu fissuré peu perméable. Identification à partir de mesures hydrochimiques le long du forage. Thèse de doctorat, spécialité Géologie, Université de Montpellier.

Biot, M.A. 1956. Theory of propagation of elastic waves in a fluid saturated porous solid. Jou. Acoust. Soc. Am., vol. 28, pp.168-191.

Cheng, C.H. and M.N. Toksöz 1981. Elastic wave propagation in a fluid filled borehole and synthetic acoustic logs. Geophysics, vol.46, pp.1042-1053.

Cheung, Ch. 1984. Fracture detection using the sonic log. 9th Formation Evaluation Symp., paper 42, Soc. Avance. Interp. Diagr. Paris.

Cornet, F.H. 1986. Stress determination from hydraulic Tests on Preexisting Fractures. The HTPF method. Int. Symp. Rock Stress and Rock Stress Measurements, Stockholm, Centek Publication, Lulea (Suede).

Cornet, F.H. and B. Valette 1984. In Situ Stress Determination from hydraulic injection test data; Jou. Geophys. Res., vol.89, B13, pp11527-11537.

Couturié, J.P., M. Binon et F. Carmier 1984. Rapport géologique sur le forage INAG 3-8, Rapport d'activité Mayet de Montagne, Inst. de Phys. du Globe, Paris.

Ducomte, C. 1984. Application d'une diagraphie de perméabilité Sealdex et Exaflo. 9th Int. Formation Evaluation. Soc. Avance. Interp. Diagr. Paris.

Lebreton, F., J.P. Sarda, E. Troquemé et P. Morlier 1978. Logging tests in porous media to evaluate the influence of their permeability on acoustic waveforms, 19th S.P.W.L.A., Logging Symp., El Paso.

Mathieu F., et P. Arditty 1986. Rapport interne Elf Aquitaine, Département Sismique de puits. Paris La Défense,

Mathieu F. and M.N. Toksöz 1984. Application of full waveform acoustic logging data to the estimation of reservoir permeability, paper BHG15, 54th SEG Meeting, Atlanta.

Mavco C.M. and A. Nur 1979. Wave attenuation in partially saturated rocks. Geophysics, vol.44, pp.161-178.

Morlier P. et J.P. Sarda 1971. Attenuation des ondes élastiques dans les roches poreuses saturées. Rev. inst. Fr. Pétrole., vol.26, pp.731-756.

Mosnier J. 1982. Détection électrique des fractures naturelles ou artificielles dans un forage. Annal. Geophys., vol.38, p.537-540.

Paillet F.L. 1980. Acoustic propagation in the vicinity of fractures which intersect a fluid-filled borehole. S.P.W.L.A., 21st well logging Symp., Lafayette Louisiana (U.S.A).

Geotechnische Aspekte der Endlagerung von Sonderabfällen in Salzkavernen
Geotechnical aspects of hazardous waste storage in salt cavities
Aspects géotechniques du stockage des déchets dans les cavités de sel

F.CROTOGINO, Kavernen Bau- und Betriebs-GmbH, Hannover, Bundesrepublik Deutschland
K.-H.LUX, Universität Clausthal-Zellerfeld, Institut für Bergbau, Bundesrepublik Deutschland
R.ROKAHR, Universität Hannover, Institut für Unterirdisches Bauen, Bundesrepublik Deutschland
H.-J.SCHNEIDER, Kavernen Bau- und Betriebs-GmbH, Hannover, Bundesrepublik Deutschland

ABSTRACT: Hazardous waste materials with high environmental risk potential are being increasingly deposited in underground repositories, because the additional geological barrier of the overlying rock ensures considerable reduction of risks to the environment in comparison with landfill depots. In the Federal Republic of Germany an underground repository in a salt mine has been in operation for a number of years. In addition to this, for capacity reasons, the State of Lower Saxony is planning the construction of an underground repository in salt caverns. Due to the differences in construction and spatial layout of salt caverns in comparison with mines, different requirements are placed on the geotechical design and the operation of salt cavern repositories. The introductory presentation of the repository concept is followed by a discussion of the geological and geotechnical preconditions for immission-free deposition in caverns, stability and safety analyses for the various operational phases and the formulation of consequent requirements placed on the nature of waste materials to be deposited as well as on the stages in repository operation. With respect to the long-term safety of salt cavern repositories, rock mechanical analyses of the long-term behaviour of the material deposited in salt rock as well as of the permanent cavern seal are being performed.

RESUME: Les déchets et résidus dangereux à haut degré de pollution pour l'environnement sont de plus en plus fréquemment stockés dans des décharges souterraines; ce, du fait que les risques de pollution de l'environnement, par comparaison avec les décharges de surface, sont notablement réduits par la barrière géologique naturelle composée par les terrains de recouvrement. Une décharge souterraine de ce type est en service depuis déjà de nombreuses années en République Fédérale d'Allemagne dans une mine de sel. En complément à cette installation existante, le gouvernement du Land de Basse-Saxe programme l'implantation d'une décharge souterraine en cavités de sel. Compte tenu des différentes conditions de constitution et de répartition spatiale, se présentent, par opposition aux mines de sel, d'autres exigences géotechniques de dimmensionnement et d'exploitation des décharges en cavités de sel. Après proposition d'un type de décharge en cavités de sel, seront étudiées les conditions géologiques et géotechniques d'un mode de stockage définitif sans immissions. Les analyses de résistance et de stabilité statique correspondant aux différentes phases d'exploitation des cavités seront réalisées en fonction des contraintes et de la structure des déchets et résidus à stocker comme du mode d'exploitation de la décharge. En ce qui concerne la sécurité à long terme du dépôt dans une cavité du sel seront réalisées les études des conditions de mécanique des roches en fonction du comportement à échéance des dépôts dans les roches salines comme des conditions de fermeture durable de la cavité.

ZUSAMMENFASSUNG: Sonderabfälle mit stark umweltbelastendem Gefährdungspotential werden zunehmend in untertägigen Deponien eingelagert, da die Risiken der Umweltgefährdung durch die zusätzliche geologische Barriere des überlagernden Deckgebirges im Vergleich zu einer obertägigen Deponie wesentlich abgemindert werden. In der Bundesrepublik Deutschland ist bereits seit mehreren Jahren eine Untertagedeponie in einem Salzbergwerk in Betrieb. Zusätzlich zu dieser bestehenden Anlage plant das Bundesland Niedersachsen aus Kapazitätsgründen die Errichtung einer Untertagedeponie in Salzkavernen. Aufgrund der unterschiedlichen Herstellung und räumlichen Anordnung von Salzkavernen im Vergleich zum Bergwerk ergeben sich abweichende Anforderungen an die geotechnische Auslegung und den Betrieb von Salzkavernendeponien. Nach einer einleitenden Vorstellung des Deponiekonzeptes in Salzkavernen werden die geologisch-geotechnischen Voraussetzungen für eine immissionsneutrale Endlagerung in Kavernen diskutiert, Stabilitäts- und Standsicherheitsanalysen für die verschiedenen Betriebsabläufe der Kavernen vorgenommen und die daraus resultierenden Anforderungen an die Beschaffenheit der einzulagernden Abfallstoffe sowie an den Ablauf des Deponiebetriebes formuliert. In bezug auf die Langzeitsicherheit der Salzkavernendeponie werden gebirgsmechanische Analysen zum Langzeitverhalten des Deponiegutes im Salzgebirge sowie zum dauerhaften Endverschluß der Kaverne angestellt.

1 EINLEITUNG

Die schlechten Erfahrungen mit Altdeponien, sogenannte Altlasten, die heute vielfach ein Sicherheitsrisiko für die Umwelt darstellen und meist mit beträchtlichem Kostenaufwand saniert werden müssen, die ständige Zunahme der Abfallmengen sowie der Nachweis von immer kleineren Schadstoffkonzentrationen in der Umwelt aufgrund der Verfeinerung analytischer Meßmethoden haben zu einer grundsätzlichen Revision der bisherigen Entsorgungspraxis sowie deren gesetzlicher Regelung geführt. Danach muß das Abfallendlager nach dem heutigen Stand der Technik einen völligen Abschluß der Schadstoffe sicherstellen.

Neben der Ertüchtigung der Barrierensysteme von Abfalldeponien kann dies nur durch eine schärfere Trennung der Abfallströme entsprechend ihrem Gefährdungspotential bzw. Elutionsverhalten und Reaktionsvermögen in obertägige und untertägige Deponien bewerkstelligt werden, da die Ablagerung in obertägigen Deponien mit größeren Sekundärproblemen, insbesondere in bezug auf den Anfall von Sickerwasser sowie die Langzeitsicherheit, behaftet ist. Abfälle, die giftig, nicht konditionierbar, wasserlöslich sind oder mit anderen Abfällen lösliche Verbindungen eingehen, sollten nicht mehr in Ober-

tage- sondern ausschließlich in Untertagedeponien eingelagert werden, da diese durch die natürliche geologische Barriere in tiefen, undurchlässigen, geologischen Formationen und durch ihre große Entfernung zum Grundwasserträger einen wirkungsvollen Abschluß zur Biosphäre sicherstellen (Abb. 1). Salzgestein ist aufgrund seiner gesteinsphysikalischen Eigenschaften für die Anlage von Untertagedeponien besonders geeignet.

2 KONZEPT UND BETRIEB DER SALZKAVERNENDEPONIE
2.1 Herstellung von Salzkavernen

Salzkavernen sind Hohlräume, die soltechnisch im Salzgebirge erstellt werden. Ihre Herstellung stellt bestimmte Bedingungen an die geologische Beschaffenheit der Salzlagerstätte, wie ausreichende Mächtigkeit und Erstreckung oberhalb 2 000 m, eine homogene Beschaffenheit des Salzes, das aus soltechnischen Gründen weitgehend frei von nichtlöslichen oder schwerlöslichen sowie hochlöslichen Bestandteilen sein muß.

Für die soltechnische Erstellung werden zunächst Bohrungen mit Hilfe der Tiefbohrtechnik bis zur vorgesehenen Endteufe der Kaverne niedergebracht, wobei die letzte zementierte Rohrtour zumindest über eine Strecke von 100 - 200 m im Salz

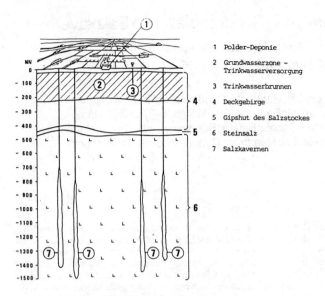

1 Polder-Deponie
2 Grundwasserzone – Trinkwasserversorgung
3 Trinkwasserbrunnen
4 Deckgebirge
5 Gipshut des Salzstockes
6 Steinsalz
7 Salzkavernen

Abb. 1. Lage der Übertagedeponie und der Salzkavernendeponie zum Ökosystem

verläuft. Für den Solprozeß werden drei ineinander hängende Rohrstränge eingebaut (Abb. 2). Über den ersten Rohrstrang wird Süßwasser oder Meerwasser in den Kavernenbereich gepumpt. Dort löst es Salz an den Kavernenwänden und sättigt sich auf. Die Sole wird über den zweiten Verbindungsweg nach Übertage abgeleitet. Über den dritten Verbindungsweg, den äußeren Ringraum, wird ein Schutzmedium in die Kaverne gepumpt, um den Lösungsprozeß in der Vertikalen, also zum Kavernendach hin, zu begrenzen.

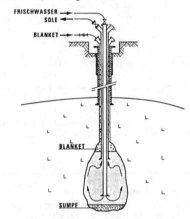

FRISCHWASSER
SOLE
BLANKET

BLANKET

SUMPF

Abb. 2. Herstellung von Salzkavernen im Lösungsbergbau nach dem direkten Solverfahren (Schemadarstellung)

Nach dem heutigen Stand der Technik ist es möglich, mit Hilfe der Soltechnik Hohlräume von gewünschten Abmessungen zu erstellen (QUAST, BECKEL, /1/). Zur Kontrolle des ausgesolten Hohlraumes wird die Kaverne während und nach der Herstellung echometrisch vermessen.

2.2 Vorbereitung des Hohlraumes für die Deponierung von Sonderabfällen

Eine soltechnisch hergestellte Salzkaverne ist am Ende des Herstellprozesses mit Sole gefüllt. Darüber hinaus befindet sich im Kavernentiefsten der sogenannte "Sumpf", in dem sich die unlöslichen Bestandteile aus dem Salz (Anhydrit und Ton) während des Solprozesses ansammeln.

Vor Beginn der Abfalleinlagerung ist die solebefüllte Kaverne mit Hilfe von Tauchkreiselpumpen oder durch pneumatische Soleverdrängung zu entleeren.

Die im Sumpf enthaltene Restsole muß bei der Deponierung von Abfällen, die mit Wasser reagieren, durch Sumpfentwässerung und/oder Abbindemittel beseitigt werden. Bei Abfällen, die leichtflüchtige, explosionsgefährdete Stoffe enthalten, ist der Kavernenhohlraum mit Inertgas zu befüllen. Die Inertgasbefüllung erfolgt in Verbindung mit der Soleentleerung.

2.3 Betrieb der Salzkavernendeponie

Die Abfälle werden über die Zugangsbohrung in die Kaverne über eine zusätzlich eingehängte Rohrtour kontinuierlich mit einer Rate von 100 000 – 200 000 m³/a eingelagert. Diese zusätzliche Rohrtour dient zum Schutz der äußeren Verrohrung vor Korrosion und Abrasion und kann bei Störfällen ausgebaut bzw. ersetzt werden.

Die Einlagerung der Abfälle kann alternativ nach zwei Verfahren
- als Feststoffschüttgut über eine Freifalleitung oder
- als pumpenunterstützte Dickstoffförderung
vorgenommen werden. Bei der Einlagerung als Suspension sind die Abfälle anschließend in der Kaverne durch Beigabe von Zuschlagstoffen zu verfestigen. Bei beiden Verfahren ist vorab eine Aufbereitung der Abfälle erforderlich (vgl. Kap. 3).

2.4 Nachbetriebsphase

Nach erfolgter Befüllung bzw. nach dem vollständigen Versatz der Kaverne mit Abfällen muß diese dauerhaft gegen die Biosphäre verschlossen werden. Hierzu werden der obere Teil des Kavernendaches und der offene Kavernenhals durch verschiedene Lagen von Salzgrus, Zement, Ton und Bitumen versiegelt. Das verrohrte Bohrloch wird, wie in der Tiefbohrtechnik üblich, mit Zement bis obertage verfüllt. Der Betriebsplatz obertage wird rekultiviert und wieder seiner ursprünglichen Bestimmung zugeführt. Eine Nachsorge und Langzeitkontrolle der Deponiekaverne ist nicht erforderlich, da die Abfälle durch die geologische Barriere des Salzgebirges und durch den Bohrlochverschluß langfristig, d.h. für geologische Zeiträume von der Umwelt abgeschlossen sind.

3 GEOTECHNISCHE FRAGEN DER SALZKAVERNENDEPONIE

3.1 Die Mehrfachbarriere als Voraussetzung für eine immissionsneutrale Lagerung

Bei der Ablagerung von Sonderabfällen ist die Umwelt generell durch Barrieren von Schadstoffen abzuschirmen. Aufgrund der geochemischen Komplexität von Sonderabfällen (HERMANN, et al. /2/) darf hierunter nicht nur ein wirksamer Abschluß gegen den Schadstoffaustrag aus der Deponie in die Umwelt verstanden werden, sondern es ist auch der Einfluß der Atmosphäre und Geosphäre auf die endzulagernden Abfälle sowie die chemische Wechselwirkung der Abfälle untereinander zu betrachten, die zu unkontrollierbaren Freisetzungen bzw. zu Neubildungen von Schadstoffen führen können, für die die Barrieren beim Bau der Deponie nicht ausgelegt wurden (GÖTTNER /3/). Aus diesem Anforderungsprofil an das Barrierenkonzept muß zwangsläufig gefolgert werden, daß bei der Anlage von Sonderabfalldeponien nicht nur auf die Wirksamkeit einer Barriere vertraut werden darf, sondern mehrere Sicherheiten bzw. Multibarrieren einzubauen sind.

Für Obertagedeponien ist ein solches Multibarrierensystem aus folgenden Einzelkomponenten aufgebaut (STIEF, 1986 /4/):
- Deponiestandort
- Deponiebasisabdichtungssystem
- Deponiekörper
- Oberflächenabdichtung
- Nutzung
- Nachsorge, Kontrollierbarkeit und Reparierbarkeit der Barrieren.
Die Einzelbarrieren einer Untertagedeponie bestehen aus:
- dem Wirtgestein
- der Entfernung zu Oberfläche und Aquifer
- dem Ausbau- und Dichtungssystem
- den Verschlußbauwerken
- der Abfallkonditionierung
- dem Versatz
- den Zugangssystemen zum Untertagehohlraum sowie
- den Kontrollmöglichkeiten und der Nachsorge.
Im Interesse nachfolgender Generationen müssen Deponiekonzepte für die Endlagerung von Sonderabfällen ausgewählt werden, die eine größtmögliche Langzeitsicherheit gewährleisten und möglichst wenig Nachsorgemaßnahmen erfordern. Hier bietet sich die Untertagedeponie für eine Vielzahl von Abfällen als einzige Lösung an.

3.2 Gebirgsmechanische Anforderungen an die Kaverne

Seit den fünfziger Jahren sind im Steinsalzgebirge weltweit hunderte von Kavernen zur Speicherung von flüssigen und gasförmigen Energieträgern angelegt worden. Dabei ergaben sich aus den unterschiedlichen Nutzungsarten und Betriebsweisen vielfältige mechanische und thermische Beanspruchungen für das die Kavernen umgebende Gebirge. Da die Kavernen aufgrund ihrer soltechnischen Herstellung nicht begehbar sind, ist ein zusätzlicher/nachträglicher Ausbau zur Sicherung des Hohlraumes gegen Abschalungen und Nachbrüche einerseits und zur Begrenzung der Gebirgskonvergenz andererseits nicht möglich. Daraus folgt, daß allein das Gebirge die mit der Herstellung und Nutzung verbundene Beanspruchung aufzunehmen hat. Erschwerend für den Entwurf dieser Tragsysteme ist ferner, daß Steinsalz bereits unter den Beanspruchungsbedingungen der Betriebszustände ein ausgesprochen inelastisches Materialverhalten aufweist, ohne dessen angemessene Berücksichtigung eine wirtschaftliche Dimensionierung nicht möglich ist.

Die inzwischen vorliegenden langjährigen Betriebserfahrungen deuten darauf hin, daß die Kavernen ein in Abhängigkeit von der Geometrie und Teufenlage, von der Qualität des umgebenden Steinsalz- und Nebengebirges sowie von der Betriebsweise sehr unterschiedliches Verhalten im Hinblick auf Standfestigkeit und Konvergenz aufweisen.

Gebirgsmechanische Analysen des im Betrieb beobachteten Tragverhaltens einzelner repräsentativer Kavernenprojekte zeigen, daß mit den heute im Rahmen der Salzmechanik zur Verfügung stehenden Entwurfsmodellen eine realistische Erfassung der charakteristischen mechanischen Phänomene möglich ist. Einzelheiten zum gebirgsmechanischen Entwurf von Speicherkavernen und zur Aussagekraft theoretischer Prognosemodelle sind in /5/, /6/, /7/ zu finden.

Im Vergleich zu den Speicherkavernen, bei denen im Rahmen der gebirgsmechanischen Analyse die Herstellphase und die Betriebsphase über einen Zeitraum von 30 bis 50 Jahren untersucht werden, ist bei Kavernen zur Deponierung von Sonderabfall neben der mehrjährigen Betriebsphase insbesondere die Jahrzehnte und Jahrhunderte umfassende Nachbetriebsphase zu betrachten. Diese Nachbetriebsphase setzt ein, wenn nach Beendigung des Befüllvorganges die Zugangsbohrung zur Kaverne verschlossen worden ist.

Einige wesentliche, mit dem gebirgsmechanischen Entwurf von Sonderabfallkavernen verbundene Fragestellungen sind in Abb. 3 stichwortartig zusammengefaßt. Dabei wird zwischen der Situation vor und nach dem Endverschluß der Kaverne unterschieden.

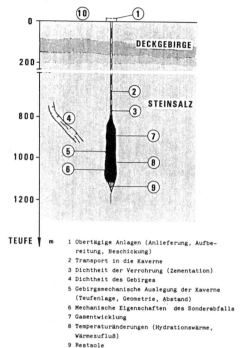

TEUFE ↓ m

1 Obertägige Anlagen (Anlieferung, Aufbereitung, Beschickung)
2 Transport in die Kaverne
3 Dichtheit der Verrohrung (Zementation)
4 Dichtheit des Gebirges
5 Gebirgsmechanische Auslegung der Kaverne (Teufenlage, Geometrie, Abstand)
6 Mechanische Eigenschaften des Sonderabfalls
7 Gasentwicklung
8 Temperaturänderungen (Hydrationswärme, Wärmezufluß)
9 Restsole
10 Meßtechnische Überwachung

Abb. 3 a. Mono-Sonderabfallkaverne
Fragen zum gebirgsmechanischen Entwurf

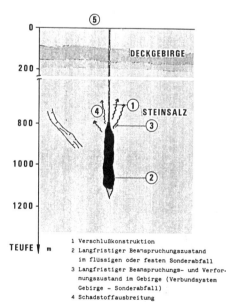

TEUFE ↓ m

1 Verschlußkonstruktion
2 Langfristiger Beanspruchungszustand im flüssigen oder festen Sonderabfall
3 Langfristiger Beanspruchungs- und Verformungszustand im Gebirge (Verbundsystem Gebirge - Sonderabfall)
4 Schadstoffausbreitung
5 Meßtechnische Überwachung

Abb. 3 b. Mono-Sonderabfallkaverne
Fragen zum gebirgsmechanischen Entwurf nach Verschluß

Aus gebirgsmechanischer Sicht bilden der in die Kaverne verbrachte Sonderabfall und das Steinsalzgebirge ein Verbundtragwerk. Dabei sind bezüglich des Sonderabfalls zumindest zwei charakteristische Situationen zu unterscheiden:

(1) Der Sonderabfall weist eine so geringe Viskosität auf, daß sich von Anfang an oder im Lauf der Zeit ein hydrostatischer Spannungszustand in der Kaverne einstellt. Schubspannungen zwischen Gebirge und Sonderabfall werden nicht übertragen (= Sonderabfall als Flüssigkeit).

(2) Der Sonderabfall ist auch langfristig in der Lage, deviatorische Spannungszustände zu ertragen. Zwischen Gebirge und Sonderabfall werden Schubspannungen übertragen (= Sonderabfall als Feststoff).

Unabhängig von einer Quantifizierung ergeben sich qualitativ gravierende Unterschiede für die gebirgsmechanische Situation in diesen beiden Fällen. Im Fall (1) ist davon auszugehen, daß sich nach dem Kavernenverschluß durch die Gebirgskonvergenz in dem gering viskosen Versatzmaterial ein über den Eigengewichtsdruck hinausgehender Innendruck aufbaut. Da das Versatzmaterial keine deviatorischen Spannungen aufnehmen kann, entspricht der Kaverneninnendruck dem Druckverlauf in einer Flüssigkeit. Das Druckniveau wird durch den auf das Versatzmaterial im unteren Kavernenbereich einwirkenden Gebirgsdruck (Radialspannung) bestimmt.

Ohne genauere Berechnungen wurde lange Zeit davon ausgegangen, daß sich im viskosen Salzgebirge langfristig im Kavernensohlbereich wieder der primäre Gebirgsdruck einstellt, so daß auch der Innendruck dort dieses Niveau erreicht. Aufgrund der geringeren Dichte des Versatzmaterials im Vergleich zur Gebirgsdichte ergibt sich dann im Kavernendach ein Innendruck, der größer ist als der dort mögliche primäre Gebirgsdruck. Infolge dieses Überdruckes, der insbesondere von der Kavernenhöhe abhängt, ist zu befürchten, daß das Gebirge in dem betroffenen Gebirgsbereich gefract wird und die Kaverne dann durch die allmähliche Rißausbreitung zum Caprock bzw. Deckgebirge hin ihre Dichtheit verliert. Damit ergibt sich zwangsläufig die Frage nach derjenigen Kavernenhöhe, bei der auch unter "Überdruck"-Bedingungen ein Aufreißen des Gebirges mit entsprechender Sicherheit noch nicht auftritt. Hierbei ist zu beachten, daß bei den derzeit betriebenen Speicherkavernen ein Innendruck zulässig ist, der nur etwa 70 - 80 % des petrostatischen Gebirgsdruckes im Bereich der letzten zementierten Rohrtour beträgt (Erfahrungsbereich).

Bei eingehender Betrachtung dieser Problematik können jedoch auch noch weitergehendere Fragen gestellt werden:

(1) Wird in der Kaverne tatsächlich langfristig überhaupt dieser maximal denkbare Überdruck erreicht?
(2) Wie groß ist die Zugfestigkeit des Gebirges, die bei dieser Beanspruchung in Ansatz gebracht werden kann? Treten überhaupt Zugspannungen auf?
(3) Wovon hängt der ertragbare Überdruck ab (Belastungsrate, Viskosität der Flüssigkeit)?
(4) Treten vor dem eigentlichen Aufreißen des Gebirges in Form von Makrorissen bereits Gefügeauflockerungen im Mikrobereich auf, die ein allmähliches Eindringen der Versatzflüssigkeit in das Gebirge ermöglichen?
(5) Wie wirkt sich die Erwärmung des Versatzmaterials durch den Wärmezufluß aus dem Gebirge auf den Innendruck aus?

Im Gegensatz zum Fall (1) ist bei dem Versatzmaterial im Fall (2) davon auszugehen, daß infolge der inneren Festigkeit des granularen oder verfestigten Versatzmaterials deviatorische Spannungen auch langfristig aufgenommen werden können und sich damit für das Gebirge nicht so ungünstige Spannungszustände im Versatzmaterial einstellen wie im Fall der flüssigkeitsgefüllten Kaverne.

Die Beanspruchung des Versatzmaterials ist in diesem Fall nunmehr insbesondere abhängig von der Wechselwirkung zwischen Steinsalzgebirge und Versatzmaterial (Verbundsystem). Daraus folgt, daß neben den mechanischen Eigenschaften des Steinsalzgebirges auch die mechanischen Eigenschaften des Sonderabfalls quantitativ ermittelt werden müssen - insbesondere das Verformungsverhalten und die Festigkeitseigenschaften. Bei granularem und verfestigtem Sonderabfall sind u.a. folgende Untersuchungen durchzuführen:

- Hydratationswärmeentwicklung
- Kompressionsverhalten
- Langzeitbeständigkeit (physikalisch-chemisch)
- Mischungsverhältnis (Sonderabfall - Zement - Wasser)
- Dichte, Porosität, Permeabilität
- Temperaturbeständigkeit
- Phasentrennung (fest, flüssig)
- Zwickel- (Brücken-)bildung
- Festigkeitseigenschaften
- Kriecheigenschaften
- Haftung Sonderabfall - Steinsalz
- Verfestigungsverhalten
- Wasser-(Sole-)aufnahmefähigkeit (chemische Bindung)
- Quellverhalten, Schwindverhalten
- Ausbreitungsvermögen

Das mechanische Verhalten des Versatzmaterials - insbesondere seine Beanspruchung und seine Festigkeit - sind auch von wesentlicher Bedeutung bei der Planung eines Sonderabfall-Kavernenfeldes, bei dem der zur Verfügung stehende Gebirgsbereich möglichst optimal ausgenutzt werden soll und die nacheinander auszusolenden und zu befüllenden Kavernen so dicht wie möglich anzuordnen sind.

Aus der Forderung, die Sonderabfallbeseitigungsanlagen wie Bauwerke zu behandeln und meßtechnisch zu überwachen, ergibt sich die Konsequenz, bei feststoffverfüllten Kavernen während der Betriebsphase den Befüll- und Aushärtungsvorgang zu überwachen. Im Vordergrund stehen dabei u.a. folgende Fragen:

- Porenraumarme Einbringung des Versatzmaterials
- Aushärtungsvorgang
- Aufbau von Enddruck (infolge Eigengewicht Versatzmaterial) und Gebirgsdruck (infolge Gebirgskonvergenz)
  im Versatzmaterial

Die Ausbildung des entstehenden Versatzkörpers und seine
mechanischen Eigenschaften könnten durch geeignete kontinuierliche oder diskontinuierliche Kontrollmethoden überprüft
werden.
   Um das prinzipielle Tragverhalten einer Sonderabfallkaverne
aufzuzeigen, wird der Fall einer mit Feststoff befüllten,
endverschlossenen Kaverne betrachtet. In Abb. 4 sind zunächst
das der Betrachtung zugrunde gelegte Berechnungsmodell und
seine Diskretisierung in finite Elemente dargestellt. Die
Diskretisierung umfaßt dabei sowohl das den Hohlraum umgebende Gebirge wie auch den Versatz (aufbereiteter Sonderabfall).
Die Kavernenform wurde vorerst aus dem Speicherkavernenbau
übernommen, um die Ergebnisse vergleichen zu können. Es ist
selbstverständlich, daß sich für Sonderabfallkavernen andere
Kavernenformen aus gebirgsmechanischen und verfahrenstech-
nischen Gründen als günstiger erweisen können. Das Kavernen-
volumen beträgt ca. 300 000 m³.

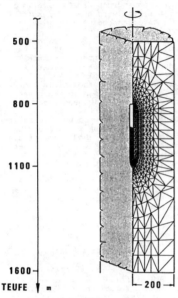

Abb. 4. Diskretisiertes Berechnungsmodell einer
        Deponiekaverne

   Das Stoffverhalten des hier betrachteten festen Versatzmaterials wird sehr vereinfacht als linear-elastisch
(HOOKEsches Stoffgesetz) angenommen. Der Verformungsmodul
wird in den Grenzen E = 50 MPa (granularer Versatz) und E =
10 000 MPa (in situ verfestigter Versatz) variiert.
   Abb. 5 zeigt den Aufbau der Spannungen im Versatzmaterial
in der Referenzteufe von 1 000 m (vollständig befüllte,
endverschlossene Kaverne). Deutlich ist zu sehen, wie sich
die Spannungen im Versatz nach dem Verschluß vom angesetzten
Ausgangszustand infolge der Gebirgskonvergenz umlagern. Dabei
nehmen zunächst die horizontalen Spannungskomponenten sehr
viel stärker zu als die vertikale Spannungskomponente. Im
Endzustand wird zuerst in horizontaler Richtung, später dann
auch in vertikaler Richtung in etwa das Spannunsniveau des
primären Gebirgsdruckes erreicht.

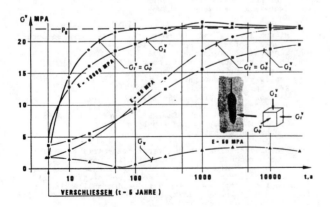

Abb. 5. Feststoffgefüllte Kaverne - Beanspruchung im Versatz

Ergänzend zu den Spannungszuständen im Versatzmaterial
zeigt Abb. 6 die Gebirgsverformungen in Kavernennähe zu ausgewählten Zeitpunkten für den Fall des steifen Versatzes.
Danach erfolgt zunächst in den ersten 5 Jahren (Solphase,
Betriebsphase mit Befüllung) eine Verschiebung der gesamten
Kavernenkontur in den Hohlraum hinein. Nach 500 Jahren hat
die Konvergenz in geringem Umfang weiter zugenommen. Nach 10
000 Jahren wird dann eine signifikante Veränderung im Verformungsverhalten erkennbar; die vertikalen Verschiebungskomponenten insbesondere im oberen Kavernenbereich haben ihre
Richtung umgekehrt und weisen einheitlich aufwärts, während
die horizontalen Verschiebungskomponenten nach wie vor in den
Hohlraum hinein gerichtet sind. Zum Zeitpunkt 20 000 Jahre
wird diese Tendenz eines "Aufwärtsschwimmens" des Versatz-
monolithen noch deutlicher.

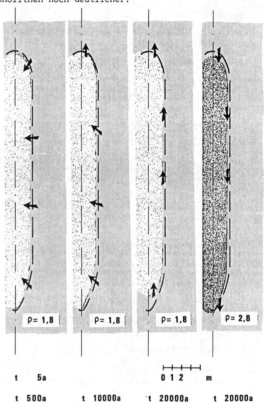

Abb. 6. Verformung der Kavernenkontur bei feststoffgefüllter
        Kaverne und langfristiges Verhalten des Versatz-
        monolithen

   Aufgrund dieses Ergebnisses liegt es nun nahe zu untersuchen,
wie sich die Dichte des Versatzmaterials auf die langfristig
zu erwartende Bewegungstendenz auswirkt. Bei Annahme einer
Dichte von $\rho_{Versatz}$ = 2,8 t/m³ ($\rho_{Gebirge}$ = 2,2 t/m³) bestätigt Abb. 6 das auch zu erwartende Ergebnis: Nach Abschluß
der mit der Verdichtung des Versatzmaterials verbundenen
Spannungsumlagerungen sinkt der Versatzmonolith im Steinsalz
ab.
   Weitere prinzipielle Ergebnisse zum Tragverhalten flüssig-
keits- und feststoffverfüllter, endverschlossener Steinsalz-
kavernen sind in /8/ beschrieben.

3.3 Anforderungen an die Abfallstoffe

Aufgrund der Einlagerung der Abfälle über ein mehrere hundert
Meter langes Befüllrohr, die behälterlose Lagerung, der in
situ Druck- und Temperaturbedingungen in der Kaverne sowie
der gebirgsmechanischen Wechselwirkung von Abfall und umgebenden Salzgestein stellen sich eine Reihe von chemisch-
physikalischen Anforderungen an das Abfallgut, die für die
Abfallstoffe eine obertägige Vorbehandlung und Aufbereitung
erforderlich machen. Diese Behandlungsverfahren müssen mit
den Eigenschaften der Abfälle im Anfallzustand abgestimmt
werden (Abb. 7).

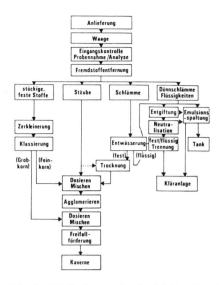

Abb. 7. Fließschema - Sonderabfallaufbereitung für
Freifallförderung

3.3.1 Chemische Anforderungen

In chemischer Hinsicht sind folgende Anforderungen an Abfallstoffe zu stellen, die sich für eine Einlagerung in Salzkavernen eignen.
- keine Reaktionen zwischen verschiedenen Abfallstoffen -
  ggfs. nach vorhergehender Inertisierung der Einzelkomponenten
- keine Reaktion mit dem umgebenden Salzgestein (diese
  Anforderung wird praktisch von allen infrage kommenden
  Abfallstoffen erfüllt)
- keine nennenswerte Bildung von Gasen und Dämpfen bei
  Umgebungstemperaturen von bis zu 60°C
- kein mikrobieller Abbau der Abfallstoffe, der zur Bildung von korrosiven und explosionsfähigen Gasgemischen
  führen kann
- möglichst geringer Gehalt an freiem Wasser
- minimale Löslichkeit von Salz auch für Flüssigkeiten
  neben Wasser (dies betrifft z.B. Alkohole und starke
  Säuren)

3.3.2 Physikomechanische Anforderungen

Bei der Definition dieser Anforderungen muß unterschieden
werden zwischen Abfall, der in einer Suspension mit einem
Bindemittel wie Zement in die Kaverne verpumpt wird, um dann
in situ zu verfestigen, und Abfall, der als Schüttgut über
eine Freifalleitung eingebracht wird. Für beide Verfahren
gilt, daß an die Festigkeit des Abfallgutes nach der Einlagerung nur geringe Anforderungen gestellt werden - verglichen
mit denen an typische Baustoffe. Der Grund liegt in der
allseitigen Belastung des eingelagerten Sonderabfalles.
  Pumpversatz: Grundlegende Forderung an einen einzulagernden
Abfall ist, daß er sich in einer Zement-, Flugasche-, Gips-
oder sonstigen Matrix binden läßt und daß der Verbund die
erforderliche Festigkeit erreicht. An die Suspension werden
u.a. die folgenden Anforderungen gestellt:
- keine nennenswerte Entmischung während des Einbringens
- ausreichende Fließfähigkeit, um eine Ausbreitung in der
  Kaverne zu gewährleisten
- geringer Anteil an Bindemittel, um das Nettospeicher-
  volumen groß zu halten.
Bezüglich der Verfahrenstechnik für das Verpumpen einer Suspension in die Kaverne und die anschließende Sicherstellung
der Verfestigung müssen noch eine Reihe von Problemen gelöst
werden.
  Einlagerung von Schüttgut: Das Fördern von Schuttgut durch
eine vertikale Freifalleitung ist Stand der Technik im Bergbau. Durch eine geeignete obertägige Aufbereitung des Abfallgutes kann ein Kornspektrum erzielt werden, bei welchem die
Gefahr von Verstopfern auf ein Minimum reduziert werden kann.
Daneben darf das Gut nicht zum Anbacken neigen. Für den
Kavernenbereich ist zu fordern, daß die Porosität nach dem
Einbringen möglichst gering ist, damit sich schnell ein Druck
im Abfall gegen das Gebirge aufbaut; hiermit kann die Gebirgskonvergenz, die auch zu Oberflächenabsenkung führt,
reduziert werden. Dies kann wiederum durch Wahl eines geeigneten Kornspektrums erreicht werden, welches eine hohe
Packungsdichte ermöglicht.

3.4 Dauerhafter Endverschluß

An die Dichtheit im Verschlußbereich der Sondermüllkaverne
werden die gleichen hohen Anforderungen gestellt, die auch
für das umgebende Salzgebirge gelten.
  Da insbesondere bei Abfall in Schüttgutform mit einer weiteren Setzung des Kaverneninhaltes zu rechnen ist, muß der
Verschluß der Kaverne im unverrohrten und im verrohrten Bohrlochsbereich vom Endlagerbereich mechanisch isoliert werden.
Hierzu wird zunächst im unteren offenen Bohrlochsbereich ein
Betonstopfen gesetzt, dessen Funktion in erster Linie in
einem Widerlager für die darüber liegende eigentliche Dich-

tung besteht. Bei dieser wird unterschieden zwischen kurz-
bis mittelfristig und mittel- bis langfristig wirkenden Dich-
tungsbereichen. Für einen kurz- bis mittelfristigen Verschluß
kommen Spezialzement und Asphalt infrage; der verrohrte Bereich wird mit Zement - wie in der Tiefbohrtechnik üblich -
verfüllt. Im unverrohrten Bereich wird bereichsweise Salzgrus
als langzeitig wirksame Dichtung eingesetzt; dieser wird
unter den in situ herrschenden Druck- und Temperaturbedingungen soweit kompaktiert, daß schließlich eine Rekristallisa-
tion und damit vollständige Verheilung von Salzstopfen und
umgebendem Salzgebirge erreicht wird.

4 SCHLUSSBEMERKUNGEN

Für Sonderabfälle, die nicht weiter mengenmäßig reduziert,
behandelt oder entgiftet werden können, kommt in Zukunft
vorrangig die Untertageendlagerung in dichten geologischen
Formationen und hier insbesondere in Salzkavernen infrage.
  Die geologische Barriere "Steinsalz" einer solchen Deponie
zeichnet sich u.a. aus durch
- absolute Dichtheit
- günstige mechanische Eigenschaften, die die Herstellung
  großer, selbsttragender unterirdischer Hohlräume zuläßt
- geringe Hohlraumerstellungskosten, da Salzkavernen von
  obertage hergestellt werden.
Der wesentliche Vorteil einer sorgfältig dimensionierten
Salzkavernendeponie liegt in der gebotenen Langzeitsicherheit
und der Tatsache, daß nach Abschluß einer Kaverne keine
weiteren Beobachtungsmessungen oder Wartungsarbeiten erforderlich sind, die selbst bei konventionellen Hochsicherheits-
deponien von vornherein mit eingeplant werden müssen.

LITERATUR

/1/ Quast, P. und Beckel, S. Derzeitiger Stand der soltech-
    nischen Planung von Speicherkavernen im Salz und die da-
    mit erzielten praktischen Ergebnisse. Erdöl-Erdgas-Zeit-
    schrift (1981), H. 6, Jg. 97, S. 213 - 217.
/2/ Hermann, A.G., Brumsack, H.J. & Heinrichs, H. (1985).
    Notwendigkeit, Möglichkeiten und Grenzen der Untergrund-
    deponie anthropogener Schadstoffe. Naturwissenschaften 72,
    S. 408 - 418
/3/ Göttner, I.J. (1985). Mögliche Reaktionen in einer Sonder-
    abfalldeponie - Folgerungen für das Deponierungskonzept.
    Müll und Abfall, Heft 2, S. 29 - 32
/4/ Stief, K. (1986). Das Multibarrierenkonzept als Grund-
    lage von Planung, Bau, Betrieb und Nachsorge von Deponien.
    Müll und Abfall, Heft 1, S. 15 - 20
/5/ Lux, K.H. und Rokahr, R.B. Dimensionierungsgrundlagen im
    Salzkavernenbau. Taschenbuch für den Tunnelbau (4), Ver-
    lag Glückauf, Essen, 1980.
/6/ Lux, K.H. Gebirgsmechanischer Entwurf und Felderfahrungen
    im Salzkavernenbau. Ferdinand Enke-Verlag, Stuttgart, 1984
/7/ Lux, K.H., Quast, P. und Rokahr, R.B. 20 Jahre Erfahrungen
    mit Speicherkavernen. Vortrag DGMK/DVGI-Tagung 1986 in
    Hamburg (Veröffentlichung in Vorbereitung).
/8/ Lux, K.H., Rokahr, R.B. und Kiersten, P. Gebirgsmechani-
    sche Anforderungen an die untertägige Deponierung von
    Sonderabfällen im Salzgebirge. Vortrag STUVA-Tagung,
    November 1985.

A displacement discontinuity model for fluid-saturated porous media
Un modèle déplacement-discontinuité pour des milieux poreux saturés de fluide
Ein Verschiebungsdiskontinuitäts-Modell für ein flüssigkeitsgesättigtes poröses Material

JOHN CURRAN, Associate Professor, University of Toronto, Ontario, Canada
JOSÉ L. CARVALHO, Graduate Student, University of Toronto, Ontario, Canada

ABSTRACT: The fundamental solutions for two-dimensional displacement discontinuities and instantaneous and continuous point fluid sources in a homogeneous, isotropic, linear elastic, fluid-saturated porous media are presented. These solutions are used to develop a two-dimensional boundary element program that computes the stress, displacement, pore pressure and fluid flux fields for problems of arbitrary geometry. Several applications are presented.

RÉSUMÉ: On présente les solutions fondamentales en deux dimensions pour discontinuités de déplacement et sources de fluide instantanées et continues dans un milieu homogène, isotrope, élastique linéaire, poreux et saturé de fluide. Nous avons développé un programme aux éléments de frontière qui permet le calcul des contraintes, des déplacements, de la pression interstitielle et de l'écoulement de fluide en problèmes de géométrie arbitraire. Quelques applications sont présentés.

ZUSAMMENFASSUNG: Die fundamentalen Lösungen zweidimensionaler Verschiebungsdiskontinuitäten und momentaner und kontinuierlicher Flüssigkeitsquellen in einem homogenen, isotropen, linear elastischen, flüssigkeitsgesättigten, porösen Material werden dargestellt. Mit Hilfe dieser Lösungen wurde ein zweidimensionales Grenzelementprogramm entwickelt, welches Spannungen, Verschiebungen, Porendrücke und Flußfelder für Probleme beliebiger Geometrie berechnet. Mehrere Anwendungen werden gezeigt.

1. INTRODUCTION

The important influence that pore pressure has on the deformation behaviour of fluid saturated porous rocks is widely recognized. The pore pressure variation in space and time controls not only the flow of fluid but also the effective stress field in the rock mass. Where the rock mass is jointed, the pore pressure distribution can strongly affect whether slip will occur along preexisting discontinuities.

A consistent theory which explicitly accounts for the coupling between the pore fluid and the rock matrix responses was developed by Biot (1941).

In this paper we use Biot's theory to develop the two-dimensional fundamental solutions (Green's functions) for the normal (aperture) and shear (ride) displacement discontinuities, as well as instantaneous and continuous point fluid sources in a poroelastic medium. The displacement discontinuity solutions are constructed using 'point force' quadrupoles together with an instantaneous point fluid source. These singular solutions form the basis of the boundary element technique developed in section 3.

2. GREEN'S FUNCTIONS

It is assumed that both the material matrix and its solid components (mineral grains) have a linear elastic behaviour and the fluid flow through the porous skeleton is laminar. The equations presented here are valid for a fluid-saturated isotropic, homogeneous material under small strain. A tension positive convention is used for the stresses while a positive pore pressure is treated as compressive. The constitutive equation for a poroelastic medium in terms of the total stresses, σ_{ij}, is given by

$$\sigma_{ij} = 2\mu e_{ij} + \delta_{ij}\lambda e_{kk} - \delta_{ij}\alpha p \qquad (1)$$

where p is the pore pressure, μ and λ Lame constants and δ_{ij} the Kronecker delta. The strain e_{ij} is given by

$$\tfrac{1}{2}\left(u_{i,j} + u_{j,i}\right) \qquad (2)$$

where u_i is the displacement in the i-direction. Throughout the paper, all subscripts take the values 1, 2, a comma denotes partial differentiation with respect to x_i and the usual summation convention is employed over repeated subscripts.

Substituting eq. (1) and (2) into the equilibrium equation

$$\sigma_{ij,j} = 0 \qquad (3)$$

results in the following differential equation

$$\mu\nabla^2 u_i + (\lambda+\mu)\frac{\partial}{\partial x_i}u_{k,k} = \alpha p_{,i} \qquad (4)$$

If it is assumed that fluid flow through the porous skeleton is governed by Darcy's law, then, by conservation of mass we obtain the following diffusion equation (Rice and Cleary, 1976)

$$\frac{\partial S}{\partial t} = \frac{k}{\gamma_f} \nabla^2 p \qquad (5)$$

where k is the permeability (L/T) of the medium and γ_f the unit weight of the pore fluid.

The variable S, which measures the change in pore fluid volume per unit volume of the poroelastic medium, is given by

$$S = \alpha e_{kk} + \frac{1}{M} p \qquad (6)$$

where $\quad \alpha = 1 - \dfrac{K_m}{K_s}$, $\quad \dfrac{1}{M} = \dfrac{n}{K_f} + \dfrac{\alpha-n}{K_s}$

and K_f, K_m and K_s are the bulk moduli of the pore fluid, the matrix and the solid components, respectively and n is the porosity of the medium (Biot and Willis, 1957).

Eq. (4) and (5) together with the appropriate boundary conditions determine the spatial and the temporal variations of u_i and p.

73

2.1. Point fluid sources

When the pore pressure field is caused by imposed fluid pressures or fluid injection rates only and the boundary pressures are zero, the displacements can be expressed in terms of a poroelastic potential of the form $u_i = \phi_{,i}$ (Goodier, 1937). Substituting $\phi_{,i}$ for u_i in eq. (4) it can be shown that

$$\nabla^2\phi = e_{kk} = \beta p \qquad (7)$$

where $\beta = \dfrac{\alpha(1-2\nu)}{2\mu(1-\nu)}$

and ν is the Poisson's ratio of the matrix.

Substituting βp for e_{kk} in eq. (6), the diffusion equation decouples from the deformation field and takes the form

$$\nabla^2 p - \frac{1}{\kappa}\frac{\partial p}{\partial t} = 0 \qquad (8)$$

where $\dfrac{1}{\kappa} = \dfrac{\gamma_f}{k}\left(\alpha\beta + \dfrac{1}{M}\right)$

The Green's function, $\bar{h}$, for an instantaneous fluid source is the solution of the diffusion equation

$$\nabla^2\bar{h} - \frac{1}{\kappa}\frac{\partial\bar{h}}{\partial t} = \delta(x_i - \xi_i; t-\tau) \qquad (9)$$

where $\delta(x_i - \xi_i; t-\tau)$ is the Dirac delta function and ξ_i and τ are the location and time, respectively, of the application of the source. The solution of eq. (9) is given as $\bar{h}$ in the appendix.

In plane strain the solution of eq. (7) is given by the logarithmic potential (Kellog, 1929)

$$\phi = -\beta\left[\ln\frac{1}{r}\int_0^r a\cdot p(a)da + \int_r^\infty a\cdot\ln\frac{1}{a}\cdot p(a)da\right]$$

which leads to the displacement field

$$u_i = \phi_{,i} = \beta\frac{x_i}{r^2}\int_0^r a\cdot p(a)da$$

where the spatial variation of p depends only on the distance from the fluid source. Differentiating the displacements and using eq. (1) yields the stress field. The displacements, stresses and fluid fluxes due to an instantaneous point fluid source are given in the appendix as $\bar{g}_i$, $\bar{t}_{ij}$ and $\bar{f}_i$, respectively.

The solution for the pressure distribution due to a continuous fluid source is obtained by integrating the Green's function for the instantaneous fluid source with respect to time (Carslaw and Jaeger, 1959), i.e.,

$$\hat{h}(r,t) = \int_0^t \bar{h}(r,t-\tau)d\tau$$

The displacements, stresses and fluid fluxes due to a continuous point fluid source are calculated in a similar manner and are given in the appendix as $\hat{g}_i$, $\hat{t}_{ij}$ and $\hat{f}_i$, respectively.

2.2. Displacement discontinuities

The solutions for displacement discontinuities (DDs) can be derived from the point force solutions for a poroelastic medium (Cleary, 1976, Wiles and Curran, 1982). The normal DD is obtained by applying to the medium two sets of double forces (dipoles). The application of the sets of double forces initiates a diffusion process which causes the magnitude of the DD to change with time. In order to keep the magnitude constant, an instantaneous point fluid sink has to be applied simultaneously with the double forces. To obtain a unidirectional DD one of the two sets of double forces has to be adjusted to prevent lateral movement at the point of application.

The shear DD is obtained by applying to the medium two sets of double forces with moment. This creates a state of pure shear, causing no change in volume, and

therefore no diffusion process, at the point of application.

The displacements, tractions, pore pressure and fluid fluxes due to a unit normal (n=k) or shear (n≠k) DD are given by

$$g^*_{ikn} = \frac{-\mu(1-\nu)}{1-2\nu}\left\{\frac{1-2\nu}{1-\nu}(g_{ik,n} + g_{in,k}) + 2\delta_{kn}\left[\frac{\nu}{1-\nu}g_{i\ell,\ell} - \frac{\nu_u-\nu}{\alpha(1-2\nu)(1-\nu_u)}\bar{g}_i\right]\right\}$$

$$t^*_{ijkn} = \frac{-\mu(1-\nu)}{1-2\nu}\left\{\frac{1-2\nu}{1-\nu}(t_{ijk,n} + t_{ijn,k}) + 2\delta_{kn}\left[\frac{\nu}{1-\nu}t_{ij\ell,\ell} - \frac{\nu_u-\nu}{\alpha(1-2\nu)(1-\nu_u)}\bar{t}_{ij}\right]\right\}$$

$$\qquad (10)$$

$$h^*_{kn} = \frac{-\mu(1-\nu)}{1-2\nu}\left\{\frac{1-2\nu}{1-\nu}(h_{k,n} + h_{n,k}) + 2\delta_{kn}\left[\frac{\nu}{1-\nu}h_{\ell,\ell} - \frac{\nu_u-\nu}{\alpha(1-2\nu)(1-\nu_u)}\bar{h}\right]\right\}$$

$$f^*_{ikn} = \frac{-\mu(1-\nu)}{1-2\nu}\left\{\frac{1-2\nu}{1-\nu}(f_{ik,n} + f_{in,k}) + 2\delta_{kn}\left[\frac{\nu}{1-\nu}f_{i\ell,\ell} - \frac{\nu_u-\nu}{\alpha(1-2\nu)(1-\nu_u)}\bar{f}_i\right]\right\}$$

where g_{ik}, t_{ijk}, h_k and f_{ik} are the displacements, tractions, pore pressures and fluxes, respectively, due to a unit point force.

The equivalences between point forces, fluid sources and DDs are shown in Fig. 1 and result in the following equalities

$$g^*_{ikn} = -t_{kni}$$

$$h^*_{kn} = 2\mu[\delta_{kn}h_{\ell,\ell} - h_{n,k}]$$

The expressions for the normal and shear unit displacement discontinuities (i.e., eq. (10)) are given in the appendix.

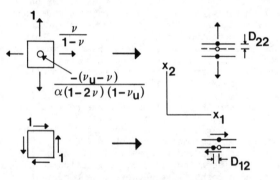

Figure 1 - Equivalences between point forces, fluid sources and displacement discontinuities

In the next section the continuous fluid source and displacement discontinuity solutions are developed into a general numerical technique for solving boundary/initial value problems in poroelastic media.

3. DISPLACEMENT DISCONTINUITY METHOD

The displacement discontinuity method for fluid-saturated poroelastic media consists of discretizing the boundary of a body into elements and determining the intensities of the normal and shear DDs and the fluid sources on the elements such that their summed effects reproduce the prescribed boundary conditions. The induced stresses, displacements, pore pressure and fluid fluxes on element 'm' due to a constant distribution of normal and shear DDs and continuous fluid sources on element 'r' are given by

$$\sigma^m_{ij}(x_1,x_2,t) = T^*_{ijk2}(x_1,x_2,t)\,D^r_{k2} + \hat{T}_{ij}(x_1,x_2,t)\,q^r$$

$$u_i^m(x_1,x_2,t) = G_{ik2}^*(x_1,x_2,t)\, D_{k2}^r + \hat{G}_i(x_1,x_2,t)\, q^r$$

$$p^m(x_1,x_2,t) = H_{k2}^*(x_1,x_2,t)\, D_{k2}^r + \hat{H}(x_1,x_2,t)\, q^r \qquad (11)$$

$$v_i^m(x_1,x_2,t) = F_{ik2}^*(x_1,x_2,t)\, D_{k2}^r + \hat{F}_i(x_1,x_2,t)\, q^r$$

where D_{k2} and q are the intensities of the displacement discontinuities and the continuous fluid source, respectively, superscripts m and r denote quantities specified in the local coordinate system of the influenced and influencing elements, respectively, and the coefficients relating the DDs and the fluid source on the influencing element to the stresses, displacements, pore pressure and fluxes on the influenced element are given in the appendix.

To allow for source strengths (both fluid and DDs) which change with time, the time marching scheme shown in Fig. 2 is used. By incrementing the source strengths at each time step and including the influences of all the previous increments, the necessity of internal discretization of the spatial domain is avoided.

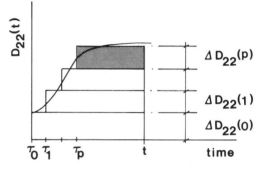

Figure 2 - Time marching scheme for the normal DD (similar schemes are used for the shear DD and the fluid source)

Representing the boundary by 's' elements the normal and tangential stresses, displacements, pore pressures and fluid flux normal to the boundary are given by

$$\sigma_{i2}^m(x_1,x_2,t) = \sum_{r=1}^{s}\left[T_{i2k2}^*(x_1,x_2,t-\tau_p)\, \Delta D_{k2}^r(p) + \hat{T}_{i2}(x_1,x_2,t-\tau_p)\, \Delta q^r(p) \right] + \sum_{\ell=0}^{p-1}\sum_{r=1}^{s}\left[T_{i2k2}^*(x_1,x_2,t-\tau_\ell)\, \Delta D_{k2}^r(\ell) + \hat{T}_{i2}(x_1,x_2,t-\tau_\ell)\, \Delta q^r(\ell) \right]$$

$$u_i^m(x_1,x_2,t) = \sum_{r=1}^{s}\left[G_{ik2}^*(x_1,x_2,t-\tau_p)\, \Delta D_{k2}^r(p) + \hat{G}_i(x_1,x_2,t-\tau_p)\, \Delta q^r(p) \right] + \sum_{\ell=0}^{p-1}\sum_{r=1}^{s}\left[G_{ik2}^*(x_1,x_2,t-\tau_\ell)\, \Delta D_{k2}^r(\ell) + \hat{G}_i(x_1,x_2,t-\tau_\ell)\, \Delta q^r(\ell) \right]$$

$$p^m(x_1,x_2,t) = \sum_{r=1}^{s}\left[H_{k2}^*(x_1,x_2,t-\tau_p)\, \Delta D_{k2}^r(p) + \hat{H}(x_1,x_2,t-\tau_p)\, \Delta q^r(p) \right] \qquad (12)$$
$$+ \sum_{\ell=0}^{p-1}\sum_{r=1}^{s}\left[H_{k2}^*(x_1,x_2,t-\tau_\ell)\, \Delta D_{k2}^r(\ell) + \hat{H}(x_1,x_2,t-\tau_\ell)\, \Delta q^r(\ell) \right]$$

$$v_2^m(x_1,x_2,t) = \sum_{r=1}^{s}\left[F_{2k2}^*(x_1,x_2,t-\tau_p)\, \Delta D_{k2}^r(p) + \hat{F}_2(x_1,x_2,t-\tau_p)\, \Delta q^r(p) \right] + \sum_{\ell=0}^{p-1}\sum_{r=1}^{s}\left[F_{2k2}^*(x_1,x_2,t-\tau_\ell)\, \Delta D_{k2}^r(\ell) + \hat{F}_2(x_1,x_2,t-\tau_\ell)\, \Delta q^r(\ell) \right]$$

where p+1 is the number of increments used to represent the DD and fluid source variation and ΔD_{k2} and Δq are their magnitudes.

Equations (12) constitute a set of linear algebraic equations which may be solved upon substitution of the boundary conditions. The boundary conditions may be of a mixed form, i.e., either the normal stress or normal displacement, the shear stress or tangential displacement and the pressure or the flow normal to the boundary may be prescribed on any of the elements.

4. VERIFICATION OF THE SINGULAR SOLUTIONS AND EXAMPLE APPLICATIONS

4.1 Layer compressed between two impermeable, frictionless, rigid plates

A two-dimensional computer program based on elements with constant distributions of displacement discontinuities and fluid sources was developed to investigate the applicability of the method. A good test case involving various types of boundary conditions was reported by Mandel (1953). The problem consists of a fluid saturated body compressed between two parallel rigid plates, and is shown in Fig. 3.

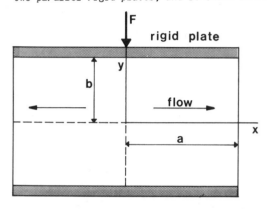

Figure 3 - Plane body compressed between two impermeable, frictionless, rigid plates

The boundary conditions for the problem are as follows. For the plate boundaries (y=±b), there is no flow in the normal direction, the normal displacement is constant (rigid plate motion) and the shear stress is zero. On the other two boundaries (x=±a), pore pressure, normal stress and shear stress are zero. Figures 4 and 5 show the pore pressure distribution along the x-axis at different times and the pressure change with time at the center point, respectively. The good agreement of the solutions and the verification of the Mandel-Cryer effect show the potential of the method.

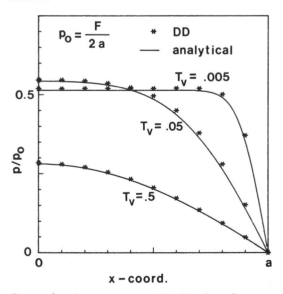

Figure 4 - Pore pressure distribution along x-axis

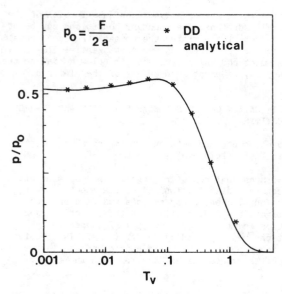

Figure 5 - Pressure change w.r.t. $T_v = \kappa t/a^2$ (dimensionless time) at center point - Mandel-Cryer effect

4.2 Pressurized crack

The flat crack shown in Fig. 6 was subjected to an internal uniform fluid pressure in a zero effective stress field.

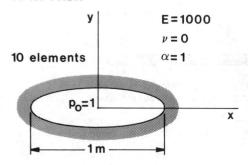

Figure 6 - Pressurized crack

Immediately after pressurization ($t=0^+$) the medium deforms elastically behaving in an undrained manner. As time progresses and fluid leaks off into the formation, the pore pressure in the vicinity of the crack increases.

A plot of the pressure contours at $T_v = 4\kappa t/L^2 = 0.8$ is shown in Fig. 7, and it can be seen that the flow path (which is normal to the pressure contours) is curvilinear near the crack tip.

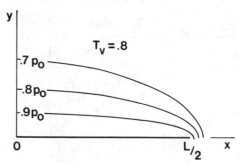

Figure 7 - Pressure contours

The maximum aperture for a flat crack under a constant pressure distribution in a poroelastic medium is approximated by (Haimson, 1968)

$$\frac{4(1+\nu)(p_s+\sigma'_{22})}{E}\left((1-\nu) - \frac{2(1-2\nu)}{9}\alpha\right) \cdot c$$

where p_s is the crack pressure minus formation pressure, c the half-length of the crack and σ'_{22} the effective stress normal to the crack. The second term inside the brackets is a consequence of fluid penetration into the formation. The DD program yields a maximum crack aperture which is within 10% of the estimation given by Haimson (1968) and a crack geometry (Fig. 8) similar to a flat elliptical crack.

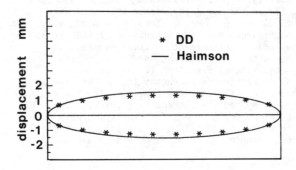

Figure 8 - Crack geometry

Another parameter that the model provides is the volume of fluid that leaks off into the formation. This allows a direct check on the 'square-root time' law (based on one-dimensional flow in a semi-infinite medium) that is generally used in oil reservoir models.

4.3 Ground subsidence

Meaningful results for ground subsidence problems are only possible using a three-dimensional model. However, it is interesting to observe the performance of the two-dimensional model in a simple example. A circular opening is cut in a semi-infinite medium. The formation has an initial pore pressure p_o and the fluid flow is prevented at the surface (fluid bearing strata are often bounded by impermeable layers). Contours of surface deflection at various times are shown in Fig.9.

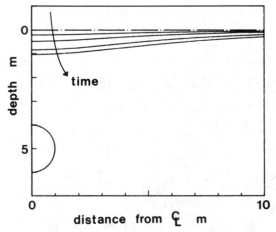

Figure 9 - Surface deflections

5. CONCLUSIONS

The two-dimensional fundamental solutions for displacement discontinuities and fluid sources were presented. The method employed to obtain the solutions for displacement discontinuities in poroelastic media is an extension of the one employed by Wiles and Curran (1982) for elastic media. The main difference is that the normal displacement discontinuity requires that an instantaneous fluid sink be applied simultaneously with the two sets of dipoles to maintain the displacement discontinuity constant in time. The solutions were used in a boundary element program that combined the displacement discontinuities with a continuous fluid source as the primary unknowns. The DD model has been found to yield good results for the problems considered to date. The two-dimensional approach has been extended to three-dimensions (Carvalho and Curran, 1986) and is presently being used to develop a three-dimensional displacement discontinuity model for fluid-saturated poroelastic media.

ACKNOWLEDGEMENT

The financial support of Dowell Schlumberger Inc. and the Natural Sciences and Engineering Research Council of Canada is greatly appreciated.

REFERENCES

Biot, M.A., "General theory of three-dimensional consolidation". Journal of Applied Physics, Vol. 12, No. 2, pp. 155-164, February 1941.

Biot, M.A. and Willis, D.G., "The elastic coefficients of the theory of consolidation". Journal of Applied Mechanics, 24, pp. 594-601, 1957.

Carslaw, H.S. and Jaeger, J.C., "Conduction of heat in solids", 2nd edition, Oxford University Press, London, 1959.

Carvalho, J.L. and Curran, J.H., "Two- and three-dimensional displacement discontinuity and fluid source fundamental solutions for poroelastic media", University of Toronto, Department of Civil Engineering, Publication 86-04, 1986.

Cleary, M.P., "Fundamental solutions for fluid-saturated porous media and application to localized rupture phenomena". Ph.D. Thesis, Brown University, 1976.

Goodier, J.N., "On the integration of the thermoelastic equations". Phil. Mag., 23(7), pp. 1017-1032, 1937.

Haimson, B., 1968, "Hydraulic fracturing in porous and nonporous rock and its potential for determining in-situ stresses at great depth". Ph.D. Thesis, University of Minnesota.

Kellog, O.D., 1929, "Foundations of potential theory". Frederick Ungar Publishing Co., New York.

Mandel, J., "Consolidation des sols (etude mathematique)". Geotechnique, Vol. 3, pp. 287-299, 1953.

Rice, J.R. and Cleary, M.P., "Some basic stress-diffusion solutions for fluid-saturated elastic porous media with compressible constituents". Reviews of Geophysics and Space Physics, 14(2), pp. 227-241, 1976.

Wiles, T.D. and Curran, J.H., "A general 3-D displacement discontinuity method". Proc. 4th Int. Conf. on Numerical Methods in Geomech., Edmonton, Canada, pp. 103-111, 1982.

APPENDIX

Influence coefficients

$$T^*_{ijk2} = a_{il}a_{jm} \int t^*_{lmk2} \, dS_r \quad ; \quad \hat{T}_{ij} = a_{il}a_{jm} \int \hat{t}_{lm} \, dS_r$$

$$G^*_{ik2} = a_{il} \int g^*_{lk2} \, dS_r \quad ; \quad \hat{G}_i = a_{il} \int \hat{g}_l \, dS_r$$

$$H^*_{k2} = \int h^*_{k2} \, dS_r \quad ; \quad \hat{H} = \int \hat{h} \, dS_r$$

$$F^*_{ik2} = a_{il} \int f^*_{lk2} \, dS_r \quad ; \quad \hat{F}_i = a_{il} \int \hat{f}_l \, dS_r$$

where a_{il} is the rotation matrix from the r (influencing) coordinate system to the m (influenced) coordinate system and the integration is carried out over the length of the influencing element.

Green's functions

$$r^2 = x_1^2 + x_2^2$$

$$\nu_u = \frac{3\nu\left(\alpha + \frac{K_m}{\alpha M}\right) + (1-2\nu)\alpha}{3\left(\alpha + \frac{K_m}{\alpha M}\right) - (1-2\nu)\alpha}$$

Ei(x) denotes the exponential integral of x.

Instantaneous point source of unit strength

$$\bar{h} = \frac{1}{4\pi\kappa t} e^{-\frac{r^2}{4\kappa t}}$$

$$\bar{f}_1 = \frac{k}{8\pi(\kappa t)^2 \gamma_f} \cdot x_1 e^{-\frac{r^2}{4\kappa t}}$$

$$\bar{f}_2 = \frac{k}{8\pi(\kappa t)^2 \gamma_f} \cdot x_2 e^{-\frac{r^2}{4\kappa t}}$$

$$\bar{g}_1 = \frac{\beta}{2\pi} \cdot \frac{x_1}{r^2}\left(1 - e^{-\frac{r^2}{4\kappa t}}\right)$$

$$\bar{g}_2 = \frac{\beta}{2\pi} \cdot \frac{x_2}{r^2}\left(1 - e^{-\frac{r^2}{4\kappa t}}\right)$$

$$\bar{t}_{11} = \frac{\mu\beta}{2\pi} \cdot \frac{1}{r^2}\left[\left(2 - \frac{4x_1^2}{r^2}\right)\left(1 - e^{-\frac{r^2}{4\kappa t}}\right) + \left(\frac{x_1^2}{\kappa t} - \frac{r^2}{\kappa t}\right)e^{-\frac{r^2}{4\kappa t}}\right]$$

$$\bar{t}_{22} = \frac{\mu\beta}{2\pi} \cdot \frac{1}{r^2}\left[\left(2 - \frac{4x_2^2}{r^2}\right)\left(1 - e^{-\frac{r^2}{4\kappa t}}\right) + \left(\frac{x_2^2}{\kappa t} - \frac{r^2}{\kappa t}\right)e^{-\frac{r^2}{4\kappa t}}\right]$$

$$\bar{t}_{12} = \bar{t}_{21} = \frac{\mu\beta}{2\pi} \cdot \frac{4x_1 x_2}{r^4}\left[-\left(1 - e^{-\frac{r^2}{4\kappa t}}\right) + \frac{r^2}{4\kappa t} e^{-\frac{r^2}{4\kappa t}}\right]$$

Continuous point source of unit strength

$$\hat{h} = \frac{1}{4\pi\kappa} \text{Ei}\left(\frac{r^2}{4\kappa t}\right)$$

$$\hat{f}_1 = \frac{k}{2\pi\kappa\gamma_f} \cdot \frac{x_1}{r^2} e^{-\frac{r^2}{4\kappa t}}$$

$$\hat{f}_2 = \frac{k}{2\pi\kappa\gamma_f} \cdot \frac{x_2}{r^2} e^{-\frac{r^2}{4\kappa t}}$$

$$\hat{g}_1 = \frac{\beta t}{2\pi} \cdot \frac{x_1}{r^2}\left[\left(1 - e^{-\frac{r^2}{4\kappa t}}\right) + \frac{r^2}{4\kappa t}\text{Ei}\left(\frac{r^2}{4\kappa t}\right)\right]$$

$$\hat{g}_2 = \frac{\beta t}{2\pi} \cdot \frac{x_2}{r^2}\left[\left(1 - e^{-\frac{r^2}{4\kappa t}}\right) + \frac{r^2}{4\kappa t}\text{Ei}\left(\frac{r^2}{4\kappa t}\right)\right]$$

$$\hat{t}_{11} = \frac{\mu\beta t}{2\pi} \cdot \frac{1}{r^2}\left[\left(2 - \frac{4x_1^2}{r^2}\right)\left(1 - e^{-\frac{r^2}{4\kappa t}}\right) - \frac{r^2}{2\kappa t}\text{Ei}\left(\frac{r^2}{4\kappa t}\right)\right]$$

$$\hat{t}_{22} = \frac{\mu\beta t}{2\pi} \cdot \frac{1}{r^2}\left[\left(2 - \frac{4x_2^2}{r^2}\right)\left(1 - e^{-\frac{r^2}{4\kappa t}}\right) - \frac{r^2}{2\kappa t}\text{Ei}\left(\frac{r^2}{4\kappa t}\right)\right]$$

$$\hat{t}_{12} = \hat{t}_{21} = \frac{\mu\beta t}{2\pi} \cdot \frac{4x_1 x_2}{r^4}\left(e^{-\frac{r^2}{4\kappa t}} - 1\right)$$

$$g^*_{122} = \frac{1}{4\pi(1-\nu)} \cdot \frac{-x_1}{r^2}\left\{(1-2\nu) - \frac{\nu_u-\nu}{1-\nu_u}\left[(1-2e^{-\frac{r^2}{4\kappa t}}) + \frac{4\kappa t}{r^2}(1-e^{-\frac{r^2}{4\kappa t}})\right] - \frac{2x_2^2}{r^2}\left[1 + \frac{\nu_u-\nu}{1-\nu_u}\left[(1+e^{-\frac{r^2}{4\kappa t}}) - \frac{8\kappa t}{r^2}(1-e^{-\frac{r^2}{4\kappa t}})\right]\right]\right\}$$

$$g^*_{222} = \frac{1}{4\pi(1-\nu)} \cdot \frac{x_2}{r^2}\left\{(1-2\nu) + \frac{\nu_u-\nu}{1-\nu_u}\left[\frac{12\kappa t}{r^2}(1-e^{-\frac{r^2}{4\kappa t}}) - (1+2e^{-\frac{r^2}{4\kappa t}})\right] + \frac{2x_2^2}{r^2}\left[1 + \frac{\nu_u-\nu}{1-\nu_u}\left[(1+e^{-\frac{r^2}{4\kappa t}}) - \frac{8\kappa t}{r^2}(1-e^{-\frac{r^2}{4\kappa t}})\right]\right]\right\}$$

$$t^*_{1122} = \frac{\mu}{2\pi(1-\nu)} \cdot \frac{1}{r^6}\left\{(x_1^4 + x_2^4 - 6x_1^2x_2^2)\left[1 + \frac{\nu_u-\nu}{1-\nu_u}\left[(1 + 2e^{-\frac{r^2}{4\kappa t}}) - \frac{12\kappa t}{r^2}(1-e^{-\frac{r^2}{4\kappa t}})\right]\right] - \frac{\nu_u-\nu}{1-\nu_u}\cdot 4x_1^2x_2^2\cdot\frac{r^2}{4\kappa t}\, e^{-\frac{r^2}{4\kappa t}}\right\}$$

$$t^*_{2222} = \frac{\mu}{2\pi(1-\nu)} \cdot \frac{1}{r^6}\left\{x_1^4\left[1 + \frac{\nu_u-\nu}{1-\nu_u}\left[(1- 4e^{-\frac{r^2}{4\kappa t}}) + \frac{12\kappa t}{r^2}(1-e^{-\frac{r^2}{4\kappa t}})-\frac{r^2}{\kappa t}e^{-\frac{r^2}{4\kappa t}}\right]\right] - 3x_2^4\left[1- \frac{\nu_u-\nu}{1-\nu_u}\left[1 - \frac{4\kappa t}{r^2}(1-e^{-\frac{r^2}{4\kappa t}})\right]\right] + 6x_1^2x_2^2\left[1 + \frac{\nu_u-\nu}{1-\nu_u}\left[(1+2e^{-\frac{r^2}{4\kappa t}}) - \frac{12\kappa t}{r^2}(1-e^{-\frac{r^2}{4\kappa t}})\right]\right]\right\}$$

$$t^*_{1222} = \frac{\mu}{2\pi(1-\nu)}\cdot\frac{2x_1x_2}{r^6}\left\{x_1^2\left[1+\frac{\nu_u-\nu}{1-\nu_u}\left[(1+5e^{-\frac{r^2}{4\kappa t}}) + \frac{r^2}{2\kappa t}e^{-\frac{r^2}{4\kappa t}} - \frac{24\kappa t}{r^2}(1-e^{-\frac{r^2}{4\kappa t}})\right]\right] - 3x_2^2\left[1 + \frac{\nu_u-\nu}{1-\nu_u}\left[(1+e^{-\frac{r^2}{4\kappa t}}) - \frac{8\kappa t}{r^2}(1 - e^{-\frac{r^2}{4\kappa t}})\right]\right]\right\}$$

$$h^*_{22} = \frac{\mu}{2\pi(1-2\nu)\alpha} \cdot \frac{\nu_u-\nu}{1-\nu_u} \cdot \frac{1}{r^4}\left\{(2x_2^2 - 2x_1^2)(1-e^{-\frac{r^2}{4\kappa t}}) + 4x_1^2 \cdot \frac{r^2}{4\kappa t}e^{-\frac{r^2}{4\kappa t}}\right\}$$

$$f^*_{122} = - \frac{k\mu}{2\pi(1-2\nu)\alpha\gamma_f} \cdot \frac{\nu_u-\nu}{1-\nu_u} \cdot \frac{4x_1}{r^6}\left\{x_1^2\left[1-e^{-\frac{r^2}{4\kappa t}} - (1 + \frac{2r^2}{4\kappa t})\frac{r^2}{4\kappa t}e^{-\frac{r^2}{4\kappa t}}\right] - 3x_2^2\left[1 - e^{-\frac{r^2}{4\kappa t}} - \frac{r^2}{4\kappa t}e^{-\frac{r^2}{4\kappa t}}\right]\right\}$$

$$f^*_{222} = - \frac{k\mu}{2\pi(1-2\nu)\alpha\gamma_f} \cdot \frac{\nu_u-\nu}{1-\nu_u} \cdot \frac{4x_2}{r^6}\left\{x_1^2\left[3(1-e^{-\frac{r^2}{4\kappa t}}) - (3 + \frac{2r^2}{4\kappa t})\frac{r^2}{4\kappa t}e^{-\frac{r^2}{4\kappa t}}\right] - x_2^2\left[1 - e^{-\frac{r^2}{4\kappa t}} - \frac{r^2}{4\kappa t}e^{-\frac{r^2}{4\kappa t}}\right]\right\}$$

$$g^*_{112} = \frac{1}{4\pi(1-\nu)} \cdot \frac{x_2}{r^2}\left\{(1-2\nu) - \frac{\nu_u-\nu}{1-\nu_u}\left[1 - \frac{4\kappa t}{r^2}(1-e^{-\frac{r^2}{4\kappa t}})\right] + \frac{2x_1^2}{r^2}\left[1 + \frac{\nu_u-\nu}{1-\nu_u}\left[(1+e^{-\frac{r^2}{4\kappa t}}) - \frac{8\kappa t}{r^2}(1-e^{-\frac{r^2}{4\kappa t}})\right]\right]\right\}$$

$$g^*_{212} = \frac{1}{4\pi(1-\nu)} \cdot \frac{x_1}{r^2}\left\{(1-2\nu) - \frac{\nu_u-\nu}{1-\nu_u}\left[1 - \frac{4\kappa t}{r^2}(1-e^{-\frac{r^2}{4\kappa t}})\right] + \frac{2x_2^2}{r^2}\left[1 + \frac{\nu_u-\nu}{1-\nu_u}\left[(1+e^{-\frac{r^2}{4\kappa t}}) - \frac{8\kappa t}{r^2}(1-e^{-\frac{r^2}{4\kappa t}})\right]\right]\right\}$$

$$t^*_{1112} = \frac{\mu}{2\pi(1-\nu)}\cdot\frac{2x_1x_2}{r^6}\left\{x_2^2\left[1+\frac{\nu_u-\nu}{1-\nu_u}\left[(1+5e^{-\frac{r^2}{4\kappa t}}) + \frac{r^2}{2\kappa t}e^{-\frac{r^2}{4\kappa t}} - \frac{24\kappa t}{r^2}(1-e^{-\frac{r^2}{4\kappa t}})\right]\right] - 3x_1^2\left[1 + \frac{\nu_u-\nu}{1-\nu_u}\left[(1+e^{-\frac{r^2}{4\kappa t}}) - \frac{8\kappa t}{r^2}(1 - e^{-\frac{r^2}{4\kappa t}})\right]\right]\right\}$$

$$t^*_{2212} = \frac{\mu}{2\pi(1-\nu)}\cdot\frac{2x_1x_2}{r^6}\left\{x_1^2\left[1+\frac{\nu_u-\nu}{1-\nu_u}\left[(1+5e^{-\frac{r^2}{4\kappa t}}) + \frac{r^2}{2\kappa t}e^{-\frac{r^2}{4\kappa t}} - \frac{24\kappa t}{r^2}(1-e^{-\frac{r^2}{4\kappa t}})\right]\right] - 3x_2^2\left[1 + \frac{\nu_u-\nu}{1-\nu_u}\left[(1+e^{-\frac{r^2}{4\kappa t}}) - \frac{8\kappa t}{r^2}(1 - e^{-\frac{r^2}{4\kappa t}})\right]\right]\right\}$$

$$t^*_{1212} = \frac{\mu}{2\pi(1-\nu)} \cdot \frac{1}{r^6}\left\{(x_1^4 + x_2^4 - 6x_1^2x_2^2)\left[1 + \frac{\nu_u-\nu}{1-\nu_u}\left[(1 + 2e^{-\frac{r^2}{4\kappa t}}) - \frac{12\kappa t}{r^2}(1-e^{-\frac{r^2}{4\kappa t}})\right]\right] - \frac{\nu_u-\nu}{1-\nu_u}\cdot 4x_1^2x_2^2\cdot\frac{r^2}{4\kappa t}\, e^{-\frac{r^2}{4\kappa t}}\right\}$$

$$h^*_{12} = \frac{\mu}{2\pi(1-2\nu)\alpha} \cdot \frac{\nu_u-\nu}{1-\nu_u} \cdot \frac{4x_1x_2}{r^4}\left\{(1-e^{-\frac{r^2}{4\kappa t}}) - \frac{r^2}{4\kappa t}e^{-\frac{r^2}{4\kappa t}}\right\}$$

$$f^*_{112} = \frac{k\mu}{2\pi(1-2\nu)\alpha\gamma_f} \cdot \frac{\nu_u-\nu}{1-\nu_u} \cdot \frac{4x_2}{r^6}\left\{x_1^2\left[3(1-e^{-\frac{r^2}{4\kappa t}}) - (3 + \frac{2r^2}{4\kappa t})\frac{r^2}{4\kappa t}e^{-\frac{r^2}{4\kappa t}}\right] - x_2^2\left[1 - e^{-\frac{r^2}{4\kappa t}} - \frac{r^2}{4\kappa t}e^{-\frac{r^2}{4\kappa t}}\right]\right\}$$

$$f^*_{212} = \frac{k\mu}{2\pi(1-2\nu)\alpha\gamma_f} \cdot \frac{\nu_u-\nu}{1-\nu_u} \cdot \frac{4x_1}{r^6}\left\{x_2^2\left[3(1-e^{-\frac{r^2}{4\kappa t}}) - (3 + \frac{2r^2}{4\kappa t})\frac{r^2}{4\kappa t}e^{-\frac{r^2}{4\kappa t}}\right] - x_1^2\left[1 - e^{-\frac{r^2}{4\kappa t}} - \frac{r^2}{4\kappa t}e^{-\frac{r^2}{4\kappa t}}\right]\right\}$$

New techniques for the determination of the hydraulic properties of fractured rock masses
Nouvelles techniques pour la détermination des propriétés hydrauliques de massifs rocheux fracturés
Eine neue Technik für die Bestimmung der hydraulischen Kennwerte für geklüfteten Fels

R.M.DE ANDRADE, Engevix S.A., Rio de Janeiro, Brazil

SUMMARY: Two new field test techniques carried out in the field are helpful for understandging how water flows in fractured rock masses and for determining their hydraulic properties. The results of these tests, performed in a borehole located at Santa Isabel Dam site be built on the Araguaia River in Central Brazil, are presented in this paper.

RESUME:On présente deux nouveaux tests au chantier dans des trous de forage qui contribuent pour meilleure compreende l'écoulement à travers les massives rocheaux de foundation et qui rendre possible determiner ses proprietées hydrauliques. On présent aussi quelques résultats obtenus avec l'utilization de ces tecniques au site de l'aménagement future de Santa Isabel.

ZUSAMMENFASSUNG:Der Verfasser beschreibt zwei neue Feldversuche, die zum Verständnis der Grundwasserströmung in einer Bauwerksgründung aus zerklüftetem Fels beitragen, eine Bestimmung ihrer hidraulischen Eigenschaften erlauben und auch die Ergebnisse, die mit der Anwendung dieser Versuche erhalten werden. Die Versuche wurden in einem Bohrloch an der Stelle des zukünftigen Wasserkraftwerkes Santa Isabel durchgeführt.

1 INTRODUCTION

In 1802 the french Engineer Charles Berigny injected for the first time in history cement grout in a rock mass in order to strengthen the foundations of the Dieppe navigation locks. In 1933 Maurice Lugeon established the first rules for the use of foundation grouting.

In 1895, after the rupture of the Bouzey Dam, in France, Maurice Levy officialy admitted in reports to the French Academy, the danger of water seepage through dam foundations.

These two events happened in different periods but the subjects are still very closely mixed due to the fact that the problems in both cases are analyzed with the help of the Water Pressure Test. In the first case, by means of injections, a solid material - the cement mortar - is introduced into the rock discontinuities.

In the second, an attempt is made to deter mine how the water flows in these discontinuities.

The Water Pressure Test consists in injecting water under pressure into a borehole at a certain depth and in measuring the volume of water that is absorbed in the ground.

This traditional but uncertain and inaccurate method has undergone very little improvement since Lugeon's time. Developing a new research line on this subject,(Andrade 1985, 1986) created two new types of tests, the Hydraulic Register Test (TRH) and the Water Injection Test under Decreasing Pressure (EIPD) that are described hereafter.

The Hydraulic Register Test (Andrade, 1986) brought answers to several questions left unanswered by the traditional Water Pressure Test, like:

-through which discontinuities does water loss occur?

-what are the water carrying discontinuities like?

-what are the attitudes of these discontinuities?

With these answers, the seepage paths can be clearly identified,remaining only the problem of determining their hydraulic properties.

The Water Injection Test under Decreasing Pressure solves this deficiency and allows the determination of hydraulic properties of fractures and discontinuities under a laminar flow condition and under any pressure gradient that may occur.

2 THE HYDRAULIC REGISTER TEST TRH

2.1 Characteristics of the TRH Test

The Hydraulic Register Test presents as a final product a document extracted from the borehole, containing information about the water carrying discontinuities. Its main characteristics are resumed as follows:

1. The test is as simple as the Water Pressure Test.

2. Information is recorded in full scale and is geographically oriented.

3. Only water carrying discontinuities are recorded.

4. Tests can be carried out during drilling or after the borehole is completed,therefore they can be repeated as necessary.

5. The record shows the attitude of the discontinuities and informs the magnitude of its width.

6. Tests can always be carried out regardless of the permeability of the rock mass.

7. Results are obtained under any pressure, provided there is water seepage through the recording fabric.

2.2 Description of the Process

The Hydraulic Register Test TRH - consists in introducing into the borehole a device provided with a permeable canvas which will register the presence and attitude of all water carrying discontinuities existing at the testing stretch.

Introducing the equipment in the borehole.

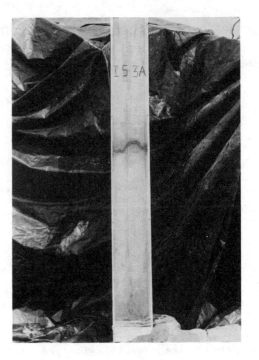

The unwrapped canvas showing the mark of the 2.41 m deep discontinuity.

The TRH equipment after the first test showing a water carrying discontinuity at the depth of 2.41 m and a 30° dip.

Identifying the discontinuity on the cores.

The device consists of a metal tube with a perforated wall which conducts the water to the testing depth. The tube has a diameter slightly smaller than the borehole's and supports the permeable canvas (see fig.2.1).The assembly is attached to a single or double packer. The single packer is used to test the lower end of the borehole. The permeable canvas is carefully wrapped around the tube and

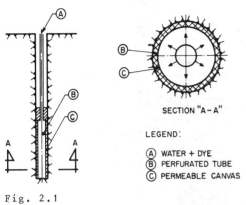

Fig. 2.1

tied to the top and bottom. The whole device is then introduced in the borehole to the desired depth, keeping a previously defined orientation in relation to geographic coordinates.

Once the device is inside the borehole the test is started with the water injection.The water pressure will press the canvas against the borehole wall. If there are any water carrying discontinuities at that stretch,the water will seep through the permeable canvas into the rock mass. A dye is then added to the water, which will mark the canvas along the discontinuities thus taking its imprints. The test takes only a few minutes to be performed and may be carried out at any desired pressure level.

The test ends when the water flow is cut and the pressure relieved. The equipment is then lifted out, care being taken not to damage the permeable canvas. On the canvas the marks of the water carrying discontinuities can then be observed. They usually take the form of a dark band about 3 cm wide with the same colour of the dye.

The discontinuity is in the center of the band. The mark is so wide due to the flow lines in the canvas converging to the discontinuity.

When there is a porous non-fractured medium at the testing level, the mark takes the shape of a shadow of the same width as the porous strech.

2.3 Some Conclusions from the Use of the TRH Test

The fact that foundation rock masses are composed of families of fractures in a spatial distribution doesn't mean that every discontinuity is a seepage path. To make this assumption without further investigation is an error.

The TRH tests performed in Cachoeira Dourada Dam, in Central Brazil, showed that most of the fractures were already sealed by oxidation, althrough they were considered by geologists as potential seepage paths.In fact only about 30% of the fractures were water carrying. The rest were sealed even under high water test pressure.

This shows that the only possible way to

learn the exact seepage path is with the help of the TRH Test, and it is a mistake to consider every family of fractures as water carrying.

Using the TRH Test the foundation mass can be mapped, furnishing new information to the initial hydrogeotechnical models.

3 THE WATER INJECTION TEST UNDER DECREASING PRESSURE-EIPD

3.1 Characteristics of the EIPD Test

The purpose of the Water Injection Test under Decreasing Pressure is to acquire the necessary data to determine hydraulic characteristics of flow in the water carrying discontinuities.

The EIPD Test has the following characteristics:

1. The test is very simple and can be carried out by non specialized labour;
2. Results are easy to obtain;
3. The tests are performed in existing boreholes and can be repeated as necessary;
4. For every type of discontinuity it is possible to perform a test under the desired flow conditions;
5. Many measuring mistakes are avoided by the fact that the test uses pressure diferences as input data.

3.2 Description of the Process

The Water Injection Test under Decreasing Pressure - EIPD - employs a water injector consisting of a cylindrical tube having a diameter that provides a controlled flow. Inside the tube a regulating system controls the water pressure. Fig. 3.1 shows the phases of the test.

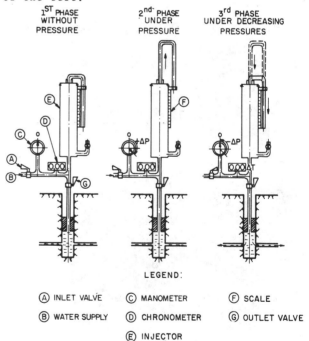

LEGEND:

(A) INLET VALVE   (C) MANOMETER   (F) SCALE
(B) WATER SUPPLY   (D) CHRONOMETER   (G) OUTLET VALVE
                   (E) INJECTOR

Fig.3.1

In the first phase water without pressure is injected in the system. In the second phase the system is under pressure and the piston is displaced to a position corresponding to the existing pressure, indicated on the scale. With the help of the manometer, the injector's

Equipment for the EIPD Test.Note the rod that indicates the water head on the scale.

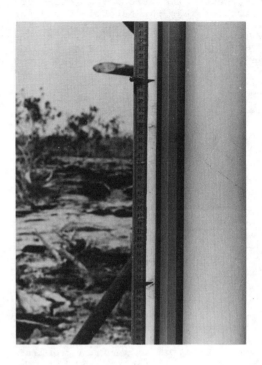

Pointer and scale for direct measuring of water heads.

Purging the equipment.

Closing the globe valve and measuring the time elapsed between pressures H_2 and H_1.

scale is calibrated and it becomes a manometer. For every position of the piston there corresponds a pressure reading on the scale.

In the third phase the water valve is closed and the volume of water contained in the tube will be slowly injected into the borehole. Duration of the injection is determined by the flow conditions in the discontinuity. The injected volume is the necessary amount to establish a laminar flow. The average seepage speed is the ratio of the difference of injected water heads to the time ΔT spent during injection. Having these values one can determine the flow factor, i.e., the conductivity of the discontinuity for a known pressure gradient under laminar flow.

The second valve is the water outlet, indicated in Fig. 3.1 and can be used to interrupt the injection for intermediate readings on the scale.

The process as described above is very simple not only in its concept but also in its operation.

3.3 Determination of Hydraulic Properties of Discontinuities

3.3.1 Vertical borehole intercepting horizontal discontinuity

As already mentioned, the EIPD Test consists in injecting into a discontinuity a volume V_o of water during a period of ΔT, submetting the water head to a variation:

$$\Delta H = \frac{\Delta P}{\gamma}$$

where ΔP is the variation of pressure and γ is the unit weight of water, from now on assumed as $1 tf/m^3$. (see Fig. 3.2).

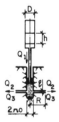

Fig.3.2

Calling D the injector's diameter and h the water head in the injector, the injection flow is

$$Q_1 = - A_1 \frac{h}{\Delta T}$$

where

$$A_1 = \frac{\pi D^2}{4}$$

but

$$V_o = h \frac{\pi D^2}{4}$$

the injection flow is

$$Q_1 = - \frac{V_o}{\Delta T} = - \frac{\pi h D^2}{4 \Delta T}$$

known

Continuity condition states that

$$Q_1 = Q_2$$

i.e., the injection flow is equal to the flow in the discontinuity. But

$$Q_2 = - A_2 \frac{\Delta H}{\Delta T}$$

where

$$A_2 = 2 \pi R \varepsilon$$

is the area of a ring of average width ε through which flows the quantity Q_2, ΔH is the variation of pressure between the initial level H_2, and final level H_1, and R is the distance between the points of pressure H_2 and H_1.
One can then write:

$$\frac{\pi h D^2}{4 \Delta T} = - 2 \pi R \varepsilon \frac{\Delta H}{\Delta T}$$

$$\boxed{R \varepsilon = \frac{h D^2}{8 \Delta H}} \qquad (1)$$

In the above expression all values of the second member are known.

Flow condition in the discontinuity is always laminar, because the injection is performed in such way that seepage velocities are kept within limits, so as to obtain this condition.

Thus,

$$Q_3 = - k_d . A_3 . i$$

or

$$Q_3 = - k_d \frac{2 \pi (R - r_o) \varepsilon}{\ln \frac{R}{r_o}} . \frac{\Delta H}{R - r_o}$$

where

$$A_3 = \frac{2 \pi (R - r_o) \varepsilon}{\ln \frac{R}{r_o}}$$

is the average cross section of flow between the points of pressure H_2 and H_1, r_o is the borehole internal radius, k_d is the hydraulic conductivity and $i = \frac{\Delta H}{R - r_o}$ the gradient.

$$k_d = \frac{g \varepsilon^2}{12 \nu}$$

where g is the acceleration of gravity and ν the kinematic viscosity.

Making

$$Q_1 = Q_3$$

and introducing the value of k_d in Q_3 and substituting ε as a function of R one comes to:

$$\boxed{R^3 \ln \frac{R}{r_o} = \frac{g}{12 \nu} \left(\frac{h D^2}{8 \Delta H} \right)^2 . \Delta T} \qquad (2)$$

Therefore, for each time interval ΔT of the test, one can determine R and ε with equation (1).

One of the conditions for a laminar flow is that Reynold's Number be less than 2,300:

$$R_e = \frac{\overline{V}.2.\varepsilon}{\nu} < 2\ 300$$

as $\overline{V} = \frac{\Delta H}{\Delta T}$ it is possible to check the flow condition for the ΔH of the test.

The gradient $i = \frac{\Delta H}{R-r_o}$ must also be computed in order to draw the curve flow x gradient.

Considering that

$$Q_3 = - f\ \frac{\Delta H}{R-r_o} = Q_1 = - \frac{\pi h D^2}{4\Delta T}$$

factor f (Andrade 1986) is:

$$f = \frac{\pi h D^2}{4\Delta H} \cdot \frac{R-r_o}{\Delta T}$$

3.3.2 Vertical borehole intercepting an inclined discontinuity

The intersection of the borehole and discontinuity is an ellipse and the injection water flows into a larger area. The seepage section is now the perimeter of an ellipse of a nominal width ε (see fig. 3.3).

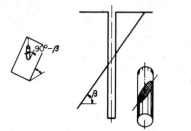

Fig. 3.3

It is then sufficient to apply the scale factor α so that

$$R = \mathcal{R}\alpha$$
$$r_o = \mathcal{r}_o \alpha$$

where

$$\alpha = \frac{1}{\cos\ \beta}\sqrt{\frac{1+\cos^2\beta}{2}}$$

and use the formulae presented in the last section.

3.3.3 Inclined borehole perpendicular to the discontinuity plane

In this case the intersection of the borehole with the discontinuity is a circle but the flow section is an ellipse. Adapting the above formulae: (see fig. 3.4).

$$R = \mathcal{R}\alpha$$
$$r_o = \mathcal{r}_o$$

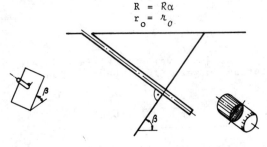

Fig.3.4

4 A CASE STUDY

4.1 - The Site

The test that shall be analyzed hereafter was carried out at Santa Isabel Dam site on the right bank of the Araguaia River in Central Brazil. The local rock is a micaschist with schistosity planes dipping 30°.
The tests were performed in SR-38A, a 10,07 m deep vertical borehole.

4.2 - Description of the Tests

Two types of tests were performed in this borehole: the Hydraulic Register Test (TRH) to identify the water carrying discontinuities and the Water Injection Test under Decreasing Pressure (EIPD) to determine the hydraulic properties of these discontinuities.
The TRH test, when carried out with a pressure of 1 kgf/cm², indicated two discontinuities. The first one was 2.41 m deep with a 30° dip as observed on the cores, and the second at 6.30 m with the same characteristics. No other water losses were detected along the borehole wall. (See Fig. 4.1).

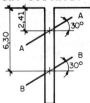

Fig.4.1

After detecting the depth and the attitude of the water carrying discontinuities, the Water Injection Test under Decreasing Pressure was carried out at the two levels.
The tests were performed with four different injection pressures varying from 0.6 kgf/cm² to 2.1 kgf/cm² as shown in Table 4.1 The equipment employed for this series of tests used a cylinder with a diameter of 9.76 cm, allowing a maximum internal pressure of 4 kgf/cm².

TABLE 4.1

	PRESSURES (kgf/cm²)	
	P initial	P final
1st Test	0,6	0,1
2nd Test	1,1	0,6
3rd Test	1,6	1,1
4th Test	2,1	1,6

The test is performed with several pressure differences, in order to be able to draw the following curves: flow (Q) x pressure gradient (i), nominal discontinuity width (ε) x gradient (i), hydraulic conductivity (k_d) x gradient (i) and flow factor (f) x gradient, thus defining all hydraulic characteristics of the discontinuities. It was observed in all cases that seepage occurred under laminar condition.
In the first test, corresponding to the discontinuity at a depth of 2.41 m the borehole was packed between depths of 1.30 m and 3.37 m.
The results obtained are shown in Table 4.2.

TABLE 4.2

P_i	P_f	ΔH	ΔT
0,6	0,1	20,3	2'07",12
1,1	0,6	19,6	48",80
1,6	1,1	19,2	18",31
2,1	1,6	19,8	4",07

In this table:

P_i, P_f - initial and final pressure in the injector (kgf/cm²);

ΔH - water head difference corresponding to the pressure decrease from P_i to P_f (in cm);

ΔT - interval of time between P_i and P_f.

The second test was performed between depths 5.32 m and 7.39 m for the discontinuity at depth 6.30 m. Results are shown in Table 4.3.

TABLE 4.3

P_i	P_f	ΔH	ΔT
0,6	0,1	19,9	6'16",27
1,1	0,6	20,2	1'57",96
1,6	1,1	20,1	26",45
2,1	1,6	21,0	4",07

Figure 4.2 shows the resulting curves for the tests. Line A corresponds to depth 2.41 m and line B to depth 6.3 m.

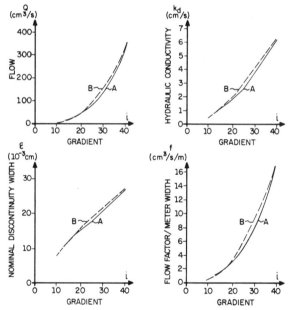

Fig.4.2

5 CONCLUSIONS

Using the two tests described in the preceding sections it is possible to determine hydraulic properties of fractured rock mass in a simple and direct way. This brings a new perspective for the understanding of water flow in discontinuities, and a sounder basis for defining foundation treatments.

6 REFERENCES

Andrade, R.M.de 1986. Propriedades Hidráulicas de Maciços Fraturados e sua Aplicação em Projetos - II Simpósio Sul-Americano de Mecânica das Rochas - Porto Alegre.
Andrade, R.M.de 1985. Contribuição nº 35 pg. 498 Q.58.35 - XV Congress on Large Dams - Lausanne.
Andrade, R.M.de 1987. Subpressão em Estruturas de Concreto - XVII Seminário Nacional de Grandes Barragens
Andrade, R.M.de 1987. Contribution for a better understanding of flow in fractured rock masses based on new field test techniques - International Workshop on Arch Dams - Coimbra
Andrade, R.M.de 1987. O projeto das estruturas hidráulicas condicionado a estudos hidrogeotécnicos - Conferência Ibero-Americana sobre Aproveitamentos Hidráulicos - Lisboa.

Three dimensional flow modeling in jointed rock masses
Modèle tridimensionnel d'écoulement dans un massif rocheux fissuré
Dreidimensionale Modelle für die Durchströmung von geklüftetem Fels

WILLIAM S. DERSHOWITZ, Golder Associates Inc., Redmond, Wash., USA
HERBERT H. EINSTEIN, Massachusetts Institute of Technology, Cambridge, USA

ABSTRACT: A number of problems involving flow through rock masses need to be analyzed by fracture flow models rather than equivalent porous medium models. The authors, and others, have developed two- and three-dimensional fracture flow models. In this paper, computational comparisons are made. They show that different fracture geometries lead to considerably different conductivities, that three diemnsional modelling leads to higher conductivities than two dimensional modelling, and that equivalent porous medium or even stochastic continuum approaches are not generally applicable to fracture flow problems. Possible approaches for modeling channeling in fractures are presented.

RESUME: C'est bien connu que les problèmes d'écoulement en masses rocheuses doivent etre analysés avec des modèles d'écoulement en fissures au lieu de modèles utilisant une masse poreuse équivalent. Les auteurs et des autres chercheurs ont développé des modèles d'écoulement en fissures en deux et trois dimensions, dont quelques-uns sont comparés sur la base de resultats calculés. Ces resultats montrent que des géometries de fissures différentes causent des conductivités très différentes, que les modèles en trois dimensions donnent des conductivités fortement supérieures au modèles en deux dimensions, et que les modèles quasi-poreux ou representant un continu stochastique ne sont généralement pas appliquable aux problèmes d'écoulement en fissures. Si, au lieu d'écoulement en fissures il y a écoulement canalisé a l'interieur des fissures, c'est possible d'utiliser des modèles plus simples qui sont présentés succinctement ci-après.

ZUSAMMENFASSUNG: Es ist bekannt, dass das Durchstroemen von Fels meistens mit Kluftdurchstroemungsmodellen anstatt von Modellen fuer ein equivalentes poroeses Medium analysiert werden muss. Die Verfasser und andere Autoren haben zwei- und dreidimensionale Kluftdurchstroemungsmodelle enwickelt, von denen einige in diesem Artikel rechnerisch verglichen werden. Die Resultate zeigen, dass verschiedene Kluftgeometrien zu sehr verschiedenen Konduktivitaeten fuehren, dass dreidimensionale Modelle wesentlich hoehere Konduktivitaeten haben als zweidimensionale Modelle, und dass Modelle fuer ein equivalentes poroeses Medium oder auch stochastische Kontinuummodelle generell nicht fuer Kluftdurchstroemungsprobeleme anwendbar sind. Wenn anstatt von eigentlichem Kluftdurchstroemen kanalisiertes Durchstroemen auftritt, koennen einfachere Modelle, die hier skizziert sind, angewendet werden.

INTRODUCTION

The conventional assumption for hydrologic analysis in rock is that flow can be modeled by a porous medium. This is a reasonable assumption where flow is dominated by the rock matrix rather than fracture or joint porosity (the terms fracture and joint will be used interchangeably), or where jointing is sufficiently intense that a representative elementary volume can be defined. However, for an increasing number of practical problems in petroleum extraction, hazardous and nuclear waste disposal, and water resources engineering, the porous medium approximation is not reasonable. Discrete fracture flow models (e.g., Long et al., 1982, Rouleau, 1983, Dershowitz, 1984) were developed to directly model flow in fractured rock masses. This paper describes the development and application of three dimensional discrete fracture flow models, and the use of those models in conjunction with conventional continuum approaches.

THEORETICAL BACKGROUND

As in many rock mechanics applications, the first problem in hydrologic modeling of fractured rock masses is to characterize fracture system geometry. Fracture geometry conceptual models developed for other rock mechanics applications by Baecher et al. (1978) and Veneziano (1978) have been applied to the problem of discrete fracture flow modeling in two dimensions by Long (1983), Rouleau (1984), and Dershowitz (1984), and more recently in three dimensional models by Long et al. (1985) and Dershowitz et al (1986).

Figures 1 through 3 illustrate three three-dimensional discrete fracture system models: the Baecher (disk) model, the polygonal fracture model by Veneziano (1978) and the further developed polygonal model by Dershowitz (1984). In all of these models, as in most rock mechanics models, the fracture system is described by distributions of fracture orientation (dip and azimuth), size (trace length or fracture area), and intensity (fractures per unit volume). The polygonal fracture models require, in addition, the

definition of an area persistence measure
defining the degree of fracture coplanarity (as
the percentage of a fracture plane consisting
of open fractures), and the Dershowitz model
also a fracture termination measure (the
percentage of fracture terminations occurring
at intersections with other fractures). Other
three-dimensional fracture system models are
the orthogonal model used by Snow (1965) and
mosaic tesselation models discussed by
Dershowitz (1984) and Dershowitz et al. (1987).
The variety of fracture system models reflect
the variety of geometries encountered in
nature. Each of the above mentioned models can
represent typical geometries and can also be
explained by particular joint genesis
mechanisms. None of the models has general
validity and, so far, no preponderance of one
model geometry compared to the others could be
established. Until evidence to the contrary is
formed, fracture flow modeling will have to be
based on all these (and possible additional)
fracture system models.

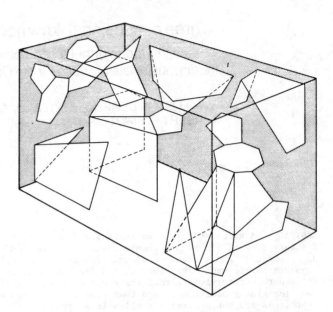

Figure 3. Generalized polygon fracture system
model.

The discrete fracture system models have been
implemented in the flow modeling package JINX
by (a) discretizing fractures into systems of
two dimensional (plate) finite elements, (b)
defining boundary conditions, and (c) solving
for flow and pressure by conventional two
dimensional finite element methods. Since the
fracture system models are stochastic, flow has
to be calculated by conducting a number of
simulations with JINX. Figures 4 and 5
illustrate the discretization of fractures used
for disk and polygonal (Dershowitz model)
fractures in the JINX package.

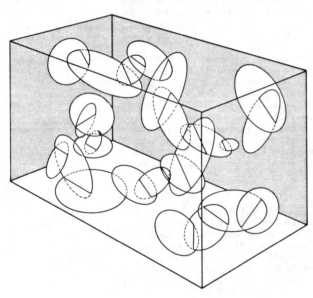

Figure 1. Baecher (disk) fracture system model.

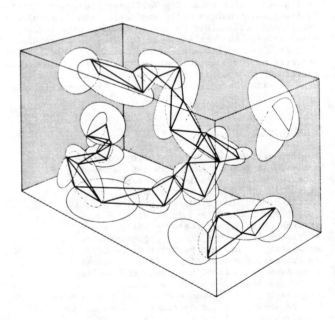

Figure 4. Discretization of disk fracture
system model.

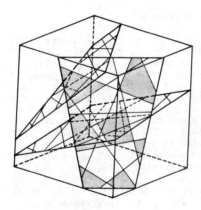

Figure 2. Veneziano fracture system model.

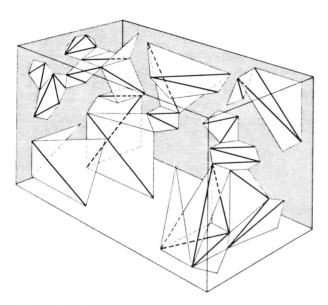

Figure 5. Discretization of polygon fracture system model.

As can be seen from the above, the development of three dimensional discrete fracture models is a fairly straightforward process, and can be carried out based upon existing, proven rock mechanics technologies. What is of interest is to investigate the usefulness and problematic aspects of three dimensional fracture flow modeling. This is done by first comparing different three dimensional fracture flow models, followed by an investigation of differences between two-dimensional and three-dimensional fracture flow models and finally some discussion of equivalent porous media compared to fracture flow models. Based on these comparisons, on reflections about field experience, and on computational considerations, some tentative conclusions on three dimensional fracture flow modeling will be made.

COMPARISON OF THREE-DIMENSIONAL FRACTURE FLOW MODELS

The Baecher and Dershowitz models are compared because they represent the extremes of fracture systems with predominant termination in intact rock (Baecher) on the one hand, and with termination at intersections with other fractures (Dershowitz) on the other hand. Flow through a cube containing the particular fracture system and with a unit gradient between two sides of the cube is computed with JINX and then expressed as equivalent hydraulic conductivity, which is the conductivity of the same size cube containing a porous medium. Figure 6 shows the thus obtained effective hydraulic conductivity in relation to the ratio of model size to fracture size. In the Baecher model, fracture size is the mean radius of the circular disks; in the Dershowitz model, it is backfigured by assuming that the polygonal fractures are represented by circles of the same areas and calculating a linear radius. The cube has edge dimensions of 100 meters and the dimensions in Figure 6 are in consistent units. Three fracture sets, each distributed according to a Fisher distribution, are assumed. The dispersion of the Baecher model is purposely made much greater (K = 5) than that of the Dershowitz model (K = 20, i.e.,

essentially subparallel and thus close to an orthogonal model) to ensure that at least some of the disk shaped joints do interconnect through the cube region. In spite of this, there is a substantially higher equivalent conductivity of the Dershowitz model and this for all joint sizes! From the comments above and from Figures 4 and 5, it appears that this difference is caused by the substantially different connectivity. This can also be shown by quantitative connectivity measures, for instance, intersection intensity or percolation probability (Dershowitz, 1984).

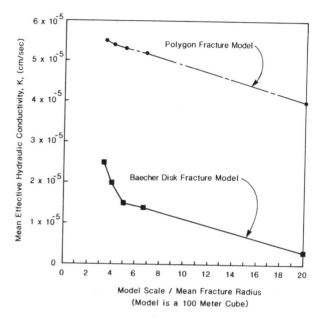

Figure 6. Scale effect on flow in three dimensional fracture systems.

COMPARISON BETWEEN THREE AND TWO DIMENSIONAL FRACTURE FLOW MODELS

In the JINX model, fracture systems and the corresponding fracture flow models can be analysed in either two or three dimensions. (In the former case, fractures are geometrically represented by fracture traces and hydraulically by "pipes"; computations of equivalent hydraulic conductivity are conducted for a square region.) Three dimensional discrete fracture network models produce substantially greater effective hydraulic conductivities than two dimensional models with comparable levels of fracture intensity.

The primary reason for the higher apparent conductivity of three dimensional models appears to be their greater connectivity. While the one dimensional fractures in a two dimensional model may intersect at a single point, two dimensional fractures in a three dimensional model intersect at a line segment, and therefore have the potential for a much greater number of intersections per fracture. This is illustrated in Figure 7. This figure shows that the number of intersections per fracture is many times greater in the three dimensional model than in the two dimensional model. This provides a feasible explanation for the greater conductivity of three dimensional models.

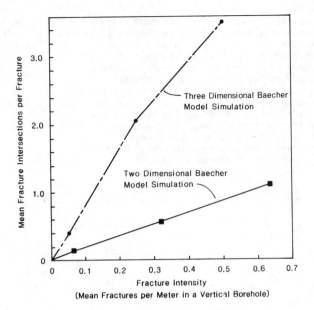

Figure 7. Fracture intersection intensity in two and three dimensional models.

EQUIVALENT POROUS MEDIA MODEL

Given the computational effort for fracture flow modeling, particularly in three dimensions, replacing the fractured by an equivalent porous medium is tempting and desirable. Extensive work in this area has been performed by Long (1983) and Long et al (1982) as well as by the M.I.T. group (Dershowitz, 1984). An illustrative method for comparing porous and fractured media approaches is to plot direction dependent conductivities (Figure 8). For porous media, this results in the so called permeability ellipses (2-D) (Figure 8a) or ellypsoids (3-D) and in irregular surfaces (Figure 8b) for fractured media. In Figure 9 permeability ellipses for two dimensional applications of different fracture flow models are presented; the resulting shapes clearly are not ellipses but irregular surfaces. As was shown in the above mentioned references, as the scale, i.e., the dimension of the region compared to the mean fracture length, increases the permeability surface approaches that of a porous medium.

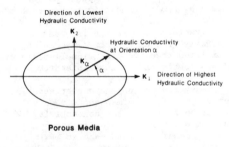

Porous Media

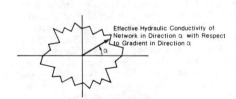

Finite Element Solution for Fractured Media

Figure 8. Permeability ellipses.

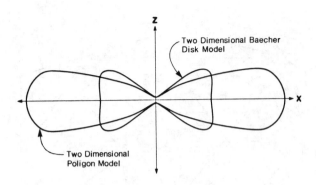

Permeability Ellipses

Figure 9. Permeability ellipses in two dimensional fracture system models.

A direct replacement of the fracture flow models by porous media models is thus generally not possible. An intermediate solution may be found by modeling small blocks with the fracture flow models and then to incorporate these blocks in a stochastic continuum model. The two dimensional stochastic continuum model STOMP (STOchastic hydrologic Modeling Program) was developed to assess the feasibility of the use to two dimensional discrete fracture flow models to develop properties for stochastic continuum models. Several approaches have been explored for relating the two models, with little success. Figure 10 shows one example in which relatively good correspondence concerning the trend (but not regarding absolute values) was found between a stochastic continuum model and a discrete fracture flow model. In this simulation, the flow anisotropy ratio K_h/K_v for fracture and stochastic continuum models was plotted as a function of the distance between constant head boundaries. In both cases, average anisotropy was substantially greater in smaller scale models, and decreased as the scale of the model being evaluated increased. This result provides some hope that a stochastic continuum model may be feasible, although it must be noted that in the majority of simulations performed, stochastic continuum models could not be developed to match discrete fracture flow model results.

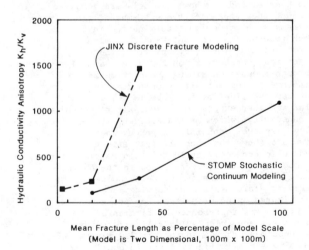

Figure 10. Comparison of continuum and discrete fracture flow models.

ASSESSMENT OF THREE DIMENSIONAL FRACTURE FLOW MODELING

The preceding presentation of results and discussion has shown that different 3-d fracture system models lead to significantly different equivalent conductivities and that 3-d fracture flow models have considerably higher conductivity than 2-d models. Both differences can be associated with the different connectivities. Also, replacing the fracture flow model by a simple porous medium model or even by a more sophisticated stochastic continuum model is not generally possible.

These comments are solely based on the comparison of computed results. Although informative, they need to be compared to results from the field. A number of efforts are under way a the time of this writing (e.g., the Canadian Underground Research Laboratory, and the Swedish Stripa Phase 3 project) to conduct field validation of fracture flow models. A number of observations can provide some indication on suitable modeling approaches:

The second author has observed in many tunnels and rock-cuts that water flowed from fractures over substantial if not the full extent of these fractures. At other occasions, concentrated flow representing that from pipes was observed. The latter can be related to "channeling" which had been mentioned by Londe (1981) as a possibility in flow through dam abutments and which has lately received increased attention in radioactive waste repository research. It is conceivable that channeling is the predominant mode of flow in undisturbed zones of the rock mass. If this is so, two-dimensional rather than three-dimensional fracture flow models can be used. A case in point is the results obtained by Schrauf et al. (1986) for the Hanford Basalt Waste Isolation Site which show good correspondence between two dimensional fracture flow modeling and observed effective hydraulic conductivities. However, whether this is in fact due to channeling type flow cannot be answered at present. Also, as shown in Appendix A, modeling of channeling flow may require a computational effort which is substantially greater than that for three dimensional fracture flow.

CONCLUSIONS

Several two and three dimensional geometric fracture models and the corresponding fracture flow models exist. These models provide the means to represent different natural conditions including the possibility to simulate fracture flow and channeling. The differences in conductivities computed with these models and the difference between these models and equivalent porous medium or stochastic continuum approaches are substantial. This points to the necessity of extensive field testing in order to obtain some idea on the actual ranges of applicability of the different models.

ACKNOWLEDGEMENTS

Research on which this paper is based has been conducted at M.I.T. under sponsorship of the Army Research Office and later at Golder Associates sponsored by the Office of Waste Technology Development. This support is gratefully acknowledged as is the involvement of a number of colleagues of the authors notably Professors Baecher and Veneziano at Massachusetts Institute of Technology, and Todd Schrauf and Scott Brown at Golder Associates.

APPENDIX A

There are several possible approaches for modeling channeling in three dimensional fractures. One is to discretize planar fractures into very small elements with autocorrelated apertures, so that flow occurs along channels defined by series of high conductivity elements (Figure A1). This approach is attractive, since it has the potential to represent the fracture roughness profiles which may actually cause channeling. However, this approach may prove to be infeasible with the current state of the art of computer technology. For example, if just 20 elements were used to define channels on each fracture, a cube 20 meters on a side with a fracture intensity of 3 fractures per meter would require almost 500,000 elements!

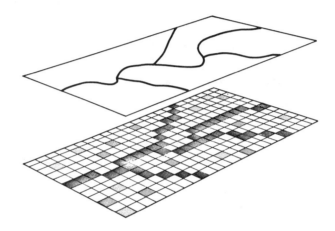

Figure A1. Two dimensional approach for modeling fracture flow channels.

Another possible approach is the direct definition of channels as one dimensional flow elements on fracture planes (Figure A2). This approach would decrease the modeling costs to a practical level by decreasing both the number of elements to be included in the model, and the connectivity of the fracture network. In this approach, no elements would need to be generated for portions of fractures which are not included in the pipe flow network defined by the one dimensional flow channels. Since the use of pipe flow elements would reduce the connectivity to the same order of magnitude as current two dimensional models, the number of elements would be reduced proportionally. For example, if 5 pipe flow channels were used for each fracture in the example above, and the connectivity were reduced by 75 percent, such that only 25 percent of fractures participated in flow, the number of elements would be reduced to 10,000, and the model would be feasible with current computer technology. This approach is currently being implemented with the JINX discrete fracture modeling package.

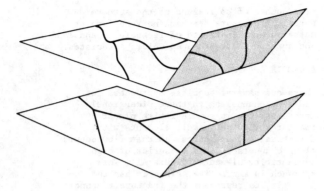

Figure A2. One dimensional approach for modeling fracture flow channels.

REFERENCES

Baecher, G.B., Lanney, N.A. & H.H. Einstein 1978. Statistical description of rock properties and sampling. Proceedings 18th U.S. Symposium on Rock Mechanics, American Institute of Mining Engineers, 5C1-8.

Dershowitz, W.S. 1984. Rock fracture systems. Ph.D. Dissertation, Massachusetts Institute of Technology. Cambridge, MA.

Dershowitz, W.S. & H.H. Einstein 1987. Characterizing rock joint geometry with joint system models. Rock Mechanics and Rock Engineering, submitted for publication.

Dershowitz, W.S., T.S. Schrauf & S. Brown 1986. Project technical report, fracture flow and solute transport modeling. Report to Battelle - Office of Crystalline Repository Development. Golder Associates. Seattle, WA.

Londe, P. 1981. Consulting discussions regarding Alta Dam, Norway.

Long, J.C.S., P. Gilhoun & P.A. Witherspoon 1985. A model for steady flow in a random three dimensional network of disc-shaped fractures. Water Resources Research 21:1105-1115.

Long, J.C.S., J.S. Remer, C.R. Wilson. & P.A. Witherspoon 1982. Porous media equivalents for networks of discontinuous fractures. Water Resources Research 18:645-658.

Rouleau, A 1984. Statistical characterization and numerical simulation of a fracture system-application to groundwater flow in the Stripa granite. Ph.D. Dissertation, University of Waterloo. Waterloo, Ontario.

Schrauf, T.S. 1986. Intermediate design of cluster injection and tracer test - fracture flow analysis. Report to Rockwell Hanford Operations. Golder Associates. Seattle, WA.

Snow, D.T. 1965. A parallel plate model of permeable fractured media. Ph.D. Dissertation, University of California. Berkeley, CA.

Veneziano, D. 1978. Probabilistic model of joints in rock. Internal report, Dept. of Civil Engineering, Massachusetts Institute of Technology. Cambridge, MA.

Physical and numerical analogues to fractured media flow
Analogies physiques et numériques pour l'écoulement en milieu fissuré
Physikalische und numerische Analogien der Grundwasserbewegung im geklüfteten Medium

DEREK ELSWORTH, The Pennsylvania State University, USA
ANDREW.R.PIGGOTT, The Pennsylvania State University, USA

ABSTRACT: The performance of an automated numerical model for the transient hydraulic analysis of three-dimensional, rigid fractured rock masses is compared against a thermal analogue to the same problem.

RESUME: Une méthode automatique et numérique a été développée pour le calcul des changements de pression interstitielle en trois dimensions de rocheaux fissurés. Cette méthode numerique a été comparé avec un étude thermique.

ZUSAMMENFASSUNG: Die leistung von automatichen numerische modellen zur reichtzeitigen hydraulischen analyse von drei-dimensionalen festen fraktierten felsmassen wird einer thermischen analyse analog zu dem gleichen problem gegen übergestellt.

1 INTRODUCTION

Accurate assessment of the transient hydraulic behaviour of sparsely fractured rock masses is of critical importance in determining the long term performance of waste containment facilities. The large body of data built up within the disciplines of hydrogeology and petroleum engineering is of limited applicability in that sparsely fractured masses behave differently to the typically highly fractured reservoirs encountered in these disciplines. More specifically, the continuum representation of discretely and discontinuously fractured masses is highly scale dependent. For steady analysis, equivalent continuum representation of the mass becomes appropriate as the sampled scale or discretization scale is increased relative to the mean trace length of the discontinuities. This facet has been observed in 'numerical testing' studies of two dimensional fracture networks, however, it is not immediately apparent how a truly three dimensional network may be reduced to two dimensions. Furthermore, in transient analysis, as the zone of hydraulic disturbance surrounding any perturbation migrates through an ever increasing volume with time, the effective scale of the analysis also changes.

In order to define the sensitivity and relative scale dependence of continuum versus discontinuum representation for the transient case, it is imperative to understand the fundamental physical factors involved. Well testing in representative rock masses is one possible approach although the attendent expense and poor definition of boundary conditions and rock structure are critical factors regulating success. Alternatively, numerical testing of a statistically equivalent rock mass is possible provided efficient and accurate numerical techniques are available. Of particular importance in this regard is efficiency. Since many hundreds of rock fissures may be required in any analysis it is important to reduce the number of equations and consequently the number of degrees of freedom of the system to a minimum while still retaining accuracy in the solution. This is of particular importance for transient analysis in that time stepping greatly increases the computational effort. The following addresses one particular approach of solving the problem of transient hydraulic analysis in three-dimensional, rigid fracture networks set within an impermeable medium. A physical heat flow analogue is used to comment on the effectiveness and accuracy of the method.

2 METHOD OF ANALYSIS

The fracture network is generated in a control volume within which the interconnection, geometric hydraulic conductivity and storativity may be evaluated. Under appropriate prescribed boundary conditions the hydraulic performance of the system may be determined with time.

2.1 Rock mass fabrication

In practical applications the rock mass may be fabricated from known and site specific statistical distributions of trace length, aperture, spacing and orientation. This generation of Poisson discs is completed for a prismoidal control volume within which relative interconnection of discrete fractures may be determined. With all mutual intersections delineated, individual constant aperture discs represent idealized two-dimensional fissure aquifers arbitrarily oriented in three-dimensional space. A direct boundary element procedure is used to further reduce the dimensionality of the problem. Discretization is required of all disc intersections and external contours as illustrated in Figure 1. An automated procedure is used to complete this stage as described by Piggott (1986).

3 FLOW ANALYSIS

The analysis is most conveniently decoupled between steady and transient states. A hybrid formulation is used to combine the inherent ease of discretization of the boundary element method (BEM) with effective transient analysis using a finite element method (FEM).

3.1 Steady analysis

For steady analysis, when an individual fissure is ascribed a constant effective hydraulic

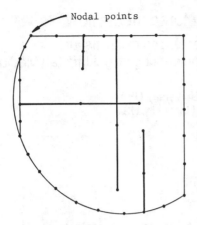

Figure 1. Discretised disc as illustrated in Figure 2(a). Note that all external boundary nodes are ultimately condensed out.

conductivity, the boundary constraint equation corresponding to the direct boundary element formulation may be represented in matrix form as

$$\underline{H} \, \underline{v} = \underline{V} \, \underline{h} \qquad (1)$$

where $\underline{H}$, $\underline{V}$ are fully populated matrices of kernel functions integrated over the domain internal and external boundaries and $\underline{v}$, $\underline{h}$ are nodal values of normal (to the boundary) velocity and total hydraulic head. Three node isoparametric elements are used to accurately represent the curved external contour of individual discs and dipole type elements are used to represent internal slits. Equation (1) may be inverted symbolically to yield

$$\underline{v} = [\underline{H}^{-1} \, \underline{V}]\underline{h} \qquad (2)$$

at which stage all equations relating to the impermeable boundary nodes are discarded. Premultiplication by the terms corresponding to the product of fissure aperture and isoparametric basis functions ($\underline{B}$) allows internal nodal velocities to be converted directly to nodal discharges as

$$\underline{q} = \underline{B}[\underline{H}^{-1} \, \underline{V}]\underline{h} \qquad (3)$$

or

$$\underline{q} = \underline{K} \, \underline{h} \qquad (4)$$

where $\underline{q}$ is a vector of nodal discharges and $\underline{K}$ is an nxn geometric conductance matrix for the fissure disc relating steady heads to discharges purely in terms of the n retained internal nodes. Although not guaranteed symmetric in this formulation, $\underline{K}$ may be easily symmetricised according to the method of Zienkiewicz et al (1977).

3.2 Transient analysis

The steady form of equation (4) may be considered as equivalent to a finite element statement obtained merely through boundary discretization. The equivalent FEM transient statement is

$$\underline{K} \, \underline{h}_t + \underline{S} \, \underline{\dot{h}}_t = \underline{q}_t \qquad (5)$$

where $\underline{K}$ is the same nxn matrix already defined, $\underline{S}$ is a diagonal matrix of storativity terms, $\underline{q}$ is a vector of prescribed nodal discharges and a superscripted dot and subscripted t refer to time derivative and current time of interest, respectively. All that remains to complete the formulation is to define the individual terms of the storativity vector ($\underline{S}$). If specific storage (S_s) is ascribed to an individual fissure to represent the volume of fluid expelled under a unit decrease head then the sum of the diagonal terms in $\underline{S}$ may be evaluated as

$$S = \int_A b \, S_s \, dA \qquad (6)$$

where S is total disc storativity, b is fissure aperture, and A is the area of the disc. This scalar value may be readily evaluated from the earlier discretization of the disc. In order to avoid internal discretization of individual discs, the total magnitude of S is distributed among the retained internal nodes according to the relative magnitude of the diagonal terms in $\underline{K}$. This procedure has proved satisfactory to date (Elsworth, 1986b).

Equation (5) may be evaluated for a single fissure disc or equivalently be assembled into the global system using standard summation convention. In global format a fully implicit and unconditionally stable time stepping sequence is used such that

$$\underline{K}^* \, \underline{h}_{t+\Delta t} = \underline{q}^*_{t+\Delta t} \qquad (7)$$

where

$$\underline{K}^* = \underline{K} + \frac{1}{\Delta t} \, \underline{S} \qquad (8)$$

and

$$\underline{q}^*_{t+\Delta t} = \underline{q}_{t+\Delta t} + \frac{1}{\Delta t} \, \underline{S} \, \underline{h}_t \qquad (9)$$

Δt is the time step increment and $\underline{q}_{t+\Delta t}$ prescribed nodal discharge at time $t+\Delta t$.

4 PHYSICAL ANALOGUE STUDY

The accuracy of the proposed model has been illustrated elsewhere for the steady (Elsworth, 1986a) and transient (Elsworth, 1986b) states. The net conclusion is that steady representation is very accurate although the early time transient response is somewhat deficient due to the sparing representation of internal pressure gradients. The effects of accelerated early response in single discs is apparent neither in the long term or for multiply connected discs in series. It is, however, important that these factors be independently corroborated if the solution technique is to be used with any confidence. In the absence of analytical solution to the multi-disc problem, two approaches are possible. The first is the effective but rather sterile possibility of checking the current formulation against another numerical scheme, say FEM. Alternatively, the physical analogue between conductive heat and fluid transfer may be used.

Analogues to porous and fractured media are reported for steady unconfined (Sharp, 1970), transient confined (Javandel, 1967), fracture network (Hudson, 1980) and tortuous fracture (Tsang, 1985) flows to mention but a few previous studies utilizing thermal and electrical analogies.

(a)

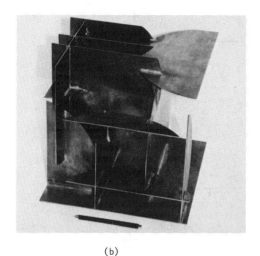

(b)

Figure 2. Two views of the disc model comprising 12 fracture discs attached to a baseplate. Pencil for scale of 140 mm.

Although representing eloquent statements in their day, the methods have been largely superceded by the rapid growth and development of digital (numerical) techniques.

In this particular application, an analogy is drawn between irrotational Darcian flow in a fracture disc of constant nominal aperture and thermal conduction in a plate, insulated perfectly above and below. The model is illustrated in Figure 2 and comprises 12 discs of 3 mm 70Cu-30Zn brass plate soldered together and attached to a baseplate. An orthogonal fracture system is chosen for purposes of convenient physical fabrication as is the consistency in fracture trace length at 300 mm. Thermal material coefficients are given for the brass plate in Table 1. The disc structure is enclosed within an insulated container and is backfilled with polystyrene pellets to minimise airborne conduction and convection. The only externally exposed surface is that of the baseplate to which the thermal perturbation is applied via an ice bath. On application of the ice bath, the equilibrated and uniform thermal regime of the disc network is disturbed as a pulse propagating through the system. This transient perturbation is monitored by 8 iron-constantan thermocouples bonded to the disc surfaces. The thermocouples are arranged as 2 on the underside of the base plate, 4 on different discs at half height within the model and the final 2 on the uppermost and most remote (from the baseplate) disc.

Table 1. Thermal properties of disc model.

Thermal property	Magnitude	Material type
Thermal conductivity (k)	120 W/m.K	Brass
Thermal conductivity (k)	46.5 W/m.K	Lead solder
Specific heat (c)	375 J/kg.K	Brass
Specific heat (c)	210 J/kg.K	Lead solder
Density (ρ)	8.90 Mg/m^3	Brass
Density (ρ)	10.20 Mg/m^3	Lead solder

K = Kelvin; J = Joule; W = Watt
Brass - 70Cu/30Zn; Lead solder - 50Pb/50Sn
Material constants over the range 20°C to 58°C
Source: Metals Handbook, 9th edition, v2, American Soc. for Metals (1979).

5 PARAMETRIC STUDIES

Since all parametric studies are completed using a single network geometry, results are most conveniently represented in dimensionless form. This allows the response of any location in the model to be uniquely defined in terms of a normalised temperature (or head) and dimensionless time. Normalised temperature is the ratio of temperature below initial ambient (T) and total change in temperature between ambient and the applied ice bath temperature (ΔT). Dimensionless time (t_D) for the thermal case is given as

$$t_D = \frac{kt}{\rho c \ell^2} \qquad (10)$$

where k = thermal conductivity, ρ = material density, c = specific heat and ℓ is any characteristic dimension, in this case chosen as the side length of the control volume cube (0.3 m). The hydraulic surrogate of dimensionless time (t_D) is obtained by replacing equation (10) with

$$k \rightarrow K \qquad ; \qquad \rho c \rightarrow S_s \qquad (11)$$

where K is hydraulic conductivity obtained from a parallel plate analogy and S_s is specific storage being some function of joint and fluid compressibility (Elsworth, 1986b). Within this framework, the results from numerical, physical and idealized analytical simulations may be directly compared.

In the numerical simulation, heads are sampled only at disc intersections. Thermocouple locations are arranged as near coincident as possible with a maximum offset of approximately 5 mm. This offset in reference to the baselength dimension (ℓ) of the model of 0.3 m is inconsequential. The thermocouples are arranged at three levels within the disc model. Thermocouples T-1 and 2 are present on the thermally perturbed baseplate, T-3,4,5 and 6 are are half height (0.5 ℓ) away from the baseplate and T-7 and 8 are at 0.85 ℓ away from the baseplate. Units T-3 and 4 are on two separate discs proximal to a single node and record identical results throughout. The readout accuracy of the thermocouple is one degree Fahrenheit.

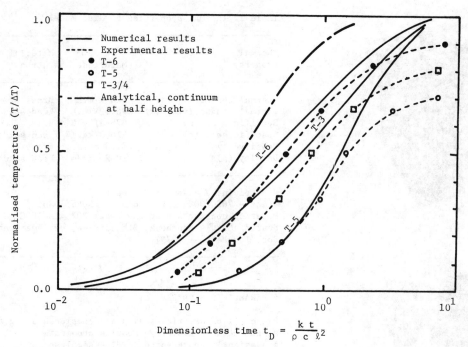

Figure 3. Thermal response of the physical model
for all thermocouples at half height compared with
numerical and analytical results.

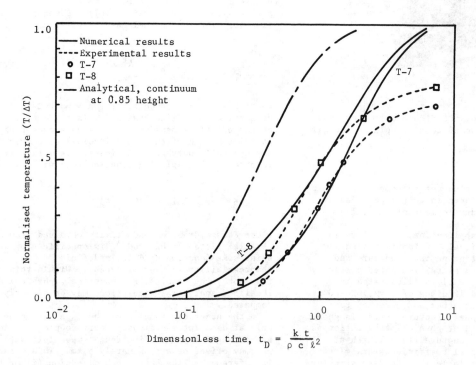

Figure 4. Thermal response of the physical model
for all thermocouples at 0.85 height compared with
numerical and analytical results.

5.1 Parametric results

The numerical simulations are completed for the 12
disc assemblage retaining a total of 45 nodal
degrees of freedom and using a total of 40 time
steps. Dimensionless time step lengths (Δt_D) of
.008, .04 and .2 are chosen to give resolution to a
steady state at $t_D > 8.0$.

The half height response of the system is
illustrated in Figure 3. It is apparent that T-6
is most directly connected to the baseplate and
response is therefore quickest. Location T-5 is
connected only via T-3/4 and this facet is borne
out in the slower numerical response of T-5. The
experimental and numerical responses maintain a
consistent pattern with rapidity of response being

in order, T-6, T-3/4 and T-5 although only T-5 bears agreeable correspondence between both the two solution techniques. It is clear that the experimental response exhibits a premature steady condition, presumably resulting from thermal radiation outside the insulated control volume. This thesis is borne out on noting that both T-5 and T-3/4 are closer to the external boundary than the internally located T-6. In the absence of external radiation and internal convection, normalised disc temperatures should clearly assymptote to unity.

For a system of continuous fractures, the boundary conditions applied to the control volume represent a one-dimensional field and the thermal response reduces to one of transient conduction in an insulated slab. It is clear that conditions of fracture continuity elicit a more rapid response than in the discontinuous system. This is true only if all fractures have constant transmissivity to storativity (Kb/S) ratios and cannot therefore be used, in general, to bound the solution. The early response of thermocouple T-6 is regulated by the analytical one-dimensional behaviour although the rapidity of equilibration is ultimately reduced.

At 0.85 height within the model, similar behavioral trends are apparent as shown in Figure 4. External thermal radiation is evident as the premature levelling of the normalised temperature. In both physical and numerical solutions T-8 responds more readily and rapidly than T-7 due to the interconnectivity of the system. The experimental and numerical results for T-7 are close in the early behaviour although diverge at larger dimensionless times as the influence of imperfect thermal insulation is felt. The effective analytical response is more rapid than that of the discrete and discontinuous system. This condition may only be guaranteed providing hydraulic short circuiting is suppressed by a fracture system with constant Kb/S ratio.

6 CONCLUSIONS

The ability to predict the hydraulic behaviour of three-dimensional fractured rock masses under varied hydraulic loads is clearly desirable. A numerical model, containing the essential physics of the problem, is illustrated to be capable of efficiently performing this task and has shown reasonable agreement with a thermal analogue. The results of numerical and experimental simulations are qualitatively in agreement with the major discrepencies emanating from the shortcomings of the thermal model.

ACKNOWLEDGMENT

The support of the Pennsylvania Mining and Mineral Resources Research Institute under Grant No. G1154142 from the Bureau of Mines, U.S. Department of the Interior is gratefully acknowledged.

REFERENCES

Elsworth, D. 1986a. A hybrid boundary element-finite element procedure for fluid flow simulation in fractured rock masses. Int. J. Num. and Analyt. Meth. in Geomechanics (in press).

Elsworth, D. 1986b. A model to evaluate the transient hydraulic response of three dimensional sparsely fractured rock masses. Water Res. Research (in press).

Hudson, J.A. and P.R. LaPointe 1980. Printed circuits for studying rock mass permeability. Int. J. Rock Mech. Min. Sci. and Geomech. Abstr. 17: 297-301.

Javandel, I. and P.A. Witherspoon 1967. Use of thermal model to investigate the theory of transient flow to a partially penetrating well. Water Res. Research, v3, 2:591-597.

Piggott, A.R. 1986. A numerical procedure for the analysis of steady state fluid flow in systems of finite discontinuities. M.Eng. thesis, Department of Civil Engineering, University of Toronto.

Sharp, J.C. 1970. Fluid flow through fissured media. Ph.D. thesis, Imperial College, London.

Tsang, Y.W. and P.A. Witherspoon 1985. The effect of tortuosity on fluid flow through a single fracture. J. Geophys. Res.

Zienkiewicz, O.C., D.W. Kelly and P. Bettes 1977. The coupling of the finite element method and boundary solution procedures. Int. J. for Num. Meth. in Eng. 11:355-375.

Underground excavations in the Muela pumped storage project
Les excavations souterraines du projet de centrale à réserve pompée de Muela
Untertagearbeiten im Muela Pumpspeicherkraftwerk

NICOLÁS NAVALÓN GARCÍA, Civil Engineer (Ph.D.). Hidroeléctrica Española, S.A., Madrid, Spain
JESÚS ALCÁZAR REINALDO, Civil Engineer, Hidroeléctrica Española, S.A., Madrid, Spain

ABSTRACT: The general characteristics of the works and of the rock in which the excavations have been carried out are described, emphasizing the importance of prior investigation of the site and planning of the final works, that have been utilised to widen knowledge on the foundation behaviour. Remarks are made on the difficulties found during excavation and how these have been solved.

RESUME: Les caracteristiques génerales de l'ouvrage et du massif où les excavations ont été réalisées sont décrites et l'on souligne l'importance de l'investigation préalable du site et l'ordonnation des travaux définitifs, lesquels ont été utilisés pour élargir la connaissance du comportement les difficultés trouvées lors de la réalisation des excavations et la manière dans laquelle elles ont été surmontées.

ZUSAMMENFASSUNG: Wir beschreiben Ihnen die allgemeinen Merkmale der Baustelle und des Massivs, wo die Ausgrabungen durchgeführt worden sind, wobei die Bedeutung der vorangegangenen Erforschungen des Ortes, sowie auch die Organisierung der endgültigen Arbeiten hervorgehoben wird, die dazu verwendet wurden, die Kenntnisse über die Merkmale des Felsmassivs zu erweitern. Es wird auf die Schwierigkeiten bei der Durchführung des Ausgrabungen und auf die Art und Weise, auf die diese gelöst wurden, hingewiesen.

I. INTRODUCTION

The Cortes-La Muela Hydro power project is located in the province of Valencia (Spain) in the river Júcar. It comprises two different functional elements: the Cortes II hydro power plant and the La Muela pumped storage project.

The Cortes II hydro power plant is a conventional installation with a hydro power station rated at 240,000 kW, replacing the previous 30,000 kW plant, built in 1922.

La Muela is a pure pumped storage system utilising favourable geomorphology, as the difference in elevation between the upper plane and the bed of the river Júcar is 600 metres, whereas the horizontal distance is only 800 metres.

The favourable topography is completed by a geological frame of great tectonic serenity, only broken by a major fault parallel to the river that, on being impermeable, has allowed the underground plant to be brought close to the lower reservoir, thus reducing the length of tunnels and galleries.

The underground pumped storage plant of La Muela houses three reversible pump-turbine units with a compound rating of 540 MW for pumping and 630 MW as turbines, requiring excavation of a complex system of shafts, tunnels, galleries and caverns.

2. INVESTIGATION AND GEOLOGICAL BACKGROUND

The underground works were carried out in a massif comprising cretacean rocks ranging from the Coniacian (C-7) to the lower Aptian (C-1), with the most important and largest excavations located in the latter.

Two blocks with different characteristics are distinguished in the rock mass separated by the Cortes fault.

The outer block, Northern block, has fallen some 200 m, following the fault plane and is only crossed by the excavations for access tunnels and draft tubes. (Fig. No. 1). The fraturing data taken therein show a high degree of dispersion.

The same is not true of the inner block, a block in which all of the major works are carried out; this appears to be tranquil with a smooth inwards dip.

In the stratigraphic column obtained in the preliminary investigation up to 36 different levels were identified within the lower Aptian (C-7), including the so-called C.1.7 comprising loose sandstone, greywackes and marly clays of 12 metres thickness. This is a highly interesting level as it is different to those predominating in the area. The petrographical analysis of the samples taken in this layer define it as sub graewackian, with fundamental components of quartz, feldspar, moscovite, sericite and chlorite. Given its special nature and location, it has been called the guide layer.

Rock fraturing shows no significant singularities and there are three main systems.

The rock paramenters deduced from the preliminary investigation are as follows, in accordance with the Beniawski classification:

Concept		Parameter Valeur
Average compressive strength of rock	647 kg/cm2	7
RQD	74,33%	13
Separation between diaclases	From 0,5 to 1 m.	15
State of the diaclases .	Closed or welded with calcite	30
Presence of water per 10 m. of tunnel	22,25 l/mim.	4
General state	Dry	
Situation of fractures with respect to excavation	Vertical main fractures	0

This represents and RMR index of 69

If the stratification were to be assimilated to a fracture and it was assumed that this was the predominant discontinuity with a dip of between 0° and 20° an RMR index of 59 is obtained.

In both cases, the rock qualification is GOOD.

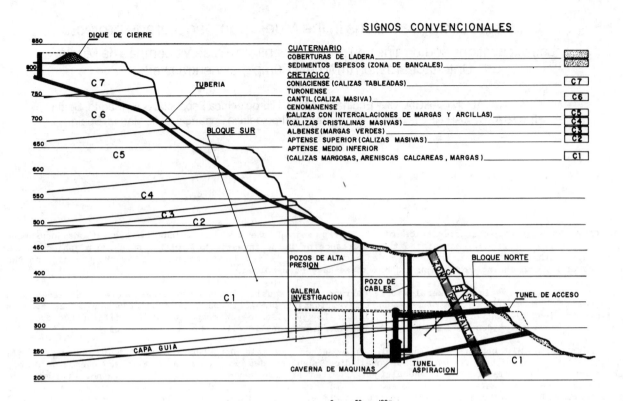

SIGNOS CONVENCIONALES

CUATERNARIO
COBERTURAS DE LADERA
SEDIMENTOS ESPESOS (ZONA DE BANCALES)
CRETACICO
CONIACIENSE (CALIZAS TABLEADAS) _____ C7
TURONENSE
CANTIL (CALIZA MASIVA) _____ C6
CENOMANENSE
(CALIZAS CON INTERCALACIONES DE MARGAS Y ARCILLAS) ____ C5
(CALIZAS CRISTALINAS MASIVAS) _____ C4
ALBENSE (MARGAS VERDES) _____ C3
APTENSE SUPERIOR (CALIZAS MASIVAS) _____ C2
APTENSE MEDIO INFERIOR
(CALIZAS MARGOSAS, ARENISCAS CALCAREAS, MARGAS) _____ C1

DIQUE DE CIERRE
TUBERIA
BLOQUE SUR
POZOS DE ALTA PRESION
POZO DE CABLES
BLOQUE NORTE
GALERIA INVESTIGACION
TUNEL DE ACCESO
CAPA GUIA
CAVERNA DE MAQUINAS
TUNEL ASPIRACION
ZONA DE FALLA

0   50   100m
ESCALA GRAFICA

Figure 1. Geological features

Even though is should be emphasized that this classification was used with the following reserves.

During the rock investigation work the presence of seven levels of confined sub-horizontal aquifers was detected, that were later shown to be intercommunicated by open preferential pathways in the vertical diaclases and fractures, which has facilitated execution of the excavations, as the filtration flows have outcropped in the excavated zones through them and have not exceeded 50 l/s.

The main conclusions of the extensive surveys carried out on the rock can be summarised as follows:
- Existence of the Cortes fault and confirmation of its impermeability.
- Detection of the so-called "guide layer", with a practically zero load bearing capacity.
- Existence of confined water tables, with minor flow rates.
- The practical absence of residual stresses in the rock, except for gravitational stresses.
- Sub-horizontal stratification, with layers of variable thickness, tending the form flat slabs on the ceilings.

3. DESIGN

The conditions described above have allowed the power station to be taken closer to the outside - thanks to the impemeability of the Cortes fault and have required the main cavern vault to be lowered in orden to provide a sufficient thickness of strata to avoid the incidence of the guide layer.

After locating the machine cavern and remaining caverns in the best quality zones, access to these zones was studied by considering a double alternative: the inclined tunnels that neccessarily would largely run through the guide layer or by means of vertical shafts and horizontal galleries or with a very slight slope, so that the guide layer would only be crossed through by the vertical shafts.

The second option was adopted and the experience obtained in the execution works confirmed that this decision was correct.

4. EXCAVATION WORK ORGANIZATION

The works were planned in two phases: phase I, comprising the access tunnels, draft tubes and corresponding portals and phase II, including four caverns (control, materials, transformers and machines), 8 vertical shafts (3 high pressure, 3 busbar, 1 materials access, 1 drainage) and an extensive intercommunication gallery network.

The rock structure made it advisable to initiate phase I excavations in three levels, corresponding to the upper zone of the high pressure shafts, the main access tunnel portal and the draft tubes portals, and delaying the start of phase II until the previous work had been completed.

Early execution of phase I allowed:
- Confirming the main rock characteristics.
- Gauging the filtration flow rates from the confined water tables.
- Studying massif behaviour, given the opening up of spaces of up to 12 metres diameter.
- Initiating and accelerating rock drainage.
- Having accesses to both ends of all shafts, which later facilitated their execution using mechanical methods for drilling, loading and removing rubble.
- Using mechanical methods for access to the cavern vault.
- Initiating the works of phase II (excavation of cavern and shafts) with a lower degree of uncertainty.

When planning the work of phase II a double objective was sought: to expand knowledge on rock behaviour and to start draining the rock mass before excavating the large machine cavern, which was achieved by the early excavation of the high pressure and busbar shafts.

On the other hand, the caverns were programmed by

excavating them from the smallest to the largest
span so that rock behaviour could be analysed as
the works progressed, particularly with regards to
the flat slabs formation.

5. EXECUTION OF THE WORKS

The underground works form a complex network (Fig.
No. 2) which can be divided into three groups from
the construction point of view:
- Horizontal tunnels and galleries
- Vertical shafts
- Horizontal caverns
The typical sections and lengths of these excava-
tions are shown in Table No. 1.

5.1 Tunnels and galleries

Tunnel and gallery excavation did not give rise to
any special difficulties, not even on passing the
Cortes fault where, of course, extra precautions
were taken.

Generally, tunnels with spans exceeding 7 m. were
executed in two stages. The first includes the upper
zones an was some 7 m. high. Depending on the local
rock characteristics, the first phase was sub-divided, bringing forward a central gallery of 7 x 5
m.

Tunnels were executed with reinforced precuts
but, due to the trend for flat slab formation, the
upper half section had to be protected by locating
fast action 5 m. long bolts and with one or two layers of shotcrete 8 cm. thick in each case, reinforced for the double layer with welded steel mesh
150x150x5 mm.

Tunnels with less than 7 m. span have a single
heading face, adopting similar protection methods
to those described above in the upper half section;

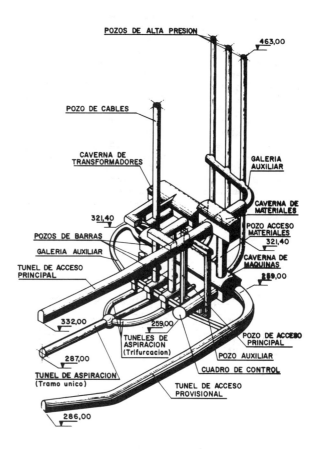

Figure 2. Axorometric view of the La Muela
underground Power Plant

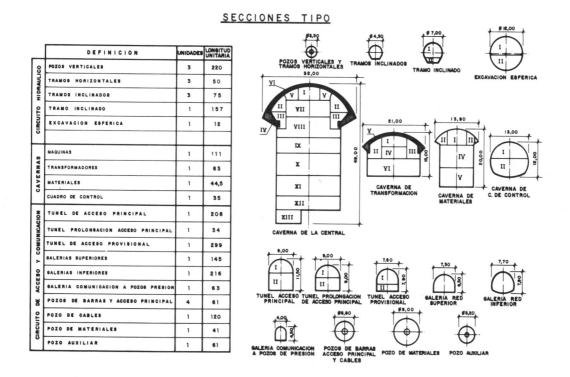

Table 1. Shafts, caverns, tunnels and galleries
cross sections

only in cases with lower curvature this strip was removed at a later stage.

The Cortes fault was crossed through by the three tunnels concerned, reinforcing the heading face protection with circular steel truses placed outside the theoretical section, type IPB 140 spaced at 1,20 to 2,00, anchored to the rock itself with 25 mm. diameter, 4 m. long bolts. Perforated Bernold type 2 mm. thick steel plates were placed between the trusses. In order to reduce ground movement, the space between the plates and the excavation profile was filled with cement mortar.

All areas covered with shotcrete were drained with a 0,50 m drilling network on a 2 x 2 m. grid.

Table n° 2 summarises the heading stages, the performance obtained and the characteristics of typical blasting for the three largest diameter tunnels.

TUNELES

	FASES	LONGITUD	SECCION	CARGA ESPECIFICA	PERFORACION ESPECIFICA	LONGITUD PERFORACION	RENDIMIENTO	
							MACHO	FALLA
		m. l.	m²	Kg/m³	m. l./m²	m. l.	m. l./dia	m. l./dia
ACCESO PRINCIPAL	I	208	41	1,44	2,21	3,70	4,80	1,54
	II	208	52	0,38	1,13	3,70	7,13	3,72
ACCESO PROVISIONAL	I	299	37	1,59	2,41	3,70	4,36	2,89
	II	299	19,20	0,408	1,35	3,70	11,20	6,87
ASPIRACION	I	157	32	1,75	2,71	3,70	3,50	2,40
	II	157	14	0,416	1,743	3,70	10,55	8,50

Table 2. Advance rates summary

5.2 Vertical Shafts

The existence of confined water tables caused by alternation of limestone-marl layers made it advisable to use an excavation system that facilitated permanent shaft bottom drainage, so the raise-drill excavation system was adopted, preparing accesses to the ends of all shafts with sufficient cross-section to allow the transit of heavy machinery.

The shaft work sequence was as follows:

1) Execution from the upper end of a guide drilling of 300 mm. diameter that allowed the remaining work to be drained.
2) Expasion of this drilling up to a diameter of 1,80 m., using the raise-drill, with a rising advance and removal of products at the bottom by loader and material transport with dumper trucks.
3) Widening up to the final design diameter by blasting of the ring, in a downwards direction, carrying out drilling with a pneumatic drill and removing rubble at the bottom of the shaft.
4) Accompanying the heading face, shotcreting of an 8 cm. thick reinforced layer on the lowest quality rock sections, employing a welded steel mesh 150x150x5 mm.
5) For shafts with a diameter of over 6 metres, protection reinforcement by placing a 5 m. long fast bolt system in the rock, located radially and separated horizontally and vertically every 2 m.
6) On completion of the excavation work, execution of the final concrete finishing layer employing an upwards slipform system.

No problems worthy of metion have arisen with these excavation methods, except those that occurred on crossing through the guide layer where a special protection system had to be adopted, as the low quality of this area became obvious when the first of the shafts crossed the guide layer and the 1,80 m. diameter left by the raise-drill spread, by fallins, and reached a bulge 5 m. in diameter.

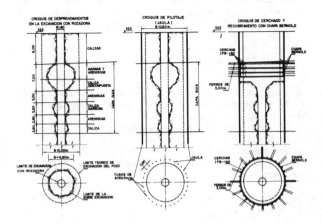

Figure 3. Shaft support system through guide layer

This served as a warning of problems that could arise in an 8 m. diameter excavation, so the following measures were adopted to limit over excavation and stabilise the zone during execution of the works:

1) Not starting the widening and stripping work until the "cage" had been completed.
2) Execution of the "cage", comprising a vertical micropile system formed of 76 mm. drillings housing a 2 mm. thick 2,5 inch outside diameter hollow steel tube driven 2 metres into the sound package on which the guide layer rests. Thus the steel tube that was essential for travelling through the guide layer with drills was utilised as a load-bearing element.
   The drillings were arranged in a circumference concentric with the shaft with a radius 0,50 m. larger and a drilling spacing of 0,50 m.
3) Injection of cement mortar through the steel tubing.
4) On completion of the cage, execution of the widening ring with a rising direction, by drilling and blasting with an advance length of 1 to 1,30.
5) Removal of rubble from each blasting, placement of horizontal steel trusses separated vertically every 1,30 m., welded to the 4 m. long 25 mm. diameter anchor bolts or, occasionally, to the cage bars.
6) Placement of 2 mm. thick Bernold plates between trusses.
7) Filling of the space between the Bernold plate and the ground with shotcrete.
8) Draining of fill by drillings located in horizontal circles and horizontally spaced at 2 m. and vertically between trusses.

Photo 1. Protection cage. Upper zone

In this way, oversize excavations were delimited by the cage limits.

The performances obtained in the different shafts are shown in Table n° 3.

FASES		LONGITUD	SECCION	CARGA ESPECIFICA	PERFORACION ESPECIFICA	LONGITUD PERFORACION	RENDIMIENTO	
		m. l.	m²	Kg./m³	m. l./m³	m. l.	MACIZO m. l./dia	CAPA GUIA m. l./dia
MATERIALES	I	41	—	—	—	—	24,40	
	II	41	3,80	—	—	—	29,20	
	III	41	62,68	0,675	1,44	7,00	1,53	0,51
BARRAS	I	61	—	—	—	—	29,60	
	II	61	3,80	—	—	—	23,40	
	III	61	32,52	0,735	2,30	2,40	3,51	1,03
PRESION	I	141	—	—	—	—	27,20	
	II	141	2,54	—	—	—	33,00	
	III	141	7,08	1,05	3,93	2,40	6,45	1,39
PRINCIPAL	I	58,30	—	—	—	—	30,60	
	II	58,30	3,80	—	—	—	19,40	
	III	58,30	32,52	0,735	2,30	2,40	3,06	0,99
CABLES	I	120	—	—	—	—	17,80	
	II	120	3,80	—	—	—	10,80	
	III	120	32,52	0,735	2,30	2,40	4,33	2,73
AUXILIAR	I	61	—	—	—	—	21,80	
	II	61	3,80	—	—	—	23,40	
	III	61	17,44	0,812	2,47	2,40	6,06	1,05

FASE I = TALADRO GUIA
FASE II = ESCARIADO
FASE III = DESTROZA

Table 3. Shafts advance rates

5.3 Caverns

As mentioned above, the smallest span ones were excavated first (control and materials, 13 and 13,80 wide, respectively) which allowed ground behaviour to be analysed for two excavations of a similar span but with a different geometry.

The treatment given to the excavation of these two caverns was similar to that described in the tunnels, so that it is not necessary to repeat it here.

The transformer cavern has required a different procedure as the free span, at the maximum point, reached 21 metres. Excavation was carried out in six stages (Fig. 4), starting with a 7 x 5 m2 central gallery in the vault, protecting the final ceiling, coinciding with that of the cavern, by means of a 5 m. long 25 mm. diameter fast bolt system placed radially on a 2 x 2 m. grid, installed immediately after each blasting operation.

The protection was reinforced with an 8 cm. thick shotcrete layer accompanied by a 150 x 150 x 5 mm. electro-welded mesh. The bolt system was completed with a new 10 m. long 25 mm. diameter bolt network injected at 3 kg/cm2 and arranged on the 2 x 2 m. grid, interposed between the 5 m. bolt grid.

Once the central gallery was sufficiently ahead, excavation commenced on the lateral galleries (stages II and III), maintaining a phase difference of some 10 metres between the advance faces of these galleries, so that the total opening of the dome was not performed until the side gallery had not been completed, with a protection system first (bolts and concrete), similar to that of the central gallery. When the excavation of the three galleries and corresponding protection was completely finished, the dome was concreted the lower bench was not removed (stage IV) until the concreting of the dome was completed.

The largest of the caverns, the machine cavern, has dimensions of 111 x 32 x 49 m. (length x width x maximum height) and was excavated in 14 stages as shown in Table No. 4.

The work was organised with the fundamental objective of completing cavern ceiling support prior to initiating bench excavation. For this, a central

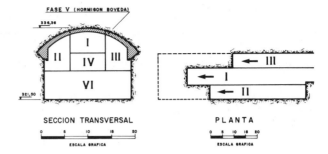

Figure 4. Transformer cavern. Excavation phases

7 x 5 m. gallery was excavated, presplitting and reinforcing the ceiling with a fast setting bolt system arranged in a 2 x 2 grid, 5 metres long and with a 25 mm. diameter, which were subjected to a compressive force as soon as they were installed by means of an ad-hoc device.

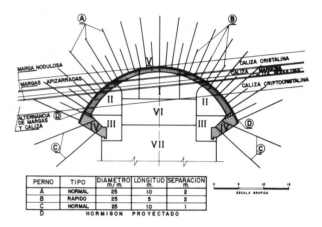

Figure 5. Machine cavern ceiling support system

On completion of the fast bolts, an 8 cm. thick layer of shotcrete was set on which a 150 x 150 x 5 mm. welded steel mesh was arranged and covered with another 8 cm. thick layer of shotcrete (Fig. No. 5). The bolt system was reinforced with a series of 10 m. long 25 mm. diameter steel bolts, grouted with cement mortar at a pressure of 3 kg/cm2 and arranged in a 2 x 2 m. grid. Finally, drainage was performed by means of 0,50 m. long drillings arranged on a 2 x 2 m. grid.

After the central gallery, the side galleries were initiated simultaneously. These, for construction reasons, were sub-divided into two phases (II and III). The protection treatment given to them is identical to that described for the central gallery (bolts, shotcrete).

On completing the side gallery excavation, the vault support beams were concreted (phase IV) on which the formwork slides and rests during concreting.

In this way, on completion of the excavation of the three galleries, the whole cavern maintains the two continuous 4 metre wide rock walls that very effectively collaborate in holding up the cavern ceiling.

Vault concreting was commenced by making partial preliminary breakages of these continuous walls over an 8 metre length, which allowed an initial 4 metre wide ring to be completed and provided sufficient space, once the initial arch was set, to break new column sections over another 4 metre length, thus obtaining sufficient space to prepare and concrete the next 4 metre wide ring. Repeating

103

Photo 2. General view of the machine cavern

Table 4. Machine cavern

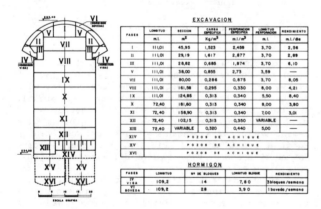

EXCAVACION

FASES	LONGITUD m.l.	SECCION m²	CARGA ESPECIFICA Kg/m³	PERFORACION ESPECIFICA m.l./m³	LONGITUD PERFORACION m.l.	RENDIMIENTO m.l./dia
I	111,01	45,95	1,523	2,459	3,70	2,56
II	111,01	29,19	1,617	2,877	3,70	2,89
III	111,01	26,82	0,685	1,874	3,70	6,10
V	111,01	38,00	0,855	2,73	3,59	—
VII	111,01	80,00	0,286	0,875	3,70	8,05
VIII	111,01	161,58	0,295	0,330	8,00	4,21
IX	111,01	124,85	0,313	0,340	5,50	8,40
X	72,40	181,60	0,313	0,340	8,00	3,80
XI	72,40	158,90	0,313	0,340	7,00	3,01
XII	72,40	102,15	0,313	0,350	VARIABLE	—
XIII	72,40	VARIABLE	0,320	0,440	5,00	—
XIV	POZOS DE ACHIQUE					
XV	POZOS DE ACHIQUE					
XVI	POZOS DE ACHIQUE					

HORMIGON

FASES	LONGITUD	N° DE BLOQUES	LONGITUD BLOQUE	RENDIMIENTO
IV VIGA	109,2	14	7,80	3 bloques/semana
VI BOVEDA	109,2	28	3,90	1 boveda/semana

BIBLIOGRAPHY

GAZTAÑAGA, J.M., LOPEZ MARINAS, J.M., (1987) "In situ stress investigation for the Muela de Cortes pumped storage project". III Intern. Cong. I.A.E.G. Madrid Vol. 10. pp. 298-301.

NAVALON, N., GAZTAÑAGA, J.M., LOPEZ MARINAS, J.M. (1979). "Muela de Cortes pumped storage project: Measurement of in situ rock stresses" Proc. International Congress on Rock Mechanics. Montreux. V. 2, pp. 467-473.

NAVALON, N. (1986) "The construction of Spain's Cortes La Muela scheme". Water Power Dam Construction February pp. 37-41.

the process, concreting was completed. Thus the maximum free ceiling span was kept below 8 m. thanks to the effective collaboration of the rock walls and the concreted arches.

For scheduling reasons, the walls were broken at 3 points, sufficiently far apart: the centre and both ends.

This execution system has avoided rock falls and practically cancels out decompression movements.

Once the vault was completed, the cavern was excavated in benches some 7 metres high, obtaining the performances shown in Table No. 3.

Prior to removing the lower bench, the side was protected with a 5 cm. thick layer of shotcrete reinforced with a series of 5 m. long bolts on a 2 x 2 m. grid.

Comportement hydromécanique d'une fracture naturelle sous contrainte normale
Hydromechanical behaviour of a single natural fracture under normal stress
Mechanisches und hydraulisches Verhalten einer Kluft unter Normaldruck

S.GENTIER, Département Génie Géologique, Bureau de Recherches Géologiques et Minières, Orléans, France

ABSTRACT : A systematical analysis of the hydromechanical behaviour of a single natural fracture under normal stress, in relation with the fracture surface morphology has been untertaken. The morphological data are got from various methods whose only the most interesting results in regard of the mechanic and hydraulic are presented hier. Study of the hydromechanical behaviour, approached in experiment and theory, leads to establish a direct relation with the fracture surface morphology.

RESUME : Une analyse systématique du comportement mécanique et hydraulique d'une fracture naturelle sous contrainte normale, en relation avec la morphologie de la surface de fracture a été entreprise. Les données morphologiques sont obtenues à partir de différentes méthodes dont seuls les résultats les plus intéressants du point de vue mécanique et hydraulique sont présentés ici. L'étude du comportement mécanique et hydraulique abordée sous l'aspect expérimental et théorique conduit à mettre celui-ci directement en relation avec la morphologie de la surface de fracture.

ZUSAMMENFASSUNG : Eine systematische Erforschung des mechanisches und hydraulisches Verhalten einer Kluft unter Normaldruck in Verbindung mit der Kluftfläschemorphologie wurde unternommen. Die morphologischen Daten werden mit verschiedenen Methoden erreicht. Nur die interressantesten Ergebnisse von einem mechanischen und hydraulischen Standpunkt sind hier präsentiert. Die Analyse des mechanisches und hydraulisches Verhalten unter eine experimentale und theorische Seite vergenommen, führt dieses zur direkten Verbindung mit des Kluftfläschemorphologie.

1. INTRODUCTION

La connaissance du comportement mécanique d'un milieu fracturé soumis à une modification du champ de contrainte, d'origine tectonique, thermique ou autre, et de ses répercussions sur la perméabilité du massif est de la plus haute importance pour tous les projets de stockage souterrain et de géothermie profonde. Les différentes façons d'aborder l'étude peuvent se résumer ainsi :
- soit le milieu fracturé est considéré dans son ensemble ; il s'agit alors de la théorie des milieux continus équivalents ;
- soit chaque fracture est considérée séparemment et le comportement hydromécanique de chacune d'entre elles est intégré sur l'ensemble du massif.

La première approche se justifie d'autant plus que la densité de fracturation est importante. Cependant, lorsque cette densité devient faible, la seconde approche pourrait se révéler plus prometteuse.

Le présent article traite du comportement hydromécanique d'une fracture unique sous contrainte normale. Il s'agit là d'une contribution à la seconde approche précédemment citée.

D'un point de vue expérimental, l'étude du comportement hydromécanique d'une fracture a été abordée par différents auteurs (Goodman 1976, Bandis et al. 1983). Pour ce qui est de la modélisation, les différentes approches tentées à ce jour relèvent principalement de l'ajustement et non de la modélisation. Dans les modèles existants il faut signaler ceux de Tsang et Witherspoon (1981) et de Zongqi (1983). Mais en général la modélisation se heurte au problème de la description quantitative de la fracture et à la variation de cette géométrie lorsque la contrainte normale augmente.

2. MATERIAU D'ETUDE

Toute l'étude morphologique, mécanique et hydromécanique a été réalisée sur un ensemble de sept éprouvettes cylindriques de 12 cm de diamètre et d'élancement 2 provenant toute d'un même bloc fracturé récolté dans la carrière du Maupuy (Creuse, France). Chaque éprouvette contient, perpendiculairement à son axe, une partie d'une seule et unique fracture. Le matériau est un granite monzonitique à grain moyen (0,5 mm) et à tendance porphyroïde (phénocristaux jusqu'à 10 mm).

Une étude pétrographique et une analyse statistique de la répartition des différentes phases minérales ont mis en évidence une légère anisotropie planaire. La direction générale de la fracture étudiée est influencée par celle-ci ; le plan moyen de la fracture étant parallèle au plan d'anisotropie. De plus, une étude statistique de la répartition des phases minérales le long de la fracture montre que le cheminement de celle-ci n'est pas aléatoire par rapport à la matrice rocheuse. On met notamment en évidence dans le plan de fracture une proportion importante de quartz et de sections longitudinales de biotites et de feldspaths ; ceci se traduisant par un caractère alternativement intra et intergranulaire de la fracture.

3. ANALYSE QUANTITATIVE DE LA MORPHOLOGIE DE LA FRACTURE

L'analyse quantitative globale de la surface de fracture est difficile voire impossible. Aussi l'étude de celle-ci a été réalisée à partir de deux séries perpendiculaires de profils sériés enregistrés sur chacune des épontes de la fracture (fig. 1). Ces profils sont obtenus à l'aide d'un "rugosimètre" avec une précision de 0,05 mm en z.

Les profils, après numérisation, sont traités pour obtenir deux types d'informations :
- les informations ponctuelles qui concernent les caractéristiques telles que la hauteur, l'angularité et la courbure. Le traitement de ces données est statistique et les résultats moyens sont présentés tableau 1,
- les informations spatiales qui prennent en compte la position relative des différents points d'échantillonnage. Différents types d'analyses, telles,

les analyses statistique, spectrale et géostatistique ont été réalisés. L'analyse géostatistique s'étant révélée la plus performante, nous présentons ici ses seuls résultats. Elle est basée sur la notion de variogramme dont les caractéristiques sont la portée et le palier. L'étude systématique des portées (fig. 2) montre que la surface de fracture peut être décrite à l'échelle de l'échantillon par la superposition de deux familles de structures respectivement de taille 4 à 6 mm et 18 à 20 mm. Celles-ci se superposent à des structures plus grandes non échantillonnables à l'échelle de l'éprouvette. A noter que la taille moyenne des structures les plus petites correspond à la courbure moyenne.

La fracture complète a été étudiée en surface et en volume à partir de pseudo-sections reconstituées à partir des profils enregistrés sur chaque éponte, et d'empreintes plastiques de la fracture à différents niveaux de contrainte. Il s'avère que l'architecture des vides est étroitement liée à la plus petite famille de structures puisque les vides peuvent être décrits par une portée moyenne de 6 mm.

L'application de méthodes stéréologiques aux pseudo-sections pour l'estimation de l'aire des zones en contact sous contrainte nulle conduit à une valeur de 10 à 15 %. L'étude des empreintes par analyse automatique d'image de la fracture sous contraintes croissantes montre que l'aire de la surface en contact augmente très rapidement pour les faibles contraintes puis se stabilise à partir d'une contrainte de l'ordre de 15 MPa à des valeurs variant entre 50 et 70 % selon les éprouvettes (fig. 3).

Tableau 1. Résultats de l'analyse statistique

No Ech.	Hauteur (mm)		Courbure (mm)		Angularité (degré)	
	$\bar{z}$	σ	$\bar{r}$	σ	$\bar{\alpha}$	σ
1	4,43	2,50	4,33	6,16	1,27	6,31
2	7,36	2,00	6,95	2,36	3,20	5,94
3	3,76	1,77	5,02	3,90	1,52	6,02
4	6,01	2,34	5,69	7,81	0,94	6,50
5	5,26	2,95	4,62	5,62	4,37	6,02
6	5,74	2,06	3,73	4,70	1,79	6,34
7	2,26	1,12	4,99	6,14	0,93	6,28

4. COMPORTEMENT MECANIQUE

Les essais mécaniques sont réalisés sur les éprouvettes précédemment citées, équipées de quatre capteurs LVDT destinés à mesurer le rapprochement des épontes de la fracture lors de l'application d'une contrainte normale au plan moyen de la fracture. La déformation de l'ensemble roche-fracture est suivie en continu pendant l'essai. La principale difficulté rencontrée lors de l'élaboration du protocole d'essai est la mise en place des épontes l'une par rapport à l'autre. Ce problème qui se pose logiquement du fait de l'histoire de l'échantillon (manipulation, enregistrement des profils) est en fait difficilement quantifiable. Une estimation de ce phénomène a été rendue possible à partir du protocole d'essai suivant : l'échantillon fracturé est soumis, après précharge, à plusieurs séries de cycles charge-décharge croissants. L'étude du retour à la précharge entre chaque cycle montre qu'il existe une fermeture résiduelle particulièrement importante à la décharge du premier cycle. Celle-ci diminue avec le nombre de cycles et de façon inversement proportionnelle à la valeur maximale de la charge lors des cycles successifs. L'échantillon est ensuite déchargé totalement.

Cette dernière opération permet de définir ce qui sera appelée fermeture résiduelle irréversible finale, par opposition à la fermeture résiduelle à la précharge qui est, au moins en partie, réversible. La répétition d'un tel ensemble de cycles, appelé expérience,

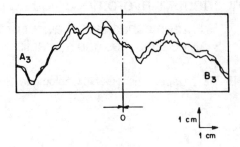

Figure 1. Exemple d'une pseudo-section reconstituée à partir de deux profils enregistrés sur chaque éponte.

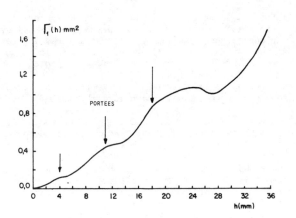

Figure 2. Exemple de variogramme mettant en évidence trois structures emboîtées de portées respectives : 4 mm, 12 mm et 18 mm.

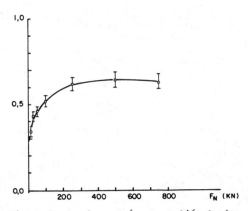

Figure 3. Courbe représentant l'évolution de la fraction de surface en contact avec la force appliquée.

montre une diminution de la fermeture résiduelle irréversible finale jusqu'à son annulation quasi complète (0 à 5 µm) à la troisième expérience (fig. 4). Au-delà on observe une réversibilité totale de la fermeture de la fracture pour des contraintes allant jusqu'à 80 MPa. Il semble donc, que pour la fracture étudiée, il soit possible de remettre en place les épontes de la fracture ou en d'autres termes, d'annuler l'histoire de l'échantillon. Il faut cependant garder en mémoire que ce protocole est valable pour des fractures qui ne se détériorent pas sous l'effet de chargements répétés. Les courbes de déplacement de l'ensemble roche-fracture en fonction de la contrainte normale confirment les résultats obtenus par ailleurs (Goodman 1976, Bandis et al. 1983, Barton et al. 1985, Trang et al. 1981), à savoir une non-linéarité pour les contraintes inférieures à 15 MPa. Au-delà, la courbe est linéaire et la pente est très proche du module élastique de la roche intacte. Il s'ensuit que c'est la fermeture de la fracture qui prédomine pour les faibles contraintes, alors que c'est le comportement élastique de la roche qui détermine le comportement de l'ensemble roche-fracture pour les contraintes élevées. A partir de 15 MPa, la fracture ne se ferme plus, sa fermeture maximale est de l'ordre de 30 à 40 % de l'ouverture moyenne de la fracture.

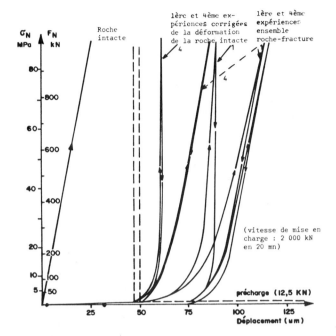

Figure 4. Courbe expérimentale contrainte-déplacement. Fermeture résiduelle irréversible finale : 32 μm pour la 1ère expérience, 2 μm pour la 4ème expérience.

Une modélisation du comportement mécanique de la fracture en contraintes normales a été proposée. Elle est basée sur la connaissance morphologique de la surface de fracture et sur les résultats expérimentaux. Le "modèle de Hertz" (Zongqi 1983) qui repose sur une assimilation de la surface de fracture à une succession de sphères, ne nous permettant pas de reproduire correctement le comportement mécanique expérimental, nous avons eu recours à un modèle dit "à dents confinées" (Billaux et al. 1984). Dans ce modèle, la fracture est assimilée à deux surfaces planes indéformables dont l'une est dotée de dents de hauteurs variables. On considère de plus que la différence de hauteur entre deux dents contigües étant faible, toute dent chargée sera toujours entourée par des dents non chargées qui exerceront sur celle-ci une pression de confinement sur la plus grande partie de sa hauteur. La pression de confinement σ_{3i} exercée sur une dent soumise à une contrainte normale σ_{1i} est fonction du rapport de la surface de la zone chargée à la surface totale qui peut s'exprimer par la quantité $1-\tau(e^+)$:

$$\sigma_{3i} = \beta \, \sigma_{1i} \text{ avec } \beta = \frac{\nu}{1-\nu}(1 - \tau(e^+))^{1/2} \, \sigma_{1i}$$

où e^+ est l'écartement (e) de la fracture normé par l'écartement maximal (e_o) ; et $\tau(e^+)$ est le degré d'ouverture de la fracture déduite de la loi de répartition des hauteurs de vide.

En introduisant le critère de rupture de Hoek et Brown (1980) on définit une contrainte normale limite σ_L à la rupture en fonction du confinement exercé sur la dent pour un niveau d'écartement donné :

$$\sigma_L = \sigma_c \, \frac{(m\beta + (m^2\beta^2 + 4(1 - \beta)^2 s))^{1/2}}{2 \, (1-\beta)^2}$$

où σ_c est la résistance à la compression simple, m et s sont les paramètres du critère de rupture.

Au-delà de la limite élastique, différents comportements post-rupture sont envisagés. La contrainte résiduelle dans la dent rompue est de la forme :

$$\sigma_R (e^+) = \sigma_L (e^+)$$

Le comportement d'une dent "i" de hauteur h_i^+ sous contrainte nulle peut se résumer comme suit (h_l^+ : hauteur limite) :

- si $h_i^+ < h_l^+$

$$\sigma_{1i} = E \, \frac{(h_i^+ - e^+)}{\alpha h_i^+} \quad \text{domaine élastique linéaire}$$

(avec $\alpha = 1-2\nu\beta$)

- si $h_i^+ > h_l^+$

$$\sigma_{1i} = \sigma_R(e^+) \quad \text{comportement post-rupture.}$$

En intégrant le comportement de chaque dent sur la surface totale de la fracture on obtient :

$$\sigma_N = \underbrace{\sigma_R(e^+) \, (1-\tau(h_l^+))}_{\text{dents en post-rupture}} + \frac{1}{\alpha} \int_{e^+}^{h_l^+} \underbrace{E \, \frac{(h^+ - e^+)}{h^+}}_{\text{dents en domaine élastique}} \tau \, (h^+) \, dh^+$$

où $1-\tau(h_l^+)$ représente la fraction de surface sur laquelle les dents ont une hauteur supérieure à h_l^+ et $\tau'(h^+)dh^+$ représente la fraction de surface sur laquelle les dents ont une hauteur comprise entre h^+ et $h^+ + dh^+$.

Pour retrouver la raideur des courbes expérimentales pour les contraintes supérieures à 15 MPa, il a fallu supposer que la contrainte résiduelle dans les dents rompues était au moins égale à la contrainte limite σ_L ($g(e^+) = 1$).

C'est ce modèle qui donne le meilleur ajustement global sur les courbes expérimentales (fig. 5). Celui-ci n'est toutefois pas tout à fait satisfaisant pour les contraintes intermédiaires (entre 5 et 10 MPa) pour lesquelles il n'a pas pu être trouvé de modèle reproduisant correctement les observations.

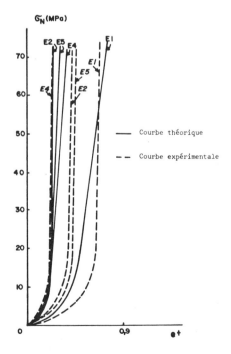

Figure 5. Courbes contrainte-fermeture théoriques (modèle à "dents confinées") et expérimentales.

107

5. COMPORTEMENT HYDROMECANIQUE

Les essais hydromécaniques sont basés sur le principe d'un écoulement radial divergent. Pour celà, une des demi-éprouvettes est forée suivant son axe en 6 mm de diamètre sur toute sa hauteur et un tube usiné collé dans le forage permet la fixation d'une tête d'injection. Le fluide est injecté à l'aide d'une pompe volumétrique à débit constant. Le protocole d'essai est établi à partir des résultats des essais mécaniques. Pour tenir compte de la mise en place progressive des deux demi-éprouvettes, l'échantillon est soumis à plusieurs séries de cycles charge-décharge. La pression d'injection pour un débit donné, et le déplacement relatif des deux éponges sont enregistrés en continu. La transmissivité intrinsèque (kf.e = $-(\mu/2\pi)$ ln (r_i/r_e) Q/P ; avec μ viscosité cinématique, r_i et r_e : rayons intérieur et extérieur, Q : débit et P : pression d'injection) diminue rapidement avec l'augmentation de la contrainte lorsque celle-ci est faible (< 10 MPa), puis se stabilise (fig. 6). A partir de la quatrième expérience, les variations sont parfaitement réversibles, ce qui corrobore les résultats mécaniques.

L'injection d'eau colorée (bleu de méthylène) lors d'un essai a permis de visualiser les points de sortie du fluide sur le pourtour de l'éprouvette. Il apparait que le fluide sort de la fracture en un certain nombre de points bien localisés ; ce qui permet de dire que l'écoulement n'intéresse pas toute la surface de la fracture, mais se concentre dans des chenaux. Le comptage systématique de ces exutoires à différentes contraintes montre que leur nombre diminue très rapidement avec l'augmentation de la contrainte pour les contraintes inférieures à 10 MPa puis se stabilise ; leur distance moyenne est de 6,5 mm pour une contrainte nulle et de 18 mm pour une contrainte de 15 MPa. Ces deux valeurs sont l'expression physique des structures géostatistiques mentionnées précédemment.

Les tentatives d'ajustement des courbes expérimentales par les relations empiriques trouvées dans la littérature (Gale 1982, Vouille 1982), reliant la transmissivité à la contrainte, montrent que l'écoulement n'est pas modélisable par celles-ci sur l'intervalle complet des contraintes étudiées. Ces relations ne sont acceptables que pour les faibles contraintes, pour lesquelles le nombre assez important d'exutoires permet d'accepter l'hypothèse d'un écoulement quasi généralisé dans le plan de la fracture, alors que ceci n'est plus le cas aux contraintes élevées.

Par ailleurs, le calcul du coefficient de rugosité C (C = 1 si $k/D_h \leqslant 0,033$ et C = 1 + 8,8 $(k/D_h)^{1,5}$ si $k/D_h > 0,033$; k/D_h est le rapport de la hauteur des aspérités et du diamètre hydraulique de la fracture), introduit par Louis (1967) dans la loi cubique, à partir des valeurs expérimentales de la transmissivité intrinsèque, de l'ouverture moyenne et du degré de séparation, conduit à des valeurs de celui-ci non cohérente avec sa signification physique.

Enfin, le couplage du modèle mécanique à dents confinées et de la loi cubique réalisé en introduisant la géométrie de la fracture non seulement par le terme en e^3 mais aussi au niveau du degré de séparation et du degré de rugosité, ne permet de retrouver le comportement expérimental que pour la moitié des échantillons. Et ceci seulement d'un point de vue qualitatif, car les valeurs calculées de la transmissivité sont systématiquement 10 à 100 fois supérieures à celles mesurées.

6. CONCLUSION

Suite à ce travail il apparait que la loi cubique même couplée à un modèle de comportement mécanique satisfaisant ne permet pas de reproduire les résultats expérimentaux. Il reste à établir une loi d'écoulement dans les fractures qui incorporerait la chenalisation progressive de l'écoulement avec l'augmentation de contrainte en relation avec la morphologie de la frac-

ture définie et quantifiée à l'aide de paramètres appropriés. Ce travail n'est de plus qu'une étape vers l'étude du comportement hydromécanique d'une fracture naturelle en cisaillement qui nécessitera une étude morphologique complémentaire notamment l'aspect angularité-courbure.

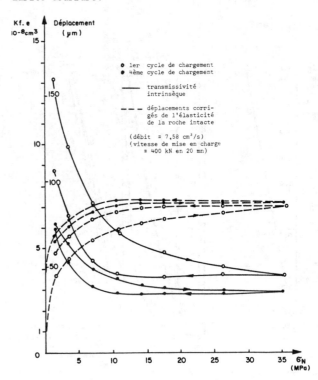

Figure 6. Courbes expérimentales contrainte-transmissivité intrinsèque.

7. BIBLIOGRAPHIE

BANDIS S., LUMSDEN A.C., BARTON N. (1983) Fundamentals of rock joint deformation. Int. J. Rock. Mech. Min. Sci., vol. 20, n° 6, p. 249-268.

BARTON N., BANDIS S., BAKTAR K., (1985) Strength, deformation and conductivity coupling of rock joints. Int. J. Rock. Mech. Min. Sci. vol. 22, n° 3, p. 121-140.

BILLAUD D., FEUGA B., GENTIER S. (1984) Etude théorique et en laboratoire du comportement d'une fracture rocheuse sous contrainte normale. Revue française de géotechnique, vol. 26, p. 21-29.

GALE J.E. (1982) The effect of fracture type on the stress-fracture closure. Fracture permeability relationship. Proc. 23 rd US Rock. Mech. Symp. Berkeley, California ; p. 290-298.

GENTIER S. (1986) Morphologie et comportement hydromécanique d'une fracture naturelle dans un granite sous contrainte normale -étude expérimentale et théorique- Thèse de l'Université d'Orléans (France) 637 p.

GOODMAN R. (1974) Les propriétés mécaniques des joints. Progrès en mécanique des roches, vol. 1, t. A (3ème congrès de la Soc. Int. Meca. Roches, Denver).

LOUIS C. (1967) Etude des écoulements d'eau dans les roches fissurées et de leurs influences sur la stabilité des massifs rocheux.Thèse Univ.,Karlsruhe,128p.

TSANG Y.W., WITHERSPOON P.A. (1981) Hydromechanical behaviour of a deformable rock fracture subject to normal stress. J. of Geophys. Res. vol. 86, n° B10, p. 9287-9298.

VOUILLE G. (1982) Etude des caractéristiques hydrauliques et thermomécaniques d'un granite fissuré. Fontainebleau, Centre d'Etudes de Mécanique des Roches (ENSMP), rapport inédit, 26 p.

ZONGQI S. (1983) Fracture mechanics and tribology of rocks and rock joints. Doctoral thesis, Lulea University (Suède), 207 p.

Contraintes et ruptures autour des forages pétroliers
Stress and rupture conditions around oil wellbores
Spannungs- und Bruchbedingungen um Erdölbohrungen

ALAIN GUENOT, Elf-Aquitaine, Pau, France

ABSTRACT : The common design methods used in underground engineering do not allow to predict with sufficient accuracy the dislocation of the wellbore walls, which leads to trouble and cost rise during oil well drilling. The different contributions to the overall loading of the rock surrounding the well, among them are thermal stresses, have been reanalysed and different rupture modes have been identified. Practical charts using normalized axes are proposed for a useful and easy comparison of these different loadings. Regarding the true rupture criteria, an important research programme involving experimental and theoretical works has been launched and the first promising results are presented.

RESUME : Les méthodes actuellement disponibles de dimensionnement des ouvrages souterrains ne permettent pas de prévoir correctement la dislocation des parois, cause majeure de difficultés et d'augmentation des coûts des puits pétroliers. On a repris en détail les chargements contribuant à ces ruptures, notamment les contraintes thermiques et ainsi mis en évidence l'existence de plusieurs modes de rupture potentiels autour du puits. Des diagrammes en axes normés permettent une analyse comparative aisée des chargements. Pour l'établissement de critères de rupture, un vaste programme de recherches expérimentales et théoriques a été lancé, dont les premiers résultats prometteurs sont présentés.

ZUSAMMENFASSUNG : Die heutigen Bemessungsverfahren der unterirdischen Hohlbauten gestatten es nicht, den Wänderzerfall -Hauptursache der den KW-Bohrlöchern anhaftenden Schwierigkeiten und Kosten- mit der erwünschten Genauigkeit vorauszusagen. Die für diese Störungen verantwortlichen Belastungen (besonders Wärmespannungen) wurden einzeln untersucht und mehrere potenzielle, die Bohrung umgebende Brucharten anschaulich gemacht. Mittels achsengenormter Schaubilder kann eine Vergleichsanalyse der Belastungen leicht ausgeführt werden. Um Bruchkriterien auszuarbeiten, hat man ein weitläufiges, theoretische und experimentelle Forschungen umfassendes Programm in Gang gesetzt, dessen erste vielversprechende Ergebnisse hier vorgestellt werden.

1 INTRODUCTION

Beaucoup de forages pétroliers, en particulier lors des premiers développements d'un nouveau champ, sont souvent sujets à des problèmes de tenue de leurs parois, qui engendrent coincements, rupture des tiges de forage, opérations de repêchage, perte d'une partie ou de la totalité du forage déjà réalisé. Toutes ces conséquences entraînent des pertes de temps extrêmement coûteuses, surtout lors des forages en mer. Même si l'expérience des hommes de terrain et de laboratoire permet, la plupart du temps, de trouver un remède adéquat, c'est toujours au prix d'essais successifs plus ou moins fructueux. C'est pour cette raison qu'a été entreprise, voici quelques années, une étude approfondie sur ce phénomène de rupture des parois, une meilleure compréhension ne pouvant qu'améliorer les procédures de préparation des sondages. Il faut également noter que cette expérience des hommes de terrain se traduit par un certain nombre de règles de l'art, qu'il convient d'expliquer.

2 POSITION DU PROBLEME

La rupture des parois est sans nul doute un phénomène mécanique. Paradoxalement, la majeure partie de la littérature sur ce sujet traite du remède et non du mal : les formulations de fluide de forage qui ont été utilisées avec succès dans des cas d'instabilité de parois.

Il existe quelques publications traitant des problèmes mécaniques de l'instabilité en forage, les plus complètes étant celles de BRADLEY (1979), MANOLESCU (1970) et BRATLI, HORSRUD et RISNES (1983). Plus récemment CHEATHAM (1984) a présenté un bon état de l'art sur ce sujet en concluant que beaucoup de progrès restaient à faire dans la connaissance de l'état de contraintes en place et de la résistance des roches en paroi. L'aspect théorique et expérimental de la question conduit à s'orienter vers les multiples approches du problème général d'instabilité des ouvrages souterrains, dans lesquelles il faut sélectionner celles qui sont extrapolables au cas des forages profonds.

La littérature sur les tunnels s'est essentiellement attachée à prévoir les déplacements de la paroi, qu'ils soient élastiques, plastiques ou visqueux et ceci d'autant plus qu'il est apparu que ce sont ces déplacements, bloqués très tôt par un soutènement même souple, qui permettent de générer les pressions de soutènement et d'assurer une auto-stabilité de la voûte. On ne s'est pas, par contre, énormément intéressé à prédire la rupture des parois, avec production d'écailles et plaques par création de nouvelles discontinuités, et avec modification sensible du profil de l'ouvrage. C'est pourtant un point essentiel dans les puits profonds où les petites déformations sont de peu d'importance dans la mesure où il est difficile technologiquement de les bloquer, mais où c'est la rupture, au sens large, des parois qui importe, soit qu'elle se produise d'une manière fragile avec élargissement de la section et production de fragments qu'il n'est parfois pas possible d'évacuer, soit qu'elle se produise d'une manière ductile avec fermeture du trou. Seuls les concepteurs d'ouvrages impensables à soutenir d'un point de vue économique (galerie de stockage) se sont posés la question, et y ont trouvé la plupart du temps des solutions par le biais de modifications dans la forme de la section.

Dans les années 70, l'approche du problème par l'introduction du modèle élastoplastique radoucissant a donné d'autant plus l'illusion d'une solution qu'elle était commode d'emploi. Il faut bien maintenant admettre que si ce modèle est adapté pour simuler a posteriori le comportement de la roche en état post-rupture, et donc très utile pour dimensionner les soutènements des parois ainsi rompues autour des ouvrages, il n'a jamais permis, par contre, de prévoir la rupture de parois, et a même été souvent mis en défaut (MAURY, 1977 - KAISER, GUENOT et MORGENSTERN, 1985).

En ce qui concerne la rupture des parois, telle que définie plus haut, le dépouillement des travaux expérimentaux montre que ces modèles élastoplastiques classiques ne permettent pas de prévoir correctement la rupture des parois de modèles de laboratoire (tubes épais, essais bi- ou triaxiaux) à partir de paramètres déterminés par des essais conventionnels (triaxial de révolution, etc.).

La table 1 rassemble un certain nombre de ces essais et présente en colonne 4, le "nombre de stabilité" N_S défini, dans ce cas, comme le rapport entre la contrainte théorique tangentielle calculée au moyen d'un modèle isotrope élastique et correspondant au chargement extérieur pour lequel on a observé la rupture des parois, et la résistance de la roche, déterminée à l'aide des essais conventionnels, Q,

$$N_S = \frac{\sigma_{\theta c}}{Q}$$

Ce nombre devrait être égal à 1 ou légèrement supérieur dans l'éventualité d'une redistribution des contraintes par plastification, indécelable à l'oeil nu. On constate, cependant, qu'il évolue entre 2 et 8 ; les cas extrêmes ont été obtenus avec du marbre.

Table 1. Essais de rupture sur cylindres creux (voir les références spécifiques)

Auteur principal	Roche	Pression interne (Mpa)	N_S
Obert	Calcaire	0	2.45
	Grès	0	1.82
	Marbre	0	2.75
Berest	Craie	0	1.8/2.6
Guenot	Charbon	0	2
Haimson, Edl	Grès	0	1.8/2.3
Simonyants	Calcaire	0/60	2/3.8
	Dolomie	0/60	2/4
	Marbre	0/60	5/8
	Grès	0/60	1.9/5.8
	Siltstone	0/60	3.2/3.55
Geerstma	Grès	?	1/16
Gay	Grès	0	2.3
Mastin	Grès	0	1.9
Haimson,Herrick	Calcaire	0	1.6/3.6
Bandis,Barton	Artificielle	0	1.5/3

Quelles que soient les explications qualitatives souvent invoquées, auxquelles on cherche à attribuer ce phénomène ("effet d'échelle", "effet de gradient", microfissuration invisible, excentricité du forage), il n'en demeure pas moins que :
- L'échantillon de roche a apparemment supporté sans rupture une charge beaucoup plus importante que celle pour laquelle il s'est rompu dans l'essai conventionnel.
- Lorsque la mesure des déformations tangentielles de la paroi a été effectuée, elle montre que l'échantillon a véritablement supporté des niveaux de déformations plus importants que ceux atteints à la rupture lors de l'essai conventionnel (Voir par exemple KAISER, GUENOT et MORGENSTERN (1985)).

Aucune théorie ne permet de quantifier ce phénomène et, pour être pratique, il n'y a pas d'autre solution que l'empirisme pour indiquer au foreur les conditions optimales, les plus proches d'un seuil qu'on ignore, lui garantissant la tenue mécanique des parois du puits.

Tous ces éléments -manque d'une réflexion complète sur l'aspect mécanique des problèmes de tenue de paroi, manque d'une théorie adéquate- ont conduit à entreprendre un programme de recherches important sur ce sujet, regroupant, outre des moyens propres, les efforts d'étudiants et de chercheurs dans différents centres de recherche en France et à l'étranger. Cet effort se poursuit encore. Le but de cet article est de rassembler les premiers résultats de ces recherches autour de l'idée initiale qui les a provoquées. Il renverra à des références plus spécifiques sur chacun des travaux dont certaines sont déjà publiées, et d'autres le seront dans un futur proche. Notons, enfin, que ce travail a bénéficié des réflexions de la Commission Internationale sur les mécanismes de rupture autour des ouvrages souterrains, à laquelle il a apporté en retour des résultats relatifs aux puits pétroliers (MAURY, 1987).

3 ANALYSE DES PARAMETRES

L'analyse des forages pétroliers diffère de celle des autres ouvrages souterrains par plusieurs points spécifiques :
- présence d'un fluide de forage exerçant sur les parois une condition à la limite mixte, mécanique et hydraulique,
- conditions de température et de pression, intermédiaires entre le génie civil/minier et la tectonophysique,
- accès à l'information (matériau d'essai, mais aussi observations en place) extrêmement limité par les considérations économiques, en retrait des possibilités de la technologie.

Il est donc utile dans un premier temps de recenser les paramètres prépondérants du problème : chargements, conditions de pression de pore.

3.1 Les chargements

Plusieurs éléments contribuent à charger ou soulager la roche en paroi ; certains sont bien identifiés et utilisés à bon escient, d'autres le sont moins.

Contraintes en place

C'est bien entendu le chargement essentiel des parois avec lequel il faut composer. Encore faut-il le connaître.

Il est possible maintenant d'estimer les directions principales du tenseur de contraintes, que cela soit par l'analyse des ovalisations des trous de forage ou des orientations de fractures hydrauliques, par les mesures de déformabilité de carottes (A.S.R, DSCA), par l'analyse du discage lorsqu'il se produit (MIGUEZ et HENRY, 1986), et récemment par l'étude de la polarisation des ondes de Stoneley (BARTON et ZOBACK, 1986). Toutes ces méthodes (sauf le discage, inexistant) ont donné des résultats consistants sur un des champs que nous opérons dans l'est du Bassin de Paris (contrainte horizontale maximum au N150°). MAURY et SAUZAY (1987) et MAURY, FOURMAINTRAUX et SAUZAY (1987) commentent d'autres résultats en champ latéral fort (Ko > 2). Il ne faut pas toutefois oublier dans ces analyses l'anisotropie mécanique du matériau en place, qui peut imposer les directions principales du tenseur des contraintes par suppression des cisaillements dans une direction, mais est-ce toujours le cas ?

En ce qui concerne les amplitudes, plutôt que des valeurs exactes qu'il n'est actuellement pas possible d'atteindre (sauf la valeur minimale estimée par la fracturation hydraulique), il est plus juste de parler d'indicateurs du champ de contraintes à partir de toutes les méthodes énoncées plus haut, permettant d'estimer une fourchette pour les valeurs des coefficients K_o de poussée latérale.

Pression de boue

Le foreur contrôle traditionnellement la stabilité des parois du puits par la densité de la boue. Un observateur extérieur pourrait parfaitement se demander l'opportunité d'un tel déploiement de forces pour analyser la tenue des parois, alors qu'il semble suffire d'augmenter la densité de la boue pour résoudre tous les problèmes. Au-delà de l'aspect économique (les "points" de densité coûtent chers), ce sont plutôt des considérations techniques qui feront privilégier la densité la plus faible possible, et ceci pour plusieurs raisons :
- détection des venues de fluide de formation : en exploration, il faut bien sûr contrôler les éruptions, mais aussi ne pas empêcher la détection de réservoirs imprégnés, qui se fait d'abord par analyse d'indices dans la boue.
- fracturation et perte de boue : il y a bien sûr une limite supérieure à la densité qui correspond à la fracturation des parois entraînant des pertes possibles de boue,
- performance en forage : on constate qu'un différentiel des pressions entre forage et formation pénalise considérablement la vitesse de pénétration instantanée de l'outil, et peut provoquer de véritables "collages" des tiges à la paroi, au droit des formations perméables,

Pour toutes ces raisons, on cherche à ne travailler qu'en légère surpression.

Pour que le soutènement exercé par la boue soit "efficace", il faut que pression de boue dans le forage, et pression de pore à la paroi soient effectivement différentes. C'est réalisé soit par la formation d'un cake de filtration en milieu perméable, soit par la faible diffusivité hydraulique en milieu peu perméable. Nous reviendrons sur ce point critique plus loin.

Contraintes thermiques

La boue qui circule dans le forage, descend très rapidement à l'intérieur des tiges, passe par l'outil, où, pour bien remplir son office de nettoyage, elle est accélérée par le biais d'orifices calibrés, remonte dans l'espace annulaire, puis passe dans des bassins de surface avant d'être réinjectée. Tous ces transits s'accompagnent de transferts de chaleur importants. Le fond du puits est refroidi, le haut du puits réchauffé : dans l'espace annulaire, il existe donc un point neutre où la température de la boue est identique à celle des terrains. Ce point neutre peut être, suivant les cas, dans la section tubée, dans le "découvert" (section non tubée), ou même, fictivement, sous le niveau de travail de l'outil : dans ce cas l'ensemble du puits est réchauffé. Notons enfin que ce point neutre évolue pendant la vie du forage ; on constate en général qu'une section située en haut de découvert, si elle est refroidie au moment où elle est forée, est par la suite moins refroidie, puis même réchauffée, au fur et à mesure de l'approfondissement (voir Fig.1 et GUENOT, 1986).

Ces échanges thermiques dans les puits en cours de forage, qui font intervenir de nombreux paramètres, sont représentés à l'aide d'un modèle numérique (CORRE, GUENOT et EYMARD, 1984).

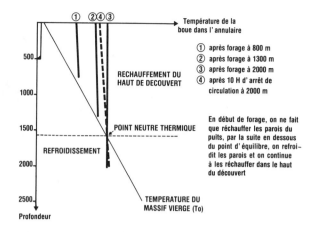

Figure 1. Evolution de la température de la boue dans l'annulaire suite à l'approfondissement du forage

Ces perturbations thermiques s'accompagnent de contraintes en paroi qui ont été pratiquement ignorées dans la littérature occidentale sur le problème de tenue de parois de puits. La littérature des pays de l'est est beaucoup plus riche sur ce sujet (exemple : KEVORKOV et al, 1974).

La figure 2 reprend schématiquement l'ensemble des paramètres intervenant dans le problème des échanges thermiques. On y retrouve essentiellement 3 termes :

1. Le refroidissement en surface ΔT_s, qui est relativement constant d'un appareil à l'autre, de l'ordre de quelques degrés pour une boue à l'eau, un peu plus pour une boue à l'huile.

2. Le terme d'échauffement par dissipation visqueuse traduisant l'effet des pertes de charges, et qui est directement proportionnel à la pression d'injection Pi. Il est lui aussi relativement constant, car, sur les appareils de forage modernes, le débit utilisé en pratique est bien souvent le maximum imposé par la valeur limite de la pression d'injection (environ 20 MPa).

L'échauffement correspondant est :

$$\Delta T_p = \frac{1}{Cp} \frac{Pi}{\rho}$$

3. Le terme d'échanges avec le terrain :

$$\Delta T_e = \frac{h \, \P \, D}{Cp \, \rho \, Q} \int_L (T_p - T_b) \, dl$$

Cette expression schématique montre bien que les échanges avec le terrain font intervenir outre le débit de boue Q, le diamètre du forage D et la capacité thermique massique de la boue ρC_p. De plus, ce flux convectif, de coefficient d'échange h, sera proportionnel à la différence de température entre la boue (T_b) et le terrain (T_R). Cela signifie que si les premiers termes sont relativement indépendants du profil de température dans le puits, ce troisième ne l'est pas. Il va donc évoluer au cours du temps et jouer un rôle de régulation du système thermique ainsi créé. Ce terme dépend également des échanges internes à travers les tiges.

La température de la boue est un paramètre de forage, généralement négligé : on ne la contrôle actuellement que pour des causes externes (pergélisol, stabilité de la boue, forage en zones géothermiques). Elle devrait être contrôlée en permanence.

Pour contrôler la température (essentiellement éviter un échauffement trop important), on jouera sur les trois termes précédents selon les modes suivants :

1. refroidir par un échangeur thermique la boue en surface,

111

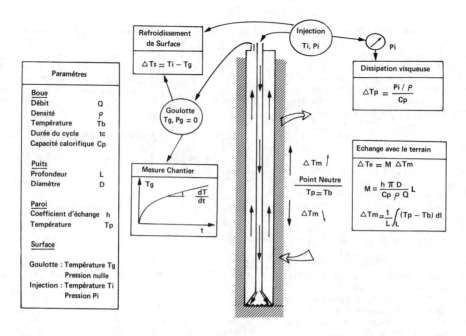

Figure 2. Bilan thermique simplifié d'un puits en forage

2. limiter les pertes de charges, essentiellement par une limitation du débit, ce qui peut aussi contribuer à la stabilité de la paroi rompue ainsi que nous le verrons plus loin,

3. limiter les échauffements de haut de découvert, par des arrêts de circulation à intervalles réguliers (c'est une règle de l'art de ne pas forer sans arrêt pendant trop longtemps), ou en diminuant les échanges par une diminution de la conductivité thermique de la boue (boue à l'huile par exemple). Il faut également envisager de limiter les échanges thermiques au travers des tiges.

C'est essentiellement le premier mode qui sera le plus efficace.

Contraintes de gonflement

L'hydratation de certains minéraux, notamment les minéraux argileux, s'accompagne d'une augmentation de volume importante. Cette variation de volume, bloquée dans son évolution, est à l'origine des contraintes de gonflement pouvant suffire à rompre la paroi (GRAY, DARLEY, 1980). Mais au contraire des contraintes thermiques, l'ampleur de cette variation de volume est fonction de l'intensité des contraintes auxquelles est soumis le matériau. S'il est possible d'évaluer au laboratoire des pressions de gonflement uniaxiale ou volumique, l'extrapolation au problème de la paroi du puits est très délicate et n'a pas été étudiée d'une manière satisfaisante à ce jour. Par contre l'origine physico-chimique du problème a été largement analysée et les boues sont traitées en conséquence (voir la synthèse de CHEATHAM, 1984).

Synthèse

Ces différents chargements doivent être analysés conjointement et les effets néfastes de l'un pourront ainsi être compensés par une action sur un autre : compenser un gonflement incontrôlable par une augmentation de densité, assurer le tenue d'un haut de découvert en jouant sur les conditions thermiques sans augmenter la densité, évitant ainsi ses conséquences sur d'autres aspects du forage. Nous avons regroupé ces chargements et leurs effets sous forme de graphiques normalisés, ce qui en simplifie l'étude paramétrique et permet une approche

comparée du mode d'action de ces paramètres, compatible avec l'incertitude actuelle sur les niveaux de contraintes en place et les conditions de rupture.

3.2 Analyse en contraintes totales

Nous ferons dans un premier temps cette analyse en contraintes totales et en ne faisant intervenir que les trois premiers chargements ci-dessus. Considérons un point en paroi d'un puits vertical dans un champ de contraintes initiales orthotrope à contrainte verticale principale, caractérisé par un coefficient de poussée latérale Ko :

$$Ko = \frac{\sigma_h}{\sigma_v}$$

Le raisonnement qui suit peut être étendu à des puits déviés dans un champ de contraintes quelconque. Nous verrons ici que le cas le plus simple peut apporter encore des éléments inattendus.

Pour la commodité des expressions, on considère que les contraintes sont normées par rapport à la contrainte verticale :

$$S = \frac{\sigma}{\sigma_v}$$

Ainsi, le chargement exercé par la boue de forage sera caractérisé par un paramètre normé d relié à sa densité par :

$$d = \frac{\gamma_b \, z}{\sigma_v} = \frac{\gamma_b}{\gamma}$$

avec γ_b : poids spécifique de la boue
γ : poids spécifique moyen des terrains sus-jacents
z : profondeur de la section considérée

Le chargement exercé par les contraintes thermiques est lui aussi caractérisé par un paramètre normé R, relié aux caractéristiques thermo-mécaniques de la roche et à la température T :

$$R = \frac{\alpha_T E \, \Delta T}{1 - \nu} / \gamma z = \frac{\Delta T}{z} \times A_T$$

avec $A_T = \frac{\alpha_T E}{\gamma (1 - \nu)}$ (m/°C) (caractéristique de la roche),

112

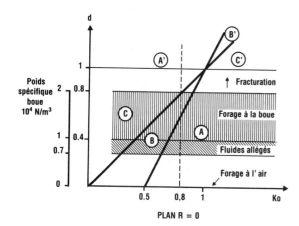

PLAN R = 0

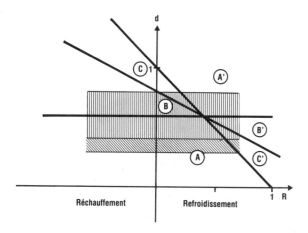

PLAN Ko = 0.8

Figure 3. Zones de contraintes avec les plages de variation possibles de Ko, d, R.

α_T étant le coefficient de dilatation thermique linéaire de la roche, E et ν, ses caractéristiques élastiques et ΔT le refroidissement de la paroi.

Nous disposons donc de trois paramètres sans dimension Ko, d, R nous permettant d'établir les graphiques évoqués au paragraphe 3.1. En élasticité, l'état de contraintes normé en paroi s'écrit, θ,r,z étant les indices relatifs à un système de coordonnées cylindriques dont l'axe est l'axe vertical du puits :

$$S_\theta = 2\ Ko - d - R$$
$$S_z = \qquad d$$
$$S_r = 1 \qquad\quad - R$$

Les analyses de contrainte citées plus haut font, la plupart du temps, mention d'état de contraintes classées dans l'ordre $S_\theta > S_z > S_r$, et considèrent que le déviateur critique résulte de la différence entre la contrainte tangentielle et la contrainte radiale. D'autres références introduisent le second invariant du tenseur des contraintes J_2 dans leur analyse et ne se préoccupent pas de cet aspect du problème.

Les équations, ci-dessus, des contraintes en paroi, permettent de déterminer trois zones selon l'ordre dans lequel ces contraintes sont rangées :

Zone A : $S_\theta > S_z > S_r$
Zone B : $S_z > S_\theta > S_r$
Zone C : $S_z > S_r > S_\theta$

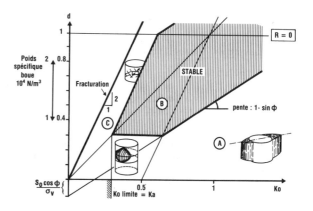

Figure 4. Polygone de stabilité et limite en fracturation dans le plan R = 0 (avec formes de rupture potentielles)

La figure 3 donne les limites de ces zones dans le plan $\Delta T = 0$ de l'espace (d,Ko,R).

La poursuite du raisonnement permet d'inclure trois autres zones correspondant aux autres combinaisons possibles :
A' $\quad S_\theta < S_z < S_r$
B' $\quad S_z < S_\theta < S_r$
C' $\quad S_z < S_r < S_\theta$

Les codes A et A', B et B', C et C', sont attribués à des zones où les contraintes extrêmes sont identiques mais inversées, et la rupture des parois devrait s'exprimer selon 3 modes différents, schématiquement représentés sur la figure 4. Le mode A est le mode de rupture classique des tunnels. Le mode B découpera des écailles de forme torique. Le mode C a une cinématique plus compliquée à saisir mais s'exprimerait à la paroi par des cisaillements le long de spirales entrecroisées. Ces trois modes de rupture sont compatibles avec les conditions rencontrées dans les forages pétroliers (Voir Fig. 3).

Il est possible, dans un premier temps, d'analyser dans ce plan la limite élastique (non pas forcément la rupture comme nous l'avons vu précédemment), définie par un simple critère de Mohr-Coulomb (cohésion So et angle de frottement ϕ). Un polygone de stabilité, tel que dessiné sur la figure 4, permet alors un certain nombre d'analyses qualitatives. Il lui est également adjoint une limite en fracturation, toujours dans l'hypothèse initiale d'un milieu non filtrant.

* On constate d'abord que le paramètre qui contrôle la taille de ce polygone est l'ordonnée à l'origine de la limite de la zone en mode A :

$$\frac{So\ \cos\phi}{\gamma.Z}$$

qui est ainsi également, une sorte de nombre de stabilité.

* Les modes de rupture B et C correspondent à des combinaisons de chargement rencontrées en forage pétroliers, notamment pour les valeurs faibles de Ko. Cependant, pour que ces deux modes de rupture apparaissent, il faut que la limite de rupture en mode B soit positive. Cela se traduit, en fait par :

$$Q = \frac{2\ So\ \cos\phi}{1 - \sin\phi} < \gamma.Z$$

Il faut donc que la profondeur Z soit suffisante et telle que la contrainte verticale soit supérieure à la résistance en compression simple du matériau, qui tient alors en place par l'effet du confinement latéral.

113

* Le coin inférieur gauche du polygone de stabilité, situé sur la limite B/C, correspond algébriquement à la valeur limite inférieure de Ko, conditionnant une stabilité en place avant forage :

$$Ko = K_a = \frac{1 - \sin \phi}{1 + \sin \phi}$$

On trouve ainsi sur le diagramme que pour des valeurs faibles de Ko, proches de K_a, associées à des valeurs fortes de C/γ.Z, il existe des cas où la plage des valeurs de densité, correspondant à la stabilité des parois, est très limitée. Les stabilisations difficiles de puits, où une augmentation de densité a un effet nul, voire néfaste, qui ont été rencontrées dans des structures géologiques particulières (ex : argiles scaglioses d'Italie), relèvent sans doute de ce cas de figure.

* La rupture en mode C se produit avant la limite en fracturation hydraulique. Elle est sans doute susceptible de l'initier à une valeur de densité inattendue. Cela implique d'inclure des essais à forte contrainte axiale (> Q) au laboratoire pour éventuellement la provoquer, alors que la plupart des essais de fracturation hydraulique sont faits actuellement à faible contrainte moyenne.

De la même manière, un polygone de stabilité, tracé dans un plan (R,d) permet de faire d'autres analyses (Fig.5).

* Une section de puits, située en haut de découvert, stable au départ avec une densité donnée, peut sous l'effet du réchauffement lié à l'approfondissement, subir un chargement excessif. Pour compenser cette augmentation de la contrainte tangentielle par effet thermique, il sera alors nécessaire d'augmenter la densité régulièrement au fur et à mesure de l'approfondissement. En outre, avec le réchauffement, la valeur de densité maximale admissible correspondant à la fracturation augmente. Il sera donc possible d'augmenter, si nécessaire, la densité de boue au-delà de la valeur critique mesurée en début de phase ("Leak-off-test"), sans constater de pertes de boue systématiques. L'effet thermique associé à l'approfondissement du forage, apporte ainsi une explication alternative à deux constatations quasi systématiques lors d'opérations sur grands découverts (GUENOT, 1986).

* Par voie de conséquence, toute mesure de gradient de fracturation doit inclure des observations sur le régime thermique du puits sous peine de dispersion incompréhensible des résultats.

L'intérêt de ces diagrammes, plutôt que de donner une réponse chiffrée, impensable dans l'état actuel de nos connaissances, tant de la contrainte en place que des conditions réelles de rupture, est de permettre, à partir d'indices probables sur les valeurs des contraintes en place (faibles ou fortes), d'ordonner les observations faites en cours de forage et sur puits, pour indiquer le sens de variation correct du ou des paramètres associés au remède à apporter.

3.3 Analyse en contrainte effective

Le même type d'analyse peut être effectué en prenant en compte la pression de pore dans la roche à la paroi u, au voisinage immédiat du trou. Considérons déjà le cas extrême, peut-être théorique, où cette pression de pore est indépendante de la pression de boue, cela ne modifie pas les limites de zones A, B, C mais cela change seulement la forme du polygone de stabilité.

Tout se passe comme si il y avait un terme de cohésion supplémentaire :

$$\Delta So = - u \, tg \, \phi.$$

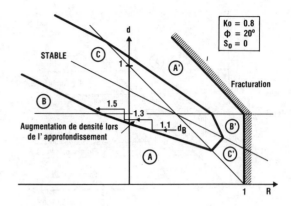

Figure 5. Polygone de stabilité et limites en fracturation dans le plan Ko = 0.8

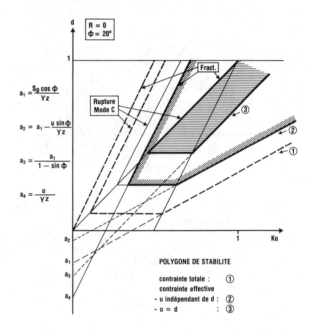

Figure 6. Effet de la pression de pore sur le polygone de stabilité et la limite en fracturation

Ce cas, où la pression de pore en paroi ne dépend pas de la densité de boue, est illustré par le polygone 2 sur la figure 6. Avec cette combinaison particulière des paramètres, la limite en fracturation (droite d'ordonnée à l'origine a_4) est située à droite de la facette en mode C du polygone de stabilité 2. En pratique, la rupture en mode C ne pourra pas apparaître dans ce cas précis.

A l'autre extrême, le cas d'une transmission intégrale de la pression de boue à la pression de pore, sans effet de gradient, est illustré par le polygone 3 indiqué sur la figure 6. Dans la réalité, il conviendrait de tenir compte du gradient d'écoulement qui donne une solution différente ; les zones A, B, C sont alors modifiées et des diagrammes de stabilité prenant en compte un cake plus ou moins efficace peuvent être tracés. Ces diagrammes avec cake peuvent être couplés également avec les contraintes thermiques. Quelques-unes des équations permettant de construire ces diagrammes sont indiquées en annexe. Elles seront reprises en détail dans une autre publication.

Le vrai problème ne réside pas tant dans l'expression analytique des contraintes liées à la filtration, qui a été développée par plusieurs auteurs, mais dans le choix adéquat du type d'analyse et, de la pression de pore en paroi, compte tenu des modes de filtration particuliers en paroi de puits pétrolier, à savoir :

* Formation d'un cake en paroi, d'autant plus rapide que le milieu permet la filtration.

* Absence de cake dans le cas de forage dans des roches argileuses saturées d'eau, (éventuellement à pression interstitielle négative, CHENEVERT, 1970) dont la faible perméabilité n'autorise pas la filtration.

* Utilisation de fluide de forage polyphasique (huile-eau) éventuellement en présence de fluide interstitiel lui aussi polyphasique (huile-eau-gaz). Il faut faire intervenir de tels types d'écoulements dans le choix du type d'analyse : il est très probable, ainsi que le mentionnaient GRAY et DARLEY (1980), que l'efficacité des boues à l'huile dans les roches argileuses soit due, non seulement à une absence de filtrat aqueux dans les zones à argiles gonflantes, mais encore au fait qu'on doit très certainement changer de type d'analyse, l'huile de la boue ne pouvant pénétrer dans le réseau poreux très fin de ces matériaux.

Compte-tenu de la différence notable entre les deux types d'analyses, le passage d'un type à l'autre exacerbe l'effet des autres chargements, dans le cas de roches peu perméables : la moindre microfissuration créée en paroi par un chargement quelconque (densité insuffisante, réchauffement) fait passer brutalement du polygone 2 au polygone 3 et peut être à l'origine de certaines déstabilisations brutales observées. Il est par ailleurs très probable que ce phénomène empêche le développement de fracture cisaillante en mode C au profit d'une fracture en extension. La moindre fissure, provoquant un changement du type d'analyse par filtration au travers de la paroi, place brutalement un point représentatif, situé sur une limite en mode C , bien au dessus de la limite en fracturation hydraulique correspondant à l'analyse en contrainte effective. C'est cet aspect du problème qui mérite encore d'être éclairci.

4 QUELQUES FAITS EXPERIMENTAUX

Tous les points développés plus haut, et notamment les problèmes de caractérisation de la rupture, ont conduit à lancer plusieurs programmes expérimentaux décrits ci-après.

4.1 Expérimentations au banc de forage Elf-Aquitaine

Ce banc permet de forer dans des conditions proches de la réalité (outil, contraintes et boue) des échantillons de 30 cm de diamètre et 50 cm de haut. Son originalité est de pouvoir soumettre des échantillons de roche naturelle à un niveau de contraintes (40 MPa axialement, 70 MPa latéralement), susceptible de provoquer des ruptures à la paroi et au front de taille du forage, et d'en analyser les conséquences sur les performances de l'outil et la tenue des parois.

Une série d'essais a été effectuée en utilisant un calcaire jurassique provenant de la bordure S.E. du Bassin Parisien (Anstrude). Les résultats sont illustrés par la figure 7, où pour chacun des essais est tracé le déplacement du point représentatif (d,Ko) lors du chargement, jusqu'à la rupture, (indiquée par une trame), détectée par une chute nette de la pression de confinement. La contrainte axiale était maintenue constante lors de l'essai ; elle était de 38 MPa pour certains essais, et de 40 pour les autres. Les essais sont effectués à drainage fermé et il semble qu'effectivement les points expérimentaux s'alignent selon une droite à 45° dans le plan (Ko,d) conformément à la limite en mode A du polygone 3 de la figure 6, correspondant à ces conditions de drainage. Les essais sont effectués

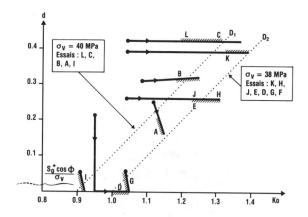

Figure 7. Chemins de mise en charge dans le plan (d, Ko), la zone probable de rupture est indiquée pour chaque essai par la trame terminant chaque trait.

à l'eau claire, mais des valeurs comparables ont été retrouvées en forant à la boue bentonitique, le drainage fermé permettant la transmission de pression sans dépôt de cake. Conformément aux observations précédentes, la roche est beaucoup plus résistante que prévue. Les deux limites en mode A (droite D_1 et D_2) correspondant aux deux valeurs de σ_v utilisées (40 et 38 MPa) fournissent, à partir de l'expression analytique de leur abscisse à l'origine, une même valeur So* de la cohésion interne ainsi mesurée (en supposant arbitrairement un angle de frottement constant). Cette valeur est égale à 2,5 fois la cohésion mesurée par les essais conventionnels. La figure 8 illustre les ruptures obtenues avec des écailles en parois d'épaisseur très régulière, qui lorsqu'elles sont détachées ont une forme voilée caractéristique. La coupe longitudinale montre que ces ruptures, qui correspondent au mode A, se développent également axialement en formant un réseau de fractures en "arête de poisson", ce qui explique la forme voilée des écailles.

La cohésion apparente obtenue, ne permet donc pas de mettre en évidence sur ce matériau une rupture en mode B, compte tenu de la limite en contrainte axiale de l'installation (40 MPa). C'est en utilisant un calcaire très tendre (calcaire lutétien de St Maximin, Q mesuré = 8 MPa) que nous avons pu obtenir des ruptures en mode B, qui se sont développées comme des portions de tores et ont formé une rupture annulaire à une distance d'environ un diamètre du front de taille.

La poursuite de ces essais, associée à des observations sur les puits, tant des formes de retombées que des profils de trou par diagraphies de diamétreurs, doit permettre de caractériser des zones de rupture en mode B, indicateur de faible Ko, et d'en tirer les conséquences pratiques pour la conduite du forage.

4.2 Autres essais sur cylindres creux

D'autres essais ont été effectués à notre instigation et avec notre aide, à l'Université de Berkeley par P.J. PERIE et R. GOODMAN qui ont utilisé des matériaux artificiels. Les premiers résultats ont fait apparaître un mode de rupture pelliculaire illustré par la figure 9. "Paradoxalement", cette rupture s'est produite pour une valeur de la contrainte en accord avec les théories usuelles. Ces essais se poursuivent, notamment par la prise en compte d'une contrainte radiale interne en essai polyaxial et donnera lieu à une publication ultérieure de la part de ces chercheurs.

DEMI-SECTION TRANSVERSALE

COUPE LONGITUDINALE

Figure 8. Ruptures obtenues dans un essai de forage sous contraintes excessives (blocs injectés de résine avant sciage)

Une autre série d'essais a été effectuée sur un grès du carbonifère anglais par F.SANTARELLI à l'Imperial College de LONDRES (article dans ce même Congrès de la SIMR) qui a montré, une fois encore, une résistance plus importante que la théorie ne le prévoit, et également des courbes de déformations tangentielles de la paroi (convergence des parois) en fonction du chargement externe présentant une courbure inverse très marquée, un peu comme un serrage, mais qui se prolonge. Ce phénomène, observé déjà par l'auteur, lors d'essais polyaxiaux sur du charbon (KAISER, GUENOT et MORGENSTERN 1985) a été expliqué par SANTARELLI, à l'aide d'une théorie qui sera décrite plus loin.

4.3 Essais polyaxiaux

Des essais sur blocs de calcaire percés ont été effectués sur un cadre biaxial au LRPC de Lyon par L. ROCHET qui a eu l'idée de mesurer par des jauges le chargement qu'il appliquait réellement au bloc.

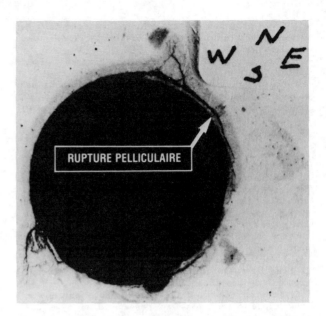

Figure 9. Essai de rupture en chargement isotrope sur matériau artificiel (reproduit avec l'accord des auteurs)

Il a constaté, d'une part, une rotation des contraintes principales par rapport à celles qui étaient censées être appliquées aux limites de l'échantillon, d'autre part, une rupture anisotrope de la paroi du trou, géométriquement en accord avec le champ de contraintes mesurées et à un niveau de contraintes anormal, mais conforme aux autres observations ($\sigma_\theta/Q \sim 3$). Les déformations tangentielles en paroi de trou mesurées par jauges ont présenté parfois des anomalies locales pour un niveau de contraintes en paroi correspondant à la compression simple, sans que la rupture ne soit visible. Ces travaux donneront lieu à publication ; ils ont déjà permis la conception d'une machine triaxiale vraie, à l'Université de Lille, permettant de charger des blocs de 50 cm d'arête jusqu'à 70 MPa par face (Voir Fig. 10). Cet équipement autorisera la réalisation de séries d'essais dans des conditions mieux définies et de mieux cerner les différents phénomènes.

4.4 Synthèse

On retiendra de ces essais, entre autres :
- qu'il existe deux types de rupture différents, l'un le long de lignes de cisaillement, l'autre beaucoup plus pelliculaire en paroi (cela correspond aux observations d'instabilité en ouvrages souterrains citées par MAURY, 1987),
- qu'il existe en cisaillement deux modes de rupture (voire trois), selon la répartition des contraintes ; ils se traduisent par des formes de retombées et de profils de trous rompus différents, qu'il faudra reconnaître in-situ.

5 QUELQUES NOUVELLES APPROCHES THEORIQUES

5.1 Elasticité dépendant de la pression de confinement

Outre l'approche expérimentale, notre effort a porté également sur le développement d'une théorie plus appropriée à décrire la rupture. SANTARELLI, BROWN et MAURY (1986) et SANTARELLI et BROWN (1987) présentent une approche originale où ils montrent que

Figure 10. Presse triaxiale vraie de l'Université de LILLE : Cadre biaxial avec passage latéral pour foreuse (communication J.P. HENRY)

la dépendance du module d'élasticité de la roche en fonction de la pression latérale (et non pas du niveau de charge axiale) peut permettre d'expliquer, par suite d'un "assouplissement" du matériau sous l'effet de la réduction progressive de la contrainte radiale en paroi, une redistribution élastique de contrainte autour du trou, avec réduction de la contrainte tangentielle en paroi et augmentation de la déformation tangentielle. La courbure inversée de la courbe (déformation tangentielle - charge externe) citée au paragraphe 4.2, s'explique aisément à l'aide de cette loi de comportement.

5.2 Approche par la théorie des bifurcations

Une approche complètement nouvelle du problème de la rupture des roches a été proposée par VARDOULAKIS (1984). Nous avons entrepris de l'appliquer au problème de la paroi de puits et elle semble intéressante malgré sa complexité mathématique. Il s'agit de considérer la rupture de la roche par la théorie des bifurcations, un peu comme on considère le flambement eulérien dans l'étude des membrures en compression. Dans le cas présent, on étudiera la bifurcation, soit par "localisation des déformations le long d'une ligne de cisaillement", soit par "instabilité de surface". Cette approche est intéressante car elle propose une interprétation indépendante des différents modes de rupture observés (sur échantillon en triaxial : colonettes, plan de cisaillement, tonnelet ; autour du trou : rupture pelliculaire, cisaillement), mais aussi elle permet d'inclure, dans l'analyse de la rupture, les conditions à la limite. En assimilant ce concept, on comprend pourquoi les lois de comportement continu ne permettent pas d'expliquer correctement la rupture. (VARDOULAKIS, SULEM et GUENOT, 1987). Ne serait-on pas, sans en avoir conscience, dans la situation d'un ingénieur du bâtiment qui interprèterait un essai de compression sur une barre d'acier périssant par flambement, à l'aide d'une loi élastoplastique avec radoucissement, et qui extrapolerait directement ce résultat à une membrure de bâtiment, sans tenir compte des conditions aux limites différentes entre laboratoire et réalité, ou de l'analyse en flambement eulérien ?

Cette approche de la rupture, par la bifurcation doit permettre de reprendre un certain nombre d'anomalies de la mécanique des roches, mais nécessite un effort théorique et numérique important.

6 CONCLUSIONS : IMPLICATIONS PRATIQUES, RECHERCHE A VENIR

Il faut convenir de manière réaliste que jusqu'à présent, la rupture des parois ne peut pas être prédite autrement que empiriquement et que les méthodes classiquement utilisées basées sur des lois de comportement plastique ne sont pas adéquates.

Les résultats, présentés ici, sont très prometteurs et la recherche doit être poursuivie dans ce sens, tant expérimentalement pour caractériser la rupture et ses différents modes, que théoriquement pour proposer une ou des analyses alternatives. L'approche par la théorie de la bifurcation, ainsi que l'introduction de l'élasticité dépendant du confinement, pourront jouer ce rôle dans un futur proche.

Au-delà de ces développements, il est impératif de pouvoir analyser correctement les difficultés rencontrées dans l'exécution des forages et pour cela, les diagrammes proposés ici pour comparer le poids relatif des différents paramètres, et aider à la décision des actions à entreprendre, sont une possibilité riche de promesses. Ils introduisent la mise en oeuvre du contrôle de la température de la boue, qui doit être considérée comme un paramètre de forage à part entière. Il faut tendre, de plus, à optimiser les trajectoires, en tenant compte des conséquences d'un angle d'inclinaison important, mais aussi de la position relative des azimuts des puits par rapport aux contraintes en place (BRADLEY, 1979 - MAURY et SAUZAY, 1987).

Il faut également distinguer en pratique deux situations dans le contrôle d'un puits, comme en travaux souterrains. La première consiste à jouer sur les chargements et la pression de pore en paroi, pour éviter la rupture ; la seconde, tout aussi importante, consiste à gérer convenablement la rupture. Des observations faites au laboratoire (KAISER, GUENOT et MORGENSTERN, 1985) et classiquement dans les ouvrages souterrains, montrent que la zone rompue (pas seulement "plastique") en paroi, constituée d'écailles frottant les unes sur les autres, stabilise la rupture ultérieure du forage. Il s'agit donc d'appliquer correctement le principe minier de la purge : c'est-à-dire évacuer les retombées qui, détachées de la paroi, ne contribuent plus à stabiliser le forage, mais peuvent provoquer des coincements des tiges avec perte de circulation, tout en n'arrachant pas les écailles en place, sous peine de mettre en route un système sans fin. La plupart des observations, faites par les foreurs lors de problèmes de ce genre : battages de tiges sur la paroi, "érosion" par un débit trop important, retombées à la suite de manoeuvres trop rapides, sont attribuables à la seconde situation : Il s'agit en particulier de trouver pour la boue, le bon débit et la bonne rhéologie, pour nettoyer le puits, sans action violente sur les parois.

Nous pouvons espérer beaucoup de la poursuite de la recherche universitaire dans ce domaine, dans la mesure où elle est confrontée en permanence à la réalité des chantiers de forage.

REMERCIEMENTS

L'auteur remercie le Groupe Elf-Aquitaine pour lui avoir permis de publier les résultats de cette recherche, Messieurs MAURY, FOURMAINTRAUX et SAUZAY pour leur aide et tous les chercheurs cités ici pour leur coopération.

REFERENCES

Barton, C.A. & M.D. Zoback 1986. Determination of in-situ stress orientation from Stoneley wave polarization in boreholes. A paraître dans J. Geophys. Res. ...

Bradley, W.B. 1979. Failure of inclined boreholes. Oil & Gas J.2.

Bratli, R.K., P. Horsrud & R. Risnes 1983. Rock mechanics applied to the region near a wellbore. Proc 5th Cong. Int. Soc. Rock Mech., Melbourne ; F1-F17.

Cheatham, J.B. 1984. Wellbore Stability. J. of Petr. Tech. June ; 889-896.

Chenevert, M.E. 1970. Shale alteration by Water Absorption. J. of Petr. Tech. 9 ; 1141-1148.

Corre, B., R. Eymard & A. Guenot 1984. Numerical Computation of Temperature Distribution in a Wellbore while drilling. 59th SPE Ann. Conf ; Houston 13208

Gray, G.R. & M. Darley 1980. Composition and properties of Oil Well drilling Fluids. Gulf Publ. Cie.

Guenot, A., 1986. Stabilité des forages profonds. Ecole d'été : Thermomécanique des roches. A paraître : Série "Manuels et Méthodes". B.R.G.M. Ed.

Kaiser, P.K., A. Guenot & N.R. Morgenstern 1985. Deformation of small tunnels -IV. Behaviour during failure. Int J. Rock Mech. Min. Sci. & Geomech. Abstr. 22-3 ; 141-152.

Kevorkov, S.A., V.N. Romashov, L.E. Simonyants, N.S. Timofeev & R.B. Vugin, 1974. Etude de l'effet de la température et des chargements thermiques cycliques sur la résistance d'une paroi de forage en trou ouvert. Neft'Khoz 6 ; 10-12.

Manolescu, G. 1970. Etat actuel des connaissances sur le problème de stabilité du trou de sonde. Traduc. CNRS de Petrol Si Gaze 21-5 ; 274-280.

Maury, V. 1977. An example of underground storage in soft rock. Rockstore 77 ; 681-689.

Maury, V., 1987. Observations, recherches et résultats récents sur les mécanismes de rupture autour de galeries isolées. Rapport de la Commission ISRM "Mécanismes de rupture". Proc. 6th Congr. Int. Soc. Rock Mech., Montreal (this volume).

Maury, V. & J.M. Sauzay 1987. Borehole instability : case histories, rock mechanics approach and results. SPE/IADC Conf. New-Orleans ; 16051.

Maury, V., D. Fourmaintraux & J.M. Sauzay, 1987. Approche géomécanique de la production pétrolière : problèmes essentiels, premiers résultats, perspectives. Proc. 6th Congr. Int. Soc. Rock Mech., Montreal (this volume).

Miguez, R. & J.P. Henry 1986. Le discage : propositions de conventions pour une étude de la morphologie. Ann. Soc. Géol. Nord Avril ; 43-46.

Santarelli, F.J., E.T. Brown & V. Maury 1986. Analysis of borehole stresses using pressure-dependent, linear elasticity. Int. J. Rock Mech. Min. Sci. & Geomech. Abstr. 23 ; 445-449.

Santarelli, F.J. & E.T. Brown 1987. Performance of deep wellbores in rock with a confining pressure-dependent elastic modulus. Proc 6th Congr. Int. Soc. Rock Mech., Montreal (this volume).

Vardoulakis, I. 1984. Rock bursting as a surface instability phenomenon. Int. J. Rock Mech. Min. Sci. & Geomech. Abstr. 21, 137-144.

Vardoulakis, I., J. Sulem & A. Guenot 1987. Stability of deep borehole as a bifurcation phenomenon. To be submitted to Int. J. Rock Mech. Min. Sci. & Geomech. Abstr.

REFERENCES RELATIVES AUX ESSAIS SUR CYLINDRES CREUX

Bandis, S.C. & N. Barton, 1986. Failure modes of deep boreholes. Proc. 27th U.S. Symposium on Rock Mech., 599-605.

Berest, P., J. Bergues & N.M. Duc, 1979. Comportement des roches au cours de la rupture : application à l'interprétation d'essais sur des tubes épais. Revue Française de Géotechnique 9 : 5-12.

Gay, N.C., 1973. Fracture growth around openings in thick-wall cylinders of rock subjected to hydrostatic compression. Int. J. Rock Mech. Min. Sci. & Geomech. Abstr. 10, 209-233.

Geertsma, J., 1978. Some rock mechanical aspects of oil and gas well completions. Proc. Europ. Offsh. Petr. Conf., London : 301-308.

Guenot, A., 1979. Tunnel stability by model tests. MSc. Th. : U of Alberta.

Haimson, B.C., & J.N. Edl, 1972. Hydraulic fracturing of deep wells. SPE Paper 4061.

Haimson, B.C., & C.G. Herrick, 1985. In situ stress evaluation from borehole breakouts, experimental study. Proc. 26th US Symp. on Rock Mech. : 1207-1218.

Mastin, L.G., 1984. The development of borehole breakouts in sandstone. Msc Th., Standford Univ.

Obert, L. & D.E. Stephanson 1965. Stress conditions under which core discing occurs. Trans. of Soc. Mining Eng. 9 : 227-235.

Simonyants, L.E., S.A. Kevorkov & N.I. Fisenko, 1970. Etude de la résistance statique de la paroi d'un trou de sonde dans un puits en découvert. Neft'i gaz 9 : 45-50.

ANNEXE : DIAGRAMME DE STABILITE

Quelques-unes des équations pour les 3 principaux domaines (ruptures et limites de zones)

A $2 K_o - d (1 + k_p + r) + u_o (\alpha (k_p - 1) + r) - S_c - R = 0$

A/B $2 K_o - d - 1 = 0$

B $R + d (k_p + r) - u_o (\alpha (k_p - 1) + r) + at - 1 = 0$

B/C $2 (K_o - d) - R + H (d - u_o) = 0$

C $2 K_o + d (v - 1) + u_o (\alpha (k_a - 1) - v) + k_a (s_c - 1) - R (1 - k_a) = 0$

C/A $R + d - 1 - H (d - u_o) = 0$

Fracture verticale : $2 K_o - d - R - e*K* (d - u_o) - \alpha u_o = 0$

Fracture horizontale : $1 - R - e*K* (d - u_o) - \alpha u_o = 0$

avec $k_p = \dfrac{1 + \sin \phi}{1 - \sin \phi} = \dfrac{1}{k_a}$

ϕ : angle de frottement
So : cohésion

$s_c = \dfrac{Q}{\gamma . Z} = \dfrac{2 So \cos \phi}{1 - \sin \phi} \dfrac{1}{\gamma . Z}$, résistance normalisée

u_o : pression de gisement normalisé = $pg/\gamma . Z$

$r = e* (K* - k_p)$ avec $e* = \alpha e$, $K* = \dfrac{\nu}{1 - \nu}$

α : coefficient de BIOT
ν : coefficient de Poisson
e : efficacité du cake : e = 0 cake parfait
                          e = 1 pas de cake

$v = e*K* (k_a - 1)$

Stress measurements at Hanford, Washington for the design of a nuclear waste repository facility

Détermination de contraintes à Hanford, Washington pour la conception d'un dépot de déchets radioactifs

Spannungsmessungen in Hanford, Washington für den Entwurf eines Lagers für radioaktive Abfälle

B.C.HAIMSON, University of Wisconsin, Madison, USA

ABSTRACT: This paper summarizes the first complete hydrofracturing stress measurements at the Hanford Site in one of the basalt flows considered for the construction of a nuclear waste repository. The results indicate a strongly anisotropic stress field suggestive of thrust faulting along east-west striking planes. These findings are supported by local geology and seismicity and by evidenceof core discing and borehole spalling. The measured in situ stresses were instrumental in the rational design of the future rock caverns with respect to site suitability, cavern orientation and shape, and support requirements.

RESUME: Cette publication présente les premiers resultats complets de mesures de contraintes par hydrofracturation au site de Hanford. Une couche de basalte y est prévue pour la construction d'un dépôt de déchets radioactifs. Les résultats suggèrent une anisotropie importante dans les contraintes. Ils sont confirmes par la géologie locale et d'autres mesures à partir de sondages. Les mesures des contraintes in situ ont été prises en compte dans l'étude des differentes cavernes prévues.

ZUSAMMENFASSUNG: In diesem Artikel sind die ersten Kompleten Hydrofractur-spannungs messungen aus einer der Basaltfluesse, die fuer den Bau einer nuclearen Abfallaufbewahrungsstelle in Hanford in Betracht gezogen wurden, zusammengefasst. Die Ergebnisse zeigen ein stark anisotropes Spannungsfeld, das eine Schubfaltung ent lang von Ost-West verlaufenden Ebenen vorschlaeght. Dieser Fund wird von der oertlichen Geologie und Seismizitaet, dem Vorfinden von Scheibenbildung an Bohrungen und Bohrlochreissen unterstuetzt. Die Gemessenen in situ Spannungen waren behilflich fuer den vernuenftigen Entwurf von zukuenftigen Felskavernen mit Ruecksicht auf Eignung des Ortes, Kavernen orientierung und-form, und Stuetzungsanforderungen.

1 INTRODUCTION

A unique requirement of underground nuclear waste repository design is a significantly longer service life than would be expected for common civil structures. The long term stability requirement makes the knowledge of local in situ stress conditions an essential element in establishing the feasibility of a site for nuclear waste disposal and in rationally designing the appropriate subterranean openings. At the Hanford Site the state of stress is a particularly critical parameter since local microseismic events (Malone et al. 1975), core discing (Moak 1981) and borehole spalling (Paillet 1985) indicate high horizontal stresses which could have considerable impact on the construction, shape, orientation, and support of excavations, as well as on the layout of waste emplacement.

Early in the site investigation at Hanford the decision was made to conduct a comprehensive series of stress measurements in the vicinity of the proposed location and depth of the planned underground waste repository. After some initial inhouse feasibility tests at an intermediate depth (343 m), we conducted the first complete set of stress measurements in a testhole that reached the depth of the lowest candidate horizon (the Umtanum flow). This paper describes these tests and their implication regarding the repository design.

2 LOCATION AND GEOLOGIC SETTING

The reported stress measurements were conducted in vertical borehole DC-12 (76 mm in diameter) located near the geographic center of the Hanford Site (lat. 46°28'6" N, long. 119°32'26"), some 24 km northwest of Richland, Washington (Figure 1). The hole is situated near the axis of the Cold Creek syncline, which is part of the Pasco Basin, one of the several

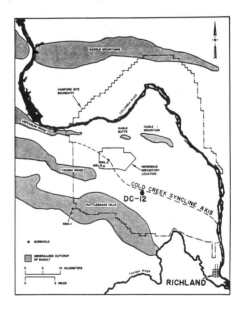

Figure 1. Map of Hanford Site, showing location of testhole DC-12 and of the Reference Repository. Also shown is the axis of the Cold Creek syncline.

basins in the western Columbia Plateau. In the Cold Creek syncline the Columbia River Basalt Group occupies much of the top 1500 m of the earth's crust, and within it several basalt flows have been identified as potential repository rock hosts (candidate horizons). At the time of our measurements the Umtanum flow was considered the preferred horizon because of its thickness (about 75 m), depth (979-1051 m in hole DC-12), lateral extent (present throughout the Cold Creek syncline), and thick uniform entablature (13 m in DC-12). All our tests were conducted in the Umtanum flow.

3 TESTING PROCEDURE

3.1 Selection of test intervals

The method used for the stress measurements was hydrofracturing (Haimson 1974). Test intervals were selected after studying the condition of the extracted core from color photographs. Six depths were identified within the Umtanum flow as being suitable for hydrofracturing (i.e. intervals, at least 1 m long, which appeared free of discontinuities).

3.2 Hydrofracturing

Hydrofracturing equipment setup and procedure were kept the same in all tests. The hydrofracturing tool consisted of two inflatable rubber packers straddling a 60 cm interval. The tool was lowered into the test hole using a slim high-pressure drillrod which also served as the hydraulic conduit to the straddled interval. The packers were independently pressurized via a slim hose strapped to the outside of the drillrod.

The hydrofracturing tool was first assembled and tested for leaks, and then lowered to the predetermined depth for test no.1. The inflatable packers were pressurized using a 70 MPa hydraulic pump to an initial level of 5 MPa which insured proper sealing of the straddled interval without creating excessive stress concentrations. The interval was then pressurized via the drillrod using an independent 70 MPa pump. The pressure in the packers was maintained at all times at 2 MPa or more higher than that in the interval. Pressurization was continued at a constant flowrate until hydrofracturing (breakdown) occurred (P_{c1}). The pump was then shut off and the pressure in the interval was allowed to stabilize at a lower level (shut-in, P_s). The pressure in the interval was later bled off, and as soon as the excess water in the fracture was backflowed the pressurization cycle was repeated several times (Figure 2). Most of the repeated cycles were conducted at the same flowrate so as to obtain meaningful secondary break-

down pressures (P_{c2}) as well as additional shut-in pressures. At least one cycle was carried out at a very slow flowrate to obtain an upper limit for P_s. At the conclusion of a test the packers were deflated and moved to the next predetermined depth without retrieving the tool to the surface.

During each entire test the packer and interval pressures were independently monitored by pressure gages and transducers and permanently recorded on time-base recorders. The flow meter monitoring was also plotted in the same recorders.

3.3 Hydrofracturing delineation

Strike and dip of the induced hydrofractures at the borehole wall were determined using an oriented impression packer, which consisted of a 1 m inflatable packer element covered by a thin layer of uncured rubber and a gyroscopic orienting tool. The impression packer was lowered to the precise depth where a hydrofracture had been induced and was pressurized to a level slightly higher than the respective shut-in pressure. This procedure enabled the soft rubber covering to penetrate the induced crack. During the one hour or so that the packer was kept inflated the gyroscopic orienting tool was lowered on a wireline until it engaged a key which was rigidly attached to the impression packer. A built-in camera took a picture of the key position as well as of an angle unit face giving the inclination and direction of the hole and the relative orientation of the key. The tool was retrieved after the picture was taken and the film was immediately developed. At the appropriate time the impression packer was deflated and brought back to surface, and the hydrofracture traces were recorded on transparent film. All fracture traces were oriented relative to the key, and thus absolutely oriented with the help of the gyroscopic tool photograph.

4 TEST RESULTS

Three important parameters are supplied by the pressure-time records obtained during hydrofracturing tests: P_{c1}, P_{c2}, and P_s. These pressures have been defined above, and some of the different ways of discerning them in a pressure-time plot have been summarized by Zoback and Haimson (1982). The other major parameter comes from the impression test in the form of the strike and dip of the induced hydrofracture.

The vertical stress (S_v) was determined independently from the measured density of the overlying strata:

$$S_v = 0.026 \ D \qquad (1)$$

where the stress is in MPa and D is the depth of the test interval measured in meters.

The least horizontal stress (S_h) was obtained directly from the mean shut-in pressure value:

$$S_h = P_s \qquad (2)$$

The major horizontal principal stress is conventionally calculated using the Hubbert and Willis (1957) expression:

$$S_H = T + 3S_h - P_{c1} - P_0 \qquad (3)$$

where P_0 is the pore water pressure (assumed to be equal to the hydraulic head of the ground water measured from the water table, which in hole DC-12 is at 33 m below the surface), and T is the tensile strength of the rock under hydrofracturing conditions. Since T was not known an alternate equation was used (Haimson 1980):

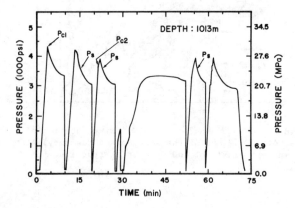

Figure 2. Typical pressure-time field record during hydrofracturing test in hole DC-12.

$$S_H = 3S_h - P_{c2} - P_o \qquad (4)$$

which is based on the assumption that under constant flowrates the value of P_{c2} is the same as that of P_{c1} minus T.

Impressions of vertical fractures were obtained in four of the six tests. The strike of each vertical fracture was taken as perpendicular to the direction of S_h. Table 1 and Figure 3 summarize the test data and the calculated stresses and their directions.

5 THE STATE OF STRESS IN THE UMTANUM FLOW

Since all six tests were conducted within a range of 40 m it is reasonable to determine the mean value of each principal stress rather than attempt to establish a relationship with respect to depth. From Table 1 the following mean principal stress components are obtained for the Umtanum flow in hole DC-12:

$$S_v = 26.2 \ (\pm 0.3) \ \text{MPa}$$

$$S_h = 34.6 \ (\pm 3.0) \ \text{MPa at N67°W} \ (\pm 20°)$$

$$S_H = 61.2 \ (\pm 7.0) \ \text{MPa at N23°E} \ (\pm 20°) \qquad (5)$$

Table 1. Summary of hydrofracturing pressures and directions and principal stress calculations (hole DC-12)

Depth	P_o	P_{c2}	P_s	Hfrac	S_v	S_h	S_H	S_H
m	MPa	MPa	MPa	direc.	MPa	MPa	MPa	direc.
1002	9.5	35.0	33.3		25.6	33.3	55.3	
1013	9.6	30.6	34.4	N40°E	25.9	34.4	63.0	N40°E
1021	9.7	30.7	31.4	N09°E	26.1	31.4	53.8	N09°E
1031	9.8	34.8	34.2	N01°E	26.4	34.2	58.2	N01°E
1036	9.8	37.0	39.4	N42°E	26.5	39.4	71.5	N42°E
1042	9.9	31.1	35.6		26.7	35.6	65.7	

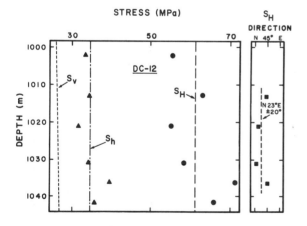

Figure 3. Summary of stress measurement results in hole DC-12. Shown are least horizontal principal stresses (triangles), largest horizontal principal stresses (circles), and the directions of the latter (squares).

The most immediate conclusion of the measurements is that the vertical stress is the overall least principal stress:

$$S_v < S_h < S_H \qquad (6)$$

suggesting potential thrust fault conditions.

Our equations (5) and (6) are in accord with focal mechanism solutions of local microseismicity (Malone et al. 1975). These indicate a nearly horizontal major compression acting roughly north-south and a minimum compression in the vertical direction, suggesting thrust faulting on east-west striking planes (Figure 4). Mapped nearly east-west folds and sub-parallel thrust and reverse faults not only support our results but also imply that the same stress regime that brought about these structural features is still active today (Caggiano and Duncan 1983). It appears that the Columbia River basalt in the Hanford area has been deforming under north-south compression for the past 14 m.y.

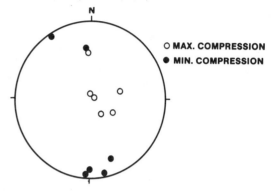

Figure 4. Principal stress directions deduced from focal mechanism solutions. showing the least overall principal stress to be vertical, and the largest principal stress to be horizontal and acting in a north-south direction (after Malone et al. 1975).

Core discing is another wide spread phenomenon at Hanford and is encountered in many of the deep holes drilled for the present site investigation. Core discing generally indicates the existence of high horizontal stresses. Finally, borehole breakouts have been observed in deep holes at Hanford that were logged with a borehole televiewer (Paillet 1985). Although breakout orientation could not always be determined because magnetic minerals in the basalt interfered with the operation of the televiewer magnetometer, all indications are that they are located on the east and west sides of the boreholes. This reconfirms the north-south trend of the maximum (horizontal) compressive stress.

If the relative magnitudes of the measured principal stresses and their directions are so well supported by other evidence, what about the absolute values determined in our tests? To answer this, among other things, Rockwell Hanford Operations undertook some additional stress measurements in the Umtanum flow in two boreholes situated within the Reference Repository Location at about 13 km from DC-12 (Kim et al. 1984). The location of the two boreholes (RRL-2 and RRL-6) is shown in Figure 1 and a comparison of results is given in Table 2. The coincidence of stress magnitudes between the two sets of tests is remarkable considering the distance between the test sites, and suggests persistence of the stress regime over a large area. The largest discrepancy between the two measurements (3.4 MPa) is in the value of S_v, and is due merely to a difference of 150 m in depth of testing. The mean S_H direction shows a difference of only 18°. It is also

Table 2. Comparison of measured principal stresses in the Umtanum flow (after Kim et al.1984)

Test hole	No. of tests	Depth m	S_v MPa	S_h MPa	S_H MPa	S_H direc.
DC-12	6	1022 ±20	26.2 ±0.3	34.6 ±3.0	61.2 ±7.0	N20°E ±20°
RRL-2 and RRL-6	4	1177 ±16	29.6 ±0.4	34.3 ±3.0	60.2 ±5.4	N05°E ±13°

Note: Location of RRL-2 and RRL-6 shown in Figure 1.

interesting to note that later stress measurements in adjacent basalt flows such as the Cohassett gave similar results (Kim et al. 1984), indicating that the stress tensor determined in DC-12 prevails over a significant three-dimensional block of the shallow crust at Hanford.

6 IMPLICATIONS FOR REPOSITORY DESIGN

6.1 Site suitability

The measured state of stress at Hanford can be used to assess the stability of the repository host rock unit. A potentially unstable formation may not be suitable as a repository site because of the uncertainty regarding the safe isolation of radionuclides.
    To study the potential for instability a Mohr diagram can be used in which are plotted the circle representing the stresses in the plane containing the maximum and the minimum principal stresses and the lines depicting the criterion of failure of the rock mass and the criterion of frictional sliding along natural fractures or joints. Since the basalt flows at Hanford are well jointed (density of 3 - 30 per meter) the test for stability is not so much whether the measured state of stress could lead to brittle failure of the rock mass, but whether it may first induce frictional sliding along favorably oriented fractures at a considerably lower stress requirement (Brace and Kohlstedt 1980). Frictional resistance of fractures has been measured in the laboratory for a variety of rock types. Significantly, Byerlee (1978) demonstrated that friction is largely independent of rock type or surface conditions. For the range of stresses that includes those measured at Hanford Byerlee found that

$$Sh_{cr} = 0.85 (S_n - P_o) \qquad (7)$$

where Sh_{cr} and S_n are the critical shear stress and the normal stress, respectively, acting on the plane of the fracture, and 0.85 is the coefficient of friction (or the slope of a straight line inclined at 40° to the normal stress axis in a Mohr diagram). Plotting this criterion (sometimes denoted as Byerlee's law) in such a diagram (Figure 5) the half circle depicting the acting effective stresses based on our measurements appears to be safely within the stable zone (entirely under the line representing equation 7). A corollary of this result is that brittle failure of the rock mass is most unlikely.

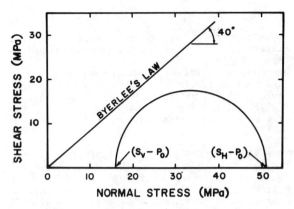

Figure 5. Mohr diagram showing the Mohr circle representing the mean state of stress in the vertical plane aligned with the direction of S_H based on our measurements in hole DC-12. Also shown is the mean criterion of slip on favorably oriented fractures as determined by Byerlee (1978).

It should be emphasized that in the above analysis we used mean values only. Were we to use the worst scenario case in which the maximum S_H and the minimum S_h measured would be employed together with the lowest coefficient of friction in Byerlee's experiments (0.6), the potential for frictional sliding would be significantly enhanced.

6.2 Orientation of rock caverns

The original design of the underground nuclear waste repository at Hanford was based on an assumed ratio of 1:1 between any two principal stresses. Such an isotropic state of stress would be ideal since it would impose no constraints on the orientation of the required rock caverns. However, our measurements in the Umtanum flow clearly indicated that such was not the case. Consequently, the design was changed to accomodate the real stress situation which may be summarized as a ratio of approximately 2:1 between the largest horizontal stress and either of the other two principal components, and a roughly 1:1 ratio between the least horizontal and the vertical stresses. Figure 6 is a sketch of a suggested solution to the orientation problem of the repository critical openings. The "storage holes", being the most crucial caverns since they will house the radioactive waste canisters, will be oriented in the horizontal stress. The "emplacement room" which leads to the storage holes will be allinged with the minimum horizontal stress (Rockwell Hanford Operations, 1982). Both caverns will require specific crosssectional shapes as described in the next section.

6.3 Shape of rock caverns

The shape of the repository caverns should minimize stress concentrations around the openings so as to optimize stability. Again, the prerequisite to the rational selection of the least destabilizing shape is knowledge of the in situ stresses. A properly conceived excavation shape can prevent large investments in reinforcement and lower the risk of rock failure around the opening. At Hanford the storage holes would have a circular crosssection , which is the desired shape for the excavation method (boring),for the function of these holes (to store circular canisters), and for maximum stability (the axes of the holes are perpendicular to both the least horizontal and the vertical stresses which are approximately equal in magnitude).

122

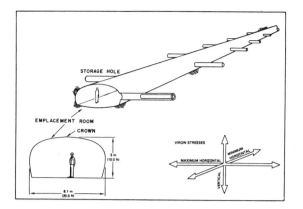

Figure 6. Proposed shape and orientation of emplacement room and storage holes at Hanford Site and their relationship to the principal stress magnitudes and directions as measured in hole DC-12 (after Rockwell Hanford Operations 1982).

The emplacement room orientation is dictated by that of the storage holes and hence is less favorable. The suggested design calls for an elongated horse shoe (quasi-elliptical) shape with the width about twice the maximum height, to correspond with the 2:1 stress ratio (between the maximum horizontal and the vertical stresses) for optimum stress concentration around the room (Rockwell Hanford Operations 1982).

6.4 Support

The measured in situ stresses at Hanford provide the basic field data required for estimating needed support and reinforcement of rock caverns. For example, the large diameter vertical shaft necessary for access to the repository could experience rock spalling, based on the highly differential horizontal stresses that would develop near the shaft wall. A suggested high density drilling fluid would reduce the amount of spalling. Corrective action strategies have been developed in the event that rock instability is encountered nevertheless (US Department of Energy 1986). Similarly, theoretical analyses of the rock stresses based on the initial measurements, suggest that rock support may be required to improve stability in some of the horizontal openings (Rockwell Hanford Operations 1982).

7 CONCLUSIONS

This paper summarizes the first complete set of deep-hole hydrofracturing stress measurements at the Hanford Site in one of the candidate repository horizons (the Umtanum flow). The results suggest a strongly anisotropic state of stress with the least principal stress vertical and the largest horizontal stress oriented roughly north-south. This stress regime favors thrust faulting along planes striking east-west, in accord with focal mechanism solutions of local earthquakes and with the type and orientation of local faults. Highly differential stresses are also indicated by borehole spalling and core discing at Hanford.

The results of our stress measurements were implemented in many aspects of the preexcavation rational design of the Hanford proposed nuclear waste repository, such as site suitability, cavern shape and orientation, and required support.

8 REFERENCES

Brace, W.F. & D.L. Kohlstedt 1980. Limits on lithospheric stress imposed by laboratory experiments. J. Geophys. Res. 85: 6248-6252.
Byerlee, J.D. 1978. Friction of rocks. Pure Appl. Geophys. 116: 615-626.
Caggiano, J.A. & D.W. Duncan (eds.) 1983. Preliminary interpretation of the tectonic stability of the Reference Repository Location, Cold Creek, Hanford Site. Report RHO-BW-ST-19P, Rockwell International, Richland, Washington.
Haimson, B.C. 1974. A simple method for estimating in situ stresses at great depths. In Field testing and instrumentation of rock, p.156-182. Am. Soc. Testing and Materials (ASTM) Special technical publication 554.
Haimson, B.C. 1980. Near surface and deep hydrofracturing stress measurements in the Waterloo quartzite. Int. J. Rock Mech. Min. Sci. & Geomech. Abstr. 17:81-88.
Hubbert, M.K. & D.D. Willis 1957. Mechanics of hydraulic fracturing. Trans. Soc. Petrol. Eng.of AIME 210:153-168.
Kim, K., Dischler, S.A., Aggson, J.R. & M.P. Hardy 1984. The state of in situ stresses determined by hydraulic fracturing at the Hanford Site. Report SD-BWI-TD-014, Rockwell International Richland, Washington.
Malone, S.D., Rothe G.H. & S.W. Smith 1975. Details of microearthquake swarms in the Columbia basin, Washington. Seism. Soc. Am. Bull. 65: 855-864.
Moak,J. 1981. Borehole geologic studies. In Subsurface geology of the Cold Creek syncline. Report RHO-BWI-ST-14, Rockwell International, Richland, Washington.
Paillet, F.L. 1985. Acoustic televiewer and acoustic waveform logs used to characterize deeply buried basalt flows, Hanford Site, Washington. Open-file report 85-419, U.S. Geological Survey, Denver, Colorado.
Rockwell Hanford Operations 1982. Site characterization for the basalt waste isolation project. Report DOE/RL 82/3, Rockwell International, Richland, Washington.
U.S. Department of Energy 1986. Nuclear Waste Policy Act, environmental assessment, Reference Repository Location, Hanford Site, Washington. Report DOE-RW-0070, vol. 2.
Zoback, M.D. & B.C. Haimson 1982. Status of the hydraulic fracturing method for in situ stress measurements. In Proceedings of the 23rd U.S. Symposium on Rock Mechanics, p. 143-156, Society of Mining Engineers, New York.

Prediction of subsidence due to groundwater withdrawal in the Latrobe Valley, Australia

Prédiction d'affaissements causés par le retrait des eaux souterraines dans la vallée de Latrobe en Australie

Vorhersage einer durch Grundwassersenkung verursachten Senkung im Latrobetal, Australien

D.C.HELM, CSIRO Division of Geomechanics, Syndal, Australia

ABSTRACT: Subsidence due to groundwater withdrawal in the Latrobe Valley, Australia, has been simulated using a one-dimensional finite difference model of nonlinear consolidation. The model uses transient changes of fluid pressure within the more permeable zones of the aquifer system(s) as input and calculates the resulting time-delayed compression and/or expansion of both the permeable sands and the more compressible but more slowly draining lenses (aquitards) and separators (semi-confining beds) as output. The standard output of a groundwater model (namely, pressure changes within the permeable zones) serves as input for the subsidence (consolidation) model at any site of interest. Examples of the predictive technique for the Latrobe Valley are presented.

RESUME: On a simulé, en employant un modèle monodimensionnel de différences finies de consolidation non linéaire, un effondrement dû au retrait de l'eau souterraine dans la vallée Latrobe en Australie. Le modèle emploie en tant que données d'entrée les changements transitoires de pression hydraulique à l'intérieur des zones plus perméables du système d'aquifère et calcule comme sortie la compression et/ou la dilatation retardée(s) des sables perméables ainsi que celle des inclusions plus compressibles mais qui s'assèchent plus lentement (aquitards) et des séparateuses (couches semi-imperméables) qui en résultent. Les données de sortie standard d'un modèle de l'eau souterraine (c.-a-d. les changements de pression à l'intérieur des zones perméables) servent de données d'entrée pour le modèle d'effondrement (de consolidation) à toute localité considérée. On présente des examples de la technique de prédiction appliquée à la vallée Latrobe.

ZUSAMMENFASSUNG: Durch Grundwasserzurückziehung verursachte Senkung im Latrobetal (Australien) ist unter Verwendung eines eindimensionellen Differenzenmodells nichtlinearer Verdichtung nachgeahmt worden. Beim Modell werden vorübergehende Veränderungen des Flüssigkeitsdruckes innerhalb der durchlässigeren Zonen des Grundwasserleitersystems als Eingabedaten verwendet und die sich daraus ergebende, verzögerte Verdichtung bzw. Ausdehnung sowohl der durchlässigen Sande als der leichter zusammendrückbaren, aber sich langsamer entwässernden Einschlüsse (Aquitarden) sowie Trennschichten (halburdurchlässiger Schichten) als Ausgabe berechnet. Die Normalausgabedaten eines Grundwassermodells (d.h. Druckveränderungen innerhalb der durchlässigen Zonen) dienen als Eingabedaten des Senkungs-(Verdichtungs)-Modells bei jeglichem in Frage kommenden Ort. Beispiele der im Latrobetal verwendeten Vorausberechnungstechnik werden vorgestellt.

1 INTRODUCTION

Open pit coal mining in the Latrobe Valley, Australia, by the State Electricity Commission of Victoria (SECV), requires depressurizing two regional confined aquifer systems in order to minimize potential heave, fracturing and flooding of the floor of the mine workings. Depressurizing the aquifer systems has in turn caused regional subsidence. In order to predict this subsidence, the mechanism of groundwater flow has been coupled through computer simulation to the mechanics of rock mass deformation.

The most compressible units in the Latrobe Valley are coal and clay which not only form lenticular interbeds within each aquifer system but also serve as regional semi-confining beds that separate one heterogeneous aquifer system from the other. These compressible units have low permeabilities and hence at any site of interest there occurs a time lag between the imposed stress due to observed drawdown and the resulting compression and subsidence. The input to the subsidence model is the fluctuating drawdown within the aquifers and the output is the partly nonrecoverable compressional response of the heterogeneous porous material.

This paper presents the strategy used to model land subsidence due to groundwater withdrawal in the Latrobe Valley as well as some examples of the prediction effort. The subsidence model COMPAC (a one-dimensional finite difference code) is calibrated at specific sites using both measured changes in hydraulic head within the aquifers and the resulting observed subsidence as functions of time.

An independent calibration of a regional groundwater model of the Latrobe Valley was carried out concurrently (Evans, 1983, 1986). The groundwater model is used to predict future drawdowns in response to anticipated scenarios for pumpage. These predicted drawdowns are then used to estimate future subsidence at key locations as a function of time. Subsidence contours can then be drawn between these key locations for any future time of interest. One output of the subsidence model is an estimate of residual subsidence which results from the fact that in the field the compressing beds almost never reach an equilibrium stress distribution. Time constants are in terms of decades to centuries.

The model COMPAC essentially uses nonlinear consolidation theory. It distinguishes between recoverable and non-recoverable compressibilities (Helm, 1975). Strain hardening of each compressibility can be introduced as well as porosity-dependent permeability values (Helm, 1976). The model can use either laboratory derived parameter values or it can be used as an independent method for estimating the parameters directly as part of the calibration process. The model has been applied as a predictive tool with excellent results to many locations in Australia (Helm, 1984), California (Helm, 1977, 1978) and Texas (Harris - Galveston Subsidence District, 1982). Application to the Latrobe Valley is the subject of the present paper. What distinguishes the Latrobe Valley problem is that (1) confined aquifers are expected to convert to unconfined aquifers and then possibly to become partly saturated aquifer systems and (2) within a unit column different parts of each system will convert at distinct times from overconsolidated to normally consolidated conditions. The aquifer systems might reconvert back again, depending on the depressurizing/dewatering/repressurizing scenario. COMPAC keeps track of these distinct physical processes automatically.

2 LOCATION AND HYDROGEOLOGY

The Latrobe Valley is located near the southeast coast of Victoria, Australia (Figure 1). It essentially follows the eastward dipping Latrobe Syncline and opens eastward towards Lake Wellington near Bass Strait. The locations for the present analysis of subsidence are south of the Latrobe River (Figure 1) near Morwell, Driffield and Loy Yang. The SECV operates a large coal winning open cut at Morwell which will be extended westward towards Driffield. A further major open cut exists at Yallourn while Loy Yang is the site of new open cut operations.

A Tertiary sequence lies unconformably on sedimentary Mesozoic basement rock. Within the syncline the Tertiary sequence reaches a maximum thickness of at least 770 m (Brumley, et al. 1981). Three main groups of coal seams of varying thickness are continuous within the Tertiary sequence and underlie a large part of the Latrobe Valley. They lie essentially within the Morwell and Traralgon Formation and have been tilted, folded and faulted.

Brumley and Reid (1982) point out that two major confined aquifer systems exist within the Latrobe Valley Group. They are called the Morwell Formation and the Traralgon Formation aquifer systems. Unconfined aquifers occur within overlying Haunted Hill Gravels and recent alluvials.

3 STRATEGY

A three-stage modelling process for subsidence prediction has been developed at the SECV for the Latrobe Valley. The first stage consists of using a quasi three-dimensional groundwater model. Its output, namely the computed changes of potentiometric head over time within the three modelled aquifer systems in the Latrobe Valley, serves at specified locations as input to the one-dimensional consolidation model COMPAC. Over the major portion of the Valley where geologic strata are flat and unbroken, the output from the consolidation model serves as the predicted subsidence at the specified point of interest.

The first two models simulate different directions of fluid flow, namely essentially horizontal flow for the groundwater model and essentially vertical flow for the consolidation model. The groundwater model primarily simulates the flow in permeable material (sands) whereas the consolidation model simulates the flow in the less permeable material (clays and coal seams).

For coupling purposes it is necessary for these two models to agree on stratigraphic interpretation of well bore data. For example, a groundwater model can afford to lump aquitards and confining beds together with negligible computational error whereas a consolidation model cannot. For vertical flow simulation, what constitutes isolated lenses (aquitards) within an aquifer system of hydraulically connected permeable material must be distinguished from what constitutes a separator (semi-confining bed) of low permeability that keeps permeable strata hydraulically separate from each other.

At specific sites in the Latrobe Valley, such as where fault zones or other geologic structures could dominate the direction and type of surface movement, a third model is being added. BITEMJ (Crotty, 1982) is a two-dimensional boundary element code which models equilibrium stresses and displacements within solids with structural discontinuities. For the present modelling effort, BITEMJ uses as input the vertical displacement output from the consolidation model. The base layer of this third-stage model has a horizontal distribution of vertical displacements imposed on it. At certain locations the land surface displacement, tilt and possibly shear are largely controlled by stratigraphy and site specific geologic structure within the passive material overlying the stressed aquifers. Examples that include this third-stage model are beyond the scope of the present paper.

4 APPLICATION OF COMPAC TO THREE SPECIFIC SITES

4.1 The Two Sites Near Driffield and Morwell

4.1.1 Field data

Settlement of the land surface versus time and in situ fluid pressures versus time data were analyzed at three sites. In addition to pressure data (Figure 2), two sites have only settlement data (Figures 3 and 4). One is near Driffield and the other is north and immediately adjacent to the Morwell Open Cut. A third site, namely the one near Loy Yang, has, in addition, vertical extensometer data (Figure 5) and will be discussed later.

Each location is assumed to have three aquifers (one unconfined and two confined). An upper separator is interpreted from stratigraphic data to lie between the uppermost confined aquifer system (Morwell Formation) and the overlying unconfined aquifer which has essentially a non-fluctuating water table at a depth of roughly ten metres beneath land surface. This convention for pressure within the unconfined aquifer conforms to the identical convention used in the groundwater model (Evans, 1983, 1986). The uppermost confined system is thin with no clay interbeds at Morwell, but at the Driffield site it is thick with five lenses of low permeability (aquitards). At each location a lower separator is interpreted to lie between the two confined aquifer systems. The lower confined system is interpreted to lie within the Traralgon Formation. DMID in Figures 3 and 4 represents the depth to the midplane of each compressing unit (two separators and two confined aquifer systems).

4.1.2 Calibration and interpretation

Table 1 summarizes the thickness estimates of aquitards and separators mentioned above. This table also indicates the average value of vertical hydraulic conductivity for the slow draining compressible beds. Near Driffield it was estimated from simulation of observed settlement at BM N94 (Figure 3) to equal about 4×10^{-4} m/yr. Near the Morwell Open Cut the simulation of observed settlement at PSM M71 (Figure 4) indicates that the

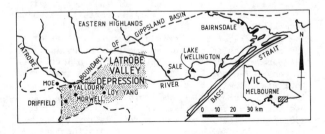

Figure 1. Location of Latrobe Valley

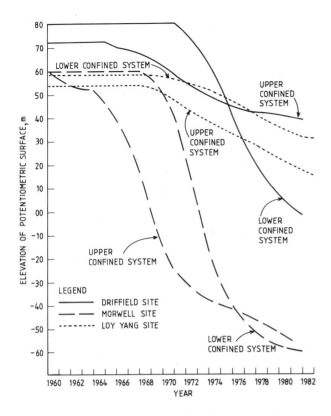

Figure 2. Potentiometric head within permeable zones versus time (measured from 1977).

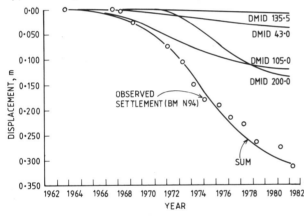

Figure 3. Computed vertical displacement versus time at the Driffield Site.

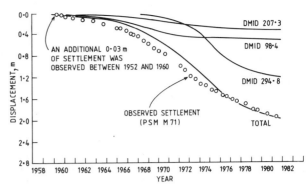

Figure 4. Computed vertical displacement versus time at the Morwell Site.

Table 1. Estimated parameter values near Driffield and Morewell.

Model Parameters		Upper Separator	Upper Confined Aquifer System	Lower Separator	Lower Confined Aquifer System	Total
		Driffield				
Thickness of Slowly Draining Beds	Cumulative Total, m	66	49	3	39.7	158
	Representative Individual Bed, m	66	14.1	3	6.3	
Vertical Hydraulic Conductivity, m/yr		4×10^{-4}	4×10^{-4}	4×10^{-4}	4×10^{-4}	
CR2		0.022	0.022	0.022	0.022	
CR3		0.2	0.2	0.2	0.2	
		Morwell				
Thickness of Slowly Draining Beds	Cumulative Total, m	163.9	0	51.8	105	321
	Representative Individual Bed, m	163.9	0	51.8	15	
Vertical Hydraulic Conductivity, m/yr		0.1	-	0.1	0.1	
CR2		0.019	-	0.019	0.019	
CR3		0.2	-	0.1	0.2	

average vertical hydraulic conductivity of the aquitards and separators may be nearly three orders of magnitude larger than the corresponding value at the Driffield site, namely 1×10^{-1} m/yr. The calibration technique selected (Helm, 1977, 1978) helps to overcome the non-uniqueness problem. The average compressibility values at the two sites, as indicated by the CR2 values in Table 1, are found to be within about 15% of each other, namely 0.022 and 0.019. The compression ratio CR is defined by the slope of equilibrium strain (namely change in thickness divided by the initial thickness) versus the logarithm of pressure to the base 10 evaluated over one log cycle of applied stress. CR3 represents compression along the virgin curve; CR2 along the recompression curve.

Actually, the average CR2 value that controls in situ behaviour at the Driffield site was found to lie between 0.019 and 0.023 with 0.022 being selected as a representative value. Similarly, the average CR2 value at the Morwell site was found to lie between 0.018 and 0.020 with 0.019 being selected as a representative value. The hydraulic

conductivity value that controls the aggregate vertical behaviour in situ at the Driffield site was estimated to lie between roughly 3×10^{-4} m/yr and 2×10^{-3} m/yr with 4×10^{-4} m/yr being selected as the most representative value. At the Morwell site, the hydraulic conductivity value that controls aggregate vertical behaviour lies between 1×10^{-2} m/yr to greater than 1 m/yr.

It is worth noting, by way of comparison that Evans (1983, 1986) independently estimated from the groundwater model that the regional vertical hydraulic conductivity of the lower separator is approximately 1×10^{-3} m/yr except beneath the Morwell Open Cut where possibly due to heave and fracturing it was required to increase with time by about three orders of magnitude.

A close look at Figure 4 shows that during the decade between 1963 and 1975, the calculated total subsidence (solid line) is smaller than the observed (circles). This might appear to suggest that the effective vertical hydraulic conductivity at the Morwell Open Cut may be larger in the field than the estimated value used in the calculations. With such an upward adjustment one would expect compression to occur earlier than it is now calculated to occur. When tried, this adjustment does not, in fact, noticeably improve the simulation.

Because fluid potential in the upper confined aquifer (Figure 2) declines earlier than it does in the lower confined aquifer, what is in fact required to improve the simulation at the Morwell site is to have the upper strata modelled as more permeable and the aquitards within the lower confined system modelled as less permeable. However, without field data from vertical extensometers to various depths at the Morwell site, it becomes a highly non-unique problem as to how to estimate and distribute any depth variation in vertical permeability. Data from two or three carefully placed extensometers at the Morwell site would have defined the vertical distribution of parameters much more neatly (see the discussion for the Loy Yang site).

4.1.3 Predictions

At the site near Driffield there are three predictions plotted in Figure 5. The first prediction (labelled "1" in Figure 5) is based on the 1982 observed potentiometric values in situ. An ultimate subsidence of nearly 0.4 m is predicted. Because about 0.32 m of subsidence has occurred (Figure 3), this implies an additional settlement of about 0.08 m of residual or latent subsidence. In other words, the COMPAC model indicates the system was about 80 percent consolidated in 1982.

The second subsidence prediction is based on SECV's groundwater model estimate of potentiometric surface elevations in the year 2020 that are associated with removing only the M1 coal seam. The third prediction requires removing also the M2 coal seam. Labels 2 and 3 in Figure 5 indicate a corresponding prediction of 0.75 m at the Driffield site associated with the removal of the M1 coal seam and 0.82 m associated with the removal of the M2 coal seam of which 0.32 m had already occurred in 1982.

An elevation of -60 m for Morwell aquifer potentiometric surface, is approximately the 1982 observed potentiometric surface at the Morwell site (Figure 2) and corresponds to the value predicted by the SECV 1983 groundwater model to occur in 2020 for the removal of the M1 coal. The groundwater model also predicted a 2020 elevation of -75 m at the Morwell site for Traralgon Formation potentiometric surface associated with removing the M1 coal. These combined stresses lead to 2.4 m of predicted ultimate subsidence of which 1.94 m had already occurred by July 1981. The eventual additional excavation of the deeper M2 seam would cause local subsidence of 4.0 m of which 1.94 m had already occurred.

In situ secondary effects and compression of sand are inherently lumped into the foregoing analyses of field data. It should be noted that all predictions in this paper require a semilogarithmic stress/strain relation. This corresponds to standard laboratory based theory. At each site, a 30 to 50 percent larger ultimate subsidence value would be predicted if a linear stress/strain relation is assumed for recompression of overconsolidated material (Helm, 1976).

4.2 The Site near Loy Yang

4.2.1 Field data

Vertical extensometer measurements of the type recommended in a previous section are available at the site near the Loy Yang Open Cut (Raisbeck, 1980). Three extensometers were installed in 1977. The deepest is to bedrock at a depth of 221 m which is assumed to be stable. Measurements of this deep extensometer are combined with subsidence measurements prior to 1977 at the survey benchmarks LY7 (1952-1975) and LY 049 (1975-1977). Two other extensometers record the aggregate changes in thickness from land surface to depths of 60 m and 130 m respectively. Figure 6 shows these field measurements using the 1977 calculated compression as the initial zero point for measured compression for the three extensometers. The stress versus time data from Loy Yang (Figure 2) used as input to the model are similar in quality to those used from Driffield and Morwell. Values after 1977 are measured. Values before 1977 are constructed from sparse data collected sporadically in the Latrobe Valley.

4.2.2 Calibration and interpretation

Figure 6 shows the computed consolidation versus time for the depth intervals 0-60 m, 60-130 m and 130-221 m. The model parameters used in the model for these calculated curves are listed in Table 2. Based on stratigraphic interpretation, the upper separator that lies between a depth of about 33 m

Table 2. Estimated parameter values near Loy Yang.

Model Parameters		Upper Separator	Lower Separator	Lower Confined Aquifer System		Total (m)
				Upper Portion	Lower Portion	
Thickness of Slowly Draining Beds	Cumulative Total, m	16.9	66.8	42.0	30	156
	Representative Individual Bed, m	16.9	66.8	7.0	30	
Total Depth Interval, m		0-60	60-130	130-191	191-221	
Vertical Hydraulic Conductivity, m/yr		5.5×10^{-4}	1×10^{-2}	3×10^{-3}		
CR2		0.059	0.021	0.00375		
CR3		0.72	0.26	0.047		

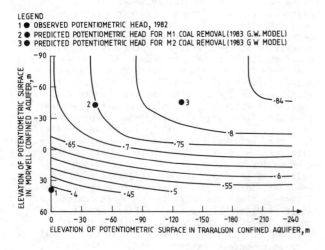

LEGEND
1 ● OBSERVED POTENTIOMETRIC HEAD, 1982
2 ● PREDICTED POTENTIOMETRIC HEAD FOR M1 COAL REMOVAL (1983 G.W. MODEL)
3 ● PREDICTED POTENTIOMETRIC HEAD FOR M2 COAL REMOVAL (1983 G W MODEL)

Figure 5. Predicted ultimate subsidence at the Driffield Site.

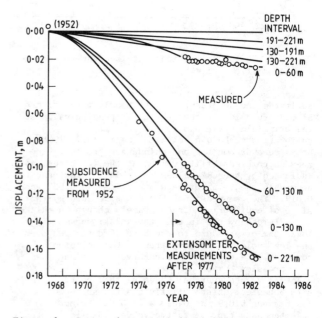

Figure 6. Computed vertical displacement versus time at the Loy Yang Site.

and 50 m is believed to be the primary compressible unit in the upper 60 metres. The lower separator that lies between a depth of about 61 m and 128 m is essentially the only compressible unit between the depth of 60 and 130 metres. The lowest depth interval (130 m - 221 m) was interpreted to comprise two types of compressible units. The upper portion (130 m - 191 m) consists of several slow-draining compressible lenses embedded within hydraulically interconnected permeable sandy material. These lenses have an aggregate thickness of 42 m. A representative lense is 7 m thick, calculated from stratigraphic data by a weight-averaging technique described by Helm (1975). This system of compressible lenses is modelled as double-draining aquitards. The lower portion (191 m - 221 m) is a single-draining compressible layer of 30 m thick, whose base rests on impermeable bedrock.

The upper separator was calibrated to be the least permeable and most compressible of all the units (Table 2). In spite of this low permeability, it was calculated to be nearly 100 percent consolidated by January 1983 (Table 2) for the anticipated long-term future stress on the system. The lower separator appears to be the most permeable as modelled whereas the compressible interbeds within the underlying confined aquifer system behave as though they are nearly incompressible. This is an unexpectedly low value for compressibility when compared to laboratory results on what is believed to be representative samples. There are several possible explanations. For example, the degree of vertical hydraulic connection of sand members within the lower aquifer system (Table 2) may possibly be less than that modelled. Correspondingly, all 72 m may not be compressing.

The computed consolidation versus time values for the three depth intervals discussed above are summed in two different ways, namely, for the total intervals, 0-130 m and 0-221 m, and are plotted in Figure 6 (solid line). Actual field extensometer data (circles) are also plotted in Figure 6 for comparison. The quality of simulation using the model with extensometer data (Figure 6) can be considered a significant improvement over the quality of simulation using only settlement data at land surface (Figures 3 and 4).

4.2.3 Prediction

Based on predicted drawdowns from the SECV's groundwater model for the year 2020, the predicted subsidence becomes slightly more than 0.28 m. Note from Figure 6 that about 0.17 m of subsidence has occurred at the Raw Coal Bunker by January 1983. This implies that only 0.11 m of additional subsidence can be expected.

5 CONCLUSIONS

The computer code COMPAC has been applied successfully to predict Latrobe Valley subsidence. It accounts for coupled hydro-mechanical transient processes within a unit column of compressible heterogeneous porous material. At the Loy Yang site where vertical extensometer measurements to different depths are available, the computer simulation was improved. If reliable early data, especially the 1960-1977 fluctuations of potentiometric head, had been collected locally for each of the regional aquifer systems at the selected key locations (namely, where computer simulations were made), the nonuniqueness problem for estimating average in situ permeability values would have been largely overcome (Helm, 1977, 1978). The estimated compressibility values at each site fell within a narrow range. The prediction of ultimate subsidence at each site can correspondingly be considered reliable.

6 ACKNOWLEDGEMENTS

The State Electricity Commission of Victoria made the present study possible, particularly John Alexander (former Chief Engineer of Fuel Department) and Robert McKenzie (former Chief Engineer of General Engineering and Construction Department). Richard Evans, Don Raisbeck, and Ian Pedler contributed significantly to the present effort. The assistance of James Jackson, Stephen Newcomb, John Brumley, Brian Newberry, Jeffery Washusen, Leo Fusinato, and Peter Learmonth of the SECV is gratefully acknowledged. The study was conducted during the author's sabbatical leave from the University of California.

REFERENCES

Brumley, J.C., C.M. Barton, G.R. Holdgate, & M.A. Reid 1981. Regional groundwater investigation of the Latrobe Valley, 1976-1981, State Electricity Commission of Victoria and the Victorian Dept. of Minerals and Energy.

Brumley, J.C. & M.A. Reid 1982. An investigation of the groundwater resources of the Latrobe Valley, Victoria, Australian Coal Geology 4, Part 2, pp. 562-581.

Crotty, J.M, 1982. User's Manual for Program BITEMJ: 2-D stress analysis for piecewise homogeneous solids with structural discontinuities. CSIRO, Australia, IEER, Div. of Appl. Geomech., Geomech. Computer Program No. 5.

Evans, R.S. 1983. The regional groundwater investigation of the Latrobe Valley, Stage 2 Mathematical Modelling Studies, Fuel Dept. Rept. DD 187, State Electricity Commission of Victoria, Melbourne.

Evans, R.S. 1986. A regional groundwater model for open cut coal winning in the Latrobe Valley, Victoria, in Groundwater systems under stress, Brisbane conference, Australian Water Resources Council, p. 459-468.

Harris-Galveston Subsidence District 1982. Water Management Study: Phase 2 and Supplement 1, Houston, Texas, 68 p. 160 exhibits.

Helm, D.C. 1975. One-dimensional simulation of aquifer system compaction near Pixley, Calif., 1) Constant parameters, Water Resources Research, V. 11, n. 3, p. 465-478.

_____ 1976. One-dimensional sumulation of aquifer system compaction near Pixley, Calif., 2) Stress-dependent parameters, Water Resources Research, V. 12, n. 3, p. 375-391.

_____ 1977. Estimating parameters of compacting fine-grained interbeds within a confined aquifer system by a one-dimensional simulation of field observations, in Land Subsidence (Johnson, A.I., ed.), Internat. Assoc. Hydrol. Sci. Publ. 121, p. 145-156.

_____ 1978. Field verification of a one-dimensional mathematical model for transient compaction and expansion of a confined aquifer system, in Verification of mathematical and physical models in hydraulic engineering, Proc. 26th Hydrau. Div. Specialty Conf., College Park, Maryland, Amer. Soc. Civil Eng., p. 189-196.

_____ 1984. Latrobe Valley Subsidence Prediction: The modelling of time-dependent ground movement due to groundwater withdrawal, 2 vols, Report Nos. DD195 and GD9, State Electricity Commission of Victoria, Melbourne.

Raisbeck, D. 1980. Settlement of Power Station structures in the Latrobe Valley, Victoria, Proceedings of the Third Australia & New Zealand Conference on Geomechanics, Wellington N.Z., V. 1, pp. 33-38.

Borehole hydraulic testing in a massive limestone
Mesure des propriétés hydrauliques des masses de roche calcaire depuis un trou de sondage
Hydraulische Prüfung in einem Bohrloch in massivem Kalkstein

R.J.HEYSTEE, Ontario Hydro, Toronto, Canada
K.G.RAVEN, Intera Technologies Ltd, Ottawa, Ontario, Canada
D.W.BELANGER, Intera Technologies Ltd, Ottawa, Ontario, Canada
B.P.SEMEC, Ontario Hydro, Toronto, Canada

ABSTRACT: A borehole hydraulic testing system has been developed to measure the hydraulic properties of relatively impermeable rock formations. It has been successfully used to measure hydraulic conductivities and to obtain estimates of formation pressures in a massive limestone. The generally low hydraulic conductivity (all values are less than 5 x 10^{-9} m/s) of this massive limestone as measured in a deep borehole is corroborated by a lack of visible ground water inflow in two nearby tunnels. The borehole test results and tunnel observations suggest that radioactive waste management underneath this rock formation or similar sedimentary rock units may enhance the safety of the repository and should be investigated further.

RÉSUMÉ: On vient de mettre au point une méthode de mesure des propriétés hydrauliques des formations de roche relativement imperméables. Cette méthode est utilisée avec succès pour évaluer le coefficient de perméabilité et la pression exercée à l'intérieur d'une masse de roche calcaire. Les coefficients de perméabilité généralement peu élevés (tous moins de 5 x 10^{-9} m/s) de cette roche calcaire obtenus lors des mesures effectuées dans un trou de sondage profond sont confirmés par l'absence de toute infiltration visible d'eau souterraine dans deux tunnels adjacents. Les mesures obtenues dans le trou de sondage et les résultats de l'observation dans les tunnels indiquent que l'enfouissement des déchets radioactifs sous cette formation de roche ou sous des formations de roche sédimentaire semblables pourrait accroître la sûreté du dépôt et que ce type d'enfouissement mérite ainsi une étude plus approfondie.

ZUSAMMENFASSUNG: Ein Verfahren zur hydraulischen Prüfung in einem Bohrloch ist entwickelt worden, womit die hydraulische Eigenschaften von verhältnissmässig undurchlässigen Felsgebilden gemessen werden können. Das Verfahren wurde erfolgreich verwendet, um die Durchlässigkeitsbeiwerte und die Bildungsdrücke in massivem Kalkstein zu messen. Die in einem tiefen Bohrloch gemessene meistens niedrige Durchlässigkeit (alle Werte sind < $5 \cdot 10^{-9}$ m/s) dieses massiven Kalksteins wird durch das Fehlen von sichtbarer Grundwassersickerung in zwei naheliegenden Tunneln bestätigt. Auf Grund der Versuchsergebnisse im Bohrloch und der Tunnelbeobachtungen ist anzunehmen, dass die Beseitigung von radioaktiven Abfällen unter diesem Felsgebilde oder ähnlichem Ablagerungsgestein die Sicherheit des Endlagers erhöhen dürfte und weiter untersucht werden sollte.

1.0 INTRODUCTION

Hydrogeologic investigations are being conducted by Ontario Hydro in thick sedimentary sequences. These investigations are part of a generic study evaluating the potential merits of disposal of nuclear fuel waste beneath a thick sedimentary sequence.

The focus of these hydrogeologic investigations has been the drilling and testing of two deep, 76 mm diameter boreholes. One of these boreholes is a 242 m long inclined (70° from the horizontal in a southerly direction) borehole (UN-2) on Ontario Hydro's Darlington Generating Station (GS) A property (Figure 1). The other borehole is a 389 m deep vertical borehole at Ontario Hydro's Lakeview GS site. The investigations that have been conducted in these boreholes includes the retrieval and logging of rock core, a television camera survey, acoustic televiewer logging and borehole geophysical logging. In addition to identifying the stratigraphy, these techniques were used to identify the location and characteristics of discontinuities intersected by the borehole. These investigations were followed by borehole hydraulic testing to determine the in situ hydraulic properties of the rock formations intersected by the borehole.

This paper focuses on the borehole hydraulic testing system, test procedures and the preliminary test results obtained to date in borehole UN-2.

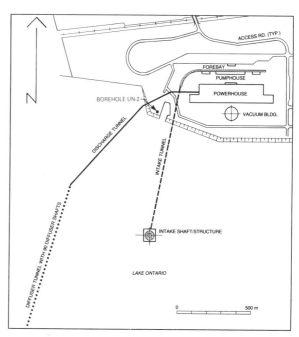

Figure 1. Location plan for Darlington GS. site.

2.0 GEOLOGY

Borehole UN-2 has intersected approximately 13.7 m of overburden, 208.3 of Paleozoic sedimentary rocks (194 m of which is limestone) and 20.0 m of Precambrian granitic gneiss. The Paleozoic limestones are overlain by a thin calcareous shale and are underlain with a very fine sandstone-siltstone unit (Semec, 1985).

The dominant joint set in the Paleozoic rocks, as determined by the aforementioned survey and logging techniques, occurs parallel to the bedding which is nearly horizontal. The spacing of these joints varies between one metre to more than three metres with the larger spacing occurring at depth. One dominant vertical joint set was found which strikes at N80°E to N90°E. All discontinuities were observed to be tight during core logging. However, several potentially permeable discontinuities were identified by the downhole survey and logging techniques. A comprehensive hydraulic testing program was conducted to quantify the permeability of these and other features in the borehole.

3.0 DESCRIPTION OF TEST EQUIPMENT AND OPERATION

A schematic diagram of the borehole hydraulic testing system is shown in Figure 2. The five main components of the system are (Waterra, 1986):

(1) Borehole straddle-packer probe;

(2) 610 m-long composite (3 hydraulic and 10 electrical conductors) cable;

(3) Trailer-mounted cable-reel hoist;

(4) Flow tank system; and

(5) Data acquisition system.

The borehole straddle-packer probe consists of an instrument pod and two steel-reinforced rubber packers (Figure 3). The packers inflate under hydraulic (water) pressure to provide a borehole seal approximately 0.78 metres in length. For the tests performed in borehole UN-2 during 1985, the two packers were joined by a spacer pipe to create a test zone length of 5.5 metres. During the 1986 testing program, the test zone length was 5.0 metres.

An instrument pod, which is connected to the top end of the upper packer contains the electronics pod. Within the electronics pod are three pressure sensors which are used to monitor the fluid pressure below (P1 Port), between (P2 Port) and above (P3 Port) the two packers. The electronics pod also protects the electrical connection for the thermistor which is situated in the test interval. The instrument pod also contains the hydraulically-operated valve and piston assemblies which are used to conduct the pressure-pulse and constant-head injection tests.

The operation of the valve and piston assemblies for both testing methods is illustrated in Figures 4a, 4b, and 4c. Figure 4a shows the valve in the relaxed (opened) position and the piston in the neutral, pulse testing position. To conduct a pressure pulse test, the valve is first closed by applying hydraulic pressure through the valve hydraulic line. This line is connected to Port B in the valve chamber (Figure 4a). When this line is pressurized to about 3 MPa at surface, the valve quickly closes to the position shown in Figure 4b. With the valve closed, the test zone is hydraulically isolated ('shut-in') from the open borehole above the upper packer and from the hydraulic lines in the composite cable.

The pressure-pulse test is conducted when the

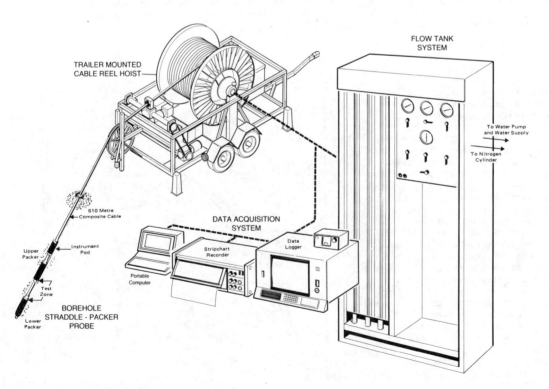

Figure 2. Ontario Hydro borehole hydraulic testing system.

132

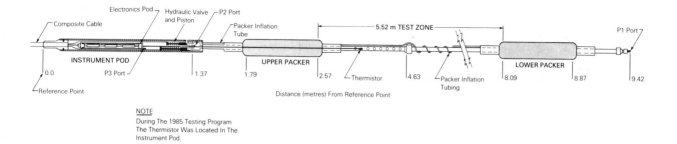

Figure 3. Borehole straddle - packer probe.

pressure in the test zone has stabilized. An instantaneous, positive or negative pressure pulse is applied to the test zone by displacing the piston. The positive pressure pulse is achieved by applying about 3 MPa pressure at surface through the piston hydraulic line. It, in turn, is connected to Port A in the piston chamber. The piston, within a period of one to two seconds, is displaced to its lower most position as shown in Figure 4b. By adjusting the piston locating screw it is possible to set the volume of water displaced by this movement at any amount from 0 mL up to 4.0 mL. A negative pressure pulse is obtained by removing the 3 MPa pressure at surface and allowing the piston to retract.

To conduct constant-head injection tests, the piston locating screw must be turned counter-clockwise several times to retract the piston above the O-ring seal as shown in Figure 4c. This operation must be done at the surface by mechanically adjusting the piston screw position. By applying pressure to the piston hydraulic line, water will flow into the piston chamber at Port A and flow past the piston to the test zone.

To measure formation (shut-in) pressures when conducting a constant-head test, the valve must be closed and the piston must be manually displaced below the O-ring seal and held in place by applying pressure to the piston hydraulic line. This will isolate the piston hydraulic line from the test zone and will permit rapid equilibration of the test zone pressure with the formation pressure. Alternatively, in more permeable test zones (hydraulic conductivities on the order of 10^{-9} m/sec or larger) this manual operaton can be avoided by simply leaving the piston in its upper most position. In this case, the shut-in procedure is performed with the piston hydraulic line open to the test zone. The hydraulic line must be closed at the surface (valve on flow tank) to permit the test zone pressure to equilibrate with the formation pressure.

4.0   DESCRIPTION OF TEST PROCEDURES

The hydraulic testing program in a borehole is completed in two phases:

(1)   reconnaissance - level testing phase, and
(2)   quality-assured (QA) testing phase.

Reconnaissance-level testing employs the pressure pulse technique and a relatively quick test procedure. The analysis of test results from this phase of testing provides an order of magnitude estimate of the hydraulic conductivity for each test interval along the entire uncased portion of the borehole. The primary objective of

reconnaissance-level testing is to identify all potentially permeable zones intersected by the borehole. These and some of the less permeable zones are then retested during the QA testing phase. QA testing differs from reconnaissance-level testing by the fact that each part of the test procedure generally has a longer duration. This permits water pressures and temperatures in the test interval to achieve a greater level of stability. This, in turn, leads to more reliable estimates of hydraulic conductivity and formation pressure. During QA testing either the pressure-pulse or constant-head injection test method may be employed. The former is typically applied in test intervals that have hydraulic conductivities of 1×10^{-9} m/s and smaller. The constant-head injection test method would be used in test intervals with hydraulic conductivities equal to 1×10^{-9} m/s and larger.

Figure 5 is a plot of the test interval pressure versus time for a pressure-pulse test conducted between 190 - 195 m depth in borehore UN-2. It is typical of many of the time-pressure plots for the pressure-pulse tests conducted in this borehole and it shows several steps in the test procedure.

Following the raising or lowering of the straddle-packer probe to the test interval, and prior to packer inflation, open-hole water-level measurements are taken for a period of thirty minutes. These measurements are needed to obtain baseline data from which pressure history effects can be determined and to compare pre- and post-test results for evaluating pressure sensor calibrations and test reliability. The time period during which open-hole measurements are taken permits thermal equilibrium to be achieved between the probe, borehole fluid and rock formation.

The packers are then inflated with the downhole valve in the open position. Approximately one hour is allowed for packer inflation. This ensures that the packers are fully inflated and that no pressure transients will be generated in the test interval due to packer 'squeeze'. This inflation time period also provides more time for the temperature equilibrium to be reached. The downhole valve is then closed and the test interval is isolated from the remainder of the borehole or 'shut-in'.

The shut-in phase of testing is the period of time after closure of the downhole valve and before hydraulic testing (by pressure-pulse or fluid injection). During this time period pressure and temperature conditions in the test interval are monitored until equilibrium conditions are obtained. Under actual test conditions a true equilibrium is seldom reached and small temporal changes in pressure and temperature are experienced. Provided that the resultant pressure

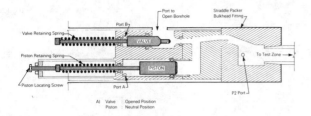

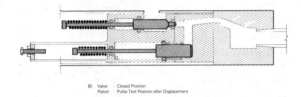

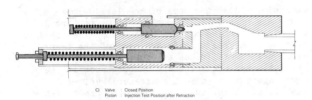

Figure 4. Operation of valve and piston assemblies.

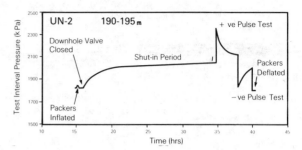

Figure 5. Typical test interval pressure history.

transients are small relative to the pressure transient generated by the pressure-pulse test, they are deemed to be acceptable. The shut-in time period is always longer than the pulse test time period. During reconnaissance-level testing the shut-in period generally lasted between 1 and 10 hours. During QA testing the shut-in period was always greater than 10 hours and often it lasted 15 to 50 hours. These time periods are generally suitable for obtaining acceptably stable pressure and temperature conditions for hydraulic testing during the reconnaissance-level and QA testing programs respectively.

During the pressure-pulse test the fluid pressure in the test interval is either increased (positive test) or decreased (negative test). The test interval pressure is than permitted to decay towards the shut-in pressure or formation pressure. During reconnaissance-level testing the pressure-pulse test normally was terminated after approximately 30 minutes if the pressure pulse decay did not exceed 10 percent of the peak pressure. For the borehole testing system used, this rate of pressure pulse decay corresponds approximately to a test interval hydraulic conductivity of less than 5×10^{-14} m/s. During QA testing the pressure-pulse decay period was longer and generally lasted several hours in relatively impermeable intervals. During QA testing the positive pressure pulse was followed by a negative pressure pulse (or vice versa) to check the repeatability of the test results.

During constant-head fluid injection tests, the fluid was typically injected from surface at three different pressures into the test interval. The volumetric flow rates were determined at surface using the flow tank system and monitored until equilibrium conditions were observed. The equilibrium fluid pressures were monitored by the downhole fluid pressure sensors.

All signals from downhole and surface pressure and temperature sensors were transmitted to a data logger and eventually stored on computer diskettes. The latter permits the easy retrieval, plotting and analysis of a large amount of test data.

5.0 TEST DATA ANALYSIS

The pressure-pulse test data were analyzed to obtain hydraulic conductivity estimates of each test interval. All pressure-pulse test results were analyzed by methods described by Cooper et al (1967) and Bredehoeft and Papadopoulos (1980). It involves the plotting of normalized test pressure values as a function of time on semi-log graph paper. This curve is fitted to a type curve to obtain match curve parameters. These are substituted into equation 8 of Bredehoeft and Papadopoulos (1980) to estimate the hydraulic properties of the rock mass at each test interval.

The results from the constant-head injection tests were analyzed assuming steady-state radial flow (Doe and Remer, 1981) to obtain an estimate of the test interval hydraulic conductivity.

Hydraulic test results from selected intervals were also analyzed by a comprehensive analytical model (Grisak et al, 1985; Intera Technologies Ltd., 1986). This analytical tool was developed to allow for variability in test conditions beyond the normal initial and boundary conditions inherent in type curve analytical methods. In the analysis of hydraulic tests, this analytical tool is able to quantitatively account for thermal induced pressure responses, borehole pressure history (resulting from a variable pressure 'skin' developed around the test interval) and dual porosity effects as well as equipment compliance. With this analytical capability it is possible to use all of the hydraulic test data (i.e. pretest, open hole, shut-in recovery, temperature, pulse test and repeat test data) to provide a complete interpretation of borehole hydrogeologic conditions.

6.0 PRELIMINARY TEST RESULTS

Figure 6 is a graphical summary of the preliminary hydraulic test results obtained to date in borehole UN-2. It represents the preliminary analysis of 56 pressure-pulse tests and two constant head injection tests that were conducted during 1985 and 1986. The hydraulic conductivity profile shown in Figure 6 accurately identifies the position of all relatively permeable zones intersected by borehole. It shows the estimated hydraulic head values for many of the relatively permeable zones. Also shown on this figure are hydraulic conductivity values obtained by testing five rock core specimens in a laboratory pressure-pulse test apparatus (Annor, 1985).

The borehole hydraulic test results indicate that borehole UN-2 intersects very low permeability rock formations. The majority of the test intervals are characterized by hydraulic conductivities below 1×10^{-13} and possibly 1×10^{-14} m/sec. However, seven intervals (square symbols on Figure 6) were identified to be relatively more permeable than the rest of the test intervals. The hydraulic conductivities in these seven test intervals range between approximately 5×10^{-9} and 5×10^{-13} m/sec. In general, these relatively permeable intervals correspond to borehole zones which had features identified by the television camera survey, borehole acoustic televiewer logging and/or borehole geophysical logging. However, it must be emphasized that all of these survey or logging techniques did not predict whether any of these seven zones were permeable. The one exception was the temperature and differential temperature logs which had anomalies that indicated potential fluid flow into or out of several borehole intervals.

Two of these relatively permeable intervals (86-91 m depth and 174-179 m depth) were subjected to 1 to 2 week slug withdrawal tests. The water-level recovery rates during these tests indicate that the permeabilities obtained by the shorter duration pressure-pulse tests may have over-estimated the hydraulic conductivities of these intervals by as much as one order of magnitude. This may be explained by the presence of a lower permeability boundary at some distance from the borehole. Alternatively, there may have been gas in the ground water which resulted in a more compressible fluid during the shorter-duration pressure-pulse test. The implications of this finding to the hydraulic conductivity estimates of these and the other relatively permeable intervals will be addressed.

The laboratory hydraulic conductivity values of rock core specimens are generally larger than the 'intact rock' borehole measurements at the same location from which the specimens were obtained. This finding is rather puzzling in light of the fact that the hydraulic conductivity measurements on rock core specimens are typically smaller than those measured in boreholes as a result of poor sampling of higher permeability rock discontinuities. This apparent discrepancy may be stress-related. The rock core specimens may have experienced stress relief when brought to ground surface. A portion of this stress relief may have been irreversible. Thus, the confining pressure applied in the laboratory test apparatus was not able to return the rock core specimen to its original in situ state and the rock core permeability may have been enhanced slightly. An alternative explanation may be that the borehole wall became plugged with drill cuttings. This, in turn, reduced the permeability of the rock adjacent to the borehole wall relative to the rock core permeability.

The computed hydraulic heads (assuming fresh water density) for three of the relatively permeable intervals below 170 m depth are generally above ground surface. The associated fluid pressures are most likely the product of increased salinity and density. As shown and discussed in Howard and Thompson (1985), brackish borehole water was indicated from borehole fluid resistivity surveys primarily below 170-175 m depth. Limited ground water sampling at 174-179 m depth indicated the presence of brackish waters with electrical conductivities greater than 50,000 $\mu S \cdot cm^{-1}$. Semec (1985) noted that rock core specimens from various depths below 175 m, had a layer of white powder form on the core surface when left exposed at ground surface. The powder had a salty taste and the analysis of the powder showed that it was

mainly calcium chloride. Its origin may have been the saline pore waters in the rock core specimens. Thus, it is most likely that many of the high fluid pressures observed below 170 m can be explained, in part, by density effects in the formation fluids. The one exception being the very large fluid pressure observed at 174-179 m depth. It is most likely that gas observed in this interval during the television camera survey (Semec, 1985) has had some effect on the measured fluid pressure.

7.0 DARLINGTON GS COOLING WATER INTAKE AND DISCHARGE TUNNELS

Nine metre diameter cooling water (CW) intake and discharge tunnels have been constructed at the Darlington GS site (Figure 1). The tunnels are 0.9 and 1.8 km in length respectively, and are located between 15 to 30 m beneath the bottom of Lake Ontario (Lo and Lukajic, 1984; Lukajic and Dupak, 1986). These tunnels are situated in the upper portion of the Middle Ordovician limestone of the Lindsay Formation (Figure 6).

As in borehole UN-2, observations of the discontinuity network during site investigations and the construction of the tunnels identified two major joint sets. One is vertical and oriented in an east-west direction and the other is oriented parallel to the near-horizontal bedding joints. Many of the joints in these two sets have been observed to be continuous over large distances. However, the joints (in particular the vertical joints) in the limestone were found to be tight at the excavation surfaces. This may be a ramification of high horizontal stresses (Haimson

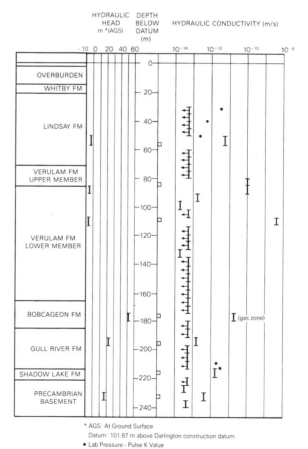

* AGS: At Ground Surface
  Datum: 101.87 m above Darlington construction datum.
● Lab Pressure - Pulse K Value
□ Hydrogeologically Significant Zone

Figure 6. Preliminary results of hydraulic tests performed in borehole UN-2.

135

and Lee, 1980) which have prevented the opening of these joints.

No major discontinuities such as faults have been observed to intersect the block of limestone rock bounded by the aforementioned tunnels and the nearby deep borings.

Visible water seepage from the limestone rock into the tunnels was not observed during the construction of the CW intake and discharge tunnels. It is most likely that the quantity of water that did enter the excavations was very small and was removed by the ventilation system as air moisture. This observation corroborates the relatively impermeable nature of the limestone as determined by hydraulic testing in the nearby borehole UN-2.

8.0   FUTURE WORK

The analysis and interpretation of hydraulic test results obtained to date in borehole UN-2 is of a preliminary nature. There will be further analysis of test results from selected intervals using the comprehensive analytical model described earlier. In addition, long-term (weeks to months) slug withdrawal tests will be conducted on the relatively permeable zones to obtain more representative values of hydraulic conductivity.

A multiple-level ground water pressure monitoring and sampling system has been installed in borehole UN-2. It will be used to measure the fluid pressure (and thus hydraulic head) distribution along the borehole and to obtain representative samples of formation fluids. The chemical and isotopic composition of the formation fluid samples will be determined. The hydrogeologic data obtained through the use of this monitoring system are necessary to interpret borehole hydraulic test results obtained to date. The data will also permit speculations to be made about the ground water flow system that may be operating in the vicinity of borehole UN-2.

9.0   CONCLUDING REMARKS

A borehole hyraulic testing system has been developed to measure the hydraulic properties of relatively impermeable rocks. The equipment consists of a straddle-packer probe which is attached to a 610 m-long composite cable. The cable, in turn, is mounted on a powered cable-reel hoist. The probe contains three pressure sensors and a thermistor, as well as hydraulically-operated valve and piston assemblies. The pressure and temperature signals are monitored by a data acquisition system. Using this equipment pressure-pulse and constant-head injection tests can be performed.

The analysis of the test results from borehole UN-2 has shown that the test interval hydraulic conductivities are generally less than 1×10^{-13} and possibly 1×10^{-14} m/s. At various depths in the borehole relatively permeable intervals in the rock have been identified. Many of these intervals correlated to apparently open discontinuities as observed in the TV camera survey and the acoustic televiewer logging results, as well to anomalies in the borehole geophysical logging results. These relatively permeable intervals have hydraulic conductivities which range between 5×10^{-9} and 5×10^{-13} m/s.

The generally low permeability of the limestone rock mass intersected by the borehole UN-2 is corroborated by observations in two nearby 9 m

diameter horizontal tunnels. No visible ground water inflow along rock mass discontinuities was observed in these tunnels. They are 0.9 and 1.8 km in length and are situated 15 to 30 m respectively beneath the bottom of Lake Ontario.[1]

The results from the borehole hydraulic tests and the observations in the tunnels indicate that further hydrogeologic studies of this and other similar sedimentary rock units may be warranted. These investigations would determine whether the safety of a radioactive waste management facility could be enhanced if it were placed beneath such a sedimentary sequence.

10.0   REFERENCES

Annor, A.B. 1985. Unpublished laboratory permeability test results from Canadian Department of Energy, Mines and Resources.

Bredehoeft, J.D. and S.S. Papadopulos 1980. A method for determining the hydraulic properties of tight formations. Water Resources Research V16(1): 233-238.

Cooper, H.H., Jr., J.D. Bredehoeft and S.S. Papadopulos 1967. Response of a finite-diameter well to an instantaneous charge of water. Water Resources Research V3(1): 263-269.

Doe, T. and J. Remer 1981. Analysis of constant-head well tests in nonporous fractured rocks. In 3rd Invitational Well Testing Symposium, March 26-28, 1980, Berkeley, CA. LBL Report 12076. p 84-89.

Grisak, G.E., J.F. Pickens, D.W. Belanger and J.D. Avis 1985. Hydrogeologic testing of crystalline rocks during the NAGRA deep drilling program. NAGRA Technischer Bericht 85-08.

Haimson, B.C. and C.F. Lee 1980. Hydrofracturing stress determinations at Darlington, Ontario. Proceedings 13th Canadian Rock Mechanics Symposium, Toronto, The Canadian Institute of Mining and Metallurgy, Montreal. p 42-50.

Howard, K.W.F. and M.J. Thompson 1985. Geophysical logging and hydrogeologic interpretation of Ontario Hydro boreholes at Niagara Falls and Darlington. Ontario Hydro Report 85295.

Intera Technologies Ltd 1986. A review and assessment of Ontario Hydro's borehole hydraulic testing program. Unpublished report to Ontario Hydro.

Lo, K.Y. and B. Lukajic 1984. Predicted and measured stresses and displacements around the Darlington intake tunnel. Canadian Geotechnical Journal V21(1): 147-165.

Lukajic, B. and D.D. Dupak 1986. Design and construction of Darlington cooling water discharge tunnel. Proceedings International Congress on Large Underground Openings, Firenze, Italy, June 8-11, 1986. p 238-243.

Semec, B.P. 1985. Darlington GS A, borehole UN-2 geological investigation. Ontario Hydro Report 85174.

Waterra Hydrogeology Inc. 1986. Development of borehole hydraulic testing system and hydraulic testing of borehole UN-2, Darlington Generating Station. Unpublished report to Ontario Hydro.

The creep characteristics of shale formation and the analysis of its loading on the oil well casing

Caractéristiques de fluage de formations schisteuses et analyse des charges conséquentes sur le tubage de forages pétroliers
Die Kriecheigenschaften einer Schieferformation und die Analyse ihres Druckes auf die Ölbohrlochverkleidung

HUANG RONGZUN, The Graduate School of East China Petroleum Institute, Beijing
ZHOU ZUHUI, The Graduate School of East China Petroleum Institute, Beijing
DENG JINGEN, The Graduate School of East China Petroleum Institute, Beijing

ABSTRACT: Improper waterflooding may increase the water content and the rate of steady-state creep of the shale formations. Based on the data of triaxial creep test, two nonlinear rheological models are recommended for the shales.To calculate the time-dependent formation load on the oil well casing a numerical method and a finite element method are studied.

RESUME: La technologie de l'injection d'eau indue peut accroître le contenu de l'eau et la vitesse du fluage de la modalité inchavirable dans les formations de la schiste.En vertu des données expérimentales du fluage sur trois axes, on présente deux modèles nonlinéaires rhéologiques qui sont applicables aux schistres. Afin de calculer la charge de la formation dependue du temps qui agit sur tubage de l'huile, on étude une solution numérique et une solution d'élément fini.

ZUSAMMENFASSUNG: Die unangemessene Wasserflut kann das Wassergehalt und die Geschwindigkeit des stabil-zuständli chen Schleichens von Schieferformationen erhöhen. Auf der Basis der Testdaten des triachsigen Schleichens werden zwei dem Schiefer entsprechende nicht-lineare rheologische Modelle empfohlen. UM die von der Zeit abhängige Formationslastung auf das Gehäuse des Ölbrunnens zu kalkulieren, hat man eine numerische Methode und eine finite Elementmethode studiert.

INTRODUCTION

Shales are composed of water-sensitive clay minerals and the change of its water content may give great influence on their mechanical properties. A steady high oil production may be achieved by water flooding operation.However if such operation had been done with too high pressure,the injected water would enter the shale formations through existing fractures, faults or weak cement rings around the casing. As a result, the stress and displacement field in the formation may be changed and the abnormal casing collapse will occur in a certain time after well completion.In this work the creep characteristics of shales with different water content were investigated under simulated in-situ stress conditions in laboratory.Based on the data of triaxial creep test, two non-linear rheological models are recommended for shales and the distribution of formation loading on the oil well casing has been studied by means of numerical analysis and finite element method for the uniform and non-uniform horizontal in-situ stresses respectively.

APPARATUS AND PROCEDURES

A triaxial creep testing apparatus was established in the ROCK MECHANICS LABORATORY of EAST CHINA PETROLEUM INSTITUTE (Fig.1). The investigated shale formations are in the depth of about 1000m of DAQING oil field. Relatively low formation temperature (about 60 °c) makes it possible to neglect its influence on the creep behavior of shales.Only the stress level and water content are considered as two important parameters in testing program.
This apparatus consists of triaxial pressure chamber 1, confining pressure control system 2, hydraulic testing machine 3, constant axial load preserving system 4 and the data acquisition system 5. The triaxial chamber may accommodate specimens with diameter of up to 100 mm and length of 250mm. The maximum operating pressure for the chamber is 1400 bar. The constant confining pressure is controlled by a given dead weight, which balances the pressure in

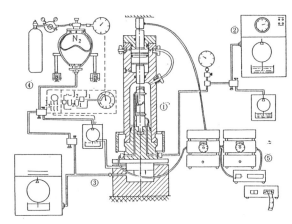

Figure 1.Apparatus for triaxial creep test

the chamber to within ± 2.5 bar.To maintain the axial load on the specimen constant, a large compensator with air dome is connected to the hydraulic system of testing machine,oil leakage of which is compensated by using an electro-contact pressure gauge to switch a supplementary pump on or off. The fluctuation of axial load may be controlled within ±2%. Axial load and confining pressure are measured by a strain gauge-type load cell outside the chamber and an in-line pressure gauge respectively. The axial and lateral strain of the specimen with less water content are measured by strain gauges,attached on the circumferential surface of specimen mid-height.For the specimens of larger water content the axial deformation is only measured by displacement transducers mounted outside the chamber. The signals of stress and strain transducers are amplified,displayed and recorded by an electric wire strain indicator and a digital volt-meter.The error of collected data is about ±1.7%.
In the laboratory,shale specimens were prepared from field cores and were saturated to a given water content w(%)with distilled water in tri-axial chamber with an additional equipment under simulated in-situ

137

stress conditions. For testing, the specimens were jacketed with 1.5 mm thick rubber tubing. The creep test data were collected after stress application at one minute interval and gradually spaced out in time as the test continued. After a constant rate of creep deformation had been observed and maintained for approximately 20 hours the test was stopped or turned to the next test with a new level of stress.

RESULTS AND DISCUSSION

The creep tests of more than three stress stages were made for each given water content of a specimen. Typical test results depicting axial creep strain as a functon of clock time are plotted in Figure 2 and 3. The rate of axial and lateral creep strain decreased with time until the steady- state creep was observed during every stage of testing program.

Shale cores were taken from 3 wells of DAQING oil field at the depth of 780-1200m.Six samples were prepared from these cores, but only the data of 3 samples are given in Fig.2 and 3. The clay mineral analysis by X-ray diffraction indicates that in accordance with the water-sensitivity the sample N.44 is the most water-senitive, N.31 is less sensitive, and the sample N.16 and N.7 are the least sensitive.

Analysis of test data indicates that for different stress stages of each specimen water content the rate of steady-state creep was proportional to the Nth power of the differential stress$(\sigma_1-\sigma_3)$. According to the form of obtained creep curves, test data were fitted to a modified non-linear Burger's model:

$$\varepsilon_1 = \frac{\sigma_1-\sigma_3}{K_2}+[\frac{1}{K_1}(1-e^{-\frac{K_1 t}{J_1}})+\frac{t}{J_2}](\sigma_1-\sigma_3)^N \quad (1)$$

where: ε_1 ----axial strain
       t----time (hour)
       σ_1 ----axial stress (MPa)
       $\sigma_2=\sigma_3$----confining pressure (MPa)
       N----constant
       K_1,K_2 and J_1,J_2 ----visco-elastic parameters of
            material

In further analysis,it has been found that for the simplicity of calculating the formation loading distribution on well casing,a modified non-linear Maxwell's model(2) may be used in place of equation(1)

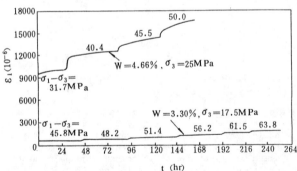

Figure 2. Creep curves for specimen N.16

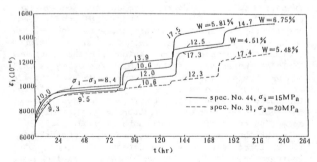

Figure 3. Creep curves for specimen N.31 and N.44

with sufficient accuracy(see Fig.4)

$$\varepsilon_1 = \frac{\sigma_1-\sigma_3}{K}+\frac{(\sigma_1-\sigma_3)^N}{J}t \quad (2)$$

where:K,J---- visco-elastic constants

Based on the research for the effect of water content on the visco-elastic properties of the specimens, parameters K and J are determined to be proportional to exp(-aw) and exp(-1/(w₀-w))respectively.So equation (2) may be written as:

$$\varepsilon_1 = \frac{\sigma_1-\sigma_3}{A}\exp(aw)+\frac{(\sigma_1-\sigma_3)^N}{B}\exp(1/(w_0-w))t \quad (3)$$

where:w----water content of shale (%)
     a,A,B----constants
     w_0----water content at which the rate of steady-
         state creep approaches infinity,w_0 is equale
         to 8.5-9.0(%) for shales tested.
The values of all the parameters N,K_1,K_2,J_1,J_2,K,J and constants a,A,B were determined from test data by means of multiple non-linear regression method and were listed in Table 1 and 2.

It is evident from Table 1 and 2 that the value of N decreases with increasing water content and it is greater for less water-sensitive shales.

Table 1.The creep parameters of shales N.16 and N.7

Shale No.	w (%)	N	In equation(1)				In equation(2)	
			K_1	K_2	J_1	J_2	K	J
16	3.30	5.74	.034 $*10^{16}$	5.2 $*10^4$	1.12 $*10^{16}$	2.44 $*10^{16}$	5.2 $*10^4$	8.32 $*10^{21}$
	4.66	3.42	.998 $*10^9$	.42 $*10^4$	2.28 $*10^9$	3.30 $*10^9$	.42 $*10^4$	1.41 $*10^8$
7	11.5	3.54	.25 $*10^9$	.12 $*10^4$	.04 $*10^6$	26 $*10^6$	.12 $*10^4$	3.09 $*10^{10}$

Table 2.The creep parameters of shales N.31 and N.44

Shale No.	w (%)	N	parameters in equation (3)			
			A	B	a	$w_0(\%)$
31	5.48	1.59				
	6.24	1.48	$1.785*10^4$	$1.23*10^7$	.2405	8.5
	7.26	1.35				
44	4.51	1.38				
	5.81	1.20	$1.005*10^4$	$2.90*10^6$	.2502	9.0
	6.75	1.11				

ANALYSIS OF FORMATION LOADING ON THE OIL-WELL CASING

In petroleum industry,the hydrostatic fluid pressure in the borehole before well completion has been considered as the only force applied on the casing in design for many years.Recently,the overburden pressure has been adopted as a formation loading on the casing in salt dome wells. The authors consider that the time-dependent external load produced by the creep deformation of rheological stratum must be taken into account in the casing design.

The theoretical solutions of formation loading on the casing may be found for linear rheological models. They were derived by using Laplace transformation method(Zhou,1984) for the uniform horizontal in-situ stress condition, and for the non-uniform in-situ stresses the solutions can be found in the reference (Yu et al.,1983).

For finding the solutions of non-linear models as equations(1)-(3),the following two methods are recommended.

For the case of non-linear Burger's model(1) and for uniform horizontal in-situ stresses,according to the principle of force equilibrium and displacement compatibility at the contact of casing with rock formation (including cement ring) and by using the convolution integral method to determine the stress history-dependent total creep strain at the moment t, an integral equation (4) can be derived for the numerical analysis of formation loading on the well casing.(Zhou et al.,1986).

$$[p_R(t)-p_i][1/K_e+1/K_r]=R \cdot A \cdot$$
$$\int_0^t (\exp(-K_i(t-s)/J_i)/J_i +1/J_2)[p_o-p_R(s)]^N \cdot ds \qquad (4)$$

Where: $p_R(t)$ ---time-dependent external pressure on the casing(MPa)

t ---time(hour)

p_i ----borehole fluid pressure before well completion(MPa)

p_o ---uniform horizontal in-situ stress(MPa)

2R---outer diameter of casing(cm)

$A=(3)^{\frac{N-1}{2}} N^{-N} (1+\mu_c)$

μ_c ----ratio of the lateral to the axial creep strain rate, $\mu_c = -\dot{\varepsilon}_3/\dot{\varepsilon}_1$,for the creep of non-volumetric change $\mu_c=0.5$

$K_r = E_r h/R/R/(1-\mu_r^2)$ defining the resistant stiffness of casing to external pressure

$K_e = E/R/(1+\mu)$ defining the resistant stiffness of the borehole wall to radial deformation

E,E_r ---Young's moduli of rock formation and casing material respectively (MPa)

μ,μ_r ---Poisson's ratios of rock formation and casing material respectively

h---thickness of well casing (cm)

For non-linear Maxwell's model(2) the integral equation can be written as:

$$[p_R(t)-p_i][1/k_e+1/K_r]=R \cdot A \int_0^t [p_o-p_R(s)]^N/J \cdot ds \qquad (5)$$

The results of computed $[p_R(t)-p_i]$ are plotted as a function of time t in Fig.4. The curves 1 and 2 in Fig.4 correspond to equation(4) and (5) respectively for the shale N.16 of water content W=4.66%. In the calculation, the values of the following parameters were taken as:R=6.985cm,h=0.772cm,E_r=2*10^5MPa, μ_c=0.5, p_i=13 MPa, p_o=24.2 MPa (depth of well=1000m), $\mu = \mu_r$= 0.3.Data of creep parameters were taken from Table 1 and 2.Good coincidence of these two curves shows that it is reliable to use the simpler non-linear Maxwell's model for further analysis.The curves 3 and 4 in Fig.4 are plotted with data of sample N.31 for w=4.58% and w=8.2% respectively.From these curves it is evident that the external pressure on the casing increases with time and approaches to a value equal to the uniform horizontal in-situ stress p_o in a certain time after well completion,which is shorter for a greater water content of the same shale formation.

For the case of non-linear Maxwell's model(3) and for non-uniform horizontal in-situ stresses, it is convenient to solve the problem by using finite element method,for which a computer program has been developed for solving plane visco-elastic creep problems by using the initial viscous strain method. In computation,the parameter values were taken as before, but the maximum and minimum horizontal in-situ stresses at the depth of 1000m σ_{h1} and σ_{h2} were assumed to be 27.5 MPa and 20.9 MPa respectively, For this case,the external casing pressure is not constant at all points of casing section.It varies as a function of cosine 2θ , where θ is the vectorial angle of the inspected point of casing section,relative to the maximum horizontal in-situ stress σ_{h1} .The computed $[p_R(t)-p_i]$ for the points of θ =0° and 90° were given in Fig.5.It can be seen from Fig.5 that at the beginning, the external casing pressure of all points increases with time,but after a specific time, which is also shorter for a greater water content, different tendencies of time-dependent pressure change occur. In the direction of minimum in-situ stress σ_{h2} (θ =90°), casing pressure tends to decrease with time and becomes even less than σ_{h2} ,and lastly remains stable at about 0.71σ_{h2} .At the same time,in the direction of maximum in-situ stress ($\theta = 0°$) the external casing pressure continues to increase,but with decreasing rate,grows even greater than σ_{h1} ,and at the

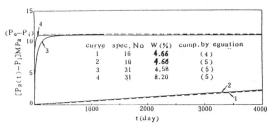

Figure 4. Formation creep loading on well casing for uniform in-situ stresses computed by numerical method

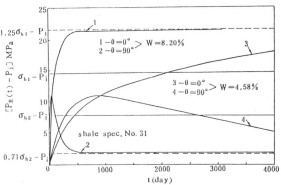

Figure 5. Formation creep loading on well casing for non-uniform in-situ stresses by finite element method

last,approaches to 1.25σ_{h1}.This fact strongly demonstrated that,as a result of formation creep the degree of non-uniformity of external casing pressure would be much more severe than that of in-situ stresses.

The load bearing capability of casing is greatly lowered by the non-uniformity of external pressure. According to the results of Nester J.H.(Nester et al., 1955),for the case investigated in Fig.5(the diameter/ thickness ratio of casing is 18.0,the final difference between the maximum and minimum external casing pressure is about 19.5 MPa),the load bearing capability under non-uniform in-situ stress is lower than that under uniform in-situ stress by a factor of 1/8. It leads to the fact that all kinds of existing API standard casing are unable to sustain the creep loading of this shale formation as shown in Fig.5.

The time to reach the stable value for either the maximum or the minimum external casing pressure is greatly influenced by the water content of shales.For w=8.20% of shale No.31 this time lowered approximately by 22 folds as compared with that for w=4.58 % (see Fig.5).Therefore the service life of an oil-well casing will be greatly lowered as a result of increasing the water content of shale formations.

All these, probably, can give a new concept for analysing the mechanism of some abnormal casing collapse in the area where an improper water flooding operation has been taken,and may provide a new method for improving the design of oil-well casing set through the creep formations.

REFERENCES

Zhou Zuhui . 1984. Creep of Shales and Primary Analysis of Formation Loading on the Casing. M.Sc. Thesis of East China Petroleum Institute.

Yu Xuefu,Zheng Yingren,Lu Huaiheng & Fang Zhengchang. 1983.The Rock Stability Analysis in Underground Engineering.Chinese Coal Industry Press.

Zhou Zuhui,Huang Rongzun & Zhuang Jinjiang.1986.The Effect of Creeping Rock Formations on the Time-dependent External Pressure of an Oil-well Casing. Chinese J.of Rock Mechanics and Engineering.5(2).

Nester J.H.,Jenkins D.R.& Simon R.1955.Resistances to Failure of Oil-well Casing Subjected to Non-uniform Transverse Loading,Drilling and Production Practice.

Analysis of ground response curves for tunnels including water flow, Hoek-Brown failure criterion and brittle-elastoplastic rock behaviour

Analyse des déformations autour d'un tunnel en fonction des infiltrations d'eau, du critère de rupture de Hoek-Brown et du comportement fragile-élastoplastique du rocher

Analyse von Tunnelreaktionen im wasserführenden Gestein mit dem Hoek-Brown Bruch Kriterium für sprödes/elastoplastisches Gesteinsverhalten

F.A.IZQUIERDO, Polytechnical University of Valencia, Spain
M.ROMANA, Polytechnical University of Valencia, Spain

ABSTRACT: The paper presents an analysis of stresses and ground response curves around a circular tunnel, introducing a seepage and using the Hoek-Brown failure criterion, a perfect elasto-brittle-plastic rock behaviour and a treatment of plastic strains according to an associated flow rule.

RESUME: L'article présent une analyse des contraintes et des curves caracteristiques autour d'une cavité circulaire, en tenant compte d'une venue d'eau et en utilisant le critére de rupture de Hoek-Brown, un comportement elastoplastique de la roche et en supossant que les déformations plastiques dérivent d'un potential plastique.

ZUSAMMENFASSUNG: Die Mitteilung präsentiert eine Analise der Spannungen und Dehnungskurven um einen Kreisförmigen Tunnel herum mit Durchströmungen, indem man das Bruckkriterium von Hoek-Brown benutzt hat ein Sprödbruch des Felsen und man angenommen hat, daß die Spannungsfunktion von einem plastischen Potential abhängt.

1 INTRODUCTION

The problem of calculating the "Ground Response Curves" or "Characteristic lines" is related with the problem of obtaining the stresses and strains around the tunnel. Due to its complexity, there are several analytical solutions in the case of circular tunnel, hydrostatic stress-field and plane strain, depending on the failure criterion, the treatment of the plastic strains and the ground behaviour model. Being a tridimensional problem, studies on the face are very scarce: Lombardi(1970) assumes it to be a "nucleus" in equilibrium before the excavation; Amberg and lombardi(1974) present a pseudo-tridimensional method to evaluate its influence; the solution proposed by Egger(1980) of evaluating the face as a semi-spherical cavity is quite simple.

With seepage the problem becomes much more complicated. Jimenez-Salas (1981) evaluated the "plastic zone radius" using the concept of "influence radius". More over, Adachi and Tamura (1978) have developed for the Seikan tunnel an elastoplastic model with a coulombian material to study the effects of water and the effectiveness of a ring of drains or/and grounting ground around a circular tunnel, admitting Terzaghi's and Darcy' laws.

This paper shows the results of an analysis made in order to evaluate stresses and strains in the tunnel cross section and in the face (Izquierdo,1984).

2 STATEMENT OF THE PROBLEM

A simple way of introducing seepage and heterogeneities consists of idealizing the real problem (figure 1) by a system of concentric rings which can have different properties. The simplest model -but allowing to study the influence of many parameters- consists of two rings: the inner one is plastic and the outer elastic (figure 2).

3 BASIC ASSUMPTIONS

* The tunnel is circular with radius R.
* The problem has cylindrical (spherical) symmetry for the tunnel cross section (face).
* The geometric heigh is not considered and the

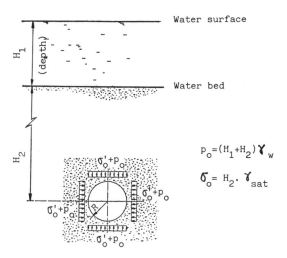

$$p_o = (H_1 + H_2)\gamma_w$$

$$\sigma_o = H_2 \cdot \gamma_{sat}$$

Figure 1. Scheme of the real problem

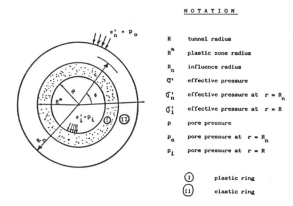

NOTATION

R	tunnel radius
R^n	plastic zone radius
R_n	influence radius
σ'	effective pressure
σ'_n	effective pressure at r = R_n
σ'_i	effective pressure at r = R
p	pore pressure
p_o	pore pressure at r = R_n
p_i	pore pressure at r = R

Ⓘ plastic ring
ⒾⒾ elastic ring

Figure 2. Scheme and notation of the model developed.

stress-field is hydrostatic (σ_o).
* The radius of the outer surface of the model is called "influence radius" and it is there where the pore pressure (p_o) is applied.

* Each ring can be admitted as a continuous medium. The material is isotropic, homogeneous, without viscosity or weight, with permeability (k) constant and it is assumed to be linear-elastic and characterised by constant Young's modulus (E) and constant Poisson ratio (ν). The failure characteristics of the material are defined by:

$$\sigma_1' = \sigma_3' + \sqrt{m\sigma_c\sigma_3' + s\sigma_c^2} \quad \text{(peak)} \quad \dots \dots \quad (1)$$

$$\sigma_1' = \sigma_3' + \sqrt{m_r\sigma_c\sigma_3' + s_r\sigma_c^2} \quad \text{(residual)} \dots. \quad (2)$$

where σ_1' and σ_3' are the major and minor principal effective stresses, respectively, m and s are constants which depend upon the properties of the rock mass and they can be esteemed following the quality ratios given by any geomechanic classification (Hoek - Brown, 1980), and σ_c is the uniaxial compressive strength.

When strength reachs its peak it falls down to the residual value (figure 3). Herein the analysis is made for a brittle-elastoplastic behaviour, being the perfect elastoplastic behaviour a particular case of this one: it is sufficient to substitute in the "brittle" equations the original ground parameters (without subindex "r") (do b'=b'' in figure 3).

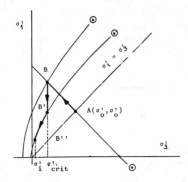

NOTATION

A : initial state of stresses

(a) : evolution of the stresses with an elastic behaviour

B : elastic stresses in the common surface

B': plastic stresses in the common surface

(b) : limit of the peak strength

(b') : limit of the residual strength

B'': stresses in the tunnel outline

Figure 3. Stresses in the brittle-elastoplastic case

* Terzaghi's and Darcy's laws are always fulfilled.
* Plastic strains derive from a plastic potential given by:

$$P = (\sigma_\theta' - \sigma_r') - (\sigma_\theta' + \sigma_r').\text{sen}\Psi \quad (3)$$

being σ_θ' and σ_r' tangential and radial effective stresses, respectively. We call "dilatation parameter", α, to:

$$\alpha = \frac{1 + \text{sen}\Psi}{1 - \text{sen}\Psi} \quad (4)$$

and it is constant.

The differential equations to be solved are:

Cross section (cylindrical case)

$$-\frac{d\dot{u}}{dr} + \alpha\frac{\dot{u}}{r} = \frac{1+\nu}{E}[\sigma_r'(1-\nu-\alpha\nu)+\sigma_\theta'(\alpha-\alpha\nu-\nu)-\sigma_o'(1-2\nu)(1+\alpha)] \quad (5)$$

Face (spherical case)

$$-\frac{d\dot{u}}{dr} + \alpha\frac{\dot{u}}{r} = \frac{1}{E}[\sigma_r'(1-\alpha\nu)+\sigma_\theta'(\alpha-\alpha\nu-2\nu)-\sigma_o'(1-2\nu)(1+\alpha)] \quad (6)$$

being $\dot{u}$ the radial displacement increment.

4 ELASTIC SOLUTIONS

For the elastic case ($R \geq R_n$), the solutions are:

Cross section (cylindrical case)

$$\sigma_r' = A - \frac{B}{r^2} - \frac{\Delta^*}{1-\nu}\ln(r/Rn)$$

$$\sigma_\theta' = A + \frac{B}{r^2} - \frac{\Delta^*}{1-\nu}[\ln(r/Rn)+2\nu-1]$$

$$\dot{u} = \frac{1-\nu-2\nu^2}{E}\{(A-\sigma_o')+\frac{B}{(1-2\nu)r}-\frac{\Delta^* r}{(1-\nu)}[\ln(r/Rn)+\nu-1]\}$$

(7)

Face (spherical case)

$$\sigma_r' = A - \frac{2B}{r^3} + \frac{\nu}{1-\nu}\cdot\frac{2\Delta^*}{r}$$

$$\sigma_\theta' = A + \frac{B}{r^3} + \frac{2\Delta^*}{(1-\nu)r}$$

$$\dot{u} = \frac{1-\nu-2\nu^2}{E}[\frac{(A-\sigma_o')}{(1+\nu)}r + \frac{B\,r^{-2}}{1-2\nu} + \frac{\Delta^*}{1+\nu}]$$

(8)

in which A, B are constants that depend on the boundary conditions; Q is the flowrate; $\Delta p = p_o - p_i$; $\eta = R_n/R$. Seepage is controlled by the parameter:

Cross section $\quad \Delta^* = \frac{Q\gamma_w}{4\pi k} = \frac{\Delta p}{\ln(\eta)}$ (9)

Face $\quad \Delta^* = \frac{Q\gamma_w}{4\pi k} = \Delta p/(\frac{1}{R} - \frac{1}{Rn})$ (10)

From these equations it is found that:
* When seepage occurs, A and B can be represented by two components, the firts one due to the excavation (A_e and B_e) and the second to the seepage (A_f and B_f):

$$A = A_e + A_f \qquad B = B_e + B_f \quad (11)$$

where:

Cross section

$$A_e = \frac{\sigma_n'\eta^2 - \sigma_i'}{\eta^2-1}$$

$$A_f = \frac{\Delta p}{2(1-\nu)(\eta^2-1)}$$

$$B_e = \frac{\sigma_n' - \sigma_i'}{\eta^2 - 1}R_n^2$$

$$B_f = \frac{\Delta p\, R_n^2}{2(1-\nu)(\eta^2-1)}$$

(12)

Face

$$A_e = \frac{\sigma_n'\eta^3 - \sigma_i'}{\eta^3 - 1}$$

$$A_f = -\frac{\nu.\Delta p.\eta.(1+\eta)}{(1-\nu)(\eta-1)(\eta^2+\eta+1)}$$

$$B_e = \frac{(\sigma_n' - \sigma_i')R_n^3}{2\eta^3 - 1}$$

$$B_f = \frac{\nu\,\Delta p}{2(1-\nu)(\eta-1)(\eta^2+\eta+1)}$$

(13)

* In figure 4 the distributions of the elastic effective stresses have been drawn (curves without seepage are represented with a thinner line). If seepage increa-

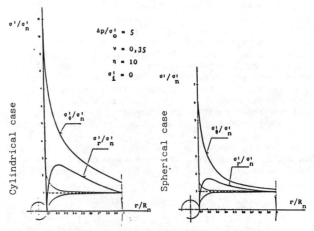

Figure 4. Distibutions of the elastic effective stresses. Seepage influence.

ses (given by the value of p_b/σ_b') the tangential stresses increase and these curves are distorted presenting a radial stress maximun.

5 PLASTIC SOLUTIONS

5.1 Stresses

If $\sigma_1' = \sigma_\theta'$ and $\sigma_3' = \sigma_r'$ the Hoek-Brown criterion is:

$$\sigma_\theta' = \sigma_r' + (m_r \sigma_c \sigma_r' + s_r \sigma_c^2)^{\frac{1}{2}} \qquad (14)$$

If we define the variable Y as the second component of this expression (the square root is defined
as positive as the tangential effective stress is the
major principal stress), the differential equations
to be solved are:

Cross section   Face

$$\frac{dY}{dr} = \frac{a1}{r}\left(1 - \frac{2\Delta^*}{Y}\right) \quad (15) \qquad \frac{Y}{a1}\frac{dY}{dr} - \frac{2Y}{r} + \frac{2\Delta^*}{r^2} = 0 \quad (16)$$

where:

$$a1 = m_r \sigma_c / 2 \qquad (17)$$

$$\Delta^* = \frac{Q\gamma_\omega}{4\pi k_r} \qquad (18)$$

For the cross section tunnel, the solution is:

$$Cr = (Y - 2\Delta^*)^{\frac{2\Delta^*}{a1}} \cdot \exp(Y/a1) \qquad (19)$$

where C is a integration constant which can be obtained making $Y=Y_i$ at $r=R$. Substitution for its value
gives the following equation:

$$\left(\frac{r}{R}\right)^{a1} = \left[\exp(Y-Y_i)\right] \cdot \left[\frac{Y-2\Delta^*}{Y_i-2\Delta^*}\right]^{2\Delta^*} \qquad (20)$$

Being $2\Delta^*/a_1$ a real number, any curve of the family (20) exists only if $Y \geqslant 2\Delta^*$. In any other case,
the factor of (20) would be a potential function with
negative base.

If there is no seepage ($Q=0$, $\Delta^*=0$) the solution
is:

$$(r/R)^{a1} = \exp(Y-Y_i) \qquad (21)$$

and substituting each term, the solution of Hoek-
Brown is obtained:

$$\sigma_r' = \frac{m_r \sigma_c}{4} \cdot \ln^2(r/R) \cdot \sigma_i' + (m_r \sigma_c \sigma_i' + s_r \sigma_c^2)^{\frac{1}{2}} \cdot \ln(r/R) + \sigma_i'$$
$$\sigma_\theta' = \sigma_i' + (m_r \sigma_c \sigma_r' + s_r \sigma_c^2)^{\frac{1}{2}} \qquad (22)$$

In figure 5, curves Y versus $(r/R)^{a1}$ are drawn for
different values of $2\Delta^*$ if $Y_i=1$. It can be concluded
that:

1. For a certain value of r, an increase in seepa-

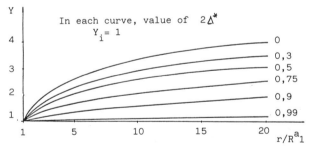

Figure 5. Cross section tunnel. Distribution of the
plastic stresses. Hoek-Brown failure criterion.

ge (Δ^*) causes Y to disminish, consequently decreasing σ_r' and increasing the plastic zone radius.
2. If $Y_i=2\Delta^*$ a critic state is produced, with
constant stresses and an infinite plastic zone radius.
3. If $2\Delta^* > Y_i$ there is no solution for stresses.
For the face without seepage, the solution is:

$$Y = 2 a1 \ln(Cr) \qquad (23)$$

where C is a integration constant. If $Y=Y_i$ at $r=R$,
the stresses are:

$$\sigma_r' = \sigma_i' + 2(2 a1 \sigma_i' + s_r \sigma_c^2)^{\frac{1}{2}} \ln(r/R) + 2 a1 \ln^2(r/R)$$

$$\sigma_\theta' = \sigma_r' + (m_r \sigma_c \sigma_r' + s_r \sigma_c^2)^{\frac{1}{2}} \qquad (24)$$

The critical effective pressure, σ_{crit}', (plastic
ring will exists only if $\sigma_i' \leqslant \sigma_{crit}'$) is given by:

$$\sigma_{crit}' = \sigma_o' - M \sigma_c \qquad (25)$$

being

$$M = 2/3 \left[\left(\frac{m^2}{9} + \frac{m\sigma_o'}{\sigma_c} + s\right)^{\frac{1}{2}} - \frac{m}{3} \right] \qquad (26)$$

The plastic zone radius, R^*, is:

$$R^* = R \cdot \exp\{\frac{1}{2}[N - \frac{2}{m_r \sigma_c}(m_r \sigma_c \sigma_i' + s_r \sigma_c^2)^{\frac{1}{2}}]\} \qquad (27)$$

where

$$N = \frac{2}{m_r \sigma_c}(m_r \sigma_c \sigma_o' + s_r \sigma_c^2 - m_r M \sigma_c^2)^{\frac{1}{2}} \qquad (28)$$

With plane strain and without seepage, Hoek-Brown
(1980) derived:

$$R^* = R \exp[N - \frac{2}{m_r \sigma_c}(m_r \sigma_c \sigma_i' + s_r \sigma_c^2)^{\frac{1}{2}}] \qquad (29)$$

where:

$$\sigma_{crit}' = \sigma_o' - M\sigma_c \qquad (30)$$

$$N = \frac{2}{m_r \sigma_c}(m_r \sigma_c \sigma_o' + s_r \sigma_c^2 - m_r \sigma_c^2 M)^{\frac{1}{2}} \qquad (31)$$

$$M = \frac{1}{4}(\frac{m^2}{16} + \frac{m\sigma_o'}{\sigma_c} + s)^{\frac{1}{2}} - \frac{m}{8} \qquad (32)$$

For the face and seepage the integration of (16) has
been made using the Runge-Kutta numerical method. Figure 6 shows the solution of Y/a_1 as a function of

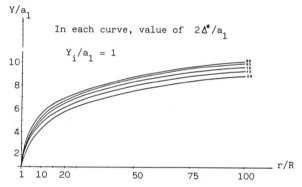

Figure 6. Face. Distribution of the plastic stresses.
Hoek-Brown failure criterion

r/R for different values of the seepage (expressed by
$2\Delta^*/a_1$) and imposing that $Y/a_1=1$ if $r=R$. With this
condition, the derivate of Y with respect to r is zero if $r=R$ and $\Delta^*=1$, and it is negative for greater va
lues. As Y has been defined as positive, it can be deduced that, as in the plane strain case, there is a
limit value for the seepage so that for values greater than this is not possible to find a solution for
the effective stresses. However, in the spherical case
such value depends on the boundary conditions imposed
at $r=R$.

5.2 Characteristic lines

Without seepage, closed-form solutions are obtained:
   Cross section (cylindrical symmetry)

$$u = DR^{-\alpha} + \frac{1+\nu}{E} \cdot R[\frac{(1-2\nu)}{2 a1}Y_i^2 + \frac{(\alpha-1)(1-\nu)}{\alpha+1}(Y_i - \frac{a1}{\alpha+1}) - (1-2\nu)(\sigma_o' + \frac{a2}{2 a1})]$$

$$(33)$$

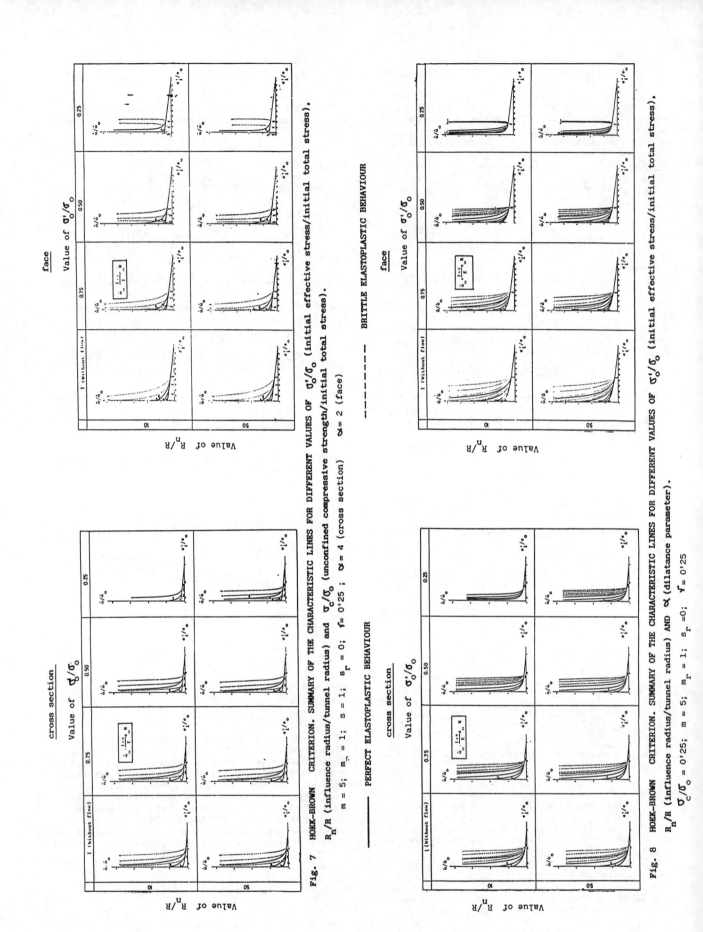

Fig. 7 HOEK-BROWN CRITERION. SUMMARY OF THE CHARACTERISTIC LINES FOR DIFFERENT VALUES OF σ_o'/σ_o (initial effective stress/initial total stress),

R_n/R (influence radius/tunnel radius) and σ_c/σ_o (unconfined compressive strength/initial total stress).

m = 5; m_r = 1; s = 1; s_r = 0; γ = 0'25 ; α = 4 (cross section) α= 2 (face)

——————— PERFECT ELASTOPLASTIC BEHAVIOUR   — — — — —   BRITTLE ELASTOPLASTIC BEHAVIOUR

Fig. 8 HOEK-BROWN CRITERION. SUMMARY OF THE CHARACTERISTIC LINES FOR DIFFERENT VALUES OF σ_o'/σ_o (initial effective stress/initial total stress),

R_n/R (influence radius/tunnel radius) AND α (dilatance parameter).

σ_c/σ_o = 0'25; m = 5; m_r = 1; s_r =0; γ = 0'25

144

where

$$a2 = s_r \sigma_c^2$$

(34)

Face (spherical symmetry)

$$u = DR^{-\alpha} - \frac{[\sigma_o' + \frac{a2}{a1}](1-2\nu)}{E}R + \frac{2.a1.R}{E(\alpha+1)}\{(1+\alpha)(1-2\nu)\ln^2(CR)-(1-\nu)(2-\alpha).$$

$$.[\ln(CR) - \frac{1}{\alpha+1}]\}$$

(35)

where D is a integration constant which can be obtained equalizing displacements at r=R.

With seepage, the integration has been made using the Runge-Kutta method. The results of the sensibility analysis are drawn in figures 7 and 8.

6 CONCLUSIONS

The analysis presented evaluates stresses and ground response curves in the tunnel cross section and in the face. We have derived some close-form solutions and results considering seepage and using the Hoek-Brown criterion (1980), a treatment of plastic volumetric strains and an elastic-brittle-residual-plastic stress-strain model with dilatation.
   Some of the more relevant conclusions are:
                Critical effective pressure
* is independent of the influence radius considered
* is independent of the dilatance parameter,α
* decreases if the unconfined compressive strength increases
* disminishes if seepage increases
                Characteristic lines
* The influence of the influence radius is not important except if it is very small and at the same time p_o is quite high.
* If seepage increases, characteristic lines are more "vertical". If Hoek-Brown criterion is used there is a limit value of the seepage so that for values greater than this it is not possible to find a solution.
* The difference between the perfect elastoplastic and brittle elastoplastic behaviour is greater as the interval between peak and residual strengths increases.
* The dilatance parameter,α , is unfavourable; if it increases, characteristic lines are more "vertical".

REFERENCES

Adachi, T.; Tamura, T. 1978. Undersea tunnel. Effect of drainage and grouting. Symp. on Soil Reforcing and stabilising Techniques. Sydney. Australia.
Amberg, W.A.; Lombardi, G. 1974. Une méthode de calcul elasto-plastique de l'état de tension et defor mation autour d'une cavité souterraine. Adv. in Rock Mech. Proc. 3th. Cong. ISRM. Vol.2, part B. Nat. Acad. Sci. Washington. 1055-1060.
Brown, E.T.; Bray, J.W.; Ladanyi, B.; Hoek, E. 1983. Ground response curves for rock tunnels. Proc. Am. Soc. Civil Eng. Geotech. Div. Vol.109. 15-39.
Egger, P. 1980. Deformations at the face of the heading and deformations of the cohesion of the rock mass. Underground Space. Vol.4. n.5 313-318.
Hoek, E.; Brown, E.T. 1980. Underground excavations in rock. The Inst. of Mining and Metallurgy. London.
Hoek, E.; Brown, E.T. 1980. Empirical strength crite rion for rock masses. J. Geot. Eng. Div. ASCE. 106 1013-1035
Izquierdo, F.A. 1984. Desarrollo del método de las líneas características en túneles circulares; con comportamiento elastoplasto-frágil del terreno; in fluencia de las heterogeneidades y de la filtración. Tesis Doctoral. Univ. Polit. de Valencia.
Jimenez-Salas, J.A. 1981. Conclusiones finales. Simp. sobre Uso Industrial del Subsuelo. Sesión 2. Madrid 127-143.
Lombardi, G. 1980. Influence of rock caracteristics on the stability of rock cavities. Tunnels and Tunnelling. Vol.2 n.1. Jan-Feb. 19-22. Mar-Apr. 104-109

Excavation of large rock caverns for oil storage at Neste Oy Porvoo Works in Finland

Excavation de grandes cavernes souterraines pour le stockage de pétrole à la raffinerie de Porvoo de la compagnie finlandaise Neste Oy
Auffahrung von Grosskavernen im Fels der Neste Oy Porvoo Werke, Finland

STIG JOHANSSON, Neste Oy, Espoo, Finland

SUMMARY

At the Porvoo Works of Neste Oy - Finland's national oil company - underground oil storage caverns of total capacity of 5.2 million m^3 are in operation.
   In April 1986 the implementation of two additional caverns totalling 150.000 m^3 in volume was initiated.
   The 37 individual caverns are divided into 24 separate plants. The caverns were constructed in 11 phases, starting in 1965.
   Cavern profile areas range from 340 m^2 to 580 m^2 and have different shapes. The strength of the Precambrian migmatite rock (a mixture of gneiss and granite) has meant that construction methods based on drilling and blasting have been most economical. Vertical shafts 2400 mm in diameter have successfully been drilled with the raise-bore method.
   Design and construction methods as well as costs for these large unlined openings are presented.

RESUME

La campagnie pétrolière finlandaise nationale NESTE OY dispose actuellement à sa raffinerie de Porvoo de 5.2 millions de m^3 de stockage souterrain de pétrole.
   En Avril 1986 deux cavernes supplémentaires dúne capacité totale de 150.000 métres cubes étaient mis en construction.
   Les 37 cavernes étant divisées en 24 dépôts détachées sont construites en 11 phases à partir de 1985.
   Le profil de la caverne varie de 340 métres carrés à 580 métres carrés étant d une forme différente. Pour raison de la structure compacte des migmatites de Precambrian (une mixture du gneiss et du granite) le méthode de forage et l explosion se sont montrées les plus économiques. Le forage des puits déxploitation verticals d' un diamétre de 2400 mm a été effectué avec succés conformément à la méthode raise-bore.
   les méthodes de planification et construction ainsi que les frais de ces stockages souterrains sont présentés.

ZUSAMMENFASSUNG

Neste Oy - die nationale Ölgesellschaft Finnlands - hat in seinen Porvoo Werken heutzutage Lagerstätten in Felskavernen mit einem Gesamtvolumen von 5.2 millionen m^3 für Öl im Betrieb.
   Im April 1986 wurde die Verwirklichung von zwei neue Kavernen mit einem Gesamtvolumen von 150.000 m^3 angefangen.
   Die 37 Einzelkavernen sind in 24 Anlagen geteilt worden. Die Bauarbeiten wurden im Jahre 1965 angefangen und die Kavernen in elf Stufen gabaut.
   Die Kavernenprofile sind von 340 m^2 bis 580 m^2 und haben verschiedene Formen. Die Festigkeit des Precambrischen Migmatite Gesteines (eine Mischung von Gneis und Granit) hat bedeutet, dass die Errichtungsmethoden, die sich auf die Bohrung und das Sprengen gründen, sehr ökonomisch gewesen sind. Die vertikalen Böhrlöcher mit dem Durchmesser von 2400 mm sind mit gutem Erfolg mit "raise-bore"-methode gebohrt worden.
   Die Projektierungs- und Errichtungsmethoden sowie auch die Kosten für diese grossen Hohlräume ohne Ausbau werden vorgestellt.

1 INTRODUCTION

Energy imports to Finland are high, because the climate is cold and many Finnish industries are energy-intensive. Consumption of oil products amounted to about 11 million tons in 1986, equivalent to 35 % of the country s primary energy consumption.
   Neste Oy, Finland's national oil company is the only company in the country involved in oil refining. The annual capacity of its two refineries in Porvoo and Naantali is 15 million tons.
   Climate is one reason why the company must store relatively large amounts of crude oil i.e. 3 to 4 million tons, for winter use; it is stored in large unlined rock caverns excavated into the bedrock at the refineries.

The crude oil storage caverns at the Porvoo Works (Fig. 1.) have a combined capacity of about 3.3 million m^3, divided between 14 separate plants. In addition the refinery also operates 8 caverns for storage of oil products, with a combined capacity of about 1.9 million m^3. The total number of individual caverns is 35.
   In late April 1986 construction of two additional cavern plants, for storage of light gas condensate and propane was initiated. The total capacity of these two pressurized caverns will be 150.000 m^3, and the caverns are scheduled to be operational by the end of 1987.
   All caverns operate according to the well-known Scandinavian method, i.e. the caverns are located well below the ground water level in bedrock that

Figure 1. The location of Neste Oy s Porvoo Works.

provides the necessary stability and tightness. No artificial linings are used.

2 GEOLOGY OF THE PORVOO WORKS AREA

Neste Oy's Porvoo Works are located about 50 km east of Helsinki on the Southern coast of the Gulf of Finland in the eastern part of the Baltic Sea.

The Precambrian rock crust in the area is a typical part of the Baltic or Fennoscandian Shield, which is the broadest shield area on the European continent. The rocks of the Porvoo area belong to the Svecokarelidic orogenic belt, the folding of which took place about 1800 million years ago. A characteristic feature of many Svecokarelidic areas is an abundance of mixed rocks with two or more components of different character. The older component is a metamorphic schist or a plutonic rock and the younger component is an igneous looking granitoid rock. The migmatite forming granites with associated pegmatites penetrate all older rocks, forming a great variety of migmatites (Simonen 1980). The principal rock types encountered are: migmatite, gneiss, granite and granite pegmatite. The migmatite is composed of mixtures of gneissose rocks, frequently microfolded mica-gneiss and microcline-granite.

Small massifs and dikes of coarsegrained granite are also common. Dikes of amphibolitic composition (metadiabase) represent less then 1 % of the rock mass, and do not show any relationship to the schistocity.

The arithmetic mean strike of the schistocity is 55° (about NE-SW). The dips are very steep (75° - 80°) towards 145° (about SE) or vertical.

The Svecokarelidic orogeny was followed by a long period of erosion and cratonization. Numerous joints, shear zones, faults and crushed zones of highly varying ages produced a mosaic structure in the crust.

The bodies of rock utilized for cavern construction are located between three major tectonic disturbance zones. The intervening rock bodies show evidence of a second order fracture tectonization, which has resulted in a number of fracture and shear zones.

The joint system of the gneissic granite rock consists of three well-defined joint sets, one set corresponding to the strike and dip of the schistose rock components. The second less developed joint set cuts the schistocity at almost right angels. Horizontal and/or gently inclined joints form the third joint set. This jointing has a tendency to decrease with depth. The types of filling material found on the fracture surfaces are mainly calcite, chlorite, epidote, hematite, "rust", talc and clayey gouge.

Considerable overbreaks in connection with blasting works have been experienced, and a great part of these overbreaks can be ascribed to moist or wet chlorite-coated or filled slickenside-type fractures. Because of their continuity on several occasions over hundreds of square metres and because of their low shear strength these slickensides affect the sliding stability of any rock resting above them, hence necessitating also additional reinforcements.

A summary of the results of laboratory testing carried out on core samples is presented in Table 1. The rock material strength can be classified as being high to very high using internationally recommended terminology.

Measured in-situ stress values indicate that a relatively "weak" horizontal compressive stress field exists in the cavern construction area, mean values ranging from about 5 MPa to about 17 MPa at depths up to 100 m in the rock.

The hydraulic conductivity values of the migmatitic rock mass, decrease with depth, about 80 % of the calculated modified Lugeon units being smaller than one Lugeon at a depth of over 30 m in the rock.

Further details of the geology of the Porvoo site have been given by Johansson (1985 b).

3 CAVERN DESIGN ASPECTS

3.1 Excavation design

The basic aim of any underground excavation design should be to utilize the rock itself as the principal structural material, with minimal disturbance being created during the excavation and as little as possible being added by way of concrete and steel support. This observation by Hoek and Brown (1980) is basic to oil storage design.

The design of oil storage caverns utilizes empirical design methods almost entirely and relates therefore to practical experience gained in previous implementations. This has in particular been the situation at the Porvoo site where the cavern plants have been constructed in 11 different phases. The influence of empirical design is for example found in the sizes and shapes of the excavation profiles used here, which have increased in size from about 340 sq. m. up to 580 sq. m.. The shapes have also differed considerably (see Fig. 2).

Table 1. Laboratory test results on rock samples

Dry density	
– granite	26.5 – 26.8 KN/m^3
– gneiss	27.1 – 28.9 KN/m^3
– migmatite	26.7 – 27.8 KN/m^3
Compr. strength	
– granite	107 – 147 MPa
– gneiss	127 – 275 MPa
– migmatite	133 – 210 MPa
Tensile strength	
– granite	6.3 – 12.5 MPa
– gneiss	8.4 – 18.9 MPa
– migmatite	7.0 – 14.6 MPa
Porosity	0.04 – 0.44 %
Young s mod.	60.1 – 103.0 GPa
Poison s ratio	0.20 – 0.40

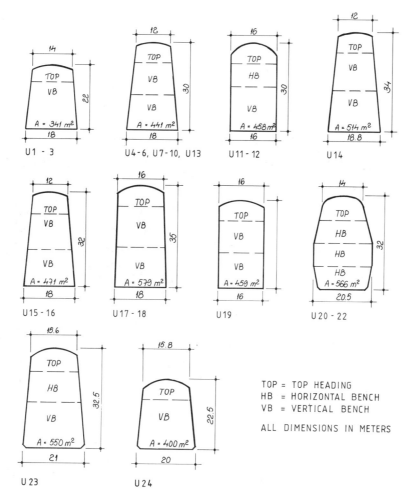

U1 - 3   U4-6, U7-10, U13   U11-12   U14

U15-16   U17-18   U19   U20-22

U23   U24

TOP = TOP HEADING
HB = HORIZONTAL BENCH
VB = VERTICAL BENCH

ALL DIMENSIONS IN METERS

Figure 2. Cavern profiles and excavation sequences.

In carrying out the design of underground rock caverns it is essential to consider construction procedures at the same time that the design is being made. If this is not done one can end up with a most sophisticated design which cannot be implemented at reasonable costs or according to a given timeschedule.

The main target in the site investigation phase has been the optimal location of the planned cavern plant and its depth levels, as the selection of these will determine the real geological conditions under which the caverns will be excavated.

The length axes of the caverns have been oriented perpendicular to the strike of the schistose components of the migmatitic rock mass, which has proved to be a beneficial solution. For example, the strike facilitates the loosening of rock when the schistocity is almost perpendicular to the advance of the excavation because the tensile strength of the rock in the direction of throw is at its minimum.

Overbreaks experienced in the cavern walls, owing to the fact that the less dominant vertical joint set parallels these have been small and have not formed any threat to the overall stability of the cavern walls. Overbreaks in the roof parts of the caverns have been insignificant as the horizontal and/or gently inclined set is not well developed. As well known horizontal fractures are a threat to roof stability in caverns with an arched vault form.

Top heading spans have varied between 12 and 17 m. and the base widths between 16 and 21 m. With the present available excavation methods the most practical cavern height has been about 30 m. Vertical bench height is limited in the design to about 16 m otherwise, deviations in drilling accuracy become too large (over 2 %). It can be

noted that the highest vertical bench blasted in one round was 23 m high and was used in the excavation of cavern U 5/6.

In order to minimize the negative impact caused by immediate reinforcements, the main principle has been that besides the main access tunnel(s) at least two overhead tunnels (top headings) should be under excavation simultaneously. Furthermore possibilities for maintaining and servicing of excavation equipment must be good. At least one vertical shaft to the ground surface should always be provided for, thereby cutting down the hours otherwise lost in ventilating only via the access tunnel.

4 EXCAVATION METHODS

4.1 Objectives

The strength and intact nature of the migmatitic rock at the Porvoo site has meant that excavation procedures based on drilling and blasting have proved to be the most economical. In addition the need to excavate a large underground space at minimal cost and in a short time makes this method feasible. Cavern design has also favoured the use of heavy mobile tunnelling and open-pit drilling equipment and the use of large-capacity loading and transportation machinery, which reduces construction time considerably (Johansson 1985a).

There are several excavation targets in a cavern project. The most important are the open-cut, the main access tunnel, the branch tunnels to different cavern excavation levels, vertical shafts for equipment to the ground surface and the actual storage caverns themselves (Fig. 3).

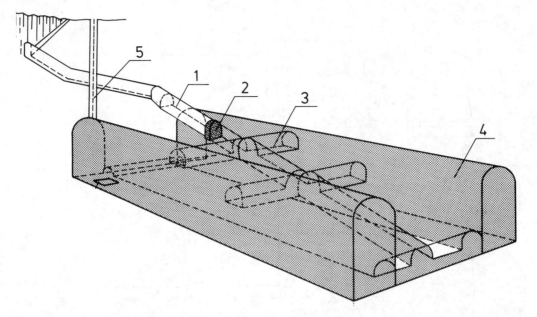

Figure 3. Principal scheme for a dual cavern oil storage facility. 1. Main access tunnel, 2. Concrete plug, 3. Branch tunnels to excavation levels, 4. Main storage caverns, 5. Vertical shaft for pumps and instruments.

4.1.2 Access and branch tunnels

The main access tunnel for the transportation of equipment and personnel as well as for that of broken rock from actual cavern space, normally has an inclination of 1:7, in extreme cases, and for very short distances an inclination of 1:5 has been used.

Depending on the size of the excavation target and the overall timeschedule, the main access tunnel is excavated entirely or partly, either for one- or two-way traffic. Two-way traffic tunnels have ranged in size from 36 to 55 sq. m.. The corresponding sizes for one-way tunnels have been 17 to 30.5 sq m. Passing points have been constructed at selected places. An example of a recently used drilling and charging pattern in an one-way traffic tunnel is presented in Fig. 4. Concrete paved access tunnels have also been introduced to increase driving speeds and to save tyre wear. Higher driving speeds have also been achieved by making the tunnels straight, which has increased total work output.

4.1.3 Vertical shafts

Vertical shafts that lead to the surface have been constructed in three main ways:
- by longhole drilling with drill and blast
- by the up-hole method by drill and blast
- by raise boring

Experience has shown that shafts with a depth of up to 40 m and a size larger than 3 m x 5 m can be excavated by the longhole drilling method, in which the entire shaft is drilled from the surface. Blasting is carried out in short rounds from the bottom upwards.

The up-hole method has been used in shafts that exceed 40 m in depth. In this case a 150 mm diameter hole is drilled from the surface through the arch of the cavern, a wire rope is sunk into the hole and a drill platform is attached to it from below. A pilot drift is excavated up to the surface by use of short rounds. After completion of the pilot drift the final contour of the shaft is excavated by smooth blasting methods from the surface towards the cavern.

The raise-boring method has, to date, been used in four shafts with diameters of 1500 mm and 2400 mm

and depths of 45 m and 112 m. At the time of writing one additional shaft with a diameter of 2400 mm and a depth of 145 m is being completed.

The following penetration rates have been achieved:
- pilot hole, 280 mm   1,8 to 3,0 m/h
- shaft, 1500 mm   0,5 to 0,9 m/h
- shaft, 2400 mm   0,3 to 0,6 m/h

The type of equipment used in these shafts were:
- Dresser Strata Borer, Model 800
- Tamrock Rhino, Model 1000
- Robbins Raise Drill, Model 71R

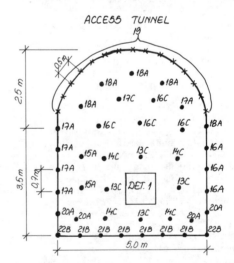

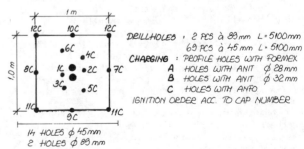

Figure 4. Typical drilling and blasting pattern in one-way access tunnel.

150

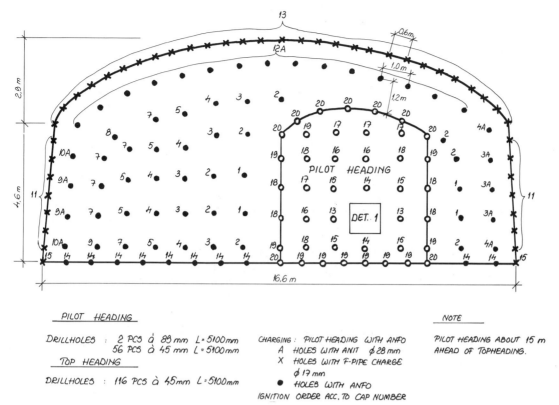

PILOT HEADING

DRILLHOLES : 2 PCS à 88 mm L = 5100 mm
56 PCS à 45 mm L = 5100 mm

TOP HEADING

DRILLHOLES : 116 PCS à 45 mm L = 5100 mm

CHARGING : PILOT HEADING WITH ANFO
A HOLES WITH ANIT ⌀ 28 mm
X HOLES WITH F-PIPE CHARGE
⌀ 17 mm
● HOLES WITH ANFO
IGNITION ORDER ACC. TO CAP NUMBER

NOTE

PILOT HEADING ABOUT 15 m
AHEAD OF TOPHEADING.

Figure 5. Typical drilling and charging pattern in top heading.

4.1.4 Cavern excavation

The excavation of large cavern profiles can be divided into two phases - tunnelling of the top heading and benching of the lower parts. The top headings have had an profile area of 80 sq. m. to 110 sq. m. (Fig. 2) and have been drilled with hand-held drills with pusher-legs and mechanized rubber-tyred drilling jumbos (both pneumatic and hydraulic). The subsequent benches have been drilled either inclined outward at 10:1 or vertically, with the use of various types of mechanized crawler rigs. This has been possible by the low arch/span ratio (average 0.16) which leaves enough space for a normal crawler rig feed boom to be placed at the contour of the profile. Vertical benching corresponds closely to normal open-pit excavation.

In two cases (U11-U12 and U23) the first bench was drilled horizontally and lifted. In cavern project U20 - U22 all benches were drilled horizontally and lifted. Horizontal benching may be considered a modified form of tunnelling.

Until the early 1970s top headings were blasted by V- and cylinder cuts, but more recently, the pilot drift method with consequent slashing has been employed (Fig. 5).

Particularly important are the contour holes of the excavation profile. These should be drilled with as small a look-out angle as possible and be kept parallel with the longitudinal direction of the cavern. Equally important is the setting out of the drillhole positions. The setting out of drill holes has been set at 0.15 m and the allowable deviation at 0.03 m per meter drill hole. This means e.g. that in a 5 m long top heading round the maximum acceptable overdrilling is 0.3 m.

An old truth is that "no rock is so good that it could not be scattered by careless drilling and blasting, nor can the blasting result be better than the drilling".

The use of explosives to remove rock requires controlled blasting to minimize damage to the remaining rock surface. The normal methods that are used in controlled blasting are smooth blasting and presplitting. Various recommendations discussed a.o. by Langefors-Kihlström (1973), Vuolio (1980,1986) have formed the basis for design considerations. Smooth blasting in which the contour charges are initiated last in the round have been applied both in top headings (Fig. 5) and in horizontal benching. Presplitting in which the contour charges are initiated before the rest of the charges in the round have also been applied in horizontal benching (Fig. 6) but mainly in vertical benching (Fig. 7).

The transient strain in rock due to blasting depends upon the linear charge concentration and the distance. For example granite is expected to fail in dynamic tension around a peak particle velocity of v = 1000 - 2000 mm/s depending on the wave type. In Fig. 8 the peak particle velocity has been shown for various distances from a 4 m long charge using different explosives manufactured in Finland (after Vuolio 1986). With the assumption that damage would occur around a peak particle velocity of v = 1000 mm/s it is possible to establish a proper blast pattern.

It is insufficient to have a low charge density only in the contour holes, the next row of holes from the contour holes towards the centre must also have a reduced charge density to prevent the formation of an extensive zone of damage (loosened zone). The zone of damage can be limited to 0.3 - 0.5 m.

Some data related to drill meters and specific charge per m^3 of solid rock is given in Table 2.

From the data presented in Table 2 it is evident that the fragmentation of the rock is related to the specific charge, the higher the charge density, the smaller the maximum block size. Generally taken rock removed from top headings have as such been used for subbase of roads, while rock from horizontal and vertical benches have served levelling and filling works.

The granitic rocks are more brittle than the gneissic rocks, which can be classified being tough. Therefore a mixture of both these rock types e.g. for crushing of aggregates has given a good product.

151

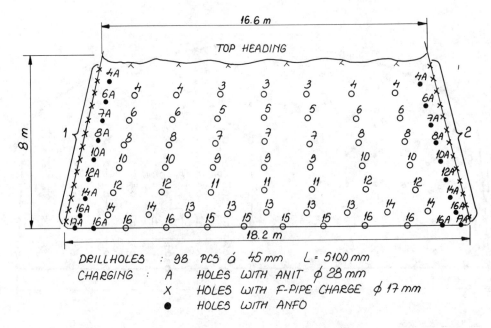

DRILLHOLES : 98 PCS ó 45mm L = 5100mm
CHARGING : A   HOLES WITH ANIT φ 28 mm
          X   HOLES WITH F-PIPE CHARGE φ 17 mm
          ●   HOLES WITH ANFO

IGNITION ORDER ACCORDING TO CAP NUMBER

Figure 6. Typical drilling and charging pattern in lifted bench (inclination 10:1).

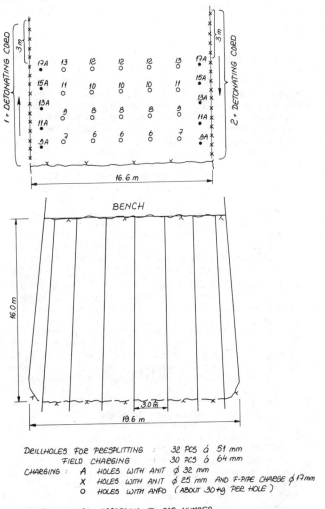

DRILLHOLES FOR PRESPLITTING : 32 PCS ó 51 mm
        FIELD CHARGING   : 30 PCS ó 64 mm
CHARGING : A HOLES WITH ANIT φ 32 mm
          X HOLES WITH ANIT φ 25 mm AND F-PIPE CHARGE φ 17mm
          O HOLES WITH ANFO (ABOUT 30 kg PER HOLE)

IGNITION ORDER ACCORDING TO CAP NUMBER

Figure 7. Typical drilling and charging pattern in vertical bench (inclination 10:1).

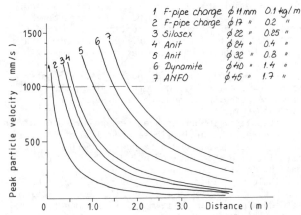

1	F-pipe charge	φ 11mm	0.1 kg/m
2	F-pipe charge	φ 17 "	0.2 "
3	Silosex	φ 22 "	0.25 "
4	Anit	φ 24 "	0.4 "
5	Anit	φ 32 "	0.8 "
6	Dynamite	φ 40 "	1.4 "
7	ANFO	φ 45 "	1.7 "

Note   Charge weight calculated in relation to dynamite

Figure 8. Peak particle velocities for different types of Finnish explosives in relation to distance from the drill hole (after Vuolio 1986).

152

Table 2. Average drilling and charging per m^3 rock in different phases in cavern excavation

Phase	Height	Drilling	Specific charge
	m	m/m^3	kg/m^3
Top heading	7.5	1.00	0.90
Horizontal bench	8.0	0.50	0.60
Vertical bench	16.0	0.35	0.50

5 COSTS

A number of factors affect the final investment costs of a cavern project and some of these factors show great variation. Based upon a detailed cost control system comprising 16 different cost items, a grouping of total costs into six major groups was carried out.

The results are presented in Table 3 in the form of minimum, average and maximum costs (from Johansson 1984). It can be seen that the major part of costs is related to excavation work. In turn the quality of the rock mass, the number of individual caverns, the profile area, the depth and length of individual caverns, as well as the excavation method have a great influence on the excavation costs.

The principal excavation sequences are shown in Fig. 2. In the company's expencience the method whereby the lower parts are excavated with vertical benching has been somewhat more economical than either a combination of horizontal and vertical benching or horizontal benching alone.

The unit cost figures for caverns U13 to U22 presented in Table 4 are all inflated to be current at 31 December 1986 by use of the official building cost index in Finland as well as Neste's own index (est. in 1948) as a basis for calculations. It should be clearly understood that the cost figures are specific for the Porvoo Works and are not, as such, universally applicable.

Table 3. Total costs, by category, for underground oil storage caverns.

Item	Percent of total cost		
	Minimum	Average	Maximum
Administration, design, construction, supervision	1.5	4.0	8.0
Excavation work	48.0	66.0	81.0
Reinforcing	4.0	5.0	8.5
Concrete structures	4.0	9.0	18.5
Mechanical equipment, piping, steel structures	5.5	9.0	14.0
Electrical installations, instruments, insulation	2.5	6.0	9.0

Table 4. Unit cost per m^3 of effective cavern volume in Finnish Marks (FIM) and U.S. dollars (USD) at constant 1986 prices.

Cavern plant	Unit cost per m^3	
	FIM	USD
U13	140	28
U14	125	25
U15 – U16	130	26
U17 – U18	140	28
U19	250	50
U20 – U21	125	25

1 USD = 5,0 FIM

Long-term operation and maintenance cost control accounting indicate that the costs for underground caverns have been approximately onesixth of the corresponding figures for steel tanks.

6 ACKNOWLEDGEMENTS

It is my pleasant duty to express my gratitude to Neste Oy and particularly to Mr Jussi Rinta, Executive Vice President, Corporate Development and Technology, and Mr Ilmo Paasi, General Manager, Neste Engineering, for permission to publish this article.

REFERENCES

Hoek, E.,Brown E.T. 1980.Underground excavations in rock.p. 527. London: The Institution of Mining and Metallurgy.
Johansson S. Costs of mined oil caverns at Neste Oy's Porvoo Works in Finland. Underground Space. Vol. 8:5-6. p. 372-380. Pergamon Press.
Johansson S. 1985(a).Twenty years experience of constructing oil cavern storage at Porvoo works in Finland. Tunnelling 85 Conf. Paper 12. 14 p. London:The Institution of Mining and Metallurgy.
Johansson S. 1985(b). Engineering geological experience from unlined excavated oil storage caverns in a Precambrian rock mass in the Porvoo area, southern Finland. Academic dissertation. 77 p. Espoo.
Langefors U.,Kihlström B. 1973.The modern technique of rock blasting.405 p.New York:J. Wiley & Sons.
Simonen A. 1980.The Precambrian in Finland.Bulletin 304.Geological Survey of Finland.58 p.Espoo
Vuolio R. 1980.Design and execution of blasting.188 p.Helsinki:SMK (Finnish text).
Vuolio R. 1986.Excavation of large rock caverns taking into account the integrity of rock structures,surrounding buildings and delicate instruments in surrounding buildings.Proc. Intl. Symp. Helsinki, Large Rock Caverns. Vol. 2,p. 1629-1646.Pergamon Press.
The Mining and Metallurgical Society of Finland.1982.Handbook of mining and excavation techniques VMY Publ.No B 29. 801 p.Helsinki. (Finnish text).

Relationships of variation of rock fissure permeability
Les régularités du changement de la perméabilité des fissures de roches
Veränderungsgesetzmässigkeiten der Wasserdurchlässigkeit klüftiger Felsgesteine

J.M.KAZIKAEV, Belgorod Technological Institute, USSR
YU.S.OSIPENKO, Belgorod VIOGEM Institute, USSR

ABSTRACT: The problems of hydromechanical parameters evaluation specifically concerning fissured rock mass filtration properties are characteristic of mining and building. They are of the greatest significance for industrial sewage and liquid toxic waste burial designing, particularly in connection with the environmental protection problems. The main principles of the report based on theoretical and experimental research allow to solve a wide variety of scientific and practical problems in this sphere.

RESUME: Les taches selon les paramètres hydrogéomécaniques, en particulier, selon les caractéristiques à filtrer des massifs des roches fissurées sont typiques pour l'industrie minière et la construction. Elles prennent la plus grande netteté au cours d'un projet de l'inhumation des écoulements industriels et des déchets liquides, toxiques surtout en rapport avec la protection de l'environnement. Les régularités présentées à ce rapport, obtenues sur la base des recherches expérimentales et théoriques permettent de résoudre beaucoup de taches scientifiques et pratiques dans ce domaine.

ZUSAMMENFASSUNG: Die Einschätzung der hydrogeomechanischen Parameter, insbesondere der Filtrationseigenschaften des klüftigen Festgesteins ist für das Bergbau- und Bauwesen kennzeichnend. Maximal akut wird sie bei der Projektierung der Begrabung von flüssigen toxischen Produktionsabfällen, besonders im Zusammenhang mit dem Umweltschutz. Die im Vortrag genannten Gesetzmäßigkeiten, die anhand der theoretischen und experimentellen Untersuchungen ermittelt worden sind, gestatten zahlreiche wissenschaftliche und praktische Aufgaben auf diesem Gebiet zu lösen.

Nowadays it is universally recognized that fissured rock filtration properties weaken with depth. However the absence of ideas on this phenomenon's mechanism makes it necessary to perform engineering geology and hydrogeology exploration and investigation work at each object which allows finally to obtain only partial empirical relationships between rock massif filtration properties and its structural and mechanical behavior.

The theoretical relationships determining values of fissured rock filtration properties on the basis of account of their individual structure establish relation between filtration coefficient on the one hand and fissure spatial orientation, fissure opening magnitude and their density on the other hand. Therefore in the light of available theoretical statements in order to make clear the nature of fissure permeability changes with depth it is necessary to establish which of the said values change with rock bedding depth and which of them are constant, i.e. are parameters. For this purpose the authors carried out a number of investigations aimed to rock fissure parameter studying and summarized and analysed the results of fissured rock studies from this point of view which were published in literature, included into reports of mineral deposit prospecting, investigations at sites of underground water intake plant construction and hydraulic structures. The analysis of rock fissuring included not only results of fissure in-situ measurements in mining excavations and rock exposures but also well core fissuring measurements were used to a maximum extent. All this gave a possibility to investigate fissured rocks characterized by essentially different geological and structural conditions, geological development history, lithological and mineralogical composition and their position in different climatic conditions. The obtained results showed that beyond the weathering influence depth which as a rule does not exceed 70-100 m and is limited by 20-50 m rock fissuring intensity does not change with depth and the same may be said about spatial orientation of main fissure systems.

The investigations at one of Ukraine ore deposits may be taken as an example which were performed on the basis of rock fissuring study of well cores. The results of these studies were confirmed by in-situ measurements of fissuring parameters in mining excavations at different mine levels.

While core characterization two gradations of metamorphized fissured rocks are adapted: 1.intensively fissured rock with medium intervals between fissures less, than 0,1 m; 2.fissured and feebly fissured rock with medium intervals between fissures more than 0,1 m. These groups fully determine metamorphized rock probable behavior with respect to degree of fissuring.

Below the zone of weathering influence the thickness of which was adopted equal to 100 m metamorphized rocks were divided along depth into 12 intervals the thickness of which was supposed to be 100 m. Thus rock fissuring degree was traced up to the 1300 m depth. In order to reveal and study the regularities of metamorphized rock fissuring change with depth dimensionless values of intensive fissuring zones were estimated the sense of which becomes clear from the following relationship:

$$\bar{d} = \frac{d_T}{d} \qquad (1)$$

where $\bar{d}$ - dimensionless zone of intensive fissuring;

d_T- thickness of intensive fissuring zone, m;

d - total thickness of metamorphized rocks in interval of observation, m.

In order to exclude the influence of rock lithological composition when fissuring pattern change establishment, processing of fissuring measurement results was done separately for each of lithological rock variety. In subsequent designs the results of fissure measurements of the well bore cores performed only along representative intervals were adopted. An interval was considered a representative one if rock lithological variety had been stripped not less than 0,7 of its extent, i.e. 70 m. A chart of changes of intensive fissuring dimensionless zone thickness in quartzite and carbonate rock with depth (Fig.1) is taken by the authors as an example. Fissuring measurements were done for 347 well bores.

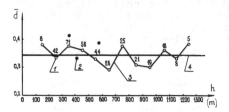

Figure 1. Change of quartzite and carbonate rock fissuring with depth.

1 - mean values of dimensionless thickness of intensive fissuring zone using well cores (figure means a number of points of observation)
2 - mean values of dimensionless thickness of intensive fissuring zone using the results of measurements in underground mining excavations
3 - regression empirical line
4 - regression theoretical line

On the basis of analysis of regression empirical line presented in the chart a hypothesis was adopted concerning the absence of relations between features being investigated, i.e. the distribution of intensive fissuring zone of dimensionless thickness does not depend on rock bedding depth. The hypothesis check is done with criterion $\bar{\chi}^2$ which was calculated according to the following formula:

$$\bar{\chi}^2 = \sum_i \sum_j \frac{(n_{ij} - \bar{n}_{ij})}{\bar{n}_{ij}} \qquad (2)$$

where n_{ij} and $\bar{n}_{ij}$ - empirical and theoretical frequences, respectively.

The calculated $\bar{\chi}^2$ value is equal to 11,1. With nine degrees of freedom (f=9) a condition $\bar{\chi}^2_{0,90;9} < \bar{\chi}^2 < \bar{\chi}^2_{0,20;9}$ is satisfied as tabulated values $\bar{\chi}^2_{0,90;9}$ and $\bar{\chi}^2_{0,20;9}$ are equal to 4,17 and 12,2 respectively. Thus it may be considered that the adopted hypothesis concerning the absence of relation between the investigated features is confirmed with sufficiently good probability.

The results of the in-situ performed studies showed that selected systems of fissures are traced at all mining levels and are not influenced by the depth of their bedding. The absence of dependence of the main fissure system orientation and fissuring

intensity on rock depth bedding on the general background of the regular diminishing of their fissure permeability give a base to state that a single value which experience change with depth rise is a fissure width. In this case if to consider rocks as deformable fissured reservoirs the reduction of their permeability may be explained by fissure opening (width) decrease under the influence of rock pressure. Consequently statistical regularity of rock fissure permeability which reflect the nature of the phenomenon being considered should be sought proceeding from the following conditions:

$$K(P_h) \to 0 \quad \text{with} \quad P_h \to \infty \qquad (3)$$

where $K(P_h)$ - function of changes of rock fissure permeability due to rock pressure

P_h - running effective pressure, P_a.

The condition of (3) is satisfied by exponential dependence, and this should be taken into account when statistical processing of fissure rock filtration properties determination results is taking place.

In order to present the statistical regularity in the more general form it is necessary to give included parameters and variables in dimensionless form. Dimensionless parameters and variables are determined in the following way:

$$\bar{K}_P = \frac{K_P}{K_0} ; \quad \bar{P}_h = \frac{P_h}{P_M} ; \quad \bar{P}_o = \frac{P_o}{P_M} \qquad (4)$$

where $\bar{K}_P$ - dimentionless coefficient of fissure permeability with running dimensionless effective pressure $\bar{P}_h$

$\bar{P}_h$ - running dimensionless effective pressure

$\bar{P}_o$ - standard dimensionless effective pressure

K_P - coefficient of fissure permeability with running effective pressure P_h, m/s

K_o - coefficient of fissure permeability with standard effective pressure P_o, m/s

P_o - standard effective pressure, P_a.

The coefficient of rock fissure permeability is equal to K_M if effective pressure at aquifer toe is equal to P_M. Let us express said values in dimensionless form as well:

$$\bar{K}_M = \frac{K_M}{K_0} ; \quad \bar{P}_M = \frac{P_M}{P_M} = 1 \qquad (5)$$

Let us limit the range of $\bar{P}_h$ change so as to put all its values within the range $\bar{P}_o$ to $\bar{P}_M$ or from P_M up to 1 as $P_o = 1$. Taking into account (4) and (5) and the exponential dependence character as well, the function $K(P_h)$ should be found in the following form:

$$\bar{K}_P = exp[a_o(\bar{P}_h - \bar{P}_o)/(1 - \bar{P}_o)] \qquad (6)$$

Analysis of relationship (6) shows that with dimensionless pressure change from $\bar{P}_o$ to 1 coefficient of fissure permeability changes from 1 to $\bar{K}_M$.

As dimensionless standard pressure $\bar{P}_o$ can always be taken as for initial level of indication then relationship (6) is simplified up to

$$\bar{K}_P = exp(\bar{a}_o \cdot \bar{P}_h) \qquad (7)$$

In this case the coefficient of fissure permeability will have values from 1 up to $\bar{K}_M$ with dimensionless pressure changes from 0 to 1.

In equation (7) there is only one unknown parameter $\bar{a}_o$, which depends on the assigned value of dimensionless coefficient of fissure permeability at aquifer toe. This conclusion allows to select values of $\bar{a}_o$, guiding by purely pragmatic considerations. It is evident that value of $\bar{a}_o$ should be selected in such a way that the error in water inflow and ground water level estimation when prediction problem solving concerning filtration due to unconfined aquifer change into the confined one does not exceed the predetermined value.

Equations (6) or (7) together with basic equations of filtration presented in dimensionless form allowed to estimate error value in determining water inflow rate into drain, completely penetrating the aquifer, by means of numerical modeling. The finite element method was used in said problem solving. Boundary conditions of the first kind (H=const) were assigned to boundaries of recharge and drainage area as it is in this case that the assumption being considered leads to major errors. The results of the set problem solution are summarized in the following table.

Table 1. Relative error in estimation of water inflow towards drain.

Dimensionless coefficient of fissure permeability at aquifer toe ($\bar{K}_M$)	Dimensionless coefficient of fissure permeability change ($\bar{a}_o$)	Relative error in estimation of water inflow towards drain (ε)
10^{-1}	-2,303	0,603
10^{-2}	-4,605	0,245
10^{-3}	-6,908	0,059
10^{-4}	-9,210	0,007

Consequently if to assume an error in the set problem solution to be a permissible one a value of dimensionless parameter of fissure water permeability change $\bar{a}_o$ can be determined which allows in its turn to calculate dimensionless coefficients of fissure permeability using equation (7) for any given value of dimensionless pressure. The absence of relationship between coefficient $\bar{a}_o$ and individual fissure rock structure of studied object reflects the unity of nature of changes in rock fissure permeability.

The relationship (7) satisfies fully all the earlier formulated conditions, i.e. it reflects the exponential character of fissure permeability coefficient decrease with depth, states inambiquously criterion and method of transition from unconfined aquifer to the confined one, reveals the nature of the process being considered and as a consequence describes the whole class of problems including fissure permeability change in deformable rocks.

The use of relationship (7) together with basic equations of underground water filtration reduced to dimensionless form gives an opportunity for simulational modelling and practical problem solution on a wide scale. In particular planning of rational positioning of test hydrogeological wells aiming to studying fissure rock filtration properties, choice of the most effective depth of position of water and drainage well intake plants, grounding the necessity and determination of principal designs of drainage and other measures directed to mining excavation protection from underground water inflow.

When solving a number of problems the obtained regularity was used in particular in order to predict perspective structures for brine burial. With the help of the relations obtained a number of the listed problems were solved and the results correspond to the values observed in in-situ conditions.

Connection between farfield and nearfield in relation to the fractures and/or flowpaths for the geological isolation of radioactive wastes

Relation entre les étendues proches et éloignées dans le contexte des fractures et/ou des chemins d'écoulement pour l'isolation géologique des déchets radioactifs

Zusammenhang zwischen Fern- und Nahfeld in Bezug zu Brüchen und/oder Flussbahnen für die geologische Isolierung radioaktiver Abfälle

K.KOJIMA, University of Tokyo, Japan
H.OHNO, University of Tokyo, Japan

ABSTRACT: The characteristics of fractures and hydrology are described here for the radioactive wastes isolation in the mobile belt. Fracture statistics, origin of highly fractured zone and sound rock blocks, and Permeability of fractured and presented, the relationship among the data from nearfield & farfield are discussed.

RESUME: Les caractéristiques des fractures et de l'hydrologie sont décrites ici pour l'isolation des déchets radioactifs dans la ceinture mobile. La présentation porte sur les statistiques de fracture, l'origine des zones extrêmement fracturées et des blocs de pierre franche, anisi que sur la perméabilité des fractures; les relations entre les données obtenues à partir des étendues proches et éloignées front l'objet de discussions.

ZUSAMMENFASSUNG: Der eigenschaften von Brüchen und Hydrologie für die Isolierung radioaktiver Abfälle in labilen zonen gegeben. Brunchstatistiken, der Ursprung starker Bruchzonen und gesunden Gesteins und die Durchlässigkeit von Bruchflächen werden erläutert sowie der Zusammenhang zwischen Nah- und Fernfeld diskutiert.

1 BACKGROUND OF SITE CHARACTERIZATION IN MOBILE BELT

Japan is located in the recent mobile belt. For any geological and long-term assessment, therefore, crustal movement, earthquake and earthquake fault, volcanic activity and hydrothermal action are important facors.

Materials of natural barrier related to geological isolation are largely classified into fractured hard rocks (granite rocks) and layered soft rocks (sedimentary rocks) in Japan. Mainly fractured hard rocks are described here.

It is often seen that many fractures are present and they contain unhardened altered clay on their surfaces. Although the presence of clay is a disadvantage from the mechanical stability standpoint, it is rather an advantage in relation to permeability and migration of nuclides.

It is not necessarily possible to specify areas where fractures are many, but they often can be seen in areas where crustal movement is noticeable, and in such areas active faults and active fold also exist.

2 DISTRIBUTION OF FAULT/SHEARED ZONE

Fractures are classified into joints and faults. The dominant direction of joints and faults are often identical.

As quantities for indicating the scale of faults, there are length, width of sheared zone and slip length. These quantities are distributed log-normally, as shown in Fig.1. And there are clear correlations between these quantities (Fig.2). Therefore, if one quantity can be measured, other quantity can be obtained. Analyses using air-photo's and surface surveys can readily measure length in farfield whitch easily appear as topography, and bore holes and investigation adits can readily measure widths in nearfield.

Another major feature of faults in terms of geometric distribution is that large faults are often surrounded by congestion of smaller faults and joints. Fig.3 shows an example case. In zones congested with fractures, sttrength is smaller compared with the peripheries and they tend to cause faults. Also, when a fault is made, the part formes the weak zone and gap

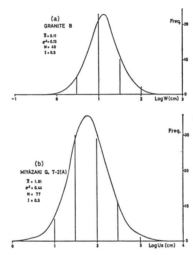

Figure 1. An example distribution of fault/sheared zone. (a)Width of sheared zone. (b)Slip length.

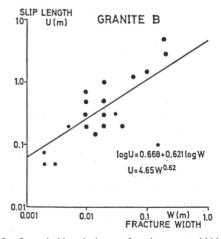

Figure 2. Correlation between fracture quantities. (a)Width of sheared zone and slip length.

159

is accumulated there. This fact can be confirmed in earthqake faults.

By combining the theory of elastic dislocations and FEM, deformative behavior of rockmass around a large fault accompanying the slip can be obtained as shown in Fig.4. Where strain of periphery and the quantity of slips of small fault correspond with each other, and thereby the fact of Fig.3 can be explained specifically (Ohtsuka & Kojima, 1979). And the fact that a "sound rock block" exsists in a part surrounded by fault larger than certain scale, can also be explained.

Fig.5 is the probability model of faults obtained by above-mentioned principles. The average scale of a block surrounded by faults having 100cm or more of gap is more or less 60m in side, and this corresponds to Fig.3.

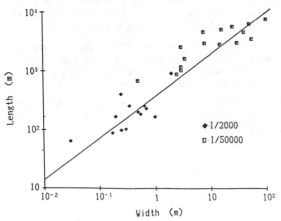

Figure 2-(b). Width and length of sheared zone.

3 FRACTURE PERMEABILITY IN NEARFIELD AND FARFIELD

The permeability of natural barrier, one of the most important factors related to the geological isolation of radioactive waste, is studied on fractured rock-masses.

Fig.6 shows the hydraulic conductivity of a cylindrical testpiece containing joints filled with various materials, and the stress (depth) dependency. The figure explains that, depending on the permeability differs, and the depth dependency decreases by the presence of unconsolidated filling materials. It is also understood that unconsolidated filling materials act as a cushion to the variation of permeability due to temperature changes (Kojima and Koike, 1984).

Fig.7 is an example showing the dispersion of permeability of single fractures in rockmass. A correlation between both was obtained by drilling boreholes in grid from into a fracture surface of 25 X 25 m located several meters below ground level, and by measurement of apertures and observation of hydraulic conductivity using the injection test. The figure explains that, even in the same fracture surface, permeability largely differs by presence of filling materials and by variability of aperture. And discoloration by weathering on fracture surface shows flow paths in this surface.

Fig.8 shows the correlation between hydraulic conductivity of rockmass and depth, the former being the mean value of those measured at 10m intervals. In sections where joints are filled with much clay shows differnt trend. it corresponds to Fig.6.

As the permeability of rockmass noticeably differ depending on the condition of fracture surface and depth as described above, compensasion and geological interpretation regarding these influential factors are needed.

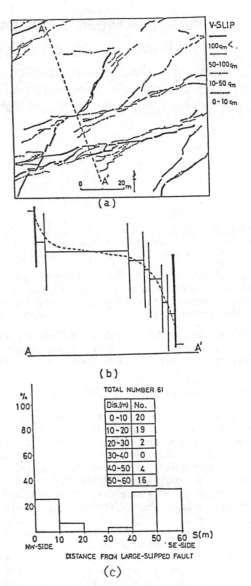

Figure 3. Aspect & fracture density of smaller faults.
 (a)A reverse fault system in the Joogashima Island.
 (b)Accumulation diagram of fault displacement along A-A' profile.
 (c)Fracture density of smaller faults in a place of the Joogashima Island (sedimentary rock in Tertiary).

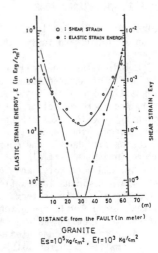

Figure 4. Deformative behavior of rock mass around a large fault accompanying the slip.

Fig.9 shows the relationship between the hydrauric conductivity and width of sheared zone obtained by applying these treatments to the result of injection tests conducted to boreholes of granite. This example is useful as a reference in conjunction with the difference of permeability between large faults appearing in farfield and samll faults appearing in nearfield, and with how to give the representative value of hydraulic conductivity in each zone.

4 CORRELATION OF FRACTURES BETWEEN NEARFIELD AND FARFIELD

Between nearfield and farfield, the size of rockmass differs noticeably. Therefore, when considering the flow of groundwater and migration of nuclides between both, data obtained from each field are noticeably different in dimension of rockmass and accuracy of measurement and are rather not applicable as are.

In several granite regions, measurements were performed on rockmasses of different sizes about the length of fault, width of sheared zone and density. An example is shown in Fig.10 & 11. The axis of abscissas corresponds to the size of rockmass by the map scale used for measurement, and the axis of ordinates gives the mean value of each. Even though the size of given fractures are different, the distribution characteristic of fractures is nearly identical between nearfiled and farfiled (Table-1). Human eyes has selected the size of fractures given on the map nearly proportionately to the scale of reduction. Also, there is an statistical self-similarity between factors of these fractures. Table-1 shows this similarity on the fractal dimension. For each factor nearly the same dimension is obtained.

Fig.12 (a) shows a forecast of geometrical distribution of faults of the nearfield scale from a 1/2000 geological map and using relations of Fig.10 only, and it corresponds to the fault model of nearfield (faults among sound blocks). From the data of farfield, the detailed fracture by using the fractal dimension. By adjusting it to the measured minimum length of nearfield, density of fracture is obtained close to the measured value. It corresponds with the joint model of nearfield (in a sound block) (Fig.12 (b)).

Fig.13 (a) & (b) correspond to Fig.12 (a) & (b) respectively. On Fig.13 (a) & (b), fractures continued to a cavern □ are only illustrated. Figure (d) are examples of simulation of the flow of water into underground caverns of (b) using FEM with linear elements and (c) in case of homogeneous flow. In Fig.13 (d) it is seen that the network of fracture congested zone of (b) forms flow paths (the black colour becomes darker as the flow rate increases).

Though the range of applicability of the similarity of physical properties of faults and joints has to wait for the results of future studies, at least in this example, it is possible to assume the pattern of flow path and trend of flow rate/velocity from the geometrical similarity of fractures.

REFERENCES

Kojima,K. Y.Ohtsuka & T.Yamada 1981. Distributions of fault density and fault dimension in rock mass and some trials to estimate them. Jour.Japan Soc.Eng.Geol. 22-1.

Kojima,K. & Y.Koike 1984. Infuluence of stress state and temperature in the permeability of rock core samples containing various joint infills. Proc.16th Sympo.Rock Mech. JSCE Japan.

Kojima,K. 1984. Geotechnical characteristics of Japanese granite for geological isolation of radioactive

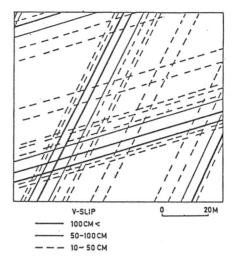

Figure 5. Probability model of faults at the Jogashima site (sedimentary rock).

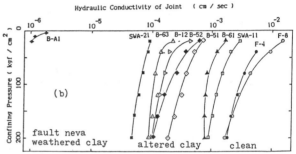

Figure 6. Relations between hydraulic conductivity of joint and confining pressure in labo. test.

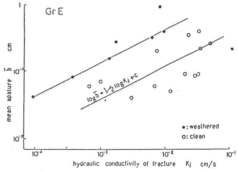

Figure 7. An exsample of the variation of aperture and permeability along the same joint plane (25 x 25 m in area) in near surface granite.

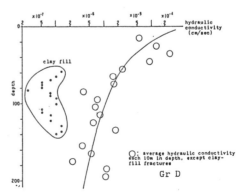

Figure 8. Permeability trend for the bore hole of Gr.D region.

waste. Annual Report of the Eng. Research Inst. Faculty of Eng. Univ.of Tokyo 43. Japan.
Nishimura,T. & K.Kojima 1986. Modelling of flow path in fractured rock mass. Proc. Ann.Meeting Japan Soc. Eng.Geol.
Ohtsuka,n. 1979. A geotechnical study of fault deformations. Dr.Thesis Faculty of Eng. Univ.of Tokyo Japan.

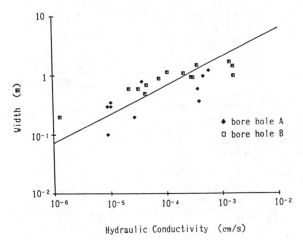

Figure 9. Relations between hydraulic conductivity of fracture and width of sheared zone in Gr.H region.

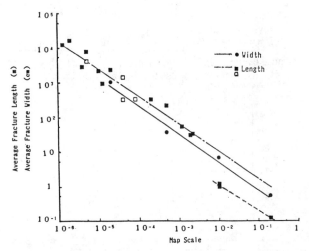

Figure 10. Relations between map scale and average fracture length, width in Gr.D & Gr.H region.
----- another method in measurment

Table 1. Average, variance and fractal dimension of fracture width in Gr.H region.

	Map Scale	Avreage (log(cm))	Variance (log)	Fractal Dimension
Field Survey	1/5	-0.248	0.439	0.931
Adit Expantion Map	1/100	0.844	0.481	0.985
Geological Map	1/2000	1.557	0.234	1.000
Geological Map	1/50000	3.011	0.487	0.969

Average Fractal Dimension 0.984±0.069

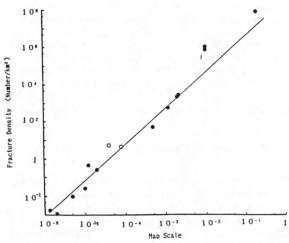

Figure 11. Relations between map scale and fracture density in various fields (Gr.D & Gr.H).

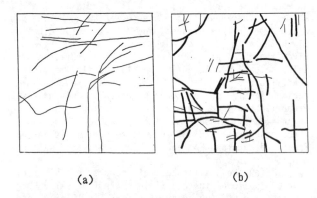

(a)   (b)

Figure 12. Geometrical distribution of fracture in Gr.H region. Quadrangle 1 x 1 m in scale.

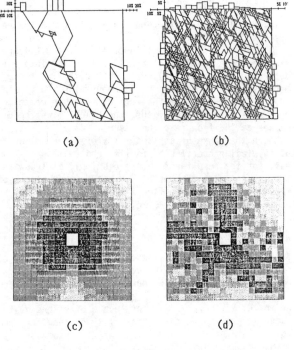

(a)   (b)

(c)   (d)

Figure 13. Network model of fracture at the site of Gr.D. Quadrangle 50 x 50 m in scale.

Prediction of hydraulic fracture extension by a finite element model
Prédiction de prolongement de la fracture hydraulique avec un modèle d'éléments finis
Die Vorhersage von hydraulischen Spaltverlängerungen mit Finite Element Modellen

H.KURIYAGAWA, National Research Institute for Pollution and Resources, Japan
I.MATSUNAGA, National Research Institute for Pollution and Resources, Japan
T.YAMAGUCHI, National Research Institute for Pollution and Resources, Japan
G.ZYVOLOSKI, Los Alamos National Laboratory, USA
S.KELKAR, Los Alamos National Laboratory, USA

ABSTRACT: Fracture extension created by fluid injection in a hot dry rock reservoir at the depth of 4 km and the temperature of 300 °C is studied using a finite-element heat and mass transfer code. In this paper, the attension is focused on the relative extension of two competing fractures with different closure characteristics.

RÉSUMÉ: On etudie prolongement d'une fissure produit dans un reservoir du roche chaud et sec per injection liquide à un fond de 4km et une temperature de 300 °C avec une code de transfer d'element fini de chaleur et de masse. On met au point la prolongement relative de deux fissures concourant avec divers caractères de clôture.

ZUSAMMENFASSUNG: Spaltverlängerung erschafft mit flussigem Einspritzen in ein heisees, trockenes gesteinreservoir bei 4km tiefe und 300 °C ist untersucht mit einem Wärme- und massen transport koden von begrenzten elementen. Hier ist die emphase auf die relative verlangerung von zwai konkurrierten bruchen mit verschiedenen verschlusseigenshaften.

1 INTRODUCTION

At Los Alamos National Laboratory, more than ten hydraulic fracturing experiments have been conducted to stimulate a hot dry rock reservoir. Two hydraulic fracturing attempts, Exp.2059 (Kelkar 1985) and Exp.2062(Robinson 1985), successfully established a large fracture system connecting an injection well,EE-3A, and a production well, EE-2. The initial closed loop test was carried out from May through June, 1986 to get information on volume, impedance and temperature of the reservoir. The fracture systems created by the two experiments show the characteristics that are different from each other. For example, higher pumping pressure was required to extend fractures in Exp.2062 than in Exp.2059. In order to understand the combined behavior of two fracture, the computer simulation techniques were used.

The parameters for the simulation are chosen to fit the experimental data. Two models are introduced to obtain a good match between the simulation and the experiment. The extension of fractures is discussed. The fracture radius is predicted when water is pumped into the combined two-layer fracture systems.

2 DESCRIPTION OF EXPERIMENT 2059 AND 2062

Experiment 2059 was conducted in May, 1985. A packer was run to 3,519m and the bottom hole was sanded back to a depth of 3,722m. In Exp.2062, which was performed in July, 1985, a packer was set at 3,653m and the total depth was 3,837m. The test intervals are illustrated in Fig.1. The pressure histories and flow rate in these tests are shown in Fig.2(a) and (b). Total injected water volume was about 1,600m^3 in Exp.2059 and about 5,722m^3 in Exp.2062. From these figures the relation of flow rate to fracture extension pressure is obtained and the results are plotted in Fig.3. The fracture extension pressure in Exp.2059 was lower than that in Exp.2062 at the same flow rate.

Temperature surveys were run in the openhole section of EE-3A. Fig.4 shows post-experiment logs of these experiments. Major depressions exist at 3,570m and 3,660m and small anomalies appear at 3,700m in Exp.2059. Fractures at 3,660m and 3,690m took the bulk of fluid during Exp.2062, and in addition, a new fracture was stimulated at 3,750m. The fracture system created by each experiment consists of several fractures. But in the simulation model, each fracture system is represented by one major fracture, because it is difficult to characterize each fracture independently for modeling.

3 MODELING OF FRACTURE SYSTEM AND RESULTS OF SIMULATION

A model of the fracture system showing the two

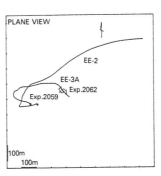

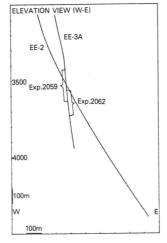

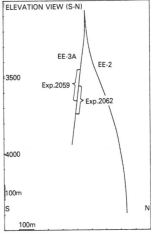

Figure 1. Test interval in Exp.2059 and Exp. 2062.

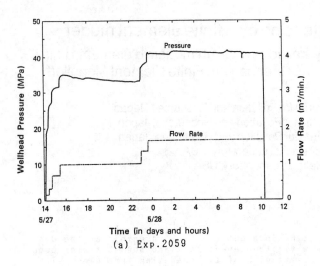

(a) Exp.2059

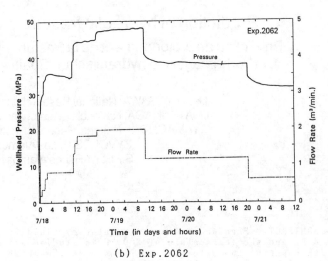

(b) Exp.2062

Figure 2. Obseved pressure response.

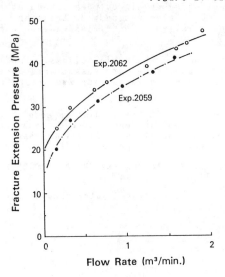

Figure 3. Relation of flow rate with fracture extension pressure.

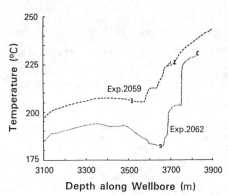

Figure 4. Temperature logs of post-experiment 2059 and 2062.

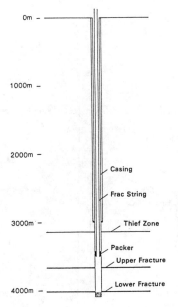

Figure 5. Concept of two-layer fracture model.

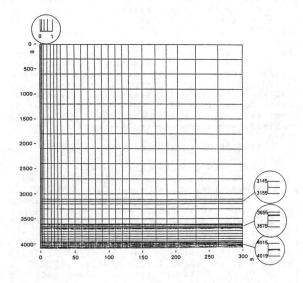

Figure 6. Finite element grid for simulation.

164

layer fractures is given in Fig.5. A packer is placed at 3,500m. The upper fracture represents those fractures stimulated in Exp.2059, while the lower fracture represents those fractures stimulated in Exp.2062. A finite element grid in a radial geometry is shown in Fig. 6. It is very important to characterize these two fracture sytems for the computer simulation. In this paper, the permeability was chosen as the main parameter to characterize the fracture system. In the first model the permeability is assumed to be constant and independent of the local pressure (linear permeability model). In the other model, the aperture is assumed to be related to the local pressure and the permeability is expressed as a fuction of the aperture (Kelkar 1986) (non-linear permeability model). The Finite Element Heat and Mass Transfer Code (FEHM) (Zyvoloski 1983) is used for the simulation. In this computer code, heat and mass transfer is allowed among the fracture, the wellbore and the strata. All boundaries are fixed as "no flow". The parameters used in this simulation are listed in Table 1. Flow rate is kept at $0.8m^3$/min for the simulation.

3.1 Linear permeability model

The permeability is assumed to be $10^{-9}m^2$ for the upper fracture and $10^{-10}m^2$ for the lower fracture, as the upper fracture representing Exp.2059 took more water. The Wellhead pressure response is dicussed for the following three cases:
1) Case 1: upper fracture alone takes water
2) Case 2: lower fracture alone takes water
3) Case 3: both fractures take water.
In the cases where one of fractures takes all water, the permeability of the other fracture is assumed to be equal to that of the surrounding rock.

The computer model results for are shown in Fig.7 by the broken line for case 1, by the dotted line for case 2 and the solid line for case 3. Note the pressure keeps increasing with the time for all cases. This tendency is somewhat different from the experimental results which show a leveling off of the pressure. We found that it is difficult to get good match with experimental data when linear permeability model is introduced. The extension pressure for the lower fracture (case 2) is about 5 MPa higher than that of the upper fracture (case 1). This pressure difference is close to that between Exp.2059 and Exp.2062. When both fractures take water, the extension pressure is about 10 MPa lower than that for case 2.

3.2 Non-linear permeability model

Non-linear permeability model is then discussed. An aperture law is introduced to consider a non-linear permeability. The fracture aperture w is related to the local pressure P_w by an emprical equation (Kuriyagawa 1986),

$$w = w_o \exp(A \cdot R_p) \qquad P_w \leq \sigma_n$$
$$w = w_1 \exp(B \cdot R_p + C) \qquad P_w \geq \sigma_n \qquad (1)$$

where w_o, w_1, A, B and C are constant. R_p is the ratio of the local pressure inside the fracture with the earth stress normal to the fracture σ_n ($R_p = P_w/\sigma_n$). And A is given by $\log(1/w_n)$. The unit of w is mm. We assume that $w_o = 0.00005$ for the lower fracture, and $w_1 = 5$, $b = 0.811$ and C=1.9096 for both fractures, σ_n is assumed to be 27 MPa for the upper fracture and 30 MPa for the lower fracture. By considering that the lower fracture is easier to open by pressurization compared with the lower fracture, we assume a larger value for w_o for the upper fracture. First of all, w_o is assumed to be 0.005. Fig.8 illustrates the relation between the local pres-sure and the fracture aperture. The permeability k is given as a fuction of w by the parallel plate law:

$$k = f w^2 / 12 \qquad (2)$$

In the simulation, constant f is assumed to be 1. Calculations were made with these aperture laws for the three cases as same as a linear permeability model.

To match the simulation result with the experimental data, w_o for the upper fracture is assumed to 0.005. The pressure is shown in Fig.9 by the broken line for case 1 and the

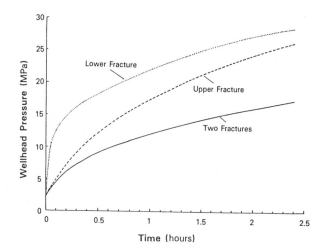

Figure 7. Calculated wellhead pressure with linear permeability model.

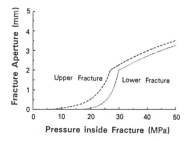

Figure 8. Fracture aperture vs pressure used for modeling.

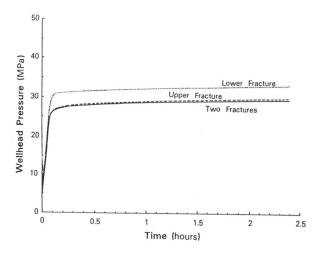

Figure 9. Calculated wellhead pressure with non-linear permeability model.

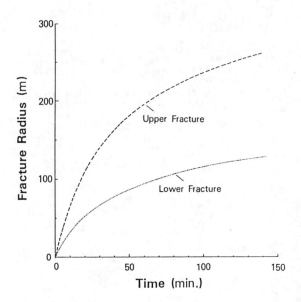

Figure 10. Fracture extension with time in case 1 and case 2.

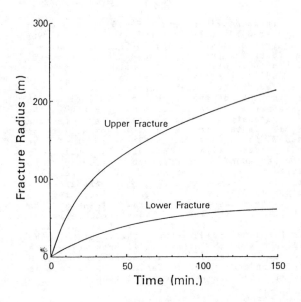

Figure 11. Fracture extension with time in case 3.

dotted line for case 2 at a flow rate. The difference of the pressure between the upper and the lower fractures was about 5MPa, which is close to the experimental result shown in Fig.3. The extension of the fracture radius during pumping is shown in Fig.10.

The computer simulation result of the pressure vs time response is illustrated in Fig.9 by the solid line when both fractures are pressurized by pumping water. The pressure showed a slightly smaller value than that obtained by pressurization of the upper fracture. In this case, as shown in Fig.11, the upper fracture extended slightly slower, while the lower fracture grows almost half as much compared to the results for pressurization of either the upper or lower fracture alone.

4 DISCUSSION

The pressurization of two-layer fractures with different properties was simulated. The fracture characteristic chosen for variation was the permeability. It was found that with linear permeability model we cannot obtain sharp leveling off of the pressure which is usually recognized in the hydraulic fracturing tests.

On the other hand, a good match of the pressure between the simulation and experimental results was obtained with a non-linear permeability model. As in the case of Exp. 2059 and Exp.2062, if the difference of the independent fracture extension pressure is less than 5MPa, both fractures grow although the upper fracture extends faster when water is pumped into both fractures.

ACKNOWLEDGMENTS

The authors express great thanks to New Energy Development Organization and Sunshine headquarters,MITI which supported us.

REFERENCE

Kelkar,S.M.1985. Exp.2059, EE-3A packer test No.3 hydraulic data. Office Memo. Los Alamos National Lab.

Kuriyagawa,M, G.Zyvoloski, S.Kelkar & Z.Dash. 1983. Simulation of fracture extension on experiment 2061. Office Memo. Los Alamos National Lab.

Kelkar,S, G.Zyvoloski & Z.Dash. 1986. Pressure Testing of a high temperature naturally fractured reservoir. 11th Workshop on Geothermal Reservoir Engineering at Stanford University.

Table 1. Parameters for simulation model

Parameter		Value
Permeability	x-direction	$5 \times 10^{-17} m^2$
	y-direction	$5 \times 10^{-19} m^2$
	Thief zone	$1 \times 10^{-12} m^2$
Porosity	Matrix	0.001
	Fracture	1.0
Thermal conductivity		$2.7 W/m^\circ C$
Initial temperature	$0 < y < 750m$	$12.35 - 0.056 y \,^\circ C$
	$750m < y$	$72.29 + 0.00996 y +$
		$1.05 \times 10^{-5} y^2 \,^\circ C$
Rock density		$2,500 \ kg/m^3$
Rock specific heat		$1,000 \ J/kg^\circ C$
Initial pressure		$0 \ MPa$

Robinson,B, Z.Dash, D.Dressen, S.Kelkar, J. Miller, B.Restine, T.Yamaguchi & G.Zyvolosli. 1985 Experiment 2062 results.Office Memo. Los Alamos National Lab.

Zyvoloski,G. 1983. Finite element methods for geothermal resevoir simulation. Int.J. for Numerical and Analytical Method in Geomechanics. 7: 75-96.

Entwurf und Dimensionierung eines Endlagerbergwerkes für radioaktive Abfälle im Salzgebirge

Design and dimensioning of a repository mine for radioactive wastes in rock salt
Plan et dimensionnement d'une mine pour le stockage des déchets radioactifs dans le sel gemme

M.LANGER, Prof.Dr., Bundesanstalt für Geowissenschaften und Rohstoffe, Hannover, Bundesrepublik Deutschland

ABSTRACT: The paper describes the work during the planning and construction of underground repositories for radioactive wastes in rock salt. In the safety assessment the geotechnical stability analysis is a critical part. Such an analysis comprises an engineering-geological study of the site, laboratory and in-situ experiments, geomechanical modelling, and numerical static calculations. Comments on new research data on the mechanical behaviour of rock salt are given. The Gorleben repository project is an example demonstrating how modern geotechnical research and investigation methods can help to secure the safe isolation of wastes from the biosphere.

RESUME: Le présent article expose les études pour la conception et la réalisation de stockages souterrains de déchets radioactifs dans le sel gemme. Pour la démonstration de la sécurité l'analyse de la stabilité géotechnique est un point essentiel. Une telle analyse comprend l'étude géologique et géomécanique du site, des essais en laboratoire et in situ, des modélisations et des calculs statiques (recherches numériques). Les derniers résultats de la recherche sur le comportement mécanique des massifs de sel sont communiqués ici. L'exemple de projet de stockage de Gorleben montre comment, par la recherche et les méthodes modernes de la géotechnique, il est possible de prouver que les déchets peuvent être éliminés de la biosphère de manière sûre.

ZUSAMMENFASSUNG: Es werden die geotechnischen Arbeiten für die Planung und den Bau von Endlagerbergwerken für radioaktive Abfälle im Salzgebirge besprochen. Im gesamten Sicherheitsnachweis nimmt die Analyse der Standsicherheit eine zentrale Stellung ein. Eine solche Analyse umfaßt die ingenieurgeologische Untersuchung des Standortes, Laborversuche, in-situ Messungen, Bildung geomechanischer Modelle und numerische statische Berechnungen. Neuere Forschungsergebnisse über das mechanische Verhalten von Salzgesteinen werden mitgeteilt. Im Projekt Endlagerbergwerk Gorleben läßt sich beispielhaft zeigen, wie moderne geotechnische Forschung und Untersuchungsmethoden dazu beitragen, den sicheren Abschluß von Abfällen gegenüber der Biosphäre zu gewährleisten.

1. Schutzziele und Sicherheitskonzept der Endlagerung radioaktiver Abfälle

Alle Überlegungen zur Sicherheit von Endlagerbergwerken haben sich auf die Schutzziele der Endlagerung auszurichten, wie sie z.B. in Deutschland im Atomgesetz, den Strahlenschutzbedingungen und Empfehlungen der Reaktorsicherheitskommission, festgelegt sind [1]. Insbesondere muß eine Störung der Langzeitstabilität des Ökosystems in der Nachbetriebsphase ausgeschlossen werden, d.h. der Transport gefährlicher Mengen von Radionukliden in die Biosphäre muß verhindert werden. Um Gesundheit und Sicherheit der Menschen über diesen Zeitraum zu gewährleisten, werden mehrere unabhängige technische und natürliche Barrieren im gekoppelten und vernetzten System "Aball/Endlagerbergwerk/geologisches Medium" zur Behinderung der Freisetzung von Schadstoffen herangezogen (multiple barrier system). Es sind dies - Technische Barrieren (Abfallform, -verpackung) -Gebirgsmechanische Barrieren (Bohrlochverfüllung/ -verschluß, Versatzmaterial, Dämme, Wirtsgestein) - Geologisches Barrieren (geologisches Umfeld als geohydraulische Barriere).

Die Entwicklung eines realistischen und prüffähigen Sicherheitskonzeptes zur Erfüllung der Schutzforderungen unter Berücksichtigung aller vorhandenen Barrieren ist äußerst schwierig und unterliegt zur Zeit großer internationaler Forschungsaktivität. Ein mögliches Konzept für eine umfassende Sicherheitsanalyse ist vom Verfasser aufgezeigt worden [2]. Dieses Konzept (Abb. 1) beinhaltet die getrennte Analyse der einzelnen Barrieresysteme (technische, gebirgsmechanische, geologische), die Analyse der physikalischen und geochemischen Prozesse im Nah- und Fernfeld des Endlagers sowie eine zusammenfassende Szenarien- bzw. Störfallbewertung für die Beruteilung technischer Barrieren steht die probabilistische Risikoanalyse zur Verfügung. Die Bewertung gebirgsmechanischer Barrieren erfolgt durch den geotechnischen Standsicherheitsnachweis. Das geologische System wird durch die Prognose zukünftiger geochemischer, hydrogeologischer und tektonischer Vorgänge analysiert ("prognostische Geologie"). In der zusammenfassenden Störfallanalyse wird das Zusammenwirken aller Barrieren bei bestimmten theoretisch denkbaren Ereig-nissen (Störfälle), die eine Gefahr der Freisetzung von Schadstoffen in die Biosphäre (Freisetzungspfade) bewirken könnten, untersucht.

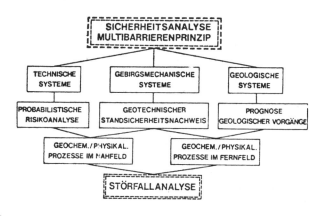

Abb. 1: Konzept der Sicherheitsanalyse

2. Prinzip des geotechnischen Standsicherheitsnachweises

Man erkennt, daß geotechnische und gebirgsmechanische Arbeiten ein Kernstück des Sicherheitsnachweises für die Deponierung radioaktiver Abfälle sind, da sowohl technische Barrieren (z.B. Versatzmaterial, Verschlußbauwerke) als auch geologische Barrieren, also das Wirtsgestein, berücksichtigt bzw. bewertet werden müssen. Da bei einer solchen Bewertung der Barrieren, als auch für den sicheren und wirtschaftlichen Entwurf des Endlagerbauwerks, sowohl das Tragverhalten als auch die Langzeitstabilität des Endlagerbereiches eine wesentliche Rolle spielen, kann ein solcher Sicherheitsnachweis nicht rein bauingenieurmäßig geführt werden, sondern muß geologische Faktoren und Prozesse integrieren; Standsicherheitsbegriffe und Sicherheitsfaktoren des normalen Ingenieurbaus reichen hier nicht aus.

Von der Bundesanstalt für Geowissenschaften und Rohstoffe, Hannover wurde deshalb für die Belange der Endlagerung, und ausgerichtet auf die zu erreichenden Schutzziele, ein komplexer geotechnischer Standsicherheitsnachweis entwickelt, der auf folgenden Grundüberlegungen beruht [3].

Wegen der Komplexität der zu berücksichtigenden Randbedingungen kann der Standsicherheitsnachweis für Endlagerhohlräume im konkreten Fall nur durch eine Kombination verschiedener Untersuchungen und Berechnungen gelingen. Ingenieurgeologische Erkundungen, geotechnische Untersuchungen, felsmechanische Messungen, statische Berechnungen, meßtechnische Überwachungen und bergbauliche Betriebserfahrung müssen zusammenwirken (Abb. 2).

Dem rechnerischen Teil des Standsicherheitsnachweises kommt im Rahmen der Gesamtanalyse der Sicherheit eine besondere Bedeutung zu, da im Zuge der Planung und Planfeststellung bereits verläßliche und überzeugende Nachweise der Sicherheit geliefert werden müssen. Darüberhinaus werden durch die zusätzliche thermische Belastung (bei wärmeentwickelnden radioaktiven Abfällen) Spannungsänderungen sowohl im Nahfeld (Bergwerk) als auch im Fernfeld (Gebirge) hervorgerufen, die sich bisherigen bergmännischen Erfahrungen und meßtechnischen in situ-Beobachtungen entziehen.

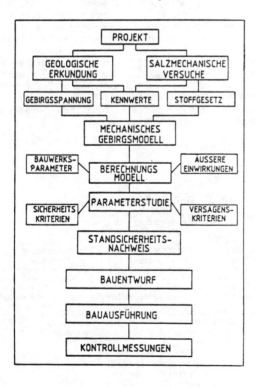

Abb. 2: Prinzip des geotechnischen Standsicherheitsnachweises

Erstes Ziel solcher Standsicherheitsberechnungen ist - den genannten Kriterien folgend - der Nachweis, daß die durch den Hohlraumausbruch hervorgerufenen Spannungsumlagerungen bruchlos einen Gleichgewichtszustand erreichen, sich keine unzulässigen Konvergenzen und Schäden während der Nutzungszeit (z. B. durch thermische Spannungen) einstellen und die Langzeitintegrität des Gebirges erhalten bleibt. Es müssen also Spannungs- und Verformungsverteilungen im Gebirge berechnet und mit der Grenztragfähigkeit der Gebirgskörper verglichen werden. Dazu gehört vor allem die Formulierung eines mechanischen Gebirgsmodells und des dazugehörigen Berechnungsmodells, Paramterstudien sowie die Festlegung von Sicherheits- und Versagenskriterien.

3. Finite-Element-Programmsystem ANSALT

Im Rahmen eines BMFT-Forschungsvorhabens wurde von der BGR in Zusammenarbeit mit CONTROL DATA GmbH, Hamburg, das Finite-Element Programmsystem ANSALT entwickelt [4]. Zielsetzung dieser Entwicklung war die Erstellung eines nach neuesten Erkenntnissen der Salzmechanik und der Rechentechnik zur Lösung nichtlinearer thermomechanischer Problemstellungen konzipierten Rechenprogramms. Damit sollten optimale technisch-wissenschaftliche Voraussetzungen für gebirgsmechanische Untersuchungen geschaffen werden.

Das Programmsystem besteht aus vier in sich geschlossenen aber miteinander korrespondierenden Teilprogrammen
- ANSPRE, einem Preprocessor zur einfachen und sicheren Aufbereitung der Eingabedaten,
- ANTEMP, einem Finite-Element-Programm zur Berechnung des Wärmetransportes,
- ANSALT, einem Finite-Element-Programm zur Lösung thermo-mechanischer Problemstellungen,
- ANSPOST, einem Postprocessor zur grafischen Auswertung und Darstellung der Rechenergebnisse.

Das Programm ANSALT enthält in seiner Elementbibliothek alle ein-, zwei- und dreidimensionalen isoparametrischen Elemente zur Diskretisierung untertägiger Hohlraumstrukturen, die zur Bearbeitung gebirgsmechanischer Problemstellungen notwendig sind. Darüber hinaus werden bestimmte Rand- und Anfangsbedingungen, die sich aus der besonderen Aufgabenstellung bergbaulicher Tätigkeit unter Tage ergeben, in geeigneter Weise simuliert.

4. Geotechnisches Stoffmodell für Steinsalz

Eine Dimensionierung von Pfeilern, Kavernen und sonstigen Grubenbauen auf der Grundlage des erläuterten Konzepts einer geotechnischen Standsicherheitsanalyse ist nur möglich, wenn das mechanische Verhalten der Gesteine mit genügender Genauigkeit erfaßt ist. Die Grundlagen der theoretischen und experimentellen Behandlung von Deformationsvorgängen im Salzgestein ergibt sich aus einer Analyse der Phänomene Kriechen, Plastizität und Kriechbruch. Darüber hat es in den letzten Jahren verschiedene zusammenfassende und bewertende Übersichten in der Fachliteratur und in Spezialsymposien gegeben [5].

Salzgestein besitzt wie alle Materialien unter mechanischer und thermischer Beanspruchung elastische und inelastische Stoffeigenschaften. Eine konstitutive Beziehung für das Materialverhalten von Salzgesteinen muß sowohl elastische Deformationen, einschließlich der thermischen Verformbarkeit, als auch Kriech- und Bruchdeformationen einbeziehen.

Kontinuumsmechanische Berechnungen erfordern außerdem eine allgemeine dreidimensionale Darstellung der Stoffgesetze. Entgegen dieser Notwendigkeit werden Stoffgesetze allerdings i.a. in eindimensionaler Schreibweise formuliert, weil sie vorwiegend auf der

Grundlage der Ergebnisse einaxialer Laborversuche entwickelt werden. Mit folgenden, versuchsmäßig bestätigten Voraussetzungen
- volumentreue Verformung,
- ein überlagerter hydrostatischer Spannungszustand ist ohne Einfluß auf das Kriechen,
- Isotropie (Kriechverformungen und Spannungen sind kollinear),
- Materialverhalten unter Kompressions- und Extensionsbeanspruchung ist gleich,

kann eine Verallgemeinerung der Materialgleichungen für die Kriechvorgänge formuliert werden. Ein aus umfangreichen Laborversuchen der BGR entwickeltes pragmatisches Stoffmodell, das nur die allgemeingültigen Anteile des Formänderungsverhaltens zusammenfaßt, stellt das in Abb. 3 vorgeschlagene Stoffgesetz dar [6]. Darin setzt sich das Gesamtdehnungsinkrement ϵ^t zusammen aus
- einem elastischen Verzerrungsinkrement ϵ^{el}
- einem thermischen Verzerrungsinkrement ϵ^{th}
- einem Kriechinkrement ϵ^{cr} (zeitabhängige irreversible Gestaltsänderungen)
- einem Bruchverformungsinkrement ϵ^f (irreversible, geschwindigkeitsunabhängige Formänderung mit Ausbildung diskret verteilter Bruchflächen und Dilatation).

Für das Kriechverzerrungsinkrement ϵ^{cr} ist nur stationäres Kriechen als wesentlicher Anteil der Langzeit-Kriechdeformation berücksichtigt. Die mathematische Formulierung basiert auf einem von Munson und Dawson [7] entwickelten Ansatz für das Kriechen als Summe dreier unterschiedlicher Kriechmechanismen
- Versetzungsklettern bei niedrigen Spannungen und hohen Temperaturen
- Übergangsversetzungsmechanismus bei Lagerstättentemperaturen und -spannungen
- Versetzungsgleiten bei hohen Spannungen.

Das Bruchverzerrungsinkrement wird bei Überschreiten des Kriteriums für die Langzeitfestigkeit in Form einer assoziierten Fließregel und einer erweiterten Drucker/Prager-Fließbedingung berücksichtigt.

$$\dot{\epsilon}_{ij} = \dot{\epsilon}_{ij}^{el} + \dot{\epsilon}_{ij}^{th} + \dot{\epsilon}_{ij}^{cr} + \dot{\epsilon}_{ij}^{f} \qquad (1)$$

$$\dot{\epsilon}_{ij}^{el} = -\frac{\nu}{E}\dot{\sigma}_{kk}\delta_{ij} + \frac{1+\nu}{E}\dot{\sigma}_{ij} \qquad (2)$$

$$\dot{\epsilon}_{ij}^{th} = \alpha_i \dot{T}\delta_{ij} \qquad (3)$$

$$\dot{\epsilon}_{ij}^{cr} = \frac{3}{2}\frac{\dot{\epsilon}_{eff}^{cr}}{\sigma_{eff}}s_{ij} \, . \qquad \dot{\epsilon}_{eff}^{cr} = \sum_{i=1}^{3}{}^i\dot{\epsilon}_{eff}^{cr}(S,\sigma_{eff},T)$$

$${}^1\dot{\epsilon}_{eff}^{cr} = A_1 \exp(-Q_1/RT)(\sigma_{eff}/\sigma^*)^{n_1}$$

$${}^2\dot{\epsilon}_{eff}^{cr} = A_2 \exp(-Q_2/RT)(\sigma_{eff}/\sigma^*)^{n_2}$$

$${}^3\dot{\epsilon}_{eff}^{cr} = 2[B_1\exp(-Q_1/RT) + B_2\exp(-Q_2/RT)] \times \sinh(D<\frac{\sigma_{eff}-\sigma_{eff}^\circ}{\sigma^*}>) \qquad (4)$$

$$\dot{\epsilon}_{ij}^{f} = \frac{1}{\eta}<F>\frac{\partial F}{\partial\sigma_{ij}}$$

$$F = \alpha (\frac{|I_\sigma|}{\sigma^*})^{m-1} I_\sigma + \sqrt{\mathbf{II_s}} - k \qquad (5)$$

Abb. 3: Stoffgesetz für rheologisches Verhalten von Steinsalz

Abhängig von der Größe des Seitendrucks bzw. der isotropen Spannung treten Brucherscheinungen erst nach Überschreiten einer bestimmten Größe der Verzerrungsgeschwindigkeit auf. Unterhalb dieser Grenze ist Steinsalz unbegrenzt tragfähig und verformt sich beliebig bruchlos.
Die untere Einhüllende aller Bruchfestigkeiten ergibt eine Minimalfestigkeit für Salzgesteine. Diese Minimalfestigkeit fällt zusammen mit der Grenze, unterhalb der Salzgestein unbegrenzt tragfähig ist und auch kein Kriechbruch mehr auftritt. Sie kennzeichnet damit den Übergang zum stationären Kriechen und stimmt mit der für Versetzungsgleiten ermittelten Beziehung überein (Abb. 4).

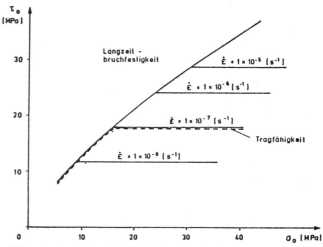

Abb. 4: Langzeittragfähigkeit von Steinsalz

Da für die Langzeitfestigkeit bei unterschiedlichen isotropen Spannungen jeweils unterschiedliche, aber bestimmte minimale Verzerrungsgeschwindigkeiten maßgebend sind, läßt sich für die Langzeitfestigkeit ein nur in Spannungen formuliertes Kriterium angeben:

$$F = \alpha (\frac{|I_\sigma|}{\sigma^*})^{m-1} I_\sigma + \sqrt{\mathbf{II_s}} - k$$

5. In-situ Spannungszustand

Die BGR führt eine Weiterentwicklung des Spannungsmeßverfahrens nach der Überbohrmethode im Felslabor Grimsel der NAGRA (Nationale Genossenschaft für die Lagerung radioaktiver Abfälle) durch [8].

Diese bekannten Verfahren konnten bisher nur in Bohrungen kleiner als 100 m eingesetzt werden. Durch gezielte Geräteentwicklung ist es nunmehr möglich, Deformations- und Spannungsmessungen in Bohrungen in 300 m und mehr Teufe durchzuführen. Das ist insbesondere dadurch erreicht worden, daß mit der Meßsonde auch ein Kleincomputer in die Bohrung eingelassen wird, der die Aufzeichnung der Meßwerte übernimmt ohne daß eine Kabelverbindung durch das Bohrgestänge hindurch notwendig wird.

Diese Untersuchungsmethode zur Bestimmung der in-situ Gebirgsspannung ist auch im Salzgebirge einsetzbar, wobei allerdings das illineare zeitabhängige Deformationsverhalten von Steinsalz berücksichtigt werden muß. Dabei zeigt sich, daß im Gegensatz zur Anwendung im elastisch reagierenden Gebirge folgende Versuchsbedingungen beachtet werden müssen [9]: Zur vollständigen Ermittlung der Entspannungsdeformation ist eine kontinuierliche Messung während des gesamten Überbohrvorgangs erforderlich. Das Verhältnis von Über-

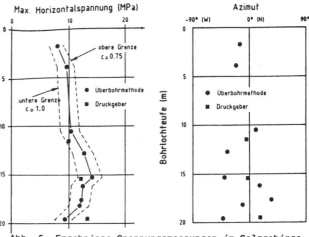

Abb. 5: Ergebnisse Spannungsmessungen im Salzgebirge

169

bohrdurchmesser zu Pilotbohrlochdurchmesser beeinflußt
die Meßergebnisse und ist unbedingt zu berücksichtigen. Die Bohrlochstandzeit von der Erstellung
bis zum Überbohren ist aufgrund der zeitabhängigen
Spannungsrelaxation in die Auswertung einzubeziehen.
Die Gegenüberstellung der mit dem BGR-Überbohrver-
fahren und mit Druckgebern ermittelten Gebirgsspannungen zeigt eine befriedigende Übereinstimmung (Abb.
5). Der parallele Einsatz dieser beiden grundsätzlich
unterschiedlichen Meßmethoden im Salzgebirge wird empfohlen.

6. Berechnung großräumiger Deformationsvorgänge

Im geplanten Endlager Gorleben ist die Einlagerung
hochradioaktiver und wärmeentwickelnder mittelaktiver
Abfälle vorgesehen. Auf die numerischen Sicherheits-
berechnungen für die Planung des Endlagerbergwerks
Gorleben haben deshalb die thermomechanischen Eigenschaften des Salzgebirges besonderen Einfluß und müssen bei der Erstellung geomechanischer Modelle unbedingt berücksichtigt werden. Die geologische Struktur
des Salzstockes Gorleben wird zwar erst nach der untertägigen Erkundung im einzelnen bestimmbar sein, jedoch sind einige Erkenntnisse zur prinzipiellen Innen-
struktur des Salzstocks in Analogie zu anderen Salz-
stöcken Nordwestdeutschlands und durch die Tiefbohrun-
gen in Gorleben vorhanden. Diese sind in einem geolo-
gisch/statischen Modell des Endlagers in Abb. 6 dargestellt.

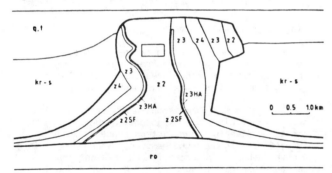

Abb. 6: Geomechanisches Modell Salzstock Gorleben

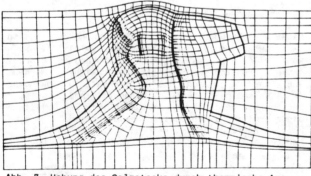

Abb. 7: Hebung des Salzstocks durch thermische Ausdehnung

Von besonderer Bedeutung sind die Anhydritschichten
im Salzstock. Nach Umsetzung dieses Modells in eine
Finite-Element-Struktur lassen sich zum Beispiel je
nach Wärmeinventar des einzulagernden Abfalles thermo-
mechanische Vorgänge im Salzstock und am Salzstockrand
berechnen. Als Beispiel zeigt Abb. 7 die Hebung des
Salzstocks durch die thermische Ausdehnung des Gebirges. Die Hebung des Salzstockes bei der hier zugrunde
gelegten modellhaften Einlagerungsgeometrie beträgt
etwa 2 bis maximal 2,5 m. Abb. 8 macht deutlich, daß
durch das Kriechen des Steinsalzes die zunächst vor-
übergehend vorhandenen Spannungsspitzen in der Umgebung des Einlagerungshohlraumes mit der Zeit abgebaut
werden und einen stabilen Zustand erreichen [10].
Mit dem Dynamik-Teil des Rechenprogramms ANSALT lassen
sich auch dynamische Vorgänge an Hohlräumen berechnen.

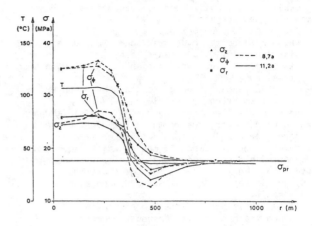

Abb. 8: Abbau von Spannungsspitzen durch Kriechen

Bewegungen an den Hohlraumwandungen beim Durchgang
einer Erschütterungswelle lassen sich so verdeutli-
chen [11]. Dabei bestätigt sich die Erfahrung, die in
Japan und USA in Erdbebengebieten gemacht wird: Un-
tertagehohlräume sind gegenüber dynamischer Beanspruchung sicherer ausgelegt als Baukonstruktionen an der
Oberfläche.

7. Schrifttum

[1] Bundesminister des Innern 1983. Sicherheitskrite-
rien für die Endlagerung radioaktiver Abfälle im
Bergwerk, Bundesanzeiger 35 2, 45-46.
[2] Langer, M. 1980. Grundlagen des Sicherheitsnachweises für ein Endlagerbergwerk im Salzgebirge,
Proc. 4th Ntl. Tagung Felsmechanik, Aachen, 1980,
Deutsche Gesellschaft für Erd- und Grundbau, Essen,
365-408.
[3] Langer, M. et al. 1986. Engineering-geological
methods for proving the barrier efficiency and
stability of the host rock of a radioactive waste
repository. - Proc. IAEA Int. Smyp. on the siting,
design and construction of underground repositories
for radioactive wastes, SM 289/23, Hannover,
S. 463-475
[4] Bundesanstalt für Geowissenschaften und Rohstoffe 1983. Entwicklung eines optimalen FE-
Programmes ANSALT zur Berechnung thermomechanischer
Vorgänge bei der Endlagerung hochradioaktiver Abfälle, Final Report to BMFT-Forschungsvorhaben,
KWA 2070/8, BGR, Hannover
[5] Hardy, H.R. jr. & M. Langer ed. 1981. The mechanical behavior of salt. Proc. 1. Conf., Penn
State Uni, Trans Tech Publ., Clausthal, p. 901
[6] Langer, M. et al. 1984. Gebirgsmechanische Bearbeitung von Stabilitätsfragen bei Deponiekavernen
im Salzgebirge. Kali u. Steinsalz, Heft 2, S. 66-
76, Essen
[7] Munson, D.E. & P.R. Dawson 1979. Constitutive
model for the low temperature creep of salt (with
application to WIPP). Sand-79-1853. Albuquerque.
[8] Pahl, A. et. al. 1986. Results of Engineering
geological research in granite. Bull. 34 of IAEG,
S. 60-65, Paris
[9] Heusermann, S. & A. Pahl 1986. Durchführung und
Interpretation direkter und indirekter Spannungsmessungen im Salzgebirge. Vortragsmanuskript Int.
Symp. Rock Stress and Rock Stress Measurements,
Sept. 1-3, 1986, Stockholm
[10] Wallner, M. 1986. Stability verification concept
and preliminary design calculations for the Gorleben high level waste repository. Waste Management
'86 (Proc. Symp. Tucson, 1986) (in print)
[11] Alheid, H.-J. et al. 1986. Stoffverhalten von
Salz bei kurzzeitigen Wechselbelastungen. Zwischen
bericht zum Forschungsvorhaben KWA 5502 8, Bundesanstalt für Geowissenschaften, Hannover

Some geostatistical tools for incorporating spatial structure in fracture network modeling

Techniques géostatistiques permettant d'utiliser la fracturation spatiale pour modéliser un réseau de fractures

Einige geostatistische Methoden für die Berücksichtigung von Raumstrukturen in Kluftnetzmodellen

JANE C.S.LONG, Earth Sciences Division, Lawrence Berkeley Laboratory, Calif., USA
DANIEL BILLAUX, Earth Sciences Division, Lawrence Berkeley Laboratory, Calif., USA
KEVIN HESTIR, Earth Sciences Division, Lawrence Berkeley Laboratory, Calif., USA
JEAN-PAUL CHILES, Bureau de Recherches Géologiques et Minières, Orléans, France

ABSTRACT: Fracture patterns observed in the field are not completely random and not completely ordered. Geostatistical techniques can be used to estimate the statistical parameters which describe the spatial variability of the fracturing. Data from the Fanay-Augères mine in France has been used as an example in developing these techniques.

RESUME: Les distributions de fractures observées dans la nature ne sont ni complètement aléatoires ni complètement ordonnées. Les paramètres statistiques qui rendent compte de la variabilité spatiale de la fracturation peuvent être estimés en utilisant les techniques de la géostatistique. Cette approche a été expérimentée sur des données provenant de la mine de Fanay-Augères, en France.

ZUSAMMENFASSUNG: Die in der Natur beobachteten Kloeftestukturen sind weder voellig zufaellig noch geordnet. Die Geostatistik-Methoden koennen benutzt werden, um die statistische Parametern abzuschaetzen, die die rauemliche variabilitaet der Kloefte beschreiben. Diese Methoden sind benutzt worden, um die Beobachtungen aus dem Bergwerk bei Fanay-Augères zu studieren.

1 INTRODUCTION

To model the behavior of a fracture system, for example the flow through a fracture network, we first develop a conceptual model for the system. The conceptual model then forms the basis of a numerical model used to calculate the flow. A fracture network model may be stochastic in that we create a random realization of a fracture system which is conceptually and statistically similar to that observed in the field. A stochastic model is not a model of the specific fractures which actually exist. However, a stochastic model may be a conditional model where we adjust the stochastic process such that features we can observe are reproduced in the model and features we can not observe are randomly generated.

To make a stochastic model, we generally specify rules for generating fractures, rules for truncating fractures and distributions for the random parameters. The rules are derived from the conceptual model, the distributions are derived from field data. For example, a rule for generating fractures in two dimensions might be that fractures are randomly located. Thus, locating fractures in space is a Poisson process with a strength, λ_A where λ_A is the areal density of the fractures (number of fractures per unit area). Thus we pick A λ_A points with random coordinates in the area, A. Now we provide an orientation distribution, f (θ) for choosing the lines through the center coordinates and a rule for truncating the lines. For instance, we might truncate the fractures where they intersect another fracture (Conrad and Jacquin, 1973; Dershowitz, 1984) or we might truncate the fractures such that they have some specified distribution of length, g(l) (Baecher et al. 1977).

Fracture networks are often conceptualized as some type of Poisson process. That is, the fractures are said to occur randomly in space. However, real systems are not usually completely random. Often we are able to identify "swarms" of closely spaced subparallel fractures, or that fracture density varies from location to location. These and other non-Poisson characteristics comprise the spatial structure of fracture networks.

The conceptual model chosen for a particular site should reflect the spatial structure of the site. For instance, one would model polygonal fracture patterns in the colonade of a basalt (Smalley, 1966). Many other conceptual-stochastic models have been proposed in the literature including Baecher (1977), Dershowitz (1984), Veneziano (1979), Conrad and Jaquin (1973) and LaPointe and Hudson (1981). The work described here is an attempt to develop a conceptual model for the fractures at Fanay-Augères, a mine in the Massif Central in France, and determine the parameters of the appropriate statistical distributions from the field data. The final goal is to model the hydrologic behavior of the fracture network.

1.1 Fanay-Augères Data

Fanay-Augères is a uranium mine owned by Cogema Co. and located in Limousin, France in the granite massif of Saint-Sylvestre. For the past 7 years this mine has been used as a test facility to develop methods and tools for investigating mass and heat transfer in granitic rocks (Barbreau et al, 1985; Lassagne, 1983). In particular we have focused on the data collected in a long section of a drift, about 3m in diameter at the 320m level. In this section fractures on the East wall have been mapped over two sections, S1 and S2, totaling 180m in length. For any fracture trace which intersected the 2m high rectangle the visible trace length, number of visible endpoints, orientation and morphology were recorded. Ten boreholes 50m long have been drilled in 3 radial patterns. Oriented core was obtained from these holes and the fractures logged. More than 220 steady-state permeability tests were performed in these holes between packers spaced at various distances. In summary, this data provided information about the location, size and characteristics of about 7000 fractures.

Gros (1982) gives a analysis of the fracturing at Fanay-Augères which identifies seven major tectonic episodes from which he identified five major sets of fractures by their orientation. However, 60% of the fractures do not fall into any of the tectonic classifications. Because this classification is not adequate for quantitative hydrologic analysis, we broadened the definition of the sets so that 93% of the fractures mapped in S2 were included. Figure 1 shows the fracture traces from S2 sorted into these five sets.

An estimate of the equivalent hydraulic aperture distribution for each set was made as follows. Each hole was studied with steady state packer tests. The core logs were studied to identify the fractures intersecting each of the test zones and the observed free aperture of these fractures, b_o. Assuming these fractures are infinite and parallel allows us to calculate an expected value of transmissivity, T_o, which can be compared to the measured transmissivity, T_m. We then adjusted the values of b_o to b_m where $b_m = Ab_o$ and $A = (T_0/T_m)^{1/3}$. This estimate accounts for heterogeneity due to apparent fracture opening but not fracture filling and roughness.

An initial study consisted of developing a two-dimensional model of the fracture traces on a vertical plane. In this model, the density and length of the fracture traces were allowed to vary spatially. We are now extending this work to three dimensions where we will include spatial variation in density and orientation as well as clustering of sub-parallel fractures. These two studies are summarized below.

2 Two-dimensional study

Our previous work on fracture networks was mainly concerned with parameter studies using the purely random Poisson model similar to that proposed by Baecher (1977). In this model fractures are randomly located, and each set of fractures has random distributions of orientation, size and equivalent hydraulic aperture (Long et al., 1982, Long, 1983, Long and Witherspoon, 1985, Long et al., 1985). In producing a two-dimensional model of the Fanay-Augères fractures, our goal was to improve this model by incorporating spatial variability in fracture density and size (Long and Billaux, 1986). To do this we assume that small subregions of the area of interest have statistically homogeneous characteristics and that a simulation of the spatial variation of the subregion characteristics can be made with geostatistical tools. Thus we are able to generate a region that is statistically heterogeneous.

We first divide the region into statistically homogeneous subregions. The dimensions of these subregions are somewhat arbitrary. Practical guidelines for the choice of subregion size include the following: (1) The subregion dimensions should be large enough to include a significant statistical sample of fractures. (2) The dimensions should be similar to those of the fracture clusters. (3) The dimensions must be significantly smaller than the length of sample available in order to get many individual measurements of the statistical parameters and thus achieve a measure of heterogeneity. (4) The dimensions of the subregions should be smaller than the range calculated in the geostatistical analysis. Perhaps the best way to describe how subregions were chosen in this case is to say we used "common sense". There is room for a more detailed and rigorous analysis of this problem in the future. For Fanay-Augères, we wished to generate a 100m by 100m area and we chose 10m by 10m subregions.

In each subregion and for each fracture set, we must now specify the areal fracture density, orientation, length, and aperture distributions. In order to develop the technique, we started by including only the density and length parameters in the geostatistical analysis. Apertures

were assumed to be statistically stationary throughout with distribution as explained above. Spatial variation of orientation was built into the model based on the observation that, for each set, fractures spaced close together tend to have similar orientations. We built this into the simulation by assigning to each set in each subregion a narrow distribution of orientation. The mean of this distribution was chosen from the distribution of mean orientation measured in 10m sections of the drift. The standard deviation was equal to the mean of the standard deviations of orientation as measured in 10m sections of the drift.

Data input to the geostatistical analysis for each set consisted of 20 values of mean length and density of fractures as measured in each 5m by 2m section of the drift wall. These values were used to produce variograms and the range of the variograms were used in turn to produce a simulation of each variable over a 100m by 100m area. The simulation technique is described in Long and Billaux (1986) and is derived from work by Matern (1960) and Alfaro (1980). The technique is different than Kriging in that the Kriged value of a regional variable is the best, linear, unbiased estimate of the variable at the given point. Thus, Kriging can be thought of as an interpolation technique which produces an estimate which is not reality. The simulated values are not the best estimates, but they do reproduce the spatial variability of the variable.

The results of the simulation are tables of values for mean length and fracture density for each of the five sets. Each table contains 100 values, one for each 10m by 10m subregion. The value of density simulated for each subregion was used directly to determine the number of fractures of that set to be generated in the subregion. The value of mean length was used as the mean of the length distribution in the subregion.

Given the local mean and the global distribution for each set, a local distribution for length was derived by letting ν_l, the local coefficient of variation, be the same in every subregion. We chose the value of this coefficient such that the combined length distribution from all the subregions and the global distribution of length for each set have the same first and second moments. We chose the form of the distribution used to generate local values of fracture length to be lognormal.

Using the data described above, we generated a 100m by 100m fracture network in a series of 100 statistically homogeneous subregions. It should be noted that a fracture generated in a given subregion may be assigned a length such that it crosses over into other subregions. In this way, communication between subregions is accommodated. As in the field measurements, a fracture is counted as belonging to a subregion if its center is in that subregion. The fracture region as it was generated has 65,740 fractures.

Within this region six square flow regions were isolated for permeability testing. Each of these was 70m by 70m and was located in the center of the generation region but oriented in different directions. Figure 2 shows one of these subregions oriented at 0°. In this figure the dead ends and isolated fractures have been removed so that it is easier to see how the network is connected. The boundaries of the other five flow regions are rotated 15°, 30°, 45°, 60°, and 75° counterclockwise from the region shown. For each of these flow regions, constant-head boundary conditions were applied as described in Long and Witherspoon (1985). This was done to each flow region in four different ways such that the gradient was sequentially perpendicular to each side. In this way a total of 24 directional measurements of flow were made.

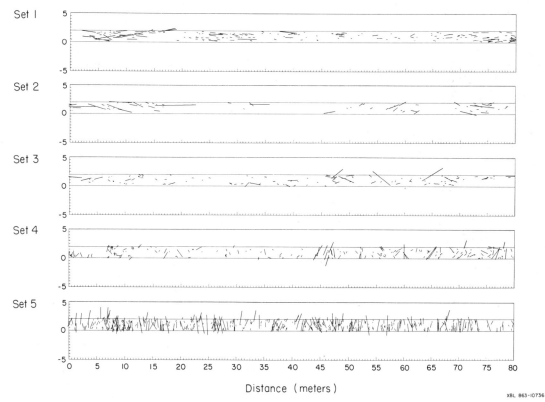

Set 1
Set 2
Set 3
Set 4
Set 5

Distance (meters)

XBL 863-10736

Figure 1.   Traces of five sets of fractures on the 80 m
section of the drift wall.

From these data the directional permeability ellipse was
calculated. These permeability data are shown in Figure
3.

Close observation of Figure 2 shows that the system is
barely connected. In fact for the orientation in Figure 3,
any flow originating on the left side of the region and
leaving on the right side of the region must all pass
through a very small proportion of the fractures, on the
order of 0.1% of the total number of fractures in the
region. Thus the permeability of these few fractures
essentially controls permeability in that direction. In the
language of percolation theory, this mesh is close to the
critical density for which infinite clusters of fractures are
found. It is not clear whether the mesh is above or below
critical, just that it is near critical.

3 Three-dimensional analysis

This extension of the two-dimensional technique
described above generates three-dimensional disc-shaped
fractures and allows regional fracture density variation
through a simulation procedure which employs the same
theory as the two-dimensional simulation. Fractures are
nucleated as "daughters" of a "parent" or seed which
allows the simulation of swarms of fractures. The simulation allows for regional variation of density and orientation within each set of fractures. The calculation of flow
through such a model has been described in Long and
Witherspoon, 1985.

Obtaining data for the two-dimensional analysis was
relatively straightforward because the vertical plane of
analysis was exposed in the drift. This made it possible to
obtain the areal fracture density by counting the fracture
traces in the plane and estimating the trace length distributions by direct measurement.

A three-dimensional analysis is necessary to achieve a
realistic picture of the hydrology. However, one can not
see into the rock to count the number of fractures per

unit volume and determine their shape. Perhaps geophysics will help us with this problem as described below, but
for now we must use statistical geometry to estimate
these parameters. For three-dimensional analysis we must
determine the disc diameter distribution from the trace
length distribution. We also must develop parameters for
the regionalized density and orientation distribution with
a parent-daughter process rather than a simple Poisson
distribution. These topics are briefly described below followed by an outline of the generation procedure. Application of this model to the site data will begin this year.

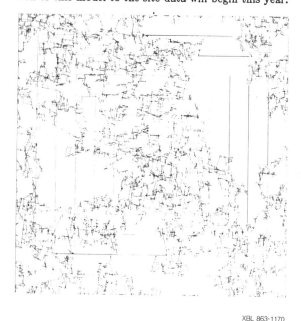

XBL 863-1170

Figure 2.   A 70m x 70m flow region isolated from the
center of the generated region with non-
conducting fractures removed.

173

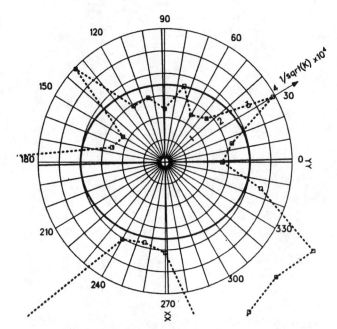

Figure 3.   Permeability results for a 70 m flow region.

3.1 Determination of the disc diameter distribution

In order to generate fractures in three dimensions from data which is essentially two-dimensional (drift wall mapping) or one-dimensional (well bore data) we need to make some assumptions about the shape and disposition of the fractures and derive the statistical relationships between the radii distribution, volumetric density, trace length distribution, frequency and areal density.

To make an estimate of fracture density we assume that the fractures are randomly located in space on the scale of the subregion (Poisson distribution). In order to estimate fracture size, it is necessary to make an assumption about the shape of the fractures and the type of distribution governing fracture sizes. We have assumed that fractures are discs. If the volume is cut by a plane, then the statistical relationship between λ_A, the areal density on that plane and λ_V, the volumetric density is:

$$\lambda_A = \lambda_V \, \overline{D} \, \overline{(\cos\theta)} \qquad (1)$$

where $\overline{D}$ is the mean fracture diameter, $\overline{(\cos\theta)}$ is a sampling orientation correction, and θ is the angle between the pole of each fracture and the plane of interest.

To determine λ_V we need to know $\overline{D}$, $\cos\theta$, and λ_A. From the drift wall mapping we can obtain the θ, and λ_A and the distribution of trace length, l. Once we have the relationship between the distribution of l and $\overline{D}$, then we can calculate λ_V.

A functional relationship between the distribution of fracture trace lengths and the distribution of fracture diameters which has been developed by Warburton, (1980) is given below. In order to determine the parameters of the diameter distribution from this relationship one must assume the form of the distribution. Assuming that the fracture discs are orthogonal to the sampling plane. Warburton (1980) shows that:

$$\tilde{g}(D) = 1/\overline{D} \; D \; g(D)$$

$$f(l) = 1/\overline{D} \int_l^\infty \frac{g(D)}{\sqrt{D^2 - l^2}} \; dD.bp \qquad (2)$$

Where

D	= diameter of a given fracture,
x	= distance between the fracture center and the plane of intersection,
l = h(x)	= the trace length of the fracture in the plane,
g(D)	= pdf (probability density function) of fracture diameters,
f(l)	= pdf of fracture traces,
$\tilde{g}(D)$	= pdf of disc diameters that intersect a given fixed plane,
$\overline{D}$	= mean of D $\equiv \int_0^\infty D \, g(D) \, dD$.

Given the above relationship one can assume the distribution of fracture diameters, g(D) and evaluate the integral to obtain the distribution of trace lengths f(l). One can then compare the computed distribution of trace lengths to that obtained from the field data. By trial and error, one can find a reasonable, but not necessarily unique distribution for fracture diameters.

To help frame the trial and error process we have extended Warburton's analysis to be able to calculate the relationship between the moments of g(D) knowing the first two moments of f(l) from the trace length distribution. This information in turn will guide initial choices of g(D) in the trial and error process. The details of this derivation are given in Long and Billaux (1986). The results are given below.

If we call μ_l and σ_l the mean and standard deviation of trace length, then this analysis gives:

$$\mu_l = \frac{\pi}{4} \frac{E(D^2)}{E(D)} \qquad (3)$$

and

$$\mu_l^2 + \sigma_l^2 = \frac{2}{3} \frac{E(D^3)}{E(D)} \; , \qquad (4)$$

If we assume the diameter distribution is lognormal with mean μ_D and standard deviation σ_D, then

$$\mu_D = \mu_l \cdot \frac{128}{3\pi^3} \frac{1}{1 + \frac{\sigma_l^2}{\mu_l^2}} \qquad (5)$$

and

$$\sigma_D = \mu_l \cdot \frac{16}{\pi^2} \left\{ \frac{2}{3[1 + (\frac{\sigma_l}{\mu_l})^2]} \left(1 - \frac{32}{3\pi^2[1 + (\frac{\sigma_l}{\mu_l})^2]} \right) \right\}^{1/2} \qquad (6)$$

3.2 Correction for trace length truncation

A problem remains in that we do not have an infinite plane on which to observe trace length, l. Thus, we are not able to measure f(l) very well due to the truncation of visible traces. Baecher et al. (1977) and Warburton (1980) have both addressed this problem and a new approach was developed by Massoud and Chiles (1986) which does not require any a priori assumption on the shape of the distribution of length. Chiles and Massoud examined the trace length data from the Fanay-Augères mine. They divided the data into three groups: no endpoint visible, fractures with one endpoint visible and fractures with both endpoints visible. For each of these groups, the relationship between the true distribution and

174

the observed one were derived assuming all the fractures are parallel, at an angle θ with x axis.

The program LOI, developed by J.-P. Chiles, uses these relationships to get the expected value of the number of fractures in Group 0, and each histogram interval in Group 1 and 2.

We then use trial and error to get λ_A and f(l). First, one makes a guess of the histogram of real length and of the number of trace centers in the survey area. The program, using the three derivations above, calculates:

- The number of traces with no endpoint, one endpoint and two endpoints
- The histogram of trace lengths with one endpoint
- The histogram of trace lengths with two endpoints

Both histograms are given in four classes.

Now to find f(l) we first fit the 4 first classes using the "two endpoints" results and then fit the tail of the distribution, knowing that the 1 endpoint histogram is more sensitive to the middle of the distribution and the number of fractures with no endpoint is more sensitive to the tail. It is easy to fit the distributions, because we can fit the tail without changing the result for Group 2, and Group 2 depends only on the first four classes of real length.

3.3 Parent-daughter model

We often observe that fractures occur in swarms or zones. In order to model fractures in swarms we use a statistical description of this type of pattern called a "parent-daughter" model. In this model, the fracture swarms, or daughters are nucleated around seeds, or parents. The location of the parents may be purely random, ie a fixed rate Poisson process, or there may be a regional variation in the density of the parents. Once the parents have been determined, the daughters are found in some distance from the parents where this distance may, for example, have a normal distribution. The location of parents, location of daughters relative to their parents, and number of daughters are taken to be independent random quantities. Thus, implementation of the model requires that we know the distribution of the density of the parents, the distribution of the number of daughters per parent, and the distribution of the daughters around the parents.

There are two major difficulties in obtaining these distributions from the field data such as that available at Fanay-Augères. The first problem is that it is not necessarily clear to which parent a fracture daughter belongs. This is further complicated by regional changes in the density of fractures. The second problem is that the parents and daughters are distributed in three-dimensional space but we can only observe traces. The centers of the traces are not the same as the centers of the fractures and the parent is a point that does not necessarily lie in the plane of observation. The first problem requires that we propose a parent daughter model and see how well it fits. Given a trial example of a three-dimensional parent daughter model with regional variation in density, Deverly (1986) worked out the statistics which would be observed on a plane. J. P. Chiles (1986) modified this derivation to account for the case that the daughters are disc shaped fractures rather than points. The result of Chiles work is that given assumed distributions for the parameters of the regionalized parent daughter model, we can calculate the theoretical

variogram of fracture trace density on the plane. Then, we can compare the theoretical variogram with that derived from the drift wall mapping. Thus, we are now able to pick an appropriate regionalized parent daughter model by trial and error. This work is summarized in Long (1986). Now, given the parameters of the regionalized parent-daughter process for discs in space we can calculate the variogram of fracture density as it would appear on a plane of observation.

3.4 Program SALVE

The preceding results have been incorporated in the program SALVE written by J.P. Chiles. To use this program we estimate the density of parents, mean number of daughters, distribution describing the dispersion of the daughters, distribution of the size of discs and the variogram of the density of parents. Given these and the size of the support sample (ie the subregion volume over which density is measured), the program calculates the variogram of the fracture density on a plane. This variogram can then be compared to the variogram derived from the field data. We can then change the estimates of the regionalized parent-daughter model until a good fit to the variogram is found. In this way we can derive a model which agrees with our data.

In implementing this trial and error process, the following guide lines may be used:

- The known number, N, of traces on the survey area is

$$N = \theta_G \cdot \theta_M \overline{D} \cdot s \qquad (7)$$

where s is the known area of the survey, and $\overline{D}$ is the known mean fracture diameter. We then know the product $\theta_G \cdot \theta_M$. To separate θ_G and θ_M, one can use the variograms of the interdistances. These should contain two peaks, one at the typical daughter interdistance and one at the typical parent interdistance.

- When the variogram is computed from values averaged over an area, the range is longer than the range of the variogram of the same variable computed from point values. The difference is approximately equal to the length of the area over which the averaging took place.

- If σ_D is the standard deviation of daughters around parents, a_p is the range of the variogram of the density of parents, and a_d is the range of the variogram of the density of all points, then

$$a_d \approx a_p + (c \cdot \sigma_D) \qquad (8)$$

where c is between 2 and 3.

- To increase the sill of the variogram of the fracture density calculated by SALVE, one can either increase the number of daughters per parent and decrease the density of parents, while keeping the produce constant, or increase the sill of the variogram of the density of parents.

When a good fit is found, we have obtained an estimate of the parameters for the parent daughter model which can be used to simulate the density of parents in each subregion. This simulation plus the other parent daughter parameters are used to generate realizations of the fracture system as described in the last section.

3.5 Orientation

Within a given set there may be a large dispersion in orientation. However locally, especially within a swarm, the orientation distribution may be quite narrow. Thus orientation has spatial structure. To account for this we can construct variograms for orientation. These variograms typically will have a nugget which represents the local random variation of orientation. If this nugget is subtracted from the variogram, the result can be used in a simulation to predict the spatial variation of mean orientation on a grid throughout the region. Now when a particular fracture is nucleated by the process described below the orientation can be found by interpolating the local mean orientation from the nearest neighbor grid points and adding to that a random component dictated by the nugget of the variogram.

The following steps are taken to create a realization of the fracture system once an acceptable parent daughter model has been obtained.

1. Using the parent-daughter model, a simulation of the spatial variations of the fracture model parameters (ie the density of the parents, distribution of the number of daughters per parent etc) is made in a three-dimensional region.

2. Define subregions on the same scale as the sample support.

3. For each subregion, read the strength of the parent Poisson process from the simulation.

4. From this Poisson process, pick the number of parents to be generated in the subregion.

5. To locate each parent, pick random values of x,y,z in the subregion.

6. Pick the number of fracture daughters from the distribution determined above.

7. Pick the random locations of the fracture centers distributed around the parent. This may be an anisotropic distribution.

8. Pick the orientation in two steps. First, pick the continuous part from the simulation described above. Then add a random component as determined by the nugget of the orientation variogram.

9. Pick the fracture length and aperture from the global distributions.

Application of this statistical model to the Fanay-Augeres data is now underway. The major outstanding problem in developing this three-dimensional model is that the observed trace pattern gives rise to a highly connected three-dimensional fracture network. However, we know from hydrologic observations that the rock does not behave like an equivalent porous medium. Therefore, it is likely that many of the observed fractures are not conductive. It remains to determine how many, and which fractures are not conductive.

REFERENCES

Alfaro, M. 1980. The random coin method: Solution to the problem of simulation of a random function in the plane. Mathematical Geology 12.

Baecher, G.B., N.A. Lanney, and H.H. Einstein 1977. Statistical descriptions of rock properties and sampling. Proceedings of the 18th U.S. Symposium on Rock Mechanics, Am. Inst. of Min. Engineers.

Barbreau, A., B. Come, S. Derlich, E. Durand, G. de Marsily, P. Peaudecerf, G. Vouille 1985. Experiments performed on granite in the underground research laboratory at Fanay-Augères, France. Proc. Internat. Symp. on Coupled Processes Affecting the Performance of a Nuclear Waste Repository, Lawrence Berkeley Laboratory, Berkeley, Ca., Sept.

Chiles, J.P. 1986. Personal Communication.

Conrad, F. and C. Jacquin 1973. Representation of a two-dimensional fractue network by a probabilistic model: Application to calculation of the geometric magnitude of matrix blocks. Univ. of Calif., Lawrence Livermore Lab. Pub. UCRL-Trans-10814, 75 pp.

Dershowitz, W.S. 1984. Rock Joint Systems, Ph.D. Dissertation, M.I.T., Department of Civil Engineering.

Gros, Y. 1982. Etude structurale de la mine de Fanay-Augères-Géométrie, cinématique et chronologie des fractures, B.R.G.M. report.

La Pointe P.R. and J.A. Hudson 1981. Characterization and interpretation of rock mass jointing patterns, Univ. of Wisconsin report, Madison, Wi.

Lassagne, D. 1983. Essai de Caractérisation du Milieu Fracturé en Massif Granitique, B.R.G.M. report 83 SGN 576 GEG, 45060 Orléans, France.

Long, J.C.S., J. Remer, C. Wilson, P.A. Witherspoon 1982. Porous media equivalents for netwos of discontinuous fractures, Water Resources Research: 18(3) 645.

Long, J.C.S. 1983. Investigation of equivalent porous medium permeability in networks of discontinuous fractures, Ph.D. Dissertation, Univ. of Ca., Berkeley.

Long, J.C.S. and P.A.Witherspoon 1985. The relationship of degree of interconnection to permeability in fracture networks, Journal Geophysical Research:90(B4) 3087-3098.

Long, J.C.S., P.A. Witherspoon, H.M. Gilmour 1985. A model for steady fluid flow in random three-dimensional networks of disc shaped fractures, Water Resources Research:21(8) 1105-1115.

Long, J.C.S., D. Billaux 1985. From field data to fracture network modeling--An example incorporating spatial structure, to be published in Water Resources Research.

Massoud, H. and J.P. Chiles 1986. La modelisation de la petite fractureation par les techniques de la geostatistique, BRGM report in preparation.

Matern,B. 1960. Spatial variation, Meddelanden Fran, Statens Skogsforskningsinstitut, Band 49, Nr 5.

Smalley, I.J. 1966. Contraction crack networks in basalt flows, Geol. Mag.:103(2).

Veniziano, D. 1979. Probabilistic model of joints in rock, Tech. Report, 4 pp. Civ. Eng. Dep. Mass. Inst. of Technol., Cambridge.

Warburton, P.M. 1980. A stereological interpretation of joint trace data, International Journal of Rock Mechanics and Mining Science and Geomechanical Abstracts:17 181-190, Pergamon Press Ltd.

Deformation of foundation and change of permeability due to fill placement in embankment dams

Déformation des fondations et variation de perméabilité provoquée par remblayage de barrages en remblai
Verformung des Fundaments und Veränderung der Durchlässigkeit mit Aufschüttung bei Schüttdämmen

NORIHISA MATSUMOTO, Public Works Research Institute, Tsukuba, Japan
YOSHIKAZU YAMAGUCHI, Public Works Research Institute, Tsukuba, Japan

ABSTRACT: The fill placement works on the foundation as overburden loading. This loading will compress the foundation and cause the decrease of permeability. Therefore, the authors actually measured the deformation of the foundation and the change of permeability due to fill placement at three damsites. The measured results indicate that the fill placement causes the compression of dam foundations and the decrease of permeability. Moreover, taking these results into consideration, the authors propose the effective designing of grouting for dam foundations, particularly weak rock foundations.

RESUME: Le remblayage produit sur les fondations l'effet d'une charge due au poids des terres. Cette charge comprime les fondations et provoque une diminution de la perméabilité. En conséquence, les anteurs ont meauré, sur le site de trois barrages, la déformation des fondations et la diminution de la perméabilité provoquées par le remblayage. Les résultats de ces mesures indiquent que le remblayage provoque la compression des fondations des barrages et la diminution de leur perméabilité. En tenant compte de ces résultats, les auteurs proposent, de plus, une conception efficace d'injection des fondations d'un barrage, plus particulierèment dans le cas de fondations à roches tendres.

ZUSAMMENFASSUNG: Die Aufschüttung wirkt auf das Fundament als Abraum-Belastung. Diese Belastung drückt das Fundament zusammen und bewirkt eine Verringerung der Durchlässigkeit. Daher haben die Autoren die Verformung des Fundaments und die Veränderung der Durchlässigkeit bei drei Dämmen gemessen. Die Meßergebnisse zeigen, daß die Aufschüttung das Zusammendrücken des Fundamentes und die Verringerung der Durchlässigkeit bewirkt. Weiterhin schlagen die Autoren unter Berücksichtigung dieser Ergebnisse die wirksame Konstruktion von Auspressung von Dammfundamenten, insbesondere schwache Felsfundamente, vor.

1 INTRODUCTION

Controlling seepage in the dam foundation is one of the most important and challenging parts of dam engineering. There are three basic methods for controlling seepage, i.e. reduction of seepage, drainage, and use of filters. As for rock foundations grouting is most frequently adopted among several seepage reduction methods. Foundation grouting at dams takes the form of curtain and blanket grouting. The curtain grouting is provided down to a comparatively deep depth at the foundation for the purpose of cutting off the underseepage. The blanket grouting is provided at a comparatively shallow depth to make the area more impermeable or to consolidate fractured or jointed rock.

Recently, in Japan, most of the sites which have satisfactory geological conditions for the construction of dams have been already developed. Therefore, dams tend to be constructed on the sites which are not always satisfactory in terms of geological conditions. In selecting the type of dams, a fill dam is often chosen for the sites which are low in bearing capacity for concrete dams. Therefore, not a few damsites for the fill dams under construction and planning are composed of weak rocks such as soft, or weathered, or fractured rocks. Since the higher grouting pressures cause fracturing or lifting the surface of weak rock foundations, the allowable pressures should be kept low in weak rocks. Furthermore, these weak rocks often have fine joints or fissures which are not groutable by cement. Therefore, it is very difficult to reduce the permeability in weak rock foundations, particularly in shallow areas, by cement grouting.

On the other hand, since weak rock foundations show large deformability, these foundations, especially shallow areas, are expected to deform due to the loading of fill placement. If the foundation rocks themselves are consolidated and the above-mentioned joints or fissures are closed, the permeability of foundations will decrease. The authors feel that investigations on the quantitative relationship between the compressive strain and the decrease of permeability are very useful for controlling seepage in dam foundations.

From the above-mentioned viewpoint, the authors try to reveal the relationship between the compressive strain and the decrease of permeability due to fill placement in fill dams through the in-situ measurements at three dmasites. Moreover, taking these results into consideration, the authors propose the effective designing method of grouting for dam foundations, particularly weak rock foundations.

Although a lot of researches on the permeability and the grouting of dam foundations have been already reported, there have been few systematic researches from the above-mentioned viewpoint. Installation of foundation galleries at the base of fill dams enabled to obtain in-situ data of consolidation effects on the permeability due to fill placement.

2 GEOLOGY OF DAMSITES

The measurements of the deformation and the change of permeability at dam foundations due to fill placement were conducted at three damsites shown in Table 1. Table 1 also shows the results of in-situ and laboratory tests for foundation rocks. Sandy tuff at Nanakita Dam has not a few clay dikes, but have only a few beddings. This is the weaker soft rock. Though tuff breccia at Shimoyu Dam is loosely consolidated soft rock, it is much harder than sandy tuff at Nanakita Dam and has a large number of open joints. The foundation of Shitoki Dam is composed of Mesozonic schist, which is jointed hard rock. Main purpose of these measurements is to grasp the relationship between the deformation of weak rock foundations such as Nanakita Dam and Shimoyu Dam foundations and the change of permeability due to fill placement.

Table 1. Damsites and testing results for foundation rocks.

Dam name	Height (m)	Measurement site	Geological age	Foundation rock	Natural water content (%)	Specific gravity	Uniaxial compressive strength (MPa)	Deformation modulus (MPa)
Nanakita	74.0	right abutment	the Miocene	sandy tuff	36.9	2.38	0.8	2.0×10^2
Shimoyu	70.0	river bed & left abutment	the Miocene to the Pliocene	tuff breccia	24.5	2.72	5 to 40	3.0×10^2
Shitoki	83.5	right abutment	the Mesozonic	green schist	0.19	3.02	96	——
				quartz schist	0.08	2.75	75	6.6×10^2

However, these measurements are also conducted on the hard rock foundation such as Shitoki Dam foundation for comparison. The reason why the authors selected the schist in Shitoki Dam among many kinds of hard rocks is that it has many regularly developed joints, or schistosity, and the change of permeability is expected to be clearly measured due to the deformation of foundation.

3 INSTRUMENTATION

Extensometers are installed and boreholes for water pressure tests are drilled in foundation rocks at three damsites in order to measure the deformation and the change of permeability due to fill placement. Permeability of foundation is evaluated by Lugeon value. Moreover, in-situ measurements are conducted after the execution of blanket grouting and before that of curtain grouting.

4 RESULTS OF MEASUREMENTS

4.1 The relationship between the height of fill placement and displacement of foundation

Figure 1 shows the measured results of the deformation of rock foundation of Shimoyu Dam. Moreover, only typical results measured at abutment are illustrated in Figure 1, because the tendency of the results measured by every pair of horizontal and vertical extensometers is similar. This figure leads to the following:

1. When the surface of fill placement reaches the vicinity of the elevation of each extensometer, the dam foundation is subjected to tension in the horizontal direction. However, this tensile displacement tends to decrease due to subsequent fill placement.
2. On the other hand, the dam foundation is always compressed in the vertical direction. And the absolute value of vertical compressive displacement is larger than that of horizontal tensile displacement at abutment.
3. The height of fill placement and vertical displacements are closely related. Moreover, displacements hardly change during the winter shut-down period of fill placement.
4. Though the vertical extensometer at river bed has the same length with that of at abutment, vertical displacement at river bed is much larger than that of at abutment. This is attributed to the fact that tuff breccia at river bed was subjected to hydrothermal alteration and is weaker and softer than that of at abutment.

4.2 The relationship between the height of fill placement and the change of permeability

Figure 2 to 5 show the relationships between the height of fill placement and Lugeon value measured at three damsites. Since there is not enough space for all measured results, only typical examples are shown in these figures. Regardless of the difference of foundation rocks or installing directions of extensometers, it is clearly evident that Lugeon values decrease with the increase of the height of fill placement from these figures. Moreover, if water pressure tests are performed by using the injection pressure larger than the critical pressure, the excess pressure produces fracturing for the rocks

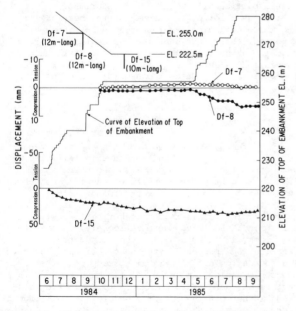

Figure 1. Displacement of foundation (Shimoyu Dam).

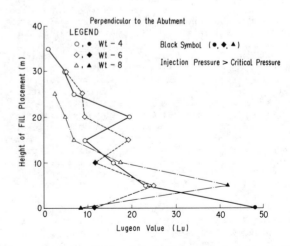

Figure 2. The relationship between the height of fill placement and Lugeon value (Nanakita Dam).

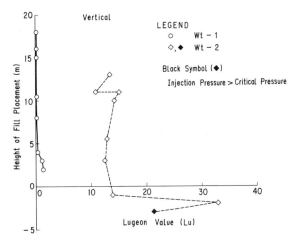

Figure 3. The relationship between the height of fill placement and Lugeon value (Shitoki Dam).

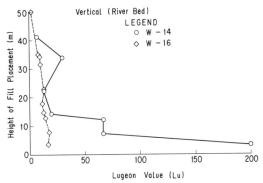

Figure 4. The relationship between the height of fill placement and Lugeon value (Shimoyu Dam).

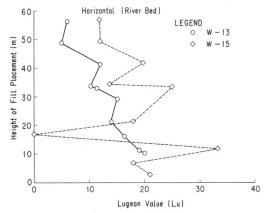

Figure 5. The relationship between the height of fill placement and Lugeon value (Shimoyu Dam).

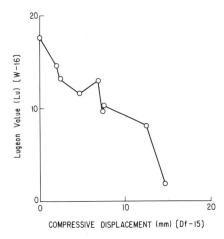

Figure 6. The relationship between Lugeon value and compressive displacement (Shimoyu Dam).

surrounding the boreholes. Therefore, some of Lugeon values increased compared to the previous tests performed at the same boreholes due to this hydraulic fracturing. However, Lugeon values tend to decrease with the increase of fill placement thereafter. It can be considered that the overburden loading consolidated the foundation, thus reducing the permeability.

4.3 The relationship between the strain in the foundation and the change of permeability

The above-mentioned results at three damsites revealed that dam foundation was consolidated due to fill placement and the permeability of foundation decreased with the increase of fill placement. In this section, the measured data are rearranged in

reference to the relationship between the strain in the foundation and the permeability. Figure 6 illustrates the relationship between Lugeon values measured in borehole W-16 and the strains measured by extensometer Df-15 at Shimoyu Dam. In this figure, the strain distribution in the vertical direction within the length of the installment is assumed to be constant. The decreasing tendency of permeability due to the increase of compressive displacement can be read from this figure.

Moreover, the authors have already explained the above-mentioned relationship by using the simple models (Matsumoto and Yamaguchi 1986 b & unpublished). However, since these models have been adequately discussed in references shown at the end of this paper and hence discussion about them is omitted.

5 DESIGNING OF GROUTING FOR WEAK ROCK FOUNDATIONS

5.1 Classification of weak rock foundations

The nature of weak rock dam foundations varies from one dam to another. Therefore, before describing the design method of grouting for weak rock foundations, the authors would like to classify weak rock foundations into four types according to the improvement degree of permeability by cement grouting.

(1) weak rock foundation I
Though this foundation is composed of tightly consolidated rock which is supposed to be impermeable, it has many large and fine joints or fissures. Therefore, this foundation has high permeability.

(2) weak rock foundation II
Though this foundation is also composed of the above-mentioned tightly consolidated rock, it has comparatively high permeability because of having many joints or fissures. However, since these joints or fissures are very fine, it is very difficult to reduce the permeability of this foundation by cement grouting. That is, it is one of the most important problems to reduce the permeability of this foundation on seepage control for dam foundations.

(3) weak rock foundation III
In this foundation the rock itself is porous and pervious because of low consolidation or grain size, and it doesn't have the fine joints nor fissures. Grouting can hardly be used for this type of rock.

(4) weak rock foundation IV
Since this foundation is scarcely consolidated, it has low strength and large deformability. However, because it hardly has joints or fissures, or its void is very small, it is supposed to be impermeable.

5.2 Designing of grouting for weak rock foundations

The authors try to summarize the grouting methods for four types of weak rock foundations classified in 5.1, hereafter.

(1) weak rock foundation I
The permeability of this foundation, like hard rock foundations, can be sufficiently improved by using portland cement (maximum grain size : approx. 100 μm) or clay-cement as main grouting materials. Clay is often utilized as an additive to cement in order to increase lubrication of grouts.

(2) weak rock foundation II
This foundation has many fine joints or fissures. Though water can flow through the above-mentioned fine joints or fissures, cement grout is hardly able to travel through these cracking system. Therefore, recently ultra fine particle cement (Blaine specific surface area : approx. 8 000 to 10 000 cm^2/g, maximum grain size : approx. 10 μm) has been developed as grouting material and the tube à manchette has been adopted for grouting to get the grout into each of the above-mentioned joints or fissures. Even though the above-mentioned grouting material or grouting method is used, it would be impossible to make a very tight curtain for this type of the foundation. Therefore, the authors recommend that grouting would be executed taking the decrease of permeability due to fill placement into consideration, in order to reduce the permeability of foundation having fine joints or fissures. Concrete method is shown in the following paragraph.

There are measured data concerning the decrease of permeability due to fill placement at three damsites. But the decreasing degree of permeability is different at each damsite. Therefore, the authors think that it is good to perform the grouting test containing excavation of foundation and fill placement (Matsumoto and Yamaguchi 1986 c). From the results of this test, the target for improvement of blanket grouting could be mitigated and the effective time of curtain grouting could be determined. However, the decreasing tendency of permeability in the main work isn't always larger than that in the trial work. Therefore, in case of blanket grouting, regrouting has to be prepared for that case. Moreover, in case of curtain grouting, since this work is critical path in the dam construction work, the effective time of this work should be determined totally taking the decrease of permeability due to fill placement, economic factor and so on into consideration. Grouting gallery should be installed at the base of fill dams in order to make the abovementioned planning of grouting.

The authors will continue to measure the decrease of permeability due to fill placement at various damsites and collect many data. After the completion of this work the authors believe that the systematic planning of grouting could be made without performing the grouting test mentioned above.

(3) weak rock foundation III
Since the permeability of this foundation depends upon the void of rock itself, it is more difficult to reduce the permeability by grouting than that of weak rock foundation II. Moreover, since this foundation hardly has even fine joints or fissures, even the decrease of permeability due to fill placement couldn't be expected. The authors have explained this phenomenon by using mathematical models (Matsumoto and Yamaguchi 1986 b & unpublished).

On the other hand, since this foundation has no joints or fissures causing high permeability, it is possible to control the seepage through foundation by installing upstream blanket, relief wells, and/or filters.

(4) weak rock foundation IV
Since the permeability of rocks which this foundation is composed of, is negligible and hardly has joints or fissures, it is basically unnecessary to execute seepage cut-off works. Occasionally it is needed to reduce the excessive uplift pressure in

special geological configurations, since impervious sedimentary layer is sometimes interbedded with previous sand layer or gravel layer. Therefore, the investigation about geological uniformity should be laid emphasis on.

6 CONCLUSIONS

The summary of this paper is as follows:
1. The authors measured the deformation of foundation and the change of permeability due to fill placement at three damsites.
2. The results measured at three damsites revealed that dam foundation was consolidated due to fill placement and the permeability of foundation decreased with the increase of fill placement.
3. Weak rock foundations were classified into four types according to the groutability.
4. The authors summarized the grouting methods for four types of weak rock foundations.
5. Specially in case of dam foundations which has fine joints or fissures, and can hardly be improved by cement grouting, the authors recommended the utilization of the consolidation effect on permeability. The authors also briefly showed the method of grouting test to establish the effective grouting procedures.

ACKNOWLEDGEMENT

The authors gratefully acknowledge the work of the many personnel and organizations involved in the projects of Nanakita Dam, Shimoyu Dam and Shitoki Dam.

REFERENCES

Matsumoto N. & Yamaguchi Y. 1986. Deformation of foundation and change of permeability due to fill placement in embankment dams (part 3), Proc. of the 18th Symposium on Rock Mechanics, JSCE, pp.376-380. (with English summary)
Matsumoto N. & Yamaguchi Y. 1986. Deformation of foundation and change of permeability due to fill placement in embankment dams, Proc. of JSCE, No. 370, pp.281-290. (with English summary)
Matsumoto N. & Yamaguchi Y. 1986. Deformation characteristics and seepage control of soft rock foundations for dams, Civil Engineering Journal, Vol.28, No. 10, pp.33-38. (in Japanese)
Matsumoto N. & Yamaguchi Y. Consolidation effects on foundation permeability, Proc. of ASCE. (unpublished)

Approche géomécanique de la production pétrolière – Problèmes essentiels, premiers résultats

Geomechanical approach of oil and gas production – Main problems, first results
Felsmechanische Annäherung der Erdöl- und Gasgewinnung – Hauptfrage, erste Ergebnisse

VINCENT MAURY, Elf-Aquitaine, Pau, France
JEAN-MICHEL SAUZAY, Elf-Aquitaine, Pau, France
DOMINIQUE FOURMAINTRAUX, Elf-Aquitaine, Pau, France

ABSTRACT: Case histories concerning difficult oilwells drillings and oilfield subsidence phenomena are detailed. Geomechanical interpretations made on the basis of carefull observations and theoretical analysis are the way to progress, in despite this method doesn't agree with some empirical practices and traditional schemes.

RÉSUMÉ: Par les exposés détaillés de cas vécus particulièrement démonstratifs de forages pétroliers difficiles et des phénomènes de compaction-subsidence dans et autour des gisements pétroliers, on montre comment les interprétations géomécaniques basées sur une observation minutieuse des faits et une analyse théorique complète fournit une méthode nouvelle d'approche des problèmes de production d'hydrocarbures, remettant en cause certaines règles de l'art et des schémas traditionnels, mais ouvrent la voie vers la compréhension des phénomènes.

ZUSAMMENFASSUNG: Wâhrend Erdoelbohrungen gestösse Schwierigkeiten und in Erdoellagern bemerkte Subsidenzerscheinungen einzeln beschrieben sind. Die auf zugenau Besbachtungen und theoretische Analyse gegründete feldmechanische Auslegunge der gute Weg nach Forschritt sind, trotzdem diese Methode nicht im vollen Einverständnis mit einigen empirischen Proxisen und herkömmlichen Abrissen ist.

1 - INTRODUCTION

Dans les diverses opérations de l'exploitation des gisements pétroliers, le comportement mécanique des roches concernées joue un rôle parfois important vis-à-vis de la quantité d'hydrocarbures susceptibles d'être extraites et sur le coût de ces opérations. Si l'on considère par exemple le domaine du gisement, c'est-à-dire de l'ensemble roche réservoir et fluides contenus, la prise en considération des propriétés des seules phases fluides (hydrocarbures et eau) se révèle insuffisante : la chute de pression interstitielle très notable provoquée par l'extraction de la phase fluide est à même d'induire une déformation de la phase solide et une réduction concommittante du volume disponible pour la phase fluide. Cette réduction de l'espace poreux peut contribuer de façon sensible à une augmentation de la quantité de fluides expulsée au cours de la récupération dite primaire par l'effet du simple différentiel de pression, imposée aux puits producteurs. Cette quantité de fluide expulsée par la réduction du volume poreux augmente énormément lorsque l'assemblage matriciel est susceptible de s'écraser sur lui-même sous l'effet des éventuelles redistributions de contraintes dues à la chute de pression interne et au maintien de la charge par le recouvrement (pore-collapse). Mais des effets nocifs se révèlent : perte de perméabilité, fermeture de drains et production de fines peuvent réduire la production d'huile ; compaction et grandes déformations des horizons producteurs peuvent induire un affaissement du recouvrement incompatible avec l'exploitation et la pérennité des installations, voire une potentialité d'évolution catastrophique selon des schémas observés en exploitations minières ou en rives de barrages lorsque les conditions de circulation des fluides et les régimes de pression dans le réseau des discontinuités du massif rocheux sont modifiés. Le comportement hydraulique et mécanique des discontinuités (fractures et fissures) du massif doit ainsi être reconnu au même titre que leur répartition dans l'espace. On peut alors soit mieux orienter les drains producteurs, que l'on sait forer maintenant jusqu'à l'horizontale au besoin, ou mieux contrôler le développement de ce champ de fractures productrices par la fracturation hydraulique en s'asssurant qu'elles peuvent conserver des caractéristiques de productivité convenable (ouverture, auto-soutènement ou soutènement artificiel).

Dans le domaine du forage, les roches ont jusqu'à présent le plus souvent été considérées en termes de résistance à la destruction c'est-à-dire de performance à l'avancement instantané du forage, malgré l'évidence du rôle premier de la tenue des parois du trou : qu'importe de forer vite à raison de plusieurs centaines de mètres par jour si on est incapable de conserver le trou sauf au prix de longues opérations de sauvetage du trou ou du train de tiges coincé, immobilisant l'appareil de forage –à raison en opérations offshore de 120 à 150 000 $US par jour- ! La méconnaissance des mécanismes d'instabilité des parois met le foreur dans l'impossibilité de se dégager d'une situation devant laquelle le fatalisme l'emporte sur le rationalisme. Les progrès possibles dans les processus de destruction de la roche au front de taille doivent être replacés dans le contexte plus général de la tenue des parois du puits, dont le fond n'est qu'une zone particulière soumise à des chargements particuliers. Comment contrôler l'état, donc la tenue des parois, sans bien connaître ce que leur fait subir l'outil lors du processus de destruction ? et corollaire : comment approcher correctement le processus de destruction sous l'outil sans bien prendre en compte toutes les conditions de chargement ?

Dans le domaine de la production sensu stricto, des difficultés apparaissent lorsque des puits forés dans des roches réservoirs faiblement cohérentes (craies, grès tendres, grès argileux) ou hyper compactées mais finalement mal consolidées (sables, silts..). Une partie de la phase solide est entraînée par l'écoulement des fluides et cet engorgement des conduites de production nécessite la pose d'équipements spéciaux toujours coûteux. Les conditions d'apparition de ces instabilités, liées aux caractéristiques géomécaniques des horizons en cause, sont à préciser de façon à optimiser la pose de ces équipements spéciaux.

L'objet de cette communication est d'illustrer la méthode d'investigation utilisée pour l'approche géomécanique des problèmes de production pétrolière principalement basée sur des analyses approfondies d'études de cas à la lumière de nombreux résultats du génie civil et minier. L'exposé ci-après des observations et des résultats obtenus sur des forages difficiles dans deux champs et un cas remarquable de subsidence, en sont des exemples démonstratifs.

2 - APPROCHE GEOMECANIQUE DES DIFFICULTES DES FORAGES PETROLIERS

Une dizaine de cas ou séries de cas de forages difficiles ont été analysés de manière approfondie. Ils sont situés dans une gamme très variée de formations géologiques et de conditions géo-structurales de massifs rocheux comme par exemple : les champs du GOLFE de GUINEE (couverture argileuse du Crabe à M'BYA ; la tectonique diapirique de Baudroie Marine au GABON ; les argilites du Malembo en ANGOLA) ; les champs d'AQUITAINE de Lacq, Meillon St Faust, CastéraLou, dans la marge nord de la chaîne pyrénéenne à tectonique compressive très marquée ; les champs du BASSIN de PARIS, dans les séries argileuses du Toarcien et argilo-gréseuses du Rhétien ; les champs d'AUSTRALIE dans le détroit de BASS et le Golfe Bonaparte ; les champs de la mer thyrénéenne italienne. Ces études ont permis de dresser un inventaire des incidents survenant au cours des forages, aussi documenté que possible sur les conditions dans lesquelles elles sont apparues et que l'on s'efforce de traduire en termes géomécaniques.

L'exposé des cas des forages dans les champs de Meillon St-Faust en Aquitaine et de Baudroie Nord Marine du Gabon servira de toile de fond à la description d'incidents de forage typiques et de la méthode employée pour cerner au mieux les paramètres géomécaniques concernés.

2.1 Forage en Aquitaine : le champ de Meillon St-Faust

2.1.1 Données qualitatives

Il est localisé (figure 1) à quelques kms à l'Est du champ de Lacq dans les coteaux du piémont pyrénéen. En 1964 le puits Meillon 1 découvrait un réservoir de gaz dans la dolomie kimméridgienne dite "dolomie de Mano" à 4335 m de profondeur. Le contexte structural est très marqué par la tectonique pyrénéenne : nombreux compartiments séparés par des contacts anormaux, chevauchements, failles et autres accidents. De manière simplifiée, le gisement est associé a une structure anticlinale faillée déjetée vers le Nord et limité au Sud par une grande faille inverse (figure 2). Le forage des puits de développement du champ de Meillon et ses satellites (Le Lanot, Rousse, Mazères, etc...)fut difficile en maintes occasions, mais principalement dans la traversée des alternances gréso-schisteuses du flysch maestrichtien (Crétacé supérieur) et le passage dans la dolomie de Mano entre 4000 et 4300 m. Dans cette dernière, gazéifère, d'importants hors profils ("cavages") du trou de forage se produisirent ; ils rendaient délicats les essais de production de gaz en trou ouvert car il était impossible d'ancrer correctement les packers de test à la paroi irrégulière du trou ; pour la même raison les diagraphies de pendagemètrie et de

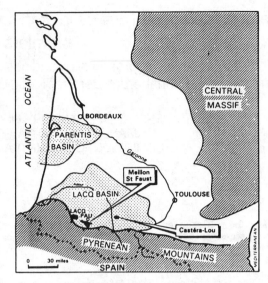

Figure 1 : Location of Meillon St-Faust gas field in Aquitaine (South West of France)

diamètre étaient mal lisibles et la cimentation des cuvelages s'est révélée délicate. A la profondeur de 4000 m environ, un cuvelage de Ø 7" (17,8 cm) est mis en place dans le trou foré en Ø 8"1/2 (21,6 cm), à la suite de quoi le forage est poursuivi dans la dolomie en Ø 5"3/4 (14,6 m) ; lors de cette phase de forage apparaissent de fortes augmentations du couple à l'outil et même des blocages de la rotation, la circulation normale de la boue de forage n'étant pas modifiée ; les remontées du train de tiges nécessitent des efforts de traction anormalement élevés allant jusqu'au coincement de la garniture de forage. Lorsque l'outil est redescendu, il faut chaque fois reforer la partie inférieure du trou tandis que parallèlement, on recueille sur les tamis vibrants du circuit de nettoyage de la boue, une grande quantité de fragments grossiers de dolomie en forme d'écailles aux bords tranchants n'excédant pas 0,5 cm de diamètre (ces écailles déta-

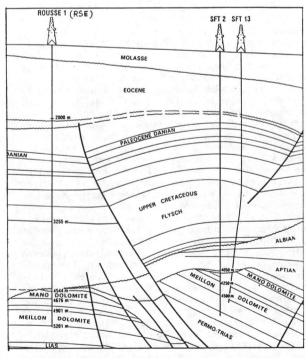

Figure 2 : South-North simplified geological cross-section in Meillon St-Faust gas field (see map figure 4)

chées de la paroi du puits et "tombées" vers le fond du puits et entraînées par le fluide de forage sont appelées des "retombées"). La mise en oeuvre de solutions traditionnelles dans l'art du foreur fut tentée : le fluide de forage -boue aux chlorures (FCl) de 1.15 à 1.18 de densité- fut changé pour une boue au gypse puis une boue à la chaux, dont on attend traditionnellement un colmatage de la paroi par une sorte d'emplâtre : à 4100 m de profondeur et une température de 140°C l'effet dépassa les espérances : la boue à la chaux subit un épaississement tel qu'elle boucha totalement l'annulaire et bloqua la circulation.

La foration stoppée ne put reprendre que 3 jours après le montage de pompes haute pression qui permirent la circulation de la boue sous 35 MPa. On revint alors à la boue d'origine, et les symptômes d'instabilités persistèrent et s'accentuèrent, en particulier la quantité de retombées. Pour nettoyer le trou de tous les dépôts et retombées, on procéda à l'envoi d'un bouchon très visqueux, qui remonta effectivement plusieurs m3 de retombées dolomitiques. Toujours dans l'espoir de renforcer les parois du trou des bouchons de ciments furent envoyés à plusieurs reprises et dans l'optique de diminuer les frottements on ajouta à la boue jusqu'à 10 % d'huile et lubrifiants divers : tout cela sans résultats positifs. Après un test de production de gaz sur la section non tubée du puits n°5, un important volume de retombées dolomitiques dut être extrait. -note : au cours de tel test la pression fluide varie de plus de 20 MPa en quelques minutes. Sur un puits proche (Lanot 1) deux diagraphies de diamètre du trou réalisées l'une juste avant et l'autre juste après un essai de production du même type montrent des différences importantes : l'essai de production provoque des instabilités de la paroi et de larges cavages augmentent très nettement le diamètre du trou. Ces cavages relevés sur les diagraphies de diamètre mènent à des sections du trou dont la forme est très ovalisée et qui atteignent sur leur grand axe des longueurs de plus de 14" (35 cm) alors que le diamètre de forage n'est que de 5"3/4 (14,6 cm), soit plus de 2 fois plus petit ; sur leur petit axe par contre des ouvertures parfois plus faibles que le diamètre nominal sont relevées. Les carottages réalisés dans ce réservoir, nécessaires pour les études de gisement, furent longs et difficiles : coincements de carottier fréquents, vitesse d'avancement très faibles, arrachement correct de la carotte difficile. La récupération est décevante et les carottes sont toujours très fragmentées ; le discage est fréquent et on constate que les zones de discage intense de la carotte semblent correspondre sur les diagraphies de diamètre à des zones de cavage plus fort de la paroi du puits.

2.1.2 Quelques données géomécanique quantitatives

La dolomie de Mano est une roche résistante (résistance en compression simple de 45 à 75 MPa voire plus, avec un module de déformation tangent de 75 000 MPa) assez homogène et isotrope. L'environnement tectonique et la profondeur dans lesquels sont observés les phénomènes décrits ci-dessus inclinent à retenir pour origine vraisemblable des instabilités le niveau et la distribution des contraintes autour du trou. Cependant les moyens d'investigation sont très limités dans ce type d'excavation souterraine. Ce sont finalement les diagraphies de pendagémétrie et de diamètre ("4-arms dipmeter") qui se révèlent l'outil d'investigation le plus performant. Les formes variées de la section du trou révélée par ces diagraphies furent analysées et répertoriées en fonction des conditions de leur apparition et de leur origine la plus probable : usure ou raclage de la paroi du trou par le train de tige (ovalisation d'usure) donnant des formes en "trou de serrure", augmentation du diamètre (isotro-

pe liée à des ruptures circulaires sous contraintes isotropes ou ovalisation par ruptures localisées par un champ de contrainte anisotrope) associé à des modifications du régime des pressions fluides (lors des tests par exemple) ou du régime thermique (lors des modifications des conditions de circulation du fluide de forage). Ensuite, les directions de ces ovalisations de la section du trou de forage furent systématiquement déterminées : la figure n°3 donne un exemple de ce traitement (un programme d'exploitation de ces diagraphies a depuis été mis au point : programme "OVAL"). Sur une projection dans le plan horizontal de la trajectoire du puits ARS-1, on a reporté l'azimut de la direction du grand axe des ovalisations du trou aux diverses profondeurs : on repère immédiatement une orientation systématique, ici de N90° à N140°, sur plus de 4000 m linéaire de forage et à travers des formations de lithologie variée rencontrées sous des pendages très divers. Réalisée sur 31 puits de ce champ de Meillon et ses satellites, cette analyse a mis en évidence une orientation systématique générale de la direction d'ovalisation vers le N 20° pour la plus grande partie du champ et vers le N110° pour les puits situés à l'Est d'un accident géologique majeur : ceci est illustré sur la figure n°4 où ne sont reportés que les résultats des mesures entre 4000 et 5000 m de profondeur ; à l'Est de la ligne AA' à droite de la figure la direction d'ovalisation est pratiquement à 90° de celle trouvée dans de la partie ouest du champ.

2.1.3 Interprétation géomécanique : approche de l'état des contraintes initiales

Le caractère systématique de l'orientation de la

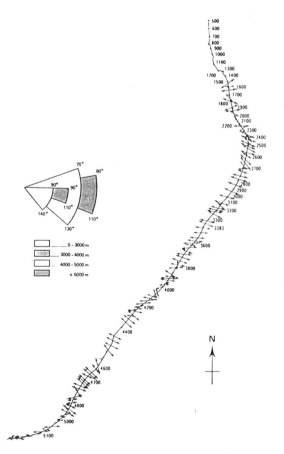

Figure 3 : Horizontal projection of borehole section elongation axis on the borehole trajectorie projection of the 5000 meter deep ARS-1 oilwell (see bottom right on map Figure 4)

direction d'ovalisation et son indépendance de celle du forage implique une cause régionale. Ceci est confirmé par les analyses semblables réalisées sur le champ de Castéra-Lou à 50 km à l'Est de celui de Meillon, pour lequel on retrouve encore une direction d'ovalisation N20°. Sur la base d'une étude théorique approfondie réalisée par A. GUENOT (1987) et à la lumière de la synthèse d'observations de ruptures autour des excavations souterraines du génie civil et minier qui remettent en cause certains résultats traditionnellement acceptés (V. MAURY, 1987), -ces deux papiers sont présentés dans ce même congrès- il est attribué un rôle majeur aux contraintes initiales, régnant dans le massif traversé par le puits, dans l'apparition des instabilités de la paroi ; les caractéristiques de la dolomie et du fluide de forage permettent d'estimer que la valeur du rapport K de la contrainte horizontale à la contrainte verticale est de 1 à 2 et que le champ des contraintes initiales horizontales doit être fortement anisotrope pour provoquer les ovalisations observées. La forme typique en "selle de cheval" des surfaces de discage à double courbure (MIGUEZ et HENRI, 1986) observées sur les carottes confirme également cette conclusion de forte anisotropie des contraintes horizontales dont la composante maximale est orientée perpendiculairement à la direction d'ovalisation. Le champ des contraintes initiales dans le champ de Meillon aurait donc une composante horizontale majeure grossièrement N110° soit NNW-SSE c'est à dire plutôt parallèle à l'axe de chaîne pyrénéenne. Ceci est en accord avec les études du mécanisme au foyer des séismes d'Arette (1967) et Arudy (1980) à 30 km de Meillon, qui concluent à une compression orientée NW-SE. Mais dans un compartiment contigü de l'autre coté d'une discontinuité tectonique, on retrouve un état correspondant au schéma plus classique de compression N-S. L'utilisation de l'interprétation du discage d'Obert et Stephenson (1965) avec l'hypothèse d'une contrainte verticale égale au poids des terres mène à une valeur de la contrainte horizontale d'au moins 150 MPa pour produire le discage centimétrique observé. Des essais de fracturation et minifracturation hydrauliques ont été réalisées dans l'optique de la stimulation de ce réservoir faiblement perméable (10^{-2} mDarcy). Au puits SFT 5 le pic de pression ne put être obtenue malgré une pression du fluide de 95 MPa à 4046 m ; plus récemment un deuxième essai (135 MPa à 4550 m) confirme la valeur élevée des contraintes horizontales. On a estimé l'état des contraintes initiales dans le champ à 4000 m de profondeur : $\sigma^{o}h$ > 100 MPa selon une direction N-S ; $\sigma^{o}H$ environ 200 MPa selon E-W ; la contrainte verticale σ_v serait de l'ordre de 100 MPa et très probablement plus faible que $\sigma^{o}h$.

2.1.4 Vers des remèdes appropriés

Le haut niveau d'anisotropie des contraintes latérales est la particularité du champ et la cause première des instabilités : un simple calcul élastique indique que sans pression interne la contrainte tangentielle à 4000 m atteint ainsi 500 MPa en N-S et plus de 100 MPa sur le diamètre E-W selon lequel des convergences ("retreints" des logs de diamètre) peuvent se produire. Cette convergence, est vraisemblablement responsable de certaines difficultés de rotation (couples élevés, coincements de garniture ou du carottier) : sur le puits plus récent de SFT 15 un coincement d'un stabilisateur de la garniture de forage avec maintien de la circulation de boue (indice d'une convergence ponctuelle et non d'une fermeture complète du trou) a été résolu par une mise en pression temporaire du trou de forage et la mise à profit de la légère irréversibilité de la déformation provoquée. Une autre solution peut être d'équiper les stabilisateurs d'éléments de coupe à l'arrière afin de pouvoir localement forer en remontant.
L'incidence de l'anisotropie du champ de contraintes horizontales sur le choix des trajectoires des puits est d'autant plus forte que les puits seront inclinés : un puits dévié rencontre des conditions de contraintes en paroi très différentes selon l'azimut de sa déviation.
Les puits de développement de champs offshore sont forés "en bouquet" à partir d'une implantation de la plateforme. Cette implantation doit être réalisée avec soin, car on peut définir pour chaque position des quadrants géomécaniquement plus ou moins favorables dans lesquels la déviation des puits sera limitée. Ce choix est encore plus sensible pour les forages horizontaux qu'il convient de placer selon l'azimut le plus favorable, leur coût étant de 2 à 5 fois plus élevé que celui d'un forage vertical.
Enfin, c'est dans ce champ de Meillon, sur le puits du Lanot 4, que l'efficacité et la faisabilité du contrôle du régime thermique des puits développée par A. GUENOT (1986) ont été illustrées pour la première fois. L'idée de base est que le refroidissement de la paroi se traduit par un relâchement de la contrainte tangentielle et réduit ainsi l'apparition des instabilités. D'autres effets sont à attendre mais pas encore explicités ni quantifiés : réduction des pressions de gonflements des argiles sensibles, des pressions interstitielles, amélioration de la longévité des matériaux (boues, additifs...) et des matériels (sondes, outils de forage, etc...), effets sur la cimentation.

2.2 Forage en massif "décomprimé" au Gabon

2.2.1 Données d'observation

Dans le champ de Baudroie Nord Marine situé dans une structure diapirique de l'offshore gabonais, les 4 premiers puits furent forés sans difficultés particulières à 2800 m de profondeur avec une inclinaison de 25° par rapport à la verticale à raison d'environ 35 jours par puits. L'objectif du 5ème puits (BDNM-5) était situé au-delà du dôme de sel au sud duquel était implantée la plateforme et la trajectoire prévue passait donc au-dessus du dôme (Figure n°5).

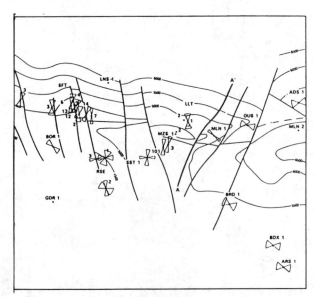

Figure 4 : Azimuth of borehole section elongations axis between 4000 to 5000 meters depth in Meillon St-Faust gasfield wells.

Le forage atteignit la profondeur de 1200 m en
∅ 14"3/4 (37,5 cm) puis 2400 m en ∅ 9"5/8 (24,4 cm)
avec une boue de forage de densité 1.25. A la suite
de la rencontre d'une digitation saline issue du
dôme et injectée dans la couverture, une recon-
version de la boue en saumure dut être entreprise
au cours de laquelle le coincement total du train
de tige obligea à en abandonner une bonne partie
et à perdre ainsi plus de la moitié du puits. Des
difficultés dues à des instabilités survenant de
plus en plus fréquemment au cours du forage après
ce premier "sidetrack" (reprise d'un forage selon
une trajectoire légèrement déviée pour s'éloigner
du trou et du "poisson" qui ont dus être abandonnés)
la densité de la boue fut portée à 1.35, sans
succès : à 1400 m au cours d'un reforage difficile
une déviation accidentelle se produisit ; les
instabilités persistèrent et cette nouvelle trajectoire dut être abandonnée. On coupa le cuvelage en
∅ 10"3/4 (27,30 cm) placé dans la première partie
du forage en ∅ 14"3/4 et il fut retiré pour pouvoir
relancer la foration à partir de 470 m de profondeur sur une nouvelle trajectoire (2ème side-track).
L'idée étant alors de pousser le plus loin possible
la foration en ∅ 14" 3/4 et poser le cuvelage
correspondant ∅ 10"3/4 le plus profondément possible
avant d'atteindre des niveaux salifères. La phase
∅ 14"3/4 à 2060 m ; poursuivi
ensuite en 9"5/8 le forage ne rencontra jamais le
sel, mais pénétra alors dans les argilites d'Aminba
de la couverture du dôme.
Leur traversée se révéla d'emblée difficile avec de
nombreuses instabilités (coincements, reforages,
mauvaise circulation...) malgré une montée de la
densité de la boue à 1.73, qui dut être rapidement
réduite à 1.60 en raison des pertes partielles et
parfois totales de la boue dans le terrain. Un
coincement inébranlable à 2400 m amena un 3ème
side-track, qui réussit à pousser le forage à
2260 m avec une boue à la saumure saturée de
densité de 1.75, dont on perdait toujours des
quantités importantes. Malgré l'envoi d'un bouchon
de ciment pour tenter de colmater les parois, mais
avec une augmentation de la densité de boue à 1.80,
les reforages systématiques nécessaires étaient
toujours de plus en plus difficiles : finalement
après 92 jours de travail et 3 side-tracks le puits
fut abandonné.

2.2.2 Analyse et interprétation

Par rapport aux 4 premiers puits, la situation de
BDNM-5 ne se distingue pas par la nature et le
comportement mécaniques des roches traversées : les
argilites rencontrées sont peu différentes et ne
contiennent pas de minéraux argileux identifiés
comme minéraux gonflants. Dans les 4 premiers puits
forés avec une boue au KCl de 1.25 de densité, les
diagraphies de diamétreurs à 4 bras ne montrent
qu'une faible ovalisation (11" à 13" pour ∅ 10"3/4
nominal) attribuable autant à l'usure par les tiges
qu'à des ruptures en paroi. Aucune orientation
systématique des directions d'ovalisation n'est
décelable, et une compilation complète de ces
données directionnelles semble plutôt indiquer une
dépendance en fonction de la localisation de la
mesure par rapport au dôme de sel : les contraintes
principales in situ apparaissent grossièrement en
position tangentielle et radiale par rapport au
dôme (un tel effet a déjà été signalé par Bradley
(1978)). De plus, on constate que la perturbation
des parois est d'autant plus accentuée que la
section du puits correspondante est proche du
dôme ; la partie au NE du dôme où les digitations
furent rencontrées semble même plus sévèrement
perturbée, la perturbation consistant en une
diminution des composantes latérales du champ de
contrainte : on aurait autour du dôme une sorte de
zone "décomprimée".

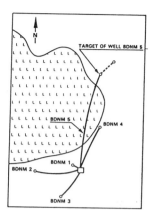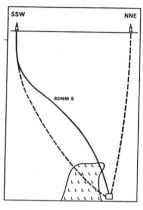

Figure 5 : (left) and Figure 6 (right). Baudroie
Nord Marine oilfield (Guinea Gulf). Simplified
map (left) and cross-section (right) showing salt
dome and platform positions and wells trajectories
(in dotted line : proposed trajectories for
BDNM-5 well).

En appliquant dans un tel schéma, la méthode
d'analyse des conditions de stabilité d'un forage
développée par Alain GUENOT (op.cit.), on constate clairement que l'augmentation des densités des
fluides de forage, en augmentant la pression
fluide interne ne peuvent qu'avoir des effets
néfastes en provoquant une rupture en paroi en
mode C (cisaillements multiples) d'où les instabilités et les pertes de fluide par un début
d'injection de la boue dans la formation.

2.2.3 Remèdes : Quelques propositions

C'est la trajectoire même du puits BDNM-5 qui est
en cause et qu'il conviendrait de modifier en se
maintenant le plus loin possible du dôme, et en
particulier en évitant le quart NE de sa couverture par exemple (figure 6) une trajectoire inclinée
à 25° à partir d'un point situé nettement à
l'E NE de la structure. La boue ne devrait pas
dépasser 1.25 à 1.30 de densité pour éviter les
ruptures en mode C et limiter les écaillages
quasi inévitables en mode A ou B. Si par contre
la plateforme de départ ne peut être déplacée, il
vaudrait mieux chercher à minimiser la trajectoire dans la zone "décomprimée" de la couverture
argileuse, traverser au plus court et passer par
l'intérieur du dôme de sel. Dans ce cas une
attention particulière devrait être apportée au
choix du fluide de forage : l'utilisation de
saumure saturée pouvant accélérer fortement le
fluage du sel en conditions de température et
pression élevées (Spiers, 1986), une boue à
l'huile sera préconisée.

2.3 La mécanique des roches au service du forage

La question de la stabilité du front et des
parois d'un forage pétrolier, comme pour un
tunnel foré au tunnelier, se pose en des termes
plus "nets" pourrait-on dire que pour une galerie
déroctée à l'explosif : la forme de départ est
parfaitement circulaire et les efforts appliqués,
-poids sur l'outil et densité de la boue- ne
varient normalement pas, brusquement. Ce sont les
variables liées au terrain qui sont bien sûr
susceptibles de varier plus rapidement : nature,
donc résistance et déformabilité de la roche ;
pressions interstitielles, contraintes (mécaniques
et thermiques). En face d'une difficulté de
forage, l'essentiel est tout d'abord de s'efforcer
d'en identifier la cause principale parmi les
multiples paramètres : semble-t-elle liée plutôt
aux caractéristiques même du matériau ? aux

contraintes en place ? aux régimes de pressions des fluides ? Ensuite on tentera d'en préciser les conditions d'apparition : OU (dans le forage) ? QUAND (au cours ou à la suite de quelle opération de forage) ? COMMENT (sous quelle forme et quels étaient les paramètres de forage) ? la rupture s'est-elle produite ? lorsqu'on fore à 5000 m de profondeur toute observation directe "au front" est impossible ; c'est le suivi de tout ce que remonte la boue et en particulier les "retombées" qui constituent l'unique source d'informations sur la rupture en association avec des mesures indirectes réalisées en cours de forage dans la région du front ainsi que celles des paramètres de forage et dont les valeurs sont transmises en temps réel par impulsions codées à travers la boue (MWD : Measurements While Drilling) : la mise en phase de ces deux types d'observations fournit les bases d'une véritable diagraphie instantanée géomécanique qui progressivement est réalisée sur nos chantiers. Les diagraphies différées serviront ensuite à mieux caler la localisation, la forme, l'extension des ruptures et autres déformations de la paroi du puits (auscultation "sonique" et résistivité multipoints, visualisation et "profilographie" de la paroi sont quelques améliorations proposées actuellement dans un domaine où des progrès nécessaires sont en cours). D'autre part, les conditions de contraintes en place sont petit à petit mieux cernées par un meilleur usage des essais de minifracturations hydrauliques. Sur la base de toutes ces informations dont la qualité ne peut que progresser, la question est alors posée au mécanicien des roches : dans de telles conditions la rupture macroscopique interviendra-t-elle, oui ou non, en paroi ? Le puits est-il alors susceptible de rester ouvert ou non ?

Des progrès sont encore indispensables pour tenir un tel challenge : les modèles actuels en déformations plastiques ne représentent pas le comportement réel à la rupture de la plupart des roches. Ce n'est que l'examen soigneux, associé à un traitement encore approximatif, des difficultés et des ruptures qui permettra de progresser à partir des ébauches entrevues actuellement.

3 SUBSIDENCE LIEE A L'EXTRACTION PETROLIERE

Le chargement par les terrains du recouvrement des roches réservoirs déplétées, souvent très poreuses et peu résistantes, provoque leur compaction. Cette compaction peut atteindre plusieurs mètres et plus dans le cas des réservoirs épais et superposés. Selon les caractéristiques géométriques et mécaniques du gisement et du recouvrement, les déformations provoquées dans le recouvrement par cette compaction du réservoir peuvent se transmettre jusqu'à la surface où elles se révèlent principalement sous la forme de déplacements verticaux analogues aux "affaissements" provoqués par les travaux miniers et par les rabattements d'aquifères ou aux "tassements" associés aux excavations du génie civil. Ces mouvements sont désignés par le terme général de SUBSIDENCE. (note : Dans la terminologie géologique française ce terme désigne l'affaissement régulier du fond d'un bassin sédimentaire autorisant l'accumulation de grande épaisseur de sédiments dans des conditions identiques). L'exploitation de certains champs pétrolifères a déjà alimenté la chronique d'exemples spectaculaires de subsidence : Au-dessus du champ de Wilmington (Californie, USA) les digues du port de Long Beach ont dû être rehaussées de 9 m (GILLULY et GRANT, 1949 ; GRANT, 1954) ; les déplacements horizontaux provoqués en surface par l'exploitation du champ d'Inglewood (en Californie également), ont été reconnus être à la base des déformations et de la rupture du barrage de Baldwin Hills en 1963 (LEE et SHEN, 1969 ; JANSEN, 1980 ; LEE, 1976). Dans la

zone de Houston-Galvestone les extractions pétrolières et les pompages en aquifères ont provoqué une cuvette de subsidence de plus de 2,50 m de profondeur et plus de 60 km de diamètre. Outre ces nuisances à l'environnement, la subsidence pétrolière peut avoir des répercussions néfastes sur les installations de production et la production pétrolière elle-même. Elles sont évoquées ci-après à travers l'exposé d'un cas actuel de subsidence d'un champ en Mer du Nord.

3.1 Subsidence et production pétrolière : un cas en Mer du Nord.

Depuis quelques années un important gisement de Mer du Nord (WIBORG & JEWHURST, 1986) s'est révélé avoir subi une compaction de plusieurs mètres et provoqué une subsidence de surface du même ordre passée jusque là inaperçue dans l'environnement marin très agité. Le réservoir dans la craie du crétacé supérieur a 300 m d'épaisseur et une superficie de 50 km2. Une épaisseur notable de ce recouvrement comporte des zones dans lesquelles les argiles n'ont pas atteint le degré de compacité correspondant normalement à leur profondeur : dans ces zones dites "sous-compactées" les pressions des fluides interstitiels sont anormalement élevées jusqu'à plus de 2 fois la pression hydrostatique. Le mur des argiles du recouvrement étant à 3000 m on relève 32 MPa à 1800 m de profondeur et des surpressions sont encore décelées à 1200 m. En forage, la valeur de ces surpressions est appréciée par la densité de la boue nécessaire pour éviter les venues de gaz. Démarrée en 1975, la déplétion atteint 20 à 30 MPa. Les principaux effets décelés à ce jour sont du haut vers le bas les suivants (Figure 7) :

3.1.1 Réduction du tirant d'air des plate-formes :

Dans la partie centrale du champ, en 1985, la subsidence du fond marin sur lequel repose les fondations a réduit de près de 3,00 m la garde vis-à-vis de la vague centennale. La sécurité n'est plus suffisante et un investissement de plus de 250 M.US $ est engagé pour rehausser de 6 m , 6 plates-formes par vérinage. En bord de cuvette, le déplacement vertical de 2 autres plates-formes est moitié moindre mais est associé à un léger basculement. Depuis 1984, des bathymétries successives indiquent une descente du fond marin de l'ordre de 0,40 m par an.

3.1.2 Dégâts sur les sea-lines :

Les conduites posées sur le fond reliant les diverses plates-formes entre elles et au continent sont le siège d'incidents et de déformations repérés par les enregistrements vidéo de télésurveillances sous-marine : mises en "portée libre" (free-spans) où sur plusieurs dizaines de mètres la conduite portée par des escarpements ne repose plus sur le fond ; flambages horizontaux correspondant à des sections de la conduite soumises à des compressions axiales, comme en provoquerait un rapprochement des appuis au sol. Les interventions de maintenance sur ces conduites sont coûteuses et le traitement des dégâts en fonction de leurs origines, supposées (par exemple des remblais pour lutter contre l'affouillement par courants sous-marins ne sont pas toujours couronnés de succès : après quelques temps, on retrouve certains remblais intacts en dessous de nouvelles portées libres.

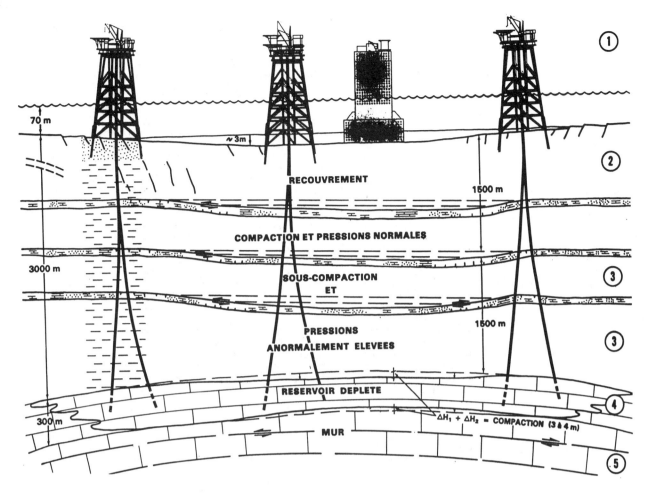

① ② ③ ③ ④ ⑤

70 m

~3 m

RECOUVREMENT

1500 m

COMPACTION ET PRESSIONS NORMALES

3000 m

SOUS-COMPACTION

ET

PRESSIONS

ANORMALEMENT ELEVEES

1500 m

RESERVOIR DEPLETE

$\Delta H_1 + \Delta H_2$ = COMPACTION (3 à 4 m)

300 m

MUR

Figure 7 : Subsidence of a North Sea oil Field.1 : at sea level, safety relativ to wave height and tilting ; 2 : in sea bottom, safety relativ to foundation because movements and stress in soils ; 3 : in the overburden, depressurization of overpressurized beds give unexpected subsidence rise ; 4 : reservoir compaction induce rock properties changes and troubles in well completion equipments ; 5 : in the underlying beds, tension cracks rise water inflow in reservoir.

3.1.3 Désordres dans les équipements des puits.

Un nombre anormalement élevé de travaux de reprise de puits ("work-over") est nécessaire depuis quelques années, pour répondre à des ruptures de cimentations, des écrasements ou ruptures de cuvelage et même de tubing de production ; ces accidents se produisent au niveau du réservoir mais aussi à différentes hauteurs dans le recouvrement.

3.2 Un remède envisagé : l'injection d'eau ou de gaz

Dans l'optique d'augmenter la "récupération primaire", il est classique d'envisager l'injection d'eau ou de gaz dans le gisement : celle-ci balaye l'huile restée piégée dans les pores. Cette technique présente également l'intérêt de limiter la déplétion et le maintien de pression pourrait efficacement éviter une partie de la compaction du réservoir et réduire, si ce n'est bloquer, la subsidence, comme cela a été obtenu à Wilmington.

3.3 Tentatives d'interprétations géomécaniques

3.3.1 Le schéma théorique le plus simple du phénomène fournit en surface l'image d'une cuvette de subsidence dont les bords extérieurs convexes sont en extension, les bords intérieurs concaves en compression et la partie centrale plate ne subit que des déplacements verticaux. Les études de cas de subsidence minière (DEJEAN et ARCAMONE, 1972 ;

ARCAMONE, 1980) ou au droit d'ouvrages souterrains du génie civil ont montré que du fait de l'hétérogénéité des terrains, on aboutit à des déformations localisées sous forme de profondes fentes de traction dans les zones en extension, de bourrelets de compression dans les zones en contraction et dans les zones intermédiaires des escarpements brutaux sous l'effet d'efforts de cisaillements localisés sur des plans subverticaux comme sur les bords des graben tectoniques. De tels mouvements du fond marin sont probablement à l'origine de certains dégâts aux sea-lines et situations délicates dans lesquelles ils se trouvent.

3.3.2 Les modifications de l'état des contraintes dans le sol sont susceptibles d'entraîner des réductions des caractéristiques mécaniques des sols de fondation des plates-formes, en particulier du coefficient de poussée latérale et de la densité en place.

3.3.3 Plus profondément dans le recouvrement, on observe dans de nombreux cas au-dessus d'exploitations minières qu'il se produit des ouvertures de diaclases et des petits déplacements tangentiels cisaillants au droit des petits niveaux plus rigides que l'ensemble (ARCAMONE, 1980) : par exemple, bancs argiliteux ou gréso-argileux dans les argiles ; bancs calcaires ou dolomitiques dans les marnes ; bancs gréseux dans les sables. Les discontinuités ainsi créées augmentent de manière sensible la perméabilité du recouvrement. Elles peuvent permettre le drainage et la dissipa-

187

tion des pressions interstitielles élevées vers le réservoir, vers la surface ou même latéralement : un deuxième mécanisme de subsidence va alors se développer, lié à la compaction des zones en surpression du recouvrement. Ses effets en seront plus forts et plus immédiats car son origine est plus proche de la surface ; il ne dépend que des conditions de pressions interstitielles et de drainage des argiles du recouvrement. Il faut noter que ce deuxième mécanisme ne peut se déclencher qu'à partir d'une certaine amplitude des déplacements induits par la compaction sur réservoir, donc après une certaine période d'exploitation du gisement. Il peut bien sûr ne pas se produire du tout.

Le drainage des horizons surpressurisés vers des niveaux déplétés ou vers la surface est susceptible d'être provoqué très probablement également par des chenaux s'installant le long des puits à la faveur des désordres produits dans les cimentations des cuvelages par les mouvements différentiels dans le réservoir et remontant dans le recouvrement.

L'apparition de ce deuxième mécanisme associé au premier par l'intermédiaire de la <u>modification des régimes d'écoulement et la variation des distributions des pressions interstitielles induites par des déplacements parfois minimes est un des problèmes-clés de la Mécanique des Roches et des sols :</u> Il est à l'origine de très nombreuses ruptures de talus, de rives et appuis de barrages et barrages proprement dits (LONDE, 1966) ou d'effondrements miniers (TINCELIN et SINOU, 1962 ; MAURY, 1979) qui comptent parmi les catastrophes majeures du génie civil ou minier.

3.3.4 Dans le réservoir, la réduction de porosité due à la compaction peut avoir elle au moins une conséquence favorable : augmenter la production d'huile expulsée (conditions dites de "compaction drive"). L'enjeu est d'importance puisque la récupération peut ainsi passer de 10 à 15 % de l'huile en place pour la simple "récupération primaire" par soutirage à 20 à 25 % et plus. Cependant, cette compaction peut s'accompagner d'une baisse de perméabilité matricielle qui compromettrait le drainage de l'huile ; si le gisement est fracturé la conductivité de fracture sera aussi modifiée, mais dans quel sens ? Si la redistribution des contrainte dans la matrice est susceptible d'amener le matériau à s'écraser sur lui-même ("pore-collapse", comportement observé couramment sur la craie au laboratoire) la part des conséquences favorables et défavorables est encore plus mal connues.

3.3.5 Enfin, en dessous du réservoir, l'analogie avec les observations en mines, fait prévoir une possible détente du mur du réservoir, accompagné ou non d'un déplacement vertical vers le haut. Cette détente horizontale pourra ouvrir ou développer une fissuration verticale et déclencher des cisaillements de discontinuités horizontales, tous mouvements capables d'augmenter très sensiblement la perméabilité du mur. Le régime hydraulique des aquifères situés sous le gisement est ainsi modifié. L'aquifère deviendrait ainsi plus actif au cours de la vie du gisement. De tels cas sont signalés d'aquifère considérés comme inactifs initialement et qui se réveillent plus tôt que ne le laissaient penser les études de gisements.

3.4 Quelques commentaires géomécaniques du cas présenté

3.4.1 La possibilité d'un relais des mécanismes de compaction-subsidence du réservoir puis du recouvrement apparaît comme la situation pratique la plus préoccupante. Dans de telles conditions où le volume des roches concerné est énorme, les données sur les propriétés des roches sont par suite insuffisantes et dispersées. Les détails des mécanismes en cause sont encore difficiles à bien évaluer pour une modélisation représentative. C'est alors l'auscultation des déformations in-situ qui devient l'outil principal d'information et de diagnostic sur le comportement du massif rocheux ; elle est seule capable à la fois de fournir immédiatement les premiers signes d'une modification de l'allure des phénomènes et donc des comportements mais aussi de fournir un moyen d'ajustements des hypothèses et des modèles.

3.4.2 Dans le cas de la craie, l'opportunité de l'injection d'eau pose un problème. De nombreuses études (DOREMUS, 1978 ; HUDSON et MORGAN, 1975 ; DESENNE, 1971 ; MAURY, 1979 ; MASSON, 1973) ont montré au cours des dix dernières années le comportement géomécanique assez particulier des craies saturées. La craie "typique" à porosité élevée (30 à 35 % et plus) sans impuretés, ni recristallisation de calcite formant des ponts cristallisés supplémentaires entre coccolithes, présente l'originalité maintenant bien connue de voir ses propriétés mécaniques (résistance, déformabilité) divisées par un facteur 2 à 3 entre l'état sec ou saturé d'un liquide apolaire (huile) et l'état saturé d'eau. L'injection d'eau risque en augmentant la déformabilité de la roche réservoir d'en accentuer les déformations et en diminuant sa résistance d'en provoquer l'écrasement ; elle concourerait ainsi à amplifier à terme les effets de la subsidence ; d'autre part l'injection sous pression associée à une réduction de la résistance peut favoriser le déclenchement de mouvement discontinus de cisaillement sur des discontinuités et provoquer de petites secousses sismiques (EVANS, 1966 ; HANDIN et RALEIGH, 1972). Or les craies typiques montrent une très forte sensibilité aux chargements dynamiques et des comportements induits très particuliers (thixotropie) (TALON, 1976 ; COMES, 1969). Une telle opération d'injection d'eau ne doit être envisagée que dans la certitude d'être en présence d'une craie atypique, insensible à la saturation en eau.

4 CONCLUSION

Des données détaillées observations et mesures replacées dans une vision géomécanique du comportement des roches autour d'une cavité permettent d'identifier les 2 ou 3 paramètres les plus importants à l'origine des difficultés de forage d'un puits pétroliers.
De la même façon un transcription en termes géomécaniques des observations disponibles sur les subsidences de certains champs pétroliers mènent à développer les dispositifs d'auscultation seuls capables à l'heure actuelle de fournir un moyen de controle et confirmation des méthodes de stabilisation ou le moyen d'alerte fournissant le plus long délai.
Un résultat principal de ces premières approches a été de souligner les nécessaires remises en cause ou approfondissements de certaines approches et hypothèses : modèles élastoplastiques, poro-élasticité, thermomécanique, compaction-subsidence des seuls réservoirs, hypothèse d'intégrité du recouvrement, etc...

REMERCIEMENTS

Les auteurs remercient la société Elf-Aquitaine au sein de laquelle ils réalisent ces travaux et en particulier les foreurs et les spécialistes "boues" et "gisements" avec lesquels ces résultats ont pu être obtenus.

REFERENCES

J. ARCAMONE. 1980. Méthodologie d'étude des affaissements miniers en exploitation totale et partielle. Thèse Doc. Ing., INPL-E.N.S. Mines, Nancy.

W.B. BRATLEY. 1978. Borehole failure near salt domes. SPE paper 7503, 53th Fall Meeting at Houston.

G. COMES & al.. 1969. La craie au laboratoire et dans un tunnel profond. C.R. 7ème Cong. Int. Mec. Sols et Fond., Mexico.

J.L. DESENNE. 1971. Etude rhéologique et géotechnique de la craie. Thèse Doc. Ing., Université de Grenoble.

C. DOREMUS. 1978. Les craies du Nord de la France : litho stratigraphie, microstructures et propriétés mécaniques. Thèse, Université Sc. et Tech., Lille.

M. DEJEAN & J. ARCAMONE. 1979. Affaissements miniers et déformations du sol. Influence de la lithologie du recouvrement. C.R. 4ème Cong. Int. S.I.M.R., Montreux.

D.M. EVANS. 1966. The Denver area earthquakes and the Rocky Mountain arsenal disposal wells. Mountain Geologist.

M. HANDIN & M. RALEIGH. 1972. Man-made earthquakes and earthquakes control. C.R. Symposium S.I.M.R., Stuttgart.

J.A. HUDSON & J.M. MORGAN. 1975. Compressive failure of chalk, TRRL. Lab Report 681, Crowthorne.

J. GILLULY & U.S. GRANT. 1949. Subsidence in the Long Beach harbor area, California. Bull. Geol. Soc. Am., 60, pp.461-530.

A. GUENOT & V. MAURY. 1986. Stabilité des forages profonds. In cours présentés à l'Ecole d'Eté de thermo-mécanique des roches du Comité Français de Mécanique des Roches, Alès. (à paraître 1987 aux éditions du BRGM, Orléans).

A. GUENOT. 1987. Contraintes et ruptures autour des forages pétroliers. à paraître 6ème Cong. Int. S.I.M.R., Montréal.

U.S.GRANT. 1954. Subsidence of the Wilmington Oil Field, Cal. Calif. Div. Mines Bull, 170, p.19-24.

R.B. JANSEN. 1980. Dams and Public Safety. US D.O.I., Water and Power Service, Water Resources Tech. Pub.

K.L. LEE & C.E. SHEN. 1969. Horizontal movements related to subsidence. J. Soil Mech. and Found. Div. ASCE, January, p. 139.

K.L. LEE. 1976. Calculated horizontal movements at Baldwin Hills, Cal.. Proc. Anaheim Symp., Int. Ass. of hydrological Sc. Publ n°21.

P. LONDE & F. SABARLY. 1966. La distribution des perméabilités dans les fondations des barrages voûtes en fonction du champ de contrainte. C.R. 1er Cong. Int. S.I.M.R., Lisbonne.

M. MASSON. 1973. Pétrophysique de la craie. Bull. Labo. P et Ch., Special V, Octobre 1973.

R. MIGUEZ & J.M. HENRY. 1986. Discage : Conventions pour l'étude de sa morphologie. Ann. Société Géol. du Nord, Lille.

V. MAURY. 1977. An Underground storage in soft rock (chalk). C.R. Rockstore 77, Stockholm.

V. MAURY. 1979. Effondrements spontanés. Revue de l'Industrie Minérale, Octobre.

V. MAURY. 1987. Observartions, recherches et résultats récents sur les mécanismes de ruptures autour de galeries isolées. Rapport du Président de la Commission Internationale de la S.I.M.R., à paraître aux C.R. du 6ème Cong. Int. de la S.I.M.R., Montreal.

L. OBERT & D.E. STEPHENSON. 1965. Stress conditions under which core disking occurs. Trans. SME, v.232, p.227-235.

J.M. SAUZAY & V. MAURY. 1987. Borehole instability : case histories, Rock Mechanics approach and results. SPE/IADC paper 16051.

C.J. SPIERS & al. 1986. The influence of fluid-rock interaction on the rheology of salt rock. Report EUR 10399 EN, Commission of European Communities. Publication N°CD-NE-86-010-EN-C, Luxembourg.

J.P. TALON. 1976. Influence de la texture sur certaines propriétés géotechniques des craies franches. Thèse Doc. Ing., Universié P. et M. Curie, Paris.

E. TINCELIN & P. SINOU. 1962. Effondrements brutaux et généralisés-coups de toit. Rev. Ind. Minerale, Avril.

R. WIBORG & J. JEWHURST. 1986. Ekofisk subsidence detailed and solutions assessed. Oil & Gas Journal (Technology) Feb. 17, pp. 47-51.

Experimental study on seepage and dispersion of seawater in fractured rock
Etude expérimentale sur l'infiltration et la dispersion d'eau de mer dans la roche fracturée
Experimentaluntersuchung über Versickerung und Dispersion von Meerwasser in brüchigem Lagergestein

HAJIME MINEO, Electric Power Development Co. Ltd, Japan
MASAYUKI HORI, Electric Power Development Co. Ltd, Japan
MASAHIKO EBARA, Electric Power Development Co. Ltd, Japan

ABSTRACT: Seepage and dispersion of seawater from the upper storage reservoir is one of the most important environmental problems for consideration in planning a seawater pumped-storage power generating project. From a phenomenological standpoint, the process consists of seawater seepage from the upper side of the groundwater table, followed by spreading and dispersion after coming in contact with the fresh groundwater zone. Field experiments were performed in the vicinity of an upper storage reservoir site to examine the differences between ideal porous media and actual fractured rock masses for the above-mentioned phenomena and to evaluate dispersion coefficients based on field experiments. In this paper, the authors describe mainly results of the field experiments and evaluation of dispersion coefficients in fractured bedrock.

RÉSUMÉ: L'infiltration et la dispersion d'eau de mer du réservoir supérieur de retenue est un des problèmes environnementaux que l'on doit prendre en ligne de compte lors de la planification d'un projet de production d'électricité à réservoir à pompage d'eau de mer. D'un point de vue phénoménologique, le phénomène consiste en une infiltration de l'eau de mer provenant du côté supérieur de la surface de la nappe d'eau souterraine, suivie par un écoulement et une dispersion après l'entrée en contact avec la zone d'eaux souterraines douces. Des expériences pratiques ont été effectuées à proximité d'un site de réservoir supérieur de retenue afin d'examiner les différences entre le support poreux idéal et les masses rocheuses réellement fracturées pour le phénomène décrit ci-dessous et pour évaluer les coefficients de dipersion fondés sur des mesures pratiques. Dans cette thèse, les auteurs décrivent essentiellement les résultats des expériences pratiques et l'évaluation des coefficients de dispersion dans de la roche de fond fracturée.

Zusammenfassung: Versickerung und Dispersion von Meerwasser im oberen Stausee ist eines der wichtigsten Umweltprobleme, das bei den Überlegungen für die Planung eines Meerwasser-Pumpspeicher-Kraftwerks berücksichtigt werden muß. Von einem phänomenologischen Standpunkt aus betrachtet, besteht der Prozeß aus Meerwasserversickerung von der Oberfläche des Grundwasserspiegels, gefolgt von Ausbreitung und Dispersion, nachdem das Meerwasser mit der frischen Grundwasserzone in Berührung kommt. In der Nähe eines oberen Stausees wurden Feldversuche durchgeführt, um die Unterschiede zwischen idealen porösen Medien und tatsächlochen Felsbruchmassen für das oben genannte Phänomen zu untersuchen und die Dispersionskoeffizienten, basierend auf Feldmessungen, auszuwerten In dieser Abhandlung beschreibt der Verfasser vornehmlich die Resultate der Feldversuche sowie die Auswertung von Dispersionskoeffizienten in brüchigem Lagergestein.

1 INTRODUCTION

A seawater pumped-storage power generating project is presently being considered in the main island of Okinawa located close to the southern tip of the Japanese Archipelago (Hashimoto 1986). From an environmental standpoint, what are of extreme importance in planning such a project are seepage and dispersion phenomena of seawater from the upper storage reservoir to the surrounding bedrock. Phenomenologically, when leakage occurs from the reservoir sealed with waterproof rubber sheets, the process will be for seawater to seep into the bedrock, and on coming into contact with the fresh groundwater zone, to be dispersed. However, there are extremely few cases of studies having been made on such dispersion phenomena in fractured bedrock, and presently the propriety of applying findings regarding the phenomena and estimated dispersion coefficients obtained in ideal porous media to actual fractured bedrock is not clear. Accordingly we have been carrying out laboratory experiments and numerical analysis concerning this problem since 1983, and as a part of a series of the studies, field experiment on seepage and dispersion in fractured bedrock has been performed from September 1985. The above-mentioned phenomena are mainly discussed in this paper based on the results of the experiments.

The objectives of the experiments are the following three items;
1. Grasping of the dispersion phenomena after contact between seawater and fresh water in fractured bedrock

2. Identification of important parameters for analyzing dispersion in the bedrock such as dispersion coefficient
3. Examination of appropriateness of analysis technique and programs comparing with the results of numerical simulation

Dispersion analysis is generally divided into a part for seepage analysis and a part for solving a diffusion type partial differential equation of the second order. In performing dispersion analysis several parameters are required, i.e. specific moisture capacity, specific storage coefficient, permeability coefficient, adsorption coefficient and dispersion coefficient etc. Permeability coefficients were obtained by in-situ permeability tests, and dispersion coefficients were estimated by fitting the theoretical solution of the one-dimensional dispersion equation to the results of field measurement of dispersion process.

2 SITE DESCRIPTION

The experiment site is at the Pacific Ocean side of Okinawa Island approximately 80 km northeast of the city of Naha, and topographically it is featured by a gently sloped flat area of elevation from 140 to 180 m and a steep slope in the vicinity of the seashore. Regarding lithological facies of the general area of the experiment site, although alternations of sandstone and mudstone can be seen in parts, black phyllite is distributed in almost all of the surrounding area, and it may be considered that as a whole the monoclinal structure of strike N26°E and dip 33°NW as shown in Fig.1 is constituted. Fig.1 was prepared from the observations and analysis using a borehole television system at Boreholes B-0, B-1, B-2, B-3, and B-4 shown in Fig.2 and there is good conformity with the results of observations of outcrops.

The phyllite at the site may be divided into four zones according to degree of weathering, and these are weathered residual soil, strongly weathered phyllite, weathered phyllite, and unweathered phyllite in order from the top. The geological profile along the observation boreholes C-1~21 shown in Fig.2 and the distribution of permeability coefficients estimated from the results of in-situ permeability tests are shown in Fig.3. Generally permeability coefficient is $1~2 \times 10^{-5}$ (cm/sec) in the unweathered phyllite zone deeper than 35 m from the ground surface. This coefficient may be slightly large for fresh bedrock, and is considered to be due to open cracks concentrated comparatively at the upper part of the unweathered phyllite at this experiment site. Permeability coefficient at this high permeable zone is estimated as approximately 2×10^{-3} (cm/sec), while at the overlying weathered phyllite zone it is of the order of roughly 10^{-4}-10^{-3} (cm/sec). As mentioned previously, the basement consists of phyllite, and generally the anisotropy due to schistosity of the rock itself is fairly great. However, cracks intersecting the schistosity are also found by observations with borehole television. Therefore it was supposed that an extreme anisotropy in seepage and dispersion would not be found in actual bedrock phenomenologically, and this has been substantiated to a certain extent by experimental results also.

The groundwater table at this site is at a depth of 20 to 25 m and located inside the weathered phyllite stratum. Although the groundwater table is inclined slightly toward the seaside, it is practically horizontal as a whole and it is considered that groundwater flow is extremely slow. Groundwater at this site has on an average a water temperature of approximately 20°C, chlorine concentration of approximately 160 mg/l, and electrical conductivity of approximately 280 μʊ/cm. On the other hand, chlorine concentration of injected seawater is approximately 18800 mg/l.

3 FACILITIES AND FIELD MEASUREMENTS

The main facilities for the experiments are seawater injection boreholes (A-1~11), observation boreholes (C-1~21), and the seawater storage pond shown in Fig.2. The direction of the row of observation boreholes was made orthogonal to the strike of schistosity discerned in geological investigations shown in Fig.1, while the row of injection boreholes consisted of 11 boreholes at 1.0 m intervals in a line to be orthogonal to the row of observation boreholes. The each injection borehole (A-1~11) had steel casings inserted excepting a part of 1.5 m from bottom of borehole and further the storage pond is also sealed with waterproof rubber sheets. Therefore seawater is injected only through a part of 1.5 m of each injection borehole. The elevations of these injection portions are from 110.34 to 111.84 m, and they are located in unweathered bedrock and below the groundwater table. The observation boreholes are arranged as shown in Fig.3 at intervals of 1.5 m, and filled with mortar excepting a part of 3 m from bottom of each borehole. This portion of 3 m is filled with gravel of grain size 0.5~3 mm, where the filter tip of BAT groundwater monitoring system (Torstensson 1984) is embedded.

The experiment consists of seawater injection from September to November 1985 and measurements of chlorine concentration, pore pressure, seepage discharge, etc. during injection and after ceasing injection. Chlorine concentration was determined by measuring electrical conductivity of sampled groundwater and converting it to the concentration. Sampling of groundwater and measurement of pore pressure are carried out by BAT system. Seepage discharge was determined by measuring the flow velocity in the injection boreholes.

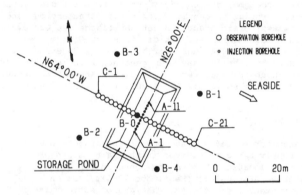

Figure 2. Layout of experiment facilities.

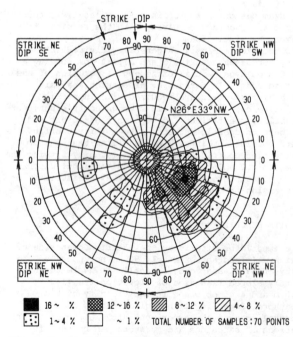

Figure 1. Crack distribution at experiment site.

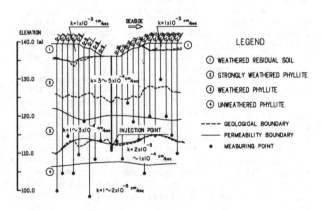

Figure 3. Geological profile and permeability coefficient distribution along observation hole row.

4 RESULTS

The discharge of seawater injected under the ground-water table showed a prominent unsteadiness at the initial stage of injection as shown in Fig.4. However, decreasing rate of discharge became small from about the time 20 days had elapsed after start of injection, and seepage of around 20 l/min came to be maintained. Seawater injection was continued for 1760.5 hours, or approximately 73 days.

The breakthrough curves obtained at C-1, C-5, C-7, C-8, and C-9 are given as examples in Fig.5. Considering the duration of injection, breakthrough curves obtained at measuring points distant from the injection borehole may be incomplete, and relative concentration C/C_0 may not become high enough. Accordingly these breakthrough curves up to the time injection was stopped are used in the evaluation of dispersion coefficients. Fig.5 shows that the decreasing rate of chlorine concentration after stop of injection is greater the higher the elevation of the measuring point. One of the reasons for this phenomenon may be the gravitational effect due to difference in density of seawater and fresh water. The seawater mass penetrated into the fresh groundwater zone even after injection had been stopped, while conversely, the interchange of seawater and fresh water at the vicinity of the groundwater table occurred at a comparatively rapid rate. The time-dependent change of the isochlors is shown in Fig.6. According to this figure, chlorine distribution apparently presents a comparatively isotropic spread in spite of anisotropic characteristics due to schistosity of the phyllite itself. As described previously it is considered that the influence of the cracks intersecting the schistosity appears in dispersion phenomena. It can also be seen that the isochlors are warped being affected by the zone of high permeability extending from the vicinity of the injection point toward the seaside.

In estimation of dispersion coefficients based on breakthrough curves, it is necessary to know seepage velocity at measuring points. However, field measurement of seepage velocity has several difficulties at present and is not used generally yet. Therefore we estimated it by the following procedure. Taking time To giving the steepest gradient on the breakthrough curve of each measuring point, this time To is assumed to be the time when the seawater front passed its measuring point. Accordingly the average rate of moving front can be calculated between two measuring points, and as a result, average seepage velocity can be obtained approximately by the distribution of the front velocity, and ranges from 5×10^{-5} to 1×10^{-2} (cm/sec) for this experiment.

5 DISPERSION COEFFICIENT

Dispersion coefficient can be estimated using the breakthrough curve obtained at each measuring point in the bedrock. The procedure adopted here is a kind of curve-fitting, where it is necessary to determine the form of the governing equation and its solution in order to curve-fit, considering experimental conditions at the site. From the macroscopic standpoint, permeability inside the ground decreases in accordance with depth from the ground surface, and such a trend can be seen at this experiment site also. Groundwater flow at such a field, caused by injection as in our experiment, can be considered to be close to a roughly horizontal one-dimensional flow except

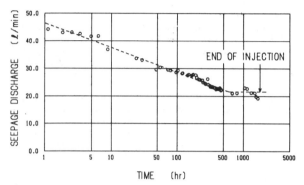

Figure 4. Discharge of seawater injected through boreholes.

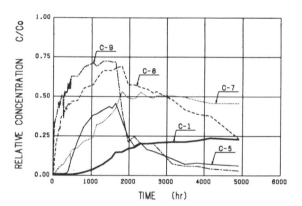

Figure 5. Time-dependent changes in chloride concentrations (C-1, C-5, C-7, C-8, C-9).

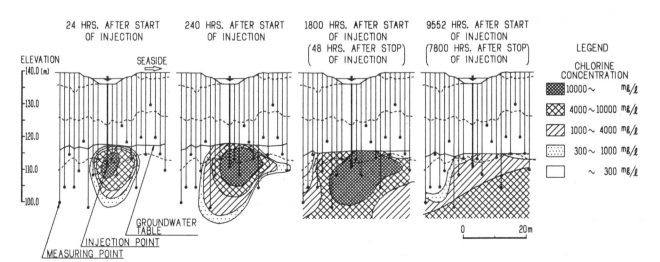

Figure 6. Time-dependent change of isochlors.

193

in the vicinity of the injection points. Based on the hypothesis above, Eq.(1) below is considered to be the governing equation of the dispersion phenomenon of this field experiment.

$$\frac{\partial c}{\partial t} + u \frac{\partial c}{\partial x} = D \frac{\partial^2 c}{\partial x^2} \quad \dots\dots\dots\dots\dots\dots\dots (1)$$

where,
  $C(x,t)$: chlorine concentration
  D : dispersion coefficient
  u : average seepage velocity

Boundary conditions and initial condition are set as follows;
$$\begin{aligned} C(0,t) &= C0 & &\text{for } t \geq 0 \\ C(x,0) &= 0 & &\text{for } x > 0 \quad \dots\dots\dots\dots (2) \\ C(\infty,t) &= 0 & &\text{for } t \geq 0 \end{aligned}$$

Based on these conditions, the solution for Eq.(1) will be as follows (Bear 1979);

$$\frac{C}{C_0} = \frac{1}{2} \operatorname{erfc}\left(\frac{x - ut}{2\sqrt{Dt}}\right) + \frac{1}{2} \exp\left(\frac{ux}{D}\right) \operatorname{erfc}\left(\frac{x + ut}{2\sqrt{Dt}}\right) \dots$$
$$\dots\dots (3)$$

Where, in case of $D/ux < 1.0$, it is considered the second term of the right-hand side of the above equation can be neglected (Rumer 1962). Consequently Eq.(3) can be simplified to the form of Eq.(5) by employing the non-dimensional parameters given in Eqs.(4).

$$\begin{aligned} \xi &= u \cdot t/x \\ \eta &= D/u \cdot x \end{aligned} \quad \dots\dots\dots\dots\dots\dots\dots\dots (4)$$

$$\frac{C}{C_0} = \frac{1}{2} \operatorname{erfc}\left(\frac{1 - \xi}{2\sqrt{\eta\xi}}\right) \quad \dots\dots\dots\dots\dots\dots\dots (5)$$

Eq.(5) can be rewritten as
$$\operatorname{erf} \Phi = 1 - 2 \frac{C}{C_0} \quad \dots\dots\dots\dots\dots\dots\dots\dots (6)$$

where,
$$\Phi = \frac{1 - \xi}{2\sqrt{\eta\xi}} \quad \dots\dots\dots\dots\dots\dots\dots\dots\dots (7)$$

The right-hand side of Eq.(6) can be evaluated by using measured concentration of chlorine, and therefore, Φ can be determined by calculating the inverse function of the error function.
Further, $\Psi = \Phi\sqrt{\eta} = (1 - \xi)/2\sqrt{\xi}$ can be calculated by Eq.(7). Therefore, each datum defining breakthrough curve can be plotted on the coordinate system with Ψ as ordinate and Φ as abscissa. If the plotted data can be approximated by a straight line, its gradient J equals $\sqrt{\eta}$, and dispersion coefficient can be calculated as $D = J^2 ux$ considering Eq.(4). The relationship between Φ and Ψ at measuring point C-1 is shown as an example in Fig.7. It shows a fairly good linear correlation, limiting to data corresponding to the period of injection, and it is possible for the dispersion coefficient to be estimated by this method. The entire data are overlaid in Fig.8. Dispersion coefficients obtained by this method are plotted against average seepage velocities in Fig.9. As is evident from this figure, these two parameters indicate fairly good correlations, and the regression equation is as shown in Eq.(8).

$$D = 569.1 \times u^{0.901} \quad \dots\dots\dots\dots\dots\dots\dots\dots (8)$$

Rumer obtained $D = 0.2u^{1.083}$ for quartz gravel of uniformity coefficient 1.26, effective diameter 1.35 mm, average grain size 1.65 mm, and porosity 0.39, and $D = 0.027u^{1.105}$ for glass beads of uniformity coefficient 1.13, effective diameter 0.35 mm, average grain size 0.39 mm, and porosity 0.39 (Rumer 1962). According to the experiments of Harleman and Rumer, $D = 0.093u^{1.183}$ was obtained for plastic beads of uniformity coefficient 1.14, effective diameter 0.86 mm, and average grain size of 0.96 mm (Harleman

and Rumer 1963).
   The above-mentioned results appear to indicate that the coefficient depends to a great extent on the geometry of the pore system of the medium, while it may be considered that the exponent of average seepage velocity assumes a value close to approximately 1.0 regardless of porous media or fractured bedrock within the limits of this field experiment.

REFERENCES

Hashimoto, T. 1986. Seawater pumped-storage scheme under study in Japan. Water Power & Dam Construction. 2: 9–11
Torstensson, Bengt-Arne 1984. A new system for ground water monitoring. Ground Water Monitoring Review. Vol.4 No.4: 131–138
Rumer, R.R. 1962. Longitudinal dispersion in steady and unsteady flow. Proc. of ASCE, J.Hyd.Div. 88, HY 4: 147–172
Harleman, D.R.F. & R.R.Rumer 1963. Longitudinal and lateral dispersion in an isotropic porous medium. J.Fluid mech. Vol.16, part 3: 385–394
Bear, J. 1979. Hydraulics of groundwater. New York: McGraw-Hill Inc.

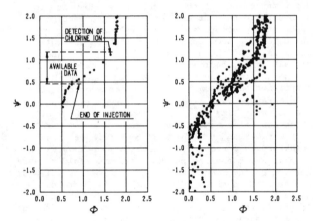

Figure 7. Φ-Ψ relationship (C-1).

Figure 8. Φ-Ψ relationship (entire data).

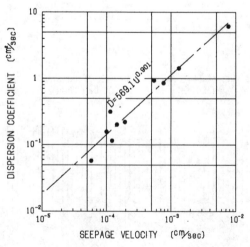

Figure 9. Dispersion coefficient versus seepage velocity.

Proposed design for high pressure isolation structure in saltrock
Plan proposé pour des structures isolantes à haute pression utilisées en milieu de roches salines
Vorgesehene Planung einer Hochdruck-Isolations-Struktur in Salzfelsen

D.Z.MRAZ, Mraz Project Consultants Ltd, Saskatoon, Canada

ABSTRACT: Recent inflows of unsaturated brines into three deep potash mines in Saskatchewan have again high-lighted the need for a design of reliable engineering barriers in saltrocks. A successful design of such iso-lation structure must not only provide an impermeable seal at the interface of rock and the dam, but it must also prevent migration of the brine through microfracture system in the vicinity of the structure. A study into the feasibility of designing such a structure has resulted in a proprietary concept described here.

RÉSUMÉ: En Saskatchewan, de récentes infiltrations de saumures non saturées dans trois mines profondes de potasse a encore mis l'emphase sur le besoin de barrières de conception sûre utilisées en milieu de roches salines. Le système approprié pour une telle structure isolante doit non seulement sceller imperméablement l'interface de la roche et du barrage mais il doit également prévenir la migration des saumures par le système de microfractures alentours de la structure. Une étude concerant la faisabilité de dessiner de telles struct-ures nous a conduits à l'élaboration originale des concepts décrits ici.

ABRISS: Zuflüsse jungeren Datums von unimprägniertem Salzwasser in drei tiefe Kali-Zechen in Saskatchewan haben wiederum das Bedürfnis nach zuverlässig konstruierten Barrieren in Salzfelsen betont. Ein erfolgreicher Plan einer solchen Isolationsstruktur muss nicht nur ein wasserdichtes Siegel an der Grenzfläche zwischen Felsen und Damm errichten, sondern auch die Wanderung von Salzwasser durch das Mikrofraktur-System in der Nähe der Struktur verhindern. Die Erforschung der Durchführbarkeit des Entwurfs einer solchen Konstrucktion hat zu den hier beschriebenen patentamtlich geschützten Planen geführt.

1 DISCUSSION

The first uncontrolled high volume flood in a deep Saskatchewan potash mine had occurred in November, 1984, at the Rocanville Division of PCS (C.M.J., 1985) and additionally, two such incidents have been experienced since. The Rocanville crisis ended in March, 1985, when a concrete bulkhead was success-fully completed in a single opening through halite rock which connected the mine with the inflow chan-nel.

Although the need for reliable engineering barr-iers in saltrocks is most dramatically demonstrated by the incidence of mine flooding, there are other applications where these structures are required. For example, the feasibility of isolation of high level nuclear waste in saltrock deposits is very de-pendent on the ability of constructing reliable eng-ineering barriers within rooms and shafts of the nuclear waste repositories. Designing of these str-uctures is a very difficult engineering problem. There are some examples of success, but there are many more examples of failures. In those deposits where the saltrock consists of more than one mineral, i.e. sylvite, carnallite, and halite, design difficu-lties seem almost insurmountable. Natural solutioning of saltrock deposits results in solutions which may be saturated with respect to NaCl, but contain only traces of other salts; such solutions are not satur-ated when they contact potash salts (Baar, 1982). It has been demonstrated that even at very low pressures, the unsaturated brine will bypass engineered barrier through the cracks and fractures in the wall of the mine opening adjacent to it. In addition to bedding planes and other lithological discontinuities, there are fractured zones in the areas subjected to stress concentrations during the excavation of the opening. Hence, prior to constructing any such barrier, all fractured or otherwise potentially discontinuous

material must be removed. However, even a complete removal of all such material will not prevent infil-tration of brine into the stress relieved zones around the opening. Scanning electron microscope photographs of potash specimens have shown that stress relieved potash rock is full of microcracks, which are the result of a stress relief and resulting difference in strains of sylvite and halite crystals (Dusseault, et al., 1985). The fracture pattern in pure halite is much less pronounced, however, some mircofractures can still be present. A successful design of the isolation structure must not only con-tain an impermeable seal at the interface of rock and the dam, but must also prevent migration of the brine through microfracture system in the vicinity of the structure.

2 ANALYSIS AND DESIGN OF EXCAVATIONS

Excavations in saltrocks undergo continuous modifi-cations of shape due to a combination of time depend-ent deformation and yield mechanisms (Mraz and Dusseault, 1986). The acting geometrical surface is called an active opening, or an equivalent opening, and its size and shape depends on the following factors:
- size and shape of initial excavation;
- sequence of excavation;
- level of stress at the time of excavation;
- extent and sequence of adjacent excavations;
- time since excavation;
- presence or absence of lithological discontinui-ties.

The shape of the active opening has a substantial effect on the stress distribution in adjacent rock. Although the number of possible variations of its shape is infinite, it is possible to estimate its

195

shape with reasonable certainty.

2.1 Shape of active opening in rock without discontinuities

Three basic cases of active opening geometry are recognized in continuum (Mraz and Dusseault, 1986):

Case 1 - The major principal stresses near the opening at the time of excavation have substantially exceeded the limiting strength criterion and the rock within the active opening fails by a brittle yield process. The limiting strength criterion can be typically expressed by an empirical expression such as the one derived by Rothenburg (1986):

$$\sigma_1 = \frac{1}{2}\sigma_c \left[(1 - sr) + \sqrt{(1 + sr)^2 + 4s\frac{\sigma_3}{\sigma_c}} \right] \quad (1)$$

$$r = \frac{\sigma_T}{\sigma_c}$$

σ_1, σ_3 ... principal stresses
σ_c ... unconfined compressive strength at long time
σ_T ... tensile strength
s ... empirical factor

Shear fracture in rectangular opening occurs along the segments which are approximately circular

$$a_v = \frac{1}{\sqrt{2}} h \quad (2)$$

$$a_H = \frac{1}{\sqrt{2}} v \quad (3)$$

a_v ... radius of segment in roof
a_H ... radius of segment in wall
h ... width of initial opening
v ... height of initial opening

After some convergence, the shape of active opening can be approximated by an ellipse with the following dimensions (Figure 1):

$$d_1 = \frac{1}{2} \left[(h_1 - C) + (\sqrt{2} - 1)(v_1 - C) \right] \quad (4)$$

$$b_1 = \frac{1}{2} \left[(v_1 - C) + (\sqrt{2} - 1)(h_1 - C) \right] \quad (5)$$

d_1 ... horizontal principal axis
b_1 ... vertical principal axis
h_1 ... width of opening after convergence
v_1 ... height of opening after convergence
C ... average total convergence of opening

Case 2 - The major principal stresses at the time of excavation have not exceeded the limiting strength criterion. Initially, there is no failure or there is only a "skin damage". However, very slow creep movement occurs nonetheless, because saltrocks creep at any differential stress level. Because the shape of openings is usually incompatible with zero volume change creep, stress concentrations will develop in corners, resulting in gradual development of fractures. Assuming that the active opening curvature can be approximated by circular arc (Figure 2):

$$a_H = \frac{v^2}{4C} - \frac{v - C}{2} \quad (6)$$

$$a_v = \frac{h^2}{4C} - \frac{h - C}{2} \quad (7)$$

Case 3 - The major principal stresses at the time of excavation are close to the yield point, causing incompletely developed latent damage. In these circumstances, continued convergence may lead to the full development of the fractured zone and, conservatively, it is appropriate to assume that the active opening takes the shape of an ellipse with conjugate diameters equal to the diagonals of the mine opening:

$$d_1 = \frac{1}{\sqrt{2}} (h - C) \quad (8)$$

$$b_1 = \frac{1}{\sqrt{2}} (v - C) \quad (9)$$

2.2 Shape of active opening in saltrock with discontinuities

The most common form of discontinuities are sedimentary features, such as seams of insoluble materials in bedded salt and potash deposits. The influence of these features on the active opening could be unpredictable. Under these circumstances, the reliable method of determining the shape of zone of fractured or yielded rock is in situ measurement of stresses and convergences.

2.3 Determination of active opening geometry by in situ measurements

The most reliable method of determining the extent of yielded zone is the in situ measurement of minor principal stresses, because the locus of zero minor principal stress coincides with active opening. It is often satisfactory to measure radial stresses along the axes of the opening, but, in some circumstances,

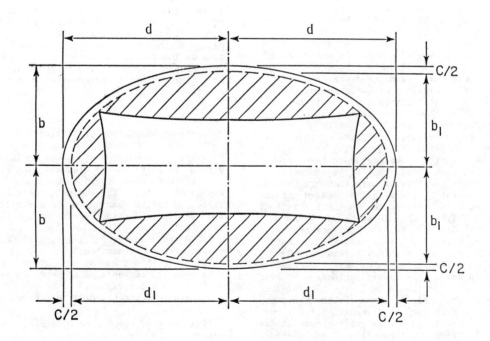

FIG. I ACTIVE OPENING GEOMETRY AFTER CLOSURE

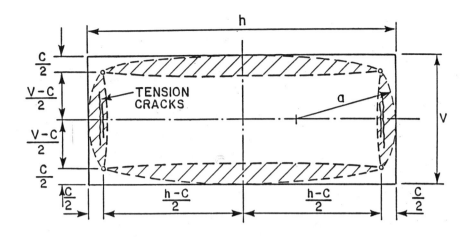

FIG. **2** ACTIVE OPENING GEOMETRY DUE TO SLOW
CLOSURE AT LOW STRESS

radial stress measurements along other directions may
be required. The measurement of radial stress can
be obtained with packer type hydraulic pressure cell
in the holes perpendicular to the opening (Mraz,1980),
or with various other stress meters in holes parallel
to the opening. Typically, the stress is measured
at several locations and the location of zero radial
stress can be determined by geometrical plot. Alter-
nately, it can be calculated by fitting stress distribution function to the measured levels of radial
stress (Mraz, 1980).

In continuous saltrock

$$\sigma_r = 2K \ln \frac{r}{a} \qquad (10)$$

$$K \cong \tfrac{1}{2}\sigma_c \qquad (11)$$

σ_r ... radial stress
K ... Prandtl Limit (Mraz and Dusseault, 1986)
r ... radial distance
a ... radius of curvature of active opening.

In rock with discontinuities

$$\sigma_r = \tau_s \ln \frac{r}{a} \qquad (12)$$

Where τ_s is shear stress limit related to the reduced shear strength along discontinuities.

2.3 Design of excavations

The extent of required excavation is very site specific. Determination of the exact location of the
active opening is one of the most important aspects
of the design. In most cirumstances, a complete
removal of the yielded and fractured rock is required,
in order to prevent leakage. The shape and the size
of the excavations is governed by the following:
- geometry of active opening;
- location of discontinuities;
- pressure and saturation of brine to be isolated;
- methods employed in host rock improvement, i.e.,
  grouting;
- methods employed in rock restressing after the completion of structure.

3 DESCRIPTION OF PROPOSED STRUCTURE

The proposed conceptual design consists of several
"lines of defense", and it is highly dependent on a
life expectancy of the structure. In other words,
the life of the structure is not necessarily consid-

ered to be infinite; it is merely designed to provide
isolation for the period of its useful service. The
components of the design are as follows:
   1. A complete removal of all the macrofractured
material around the opening;
   2. Construction of barriers, i.e., concrete dams;
   3. Sealing of the interface between the barrier
and the host rock;
   4. Sealing the host rock on the pressure side of
the barrier;
   5. Sealing of the microfracture system in the
host rock around the barriers through rock repressurization;
   6. Impregnating of the microfracture system
within the host rock with hydrophobic medium;
   7. Installation of monitoring instrumentation;
   8. Supporting the host rock on both sides of the
structure by backfill.

3.1 Construction of barriers

The shape of barriers required in any given situation
is again site specific. These structures must be
keyed into the host rock and the shape of initial
excavation discussed above will dictate their shape
and size. In homogeneous rocks, such as halite, a
plug-like bulkhead may be feasible (Herbert and
Stöver, 1986). However, in lithologically and geotechnically complex environments, the shape and size
of these structures must conform to the local requirements. For example, clay seams capable of serving
as a water conduit may have to be removed and replaced with impermeable barrier, and local conditions
will also dictate the type of construction materials
used.

3.2 Proposed design of isolation barrier

The proposed proprietary conceptual design (patent
pending) is shown in Figure 3. It requires two dams
(5, 6) constructed of suitable material, such as
concrete, and designed to suit local conditions as
described above. The space between the two dams is
equipped with sump (10), which is connected to a
sump withdrawal pipe (11). The wet side of dam (5),
as well as adjacent rock surfaces, are coated with
impervious liner (3), such as spray-on elastomer.

Optionally, the space (1) can be backfilled with
local material to provide additional support to the

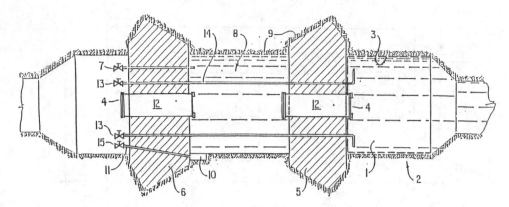

FIG.**3** PROPOSED DESIGN FOR HIGH PRESSURE ISOLATION
STRUCTURE IN SALT ROCK.

rock and the liner. Next, the space between the dams is filled with hydrophobic liquid lighter than inclosed brine, such as diesel fuel, which is pressurized. The level of pressure required depends on the local geophysical and hydrological conditions and it must be maintained over the lifetime of the structure. This will cause repressurization of the adjacent saltrock and a closure of the microcracks in the rock, as well as, a sealing of the interfaces between the dams and the rock through creep. In addition, any remaining microcracks or fissures will be filled with the hydrophobic liquid, thereby resisting the entry of inclosed brine through the cracks. If necessary, pressure of the hydrophobic liquid can be regulated at will to achieve exactly the rock stress conditions desired. For this purpose, both the host rock and the structure would be sufficiently instrumented and monitored.

The enclosed brine can be monitored and sampled through monitoring and sampling instrumentation (13). If any brine enters into the area (8) between the dams, its higher specific gravity will cause it to fall to the sump (10) and it can be withdrawn through valve (15). Withdrawn fluid will be replaced through port (7). Thus, even if a leak occurs, the integrity of the structure could be maintained until a corrective action, such as grouting, can be implemented. As well, pressure between dams can be somewhat manipulated, which may assist in improving the integrity of the barrier.

3.3 Alternate design with expansion joint

One of the most serious concerns in installing the isolation dams is the period of time it might take to repressurize the rock and close the microfractures. The solution suggested above may not be always acceptable. An alternate method of accelerated rock repressurization is shown in Figure 4.

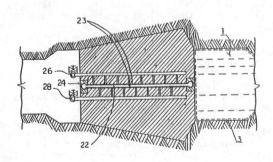

FIG.**4** EXPANSION JOINT FOR HIGH PRESSURE
ISOLATION STRUCTURE

In this case, the proposed structure includes a proprietary design of expansion joint (22), consisting of liner plates (23) connected with flexible joint (24) designed for required pressure. The flexible joint will allow the plates to separate upon application of the hydraulic medium through injection manifold (26). Once required rock stress is achieved, the hydraulic medium between the plates can be displaced with grout, utilizing drainage manifold (28). If required, both horizontal and vertical expansion joints can be utilized in one dam, achieving improved rock integrity and dam performance.

4 CONCLUSION

New types of engineering barriers for inclosure of unwanted liquids in the mines, or other underground spaces, within incompetent or soluble rocks are proposed. They may be useful in situations involving flooding of mines, as well as, in the design and construction of underground repositories for radioactive wastes.

REFERENCES

Canadian Mining Journal, Anatomy of a crisis, December, 1985, pps. 10-11.

Baar, C.A., Effect of salt removal from underground deposits and their control, Energy, Mines and Resources, Canada, CANMET No. OSQ81-00063, March, 1982.

Dusseault, M.B., et. al., Test procedures for saltrock, 26th U.S. Symposium on rock mechanics, June, 1985.

Mraz, D.Z., Dusseault, M.B., Effects of geometry on the bearing capacity of pillars in saltrock, CIM RMSCC Workshop, Saskatoon, November, 1986.

Mraz, D.Z., Plastic behaviour of salt rock utilized in designing a mining method, CIM Bulletin, 1980.

Hebert, H.J., Stöver, W.H., Geochemical, geomechanical and geophysical measurements and test of an underground dam during the flooding of a salt mine, International Symposium on the siting, design and construction of underground repositories for radioactive wastes, Hanover, March, 1986.

Study on compressed air storage in unlined rock caverns
Etude de conservation de l'air comprimé dans des poches rocheuses non revêtues
Untersuchung über Druckluftlagerung in Felskavernen ohne Auskleidung

K.NAKAGAWA, Central Research Institute of Electric Power Industry, Japan
H.KOMADA, Central Research Institute of Electric Power Industry, Japan
K.MIYASHITA, Simizu Construction Co. Ltd
M.MURATA, Kaihatsu Computer Center Co. Ltd

ABSTRACT: The laboratory experiments were performed to show that the air can be stored in an unlined rock cavern if the air pressure is lower than the hydrostatic pressure at the depth of the cavern. The finite element analysis, considering the gas-liquid two phase flow in the ground, was also applied to the investigation. The result affirmed this condition and indicated the numerical procedure proposed here is effective to estimate the leakage of the air.

RESUME: Des expériences ont été conduites en laboratoire pour monter qu'il est possible de stocker l'air dans une poche rocheuse non revêtue si la pression del'air est inférieure à la pression hydrostatiqu au fond de la poche. Une analyse de l'élément fini, tenant compte du debit biphasé gaz-liquide dans le sol, a également été appliquée à l'étude. Les résultats confirment cett condition et indiquent que la procédure numérique proposée ici permet d'estimer de maniere efficace les fuites d'air.

ZUSAMMENFASSUNG: Die Laborversuche wurden durchgefuhrt, um zu zeigen, daß sich Luft in unverrohrten Felsenbetthölen lagern läßt, wenn der Luftdruck niedriger ist als der hydrostatische Druck am Höhlentiefpunkt. Die Finit-Element-Analyse, unter Berücksichtigung des zweiphasigen Gas-Flüssigkeits-Flusses, wurde ebenfalls auf die Untersuchung angewandt. Das Ergebnis bestätigt die Voraussagen und zeigt, daß das hier vorgeschlagene numerische Verfahren für die Abschätzung der entweichenden Luft wirksam ist.

INTRODUCTION

The energy storage in the form of compressed air is considered to be one of the advandageous systems to transfer the off-peak electric power to the peaking period. In this system the pressurized air is accumulated into underground rock caverns during slack hours and is used to generate electricity for peak hours. Since the efficiency of the system is supposed to be influenced by the leakage of the air, it is an important subject to study the prevention of the air leakage from the storage caverns. Especially in the case that the compressed air is stored in unlined rock caverns by means of underground water flow, it is required to investigate the hydraulic concerning the prevention of the air leakge.

Researches hitherto attempted to determined the necessary conditions based on a critical hydraulic gradient at which the downward water flow would prevent bubbles from moving upwards in rock fractures. The elementary concept requires the underground water to flow downwards with vertical hydraulic gradient larger than 1 (Aberg 1977). In order to satisfiy this condition, unlined reservoirs should be located in awful depth unless the pressurized water be injected from an array of boreholes above the caverns.

This paper describes the works to investigate the rational condition necessary for the underground water to prevent the air leakage from unlined rock caverns, and to propose the numerical procedure to estimate the leakage of the air from the storage caverns in the field.

ELEMENTAL STUDY WITH MODELS OF ROCK FRACTURE

The elemental experiment using models of rock fracture was carried out to examine the critical conditions at which the air escapes into the rock fracture filled with water (Nakagawa 1981). Fig.1 shows the layout of the experimental apparatus. The crack models, one of which is shown as Fig.2, were made of parallel acryl plates with 1mm aperture. The model had a notch with an angle α at the mouth of the simulated fracture to investigate the effects of irregularities of the ceilings of the caverns. Five different angles ($\alpha=30°$, $60°$, $90°$, $120°$ and $180°$) were used in the experiment. $\alpha=180°$ means the case with a flat bottom model. The upper and lower chambers of the apparatus were connected with theouter two reservoirs respectively. The height of each reservoir was changeable to control the difference of the pressure head between the top and the bottom of the model. The pressure was measured with a transducer in each chamber.

At the beginning of the test the air was injected into the chamber beneath the crack model under the condition that the flow reate of water through the crack was kept large enough to prevent the air from rising. And then the pressure at the top of the fracture was reduced until the air began to leak. Table 1 shows the water pressure, Pu, at the top of the crack, and the pressure, Pa, of the air stored in the lower chamber, Which were measured just before the air leakage began. Table 1 also shows the ratio of Pws

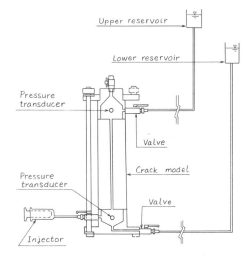

Fig. 1. Layout of apparatus

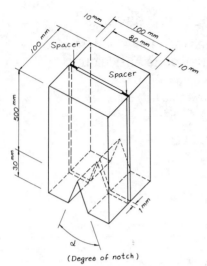

Fig. 2. Acryl crack model

(the hydrostatic pressure at the top of the notch) to Pa. Pws is calculated as $Pws=Pu+\rho_w \cdot g \cdot L$, where ρ_w is the density of water, g is the acceleration of the gravity and L (=0.5m) is the length between the top of the crack model and the top of the notch. This result shows that the stored air pressure is almost equal to the hydrostatic pressure at the top of the notch in the critical state of the prevention of the air leakage and this indicates that the critical hydraulic gradient to prevent the air leakage, In, is nearly equal to zero (In=0). From the above it is inferred that the pressurized air can be stored if the ground water flows with positive hydraulic gradient normal to the surface of the cavern, or, if the air pressure is lower than the hydrostatic pressure equivalent to that of the depth of the ceiling of the cavern. Even in the case that the irregularities of the surface of the cavern be found, this conclusion would be applicable to the envelope surface of the irregurlarities.

MODEL EXPERIMENT OF COMPRESSED AIR STORAGE

Another laboratory experiment was performed using the model of unlined cavern in order to study the phenomena of the air leakage as well as to affirm the condition to prevent the air leakage (Komada 1985). A schematic diagram of the apparatus for the model experiment is shown as Fig. 3. The model of an unlined rock cavern is shown as Fig.4. The model was made of the porous plate which was thrmally compacted assemblage of plastic beas. The clear acryl plate was attach to the front of the model in order to observe the behavior of the surface of water. Three different materials were used in the experiment. Permeable properties are shown in Table 2.

Table 1 Critical pressures to prevent air penetration into the crack

Degree of notch	Water pressure at the crack	Stored air pressure	Estimated static water pressure	
α (Deg)	Pu (kPa)	Pa (kPa)	Pws (kPa)	Pws/Pa
30	7.93	12.54	12.93	1.03
60	8.06	13.02	13.06	1.00
90	6.48	11.03	11.48	1.04
120	8.46	13.66	13.46	0.99
180	2.57	8.05	7.57	0.94

Table 2 Permeabilities of materials

Material	Water permeability (m/s)	Air permeability (m/s)
B	2.60×10^{-3}	4.77×10^{-3}
C	1.58×10^{-3}	4.72×10^{-3}
D	4.85×10^{-4}	4.42×10^{-3}

Table 3 Test conditions

Case No.		Total head (mm) *	
		Left reservoir	Right reservoir
	0	760	760
	1	700	700
I	2	700	500
	3	700	300
	4	700	200
	1	500	500
II	2	500	400
	3	500	300
	4	500	200
	1	400	400
III	2	400	300
	3	400	200
IV	1	300	300
	2	300	200
V	1	200	200

* Total head is 0 at the height of the crown of the cavern

The test conditions are listed in Table 3. In each case of the tests the waterlevels of the right and left reservoirs were kept at the appointed height respectively. The pressure of the air in the cavern was gradually increased by the air injection until the air leakage began. And then the pressure was gradually decreased until the leakage ceased. The air leakage could not be observed through the front acryl plate but be certified by the measurement of water flow rate of the air injection and by the observation of the bubbles yielding at the top or the side of the

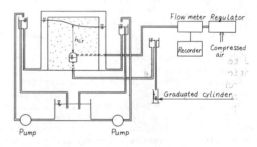

Fig. 3. Layout of apparatus for model experiment of air storage

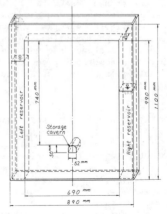

Fig. 4 Dimensions of model for compressed air storage

model.

In every case the free surface of water gradually moved upwards according to therise of the pressure of the air in the reservoir. Fig. 5 shows examples of the free surfaces of water which were observed when the air pressure was atmospheric pressure and when the air began to leak.

Let the vertical hydraulic gradient in the critical state defined as $I_{cr}=(h_{cr}-(P_{cr}/\rho_w \cdot g))/h_{cr}$, where P_{cr} is the critical air pressure when the air begins to leak or ceases to leak the stored air, and h_{cr} is the depth of the cavern from the free surface of water at that time. From the result of this experiment, I_{cr} varied from -0.4 to 0.3. Especially at the time when the air began to leak while the air pressure was increased, the values of I_{cr} gathered around 0. This means that the storage pressure of the air in the cavern is possible to be raised without leakage up to the hydrostatic pressure corresponding to the depth of water at the crown of the cavern.

TWO-PHASE FLOW ANALYSIS

In order to estimate the amount of the leakage of the air from the unlined storage caverns in the field, a numerical procedure was studied. The procedure was derived from the concept of the gas-liquid two phase flow in the ground around the caverns.
The fundamental equations of the gas-liquid two phase flow in the ground were derived from Darcy's law and the continuity condition as follows (Meiri 1982) (the equations are described with the conventional tensor notaion, that is to say, the repetition of the indices implying summation):

$$\frac{\partial}{\partial X_i}\left[\frac{K_{ra}}{\mu_a \beta_a}K_{ij}\left(\frac{\partial P_a}{\partial X_i}+\rho_a g\frac{\partial X_3}{\partial X_j}\right)+R_s\frac{K_{rw}}{\mu_w \beta_w}K_{ij}\left(\frac{\partial P_w}{\partial X_j}+\rho_w g\frac{X_3}{X_j}\right)\right]$$
$$=\frac{\partial}{\partial t}\left[f\left(\frac{S_a}{\beta_a}+\frac{R_s S_w}{\beta_w}\right)\right]-q_a \quad (i,j=1,2,3) \quad \cdots\cdots\cdots(1)$$

$$\frac{\partial}{\partial X_i}\left[\frac{K_{rw}}{\mu_w \beta_w}K_{ij}\left(\frac{\partial P_w}{\partial X_i}+\rho_w g\frac{\partial X_3}{\partial X_j}\right)+R_v\frac{K_{ra}}{\mu_a \beta_a}K_{ij}\left(\frac{\partial P_a}{\partial X_j}+\rho_a g\frac{\partial X_3}{\partial X_j}\right)\right]$$
$$=\frac{\partial}{\partial t}\left[f\left(\frac{S_w}{\beta_w}+\frac{R_v S_w}{\beta_w}\right)\right]-q_w \quad (i,j=1,2,3) \quad \cdots\cdots\cdots(2)$$

where P_a and P_w are the pressures of air and water, ρ_a and ρ_w are the densities of air and water, μ_a and μ_w are the viscosities of air and water, β_a and β_w are the ratios of the volume of air and water to each volume in the standard condition, K_{ra} and K_{rw} are the relative permeabilities of air and water, K_{ij} is the

absolute permeability tensor, S_a and S_w are the ratios of saturation of air and water, R_s is the solubility of air into water, R_v is the volatility of water, q_a and q_w are the rates of inflow of air and water, f is the porosity, X_i is the coordinate (X_3 is the vertical coordinate) and t is time, respectively.

These governing equations were formulated into numerical manner by the finite element method with Galerkin's process. This finite element procedure was applied to the model experiment of the compressed air storage presented in the above. The relative permeability and the capillaru properties were given by the test as shown if Fig. 6 and Fig. 7, respectively. Other parameters were assumed.

In Fig.5 the free surfaces of calculated result were compared with those of Experimental result. Fig. 8 and Fig. 9 show the calculated relations of the water vs. the storage air presuure and theair leakage vs. the air pressure, respectively. Those figures show that the numerical solutions agree well with the measured values. As a result it is indicated that the numerical procedure proposed here is effective to simulate the behavior of the ground water and the air around the reservoir.

The calculation was carried out to study the compressed air storage system. The model shown in Fig. 10 represents half the region around the storage system which consists of four unlined rock caverns in 334m depth of water. The ground was regarded as homogeneous. And the material properties were assumed as follows: $K_w=1\times10^{-7}$ m/s, $K_a=7.4\times10^{-6}$m/s, $\rho_w=1\times10^3$ kg/m^3, $\rho_a=1.247$ kg/m^3, $\beta_w=1$, $\beta_a=0.105$/Pa (Pa is in MPa), $R_v=9.28\times10^{-5}$/(Pa-1.23×10^{-3}) and $R_s=0.02$.

The free surfaces of underfround water were obtained as Fig. 10. This figure shows that the free surface moves upwards as the air pressure increases in the reservoirs. Fig. 11 shows that the calculated results of the water inflow and the air leakage. This figure shows that the water inflow into the caverns linearly decreases according to the rise of the air pressure, and becomes zero when theair pressure is 35 atm (3.55 Mpa). The figure also indecates that the air leakage is almost zero when the storage pressure is lower than 35 atm and that the air leaks when the air pressure becomes larger than this value. This analysis leads tha conclusion that the most rational pressure of the air is 35 atm or 3.55 MPa in the reservoirs in 334m depth of water.

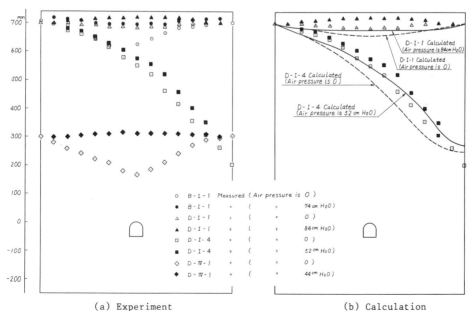

(a) Experiment   (b) Calculation
Fig. 5. Free surfaces of water

CONCLUSION

This paper has described the experimental and numerical works to investigate the ground water conditions concerning to the storage of the compressed air in an unlined rock cavern. The conclusions of this study are summarized as follows:

1. The compressed sir can be stored without leakage on condition that the air pressure is lower than the hydrostatic pressure equivalent to the depth of water at the ceiling of the storage caverns.

2. The finite element procedure, which is derived from the concept of gas-liquid two phase flow in the ground, is an effective means to estimate the amount of leakage of the air or gases from the stroge in the field.

REFERENCES

Aberg,B 1980. Prevention of gas leakage from unlined reservoirs in rocks. Symp. on storage in excavated rock caverns, Rock store 77,pp.399-414. Stockholm,Sweden.

Komada,H. and K. Nakagawa 1985. Study on compressed air storage in rock caverns by model experiment. CRIEPI Report No. 384044. Japan.

Meiri,D. and G.M. Karadi 1982. Simulation of air storage aquifer by finite element model. Int.J.Numer. Anal.Methods Geomech.,Vol.6,pp.339-351

Nakagawa,K. and H. Komada 1981. Fundamental study on prevention of gas leakage from unlined storage cavern under water table. CRIEPI Report No.380056

Nakagawa,K. and H. Komada 1985. Study on compressed air storage in rock caverns by Two-phase flow analysis. CRIEPI Report No. 384045

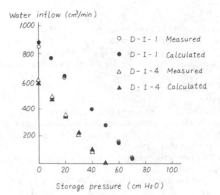

Fig. 8. Relations between water inflow and storage pressure

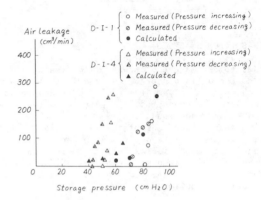

Fig. 9. Relations between air leakage and storage pressure

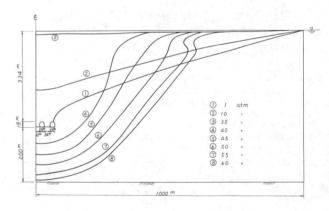

Fig. 10. Free surfaces of water

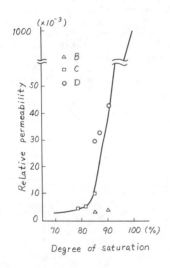

Fig. 6. Rerative permeability

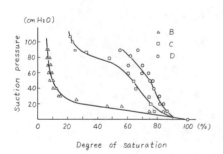

Fig. 7. Capillary property

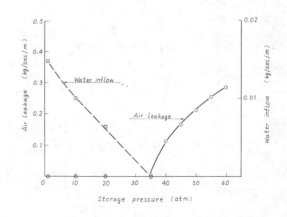

Fig. 11. Calculated air leakage and water inflow

Permeability tensor of jointed granite at Stripa Mine, Sweden
Tenseur de perméabilité du granit jointif dans la mine de Stripa en Suède
Durchlässigkeitstensor von klüftigem Granit im Bergwerk von Stripa in Schweden

M.ODA, Saitama University, Japan
Y.HATSUYAMA, Saitama University, Japan
K.KAMEMURA, Taisei Corporation, Kanagawa, Japan

ABSTRACT: Rock masses containing a large number of geological discontinuities (called cracks) are treated as homogeneous, anisotropic porous media. A permeability tensor, hydraulically equivalent to a flow network formed by cracks, is formulated in terms of two crack tensors and a non-dimensional scalar, both depending only on the crack geometry. The permeability tensor is calculated by using the information about the crack geometry which is obtained from the ventilation drift at Stripa mine, Sweden, and is compared with the result of the large scale hydraulic conductivity test.

RÉSUMÉ: Les masses rocheuses, qui renferment une grande quantité de discontinuités géologiques (dites fissures), sont considérées comme des milieux homogènes, anisotropes et poreux. Un tenseur de perméabilité, équivalent hydrauliquement à un réseau d'affluence formé par des fissures, est formulé sur la base de deux tenseurs de fissures et d'un scalaire sans dimensions dont tous les deux ne dépendent que de la géométrie de fissures. On calcule le tenseur de perméabilité en utilisant l'information sur la géométrie de fissures obtenue à partir de la dérive de ventilation dans la mine de Stripa en Suède et il est comparé avec le résultat de l'essai de conductivité hydraulique à grande échelle.

ZUSAMMENFASSUNG: Gesteinsmassen, welche eine grosse Menge von geologischen Ungleichmässigkeiten (die sogenannten Bruchrisse) enthalten, werden als gleichmässige anisotrope luckige Mittel betrachtet. Ein Durchlässigkeitstensor, welcher einem aus Bruchrissen gestalteten Strömungsnetz hydraulisch gleichwertig ist, wird auf der Grundlage von zwei Risstensoren und von einem dimensionslosen Skalar festgelegt und beiden hängen nur von der Rissgeometrie ab. Man berechnet den Durchlässigkeitstensor durch die Verwendung der Angaben über die Rissgeometrie, welche aus der Lüftungsversetzung im Bergwerk von Stripa in Schweden bekommen wurden, und er wird mit dem Ergebnis einer umfangreichen hydraulischen Leitfähigkeitsprüfung verglichen.

1 INTRODUCTION

In relation to the deep underground disposal of high level nuclear waste, extensive research works are going on to make clear the hydraulic properties of crystalline rock masses. It is no wonder that much attention is focused on the role of geological discontinuities (called cracks) in the migration mechanism of radionuclides. In fact, it can be a crucial problem for designing the nuclear waste repositories with high safety. Oda (1985 & 1986) has proposed a theory in which discontinuous rock masses are treated as homogeneous, anisotropic porous media. The corresponding permeability tensor was formulated in terms of in situ measureable quantities such as orientation data of cracks projected on a Schmidt's equal area net and maps of crack traces visible on excavated walls. The purpose of the present paper is to check the validity of the theory using the field measurements at Stripa mine.

2 PERMEABILITY TENSOR OF DISCONTINUOUS ROCK MASSES

If a rock mass is assumed to be an anisotropic porous medium, it obeys Darcy's law in which the apparent velocity is related to the gradient of total head through a linking coefficient k_{ij} called the permeability tensor. Oda (1985) has obtained a permeability tensor of rock mass on the followings assumptions: 1) Any crack has a shape similar to a penny, with a diameter r and a uniform hydraulic aperture t, whose orientation is identified by a unit vector $\underset{\sim}{n}$ normal to the principal plane. (Here, r, t and $\underset{\sim}{n}$ are are random variables. Any crack is characterized by a set of these variables. So, $(\underset{\sim}{n},r,t)$-crack is used if the unit normal vector $\underset{\sim}{n}$ is oriented inside a small solid angle $d\Omega$ around $\underset{\sim}{n}$ and if the diameter and aperture are within r to

r+dr and t to t+dt, respectively. A density function $E(\underset{\sim}{n},r,t)$ is introduced in such that $2E(\underset{\sim}{n},r,t)d\Omega$ drdt gives the probability of $(\underset{\sim}{n},r,t)$-cracks among others (see Oda, 1985)). 2) The hard matrix is impermeable. 3) Hydraulic gradient J is uniformly distributed over the elementary volume. 4) Seepage flow through a crack can be treated as a laminar flow between parallel planar plates with a uniform aperture t. 5) There is no head loss at intersections between cracks.

$$k_{ij} = \lambda (P_{kk} \delta_{ij} - P_{ij}) \qquad (1)$$

Here, P_{ij} is a symmetric, second-rank tensor defined by

$$P_{ij} = (\pi\rho/4)\int_0^{t_m}\int_0^{r_m}\int_\Omega r^2 t^3 n_i n_j E(\underset{\sim}{n},r,t)\,d\Omega\,drdt \qquad (2)$$

where ρ is the number of cracks per unit volume, t_m and r_m are the maximum aperture and the maximum crack size respectively, δ_{ij} is Kronecker's delta and Ω is the entire solid angle equal to 4π. (The summation convention is adopted if any subscript appears twice. This is not the case, however, when superscripts are in parentheses. Note also that $\pi/4$ in Eq.(2) comes from the shape of cracks. If cracks are squares in shape, then it is omitted.) In Eq.(1), λ is a non-dimensional coefficient satisfying an inequality of $0 \leq \lambda \leq 1/12$. If cracks are infinite in size, λ can be set to 1/12. On the basis of numerical experiments on fluid flow through crack systems, Oda (1986) proposed the following relation

$$\lambda = \lambda(F_{ij}) \qquad (3)$$

where $F_{ij}=(\pi\rho/4)\int_0^{t_m}\int_0^{r_m}\int_\Omega r^3 n_i n_j E(\underset{\sim}{n},r,t)d\Omega drdt \qquad (4)$

is a dimensionless, second-rank tensor.
As shown in Fig.1, the dimensionless scalar λ is

given in terms of F_o and $A^{(F)}$ for two-dimensional cases.

$$\lambda = \lambda (F_o, A^{(F)}) \qquad (5)$$

where $\quad A^{(F)} = (1/F_o)(3F'_{ij}F'_{ij})^{1/2}$

is an index measure to show the anisotropy due to the preferred alignment of cracks (Oda et al., 1986). F_o is the trace of F_{ij} and F'_{ij} is the deviatoric tensor of F_{ij}.

3 CRACK GEOMETRY OF THE VENTILATION DRIFT AT STRIPA MINE.

Stripa mine is located at Central Sweden, where extensive research works dealing with the thermo-hydro-mechanical behaviour of a jointed granite are going on in relation to the high level nuclear waste disposal. Recently, Rouleau and Gale (1985) reported complete data on rock joints collected from the ventilation drift at the 338 m level of the mine. Using the joint data in Eq.(1), a permeability tensor for the mine is estimated, and it is compared with the hydraulic conductivity reported by Wilson et al.(1983).

All data used in this report were obtained either from Rouleau and Gale (1985) or from Gale and Rouleau (1985). They collected the joint data from three inclined surface boreholes, thirteen sub-surface boreholes drilled from the ventilation drift and the crack trace maps of the walls and the floor of the ventilation drift.

1) Orientaion : All orientaion data were compiled in the lower-hemisphere, equal area Schmidt's net of Fig.2 (Rouleau and Gale, 1985). Every cracks were classified into one of the four joint sets (1), (2), (3) and (4) according to the concentrations of poles to joints. Let us assume that the statistical variables n, r and t are mutually independent if each joint set is separetely considered. Then, the density function $E(n,r,t)$ is given by

$$E(n,r,t) = E(n)f(r)g(t) \qquad (6)$$

where $E(n)$, $f(r)$ and $g(t)$ are the density functions of n, r and t, respectively.

2) Trace length : The maps of crack traces visible on the walls and the floor of the ventilation drift were used to give histograms of the trace lengths for each joint set. Correction for sampling errors (censoring and truncation) was made to provide unbiased trace length data by Rouleau and Gale. Here, a negative exponential is accepted as a possible density function $H(\ell)$ of the trace lengths.

$$H(\ell) = ae^{-a\ell} \qquad (7)$$

The means $1/a$ of the distributions were reported as follows; for set(1), $1/a$ =2.16 m; for set(2), $1/a$ =0.86 m; for set(3), $1/a$ =1.51 m; for set(4), $1/a$ =1.03 m.

3) Spacing : Gale and Rouleau (1985) reported the distribution of spacings between every pair of consecutive cracks belonging to the same joint set. Let $N^{(q)}/h$ be the mean number of the cracks intersecting the unit length of drill cores. Their results were summarized as follows; for set(1), $N^{(q)}/h$ =1.03 m; for set(2), $N^{(q)}/h$ =2.64 m; for set(3), $N^{(q)}/h$ =0.91 m; for set(4), $N^{(q)}/h$ =1.83 m. The superscript q, which denotes a direction q (unit vector) of a scanline, was selected parallel to the maximum concentration of poles for each joint set.

4) Aperture : Since the determination of apertures of natural cracks requires tedious work, the complete knowledge has not been established yet in spite of the important role in rock hydraulics. Hydraulic aperture, instead, can be determined from packer injection tests. From the injection tests installed at a very short packer spacing, in which one crack is expected to be isolated from others, Gale and Rouleau (1985) reported the mean hydraulic aperture to be 5.43 microns.

Bianchi and Snow (1968) determined the crack aperture by means of the macrophotography and fluoresent liquid penetrants. They sampled cracks from various levels of depth at ten different dam-sites and tunnels composed of granite and gneiss, and reported their result as the change of the aperture with depth.

A theoretical study, if supplemented by experiments in a laboratory, can provide one of the ways to determine a crack aperture at a depth. For

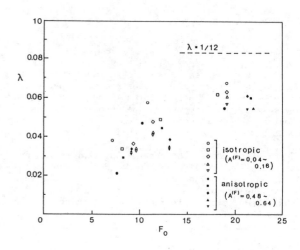

Fig.1 Dimensionless scalar λ in terms of F_o and $A^{(F)}$ for two-dimensional crack systems. (According to the numerical experiments by Oda (1986), water flow through two-dimensional crack systems is well simulated by using the dimensionless scalar of Fig.1 in Eq.(1). If F_o is less than about 6, however, the crack system is not percolative.)

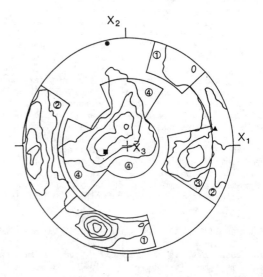

Fig.2 Schmidt's equal area projection (lower hemisphere) of 2648 poles to joints (Rouleau and Gale, 1985). (Density of the poles is given by contour lines of 1 , 2 , 3 , 4 and 5 per 1 percent area. If these percents are divided by 4π, the contour lines becomes equivalent to $E(n)$. The encircled numbers indicate the corresponding sets of joints. Symboles ● , ■ and ▲ show the directions of the major, intermediate and minor principal axes of the tensor P_{ij} respectively.)

example, Oda (1986) has proposed an equation, which basically relies on the experimental study on stiffnesses of joints reported by Bandis, et al.(1983).

$$t = t_o\{1 - z/(H + z)\} \qquad (8)$$

where z is the depth, t_o is the aperture at ground surface ($z=0$), and H is a constant depending on both of the stiffness and overall shape of cracks. The predicted value was compared with the measured one, with the conclusion that there is good agreement between them if $t = 200$ microns and $H = 20$ m are used in Eq.(8) (Fig.3).

4 CALCULATION OF PERMEABILITY TENSOR

Consider that the crack aperture is constant ($t=t_o$ =5.43 microns). The variables n and r are statistically independent when each joint set is individually treated. Accordingly, the crack tensor P_{ij} is given by

$$P_{ij} = t_o^3(\pi\rho/4)\int_0^{r_m} r^2 f(r)dr\int_\Omega n_i n_j E(\underset{\sim}{n})d\Omega$$

$$= (\pi\rho/4)t_o^3 <r^2> N_{ij} \qquad (9)$$

where $N_{ij} = \int_\Omega n_i n_j E(\underset{\sim}{n})d\Omega$

is a symmetric, second rank tensor. Note that N_{ij} can be calculated only if the density of poles normal to cracks is available (Fig.2).

Stereology, based on geometrical probability, provides a powerful tool to investigate the crack geometry. Oda (1985) has shown that $\rho<r^2>$ is related to $N^{(q)}/h$, the number of cracks crossed by a unit length of a scanline with a direction parallel to $\underset{\sim}{q}$.

$$\rho<r^2> = \frac{4N^{(q)}/h}{<|\underset{\sim}{n}\cdot\underset{\sim}{q}|>} \qquad (10)$$

where $<|\underset{\sim}{n}\cdot\underset{\sim}{q}|> = \int_\Omega |\underset{\sim}{n}\cdot\underset{\sim}{q}|E(\underset{\sim}{n})d\Omega$

is a correction term with respect to the choice of the direction $\underset{\sim}{q}$. Substituting Eq.(10) in Eq.(9), we finally have

$$P_{ij} = \frac{t_o^3 N^{(q)}/h}{<|\underset{\sim}{n}\cdot\underset{\sim}{q}|>} N_{ij} \qquad (11)$$

To calculate the tensor actually, reference axes x_1, x_2 and x_3 are set parallel to the eastward, northward and upward respectively (Fig.2). All terms in Eq.(11) are self-explaining. However, a few comments will still be helpful. 1) Orientation of cracks is distributed even if they belong to a joint set. Here, the orientation diagram (Fig.2) is simplified by replacing it by four poles corresponding to the maximum concentrations of the four joint sets. This simplification may cause some error especially when the distribution of $\underset{\sim}{n}$ is extremely biased around the maximum concentrations. Otherwise, as in the case of Stripa mine, the error is negligibly small. (If strikes and dips of all joints are known, no difficulty arises to take into account the distribution.) 2) In the report by Gale and Rouleau (1985), scanlines were chosen parallel to the maximum concentrations of joint sets. Then, the correction term $<|\underset{\sim}{n}\cdot\underset{\sim}{q}|>$ can be set to unity. 3)

Crack tensors $P_{ij}^{(1)}$, $P_{ij}^{(2)}$, $P_{ij}^{(3)}$ and $P_{ij}^{(4)}$ corresponding to joint sets (1),(2),(3) and (4) respectively, are individually calculated and they are summed up to give the final one P_{ij}. The components of P_{ij} with respect to the principal axes are given in matrix form as follows;

$$P_{ij} = P_{ij}^{(1)} + P_{ij}^{(2)} + P_{ij}^{(3)} + P_{ij}^{(4)}$$

$$= \begin{pmatrix} 5.553 & 0 & 0 \\ 0 & 1.374 & 0 \\ 0 & 0 & 3.332 \end{pmatrix} \times 10^{-17} m^2 \qquad (12)$$

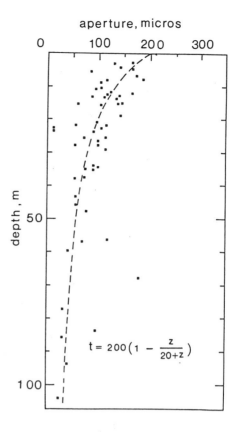

Fig.3 Change of crack aperture with depth. (Data in this figure were taken from Bianchi and Snow (1968). Since the ventilation drift at Stripa mine is located 338 m below the surface, using Eq.(8), the aperture is estimated as 11.2 microns, about two times larger than the hydraulic aperture.)

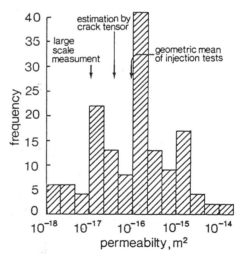

Fig.4 Comparison between the permeabilities of the ventilation drift at Stripa mine, Sweden. (The histogram with a geometric mean of 8.9×10^{-17} m^2 shows the variation of permeabilities obtained from 148 packer injection tests by Gale and Rouleau (1985). The permeability of 1.0×10^{-17}m^2 was from Wilson et al. (1983).)

To calculate the tensor F_{ij}(Eq.(4)), the term $<r^3>$ must be rewritten by some $N^{(q)}$ in situ measureable quantity. Unfortunately, $N^{(q)}/h$ cannot be used, because it is only related to $\rho<r^2>$. Again, stereological study is useful for the present purpose. Oda (1985) related the nth moment $<\ell^n>$ to the moments of r as follows:

$$< \ell^n > = \int_0^\infty \ell^n H(\ell)d\ell = \frac{<r^{n+1}>}{<r>} \int_0^{\pi/2} \sin^{n+1}\theta d\theta \qquad (13)$$

Eq.(10) and (13), together with Eq.(6), leads to the following equation.

$$F_{ij} = \frac{3\pi}{8}\frac{<\ell^2>}{<\ell>}\frac{N^{(q)}/h}{|\underset{\sim}{n}\cdot\underset{\sim}{q}|} N_{ij} \qquad (14)$$

If $H(\ell)$ is given by a negative exponential, then the nth moment of becomes

$$< \ell^n > = \int_0^\infty ae^{-a\ell}d\ell = (1/a^n)\Gamma(n+1) \qquad (15)$$

where $\Gamma(n+1)$ is the gamma function and $1/a$ has been already given for each joint set. Finally, F_{ij} is calculated as

$$F_{ij} = \begin{pmatrix} 7.762 & 1.589 & 0.150 \\ 1.589 & 4.348 & 0.816 \\ 0.150 & 0.816 & 5.940 \end{pmatrix} \qquad (16)$$

If these components are rewritten with respect to the principal axes, the corresponding principal values are 8.491, 6.071 and 3.518 which give $F_0 = 18.1$ and $A^{(F)}=0.337$.

To calculate the permeability tensor using Eq.(1), the non-dimensional coefficient λ must be determined. Fig.1 is not adequate because they were prepared for two-dimensional crack systems. Oda (1986) has shown, for two-dimensional cases, that $F_0=6$ gives the threshold value under which the crack system is not percolative. According to Robinson (1984), the threshold values for three-dimensional crack systems are: $F_0=1.56$ for the isotropic system composed of three orthogonal sets of square cracks and $F_0=1.05$ for the isotropic system composed of square cracks with uniform orientation. Note that these values are quite small as compared with those for the two-dimensional cases. In other words, the three-dimensional crack systems are more percolative than the two-dimensional ones for a given F_0. Since F_0 is 18.1 in the ventilation drift, the related joints are considered to be highly connected to make many flow paths. Accordingly, it is reasonable to accept tentatively $\lambda=1/12$, keeping in mind that the choice tends to give an overestimation of the permeability tensor.

All necessary information is now ready. Substituting Eq.(12), together with $\lambda=1/12$ and $t_0=5.43$ microns, in Eq.(1), we finally have the following permeability tensor for the ventilation drift at Stripa mine:

$$k_{ij} = \begin{pmatrix} 3.92 & 0 & 0 \\ 0 & 7.40 & 0 \\ 0 & 0 & 5.77 \end{pmatrix} \times 10^{-17} m^2 \qquad (17)$$

where the components are given with respect to the principal axes of P_{ij}(Fig.2). The permeability in the major direction is about 2 times larger than that in the minor one. Anisotropy is obvious, but is not so serious. Then, the mean permeability k_0 ($=k_{ii}/3$) is considered here. From Eq.(17), k_0 equals $5.7\times10^{-17}m^2$.

The hydraulic conductivity for the ventilation drift was determined by measuring the ground water flow into the 33m long, 5 m diameter drift by Wilson et al. (1983). The volume seems to be large enough to get information about the hydraulic properties of the rock mass. They have concluded that "the average hydraulic conductivity of the monitored rock mass, exclusive of a zone of lower conductivity immediately surrounding the drift, is approximately 9.8×10^{-11} m/s". Since the temperature is about $20^\circ C$, the conductivity is converted to the permeability of 1.0 $\times 10^{-17}$ m^2.

In addition to the large scale hydraulic conductivity test, many packer injection (and also withdraw) tests were performed with packer spaces ranging from 1 m to 3 m (Gale and Rouleau, 1985). In Fig.4, 148 permeabilities, which were measured using the boreholes drilled from the ventilation drift, are compiled. The individual values show wide variation from place to place, and they therefore cannot be representative. This is because they are seriously affected by local inhomogeneities. Gale and Rouleau used the geometric mean as a representative value. The three values of permeabilities are now available. Note that the permeability k_0, which was predicted through the stereological study on the crack geometry, is just intermediate between the permeabilities measured in site.

REFERENCES

Bandis, S.C., A.C. Lumsden and N.R. Barton 1983. Fundamentals of rock joint deformation. Int. J. Rock Mech. Min. Sci. & Geomech. Abstr. 20 (6): 249-268.

Bianchi, L. and D.T. Snow 1968. Permeability of crystalline rock interpreted from measured orientations and apertures of fractures. Annals of Arid Zone, 8 (2): 231-245.

Gale, J.E. and A.rouleau 1985. Hydrogeological characterization of the ventilation drift area. NEA Information Symposium on In Situ Experiments in Granite Associated with the Disposal of Radioactive Waste, Stockholm, Sweden.

Oda, M 1985. Permeability tensor for discontinuous rock masses. Geotechnique. 35 (4): 483-495.

Oda, M 1986. An equivalent continuum model for coupled stress and fluid flow analysis in jointed rock masses, submitted to Water Resour. Res.

Oda, M., T.Yamabe and K. Kamemura 1986. Crack tensor and its relation to anisotropy of longitudinal wave velocity in jointed rock masses. To appear in Int. J. Rock Mech. Min. Sci. & Geomech. Abstr.

Robinson, P.C. 1984. Connectivity, flow and transport in network models of fractured media. Ph. D. disertation, Oxford University.

Rouleau, A. and J.E. Gale 1985. Statistical characterization of the fracture system in the Stripa granite, Sweden. Int. J. Rock Mech. Min. Sci. & Geomech. Abstr., 22 (6): 353-367.

Snow, D.T. 1968. Rock fracture spacings, opening, and porosities. J. Soil Mech. Found. Divi., Proc. ASCE. 94 (SM1): 73-79.

Wilson, C.R., P.A. Witherspoon, J.C.S. Long, R.M. Galbraith, A.O. Dubois and M.J. Mcpherson 1983. Largescale hydraulic conductivity measurements in fractured granite. Int. J. Rock Mech. Min. Sci. & Geomech. Abstr., 20 (6): 269-276.

Thermo-hydro-mechanical behavior of rocks around an underground opening

La conduction thermo-hydro-mécanique des roches autour d'une ouverture souterraine

Thermo-hydraulisch-mechanisches Verhalten von Fels in einem unterirdischen Hohlraum

YUZO OHNISHI, Kyoto University, Japan
AKIRA KOBAYASHI, Hazama-gumi, Tokyo, Japan
MAKOTO NISHIGAKI, Okayama University, Japan

ABSTRACT: The assessment of the performance of a nuclear waste repository involves the evaluation of the combined effects of many different processes in geological systems. The coupled interaction between heat transfer and natural water in rock systems is one of the most important subjects to be studied not only with experiments and measurements, but also with powerful numerical analyses. This paper describes the coupled thermo-hydro-mechanical analytical method with application to the Stripa project. The buffer mass test and the macropermeability experiment are chosen to evaluate the capability of the method. The numerical results indicate that the complex nonlinear behavior of the rock mass with water and heat must be considered to interpret the field data at the site.

RESUME: L'éstimation d'un dépositoir de déchets nucléairs inclut l'évaluation des effets combinés de nombreux procéssus en systèmes géologiques. L'interaction accouplée entre le transfert de chaleur et de l'eau naturelle en système roc est l'un des plus importants sujets devant être étudiés non seulement avec des expériences, des mesures mais aussi avec de puissantes analyses numériques. Ce papier descrit la méthode accouplée thermo-hydro-mécanique et analytique avec une application dans le projet Stripa. Un examen en tampon de masse et une macroperméabilité en expérience sont choisis pour évaluer la capabilité de la méthode. Les résultats numériques indiquent que la complexe conduite nonlinéaire de la masse roc avec l'eau et la chaleur doit être considérée pour interpréter le champ des données dans le site.

ZUSAMMENFASSUNG: Die Einschaetzung der Ausfuehrung eines Aufbewahrungsortes fuer Atommuell macht eine Abschaetzung der kombinierten Wirkungen verschiedenster Prozesse geologischer Systeme erforderlich. Eines der wichtigsten zu studierenden Themen ist die gekoppelte Interaktion zwischen Hitzeuebertragung und natuerlichem Wasser in Felssystemen, und dies nicht nur durch Experimente und Messungen, sondern auch mit leistungsfaehigen numerischen Analysen. Dieser Artikel beschreibt die kombinierte thermo-hydro-mechanisch analytische Methode mit Anwendung auf das Stripa-Projekt Der Puffermasse-Test und das Makrodurchlaessigkeitsexperiment werden gewaehlt, um die Tauglichkeit der Methode zu evaluieren. Die zahlenmaessigen Resultate weisen darauf hin, dass das komplexe nichtlineare Verhalten der Felsmasse mit Wasser und Hitze zur Interpretation der Felddata an Ort in Betracht gezogen werden muss.

1. INTRODUCTION

Nuclear power production creates wastes that have to be managed and ultimately disposed of in a safe manner. Some of these wastes are highly radioactive and must be isolated. Most possible way to isolate nuclear waste seems to deposit in underground caverns built in hard, competent rock. Once emplaced in these caverns, the radioactive decay of high-level radioactive waste will affect the host rock mechanically, hydraulically and thermally in these deep caverns.

The assessment of the performance of a nuclear waste repository involves the evaluation of the combined effects of many different processes in complex geological systems. The coupled interaction between heat transfer and natural water in rock systems is one of the most important subject to be studied not only with experiments, measurements but also with powerful numerical analyses.

To shed light on the response of rock to such coupled interactive loading, a series of experiments has been conducted in the Stripa Mine, Sweden, as the international cooperative program. Some of the typical experiments are the macro-permeability experiment (MPE) and the buffer mass test (BMT). MPE is attempted to determine flow parameters for large-scale low permeability fractured rock mass. BMT is to check the suitability and predicted functions of certain bentonite-based buffer materials under real conditions on site.

In parallel with experimental developments, the analyses have been performed with the aid of advanced computers. Recent studies are involved in the problems of coupled hydro-thermo-mechanical phenome-na. Bear and Carapcioglu (1981) derived the basic equations which describes thermo-elastic behavior of a ground. Hart (1981) presented a model which described fully coupled thermal-mechanical-fluid flow behavior of highly nonlinear porous geologic systems and analysed the model by an explicit finite difference method. Noorishad, et al. (1984) applied a finite element method with joint elements to fully coupled phenomena for a saturated fractured porous rock mass. Ohnishi et al.(1985) developed a finite element code to handle the problem of coupled hydro-thermo-mechanical behavior of saturated-unsaturated geologic medium.

This paper describes the numerical model which is applied to interpret the results obtained at the MPE and BMT of the Stripa project. The model is based upon the code by Ohnishi et al.(1985) and the computed results are used to investigate the possible behaviors of the buffer materials and the host rock mass under complex coupled loading conditions.

2. NUMERICAL PROCEDURE

2.1 Governing equation

The governing equations are derived under the following assumptions: 1)The medium is isotropic and poro-elastic in the plane strain and axisymmetric conditions. 2)Darcy's law is valid in the saturated-unsaturated medium. 3)Energy transfer by gas, phase change between liquid and gas phase are not considered. 4)Heat transfers among three phases(solid, liquid and gas) are neglected. 5)Fourie's law holds for heat flux. 6)Water density varies depending upon the temperature and pressure of water.

Finite element formulation is performed by using

the following governing equations. Detailed derivation of the equations are given in Ohnishi et al.(1985).

$$[\frac{1}{2}C_{ijkl}(u_{k,l}+u_{l,k})-\beta\delta_{ij}(T-T_0)+\chi\delta_{ij}\rho_f gh]_{,j}$$

$$+\bar{\rho}_s b_i=0 \tag{1}$$

$$[\rho_f k(\theta)h_{,i}]_{,i}-\rho_{f0}nS_r\rho_f g\beta\frac{dh}{pdt}-\rho_s C(\psi)\frac{dh}{dt}-\rho_f S_r\frac{du_{i,i}}{dt}$$

$$+\rho_{f0}nS_r\beta\frac{dT}{Tdt}=0 \tag{2}$$

$$(\rho C_v)\frac{dT}{mdt}+nS_r\rho_f C_{vf}v_{fi,i}T_{,i}-K_{Tm}T_{,ii}+nS_r T\frac{\beta_T}{\beta_p}k(\theta)h_{,ii}$$

$$+\frac{1}{2}(1-n)\beta T\frac{d}{dt}(u_{i,j}+u_{j,i})\delta_{ij}=0 \tag{3}$$

$$(i,j,k,l=1,2,3)$$

where Cijkl: elastic constants, u: displacement, T: temperature, χ: Bishop's parameter for unsaturated state, ρ: density, h: total head, b: body force, k: permeability, θ: volumetric water content, n: porosity, Sr: degree of saturation, g: gravity β_p, β_T: compressibility and expansivity for fluid, respectively, $C(\psi)$: specific water content, C_v: specific heat, K_T: coefficient of heat conduction.

The interaction parameter which defines the effect of thermal expansion on stress is expressed as follows:

$$\beta=(3\lambda+2\mu)\alpha_s$$

where λ and μ are the Lame's elastic constants, and α_s is a coefficient of thermal expansion of solid.

2.2 Permeability of Rock Mass

In order to interpret the complex rock mass behavior, it is necessary to develop an analytical model for fractured rock characteristics.

Kelsall et al.(1984) proposed a simple method to take into account the effects of stress on rock mass hydraulic conductivity. Modifying their method, the following stress dependent permeability function is implemented in the finite element code.

$$\frac{k_e}{k_{od}}=[1+A(\frac{\sigma_{eo}}{\zeta})t']^3/[1+A(\frac{\sigma_e}{\zeta})t']^3 \tag{4}$$

where k_e is the intrinsic permeability at the stress state σ_e and k_{od} corresponds to the reference stress state σ_{eo}. Since anisotropy of k_e is considered in this paper, σ_e is acting in the principal direction of k_e. Effective stresses should be used for the stress expression. A, t' and ζ are material constants for a rock fracture.

This empirical equation is derived from the results for a single fracture and neglects the effect of rock matrix blocks. It is applied only to an equivalent porous medium rock mass conductivity.

2.3 Permeability of Highly Compacted Clay

In order to evaluate the expected function of the highly compacted clay (HCC) as an artificial barrier, it is necessary to know the permeability change in wetting process. However, the unsaturated flow properties of HCC is not investigated intensively in the Stripa program. Reviewing the BMT reports by Push et al.(1985), we proposed the following water retention curve which was dependent on the change of void ratio e

$$k=10^{-13}\exp(\frac{e-0.26}{1.09/2.30})\times S_e^{14.27} \quad (cm/s) \tag{5}$$

$$\psi=\exp(-1.76e+7.0)\times S_e^{-2.7} \tag{6}$$

where Se is the degree of saturation and ψ denotes suction.

It should be noticed that these equations do not consider the evaporation effects even when the temperature of water goes up.

Swelling pressure Ps is determined as an external force for the finite element computation. While Sr is less than 100%, Ps = ψ and while Sr equals 100%,

Ps is obtained from Table 1 which is a function of the temperature and the density.

Kinematic viscosity is considered to be a function of the temperature. Table 2 shows the material constants for the rock mass at the initial state. Table 3 is the material properties of HCC.

3. MACRO-PERMEABILITY TEST (MPT)

At Stripa, macroscopic permeability test has been carried out. A 33 m length of the ventilation drift has been sealed off and equipped with a ventilation system whose temperature can be controlled to evaporate all water seeping into the room. The water seepage is determined from measurements of the mass flow rate and the difference in the humidities of entering and existing air streams. The pressure gradients in the rock walls are measured in holes that radiate out from the sealed room.

For a coupled finite element analysis, the rock cavern was simply modeled as shown in Fig.1. Excavation of the underground space was simulated with the nonlinear equation Eq.(4). Fig.1 also shows the comparison of the water table between field measurement and computed result.

Computed distribution of the horizontal permeability at the drift wall is shown in Fig.2 in comparison with in situ measurements. Reduction of permeability in the figure seems to indicate the closure of fracture due to circumferencial stress increase. When the room temperature goes up to 30, the rock matrix block expands and the fractures close. As a result, the permeability of the rock mass reduces as shown in Fig.2. It also causes the higher water level as shown in Fig.1.

The calculated amount of water seeping into the room due to the temperature change is shown in Table 4. Since the hydraulic boundary condition is not certain, the quantitative analysis is difficult. However, the numerical results indicate that the calculated water inflow with nonlinear properties of k and μ reduces 33% comparing to the linear analy-

Table 1. Swelling pressure P_s versus bulk density ρ of HCC at different temperatures.

$\rho(tf/m^3)$	P_s (tf/m²)		
	20°C	70°C	90°C
2.15	4592.	4082.	3571.
2.10	3061.	2041.	1735.
2.05	1531.	1020.	816.
2.00	714.	510.	408.
1.95	459.	306.	255.

Table 2. Data of rock used for analysis

Property	Value
mass density	2.65 tf/m³
porosity	0.25
Young's modulus	5.0×10^6 tf/m²
Poisson's ratio	0.33
thermal expansion coef.	8.0×10^{-6} °C^{-1}
specific heat	837. J/kg°C
thermal conductivity	3.0×10^{-3} kJ/ms°C
permeability	1.0×10^{-17} m²
experimental constant A	0.0276
experimental constant ζ	0.00217
experimental constant t'	0.728

Table 3. Data of HCC used for analysis.

Property	Value
mass density	2.15 tf/m²
porosity	0.42
Young's modulus	1.0×10^6 tf/m²
Poisson's ratio	0.3
thermal expansion coef.	6.0×10^{-6} °C^{-1}
specific heat	1220. J/kg°C
thermal conductivity	1.46×10^{-3} kJ/ms°C
permeability	1.0×10^{-19} m²

Table 4. Variation of fluid inflow to the drift by the increase in temperature.

Nonlinear properties	Fluid inflow(ml/min)	
	20°C	30°C
k_0 only	31.0	18.8
μ only	58.5	63.3
k_0 & μ	31.0	21.0
observed at MPE	50.0	42.0

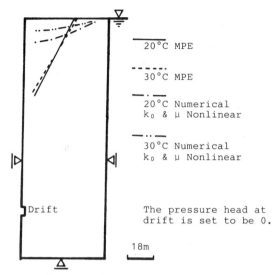

20°C MPE

30°C MPE

20°C Numerical
k_0 & μ Nonlinear

30°C Numerical
k_0 & μ Nonlinear

The pressure head at drift is set to be 0.

18m

Fig.1 Schematic model of MPE, and comparison of the water tables calculated by numerical analysis and inferred by MPE.

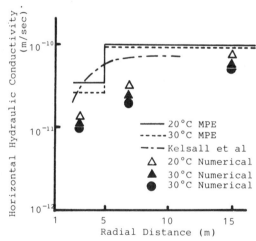

— 20°C MPE
---- 30°C MPE
—·— Kelsall et al
△ 20°C Numerical
▲ 30°C Numerical
● 30°C Numerical

Fig.2 Comparison of horizontal hydraulic conductivities calculated by numerical analysis and inferred from MPE. Triangles are calculated by the use of k_0 and μ nonlinearity, circle is calculated by the use of k_0 nonlinearity only.

sis. The rate of reduction is 39% when the non-linearity of k only is taken into account. The analyses concluded that the change of inflow rate was strongly influenced by the temperature dependency of the kinematic viscosity μ.

Fig. 3 shows the calculated vertical permeability distribution on the horizontal section. It means that the high permeable zone which corresponds to a loosened area is recognized up to the depth of 7m from the wall surface.

4. BUFFER MASS TEST

Canisters of high level waste will be stored in shallow vertical boreholes in the floor of hori-

zontal tunnel. The annulus between the canisters and the borehole will consist of a highly compacted clay (HCC). The buffer material has a very low permeability and high swelling capacity when absorbing water. The objective of the buffer mass test is to verify the barrier function of highly compacted bentonite and sand under real conditions.

Numerical analyses have been performed with the finite element mesh as shown in Fig.4. Material properties of HCC are obtained from the nonlinear equations Eq.(5)(6). The electric heater intensity is assumed to be 600W. A bentonite powder is placed between HCC and the host rock.

Figs.5a)b)c)d) show the distributions of n, Sr, K_T and permeability k in the borehole, respectively. The swelling HCC causes the increase of porosity. Sr goes up to 100% once after reduction of its value, depending on the combination of the conditions of porosity and swelling. The heat conductivity

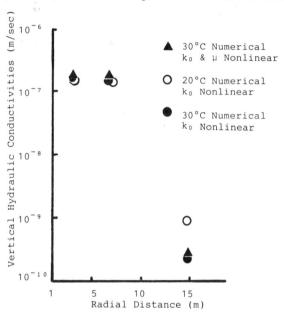

▲ 30°C Numerical
k_0 & μ Nonlinear
○ 20°C Numerical
k_0 Nonlinear
● 30°C Numerical
k_0 Nonlinear

Fig.3 Vertical hydraulic coductivities calculated by numerical analysis.

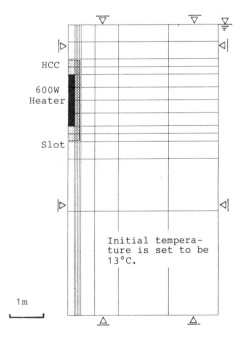

HCC

600W Heater

Slot

Initial temperature is set to be 13°C.

1m

Fig.4 Finite element model of BMT. Right, top and bottom boundaries are set to be fixed temperature condition.

gradually decreases depending upon the porosity increase.

The comparison between the numerical results and field test results is shown in Fig. 6a)b)c) for temperature, water content and swelling pressure, respectively. The temperature distribution obtained at the field measurement is well predicted by the computation as shown in Fig.6a). The distribution of water content disagrees with the measurement. This is possibly because the evaporation of water is neglected. It is concluded that the water content may strongly be influenced by water movement due to evaporation. Fig.6c) indicates that the initial swelling pressure calculated is larger than the measured one. This may be caused by the overestimation of suction in Eq.(6) at the initial state.

5. CONCLUSIONS

The work described in this paper is the application to the Stripa experiments of the coupled thermo-hydro-mechanical analytical method. The macro-permeability experiment and buffer mass test are chosen to evaluate the capability of the method. Some of the nonlinear coupling effects have been handled by introducing the nonlinearity of material parameters. It is concluded that the qualitative interpretation of the behavior of the fractured rock mass subjected to complex loading can be done by the powerful numerical tool. For the better estimation, it is necessary to know more the complex nonlinear behavior of the rock mass and buffer materials under hydraulic and thermal conditions.

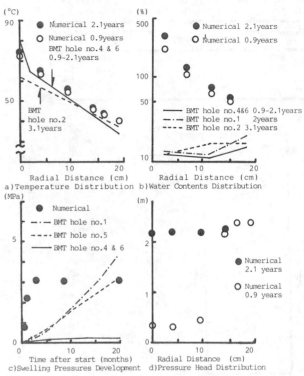

Fig.6. Comparison of recorded values at mid-height heater in holes of BMT and calculated ones by numerical analysis. a)Temperature distribution as a function of radial distance from heater. b)Water contents distribution as a function of radial distance from heater. c)Swelling pressures development at rock/HCC interface as a function of time after test start. d)Pressure head distribution as a function of radial distance from heater.(no recorded values)

REFERENCES

Bear,J. and Corapcioglu,M.Y.(1981): A mathmatical model for consolidation in a thermoelastic aquifer due to hot water injection or pumping, Water Resor. Res., Vol.17,No.3,pp.723-736.

Hart,R.D. (1981): Afully coupled thermal-mechanical-fluid flow model for nonlinear geologic system, PH.D.thesis, Univ. of Minesota.

Kelsall,P.C., Case,J.B. and Chabaaes,C.R.(1984): Evaluation of excavation - induced change in rock permeability, Int. J. Rock Mech. Min. Sci. & Geomech. Abst. Vol.21,No.3,pp.123-135.

Noorishad,J.,Tsang,C.F. and Witherspoon,P.A.(1984): Coupled thermal-hydraulic-mechanical phenomena in saturated fractured porous rocks, numerical approach, J.G.R. Vol.189,No.B12,pp.10365-10373.

Ohnishi,Y.,Shibata,H. and Kobayashi,A.(1985): Development of finite element code for the analysis of coupled thermo-hydro-mechanical behaviors of saturated-unsaturated medium, Int. Symp. on Coupled Process Affecting the Performance of a Nuclear Waste Repository.

Push,R. and Borgesson,L.(1985): Final report of the buffer mass test - Volume ll; test results, SKB Technical Report 85-12.

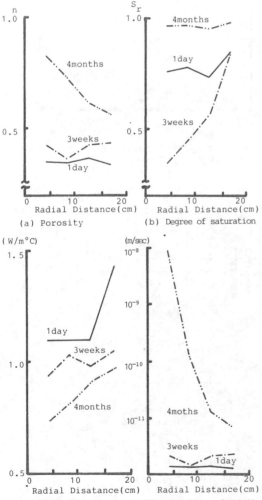

Fig.5. Calculated properties profiles as a function of radial distance from heater. (a)Prosity (b)Degree of saturation (c)Thermal conductivity (d)Hydraulic conductivity

Untersuchungen von Durchlässigkeit und Spannungszustand im kristallinen Gebirge zur Beurteilung der Barrierewirkung

Investigation of permeability and stress field in crystalline rock to assess the barrier efficiency
Recherches sur la perméabilité et sur l'état des contraintes dans la roche crystalline pour l'estimation de son rôle de barrière

A.PAHL, Bundesanstalt für Geowissenschaften und Rohstoffe, Hannover, Bundesrepublik Deutschland
L.LIEDTKE, Bundesanstalt für Geowissenschaften und Rohstoffe, Hannover, Bundesrepublik Deutschland
B.KILGER, Bundesanstalt für Geowissenschaften und Rohstoffe, Hannover, Bundesrepublik Deutschland
S.HEUSERMANN, Bundesanstalt für Geowissenschaften und Rohstoffe, Hannover, Bundesrepublik Deutschland
V.BRÄUER, Bundesanstalt für Geowissenschaften und Rohstoffe, Hannover, Bundesrepublik Deutschland

ABSTRACT: The assessment of deep-seated rock in terms of geotechnology with respect to prove the barrier efficiency of crystalline rock for a final repository for radioactive waste requires the knowledge of the in-situ behavior of the rock, in particular the stress and deformation behavior as well as permeability. The waterpercolation through fissured granite and the stress field is being studied in a Swiss German cooperation by the BGR in the Grimsel Rock Laboratory in Switzerland. This laboratory is operated by the NAGRA, Switzerland. German participants are the BGR and the GSF. BGR is responsible for the Fracture System Flow Test, a modified water injection test and Rock Stress Measurements in deep boreholes. Target of both projects is the development of methods and instruments. Some of the results of the permeability and stress measurements are used to discuss the methods.

RESUME: L'estimation géotechnique d'une roche située en profondeur en ce qui concerne sa qualification en tant que barrière pour le stockage final de déchets radioactifs exige la connaissance du comportement de la roche in-situ, particulièrement le comportement mécanique et la connaissance de la perméabilité. La percolation de l'eau à travers du granite fissuré et l'état des contraintes primaires sont examinés dans le cadre d'une coopération suisse-allemande par le BGR dans le laboratoire souterrain du Grimsel/Suisse. Le laboratoire est exploité par la Cédra, Suisse. Les participants allemands sont le BGR et la GSF. Le BGR est responsable des essais de perméabilité, un essai modifié d'injection d'eau et des mesures de l'état des contraintes primaires dans des forages profonds. L'objectif des deux projects est le développement non seulement des techniques de mesures mais aussi des dispositifs d'essais. Les techniques appliquées et les résultats des mesures de l'état des contraintes primaires aussi bien que les essais de perméabilité sont expliqués.

ZUSAMMENFASSUNG: Die geotechnische Beurteilung tiefliegender Felsarten hinsichtlich ihrer Barrierewirkung für die Endlagerung radioaktiver Abfälle erfordert die Kenntnis der Gebirgseigenschaften in situ, insbesondere der Durchlässigkeit und des Spannungs-Verformungsverhaltens. Die Wasserzirkulation im geklüfteten Granit und der Gebirgsspannungszustand sind daher Gegenstand von Untersuchungen, die in einer deutsch-schweizerischen Zusammenarbeit von der BGR im Felslabor Grimsel/Schweiz derzeit durchgeführt werden. Dieses Felslabor wird von der NAGRA, Schweiz, betrieben. Deutsche Partner sind die BGR und die GSF. Die BGR ist verantwortlich für den Bohrlochkranzversuch zur Durchlässigkeitsuntersuchung und für Gebirgsspannungsmessungen in tiefen Bohrlöchern. Ziel beider Forschungsvorhaben ist die Entwicklung und Erprobung von Untersuchungsmethoden im Granit. An einigen Ergebnissen von Durchlässigkeits- und Spannungsmessungen werden die eingesetzten Methoden erläutert.

1 EINFÜHRUNG

Im Zuge einer 1983 vereinbarten deutsch-schweizerischen Zusammenarbeit werden ingenieurgeologische, felsmechanische, felshydraulische und geophysikalische Untersuchungsverfahren im kristallinen Gebirge getestet und weiterentwickelt. Partner sind die Nationale Genossenschaft für die Lagerung radioaktiver Abfälle (NAGRA), die Bundesanstalt für Geowissenschaften und Rohstoffe (BGR) und die Gesellschaft für Strahlen- und Umweltforschung mbH (GSF) (Brewitz & Pahl 1986).

Die NAGRA betreibt das Felslabor Grimsel, das im Aaremassiv der Schweizer Zentralalpen in der Nähe des Grimselpasses liegt. Der Hauptzugangsstollen zur Zentrale Grimsel II der Kraftwerke Oberhasli AG ist von der NAGRA untersucht und ein Gebiet unter dem Juchlistock in ca. 400 - 500 m Tiefe für das Felslabor ausgewählt worden (NAGRA 1985). Das Gebirge besteht hauptsächlich aus Granit und Granodiorit. Der Laborstollen und die Versuchsorte wurden 1983/84 aufgefahren (Abb. 1).

Die ingenieurgeologische und geotechnische Beurteilung des tiefen Gebirges in Hinsicht auf seine Eignung als Wirtgestein für die Endlagerung radioaktiver Abfälle erfordert die Kenntnis des in-situ-Verhaltens des Gebirges, insbesondere der Spannungs- und Verformungseigenschaften sowie der Permeabilität. Von der BGR werden dazu im Felslabor Grimsel die folgenden Forschungsprojekte durchgeführt:

Bohrlochkranzversuch (BK): Ein modifizierter felshydraulischer Wasserinjektionstest mit einer zentralen Bohrung für die Injektion und mehreren Beobachtungsbohrungen. Er hat zum Ziel, Richtung und Geschwindigkeit der Wasserströmung in Klüften und Rissen zu bestimmen und die Permeabilität in situ zu messen.

Gebirgsspannungen (GS): Das Projekt dient der in-situ Bestimmung felsmechanischer Eigenschaften wie Spannungsfeld und Deformationsverhalten. Die in-situ-Messungen werden in einer tiefen Bohrung im Gebirge durchgeführt.

Beide Projekte haben außerdem zum Ziel, Methoden und Geräte weiterzuentwickeln. Die felsmechanischen und -hydraulischen Tests beinhalten die folgenden Methoden:

Direkte Methoden:

- in-situ-Spannungsmessungen in tiefen Bohrungen mit Hilfe der Überbohrtechnik, hydraulischer und mechanischer Frac-Versuche (GS),
- Belastungs- und Kompensationsversuche an Bohrkernen im Labor zur Bestimmung des Spannungs- und Verformungsverhaltens sowie zur Simulation des Spannungsfeldes (GS),
- in-situ-Durchlässigkeitsversuche mit einem Injektionsbohrloch und mehreren Beobachtungsbohrlöchern; die räumliche Lage der Bohrungen wird dabei anhand der geologischen Struktur festgelegt (BK),

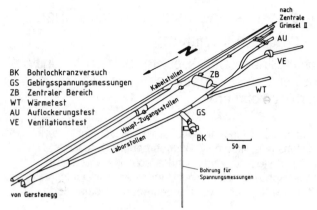

BK Bohrlochkranzversuch
GS Gebirgsspannungsmessungen
ZB Zentraler Bereich
WT Wärmetest
AU Auflockerungstest
VE Ventilationstest

Abb. 1: Felslabor Grimsel

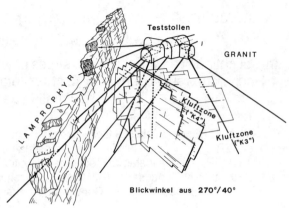

Abb. 2: Übersicht des BK-Versuchsortes

- Laboruntersuchungen an Felsblöcken zur Bestimmung
  der hydraulischen Leitfähigkeit von Kluftflächen
  (BK).

Indirekte Methoden:

- Spezielle ingenieurgeologische Aufnahmen des Versuchsortes, Bohrkernbeschreibung und graphische Darstellung (BK, GS),
- Charakterisierung von Gebirgsbereichen auf der
  Grundlage der geologischen Struktur, des Spannungsfeldes und der Wasserzirkulation auf Klüften,
- Simulation von Versuchen mit Hilfe numerischer Methoden (GS) einschließlich der Berechnung der hydraulischen Leitfähigkeit (BK),
- Verifizierung der Simulationsmodelle (BK).

Anhand einiger Ergebnisse von Durchlässigkeitsuntersu-
chungen und Spannungsmessungen werden die eingesetzten
Methoden erläutert.

2 UNTERSUCHUNGSERGEBNISSE

2.1 Durchströmungsversuche in Kluftsystemen

Im harten Gestein, z.B. Granit, sind Klüfte, Risse und
Spalten die Haupteinflußfaktoren auf die Durchlässigkeit eines Gebirges, wobei die Durchlässigkeit eines
intakten Gesteines ohne Klüfte von untergeordneter
oder sogar vernachlässigbar kleiner Bedeutung ist.

Die hydraulische Leitfähigkeit hängt dann vor allem
von den vorhandenen Klüften, ihrer Öffnungsweite, dem
Füllmaterial, der räumlichen Klufterstreckung, ihrer
Entwicklung, dem Durchtrennungsgrad und den Spannungen
sowie ihrer Orientierung zu den Trennflächen ab. Die
BGR hat eine Durchströmungsmeßapparatur für Forschungsarbeiten im Zusammenhang mit dem Bau unterirdischer Kernkraftwerke entwickelt (Pahl & Schneider
1982).

Das Ziel der Durchströmungsversuche in Kluftsystemen
ist einerseits die Weiterentwicklung von Geräten für
die Bestimmung der Fließwege und hydraulischer Leit-
fähigkeiten des kristallinen Gesteins in Trennflächen,
wie Klüfte, Schieferung, Risse und Störungen, andererseits die Entwicklung eines Modells, das auf den oben
beschriebenen direkten oder indirekten Methoden basiert. Dieses Modell soll es ermöglichen, mit Hilfe
numerischer Berechnungen nach der Finite-Element-Methode die Ergebnisse von Durchlässigkeitsuntersuchun-
gen auf größere, geologisch bekannte Bereiche zu übertragen. Dazu müssen zunächst die gemessenen Daten mit
den aus dem Modell berechneten Daten verglichen werden. Unter der Voraussetzung, daß ein direkter Zusammenhang von geologischer Struktur und Wasserdurchlässigkeit auf Diskontinuitäten besteht, kann mit dieser
Methode die Qualität und Schutzwirkung der geologischen Barriere bestimmt werden.

Ein vereinfachtes Blockbild des Untersuchungsortes
und der Anordnung der Bohrungen ist in Abb. 2 zu
sehen.

Die in-situ-Meßapparatur besteht im wesentlichen aus
Sonden und Packern, aus Pumpaggregaten zur Erzeugung
der Injektionsdrücke und aus Computern zur Meßwerter-
fassung und -auswertung (Liedtke & Pahl 1984). Die
Sonden messen Wasserdruck, Temperatur und elektrische
Leitfähigkeit und werden zwischen zwei pneumatischen
Packern in der Beobachtungsbohrung (86 mm ø) installiert. Im allgemeinen werden bis zu zehn Sonden, jeweils durch einen Packer voneinander getrennt, hin-
tereinander eingesetzt. Das Kabel für die Datenüber-
mittlung verläuft zwischen den Sonden und dem Auf-
zeichnungsgerät in einem wasserdichten, nichtrosten-
den Stahlrohr.

Von jeder Bohrung werden die Petrographie, Klüfte mit
Kluftöffnung, deren Öffnungsweite, Kluftorientierung,
Kluftfüllung, Störungen und Schieferung im einzelnen
erfaßt. Die Darstellung der Kluftsysteme und die
räumliche Extrapolation von einer Bohrung zu anderen
Bohrungen und zum Test-Stollen erfolgt mit speziellen
Computerprogrammen. In Abb. 3 ist die Anordnung der
Bohrungen im Grundriß dargestellt. Die Schraffuren
geben die Klufthäufigkeit an. Die Injektion, in Stufen bis ca. 50 bar durchgeführt, erfolgt bevorzugt in
den stark klüftigen Bohrlochabschnitten.

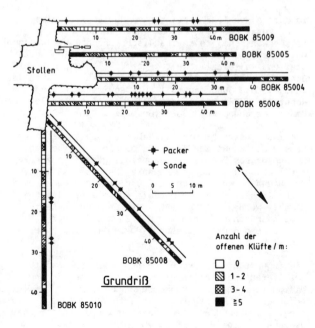

Abb. 3: Bohrlochanordnung im BK-Versuch

In dem umgebenden Gebirge wird in abgepackerten Bohrabschnitten gemessen, wie sich der Druck verändert, insbesondere in der Richtung des injizierten Kluftsystems.

Die Wasserzirkulation in geöffneten Klüften wird u.a. von der Kluftöffnungsweite beeinflußt, die sich in Abhängigkeit vom Wasserdruck und vom Verformungsverhalten des Gebirges ändern kann. Der natürliche Wasserdruck beträgt in den vorliegenden offenen Kluftsystemen z.B. bis 40 bar, entsprechend der Überlagerungshöhe des Gebirges. In geschlossenen Kluftsystemen kann der Wasserdruck jedoch durch Wärmeausdehnung, z.B. durch Einlagerung wärmeproduzierender Abfälle hervorgerufen, oder durch Kriechvorgänge im Gebirge bis zur Höhe des Gebirgsdrucks ansteigen.

Daher wurde eine numerische Berechnung der Kluftöffnungsweite in Abhängigkeit vom Kluftwasserdruck unter Verwendung des Finite-Element-Programms ADINA am Beispiel einer Kluft unterhalb eines Stollens durchgeführt (Liedtke & Pahl 1984). Die Gebirgssteifigkeit wurde dabei entsprechend den Ergebnissen aus Dilatometermessungen mit E = 40 GPa angenommen (Pahl & Heusermann 1986).

Zur Simulierung von Fließvorgängen in Klüften wurde das FE-Programm ADINAT herangezogen und ein neues spezielles FE-Programm DURST entwickelt. Anhand eines dreidimensionalen Modells mit zwei Klüften und einer Öffnungsweite von 1 mm wurden die Kontinuitätsbedingungen verifiziert. Ein Vergleich der mit beiden Programmsystemen ermittelten Ergebnisse zeigt hinsichtlich Fließvolumen, Äquipotentiallinien und Fließgeschwindigkeit, daß die Unterschiede geringer als 3 % sind.

In Abb. 4 sind auf der Grundlage der in der ersten Untersuchungsphase erzielten Meßergebnisse Zonen mit annähernd gleicher Wasserdurchlässigkeit des Gebirges dargestellt. Der untersuchte Gebirgsbereich umfaßt etwa 50 x 50 x 50 m. Diese Zonen decken sich im Prinzip mit unterschiedlichen Kluftsystemen, die sich aus den speziellen ingenieurgeologischen Untersuchungen ergeben. Aus den strukturgeologischen Aufnahmen und den Versuchsergebnissen lassen sich Einzelbereiche ableiten, die durch unterschiedliche Trennflächengefüge und tektonische Genese charakterisiert sind. Danach können unterschiedliche Homogenbereiche identifiziert werden. In der nächsten Untersuchungsphase wird zu klären sein, ob sich diese felshydraulischen Homogenbereiche auf der Grundlage einzelner Versuchsergebnisse und geologischer Untersuchungen zu größeren Einheiten zusammenfügen lassen und ob eine Extrapolation auf größere Gebirgsabschnitte, d.h. ca. 500 x 500 x 500 m, möglich ist.

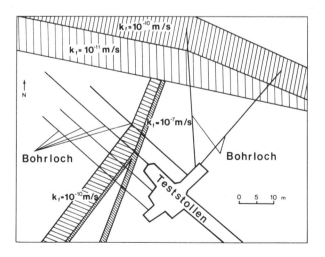

Abb. 4: Bereiche konstanter Permeabilität

2.2 Gebirgsspannungen und geologische Struktur

Spannungsmeßgeräte und Dilatometer werden zur Untersuchung des Spannungsfeldes und des Gebirgsverformungsverhaltens benutzt. Diese Geräte sind bisher jedoch selten in tiefen Bohrungen (d.h. tiefer als 100 m) angewandt worden. Es soll daher in diesem Untersuchungsprojekt festgestellt werden, ob die o.a. Geräte und Verfahren in großen Tiefen verwendet werden können und welche Gerätemodifizierungen notwendig sind.

Die Überbohrversuche erfolgten während des Niederbringens der Kernbohrung in einer Bohrtiefe bis 171 m unterhalb des Laborstollens, entsprechend einer Tiefe von ca. 600 m unter der Oberfläche (Pahl & Heusermann 1986).

Im ersten Abschnitt der Untersuchungen lag der Schwerpunkt in der Geräteentwicklung, insbesondere der Weiterentwicklung des von der BGR angewandten Überbohrverfahrens. Für den Versuch wird eine induktiv messende Sonde in einer Vorbohrung mit einem Durchmesser von 46 mm installiert und anschließend mit einem Durchmesser von 146 mm überbohrt. Besonders hervorzuheben ist die neue 3-D-Sonde der BGR und der Einsatz eines Computers in der Bohrung zusammen mit der Sonde. Dieses Verfahren ermöglicht insbesondere in tiefen Bohrungen das Speichern der Meßwerte ohne das sonst übliche Kabel im Bohrgestänge. Ein Beispiel der kontinuierlichen Messung der Entspannungswege während des Überbohrens zeigt Abb. 5.

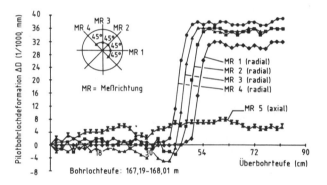

Abb. 5: Bohrlochdeformationen im Überbohrversuch

In Abb. 6 wird deutlich, daß die maximalen horizontalen Spannungen zwischen 15 und 30 MPa liegen. Damit liegen diese Spannungen über dem teufenbezogenen gravitativen Gebirgsdruck. Obgleich der Granit in der Bohrung den Eindruck einer gleichförmigen Ausbildung macht, zeigen sich in vier Abschnitten der Bohrung abweichend hohe Horizontalspannungen und zwar in dem Abschnitt unterhalb des Laborstollens bis 40 m, darunter bis 100 m, dann bis etwa 170 m und schließlich im tiefsten Teil der Bohrung bis etwa 180 m. Diese Abschnitte sind auch durch unterschiedliche Richtungen der Horizontalspannungen gekennzeichnet.

Die gleichzeitig mit den Spannungsuntersuchungen vorgenommenen detaillierten geologischen Aufnahmen des Versuchsortes und der Bohrkerne, die orientiert entnommen wurden, hatten folgende Ziele: An jedem Meßpunkt mußte festgestellt werden, ob Beeinflussungen der Messung durch einzelne Klüfte, Störungen oder petrographische Besonderheiten vorliegen. Abschnittsweise war zu klären, ob ein Zusammenhang zwischen Kluftsystemen, ihrer Ausbildung und Richtung mit den Spannungen erkennbar ist. Schließlich war zu prüfen, ob sich geologische Homogenbereiche mit Bereichen gleicher Spannung decken. Wenn das der Fall ist, könnten Ergebnisse einzelner, punktueller Spannungsmessungen auf größere geologische Homogenbereiche übertragen werden.

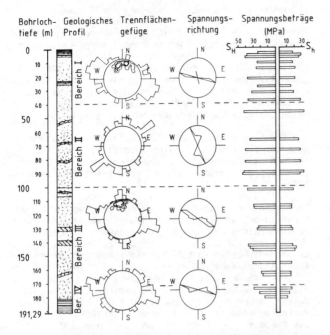

Bohrloch-tiefe (m)	Geologisches Profil	Trennflächen-gefüge	Spannungs-richtung	Spannungsbeträge (MPa)

Abb. 6: Gebirgsstruktur und Spannungsfeld

Das schematisierte geologische Profil der Bohrung (Abb. 6) zeigt den Zentralen Aare Granit (ZAGr), der petrographisch und texturell relativ gleichförmig ausgebildet ist, abgesehen von einigen Zonen mit ausgeprägter Paralleltextur bzw. starken Zerklüftungen. Der obere Abschnitt bis 40 m Tiefe ist durch eine Schieferungszone und relativ große Kluftdichte gekennzeichnet. Darunter folgt bis 100 m gleichförmig ausgebildeter kompakter Granit mit sehr wenig Klüften (RQD meistens 100 %). Unterhalb einer annähernd horizontal verlaufenden Zerrkluft mit hydrothermaler Zersetzungszone folgt wieder ein gleichförmiger Granitabschnitt und schließlich im Bohrlochtiefsten ein Lamprophyr mit Schieferungszone.

Die Darstellung der Orientierung von Klüfte und Spannungen läßt den postulierten Zusammenhang zwischen Kluftsystemen und Spannungsrichtungen deutlich hervortreten. In Abb. 6 sind für den Bohrabschnitt bis 40 m offene Klüfte als Polhäufung durch Isolinien dargestellt (Schmidt'sches Netz, Projektion der unteren Halbkugel) und außen die Richtung geschlossener Klüfte. Die horizontale Hauptspannung (WNW-ESE) deckt sich ungefähr mit der Richtung der steil einfallenden Klüfte. Ein ganz anderes Bild zeigt der darunterliegende Bereich bis 100 m. Hier ist die Klufthäufigkeit wesentlich kleiner, es treten keine offenen Klüfte auf, und die horizontale Hauptspannung verläuft mehr NNW-SSE, nicht im Kluftstreichen. Unterhalb der alpidischen Zerrkluft, bei 100 bis 170 m, sind wieder offene Klüfte zu beobachten, und die Spannungsrichtung liegt bei WNW-ESE, wie in den obersten 40 m. Mit Annäherung an einen Lamprophyrgang im Bohrlochtiefsten folgt ein Abschnitt mit deutlich abnehmenden Spannungen, gemessen durch Hydrofrac-Versuche.

So läßt sich durch detaillierte geologische Untersuchungen und intensive Spannungsmessungen ein Zusammenhang zwischen der geologischen Struktur und den Gebirgsspannungen nachweisen, insbesondere durch die Klufthäufigkeit und offene oder geschlossene Kluftsysteme, besonders auffallend an einer alpidischen, also geologisch jungen Zerrkluft. Außerdem scheinen Gänge anderer Gesteinsarten mit begleitenden Diskontinuitäten eine Beeinflussung der Spannungen im Gebirge zu bewirken.

Die Ergebnisse der Spannungsmessungen und der geologischen Aufnahme führen zu einer Bewertung, die für zukünftige Planungen wie auch für die Übertragbarkeit

von Meßergebnissen wichtig ist. Spannungsmessungen werden punktuell angesetzt und für größere Gebirgsabschnitte in Ansatz gebracht. Dieses Vorgehen ist nur sinnvoll, wenn, wie hier nachgewiesen, für geologische Homogenbereiche aussagefähige Ergebnisse in ausreichender Zahl zur Verfügung stehen. Die Planung von Spannungsmessungen sollte deshalb nur im Zusammenhang mit der detaillierten geologischen Aufnahme und im entsprechenden Umfang erfolgen.

3 ZUSAMMENFASSUNG UND ZUKÜNFTIGE AUSSICHTEN

In der ersten Testphase des Bohrlochkranzversuches ließen sich bereits gute Ergebnisse erzielen. Ingenieurgeologische Spezialkartierungen verbunden mit Wasserinjektionen sowie Beobachtungen in acht Bohrungen lassen die bevorzugte Wasserdurchlässigkeit des kristallinen Gebirges auf bestimmten Kluftsystem erkennen. Zonen gleicher Durchlässigkeit können geologischen Homogenbereichen zugeordnet werden. Für die numerische Modellierung des Gebirges werden Finite-Elemente-Programme entwickelt, um die gemessene Durchlässigkeit mit Berechnungen vergleichen zu können. Mit dieser Methode soll versucht werden, die Ergebnisse auf größere Gebirgsbereiche zu übertragen. Für die nächste Testphase sind höhere Injektionsdrükke, längere Bohrlöcher und die Injektion von aufgeheiztem Wasser und Salztracern vorgesehen. Zudem sind Tests an Granitblöcken mit Schnittlängen von ca. 1 m im Labor geplant. Diese Tests haben den Vorteil, daß jede künstliche oder natürliche Kluft exakt gemessen werden kann. Die klaren Randbedingungen erlauben dann auch die bestmögliche Anpassung des FE-Modells. Die notwendige Extrapolation auf große Gebirgsbereiche ist dadurch verläßlicher.

Die Untersuchungen der Gebirgsspannungen konnten einen Zusammenhang mit den geologischen Strukturen nachweisen. Im allgemeinen werden Spannungsmessungen punktuell ausgeführt und auf größere Gebirgsabschnitte übertragen. Dieses Vorgehen ist sinnvoll, wenn für geologische Homogenbereiche genügend aussagefähige Ergebnisse vorliegen. Außerdem muß gewährleistet sein, daß in Bohrungen ausgewiesene Homogenbereiche die angenommene räumliche Ausdehnung haben. Die Gebirgsspannungsmessungen mit der BGR-Überbohrapparatur sind bisher bis ca. 170 m Tiefe durchgeführt worden. Mit dem neu entwickelten Bohrlochcomputer sind aber auch Messungen in tieferen Bohrungen (1000 m) möglich. Im Zuge der weiteren Arbeiten ist die Entwicklung einer mechanisch arbeitenden Frac-Sonde vorgesehen. Außerdem sind Versuche an Überbohrkernen im Labor geplant. Schließlich soll neben dem Makrogefüge auch das Mikrogefüge im Hinblick auf Zusammenhänge mit Gebirgsspannungen untersucht werden.

LITERATUR

Brewitz, W. & A. Pahl 1986. German participation in the Grimsel Underground Rock Laboratory in Switzerland - targets and methods of in situ experiments in granite for radioactive waste disposal, IAEA-SM-289/24, Int. Symp., Hannover.
Liedtke, L. & A. Pahl 1984. Water injection tests and finite element calculations of water percolation through fissured granite. CEC/NEA Workshop on Design and Instrumentation of In-Situ Experiments in Underground Laboratories for Radioactive Waste Disposal. Brussels.
NAGRA 1985. Felslabor Grimsel - Rahmenprogramm und Statusbericht. NTB 85-34, Baden.
Pahl, A. & S. Heusermann 1986. Stress measurements at the Grimsel Rock Laboratory. Geol. Jahrbuch, C 45, Hannover.
Pahl, A. & H.J. Schneider 1982. Standortmöglichkeiten für unterirdische Kernkraftwerke im Fels aus ingenieurgeologisch-felsmechanischer Sicht. Symp. Underground Siting of Nuclear Power Plants. Hannover.

The influence of the uniaxial compression stress intervals on the rock deformability anisotropy

L'influence des intervalles de contrainte de compression uniaxiale sur l'anisotropie de déformabilité des roches
Einfluss der Uniaxialdruckspannungsintervalle auf die Verformungsanisotropie der Gesteine

F.PERES-RODRIGUES, Laboratório Nacional de Engenharia Civil, Lisboa, Portugal

ABSTRACT: This paper discusses the influence of stress intervals on the characteristics of the anisotropy surfaces of six types of rock, as obtained in a first loading-unloading quick test. Some tendances observed, thought relevant, are briefly described.

RÉSUMÉ: Dans cette communication on étudié l' influence des intervalles de contrainte, sur les caractéristiques des surfaces d'anisotropie de six types de roche, obtenues pendant un premier essai rapide de mise en charge-déchargement. On indique quelques tendances observées qui sont considérées importantes.

ZUSAMMENFASSUNG: Diese Arbeit untersucht den Einfluß der Spannungsintervalle auf die Characteristika der Anisotropieflächen von sechs Gesteinsarten, die während des ersten schnellen Be - und Entlastungsversuchs erhalten wurden. Es wird auf einige beobachtete Tendenzen, die als wichtig erachtet werden, hingewiesen.

1 INTRODUCTION

As is known, most rocks, mainly fractured and decomposed rocks display a mechanical behaviour that is far from linear and elastic. In rather simplistic terms one may say that their behaviour is closer to being viscuous-elastic even in quick tests when the influence of the time factor can be neglected. Their behaviour will also differ in loading or in unloading, chiefly in the first test cycle, and as a rule significant irrecuperable deformation occurs.

Fig. 1 presents the typical behaviour of fractured and/or decomposed rock, in a first loading-unloading quick test. The loading curve may be divided into three phases: the first corresponds to reclosing of the fissures and decreasing of porosity, during which the moduli of deformability, tangent, by points, or by intervals do increase; in the second phase the behaviour is practically linear; lastly in the third phase the viscosity effect is markedly felt, which ordinarily does not occur in tests when the maximum stress applied does not exceed one fifth of the ultimate strength. The unloading curve can also be divided into three phases. In the first the moduli of deformability already referred to are comparatively high, mainly due to a kind of mechanical inertia; there follows a practically linear phase of small or even null amplitude, and finally a phase with irrecuperable deformation to a smaller or larger extent.

2 ROCKS TESTED

For this study advantage was taken of anisotropy tests carried out on rocks from several sites where LNEC studied the rock mass characteristics of concrete dams foundations, as follows: from the Albarellos site, in Spain, laboratory anisotropy tests carried out on specimens of gneissoid granite, coarse-grain granite and micaceous schist [1]; from the Torrão site, tests on specimens of sound granite and weathered granite [2]; and lastly from the Funcho site, tests on graywacke [3].

3 TESTS

A cube about 40 cm in edge was taken in situ for the six types of rocks referred to in 2. As a result 9 to 12 prismatic specimens, 5 x 5 x 12 cm^3 in size, could be prepared in laboratory with such orientations and nature as described in Fig. 2. Each specimen was submitted to a simple compression test with several loading-unloading cycles up to a maximum normal stress about one fourth of ultimate compression stress. In this study only the first loading-unloading cycle was considered since then the material was still intact without any disturbances due to mechanical actions. The application of stresses is quick enough for the influence of factor time not be significant. All cubes extracted are referred to the ground by the attitude of plane Oxy and the direction of the semi-axes Ox, Fig. 2. The two loading-unloading curves were divided into the same number of intervals at constant stress variation; in each interval the modulus of deformability between extreme points was determined. We thus obtain the evolution of the modulus of deformability with the stresse interval, either for loading or for unloading, in each specimen referred to a given direction, Fig. 1. Hence it was possible to determine the most probable anisotropy surfaces of the moduli

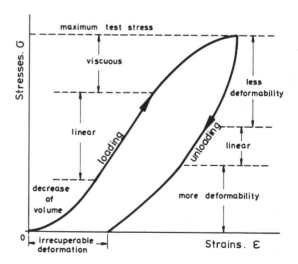

Fig. 1 - Behaviour of a fracture and/or decomposed rock

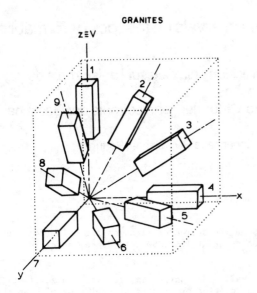

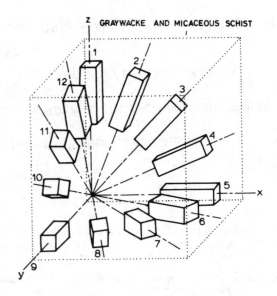

Fig. 2 - Extraction of specimens

of deformability, for each stress interval, by using the least-square method [4], [5] and [6].

4 RESULTS

Of the rocks studied, only results referring to the Albarellos gneissoid granite are presented to keep the paper within the page limits prescribed.

Table I successively indicates for each of the eight stress intervals of 25 kg/cm^2, on the curves of the first loading-unloading cycle: the most probable semi-axes a, b, c, of the normal equation of ellipsoid that defines the anisotropy, obtained on basis of the least square method; the value of the geometrical mean $\overline{E}$ of the three semi-axes, equivalent to the radius of the sphere whose volume equals that of the ellipsoid

obtained, the direction angles that each of the three principal axes of the ellipsoid, X, Y, Z make with the reference axes x, y and z; the coefficients of the mass anisotropy, a_m and maximum anisotropy, a_M, [6] defined by

$$a_m = \frac{R}{E} \qquad a_M = \frac{R}{r} \qquad (1)$$

where R is the major semi-axes and r the minor. Lastly the coefficient of variation [6] is given by

$$\delta_v = \sqrt{\frac{\sum\limits_{i=1}^{n} \left(\dfrac{t_i - e_i}{t_i}\right)^2}{n-1}} \qquad (2)$$

Table I. Modulus of deformability (10^2 MPa) in simple compression

CURVE	σ_a (MPa)		Normal equation $\dfrac{x^2}{a^2}+\dfrac{y^2}{b^2}+\dfrac{z^2}{c^2}=1$			$\overline{E}$	Direction angles ($^\circ$)									Coefficient of		
							X			Y			Z			Anisotropy		Variation (%)
	From	To	a	b	c		x	y	z	x	y	z	x	y	z	a_m	a_M	
LOADING	0	2.5	185	244	125	178	47	53	64	135	46	81	102	114	27	1.37	1.95	2.7
	2.5	5.0	197	229	142	186	35	66	66	122	39	72	101	119	31	1.23	1.61	1.4
	5.0	7.5	199	212	171	193	173	86	84	84	22	69	86	112	22	1.10	1.24	2.0
	7.5	10.0	196	234	173	199	158	69	83	71	33	66	93	114	25	1.17	1.35	2.6
	10.0	12.5	228	233	183	213	131	135	75	119	49	55	55	105	39	1.09	1.27	2.2
	12.5	15.0	206	268	183	216	157	78	71	69	36	62	80	123	35	1.24	1.46	3.6
	15.0	17.5	207	259	237	233	47	110	50	126	142	79	116	59	42	1.11	1.25	3.7
	17.5	20.0	235	271	211	237	150	84	61	72	28	69	67	117	37	1.14	1.28	3.4
UNLOADING	20.0	17.5	402	478	431	436	135	133	81	130	44	74	73	96	18	1.10	1.19	5.5
	17.5	15.0	429	396	329	382	141	98	52	92	9	81	51	95	39	1.12	1.30	4.0
	15.0	12.5	337	402	314	349	33	58	80	122	49	57	80	123	34	1.15	1.28	2.3
	12.5	10.0	316	320	266	300	148	59	84	59	40	66	97	113	25	1.07	1.20	3.6
	10.0	7.5	265	302	230	264	10	89	80	95	22	68	99	113	24	1.14	1.31	1.4
	7.5	5.0	223	261	184	220	21	75	76	110	30	68	97	116	27	1.18	1.42	2.8
	5.0	2.5	187	212	144	179	38	52	85	126	45	67	80	111	23	1.19	1.47	1.0
	2.5	0	190	139	103	140	136	78	49	79	147	59	48	61	56	1.36	1.84	9.1

where t_i is the theoretical modulus of deformability obtained from the ellipsoid and e_i is the corresponding experimental value for each of the n directions tested. This coefficient thus measures fitting between the most probable ellipsoid and the experimental values, and is considered stable provided it does not exceed 15%.

For three stress intervals in loading and in unloading, Fig. 3 shows the most probable ellipsoids, test points, the major referencial OXYZ and that of reference Oxyz.

In Fig. 4 are shown the diagrams representing the evolution of the maximum and minimum moduli of deformability of the ellipsoids and of the mean value $\bar{E}$ already defined as a function of stress intervals and of loading and unloading curves (see Table I).

5 CONCLUSIONS

With the results obtained on the six rocks tested, it was possible to infer some tendencies, to be confirmed or adjusted as future tests may indicate. The most relevant tendencies detectec are as follows:

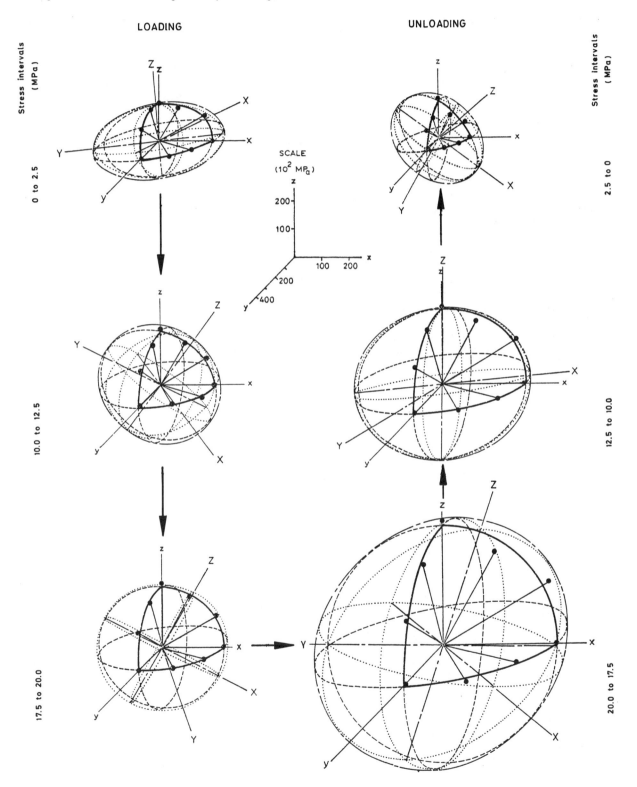

Fig. 3 - Anisotropy surfaces (ellipsoids).

217

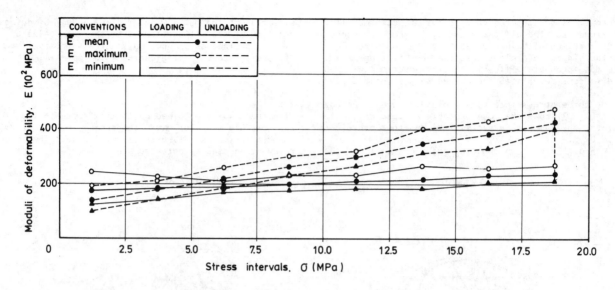

Fig. 4 - Evolution of the moduli of deformability at different stress intervals.

a) anisotropy is influenced by stress intervals;

b) anisotropy is influenced by loading and unloading curves, the unloading curve being the one that exerts more influence;

c) both the largest and the smallest values of the modulus of deformability are obtained on the unloading curve;

d) coefficients of anisotropy tend to decrease with loading and increase with unloading;

e) $\bar{E}$ values systematically increase with loading and decrease with unloading;

f) it does not seem to exist a law of variation of the major axes of the anisotropy surfaces with stress intervals.

Those conclusions give an incitement to pursuing this study as regards the modulus of deformability and also other properties.

6 ACKNOWLEDGEMENT

The author is indebted to LNEC's experimental assistant Manuel Reis e Sousa, for his valuable cooperation both in testing and in drawing of figures.

7 REFERENCES

[1] LNEC, Estudo das fundações da barragem de Albarellos (Espanha), Lisbon (1971).

[2] LNEC, Estudo das fundações da barragem do Torrão, Lisbon (1971).

[3] LNEC, Estudo das fundações da barragem do Funcho, Lisbon (1977).

[4] PERES-RODRIGUES, F., Anisotropy of granites-modulus of elasticity and ultimate strength ellipsoids, joint systems, slope attitudes, and their correlations, Proc. 1st Cong ISRM, Vol. I, pp. 721-732, Lisbon (1966).

[5] PERES-RODRIGUES, F., Anisotropy of rocks-most probable surface of the ultimate stresses and of the moduli of elasticity, Proc. 2nd Cong. ISRM, Vol. I, pp. 1-20, Beograd (1970).

[6] PERES-RODRIGUES, F., Anisotropia das rochas e dos maciços rochosos — aplicação de quárticas fechadas ao estudo da deformabilidade e da rotura, memória 483 — Vol. I, LNEC, Lisbon (1977).

Borehole creep closure measurements and numerical calculations at the Big Hill, Texas SPR storage site*

Mesures de convergence par fluage d'un forage et calculs numériques au site de stockage SPR de Big Hill, Texas

Messungen der Bohrlochkonvergenz und numerische Berechnungen für das Big Hill, Texas SPR Lager

DALE S.PREECE, Applied Mechanics Division I, Sandia National Laboratories, Albuquerque, N.Mex., USA

ABSTRACT: The Big Hill Texas salt dome is being developed as an additional site for the U.S. Strategic Petroleum Reserve (SPR). Ten deep (1433 m) boreholes were drilled in 1983 for leaching the first five storage caverns. Leaching has been delayed and the ten wells have been used to study the in-situ creep behavior of rock salt. Creep closure has been quantified by measuring wellhead pressure rise, volume of brine on pressure bleed-down and by performing caliper logs several years apart. Finite element calculations of borehole creep closure have also been performed. The calculated creep displacement of the borehole wall was used to compute volume change which could be used to compute wellhead pressure rise. The calculated volume change and wellhead pressure rise were found to underpredict the field data by a factor of 2.5. The calculated displacements underpredicted measured borehole closures by a factor of 4.

RESUME: Des travaux sont en cours pour faire du dôme salin de Big Hill, Texas, un emplacement supplément-aire de la U. S. Strategic Petroleum Reserve (SPR). Dix trous de sondage profonds (1433 m) ont été forés en 1983 pour lessiver les cinq premières cavernes de stockage. Le lessivage a été retardé et les dix puits ont été utilisés pour étudier le comportement de fluage sur place du sel gemme. La fermeture par fluage a été établie quantitativement en mesurant l'augmentation de pression à la tête du puits, le volume de saumure à la décompression et en effectuant des profils de calibrage à quelques années de distance. On a procédé également à des calculs par éléments finis de la fermeture des trous de sondage par fluage. Le déplacement par fluage calculé de la paroi du puits a été utilisé pour calculer le changement de volume qui pourrait être utilisé pour calculer l'augmentation de pression à la tête du puits. On a trouvé que les changements de volume et l'augmentation de pression à la tête du puits calculés sous-évaluaient, d'un facteur de 2,5, les données observées sur le terrain. Les déplacements calculés sous-évaluaient, d'un facteur de 4, les fermetures mesurées des trous de forage.

ZUSAMMENFASSUNG: Der Salzdom in Big Hill, Texas, wird zur Zeit als zusätzlicher Lagerort für die U.S. Strategic Petroleum Reserve (SPR) ausgebaut. Zehn 1433 m tiefe Bohrlöcher wurden 1983 zur Auslaugung der ersten fünf Speicherkavernen abgesenkt. Die weitere Auslaugung wurde aufgeschoben, und die zehn Bohrlöcher wurden dazu benutzt, das Kriechverhalten von Steinsalz an Ort und Stelle zu beobachten. Die Kriechkonvergenz wurde durch Messungen des Druckanstiegs am Bohrlochkopf, des Salzlaugenvolumens beim Druckablassen sowie mittels Kalibermessungen in Abständen von mehreren Jahren quantifiziert. Bohrloch-kriechen wurde auch mit Hilfe von Finite Element Analysen ermittelt. Die errechnete Kriechverschiebung der Bohrlochwand wurde zur Bestimmung von Volumenänderungen herangezogen, die wiederum zur Errechnung des Druckanstiegs am Bohrlochkopf verwendet werden konnten. Die analytisch bestimmte Volumenänderung und der Druckanstieg am Bohrlochkopf erwiesen sich um den faktor 2,5 niedriger als die Messungen in situ. Die theoretisch errechneten Verschiebungen lagen um einen Faktor von 4 unterhalb der gemessenen Bohrloch-konvergenz.

1 INTRODUCTION

The Big Hill salt dome is located in southwestern Jefferson County, Texas, approximately 32 km southwest of Port Arthur, Texas and 8.5 km north of the Intracoastal Waterway. The dome created a hill on the ground surface which is approximately 1.6 km in diameter and 8.2 m above the surrounding flat terrain. The Big Hill site characterization is documented by Hart et al 1981.

The U. S. Department of Energy (DOE) obtained the major portion of the site for use in the U. S. Strategic Petroleum Reserves (SPR). Big Hill is planned as an additional site where 14 caverns will be leached, each with a crude oil storage volume of 1.59×10^6 m³. Ten wells for the first five caverns were drilled in 1983. Sandia National Laboratories has monitored the wells and accurately measured the wellhead pressure increase due to creep closure. The volume of brine produced on pressure bleed-down has been carefully measured and caliper logs have also been obtained two years apart. The wells at Big Hill were drilled into virgin salt so the in situ stresses are assumed to be equal in all directions and vary linearly with depth. The geometry of a well is simple and the data acquisition has been high quality.

This paper contains the field data from a number of the wells and describes the finite element analyses that were performed to test current predictive capabilities for underground structures in rock salt.

* This work performed at Sandia National Laboratories supported by the U.S. Department of Energy under contract number DE-AC04-76DP00789.

2 WELL DESCRIPTIONS

Figure 1 shows a cross-section of the dome through the first five of the proposed storage caverns. The caverns to be leached at Big Hill are labeled 106 through 110 with wells labeled 106A and B through 110A and B. Each well is cased to a depth of 640 m leaving an open borehole in salt from there to 1433 m. The A wells have a single 0.273 m diameter hanging string to a depth of 817 m and the B wells have two hanging strings, the outer one being 0.273 m diameter to a depth of 1330 m and the inner one being 0.178 m diameter to a depth of 1422 m. The hanging strings are configured for the start of leaching.

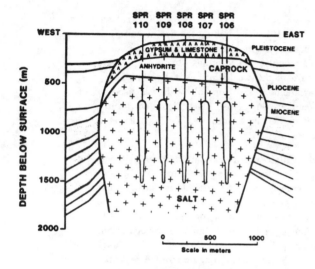

Figure 1. Cross-section of the Big Hill Salt Dome showing proposed SPR storage caverns.

3 EXPERIMENTAL PROCEDURES

3.1 Wellhead Pressure and Brine Volume Bleed-Off Measurements

Pressure testing was performed on each well as soon after well completion as possible, the elapsed time ranged from 20 to 150 days. After pressure testing the wells were shut-in which results in wellhead pressure rise as the open salt portion of the borehole creeps inward reducing the borehole volume. Measured brine bleed-off from a well was done at wellhead pressures just under 3.45 MPa which was the pressure at which the relief valve is set to open. The brine was bled from the well through a flexible hose into a volume calibrated container. During the first few pressure cycles the wellhead pressure was bled to almost atmospheric pressure to increase the length of the cycle for convenience. After the first seven months the volumes bled were limited to prevent the wellhead pressure from dropping below approximately 2.0 MPa to reduce the closures.

Pressures were read daily from dial gauges on each wellhead by on-site personnel. The wellhead pressure and bleed-off brine volume measurements were made continuously from February 1984 to December 1985. The pressure-versus-time data for each well has been computerized and is given for the first few cycles of wells 106B, 107B and 109A in Figure 2. More detailed pressure cycle information on all the wells is available in Beasley et al 1986.

It is important to point out that half of the wells at Big Hill have consistently produced gas since they were completed.

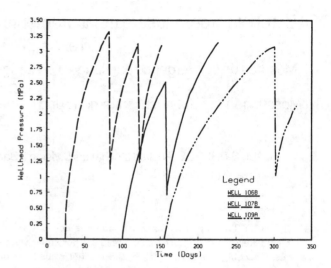

Figure 2. Measured pressure versus time for wells 106B, 107B and 109A. Data has been smoothed.

Two of the wells included in this study, 106B and 107B, are gassy which causes the wellhead pressure to increase at rates approximately 1.8 times higher than the rates of the other wells. Well 109A of this study produces negligible gas.

3.2 Borehole Caliper Logs

A portion of each uncased borehole between the bottom of the hole and the bottom of the hanging strings was available for caliper logging. Logs were performed in February 1984 and again in December 1985. The logging tool was a Micro Gage, Inc. 4-arm caliper tool. Closure results from wells that were caliper logged in both February 1984 and December 1985 are given in Table 1.

Table 1. Results of borehole caliper logs at 1433 m depth.

Well Number	Avg. Dia. (mm)		Radius Change (mm)
	Feb-84	Dec-85	
106A	394.72	371.88	-11.43
106B	318.34	302.36	-7.98
107B	346.86	336.91	-4.98
110A	373.25	368.78	-2.24
110B	358.17	331.29	-13.44
Average Radius Change = -8.0 ± 4.6 mm			

3.3 Borehole Temperature Measurements

Temperature logs were performed on wells 106A, 107A and 110A in February 1984 and on wells 106B and 110A in December 1985. The temperature logs are given in full detail by Beasley et al 1986. Since the volume of brine in the borehole is relatively small, the temperature in a well comes to thermal equilibrium in a matter of days and the well has very little thermal influence on the surrounding formation.

Proof of this is provided by the fact that the February 1984 and December 1985 logs were almost identical below 300 m. Thus the temperatures obtained from the brine well logs are considered representative of the formation temperatures. The logs show that the temperatures below 300 m can be represented by equation (1)

$$T = 25.90 + 0.0245D \qquad (1)$$

where T is temperature in $°C$, and D is depth in meters.

4 FINITE ELEMENT ANALYSIS

Calculations of borehole closure due to salt creep were made using the finite element computer program named SANCHO (Stone, Krieg and Beisinger 1985). The material model for creep is a power law model for secondary (steady state) creep. The creep model is integrated semi-analytically which has been shown to be accurate for any strain step size. This method has no stability or time step restrictions as are usually associated with classical Euler integration. The only restriction is that the strain rate should be approximately constant during the time step (Krieg 1983). A few references documenting finite element calculations that have employed this secondary creep formulation are: (Krieg et al 1981), (Preece 1986), and (Morgan et al 1986).

4.1 Material Properties

At the time this study was initiated, salt from Big Hill had not been tested to obtain elastic or creep material properties. The creep model parameters derived from extensive triaxial testing of West Hackberry core (Wawersik and Zeuch 1984) have been used in this study. Subsequent to these calculations and prior to the writing of this paper, creep testing of Big Hill core was completed and the model was found to fit well with the West Hackberry creep model (Wawersik 1986).

As mentioned previously, the program uses a secondary creep model where creep strain rate magnitude, $\dot{\bar{e}}$, is a function of effective stress, $\bar{\sigma}$ expressed as

$$\dot{\bar{e}} = D exp(-Q/RT)\bar{\sigma}^n \qquad (2)$$

where D and n are constants and Q is an activation energy, R is the universal gas constant and T is the material temperature. The laboratory determined creep coefficients for West Hackberry salt are given in Table 2 along with the elastic constants obtained from quasi-static tests of West Hackberry core (Price 1986).

Table 2. Material properties of West Hackberry salt

D	=	9.5234E-6 $\frac{1}{(s)(MPa)^n}$
Q	=	13.12 $\frac{kcal}{(mole°K)}$
n	=	4.73
Young's Modulus	=	38.4 GPa
Poisson's Ratio	=	0.30

The magnitude of the creep strain rate, $\bar{e}$, and the effective stress, $\bar{\sigma}$, can be calculated from the creep strain rate tensor, $\dot{\mathbf{e}}^c$, and the deviatoric stress tensor, σ', respectively as follows.

$$\dot{\bar{e}} = \sqrt{\frac{2}{3}\dot{\mathbf{e}}^c \dot{\mathbf{e}}^c} \qquad (3)$$

$$\bar{\sigma} = \sqrt{\frac{3}{2}\sigma' \sigma'} \qquad (4)$$

The creep strain rate tensor is related to the deviatoric stress tensor by

$$\dot{\mathbf{e}}^c = |\dot{\mathbf{e}}^c|\frac{\sigma'}{|\sigma'|} \qquad (5)$$

The components of the total strain rate tensor, $\dot{e}_{ij}$, are obtained by summing the components of creep strain rate tensor, $\dot{e}_{ij}^c$, along with the components of the bulk, elastic and thermal strain rate tensors as follows.

$$\dot{e}_{ij} = \dot{e}_{ij}^c + \frac{\nu}{E}\dot{\sigma}_{kk}\delta_{ij} + \frac{1+\nu}{E}\dot{\sigma}_{ij} + \alpha\dot{T}\delta_{ij} \qquad (6)$$

Where σ_{ij} are the components of the stress tensor, ν is Poisson's ratio, E is Young's modulus, T is absolute temperature, α is the coefficient of linear thermal expansion, δ_{ij} is the Kronecker Delta.

4.2 Finite Element Model

A borehole that is 38.1 cm in diameter and 518 m high has an aspect ratio that results in very large numbers of elements if it is modeled with a single typical finite element mesh. One method that has been used before (Prij and Mengelers 1980) employs a finite element model of a thin horizontal slice such as that shown in Figure 3. This model is two-dimensional axisymmetric with brinehead pressure on

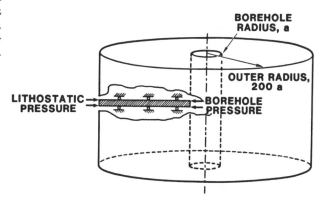

Figure 3. Two-dimensional axisymmetric finite element model of borehole horizontal slice.

the left end and lithostatic pressure on the right end. The top and bottom of the model are constrained to prevent vertical movement but allow horizontal movement. A single borehole is represented by four of these models each of which corresponds to a different depth. The depths are given in Table 3 which also lists the temperatures used to define the creep model and pressures for each depth. The temperatures were obtained from equation (1), the brinehead pressures (assuming zero wellhead pressure) are based on a gradient of 0.0118 MPa/m and the lithostatic pressures are based on a gradient of 0.0226 MPa/m.

Table 3. Temperature and loading variation with depth

Depth (m)	Temp. (°C)	Brinehead Pres. (MPa)	Lithostatic Pres. (MPa)
610	41	7.18	13.79
914	48	10.78	20.68
1219	56	14.37	27.58
1433	61	16.88	32.41

A cavern or borehole in rock salt is typically cycled by allowing the pressure in the cavern or borehole to build and then reducing the pressure by bleeding off brine (see section 3.1). The pressure increase is the result of creep closure of the borehole with the wellhead closed. Pressure cycles were accurately measured for a year after borehole creation. The first few cycles given in Figure 2 are representative of the first year of pressure data for wells 106B, 107B and 109A. Piece-wise linear approximations of these curves were added to the brinehead pressure as a function of depth given in Table 3 and used to determine the pressure as a function of time on the borehole wall in the finite element analyses. The result is an accurate representation of the brine pressure on the borehole wall as a function of time. It is important to point out that the pressure variation with time was part of the input data for each calculation.

4.3 Finite Element Results

The results from the finite element calculations are typically stresses, strains and displacements. Radial closure versus time calculated at the four different depths are shown in part A of Figures 4, 5 and 6 for wells 106B, 107B and 109A respectively. It is interesting to note that radial closure essentially doubles for each 300 m of increased depth. The increase is logical because creep strain (equation 2) is an exponential function of temperature and a power function of effective stress, both of which increase linearly with depth.

Total borehole volume is calculated at any point in time using the instantaneous radius computed at the four different levels and the assumption of linear variation between levels. Volume versus time curves are shown in part B of Figures 4, 5 and 6. The fact that the radial closure increases significantly with depth indicates that the bottom portion of a borehole dominates the volumetric response. The total borehole volume is used along with the instantaneous fluid volume and the compressibility to compute the wellhead pressure. The computed and measured wellhead pressure versus time curves are shown in part C of Figures 4, 5 and 6. The initial (zero pressure) portion of each pressure curve represents the time between well completion and wellhead shut-in. The calculated wellhead pressure increases more slowly than the measured wellhead pressure for the three wells. At the highest pressure peaks the calculated pressure is below the measured pressure by a factor of approximately 2 or 3 for wells 106B and 107B and a factor of 1.5 for well 109A. Recall that 106B and 107B are gassy while 109A is not.

Figures 4, 5 and 6 have been arranged so that the influence of pressure changes, particularly large pressure drops due to bleed-down, can be seen in the calculated radial closure and volume curves. Large drops in pressure lead to temporary increases in radial and volumetric closure rates.

Additional finite element calculations were made to cover a longer period of time (2000 days). These calculations did

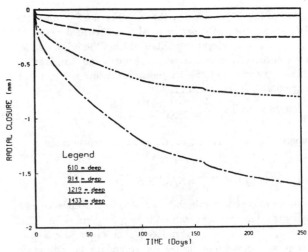

A. Radial Closure.

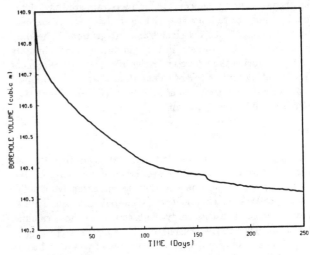

B. Volume.

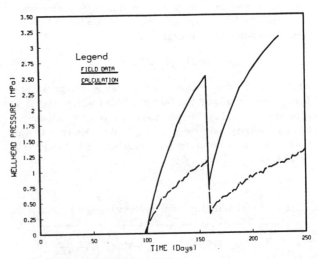

C. Wellhead brine pressure.

Figure 4. Finite element results for Big Hill Well 106B.

not include any pressure cycling of the borehole boundary conditions but were made at constant wellhead pressures of 0.0, 1.379 and 2.758 MPa. These pressures cover the operating range seen in the three wells. The field data used for comparison with these calculations came from all of the wells. The influence of wellhead pressure on the borehole volume is illustrated in Figure 7. Over a 22 month (669 day) test period the non-gassy wells (section 3.1) produced an average of 1.489 m³ of brine at an average pressure of 2.668 MPa. The finite element calculations over the same period of time and at approximately the same wellhead pressure (2.758 MPa) indicated a well volume reduction of 0.595 m³.

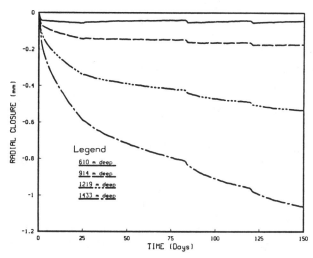

A. Radial Closure.

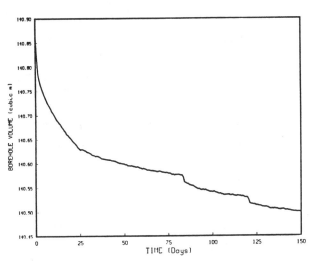

B. Volume.

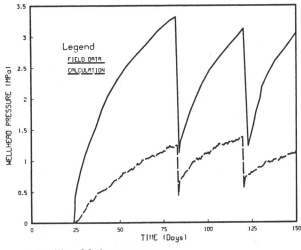

C. Wellhead brine pressure.

Figure 5. Finite element results for Big Hill Well 107B.

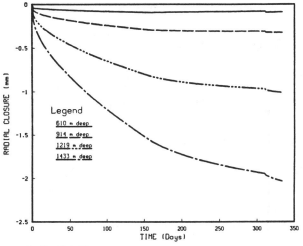

A. Radial Closure.

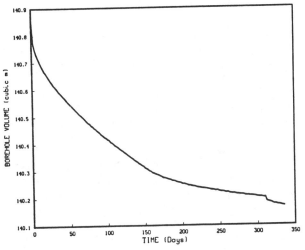

B. Volume.

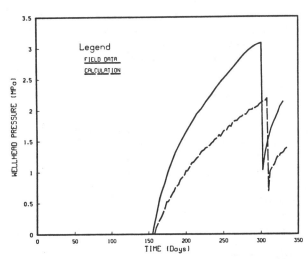

C. Wellhead brine pressure.

Figure 6. Finite element results for Big Hill Well 109A.

Thus, the finite element calculated volume is a factor of 2.5 less than the measured volume.

The volume change results of Figure 7 were also used to calculate pressure increase rates at times between 500 and 660 days. This period of time corresponds to the last six

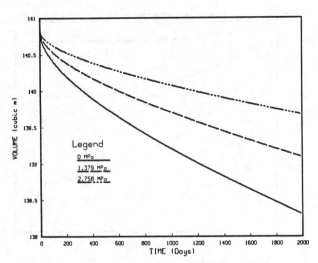

Figure 7. Calculated volumetric closure at brine pressures of 0.0, 1.379 and 2.758 MPa.

months during which pressures were measured. The calculations with 0.0, 1.379 and 2.758 MPa wellhead pressure gave pressure increase rates of 24.13, 13.10 and 8.96 KPa/day respectively. The average measured pressure increase rate in the non-gassy wells over the same period of time and at an average pressure of 2.668 MPa was 17.93 KPa/day. Thus, the measured pressure increase rate of 17.93 KPa/day is a factor of 2.0 higher than the calculated value of 8.96 KPa/day where the average wellhead pressures (2.668 and 2.758 MPa) were approximately the same. The factor of 2.0 difference between measured and calculated pressure increase rates is generally consistent with the factor of 2.5 difference in measured and calculated volume.

The caliper log results of Table 1 indicate a large range of reduction in the measured borehole radius over the 22-month test period. The average radius reduction indicated is 8.0 ±4.6 mm. The finite element results for the same time period at 2.758 MPa wellhead pressure gave a radius reduction of 2.03 mm, a factor of 4 lower than the measured value. It should be noted that this discrepancy is twice that obtained between measured and calculated pressure and volume. The most likely explanation for this difference is inaccuracies in the borehole caliper results. The standard deviation on the mean radial closure in Table 1 was half the average closure value. The caliper tool was used to measure values that are at the very edge of its accuracy limit and the measurments obtained, though useful, should not be taken as absolute. Morgan et al 1985 and Morgan et al 1986 also experienced underprediction of creep displacements in salt. They were calculating the creep closure of drifts in bedded rock salt at the WIPP site in southeastern New Mexico and observed a factor of approximately 3 between measured and calculated closure.

CONCLUSIONS

An experimental program was carried out with several Big Hill wells to determine creep behavior in situ. The quantities measured were wellhead pressure, bleed-down volume, and radial closure through caliper logging. Numerical calculations of borehole closure using a finite element program with a secondary creep model were also performed. The calculated wellhead pressure underpredicted the measured pressure by a factor of 2.0 and the measured volume loss was below that calculated by a factor of 2.5. The calculated

radial closure was a factor of 4 below the average measured with caliper logs. The caliper measurements are uncertain because the values measured are very close to the accuracy limit of the caliper tool. It was also determined that the bottom portion (approximately 300 m) of a borehole tends to dominate its volumetric response since the closure increases exponentially with depth.

BIBLIOGRAPHY

Beasley, R. R., K. L. Goin and D. S. Preece, 1986, Experimental and Theoretical Studies of Salt Creep Closure of the SPR Big Hill Site Wells 106 Through 110, SAND86-0190, Sandia National Laboratories, Albuquerque, NM.

Hart, R. J., T. S. Ortiz and T. R. Magorian, 1981, Strategic Petroleum Reserve (SPR) Geological Site Characterization Report Big Hill Salt Dome, SAND81-1045, Sandia National Laboratories, Albuquerque, NM.

Krieg, R. D.,C. M. Stone and S. W. Key, 1981, Comparisons of the Structural Behavior of Three Storage Room Designs for the WIPP Project, SAND80-1629, Sandia National Laboratories, Albuquerque, NM.

Krieg, R. D., 1983, Implementation of Creep Equations for Metal Into a Finite Element Computer Program, Computer Methods for Nonlinear Solids and Structural Mechanics, ASME, AMD-Vol 54, pp 133-144.

Morgan, H. S., C. M. Stone and R. D. Krieg, 1986, An Evaluation of WIPP Structural Modeling Capabilities Based on Comparisons With South Drift Data, SAND85-0323, Sandia National Laboratories, Albuquerque, NM.

Preece, D. S., 1986, Physical and Numerical Simulations of Fluid- Filled Cavities in a Creeping Material, PhD Dissertation, Civil Engineering Dept., University of New Mexico, Albuquerque, NM.

Price, R. H., 1986, Uniaxial and Triaxial Mechanical Experiments on Rock Salt Samples from Three Gulf Coast Domes, Geophysical Research Letters, October 1986, Vol. 13, No.10.

Prij, J and J. Mengelers, 1980, On the Derivation of a Creep Law From Isothermal Bore Hole Convergence, Stichting Energieonderzoek Centrum Nederland (ECN), ECN-80-169.

Stone, C. M., R. D. Krieg and Z. E. Beisinger, 1985, SANCHO - A Finite Element Computer Program for the Quasistatic Large Deformation, Inelastic Response of Two-Dimensional Solids, SAND84-2618, Sandia National Laboratories, Albuquerque, NM.

Wawersik, W. R. and D. H. Zeuch, 1984, Creep and Creep Modeling of Three Domal Salts - A Comprehensive Update, SAND84-0568, Sandia National Laboratories, Albuquerque, NM.

Wawersik, W. R., 1986, Creep Measurements and Microstructural Observations on Rock Salt From Big Hill, Texas, SAND86-2009, Sandia National Laboratories, Albuquerque, NM.

Hydraulic and mechanical properties of natural fractures in low permeability rock
Propriétés hydrauliques et mécaniques de fractures naturelles dans une roche peu perméable
Hydraulische und mechanische Eigenschaften der natürlichen Bruchflächen in einem wenig durchlässigen Fels

LAURA J.PYRAK-NOLTE, Earth Sciences Division, Lawrence Berkeley Laboratory, and Department of Materials Science and Mineral Engineering, University of California, Berkeley, USA
LARRY R.MYER, Earth Sciences Division, Lawrence Berkeley Laboratory, and Department of Materials Science and Mineral Engineering, University of California, Berkeley, USA
NEVILLE G.W.COOK, Earth Sciences Division, Lawrence Berkeley Laboratory, and Department of Materials Science and Mineral Engineering, University of California, Berkeley, USA
PAUL A. WITHERSPOON, Earth Sciences Division, Lawrence Berkeley Laboratory, and Department of Materials Science and Mineral Engineering, University of California, Berkeley, USA

ABSTRACT: The results of a comprehensive laboratory study of the mechanical displacement, permeability, and void geometry of single rock fractures in a quartz monzonite are summarized and analyzed. A metal-injection technique was developed that provided quantitative data on the precise geometry of the void spaces between the fracture surfaces and the areas of contact at different stresses. At effective stresses of less than 20 MPa fluid flow was proportional to the mean fracture aperture raised to a power greater than 3. As stress was increased, contact area was increased and void spaces become interconnected by small tortuous channels that constitute the principal impediment to fluid flow. At effective stresses higher than 20 MPa, the mean fracture aperture continued to diminish with increasing stress, but this had little effect on flow because the small tortuous flow channels deformed little with increasing stress.

RESUME: Les résultats d'une étude complète en laboratoire du déplacement mécanique, de la perméabilité, et de la géometrie des vides d'une fracture unique dans un quartz monzonite sont résumés et analysés. Une technique d'injection de métal est présentée, elle fournit des résultats quantitatifs sur l'exacte géométrie des vides compris entre les surfaces de la fracture et les zones d'aspérités en contact. Une augmentation de la contrainte effective entraine une diminution non linéaire de l'ouverture moyenne des vides, et une augmentation non linéaire de la surface de contact moyenne. Sous une contrainte effective de moins de 20 MPa, l'écoulement des fluides est proportionnel à une puissance supérieure à trois de l'ouverture moyenne de la fracture. L'augmentation de la contrainte provoque le changement de la surface de contact, les vides deviennent alors reliés par d'étroits canaux tortueux qui constituent le principal obstacle à l'écoulement. Sous une contrainte effective supérieure à 20 MPa, l'ouverture moyenne de la fracture continue de diminuer avec l'augmentation de contrainte, cela a peu d'influence sur l'écoulement parce que les canaux étroits et tortueux se déforment peu quand la contrainte augmente.

ZUSAMMENFASSUNG: Es werden die Ereignisse eines ausführlichen Laborversuchs sur Bestimmung der mechanischen Derschiebung, der Permeabilität und der Kluftgeometrie einselne Brüche in einem Quars monsonit zusammen gefasst und untersucht. Es wird ein Metalleinspritz-verfahren beschrieben, das in der Lage ist, quantitative Messwerte über die genaue Geometrie des Porenraums zwischen den Bruchflächen und unebenen Kontaktflächen zu liefern. Mit effektivem Druck nimmt die Poren-öffnung in nicht linearem Masse ab und die durchschnittliche Kontaktfläche in nicht linearem Masse zu. Bei effektiven Spannungen von weniger als 20 MPa fliesst die Flüssigkeit proportionel zur mittleren Poren-öffnung zu einem Exponenten grösser als drei. Mit ansteigendem Druck wird die Anzahl der Flächen in Kontakt verändert und Porenräume werden über kleine gewundene Kanäle verbunden, welche das Haupt hindernis für die Flüssigkeit bilden. Bei effektiven Spannungen grösser als 20 MPa verringert sich die mittlere Bruch öffnung weiter mit ansteigender Spannung, aber dies hat wenig Einfluss auf den Fluss, weil die gewundenen Kanäle nur wenig mit ansteigender Spannung verformt werden.

1 INTRODUCTION

Fractures, including joints and faults, are a major concern in the geologic isolation of nuclear wastes in low-permeability rock. Fractures are the principal conduits along which potentially contaminated groundwater can flow in rock masses with low permeability. Knowledge of the mechanical and hydraulic behavior of fractures in low-permeability rock masses is fundamental to the study of nuclear and toxic waste isolation, oil and gas recovery, and fault mechanics.

Fluid flow through fractures in low-permeability rocks depends on the state of stress in the rock mass. A fracture can be thought of as two surfaces in partial contact. When a fracture is stressed, the void space deforms and changes in contact area occur, affecting the hydraulic and mechanical properties of the rock. Several researchers have investigated the displacement of fractures and the increase in contact area as functions of applied normal stress. Goodman (1976) measured the deformation of fractured and whole samples and developed an empirical linear relationship between fracture displacement and the logarithm of effective stress. Goodman suggested that the nonlinear behavior of fracture permeability (which depends on aperture) under stress is accounted for by the nonlinear and inelastic behavior of a fracture under compression. Swan (1983) measured fracture surface topography and normal

stiffness in slate joints. For certain conditions and assumptions, he observed that hydraulic conductivity, normal stiffness, and true contact area are simple function of pressure and initial aperture and that the surface roughness properties appear to be irrelevant. Brown and Scholz (1985), however, determined from theoretical and experimental results that fracture displacement is not only a function of the elastic properties of the rock but also depends critically on the way in which the surface topography affects the distribution of contact area.

Measurements of fracture contact area have been made using sheets of pressure-sensitive paper (Duncan and Hancock 1966) and deformable film (Iwai 1976; Bandis et al. 1983). Both methods have shown that contact area increases with applied stress and is dependent on rock type. However, these methods are subject to considerable error, especially when the apertures are smaller than the thickness of the sheet.

Several investigations of fluid flow through single fractures have been carried out in the laboratory under varying conditions. Iwai (1976) studied flow through single induced fractures as functions of displacement, contact area, and stress up to 20 MPa. He found that there is always a residual flow, even at high stresses, and that the cube of the aperture is proportional to flow when corrected by the amount of the residual flow. From an experiment in which the contact area was varied artificially, Iwai determined that flow decreases hyperbolically as the contact area increases. For real fractures, however, he observed that contact area has a small effect on the flow for some samples but a greater effect in other samples, especially those with apertures less than 15 μm. Gale and Raven (1980) studied the effect of stress on radial fluid flow through natural fractures for samples of various sizes and stresses up to 30 MPa. Finding that their results did not follow the cubic law model, they suggested that the change in contact area with stress plays a major role in decreasing the flow. Kranz et al. (1979) measured the permeability of artificial and induced joints in granite for stresses up to 200 MPa. They observed that permeability is not directly related to effective stress but is a function of confining pressure and pore pressure, each multiplied by a pre-factor that depends on surface roughness of the joint and on ambient pressure. Engelder and Scholz (1981) studied fluid flow along artificial fractures using effective stresses up to 200 MPa. They found that changes in confining pressure have a larger influence than changes in pore pressure and that their data agree well with a "cubic" law modified for variable cross section as a result of applying stress. Walsh (1981) derived a relationship for fluid flow as a function of confining pressure and pore pressure. He concluded that the flow rate is proportional to the product of two factors: tortuosity and aperture. Examining the data of Kranz et al., he observed that the effect of tortuosity could be neglected for fluid flow, since aperture is raised to the third power and dominates the tortuosity term, which is only raised to the first power. In contrast, a theoretical study by Tsang (1984) concludes that tortuosity and surface roughness greatly affect the flow, especially when contact area is greater than 30%.

This brief review points out several inconsistencies, such as the applicability of the cubic law and the effect of void geometry (contact area, tortuosity) on flow. This paper reports the results of experiments and analyses of the hydraulic and mechanical properties of natural fractures as a function of stress and discusses the relationship of these properties to changes in the fracture contact area and void geometry. Hydraulic and mechanical properties of three fractures were measured at effective normal stresses up to 85 MPa. Using a nondestructive metal-injection technique, actual casts of fracture void space were obtained at stresses up to 85 MPa, providing quantitative data on the changes in fracture contact area and void geometry with increasing stress.

2 EXPERIMENTAL PROCEDURE

Tests were performed on three core samples (E30, E32, and E35) of quartz monzonite (Stripa granite) measuring 52 mm in diameter by 77 mm in height. Each sample contained a single natural fracture orthogonal to the long axis of the core. Hydraulic conductivity and mechanical displacement were measured for each of these samples, and changes in fracture void geometry and contact area were evaluated for samples E30 and E32.

A linear flow technique (quadrant flow) was used to measure hydraulic conductivity. In this technique the intersection of the fracture plane with the circumference of the sample is sealed everywhere except along two diametrically opposed quadrants, as shown schematically in Figure 1. A manifold placed on the sample permits fluid to flow through the fracture from one unsealed quadrant to the other. This technique was employed because of sample size; if a radial flow method had been used, the central borehole would have been very small and the results would have reflected the loss of energy predominantly due to flow adjacent to the borehole rather than the loss of energy due to flow through the whole fracture.

A constant effective stress was maintained in the fracture by applying an axial load to the sample in a test machine. An upstream pressure of 0.4 MPa was applied, and flow measurements were made for one complete loading and unloading cycle and an additional loading cycle up to a maximum effective stress of 85 MPa.

Fracture displacement was also measured on the same samples used for fluid flow measurement. Three annular collars were attached to the sample. The inside surface of

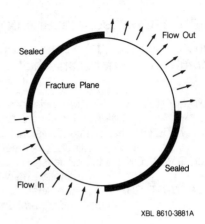

Figure 1. Flow path in quadrant flow technique.

the annulus did not touch the rock surface; each collar was secured to the rock by three pointed screws. The collars were carefully placed to ensure that they were equally separated and parallel. The assembly was placed in the test machine, where axial stresses up to 85 MPa were applied normal to the fracture plane. Four precision linear variable differential transformers (LVDTs, rated repeatability of 1.0×10^{-7} m) were attached in pairs on diametrically opposed sides of the sample. One set of LVDT's spanned the fracture and measured the displacement across both the fracture and the intact rock adjacent to the fracture. The second set measured the displacement of an equal length of the intact rock. The fracture displacement was calculated from the difference between the displacements measured by the first set of transducers and those of the second set. Performance of the measurement system was evaluated by duplicating the experiment using a solid aluminum cylinder of the same dimensions as the rock sample.

A metal injection technique was used to study the fracture void geometry and contact area as a function of stress. The metal used for injection is one of a family of bismuth-lead-tin alloys of which Wood's metal is the most commonly recognized. In the liquid phase, these metals are nonwetting with an effective surface tension of roughly six-tenths that of mercury (Swanson 1979), namely, 0.282 N/m for the Wood's metal. The alloy (Cerrosafe) used in these experiments has a melting point of 160 °F to 190 °F. The Wood's-metal injection technique is similar to mercury porosimetry methods, but it has the advantages of yielding actual metal casts of the voids for the same fracture in experiments at different stresses. These casts can then be studied in detail as described below. Wood's-metal injection techniques have been used by other investigators (Swanson 1979; Yadav et al. 1984) to study the pore geometry of sandstone cores.

For the study of fracture void geometry, the fractured sample was held in a triaxial test vessel maintained at a temperature just above the melting point of the alloy. The vessel was placed in the test machine and an axial load applied normal to the fracture surface. To perform an injection test, the triaxial vessel was evacuated and molten metal pumped into the test vessel until the desired pore fluid pressure was obtained. Both the fluid pressure and the axial load were maintained until the metal in the test vessel had solidified. When the sample was removed from the vessel, the two halves of the sample were separated to reveal metal casts of the void geometry corresponding to the effective stress of the test. Some casts adhere to one surface, some to the other.

The distribution of metal on the two fracture surfaces was examined using both a scanning electron microscope (SEM) and photographic techniques. Images of each of the two fracture surfaces were superimposed to form a composite image of the contact area and void geometry. A Zeiss image analyzer was used in quantitatively evaluating the composite images.

Injections were made on samples E30 and E32 at effective stresses of 3, 33, and 85 MPa.

3 RESULTS

Fracture displacements for two complete loading/unloading cycles for one sample (E30) are shown in Figure 2. Though nonlinear, there was very little hysteresis and, within the scatter of the results, no difference in displacement between loading cycles. Deformation of asperities and voids within the fractures must therefore have been elastic. It should be noted that the fracture had been loaded to 85 MPa on several occasions for different tests prior to collection of the data shown in Figure 2. Typically, some hysteresis is observed in the initial loading cycle for natural fractures. However, upon further cycling, hysteresis decreases, and hence so does the difference in displacement between cycles (Bandis et al. 1983; Gale and Raven 1980).

Figure 3 shows fracture displacement versus applied normal stress for the three different fractures, E30, E32, and E35. Though the magnitude of the total displacement varies for the different samples, all the samples exhibit a rapid increase in displacement up to about 10 MPa. Above 10 MPa, the increase in fracture displacement becomes more gradual but does not reach zero, even at 85 MPa. This suggests that even at the highest stresses, numbers of voids must still have been open in all these fractures, least of all for E32 and most of all for E35. The maximum fracture displacement values measured at 85 MPa for samples E30, E32, and E35 are listed in Table 1.

Table 1. Values of experimentally determined displacement at 85 MPa and maximum displacement, d_{max}, determined from equation (1).

Specimen	Displacement at 85 MPa	d_{max}
E32	4.5 μm	6.6 μm
E30	9.5 μm	12.5 μm
E35	28.1 μm	46.0 μm

The inverse of the tangent slopes to the displacement-versus-stress curve is defined as the specific stiffness of the fracture. Values of specific stiffness for these fractures plotted in Figure 4, were determined graphically from the data in Figure 3.

Specific stiffness for all of the samples is nonlinear with stress, in disagreement with the linear relationship assumed by Walsh (1981). Specific stiffness increases rapidly with stress up to about 10 MPa and then increases at a decreasing rate. Though the trend is less pronounced for sample E30, the specific stiffness for these fractures appears to approach a constant value at high stress levels. Ultimately, of course, when all the void spaces between the fracture surfaces close, the specific stiffness must become infinite for each fracture. However, the stresses needed to achieve complete closure would have to be very high.

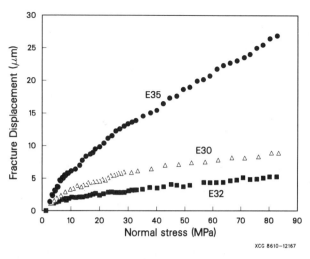

Figure 3. Fracture displacements as a function of normal stress on fracture for samples E35, E30, and E32.

Fracture displacement tests have also been performed on E30 under dry conditions at 100 °C and under saturated conditions at room temperature and at 95 °C. No significant difference in displacement behavior is observed. The difference between the best-fit curves for tests under each of these conditions and that for the test at room-temperature dry conditions is less than one micron at any stress level.

Figure 5 compares fluid flow data for all three samples. Data for E30 and E32 are obtained from the first unloading cycle. Though the magnitude of the flow varies between loading and unloading cycles, the trend of a rapid decrease in flow at low stresses and a much more gradual decrease in flow at higher stresses is observed to be independent of the load cycle. Similar behavior has also been observed in flow measurements made by other investigators (Iwai 1976; Engelder and Scholz 1981; Raven and Gale 1985). Data presented for E35 are obtained from the first loading cycle because that set is the most complete. For samples E30 and E32, a rapid decrease in flow is observed as stress is increased to about 20 MPa. This change in flow corresponds to the rapid increase in fracture closure observed in the displacement measurements. Flow in sample E35 continues to decrease with increasing stress above 20 MPa, whereas the flow for E30 and E32 appears to approach a constant value. The small but finite flow in samples E30 and E32, even at 85 MPa, is significant in that it suggests that flow approaches an irreducible level at high stresses. At these high stresses, mechanical displacement of the fracture continues as demonstrated by the finite values

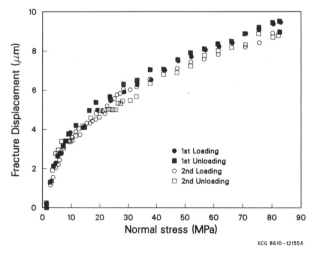

Figure 2. Comparison of fracture displacements for two sequential loading/unloading cycles for sample E30.

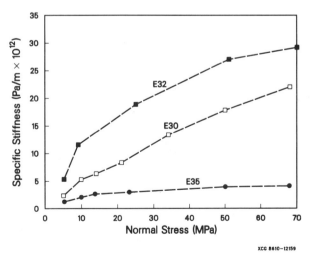

Figure 4. Specific stiffness for the three mechanical displacement tests shown in Figure 3.

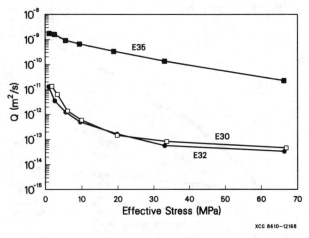

Figure 5. Comparison of flow per unit head drop as a function of effective stress for samples E35, E30, and E32.

of specific stiffness (Figure 4). However, this deformation, which must be a result of void closure, has very little effect on the fluid flow for samples E30 and E32. Iwai (1976), and Raven and Gale (1985) have also observed irreducible flow at the highest stresses, even though fracture displacement continued.

In Figure 6, the flow test data are plotted in the conventional format of the logarithm of average fracture displacement versus logarithm of flow. The effect of the irreducible flow is seen as a departure from linearity in this plot. At the lowest values of flow, mechanical displacement continues with virtually no change in the flow. The slope of the linear portion of the curve is not equal to one-third as would be expected if the flow followed the commonly accepted "cubic" law relationship derived for a parallel plate model.

In the analysis of fracture displacement and fluid flow data, a cubic relation is not assumed. Instead, the model consists of a general power relationship between fracture closure and fluid flow as well as a constant term representing the irreducible flow:

$$Q = Q_\infty + C (d_{max}-d)^n \qquad (1)$$

where:
Q = measured value of flow
Q_∞ = irreducible flow
C = fitted constant
d_{max} = maximum value of displacement
d = measured value of displacement
n = power relation.

A linear least-squares (Press et al. 1986) fitting routine applied to the logarithm of these quantities in equation (1) is used to determine Q_∞, d_{max}, n and C.

Table 1 lists the measured values of displacement at a stress of 85 MPa and the values of the fitted maximum displacement, d_{max}. Physically, the parameter d_{max} corresponds to the fracture displacement which would result if stress were increased until all voids in the fracture closed. Though obtained by least-squares fitting to flow data, values of d_{max} are plausible when compared to the mechanically measured values at 85 MPa and the slope of the displacement-versus-stress curves. For example, it is plausible that E32, the stiffest fracture, would close to a maximum of 6.6 μm if sufficient stresses were applied, whereas E35, the most compliant fracture could close an additional 20 μm.

When equation (1) is applied to the data, the result, as shown in Figure 7, is a power law relationship between flow and displacement. However, the values of the exponents in equation (1) of 8.3, 9.8, and 7.6, for E30, E32, and E35, respectively, differ greatly from a cubic law representation.

Figure 8 shows composite images made from SEM micrographs of the Wood's metal in the fracture voids of sample E30 at effective stresses of 3, 33, and 85 MPa. Areas of contact are depicted as black areas and the flowpaths are in white. At 3 MPa (Figure 8a), the contact area appears as isolated "islands" in "oceans" of metal. The void area in the fracture is much greater than the contact area and is freely interconnected. The islands of contact area range in size from 5 μm to 0.1 mm. As the stress increases (Figure 8b & 8c), the "islands" of contact areas become "continents," with "lakes" of metal connected by filamentary, tortuous "streams" of metal. Although the lakes of metal may be measured on a scale of millimeters, many of the streams are measured in micrometers. The tortuosity of the interconnected flowpaths (white) is quite noticeable in Figure 8c.

Figure 9 shows composite images of the Wood's metal between the fracture surfaces for samples E30 and E32 for effective stresses of 3, 33, and 85 MPa. Each image in the figure is a composite of two photographs, one of each fracture surface. Areas of contact are white; flowpaths or voids are black. A difference in the shape, size, and distribution of contact area is observed between E30 and E32 at all stresses. Sample E30 possesses many large areas of contact, with more and more smaller areas of contact being generated near these large areas as the stress is increased. The contact areas of E30 tend to be large and clustered, whereas those of E32 are smaller, more numerous, and elongate.

An image analyzer was used to obtain percent contact area at different stress levels for E30 and E32 from the composite images shown in Figure 9. Results are plotted in Figure 10. At 3 MPa, the contact area of sample E30 is about 8%; it increases to about 15% at 33 MPa and to 30% at 85 MPa. For sample E32, the contact area was higher at all stress levels, beginning at about 15% at 3 MPa, increasing to 42% at 33 MPa, and remaining essentially constant at 85 MPa. Greenwood and Williamson (1966) postulated that contact area is porportional to load.

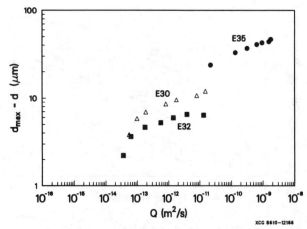

Figure 6. Fracture displacement versus flow per unit head drop for samples E35, E30, and E32.

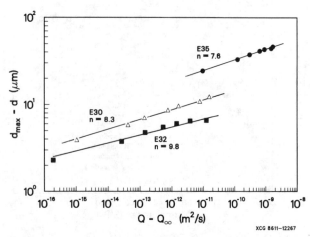

Figure 7. Fracture displacement versus flow per unit head drop after subtraction of irreducible flow for samples E35, E30, and E32.

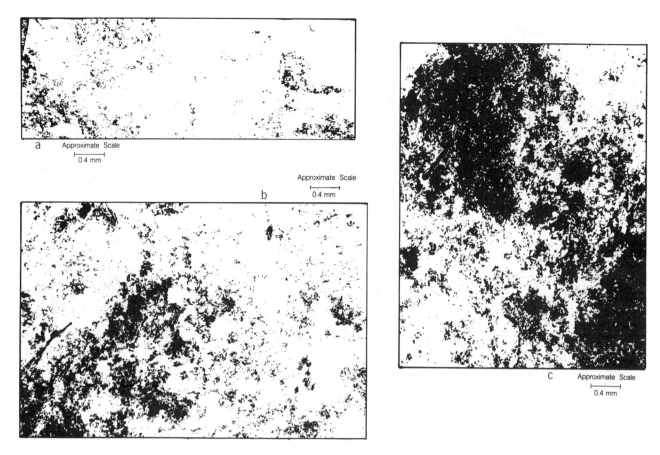

XBL 8612-4934

Figure 8. Compositve from micrographs of portion of fracture surfaces of sample E30 at three effective stresses: (a) 3 MPa; (b) 33 MPa; (c) 85 MPa (white regions are metal).

Although E30 shows a more or less linear relationship between stress and contact area, E32 does not; our experimental results show a nonlinear relation between specific stiffness and stress.

The increase in contact area with stress for sample E30 correlates well with observed increases in stiffness with stress. Though the stiffness of sample E32 gradually increases at higher stresses, the contact area appears to be almost constant. This suggests that the gradual increase in stiffness is due to contact at many small points not discernible on the large composite images (Figure 9).

4 DISCUSSION

Wood's metal injection tests, in combination with mechanical and hydrologic measurements, have provided clear indications of how changes in these properties with stress are related to changes in contact area and void geometry.

At low normal stresses the mechanical behavior of the fractures is characterized by nonlinear displacement and rapidly changing specific stiffness and contact area. Though the observed mechanical response could be attributed to nonlinear elastic properties, there is no evidence to suggest that hard rocks exhibit significant nonlinear elasticity. The rapid increase in specific stiffness must be caused primarily by an increase in the number of asperities coming into contact. The higher specific stiffness values for E32 than in E30 suggest a higher contact area, which is confirmed by the injection tests.

At higher stresses in samples E32 and E35, the nearly constant values of specific stiffness reflect continued elastic deformation of voids, with negligible changes in geometry or contact area. Therefore, the voids that remain open at high stresses must have aspect ratios smaller than those that would close at a stress of 85 MPa. The lower stiffness

of sample E35 compared with that of the other samples is probably a result of the presence of larger voids.

For laminar flow between parallel plates, a cubic relationship exists between flow and aperture. The flow in our experiments is determined to be laminar, both from estimations of the Reynolds number and from flow experiments in which the viscosity of the fluid was changed from 1.0 cP through 1000 cP. In these experiments, flow varied inversely as the viscosity to within a few percent. Tsang (1984) determined that flow will depend on the number of small apertures and that for contact area greater than 30%, flow is greatly affected by tortuosity.

If a parallel plate model is assumed to represent the fluid flow through the fracture, the large values of n (exponent in equation (1)) determined from our experiments suggest a nonlinear relationship between mechanical displacement and hydraulic aperture. Figure 11 presents the changes in hydraulic aperture, b, calculated from the cubic law and the measured values of the mechanical displacement, d.

At low stresses, the effective hydraulic aperture changes more rapidly than is reflected by the mechanical displacement, whereas at high stresses relatively large mechanical displacements occur with little change in hydraulic aperture. This implies that at low stresses the voids through which the fluid flows close preferentially. At high stresses, the principal impediments to fluid flow must consist of tubes or orifices with equidimensional cross section that do not diminish significantly with increasing stress. Moreover, there must be significant areas of voids with apertures large enough to accommodate the mechanical displacement up to 85 MPa without closing. Presumably, these voids must be very poorly interconnected hydraulically, or else they would constitute relatively free flow paths for the fluid. Evidence to support this interpretation was provided by the metal-injection tests. At low stresses, contact area was isolated (Figure 8a) so that fluid flowed through large void areas,

229

Approximate Scale

|—————|
10 mm

XBL 8612-4933

Figure 9. Composite photographs of entire fracture surface of sample E30 (upper row) and E32 (lower row) at three effective stresses: (a) 3 MPa; (b) 33 MPa; (c) 85 MPa (dark regions are metal).

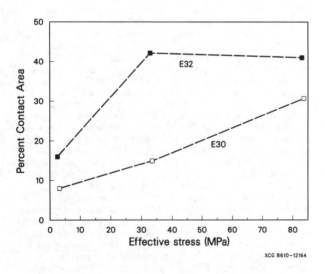

XCG 8610-12164

Figure 10. Contact area as a function of effective normal stress for samples E30 and E32; results of image analysis of photographs in Figure 9.

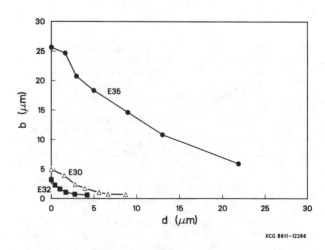

XCG 8611-12266

Figure 11. Calculated value of hydraulic aperture versus measured mechanical aperture for samples E35, E30, and E32.

230

which would also be subject to mechanical closure. At high stresses, large void areas still remained (Figure 8c) to accommodate further mechanical closure. However, the contact area was distributed such that the hydraulic connections between the voids were limited to small tortuous channels.

5 CONCLUSIONS

The hydraulic and mechanical properties of low-permeability rock are dominated by fractures. A comprehensive study of the mechanical and hydraulic behavior of natural fractures in crystalline rock under normal stress has been performed. From measurements of the mechanical displacement of a fracture, contact area of a fracture, and fluid flow through a fracture as a function stress, a clear, though qualitative, indication is given of how changes in asperity contact and void geometry affect both the mechanical and hydraulic behavior of a fracture. These results illustrate the important influence of changes in fracture geometry with increasing stress; i.e., changes in size and distribution of asperities and voids.

The cubic law relating the fluid flow through fractures in rock to the mean aperture between the fracture surfaces has been proposed on theoretical grounds. Most of the experiments that have substantiated this cubic relationship have been done either on surfaces prepared by machine or artificially induced tensile fractures. In the former case, the variability in the actual aperture because of the imperfectly planar nature of the surfaces can be expected to be small. In the latter case, the topographies of the two surfaces may be quite pronounced but the two surfaces register almost exactly, so that the apertures are everywhere very similar despite their tortuosity in the out-of-plane direction.

Natural rock fractures that have been subjected to any degree of shear motion or alteration are not only far from perfectly planar but the topographies of the opposing surfaces fail to register. Therefore, contact between natural fracture surfaces will occur at asperities, corresponding to mutual topographic highs. Surrounding the asperities will be voids, corresponding to mutual lows. Both asperities and the adjacent voids deform under increasing stress, increasing the numbers of asperities in contact and diminishing the volumes of the voids. It is these geometrical changes that result in the nonlinear displacement of a fracture with stress and the change in specific stiffness with stress. These changes can also be expected to change the apertures through which fluid flows.

The experiments reported here show that the cubic relationship between fixed flow and mechanical fracture displacement does not hold at either high or low stresses for natural fractures, as has also been found by Raven and Gale.

It is clear that the observed behavior of natural fracture surfaces must be the result of the complex relationship between the geometries of asperities in contact and the adjacent voids. This geometry is a result of the interaction between the topographies of the two fracture surfaces and their deformation under stress. This results in highly non-linear changes in the plan areas of the voids, their apertures, and the way in which they are interconnected. Accordingly, it will be necessary to study the detailed distribution of the apertures of voids and their interconnections at different values of applied normal stress.

6 ACKNOWLEDGEMENTS

This work was supported by the Office of Civilian Radioactive Waste Management through the Office of Geologic Repositories and the Crystalline Repository Program and by the Director, Office of Basic Energy Sciences, Division of Engineering, Mathematics and Geosciences of the U.S. Department of Energy under Contract No. DE-AC03-76SF00098.

7 REFERENCES

Bandis, S.C., A.C. Lumsden and N.R. Barton 1983. Fundamentals of rock joint deformation. Int. J. Rock Mech. Min. Sci. Geomech. Abstr. 20(6):249—268.

Brown, S.R. and C.H. Scholz 1985. Closure of random elastic surfaces in contact. J. Geophys. Res. 90(B7):5531—5545.

Duncan, N. and K.E. Hancock 1966. The concept of contact stress in assessment of the behavior of rock masses as structural foundations. Proc. First Cong., Int. Soc. Rock Mech., Lisbon 2:487—492.

Engelder, T. and C.H. Scholz 1981. Fluid flow along very smooth joints at effective pressures up to 200 megapascals. In Mechanical behavior of crustal rocks. Am. Geophys. Union Monogr. 24:147—152.

Gale, J.E. and K.G. Raven 1980. Effects of sample size on the stress-permeability relationship for natural fractures. Lawrence Berkeley Laboratory Report, LBL-11865 (SAC-48), Berkeley, California.

Goodman, R.E. 1976. Methods of geological engineering in discontinuous rocks, p. 172. New York: West Publishing.

Greenwood, J.A. and J.B.P. Williamson 1966. Contact of nominally flat surfaces. Proc. R. Soc. London A295.

Iwai, K. 1976. Fundamentals of fluid flow through a single fracture. Ph.D. thesis, 280 p. Univ. of Calif., Berkeley.

Kranz, R.L., A.D. Frankel, T. Engelder and C.H. Scholz 1979. The permeability of whole and jointed Barre granite. Int. J. Rock Mech. Min. Sci. Geomech. Abstr. 16:225—234.

Press, W.H., B.P. Flannery, S.A. Teukolsky and W.T. Vettering 1986. Numerical recipes, p. 509. New York: Cambridge Univ. Press.

Raven, K.G. and J.E. Gale 1985. Water flow in a natural rock fracture as a function of stress and sample size. Int. J. Rock Mech. Min. Sci. Geomech. Abstr. 22(4):251—261.

Swan, G. 1983. Determination of stiffness and other joint properties from roughness measurements. Rock Mech. Rock Eng. 16:19—38.

Swanson, B.F. 1979. Visualizing pores and non-wetting phase in porous rock. J. Petrol. Tech. 31:10—18.

Tsang, Y.W. 1984. The effect of tortuosity on fluid flow through a single fracture. Water Resour. Res. 20(9):1209—1215.

Walsh, J.B. 1981. Effect of pore pressure and confining pressure on fracture permeability. Int. J. Rock Mech. Min. Sci. Geomech. Abstr. 18:429—435.

Yadav, G.D., F.A.L. Dullien, I. Chatzis and I.F. MacDonald 1984. Microscopic distribution of wetting and non-wetting phases in sandstones during immiscible displacements. SPE 13212, Society of Petroleum Engineers of AIME, Dallas, Texas.

Measurement and analysis of laboratory strength and deformability characteristics of schistose rock

Mesure et analyse en laboratoire de la résistance et de la déformation des roches schisteuses
Messung und Deutung der im Labor bestimmten Druckfestigkeit und Deformierbarkeit schiefriger Gesteine

S.A.L.READ, New Zealand Geological Survey, Lower Hutt
N.D.PERRIN, New Zealand Geological Survey, Lower Hutt
I.R.BROWN, Ian R.Brown Associates Ltd, Wellington, New Zealand

ABSTRACT: The results from laboratory testing of anisotropic schistose rocks to determine their uniaxial compressive strengths are influenced by the orientation of the foliation with respect to the direction of load application. The tangent moduli of elasticity values determined from strain gauged uniaxial tests are not greatly affected by either foliation orientation or by the positioning of the deformation measuring points relative to it. However, Poisson's ratio values appear to be affected by these factors. Recommendations are made for the placing of strain gauges in positions which are different from those in published test methods. The test results are applied to a model of schist as a transversely isotropic rock defined by five independent elastic constants.

RESUME: Les résultats des tests effectués en laboratoire sur des roches schisteuses anisotropes pour déterminer leur résistance à la compression uniaxiale sont influencés par l'orientation de la foliation par rapport à la direction d'application de la charge. Les valeurs tangentes des modules d'élasticité, déterminées par des tensiomètres travaillant en contrainte uniaxiale, ne sont influencées de manière sensible ni par l'orientation de la foliation, ni par la position des points de mesure de déformation par rapport à la foliation. Cependant, les valeurs du coefficient de Poisson semblent influencées par ces deux facteurs. Des recommandations sont faites pour le placement des tensiomètres, qui diffèrent de celles habituellment publiées dans les méthodes de test. Les résultats des tests sont appliqués à un modèle selon lequel les schists sont considérés comme des roches transversalement isotropes pouvant être définies par cinq constantes élastiques indépendantes.

ZUSAMMENFASSUNG: Die Resultate von Experimenten an anisotropen schiefrigen Gesteinen zur Bestimmung der einachsigen Druckfestigkeit hängen von der Orientierung der Schieferung relativ zur Druckrichtung ab. Die Tangentialmodule von Elastizitätswerten, die auf einachsigen Dehnungsmessungen beruhen, werden weder von der Orientierung der Schieferung, noch von der Lage den Messpunkte relativ dazu, stark beeinflusst. Dagegen scheinen die Werte der Querzahl von diesen Faktoren abzuhängen. Es wird empfohlen die Dehnungsmessstreifen anders als üblichwerweise zu plazieren. Unsere Resultate werden im Lichte eines transvers-isotropen Modellgesteins interpretiert, das durch fünf unabhängige elastische Konstanten bestimmt ist.

INTRODUCTION

The unconfined uniaxial compressive strength (q_u) of rock is one of the three major strength parameters which can be determined in the rock mechanics laboratory, and is probably the most widely quoted parameter in rock mechanics (Hoek 1983). Although useful as an index to compare different rock types, the q_u value on its own is of little value to a geotechnical design engineer, unless deformation characteristics are measured during the test and other strength parameters (i.e. tensile or shear strength) are determined in an accompaning testing programme. The modulus of elasticity, or Young's modulus (E) and Poisson's ratio (ν) may be determined from strain measurements made axially and circumferentially in an unconfined compression test.

Several test methods are used to determine q_u, E and ν including International Society for Rock Mechanics (ISRM) (Bieniawski co-ord 1978), Canada Centre for Mineral and Energy Technology (CANMET) (Gyenge & Herget 1977) and American Society for Testing and Materials (ASTM) (1980). In these methods the modulus of elasticity, which is the slope of the plot of axial stress against axial strain, may be determined in three ways: average, tangent and secant. The tangent modulus of elasticity E_t (at 50% of the ultimate strength, q_u) is the most commonly used and is adopted in this paper. In the test methods Poisson's ratio (ν) is found either by dividing the axial tangent modulus (at 50% of q_u) by the circumferential tangent modulus at the same stress level, or alternatively

from the gradient of the plot of circumferential strain against axial strain at 50% of q_u. For the testing of hard rocks (i.e. $q_u > 50$ MPa) strain is generally measured using electrical resistance strain gauges mounted at the mid-height of the specimen.

The recommended positions of the strain gauges vary between the methods: CANMET recommends a pair of axial gauges mounted diametrically opposite one another, and a pair of circumferential gauges mounted diametrically opposite one another and at 90° around the specimen circumference from the axial gauges. ASTM recommends the use of at least two axial gauges equally spaced and at least two circumferential gauges equally spaced. ISRM recommends two axial gauges and two circumferential gauges equally spaced.

All three methods call for the average of the strain measured by the axial and the circumferential gauges to be determined. Whether this is achieved by connecting the gauges in series to directly record an average or by calculation using the outputs of the individual gauges is not explicitly stated.

Tests performed in the New Zealand Geological Survey Geomechanics Laboratory to determine E_t and ν of homogeneous hard sandstone ("greywacke") using pairs of gauges connected in series and mounted opposite each other produced consistent results (Figure 1). Elastic behaviour was indicated by the linear plots of circumferential strain (ε_c) against axial strain (ε_a) (Figure 1b) with a Poisson's ratio value of 0.27. However, similar tests carried out on foliated anisotropic schist did not give consistent

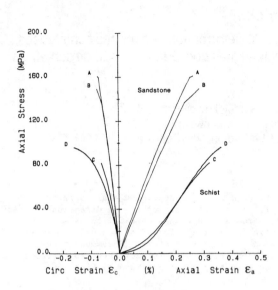

Fig. 1a. Axial stress against axial and
circumferential strain

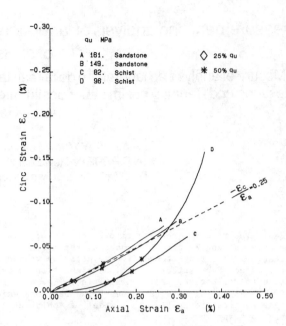

Fig. 1b. Circumferential strain against axial strain

Figure 1. Uniaxial compression test results for "greywacke" sandstone and schist

results. The plots of circumferential strain against axial strain were often non-linear, with Poisson's ratio values at 50% q_u commonly greater than 0.30 (Figure 1b).

The results of a testing programme to investigate the effects of the orientation of foliation and strain gauge position on the derived values of tangent modulus of elasticity and Poisson's ratio in schistose rocks are given in this paper. The implications of the anisotropy for modelling are discussed.

The rock samples tested are from the Haast Schist Group of the schist belt of New Zealand and were collected from the Central Otago region (Figure 2). The samples are textural zones III and IV schists which generally exhibit a well developed segregation foliation (Suggate et al. 1978). For the purposes of this paper the samples are considered to be of a similar lithology.

SPECIMEN PREPARATION AND TESTING

The samples were taken either from drillholes or from blocks of rock cored in the laboratory at selected orientations. All cores were HQ size (62mm nominal diameter) and were prepared to give specimens with a nominal length to diameter ratio of 2.5:1.

All specimens were loaded to failure in a 120 tonne compression frame. Axial and circumferential deformations were measured using two 30 mm strain gauges mounted axially and two 10 mm strain gauges mounted circumferentially. Each gauge was monitored individually using digital strain indicators in half-bridge configuration, with a dummy gauge mounted on an identical core of schist for temperature compensation (Read et al 1986). Specimen preparation and loading were performed in accordance with the ISRM recommended method.

The strain gauge configuration used is shown on Figure 3. Circumferential gauges were mounted diametrically opposite axial gauges with each mounting position being either in the direction of the strike or the dip of foliation. This configuration results in one axial and one circumferential gauge being mounted in each of the "strike" and "dip" positions enabling the comparison of E_t and ν values determined for each position. The angle of dip of foliation from horizontal (α) shown on Figure 3 differs from the angle of inclination from the core axis (β) that is used by other authors (e.g. Hoek & Brown 1980) where $\alpha = 90 - \beta$.

Figure 2. Location map

234

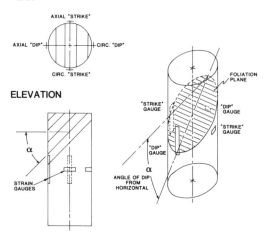

PLAN

ELEVATION

Figure 3. Strain gauge configuration with respect
to foliation attitude

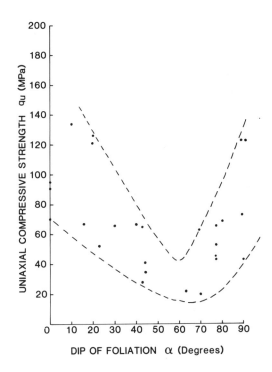

Figure 4. Uniaxial compressive strength against
dip of foliation

RESULTS

Uniaxial Compressive Strength

The uniaxial compressive strengths (q_u) determined
are plotted against the dip of foliation on Figure
4. At shallow angles of dip (0-20°) the strengths
are high and failures were generally not influenced
by foliation but occurred diagonally or axially
across it. With increasing angles of dip the
strengths decrease reaching a minimum in specimens
with dips of 65-70° where the foliation forms an
end-to-end diagonal plane through the specimen.
Failures were along the foliation planes, which
"daylighted" along the core sides. Specimens with
dips greater than 70° show higher strengths as the
foliation planes are constrained (i.e. do not
"daylight" on the sides) between the testing machine
platens. Failure was axial and parallel to
foliation.

This variation in strength with dip of foliation
has been well documented e.g. Hoek & Brown (1980)
and Donath (1972) and for this reason is not
discussed further.

Modulus of Elasticity (E_t)

Of the tangent modulus of elasticity (E_t) values
determined, the majority of the "strike" and "dip"
position values agree to within ± 10%. The averaged
values are plotted against dip of foliation on
Figure 5. Little variation is apparent for angles
of dip of foliation between 0 and 70°. The values
parallel to foliation (i.e. α = 80-90°) are about 1½
times greater than those perpendicular to foliation,
thus reflecting the anisotropy of the schist.

For dips up to 70° the change in mode of failure
from across foliation to along the foliation plane
appears to have little influence on the E_t values.
For dips greater than 70° higher values occur as the
foliation is constrained between the machine
platens.

The test results show that the E_t values
determined do not appear to be influenced by the
position of the axial gauges with respect to
foliation.

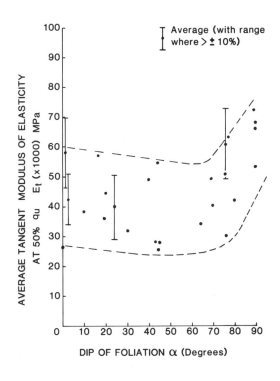

Figure 5. Tangent modulus of elasticity against
dip of foliation

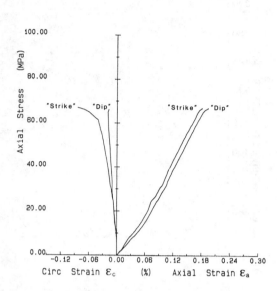

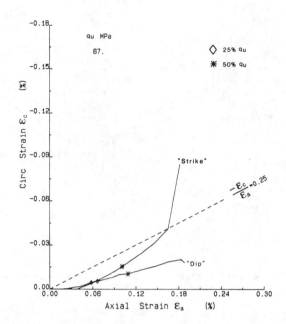

Fig. 6a. Axial stress against axial and
circumferential strain

Fig. 6b. Circumferential strain against axial strain

Figure 6. Uniaxial compression test result for schist. (Dip of foliation 40°)

Poisson's Ratio

Many of the Poisson's ratio (ν) values that were
determined show marked differences depending on
whether they were measured in the "dip" or "strike"
positions. Because the axial modulus values are not
greatly influenced by the gauge positions,
variations in Poisson's ratio values are therefore
inferred to be due to the different deformations
measured by the two circumferential gauges.

The method considered most appropriate to assess
the Poisson's ratio values is to plot
circumferential strain against axial strain. For
elastic behaviour (such as that shown by sandstone
in Figure 1b) the plot is linear with a constant
gradient. Anomalous or inelastic behaviour can be
seen by variations in the gradient (such as
curvature or sharp changes) as load increases (e.g.
schist in Figure 1b).

At low angles of dip (0-20°) many of the plots of
circumferential strain against axial strain are
curved; in some cases this is true for both the
"dip" and "strike" positions. This curvature occurs
even though, after bedding in, the plot of axial
stress against axial strain is linear. The
curvature is considered likely to be due to plastic
deformation of more micaceous layers within the
schist and is recorded by the circumferential
gauges. If the test results that produce the most
curved or irregular plots are disregarded there
often remains a reasonable agreement between the
"dip" and "strike" position Poisson's ratio
values, which range between 0.20 and 0.35.

For moderate angles of dip (30-65°) failure is
along foliation. Generally linear plots of
circumferential strain against axial strain are
shown for both "dip" and "strike" positions until
the influence of deformation along foliation
dominates. This influence is particularly evident
for the "strike" position circumferential gauge, as
it is positioned across the foliation plane and
records excessive deformation generally due to
shear. This behaviour is illustrated on Figure 6b

where the "strike" position circumference gauge
records significantly higher strains at stress
levels greater than 30% of q_u. The linear plot of
the "dip" position gauges gives a Poisson's ratio
value which may be accepted with confidence.
However, in some cases for the lowest strength
materials the deformation during loading may be
preferentially taken up in the plane of foliation.
This may result in very low Poisson's ratio values
for the "dip" position (significantly < 0.10) and
very high "strike" position values (> 0.50). It
should be noted that although apparently reasonable
results could be obtained by averaging these "dip"
and "strike" position values, this is not
appropriate.

For steep angles of dip the ultimate strength
failure mode is dominated by axial splitting, and
consequently the two circumferential gauges may
record widely differing strains. The "strike"
position gauge may record excessive strains due to
the splitting even at early stages of loading, while
the "dip" position gauge may record smaller strains
depending on its proximity to incipient splitting
and whether any tilting or rotation of the specimen
is occurring. As for moderate angles of dip, the
reliability of the Poisson's ratio values obtained
for the two positions can be assessed by inspection
of the plots of circumferential strain against axial
strain. More reliance is placed on the value from
the "dip" position gauges.

The Poisson's ratio values, which have been
generally determined from the "dip" position gauges
are plotted against dip of foliation, on Figure 7.
There is a general trend for the values to decrease
with increasing angle of dip. For angles of dip
greater than 70° the scatter of results is greatest
and it is uncertain whether the downward trend
continues or the values remain constant.

It is clear that the Poisson's ratio values
obtained depend on the position of the
circumferential strain gauges. The use of strain
gauges mounted in opposite pairs will result in both
gauges of one type being in either the "dip" or

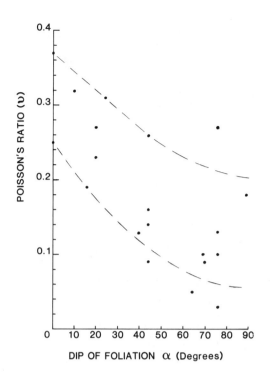

Figure 7. Poisson's ratio against dip of foliation

"strike" position (assuming they are mounted with a consideration for the orientation of foliation). In the authors' opinion this should be avoided and gauges should be mounted with a consideration of the orientation of foliation. The positioning of gauges in both the "dip" and "strike" positions as illustrated in Figure 3 allows the comparison of results in the different directions, a necessary feature when testing anisotropic rocks such as schist. Testing of isotropic rocks such as "greywacke" sandstone using this configuration of gauges has given good agreement between the two gauge positions, and similar Poisson's ratio values.

APPLICATION OF RESULTS TO DETERMINE THE CONSTITUTIVE RELATION FOR SCHIST

The general form for the constitutive relation $(\varepsilon_{ij}) = (c_{ijkl})(\sigma_{ij})$ can be written for an orthotropic material, i.e. with planes of symmetry normal to the co-ordinate axes, as:

$$
\begin{bmatrix} \varepsilon_x \\ \varepsilon_y \\ \varepsilon_z \\ \gamma_{yz} \\ \gamma_{xz} \\ \gamma_{xy} \end{bmatrix} =
\begin{bmatrix}
\frac{1}{E_x} & \frac{-\nu_{yx}}{E_y} & \frac{-\nu_{zx}}{E_z} & 0 & 0 & 0 \\
\frac{-\nu_{xy}}{E_x} & \frac{1}{E_y} & \frac{-\nu_{zy}}{E_z} & 0 & 0 & 0 \\
\frac{-\nu_{xz}}{E_x} & \frac{-\nu_{yz}}{E_y} & \frac{-1}{E_z} & 0 & 0 & 0 \\
0 & 0 & 0 & \frac{1}{G_{yz}} & 0 & 0 \\
0 & 0 & 0 & 0 & \frac{1}{G_{xz}} & 0 \\
0 & 0 & 0 & 0 & 0 & \frac{1}{G_{xy}}
\end{bmatrix}
\begin{bmatrix} \sigma_x \\ \sigma_y \\ \sigma_z \\ \tau_{yz} \\ \tau_{xz} \\ \tau_{xy} \end{bmatrix}
$$

... 1

where E_x, E_y, E_z are Young's moduli in the x, y, and z directions; G_{xy}, G_{yz} and G_{xz} are shear moduli in planes parallel to co-ordinate planes xy, yz, and xz. The Poisson's ratios ν_{ij} refer to strain responses in the j direction to a uniaxial stress acting in the i direction.

The twelve elastic constants in equation 1 are not all independent, and because of symmetry conditions only five independent elastic constants are needed to describe the compliance matrix for a transversely isotropic material such as schist (Pinto 1970; Brown et al 1980; Savage et al 1986).

The direction of foliation in schist corresponds to the plane of transverse isotropy. When this is parallel to the xy co-ordinate plane (Figure 8), then

$$E_y = E_x = E_1$$

$$E_z = E_2$$

$$\nu_{xy} = \nu_{yx} = \nu_1$$

$$\nu_{zx} = \nu_{zy} = \nu_2$$

$$G_{xz} = G_{yz} = G_1$$

Also, $\quad G_{xy} = \dfrac{E_1}{2(1+\nu_1)} \qquad \qquad \ldots 2$

and $\quad \nu_{xz} = \nu_{yz} = \nu_2 \dfrac{E_1}{E_2} \qquad \ldots 3$

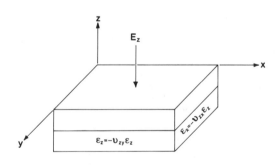

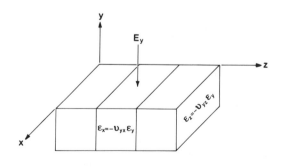

Figure 8. Definition of elastic constants for schist with foliation parallel to the xy plane

237

In terms of the laboratory testing results reported, E_1 corresponds to the modulus measured when angle of dip of foliation (α) is 0°, and E_2 corresponds to the modulus measured when α = 90°. Similarly ν_1 is the Poisson's ratio measured by both the "strike" and "dip" position gauges when α = 0°, and ν_2 is the Poisson's ratio measured by the "dip" position gauges when α = 90°.

Using data given in Figure 5, and taking an average of the results, we find

$$E_1 = 45\ 000 \text{ MPa}$$
$$E_2 = 65\ 000 \text{ MPa}$$
$$\nu_1 = .3$$
$$\nu_2 = .15$$

then using equation 2

$$G_{xy} = 17\ 000 \text{ MPa}$$

The testing programme has not allowed the measurement of G_1, the shear modulus in planes normal to the plane of tansverse isotopy. A suitable torsion test is required to provide this data.

CONCLUSIONS

Uniaxial compressive strength testing of anisotropic schistose rocks from Central Otago, New Zealand, confirmed their strengths are dependent on the dip of the foliation with respect to the loading direction. The lowest strengths are found when the dip of foliation is closest to the diagonal through the tested specimen (68°). For angles of dip of up to 70° the tangent modulus of elasticity does not appear to be affected by the orientation of foliation, although the anisotropy of the schist is reflected by an increase in values at dips greater than 70°. Poisson's ratio values are affected not only by the dip of foliation, but also by the position at which the circumferential strain is measured.

It is recommended that strain gauges are mounted so that circumferential gauges are positioned diametrically opposite axial gauges with each mounting position being either in the direction of the strike or the dip of the foliation plane. This arrangement results in one of each type of gauge in both the "dip" and "strike" positions. Such an arrangement should allow consistent values of Poisson's ratio to be obtained as long as all four gauges are monitored individually (i.e. not averaged electronically by connecting them in series).

Generally the "dip" position gauges yield more consistent Poisson's ratio values. This should be confirmed by the examination of the plots of circumferential strain against axial strain for both gauge positions.

The results obtained have been applied to a model of schist as a transversely isotropic rock defined by modulus of elasticity and Poisson's ratio values normal and parallel to the plane of foliation, and the shear modulus in planes normal to the plane of transverse isotropy.

ACKNOWLEDGEMENTS

The authors gratefully acknowledge the contribution made by David Wong during the testing programme and preparation of the paper. Thanks are given to G.T.Hancox, C.A.M.Franks and R.D.Beetham who reviewed the manusrcript.

REFERENCES

ASTM 1980. Standard test method for elastic moduli of intact rock core specimens in uniaxial compression. ASTM (American Society for Testing and Materials) Designation D3148-80. In 1984 annual book of ASTM standards, p. 501-507.

Bieniawski,Z.T. (co-ord) et al 1978. Suggested methods for determining the uniaxial compressive strength and deformability of rock materials. In E.T.Brown (ed.) 1981. Rock characterisation testing and monitoring. ISRM (International Society for Rock Mechanics) suggested methods, p. 111-116. Pergamon Press.

Brown,I., M.Hittinger & R. Goodman 1980. Finite element study of the Nevis Bluff (New Zealand) rock slope failure. Rock Mechanics 12: 231-245.

Donath,F.A. 1972. Effects of cohesion and granularity on deformational behaviour of anisotropic rock. Geol.Soc.Amer.Memoir 135: 95-128.

Gyenge,M. & G.Herget 1977. Determination of elastic modulus and determination of Poisson's ratio. In Pit slope manual supplement 3-2 - laboratory tets for design purposes. In CANMET (Canada Centre for Mineral and Energy Technology, formerly Mines Branch, Energy, Mines and Resources Canada) report 77-26.

Hoek,E. 1983. Strength of jointed rock masses. Geotechnique 33(3): 187-223.

Hoek,E. & E.T.Brown 1980. Underground excavations in rock. London: Inst. Mining & Metallurgy.

Pinto,J.L. 1970. Deformability of schistose rocks. Proc. of Sec. Cong., Int. Soc. Rock Mechanics, Belgrade: 2-30.

Read, S.A.L., N.D.Perrin & D.Wong 1986. Uniaxial compression testing of highly anisotropic rocks. New Zealand Geological Survey geomechanics laboratory report 025.

Savage,W.Z., B.P.Amadei & H.S.Swolfs 1986. Influence of rock fabric on gravity induced stresses. Proceedings of the International Symposium on Rock Stress Measurements, Stockholm: 99-110.

Suggate, R.P., G.R.Stevens & M.T.TePunga (eds.) 1978. The geology of New Zealand. 2 vol. Wellington: Government Printer.

Monitoring the internal microcracking of heated rocks by remote sensing procedures, AE/MS

Contrôle de la micro-fissuration d'un rocher chauffé par des méthodes de télé-détection AE/MS (émission acoustique/activité microséismique)

Beobachtung des internen Mikrobruchs erhitzter Gesteine durch Fernaufnahme von akustischen Emissionen und Mikroseismik

V.G.RUIZ DE ARGANDOÑA, Department of Geology, University of Oviedo, Spain
L.CALLEJA, Department of Geology, University of Oviedo, Spain
L.M.SUÁREZ DEL RÍO, Department of Geology, University of Oviedo, Spain
M.MONTOTO, Department of Geology, University of Oviedo, Spain
A.RODRÍGUEZ-REY, Department of Geology, University of Oviedo, Spain

ABSTRACT: Cracks of thermal origin can be developed in the vicinity of the canisters (in the heated intact rock of a HLW repository) increasing its "flow porosity", that is allowing an easier water flow through the intact rock. As it is proved in this paper acoustic emission / microseismic activity (AE/MS) can be considered as the most appropriate remote sensing method for monitoring stressed rocks. This technique provides very valuable information about the stress level affecting the rock and the internal development of new microcracks. Moreover the significant "thermal microfissuration threshold", T.M.T., of the intact rock can be so determined in such a way. The petrographic interpretation of AE and its relation to the progressive evolution of microcracking in the stressed intact rock, provides a powerful tool for understanding the mechanical behaviour of the heated rock.

Experimental tests have been carried out with three different types of Spanish "intact rock" (granodiorite, epidiorite and serpentinite) subjected to uniform heating from laboratory temperature up to 430°C, under different heating rates (0.7, 2 and 6°C/min). The AE/MS so generated was continuously monitored. The variations of some physical properties after testing were also evaluated. The registered AE/MS in the three heated rocks exhibit a clear correlation with temperature and with the variations in porosity. The acoustic emission is assigned to the development and propagation of cracks generated by the differential thermal expansion of the rock-forming minerals. The T.M.T. has been determined according to the AE and porosity analysis: granodiorite 110-115°C, epidiorite 130-135°C and serpentinite 230-235°C.

RESUME: Les fractures d'origine thermale peuvent se développer à proximité des boites (dans la "intact rock" chauffée d'un répositoire de HLW) en augmentant leur "porosité de flux", c'est-à-dire, elles permettent un flux d'eau plus facile à travers la "intact rock". Comme on le prouve dans ce travail, l'emission acoustique/activité microsismique, peut être considerée la methode de contrôle à distance la plus apropiée pour monitoriser les roches sous effort. Cette technique fournit des rasseignements très utiles sur le niveau d'effort qui affecte la roche et le développpment interne de nouvelles microfractures. Plus, le seuil de microfissuration thermale (T.M.T.) de la "intact rock" peut être determiné de cette manière. L'interpretation pétrographique de la AE/MS et sa relation avec l'evolution progressive de la microfracturation dans la "intact rock" sous effort, proportionne un instrument magnifique pour la compréhension du comportement mécanique de la roche chauffée.

Nous avons developpé des tests experiméntaux avec trois types différents de "intact rock" espagnole (granodiorite, epidiorite et serpentinite) soumises à un chauffage uniforme à partir de la température du laboratoire jusqu'à 430°C, sous des niveaux de chauffage differents. (0.7, 2 et 6°C/min). Nous avons monitorisé la AE/MS generée en cette forme. Les variations de certaines proprietés physiques après le testing ont eté evaluées aussi. La AE/MS enregistrée à les trois roches chauffées présente une corrélation claire avec la temperature et avec des variations de porosité. L'emission acoustique s'assigne à le développment et propagation des fractures generées par la differente expansion thermale des mineraux qui forment la roche. Le T.M.T. a eté determiné d'accord avec les analyses de AE/MS et porosité: granodiorite 110-115°C; epidiorite 130-135°C et serpentinite 230-235°C.

ZUSAMMENFASSUNG: Risse von thermischem Ursprung können sich in der Nähe von Kanistern (in dem erhitzten intakten Gestein von einem HLW Speicher) entwickeln und so die "Strömungsporosität" erhöhen, das heißt, sie erleichtern das Fließen von Wasser durch das intakte Gestein. Wie in der vorliegenden Arbeit erwiesen wird, kann die akkustische Emission/mikroseismische Aktivität (AE/MS) als die geeignetste Methode von ferngesteuerten Meßverfahren zum Überwachen von Gestein unter Druck betrachtet werden. Diese Technik liefert wertvolle Informationen über den Druckpegel des Gesteins und über die interne Entwicklung von neuen Mikrorissen. Außerdem kann die bedeutende Schwelle thermischer Mikrorisse (thermal microfissuration threshold, T.M.T.), des intakten Gesteins auf diese Weise bestimmt werden. Die gesteinskundliche Interpretation von AE und seine Beziehung zur fortschreitenden Entwicklung von Mikrorissen im intakten Gestein unter Druck stellt ein wertvolles Instrument zum Verständnis des mechanischen Verhaltens von erhitztem Gestein dar.

Es wurden Versuche mit drei verschiedenen Arten von spanischem "intakten Gestein" durchgeführt (Granodiorit, Epidiorit und Serpentinit), bei gleicher Erhitzung bei einer Labortemperatur bis zu 430°C bei unterschiedlichem Erhitzungsgrad (0.7, 2 und 6°C/min). Die so entstandene AE/MS in den drei erhitzten Gesteinsarten weisen ein klares Verhältnis zur Temperatur und zu den Veränderungen der Porosität auf. Die akkustische Ausstrahlung beschäftigt sich mit der Entwicklung und Verbreitung von Rissen die durch die unterschiedliche thermische Expansion des gesteinsbildenden Minerals entstehen. Die T.M.T. wurde der AE und der Porositätsanalyse entsprechend festgelegt: Granodiorit 110-115°C; Epidiorit 130-135°C und Serpentinit 230-235°C.

1 INTRODUCTION

Stressed rocks exhibit acoustic emission/ microseismic activity signals, AE/MS; this remote sensing method provides very valuable information about the stress level affecting the rock and the internal development of new microcracks.

The available bibliography is full of convincing examples supporting such assertion.

Laboratory experiments, carried out monitoring the AE/MS, clearly determined the development of new cracks in the microstructure of the rock, which obviously allows the increase of the "flow porosity".

AE/MS can be considered as the most appropiate remote sensing method for monitoring the development of new cracks in the heated "intact-rock" of a HLW repository. In such repositories, the "intact-rock" in the vicinity of the canisters is subjeted to thermal stresses, sometimes high enough to induce thermal cracks, allowing an easier water flow through the rock. Thus, it is important to know the "thermal microfissuration threshold" T.M.T., (i.e. the temperature at which new cracks start to be formed), in order to guarantee the watertightness of the "intact rock" and avoid the waterflow to the biosphere.

To locate the newly formed cracks is not the main problem to be solved, but rather to verify that this phenomenon has undergone in the intact rock, in the vicinity of the canisters, contributing to a higher porosity of the rock and, consequently, allowing an easier water-flow through the intact rock.

In one experiment at the Stripa iron-ore mine in Sweden, conducted by the Lawrence Berkeley Laboratory (LBL) of the University of California and the Swedish Nuclear Fuel Safety Agency (KBS), electrical heaters in granite were employed to study thermal stress effects. AE studies were started well after the heaters were turned on, and only one sensor was used. However, considerable AE activity was still observed during the later stages of the heating phase, and activity increased during the initial stages of the cool-down period (Paulsson et al., 1980).

2 PETROGRAPHY

2.1 Granodiorite

Its modal composition is: quartz (31.5%), microcline (14.6%), plagioclase (An_{25-40}) (41.5%) biotite (9.3%) and muscovite (2,6%). The minerals are distributed in an allotriomorphic, heterogranular texture of a medium grain size. From the fractographic point of view, it has transgranular fissures connected to intergranular cracks.

2.2 Epidiorite

It exhibits an subidiomorphic granular texture with subophitic tendency. It is medium grain sized. The mineralogical composition is: plagioclase (An_{35}) (66.5%), pigeonite (11.3%), chlorite (15.5%) amphiboles (2.3%) and iron oxides (3.4%). It has almost no microfissuration, and the pore porosity is mainly located inside the feldspars.

2.3 Serpentinite

Its mineralogy is: serpentine (77.2%), magnesite (13.5%) and iron oxides (9.3%) as well as talc and sericite. The serpentine has a cryptocrystalline texture, and the magnesite is located in subparallel bands with a polygonal granular texture. The fractography is represented by open fissures associated to the magnesite veins.

A most exhaustive description of these rocks appears in Ruiz de Argandoña (1985).

3 PHYSICAL PROPERTIES

In order to measure the variation of some physical properties with temperature, rock samples were heated in an electrical oven up to 130, 230, 330 y 430°C. The heating was performed with a maximum heating rate of 2°C/min,

and the cooling rate was about 1°C/min.

After cooling to room temperature, measurements of elastic wave propagation (Vp and Vs) were taken on cylindrical cores (length: 110 mm; diameter: 50 mm) with an OYO SONICVIEWER equipment. Two couples of transducers were used, one of them to measure Vp (nominal frequency: 200 kHz) and the other for Vs (nominal frequency: 100 kHz). The pulse voltage was 200 V and the pulse frequency 20 microseconds. The dynamic Young's modulus (E_{dyn}) was evaluated following the A.S.T.M. standard (D.2845, 1976).

The total porosity, following the procedure suggested by Belikov et al. (1967), was measured in other heated specimens in order to use the data for the petrophysical interpretation of the elastic wave propagation and the AE/MS.

The values of Vp, Vs, E_{dyn} and total porosity (n) for the three tested rocks after heating to different temperatures, are presented in Table I.

Table I.- Physical properties of the tested rocks heated up to different temperatures (T). n: total porosity; Edyn: dynamic Young's modulus; Vp: P-wave velocity; Vs: S-wave velocity.

	T	n	E_{dyn}	Vp	Vs
	°C	%	MPa.10⁴	m/s	m/s
GRANODIORITE	20	1,37	4,73	4520	2694
	130	1,37	4,68	4463	2688
	230	1,55	4,10	4194	2517
	330	1,74	3,65	3925	2386
	430	2,10	3,25	3685	2268
EPIDIORITE	20	1,78	7,05	5974	3046
	130	1,43	7,24	5948	3093
	230	1,78	6,88	5596	3050
	330	1,95	6,43	5385	2957
	430	2,12	6,18	5311	2896
SERPENTINITE	20	3,78	2,83	3421	2644
	130	3,60	2,71	3377	2627
	230	4,32	3,18	3543	2672
	330	4,50	2,65	3427	2696
	430	4,50	2,88	3510	2735

4 TEST PROCEDURE

Experimental thermal cycles have been conducted with cubic specimens of 50 mm edge. All the specimens were heated in a programmable electric oven. The ovenheating rate was regulated by an optoelectronic regulator at 0.7, 2 and 6°C/min and the cooling rate was always under 1°C/min. The temperature in the specimen interior and exterior was continuously monitored with type J termopars (iron-constantan).

So, maximum temperatures were 130, 230, 330 and 430°C for different samples of the three rock types.

When all the rock specimen was stabilized at the maximum temperature, this was maintained till acoustic emission activity appeared to have ceased.

The acoustic emission activity was monitored with an A.E.T. 204GR equipment. A transducer, 100-300 kHz sensitive, placed out of the oven and connected to the rocksurface through a glass wave guide, picked up the AE/MS. (Fig. 1). The rock was attached to the glass wave guide using an Ag-glue interface.

The acoustic emission activity was conti-

Fig. 1. Overall view of the experimental facilities. (1) Electric oven, (2) Glass wave-guide, (3) AE transducer, (4) Preamplifier, (5) AE monitoring system, (6) Oven regulator, and (7) Dual-Channel strip chart recorder.

nuously registered (in counts/10s.) under a gain of 92 dB.

5 RESULTS

5.1 Granodiorite

In Fig. 2, the evolution of the Acoustic Emission during the heating is plotted (heating rate: 2°C/min). It can be seen that the AE starts to have significant values when the temperature in the oven reachs 110-115°C, as it was previously stated by Ruiz de Argandoña and Calleja (1984) and Montoto et al. (1985).

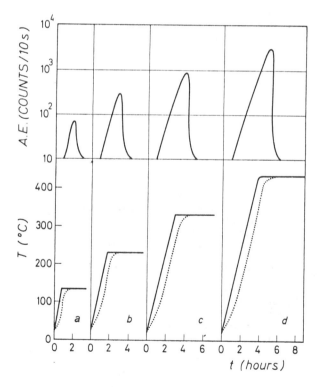

Fig. 2.- Evolution of the AE rate in the granodiorite during heating up to different temperatures: a) 130°C; b) 230°C; c) 330°C; d) 430°C. Heating rate 2°C/min. Solid line: oven temperature; dashed line: temperature at the rock specimen center.

This temperature, at which the Acoustic Emission has a significant increase, is called "thermal microfissuration threshold" (T.M.T.).

Therefore, before reaching this temperature, some AE events (with low AE rate) are monitored, some authors attributing them to textural rearrangements inside the rock (Cooper and Simmons, 1977).

The highest AE values for each maximum temperature are reached when the temperatures inside and outside of the rock specimen become the same.

Fig. 3 shows that the T.M.T. is not affected by the heating rate, but this influences the values of AE and the shape of the AE curve.

The maximum values of AE rate for each maximum temperature and for the different heating rates are summarized in Table II.

This rock, as well as the other tested rock types, shows a decrease of the AE rate once the maximun temperature was reached and the temperatures inside and outside of the specimen are very similar. The shape of the AE rate decrease, when maintaining the temperature, was similar for all the rock types tested.

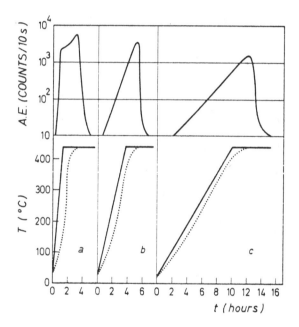

Fig. 3.- AE rate generated during heating up to 430°C with different heating rates; a) 6°C/min; b) 2°C/min; c) 0.7°C/min. Granodiorite.

Table II.- AE rate peaks (counts/10s) at each temperature (T, °C) and different heating rates (HR, °C/min.). Granodiorite.

HR\T	130	230	330	430
0.7	43	145	550	1598
2.0	70	298	972	3219
6.0	116	483	1512	5760

5.2 Epidiorite

The variation of the AE rate with the temperature for a heating rate of 2°C/min is plotted in Fig. 4. The T.M.T. is located at a temperature of 130-135°C.

When the temperature reachs 200°C there is a decrease of the AE rate although the temperature is going up, and at 300°C the AE rate starts to increase again, reaching the maximum value when the temperature equals in all the specimen volume.

In Fig. 5 it can be seen that the evolution of the AE rate for different heating rates follows the same tendency.

The maximum values of the AE rate for the different maximum temperatures and heating rates are expressed in Table III.

5.3 Serpentinite

The evolution of the AE rate against the temperature (heating rate: 2°C/min) is presented in Fig. 6. The microfissuration starts at 230-235°C for all the heating rates (0.7, 2 and 6°C/min), and, as in the

other tested rocks, the maximum AE rate appears when the temperature is equal in all the specimen (Fig. 7).

Table IV summarizes the maximum values of the AE rate for all the temperatures and heating rates.

Table III.- AE rate peaks (counts/10s) at each temperature (T, °C) and different heating rates (HR, °C/min.). Epidiorite.

HR\T	130	230	330	430
0.7	18	23	30	53
2.0	22	30	40	70
6.0	30	37	56	82

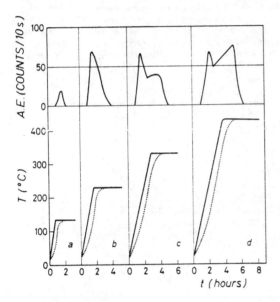

Fig. 4.- AE rate during the heating of the epidiorite up to 130, 230, 330 and 430°C. Heating rate: 2°C/min.

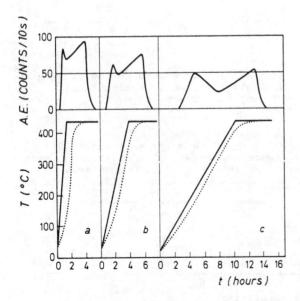

Fig. 5.- AE rate generated during heating up to 430°C with different heating rates; a) 6°C/min; b) 2°C/min; c) 0.7°C/min. Epidiorite.

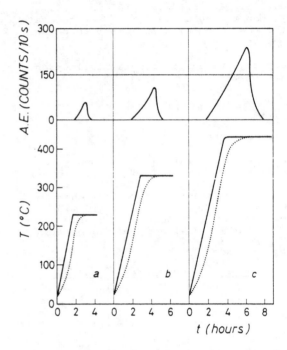

Fig. 6.- AE rate during the heating of the serpentinite up to 230, 330 and 430°C. Heating rate: 2°C/min.

6 PETROPHYSICAL INTERPRETATION

Comparing the porosity values for each maximum temperature (expressed in Table I), Ruiz de Argandoña and Calleja (1985) have stated that the porosity is exponentially related with the maximum temperature, coinciding with the conclussions of Simmons and Cooper (1978) for other rocks.

Acoustic emission is mainly generated by development and growth of microcracks (Hardy, 1972). This microfissuration increase is due to the different thermal expansion of the rock forming minerals and, even in the same mineral group, to the anisotropy of such an expansion.

Fig. 8 shows the relationship between the maximum values of AE rate (heating rate: 2°C/min) and the porosity, for each maximum temperature.

Below the T.M.T. a partial closure of the pre-existing porosity is produced, due to mineral expansion phenomena (Calleja and Ruiz de Argandoña, 1985); the acoustic emission monitored below the T.M.T. can be interpreted to the closure of voids. A similar behaviour

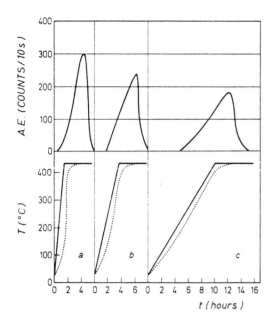

Fig. 7.- AE rate generated during heating up to 430°C with different heating rates; a) 6°C/min; b) 2°C/min; c) 0.7°C/min. Serpentinite.

Table IV.- AE rate peaks (counts/10s) at each temperature (T, °C) and different heating rates (HR °C/min). Serpentinite.

HR\T	230	330	430
0.7	45	85	180
2.0	52	108	240
6.0	60	132	296

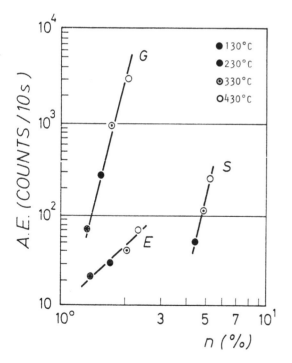

Fig 8.- Relationship between the AE peaks at 430°C and the porosity of the rocks after cooling. G: Granodiorite; E: Epidiorite; S: Serpentinite.

has been clearly showed by Montoto et al. (1983), in weathered granitic rocks with high porosity (n= 5.21%) when subjected to uniaxial compressive loads.

In all the materials tested and for all the heating rates, the AE rate peaks have a log-normal relationship with the maximun temperatures. Thus, it can be stated that, above the T.M.T., the AE rate increases with the temperature and the heating rate.

Also, once the maximum AE rate is reached, this gradually starts to decrease, even when the temperature is maintained constant. The acoustic emission so generated after reaching the maximum temperature in all the sample volume can be attributed to the fact that crack equilibrium configuration for that temperature has not been achieved (Johnson et al., 1978; Kurita and Fujii, 1979).

In the granodiorite, the fissuration is mainly due to the different thermal expansion coefficients of quartz and feldspars.

In the epidiorite, the difference between the volumetric expansion coefficients of pigeonite and plagioclases is the main source of thermal microfissuration (Skinner, 1966), generating high stresses mainly located in intergranular positions.

Looking at the evolution of these coefficients along with the temperature, a higher increase can be observed between 20 and 200°C than between 200 and 400°C for the pigeonite and plagioclase, explaining the AE rate evolution of this rock.

In the serpentinite, the thermal microfissuration is due to the anisotropic behaviour of the magnesite, the cracks being located in intergranular positions inside the veins of this mineral and, in a lesser amount, in the contact between the magnesite and serpentinite (Ruiz de Argandoña, 1985).

All these increases in porosity induce a decrease of the Vp and Vs (and therefore E_{dyn}) (see Table I) in the granodiorite and epidiorite, showing a degradation of the rock matrix. In the serpentinite, the elastic characteristics are almost not affected, due to the particular location and orientation of the cracks.

7 CONCLUSIONS

The Acoustic Emission technique is a good procedure to monitor and evaluate the microfissuration processes developed in the intact rock by thermal stresses.

Under similar heating conditions, the three intact rock types have shown a clear relationship between the maximum temperature at whicn they were subjected and the AE rate monitored, for all the heating rates used in the tests.

The acoustic emission is mainly originated by the generation and propagation of microcracks, due to the different thermal expansion coefficients of the rock-forming minerals. The T.M.T. have been evaluated by AE/MS: granodiorite, 110-115°C; epidiorite, 130-135°C and serpentinite, 230-235°C. These values have been later corroborated by porosity and Vp-Vs measurements.

The heating rate does not influence the temperatures at which the thermal microfissuration starts (T.M.T).

The peak of AE rate appears when the temperature becomes uniform in all the sample volume, once the maximum temperature was reached. After this, a gradual decrease of the AE rate starts, even when the temperature is maintained. The acoustic emission ceases when the crack equilibrium configuration is attached.

8 ACKNOWLEDGMENTS

We wish to thank "Comision Asesora de Investigación Científica y Técnica (C.A.I.C.Y T.), Spain, by supporting this research.

9 BIBLIOGRAPHY

A.S.T.M. 1976. Standard method for laboratory determination of pulse velocities and ultrasonic elastic constants of rock (D-2845). In A.S.T.M. Standards (eds.).

Belikov, B.P.; B.V. Zalesskii; Y.A. Rozanov; E.A. Sanina & I.P. Timchenko, 1967. Methods of studying the physicomechanical properties of rocks. In Zalesskii, B.V. (ed.) Physical and mechanical properties of rocks. Academy of Sciences of the USSR. p. 1-58. Jerusalem: Israel Program for Scientific Translations.

Calleja, L. & V.G. Ruiz de Argandoña, 1985. Variación de la expansión térmica en rocas cristalinas. Trabajos de Geología. Univ. de Oviedo, 15: 307-313. Oviedo. Spain.

Cooper, H.W. & G. Simmons,1977. The effect of crack on the thermal expansion of rocks. Earth and Planetary Sci. Letters, 36: 404--412.

Hardy, H.R., Jr. 1972. Application of acoustic emission techniques to rock mechanics research. In A.S.T.M, Sp. Tech. Publ. 505. Acoustic Emission, p. 41-83.

Johnson, B.; A.F. Gangi & J. Handin, 1978. Thermal cracking of rock subjected to slow, uniform temperature changes. Proc. 19th U.S. Symp. on Rock Mech., p. 259-267.

Kurita, K. & N. Fujii, 1979. Stress memory in crystalline rocks in acoustic emission. Geophys. Res. Lett., 6, 1: 9-12.

Montoto, M.; L. M. Suárez del Río; A. W. Khair & H.R. Hardy, Jr., 1983. Acoustic emission in uniaxially loaded granitic rocks in relation to their petrographic character. In Hardy, H.R., Jr. and Leighton F.W. (eds). Proc. 3th. Conf. on AE/MA in Geologic Structures and Materials. The Pennsylvania State University. U.S.A., p. 83-100. Clausthal, Germany: Trans Tech Publications.

Montoto, M.; V.G. Ruiz de Argandoña; L. Calleja & L.M. Suárez del Río, 1985. Kaiser effect in thermo-cycled rocks. In Hardy, H.R., Jr. and Leighton, F.W. (eds.). Proc. 4th Conf. on AE/MA in Geologic Structures and Materials, The Pensylvania State University. U.S.A., (In press). Clausthal, Germany: Trans Tech Publications.

Paulsson, B.N.P.; M.S. King & R. Rachiele, 1980. Ultrasonic and acoustic emission results from the Stripa heater experiments. Lawrence Berkeley Laboratory, Report SAC-32, LBL-10975.

Ruiz de Argandoña, V.G. 1985. Estudio de la microfisuración térmica mediante emisión acústica: interpretación petrográfica. Ph. D. Thesis. Fac. Geología. Univ. de Oviedo, Spain, 253 p.

Ruiz de Argandoña, V.G. & L. Calleja, 1984. Determinación del umbral de microfisuración térmica mediante emisión acústica/actividad microsísmica. In: S.E.M.R. (ed.). VIII Simp. Nac. de Mecánica de Rocas: Reconocimiento de Macizos Rocosos. 1: 2-7, 5 p., Madrid.

Ruiz de Argandoña, V.G.; L. Calleja & M. Montoto, 1985. Determinación experimental del umbral de microfisuración térmica de la "roca matriz" o "intact rock". Trabajos de Geología. Univ. de Oviedo, 15: 299-306. Oviedo.

Simmons, G. & H.W. Cooper, 1978. Thermal cycling cracks in three igneous rocks. Int. J. Rock Mech. Min. Sci. & Geomech. Abstr. 15: 145-148.

Skinner, B.J. 1.966. Thermal expansion. In: Clark, S.P. (ed). Handbook of physical constants. Geol. Soc. Am., Memoir 97: 78--96.

Applied study on transfer of radioactive nuclides and heat diffusion around radioactive waste disposal cavern in deformable fractured rock mass

Etude appliquée sur le transfert de soluté nucléaire et la diffusion de chaleur autour d'une caverne de disposition de déchets radioactifs dans la roche mère

Angewandte Forschung über Transport der Nuklearlösung und Diffusion der Wärme in der Nähe des Hohlraums für Aufspeicherung der radioaktiven Abfälle im Fels

K.SATO, Saitama University, Japan
Y.ITO, Kumagai Gumi Co. Ltd, (Research member of Saitama University), Tokyo, Japan
T.SHIMIZU, Kumagai Gumi Co. Ltd, Tokyo, Japan

ABSTRACT: This paper presents the numerical analyses of the heat diffusion and nuclide migration around a disposal cavern of radioactive wastes in rock mass. The rock mass model used in the present study is deformable, and it is consisted of a number of elastic blocks and intersticial materials in fractures. The coupled analyses of a set of fundamental equations are done by the Finite Element Method and Fluid in Cell Method.

RESUME: Ce rapport présente les analyses numériques de la diffusion de chaleur et la migration des nucléons autour de une caverne de disposer les matières radioactives des rebuts dans la roche mère. Le modèle de la roche du present travail est déformable, et C'est consisté á beaucoup de blocs élastiques et màtériels dans les fractures. Les analyses de la combination sur l'emploi d'un equations fondamentales ont été effectuées par la méthode d'element fini et la méthode du fluide a la cellule.

ZUSAMMENFASSUNG: Diese abhandlung zeight die zählenmäβige analysis der warmer diffusion und konvektiven migration bei der höhle für aufspeicherung des radioaktiven abfalle stoffs im felsboden. Das modell des felsboden in dieser forschung ist deform, und es ist zusammengesetzt aus vieler felsenmasse der elastik und der kleine materials im spalt. Die verbunden analysis mit ein satz gleichung wird durch dem Finite Element Methode und der Flüssigkeit in Zelle Methode gemacht.

1. INTORODUCTION

Recently, it attracts public attention how to dispose the radioactive nuclide waste. The usage of underground cavern in rock mass is considered as one of repository ways. In safety evaluation of this system, the analyses on geohydraulic migration of nuclides and heat conduction are essential. The development of coupled analysis may afford the required solution. Some of the existing studies led to success in solving the coupling problems between groundwater flow and heat transfer in fractured rock system of double porosity (Sato et al. 1985a). With respect to coupled analysis on stress, heat and seepage flow, Ohnishi et al(1986) tried to solve the coupled problem by a simple rock mass. The analyses of heat transfer and nuclide migration around a disposal canister were reported by Sato et al(1985b, c). They are, however, not sufficient to explain not only the dynamic change of permeability around the cavern but also the groundwater behavior.

The authors aim at development of numerical technique for analyzing coupled seepage-heat conduction and solute transfer around disposal cavern in deformable rock mass. The governing equations of groundwater flow are derived from a concept of rock block model which is composed of a number of rock blocks and intersticial materials, and which is subjected to an elastic deformation by dynamic change of water pressure.

Groundwater and stress analyses in rock mass are based on the finite element method in applying of Galerken technique, and the heat conduction and solute transfer are calculated numerically by the modified fluid in cell (FLIC) method for solving the dispersion problems at a large Peclet number flow.

2. PERMEABILITY FORMULATION AND GOVERNING EQUATIONS

2.1 Permeability equations

In the present study, the rock block model equivalent to fractured rock mass is shown in Fig.1. The concept of the model has been originally proposed by Snow(1968) and Sato (1982). A small column block in fracture space is adopted as a model of contact points in irregular fracture wall, and it behaves as an elastic body in response to a change of effective stress. According to the rock block model,

the equations of permeability and porosity can be formulated as follows(Sato et al, 1986):

$$k_i = k_{oi} \frac{\left\{ 1 + \frac{1}{a_i e_c} \left(\frac{\Delta P_f}{E_r} - \frac{\Delta P_{tj}}{E_r} \right) \right\}^3}{1 + \frac{S_{oi}'}{S_{oi}} \left\{ \beta_i \frac{\Delta P_f}{E_r} - (1 + \beta_i) \frac{\Delta P_{tj}}{E_r} + M_r \frac{\Delta P_{tj}}{E_r} + M_r \frac{\Delta P_{tk}}{E_r} \right\}} \quad (1)$$

$$k_{oi} = \frac{2}{3} \frac{g}{\nu} \frac{d_{oi}^3}{S_{oi}}, \quad \beta_i = \frac{1}{a_i e_c} \frac{2d_{oi}}{S_{oi}'}, \quad e_c = \frac{E_c}{E_r}, \quad i, j, k = 1, 2, 3$$

$$k_x = k_1 + k_2, \quad k_y = k_1 + k_3, \quad k_z = k_2 + k_3, \quad i \neq j \neq k$$

$$n \fallingdotseq \sum_{i=1}^{3} \frac{2d_{oi}}{S_{oi}} \frac{1 + \frac{1}{a_i e_c} \left(\frac{\Delta P_f}{E_r} - \frac{\Delta P_{tj}}{E_r} \right)}{1 + \frac{S_{oi}'}{S_{oi}} \left\{ \beta_i \frac{\Delta P_f}{E_r} - (1 + \beta_i) \frac{\Delta P_{ti}}{E_r} + M_r \frac{\Delta P_{tj}}{E_r} + M_r \frac{\Delta P_{tk}}{E_r} \right\}}$$

$$\cdots\cdots\cdots (2)$$

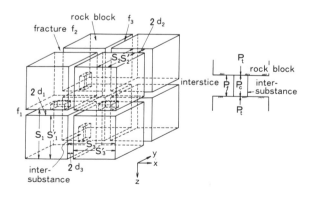

Fig.1 Rock mass model

where
k: permeability,
g: acceleration of gravity,
2d:width of interstice,
s: spacing of rock block model,
a: ratio of contact area to flat area of rock block($a < 1$),
E_r: elasticity modulus of rock block,
M_r: Poisson's ratio of rock block,
P_t: total stress($P_t = P_f + P_c$, P_c : effective stress acting on directional fractures),
n: porosity,
ν :kinematic viscosity of fluid,
s':size of block,
Δ:differnce of dynamic quantities from an initial state,
E_c:elasticity modulus of inter-substance,
M_c:Poisson's ratio of inter-substance,
P_f:fluid pressure of void,
x,y,z: coordinates,
and suffix o indicates an initial value.

The comparison between field measurments and theoretical permeabilities with depth is shown in Fig.2. The parameter aE_c in Eq.(1) is $2.86 \sim 6.47$ MN/m² for $K = 0.7 \sim 1.5(\Delta P_{ti} = K\Delta P_{ti}, i = 2, 3)$.

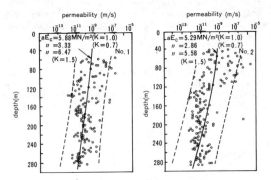

Fig.2 Comparison between theoretical and measured permeabilities with depth by Gale et al(1982)

2.2 Governing equations of groundwater flow

The governing equation of groundwater flow are given by,

$$\frac{\partial}{\partial x}\left(k_x \frac{\partial h}{\partial x}\right) + \frac{\partial}{\partial y}\left(k_y \frac{\partial h}{\partial y}\right) + \frac{\partial}{\partial z}\left(k_z \frac{\partial h}{\partial z}\right)$$

$$= S_f \frac{\partial h}{\partial t} - S_{ta}\frac{\partial h_{ti}}{\partial t} - S_{tb}\frac{\partial h_{tj}}{\partial t} - S_{tc}\frac{\partial h_{tk}}{\partial t} \qquad (3)$$

where,
$S_f = \{ n/e_w + \sum_{i=1}^{3} \beta_i (s_{oi}'/s_{oi})(1 + n\zeta_{ia} \cdot (s_{oi}'/s_{oi})\} \gamma_w/E_r$,

$S_{ta} = \{ \sum_{i=1}^{3} (s_{oi}'/s_{oi})(\beta_i s_{oi}'/s_{oi} \cdot \zeta_{ib} + n + n\beta_i \} \gamma_w/E_r$,

$S_{tb} = S_{tc} = \{ \sum_{i=1}^{3} (s_{oi}'/s_{oi})(\beta_i s_{oi}'/s_{oi} \cdot \zeta_{ic} + n + nM_r \} \gamma_w/E_r$,

$e_w = E_w/E_r$,

$\zeta_{ia} = (s_{oi}'/s_{oi} - aie_c\beta_i - \gamma_w/E_r \Delta h_{ti} + M_r\gamma_w/E_r \cdot \Delta h_{tj} + M_r\gamma_w/E_r \cdot \Delta h_{tk})/\eta$

$\zeta_{ib} = \{ s_{oi}'/s_{oi} - aie_c(1 + \beta_i) - \gamma_w/E_r \cdot \Delta h_f + M_r\gamma_w/E_r \cdot \Delta h_{tj} + M_r\gamma_w/E_r \cdot \Delta h_{tk} \}/\eta$

$\zeta_{ic} = M_r(aie_c + \gamma_w/E_r \cdot \Delta h_f - \gamma_w/E_r \cdot \Delta h_{ti})/\eta$

$\eta = [1 + s_{oi}'/s_{oi} \{ \beta_i\gamma_w/E_r \cdot \Delta h_f - (1 + \beta_i)\gamma_w/E_r \cdot \Delta h_{ti} + M_r\gamma_w/E_r \cdot \Delta h_{tj} + M_r\gamma_w/E_r \cdot \Delta h_{tk} \}]^2$

$i, j, k = 1, 2, 3, \quad i \neq j \neq k$,

where
t : time,
E_w: compressibility of fluid,
h_f: pressure head($h_f = P_f/\rho_w g$),
h_t: hydraulic head equivalent to total stress,
γ_r: unit weight of rock block,
γ_w: unit weight of fluid,
ρ_w: density of fluid,
h : piezometric head($h = h_f + z$).

2.3 Heat and solute transfer equations

The equation of heat conduction is obtained from the energy conservation, and the transfer equation of solute including a linear interaction between solute and solid phases in porous media is given by the formulation of mass conservation as follows:

$$(\rho c)\frac{\partial T}{\partial t} = \frac{\partial}{\partial x}\left(\kappa_x \frac{\partial T}{\partial x}\right) + \frac{\partial}{\partial y}\left(\kappa_y \frac{\partial T}{\partial y}\right) + \frac{\partial}{\partial z}\left(\kappa_z \frac{\partial T}{\partial z}\right)$$
$$- (\rho c)_f \frac{\partial (U_x T)}{\partial x} - (\rho c)_f \frac{\partial (U_y T)}{\partial y} - (\rho c)_f \frac{\partial (U_z T)}{\partial z} \qquad (4)$$

$$\frac{\partial C}{\partial t_R} = \frac{\partial}{\partial x}\left(D_x \frac{\partial C}{\partial x}\right) + \frac{\partial}{\partial y}\left(D_y \frac{\partial C}{\partial y}\right) + \frac{\partial}{\partial z}\left(D_z \frac{\partial C}{\partial z}\right)$$
$$- \frac{\partial (V_x C)}{\partial x} - \frac{\partial (V_y C)}{\partial y} - \frac{\partial (V_z C)}{\partial z} - \lambda C \qquad (5)$$

$$t_R = t/R_f, \quad R_f = 1 + (1-n)/n \cdot \rho_s \cdot K_d$$

where
ρ_s: solid phase density,
V : seepage velocity vector,
D : dispersion coefficient,
K_d : adsorption coefficient,
$(\rho c)_f$: heat capacity of fluid,
T : temperature,
κ : equivalent thermal conductivity,
C : solute concentration of radioactive nuclide,
λ : decay parameter,
(ρc): equivalent heat capacity of porous medium,
c : specific heat,
U : average seepage velocity vector.

2.4 Computational technique and procedure

Fig.3 shows computational procedure of analyses in this study. The numerical solutions of groundwater flow are obtained from Eqs.(1),(2),(3) and linear elastic analysis by means of the finite element method as a coupling problem. Piezometric head and total stress are computed in a number of finite elements under boundary conditions, and the permeability distributions depend dynamically on fluid pressure and total stress.

On the other hand, solute concentration and temperature are obtained from combination of Eq.(3) with Eq.(4), and Eqs.(3) and (5) as a coupling problems of two equations. The computing schemes of Eqs. (4) and (5) are based on the modified fluid in cell method for saving the computation time and for maintaining the computational stability(Ito et al,1985a).

The FLIC method is a kind of integrated finite difference schemes. The calculation of FLIC method can be done by the following two steps. In the first step, an equation ignored the convective term of original equation is solved by the Eulerian way. Next, the moving locus of each particle can be computed by the Lagrangian way. The FLIC method provides both Lagrangian and Eulerian merits. For the sake of the treatment of boundary as well as saving economy of computation time, the modified FLIC method is applied to the triangular elements. The flow domain is divided into a number of small triangular elements like the finite element method.

3 MODEL OF CAVERN AND NUMERICAL CONDITIONS

The numerical analyses are done in a rock mass below the alluvial layer as shown in Fig.4. The cavern is B wide and Hc high, and its depth is H_o below the ground surface. The canister for reposing the radioactive nuclides is located at the bottom of cavern, and the length is H_b.
The groundwater flows uniformly under hydraulic gradient i=1/100. The filler in the cavern is packed after excavating the cavern. The two-dimensional stress analysis is carried out by assuming a state of plane strain.

All given physical quantities in the present computation are summarized in Table-1. The coefficient of vertical and horizontal acting pressure in rock mass is K=1.0 and the permeability at the non-stress condition is $k_{ox} = k_{oz} = 2.0 \times 10^{-8}$m/s. The values of inter-substance parameter are aE_c

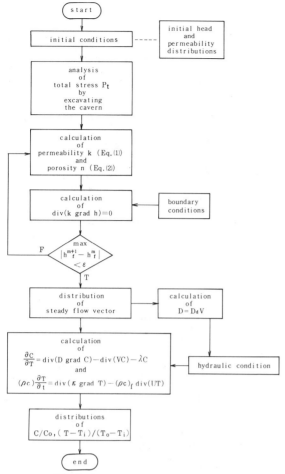

Fig.3 Computational procedure

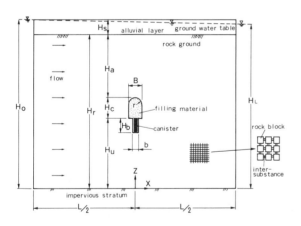

Fig.4 Cavern model with repository canister
for reposing radioactive nuclides

=3.5 MN/m². Other values correspond to given notations in Table-1. The number of elements is 705 and the nodes are 392.

4 COMPUTED RESULTS AND DISCUSSION

Some of typical results concerning with the permeability, heat transfer and nuclide migration around the cavern are demonstrated in Figs.5~9.

Fig.5 shows the space distribution of total stress in rock mass after excavating the cavern with canister, and Figs.6 and 7 are the non-dimensional permeability distributions of k_x/k_{ox}, k_z/k_{oz} and porosity ratio n/n_o (suffix o indicates the value at the initial state), respectively.

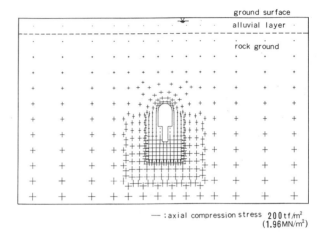

— : axial compression stress 200 tf/m²
(1.96 MN/m²)

Fig.5 Distribution of axial compression stress
P_{tx} and P_{tz}

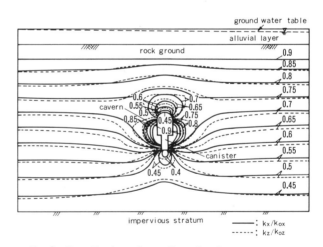

Fig.6 Distribution of non-dimensional
permeability k_x/k_{ox} and k_z/k_{oz}

Table-1 Computational conditions

aE_c (MN/m²)	block size s'_o (m)	interstitial width $2d_o$ (×10⁻⁵m)	porosity n_o (%)	initial permeability $k_{ox}(=k_{oz})$ (m/s)	thermal conductivity κ_r (W/mK)	hydraulic conditions and parameters
3.5	1.0	2.3	6.9×10^{-3}	2.0×10^{-8}	2.42	$H_o=60$m, $H_L=59$m, $L=100$m, $H_u=25$m $H_r=55$m, $H_s=5$m, $H_a=22.5$m, $H_c=7.5$m, $H_b=5$m, $b=2$m, $B=5$m $n_w=30\%$, $E_r=14.7$ GN/m², $M_r=0.3$ $K_w=1.0\times10^{-4}$m/s, $\gamma_s=19.6$kN/m³ $\gamma_w=9.8$kN/m³, $\gamma_r=25.5$kN/m³ $D_d=5.0\times10^{-4}$m, $\lambda=1.4\times10^{-15}$1/s $(\rho c)=2.2$MJ/m³K, $(\rho c)_f=4.2$MJ/m³K

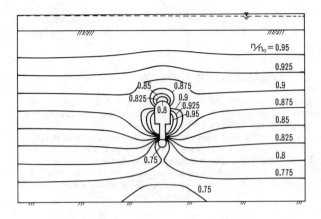

Fig.7 Distribution of non-dimensional porosity
n/n_0

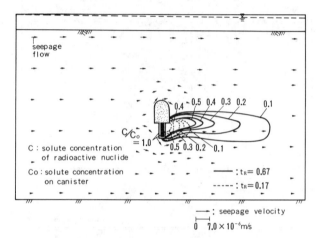

Fig.8 Flow pattern and distribution of
relative concentration C/C_0

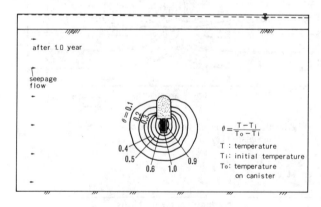

Fig.9 Distribution of non-dimensional
temperature θ after 1.0 year

The permeability and porosity distributions decrease with the depth because the total stress in the vertical direction increases except in the neighbor of cavern. This is caused by the deformation of rock mass in the vertical direction. Both distributions near cavern are complicated in accordance with those of total stress profiles. The changes of local distributions appear at the top and bottom of cavern and in the both sides of cavern.

Fig.8 shows the seepage velocity vectors in the steady state for the same condition as in Fig.5, and the relative concentration of nuclide solute C/C_0 around the canister at the non-dimensional times $t_R=0.17$, 0.67 (real time t= 6700 years, $1+(1-n)/n \cdot \rho_s \cdot K_d=10000$). The flow vector tends towards the downstream, but the magnitude of vector decreases with depth. The distribution of concentration is reasonable from inferring the flow pattern.

Fig.9 demostrates the relative temperature profiles $\theta = (T-T_i)/(T_0-T_i)$ after t=1.0 year. The profiles are partially distorted by the existence of flow. Judging from the computed results, the validity of computation may be recognized.

5. CONCLUSIONS

The heat transfer and nuclide migration around a disposal cavern of radioactive wastes in deformable rock mass have been studied by means of a computational model based on the FEM and FLIC method in a two-dimensional flow. Main results in this study are concluded as follows:

(1) The extensive possibility of coupled analysis in the stress and seepage is verified, and groundwater flow in responese of permeability change in deformable rock mass is clarified by the rock block model proposed in this study.

(2) The permeability in the direction of depth decreases as the depth increases, and the local distribution of permeability becomes larger at the side of cavern.

(3) The migration of nuclide may be largely dependent on groundwater flow, but the heat transfer is less affected by it because the heat conduction prevails.

REFERENCES

Snow,D.T.1968. Fracture deformation and changes of permeability and seepage upon changes of fluid pressure, Quarterly of Colorado School of Mines, Vol.63, No.1, Janu., pp.201-244.

Sato,K. and M.Iizawa 1982. Groundwater analysis of underground cavern by means of rock block model, Soils and Foundations, Vol.22, No.4, Dec., pp.103-117.

Gale J.E. et al. 1982. Hydrogeologic characteristics of a fractured granite, AWRC Conf. Groundwater in Fractured Rock, Canberra, pp.95~108.

Ito,Y., K.Sato and T.Shimizu 1985a. Analysis of heat and pollutant diffusions by means of the Modified Fluid in Cell(FLIC) method in fractured rock models, Proc. of the 29th JCH, Feb., pp.893-898.

Sato,K., T.Shimizu and Y.Ito 1985b. Applied study of pollutant and heat diffusion around underground cavern, 21th IAHR, Vol.1, Melbourn.

Sato,K. and Y.Ito 1985c. A study on heat conduction around caverns, JSCE, No.363/II-4, Nov., pp.97-106.

Sato,K. and Y.Ito 1986. Groundwater analyses of rock cavern by means of rock block model, JSCE, No.369/II-5, May, pp.51-60.

Ohnishi,Y., H.Shibata and A.Kobayashi 1986. Numerical technique for analysis of coupled thermal-hydraulic-mechanical problem by finite element method, JSCE, No.370 /III-5, Janu., pp.151-158.

Konzeption, Bau und Überwachung eines Abschlussbauwerkes gegen hohe Flüssigkeitsdrücke in Salzformationen

Design, construction, and monitoring of a bulkhead against high water pressures in salt formations
Conception, construction et contrôle d'un bouchon contre les hautes pressions d'eau dans une formation saline

H.H.SCHWAB, Institut für Grundbau, Boden- und Felsmechanik, TH Darmstadt, Bundesrepublik Deutschland
W.R.FISCHLE, Gesellschaft für Strahlen- und Umweltforschung mbH, Abteilung Tieflagerung, Schachtanlage Asse, Bundesrepublik Deutschland

ABSTRACT: It is reported on a bulkhead construction which seals tightly a gallery in a salt stock against single acting brine pressure of 6 MN/m2. About 30 m2 in cross section it consists of two supporting elements of salt concrete and in between a mastic core and an impervious blanket on the pressure side. Aspects for the selection of building materials are presented as well as measuring results.

RESUME: Il s'agit d'un cloison étanche que ferme étanchement une galerie au noyau de sel contre la pression de 6 MN/m2 de l'eau salée agissant d'un côté. Environ 30 m2 en travers il consiste en deux supports de béton de sel et entre les deux un noyau étanche de mastic et un revêtement d'étanchéité à côté de l'eau salée. Critères de sélectionner les materiaux de constuction sont présenté ainsi que les résultats des mesures.

ZUSAMMENFASSUNG: Es wird über ein Abschlußbauwerk berichtet, das eine Strecke im Salzgebirge gegen einseitig wirkenden Laugendruck von 6 MN/m2 dicht verschließt. Es hat einen Querschnitt von 30 m2 und besteht aus zwei Stützteilen aus Salzbeton, einer dazwischen liegenden Kerndichtung aus Sandasphalt und einer laugenseitigen Oberflächendichtung. Kriterien zur Auswahl der Baustoffe werden vorgestellt, Meßergebnisse berichtet.

1 EINLEITUNG

Entwurf und Bau eines Abschlußbauwerkes erfordern beides, das Können der beteiligten Ingenieure und den Einsatz der Wissenschaft. Theorie und praktische Erfahrung arbeiten Hand in Hand, einen sicher arbeitenden Streckenverschluß zu konzipieren und zu bauen, der auch bei dem größten anzunehmenden Unfall den Austritt von Radionukliden in die Umwelt verhindert.

Von der Gesellschaft für Strahlen- und Umweltforschung mbH und der Kavernen Bau- und Betriebsgesellschaft mbH wurde mit Unterstützung des Bundesministers für Forschung und Technologie ein umfangreiches Forschungs- und Entwicklungsprojekt zu Fragestellungen im Zusammenhang mit der Einlagerung radioaktiver Abfälle in Salzformationen durchgeführt. Neben anderen Teilprojekten wurde im stillgelegten Salzbergwerk Hope auf der 500 m Sohle ein Abschlußbauwerk als Streckenverschluß konzipiert und gebaut.

2 ZIELSETZUNG

Als größter anzunehmender Störfall bei der Lagerung radioaktiver Abfälle im Salzgebirge wird der Einbruch von Wasser bzw. Lauge in ein Endlagerbergwerk betrachtet. Bei einem derartigen Störfall muß ein Abschlußbauwerk wirksam, d.h. dicht sein. Der Zeitpunkt eines möglichen Störfalls ist nicht vorherzubestimmen. Deshalb muß während der gesamten Nachbetriebsphase eines Endlagers sichergestellt sein, daß das Bauwerk ohne Wartung ständig funktionsfähig ist.

3 KONZEPTION DES ABSCHLUSSBAUWERKES

Das in Abbildung 1 dargestellte Konzept für das Absperrbauwerk besteht aus insgesamt vier Funktionsteilen, dem luftseitigen Stützkörper (B), der Kerndichtung aus Sandasphalt (C), dem laugenseitigen Stützkörper (A) und der laugenseitigen Oberflächendichtung (D). Es ist ausgelegt für Laugendrücke bis 6 MN/m2. Kern- und Oberflächendichtung binden ca. 0.5 m in das Gebirge ein und dichten somit die oberflächennahen Auflockerungszonen ab.

Das Abschlußbauwerk wirkt als 'selbstregelnder' Verschluß. Eine Erhöhung des Druckes auf die Stirnfläche führt zu einer Verbesserung der Dichtwirkung. Verschiebungen des Bauteils A infolge Laugendruck führen zu einem erhöhten Druck in der Kerndichtung und damit zu einer Verringerung der Durchlässigkeit der Kontaktfuge und des umgebenden Salzgebirges. Das Bauwerk ist redundant. Wasserwegigkeiten um die Stirnflächendichtung herum werden durch die Kerndichtung abgesperrt. Der luftseitige Stützkörper (Bauteil B) reicht aus, um die Lasten aus dem Laugendruck in das Gebirge abtragen zu können.

FE-Berechnungen von Nipp (1986) haben gezeigt, daß durch den Laugendruck die Kontaktfuge zwischen Gebirge und Bauwerk geöffnet werden kann, dies besonders dann, wenn der Lastfall Laugendruck zu einem Zeitpunkt eintritt, zu dem noch kein ausreichend großer Konvergenzdruck gegeben ist. Die Berechnungen haben auch gezeigt, daß eine mögliche Relativverschiebung

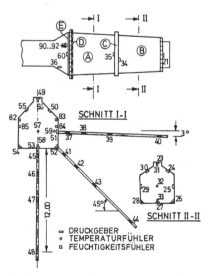

Bild 1. Abschlußbauwerk, Querschnittsfläche ca. 30 m² .

zwischen Bauwerk und Gebirge besondere Beachtung erfordert z.B. auch im Hinblick auf die Einspannung der
Kerndichtung zwischen den Stützteilen und dem Gebirge.
Über die Größe der zu erwartenden Konvergenz-
drücke und die dadurch hervorgerufene Beanspruchung
der einzelnen Bauteile bestehen bisher nur auf der
Grundlage von theoretischen Überlegungen bzw. von
FE-Berechnungen entwickelte Vorstellungen. Überbe-
anspruchungen der Baustoffe, verbunden mit örtlichen Rißbildungen sind demnach nicht auszuschließen,
sie dürfen jedoch die Funktionsfähigkeit des Absperr-
bauwerkes nicht nachteilig beeinflussen.
Die Verwirklichung der vorgeschlagenen Konzeption
setzt eine sorgfältige Auswahl der Baustoffe voraus.
Umfangreiche Voruntersuchungen im Labor waren notwendig, z.T. mußten neue Untersuchungsmethoden, besonders zur Überprüfung der Eignung des Sandasphaltes,
entwickelt werden.

4 AUSWAHL DER BAUSTOFFE

Die Baustoffe für das Abschlußbauwerk waren der Aufgabenstellung entsprechend zu wählen. Darüber hinaus
mußten die vorhandenen Transportmöglichkeiten und
die Bedingungen an der Einbaustelle berücksichtigt
werden. Für die Stützkörper (A) und (B) wurde Salzbeton verwendet, ein Gemisch aus Zement, Wasser und
Salz im Verhältnis 2:1:6. Umfangreiche Voruntersuchungen wurden von Fischle (1986) durchgeführt.
Verwendet man Steinsalz als Zuschlagstoff für einen
Beton, so erhält man einen Baustoff, der sich beim
Einbau den unregelmäßigen Oberflächen des Gebirges
sehr gut anpaßt, ausreichende Festigkeit besitzt und
dennoch dem Salzgestein vergleichbare Verformungseigenschaften hat, insbesondere auch hinsichtlich
seiner Kriecheigenschaften. Bei entsprechender Zusammensetzung besitzt er gegenüber gesättigter Salz-
lauge eine sehr niedrige Durchlässigkeit. Der Zuschlagstoff kann vor Ort gewonnen werden, zur Herstellung und zum Einbau lassen sich bewährte Technologien anwenden.

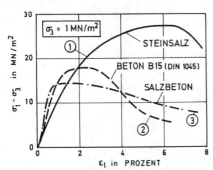

Bild 2. Triaxialversuch mit konstantem Seitendruck

In Bild 2 ist beispielhaft die Arbeitslinie des an
der Sperrenstelle angetroffenen Steinsalzes derjenigen des eingebauten Salzbetons gegenübergestellt.
Die Arbeitslinien wurden im Triaxialversuch bei einem
Seitendruck von 1 MN/m2 ermittelt. Zusätzlich aufgetragen ist die entsprechende Arbeitslinie eines Betons B 15, bei dem als Zuschlagstoff Flußkies verwendet wurde. Man kann erkennen, daß in dem hier in-
teressierenden Bereich kleiner Dehnungen der Salzbeton in seinen Verformungseigenschaften zufriedenstellend mit denjenigen des Salzgesteins übereinstimmt.
Bild 3 zeigt im oberen Teil Arbeitslinien bei unterschiedlichen Seitendrücken für Salzbeton und
Steinsalz. Darunter sind in gleicher Weise die zuge-
hörigen Volumenänderungskurven dargestellt. Daraus
ist zu ersehen, daß der Salzbeton durch Schubbean-
spruchung bei niedrigen Seitendrücken stärker zur Vo-
lumenvergrößerung neigt als das Steinsalz. Mit
zunehmendem Seitendruck kehrt sich diese Tendenz um,

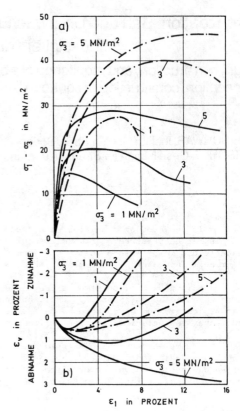

Bild 3. Triaxialversuch mit Salzbeton (ausgezogene
Linien) und Steinsalz (strich-punktierte Linien).

bei einem Seitendruck von 5 MN/m2 erfährt der Salzbeton beim Abscheren nur noch Verdichtung. Dies bedeutet, daß das Volumen des Salzbetons mit zunehmender
Ausnutzung der Scherfestigkeit ab- und somit seine
Dichtigkeit zunimmt. Je größer die Einspannung z.B.
durch Konvergenz oder durch Laugendruck ist, um so
günstiger wirkt sich dies auf die Dichtigkeit aus.
Auch örtlich auftretende Überschreitungen der Festigkeit des Salzbetons und damit einhergehende örtlich begrenzte Rißbildungen führen nicht zu einer Erhöhung der Durchlässigkeit, sofern eine ausreichende
Einspannung gegeben ist. In diesem Zusammenhang ist
an das Phänomen der 'Selbstheilung' zu erinnern. Darunter versteht man in der Bodenmechanik (Staudammbau
und Deponieabdichtungen) die Eigenschaft minerali-
scher Dichtungsschichten, vorhandene Risse durch den
Einspannungszustand wieder zu schließen, wobei ein
großer Teil der verlorengegangenen Scherfestigkeit
und Dichtigkeit wiedergewonnen wird. Derartige
Selbstheilungsvorgänge von mineralischen Abdichtungen, Salzgestein und Salzbeton sowie von bitumengebundenen Dichtungsmaterialien werden z.Zt. an der
Technischen Hochschule Darmstadt, FRG, untersucht.
Das Verhalten der Kontaktfuge zwischen Gebirge und
Bauwerk wurde im Scherversuch untersucht. Hierzu
wurde Salzbeton auf eine 'rauhe' Salzoberfläche auf-
betoniert und so abgeschert, daß die Scherung in der
Kontaktfuge erzwungen wurde. Das Ergebnis eines Ver-
suches mit unterschiedlichen Normalspannungen in der
Kontaktfuge ist in Bild 4 dargestellt. Die Darstel-

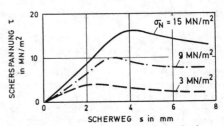

Bild 4. Scherfestigkeit in der Kontaktfuge
Salz/Salzbeton.

lung zeigt, daß in den am Absperrbauwerk zu erwartenden Normal- und Scherspannungsbereichen nur sehr geringe Relativverschiebungen zwischen Bauwerk und Gebirge auftreten werden.

Aufgrund vorliegender Erfahrungen mit bitumengebundenen Dichtungen im Staudammbau (Breth/Schwab 1979, Schwab 1983) wurde zur Herstellung der Kerndichtung die Verwendung von Sandasphalt, bestehend aus Sand, Kalksteinmehl und Bitumen B 80 im Gewichtsverhältnis 8:2:1 vorgeschlagen. Übertage hergestellte Blöcke mit den Abmessungen 50/20/20 cm wurden an der Einbaustelle mit einem gefüllten Bitumen heiß verklebt. Der Anschluß des 'Mauerwerks' an das Gebirge erfolgte durch Ausstampfen des Zwischenraums mit auf 150° C erwärmtem Sandasphalt. Der Aufbau der Sandasphaltdichtung erfolgte abschnittsweise. Nach Fertigstellung des Stützteils (B) wurde zunächst die Kerndichtung auf eine Höhe von ca. 1 m 'aufgemauert'. Dieser Abschnitt mußte solange frei stehen, bis ein ca. 1 m breiter Stützteil aus Salzbeton vorbetoniert worden war. Nach dem Abbinden des Salzbetons konnte der nächste Abschnitt der Kerndichtung in gleicher Weise erstellt werden. Durch diese Vorgehensweise wurde gewährleistet, daß sich bereits während der Herstellung der Dichtung diese durch ihr Eigengewicht vollflächig sowohl an das Gebirge als auch an den Salzbeton der Stützteile anpreßte. Damit war eine vollständige Ausfüllung aller Hohlräume im Bereich der Kerndichtung gegeben. Ein Ablösen der Dichtung im Bereich der Firste war bei der vorgegebenen Form ausgeschlossen. Der Sandasphalt mußte einerseits bis zu seiner Stützung ausreichend standfest, andererseits zur Anpassung an die Oberflächenrauhigkeit ausreichend verformbar sein. Dies wurde in umfangreichen Voruntersuchungen geprüft. Bild 5 zeigt das Ergebnis von einaxialen Standversuchen mit Sandasphalt unterschiedlicher Zusammensetzung. Dabei wurden sowohl die

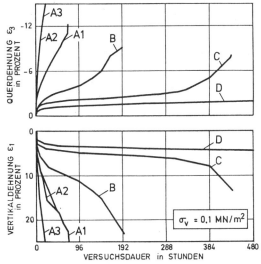

Bild 5. Prüfung der Standfestigkeit von Sandasphalt bei einaxialer Belastung (A - B - C - D = abnehmender Bitumenanteil)

Vertikaldehnung (unteres Diagramm) als auch die Querdehnung (oberes Diagramm) der vertikal mit 0.1 MN/m2 belasteten Prüfkörper über die Zeit beobachtet. Eingebaut wurde die Mischung C. Sie zeigte während einer Prüfdauer von 10 Tagen ausreichende Standfestigkeit. Andererseits konnte im Kompressionsversuch (siehe Bild 6) nachgewiesen werden, daß der Sandasphalt der Mischung C ausreichend verformbar ist. Durch die bei Belastung auftretenden Querdehnungen füllen sich auch größere Unebenheiten und Hohlräume relativ schnell.

Sowohl mit dem Salzbeton als auch mit dem Sandasphalt wurden eine Fülle weiterer Untersuchungen und Eignungsprüfungen ausgeführt, auf die hier nicht eingegangen wird. Mit den Untersuchungen konnte nachge-

wiesen werden, daß es möglich ist, auf der Grundlage von Laborversuchen Rezepturen zu entwickeln, die es ermöglichen, die Baustoffe an die im Einzelfalle gegebenen Randbedingungen anzupassen. Im vorliegenden Fall war es gelungen, sowohl die Forderungen nach Dichtigkeit und nach angepaßten Festigkeits- und Verformungseigenschaften zu erfüllen, als auch Herstellungs-, Transport- und Einbaubedingungen zu berücksichtigen.

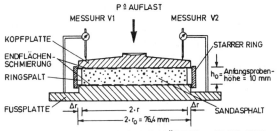

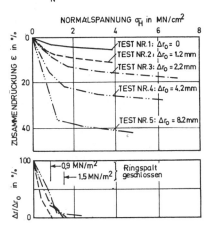

Bild 6. Prüfung der Verformbarkeit des Sandasphalts (Anpassung an Oberflächenrauhigkeiten).

Als Oberflächendichtung wurden laugenseitig abwechselnd je zwei Lagen aus Bitumenbahnen und aus 0.05 mm dicken Edelstahlbändern mit Heißbitumen verklebt. Ein vorgesetztes Mauerwerk dient zur Stützung und zum Schutze der Dichtung. Der Zwischenraum ist mit Mörtel verfüllt.

5 INSTRUMENTIERUNG UND MESSUNG

Vor, hinter und im Abschlußbauwerk sowie im umgebenden Gebirge wurde eine Vielzahl von Gebern zur Messung von Spannungen, Flüssigkeitsdrücken, Temperatur, Feuchtigkeit und Verschiebungen installiert. Bild 1 zeigt Lage und Art eines Teils der insgesamt 89 eingebauten Meßwertgeber. Ausführlicher wird hierüber von Fischle (1986) an anderer Stelle berichtet.

Alle Meßstellen können automatisch abgefragt und auf zwei voneinander unabhängigen Kabelwegen nach Übertage zur ca. 5 km entfernten Meßstation übertragen werden. Sämtliche Daten werden auf Magnetband aufgezeichnet und für die Weiterverarbeitung bereitgestellt.

6 AUSGEWÄHLTE MESSERGEBNISSE

Nach Fertigstellung des Bauwerkes im November 83 dauerte es bis zum April 85 bis die Lauge das Abschlußbauwerk erreichte. Der Flutungsverlauf ist im oberen Teil des Bildes 7 dargestellt. Im unteren Teil sind, ebenfalls über der Zeit, die Ergebnisse der Messungen von drei Absolutdruckaufnehmern aufgetragen. Der Geber Nr. 50 ist in der Firstschräge, der Geber Nr. 49 in der Firste angeordnet und zwar in der Mitte des vorderen Stützteils (A) (siehe auch Bild 1). Mit dem Geber Nr. 51 wird der Absolutdruck im Bereich der Ulme gemessen. Aus der Darstellung im Bild 7 ist zu erkennen, daß über einen Zeitraum von ca. 1 Jahr nach Herstellung des Bauwerkes keine nennenswerten Druckveränderungen gemessen werden konnten. Ab Oktober 84 bis zum beginnenden Einstau am 18.4.85 zeigten alle drei Geber eine zwar unterschiedliche, aber deutliche Zunahme des Druckes. Die unmittelbare Auswirkung des Einstaus auf alle drei Geber in der darauffolgenden Zeit bis Ende 1985 ist offensichtlich.

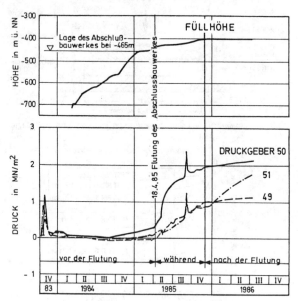

Bild 7. Füllkurve und Anzeige der Absolutdruckgeber 49, 50 und 51

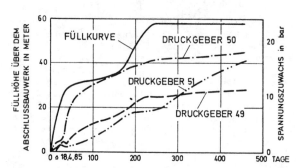

Bild 8. Füllkurve und gemessener Spannungszuwachs an den Meßstellen 49, 50 und 51 bezogen auf den Einstaubeginn.

Seit Ende 1985 erhöht sich der Laugendruck nicht mehr. Trotzdem steigen die Absolutdrücke an den Gebern weiter an. Bild 8 zeigt einen Ausschnitt der Füllstandskurve und des zeitlichen Verlaufs der Meßwerte der ausgewählten Geber, wobei nur noch der seit dem 18.4.85 gemessene Zuwachs aufgetragen wurde. Zum einen wird hier nochmals die unmittelbare Abhängigkeit der Druckanzeige von der Füllstandshöhe deutlich. Zum anderen ist zu erkennen, daß sich bei konstanter Füllhöhe der Druckanstieg an den ausgewählten Gebern fortsetzt und zwar z.T. mit einer deutlich größeren Geschwindigkeit. Wie die Scherversuche in der Kontaktfuge Salzgestein/Salzbeton gezeigt haben (Bild 4), werden schon bei sehr geringen Relativverschiebungen sehr große Scherkräfte geweckt. Es ist deshalb nicht zu erwarten, daß bei dem bisher am Abschlußbauwerk anstehenden Laugendruck große Verschiebungen des Bauwerkes relativ zum Gebirge eingetreten sind. Die bisher vorliegenden Verschiebungsmessungen lassen den gleichen Schluß zu. Dies bedeutet jedoch, daß der Druckanstieg nach Beendigung der Einleitung auf Konvergenz zurückzufüh-

ren ist. Die Geschwindigkeit, mit der der Absolutdruck an den ausgewählten Gebern zunimmt, ist in der Tabelle 1 zusammengestellt. Dort wird unterschieden in den Zeitraum vor und nach der Flutung des Abschlußbauwerkes. Geber 49 und 50 sind von der Lage her miteinander vergleichbar. Sie haben in der Zeit nach der Flutung in etwa die gleiche Anstiegsrate. Der seitlich im Bereich der Ulme angeordnete Geber 51 hat nach der Flutung eine zehnfach größere Anstiegsrate als vorher. Ähnliches Verhalten zeigen auch die übrigen, hier nicht dargestellten Geber. Diese Ergebnisse stehen zunächst im Widerspruch zu den Berech-

Tabelle 1. Druckanstiegsrate an den Meßstellen 49, 50 und 51

Tabellenwerte in MN/m² je 100 Tage	FIRSTE GEBER 49	SCHRÄGE GEBER 50	ULME GEBER 51
vor der Flutung	0.27	1.38	0.43
nach der Flutung	0.63	0.69	4.43

nungsergebnissen, wonach die konvergenzbedingte Druckrate mit der Zeit abnimmt.

Die bisher vorliegenden Meßergebnisse lassen sich wie folgt interpretieren: (i) Konvergenz findet statt, (ii) bis Anfang IV/84 wirkt sich diese wegen fehlendem Kraftschluß an den Druckgebern nicht aus, (iii) ab diesem Zeitpunkt baut sich der Kraftschluß infolge der Konvergenz erst auf, (iv) der durch die Flutung bedingte Laugendruck wirkt sich unmittelbar auf den Kontaktdruck aus und bewirkt (v) den besseren Kraftschluß an den Gebern, dies hat (vi) zur Folge, daß nach der Flutung die konvergenzbedingte Druckanstiegsrate zunächst größer ist als vor der Flutung. Aus dem Verlauf läßt sich jedoch bereits jetzt erkennen, daß (vii) die konvergenzbedingte Druckanstiegsrate, wie in den Berechnungen vorhergesagt, abnehmen wird. Desweiteren ist offensichtlich, daß (viii) bisher keine Relativverschiebungen zwischen dem Abschlußbauwerk und dem Gebirge stattgefunden hat, was auch nach den Ergebnissen der Laborversuche nicht zu erwarten war. Es muß somit bis jetzt vom Grenzfall des schubfesten Verbundes ausgegangen werden. Alle bisher vorliegenden Meßergebnisse geben (ix) keinerlei Hinweise auf eine Undichtigkeit des Abschlußbauwerkes.

7 SCHLUSSBEMERKUNG

Das in dem vorliegenden Beitrag vorgestellte Konzept für ein Abschlußbauwerk erfüllt die hohen Anforderungen an einen Streckenverschluß eines Endlagers für radioaktive Abfälle. Die Herstellung konnte unter in-situ Bedingungen erfolgen, die Arbeitsweise und die Funktionsfähigkeit des selbstregelnden Verschlusses wurde im Maßstab 1:1 geprüft.

REFERENCES

NIPP, H.K. 1986. Versuchsbegleitende Rechnungen zur Geomechanik und zum Abschlußbauwerk. Statusseminar zum FE-Projekt Hope, GSF-Bericht 20/86, Gesellschaft für Strahlen- und Umweltforschung, München, p.253-279.
Fischle, W.R. & Schwieger, K. & Stöver, W.H. 1986 Konzeption, Bau und Überwachung des Abschlußbaubauwerkes. Veröffentlicht wie vor, p.230-252.
Fischle, W.R. 1986. Konstruktion eines Streckenverschlusses und Messung unter Laugendruck. Vortrag zum SMRI Meeting 1986.
Breth, H. & Schwab, H.H. 1979. Zur Eignung des Asphaltbetons für die Innendichtung von Staudämmen. Wasserwirtschaft 69 (1979), Heft 11, p.348-351.
Schwab, H.H. 1983. Asphaltbeton für die Innendichtung von Staudämmen - Ein Beitrag zur Frage nach den zeitabhängigen Verformungen. Beiträge zum Staudammbau und Bodenmechanik. Festschrift zum 70. Geburtstag von H. Breth, Inst. f. Grundbau, Boden- und Felsmechanik, TH Darmstadt.

Felskavernen mit Innendichtung zur Lagerung von Sonderabfällen

Lined rock caverns for the storage of hazardous waste
Cavernes avec étanchement intérieur pour le stockage de déchets toxiques

STEPHAN SEMPRICH, Bilfinger & Berger Bauaktiengesellschaft, Mannheim, Bundesrepublik Deutschland
SEPP-RAINER SPEIDEL, Bilfinger & Berger Bauaktiengesellschaft, Mannheim, Bundesrepublik Deutschland
HANS-JOACHIM SCHNEIDER, Kavernen Bau- und Betriebs GmbH, Hannover, Bundesrepublik Deutschland

ABSTRACT: For reasons of environmental protection the storage of hazardous waste in unlined rock caverns is possible to a very limited extent only. Therefore, the authors have recently developed technologies for the lining and sealing of rock caverns. In the process, sealing systems of synthetic materials or metals have proved suitable. Synthetic materials can be used in the form of either sheets or coatings with various materials such as epoxy resins, polyethylenes etc. being used. Metal sealings consist of thin sheets or foils which are either welded or bonded. In either case, the structural design must provide for a leakage control possibility. The article describes the design principles, the structural and operational aspects as well as the control measures with regard to the planning and execution of lined rock caverns for the storage of hazardous waste.

RESUME: Pour des raisons de la protection de l'environnement, le stockage de déchets toxiques dans des cavernes sans étanchement intérieur n'est possible que dans une mesure restreinte. C'est pourquoi les auteurs ont récemment développé des technologies pour le revêtement et l'étanchement de cavernes. Dans le cadre de ces travaux des systèmes d'étanchement sur base de matières plastiques ou de métaux se sont revélés très efficaces. Les étanchements plastiques peuvent être employés sous forme de pans ou d'enduits protecteurs, en utilisant des matières plastiques, telles que la résine époxyde, les polyéthylènes etc. Des étanchements métalliques sont utilisés sous forme de tôles ou de feuilles métalliques qui sont soudées ou collées. En tout cas, pendant la construction, une possibilité de contrôle de fuites doit etre prévue. L'article décrit les principes du projet, les aspects relatifs à la construction et l'exploitation ainsi que les mesures de contrôle en vue de l'établissement de l'étude et de la construction de cavernes avec étanchement pour le stockage de déchets toxiques.

ZUSAMMENFASSUNG: Aus Gründen des Umweltschutzes ist die Lagerung von Sonderabfällen in nicht ausgekleideten Felskavernen nur sehr begrenzt anwendbar. In jüngster Zeit wurden daher von den Autoren Technologien für ausgekleidete und abgedichtete Kavernen entwickelt. Dichtungssysteme in Verbindung mit Kunststoffen oder Metallen haben sich dabei als geeignet erwiesen. Kunststoffabdichtungen lassen sich in Form von Bahnen oder Beschichtungen einsetzen, wobei verschiedene Kunststoffarten wie Epoxidharze, Polyäthylene u.a. Verwendung finden. Metallische Dichtungen werden in Form von dünnen Blechen oder Folien eingesetzt, die miteinander verschweißt oder verklebt werden. In jedem Fall ist konstruktiv eine Leckagekontrollmöglichkeit vorzusehen. Der Beitrag beschreibt Entwurfsgrundlagen, bautechnische und betriebliche Aspekte sowie Überwachungsmaßnahmen im Hinblick auf die Planung und Ausführung von ausgekleideten Felskavernen zur Lagerung von Sonderabfällen.

1 EINLEITUNG

Sowohl bei der Zwischen- als auch bei der Endlagerung von Sonderabfällen ist ein Abschluß der Schadstoffe von der Umwelt eine unabdingbare Voraussetzung an ein Deponiekonzept.

Bei einer Zwischenlagerung von Sonderabfällen ist das Auskleidungssystem als die technische Barriere auf die geplante Betriebsdauer auszulegen und ihre Funktionsfähigkeit durch geeignete Kontrollen zu überwachen. Handelt es sich jedoch um ein Endlager, so kann aufgrund der zeitlich begrenzten Lebensdauer der technischen Barriere der Abschluß der Schadstoffe von der Biosphäre nur durch die zusätzliche Nutzung von geologischen Barrieren sichergestellt werden. Diese Forderungen lassen sich im Hinblick auf eine untertägige Lagerung nicht nur mit Kavernen im Salz sondern auch mit Felskavernen erfüllen. Das setzt allerdings voraus, daß das die Kaverne umgebende Gebirge die Schadstoffe vom Niederschlags- und Grundwasserbereich sowie von der Atmosphäre abzuschirmen vermag.

Nachfolgend werden verschiedene Baukonzepte solcher Felskavernen vorgestellt. Im einzelnen wird dabei auf die Entwurfsgrundlagen, die bautechnische Ausführung und den Betrieb der Kavernenbauwerke eingegangen.

2 ENTWURFSGRUNDLAGEN

Felskavernen zum Zweck der Lagerung von Abfallstoffen lassen sich nur verwirklichen, wenn bezüglich der geologischen und der technischen Barriere bestimmte Voraussetzungen erfüllt sind.

Wesentliche geologische Voraussetzungen hängen mit der petrographischen und der stratigraphischen Beschaffenheit des Gebirges sowie mit Einflüssen aus der Tektonik zusammen (Tab. 1).

Die Durchlässigkeit des Gesteins bzw. des Gebirges, die Entfernung und die Höhenlage in Bezug zu einem durchströmten Grundwasserbereich sowie mögliche Inhomogenitäten mit einer Verbindung zum Grundwasser sind die maßgeblichen hydrogeologischen Parameter, nach denen ein Standort zu bewerten ist (Tab. 1). Als geeignete Felsformationen dürften z.B. Anhydrit, Tonsteine und Granit in Frage kommen.

Die Anforderungen an das Auskleidungssystem können aus den Beanspruchungen, die auf die Auskleidung einwirken, und denen sie widerstehen muß, abgeleitet werden:

- Stützung des Hohlraums im Bau- und Endzustand

- Undurchlässigkeit des Dichtungssystems gegenüber den Abfallstoffen und dem Bergwasser

- Leckagekontrollmöglichkeit

Tabelle 1. Geologische und hydrogeologische Faktoren zur Standortbewertung von Felskavernendeponien

Petrographie	Stratigraphie	Tektonik	Hydrogeologie
Gesteinseigen-schaften	Lithologische Diskontinuitäten	Lagerung und Trennflächengefüge	Anbindung an die Biosphäre
Physikalisch-chemische Zusammensetzung	Schichtmächtigkeit	großräumige Struktur	Gebirgsdurchlässigkeit GW-Zutrittswege
Porosität	Wechsellagerung versch. Gesteine	Verfaltung	GW-Strömung
Permeabilität	Abfolge im Hangenden/Liegenden	Schieferung	GW-Mengen
Lösungsverhalten		Klüftung	Entfernung zu Aquiferen
Sorptionsvermögen	Schichtgrenzenphenomäne	Störungen	
Reaktions-/Alterations-verhalten		Alterations- und Verwitterungsgrad	Verbindung zu Trinkwasser-einzugsgebieten
Festigkeit			

- Ausreichende Dehnfähigkeit der Dichtungsmateriali-en, um Verformungen der Kavernenwandung schadlos aufnehmen zu können

- Widerstand gegen die mechanische Beanspruchung bei Be- und Entladevorgängen

- Schadlose Aufnahme der Beanspruchung aus dem Ei-gengewicht der Abfallstoffe

- Chemische Widerstandsfähigkeit der Dichtungsmate-rialien gegenüber dem Lagergut und dem Bergwasser

- Alterungsbeständigkeit

Standsicherheit der Kaverne und Dichtigkeit des Auskleidungssystems sind demnach die wesentlichen Elemente bezüglich der Funktionsfähigkeit der tech-nischen Barriere.

Felskavernendeponien können nach verschiedenen Konzepten angelegt werden. Dabei wird unterschieden zwischen Stollenbauweise, Kavernenbauweise und Schachtbauweise. Entscheidend für die Auswahl ist dabei die Funktion als Zwischen- oder Endlager, die Art der Abfälle, die Lagertechnik und die Eigen-schaften des Gebirges.

Die Stollenbauweise (Abb. 1) hat beispielsweise den Vorteil, daß sich auch noch in Felsarten mit größerer Verformbarkeit kostengünstige Hohlräume erstellen lassen. Die Möglichkeit einer gleichzei-tigen Bauausführung an mehreren Betriebspunkten

bewirkt kürzere Bauzeiten. Eine Unterteilung der einzelnen Stollen in einzelne Kammern erlaubt eine flexible Betriebsweise. Nachteilig gegenüber den anderen Bauweisen wirkt sich dagegen das ungünstige Volumen/Oberflächen-Verhältnis der Stollen aufgrund der höheren Kosten für das Dichtungssystem aus.

Die Felsausbrucharbeiten erfolgen i.a. im Bohr-und Sprengbetrieb. Um an der Hohlraumwandung auf-tretende Auflockerungen und damit verbundene erhöhte Wasserwegigkeiten möglichst weitgehend zu vermeiden, ist der Außenbereich des Ausbruchquerschnittes gebirgsschonend zu schießen. In weniger festen Gesteinen kann darüberhinaus der vollständige Aus-bruch mittels Fräse die wirtschaftlichste Lösung darstellen. Ausgerundete Querschnittsformen sind vorzuziehen.

3 AUSKLEIDUNGSSYSTEME

Die Auskleidungssysteme von Felskavernen bestehen aus mehreren Komponenten, die in geeigneter Weise kombi-niert werden, um den standortspezifischen und den sich aus dem eingelagerten Sonderabfall ergebenden Anforderungen optimal zu entsprechen. Auf die einzel-nen Systemkomponenten und ihre Kombinationsmöglichkei-ten wird nachfolgend eingegangen.

Zur Aufnahme statischer Beanspruchungen aus Ge-birgsdruck sowie Lasten aus den eingelagerten Abfall-

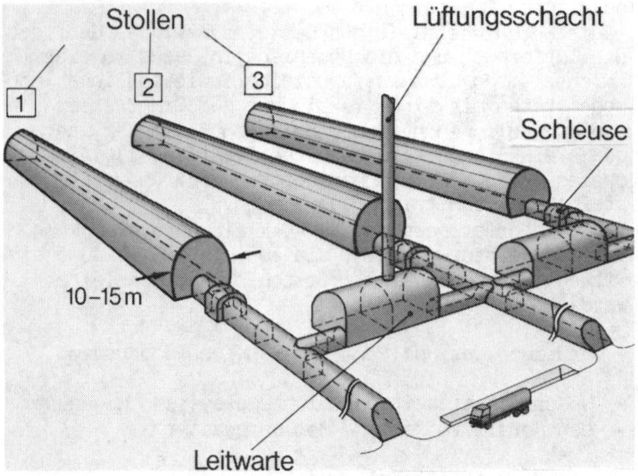

Abb. 1. Beispiel für die Tunnelbauweise

Abb. 2. Einbau der Auskleidung in Spritztechnik

254

stoffen dient in der Regel eine Kavernenauskleidung aus Stahlbeton. Aus wirtschaftlichen Gründen empfiehlt sich der Einbau einer solchen Auskleidung in Spritztechnik (Abb. 2). Stahlbetonkonstruktionen sind jedoch nicht diffusionsdicht und i.a. nicht wasserdicht. Darüberhinaus ist Beton, wenn überhaupt, nur gering säurebeständig, so daß ergänzende Dichtungsmaßnahmen notwendig sind.

Zur Abdichtung werden in der Regel Kunststoff- oder Metallauskleidungen verwendet.

Kunststoffabdichtungen werden in Form von Beschichtungen oder Auskleidungen mit vorkonfektionierten Bahnen vorgenommen.

Kunststoffbeschichtungen können auf trockene, in der Regel kantenfreie Oberflächen unregelmäßiger Geometrie relativ problemlos, z.B. durch Spritzen, aufgebracht werden. Die rißüberbrückenden Eigenschaften können durch Gewebeeinlagen verbessert werden. Bergseits anstehender Wasserüberdruck kann nur in geringer Größe aufgenommen werden. Bereits ein Druck von wenigen Metern Wassersäule führt zu flächigen Ablösungen mit daraus resultierenden Undichtigkeiten der Beschichtung. Bergwasser muß daher durch Drainageschichten oder andere Maßnahmen abgeführt werden.

Neben den Beschichtungen am weitesten verbreitet sind heute Abdichtungen mit vorgefertigten Folien, die auf die Betonoberfläche aufgeklebt oder mechanisch befestigt werden. Untereinander werden die einzelnen Bahnen ebenfalls verklebt oder verschweißt. Verklebte Bahnen haben den Vorteil, daß sich eine Wasserwegigkeit hinter den Folien nicht ausbilden kann. Mechanisch befestigte Bahnen haben dagegen gute rißüberbrückende Eigenschaften und passen sich Verformungen in weiten Grenzen an. Praktisch alle Folien sind gegen mechanische Beschädigungen empfindlich. Dagegen sind sie gegen eine Vielzahl von Chemikalien beständig. Eignungstabellen für die meisten gängigen Folien liegen vor.

Zum Schutz der Dichtung gegen mechanische Beschädigungen von innen ist eine innenliegende bewehrte Betonschale geeignet. Nachteil einer inneren Betonschale ist jedoch, daß undichte Stellen in der Kunststoffdichtung nur mit erheblichem Aufwand geortet und repariert werden können. Als zusätzliche Maßnahme sollte deshalb grundsätzlich eine Außendrainage vorgesehen werden, um örtlich auftretenden Wasserüberdruck auf möglichst niedrige Werte zu begrenzen. In Abb. 3 ist ein solches System mit einer flächigen Drainageschicht aus Porenspritzbeton dargestellt.

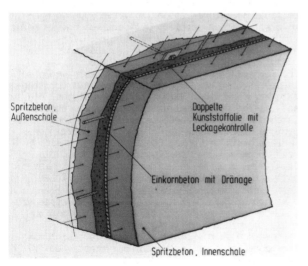

Abb. 3. Auskleidungssystem mit Dichtungsfolien aus Kunststoff

Von ganz besonderer Bedeutung sind bei der Herstellung von Abdichtungen die qualitätssichernden Maßnahmen. Sämtliche Nähte sowie das Dichtungsmaterial selbst müssen auf Fehl- und Leckstellen überprüft

werden. Bei Kunststoffsystemen wird dies in der Regel durch Hochspannungsprüfung vorgenommen. Die Kosten für Ortung und Beseitigung nur einer Fehlstelle im Dichtungssystem einer in Betrieb befindlichen, mit Sonderabfällen beladenen Kaverne, stehen in keinem Verhältnis zu den Aufwendungen für die Qualitätssicherung während der Bauzeit.

Neben den Kunststoffen bieten sich auch metallische Auskleidungen (sogenannte Liner) als Dichtungssysteme an. Durch die Entwicklung geeigneter Klebe- und Schweißtechniken ist es möglich geworden, extrem dünne Folien oder Bleche mit ausgezeichneten rißüberbrückenden Eigenschaften zu verarbeiten. Nachdem der Materialpreis bei den zur Anwendung kommenden hochwertigen Edelstählen eine erhebliche Rolle spielt, konnte durch die Reduzierung der Wandstärken eine bedeutende Einsparung erzielt und diese Art der Auskleidung für viele Anwendungen erst preislich akzeptabel gestaltet werden. Die Forderung nach Diffusionsdichtigkeit kann durch metallische Auskleidungen voll erfüllt werden. Vorbehaltlich spezieller Anforderungen an die Beständigkeit des Liners aus dem Lagergut sind für den Regelfall austenitische Stähle wie etwa X 5 CrNiMo 17 12 2, Werkstoff-Nr. 1.4401, eine Materialgüte für schwache bis mäßig korrosive Umgebung oder X 1 CrNiMoCu 25 2 05, Werkstoff-Nr. 1.4539, ein hochkorrosionsbeständiges, weitgehend säurefestes Material geeignet.

Mit verklebten Kunststofffolien vergleichbar ist ein Edelstahlliner, der aus einer genoppten Folie mit einer Dicke bis zu 0,2 mm besteht. Die Folie wird mit einem gegenüber vielen Chemikalien beständigen Zweikomponenten-Polyurethan-Kleber auf die Betonoberfläche geklebt. Von besonderem Vorteil ist, daß die Kombination aus Noppenfolie und Kleber lokal auftretende Dehnungsspitzen ausgleichen und so mehrere millimeterbreite Risse in der Betonoberfläche sicher überbrücken kann (Abb. 4). Die maximal zulässige Riß-

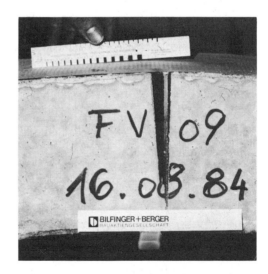

Abb. 4. Versuch zur Bestimmung der rißüberbrückenden Eigenschaften von Edelstahlfolien

weite hängt im wesentlichen von dem Druck ab, der über dem Riß auf die Folie einwirkt. Der Folienliner wird in 1000 mm breiten Bahnen angeliefert und durch überlappte Verklebung gestoßen. Eine gespachtelte Spritzbetonoberfläche ist als Untergrund ausreichend; besondere Vorbehandlungsmaßnahmen sind in der Regel nicht erforderlich, jedoch darf ebenso wie bei den Kunststoffsystemen der Untergrund nicht durchfeuchtet sein. Beschädigungen des Folienliners können durch einfaches Überkleben repariert werden.

Wird eine hohe mechanische Belastbarkeit gefordert oder ist die Kavernenoberfläche zum Beispiel wegen permanenter Feuchtigkeit zur Verklebung nicht geeignet, bietet sich die Verwendung von Edelstahldünnblech-Linern an. Hierbei werden vorkonfektionierte

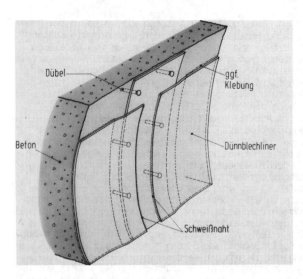

Abb. 5. Abdichtung mit Hilfe von Dünnblechlinern aus Edelstahl

Abb. 7. Geotechnische Messungen zur Überwachung des Bauwerks

Dünnbleche von etwa 0,5 bis 2 mm Dicke mit Kehlnähten auf vorab befestigte Blechstreifen des gleichen Materials aufgeschweißt. Ist Verkleben möglich, können die Dünnblechtafeln zwischen den Blechstreifen zusätzlich aufgeklebt werden, um eine Wasserwegigkeit hinter den Blechen im Fall von Leckagen zu vermeiden. Die Haltebleche werden angedübelt (Abb. 5).

Wie bei den Kunststoffsystemen besteht auch bei den Edelstahllinern die Möglichkeit, eine innere Betonschale vorzublenden. Die Möglichkeiten, den Liner im Falle von Leckagen zu reparieren, werden hierdurch allerdings erschwert.

In den bisherigen Ausführungen wurden Auskleidungen aus verschiedenen Materialien vorgestellt, bei denen jeweils eine Schicht die tragende (Spritzbetonschale) und eine andere die Dichtfunktion (Kunststoffbahnen, Kunststoffschichten, Edelstahlliner) übernimmt. Für die praktische Anwendung wird jedoch aus Sicherheitsgründen die Dichtigkeit durch zumindest zwei unabhängige Barrieren gewährleistet. Solche Systeme können durch zweckmäßige Kombination der vorgestellten Lösungen aufgebaut werden. In Abb. 6 ist als Beispiel das Modell eines zweischaligen Linersystems gezeigt. Auf die direkt auf den Beton geklebte Edelstahlnoppenfolie folgt eine begrenzt druckfeste, poröse Schicht, die als Abstandshalter für die zweite Schale dient, zugleich jedoch auch wegen der Porosität den Anschluß eines Leckdetektionssystemes etwa auf Unterdruckbasis

ermöglicht. Auf der Innenseite ist ein gas- und flüssigkeitsdicht verschweißter Edelstahldünnblechliner angeordnet.

Das beschriebene zweischalige System dient als Beispiel für eine Vielzahl möglicher Lösungen, die durch Kombination der vorgestellten Komponenten gewonnen werden können. Die Kavernenauskleidungen können damit in weiten Grenzen und äußerst flexibel den standortbedingten, geologischen und aus den eingelagerten Sonderabfällen resultierenden Beanspruchungen angepaßt werden. Bei Einbeziehung weiterer Dichtungsschichten und geologischer Barrieren können auch vielfach redundante Systeme, sogenannte "Multi-Barrier-Systems" für höchste Sicherheitsanforderungen realisiert werden.

4 BETRIEB UND KONTROLLMASSNAHMEN DER FELSKAVERNEN

Felskavernen eignen sich sowohl zur Zwischenlagerung flüssiger wie fester Sonderabfälle.

Feste Abfälle sind unter dem Gesichtspunkt der Rückholbarkeit grundsätzlich in Gebinden zu lagern. Die Abfälle können dabei nach Gefahrenklasse und chemischem Reaktionsvermögen in verschiedenen Kammern bzw. Kavernen sortiert gelagert werden. In bezug auf mögliche Störfälle sind Wettertore, Sicherheitsschleusen, Rückhaltesperren u.ä. vorzusehen.

Endzulagern in Felskavernen sind grundsätzlich nur feste Sonderabfälle. Werden die Abfälle in Gebinden gelagert, so ist der verbleibende Hohlraum vollständig zu versetzen. Der Versatz hat die Aufgabe, die Standsicherheit des Hohlraumes langzeitig zu gewährleisten und die Abfallbehälter als zusätzliche Barriere dicht einzuschließen. Aufgrund ihrer geringen Durchlässigkeit und hohen Festigkeit eignen sich hierzu besonders Reagips-Flugaschengemische, die hydraulisch versetzt werden können und somit eine hohlraumfreie Verfüllung der Kaverne ermöglichen.

Selbstverständlich erfordert eine Felskavernendeponie Maßnahmen zur Überwachung im Bau- und Betriebszustand. Das gilt sowohl für die geologische als auch für die technische Barriere.

Das den Hohlraum umgebende Gebirge läßt sich mit Hilfe geotechnischer Messungen überwachen (Abb. 7). Gegebenenfalls sind darüberhinaus zusätzliche Inspektionsstollen erforderlich. Der Nachweis der Dichtigkeit des Auskleidungssystems ist erstmalig durch sorgfältige Dichtigkeitsprüfungen im Zuge der Bauausführung zu erbringen. Der entsprechende Nachweis im Betriebszustand kann durch den Einbau von Leckagekontrollsystemen geführt werden.

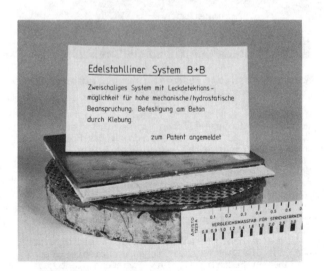

Abb. 6. Modell eines zweischaligen Linersystems mit Leckdetektionsmöglichkeit

Models of quasi-static and dynamic fluid-driven fracturing in jointed rocks

Modèles quasi-statiques et dynamiques de fracturation produite par des fluides dans un milieu rocheux fracturé

Modelle für einen quasi-statischen und dynamischen flüssigkeitsgetriebenen Bruchmechanismus im geklüfteten Fels

R.J.SHAFFER, Lawrence Livermore National Laboratory, Calif., USA
F.E.HEUZE, Lawrence Livermore National Laboratory, Calif., USA
R.K.THORPE, Lawrence Livermore National Laboratory, Calif., USA
A.R.INGRAFFEA, Cornell University, Ithaca, N.Y., USA
R.H.NILSON, S-CUBED, La Jolla, Calif., USA

ABSTRACT: We describe the development and the applications of a numerical model to simulate fluid-driven fracturing in rock masses. It is a finite element computer program which couples solid mechanics, fracture mechanics and fluid mechanics. The fractures are driven either in a quasi-static fashion by conventional hydrofracturing liquids, or in a dynamic fashion by gases from burning solid propellants. Fracture propagation can be arbitrarily modeled in a mixed-mode, in media which already contain discontinuities. The code is applied to the analysis of stimulation of tight gas reservoirs such as the Gas Sands in the Western U.S. and the Coal Beds in the Southern U.S.

RESUME: On présente un nouveau modèle numérique qui simule la propagation de fractures par des fluides, dans les milieux rocheux. Ce programme d'éléments finis intègre de facon couplée la mecanique du milieu continu, la mecanique de la fracture et la mecanique des fluides. Les fractures se propagent de façon quasi-statique, comme en hydrofracturation classique, ou de façon dynamique, comme lorsque poussées par la déflagration de propergols. Le cheminement des fractures est arbitraire, en mode mixte, dans un milieu qui peut déjà être discontinu. On présente des applications à l'analyse de la fracturation dans les réservoirs de gaz à roches très peu perméables tels que les grès lenticulaires de l'ouest des U.S.A; on décrit aussi une analyse de fracturation pour dégazage de bancs de charbon dans le sud des U.S.A.

ZUSAMMENFASSUNG: Wir berichten über die Entwicklung und die Anwendung eines numerischen Modelles für Flüssigkeit betriebenen Bruch in geklüftetem Fels. Das Modell ist ein finites Element Computermodell das die Mechanik der Festkörper, der Flüssigkeiten und der Brüche gleichzeitig berechnet. Die Brüche werden entweder quasistatisch oder dynamisch durch Brennabgase von Festantriebsstoff betrieben. Das Wachstum von Brüchen kann beliebig in Körpern berechnet werden die schon Diskontinuitäten enthalten. Das Computermodell wird auf die Anregung von Gas Reservoirs wie die Gas Sande in den westlichen Vereinigten Staaten und wie die Kohlelagerungen in den südlichen Vereinigten Staaten angewendet.

1. INTRODUCTION

To recover gas from low permeability formations it is necessary to "stimulate" the rock reservoirs, that is, to produce large, man-made fractures that penetrate the reservoirs and drain the gas toward wells. Fractures can be created either by injecting fluids under presure (hydrofracturing) or by burning solid propellants in the wells. To use either technique effectively, one must be able to predict how the rock will fracture. Rock reservoirs already contain natural fractures and joints, making it difficult to predict the effects of induced fracturing. An analysis of fracturing in such a medium must include models for the joint systems, and for the preexisting cracks packed with infill material. The joints are subject to compressive and shear forces; if compression becomes large, the joints become very stiff, and if shearing becomes too large, slippage occurs. Consequently, in a useful model, joint properties must change with stress conditions. In the Western gas sands, interfaces are the regions where shales and sandstones meet, and these interfaces behave much like joints.

We have developed a computer code, FEFFLAP (Finite Element Fracture and Flow Analysis Program), that has enabled us to make great progress in describing the complex physics of fluid-driven fractures propagating in jointed media [1,2]. The coupled FEFFLAP model includes solid mechanics, fracture mechanics, and fluid mechanics. From the quasi-static to the dynamic regime this has applications to fluid-flow in jointed rocks, hydraulic fracturing for hydrocarbon recovery, and comminution of rock masses. For dynamic analyses, the steady-state FEFFLAP was coupled to the FAST fluid dynamics module [3].

2. DESCRIPTION OF THE QUASI-STATIC FRACTURING MODEL

2.1 The Solid Fracture Model (FEFAP)

Our FEFFLAP code represents the coupling of the FEFAP discrete fracture propagation code [4] and of the JTFLO program, a LLNL-enhanced version of an earlier code for analysis of fluid flow in rock fractures [5]. FEFAP analyzes planar and axisymmetric structures for crack initiation and growth. The program combines fracture mechanics theory, the use of interactive computer graphics, and a unique, automatic remeshing capability to allow the user to initiate and propagate up to ten discrete cracks simultaneously. The salient capabilities available in FEFAP were:

o complete interactive-graphical execution of the program. One is not locked into a batch-produced result via the initial data input.
o automatic, discrete crack nucleation at arbitrary points and angles on an edge, as specified by the analysis.
o automatic, discrete crack propagation capability with optional interactive mesh adjustment along the propagating crack.
o automatic nodal adjustment for singular elements, and direct, automatic extraction of the stress intensity factors.

o automatic bandwidth minimization and nodal re-
numbering.

This automatic crack extension is an intricate process, the logic of which is unique to the FEFAP
code.

FEFAP is built under the assumption of linear
elastic fracture mechanics (LEFM); stress intensity
factors control crack stability and trajectory. The
logic of FEFAP is:

1. compute stress intensity factors, K_I and K_{II},
   for present structure and loading.
2. substitute K_I and K_{II} into any of three
   mixed-mode interaction formulas. Compute new
   crack direction and assess crack stability. If
   the crack is unstable, continue. If stable, go
   to step 4.
3. mesh for a selected increment of propagation.
   Repeat steps 1 through 3 until the crack is
   stable or fracture occurs.
4. if the crack is stable, raise the load level
   until instability is predicted by the chosen
   interaction formula. Continue with step 3.

2.2 The joint models (JPLAXD)

To apply FEFAP to the problem of fluid-driven
crack propagation in jointed rock, two major extensions of its capabilities were required. First a
rezoning was implemented which can handle crack
propagation into, across, or from discrete joints.
Second, the joint constitutive models were enhanced
to accommodate nonlinear shear and normal behavior,
such as tension cut-off, maximum closure, and
shear-strain softening. The non-linear algorithms
were those developed earlier in the JPLAXD code [6].
Then FEFAP was coupled to the flow model described
next.

2.3 The flow model (JTFLO)

The fluid flow model in FEFFLAP is that of the
JTFLO finite element program. JTFLO is a LLNL-
enhanced version of a coupled stress and flow
analysis model developed by Noorishad [5]. It provides for the steady-state solution of flow in
parallel or tapered channel such as joints, cracks,
and interfaces in rock. Both flow rate and pressure
boundary conditions can be specified. The fluid
conductivity of individual flow elements is described by

$$k_p = \rho g b^2 / 12 \mu$$

where b is the mean element aperture, ρ is the
specific gravity of fluids, g the acceleration of
gravity and μ the dynamic viscosity. Then, the
areal permeability along a family of parallel such
fractures is proportional to b^3. The merits and
limitations of this so-called "cubic law" model have
been discussed at length by others [7,8].

2.4 Coupling of JTFLO with the enhanced FEFAP: The
FEFFLAP code

FEFFLAP allows for flow and pressurization in
existing and propagating cracks as well as pressurization and fluid flow in joints. In the general
operation of the model one initiates or extends a
crack, obtains the new pressure distribution due to
the new crack geometry, and follows any non-linear
joint behavior by secant iterations on interface
element properties, as required. Since the joints
can behave non-linearly, care must be exercised in
selecting the sequence of events. Figure 1 shows
the logic of the coupled structural, fracture, and
flow analyses.

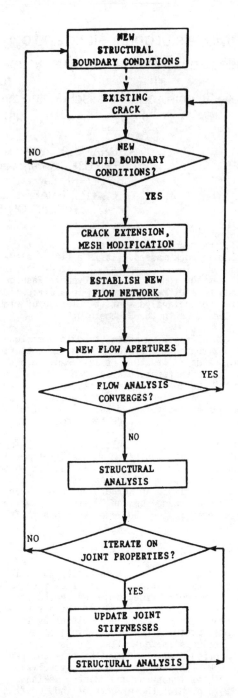

Figure 1. Logic of the FEFFLAP Code for Fluid-
Driven Fracture Analysis.

3. VERIFICATION OF THE QUASI-STATIC FRACTURING MODEL

3.1 Comparison of FEFFLAP Analyses with Hydrostone
block experiments

Sixteen hydrostone block experiments were performed at LLNL to provide physical test data related
to hydrofractures crossing interfaces [9]. The
basic test layout is shown in Figure 2. The problem
involved two types of hydrostone separated by an
interface, and also included the steel platens that
were used to load the block. Thus three different
solid material types were used in the analysis.
Four joint-interface types were required: (1) the
interface between the two hydrostone materials,
(2) the interfaces between steel platens and the
hydrostone, (3) the joint elements that are inserted into the crack as it propagates, and (4) a
set of joint elements around the interior of the
borehole, which provides a convenient way to

pressurize the hole. The last two joint types are necessary for the fluid flow part of the analysis.

In order to determine the adequacy of FEFFLAP, a 2-D code, to handle the 3-D geometry of Figure 2, the stresses in the mid-vertical section of the block were calculated both with a plane stress FEFFLAP solution, and with JROC3D, a 3-D jointed block code developed by C. St. John at Imperial College [10] and enhanced by F. Heuze, at LLNL. Results agreed to better than 1% [11].

Then, two of the tests were analyzed with FEFFLAP. Figure 3 shows the results of a FEFFLAP analysis of one experiment in which the crack stopped at the interface; vertical and horizontal loading stresses were 4.8 and 0.7 MPa, respectively, and the peak pressure in the borehole was 19.3 MPa. The FEFFLAP analysis of another experiment shows that a crack reinitiates from the interface; for this case the vertical and horizontal loads were 12.4 and 5.2 MPa respectively, and the peak borehole pressure was 23.5 MPa. The borehole pressure required in FEFFLAP to reinitiate the crack was higher than that in the experiment. We suspect the cracks crossed the interface dynamically in the experiments because the cracks went straight through the interfaces.

3.2 Comparison of FEFFLAP with problems for which solutions are known

FEFFLAP was tested on a cracked borehole problem (standard hydrofrac geometry) by calculating Mode I stress intensity factors for two types of loading: a remote biaxial tensile stress, and uniformly pressurized borehole and cracks. The results were compared to established values [12] to obtain an estimate of the code's accuracy. For both types of loading each crack length was 1.5 times the borehole radius. The Mode I stress intensity factor calculated in FEFFLAP was 7 percent higher than the established value for both cases. These results are quite good when one considers the coarse finite element mesh. In addition, the mesh is truncated at 10 times the borehole radius while the established values correspond to an infinite medium.

The multicrack capability of FEFFLAP was verified against an analytical solution for six pressurized cracks emanating from a borehole. The analytic results are due to Ouchterlony [13]. The geometry of the cracks in half plane symmetry are shown in Figure 4; it has 6-fold symmetry. The borehole and the cracks are subject to a constant pressure P. Table 1 lists the values for the nondimensional Mode I stress intensity factors for each crack tip. The quantity μ is a multiplicative factor of the borehole radius and its value yields the distance from the center of the borehole to the crack tip. Thus μ is a measure of crack length + borehole radius. There is a slight varition in the values of μ for the three cracks. This is due to the fact that the crack tip locations are identified by positioning a cursor on the computer screen. Every crack tip should have the same value of stress intensity. Table 1 shows the FEFFLAP values, which vary by less than 4% from the analytical one.

Table 1. Comparison of FEFFLAP results with analytical results from Ouchterlony.

Crack Number	μ	$K_I/P\sqrt{\pi\mu R}$
1	2.679	0.779
2	2.678	0.781
3	2.676	0.775
Ouchterlony [13]	> 1.5	0.743

Figure 2. Physical Tests of Hydrofracturing in Jointed Blocks Made of Two Different Materials.

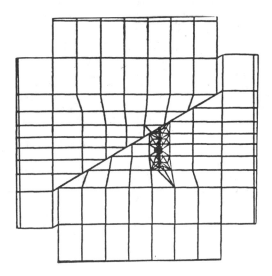

Figure 3. Fracture Does Not Cross the Interface (σ_v = 4.8 MPa, σ_h = 0.7 MPa).

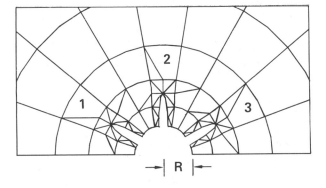

Figure 4. FEFFLAP Analysis of Multiple Crack Propagation, for Comparison with Ouchterlony's Analytical Results.

4. QUASI-STATIC APPLICATIONS

4.1 Fracturing of a jointed sandstone reservoir

Figures 5 and 6 show the kind of realistic field problem FEFFLAP was designed to solve. In this case, there is a joint system (heavy lines) around a borehole, and the ends of the two discontinuous joints above the borehole are treated as points of possible crack extension. The sequence in Figure 5 illustrates successive steps in the analysis. In this example, the fluid flows into a preexisting joint system; in other cases, the induced crack may cross the first interface. This numerical calculation also showed, for the first time, a phenomenon that had been predicted only analytically: the advancing crack front tends to open the natural fracture by inducing tensile stresses ahead of itself. In Figure 5b the crack has proceeded from the borehole and intersected a joint. The flow can now go out into the joint network. For this particular problem the flow was apportioned as shown in Figure 6. Notice how lubrication (some flow) occurs in almost every joint. If the medium is under stress one could expect slippage from these joints and therefore some seismic noise.

Sandia National Laboratories, Albuquerque, is conducting the Multi-Well Experiment in the Western gas sand formations near Rifle, Colorado, at a depth of several thousand feet [14]. Three proximate wells have been drilled, and one has been hydraulically fractured. The progress of a hydrofracture in one well was monitored from a second well by passive seismic means as shown in Figure 7. Sandia found that the seismic noises originated from a zone between 6 and 24 m wide, considerably greater than the original fracture width (about 2.5 cm). FEFFLAP can demonstrate (see Fig. 6) how a large volume of rock may react to a single hydrofracture injection, and we are using it to help explain the above observations at the Multi-Well Experiment.

4.2 Hydrofracturing in coal for methane recovery

Figure 8 shows a plan view of a hydrofracture in the Oak Grove Mine near Birmingham, Alabama. An analysis was done to determine why the fracture went into the roof rock [15]. The results indicate that no net tension would be created in the shale roof rock due to the coal hydrofracture. However, inflation of an existing flaw in the shale roof could occur if fluid got into it, and this new crack then could propagate if it were as small as a centimeter, to start.

The geometry that was constructed to model the Blue Creek coal seam included the one foot thick Marylee seam for completeness. There were five horizontal joint systems; all of them were coal-shale interfaces except for a coal-coal interface at the bottom of the hydrofrac. This geometry represents a cross section of about 12 m x 12 m. The hydrofracture in the coal was inflated and an analysis was done to obtain the stresses and determine if there was any nonlinear behavior of the slickensided joints near the hydrofracture. There was no joint inelastic slip.

Figure 9 represents a "blow up" of the hydrofracture - shale intersection. This grid corresponds to an area 60 cm wide by 52 cm high. This analysis was done for the same reasons as above, except here the details are fine enough to look at five face cleats and treat the fracture as being 5 cm wide. The boundary conditions for this analysis were obtained from the previous analysis, i.e., the stresses corresponding to the perimeter of Figure 9 were extracted from the results of the larger grid stress analysis. As before, there was no joint inelastic slip at these loads. Note that this slip,

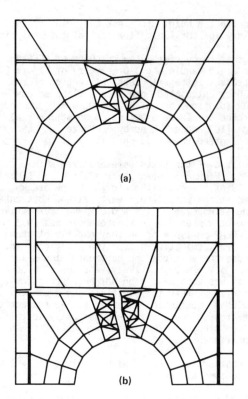

Figure 5. FEFFLAP Analysis of Fluid-Driven Crack Propagation in a Jointed Block System.

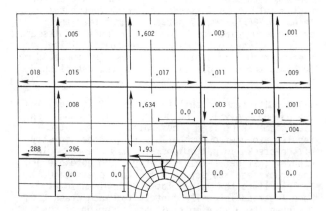

Figure 6. Coupled Fracturing and Flow Analysis in a Jointed Medium.

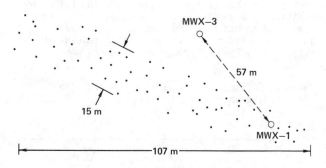

Figure 7. Microseismic Activity Due to Massive Hydraulic Fracturing at MWX Well No. 1, as Monitored from Well 3.

if it occurred, would deconcentrate stresses and for our particular problem it would tend to retard or diminish the creation of tensile stress across the interface in the shale. The stresses just across the interface did not show tension above the 5 cm crack. Indeed, there was compression, with the horizontal component in the 2.1 to 3.5 MPa range. What happened, then? We know from field observations that the shale had a dilated proppant-filled crack.

A possible explanation is the following: suppose that there is a flaw of some height h in the shale, that intersects the coal-shale interface. Suppose further that the 5 cm wide hydrofracture comes in contact with this flaw. If fluid can get into this crack under compression, the crack could extend. There is a way for this to occur - namely - a mis-match of the flaw's surfaces. Assuming inflation due to mismatch, the problem to solve with FEFFLAP is the minimum height h of the pre-existing crack to allow the fluid to extend it. Five calculations were run: crack lengths of 1/2 cm and 2 cm with the mesh of Figure 9, and crack lengths of 7, 12, and 62 cm with the larger mesh. Table 2 summarizes the results of the calculations in terms of stability. It shows the Mode I stress intensity factor K_I and the load factor, F, for each calculation. A load factor F < 1 means crack instability. If the load factor is greater than 1, the crack will become unstable when the load vector is multiplied by F. The shale fracture toughness was taken as K_{IC} = 1.2 MPa√m.

Table 2: Summary of 5 Crack Stability Calculations.

Crack Length (cm)	K_I(MPa√m)	Load Factor, F
0.5	0.63	1.065
2.0	1.16	1.003
7.0	1.26	< 1
12.0	1.54	< 1
62.0	2.47	< 1

The calculations show that a crack as short as 2 cm is incipiently unstable if the 7.6 MPa fluid pressure can get to it. In fact, the shortest crack is also potentially unstable if the horizontal in-situ stresses were slightly less than the assumed value of 6 MPa, or if the fluid pressure were slightly in excess of 7.6 MPa. Additionally, it is well established that K_{IC} decreases with very short crack length so that we may be over estimating the load factor, F, for the shorter cracks. Thus, the pressurizing of a small flaw in the roof shale is proposed as the most likely explanation for the observed hydrofrac vertical propagation.

5. DEVELOPMENT OF THE DYNAMIC FEFFLAP MODEL

5.1 The FAST/FEFFLAP coupled model

The new developments build upon the existing steady state FEFFLAP model and a fluid propagation model called FAST [16]. In the FAST model, the rate of fluid advance and the pressure variation along each individual crack is determined by solving the conservation equations for mass and momentum of the fluid. The one-dimensional form of the equations is appropriate, since the flow channels are very long and narrow. The friction factor is approximated by a simple additive expression, which includes both laminar and turbulent components. The lateral seepage velocity, v, of fluid into the walls of the fracture is estimated with Darcy-flow analysis normal to the fracture plane. The temperature distribution along the crack is determined from conservation of energy.

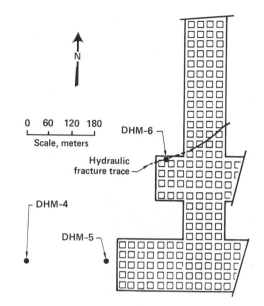

Figure 8. Hydrofracturing at the Oak Grove Mine, near Birmingham, AL, in the Blue Creek Coal Seam.

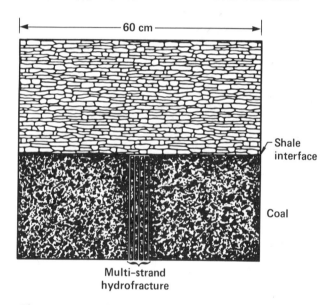

Figure 9. Detailed FEFFLAP Analysis of the Fracture at the Interface.

In terms of coupling with FEFFLAP, the FAST module essentially replaces the steady state flow module. FEFFLAP provides crack and joint apertures to FAST and FAST returns pressure profiles due to those apertures and the flow boundary conditions.

5.2 Application problems and initial verification

Figure 10 typifies a geometry of interest for gas driven fractures from a borehole, where two fractures have begun to dominate. The interaction of these expanding cracks with joint systems (not shown) is also of interest.

An "exact" similarity solution for fracture tip velocity is applicable under some restrictions. In comparison with the similarity solution we assumed that the rock permeability was negligible. As shown in Figure 11, the numerical results are in good agreement with the analytical results at larger crack lengths; this is consistent with the fact that the similarity solution requires the cracks to be long, compared to the borehole diameter.

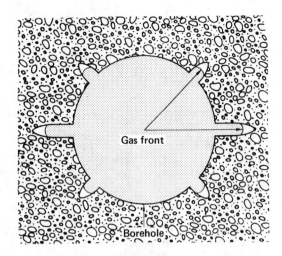

Figure 10. Geometry for FEFFLAP/FAST Analysis of Multiple Gas-Driven Fractures.

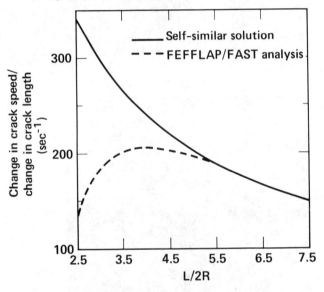

Figure 11. Comparison of FEFFLAP/FAST Results with a Self-Similar Solution.

REFERENCES

1. Ingraffea, A.R., Shaffer, R.J., and Heuze, F.E., (1985) "FEFFLAP: A Finite Element Program for Analysis of Fluid-Driven Fracture Propagation in Jointed Rock, Vol. 1: Theory and Programmer's Manual," Lawrence Livermore National Laboratory, UCID-20368, March.

2. Shaffer, R.J., Ingraffea, A.R., and Heuze, F.E., (1985) "FEFFLAP: A Finite Element Program for Analysis of Fluid-Driven Fracture Propagation in Jointed Rock, Vol. 2: User's Manual and Model Verification," Lawrence Livermore National Laboratory, UCID-20369, March.

3. Nilson, R. H., (1986) "An Integral Method for Predicting Hydraulic Fracture Propagation Driven by Gases and Liquids," Int. J. Num. and Analyt. Methods in Geomechanics, Vol. 10, pp 191-211.

4. Ingraffea, A. R., and Saouma, V., (1984) "Numerical Modeling of Discrete Crack Propagation in Reinforced and Plain Concrete," Application of Fracture Mechanics to Concrete Structures, Martinus Nijhoff Publishers.

5. Noorishad, J., Witherspoon, P. A., and Brekke, T. L. (1971) "A Method for Coupled Stress and Flow Analysis of Fractured Rock Masses," Publ. No. 71-6, University of California, Berkeley, March.

6. Heuze, F. E. (1981) "JPLAXD: A Finite Element Program for Static, Plane and Axisymmetric Analysis of Structures in Jointed Rock," Lawrence Livermore National Laboratory, UCID-19047.

7. Witherspoon, P. A., Wang, J. S. Y., Iwai, K., and Gale, J. E. (1979) "Validity of Cubic Law for Fluid Flow in a Deformable Fracture," Lawrence Berkeley Laboratory, LBL-9557.

8. Gale, J. E. (1982) "The Effect of Fracture Type (Induced vs. Natural) on the Stress-Fracture Closure-Fracture Permeability Relationships," Proc. 23rd U.S. Symp. on Rock Mechanics, Berkeley, CA (AIME, Littleton, CO).

9. Thorpe, R. K., Heuze, F. E., and Shaffer, R. J., (1984) "An Experimental Study of Hydraulic Fracture-Interface Interaction," UCID-20114, Lawrence Livermore National Laboratory, Livermore, July.

10. St. John, C. M., (1971) "Three-Dimensional Analysis of Rock Slopes," Proc. Symp. Int. Soc. Rock Mech., Nancy, France, Sept.

11. Shaffer, R. J., Thorpe, R. K., Ingraffea, A. R., and Heuze, F. E., (1984) "Numerical and Physical Studies of Fluid-Driven Fracture Propagation in Jointed Rock," 25th Rock Mechanics Symposium, (SME, Littleton, CO), p 117-126.

12. Sih, G. C. (1973) Handbook of Stress Intensity Factors, Institute of Fracture and Solid Mechanics, Lehigh Univ., Bethlehem, PA, p 1.2.8.

13. Ouchterlony, F., (1982) "Analysis of Cracks Related to Rock Fragmentation," Part 1 of Lectures at the International Center for Mechanical Sciences (CISM), Udine, Italy, Report DS 1982:4.

14. Northrop, D. A., (1986) "Second Technical Poster Session for the Multiwell Experiment," SAND 86-0945, Sandia National Laboratories, Albuquerque. Also, Proc. SPE/DOE Symposium on Low Permeability Reservoirs, May, 1986. Denver, CO. (Soc. Pet. Eng., Richardson, TX).

15. Shaffer, R. J., Nilson, R. H., Heuze, F. E., and Swift, R. P. (1986) "Quasi-static and Dynamic Arbitrary Fracture Propagation in Jointed Rock," Proc. 27th U.S. Symposium on Rock Mechanics, University of Alabama, June 13-26. Also, Lawrence Livermore National Laboratory report UCRL-93429.

16. Nilson, R. H., and Peterka, D. L. (1986) "User's Manual for the FAST Hydrofracture Code," Report SSS-R-86-7590 to Lawrence Livermore National Laboratory, by S-CUBED, La Jolla, CA, August.

ACKNOWLEDGMENTS

This research was performed under the LLNL Unconventional Gas Program - F. Heuze, Manager. Program monitoring is performed by Morgantown Energy Technology Center under contract W-7405-ENG-48 with the U.S. Department of Energy. The careful typing of Lydia Grabowski is gratefully acknowledged.

An investigation of the material parameters that govern the behavior of fractures approaching rock interfaces

Investigation des paramètres d'un matériau affectant le comportement de fractures à l'approche d'interfaces rocheuses
Eine Untersuchung der Stoffparameter, die die Ausbreitung von Bruchflächen bestimmen

M.THIERCELIN, Dowell-Schlumberger, Tulsa, Okla., USA
J.C.ROEGIERS, Dowell-Schlumberger, Tulsa, Okla., USA
T.J.BOONE, Cornell University, Ithaca, N.Y., USA
A.R.INGRAFFEA, Cornell University, Ithaca, N.Y., USA

ABSTRACT: Fractures approaching interfaces in rock are analyzed in terms of nonlinear fracture mechanics concepts. The results of some Brazilian-type fracture tests, which include bi-material interfaces, are presented. The tests have been simulated using a finite element code. It is concluded that the two primary material parameters that govern fracture containment at interfaces are the shear capacity of the interface and the tensile strength of the material into which the fracture could potentially propagate.

RESUME: Dans cet article, la mecanique nonlineaire de la rupture est utilisee pour interpreter la propagation des fractures au travers d'interfaces entre des materiaux de proprietes differentes. Des essais experimentaux de type Breselien ont ete conduits. Ces essais ont ete simules par une methode aux elements finis. Cette etude montre que les principaux parametres gouvernant l'arret de la fracture a l'interface sont la resistance au cisaillement de l'interface et la resistance a la traction du milieu non encore penetre.

ZUSAMMENFASSUNG: Das Verhalten von Rissen im Fels bei der Interaktion mit Trennflaechen wird mit nichtlinearen Konzepten der Bruchmechanik untersucht. Die Ergebnisse aus Testversuchen an geschichteten Spaltzugproben mit unterschiedlichen Materialien und die entsprechende analytisch-numerische Simulation mit einem speziellen FE-code werden vorgestellt. Die durchgefuehrten Untersuchungen zeigen, dass die Scherfestigkeit entlang der Trennflaeche und die Zugfestigkeit der Materialien wesentliche Parameter fuer das Rissverhalten im Bereich von Trennflaechen darstellen.

1 INTRODUCTION

The design of fractures in rock is of prime importance to the oil industry and also may be of importance for projects involving geothermal energy extraction, waste containment and rock excavation. The effect of material interfaces on the propagation of fractures in rock is an area of common interest. Figure 1 illustrates four distinct events that can occur when a fracture intersects an interface: the fracture may penetrate, arrest (or be contained), jog or divert. All of these phenomena have practical significance. For example, when a hydraulic fracture is employed for the enhancement of hydrocarbon production, it is desirable to contain the fracture within the portion of the rock formation that bears the desired fluids. Material interfaces could potentially produce the desired containment. However, joints and interfaces also can be detrimental if encountered within a zone where propagation is desirable.

An understanding of the factors promoting the arrest of fractures at and propagation through interfaces therefore is of great significance in designing the fracture in rock. The concepts of linear elastic fracture mechanics (LEFM) have previously been applied to this task, but only with marginal success. The intent of this work is to apply nonlinear fracture mechanics concepts to interface phenomena, to produce a more effective model of the physical processes involved.

The bulk of the research into the effects of interfaces on fracture propagation in rock has centered on hydraulic fracture as it pertains to oil production. However, the spacing of joints and bedding planes is known to have a great influence on blasting efficiency as well [1]. Experiments which have studied explosivly driven fractures monitored by dynamic photo-elasticity have produced similar conclusions to quasi-static hydraulic fracture experiments [2]. Fracture-interface interaction is studied, in this paper, in a quasi-static sense which is appropriate for hydraulic fracture. Although, the methods also can be applied to dynamic fracture

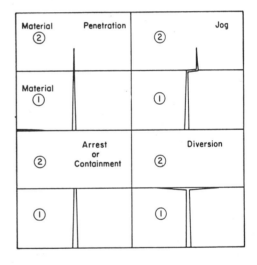

Figure 1. Four potential types of fracture propagation at an interface.

problems.

Within the paper the term "first material" shall be used in reference to the material in which the fracture is initially found. The term "second material" shall be used to identify the material which forms the interface with the first material and into which the fracture propagates.

2 PREVIOUS RESEARCH

2.1 LEFM, Experimental

Analytical studies [3,4,5] which have used LEFM as their basis in considering cracks approaching bi-material interfaces have shown that as a crack progresses toward an interface from a material of

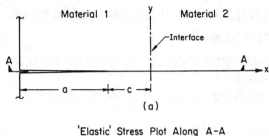

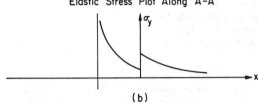

'Elastic' Stress Plot Along A-A

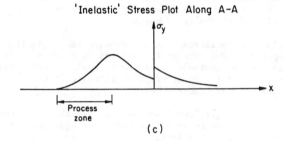

'Inelastic' Stress Plot Along A-A

Figure 2. (a) Idealized fracture in the vicinity of a material interface. (b) Linear elastic concept of the stress (σ_y) along the line of fracture propagation. (c) Nonlinear concept of the stress (σ_y) along the line of fracture propagation.

higher stiffness into a material of lower stiffness, the stress-intensity approaches zero. The conclusion drawn by Simonson et al [3] was that a crack would arrest close to the interface without progressing through it. However, experimental results [6-9] have clearly shown that cracks in rock are not contained solely by the effects of a contrast in the material stiffness. It has been concluded by Teufel and Clark [7] that elastic stiffness properties may not be the most significant factor in the containment of hydraulic fractures. Rather, the important factors are a weak interfacial shear strength of the layers and an increase in the minimum horizontal in-situ stress in the bounding layers. The dependence of fracture containment on the interfacial shear strength has been established in the other cited references, as well. Additionally, Teufel and Clark have shown that the tensile strength of the material on the opposite side of the interface (to the fracture) is important in determining if a fracture will penetrate through the interface.

It is apparent from the contradictory results of the experimental and the analytical work that LEFM principles are inadequate in this case. A premise of LEFM is that the inelastic zone near the crack tip must be small compared to all other dimensions of the problem. In the case being studied the critical dimension, labeled "c" in Figure 2(a), is the distance from the crack tip to the interface. This critical dimension approaches zero as the crack approaches the interface, which violates the previously cited premise of LEFM.

2.2 Nonlinear Fracture Mechanics

In recent years, numerous researchers have directed considerable effort toward developing nonlinear fracture mechanics concepts for application to rock

structures [10-14]. The concepts of strain softening and a fracture process zone are well established in the fields of rock and concrete mechanics. Figures 2(b) and 2(c) contrast the stress fields which are assumed to exist when applying linear and nonlinear fracture mechanics concepts. In this paper, it is shown that the nonlinear model can be used to describe and predict the behavior of cracks approaching interfaces. It is demonstrated that these principles are consistent with and aid in the interpretation of the results of previous experimental work. New experiments are reported herein which were conducted to assist in verifying the model and illustrating the behavior. A finite element code which incorporates a nonlinear interface and a nonlinear fracture mechanics model has been used to simulate the experiments [15].

The term "fracture" shall be used herein to describe the full crack length inclusive of the stress-free crack length and the fracture process zone where the material can still carry stresses across the crack. In fact, a fracture may be entirely a process zone. The fracture process zone also is considered to be a zone of strain localization, and fracture initiation is considered synonomous with initiation of strain localization. Refer to Ref. 12 for a more detailed discussion of these terms and phenomena.

2.3 Interface Slippage

Prior discussion assumed that the interface was bonded and, therefore, no slippage could occur along it. However, rock interfaces often do not exhibit perfect bonding and, therefore, have a limited shear capacity. The shear capacity is typically modeled using a Coulomb friction equation that accounts for the effect of normal stresses. The pertinence of the shear capacity of fractures approaching interfaces has been studied in some detail, theoretically [16,17,18] and experimentally [7,8]. The typical theoretical concept of a fracture which has intersected an interface is shown in Figure 3. This concept assumes that a crack can be blunted as illustrated in Figure 3 and thus arrested at the interface. Experimental work has clearly shown that interfacial slip is a mechanism by which a fracture can be arrested.

A relationship between the tensile strength of rock materials forming an interface and the shear capacity of the interface has been reported by Teufel and Clark [7]. Figure 4 shows a plot taken after Figure 7 of Reference 7. The observed linear relationship would seem to be founded in the concept that the tensile stress that can be generated in the second material by

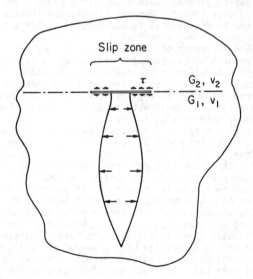

Figure 3. Idealized model of fracture at an interface where slippage has occurred.

a hydraulically loaded fracture in the first material is proportional to the shear capacity of the interface.

Essentially, the problem of fracture propagation through an interface where slip can occur reduces to a problem of fracture initiation at the interface in the second material. It has been suggested by Cleary that crack initiation in the second material is dependent on the inherent flaw size in the second material and the tensile stresses generated in the second material [18]. However, LEFM approaches are not applicable at the scale of the typically inherent flaws in rock. An alternative approach to fracture initiation in rock has been discussed in Reference 12 by two of the authors and is used subsequently in this work. The approach provides a transition from a strength of materials methodology which predicts initiation of strain localization to large cracks where LEFM may be applicable.

3 EXPERIMENTS

A series of experiments has been conducted on modified Brazilian-type specimens of Berea sandstone and Indiana limestone. The specimen geometries are shown in Figure 8. Compressive loads were applied to the truncated edges of the cylindrical sections by the platens of a 100-kN MTS load frame. The applied loads produce a tensile stress concentration at the crown and invert of the hole in the center of the specimen. Natural cracks form at these points and develop as shown in Figure 5(a). The compression in the region of the platens subsequently arrests the crack growth. There are three important characteristics of this geometry: (1) normal stress is applied to the interface during the process of loading; (2) there is a discontinuity in the tensile stress across the interface between the limestone and the sandstone; and (3) the specimen can be varied such that crack growth is stable throughout the process of crack formation and propagation [19].

The specimens were instrumented with 6 mm foil strain gages along the expected crack path. It was found that strain gages could be used to give reliable qualitative information about crack formation, and the development of strain softening across the gage. The diametric expansion of the hole in the center of the specimen was monitored with a MTS clip gage which had been fitted with a pair of inner caliper-type fixtures.

The elastic material properties of the two rock types are listed in Table 1. These properties were determined from compression tests on standard 25.4 mm diameter by 50.8 mm cylinders. Also included in Table 1 is the material tensile strength determined from direct tension tests on notched 50.8 mm diameter by 50.8 mm cylindrical specimens. The tension test on the Berea sandstone produced a complete strain softening, which is shown in Figure 6. The Indiana limestone tensile test specimens failed in an uncontrollable manner.

3.1 Single-Material Tests

Tests were first conducted on specimens composed solely of Berea sandstone or Indiana limestone with the geometry illustrated in Figure 5(a). The objective of these tests was to provide verification of the material models used for analysis, and to determine the relationships between strain gage measurements and crack opening. The latter objective was achieved by placing a strain gage immediately above and below the hole. It was postulated and verified analytically that the lateral diametric expansion of the hole, which was measured by the clip gage, approximates the crack opening displacement at the crack mouth (CMOD). These and the other tests

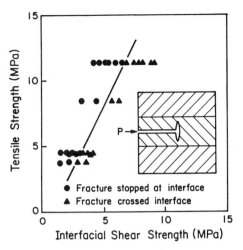

Figure 4. Relation of interfacial shear strength to tensile strength on the extent of fracture propagation in three-piece specimens composed of a single rock type (after Figure 7 of Reference 7).

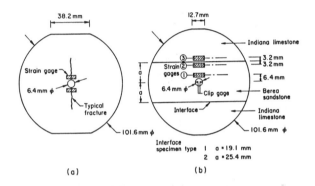

Figure 5. Typical specimen dimensions for (a) a single material specimen and (b) two types of bi-material specimens.

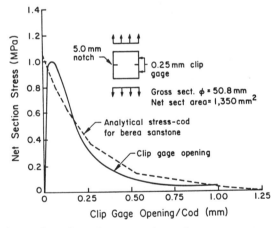

Figure 6. Plot of an experimental stress vs gage opening relationship and an analytical stress vs COD relationship for Berea sandstone.

were conducted under displacement control at a rate of 0.028 μm/sec. The specimens were tested in an air-dried condition, at room temperature, and at a humidity of approximately 70%.

The results of these tests are illustrated in Figure 7, which shows that the relationship between the clip gage measurements and the strain measurements is approximately bilinear for the sandstone and limestone. These regions of linearity can be associated (1) with primarily elastic deformation and (2) with deformation due to crack opening.

265

TABLE 1

MATERIAL PROPERTIES

Rock Type	Young's Modulus E^1 (MPa)	Poisson's Ratio v^1	Tensile Strength σ_t^2 (MPa)	Fracture Energy $G^{}$ (J/m^2)	Constant Eq. (1) k^5 (m^{-1})	Interface Friction ϕ^6
Indiana Limestone	20,000	.26	3.8	60^3	.063	
						0.75
Berea Sandstone	8,000	.24	1.0	120^4	.0053	

NOTES:
1) Determined from standard cylinder compression tests.
2) Assumed to be 5% greater than the net stress at failure of notched specimens.
3) Interpreted from an assumed K_{Ic} of 1.1 MPa$\sqrt{m}$.
4) Determined from the area under the stress-displacement curve for the notched specimen (Fig. 2).
5) Interpolated using material properties and Eq. (1).
6) Determined from friction tests of Berea sandstone on Indiana limestone.

3.2 Bi-Material Tests

The two different geometries of the interface test specimens are shown in Figure 5(b) along with the location of the strain gages. The only variation in the specimens is the position of the interfaces. The two specimen geometries are referred to as type 1 and type 2 interface tests as labeled in Figure 5(b). Behavioral differences can be expected to arise at the interface from differences in end effects and slip effects. It should be noted that these and the other tests described were repeated to verify the measurements. However, typical results from only one test are presented herein. The specimens were cut and ground smooth with dimensions typically within 1 mm of those shown in Figure 5(b). The sandstone and limestone pieces were simply placed in contact without applying an adhesive. The compressive force across the interface obviously would restrict slip along it, as will be discussed subsequently in the paper.

The strain gage measurements are plotted as a function of the controlled displacement in Figures 8 and 9. for the two specimen types. The authors' interpretation of these results is that strain softening commenced at the hole, as expected and progressed towards the interface. In both cases, the crack progressed through the interface without apparent restriction.

Inspection of the specimen after testing showed the crack to be clearly visible; however, all three rock

sections remained intact. The width of the crack on the Berea sandstone side of the interface was visibly wider than that on the Indiana limestone side of the interface, for all tests. The cracks followed tortuous paths that varied by about 3 mm from the center line of the specimen. The crack generally jogged at the interface 1 to 2 mm from its mate in the adjacent material.

4 COMPUTER SIMULATION

The experiments have been simulated using a dynamic, explicit, finite element code developed by Swenson [15]. A typical finite element mesh is shown in Figure 10. The code employs interactive computer

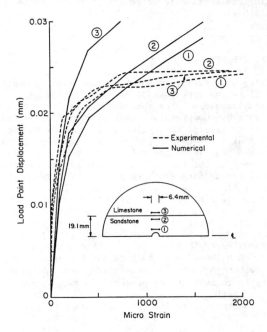

Figure 8. Experimental and analytical plots of load point displacement vs microstrain for bi-material test type 1.

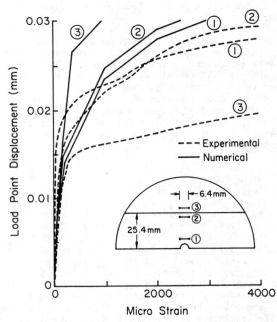

Figure 9. Experimental and analytical plots of load point displacement vs microstrain for bi-material test type 2.

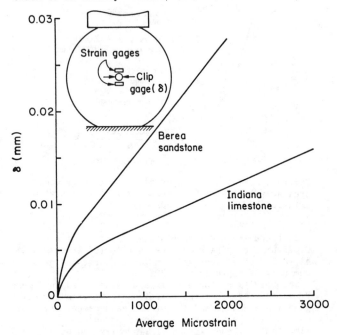

Figure 7. Clip gage opening (δ) vs microstrain for single-material specimens.

266

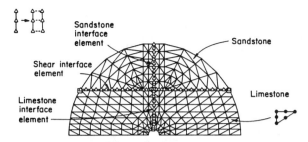

Figure 10. Typical finite element mesh used for the numerical simulations.

graphics which allow the user to monitor displacements, strains and stresses throughout the simulation. In the simulation displacements were applied at constant rates as was done in the test situation. However, a displacement rate of 50 mm/sec was used in the computer analyses to minimize the computation time. While this rate is significantly faster than the rate applied to the specimens during testing, it was found that dynamic effects were minimal and did not influence the results. An important advantage of the dynamic, explicit, finite element code is that strain softening and friction slip models could easily be integrated into it.

4.1 Nonlinear Fracture Mechanics

The strain softening models were incorporated into interface elements which modeled the progressing fracture and the fracture process zone. Similar analyses have been described in detail elsewhere [12]. A simple exponential decay model was used to characterize the strain softening [20,21] such that

$$\sigma = \sigma_t e^{-kw} \qquad (1)$$

where σ = stress across the crack, w = crack opening, k is a decay constant and, σ_t = tensile strength of the rock.

This decay function has been found to be typical of mortar and concrete, and can be seen to describe rock as well [13]. The material parameters used in these analyses are listed in Table 1. The strain softening model described by Equation 4 is plotted in Figure 6 for Berea sandstone.

4.2 Interface Slip Model

The bi-material interface has been characterized by a Coulomb friction model assuming no cohesive strength. The relationship between normal stress and shear resistance is given in Equation 2.

$$\tau \leq \pm\mu\sigma \qquad (2)$$

where τ is the local shear stress at the interface, μ is the friction coefficient and, σ_n is the compressive stress across the interface.

The friction model is simple and has been shown to be reasonable for a wide range of normal stresses [7].

4.3 Analysis Results

The results of an analysis of the interface test type 1 and 2 are shown in Figures 9 and 10 respectively. Figure 11 shows a series of displaced mesh plots and Figure 12 shows the same temporal series of plots for

stress along the crack line. (The stress plots presented in this paper are not averaged across the elements so discontinuities between elements are apparent.)

Strain softening can be seen to initiate at the center of the specimen once the tensile strength of the Berea sandstone is exceeded, which is just prior to the stress state shown in Figure 12(a). This point can be considered the beginning of fracture initiation. The stress at the interface in the limestone increases until the tensile strength of the limestone is reached as shown in Figure 12(b). A fracture then begins to initiate at the interface in the limestone. At this point, strain softening has progressed through the sandstone to the interface and noticeable slip has occurred at the interface as well.

It can be seen that fracture initiation in the sandstone and the limestone are separate and distinct events. If the tensile strength in the limestone had been only 2 Mpa, then strain softening would have initiated in the limestone before it progressed through the sandstone. This will be discussed subsequently as a possible mechanism for jogs at interfaces.

It can be seen that throughout this fracture process, stress is carried across the line of strain softening or process zone in the Berea sandstone. In fact, a true stress-free crack never forms. This is confirmed by observation of the specimens after testing. A crack is clearly visible through the sandstone section while the section remains firmly intact. Also, it can be seen from the analysis results that the stress in the process zone of the limestone decreases at a much faster rate than in the sandstone. This effect is associated with the material fracture energy (G_c) through Equation 3. It can be important if the fracture energy is high, since there could be considerable stress redistribution as the process zone develops.

As noted previously, considerable slip occurred along the interface. It can be seen from Fig. 11(d) that the interfacial slip has a form similar to that shown in Figure 3. The shear along the interface transfers tensile stresses from the first material into the second material. This leads to the conclusion that the shear strength of the interface is important in determining if, or at what load level, a fracture will progress through an interface.

A limitation of the model is that it does not completely capture the ultimate failure mode of the test specimens as shown in Figure 13. This mode is asymmetric and involves a form of shear driven fracture in the vicinity of the load platens, which is beyond the scope of this simulation. Once the fracture reaches the exterior edge of the specimen, it can be expected that its load carrying capacity will have peaked and that the strain measured across the fracture path would increase rapidly, as can be observed in Figures 9 and 10. Prior to the fracture extending in the asymmetric manner, the simulation is expected to aproximate the experimental results.

5 DISCUSSION

There exist two limiting cases in the analysis of fractures approaching interfaces: the case where the interface acts in an effectively bonded manner, and the case where the interface is unbonded and there is partial slip. The first instance is typical of rock under moderate to high confining pressures. In this case, it can be expected that the fracture will pass through the interface with no apparent resistance. The relative toughness and stiffness of the materials may influence the rate of fracture growth and the crack opening profile, but the interface per se will not be an obstruction to fracture growth. This conclusion has been drawn by several experimentalists based on small-scale experiments and some mine-back operations. It is supported by the scale arguments presented herein, which refute LEFM concepts that suggest that the stress intensity should approach zero for fractures approaching some interfaces.

If the interface is only partially bonded or in-situ stresses are relatively low, then slip can occur at an interface prior to the fracture intersecting it. Slippage reduces the stresses in the second material and blunts the crack. The problem then becomes one of crack initiation in the second material. If there is sufficient stress produced in the second material to exceed the tensile capacity of that material, then strain softening will be initiated in the second material as well. The fracture will effectively pass through the interface. The processes involved have been illustrated by the experiments and analyses presented herein. Most importantly, the experimental results of other researchers have also been shown to be consistent with the nonlinear fracture mechanics models.

In summary, it can be stated that the condition necessary for a crack to cross an interface is:

> The interface must have sufficient shear capacity to develop a tensile stress in the second material equal to the tensile capacity of that material.

This concept accounts for experimentally observed phenomena, such as the dependence on the friction coefficient of the interface and the dependence on the normal stress across the interface, both of which affect the shear capacity of the interface. Furthermore, the concept is consistent with the conclusions drawn by Teufel and Clark [7] who have shown experimentally that the shear capacity required for a fracture to penetrate an interface is dependent on the tensile strength of the second material.

The nonlinear model allows us to induce a mechanism by which a fracture may jog at an interface. If a crack is to cross an interface without jogging, a local tensile stress concentration must be created immediately ahead of the progressing crack in the second material. This situation is analogous to notching a tensile test specimen so that a crack initiates at the notch which is a point of stress concentration. The stress field ahead of a fracture can produce a tensile stress concentration at an interface which increases in magnitude as the crack progresses toward the interface. If the tensile strength of the second material is not exceeded until the fracture has progressed within close proximity of the interface, then the tensile stress will be concentrated over a small region and strain localization will initiate in the second material in line with the fracture in the first material. Thus, the crack will not jog at the interface.

However, if strain localization initiates in the second material before the fracture is in close proximity to the interface the stress perturbation is smaller in magnitude and the tensile stress is less concentrated. This provides a much broader range in the length along the interface where strain localization could initiate and a jog in the fracture is likely to occur. This argument leads to the conclusion that conditions which promote jogging at the interface are (1) a low tensile strength in the second material relative to the first material, (2) a high stiffness in the second material relative to the first material, and (3) a weak shear capacity along the interface. All of these conditions produce a broader region of tensile stress along the interface.

6 CONCLUSIONS

The application of nonlinear fracture mechanics concepts to fractures at interfaces leads to the

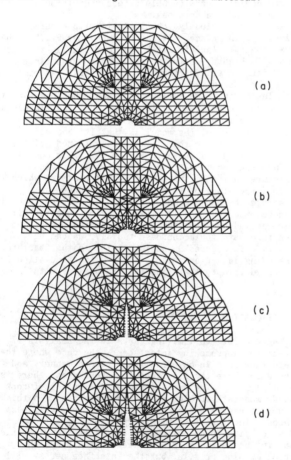

Figure 11. Series of displaced mesh plots for bi-material test type 1. Plots correspond to experimental load point displacements of (a) 0.013 mm, (b) 0.020 mm, (c) 0.025 mm, and (d) 0.030 mm.

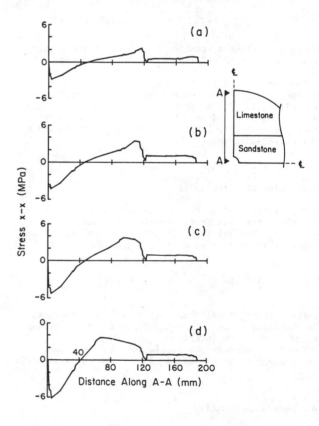

Figure 12. Series of stress plots for bi-material test type 1. Plots correspond to experimental load point displacements of (a) 0.013 mm, (b) 0.020 mm, (c) 0.025 mm, and (d) 0.030 mm.

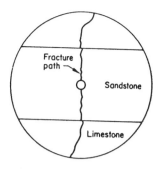

Figure 13. Ultimate failure mode of the test specimens.

following conclusions.

(1) LEFM concepts are not applicable to fractures approaching interfaces in rock due to scale effects.

(2) nonlinear fracture mechanics concepts which include strain softening characteristics and a fracture process zone can be applied to fractures approaching rock interfaces, and are consistent with observed experimental behavior.

(3) Bonded rock interfaces cannot be expected to provide significant impedance to fracture propagation.

(4) Fractures may penetrate partially bonded interfaces, if the interface has sufficient shear capacity to develop a tensile stress in the second material equal to its tensile strength.

(5) The material parameters that determine if a fracture will penetrate an interface are the tensile capacity of the second material, the relative stiffness of the two materials (which determines the stress contrast across the interface), the fracture energy of the materials (which influences the amount of stress redistribution that can take place), and the shear capacity of the interface (which is a function of the normal stress across the interface).

(6) Conditions which promote jogs at interfaces are a weak shear capacity, a low tensile strength in the second material, and a high stiffness ratio between the second and first materials.

Acknowledgments: The experimental work in this paper was conducted at Dowell Schlumberger laboratories in Tulsa, Oklahoma. The computational work was done at the Program for Computer Graphics, Cornell University in Ithaca, New York. The work has been sponsored in part by grants from the National Science Foundation, No. PYI 8351914 and No. CEE 8316730. Dr. Daniel Swenson's assistance and comments also are gratefully acknowledged.

REFERENCES

1. Singh R.N., Elmherig A.M. and Sunu M.Z. Application of rock mass characterization to the stability assessment of blast design in hard rock surface mining excavations. Rock Mechanics: Key to Energy Production, 27th U.S. Symposium on Rock Mechanics, Edited by H.L. Hartman, pp. 471-478 (1986)
2. Shukla A. and Fourney W. L. Explosively driven crack propagation across an interface. Int. J. Rock Mech. Min. Sci. & Geomech. Abstr. 22, 6, 443-451 (1985)
3. Simonson E.R., Abou-sayed A.S. and Clifton R.J. Containment of massive hydraulic fractures. Soc. Pet. Eng. J. 18, 1, 27-32 (1978)
4. Cook T.S. and Erdogan F. Stresses in bonded materials with a crack perpendicular to the interface. Int. J. Eng. Sci. 10, 677-697 (1972)
5. Hanson M.E. and Shaffer R.J. Some results from continuum mechanics analyses of the hydraulic fracturing process. Soc. Pet. Eng. J. 20,4, 86-94 (1980)
6. Daneshy A.A. Hydraulic fracture propagation in layered formations. Soc. Pet. Eng. J. 18, 1, 33-41 (1978)
7. Teufel L.W. and Clark J.A. Hydraulic fracture propagation in layered rock: experimental studies of fracture containment. Soc. Pet. Eng. J. 24, 2, 19-32 (1984)
8. Anderson G.D. Effects of friction on hydraulic fracture growth near unbonded interfaces in rocks. Soc. Pet. Eng. J. 21, 2, 21-29 (1981)
9. Warpinski N.R., Schmidt R.A. and Northrop D.A. In-situ stresses: The predominant influence on hydraulic fracture containment. J. Pet. Tech. 653-664 (March 1982)
10. Labuz J.F.,Shah S.P. and Dowding C.H. Experimental analysis of crack propagation in granite. Int. J. Rock Mech. Min. Sci. & Geomech. Abstr. 22, 85-98 (1985)
11. Bazant Z.P. Crack band model for fracture of geomaterials. Proc. 4th Int. Conf. on Numerical Methods in Geomechanics, pp. 1137-1152. Edmonton, Canada (1982)
12. Boone T.J., Wawryznek P.A. and Ingraffea A.R. Simulation of the fracture processes in rock with application to hydrofracture. Int. J. Rock Mech. Min. Sci.& Geomech. Abstr. 255-265 (1986)
13. Krech W.W. The energy balance theory and rock fracture energy measurements for uniaxial tension. Advances in Rock Mechanics, Proceedings of the 3rd ISRM Conference. V.II-A,pp. 167-173 (1974)
14. Perkins T.K. and Krech W.W. The energy balance concept of hydraulic fracturing. Soc. Pet. Eng. J. 18, 1-12 (1978)
15. Swenson D.V. Modelling mixed-mode dynamic crack propagation. PhD thesis, Cornell University, (1986)
16. Keer L.M. and Chen S.H. The intersection of a pressurized crack with a joint. J. Geophys. Res. 86, 1032-1038 (1981)
17. Weertman J. The stopping of a rising, liquid-filled crack in the earth's crust by a freely slipping horizontal joint. J. Geophys. Res. 85, 967-976 (1980)
18. Lam K.Y. and Cleary M.P. Slippage and re-initiation of (hydraulic) fractures at frictional interfaces. Int. J. Num. Anal. Meth. Eng. 8, 589-604 (1980)
19. Thiercelin M. and Roegiers J.-C. Toughness determination with the modified ring test. 27th U.S. Symposium on rock mechanics, pp. 615-622, Alabama (1986)
20. Gopalaratnam V.S. and Shah S.P., Softening response of plain concrete in direct tension, ACI J. 82, 3, 310-323 (1985)
21. Boone T.J., Wawryznek P.A. and Ingraffea A.R. Discussion 82-27 (of Ref. 20), ACI J. 83, 2, 316-318 (1986)

Calculation-experimental method of evaluating mechanical characteristics of scale heterogeneous rock masses

La méthode de calcul et d'expérimentation pour déterminer les caractéristiques mécaniques de la roche d'envergure hétérogène

Berechnungs- und Experimentalverfahren zur Bestimmung der mechanischen Parameter grössenordnungsmässig heterogener Gesteine

S.B.UKHOV, MICI named after V.V.Kuibyshev, Moscow, USSR
V.V.SEMENOV, MICI named after V.V.Kuibyshev, Moscow, USSR
E.V.SCHERBINA, MICI named after V.V.Kuibyshev, Moscow, USSR
A.V.KONVIZ, MICI named after V.V.Kuibyshev, Moscow, USSR

SYNOPSIS: Here a new method for determining characteristics of mechanical properties of scale heterogeneous rocks, based on mathematical modelling of the experiment is suggested. The method proves to be effective when traditional methods are unacceptable due to large heterogeneity elements dimensions. The method makes it possible to fix deformation processes and local failures in heterogeneous material samples all over the way of loading. The method has got experimental verification. Basic results obtained with loam, wasteland and crushed stone compositions are presented.

RESUME: On propose la nouvelle méthode de détermination les caractéristiques mécaniques de la roche d'envergure nonhomogène, basée sur la simulation mathématique d'expérience. La méthode est efficace en cas où les méthodes d'étude traditionelle ne sont pas inadmissible, à cause de grandes dimensions des élèments nonhomogénéité. Le terme spécial "la roche d'envergure nonhomogène", signifie la roche avec plusieures niveau de la nonhomogénéité. Cette méthode permet fixer les processus de la déformation et des destructions locales dans les modèles des materiaux nonhomogènes au cours de toute l'opération de chargement. La méthode a été confirmée expérimentalement. On donne les résultats essentiels, récu dans l'exemple des mélange de limon et d'arène, d'une côté, et du blocage de l'autre côte.

ZUSAMMENFASSUNG: Der Vortrag behandelt das neue Bestimmungsverfahren der mechanischen Parameter grössenordnungsmäßig heterogener Gesteine, das auf der mathematischen Modelliering von Experimenten beruht. Das Verfahien erweist sich in den Fällen als effektiv, wenn aufgound des groBen Abmessungen der heterogenen Elemente, traditionelle Berechnungsverfahren versagen. Es gestallet, die Verformungsprozesse und lokalen Brüche in emer Probe aus heterogeneren Material entlang des gesamten Belastungspfades zu fixieren. Das Vestahren ist experimentell bestätigt. An den Beispielas der Gemische "Lehm mit Felsbruch" und "Schottes" werden die grundlagender Ergebnisse erläutert.

In many cases rock masses may be regarded as bodies of composite structure. They include, for example, boulder-clod deposits with loam and sandy loam filler, widely spread in Northern regions of the USSR and Canada. Large rock masses of this type incorporate fissured clods of primary material (limestones, dolomits marls) of various configuration and orientation in space with sizes varying from tens of meters to tens of centimetres across and interclod filler in the form of sandy loam and crushed loam. Rock masses slip zones made by crushed loam and sandy loam with boulder inclusions of primary material, eluvial soil strata in the weathered rock zones, etc. may also serve as an example of such rock masses.

Rocks of this type possess scale heterogeneity, i.e. small bulks of geological body lack some elements of heterogeneity, possessed by large bulks. As for large bodies they have heterogeneity elements of different size, composition, structure possessing appreciably differebt mechanical properties characteristics.

Traditional methods of determining such rock masses mechanical properties are of little effect: customary field and laboratory experiments fail to provide the condition of representability of the studied rock bulk with respect to a large rock mass area. That is why in practice design characteristics of such rock masses mechanical proper-ties are based on test findings of the weakest fractions (the filler), disregarding the effect of stronger elements (boulder and clod inclusions) on these characteristics. This inevitably leads to design characteristics underrating and as a result to making unoptimum engineering decisions.

To eliminate these defects the calculation-experimental method of determining mechanical characteristics of highly heterogeneous rock masses was worked out in MICI named after V.V. Kuibyshev. Basic propositions presented in (Ukhov et al. 1985; Ukhov et al. 1986) are the following. "Typical structures", characteristic of different heterogeneity levels of the rock area under study are worked out according to the data of rock mass engineering-geological analysis. It is necessary that "typical structures" of small bulks were the ingredients of "typical structures" of large bulks. So the schematized rock areas, containing rock blocks and loam-crushed stone filler between them, may be considered as "typical structures" of the highest heterogeneity level. Separate blocks schemes reflecting their composition, jointing and the schemes of the filler, containing loamy material and solid inclusions are "typical structures" of lower level of heterogeneity. Such schematization may be practiced on still lower levels, if necessary.

Experimental methods are used to study mechanical properties of the fractions, corres-

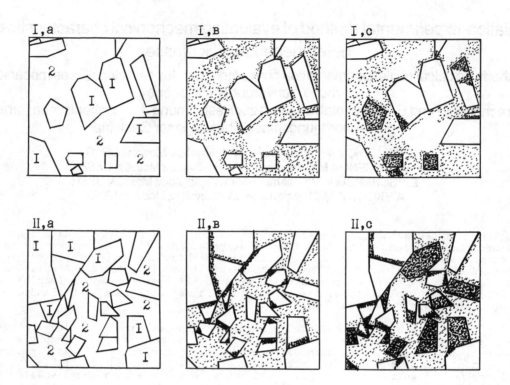

Figure 1. "Typical structures" of the samples (I,a - n = 35%; II,a - n = 57%) and plastic deformation area development in limiting state (I,b; II,b - inclusions strength is maximum; I,c; II,c - inclusions strength is minimum); 1 - crushed marl; filler (landwaste loam).

ponding to the "typical structures" of the smallest bulks, which can be investigated by these methods. The models of their mechanical behaviour and models' parameters are also established by experimental methods. Then the "typical structure" of larger bulk, being the composition of already studied elements is investigated by the method of mathematical modelling of experiment and mechanical properties effective characteristics of this structure are determined. Since "typical structure" of every next heterogeneity level consists of finite number of smaller bulks "typical structures", which effective mechanical characteristics can be determined by calculation method, mathematical modelling can be spread to rock masses of any size.

As an example, the application of this method is demonstrated by "typical structures" of crushed loam samples, represented in Fig. 1 (I,a; IIa). These schemes show some of the compositions of strong marl inclusions and less strong loam filler, typical for the material under study. They differ in percentage of inclusions, in total mass - n, in size, shape and orientation of inclusions, which will be denoted by the general term "architecture". Experimental analysis of the filler and inclusion material made it possible to establish the regularities of their deformation and failure. To describe filler mechanical properties in pre-limiting state non-linear deformation model was used, the model of ideally plastic body being applied for limiting state.

The research programme involved mathematical modelling of samples loading in corresponding conditions and determination of effective mechanical characteristics of the compositions. For this purpose the authors used the method of final elements. Iteraction procedure of calculation is realized using the method of variable rigidity, ac-

cording to the scheme presented in papers (Ukhov et al. 1985; Semenov 1986).

While working out the technique of numerical experiment some problems were specially studied, which permitted to set some principles, providing reliable enough (at least not overrated as compared to physical experiment) data:

- numerical modelling may be carried out according to the scheme of plane deformation, disregarding volumetrical stressed state of the sample. The resulting strength and deformation characteristics proved to be practically equal to or slightly less than those obtained in physical experiment at volumetrical stress state. Theory of composites gives corroboration to this idea (Christensen 1982).;

- the dimensions of the area under study should be at least 4-6 times larger than characteristic linear dimension of the largest inclusion. It provides the representability of the composition bulk and agrees with the notion of quasicontinuity and quasihomogeneity criterion proposed in paper (Ukhov 1975);

- final elements approximation detailing of the scheme under study should be high enough. Special attention should be payed to modelling of interaction conditions at "inclusion-filler" and "inclusion-inclusion" contacts.

The investigation made resulted in obtaining compositions effective characteristics, which are close to and never overrating the data of physical experiments. The investigation also permitted to study the processes of plastic failure areas formation in the samples, both in the course of loading and in limiting state (blackened zones in schemes I,b,c; II, b,c - Fig. 1). It should be noted that plastic deformation zones development character agree both with qualitative representations of the theory of composites (Christensen 1982) and with quantitative regularities of strength deformability of samples observed in mathematical and physical experiment.

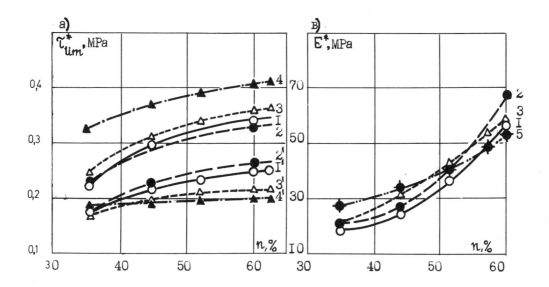

Figure 2. The effect on bulky fragments inclusions volumetrical content on strength (a) and deformability (b) of the studied soil (σ_m = 0.5 MPa; τ_3 = 0.168 MPa; E_3 = 13 MPa) with maximum (curves 1-4) and minimum (curves 1'-4') strength of bulky fragments inclusions: 1, 1'- mathematical modelling; 2,2' - physical experiments; 3,3' - calculation by interpolation formulas (2) and (3); 4,4'- calculation by V.I. Fedorov's method (Fedorov 1984); 5 - calculation by formula (Compositional... 1978).

To study the effect of inclusions percentage in the total rock mass, composition's "architecture", proportion of strength and deformation characteristics of inclusions and filler on effective mechanical characteristics of the composition, numerous series of numerical computations were carried out accompanied by control physical experiments. The analysis was carried out with n changing in the interval from 35 up to 60%. The reinforcing effect was negligible with n < 35% for relatively not strong and not rigid enough inclusions material. With n > 60% discordances between the computation and experimental results started to appear, which gained in magnitude with n increasing. It is connected with filler continuity disturbance in the pores between inclusions and appearance of unfilled cavities.

The following indices were used as basic characteristics of the filler and whole composition mechanical properties: τ - limiting shearing resistance, E - modulus of total deformation, J - lateral deformability ratio, defined as characteristics of corresponding non-linear dependences at predetermined stress intervals. The indices used for inclusions were the following: R_t, R_c - uniaxial compression and tension resistance, E and ν. Further on coefficient's indices show that they refer to "K"-composition, "B"-inclusion, "3"-filler.

The experiments conducted resulted in establishing a number of important regularities.

1. Compositions mechanical behaviour is qualitatively similar to that of the filler: they are characterized by inelastic bulk and shape deformation, dependence of shape deformation on hydrostatic constituent of the load, exhibition of contraction-delatation properties.

2. In all cases effect on inclusions is registed, i.e. when n increases, composition strength increases too and deformability decreases.

3. "Architecture" of the samples does not have any noticeable influence on the effective characteristics, when the arrangement of inclusions is chaotic, all other conditions being equal.

4. Inclusions strength affects τ * value considerably which can be compared with n effect and produces much less effect on value. With $\tau_b/\tau_3 > 10$ inclusions may be considered to be absolutely strong as compared to the filler. In this case they are practically undestructible in the process of loading, plastic deformation area is formed only in the filler (Fig. 1, schemes I,b; II,b), and inclusions rigidity practically does not affect shearing resistance value.

5. When inclusions strength values are low ($\tau_b/\tau_3 \approx 3.5$) the increase of their rigidity leads to τ * decrease due to stresses concentrations in the areas of contact. Plastic deformation area in limiting state covers not only the filler but also separate destructing inclusions (Fig. 1, schemes I,c; II,c).

6. In all cases E_b/E_3 increase leads to E* increase and some J * decrease. When E_b/E_3 = 100 inclusions may be considered to be absolutely rigid with respect to the filler. This parameter further increase does not cause any noticeable change of effective characteristics.

7. ν_b / ν_3 relationship influence on mechanical properties effective characteristics is negligible.

Multifactor analysis with the application of the theory of experiment planning was carried out on the basis of received regularities. The task of getting the following relationships was set:

$$\left. \begin{array}{l} \tau^*/\tau_3 = f_1(\, n, \ \tau_b/\tau_3 \, , \ E_b/E_3 \, , \ \nu_b/\nu_3) \\ E^*/E_3 = f_2(\, n, \ \tau_b/\tau_3 \, , \ E_b/E_3 \, , \ \nu_b/\nu_3) \\ \nu^*/\nu_3 = f_3(\, n, \ \tau_b/\tau_3 \, , \ E_b/E_3 \, , \ \nu_b/\nu_3) \end{array} \right\} \qquad (1)$$

within the limits of factors variation: n - from 35 up to 60%, τ_b / τ_3 from 3.5 up to 10, E_b/E_3 - from 5 up to 100; ν_b / ν_3 from 0.5 up to 1.0.

Regression equations in actual sizes are rather awkward. Below interpolation formulas for limiting cases are given:

$$1. \quad \begin{cases} \tau_b/\tau_3 \geqslant 10 \; ; \; E_b / E_3 \geqslant 100 \\ \tau^*/\tau_3 = -1.57 + 0.12 \cdot n - 9.60 \cdot 10^{-4} \cdot n^2 \\ E^*/E_3 = 1.20 - 4.66 \cdot 10^{-2} n + 1.73 \cdot 10^{-3} n^2 \end{cases} \qquad (2)$$

$$2. \quad \begin{cases} \tau_b/\tau_3 = 3.5 \; ; \; E_b/E_3 = 5 \\ \tau^*/\tau_3 = -1.26 + 9.84 \cdot 10^{-2} n - 9.60 \cdot 10^{-4} n^2 \\ E^*/E_3 = 3.10 - 0.12 n + 1.73 \cdot 10^{-3} n^2 \end{cases} \qquad (3)$$

For both cases: $\nu^*/\nu_3 = 1.92 - 3.68 \cdot 10^{-2} n + 3.27 \cdot 10^{-4} n^2$ \qquad (4)

Relationships corresponding to formulas (2) and (3) are represented in Fig. 2,a,b. They agree both with the results of mathematical modelling and physical experiments made by the authors and sith other author's data, published in technical literature (Fedorov 1984; Compositional... 1978).

In our opinion, the data presented in this paper convincingly corroborate the validity of the calculation-experimental method of determining mechanical characteristics of scale heterogeneous rock masses, worked out in MICI named after V.V. Kuibyshev. The proposed research method may be used in practice and interpolation formulas given here - for preliminary evaluation of inclusions effect on mechanical properties of the rocks of composite structure.

BIBLIOGRAPHY

Ukhov, S.B., Chernyshov, S.N. & V.V. Semenov 1985. The determination of effective charactistics of highly heterogeneous rock masses. Int.Symp. on the role of rock mechanics in excavation for mining and civil works, Mexico: 64-68.
Ukhov, S.B., Semenov, V.V., Scherbina, E.V. & A.V. Konviz 1986. Calculation-experimental method of evaluating mechanical characteristics of scale heterogeneous rock masses. In the book "Application of numerical methods to geomechanic problems", MICI, Moscow: 6-21.
Semenov, V.V. 1986. Joint static and filtration calculations of concrete dams rock footings. In the book "Application of numerical methods to geomechanic problems", MICI, Moscow: 78-90.
Christensen, R. 1982. Introduction into mechanics of composites. Moscow: "Mir", 334.
Ukhov, S.B. 1975. Rock footings of water power structures. Moscow: "Energia", 263.
Fedorov, V.I. 1984. Methods of strength and compressibility evaluation of disintegrated-clay soil. "Footings, foundations and soil mechanics", 3: 18-21.
Compositional materials. Vol. 5. Failure and fatigue. Moscow: "Mir", 484.

High-altitude monitoring of rock mass stability near the summit of volcano Nevado del Ruiz, Colombia

Surveilance à haute altitude de la stabilité du rocher près du sommet du volcan Nevado del Ruiz, Colombie

Die Überwachung aus grosser Höhe der Felsstandfestigkeit am Gipfel des Vulkans Nevado del Ruiz, Colombien

BARRY VOIGHT, Department of Geosciences, Pennsylvania State University, University Park, USA
MARTA LUCIA CALVACHE, Observatorio Volcanologico de Colombia, Manizales
OSCAR OSPINA HERRERA, Observatorio Volcanologico de Colombia, Manizales

ABSTRACT: An investigation to resolve questions of gravitational stability at Nevado del Ruiz include installation of robust EDM reflector stations at rock sites above 5000 m altitude. Line lengths periodically measured from helicopter-supported instrument stations 6.7 to 9.3 km distant suggest that the movement of the summit area dominantly reflects movements in an ice cap underlying volcanic deposits. Motions of glacier ice and underlying bedrock are decoupled. These data place constraints on the volume of rock susceptible to massive gravitational failure. The case history provides estimates of the precision and accuracy of long range EDM monitoring in high altitude terrain and some useful lessons for geotechnical monitoring.

RESUME: Une etude faite pour resoudre des problemes de l'instabilite gravimetrique a Volcan Nevado del Ruiz a commence par l'installation de reflecteurs EDM dans des sites audessus de 5000 m d'altitude. Les mesures faites periodiquement par des instruments installes a des distances de 6,7 - 9,3 km indiquent que le movement du sommet est surtout provoque par une carapace de glace en-dessous des debris volcaniques. Il existe un decollement entre le glacier et le substratum rocheux. Ces donnees mettent des contraintes sur le volume susceptible a un glissement gravimetrique massif. Cette etude donne des estimations de la precision et l'exactitude des methodes de surveillance par EDM a haute altitude et quelques lecons utiles pour la geotechnique.

ZUSAMMENFASSUNG: Wir haben die Gravitationsstabilitat bei Vulkane Nevado del Ruiz studiert mit EDM-Messungstechnik in Felsstationen hoher als 5000 m. Messungen wurden periodisch gemacht von Stationen auf 6,7 - 9,3 km. Die Ursache der Bewegungen der Spitze ist im Eis unter die pyroklastische sedimente. Es gibt ein Bewegungsdiskontinuitat zwischen Eis und unterliegenden Fels. Diese Methode gibt eine Schatzung der Praxision von EDM-Messungen in hohere Gebirge und einige brauchsame vorhafte fur geotechnische Praxis.

1 INTRODUCTION

Nevado del Ruiz, the most northern of active volcanoes of the Andes and the largest of volcanoes forming the crest of the Cordillera Central of Colombia (alt. 5400 m), is located 30 km southeast of Manizales (population 350,000). A small magmatic eruption on 13 November 1985 melted snow and ice from its summit plateau to mobilize the deadliest lahars of historic time, with the losses exceeding 23,000 fatalities and 200,000 people directly affected. Ruiz experienced a major eruption in 1595, and in 1845 was reportedly the source of a large gravitational slope failure of rock and ice that transformed into a lahar, killing at least 1,000 persons below the eastern flank of the mountain (Ramirez, 1975).

On 13 November 1985, pyroclastic flows and surges were deposited onto the summit ice cap during the paroxysmal phase of the eruption. In the following weeks, large cracks were observed in these pyroclastic deposits and in subjacent ice on the summit plateau. Many cracks progressively opened, and additional fractures developed throughout December (R.J. Janda, 1986, personal communication). Electronic distance measurement (EDM) radial lines had been installed to monitor inflation and deflation events of the volcano in order to aid eruption forecasts (Fig. 1). Systematic radial line shortening of 5-10 cm/day was recorded in December for targets on the summit plateau surface near the head of the Rio Azufrado, compared to both shortening and lengthening motions of a few mm to 5 cm/day elsewhere (Fig. 2; cf. SEAN Bulletin, v. 10, no. 12).

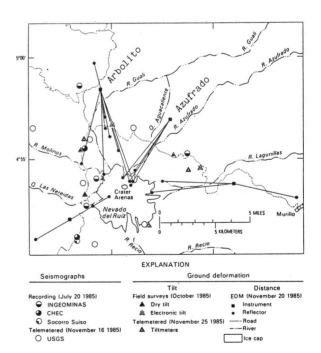

EXPLANATION

Seismographs	Ground deformation		
	Tilt		Distance
Recording (July 20 1985)	Field surveys (October 1985)		EDM (November 20 1985)
◐ INGEOMINAS	▲ Dry tilt		■ Instrument
◑ CHEC	⚭ Electronic tilt		● Reflector
◎ Socorro Suiso	Telemetered (November 25 1985)		------ Road
Telemetered (November 16 1985)	△ Tiltmeters		—— River
○ USGS			▭ Ice cap

Fig. 1. Volcano monitoring network. Dates refer to initial operations of parts of network. Stability investigations used EDM instrument stations at Arbolito and Azufrado.

Because of complicated geology and difficulty of access, the specific causes of fracturing and associated ground movements could not be immediately and unambiguously determined. Among the various possibilities were glacier flow adjustments, deep-seated gravitational creep and rupture of bedrock, tectonic processes, or deformation produced by magmatic intrusion. Concern was expressed by the Comite de Estudios Vulcanologicos -- the Ruiz hazard evaluation team staffed by Colombians and international scientists -- that the fractures might be a precursor to a large gravitational failure that could transform downstream into a catastrophic lahar, a sequence of events that recalled the aforementioned deadly scenario of 1845.

The investigation described here was carried out in response to this concern (Voight, 1986). Resolution of this problem required field study and a program of precautionary monitoring. This mission was initiated in January 1986 under the auspices of the Comite de Estudios Vulcanologicos, Manizales. The participation of Voight was supported by the Office of Foreign Disaster Assistance, U.S. State Department, in cooperation with the U.S. Geological Survey. Contribution 872, Observatorio Vulcanologico de Termales del Ruiz.

ARBOLITO – RIM

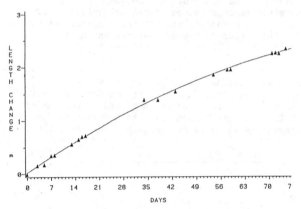

Fig. 2. Shortening of Arbolito-Rim EDM line from December 10.

2 GEOLOGIC SETTING OF SUMMIT PLATEAU

The summit crater Arenas perforates the ice cap near the northern margin of the summit plateau. The crater area consists mainly of a succession of older andesite lavas, overlain by glacier ice and the 1985 pyroclastic sequence (Herd, 1974; Thouret et al., 1985; Calvache, 1986). The structure of the volcano interior is revealed in the walls of the glacially-incised valley of the Rio Azufrado, where hydrothermally-altered andesite flows dip 12°-18° northward, approximately radial to the crater. The bottom of the crater is kept free of ice by fumrolic activity. In February some standing water existed in the crater bottom but there was no lake of appreciable volume. Ice is <10m thick on the north crater rim facing the Rio Azufrado, but thickens considerably toward the south.

Airphotos indicate that in 1959, streams of glacier ice extended continuously over the summit plateau rim into the Rio Azufrado valley. These lobes have now been decapitated as a result of events of November 1985. In the Rio Azufrado, a beheaded glacier segment rests on rock slopes between approximate elevations 4900 to 4600 m. Smaller glacier slabs are perched on the upper valley walls, some on slopes that average more than 30°. Scars of bare rock surrounded by ice cover

locally indicate where large slab avalanches of glacier ice were released during November 1985. These avalanches were probably triggered by basal water pressures generated by percolating meltwater during the eruption. Ground motion associated with earthquakes and tremor may have been a contributing factor.

3 COMPLEX FRACTURE PATTERNS OF THE SUMMIT PLATEAU

The following considerations apply to the interpretation of fracture systems in the vicinity of the crater. A polygonal crack pattern has developed within the 13 November 1985 sequence of pyroclastic deposits. Such cracks form in response to thermal tension set up by the tendency of cooling pyroclastic deposits to contract. Crack polygons range from a few meters to tens of meters in diameter. The formation of fractures obvious on the ground surface was commonly preceded by the development of a network of linear pale hydrothermal alteration zones reflecting concentrated vapor and heat flow. Apparently the fractures initiated at depth, and propagated both laterally and toward the ground surface. Fractures progressively opened to produce gaps greater than 1 m wide and 6 m deep. The polygonal cracks perhaps were most strongly influenced by a 6 m thick welded pyroclastic flow unit described by Calvache (1986), which was emplaced at a temperature of approximately 900°C (W.G. Melson, 1986, oral communication). The boundary of this unit, thus far only approximately defined, perhaps coincides with the boundary of the polygonal cracking zone.

Differential vertical displacements occurred along some of these cracks, and multiple crack intersecions tended to localize dome-shaped features with outward-dipping slabs. These features probably represent areas of lagging subsidence, rather than true upward displacement, with subsidence caused by progressive ice melting under pyroclastic deposits. Pyroclastic slabs are warmest in polygon interiors, and cooler at boundary cracks where heat is vented to the atmosphere. Intensity of ice melting and differential subsidence should reflect the heat loss and temperature distribution in the pyroclastic deposits. The hypothesis remains to be tested by direct observation of the pyroclastic deposit-ice contact in the vicinity of features displaying lagging subsidence.

Cracks are also associated with movement of the ice cap under the pyroclastic veneer. The fundamental crevasse pattern in the vicinity of the crater is due to longitudinal, extending, northward flow, with the cracks opening up approximately at right angles to the flow direction. Down-valley curvature occurs near valley walls due to the influence of drag. This characteristic pattern also appears on 1959 aerial photographs (Fig. 3), where the crevasse field intercepts the north margin of Arenas crater. The dominant pattern reflects the convex longitudinal profile at the head of the Rio Azufrado valley. The basic pattern of transverse crevasses above an ice-fall continues despite decapitation of segments of the glacier snout by ice avalanches in November 1985.

A circumferential fracture pattern surrounds the Arenas crater, reflecting local radial ice flow into the crater, and possibly also deep-seated differential subsidence related to magmatic processes.

The summit plateau is also crossed by fractures and faults of volcano tectonic origin. The most-pronounced comprise a series of north-south oriented features that cross virtually the entire volcano, passing the Arenas crater on the east (Fig. 3). Other orientations of fracture zones are present (Fig. 3). Several fractures cross the Arenas crater, with the most prominent orientations

north-south and west-northwest. Most of these
fractures were present prior to the present eruption cycle, but recurrent movement presumably has
occurred on some. Locations of some recent epi-
centers suggest that the main north-south set may
be in extension in association with volcano inflation. Bulged areas under the ice sheet seem
to reflect old eruptive centers (Fig. 3).

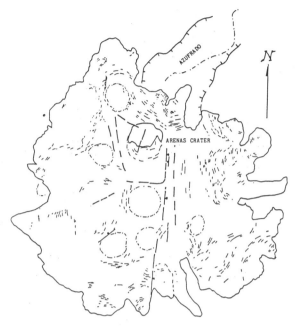

Fig. 3. Volcano-tectonic linear features and
crevasse patterns at Nevado del Ruiz, based
on USAF photos, 10 February 1959.
Circular features represent old domes and
vents under glacial cover.

Fracture patterns mapped by M.L. Calvache and B.
Voight show combinations of the fracture elements
previously discussed. Random polygonal cracks in
pyroclastic flows of the summit plateau give way to
less random, more structured, and broadly arcuate
patterns as either the Arenas crater or the Rio
Azufrado headwall are approached. This is interpreted to reflect the influence of crevasse
patterns in glaciers underlying the pyroclastic
flows. At the Nareidas glacier, pyroclastic flows
have filled or partly filled glacier crevasses. A
similar style of crevasse filling has occurred on
the summit plateau.

Progressive melting, continued opening of
crevasses, and collapse of old crevasse bridges
could explain the recurrent movement on and progressive enhancement of surface fractures that originated as cooling cracks in pyroclastic flows.
Clusters of collapse pits are most easily interpreted in this manner, i.e. as zones of profound
local subsidence in the pyroclastic sequence that
reflect underlying crevasses.

Fractures associated with the above mechanisms
could be distinguished, at least in principle.
Our more pertinent question was to identify the
volume of rock and ice that could potentially be
associated with a mass movement. The resolution of
this question was more problematic, because rock
gravity movements can produce fractures similar to
those caused by glacier mechanisms. Transverse
cracks and grabens develop in regions of extending
rock mass deformation whereas thrust faults and
buckling can occur near the slide toe. Tension
cracks near the head of a slide, and internal
shearing, often precede toe rupture. Therefore
deformational patterns that resemble the effects of

glacier motions could also be produced by rock mass
movements.

A large fracture east of the crater, that had
progressively opened since November, appeared to
connect to a prominent fracture in rock inside the
crater. This observation raised the possibility of
deep-seated bedrock failure, with the surfaces of
detachment in part consisting of pre-existing,
appropriately oriented fracture segments of volcano-
tectonic origin. The fracture patterns did not rule
out the possibility of a significant mass movement,
one that could remove a large segment of the Arenas
crater and possibly cause depressurization of the
underlying magma chamber. Monitoring was necessary.

4 PHILOSOPHY AND METHODOLOGY FOR MONITORING

Instability and failure of rock masses always involve movement. Larger movements are almost always
preceded by smaller displacements and accelerations
that may be detected by precautionary monitoring,
provided that instrumentation of sufficient sensitivity is used and adequately located. But in this
instance comprehensive and complex investigations
were ruled out. A hazard evaluation was urgently
needed. The time available for hazard investigation was therefore short. Furthermore, funds were
limited, and logistical problems were appreciable.
The summit plateau at Ruiz soars above the service
ceiling of the available helicopter. Urgency pre-
cluded adequate acclimatization to high altitude
endeavor, and gas emissions required the intermittent use of masks that further reduced the efficiency of field efforts. Field operations had to be
based on over glacier travel on foot under a perceived explosive eruption alert, with prevailing
weather uncertain and permitting at best a narrow
daily time window. Given the environment of an
active volcano, such conditions did not seem surprising. But, in retrospect these obstacles
proved barely surmountable.

The crux to overcome obstacles lay in planning the
hazard investigation as an experiment, in order to
discover what Ralph Peck (1972) termed the "critical
question": Could one question, capable of being
answered negatively or affirmatively by an efficient
investigative program, decide the issue? Not
surprisingly, the question so identified was: Is
there appreciable downward and outward movement of
the bedrock mass under the cap of glacial ice in the
Azufrado sector?

To resolve this question, a program of geotechnical monitoring was devised that involved the adapt-
ation of an EDM-based geodetic net already established for the purposes of inflation monitoring and
eruption forecasting (Banks et al., 1986). The instrument stations to be employed were those located
at "Arbolito" and "Azufrado", respectively about
9.3 and 6.7 km from the head of the Azufrado canyon.
Because these stations were located at elevations
well below the summit plateau, access by helicopter
was possible during good weather conditions. Line-
of-site visibility to various reflector points also
required fair weather, a problem at Ruiz where a
cloud cap commonly masked the summit, even on fair
weather days. Variations in pressure, temperature
and humidity could normally be recorded only at the
instrument site.

Line lengths between instrument and reflector
stations were established by phase comparison with a
Ranger VA instrument, using a directly-modulated, 3
milliwatt helium-neon (red) laser as a light source.
Modulating frequencies were generated by an internal
quartz crystal oscillator with a stability of 1 ppm.

Reflector stations consisted of weatherproof
hollow precision optical retroreflectors of 6.4 cm
aperture bolted on steel rods driven into volcanic
debris, or mounted on spikes driven into the
weathered rind on the bedrock surface. Where poss-

ible at a given station the retroreflectors were
installed in pairs.

The key reflector stations are the following
(Fig. 4):

RIM and CIMA, mounted in deposits of the November
1985 eruption that overlie glacier ice on the
summit plateau.

MARTICA and FINGER, mounted in bedrock midway
across the Rio Azufrado headwall, appproximately at
elevation 5000 m below beheaded ice cliffs on the
summit plateau adjacent to the crater.

The two bedrock stations were intentionally
located within about 100 m of each other, for redundancy of measurements was desired. A system for
precautionary monitoring should be such as to
generate confidence in the results. We did not
wish to risk misinterpretation owing to an isolated
defective station.

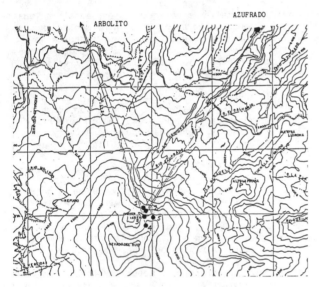

Fig. 4.   EDM stations in Rio Azufrado Sector
1, Finger; 2, Martica; 3, Rim; 4, Cima;
5, Peasoup.

5 RESULTS AND CONCLUSIONS

The summit plateau stations were installed with the
volcano monitoring study in December. The bedrock
stations were installed by roped descent as part
of the geotechnical investigation on February 12.
Two days later, initial line lengths to MARTICA and
FINGER were recorded from the Azufrado instrument
station. Within a week three additional measurements were made.

Initial results, to late February, were suggestive of slow movement of bedrock (Fig. 5 ; Voight
et al., 1986). By March 7 the trend of bedrock
measurements had stabilized (Fig. 6). It was now
realized that this initial trend was misleading,
probably because of the influence of atmospheric
factors on apparent line length. Temperature
control was lacking over the line of measurement,
and absorption, scattering and refraction were
influenced by particles and air turbulence.

The data points from the two bedrock stations
showed nearly identical apparent movement trends
(Fig. 6), with the difference between the line
length changes at the two nearby stations representing a measure of precision of measurement--
commonly about 2 ppm. Accuracy--that is, the departure of a measurement from the presumed actual
value--is more than this, as much as 5 ppm.

MARTICA AND FINGER TO FEBRUARY 21

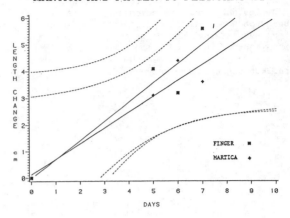

Fig. 5.   Martica and Finger reflectors, Azufrado
instrument station. Line shortening,
February 14-21.

MARTICA AND FINGER TO MARCH 10

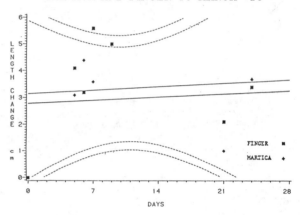

Fig. 6.   Martica and Finger reflectors, Azufrado
instrument station. Line shortening,
February 14-March 10.

RIM AND MARTICA TO MARCH 10

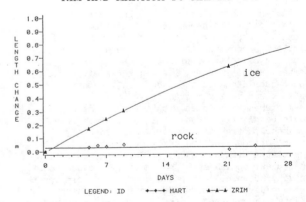

Fig. 7.   Martica and Rim reflectors, Azufrado
instrument station. Line shortening,
February 14-March 10.

The data permitted the following conclusions to be drawn on March 8 (Voight, 1986):

(1) Motions of ice and bedrock are decoupled on the summit plateau at the head of the Azufrado canyon (Fig. 7).

(2) The steady outward motion of the summit deposits reflects motion in ice, not rock.

(3) The rate of ice movement has systematically decreased over the period of measurement (Figs. 2, 7). This suggests the progressive diminishment of the strong influence on basal ice motion of copious quantities of meltwater generated in the November 1985 eruption.

(4) The difference in rates of motion for stations RIM and CIMA indicate extensional ice flowage commensurate with the development of observed surface extension fracturing.

(5) Ice avalanches of small to moderate size (10^5-10^6 m^3) may be expected periodically due to collapse of cantilivered ice blocks at the Azufrado headwall rim.

(6) The data do not suggest significant discernible movement of rock.

(7) Constraints can therefore be placed on the volume of rock potentially susceptible to gravitational failure under existing conditions of loading. The possibility of massive gravitational failure of the Rio Azufrado headwall now seems remote unless loading conditions change, and the same conclusion applies to rockslide-generated depressurization and explosive triggering of magma or hydrothermal fluid chambers.

These results, which provided a measure of relief to those responsible for the hazard response at Ruiz, were sustained by additional measurements carried out in late spring and early summer.

Lessons that might be applied elsewhere include information on the accuracy to be expected for EDM instrumentation. The practical limitations of site investigation in poorly accessible, inhospitable terrain imply uncertainties in line length of the order of 5 ppm, significantly greater than the 1 ppm accuracy available with expensive and time consuming techniques theoretically available for measurement of the relevant meteorological parameters. We emphasize the term theoretically -- at Ruiz, for example, operation of helicopters to the altitude of the instrument stations stretched their operational limits. Helicopter operation for meteorological data acquisition at the altitude of the reflector stations, or along the line connecting reflector and instrument stations, was simply not feasible. Nevertheless, some sacrifice of precision and accuracy is often appropriate in geotechnical hazard investigations. This sacrifice can be justified in order to increase the speed of completing the measurements, increase the frequency of measurements, and keep the costs economical. Neither engineering nor scientific decisions necessarily demand that the theoretical limits of accuracy be approached.

The hollow 6.4-cm aperture retroreflectors employed at Ruiz were successfully used at distances in excess of 9 km, under complex and often presumably unfavorable meteorological conditions. Far in excess of the manufacturer's stated limit of 800 m, this range may reflect an increased efficiency of EDM operation in the thin air at high altitudes.

On the other hand, the spike mounting system used at Ruiz seems only marginally robust and marginally reliable for hard rock. An improved system is needed. A mounting system involving technical climbing pitons might be investigated for future efforts of this kind.

Similar monitoring systems could be effectively applied to other volcanoes and to other high-altitude mountain sites where questions of rock and ice stability are required for hazard analysis.

Finally, Peck (1972) has suggested that more than a few projects seem to be provided with elaborate instrumentation at considerable expense, on the hopeful premise that the results of such endeavor will surely be capable of answering some question that will turn out to be of interest. With data in hand, the investigator then searches for a question that the findings can somehow resolve. In contrast, in the Ruiz project we attempted to emulate the working methods of the classical physicists--to concentrate on asking the critical question, and to seek its resolution by an unambiguous experiment employing sparse though robust instrumentation. In this instance our approach was conditioned by necessity, more by the several obstacles presented by Nevado del Ruiz than an exemplary professional attitude. Nevertheless the study served to remind us anew that instrumentation should not be a blind requirement, and that effective observational monitoring depends more on the questions to be answered than on the number of instruments used.

REFERENCES

Banks, N.G. Van der Laat, R. Carvajal, C. & Serrano, T. 1986. Deformation monitoring at Nevado del Ruiz, Colombia. EOS, 67(16):403.

Calvache, M.L. 1986. The eruption of the Nevado del Ruiz Volcano, Colombia. EOS, 67(16):405.

Herd, D.G. 1974. Glacial and volcanic geology of the Ruiz-Tolima volcanic complex, Cordillera Central, Colombia. Pub. Geologicas Especiales del Ingeominas, Bogota. 8:1-48.

Peck, R.B. 1972. Observation and instrumentation. Highway Focus, U.S. Dept. Trans., Fed. Hwy. Admin., Pub. 131. 4(2):1-5.

Ramirez, J. 1975. Historia de los terremotos en Colombia. Inst. Geographicos "Agustin Codazzi," Bogota.

Thouret, J.C. Murcia, A. Vatin-Perignon, N. & Salinas, E. 1985. Cronoestratigrafia mediante dataciones K/Ar y ^{14}C de los volcanoes compestos del Complejo Ruiz-Tolima y aspectos volcano-estructurales del Nevado del Ruiz. Ingeominas, Bogota.

Voight, B. 1986. Gravitational failure hazards, Nevado del Ruiz, Colombia--Report of a mission from 30 January-16 February 1986. Comite de Estudios Vulcanologicos, Manizales. p. 20.

Voight, B. Calvache, M.L. & Ospina Herrera, O. 1986. Is the Arenas Crater at Nevado del Ruiz endangered by gravitational slope collapse? EOS. 67(16):406.

Thermal diffusivity measurements of rocks by difference calculus of the heat conduction equations

Détermination de la diffusibilité thermique de roches à partir des formules de l'état thermique
Messung des Wärmeausbreitungsvermögens von Gestein durch diffuse Berechnung der Wärmezustandsgleichungen

NARUKI WAKABAYASHI, Institute of Technology, Shimizu Construction Co. Ltd, Tokyo, Japan
NAOTO KINOSHITA, Institute of Technology, Shimizu Construction Co. Ltd, Tokyo, Japan
TADASHI HANE, Institute of Technology, Shimizu Construction Co. Ltd, Tokyo, Japan

ABSTRACT: A new method for measuring thermal diffusivity of rocks is presented. The theory is obtained from linear and 3-dimensional heat conduction equations by discrete approximation. This method has the advantages that thermal diffusivity can easily be calculated by measuring the temperatures of three points and the measurement can be performed at a site as well as in the laboratory. The diffusivities of three kinds of rock specimens and an in-situ rock mass are measured. According to the investigation on measurement errors, this method shows sufficient accuracy.

RESUME: Une nouvelle méthode de mesure de la diffusibilité thermique de roches est présentée. La théorie est obtenue à partir des formules de l'état thermique linéaires et à 3-dimensions par approximation discrète. Cette méthode a pour avantage de permettre de calculer facilement la diffusibilité thermique en mesurant la température à trois points. La mesure de température est possible aussi bien in situ que dans un laboratoire. La diffusibilité de trois types d'échantillons de roches fait l'objet de cette mesure. L'étude sur l'érreur de mesure a démontré que cette méthode offre une haute précision.

ZUSAMMENFASSUNG: Diese Arbeit behandelt ein neues Verfahren zur Messung der Wärmeausbreitungsvermögen von Gestein. Die Theorie wird auf dem Weg der diskreten Näherung von linearen und dreidimensionalen Wämeleitungs-gleichungen abgeleitet. Die Vorteile dieses Verfahrens liegen darin, daß die wärmeausbreitungsvermögen durch Messung der Temperatur an drei Punkten leicht berechnet werden kann und die Messung sowohl vor Ort als auch im Labor möglich ist. Es werden die an drei Arten von Gesteinsproben sowie die vor Ort an Gestein gemessenen Werte eingeführt. Die Bestimmung des Meßfehlers ergab eine ausreichende Genauigkeit für das hier untersuchte Verfahren.

1 INTRODUCTION

In this paper, a new method for measuring thermal diffusivity of rocks is presented. In designing an underground high-level radioactive waste repository or an underground nuclear power plant, it is important to evaluate the thermal properties of in-situ rock mass over a wide temperature range. Nearly all measurement methods for thermal properties are based on the rigorous solution of the heat conduction equation. The rigorous solution, however, can be obtained only under certain specific conditions. For example, the following conditions are required.
1. The initial temperature distribution must be uniform.
2. Thermal properties must be temperature independent.
As a result, it is very difficult to measure the in-situ thermal properties of rock mass.

Tews et al.(1977) and Jeffrey et al.(1979) determined thermal conductivity and diffusivity by using the least squares method in the case of a finite line source conduction. They also assumed that the initial temperature distribution was uniform and the thermal properties were temperature independent.

Vost(1976) measured the temperatures of three points in the rock mass, and determined the thermal diffusivity by comparing the measured midpoint temperature with the midpoint temperature calculated from the both ends. This method has the advantage that the initial temperature distribution is not required to be uniform. However, it must be assumed that the thermal properties are temperature independent due to the difficulties in numerical calculations.

However, it is well known that the actual thermal properties of rocks show temperature dependence as Lindroth (1974) and Hanley et al.(1978) point out. The new measurement method of thermal diffusivity proposed here is obtained from the difference equation method for a linear heat conduction equation and a 3

dimensional(3-D) heat conduction equation in terms of spherical polar coordinates. By this method, the thermal diffusivity of rocks can be measured as a function of temperature at a site as well as in the laboratory. It is not necessary that the initial temperature distribution of rocks is uniform. In this paper, the theory of this method, the results of in-situ and laboratory measurements, and the measurement error are discussed.

2.THEORY OF MEASUREMENT METHOD

There are two measurement theories for plane and point heat sources. The first theory is derived from the following linear heat conduction equation.

$$a \frac{\partial^2 \phi(x,t)}{\partial x^2} - \frac{\partial \phi(x,t)}{\partial t} = 0 \qquad (1)$$

where a, x, t and $\phi(x,t)$ are thermal diffusivity, location, time and the temperature measured at the point (x,t) respectively. Using nine temperatures at the three points x_1, x_2, x_3 (which is spaced at the distance Δx respectively) at three different times t_1, t_2, t_3 (which has a time interval of Δt respectively), the two terms in Eq.(1) can be expressed as follows.

$$a \frac{\partial^2 \phi(x,t)}{\partial x^2} = a \frac{\phi(x_1,t_2) - 2\phi(x_2,t_2) + \phi(x_3,t_2)}{\Delta x^2} \qquad (2)$$

$$\frac{\partial \phi(x,t)}{\partial t} = \frac{\phi(x_2,t_3) - \phi(x_2,t_1)}{2\Delta t} \qquad (3)$$

Substituting Eq.(2) and (3) into Eq.(1), Eq.(4) is obtained.

$$a \frac{\Delta t}{\Delta x^2} = \frac{\phi(x_2,t_3) - \phi(x_2,t_1)}{2\{\phi(x_1,t_2) - 2\phi(x_2,t_2) + \phi(x_3,t_2)\}} \qquad (4)$$

where $a\Delta t/\Delta x^2$ is a Fourier number. Taking each coefficient of temperature $\phi(x_i,t_j)$, Eq.(4) is diagrammatically rewritten as follows.

$$a\frac{\Delta t}{\Delta x^2} = \begin{bmatrix} 0 & 0 & 0 \\ -1 & 0 & 1 \\ 0 & 0 & 0 \end{bmatrix}\phi(x_i,t_j)\Bigg/2\begin{bmatrix} 0 & 1 & 0 \\ 0 & -2 & 0 \\ 0 & 1 & 0 \end{bmatrix}\phi(x_i,t_j) \quad \begin{pmatrix} i=1\sim3 \\ j=1\sim3 \end{pmatrix} \quad (5)$$

By using Sympson's mean formula, Eq.(6) is obtained.

$$a\frac{\Delta t}{\Delta x^2} = \begin{bmatrix} -1 & 0 & 1 \\ -4 & 0 & 4 \\ -1 & 0 & 1 \end{bmatrix}\phi(x_i,t_j)\Bigg/2\begin{bmatrix} 1 & 4 & 1 \\ -2 & -8 & -2 \\ 1 & 4 & 1 \end{bmatrix}\phi(x_i,t_j) \quad \begin{pmatrix} i=1\sim3 \\ j=1\sim3 \end{pmatrix} \quad (6)$$

Eq.(6) is the fundamental measurement equation of thermal diffusivity. Furthermore, in order to improve the accuracy, the numerical interpolation along in the space direction is performed. Three types of Eq.(6), in which the midpoints x_1, x_2, x_3 are added up, is used to obtain Eq.(7).

$$a\frac{\Delta t}{\Delta x^2} = \begin{bmatrix} -1 & 0 & 1 \\ -6 & 0 & 6 \\ -10 & 0 & 10 \\ -6 & 0 & 6 \\ -1 & 0 & 1 \end{bmatrix}\phi(x_i,t_j)\Bigg/2\begin{bmatrix} 1 & 4 & 1 \\ 0 & 0 & 0 \\ -2 & -8 & -2 \\ 0 & 0 & 0 \\ 1 & 4 & 1 \end{bmatrix}\phi(x_i,t_j) \quad \begin{pmatrix} i=0\sim4 \\ j=1\sim3 \end{pmatrix} \quad (7)$$

The temperatures at x_1, x_3 are approximated from the other temperatures at x_0, x_2, x_4.

$$\phi(x_1,t_j) = \{3\phi(x_0,t_j) + 6\phi(x_2,t_j) - \phi(x_4,t_j)\}/8 \quad (8)$$

$$\phi(x_3,t_j) = \{-\phi(x_0,t_j) + 6\phi(x_2,t_j) + 3\phi(x_4,t_j)\}/8 \quad (9)$$

Substituting Eq.(8) and (9) into Eq.(7) and replacing $2\Delta x$ to Δx, the final equation in the case of a plane heat source is obtained as follows.

$$a\frac{\Delta t}{\Delta x^2} = \begin{bmatrix} -5 & 0 & 5 \\ -38 & 0 & 38 \\ -5 & 0 & 5 \end{bmatrix}\phi(x_i,t_j)\Bigg/16\begin{bmatrix} 1 & 4 & 1 \\ -2 & -8 & -2 \\ 1 & 4 & 1 \end{bmatrix}\phi(x_i,t_j) \quad \begin{pmatrix} i=1\sim3 \\ j=1\sim3 \end{pmatrix} \quad (10)$$

The second measurement equation can be obtained from the 3-D heat conduction equation in terms of spherical polar coordinates. Supposing the temperature is constant regardless of the angle, the equation is written as follows.

$$a\left(\frac{\partial^2 \phi(r,t)}{\partial r^2} + \frac{2}{r}\frac{\partial \phi(r,t)}{\partial r}\right) - \frac{\partial \phi(r,t)}{\partial t} = 0 \quad (11)$$

where r is the radius from the origin point. By the similar procedure which Eq.(10) is derived by, the final equation for thermal diffusivity in the case of a point heat source is obtained as follows.

$$a\frac{\Delta t}{\Delta x^2} = \frac{\begin{bmatrix} -5 & 0 & 5 \\ -38 & 0 & 38 \\ -5 & 0 & 5 \end{bmatrix}\phi(r_i,t_j) - 3\frac{\Delta r}{r_2}\begin{bmatrix} -1 & 0 & 1 \\ 0 & 0 & 0 \\ 1 & 0 & -1 \end{bmatrix}\phi(r_i,t_j)}{16\left(\begin{bmatrix} 1 & 4 & 1 \\ -2 & -8 & -2 \\ 1 & 4 & 1 \end{bmatrix}\phi(r_i,t_j) - 3\frac{\Delta r}{r_2}\begin{bmatrix} 1 & 4 & 1 \\ 0 & 0 & 0 \\ -1 & -4 & -1 \end{bmatrix}\phi(r_i,t_j)\right)} \quad \begin{pmatrix} i=1\sim3 \\ j=1\sim3 \end{pmatrix} \quad (12)$$

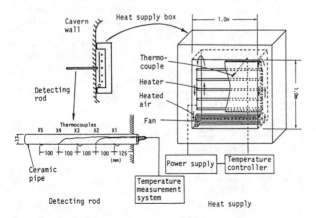

Fig.1 The schematic diagram of the measurement apparatus for in-situ thermal diffusivity.

where r_1, r_2, r_3 are the radii with the distance Δr respectively.

The thermal diffusivity can be easily calculated from the temperature at three points and at three different times by using Eq.(10) and (12).

3. IN-SITU MEASUREMENT

The in-situ thermal diffusivity measurement based on Eq.(10) was carried out on a cavern wall in "Ohya tuff" at Tochigi prefecture, Japan. "Ohya tuff" is a very porous rock having a porosity of about 45%. The rock mass was nearly saturated and its degree of saturation was about 95% except in the cavern surface. "Ohya tuff" contains a lot of montmorillonite clods ranging from several millimeters to sometimes tens of millimeters in diameter. However, the rock mass is generally homogeneous and continuous. And there was also no discontinuity in the measurement area.

The schematic diagram of the measurement apparatus for in-situ thermal diffusivity is shown in Fig.1. The heat supply box(1m×1m) shrouded with asbestos was attached to the cavern wall after the rough surface was made smooth. The box has five line heaters and a fan to keep the temperature of the wall surface uniform. The wall surface was heated up to 150°C in 10 hours and heated smoothly up to 200°C after 20 hours. The temperature was held constant after up to 40 hours. The temperature detecting rod was buried into the bore hole(Φ25mm×650mm). The five thermocouples(X1~X5) were fixed on a ceramic pipe(Φ17mm× 600mm) at 100mm intervals. The temperatures were measured every 5 minutes. The initial surface temperature was 5.6°C and that at 52cm depth was 3.6°C. The initial temperature distribution was not uniform.

The thermal diffusivity was calculated in combinations of X1-X2-X3, X2-X3-X4 and X3-X4-X5. In each case, the distance(Δx) was 100mm and the time interval (Δt) of 8 hours was selected in order to satisfy the condition that the Fourier number($a\Delta t/\Delta x^2$) was about 1.0. The results are shown in Fig.2, where the abscissa is the temperature at the midpoint of each combination. The thermal diffusivities of the three cases are almost temperature independent and range within $0.33\sim0.37\times10^6$m²/s.

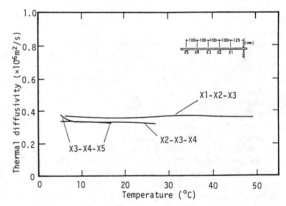

Fig.2 The in-situ thermal diffusivity of "Ohya tuff".

4. LABORATORY MEASUREMENTS

The thermal diffusivities of three kinds of rocks, "Inada granite", "Akiyoshi marble" and "Sanjome andesite", were measured in the laboratory using the same method as the in-situ measurement. The specimen size was 200mm×200mm×150mm. The density and porosity of specimens are shown in Table 1. A ceramic bar(Φ6mm ×150mm) attached with six thermocouples(X1~X6) was buried into a hole(Φ8mm×145mm) drilled in the rocks.

Measurement was performed under dry and saturated rock conditions. Rock specimens wrapped with aluminum foil to avoid absorption or evaporation of moisture,

Table 1 The density and porosity of rock specimens.

Properties	Inada granite	Akiyoshi marble	Sanjome andesite
Density (kg/m³)	2620	2720	2140
Porosity (%)	1.3	0.7	15.8

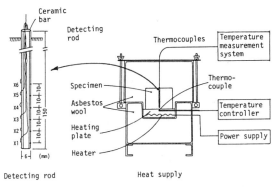

Fig.3 The schematic diagram of the apparatus for measuring thermal diffusivity in the laboratory.

were put on the heating plate and shrouded with asbestos wool. The schematic diagram of the measurement apparatus is shown in Fig.3. The specimen was heated at a rate of 30°C/h for 7 hours and the temperatures were measured every 1 minute. The thermal diffusivities were calculated in the two cases $\Delta x=20mm$(X1-X3-X5,X2-X4-X6). And the time interval Δt was selected on condition that $a\Delta t/\Delta x^2 \fallingdotseq 1.0$.

These results are shown in Fig.4, 5 and 6. The thermal diffusivities of "Inada granite" and "Akiyoshi marble" are remarkably temperature dependent. As the temperature rises from 30°C to 150°C, the diffusivity under dry condition of the former decreases from $1.7 \times 10^6 m^2/s$ to $1.2 \times 10^6 m^2/s$ and the latter decreases from $1.1 \times 10^6 m^2/s$ to $0.7 \times 10^6 m^2/s$. However, the diffusivity of "Sanjome andesite" is not temperature dependent and ranges around $0.8 \times 10^6 m^2/s$. Under saturated condition, thermal diffusivity over 70~80°C can not be determined because of the influence of evaporation. The saturated thermal diffusivities of "Inada granite" and "Akiyoshi marble" are larger than those under dry condition. However, the saturated diffusivity of "Sanjome andesite" is a little smaller than that under dry condition.

Both "Inada granite" and "Akiyoshi marble" have very low porosities as shown in Table 1, and their pore shapes mainly belong to the crack-type. However, "Sanjome andesite" shows high porosity, and its dominant pore shape is spherical-type. Walsh and Decker(1966)

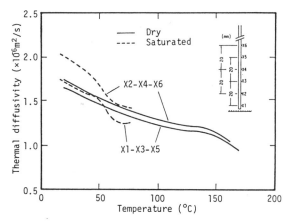

Fig.4 The thermal diffusivity of "Inada granite".

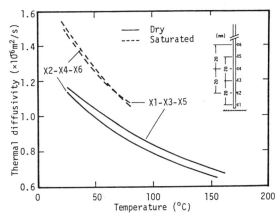

Fig.5 The thermal diffusivity of "Akiyoshi marble".

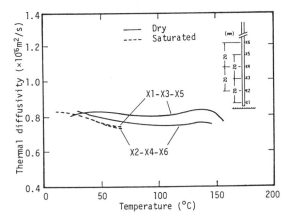

Fig.6 The thermal diffusivity of "Sanjome andesite".

pointed out that the effects of pore water on the thermal conductivity highly related to the pore shape. In other words, even if the porosity of crack-type rock is low below 1%, its conductivity remarkably increases by saturation. However, the conductivity of a spherical-type rock does not increase a lot by saturation in spite of its high porosity. These behaviors can also be stated in terms of the thermal diffusivity. The diffusivity of a crack-type rock increases by saturation, while that of a spherical-type rock slightly decreases due to the increase of heat capacity. Hence, it is found that the measurement results of these three kinds of rocks show reasonable correlation with the results given by Walsh et al.

5. DISCUSSION ON MEASUREMENT ERROR

The measurement errors of thermal diffusivity are pointed out as follows.
1. The error due to using the difference equation method in the theory.
2. The error due to using the finite size heat source instead of the infinite plane heat source.
3. The error due to mislocation of the thermocouples.

According to the numerical analysis, it is realized that the first error is within 1% if Δx and Δt are selected in order to satisfy that the Fourier number is between 0.5 and 3.0. Therefore, the first error of the thermal diffusivities measured at a site and in the laboratory on condition that Fourier number is about 1.0, must be within 1%.

The second error is estimated by using finite difference analysis. For in-situ measurement, the thermal diffusivity of $0.33 \times 10^6 m^2/s$ is assumed in the analysis. As shown in Fig.7, the error increases up to about -10% after 30 hours of heating. Similarly,

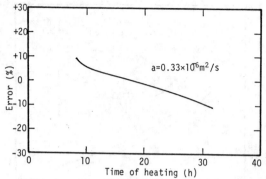

Fig.7 The error due to the finite size heat source for in-situ measurement (a=0.33×10⁶m²/s).

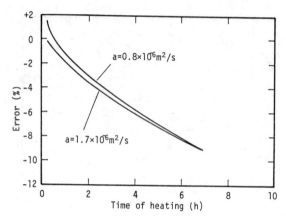

Fig.8 The error due to the finite size heat source for the laboratory measurement (a=0.8 and 1.7×10⁶m²/s).

the error of laboratory measurements is estimated on assumtion that the diffusivities are 0.8×10^6m²/s and 1.7×10^6m²/s. As shown in Fig.8, the errors in cases of 0.8×10^6m²/s and 1.7×10^6m²/s are also -9.2% after 7 hours of heating.

The mislocation of the midpoint where the distance between the points is small, tends to influence the third error to a great extent. The maximum mislocation is ±0.05mm in the laboratory measurement. The results on assumption that the diffusivities are 0.8×10^6m²/s and 1.7×10^6m²/s, are shown in Fig.9. If the midpoint shifts to the heat supply side, the calculated diffusivity becomes larger. On the contrary, shifting far away from the heat supply side, it becomes smaller. The estimated third errors are about ±4.5% for 0.8×10^6m²/s and about ±7.0% for 1.7×10^6m²/s after 7 hours of heating.

According to the measurement results of the three kinds of rocks, this error is observed to some extent in the case of dried "Sanjome andesite". The difference of diffusivity between the combinations of (X1-X3-X5) and (X2-X4-X6) tends to become larger with time because of the mislocation of any of the six points.

Whereas, the third error of in-situ measurement can be disregarded because of a large distance Δx.

6.CONCLUSION

In this syudy, a measurement method was proposed, the thermal diffusivities were measured at a site and in the laboratory, and the measurement errors were discussed. The conclusions are summarized as follows.

1. New measurement equations of thermal diffusivity are obtained by using the difference equation method. Thermal diffusivity can be easily estimated from the temperatures of the three points at three different times.

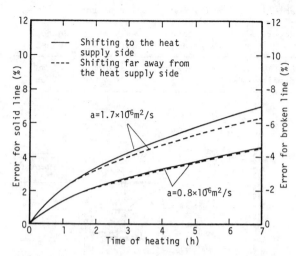

Fig.9 The error due to mislocation of the midpoint thermocouple (mislocation +0.05mm, x=20mm).

2. This method has the characteristic that it is possible to measure the thermal diffusivity at a site as well as in the laboratory even if the initial temperature distribution of rocks is not uniform.

3. In this method, it is also possible to obtain the thermal diffusivity as a function of temperature over a wide range.

4. The main measurement errors are due to using the finite size heat supply plate and the mislocation of thermocouples. These errors become larger with time.

REFERENCES

Toews N.,Larocque G.and Wong A.S. Estimation of in-situ thermal properties assuming a linear, homogeneous and isotropic rock. Mining Research Laboratories Report MRP/MRL 77/8 (TR), Ottawa, Canada, 1977.

Jeffry J.A.,Chan T.,Cook N.G.W.and Witherspoon P.A. Determination of in-situ thermal properties of Stripa from temperature measurements in the full-scale heater experiments, method and preliminary results. LBL-8423, 1979.

Vost K.R. In-situ measurements of thermal diffusivity of rock around underground airways. Trans.I.M.M., A57-a62, 1976.

Lindroth D.P. Thermal diffusivity of six igneous rocks at elevated temperatures and reduced pressures. U.S.Bureau of Mines Report of Investigations, R1-7954, 1974.

Hanley E.J.,DeWitt D.P. and Roy R.F. The thermal diffusivity of eight well characterized rocks for the temperature range 300-1000K. Eng.Geol.12, 31-47, 1978.

Walsh J.B.and Decker.E.R. Effect of pressure and saturating fluid on the thermal conductivity of compact rock. J.Geophys.Res.71, 3053-3061, 1966.

Instability of the interface between gas and liquid in an open fracture model
Instabilité de l'interface entre gaz et liquide dans le modèle d'une fracture ouverte
Die Unstabilität der Zwischenphase zwischen Gas und Flüssigkeit an Hand eines Beweismodells

KUNIO WATANABE, Hydroscience and Geotechnology Laboratory, Saitama University, Urawa, Japan
KEJI ISHIYAMA, Hydroscience and Geotechnology Laboratory, Saitama University, Urawa, Japan
TAKASHI ASAEDA, Department of Civil Engineering, University of Tokyo, Japan

ABSTRACT: Two phase flow in a model of a vertical open fracture developing over an underground cavern is experimentally studied by the use of a Hele-Shaw cell. Finger type infiltration occurs in the large gap width part
of the cell. The influence of fingers, which are formed around the cavern, on the initiation of gas leakage
is mainly investigated. It is found that fingers tend to decrease the critical gas pressure at which gas in the
cavern starts to leak into groundwater.

RESUME: Un écoulement à deux phases dans le modéle d'une fracture ouverte au dessus d'une caverne a été étudié
de maniére experimentale au moyen d'une ellule Hele-Shaw. L'influence des filets d'eau souterraine qui se
forment autour de la caverne lors de fuites de gaz et particuliérement analysée. On constate que les filets
d'eau tendent à diminuer la pression critique, pour laquelle le gaz de la caverne commence à s'infiltrer dans
l'eau souterraine.

ZUSAMMENFASSUNG: Zweierphasen fluß bei einen Modell von einer vertikal geöffneten Spalte die sich über einer
unterirdischen Kammer wird studiert mittels der Hele-Shaw zelle. Die Fingertypinfiltration tritt in der grossen
öffnungsweite als Teil der zelle auf. Der Einfluß der Finger, welche sich um die Höhlen herum bildem und des
Eintreten von der Gasdurchdringunf sollte hauptsächlich überprüft werden. Es wurde entdeckt, dass die Finger
dazu neigen den kritischen Gasdruck herabzusetzen bei dem das Gas in den Kammer beginnt in das Grundwasser
durchzudringen.

1 INTRODUCTION

Two phase flow in an open fracture is closely
related to many problems in the field of rock
hydraulics, e.g., rain infiltration in unsaturated
part of rock mass and gas leakage into groundwater
from an underground cavern which is storing oil and
gas. The motion of the interface between gas and
water in an open fracture should be detaily studied
to make clear the mechanism of the flow. An open
fracture can be simplified as a narrow space between
two parallel plates (Hele-Shaw cell). Therefore,
many previous authors have studied the characteristic nature of this motion on the basis of the
stability analysis of flow in the cell to a small
disturbance given on the interface (Saffman and
Taylor (1958), Phillip (1975), White et al. (1977),
Park and Homsy (1984)). Recently, Tryggvasson and
Aref (1983) suggested a technique to numerically
simulate the deformation of the interface. Åberg
(1977) proposed a practical technique to prevent the
gas leakage from an underground cavern, based upon
the theoretical study of the initiation of the
upward movement of a small gas bubble in an open
fracture. However, many unsolved problems on the
nature of the two phase flow are remained yet. The
present authors tried to basically clarify the
pattern of this flow around an underground cavern by
the use of a new type of the Hele-Shaw cell.

2 FlOW CONDITION CONSIDERED AND EXPERIMENTAL
APPARATUS

A vertical open fracture which connects between the
surface layer overlying a impervious rock mass and
an underground cavern is considered (see Figure
1(a)). This Figure presents the crosssection
cutting at right angle to the longest axis of the
cavern. The surface layer is composed of high
permeable soil and highly weathered part of rock
mass. It is assumed that the strike of this
fracture is identical with the direction of the

longest axis of the cavern and that the groundwater
table exists in the surface layer. The cavern is
filled with air. In usual, the gap width of an open
fracture is not uniformly distributed along the
fracture and differs from place to place. This gap
width in the part near the cavern may be larger than
it in the other part because of the stress release
occuring around the cavern. Now, an open fracture
of which gap width stepwise increases from d_1 to d_2
at a certain depth of L_1 measured downwardly from
the surface of the rock mass is considered (Figure
1(b)). In this Figure, H_S is the height of the
groundwater table measured from the bottom of the
surface layer and H_0 is the sum of L_1 and H_S . The
two phase flow in this type of an open fracture is
fundamentally studied.

The experimental apparatus used in the study is
schematically shown by Figure 2. This apparatus is

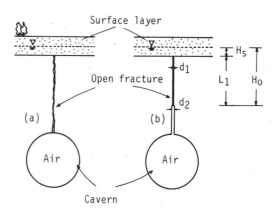

Figure 1. Schematic of the hydraulic condition and
the shape of an open fracture assumed.

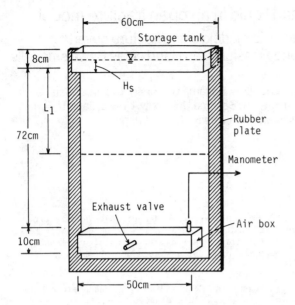

Figure 2. Schematic view of the experimental apparatus

Table 1. Experimental conditions.

Exp. case	L_1 (cm)	H_s (cm)	γ_g*	Exp. case	L_1 (cm)	H_s (cm)	γ_g*
Exp-1.1	16	2.5	1.258	Exp-3.1	56	2.5	1.252
" -1.2	"	4.25	"	" -3.2	"	4.25	"
" -1.3	"	6.0	"	" -3.3	"	6.0	1.250
" -1.4	"	8.0	1.256	" -3.4	"	8.0	1.258
Exp-2.1	40	2.5	1.256	Exp-4.1	72	2.0	1.252
" -2.2	"	4.25	"	" -4.2	"	4.25	"
" -2.3	"	6.0	"	" -4.3	"	8.0	"
" -2.4	"	8.0	1.252				

* Specific gravity of glycerol (gf/cm³)

a vertical Hele-Shaw cell having a storage tank on it and an air box at the lowest part of it. The cell and the air box simulate an open fracture and a cavern, respectively. This apparatus is made of acrylic plate. The gap width of the cell is 2mm. The leakage of air and liquid in the cell through both sides and the bottom of the cell is prevented by rubber plate as shown by this Figure. Two thin aclylic plates of 0.5mm thick, L_1 long and same wide as the cell are pasted on the upper part of inner sides of both walls of the cell to partly decrease the gap width. Consequently, the gap width discretely changes from 1 mm to 2 mm at the depth of L in the cell. High viscous glycerol instead of water is downwardly flowed from the storage tank to the air box. The specific gravity and the viscosity of glycerol is about 1.26 gf/cm³ and 600 cp, respectively. The depth of the glycerol in the storage tank is held constant H_s in the course of each experimental run.

The air box has an exhaust valve. When the valve is opened, the air pressure in this box is equal to the atmospheric pressure. On the other hand, if the valve is closed, the air pressure gradually increases because grlycerole flowing into the box decreases the volume of air in the box. The change of the air pressure is measured by the use of a manometer.

3 EXPERIMENTAL RESULT

3.1 Experimental cases

15 cases of experimental runs were carried out with different L_1 and H_s values. The procedure of each experimental run is devided into two steps. At the first step, the exhaust valve is opened and glycerol in the storage tank is downwardly flowed toward the air box. The transient change and the steady state of the flow under this condition is firstly observed. Then, as the second step in each run, the exhaust valve is closed after when the flow in the first step reaches to its steady state. The air pressure gradually increases under this condition, and when the pressure exceeds a certain value, air starts to break into glycerol in the cell. The critical pressure at which air starts to leak is measured and the transient change of the motion of the leaked air blob in glycerol is also observed.

Experimental conditions fixed in every runs

are summarized in Table 1. All runs are firstly classified into four experimental series and then every series are subdivided into three or four cases, based on L_1 and H_s values, respectively. Three runs were carried out in each experimental case. The gap width of the cell in the experimental series of Exp-4 is constant of 1.0 mm because L_1 in these cases are equal to the height of the cell. The surface tension force acting on the interface between glycerol and air is about 65.8 dyn/cm. The capillarly height of glycerol in the space of 1.0 mm wide is not so large (about 2.0 mm high). Every experimental runs were carried out under the constant temperature condition of 20°C.

3.2 Experimental result

White et al. (1977) already pointed out that the finger type infiltration takes place below the level at which the gap width of the cell stepwise increases if the similar condition to the first step of each experimental run is given. As an example, the comb-like fingers observed in Exp-2.1 is shown in Figure 3. It can be seen that every intervals

Figure 3. Comb-like fingers observed in Exp-2.1.

between two adjacent fingers are almost constant. Hereinafter, the average value D of these intervals observed in the first step is refered as D_{in}. The upper part of the cell in which the gap width is 1.0 mm is completely filled with glycerol.

The average interval D increases after closing the exhaust valve because of the increment of air pressure in both the air box and the cell. As the air pressure reaches to its maximum value just before the start of the leakage of air, so that the D value also takes its maximum value D_{max} at this moment. Figure 4 shows four examples of the difference between the D_{in} and the D_{max} observed in the series of Exp-3. However, all fingers were completely vanished and D_{max} became infinite before the start of the leakage in the series of Exp-1.

Figure 5 shows the transient change of the shape of the air blob leaking through the upper part of the cell which is filled with glycerol. As shown in

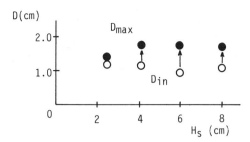

Figure 4. Differences between D_{in} and D_{max} observed in the series of Exp-3.

Figure 5. Leakage of an air blob from an interval between fingers (Exp-3). The shape of the blob is emphasized with black line.

this Figure, the leakage of air breaks out from only one of many intervals between fingers. The air blob deforms its shape and some branches are created from the main body of the blob in the course of its upward movement. This type of the deformation of the interface between air and glycerole is also caused from the instability of flow as Paterson (1981) pointed on the radial infiltration problem.

The relation between the critical air pressure head (H_a) in the box and the hydraulic head (H) of glycerol in the storage tank observed in every experimental runs is shown by Figure 6. H_a and H are defined as follows;

$$H_a = P_a/\gamma_w \qquad (1)$$
$$H = (H_s + L_1)/(\gamma_g/\gamma_w) \qquad (2)$$

where, P_a is the critical pressre at which leakage starts, g is the gravity acceleration and γ_w and γ_g are specific gravities of water and glycerol in gf/cm^3, respectively. It is found that the H_a tends to be smaller than H in the series of Exp-2 and Exp-3. On the other hand, H_a is almost equal to H in Exp-1 and in some cases of Exp-4. When the downward flow of glycerol is stopped by the increase of the air pressure, in the other word, all fingers are vanished, H_a must be larger than the sum of H and the head of the capillary force H_c acting on the interface at the begining of the leakage. Because the capillary head is not so large and all fingers are vanished in the series of Exp-1 as mentioned above, so that H_a takes almost same value as H in each case of this series. On the contrary, under the condition that many fingers are remaining as shown by Figure 5, the pressure head immediately above the interface should be considerably smaller than H. Consequently, the leakage starts with lower

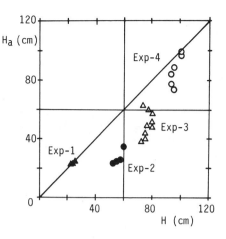

Figure 6. Relation between the critical air pressure (H_a) and the hydraulic head (H) in the storage tank.

pressure head than H as in the series of Exp-2 and Exp-3 because many fingers were remaining in these series. It can be said that fingers considerably influence on the critical air pressure. However, the reason why many fingers were remaining in the series of Exp-2 and Exp-3 is not clarified yet.

4 THEORETICAL CONSIDERATION ON THE INTERVAL D BETWEEN FINGERS

Phillip (1975) and White et al. (1977) theoretically studied the interval D between fingers on the basis of the analysis of the stability of flow in the cell to a small disturbance of the interface between flowing liquid and air. These authors found that the small disturbance of which the wave number is M grows and makes fingers if the value defined by following equations is positive.

$$\sigma = C M (G - M^2 T /\gamma_g) \qquad (3)$$
$$M = 2 \pi / \lambda \qquad (4)$$
$$G = \{-(H_s \gamma_g/\gamma_w - H_c - H_a) + L_1(\kappa-1)\}/L_1 \kappa \qquad (5)$$
$$\kappa = k_2/k_1 \qquad (6)$$

where, C is the constant, T is the surface tension force, γ_g is the specific gravity of liquid (this case; glycerol), λ is the wave length of the disturbance (identical to D), G is the pressure gradient immediately above the interface and k_1 and k_2 are hydraulic conductivities of the small and large gap

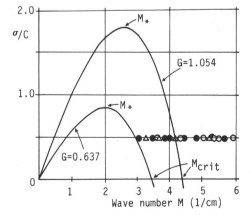

Figure 7. Relation between σ/C and wavenumber M. Solid and open circles are corresponding to D_{in} and D_{max} respectively. Open triangles are results of White et al. (1977).

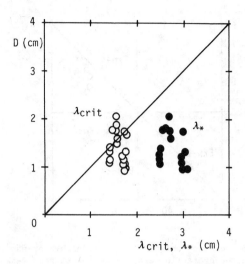

Figure 8. Comparison of observed intervals D with two theoretical wave lengths λ_{crit} and λ_*. Open and solid circles are corresponding to λ_{crit} and λ_*, respectively.

width parts, respectively. The M value which corresponds to $\sigma = 0$ is the critical wave number (M_{crit}) of disturbance which can grow. On the other hand, the M_* which makes the σ value maximum is defined. M_{crit} and M_* can be calculated by the following equations;

$$M_{crit} = (\nu_g \, G/T)^{1/2} \qquad (7)$$
$$M_* = (\nu_g \, G/3T)^{1/2} \qquad (8)$$

Figure 7 shows two curves which represent the theoretical relation between σ/C and M corresponding to the maximum and minimum values of G in every experimental runs. Relations for other runs are situated between these curves. M value can be also calculated by putting the measured D value into equation (4) as λ. M values obtained from measured D_{in} and D_{max} values in each experimental run are also shown on the M axis of $\sigma/C = 0.5$ in this Figure by open and solid circle, respectively. Open triangles are results of White et al. (1977). Although measured M values are scattered, it can be seen that these measured values are plotted in or near the region of M_{crit} value. Wave length λ_{crit} and λ_* corresponding to M_{crit} and M_*, respectively are reversely calculated by the use of equation (4). Figure 8 shows the comparisons among measured D values, λ_{crit} and λ_*. Open circles represent the relation between the measured D values and λ_{crit} . Solid circles are the relations between measured D and λ_*. It can be concluded that λ_{crit} is better approximation of measured value than λ_*.

5 DISCUSSION

The gap width is not constant along an open fracture. In usual, it can be thought that the gap width tends to be enlarged near an underground cavern excavated in rock mass. Consequently, the finger type infiltration takes place and comb-like fingers of groundwater is formed near the cavern in each open fracture. When the gas pressure in the cavern increses, gas starts to upwardly leak from one of intervals between these fingers. The critical pressure of gas at which the leakage begins may be essentially determined by the distribution of water pressure immediately above the interface between gas and water. When all fingers are not vanished before the initiation of gas leakage, this critical pressure is less than the sum of hydrostatic pressure and capillary force. It can be said that fingers play a very important role on the initiation of gas leakage. More detailed study on the flow in fingers should be needed to precisely evaluate the critical pressure.

These types of fingers may also develop in unsaturated zone in rock mass because of the ununiformity of the distribution of gap width in each open fracture.

6 CONCLUDING REMARKS

Two phase flow in an vertical open fracture in which its gap width is stepwise increases from a certain depth is experimentaly studied. Experiments were carried out by the use of a type of the Hele-Shaw cell. At the lower part of the cell, an air box that simulates an underground cavern is prepared. Glycerol is used as a flowing liquid. It is clearly found that comb-like fingers are formed under the level at which the gap width stepwise increases. The intervals between fingers are well evaluated by the theory proposed by White et al. (1977). When the air pressure rises, air leaks from only an interval between fingers. When all fingers are not vanished, the leakage starts at lower level of the air pressure than the sum of the hydrostatic pressure and the capillary force.

6 REFERENCES

Åberg, B. 1977. Prevention of gas leakage from unlined reservation in rocks. Symp. on Storage in Excavated Rock Cavern, Rock Store 77, p.399–414.
Park, C.W. & Homsy, G.M. 1984. Two-phase displacement in Hele-Shaw cell. J. Fluid Mech., 139: p. 291–308.
Paterson, L. 1981. Radial fingering in a Hele-Shaw cell. J. Fluid Mech., 113: p.513–529.
Philip, J.R. 1975. Stability analysis of infiltration. Soil Sci. Soc. Am. Proc., 39: p.1042–1049.
Saffman, P.G. & Taylor, G.I. 1958. The penetration of a fluid into a porous medium or Hele-Shaw cell containing a more viscous liquid. Proc. R. Soc. London, A245: p.312–329.
Tryggvason, G & Aref, H. 1983. Numerical experiments on Hele-Shaw flow with a sharp interface. J. Fluid Mech., 136: p.1–30.
White, I., Colombera, P.M. & Philip, J.R. 1977. Experimental studies of wetting front instability induced by gradual change of pressure gradient and by heterogeneous porous media. Soil Sci. Soc. Am. Proc., 41: p.483–489.

2

Rock foundations and slopes
Fondations et talus rocheux
Felsgründungen und Böschungen

Shear behaviour of natural filled joints

Comportement au cisaillement des joints naturels remplis
Schubverhalten von natürlichen ausgefüllten Trennungsflächen

G.BARLA, Polytechnic of Turin, Italy
F.FORLATI, Polytechnic of Turin, Italy
P.BERTACCHI, ENEL-DSR-CRIS of Milan, Italy
A.ZANINETTI, ENEL-DSR-CRIS of Milan, Italy

ABSTRACT: A research program on the shear behaviour of filled discontinuities was initiated jointly by the Polytechnic of Turin and the ENEL-DSR-CRIS of Milan. The relevant steps of this research program are described in the present paper, with emphasis being placed on: a) the design and construction of a special shear testing equipment, developed so as to simulate the failure behaviour of natural joints; b) the performing of a number of shear tests on both artificial and natural discontinuities. The following influencing factors are considered: different infilling materials of varying thickness; various surface profiles (roughness and asperities); different rock types (Limestone and Sandstone). After a detailed characterization of the unfilled discontinuities and of the infilling materials, the most attention is devoted to the study and the interpretation of the first shear tests performed on filled joints.

RESUME': Un programme de recherche sur le comportement au cisaillement de discontinuités remplies a été commencé par l'ENEL-DSR-CRIS de Milan conjointement avec l'Ecole Polytechnique de Turin. Dans ce mémoire on décrit les phases les plus importantes de ce programme de recherche, en mettant en évidence particulièrement les points suivants:a) l'étude de projet et la construction d'un appareillage spécial d'essai au cisaillement, développé pour simuler le comportement à la rupture de joints naturels; b) l'exécution d'un certain nombre d'essais de cisaillement sur les discontinuités soit naturelles soit artificielles, en prenant en considération les facteurs d'influence suivants: différents matériaux de remplissage d'épaisseur variée; profils de surface différents (rugosités et aspérités); divers types de roche (Calcaire et Grès). Après une caractérisation détaillée des discontinuités vides et des matériaux mêmes de remplissage, la plus grande attention a été consacrée jusqu'à présent à l'étude et à l'interpretation des premiers essais de cisaillement effectués sur des joints remplis.

ZUSAMMENFASSUNG: Das Polytechnikum von Turin und die ENEL-DSR-CRIS von Mailand haben gemeinsam ein Forschungsprogramm über das Scherbedingung von ausgefüllten Trennungsfugen eingeleitet. Die wichtigsten Schritte dieses Forschungsprogramms sind in diesem Artikel beschrieben, wobei die folgenden Punkte hervorgehoben werden: a) Entwurf und Konstruktion einer Spezialausrüstung für den Scherversuchen, die so entwickelt ist, daß sie die Bruchbedingung von natürlichen Fugen nachahmen kann; b) die Ausführung einer Reihe von Scherversuchen sowohl an künstlichen als auch an natürlichen Trennungsfugen bzw. Unstetigkeitsstellen und zwar unter einer gewissen Anzahl von Versuchsbedingungen, wobei die nachstehenden Einflußfaktoren zu berücksichtigen sind: verschiedene Füllmaterialien mit variierender Dicke; unterschiedliche Oberflächenprofile (Rauhigkeit und Unebenheiten); verschiedene Gesteinstypen (Kalstein und Sandstein). Nach einer gründlichen Prüfung der Eigenschaften von nicht gefüllten Trennungsfugen und den Füllmaterialien selbst haben die Studie und die Auswertung der ersten Scherversuchen an ausgefüllten Fugen bis heute die größte Aufmerksamkeit auf sich gelenkt.

1 INTRODUCTION

The presence of discontinuities in a rock mass is known to influence greatly strength and deformability. As a consequence, the characterization of natural discontinuities is a very relevant step in rock engineering design of slopes and foundations.

A significant aspect of characterizing rock joints relates to the infilling material and to the assessment of its influence on shear behaviour. The presence of filled discontinuities is a peculiar characteristic of "Complex Rock Formations" (A.G.I., 1977), which are often met during excavation workings in Italy.

With the main purpose to increase the present knowledge on the subject, the Polytechnic of Turin and the Hydraulic and Structural Research Centre of the Italian Electricity Board of Milan (ENEL-CRIS) are involved in a joint research program on the shear behaviour of filled discontinuities (Barla et al., 1986).

Given the complexity and the extent of this research program, the work has been subdivided in a number of steps as follows:

a) design and construction of the shear equipment (i.e. the shear box used for testing filled discontinuities);

b) choice of typical filled discontinuities;

c) preparation of artificial discontinuities, with various roughness profiles;

d) physical and mechanical characterization of the

rock material, forming the surfaces of discontinuity, of the joints and of the infilling;

e) set up and calibration of the shear equipment; choice of the most appropriate testing procedures to be used;

f) performance of the shear tests on artificially filled joints;

g) performance of the shear tests on naturally filled joints;

h) analysis and interpretation of experimental data.

The shear tests on both artificial and natural discontinuities consider the following influencing factors:

- different infilling materials of varying thickness;
- various surface profiles (roughness and asperities);
- different rock types;

for two typical conditions which are usually met in situ:

- infilling material, in a joint within a porous rock (Sandstone), drained during the consolidation phase, and drained or undrained during the shear displacement;
- infilling material, in a joint within a nearly impervious rock (Micritic Limestone), drained radially during consolidation, and drained or undrained during shearing.

For each condition above, the tests are being carried out on rock specimens taken from the same block, containing either natural or artificial joints.

In the present paper, particular attention is devoted:

- to describe the peculiar characteristics and performances of the new shear device, by showing the experimental procedures being adopted and discussing a few methodological aspects;
- to determine the shear strength characteristics for both the rock and the infilling: the shear behaviour of filled discontinuities is to be defined as intermediate between that of a joint with no infilling and of the infilling itself;
- to describe some typical results with a few preliminary remarks on the shear behaviour of filled joints.

2 FUNDAMENTALS OF THE RESEARCH WORK

The experimental investigation on the shear behaviour of filled rock joints has been initiated on the basis of the following assumptions:

1) the shear strength of filled joints is to be defined as intermediate between that of a joint with no infilling and of the infilling itself;

2) the shear behaviour of thick, non-interfering joints, may be approached in the framework of Soil-Mechanics. Nevertheless, the asperities on the joint surface induce normal "confining" stresses, during the loading phase, which are not negligible with respect to the global shear behaviour;

3) consideration is to be given to the following influencing factors:
- physical and mechanical properties of the

infilling material and of the rock type;
- state of strain and stress history;
- drainage conditions;
- roughness and strength of the discontinuity surfaces;
- aperture of the joint and thickness of the infilling material;
- stress level and rate of shear displacement.

In order to make the results of significance and well accepted, the specimen size is to be such as to represent appropriately the natural joints, mostly with respect to the roughness profiles, and other important factors such as waveness, angle of asperities and infilling structure;

4) to simulate, as closely as possible, in the laboratory, the horizontal-failure of a natural joint, under in situ shear conditions, the following most important aims are to be achieved:
- during the consolidation process, the specimen is to be loaded with the same effective stresses corresponding to the in situ depth of sampling. The assumption for an infinitely wide joint is that the consolidation occurs under conditions of no lateral strains;
- during the shearing phase, the specimen is subjected to gradually increasing stress and strain up to failure, so as to allow that the principal axes are free to rotate without lateral strains: the assumption is that the infilling material is subjected to a strain state of "simple shear".

5) Tests are to be carried out under undrained and drained conditions: drainage in a vertical direction, both during consolidation and shearing, within a porous rock; drainage in a horizontal direction, within a nearly impervious rock;

6) saturated infilling materials and rock joints fitted together are to be used;

7) for the characterization of the shear strength of rock joints, the Barton's relationship (1974) is used; the Joint Roughness Coefficient (JRC) is also determined by means of the correlation suggested by Tse and Cruden (1979).

3 THE SHEAR EQUIPMENT

The shear testing in the laboratory does not allow the in situ behaviour of filled discontinuities to be well represented, due to a number of factors such as specimen disturbance, original state of stress, scale effects, etc. However, it is believed that, by using a well developed experimental equipment, significant steps in the understanding of this same behaviour can be made.

This is possible if the equipment is such as to reproduce in the laboratory the strain conditions which arise in situ during failure. Also, a physical interpretation of the phenomena which occur during testing is to account for different influencing factors.

In recent years, a number of experimental studies have been developed on the shear behaviour of filled joints, by considering direct shear tests carried out on small specimens, often artificial,

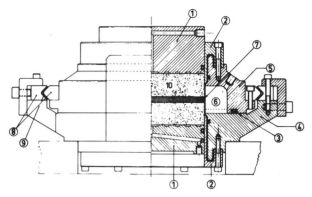

Figure 1. Cross section of the shear box: 1)upper and lower pistons;2)loose flanges; 3) O-ring;4) bottom platen support;5) head protection; 6)lateral chamber;7) membrane;8) V-guides; 9) no-friction roller;10) rock joint specimen.

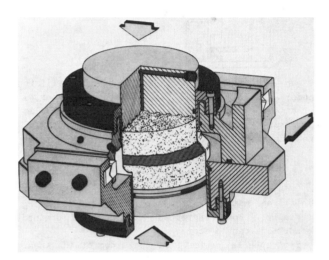

Figure 2. A partly sectioned, isometric view of the shear box.

at low normal stress levels (Kanji, 1974; Ladanyi et al., 1977; Lama, 1978; Kutter and Rautenberg, 1979).

With the main purpose to increase the present knowledge on the subject, a new shear box has been designed and constructed at the ENEL-CRIS of Milan.

The following is achieved:
- the specimen is tested in a strain state of simple shear;
- the horizontal total stress is controlled both during consolidation and shearing;
- the pore pressure is measured during the test;
- both drained and undrained conditions are taken into account;
- the box is to allow for testing cylindrical specimens of rock of 150 mm in diameter, with a height ranging from 65 to 80 mm, according to the infilling thickness to be reproduced (5÷25 mm);
- the dilatancy of the specimen is allowed to take place vertically;
- the upper part of the specimen moves in the horizontal plane, parallel to the shearing direction;
- the maximun normal stress level is in the range 3 to 5 MPa.

Figure 3. A view of the shear box.

A cross section and an isometric view of the shear box are shown in figures 1,2 and 3. The same box is to be used in a 1 MN shear machine, available for direct shear testing purposes (figure 4).

The specimen is prepared by filling the joint with the infilling material, sampled in situ and remoulded with a water content equal to two times the liquid limit (figure 5 and table 1 a, b).

The most attention is to be devoted, during this phase, to obtain desirable testing conditions. The complete absence of air is to be achieved; the rock joint and the infilling material need be saturated before testing under vacuum conditions.

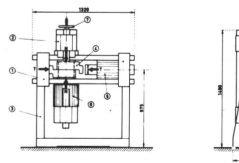

Figure 4. 1 MN shear equipment used for testing: 1) horizontal frame; 2) vertical frame; 3) machine support; 4) shear box; 5) shear ram; 6) normal ram; 7) displacer wheel.

293

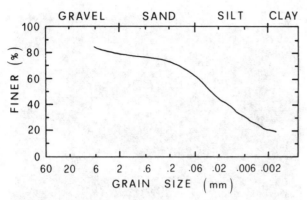

Figure 5. Grain size curve of the infilling material in Micritic Limestone.

Table 1a. Results of the diffrattometric analyses on the infilling material.

Mineral	Percental value
Calcite	77.6
Quartz	9.92 ± 2.07
Illite	12.50 ± 3.37
Chlorite	0.2
Smectite	< 0.2
Kaolinite	< 0.2

Table 1b. Geotechnical parameters obtained for the infilling material.

Principal properties			Mean value
Unit weight of solid particles	γ_s	kN/m^3	26.64
Liquid limit	w_L	%	51
Plastic limit	w_P	%	31
Plasticity index	I_P	%	20
Coefficient of consolidation *	c_v	m^2/s	$3 \times 10^{-8} \div 9 \times 10^{-8}$
Rate of secondary consolidation *	C_α	%	$0.3 \div 0.4$
Coefficient of permeability *	k	m/s	$1 \times 10^{-10} \div 3 \times 10^{-11}$
* Normal stress range	σ_n	MPa	$0.2 \div 3.2$

The "sandwich" specimen, protected by a highly deformable membrane, is enclosed in an anular chamber all around it. This hydraulic chamber, movable during shearing, at constant volume conditions, is able to carry out a shear displacement up to 25÷30 mm.

The chamber is then saturated with deaerated water by means of the special devices shown in figure 6. After assembling and preparing the specimen, the measurement and control of the lateral pressure change during testing are carried out by means of a special transducer, as the

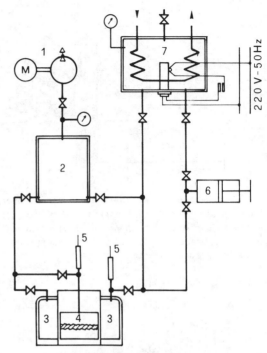

Figure 6. Scheme of the hydraulic circuits: 1) air pump; 2) air tank; 3) confining chamber; 4) specimen; 5) pressure transducers; 6) hydraulic plunger; 7) device for water deaeration by means of heating-cooling process.

infilling material is prevented from spreading radially. Both undrained and vertical drained conditions can be achieved.

Two opposite guides, on the sides of the shear box, make the upper half of the specimen move in the shearing direction, while keeping the lower and upper parts parallel and allowing for dilatancy to take place vertically.

In conjunction with the usual monitoring and control already taking place with the 1 MN shear equipment, additional displacement and pressure transducers are used, to measure continuously all the testing quantities.

According to the main features of the shear box, as described above, this is to operate as follows:
a) during consolidation:
  - as a cell instrumented for the lateral stress application or measurement;
b) during shearing:
  - as a simple shear testing device, where the lateral pressure is being continuously monitored.

The cell allows for pore pressure measurement when changes of it occur during the consolidation and shearing phases. Pressure transducers enclosed in a porous stone are used for this purpose.

Due to the great difficulty in inserting the transducers at the joint-infilling interface, in all the specimens tested, the following procedures were applied:
- "undrained tests" were performed under constant volume conditions, by varying the normal load for any pore pressure change (height of the specimen = constant);

- "drained tests" were performed, under very slow shearing velocity, nearly 0.003 mm/min (based on consolidation time), in order to permit the specimen to drain completely.

It is necessary to note that the consolidation times are generally long (function of the infilling thickness). For this reason, it is preferred to perform the pre-consolidation phase in a simple cylindrical cell, with stiff walls, up to a certain load level.

Loading is to continue, when the specimen is positioned in the shear box, so as to complete consolidation before shearing. In this way, as a specimen is being tested in the shear box, another one is being consolidated.

4 DISCONTINUITIES

The choice of "Natural Discontinuities" was made accounting for the need to use:
- homogeneous rock types, with mating surfaces characterized by well defined and repeatable roughness profiles;
- discontinuities easy to sample and rock easy to machine (with the purpose to obtain the artificial joints);
- homogeneous infilling materials, without swelling minerals and organic elements.

For testing purposes, the following rock formations, differing for permeability and strength properties (table 2), with natural infilled discontinuities, were selected: Micritic Bioclastic Limestone (Low Cretaceous) (figure 7), and Quartzitic Sandstone (Low Cretaceous) (figure 8).

Discontinuities were also obtained by shearing of Sandstone samples in the direct shear equipment (figure 9), under different stress levels

Figure 8. A typical discontinuity in Quartzitic Sandstone.

Figure 9. A view of natural joints specimens of Sandstone.

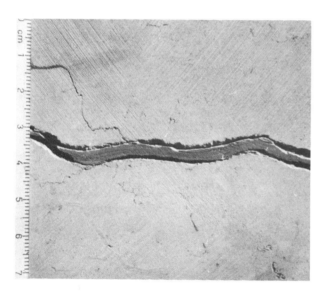

Figure 7. A view of the infilling material in Micritic Limestone.

Table 2. Material properties measured on the intact rock.

Principal properties mean value ± standard deviation (%)			Limestone Dry	Sandstone	
				Dry	Saturated
Unit weight	γ	kN/m^3	27.04	27.08	26.04
Dry unit weight	γ_d	kN/m^3	26.26	25.47	–
Porosity	n	%	2.7	5.8	–
Permeability	k	m/s	6×10^{-13}	5×10^{-10}	–
Tensile strength	σ_t	MPa	10 ± 11	5 ± 10	2 ± 12
Compressive strength	σ_c	MPa	214 ± 17	76 ± 4	44 ± 9
Modulus of elasticity	E_t	MPa	77650 ± 6	13300 ± 8	7170 ± 16
Cohesion	c	MPa	–	4.07	–
Peak friction angle	$\emptyset_p$	(°)	–	67 ÷ 51	–
Ultimate friction angle	$\emptyset_u$	(°)	–	37	–

Figure 10. Artificial discontinuities obtained on Sandstone.

(σn = 0.5 ÷ 10 MPa, on 150 mm diameter specimens), and by splitting natural joints, during drilling of the rock material.

The "Artificial Discontinuities" were obtained from rock blocks collected in the field. In order to gather a number of samples which are sufficiently representative of the rock types, a series of measurements of surface hardness by the Schmidt hammer were carried out in situ, in conjunction with petrographic and structural analyses.

Three different artificial joints are being used, obtained by appropriate preparation procedures with an automated machine, which allows a number of joints to be created with different roughness profiles (i.e. by using various mechanical tools for different values of the applied pressure at the machine),(figure 10).

5 JOINTS AND INFILLING MATERIAL CHARACTERIZATION

5.1 Sandstone joints

The Joint Roughness Coefficient (JRC) was determined on the basis of measurements of the roughness profiles and by using the relationships due to Tse and Cruden (1979) and to Barton (1974).

A typical scheme of joint description and testing results is shown in figure 11.

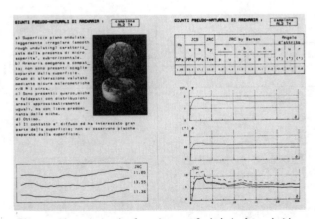

Figure 11. A typical scheme of joint description and testing data collection.

The JRC values of the artificial discontinuities and of the joints, obtained by shearing the intact rock in the shear equipment, were studied under different stress levels (figure 12). The variation in the JRC values of these joints is small (8.5 ≤ JRC ≤ 12.5), for σn ranging from 0.5 to 10 MPa. The JRC values are in the same range (mean value equal to 11.9 ± 1.3) as obtained for joints which were splitted when drilling the rock material.

On the basis of the results obtained when testing the saw cut surfaces and the natural joints, the following is observed:
- the base friction angle $\emptyset$b (figure 13) is shown to increase significantly with σn (expecially for dry specimens); it is smaller for dry specimens than for saturated ones. A possible

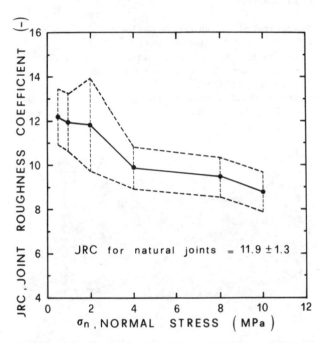

Figure 12. Change in the JRC coefficient, obtained by shearing Sandstone specimens at different stress levels.

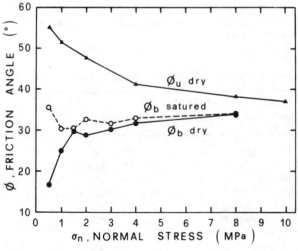

Figure 13. Change in the friction angle during shearing at various normal stress levels (ultimate and base values).

explanation for this is offered by the role of the quartz grains on the saw cut surfaces: they appear to interlock gradually, with higher σ_n stress values being reached.

- The JCS coefficient is shown to decrease during shearing, with σ_n increasing as follows:

$$\sigma_n = 0.278 \, (\Delta JCS/JCS)^{0.479} \, , \, r^2 = 0.86 \quad (1)$$

The decrease in JCS is negligible for lower normal stress levels.
- The minimum error ($\sim$ 5 percent) in determining JRC by the two methods above is obtained when using the JCS values in Barton's equation, as measured at the end of the shearing tests.
* The results obtained up to the present, when testing the sandstone joints, allow us to draw the following conclusions:
- the friction angle determined on a natural joint with zero roughness (JRC = 0), following a large shear displacement, is near to the base friction angle ϕ_b.
- The JCS coefficient, determined after shearing, is a measure of the the wall compressive strength mobilized when peak conditions occur.
- When the friction angle, measured subsequent to peak strength, is found to be smaller than ϕ_b, it is to be referred to residual strength conditions.
- The friction angle, following the attainment of peak strength, may be predicted by using the "mobilized" JRC and JCS coefficients. It is therefore stated that the surface geometry of a joint is changed during shearing. This holding true, the "mobilized" shear strength of a given joint may be obtained from the results of testing another joint (of the same rock type and at the same σ_n stress level), as long as this is characterized by the same JRC and JCS values, even if reached under different horizontal displacements. It is concluded that, given a set of experimental data for a rock joint, the τ vs. σ_n curves can be drawn for each "JRC mobilized" value. Therefore, the shear behaviour of a joint family pertaining to the same rock type can be anticipated on the basis of a limited number of specimens being tested.

The shear test results obtained for sandstone unfilled discontinuities may be summarized by reporting the τ vs. σ_n curves, as shown in figure 14, where three different curves are given:
a) peak curve, obtained by using Barton's shear envelopes for JRC ranging from 2 to 12, JCS decreasing according to equation (1), ϕ_b being kept constant, under saturated conditions;
b) base friction line, for saw cut surfaces under saturated conditions, $\phi_b = 31.95 \pm 1.06$ (°);
c) residual friction line, for rock joints under saturated conditions, $\phi_r = 26.35$ (°).

5.2 Infilling materials

A set of results, as obtained when testing the infilling materials under different laboratory conditions, are reported in figure 15, by giving:
d) the failure envelope obtained by CIU triaxial compression test: $\bar{\phi} = 32.93$ (°), $\bar{c} = 0.0026$ MPa

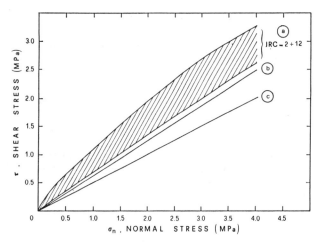

Figure 14. τ vs. σ_n envelopes obtained on rock discontinuities.

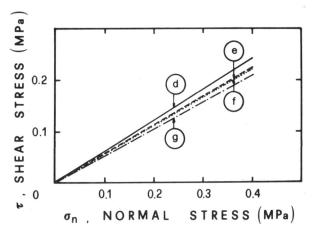

Figure 15. τ vs. σ_n envelopes obtained on infilling material.

at $(\bar{\sigma}_1 - \bar{\sigma}_3)_{max}$ and $\bar{\phi} = 33.16$ (°) at $(\bar{\sigma}_1/\bar{\sigma}_3)_{max}$;
e) the failure envelope obtained by CID triaxial compression test: $\bar{\phi} = 31.34$ (°), $\bar{c} = 0.0012$ MPa at $(\bar{\sigma}_1 - \bar{\sigma}_3)_{max}$ and $\bar{\phi} = 31.41$ (°) at $(\bar{\sigma}_1/\bar{\sigma}_3)_{max}$;
f) direct shear test results under drained conditions: $\bar{\phi} = 28.83$ (°), $\bar{c} = 0.0010$ MPa;
g) direct shear test results under drained conditions at the soil-rock interface: $\bar{\phi} = 28.01$ (°), $\bar{c} = 0.0009$ MPa.

A number of specimens with a height to diameter ratio of 0.16 were also tested as follows (the results are not shown in figure 15):
- CKoU direct-simple shear test (Geonor device); $\bar{\phi} = 28.24 \pm 1.71$(°) at τ_{hmax} (when it is assumed that the horizontal plane is the plane of the maximum shear stress) (Wroth, 1984; Ladd and Edgers, 1972); at τ_h/σ_{vmax} (not always reached during the test) $\bar{\phi} = 34.57 \pm 0.7$(°).
- CKoD direct-simple shear test (Geonor device); $\bar{\phi} = 28.75 \pm 1.25$(°) at τ_h stable for $\gamma \geqslant 35\%$ (the same assumption as above holding true).

297

A number of tests have been carried out by using the newly developed shear box. The following testing conditions were considered:
- drained conditions on the filled, natural and artificial joints: two thicknesses of the infilling material, one normal stress level (0.5 MPa) and, for natural joints, JRC values ranging from 8.19 to 11.96;
- undrained conditions on the infilling material: one single value of the normal stress and of the specimen thickness.

To interpret the results obtained during testing, the following assumptions were made:
- the radial, measured stress σr represents the average lateral stress σx acting on a vertical plane in the infilling material;
- the onset of failure occurs when the ratio $\bar{\sigma}_1/\bar{\sigma}_3$ reaches its maximum value.

✱ The following conclusions are drawn:
- the consolidation parameters measured for the infilling (figure 16) are generally found to be in good agreement with the results obtained in the conventional oedometric cell.
- The average ratio $\bar{\sigma}h/\bar{\sigma}v$ obtained during consolidation is equal to 0.450 ; this value is smaller than the 0.495 value determined for the same infilling material in the oedometric cell, instrumented for lateral pressure measurements.
- A "slippage" is observed to occur at the joint-infilling interface in specimens where JRC = 8.19 ÷ 8.37 and H (infilling thickness) = 3.36 cm. No-slippage is however experienced for specimens characterized by a smaller thickness. The internal friction angle of the infilling material is in the range 27.6 to 34.2 (°).
- The measured shear strength τh is in inverse proportion to JRC (when slippage occurs, the measured shear strength is much smaller than the same strength for no-slippage occurrence). Also, for non-interfering joints, the shear strength increases with the infilling thickness decreasing.
- The angle of inclination of the failure plane, as measured on a filled joint specimen after shearing (figure 17), is well predicted on the

Figure 17. A filled joint after the shear test.

basis of the state of stress at failure and according to Coulomb's shear strength criterion.
- Under drained conditions, the ratio $\bar{\sigma}h/\bar{\sigma}v$ at failure varies with the infilling thickness ($\bar{\sigma}h/\bar{\sigma}v$ = 0.54 ÷ 0.69 for H ranging from 3.36 to 1.97 cm),(figure 18); this same ratio under undrained conditions (in a few preliminary tests) is found to increase significantly ($\bar{\sigma}h/\bar{\sigma}v$ = 0.87 for H = 2.93 cm).

The results obtained may be interpreted by accounting for the following:
- due to the presence of roughness on the joint surface, the infilling will be restrained near its ends, and the stresses in it will be prevented from being uniformly distributed; this stress distribution varies throughout the infilling and it is more influenced as the infilling thickness decreases and/or the joint roughness increases.
- The measured $\bar{\sigma}v$ stress is expected to be greater than the axial stress acting within the infilling; also, the measured $\bar{\sigma}h$ stress represents the average lateral stress acting on the specimen lateral surface; additionaly, these stresses could be further influenced by specimen geometry and testing conditions.

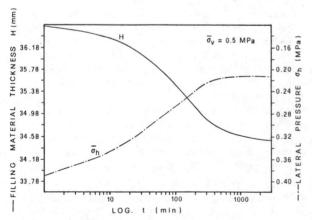

Figure 16. Consolidation phase: filling thickness and $\bar{\sigma}h$ vs. log t diagrams.

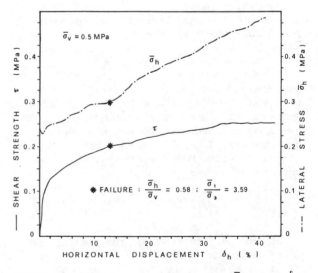

Figure 18. Shearing phase: τh and $\bar{\sigma}h$ vs. δh diagrams.

As a consequence of the above effects, the observed behaviour of the infilling material during consolidation and shearing may be interpreted as follows:

** Consolidation Parameters

- The influence on the $\bar{\sigma}_h / \bar{\sigma}_v$ ratio, obtained with the new shear box, as mentioned above, is dependent on the roughness at the joint-infilling interface.

** Shear Behaviour

- The shear strength of the infilling material is underestimated in the new shear box, as the interpretation is performed on the basis of Coulomb's shear strength criterion and by using the measured value for the $\bar{\sigma}_v$ stress, which is known to be greater than that effectively induced during testing.
- The decrease of the friction angle ϕ with increasing JRC is due to the marked differences between measured and acting stresses within the infilling material, as the roughness at the joint-infilling interface increases. Also, the decrease in shear strength with decreasing infilling thickness, is due to this same effect.

Finally, the occurence of "slippage" at the joint-infilling interface seems to be dependent on the geometry of the surface of asperity. A critical JRC is therefore defined when no-slippage takes place. This critical coefficient is found to be dependent on:

- stress level;
- frictional resistence at the joint-infilling interface;
- thickness and nature of the infilling material.

7 CONCLUDING REMARKS

The most important aspects of a research program devoted to the study of filled joints under shear conditions have been reported. The results obtained up to the present time, regarding the behaviour of "non-interfering" joints, may be summarized as follows:

1) The shear strength of "non-interfering" joints is not dependent on the shearing direction; however, the presence of roughness on the discontinuity surfaces makes the stresses within the infilling be non-uniformly distributed.

2) This roughness effect may vary according to:
   - the thickness of the infilling material;
   - the stress level;
   - the frictional resistance at the joint-infilling interface.

3) Any filled joint (type and thickness of the infilling material, morphological characteristics of the discontinuity surfaces, etc. being defined) is cheracterized by a "critical" value of the Joint Roughness Coefficient (JRC), below which the infilling material is not restrained and "slippage" will occur at or near the joint-infilling interface.

4) For all practical purposes, the shear strength of a filled joint, below the critical JRC, is nearly the same as the frictional strength at the soil-rock contact, for "slippage" occurring along the entire interface. Above the critical JRC, the shear strength is defined by the strength of the infilling material.

The experimental results obtained so far are encouraging; a number of aspects need be investigated, with emphasis being placed on the following:

a) the measurement of the $\bar{\sigma}_v$ stress induced within the infilling material, by using pressure transducers at the joint-infilling interface;

b) the valuation of critical JRC coefficients, depending on the thickness of the infilling material;

c) the study of the shear behaviour of interlocking and interfering joints;

d) the presence of infilling materials of various nature and characteristics.

REFERENCES

A.G.I. 1977. Proc.Int.Congr. on the Geotechnics of Structurally Complex Formations, Capri - Italy.

Barla G., Bertaccchi P.,Forlati F. and Zaninetti A. 1986. Laboratory tests on the shear behaviour of filled discontinuities. Proc.Int.Symp.on Engineering in Complex Rock Formations. Beijing, China.

Barton N. 1974. A review of the shear strength of filled discontinuities in rock. N.G.I., Oslo 105.

Kanji M.A. 1974. Unconventional laboratory tests for the shear strength of soil-rock contacts. Proc. of the 3rd Congress of I.S.R.M. Denver 2, 241-247.

Kutter H.K. and Rautenberg A. 1979. The residual shear strength of filled joints in rock. Proc. of the 4th Congress of I.S.R.M. Montreux 1, 221-227.

Ladanyi B. and Archanbault G. 1977. Shear strength and deformability of filled indented joints. Proc.A.G.I. - The Geotechnics of Structurally Complex Formations. Capri 2, 317-326.

Ladd C.C. and Edgers L. 1972.Consolidated-undrained direct-simple shear tests on saturated clays. Research Report. Soils Publications 284-MIT.

Lama R.D. 1978. Influence of clay filling on shear behaviour of joints. Proc. of the 3rd Congress of Int.Assoc.Eng.Geol. Madrid 2, 27-34.

Tse R. and Cruden D.M. 1979. Estimating Joint Roughness Coefficients. Int.J.Rock Mech. Min.Sci. 16.

Wroth C.P. 1984. The interpretation of in situ soil tests. Geotechnique 34, 4 , 449-489.

A probabilistic approach of slope stability in fractured rock

Approche probabiliste de la stabilité d'une pente dans un rocher fracturé

Wahrscheinlichkeitsrechnung für die Bestimmung der Böschungsstabilität im geklüfteten Gebirge

ALBERT BOLLE, State University of Liège, Belgium
FRANÇOIS BONNECHÈRE, State University of Liège, Belgium
RAYMOND ARNOULD, State University of Liège, Belgium

ABSTRACT : On account of the complexity of the problem, the use of a conventional method presented with nume-rous questions concerning the choice of the geometrical and mechanical characteristics to put into the analysis. The probabilistic approach appeared as the only possible one. It allows to take into account parameters with a considerable variation, as well as random actions like earthquakes. The computing technique uses the Rosen-blueth's multivariate point estimate method. The probability of failure was computed before, during and after the works, as well as the anchor forces required to keep this probability below a reasonable level.

RESUME : En raison de la complexité du problème, l'application d'une méthode traditionnelle posait de nombreuses questions quant au choix des caractéristiques géométriques et mécaniques à introduire dans les calculs. Il est ainsi apparu que la seule approche possible était l'approche probabiliste, qui permet de prendre en compte des paramètres présentant une variation appréciable, ainsi que des sollicitations ayant un caractère aléatoire comme les séismes. La technique utilisée est la méthode d'estimation ponctuelle de Rosenblueth avec variables multi-ples. La probabilité de rupture a été calculée avant, pendant et après les travaux, ainsi que les ancrages né-cessaires pour maintenir cette probabilité à un niveau acceptable.

ZUSAMENFASSUNG : Die Anwendung einer traditionellen Methode musste in Frage gestellt werden, da sich durch die Komplexität des Problems bei der Wahl der geometrischen und mechanischen Eingabedaten Schwierigkeiten ergaben. Eine probabilistische Methode erwies sich als die einzige Lösungsmöglichkeit. Sie erlaubt, sowohl Parameter, die beträchtlichen Schwankungen unterliegen, als auch seltene Beanspruchungen wie Erdbeben, zu berücksichtigen. Zur Anwendung kam die multivariate Punktapproximierung nach Rosenblueth. Die Versagenswahrscheinlichkeit wurde vor, während und nach dem Bau berechnet. Zudem wurden diejenen Ankerkräfte ermittelt, bei denen diese Wahrschein-lichkeit ein vernünftiges Niveau nicht überschreitet.

1 GENERAL DESCRIPTION

The building of the Northern mouth of the motorway tunnels, described by Delapierre et al. in the here-above communication "Méthodologie pour le calcul de la stabilité des tunnels autoroutiers", requires the opening of a large excavation at the toe of a slope. A 50 m wide and 20 m high piles wall will be bored and anchored in the hill.

The weathered rock mass exhibits a rather complex geometry of fractures, faults and bedding planes, with a very large scatter in the mechanical proper-ties. Furthermore, several slides occured in the vicinity during the last years and the actual seems to be rather unstable.

The site is located in an active seismic area and a seismic action cannot be ignored.

2 STABILITY ANALYSIS METHOD

The "block theory" (Goodman and Shi, Hoek and Bray) method was modified to include the effect of the piles wall and its anchors. The shear resistance of the wall has been taken in account, and the geometry of the moving volume was adapted to conform with the geometry of the wall and with the increase of the strength with depth. The moving volume is therefore defined by the wall itself, the natural slope and any valid combination of the planes of discontinuity inside the rock mass.

The acting forces are the weight, the seismic actions and water pressures along discontinuity planes.

The resisting forces include cohesion and friction along the sliding planes, and several external actions (shear strength in the wall, anchors, K_o-pressures).

The soil of the uppermost layer have similar pro-perties as the heavy weathered rock mass and is not considered separately.

3 PROBABILISTIC APPROACH

A deterministic method requires the choice of "pro-bable" or "likely" values for each parameter involved in the computing method, and the result is expressed as a "safety factor" which has to be compared with allowable values. In this particular case, this choice of the parameters, particulary the geometry and the mechanical strength of the discontinuity planes, is almost impossible due to the very large scatter of the observed values.

The probabilistic method allows the use of parame-ters given by their probability distributions, and the results of such a method are also probability dis-tributions. A failure is defined as, for instance, the occurence of negative values for the safety margin (difference between acting and resisting forces), and the "probability of failure" is defined as the proba-bility of such an occurence.

Numerous applications were carried out in the past using the classical methods :
    the Taylor-series expansion,
    the Monte-Carlo simulation.
Both methods have limitations, particularly when the algorithm becomes complex and with correlated random variables.

The point estimate method of Rosenblueth was pro-posed by Harr since 1981 as an efficient tool for the solutions of such problems. It allows the use of several correlated random variables given by their two or three first statistical moments (mean, stan-dard deviation, skewness) with any design method. The results are expressed as the first statistical moments of any desired parameter.

The original method, according to Rosenblueth's first paper, was limited to three correlated variables. Its extension to any number of correlated or indepen-dant random variables has been critized by Rosenblueth himself and by the first author.

The distribution of one ramdom variable x_i is concentrated at two particular points, located at $x_{i+} = m_i + S_i$ and $x_{i-} = m_i - S_i$, with :

m_i = mean value of x_i,

S_i = standard deviation of x_i.

The probabilities associated with these points are P_+ and P_-, functions of the skewness of the distribution.

For N correlated variables, the 2^N points of estimation are located on any combination of the 2^N x_{i+} and x_{i-} values, and the 2^N associated probabilities are functions of the partial correlations of the variables. To now, the use of skewed distributions is not allowed for any value of N.

These 2^N sets of values can be introduced in the design procedure and the 2^N sets of resulting values, weighted by the probabilities associated with each combination of the input values, allows the estimation of the first statistical moments of the resulting parameters.

In this particular case, for each set of the input variables, an iterative procedure was applied to obtain the most critical situation, by varying the dimensions (not the angles, given as input variables) of the sliding volume. The most critical situation was assumed to correspond to the lowest deterministic safety factor.

The main resulting parameter, used for the probabilistic analysis, is the "safety margin", defined as the difference between stabilizing and acting forces along the sliding direction. The complete geometry and all the separate force were also stored for further computation of the additionnal anchors.

4 RANDOM VARIABLES

The geometry of the discontinuity planes was expressed as correlated random variables through the analysis of a great amount of observations in borings and in observation pits and galleries, completed by geological considerations concerning the variations of the discontinuities throughout the rock mass.

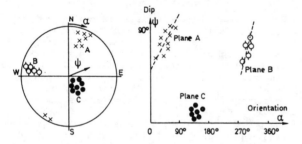

Figure 1. Definition of the geometry of the discontinuity planes.

The strength characteristics of the actual and potential discontinuities were measured by more than hundred laboratory shear tests on large samples. Both peak and residual values of the shear strength were expressed in terms of apparent cohesion and friction angle.

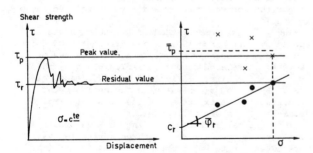

Figure 2. Definition of shear strength characteristics.

Each random variable was assumed to be symmetricaly distributed, and the observed correlations between dip angle and orientation of the discontinuities were tested for their confidence level.

The seismic actions were also considered as random variables, with a Poisson distribution, but only two particular cases were examined.

5 RESULTS

A first analysis was carried out following the described method for the existing slope before any work. It shows a probability of failure PF = 2 to 5 %, depending of the seismic hypothesis.

The 50 m wide excavation, including the resisting forces of the piles wall and its anchors, increases PF up to 10 %. The additionnal anchor forces necessary to reduce PF to 0.5 % (design value) are then 60 to 100 MN (at failure), depending of the orientation and dip of the anchors. Such a quantity of additionnal anchors is almost impossible to install, based on technical and economical considerations.

An excavation procedure by halves was then examinated, taking in account a K_O pressure of the soil remained in place during the first phase, and the resisting action of the reinforced concrete half-mouth during the second phase.

In this case, the maximum value falls below 0.5 %, resulting in a better stability thant for the natural slope, due to the wall anchors. No additionnal anchor is therefore necessary for the design value PF = 0.5 %.

Finally, the stability of the whole mouth structure after releasing of the anchors was examined, using the results of the prior study and assuming the piles wall was replaced by the mouth structure. The soil-structure shear resistance required in accordance to the allowed PF remains below the estimated values and the long-term stability is achieved without permanent anchors.

6 REMARKS AND CONCLUSIONS

Both peak and residual shear strength characteristics were used for the analysis. Although the mean values of the safety margin were found higher with peak characteristics, the estimated PF's remain below the PF value obtained by the residual ones, due to the very large scatter of the peak resistances. As the shear strength is lower-bounded by the residual values, the higher PF's obtained by use of the peak characteristics were discarded and the residual values were used for the whole study. A conservative deterministic approach should use the same values.

The PF of the natural slope (2 to 5 %) is consistent with the occurrence of several recent slidings.

Moderate seismic actions (0.05 g and 0.10 g) or a partial water pressure applied into the sliding planes gives reduced safety factor in a classical approach. In the probabilistic approach, the probability of occurrence of such events is taken in account ant the resulting probability of failure is not increased.

As skewness was observed for the distribution of the safety margin. It could be interpreted as a "functionnal skewness", resulting from the strongly non-linear behaviour of the rigid-plastic (Mohr-Coulomb) model. A gamma distribution was adopted rather than a normal one to describe the distributions of the safety margin and the additionnal anchor forces. The gamma distribution was chosen because it requires only one more parameter. It links continuous by the normal (symmetric) and the exponential (the skewest) distributions.

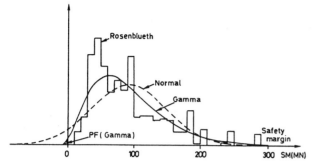

Figure 3. Distribution of the safety margin.

The probabilistic approach of civil engineering problems by the Rosenblueth's point estimate method gives the possibility to analyse any kind of structure including the variability and the correlation of the variables, as well as random sollicitations. It uses classical design methods as tools to obtain a "statistical sample" of any desired result. These results can always be expressed in terms of probability, avoiding the choice of allowable design value of the safety factor, depending of the kind of structure and of the design method itself.

REFERENCES

Rosenblueth, E. 1975. Point estimates for probability moments. U.S.A. Proc. Nat. Acad. Sci. 72.
Harr, M.E. 1977. Mechanics of particulate media - A probabilistic approach. New York : McGraw Hill.
Hoek, E. & Bray, J.W. 1977. Rock slope engineering. London : The Institution of Mining and Metallurgy.
Rosenblueth, E. 1981. Two-point estimates in probabilities. Appl. Math. Modelling. 5.
Grivas, D.A. 1983. Risk analysis in geotechnical engineering. Lausanne : E.P.F.
Goodman, R.E. & Shi, G.-H. 1985. Block theory and its application to rock engineering. Prentice-Hall.
Bolle, A. & Bonnechère, F. 1986. Etude de la stabilité de la tête Guillemins. Procès-verbal 39477. G.C. Liège.
Bolle, A. & Bonnechère, F. 1986. Etude de la stabilité de la tête Guillemins. Procès-verbal 40933. G.C. Liège.
Bolle, A. 1986. Etude probabiliste de la stabilité d'une pente rocheuse. Lausanne : E.P.F.

Un modèle tridimensionnel non linéaire aux éléments finis contribue à évaluer la convenance du site géologiquement difficile de Longtan en Chine au support d'une voûte de 220 mètres de hauteur

The contribution of a non-linear, three dimensional finite element model to the evaluation of the appropriateness of a geologically complex foundation for a 220 m high arch dam in Longtan, China

Der Beitrag eines nichtlinearen FEM-Modells zur Bewertung einer geologisch komplizierten Baustelle für eine 220 m hohe Bogenstaumauer bei Longtan, China

ALAIN CARRERE, Coyne et Bellier, Paris, France
CHRISTIAN NURY, Electricité de France, Chambéry
PHILIPPE POUYET, Coyne et Bellier, Paris, France

ABSTRACT : The site for the Longtan arch dam (220 m high) in Southern China is criss-crossed by a number of faults containing compressible gouge which have a low angle of shearing resistance. The behaviour of the foundation under arch thrust and with uplift in the abutments was evaluated with a large finite-element computer model. The faults were represented with joint elements constituted with the Mohr-Coulomb law of elastoplasticity. The simulation of the full history of foundation loading necessarily used simplifying assumptions with respect to the distribution of stresses, but these were taken on the pessimistic side. The major element of instability proved to be the range of uplift pressures in the upstream part of the banks when the reservoir was full. Displacements and stresses in the dam were seen to be acceptable on condition that certain faults are stiffened near the abutments. A system of mechanical reinforcement composed of concrete-filled galleries was therefore designed to cater for this.

RESUME : Le site du projet de voûte de Longtan (h = 220 m) en Chine du Sud est affecté par plusieurs failles remplies de matériaux compressibles et faiblement résistants au cisaillement. Le comportement de la fondation sous la poussée de la voûte et des sous pressions dans les rives a été évalué grâce à un large modèle mathématique aux éléments finis. Les failles ont été modélisées avec des éléments joints suivant la loi élastoplastique de Mohr-Coulomb. L'historique du chargement de la fondation a été simulé entièrement, avec des hypothèses nécessairement simplificatrices mais pessimistes sur le champ des contraintes naturelles. L'élément destabilisateur majeur s'est avéré être le champ des sous pressions dans l'amont des rives à retenue pleine. Les déplacements et les contraintes ont été reconnues acceptables dans le barrage, à condition de rigidifier certaines failles à proximité des appuis. Un renforcement mécanique, constitué de galeries bétonnées, a été projeté en conséquence.

ZUSAMMENFASSUNG : Die Stelle für das Bogenstaudammprojekt von Longtan (Hohe 220 m) in Südchina weist mehrere Verwerfungen auf, die aus kompressiblem und schwach scherfestem Material bestehen. Das Verhalten des Dammfundaments unter dem Druck der Staumauer und des Sohlwasserdrucks in den Ufern wurde durch ein umfassendes mathematisches Modell mit finiten Elementen berechnet. Die Verwerfungen wurden mit Fugenelementen nach den elastisch-plastischen Kriterien von Mohr-Coulomb dargestellt. Die Belastungsetappen des Fundaments wurden mit gezwungenermassen vereinfachten, aber pessimistischen Hypothesen für das natürliche Spannungsfeld völlig simuliert. Das Auftriebsfeld flussaufwärts in den Ufern bei vollem Stausee erwies sich als der Hauptstörfaktor. Die Verschiebungen und Spannungen im Staudamm wurden als annehmbar angesehen, unter der Bedingung, einige Verwerfungen in der Nähe der Widerlager zu versteifen. Eine mechanische Verstärkung aus Betongalerien wurde folglich geplant.

1 LE CADRE DE L'ETUDE

Le Ministère de l'Eau et de l'Energie Electrique de la République Populaire de Chine a prévu de commencer dans les 5 prochaines années la réalisation d'un gros aménagement hydroélectrique à LONGTAN, sur la Hong-Shui River qui traverse le Sud de la Chine d'Ouest en Est (fig. 1). La puissance installée de 5 000 MW devrait contribuer à satisfaire la demande en forte croissance dans les provinces du Sud et jusqu'à Hong-Kong.

Pour régulariser les 55 milliards de mètres cubes fournis par le bassin versant de 100 000 km², une retenue de 27 milliards de mètres cubes doit être constituée, grâce à un barrage de 220 mètres de hauteur et 700 mètres de long environ au couronnement. Avec des crues décennales de l'ordre de 16 000 m³/s, des crues de projet supérieures à 35 000 m³/s, un fond de vallée relativement étroit (200 mètres), les conditions sont particulièrement défavorables au choix d'un barrage en remblai, dont les exigences en matière de protection contre les crues pendant la construction sont lourdes donc onéreuses.

C'est la raison pour laquelle les Autorités chinoises ont tout d'abord imaginé sur ce site un barrage de type poids en béton. Devant les grandes proportions de l'ouvrage (environ 9 millions de mètres cubes de béton), ils ont demandé aux ingénieurs français, dans le cadre d'une coopération intergouvernementale, d'étudier une variante de type poids-voûte susceptible d'apporter une économie appréciable.

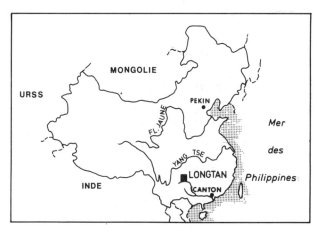

Figure 1. Localisation du projet de Longtan en Chine

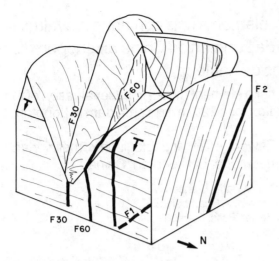

Figure 2. Schéma géologique sommaire du site

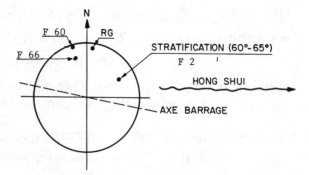

Figure 3. Direction des principales discontinuités

Toutefois, les conditions géologiques du site sont a priori peu favorables, ce qui a conduit à approfondir les études de faisabilité technique jusqu'à un niveau assez peu commun.

2 CONTEXTE GEOLOGIQUE ET GEOTECHNIQUE

Le site de LONGTAN est situé dans une zone montagneuse sur les flancs Sud-Est du plateau de YUNNAN-QUIZHOU. Les affleurements datent du cambrien au trias, avec dominance des grès et argillites du trias. La tectonique régionale est complexe, avec quatre directions principales de failles. La séismicité est modérée, avec des intensités maximum probables ne dépassant pas VI sur l'échelle chinoise soit 7 M.S.K.

Sur le site, la rivière occupe en temps normal une centaine de mètres de large avec une hauteur d'eau de 15 à 20 mètres. Elle coule entre des rives pentées à 35-40° environ, entaillées par de profonds thalwegs (fig. 2). L'altération superficielle est développée jusqu'à une cinquantaine de mètres horizontalement.

La matrice rocheuse est en général très saine, avec en laboratoire des résistances à la compression simple de 150 MPa en moyenne, peu dispersées et peu influencées par la saturation. Les modules élastiques sont aux alentours de 80 GPa. Signalons que les reconnaissances, essentiellement par galeries (près de 5 km) et essais in-situ, sont très denses et de bonne qualité. Elles autorisent une appréciation valable des qualités du site.

Les faiblesses du site tiennent d'une part à la morphologie des rives découpées par les thalwegs (ce qui limite les possibilités d'implantation), et surtout à la présence de plusieurs familles de discontinuités bien développées (fig. 3). La stratification, très régulière, pend à 60-65° vers l'Est c'est-à-dire vers l'aval et légèrement vers la rive gauche. On y trouve un grand nombre de joints, fermés, minces mais très plans, à une distance plurimétrique donc faible à l'échelle du barrage. Peu ou pas compressibles, ces joints sont susceptibles d'entraîner une anisotropie du massif à grande échelle, du moins sous fortes contraintes, du fait de leur résistance au cisaillement modérée (évaluée entre 30 et 38°). Toutefois, l'anisotropie du module de déformation mesuré du rocher est faible.

Il existe d'autre part un certain nombre de failles qui intéressent directement la fondation du barrage. Elles sont classées en 4 familles, dont l'une correspond à la stratification.

Au-delà de la couche d'altération superficielle, ces failles présentent des faciès d'altération variés, allant de la roche fracturée à la mylonite, rarement pure il est vrai. Leur compressibilité est forte et leur résistance au cisaillement a été estimée, en grand, à 30° au moins.

3 FORMULATION DES PROBLEMES DE STRUCTURE

Les questions liées à la faisabilité technique d'une grande voûte sur ce site étaient les suivantes :

a) Sous les actions conjuguées de la poussée du barrage et des sous pressions induites par la retenue, la stabilité des rives, c'est-à-dire des divers blocs découpés par les failles, est-elle assurée ?

b) Si oui, les déplacements des divers éléments de la fondation sous les mêmes poussées sont-ils compatibles ou non avec la structure de la voûte, et avec l'intégrité de ses dispositifs annexes, notamment du voile d'injection ?

c) Si cette dernière condition n'est pas satisfaite naturellement, quels travaux faut-il faire pour qu'elle le devienne et à quel prix ?

Les premières étapes de cette démarche, courantes de nos jours et nécessaires, ne sont rappelées ici que pour mémoire :

. choix de l'implantation et de la forme de la voûte en tenant compte de la topographie du rocher (abstraction faite de la couche d'altération), de la position des failles les plus importantes, et de la présence d'une importante zone fauchée en amont rive gauche ;

. calculs simples de vérification de la forme, avec un programme d'éléments finis en coque épaisse, dans lequel la fondation est simplement représentée par des modules globaux qui tiennent compte grossièrement des discontinuités à toutes les échelles ;

. analyse de la stabilité des blocs de la fondation isolés par les failles, au moyen d'un programme de calcul utilisant la méthode dite "des coins de LONDE".

On a engagé ce processus à plusieurs reprises successives en adaptant la forme de la voûte à chaque fois, pour aboutir ainsi à une voûte de 5 000 000 m³, soit 58 % du volume du barrage-poids. Par rapport à la géométrie de la vallée, et entre autres sa largeur relative (L/H = 3.9), ce volume n'est pas particulièrement bas. Si l'on considère la forme plus en détail, il apparaît que la voûte est plutôt mince en son centre, et assez abondamment épanouie le long de toute la surface de contact avec la fondation.

Pour franchir l'étape suivante, qui consistait à évaluer les déplacements sous les effets hydrostatiques, leur répartition dans l'espace, et les contraintes dans la voûte, les moyens de calculs courants étaient trop éloignés de la réalité ; il fallait en effet modéliser la voûte et la masse de la fondation, bien sûr, mais aussi les principales failles avec leurs propres caractéristiques mécaniques, au moins élastiques. En fait, une évolution récente des programmes de calcul aux éléments finis

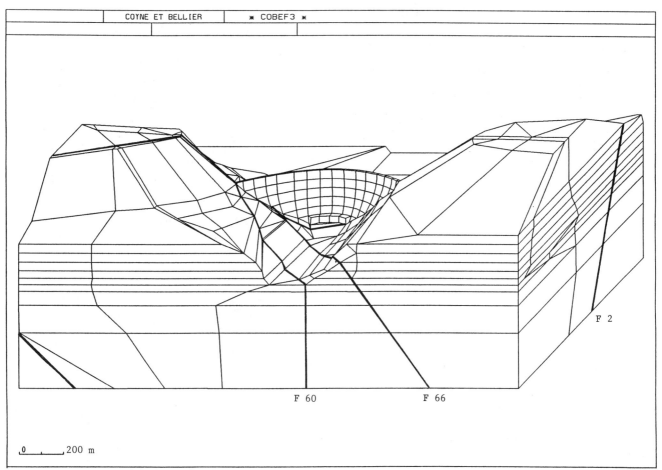

0 ⌐___⌐ 200 m

Figure 4. Maillage général vu d'aval

permettait d'introduire, notamment dans les éléments
joints, des lois de comportement non linéaires
élasto-plastiques de type MOHR-COULOMB. Le modèle
convenait parfaitement pour les failles du site de
LONGTAN et il fut décidé de l'utiliser.

4 REALISATION DU MODELE

Le maillage (fig. 4 et 5) représente le barrage et
la masse de la fondation au moyen d'éléments de
volume isoparamétriques quadratiques : briques à
20 noeuds, prismes à 15 noeuds et tétraèdres à
10 noeuds. La topographie des rives a été respectée,
abstraction faite du rocher décomprimé de surface
dont la rigidité est négligée.

Les trois principales failles du site sont représentées par trois plans d'éléments joints (fig. 6)
dont l'épaisseur est de 50 cm ; ce sont des quadri-
latères à 16 noeuds ou des triangles à 12 noeuds.
L'incorporation des failles dans le maillage, et
surtout leurs intersections avec la surface d'appui
du barrage, ont considérablement compliqué le modèle
et ont parfois conduit à des formes d'éléments
inhabituelles. Il n'y a toutefois pas moins de
4 000 noeuds et 929 éléments, dont 639 pour la masse
de la fondation, 218 pour les failles et seulement
72 pour le barrage. Le tout s'inscrit dans un
parallélépipède de 860 m de haut, 2 300 m de large
et 1 600 m de long. Ces dimensions peuvent sembler
importantes au regard de celles du barrage (220 m de
haut et 860 m de développement) ; on a voulu ainsi
être sûr que les frontières n'influenceraient pas
les déplacements, surtout les déplacements
plastiques (fig. 7 et 8).

Les chiffres ci-dessus montrent clairement
l'importance exceptionnelle du modèle par rapport à
ceux qu'on utilise pour les études courantes de

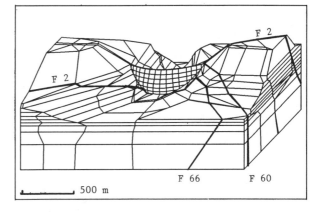

Figure 5. Maillage général vu d'amont

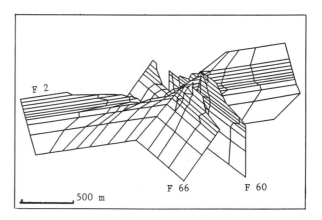

Figure 6. Maillage des failles vu d'amont

307

voûtes, qui dépassent rarement 300 éléments.

5 LES LOIS DE COMPORTEMENT DES MATERIAUX (fig. 9)

La base des lois de comportement dans les éléments
de volume est l'élasticité linéaire isotrope caractérisée par un module d'Young et un coefficient de
Poisson. Dans les joints, seules les déformations
normales εn et tangentielle εt sont considérées ; le
matériau des joints se caractérise, en comportement
élastique, par les raideurs normale et tangentielle
KN et KT :

$$KN = \frac{\Delta\sigma n}{e.\Delta\varepsilon n} \quad \text{et} \quad KT = \frac{\Delta\sigma t}{e.\Delta\varepsilon t}$$

soit KN = E'/e et KT = G/e, avec :

$$E' = \frac{E.(1-v)}{(1+v).(1-2v)} \quad \text{(module oedométrique)}$$

$$G = \frac{E}{2.(1+2v)} \quad \text{(module de cisaillement)}$$

Dans certains éléments de volume, on a utilisé la
loi élastique plastique de "non traction" selon
laquelle le matériau ne présente aucune résistance
dans les directions des contraintes principales
lorsque celles-ci sont des tractions. La méthode
numérique employée est exposée plus loin. Elle a été
mise au point dès 1980 pour représenter le comportement de la partie de fondation mise en extension
lors du remplissage sous le pied amont au centre
d'une voûte (réf. 1, 2). Signalons que dans ce
modèle, il n'y a pas mémorisation des déformations
irréversibles ; on représente ainsi des fissures
diffuses qui s'ouvrent dans un massif tendu mais
sont fictivement injectées au fur et à mesure, ne
pouvant ainsi se refermer.

La loi élastoplastique utilisée dans les joints
est beaucoup plus riche. Mise au point à l'origine
pour permettre d'analyser correctement le
comportement d'ouvrages anciens en béton divisés par
des fissures, en nombre limité mais de grande
extension, elle s'applique évidemment aussi bien aux
failles d'un massif rocheux. La seule condition est
que l'on puisse considérer que l'ensemble est
influencé par des discontinuités planes en nombre
limité. C'est souvent le cas des rives rocheuses
servant d'appuis aux barrages.

Le modèle de joints obéit au critère de MOHR-
COULOMB : si σn et t sont les contraintes normale et
tangentielle en un point du joint, le comportement
est élastique à condition que :
a) σn soit une compression ;
b) t/σn soit inférieur à une limite fixée tan Ø ;
c) le joint ne soit pas déjà ouvert.

Si la condition a) ou la condition c) n'est pas
satisfaite, on simule une résistance du joint
(normale et tangentielle) nulle.

Si la condition b) n'est pas satisfaite alors que
les autres le sont, le joint subit une déformation
de cisaillement parallèlement à l'effort tangentiel.

Les déplacements relatifs normaux non linéaires
sont mémorisés, ce qui permet dans des chargements
non monotones de voir les joints s'ouvrir, puis se
refermer.

Dans l'application au problème de LONGTAN, les
valeurs numériques suivantes ont été adoptées :
- E = 13 000 à 8 000 MPa, v = 0,27 dans le rocher de
la fondation (zonage des modules d'une rive à
l'autre) ;
- E = 20 000 MPa et v = 0,20 dans le béton
(sollicitation lente) ;

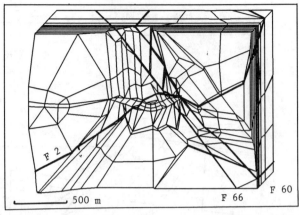

Figure 7. Maillage vu de dessus, sans barrage

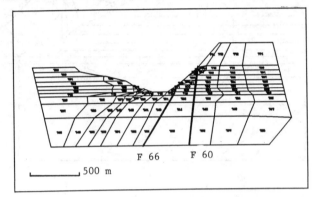

Figure 8. Maillage de la faille F2

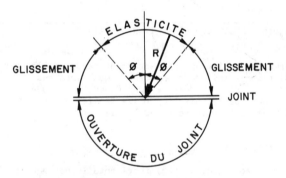

Figure 9. Loi élastoplastique dans les joints

- KN = 1 000 MN/m³ et KT = 400 MN/m³ dans les
failles, ce qui correspond à un module
oedométrique de 500 MPa sur 50 cm d'épaisseur ;
- dans les parties des failles proches du barrage et
supposées traitées, ces valeurs ont été localement
multipliées par 6.
- tan Ø = 30°, limite retenue dans les failles.

6 METHODE NUMERIQUE DE SIMULATION DES NON LINEARITES

Le code support des calculs est COBEF3 (réf. 3), un
programme d'éléments finis tridimensionnels d'usage
général mis au point en 1971 et régulièrement
utilisé depuis pour toutes sortes de calculs de
structure. Sa bibliothèque d'éléments est très
complète et permet d'aborder toutes sortes de
représentations particulières. Pour les études de
barrages et de fondation, les éléments les plus
courants sont :
- les éléments de volume, isoparamétriques
quadratiques à 20, 15 ou 10 noeuds ;
- les éléments joints 16 ou 12 noeuds.

308

Les non linéarités sont introduites sous forme d'un programme complémentaire NOTEN3 (réf. 1, 2), à l'intérieur duquel existent plusieurs sous-programmes relatifs à chaque non linéarité :

NOTEN : non résistance à la traction dans un élément de volume ;

NOCONT : non résistance à une contrainte, quel que soit son signe, dans un élément de volume ;

ORTH : relaxation des contraintes normale et tangentielle selon une direction donnée, dans un élément de volume.

JNOT : loi unilatérale de MOHR-COULOMB dans un élément joint, avec T/N limite de tan Ø fixé.

Dans tous les cas, le principe numérique est le même (réf. 4, 5). Le calcul procède par itérations correctives après un premier calcul élastique. A chaque itération, les déséquilibres constatés entre les contraintes obtenues à l'étape précédente d'une part, et les critères limites fixés (non-traction, MOHR-COULOMB, ...) d'autre part, sont intégrés dans les éléments sous forme de forces nodales. L'ensemble de ces forces constitue un cas de charge additionnel, dont les résultats (déplacements et contraintes) fournissent après résolution l'incrément non-linéaire. Les incréments de déplacements et de contraintes sont ajoutés aux déplacements et contraintes de l'étape précédente ; on obtient ainsi à la fin de l'itération un nouvel état en général plus proche du critère fixé. Il suffit alors de répéter l'opération suffisamment pour atteindre la convergence c'est-à-dire une satisfaction de tous les critères élasto-plastiques avec des déplacements finis. Il peut ne pas y avoir convergence, ce qui traduit alors l'instabilité du modèle donc la rupture en grand.

7 APPLICATION AU SITE DE LONGTAN

Trois étapes successives ont été étudiées :

- l'état initial du site, c'est-à-dire le champ des contraintes naturelles ;
- l'état de contraintes quand le barrage est achevé ; effet du poids propre ;
- l'état de contraintes avec retenue pleine, et les déplacements engendrés par le remplissage.

Du fait que l'attention des projeteurs devait se porter essentiellement sur le comportement des matériaux remplissant les failles, caractérisé par des contraintes effectives (contraintes totales moins pression interstitielle), l'ensemble du calcul a été conduit en contraintes effectives pour tout ce qui concerne la fondation.

7.1 Constitution de l'état de contraintes naturelles initiales

En l'absence de données précises sur l'état réel des contraintes dans la fondation, il a fallu ne considérer comme éléments générateurs de contraintes que le poids du rocher de la fondation. C'est une hypothèse, ni plus ni moins arbitraire qu'une autre, mais c'est la plus simple ; ajoutons que la géométrie très découpée des rives rend peu vraisemblable l'existence de contraintes tectoniques importantes, du moins au-dessus du lit de la rivière.

Le chargement correspondant a été réalisé dans les conditions suivantes :
- conditions aux limites : base du modèle fixe ; limites latérales libres seulement selon la verticale ;
- application d'une densité 2.7 sur le rocher au-dessus de la nappe actuelle, supposée horizontale à la cote 218 ; application d'une densité 1.7 en dessous ;

- application sur la surface du modèle d'une surcharge représentant le poids du mort-terrain non modélisé par des éléments.

Dans la première partie de ce calcul, avec un comportement purement élastique, il est apparu sur la faille parallèle à la stratification des résultantes inclinées à près de 30°, avec quelques dépassements locaux. Quelques itérations de non-linéarité avec limitation du frottement à 30° sur les failles ont fait disparaître ces excès, sans déplacements importants (fig. 10). On a donc constitué ainsi un champ initial de contraintes assez défavorables puisque la réserve de résistance au cisaillement mobilisable pour les charges ultérieures était minimisée.

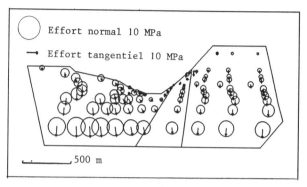

Figure 10. Contraintes sur F2 - Etat initial

7.2 Simulation de la construction du barrage

Le poids du barrage a été appliqué d'une manière courante pour les voûtes qui consiste à considérer que les différents plots qui constituent le barrage sont indépendants, le monolithisme de la voûte n'étant réalisé qu'à la fin de la construction, au moment des injections de clavage.

Numériquement, la construction par plots a été simulée par une loi non-linéaire qui annule les contraintes sur les joints du barrage. Peu de remarques sur l'évolution des contraintes dans la fondation pendant cette étape, si ce n'est l'éventualité de glissements plastiques millimétriques sur les failles si celles-ci ne sont pas renforcées à proximité des appuis.

7.3 Remplissage de la retenue

Les effets hydrostatiques ont été introduits par :
- application de la pression d'eau sur le parement amont du barrage ;
- application de la pression d'eau amont sur une surface continue transversale censée représenter le couple des voiles d'injection et de drainage. Au-dessous de la cote 218 (niveau initial de la nappe), la pression appliquée est constante ; ainsi le gradient hydraulique dans la fondation a-t-il supposé concentré sur le voile ce qui est sévère ;
- application d'une densité - 1 sur le rocher situé dans la zone de marnage de la retenue (218-405) à l'amont du voile, pour simuler le déjaugeage.

La résolution de ce cas de charge a montré des contraintes additionnelles (fig. 11) de faible amplitude par rapport aux contraintes initiales, sauf à proximité immédiate des appuis. Les contraintes résultantes (somme des cas de charge) ont été également calculées (fig. 12).

Les déplacements pendant le remplissage de la retenue (fig. 13, 14) montrent quelques phénomènes locaux, en particulier à la traversée de la surface d'appui par la faille F 66, a mi-hauteur en rive droite : un mouvement de compression et de cisaillement de la faille y est visible (fig. 14) ; cela

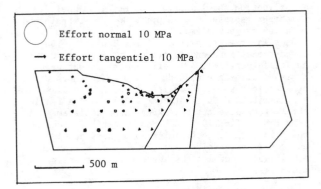

Figure 11. Contraintes sur F2 - Effet de la retenue

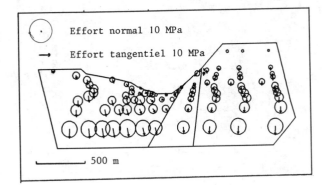

Figure 12. Contraintes sur F2 - Etat final

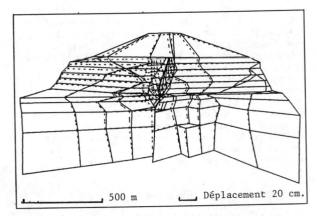

Figure 13. Déplacement élastique sous l'effet hydrostatique - Vue demi-coupe

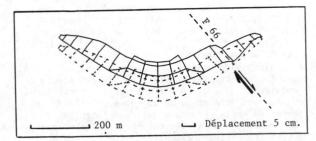

Figure 14. Déplacement élastique sous l'effet hydrostatique - Appui du barrage

se confirme par une concentration de contraintes dans le béton de la voûte à cet endroit (fig. 15), pratiquement doublées localement jusqu'à 13.5 MPa. Cette situation, susceptible d'engendrer des désordres locaux dans la voûte et de désorganiser les gradients hydrauliques sous le barrage, a été

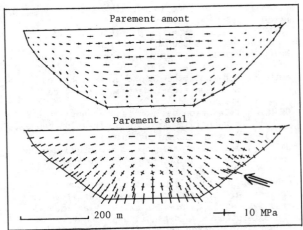

Figure 15. Contraintes sur les parements du barrage (fondation non renforcée)

considérée inacceptable ; elle a conduit à projeter (fig. 16 et 17) un traitement mécanique constitué de puits et de galeries creusés à cheval sur la faille, bétonnés et collés au terrain par injections, dans le but de rigidifier la faille. Avec un entraxe entre galeries de 25 à 30 m, la compressibilité de la faille se trouve annulée ; accessoirement sa résistance au cisaillement se trouve accrue.

Un phénomène particulier est apparu sur les surfaces des failles situées à l'amont du voile d'injection : à cause de la distribution des gradients hydrauliques, les contraintes normales effectives sont diminuées et l'inclinaison des résultantes augmente. Après avoir limité le frottement à 30° au cours d'une série d'itération non linéaires, on a constaté (fig. 18) un glissement de plusieurs centimètres entre les faces de la faille. Toutefois, les mouvements ne dépassent pas 2 mm au niveau de l'appui du barrage.

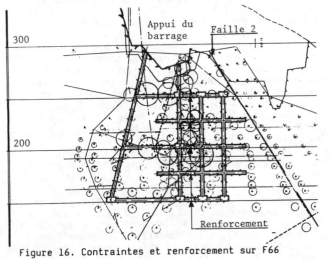

Figure 16. Contraintes et renforcement sur F66

8 CONSEQUENCES POUR LE PROJET

Tout d'abord, l'examen de coupes diverses dans le modèle (fig. 19 à titre d'exemple) a confirmé les études préalables de stabilité des rives en montrant qu'on ne pouvait pas trouver de blocs rocheux en limite de stabilité sur toutes leurs faces au contact avec le reste de la fondation.

Il a été démontré que les failles situées en rive droite sous l'appui du barrage devaient être traitées, d'un point de vue mécanique, pour limiter leur jeu près du barrage et éviter ainsi les

310

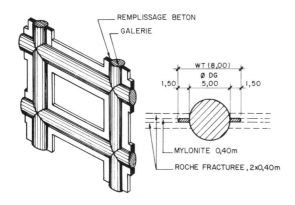

Figure 17. Principes de renforcement des failles

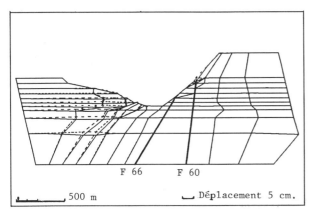

Figure 18. Déplacements plastiques relatifs des faces de la faille F2

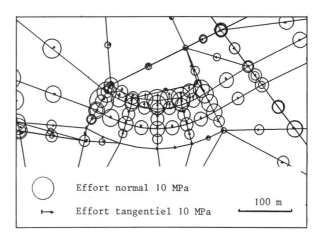

Figure 19. Contraintes sur une coupe horizontale

concentrations excessives de contraintes dans l'ouvrage.

La mise en évidence de petits mouvements possibles sous l'effet des gradients hydrauliques a conduit à renforcer le voile d'injection et le voile de drainage sur les failles, grâce à des puits bétonnés et visitables creusés à cheval sur celles-ci.

Le montant des sommes totales à investir dans la fondation est évalué à 15 % environ du prix du barrage : ces 15 % se répartissent en trois groupes de travaux sensiblement égaux en coût :
- réalisation des voiles d'injection et de drainage courants ;
- renforcement de ces voiles à la traversée des failles ;
- renforcement mécanique des failles sous les appuis.

Dans ces conditions, il a été possible de répondre positivement à la question de la faisabilité technique d'un barrage de type voûte sur le site étudié, et d'avancer une estimation des surcoûts nécessaires pour pallier aux problèmes particuliers de la fondation. D'un point de vue économique, la solution voûte apparaît alors attrayante, d'autant plus qu'une bonne partie des traitements de fondation projetés pour la voûte, a) et b) notamment, serait aussi nécessaire pour les autres types de barrage.

9 CONCLUSIONS

Le modèle utilisé pour représenter finement la voûte et la fondation de LONGTAN est un outil de calcul de type vérificatif et non pas de dimensionnement. C'est pourquoi il est prévu, au fur et à mesure de l'étude de détail du projet et tandis que s'accumuleront les résultats des reconnaissances (mesures

in-situ des contraintes naturelles ; mesures in-situ des résistances au cisaillement), de réutiliser le même modèle avec des hypothèses de traitement affinées chaque fois, jusqu'à obtenir un dimensionnement optimal des renforcements mécaniques des failles.

Il faut avoir conscience des limites des calculs même très sophistiqués, comme l'a justement fait remarquer PECK (réf. 6) : "I even rely on judgment to tell me whether I should believe the results of a finite element study or a computer calculation". Le choix des paramètres et hypothèses apparaît de plus en plus le point crucial, au fur et à mesure que les modèles se perfectionnent. C'est justement un point positif de tels calculs que de mettre l'accent sur les paramètres les plus importants et d'optimiser ainsi l'usage des reconnaissances.

On vient d'entrevoir la possibilité nouvelle de calculer les appuis d'un barrage au même titre qu'une structure faite de main d'homme. La prise en compte de la topographie réelle des rives, puis propriétés de la roche, et enfin l'introduction des discontinuités frottantes, constituent certainement un progrès sensible dans la représentation et dans la compréhension de la physique réelle de nos barrages.

REFERENCES

1. COYNE ET BELLIER, 1981. Barrage de Laparan - Dossier Définitif au Comité Technique Permanent des Barrages - Paris.

2. HAMON, M., POUYET, P., CARRERE, A., 1983. Three-dimensional finite element analysis of the Laparan arch dam by various methods. Water Power and Dam Construction, Août 1983. Londres.

3. TARDIEU, B., GUELLEC, H., 1971. Les éléments finis tridimensionnels. Congrès ITBTP. Paris

4. ZIENKIEWICZ, O.C., 1977. The finite element method, 3rd edition. Londres.

5. ZIENKIEWICZ, O.C., VALLIAPAN, I.P., 1968. Stress analysis of rock as a "no-tension" material. Geotechnique, 18, 56-66.

6. PECK, R.B., 1980. Where has all the jugment gone? The 5th L. Bjerrum memorial lecture. Canadian Geotechnical Journal, Vol. 17 N° 4, Nov. 1980.

REMERCIEMENTS

Les Auteurs tiennent à exprimer leurs remerciements pour leur coopération aussi amicale que compétente aux Responsables, Ingénieurs et Projeteurs chinois du M.W.R.E.P., C.W.H.D.C., et M.S.D.I.

311

Shaft friction of drilled piers in weathered rocks
Friction le long de la paroi de pieux forés dans des roches altérées
Mantelreibung an Bohrpfählen im verwitterten Fels

M.F.CHANG, Nanyang Technological Institute, Singapore
I.H.WONG, Ebasco Services Inc., N.Y., USA

ABSTRACT: This paper describes a number of load tests on drilled piers installed in weathered rocks and residual soils derived from the Bukit Timah granite and the sedimentary Jurong Formation in Singapore. Both full-scale drilled piers and miniature bored piles were tested. The full-scale piles were instrumented with both strain gauges and tell-tales. The full-scale tests indicate that, under normal working loads, drilled piers installed in the weathered rocks of Singapore behave predominately as friction piles and that the end bearing resistance is not developed to any significant extent. The current practice among many engineers of designing drilled piers in weathered rocks as end bearing piles is very conservative. The shaft friction factor α was found to have values of around 0.20 to 0.25 for the weathered rocks of the Jurong Formation. Pull-out tests on miniature bored piles indicated that the average α - values were around 0.78 for the completely decomposed Bukit Timah granite and 0.46 for the completely weathered Jurong Formation. The average shaft displacement required to fully mobilize the shaft friction for the miniature bored piles was around 6 to 10% of the pile diameter.

RESUME: Cet article présente les resultats des charges d'essai sur des pieux forés et coules en place installés dans les roches altérées et les sols résiduels, des derivées de le Granit de Bukit Timah et la formation sédimentaire de Jurong dans Singapour. Les essais sont effectués sur les prototypes et modèles à échelle des pieux forés. Les prototypes sont armés des jauges contraintes et des extensomètres. Les essais de prototype montrent que, sur la charge de fonctionnement, les pieux forés installés dans les roches altérées de Singapour se condirent comme des pieux a frottement et que le pointe portante est négligeable. L'hypothèse que les pieux forés dans les roches altérées se conduirent comme des pieux à pointe portante et qui est adoptée couramment par les ingenieurs dans la conception des pieux forés est trop conservative. Le facteur de frottement de l'arbre α se trouvent entre 0,20 et 0,25 pour les roches altérées de Jurong. Les essais d'arrachement axiaux sur les modèles indiquent que le α moyen est environ 0,78 pour le Granit de Bukit Timah et est, environ 0,46 pur les roches altérées de Jurong. Le déplacement moyen de l'arbre nécessaire à la mobilization complète de frottement de l'arbre pour les modèles est entre 6 à 10 pouraut du diamètre de pieux.

ZUSAMMENFASSUNG: Die Ergebnisse von Grossbohrpfäle in verwitterte Fels und Residuelboden sind in dieser Veroffentlichung erleutert. Diese Bodenarten send Abkömmling von Zwei haupt Felsarten in Singapore, nämlich der Bukit Timah Granit und die Jurong Sedimentärformation. Die Versuchsbelastungen waren an Grossbohrpfähle und auch an Modelpfähle durchgeführt. Electrische Dehnungsmessern und selbstätige Anzeigevorrichtungen waren in Grossbohrpfähle eingebaut. Durch die Belastungen, es werde festgestellt, dass unter Gebrauchslast, fur die Grossbohrpfähle in verwitterte Fels in Singapore, die Traganteile der Mantelreibung eine entscheidente Bedeutung sind. In diesem Fall, das Spitzendruck sehr nachlassigen Wert erreichte. Der Entwurf der Grossbohrpfähle bei. Vielen Ingeniuere in Singapore als Endtragend, ist übekonservativ. Bei der Beurteilung der Ergebnisse von Versuchsbelastungen an Grossbohrpfähle in verwitterte Jurong Formation, es war die Werte von Mantelreibungsfactor α zwischen 0.20 bis 0.25 gefunden. Von den Zugversuchen an Modelpfähle in völlig verwitterte Bukit Timah Granit und auch in völlig verwitterte Formation in Jurong, die respective ermittelte durchschnittwerte von α fur beide Bodenarten waren 0.78 und 0.46. Für die Modelphfähle, eine Spitzensenkung von eswa 6 bis 10% der Pfahldurchmessern fur die voll Einwirkung der Mantelreibungskräfte benötigt war.

1 INTRODUCTION

A large number of drilled piers or cast-in-place bored piles are installed in Singapore each year in conjunction with housing, commercial, and industrial developments and construction of roadways and the associated facilities. The Mass Rapid Transit system of Singapore currently under construction further increases the use of drilled piers in suburban areas where the two most important Singapore rock formations, the Bukit Timah granite and the sedimentary rocks of the Jurong Formation, are present. Over 70% of the drilled piers constructed in Singapore are embedded in weathered materials derived from these two rock types, which occupy two-thirds of the total land area of the Singapore island. The weathered rocks are generally covered by residual soils of varying thicknesses ranging from less than a meter to as great as 35 meters.

Drilled piers are often used in Singapore in weathered rock mainly because of their high load carrying capacity and the ease with which the required lengths can be adjusted. The local practice of drilled pier design is to determine the pile dimensions primarily from the standard penetration resistance N of the geological stratum and the loading condition assuming that the total loads are carried by end bearing and neglecting shaft friction. An overall factor of safety of 2.5 to 3 is used. The design is sometimes supplemented with proof testing of piles using a maximum applied load of 1.5 to 2 times of the design load. Occassionally, the piles are loaded to failure. Some projects require that the piles be socketed into a competent rock irrespective of the strength of the material above the competent layer and the shaft resistance that it can offer. As a result, the piles are often overdesigned.

This paper presents results from load tests on three instrumented full-scale drilled piers installed in the weathered rocks of the Jurong Formation for actual construction projects in Singapore. The load transfer characteristics of these piles under the test loads are discussed and used as a basis for evaluating the current practice of drilled pier design in Singapore. Results from load tests on a number of miniature bored piles installed in both completely decomposed Bukit Timah granite and completely weathered rocks of the Jurong Formation for the study of the shaft friction are also presented in the paper.

2 LOAD TESTS ON FULL-SCALE INSTRUMENTED DRILLED PIERS

Three instrumented drilled piers, numbered as M1, J1 and J2, installed in the weathered rocks of the Jurong Formation were investigated. Test pile M1 was 900 mm in diameter and 24 meters in length. It was designed to carry an axial load of 2,500 kN. The pile was tested to a maximum load of 10,000 kN. Test piles J1 and J2, both 1000 mm in diameter, were designed to carry an axial load of 3,500 kN. These two piles were tested to 2 1/2 times the design load. The lengths of Pile J1 and Pile J2 were 14 meters and 28 meters, respectively. The following paragraphs describe these instrumented piles and the test results.

2.1 Test Pile M1

The test site was covered by a residual soil consisting of 11 meters of medium stiff to hard silty clay with an undrained shear strength, c_u, of 40 to 200 kPa. The residual soil was underlain by a highly weathered siltstone (N = 50-145). The 24 m long test pile, 900 mm in diameter, was drilled into the siltstone. The pile was instrumented with five vibrating wire strain gauges at levels 7.5, 11.0, 15.5, 20.5 and 24.0 meters below the ground surface. The tremie method was used to cast the lower half of the pile. The pile was tested one month after its installation using maintained load method.

Figure 1 shows the load-settlement curve for the test pile. At four times the design load, there were no signs of failure, even though the test was originally intended to test the ultimate capacity of the pile. Also shown in the same figure is the elastic compression of the pile calculated based on the strain gauge readings and a Young's modulus (E_c) of 31 kN/mm^2 for the pile material. It is interesting to note that the axial displacement of the pile head corresponded to the elastic compression

of the pile shaft, indicating that the pile tip movement was negligible throughout the entire loading range. Probably because a constant Young's modulus was used in the analysis, the calculated elastic compression was found to exceed the measured displacement of the pile head for loads below 3,000 kN. A sudden increase of the displacement of the pile head was observed at 3000 kN. The reason for this sudden jump is not clear.

Figure 2 shows the load transfer characteristics of the test pile. Although, the pile was designed as primarily an end bearing pile, the mobilized point resistance accounted only for 4% of the total applied load at the maximum test load of 10,000 kN. The majority of the applied load was transferred to the weathered siltstone by shaft friction. At the design working load of 2,500 kN, the tip resistance, Q_p, accounted for only 2% of the total applied load. On the contrary, shaft friction, Q_s, in the residual soil accounted for 28% and that in the weathered siltstone for 70% of the total applied load. The mobilized tip resistance was small mainly as a result of the negligible tip movement as shown in Figure 1.

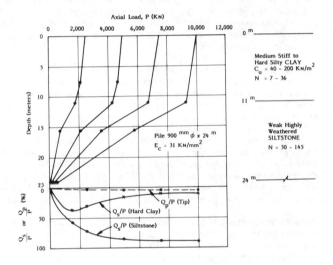

Figure 2 Load transfer characteristics of Test Pile M1

Figure 3 shows the mobilization of the average unit shaft resistance with the increase in average shaft displacement, represented by the displacement at the middle level of each stratum, for both the residual soil and the weathered siltstone in which Pile M1 was embedded. It is noted that the shaft friction in the residual soil was nearly fully mobilized at a relatively small shaft displacement of around 0.2% of the pile diameter. The shaft friction in the weathered siltstone, however, was still increasing at the maximum test load (10,000 kN).

2.2 Test Pile J1

At the site where Test Pile J1 was installed, the weathered rocks consisted of 3.9 meters of highly weathered silty sandstone (N_{avg} = 115) and a highly to moderately weathered siltstone bedrock (N_{avg} = 180). The rocks were covered by a 3.2 m thick layer of medium stiff silty clay (N_{avg} = 7) and a 3.4 m layer of dense to very dense silty sand (N_{avg} = 56) derived from a complete weathering of silty sandstone. The pile, 1000 mm in diameter and 14 meters in length, was instrumented with eight vibrating wire strain gauges, one 15 mm diameter and three 50 mm diameter galvanized iron pipes for coring and sonic

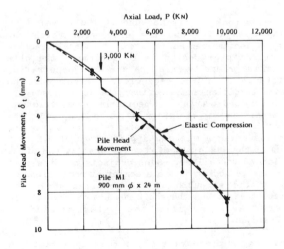

Figure 1 Load-settlement curve for Test Pile M1

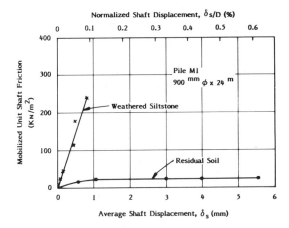

Figure 3· Mobilized unit shaft friction versus shaft displacement for Pile M1

logging and two 30 mm diameter PVC pipes for housing of the tell-tales. The strain gauges were installed in pairs at depths of 0.5, 7.5, 10.5 and 13.5 meters below the ground surface. The tell-tales were terminated at depths of 8.0 and 13.5 meters. The pile was cast using a tremie pipe and tested three weeks after its installation using the maintained load method.

Figure 4 shows the load-settlement curve of Pile J1 together with the elastic compression of the pile as determined from strain gauge readings. The pile did not fail at the maximum test load, 2 1/2 times the design load, although the rate of pile head movement per unit load increment had increased. A comparison between the pile head movement and the elastic compression indicates that there was a significant amount of tip movement. The tip movement calculated was around 4.4 mm, which compares very well with that deduced from the tell-tales.

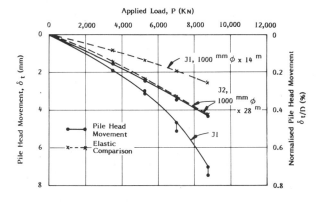

Figure 4 Load-settlement curves for Test Piles J1 and J2

Figure 5 shows the load transfer characteristics of the test pile. The E_c-values, ranging from 40 to 49 kN/mm^2, were deduced from the strain gauge readings at 0.5 m depth. The relatively high E_c-values are partly due to the main reinforcement bars and the galvanized iron pipes present in the test pile. It is noted that around 87% of the applied load was supported by shaft friction even at the maximum applied load of 8,750 kN. The shaft friction in both the weathered sandstone and siltstone layers

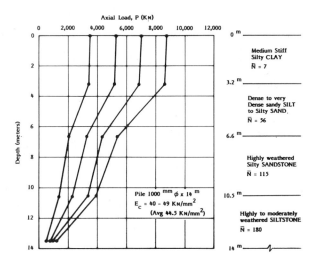

Figure 5 Load transfer characteristics of Test Pile J1

was not fully mobilized under the maximum test load as can be seen from Figure 6 in which the mobilization of shaft friction is shown as a function of the shaft displacement.

2.3 Test Pile J2

This test pile was located next to a river. The upper 14.5 meters of the ground consisted of soft peaty clay (N = 2 - 3). The pile was cast in a cased hole drilled through the soft clay and embedded in a weak, highly weathered shale with an average N-value varying between 100 and 150. The pile was terminated at a depth of 28 meters below the ground surface in weathered shale (N_{avg} = 100).

The pile, 1000 mm in diameter, was instrumented with six pairs of vibrating wire strain gauges at 0.5, 14, 18, 22, 25 and 27.5 m depths. In addition, one 150 mm diameter and three 50 mm diameter galvanized iron pipes for coring and sonic logging and three 30 mm diameter PVC pipes for housing of the tell-tales were installed. The tell-tales were terminated at depths of 14.0, 20.0 and 27.5 m. The pile was cast using a tremie pipe. It was tested two weeks after its installation.

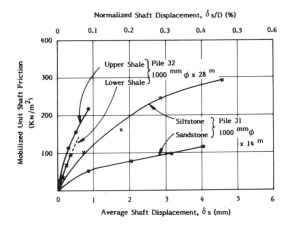

Figure 6 Mobilized unit shaft friction versus shaft displacement for Piles J1 and J2

The pile was tested to a maximum load of 2 1/2 times the design load. It is apparent, from the load-settlement and the elastic compression curves shown in Figure 4, that the pile was subjected primarily to elastic compression. There was hardly any movement at the pile tip even at the maximum test load of 8,750 kN.

Figure 7 shows the load transfer curves of Test Pile J2 calcualted from E_c-values, ranging from 50 to 56.5 kN/mm^2, and the strain gauge readings at various levels. It is apparent that the applied load was mainly (95%) carried by shaft friction along the pile-rock interface. The contribution from end bearing was approximately 5% of the total applied load at the maximum test load of 8,750 kN. The shaft friction along the pile-rock interface, although constituting more than 80% of the total resistance, was still not fully mobilised, as illustrated in Figure 6.

3 FINDINGS FROM FULL-SCALE LOAD TESTS AND SINGAPORE PRACTICE

The full-scale load tests on instrumented drilled piers discussed above illustrate the importance of shaft friction for drilled piers and the conservative design of drilled piers in Singapore.

All the three load tests were intended to be an ultimate load test. However, none of the pile failed at the maximum test load, which was four times the design load for Pile M1 and 2 1/2 times the design load for Piles J1 and J2. For Piles M1 and J2, which were 24 and 28 meters long, respectively, the settlement of the pile head was approximately equal to the elastic compression of the pile. The tip movements were very small. For this reason, the pile tip carried less than 5% of the total applied load even at the maximum test load for both piles. At the design load, the point resistance was less than 2%, as the load transfer curves in Figures 2 and 7 show. It is clear that both piles were heavily over-designed. The extrapolated ultimate load of Pile M1 based on Chin (1970) is approximately 67,300 kN. This value, which represents the upper bound of the most probable ultimate capacity, is more than 25 times the design load of 2,500 kN. The extrapolated ultimate load for Pile J2 is approximately 60,000 kN, which corresponds to 17 times of the design load of 3,500 kN.

Test Pile J1, which was only 14 meters in length, was designed to carry the same load as the 28 m long Pile J2. Because the tips of both piles were embedded in similar rocks (see Figures 5 and 7) and the shaft friction was neglected, the design load was, therefore, the same. These two piles behaved very differently when they were tested. This can be clearly seen in Figure 4 when the two load-settlement curves are compared. Pile J1 shows a much larger pile head movement than Pile J2 even though its length is a half of that of Pile J2. For Pile J1, the tip movement was 4.5 mm at the maximum test load of 8,750 kN. The pile tip, as the load transfer curves in Figure 5 show, carried 13% of the total applied load at the maximum test load. At the design load, the load distribution was about the same. For this pile, the extrapolated ultimate load of 22,000 kN still represents an over-all factor of safety of around 6 when compared with the design load of 3,500 kN.

It should be noted that the shaft resistance for Pile M1 was also significant in the medium stiff to hard silty clay (N = 7 - 36 and c_u = 40 - 200 kPa), although it is often considered as a 'weak' material. At the design load, the total shaft resistance along the 11 meter long pile section accounted for as much

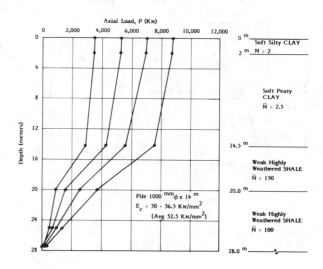

Figure 7 Load transfer characteristics of Test Pile J2

as 28% of the total mobilized resistance, whereas the tip resistance contributed less than 2% to the total resistance (See Figure 2).

Figure 3 shows that the unit shaft friction, f_s, for Pile M1 was fully mobilized at a relatively small shaft displacement. The corresponding shaft friction or adhesion factor, $\alpha = f_s/c_u$, is about 0.6 if the lower c_u-value is considered. Using an N-value of 7 and the relationship between N and c_u proposed by Stroud (1974), the α-value should be around 0.7. These values represent probably an upper limit. The most probable α-value could be close to the commonly accepted value of 0.45 as recommended by Skempton (1955) for bored pile design in London clay. Based on the measured shaft friction along Pile J1, the α-value for the upper 3.2 meters of medium stiff silty clay (N = 7), which is a very similar material, is estimated at around 0.43. The relatively high α-value of 0.84 observed by Chin (1982) for a pile installed in a similar material of the same formation with c_u = 164 kPa could not be verified.

Figures 3 and 6 show that, for all the three test piles, the shaft friction was not completely mobilized along the section of pile embedded in the weathered rock. This was particularly the case for Piles M1 and J2. Thus, the α-values of these piles could not be calculated. Nevertheless, the measured shaft friction values along Piles J1 and J2, when compared with their respective average standard penetration resistance (N_{avg} = 100 to 180), indicated a minimum α-value of between 0.14 and 0.18, using the correlation between N and c_u recommended in CP2004 of British Standard Institution (1972). A crude extrapolation of the shaft friction versus displacement curves for Pile J1 shown in Figure 6 indicates an ultimate unit shaft friction, f_s, of 150 kPa for the weathered sandstone and 400 kPa for the weathered siltstone. These correspond to an α-value of 0.20 for the weathered sandstone and 0.25 for the weathered siltstone. These values compare fairly with the ranges of shaft friction ratio, m = f_s/σ_f (σ_f = unconfined compressive strength) as reported by Meigh and Wolski (1979), although they are much lower than the α-values observed by Davies et al (1979) for bored piles in a completely weathered clayey silt-stone and a highly weathered sandstone. Horvath and Kenney (1979) indicated that the m-value were, in general, larger than 0.2, corresponding approximately to α = 0.4, for rocks with σ_f < 3,500 kPa. The extrapolated α-values are also lower than that of

0.70 reported by Chin (1982) for a similar material of the same formation with an undrained strength c_u of 350 kPa as obtained from a borehole plate load test.

It is clear, from the above discussions, that shaft friction is significant for drilled piers in weathered rocks and residual soils. The current practice in Singapore of neglecting the shaft friction and designing of drilled piers as primarily end-bearing piles is very conservative.

4 SHAFT FRICTION AS DETERMINED FROM MINIATURE BORED PILE TESTS

Some studies were carried out to investigate the skin resistance of miniature bored piles in the completely decomposed Bukit Timah granite and the highly weathered sedimentary rocks of the Jurong Formation in Singapore.

4.1 Tests in Bukit Timah granite

Five 76 mm diameter miniature bored piles were installed at one site (Site A), while seven 76 mm and five 152 mm diameter miniature bored piles were installed at another site (Site B) at a bottom of a cut for a roadway. The pile lengths ranged from 0.5 m to 2.9 m, and the depths of the pile tip levels varied from 0.7 m to 3.0 m below the ground surface.

The degree of weathering of the Bukit Timah granite varies with depth. The completely decomposed granite or residual granitic soil layer is 15 to 30 m thick when exposed. It retains the original structure of the parent rock. The residual soil at Site A was an inorganic, highly plastic, silty to sandy clay. Its fines content was 47% to 56%. At Site B, it consisted of inorganic silty and sandy clay with low plasticity. The content of fines was 53% to 70%. The average undrained strength as determined from UU-triaxial tests was around 106 kPa for Site A and 116 kPa for Site B. The N-value within the top four meters varied in general between 5 and 13.

Pull-out tests were carried out on the miniature piles. The majority of the piles were tested in tension either at maintained loading or at a constant rate of penetration (or uplift). The tensile reinforcement of one of the piles was connected to a basal plate and the pile was therefore subjected to compression during the pull-out test. The test results show that the unit shaft friction did not vary with the type of loading or testing.

A summary of the pull-out test results for the piles at both sites is shown in Table 1. It can be seen that the shaft friction factor, α, calculated from the measured shaft resistance and the undrained shear strength as obtained from UU-triaxial tests on undisturbed block samples, ranged from 0.72 to 0.91. The high values of α are likely to be due to the highly weathered parent rock that has been converted into a stiff clay. They could also be a result of the relatively high sand content of the granitic soils. The small diameter of the miniature bored piles could also be responsible for the high α-values obtained (Horvath and Kenney, 1979).

Table 1 also indicates that an average shaft displacement, of about 10% of the pile diameter was generally necessary to fully mobilize the skin friction. Probably because of the small diameter of the miniature piles, this is significantly larger than the 0.5 to 2% normally required for conventional full-scale bored piles. For the pile subject to compression during the pull-out test, the average shaft displacement was found to be around 15 mm or 20% of the pile diameter when the shaft friction was fully mobilized.

4.2 Tests in sedimentary rocks of the Jurong Formation

Eight miniature bored piles were installed in the Nanyang Technological Institute campus at depths ranging from 0.8 m to 2.8 m. The length of the 76 mm diameter miniature piles ranged from 0.77 m to 2.5 m. One of the piles was installed with a soft toe. Another two piles had their reinforcement cables connected to a basal plate located at the pile tip. The upper few meters at the site were primarily residual soils consisting of clayey silt, clayey sand, silty clay and sandy clay. The fines contents of the soils ranged from 56% to 90%. The undrained shear strength based on UU triaxial tests carried out on U4 samples ranged from around 70 to 110 kPa.

All the piles except the one with a soft toe were subjected to pull-out tests. The concrete in the piles with basal plates was in compression during the pull-out tests. The pile with a soft toe was subjected to a compression test. Maintained loads were used in all the tests.

Table 2 shows a summary of the test results. The average α-values varied between 0.43 and 0.49 based on three different types of tests. These values agreed with those recommended by Skempton (1959) for stiff clays. The α-value did not seem to vary with

Table 1 Summary of pull-out tests in weathered Bukit Timah granite

Test site	Pile diameter D (mm)	Pile length L (m)	No. of piles tested	Shaft friction factor, α	Mid-level shaft disp. δ_m (mm)	Normalized shaft disp. δ_m/D (%)
A	76	1.0 - 2.5	5	0.75 - 0.91 (Avg = 0.82)	5.6 - 9.4 (Avg = 7.3)	7.4 - 12.4 (Avg = 9.6)
B	76	0.7 - 2.0	6	0.75 - 0.82 (Avg = 0.79)	3.8 - 9.7 (Avg = 7.5)	5.0 - 12.7 (Avg = 9.9)
B	152	0.5 - 2.9	5	0.72 to 0.79 (Avg = 0.75)	7.9 - 22.2 (Avg = 16.8)	5.2 - 14.6 (Avg = 11.0)
B	76	2.0	1*	0.73	15.2	20.0

* The concrete is loaded in compression during the pull-out test

Table 2 Summary of miniature bored pile tests in Jurong Formation

Type of test	No. of piles tested	Pile length L (m)	Undrained strength c_u (kPa)	Shaft friction factor, α	Mid-level shaft disp. δ_m (mm)	Normalized shaft disp. δ_m/D*
Pull-out (Tension)	5	0.77 - 2.50	69 - 110	0.34 - 0.51 (Avg = 0.43)	2.5 - 7.9 (Avg = 4.8)	3.3 - 10.4 (Avg = 6.3)
Pull-out (Comp.)	2	1.50	69 - 113	0.37 - 0.55 (Avg = 0.46)	6.4 - 8.1 (Avg = 7.3)	8.4 - 10.7 (Avg = 9.6)
Compression	1	1.50	113	0.49	5.9	7.8

* D = 76 mm (pile diameter)

the type of test.

The average shaft displacement required for a complete mobilization of the shaft friction was 6 to 8% of the pile diameter based on both pull-out (tension) and compression tests. For piles subject to compression during the pull-out test, an average shaft displacement equal to about 10% of the pile diameter was required to fully mobilize the shaft friction. Similar to what was found from the tests in Bukit Timah granite, this required shaft displacement was slightly larger than that calculated from the pull-out (tension) and compression tests. One possible reason could be that part of the upward shaft displacement was compensated by the elastic shortening of the pile as a result of compression in the pull-out test.

5 CONCLUSIONS

Three full-scale drilled piers of 900 mm to 1000 mm in diameter and 14 to 28 meters in length which had been installed in highly weathered sedimentary rocks in Singapore were tested. The shaft friction resistance along a number of miniature bored piles installed in a completely decomposed granite rock and a completely weathered sedimentary rock formation in Singapore have also been investigated. The following tentative conclusions are drawn from these tests:

(1) The current design practice for drilled piers or cast-in-place bored piles in weathered rocks in Singapore is very conservative.

(2) The contribution of shaft friction is very important for piles well socketed in weathered rocks. The end bearing resistance, although it can be very large is, in most cases, not significant under normal working loads.

(3) The shaft friction factor α, as extrapolated from the full-scale tests, was found to be around 0.20 for the weathered sandstone and 0.25 for the weathered siltstone of the Jurong Formation in Singapore.

(4) The average α-values for the completely decomposed granite and the completely weathered sedimentary rocks investigated were found to be about 0.78 and 0.46, respectively, based on the pull-out tests on miniature bored piles. The shaft displacement required for a complete mobilization of the shaft friction was found to be on the order of 6 to 10% of the pile diameter. Because of the small diameter of the miniature piles, the ratio of the shaft displacement and the pile diameter was much larger than 0.5 to 2% derived from conventional full-scale bored piles tested in compression.

6 ACKNOWLEDGEMENTS

The Authors wish to thank the Nanyang Technological Institute (NTI) of Singapore for granting an applied research fund for studying the behavior of bored piles in the residual soils and weathered rocks of Singapore. The Authors also wish to thank both Bored Piling Pte. Ltd. and Cosmos Piling Construction Pte. Ltd. of Singapore for their assistance in acquiring the results of full-scale load tests, and Public Works Department of Singapore for assisting in the installation of some of the miniature bored piles. They wish to thank Dr. Anthony Goh of NTI for his kind assistance in the preparation of the maniscript. Special thanks are due to Professor Bengt B Broms of NTI for his careful review of the manuscript. Thanks are also due to Jamillah Sa'adon and Sherlene Lim for typing the manuscript and Mr Victor Cheah for the illustrations.

7 REFERENCES

British Standard Institution 1972. Code of Practice 2004 (1972), Foundations. BSI, London.

Chin, F. K. 1970. Estimation of the ultimate load of piles from tests not carried to failure, Proc. 2nd Southeast Asian Conf. on Soil Engineering, Singapore, pp. 81-90.

Chin, Y. K. 1982. Instrumented ultimate load tests on bored piles. Fifth PWD Technical Seminar. Public Works Department, Singapore.

Davies, P., Webb, D.L., Hooley, P. and Yeats, J.A. 1979, Geotechnical parameters for design of bored piles founded in weathered siltstone rock, Proc. 7th ECSMFE (Design Parameters in Geotechnical Engineering), BGs, London, Vol 3, pp. 43-50.

Horvath R. G. and Kenney, T.C. 1979, Shaft resistance of rock-socketed drilled piers, Proc. Symp on Deep Foundations, ASCE National Convention, Atlanta, GA., pp. 182-214.

Meigh, A. C. and Wolski, W 1979. Design parameters for weak rocks. Proc. 7th ECSMFE (Design Parameters in Geotechnical Engineering), BGS, London, Vol. 5, pp. 59-79.

Skempton, A. W. 1959. Cast-in-situ bored piles in London clay, Geotechnique, Vol. 9, No. 4, pp 153-173.

Stroud, M. A. 1974. The standard penetration test in insensitive clays and soft rocks, Proc. 1st European Symposium on Penetration Testing, Stockholm, Vol. 2.2, pp 367-375.

New development in seam treatment of Feitsui Arch Dam foundation
Nouveau développement dans le traitement de certaines couches de la fondation du barrage voûte de Feitsui
Neue Entwicklung in der Kluftverbesserung für die Feitsui Staumauergründung

Y.CHENG, member SEAGS, President Sinotech Engineering Consultants Inc., and Director of Feitsui Reservoir Project, Taiwan

ABSTRACT: To meet with the extraordinary safety requirement of the project, bedding parallel clay seams found in the competent sandstone and siltstone foundation of the Feitsui arch dam were treated by excavating adits, washing weak seams with high pressure water jets in adits and backfilling the washed seams with non-shrinking mortar. As a consequence, an integrated strong foundation was formed to resist the arch thrust. The treatment work was smoothly completed on time. During construction, systematically drilled check cores showed a good shear strength of the contact surface between the rock and the mortar. This new seam treatment method has thus proved itself to be practicable, reliable and comparable in cost.

RESUME: Pour satisfaire aux conditions extraordinaires de sécurité de ce projet, les traitements des couches argileux parallèles aux stratifications, découvertes dans la fondation de grès et de microgrès siliceux du barrage-voûte de Feitsui, sont de creuser des galeries d'écoulement, de laver les couches faibles avec un jet d'eau à haute pression, et de remblayer les couches lavés avec le mortier non-contractile. Une fondation solide et intégrale est alors formée, qui peut résister à la poussée du barrage, des travaux d'aplanissement ont été achevés à temps. Pendant la construction, les forages de recherche effectués systématiquement ont démontré une bonne résistance au cisaillement de la surface de contact entre la roche et le mortier. Il est prouvé que cette nouvelle méthode de traitement des couches argileux est pratique, sûre, et d'un prix comparable aux autres méthodes.

ZUSAMMENHANG: Um den aussergewohnlichen Anspruch auf die Sicherheit des Projekts zu erfüllen, werden die im Bereich des Fundaments von Sandstein und Siltstein des FEITSUI Gewölbemauer mit Schieferung parallel liegende Kluftfüllung behandelt durch Ausgraben und Ausspülen der schwachen Klüfter mit dem Hochdruck-Wasser-Einspritzung und Zurückfüllung mit dem nichtschwind Mörtel. Dadurch wird die Festigkeit der Felsgeklüfter verbessert und erhält man einen stabilen, kräftig integrierten Fundament, das die Belastung des Damms sicherlich tragen kann. Die feine Bearbeitung für die ausgespülten Klüfter wird auch gleichzeitig fertig gemacht. Während der konstruktion zeigt die systematische Prüfung mit Kembohren gute Schubfestigkeit zwischen Felsen und gefüllten Mörtel. Diese neue kluftfüllung Verbessern Methode scheint sehr praktikäbel, zuverlässig und ist vergleichbar mit anderen Methoden im Kost.

1 INTRODUCTION

1.1 Description of project

The Feitsui Reservoir Project is located at Peishih Creek, about 30 km away from Taipei (see Fig. 1), Taiwan, R.O.C. The primary function of this project is to provide long-range water sources to meet the increasing demand of water supply for the metropolitan Taipei area till the year 2030. The gross reservoir capacity is 406,000,000 m³. To fully utilize the water resources and the head, a power plant with an installed capacity of 70 MW is also constructed as an associated function.

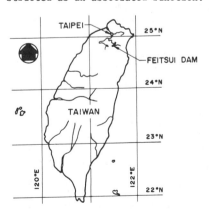

Figure 1. Project location

The Feitsui Dam is a three-centered, double curvature and variable thickness arch type structure having a height of 122.5 m, a crest length of 510 m, and a concrete volume of 700,000 m³. The dam is divided into 29 blocks. The contraction joints between the blocks are grouted after the dam concrete has been cooled down to 14.4°C. The appurtenant structures are briefly described as follows:

The crest spillway is located at the center portion of the dam crest with eight bays, each controlled by a 14 m wide and 9.3 m high radial gate. The total design discharge of the spillway is 7,670 cms which is capable of handling once in 5,000 years floods.

Three sluiceways are provided at the mid-height of the main dam with a total design capacity of 700 cms. The functions of the sluiceways are for release of flood, flushing of sediment and evacuation of reservoir water in case of emergency.

The 1.6 mφ river outlet, having a maximum design discharge of 45 cms, is provided below the left sluiceway. The function of the river outlet is to release water for downstream use when powerplant is not in operation.

The auxiliary dam, located at 170 m downstream of the main dam, consists of two parts. The center portion is an arch overflow section having a height of 25 m and a crest length of 120 m. On both sides of the arch are two gravity type non-overflow sections having a height of 36.5 m and a total length of 138 m. The purpose of the auxiliary dam is to create a plunge pool to dissipate the energy of water released from the crest spillway and the sluiceways.

The river diversion plan comprises a 29 m high cofferdam of arch type and a 10 m diameter concrete lined diversion tunnel with a design discharge of 1,190 cms which is the maximum recorded flood in the non-typhoon season. The downstream part of the diversion tunnel is converted to form a part of the tunnel spillway after final closure of the diversion tunnel. Of course, the conjunction and inclined shaft and the new intake of the tunnel spillway shall be newly constructed. The design capacity of the tunnel spillway is 1,500 cms.

The semi-underground R.C. powerhouse and power waterway are located at the right bank, with the power intake being attached to the dam body.

Figure 2 shows the plan, elevation and profile of the project.

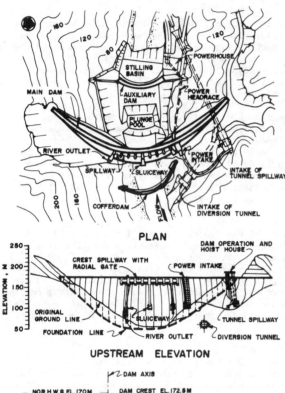

PLAN

UPSTREAM ELEVATION

PROFILE

Figure 2. General layout of the Feitsui Dam

The main construction work for the project were started in July 1981. The total progress completed was 99.54% at the end of October 1986. The project will be one hundred percent completed in April 1987 after installation of the spillway radial gates and completion of the conjunction construction work for the tunnel spillway.

1.2 Damsite geology

The project area is in a folded Oligocene sedimentary sequence in the tectonic inactive zone of northern Taiwan. The damsite is situated on the north limb of the Houshaochang Anticline, and is composed of massive and indurated sandstone and siltstone with thin alternated beds in parts. The bedding strike is generally parallel to the river channel and dips approximately 40 degrees toward the right bank, thereby creating a dip-slope left abutment and a scarp-slope right abutment (see Fig. 3).

Being a homoclinal structure, the rock strata at the dam foundation are very regular in their strike and dip, and are essentially not offsetted by fault. Both the sandstone and siltstone are strong and hard, and joints of the longitudinal and diagonal types are discontinuous and widely spaced. However, bedding shears due to flexural slippage during folding and tilting of the rock strata in the geologic time appear to have occurred occasionally, resulting in formation of unfavorable weak planes in the abutments and riverbed foundation. This is especially true with regard to the shallow strata in the left abutment which also have been affected by stress relief and are more complicated. These bedding parallel clay seams either of bedding shears or stress relief seams are rather thin (most of them range from less than 1 cm to about 15 cm). They were called 'bedding seam' in the project and will also be so named in this paper for easy description.

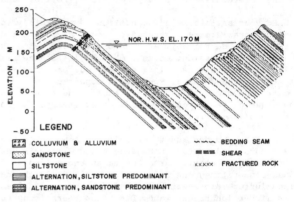

Figure 3. Typical geologic section of the Feitsui damsite

Shears transverse to bedding and clay-filled joints are also found at the damsite. However, these are discontinuous and are in the minority in comparison with the bedding seams. Therefore, they were not regarded as a control factor in the foundation design, although the bedding seams were. After proper treatment of the bedding seams, the damsite has been evaluated and is considered to be suitable for the construction of a high arch dam.

1.3 Need of seam treatment

The Feitsui Reservoir is only 30 km away from the metropolitan Taipei area in which the total population amounts to almost five million people. It has therefore been decided, right from the beginning of planning and design, that the utmost safety concern should be given to both the dam and its foundation. Because the presence of bedding seams at the Feitsui damsite could be detrimental to the deformation and stability of the dam abutments, especially concerning the shear detrusion around the seams, a sound and reliable seam treatment measure had to be developed to enable the foundation to safely transmit the arch thrust into the deep.

There are several methods which are commonly used for dam foundation treatment program, such as: deep foundation excavation, dental treatment, washing and grouting in drill holes, anchoring rock layers with pre-stressed tendons, anchoring the rock layers by backfilling a series of foundation tunnels with concrete, fault treatment by mining and backfilling with concrete or fault treatment by excavating parallel tunnels and excavating thick fault by high pressure water jet (used at the Nagawado Dam in Japan). However, after a careful study of the results, these treatment methods were found to be either inadequate or not suitable for application to the Feitsui arch dam foundation because of the special geologic conditions of the thin clay seams encountered and of the extraordinary safety requirement of the project. A more complete treatment by excavating a series of adits and washing out the thin clay seams in these adits by high pressure water jets and then backfilling them with non-shrinking was finally envisaged.

2 EXPERIMENTS OF SEAM TREATMENT

2.1 Purpose of experiments

The purpose of the following experiments of seam treatment was to find out some information pertinent to the best treatment method, working rate and cost of treatment. Based on these data, the most desirable method for treatment of bedding seam was derived.

2.2 Pilot experiment

In the Definite Plan Study stage of the Feitsui Reservoir Project in 1977, an experimental treatment of bedding seam was carried out along seam F-4 in the newly excavated inclined working shaft (2.0x1.4 m) between the existing exploratory adits TA-7 and TA-19 of which the difference in elevation was 15 m. A wooden ladder was installed in the inclined shaft to enable the workers to wash the bedding seams. The washing process was performed from lower to higher elevation and from shallower to deeper location to wash out weak materials. After completion of the washing process, the void was filled with cement mortar. The grout was a mixture of cement, fly ash and sand at the proportion by weight of 1:0.4:2.0. The water cement (cement plus fly ash) ratio used was 0.6. Subsequently, after the cement mortar had solidified and shrunk, pressure grouting was applied through the embedded pipes to ensure that no void would be left. Twenty-one days after grouting, NX core drilling of the treated seam and in-situ shear test were carried out. The major equipment used for the treatment work consisted of a high pressure pump which had a maximum value of 10 MN/m², flushing joint type steel rods of different sizes and 1 m long with a center hole of 10 mm in diameter, and nozzles with forward and sideward holes ranging from 5 to 15 mm² in cross sections, and a set of grouting equipment.
The results of this pilot experiment were evaluated and summarized as follows:
1. Although the washing depth of this test was almost 6 m, considerable reduction of washing efficiency was observed when the depth was over 3 m. For bedding seam of the thickness of around 10 cm, the optimum washing depth was 3 m with the washing efficiency registered at 0.2 m²/hr. When the thickness of bedding seam decreased, the washing depth would reduce to 2 m for seams whose thickness was between 5 and 10 cm, and to 1 m for seams whose thickness was less than 5 cm. It was observed that higher pressure water jet washing would be required for greater washing depth and higher washing efficiency.
2. Core drilling of treated seam showed that all weak materials in the seam had been completely washed out. However, breakage of the upper contact face between mortar and rock indicated that the use of non-shrinking mortar would be advisable.
3. In-situ shear test showed that the shear strength of the contact face of the treated seam would be c(apparent cohesion)= 0.08 MN/m², ϕ=48°, which had considerably increased, compared with the strength before treatment (c=0.07 MN/m², ϕ=19°).

This result indicated the methodology of seam treatment is feasible and reliable if washing pressure could be further increased.

2.3 Experiments in Basic Design stage (1979)

The objectives of the experiments carried out in this stage were: 1) to increase washing efficiency through increase of the water jet pressure; 2) to find out an effective way for treatment of bedding seams underneath the riverbed foundation; 3) to apply waterdrill and waterknife for treatment of papery thin seams; and 4) to develop a suitable mixture of non-shrinking grout for backfill grouting use.
- Experiment for treatment of bedding seams thicker than 1 cm:
This experiment was carried out in adit TAB-1 at the left abutment. Four seams each having a different thickness of 1-2 cm, 2-5 cm, 5-10 cm and 10-20 cm were selected for the test. Most of the seams contained rock fragments and thin clay layers. Undulation of seam was prominent. The equipment used for the experiment were:
1. High pressure water pump -- maximum water pressure, 21 MN/m²; discharge rate, 0.105 m³/min at the pressure of 20 MN/m².
2. Washing rod -- flush joint type steel pipes of different diameters of 38 mmϕ, 25 mmϕ, 19 mmϕ, and 9.5 mmϕ.
3. Nozzles -- conical shaped alloy steel nozzles of different opening sizes of 1 mmϕ, 2 mmϕ and 3 mmϕ.
4. Grouting pump -- maximum pressure, 3 MN/m²; grouting rate, 0.1 m³/min at the pressure of 1 MN/m².
Before washing, the existing adit at the location of washing should be enlarged to a cross section of 3 m by 3 m with a length of 2 m. The washing operation was started at the lower side and moved upward. Different washing rods, nozzles, washing pressure and flow rates were used for different thickness of seams to identify the best washing performance. After washing, the seam opening was sealed with cement mortar and cobbles. Meanwhile, 32 mmϕ grouting pipes were installed at the interval of every 1.5 m and a vent pipe was installed to reach the top of the washed seam. The grout mix by weight of cement, sand, and non-shrinking agent (DENKA TASCON) was 1:1.123:0.126. The water cement ratio was 0.43. After the grout had gained a sufficient strength, cores were drilled through the treated seam to examine the bonding condition and for direct shear test. In addition, an in-situ test was also conducted. Following are the test results:
1. All bedding seams whose thickness was greater than 1 cm could be thoroughly washed in this experiment. The washing rate and washing depth were as follows: For thickness greater than 10 cm, 0.87 m²/hr, with a depth up to 3.5 m; for thickness of 5-10 cm, 0.75 m²/hr, depth up to 3.2 m; for thickness of 2-5 cm, 0.71 m²/hr, depth up to 3.0 m; for thickness of 1-2 cm, 0.65 m²/hr, depth up to 3.0 m.
2. The use of non-shrinking mortar to fill the washed seam led to a very satisfactory result. The fact that breakage of the core specimen under the laboratory tension test was not at the contact face between rock and mortar but within the grout showed that excellent bonding had been achieved.
3. The in-situ shear strength of treated seams before and after treatment were registered as follows:
Before treatment -- c=0.03-0.2 MN/m²; ϕ=17.5-30°.
After treatment -- c=2.2 MN/m²; ϕ=38.7°.
4. The shear strength of the contact face of the treated seam of NX cores had an apparent cohesion c=2.2 MN/m² and ϕ=38°.
From the evaluation of the test results mentioned above, it could be realized that bedding seams having a thickness of greater than 1 cm can be

satisfactorily treated. The improved washing rate and greater shear strength indicate that this method can be adopted for future construction. The washing method used for this experiment is designated as 'Method A'.

- Experiment for treatment of bedding seams underneath the riverbed:

This experiment was aimed at treating the seams in the foundation below the riverbed elevation where excavation of tunnels and treatment operation would not be practical. The washing process was done by high pressure water jet with air revolving in the drill holes drilled downward along the seam in adit located above the riverbed. The experiment was carried out along a seam which had a thickness of about 10 cm in adit TAB-1. The procedures of the experiment were:

1. Enlargement of the adit to a cross section of 2 m by 2.3 m to provide adequate working space for the treatment operation. A grout cap, 0.5 to 1 m thick, was provided with pipes installed in the cap at the invert of the adit spaced at 80 cm and 40 cm apart.

2. Two BX test holes spaced at 80 cm apart were firstly drilled along the seam by using a pilot type bit. Then, the Lugeon test was performed.

3. After the Lugeon test, the double tube rod with a nozzle head attached was mounted on the drilling machine and then washing of the seam was carried out by revolving the rod with high pressure water at 20 MN/m^2, 0.105 m^3/min and compressed air at 0.6 MN/m^2, 4.5 m^3/min, jetting from two side nozzles. The washing rod revolved at a speed of 10 rpm with an upward or downward movement at the rate of 2.5 cm/min. The washing operation was continued by moving the rod up and down until the return water became clear.

4. Grouting the washed space with a non-shrinking cement paste was carried out through the drilling rod with a grouting pressure of 0.2 MN/m^2. The ratio of cement to TASCON (C/T) and of water to cement (W/C+T) used were 1:0.126 and 1:2 respectively.

5. When the strength of the grout became sufficient, two NX check holes, one parallel and the other perpendicular to the treated seam, were drilled and the Lugeon tests were conducted again to determine the effectiveness of the treatment.

6. Procedures 2 to 5 mentioned above were repeated for two test holes spaced at 40 cm apart.

The major equipment used were:

1. A high pressure water pump similar to the one used for Method A.

2. A rotary drilling machine of the hydraulic operated type, equipped with automatic lifting and timing devices.

3. A 'BX' double tube washing rod with the inner tube for water and the outer tube for air.

4. A 3 mmφ double ring nozzle attached to the rod with water nozzle in the outer ring and air nozzle in the inner ring.

Examining the cores and the test results, it could be concluded that:

1. The method would generate satisfactory and practical results with the wash hole spacing at 50 cm for seams with a thickness of less than 50 cm.

2. Rock fragments left in the seam during the course of the washing operation were well cemented with the grout so that the core looked like a core drilled from prepacked concrete.

3. Some clean fine sand was found at the bottom of the washed seam. This defect can be corrected by using the jet grouting method during the course of grouting of mortar.

4. Bonding condition of the contact face between rock and grout was very satisfactory, and that the Lugeon value dropped from 7.13 to 1.1 Lugeon after treatment.

It is evident that this type of treatment is applicable to seams below the riverbed elevation. The washing method used for this experiment is designated as 'Method B'.

- Experiment for treatment of papery thin seam:

In 1981, Sinotech Engineering Consultants, Inc.

assigned the Ret-Ser Engineering Agency (RSEA) and the Flow Industries, Inc. (USA) to carry out seam treatment experiments in adit LAT-5. The main equipment used for the experiments included:

1. An intensified type high pressure pump with a maximum pressure of 240 MN/m^2 and a water discharge rate of 0.011 m^3/min.

2. A drill set consists of a drill stem and a drill bit mounted on the stem. The drill stem is made of 14 mmφ stainless steel tubing and rotates at approximately 500 rpm by air powered drill motor. The drill bit for waterdrill has three jets each of a 0.38 mm diameter shooting from sapphire nozzles mounted on the tip of the bit at a velocity of about 610 m/s. The bit for waterknife is almost the same as that of the waterdrill except two side nozzles are used.

The aforementioned equipment are manufactured by the Flow Industries, Inc. Washing of papery thin seams was started using waterdrill along the seams with a hole spacing of 13 cm to 20 cm. Due to limited working space inside the adit, the washing depth was set at about 1.8 m. After drilling of holes, the drill tip was replaced with a waterknife to wash out clay from the seam between the holes. During this experiment, washing of thicker seams between adits LAT-5 and LAT-6 by the use of a waterdrill was also performed. According to the washing test, good washing result of papery thin seams can be obtained by using waterdrill and waterknife with hole spacing set at about 20 cm and hole depth at less than 2 m. The washing rate was about 1.0 m^2/hr. The washing depth could be deeper than 2 m if longer washing rod was used. The washing rate for thicker seams by using waterdrill was about 1.87 m^2/hr which is much higher than that of Method A. Use of waterdrill and waterknife for washing bedding seams can not only remove all the clay and weathered rock fragments in the seam, but also roughen the washed rock faces. It shows that better washing efficiency can be achieved. Therefore, backfill grouting of the washed seam for testing the shear strength was omitted. The washing method using waterdrill and waterknife is designated as 'Method C'.

- Test of non-shrinking grout:

For determination of the most suitable brand of admixture and its optimum dosage to be blended in the mortar or in the paste for backfilling of washed seams, a comprehensive program for testing the grout had also been conducted. Selection of a proper kind of admixture and dosage was based on the following factors: 1) better workability; 2) non-shrinking characteristics; 3) adequate strength; and 4) economy. Among the six admixtures and the different mix designs tested, the PLA series of the Five Star Brand provided better results. To determine the amount of PLA required, two important factors were considered: 1) since the bedding seams have a dip angle of 40°, any shrinkage occurred before the mix starts to develop strength will not create gap along the rock walls, and 2) the mix should have an expansion value of 0.07% to 0.10% so that adequate confinement pressure acting on the rock walls can be achieved. Based on these considerations, an optimum PLA dosage was recommended in the following mix design.

For mortar:
  Cement: Type II cement with 20% fly ash
  Sand-cement ratio: 1.1
  Water-cement ratio: 0.53
  PLA dosage: 15 kg/m^3 of grout
For paste:
  Cement: Type II cement
  Water-cement ratio: 0.40
  PLA dosage: 25 kg/m^3 of grout.

3 GENERAL LAYOUT AND CRITERIA FOR SEAM TREATMENT

Results of the experiment program for bedding seam treatment indicated that all kinds of clay seams could be effectively treated and could be applied for actual construction. Therefore, a seam treatment program was adopted for design of the Feitsui Reservoir Project.

For construction, the formulation of a set of practical working criteria was deemed necessary. The seam treatment program requires firstly the excavation of a series of access tunnels and working adits in the dam loaded zone. Then, the seams can be treated through the tunnels and adits according to the working criteria. After treatment, most of the tunnels and adits are backfilled with concrete and finally backfill grouting is applied to fill all possible gaps outside the backfilled concrete.

3.1 Methods adopted

The methods adopted for washing of clay seams were:
For seams thicker than 1 cm: Method A or Method C.
For seams below the riverbed elevation: Method B.
For seams thinner than 1 cm: Method C.

3.2 Design shear strength of treated seam

As described in section 2.3 above, the shear strength of the contact face between the rock and the grout of the treated seam had an apparent cohesion ($c=2.2$ MN/m^2) and internal friction angle ($\phi=38.7°$). Through the application of a scale factor of 2.0, the adopted design shear strength for the treated seam is: $c=1.1$ MN/m^2 and $\phi=38°$.

3.3 Layout of access tunnels and working adits

Two factors were considered for layout of the access tunnels and working adits for seam treatment: 1) the condition and the continuities of the bedding seams, and 2) the suitable working depth for treatment of various types of seams. Since the left abutment is a dip-slope, and also because the bedding seam LS2 is a 100% continuous seam located right below the dam base, therefore this particular seam is a main seam which requires full treatment within the dam loaded zone. To ensure the efficiency of the treatment work, 19 horizontal grids of the access tunnels and working adits were laid out at various levels along the left abutment. The vertical spacing between two levels is about 7.5 m. Each grid consists of: 1) a main access tunnel which is normal to the ground line of the left abutment and ends at main seam LS2, 2) an access tunnel which is parallel to LS2 and is used for full treatment of LS2, and 3) working adits which are normal to the access tunnel with a spacing of 9 m and are used for treatment of other seams in the loaded zone. Figure 4 shows a typical plan of the access tunnel and the working adit grid.

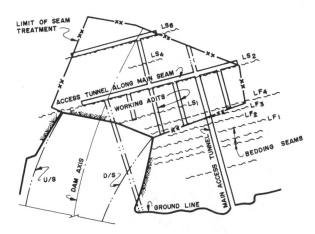

Figure 4. Typical plan of the access tunnel and the working adit grid at the left abutment

Because the bedding seams at the right abutment dip into the bank slope, each seam can be treated by a grid of an access tunnel and working shafts constructed along the seam and spaced at 9 m. Figure 5 shows the typical layout of the access tunnel and the working shaft grid for each seam.

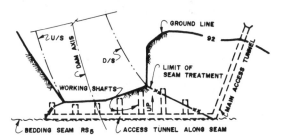

Figure 5. Typical layout of the access tunnel and the working shaft grid at the right abutment

3.4 Working criteria for seam treatment

- Basic considerations:
The working criteria to be developed should fulfill the design requirement and should also be practical for construction use. The following basic considerations were made for formulation of the working criteria for seam treatment:
1. The treated portion of the dam foundation should have sufficient strength to sustain the arch thrust from the dam so that no shear detrusion along the seam would occur.
2. The dam foundation would be more uniform in effective deformation modulus.
3. The treatment method applied and the treatment depth specified should be practical and not overconservative.
4. The complexity of seam condition should be taken into account.
- Characteristics of bedding seam:
By detailed geological mapping in tunnels and adits, the characteristics of seams inside the dam foundation can be summarized as follows:
1. At the left abutment, the bedding seams are parallel to the slope surface. Bedding seam LS2, which is located right below the dam base and has a 100% continuity, is the main weak plane in the left dam foundation and requires full treatment. Other bedding seams in the dam foundation are less continuous and are more complicated in their internal structure. They usually contain sheeted rock pieces and blocks with infilling clay. The seams usually swell and pinch and may become rock to rock contact at certain locations, and may also be replaced by branches. Therefore, only local treatment of these seams was required through the adoption of different methods and treatment depths depending on the specific conditions of the seam.
2. At the right abutment, the bedding seams dip into the slope and are rather thin (most of them ranging from 0.1 cm to 5 cm). In an early stability study of the dam foundation, the combination of a longitudinal joint and a bedding seam may form a potential sliding wedge, but it was found after the keyway had been excavated that these longitudinal joints were poorly developed, usually clean, tight and of very low continuity. Therefore, the formation of a potential wedge was less possible. According to the results of foundation stability analyses, these bedding seams and clay filled joints require only local treatment.
3. As to the riverbed foundation, bedding seams are running in the upstream-downstream direction with a dip toward the right bank. Because there is no unfavorable joint set developed at the riverbed in combination with the bedding seams to form a potential sliding wedge, treatment of the bedding seams in the riverbed foundation is primarily for

improvement of the deformation behavior as well as the seepage cut-off. The treatment depth required is shallow.
- Working criteria:

According to the basic considerations mentioned above and the characteristics of the seams, the working criteria to serve as a general guideline to define the extent and methods of seam treatment were derived as listed in Table 1.

Table 1. Working criteria for seam treatment

Zone of treat-ment	Seam condition			Extent of treatment	Method of treat-ment
	Type	Thick-ness (cm)	Conti-nuity (%)		
LEFT ABUT-MENT	Bedd-ing seams	>3	>50	completely treated	A
		1-3	>50	4 m depth radially above invert of adit	A
		<1	>50	2 m depth radially above invert of adit	C
		>1	<50	- ditto -	A
		Others		No seam treatment	
RIVER-BED FOUND-ATION	Bedd-ing seams	>3	>50	treated to the required depth	B
		1-3	>50	6 m downward	B
		<1	>50	3 m downward	C or B
		>1	<50	3 m downward	C or B
		<1	<50	2 m downward	C
RIGHT ABUT-MENT	Bedd-ing seams	Shaft extension, over-excavation of seam in tunnel and on keyway and backfilled with concrete.			
	Clay-filled joints on keyway	>3	>50	4 m depth	A
		1-3	>50	3 m depth	A
		<1	>50	3 m depth	C
		>1	<50	3 m depth	A

4 CONSTRUCTION

Excavation of seam treatment tunnels and adits at both abutments of the Feitsui Dam was started in March 1981 to inaugurate the seam treatment work, and the treatment work was completed after the last treatment tunnel had been backfilled in March 1985. During the construction period, progress of the seam treatment work was smooth and no delay of the dam concrete placement work was caused.

4.1 Aditing

Excavation of a large number of treatment tunnels and adits should be started well in advance of the excavation of the dam keyway so that the excavated keyway would not be further shattered. The tunnel excavation work started from top elevation and the construction specifications stipulated that the construction of tunnels and adits should at least be 10 m ahead of the excavation of the dam keyway. The smooth wall blasting method was applied for excava-tion of the tunnels and adits under very strict inspection. Examination of the cores drilled from the blasted walls revealed that shattering of the surrounding rock was very shallow.

4.2 Seam treatment

Seam treatment work started from lower elevation and placement of dam concrete could only be carried out after the treatment tunnels and adits at that loca-tion had been completely backfilled.
- Seam washing:

Before the commencement of seam washing work,

geologic mapping was carried out. Examination of seam conditions in greater details was also performed, and based on the information thus collected, the washing method and the washing depth required for various locations along the seam were determined in accordance with the working criteria for seam treatment.

For washing of seams in the left abutment above the riverbed elevation, either Method A or Method C was adopted. The main seam LS2 was washed in paral-lel access tunnels having a cross section of 2.0 m by 2.3 m. The seam was divided into many washing blocks each having a width of about 4 m. To avoid loosening the top rock layer during the course of treatment, it was specified that washing of the neighboring blocks should not be started until after the blocks under treatment had been completely grouted. Other than LS2, seam washing operations were carried out in working adits. At the washing point, the adit was enlarged to 2.5 m by 1.8 m with a length of 1.3 m to facilitate the washing operation.

For washing of seams in the right abutment, either Method A or Method C was adopted. Downward washing operation by Method C was carried out from the ex-cavated keyway surface.

For washing of seams below the riverbed elevation, either Method B or Method C was adopted. The downward washing operation was carried out along the seam either from the lowest tunnels and adits or from the excavated dam foundation.

Figure 6 through Figure 8 show the washing opera-tion using Method A, Method B and Method C respec-tively.
- Backfill grouting:

After washing of a block had been completed and thoroughly inspected, all the openings exposed to the working tunnel or adit were sealed with cement mortar and concrete blocks with grouting and vent pipes installed. Grouting with non-shrinking grout was started from the bottom and continued upward to fill the void. When the grout flowing from the vent pipe showed the same consistency as that of the incoming grout, the vent pipe could be closed. However, the grouting operation was maintained at a pressure of 0.05 MN/m² measured at the vent until the grout take was less than 1 liter for 15 minutes.
- Check hole drilling:

After the grouting operation had been carried out to a certain extent, a systematic drilling program to check the effectiveness of the treatment was started. A total of 632 check holes had been drilled and were examined. Of these holes, 520 were drilled perpendicular to the treated seam and 112 were parallel to the treated seam. If any defect appeared in the core, re-washing and re-grouting at the loca-tion of the core would be carried out until the re-drilled core showed that the defect had been completely eliminated. A part of these cores were also sent to the laboratory for direct shear test to check the shear strength of the contact face between the rock and the mortar.

Figure 6. Washing operation using Method A

Figure 7. Washing operation using Method B

Figure 8. Washing operation using Method C

4.3 Tunnel backfilling

After treatment work in a tunnel had been completed, the tunnel was backfilled section by section with concrete with each section having a length of about 15 m. Before concrete was placed, grout pipes and vent pipes were installed at the top of the tunnel. Backfill grouting at the pressure of 0.5 MN/m² was carried out, after the concrete had been placed for 14 days, to fill the voids outside the backfilled concrete.

4.4 Quantities and construction cost of seam treatment

The quantities and the cost of the various items of seam treatment work are summarized in Table 2. From this table, it can be seen that total cost for seam treatment for the Feitsui Arch Dam foundation was about US$ 7,575,174, or only 4.1 % of the total project direct cost of US$ 183,580,000.

5 EVALUATION OF SEAM TREATMENT RESULT

To verify the effectiveness of seam treatment work, the following evaluation programs were carried out:
1. Systematic drilling of cores of the treated seams to check the bonding condition and the shear strength of the contact face between rock and mortar.
2. Testing the permeability of the cores (for Method B treatment).
3. Installing a series of extensometers to monitor the displacement of treated seams.

Table 2. Quantities and construction cost of seam treatment

Work item	Location and method		Quantity	Construction cost (US$)
EXCA-VATION*	Left abutment		3,394 m	1,448,161
	Right abutment		1,149 m	490,259
	Subtotal:		4,543 m	1,938,420
SEAM TREAT-MENT	L. abut.	Method A	8,922 m²	1,327,910
	"	Method C	3,161 m²	724,128
	Riverbed	Method A**	84 m²	12,502
	"	Method B	4,752 m²	790,675
	"	Method C	22 m²	5,040
	R. abut.	Method A	53 m²	7,888
	"	Method C	1,600 m²	366,531
	Subtotal:		-	3,234,674
BACKFILL OF TUNNELS & ADITS	Left abutment		3,865 m	1,748,077
	Right abutment		1,446 m	654,003
	Subtotal:		5,311 m	2,402,080
Grand total:			-	7,575,174

*　Excavation for seam treatment only.
**　Method A is used for grouting cap.

5.1 Bonding and shear strength

The bonding condition between the rock and the mortar was visually inspected on rock cores drilled at a uniform distribution pattern from treated seams. Figure 9 shows the sample check cores for Method A, Method B and Method C. A total of 520 check cores were drilled perpendicular to the treated bedding seams in the left abutment. Some 95% of these check holes showed very good results. This indicated that the seamy materials had been thoroughly washed out and the grout fully backfilled the seam openings and was tightly bonded to the sound rock. Even most of the fine fissures subparallel to the seams were also treated during the course of treatment operation. Inspection of some 5% of unsatisfactory cores showed that the failure was most likely caused by incomplete washing or grouting defects. Re-washing and re-grouting of the location of the unsatisfactory cores were carried out and examination of the check holes of these further treated locations showed that the indications of unsatisfactory treatment had virtually been eliminated.

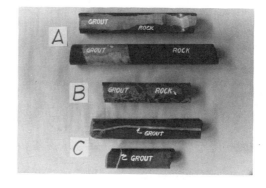

Figure 9. Sample check cores for Method A, Method B and Method C

To evaluate the shear strength of the contact face of the treated seams, direct shear tests of 192 rock core samples for Method A and 108 rock cores for Method C were performed. Figure 10 shows that the average peak shear strength of seams treated by Method A is far greater than the design value of Cp=1.1 MN/m², φp=38°. The average peak shear strength for Method C is even much higher. In addition, two sets of in-situ direct shear tests with a specimen of the size of 70 cm square were also performed for the main seam LS2 in the left abutment after treatment had been conducted. The average peak shear strength obtained is Cp=1.9 MN/m², φp=38°, which is still much higher than the design value.

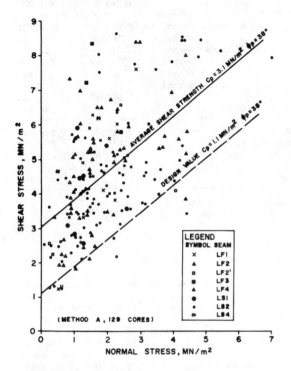

Figure 10. Peak shear strength of the contact face of the treated seams

5.2 Permeability

For seams treated through application of Method B, only seam-parallel cores were drilled. The in-situ Lugeon tests of 18 check holes were also conducted to check the permeability after treatment. The results of the tests showed that the Lugeon values obtained from most of the check holes were zero. The remainder was only about 0.4-1.3 Lugeon. This indicated that seams treated through application of Method B were sufficiently water-tight for effective control of seepage and piping.

5.3 Monitoring results

A total of 32 mechanical extensometers were installed in the abutments to monitor the displacement of treated seams. To date the reservoir water has risen to about 90% of the full level. The results, as revealed by these extensometers, showed that there is no relative movement along the treated seams.

6 CONCLUSION AND RECOMMENDATION

To meet with the need of treating seams inside the Feitsui Dam foundation, the method of washing the seams with high pressure water jets in adits and backfilling the washed seams with non-shrinking mortar was developed and applied with highly satisfactory results. The fact that the treatment work was smoothly carried out and was completed on time verifies the practicality of the method. During the course of the treatment work, the effectiveness of the work could be visually inspected and identified.

Boring of check holes after the seam treatment work also indicated sound bonding condition and good shear strength of the contact face between the mortar and the rock. These findings gratifiably reflect the reliability of this seam treatment method.

In addition to the above, the cost of the seam treatment work for the Feitsui Dam accounts only for 4.1 percent of the total direct cost of the project. Application of the new seam treatment method has therefore been proved to be economically justifiable, compared with the other traditionall treatment methods. At the current reservoir level of 90% full the Feitsui Dam has been observed and measured with various monitoring instruments and the results show that the structural behavior of the dam has been quite normal.

Based on the aforementioned facts, it may be concluded that the new seam treatment method employed for the Feitsui Reservoir Project is very successful and well deserves consideration for application to the construction of other dam and geotechnical projects. Particular attention should be given to 'Method C' which employs waterknife and waterdrill with washing pressure going up to 240 MN/m² and attainment of the highest washing efficiency, especially for treatment of papery thin seams. For construction of structures of high safety requirements, consideration of the application of this new seam treatment method will be highly desirable.

REFERENCES

Sinotech Engineering Consultants, Inc., Taipei Regional Water Supply Fourth Stage Development Project, 1978. Definite Plan Study for Water Source Development Sub-Project, Technical Report E: Foundation rock tests at damsite.
Sinotech Engineering Consultants, Inc., Feitsui Reservoir Project, Basic Design Reports:
- BDR-13, 1980. Experiments of bedding seam treatment for Feitsui Dam.
- BDR-23, 1980. Determination of the extent of seam treatment.
- BDR-28, 1980. Test on non-shrinking mortar for bedding seam treatment.
Sinotech Engineering Consultants, Inc., Feitsui Reservoir Project, Detailed Design Notes:
- DDN-03, 1981. Additional seam washing test by using waterdrill and waterknife.
- DDN-04, 1981. Working criteria for seam treatment.
- DDN-06, 1981. Controlled blasting tests for foundation treatment.
- DDN-08, 1982. Test of non-shrinking grout for bedding seam treatment.
- DDN-19, 1982. Revision of working criteria for seam treatment in right abutment.
- DDN-40, 1986. Progress report on seam treatment.
- DDN-41, 1986. Progress report on evaluation of treated seams and clayfilled joints.
ELC-Electroconsult, Tachien Dam Project, 1972. Dam left bank, facing concrete zone, seam treatment study and tendons design for actual excavation geometry.
Su, D.T., 1984. A brief introduction for the fault treatment at the foundation of Minghu Reservoir dam. Engineering Proc., Chinese Institute of Engineers. 57-5:28-33.
Taiwan Power Company, Kukuan Hydro Project, 1963. Construction report, chapter 5, section B: Strengthening of river banks. p.89-96.
Toshio, Fujiio 1970. Fault treatment at Nagawado dam. Commission Internationale Des Grands Barrages. p.371-393.

Geotechnical engineering work for the restoration of the Temple of Apollo Epicurius, Bassae

Travaux de génie géotechnique pour la restauration du temple d'Apollo Epicurius, Bassae
Die geotechnischen Arbeiten für die Restaurierung des Tempels von Apollo Epicurius, Bassae

S.D.COSTOPOULOS, Consulting Geotechnical Engineer, Athens, Greece

ABSTRACT: Extended disorder in the Temple of Apollo Epicurius, Bassae, is thought to be due to joint deformability of the rock foundation, triggered by transient ground water flow. This view is supported by in-situ measurements. Control of rock displacements is proposed to be done by an appropriate underground drainage system.

RESUME: Le désordre enregistre au Temple d'Apollo Epicurius, Bassae, a été attribué à la déformabilité des joints de la masse rocheuse excitee par le mouvement de l'eau souterrainne. Cette thèse a été confirmée par des mesures in-situ. L'auteur a proposé un système de drainage souterrain pour contrôler les mouvements du Temple.

ZUSAMMENFASSUNG: Die ausgedehnte Verdrangung des Tempels von Apollo Epicurius, Bassae, ist auf eine Fuge-Verlegung zuruckzufuhren, die aus einem unstaten Grundwasser-Lauf veranlasst wird. Die controlle dieser Verlegung wird, nach Vorschlag, mittels einer geeigneten unterirdischen Kanalisation ausgefuhrt.

1 INTRODUCTION

The need for restoration of the Temple of Apollo Epicurius, a chef-d'ouevre of Ictinos built in the midwest of Peloponnese, Greece, during the latter fifth century BC, has long been recognized as a task of primary national importance. When the Temple was discovered by Bocher in 1765 AD, 36 of the 38 columns of the peristyle were still standing; however, the rest of its structural elements (ceiling, entablature, walls of cella, coffers of pronaos and opistodomos etc) have all collapsed. Fifty years later, Coquerelle faced a nearly similar situation. Nowadays, things appear to be worse, as horizontal displacements at the top of most columns are recorded as large as 250 mm and floor differential settlements approximate 100 mm; hense,the stability of the Temple seeks to question (fig 1).This disorder was thought to be due to inadequate foundation on the underlying limestone (fig 3).

Certain restoration works have been performed within the first decade of the 20th century; however, on account of the Ministry of Culture, systematic investigations were only put forward after 1975. They included archaeological trenches, levelling measurements, seismic risk evaluation of the major area, several verifications of the structural integrity and stability of the monument, as well as an investigation programme of geological - geotechnical nature. This programme involved geological mapping, geoelectrical sounding, five exploratory boreholes, laboratory tests, Lugeon permeability tests, the installation of three piezometers and two in-situ direct shear tests (fig 2).

Critical evaluation of the above data was of considerable significance in identifying confidently the role of groundwater as a triggering mechanism in the evolution of differential ground and foundation movements, as well as in composing an appropriate geotechnical proposal for their control. This paper presents and discusses the basic points of this proposal followed by the first set of geodetic levelling results that support the proposal made by the author (Costopoulos 1984).

2 GROUND CONDITIONS

2.1 Geologic setting

The Temple has been founded on the top of an anticline with a N.NE-S.SW direction. The wider area is composed of thin bedded (100 mm), platy, white to gray-white limestones of the Upper Cretaceous, laminated with thin (10 mm) bands of red clay-shales and marls. The rock material appears semi-crystalline with local calcite or pyrite vanes and signs of incomplete fissility. Bed-

Figure 1. The Temple of Apollo Epicurius, Bassae

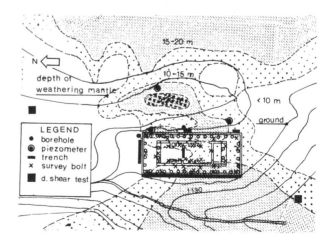

Figure 2. Geology and investigation zones on the site

ding appears fairly normal with dips 25 to 40 degrees East. The rock mass is intensively folded and strongly fractured or brecciated, often exhibiting karstic cavities and minor displacements. The bedrock is normally covered by up to 4 m of topsoil consisting of soft clayey material mixed with limestone fragments.

Over the narrow area of the Temple, the relief of the calcareous bedrock has been investigated by 88 geoelectrical soundings using the Sclumberger electrode array. The method revealed a strongly uneven and saddled weathering mantle of thickness fluctuating between 5 and 20 m (fig 2).

A series of seismographic and microtremor measurements over the site indicated the occurence of shallow earthquakes and a very short predominant period of the ground (0.13 sec).

2.2. Geotechnical data

The clayey material filling the joints - and used normally as earthfill - was classified as a highly plastic silty clay, with low sand content (13.5%) and occasional organic material (2.5%). Atterberg limits varied between 55% and 66% (w_L), 22% and 24% (w_p), water content was almost 33%, while the specific gravity of solids was 27 MN/m3. Consolidation tests indicated a range of 3×10^{-4} to 34×10^{-4} cm2/sec for the coefficient of consolidation and a range of 0.26 to 0.36 for the compression index. Initial void ratio was found to be as large as 1.34 in the consolidation tests.

Index tests on the limestone matrix displayed an unconfined compressive strength within the range of 5.6 to 48.2 MN/m2, a unit weight ranging between 25.6 and 26.8 MN/m3, while natural water content was measured as low as 0.48% up to 6.1%. Mohs hardness was found to be between 4 and 5.

In situ direct shear tests (fig 2), vertical to bedding on a 0.70x0.70 m rock mass, have shown that the overall shear strength parameters were assigned residual values between 25 and 40 degrees for the angle of shearing resistance and between 0.24 and 0.35 MN/m2 for the cohesion. Initial deformation modulus was shown to be between 3500 and 5000 MN/m2.

Core recovery varied between 20% and 50%, while RQD was very close to zero in most cases.

In situ permeability Lugeon tests were made every 3 m and indicated losses ranging between 100 and 800 lt/

order throughout the foundation of the Temple, which was in full contrast with the isodomic order of the superstructure. Crepidomas (fig 1,3) had been founded partly on narrow pinnacles and wide slots of weathered limestone, and partly on artificial earthfill of silty clay, with thickness varying between a few centimeters up to 3.2 m in the southwestern zone. The arrangement of the foundation stones was seen to be similar to that of the underlying bedrock (fig 3). On the contrary, the walls of cella had been founded on the calcareous bedrock, either through well cut stones, either directly on the excavated limestone, mainly in the northeastern zone.

Levelling measurements, albeit of relatively low precision, indicated that the columns of the peristyle were generally leaning towards the east; this finding opposes markedly to the recorded differential settlements of the lower parts of the crepidoma (euthynteria), seen in fig 3. On the stylobate level (fig 1), these settlements were of the order of 22 mm from east to west, while on the euthynteria level local differential settlements of the western side were approximating 10 mm. Of the same order of magnitude as above were the differential settlements between the east and the west walls of the cella. As a general rule, differential settlements were greater wherever the weathering mantle was deeper.

Under seismic loading of 12%g, pseudo-static evaluation of the structural integrity and stability of the monument provided evidence of low safety factors (below 1) against overturning in almost all of the columns. However, most of the architraves (fig 1) were found safe against bending rupture.

4 EVALUATION OF THE COLLECTED DATA

Due to harsch topography and intense fracturing of the limestone drainage is almost complete, such as no permanent ground water level appears, although high precipitations (1130 mm) are present in the area. However, folding and fracturing facilitate infiltration and transient flow of the rain water inside the rock mass, giving rise to mechanical and chemical erosion followed by karstic solution and dislodging of the infilled material. In fact, archaeological trenches revealed an extremely irregular bedrock·surface with sharp pinnacles and deep karst depressions (fig 3). On the o-

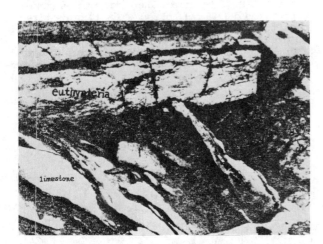

Figure 3. Foundation of the NW euthynteria

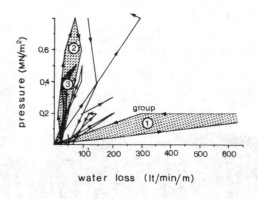

Figure 4. Results of Lugeon tests

min/m for pressures of 0.8 to 0.2 MN/m2 respectively. Significant up to total losses of water have also been recorded during the drilling procedure. No permanent ground water level has been found.

3 OTHER COLLECTED DATA

Archaeological trenches disclosed the occurence of dis-

ther hand, the results of in-situ permeability tests inside the weathering mantle infer the strong brecciation of the bedrock and the pertained conditions of a readily circulating groundwater. As seen in fig 4, the obtained Lugeon diagrammes can be classified into three groups : Group (1), detected in almost all tests at depths greater than 6 m, suggest filling of the cracks with fine material (Louis 1975), probably due to the washing out of the clayey topsoil; average water loss was 250 lt/min/m for a pressure of 0.25 MN/m2. In the same zone, Group (2) diagramme implies the opposite phenomenon, with an average water loss of about 100 lt/min/m for a pressure of 0.4 MN/m2. In other words, emptying the joints from the clayey material is likely to occur at heavy precipitations. Classical-form Group (3) diagramme appeared occasionally at depths between 7 and 10.5 m. It is worthnoting that points located on the horizontal axis insinuate cavities inside the rock mass and appear in almost all of the boreholes.

Dislodging of the infilled material is thus fairly common at depths greater than 6 m. In the upper zone of the weathering mantle brecciation of the limestone is so extended that no valuable tests could be performed. As a matter of fact, transient ground water flow is the main cause of rock mass deformability. Due to its inherent anisotropy - indicated by the wide range of the unconfined compressive strength - changing conditions of ground water flow often occur inside the mantle, pertaining the infilling or the washing out of the clayey material in the joints; two adjacent limestone plates are thus brought near or withdrawn, giving rise to joint deformability. However, local shear displacements of the plates and/or densification of the loose brecciated material could also be the cause of minor movements during an earthquake. No detectable settlements due to creep of the rock mass should be expected.

Previous levelling made evident that settlements do occur inside and under the foundation of the Temple. As the walls of cella had been "well" founded on bedrock, recorded settlements of this part can be attributed solely to the deformation of the rock mass. However, the far greater displacements of the crepidomas are normally due to the deformation of both the "poor" foundation and the rock mass. In the southern part, the occurence of a considerably thick artificial earthfill under the euthynteria aggravated the phenomenon, in a way that very high settlements are produced. Consequently, the differential settlements between the stylobate and the walls of cella are due to the foundation built up.

Settlement analysis under full loading that should had been present in the working lifetime of the monument has shown that, under a uniformily distributed stress on the foundation level ranging between 0.05

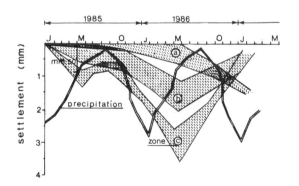

Figure 5. Monitoring results (first set)

and 0.20 MN/m2, the change in joint thickness (Duncan 1969) is able to produce overall settlements of the order of 1 to 2 mm (Costopoulos 1984). Intensity of joints was thought to be as large as one joint every 30 mm, according to the RQD values (Goodman & Smith 1980). On the other hand, analysis made evident that the consolidation settlements of the earthfill do not exceed 7% of its thickness and should have been produced during the construction period of the Temple. Movements of this magnitude are easily detectable by naked eye and it is more than certain that Ictinos himself had given the order to rectify such "constractional defects". As a matter of fact, recorded differential movements of the southwestern zone of the Temple is unlikely to be due to the consolidation of the earthfill under the euthynteria.

The estimated catchment area around the Temple is about 85000 m2, such as maximum annual precipitation (1130 mm) is able to produce an average dicharge of approximately 13 m3/hour. Given the mean site temperature of 16°C, the annual water loss due to evaporation and transpiration is estimated as high as 790 mm (Buttler 1957). As a consequence, annual dicharge due to surface runoff and infiltration of the rain water inside the rock mass should be 510 mm. However, the capacity of infiltration of the rock mass is several orders of magnitude greater than the amount of infiltrating water; in case of vertical streamlines, the rate of infiltration is the coefficinet of permeability, which was found as large as 10^{-}mm/sec from the executed Lugeon tests. Nevertheless, surface flow is thought to be three times less than the infiltration (Recordon 1974); as a matter of fact, the total amount of infiltrating water into the weathering limestone mantle is estimated 3.8 m3/hour, on an annual basis. Such a quantity of water is considerably large to be ignored and supports the author's view concerning the rock mass deformability due to dislodging of the infilled material. It is thus necessary to catch the infiltrating amount of water in order to control the deformability of the rock mass under the Temple.

5 GEOTECHNICAL PROPOSAL

Author's proposal consisted in the setting up of a drainage system formed of surface drains and vertical wells down to a depth of 20 m, discharging to an underground subhorizontal gallery along the outer perimeter of the Temple. On the other hand, surface runoff inside the Temple should be controlled by sheltering the monument; structural consultants designed a light acrylic membrane roofing suspended from 20 girder steel columns and hold down with the aid of 20 permanent 50 t rock anchors. Vertical well diameter was 300 mm giving an overall group dicharge capacity of 1.88 m3/hour (Harr 1962). The draining gallery was designed with an section of 4 m2 in order to provide a maximum dicharge capacity of 5.5 m3/hour. The vertical wells were put every 5 m and filled with gravel.

It is believed that a geotechnical system of this kind should restrict considerably the settlements of the weathering mantle and could probably stop them in the long run. Restoration works, whichever would be, could thus be easily performed. The proposal was considered to be improved by suggesting monitoring of the rock mass settlements.

6 MONITORING RESULTS

With the purpose of field evaluating the basic design assumption it was decided to implement an instrumentation programme to observe movements of the Temple's foundation. Precise levelling was used to detect settlements of the stylobate and the walls of cella on all sides; a grid of 35 survey bolts (fig 2) were attached to 11 points of the peristyle and to 24 points of the cella in both east and west sides. The plan coordinates of the bolts were established relative to a reper point, on a sound rock outcrop away of the Temple. Readings were made every five months using a WILD N3 apparatus; accuracy of the measurements were of the order of 5×10^{-4}m. Six inclinometer tubes and eleven stand pipes are to be installed in the near future to monitor rock mass displacements and ground water regime in the immediate viscinity of the Temple.

Besides the fact that the measurements made up to

now (January 1987) cover a very short period of time relative to the time required for the evolution of detectable rock mass displacements, certain discernible trends can be valuable. It is to be noted first that the settlement rate is generally aggravated during the winter period, where the precipitations increase (fig 5). On the other hand, major settlements appear in the southwestern zone of the Temple (zone c in fig 5), where "poor" foundation of the euthynteria is fairly extended. As a consequence, the walls of cella in that zone have also exhibited settlements, albeit of minor importance, due to the deformation of the rock mass (zone b in fig 5). These settlements apear to be greater than those of the eastern stylobate and the main part of the interior floor (zone a in fig 5), probably because of the greater depth of the weathering mantle in that zone.

These facts support the author's view concerning the rock mass deformability due to dislodging of the infilled material, which is much more pronounced in the zones of deeper weathering mantle.

It is believed that continuation of monitoring would give a clear picture of the response of the weathered rock mass to transient ground water regimes.

6 CONCLUSIONS

The restoration of the Temple of Apollo Epicurius, Bassae, has long been recognized as a task of primary importance in Greece. Based on extensive investigation data, the author proposed an underground drainage system to control the rock foundation movements that seeked to question the stability of the monument. The basic assumption was that joint deformability due to dislodging of the infilled material was the main cause of the recorded settlements. This was the result of critical evaluation of the collected data from archaeological trenches, levelling measurements, geological mapping, geoelectrical sounding and geotechnical insitu and laboratory investigations. Lugeon tests gave a clear picture of the triggering mechanism that provoqued ground settlements, while settlement analysis based on conventional models provided the order of magnitude of these settlements, mainly appeared inside the weathering mantle of the limestone bedrock. The drainage sytem, consisting of vertical wells and a subhorizontal gallery,was designed on the basis of simplified hydraulic models of the subsurface water regime. A first set of precise levelling,covering a period of approximately two years, provided valuable trends that support the basic design assumption.

It is believed that reasoning in the evaluation of collected data of various nature , supported by monitoring,is the key factor in making geotechnical decisions concerning delicate restoration works of monuments.

7 BIBLIOGRAPHY

Buttler,S.S. 1957. Engineering Hydrology. Prentice - Hall, p 356. London
Costopoulos,S.D. The restoration of Apollo Epicurius, Bassae. Geotechnical report (in greek). 1984. p 56, Athens.
Duncan,N. 1969. Engineering Geology and Rock Mechanics. Vol 2, p 270. L.Hill Pub. London
Goodman,R.E. 1980. Introduction to Rock Mechanics. p 468. J.Wiley & Sons. New York.
Harr,M.E. 1962. Groundwater and Seepage. p 315. Mc Graw-Hill. New York.
Louis,C. 1975. A study of groundwater flow in jointed rock and its influence on the stability of rock masses. Doctorate Thesis, University of Karlsruhe (english translation).
Recordon,E. 1971. Ecoulements souterrains. Deuxième partie. Notes de cours de 3ème cycle. Dépt de Génie Civil, EPFL, Lausanne.

Dimensioning and performance of a deep cutting for an express highway
Dimensionnement et performance d'une coupure profonde pour une autoroute
Dimensionierung und Verhalten eines tiefen Einschnitts für eine Autobahn

G.DENZER, Autobahnamt Baden-Württemberg, Stuttgart, FRG
L.WICHTER, Forschungs- und Materialprüfungsanstalt Baden-Württemberg, Otto-Graf-Institut, Stuttgart, FRG

ABSTRACT: Problems which occured during the opening of a deep cutting for a new express highway are described. Due to changes in the gradient the cutting is 20 m deeper than originally planned. Stability problems of the cutting slopes occured from the start. These were caused by a disadvantageous joint set which was not known from the exploration works. The stability problems forced the highway authorities to change the concept of supporting the slopes. In addition to the planned weathering protection of the slope surfaces (40000 m² of plantable revetment walls) around 2500 heavy prestressed rock anchors were necessary. The geotechnical situation in the cutting, the stability considerations, and the execution of construction are described. Some observations concerning the relaxation behaviour of the rock slopes are discussed.

RESUME: Des problèmes qui sont survenus pendant le creusement de deux déblais pour une autoroute nouvelle sont décrits. Par des modifications de la planification du tracé les déblais sont devenus 20 m plus profonds que projeté d'abord. Pendant le début des travaux de terrassement des problèmes de stabilité des escarpements sont survenus. Ils ont été causés par une bande des fissures très désavantageux qui n'ont pas été découvrés pendant les travaux d'exploration. Les problèmes ont obligé le maître d'oeuvre à changer la méthode de stabilisation des falaises. Environ 2500 ancres précontraints ont été nécessaires pour stabiliser les falaises. A l'origine seulement une protection contre la désagrégation (environ 40000 m² des murs de revêtement plantables) a été prévu. La situation géotechnique dans la coupure, les considérations sur la stabilité et la réalisation de la construction sont décrits. Quelques observations concernant le déplacement de la surface du rocher sont décrits.

ZUSAMMENFASSUNG: Es wird über die Schwierigkeiten berichtet, die sich während der Herstellung eines tiefen Einschnitts für eine neue Autobahn ergaben. Durch Änderung der Gradiente während der Planung wurde der Einschnitt 20 m tiefer als ursprünglich vorgesehen. Bereits zu Beginn der Aushubarbeiten ergaben sich Standsicherheitsprobleme bei den Böschungen. Sie wurden durch eine in bezug zur Böschungsoberfläche sehr ungünstig verlaufende Kluftschar hervorgerufen, die aus den Erkundungsbohrungen nicht bekannt war. Die Standsicherheitsprobleme zwangen den Bauherrn, das Konzept der Böschungssicherung zu ändern. Zusätzlich zum vorgesehenen Verwitterungsschutz (ca. 40000 m² begrünbare Futtermauern) wurde der Einbau von rd. 2500 schweren Felsankern erforderlich. Die geotechnische Situation im Einschnitt, die Stabilitätsbetrachtungen und die Ausführung der Bauarbeiten werden beschrieben. Einige Beobachtungen über das Entspannungsverhalten der Felsböschungen werden mitgeteilt.

1. INTRODUCTION

The new section of the federal express highway A 7 between the cities of Ulm and Würzburg will close a gap in the second long-distance highway connection between Scandinavia and the mediterranean countries. It ascends the escarpment of the Schwäbische Alb, which is around 190 m high, in two parallel tunnels and two cuttings. The cuttings, which are the subject of this report, are 55 m and 40 m deep. They are divided by a 50 m high rockfill dam. The shallower cutting has a length of 850 m and the deeper one is 1050 m long. In the early planning stages it was proposed to keep a constant gradient of 4 % in the tunnels and the cuttings. New findings about the accident risks in descending high-

way tunnels caused the planners to decrease the gradient in the tunnels to 3 %. This had the consequence that the following cuttings became much deeper than it had been originally planned. Fig. 1 shows a view of the deeper (northern) cutting which made the problems described in this report.

2. GEOLOGICAL SITUATION

The escarpment of the Schwäbische Alb is the most remarkable scarp in the South West German landscape. It consists of the limestone and marlstone series of the

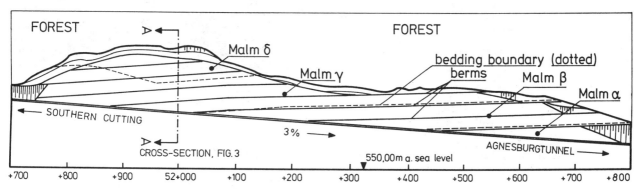

Fig. 1. View (longitudinal section) of the A 7 ascent of the Schwäbische Alb escarpment

Upper Jurassic (Malm, "White" Jurassic), which are in general much more resistant to weathering processes than the subjacent claystones and marls of the Middle and Lower Jurassic (Dogger, Lias). The good resistance to weathering, especially of the limestones which form the main part of the strata, is the reason for the existence of the escarpment. On the other hand its calcareous rocks have undergone karstification processes, and karst caves of different sizes can be found, mainly in the Malm β and Malm δ series. The highway cuts through the whole stratigraphic sequence of the Upper Jurassic. In fig. 1 the sequence of bedding relative to the highway can be seen. The dip of stratum is, in general, nearly horizontal, but may be locally 10° to 20° due to scattered fossile coral reefs.

3. ORIGINAL DESIGN OF THE CUTTING SLOPES

The geological forecast of the stability of the cutting slopes was based on the observations made in many quarries where the respective limestones and marlstones are quarried for building stones and as a raw material for the cement industry. The quarry slopes, which are nearly vertical, are stable over many decades with heights of 50 m and more. Weathering causes some rockfall, but the overall stability normally is not in question. With regard to the stability of the quarry slopes the highway authorities geological advisors proposed the following concept: The slopes of the cuttings should be divided into single slopes each of 11 m height and each with a dip of 79°. Between these slopes 5 m wide berms were proposed to allow the cultivation of the slope surfaces. The single berms and slopes should follow the bedding boundaries (which meant that they would not be parallel to the gradient of the highway). Fig. 2 shows a view of the northern cutting during the excavation measures (seen from the south), fig. 3 shows two cross-sections of the cuttings.

Fig. 2. View of the northern cutting, seen from the south

The strata resistant to weathering should be covered only by rockfall wire nets (Malm β, Malm δ). Completely weathered zones (near the surface of the terrain) should be supported using a system of stackable concrete parts as shown in fig. 4, or gabbion walls (fig. 5). Weatherable series of marlstones and limestones (Malm α/Malm γ) should be protected using retaining elements as shown in fig. 6. The vertical concrete piers which carry the soil-filled horizontal beams should be fixed at the rock mass using 8 m long single rod rock anchors. These anchors each with a nominal load of 490 kN should acquire their full load during the lifetime of the construction due to the increasing pressure caused by progressive weathering behind the retaining elements. They should be prestressed to only 50 kN.

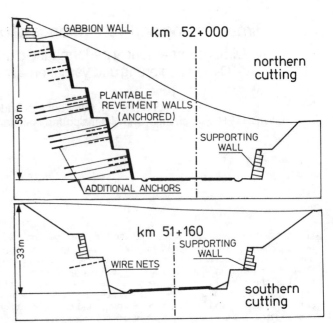

Fig. 3. Characteristic cross-sections of the cuttings

Fig. 4. Disintegrated supporting wall of 11 m height (southern cutting)

Fig. 5. Gabbion wall at the top of the northern cutting (view to the south)

4. NEW FINDINGS AT THE START OF THE CONSTRUCTION AND MODIFIED STABILIZATION CONCEPT

Shortly after the start of the excavation slope stability problems arose when the first 79°-slopes were worked. A joint system not known from the exploration

Fig. 6. Plantable retaining wall

borings was discovered. It had a strike nearly parallel to the longitudinal axis of the highway, and dipped
50° to 70° out of the slope. Fig. 7 shows these joints,
on which wedge-shaped rock bodies slipped into the
foundation-trenches.

Fig. 7. Slope-parallel rock joints

Besides the problems concerning the stability of the
partial slopes, the question of the total stability
of the northern cutting slope was raised. Unexpected
water inflow and a karst cavity in the highest section
(which had to be filled with about 1000 m³ of concrete), additionally complicated the construction. Because the earth works were being carried out with a great
speed the stabilization concept had to be changed
within a very short time. It was decided that the
stability of the individual 11 m high slopes should
be achieved by using prestressed multiple-rod rock
anchors. The anchor forces and anchor lengths should
be selected taking into consideration the different
cutting depths and geological structures in such a
way that the total stability of the cutting could also
be guaranteed by the anchoring.

5. ROCK MECHANICS INVESTIGATIONS

For the dimensioning of the additional anchoring, in
situ and laboratory rock mechanics tests were carried out. The shear parameters along the horizontal
marly beds were determined in situ. Three direct
shear tests on 80 cm diameter samples (see fig. 8)
gave the following values:

Table 1. Shear strength parameters

Test no.	Friction angle	Cohesion
1	34 °	80 kN/m²
2	31 °	95 kN/m²
3	36 °	35 kN/m²
Average	34 °	70 kN/m²

Before carrying out the shear tests slip planes were
induced in the samples by shearing them under a low
normal load and then pushing them back to the original
position. This was carried out in order to simulate
deformations of the rock masses due to unloading which
could have already caused some softening effects on the
marly layers.

Fig. 8. Direct shear test in situ

The shear parameters on the inclined joints were
checked using laboratory shear tests on completely separated rock joints sampled at the site. These tests
were carried out in a Robertson type shear apparatus
on samples with about 50 cm² shear plane area. The
average value of friction angle from 8 shear tests was
43°. In addition direct shear tests on clayey and silty joint fill materials were carried out. Friction
angles of 30° and cohesion values around 50 kN/m² were
obtained for these soils. In order to determine the
compressive strength of the jointed and layered rock
at the foot of the cutting two triaxial tests on rock
cores of 60 cm diameter were carried out using a
technique described by Wichter and Gudehus (1976). The
uniaxial compressive strength was determined to be
1,7 MN/m². The friction angle was 45°, and the cohesion value 350 kN/m². Fig. 9 shows the sampling of the
drilling cores and a specimen before the test.

6. STABILITY ANALYSIS

The stability analysises were carried out for 3 possible failure mechanisms:
  Sliding of a wedge-shaped rock body on the outward
  dipping joints (stability of the particular slope)

  Failure of the total slope on an inclined slip plane
  behind the grouted sections of the anchors, and on

Fig. 9 a. Sampling of a 0,6 m diameter rock core

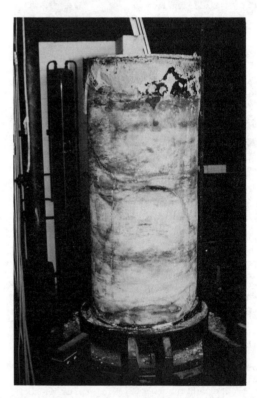

Fig. 9 b. Specimen before test

a horizontal layer

Local failure at the foot of the slope caused by exceeding the compressive strength of the jointed and fissured rock

Fig. 10 shows these mechanisms for a characteristic cross section in the northern cutting. Because of the possibility of finding large unknown karst cavities, locally very disadvantageous joint directions, or unexpected heavy water inflow, a very conservative estimate of the shear strength parameters and the joint plane structures was made.

The stability analysis for the individual slopes had to take into account that the rock structure near the

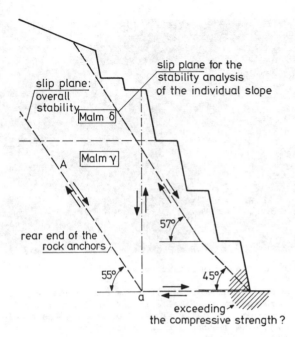

Fig. 10. Failure mechanisms

slope surface might have undergone some disaggregation due to blasting. Also any error in the estimate of the stability against sliding on the regionally clay-filled and flat joints could have endangered the lives of the workers. With the construction method (see chapter 7) and the type of retaining walls used it would have been very difficult to place any additionally required anchors in a height of some 10 m. So the anchor forces were determinated using the shear parameters $\phi = 30°$ and $c = 50$ kN/m² as characteristic values for the inclined joint surfaces.

The stability of the total slope, which had not been discussed before the beginnung of the works, determined the anchor lengths. It was assumed that failure would take place along the outward dipping joints and a marl layer at the foot of the cutting. The failure mechanism therefore was confined by these planes. The inclined slip plane A (see fig. 10) was moved hillwards until the stability calculations gave a stability factor of $\eta = 1,3$ (definition after Fellenius). Kinematic reasons require for this type of mechanism an intermediate slip plane through the lower edge (a) of the failure figure (Goldscheider, 1979). The shear parameters on this plane were set as 0. Using this method anchor lengths between 15 m and 22 m depending on the different slope heights were determined. In the highest section of the northern cutting slope, where a large karst cave had been found early after the start of the excavations, the anchor lengths were chosen without calculation as 30 m. The question of exceeding the compressive strength of the rock at the foot of the slope has been the subject of a finite element study and of some estimates based on the former geological overburden and stress distribution after the excavation. All these estimates showed that the stresses in the wedge were essentially lower than the compressive strength determined on large rock cores. After the excavation no sign of overstress could be seen even in the heavily karsted areas.

7. EXECUTION OF CONSTRUCTION

The excavation measures were carried out in such a way that on the downhill side of the cutting a trench was dug out down to the level of the highway as soon as possible. This trench was needed as a site access road and gave important informations about the geological structures. Then the single slopes were constructed in three steps of around 3 m height each. In the areas where additional anchoring was necessary the rock anchors were placed and stressed before the next step was started. Every three anchors in a horizontal row were stressed simultaneously and anchored on a horizontal concrete anchor beam. In order to drain the rock masses in some water-bearing sections drainage borings were installed from each excavation level ascending with about 10°. About 2/3 of the borings were 35 m long and reached behind the grouting sections of the anchors. Dependent on the quality of rock the slopes were then supported using the retaining structures as shown in the figures 4 and 5, or were only covered with a rockfall wire net. Karst cavities were filled with concrete if they were large enough to adversely affect the local stability. Zones extremely disintegrated were stabilized using rock bolts and shotcrete. In the southern cutting an additional row of anchors was necessary below the berm shoulder. They had to stabilize the heavily karsted and blast-disintegrated rock below the foundation of the retaining walls above, before the lower part of the cutting with a maximum height of 14 m was excavated.

8. ROCK SURFACE DISPLACEMENT DURING CONSTRUCTION

One of the main questions at the beginning of the anchoring measures was the amount of outward movement of the slope surface due to unloading in horizontal direction. This outward movement would have stressed the anchors in addition to the regular load, and would probably have lead to an overstressing which could not be tolerated. The multiple rod anchor type used does not allow a simple unloading of the anchors because the individual rod is fixed by wedges in the anchor head. There were two possibilities of keeping the anchor forces within the range of the allowed loads:

a) Overdimensioning of the anchors on the basis of an estimated horizontal outward movement, which will (probably) stress them to their nominal load after excavation.

b) Mounting of special half-circle shaped base-plates (with a thickness of the expected displacement due to unloading) under the anchor heads and removal of the plates, if the displacements occur.

The highway authority decided that both methods should be used. In cutting sections where the overall stability was discussed the method a) should be used to avoid any unloading over a large area. In cuttings where only local displacements due to karst holes or blasting shocks were expected the method b) was used. The extensometers distributed over the whole cutting surface showed a maximum horizontal displacement of about 10 mm outward in the areas where anchors were installed. Even taking into account that these extensometers cannot measure the total amount of displacement due to unloading (because they cannot be installed before the cutting is partially opened) the measured displacements were remarkably lower than the estimates (about 1/10). This might be probably an effect caused by the morphological situation (cutting in a scarp position).

9. EXPERIENCES AND RECOMMENDATIONS

The excavation of the two cuttings was carried out in the years 1984 to 1986. This means that the displacement behaviour of the slopes, and the stress development in the anchors, which will probably show some long-term effects, are not completely known at the date of this short report. A detailed report on the long-term behaviour will therefore be given in about two years.
  The experiences up to now may be summarized as follows:
a) Whenever possible the geological conditions for deep cuttings should be investigated using additional test pits which should open up the whole rock

mass of the cutting. The estimation of the rock quality and fabric using only drill cores may lead to wrong conclusions. Rock mechanics parameters as well as joint fabric can only be reliably determined, for cutting boundary conditions, in test pits. Large cuttings need test pits as well as tunnels, in many cases these pits can be made only during construction.
b) Multiple rod rock anchors allow high anchor forces and can be easily adapted to the geological situation (anchor forces as well as anchor lengths). Because of the mostly unknown behaviour of the rock mass due to unloading the possibility of reducing the anchor forces should be possible at least until the end of the earth works, even though this is more difficult than when using single rod anchors.
c) The research work concerning the displacements of cutting slopes due to unloading should be intensified to allow more exact predictions of these displacements.

REFERENCES

Goldscheider, M. 1979: Standsicherheitsnachweis mit zusammengesetzten Starrkörper-Bruchmechanismen (Stability analysis using combined rigid-body failure mechanisms). Geotechnik No. 2, p. 130-139

Wichter, L., G. Gudehus 1976. Ein Verfahren zur Entnahme und Prüfung von geklüfteten Großbohrkernen (A method for sampling and testing jointed large drill cores). Ber. 2. Nat. Tag. Felsmech. Aachen

Three-dimensional dynamic calculation of rockfalls
Calcul dynamique tridimensionnel de la chute de blocs rocheux
Räumlich dynamische Berechnung von Felsstürzen

F.DESCOEUDRES, Federal Institute of Technology, Laussane, Switzerland
TH.ZIMMERMANN, Federal Institute of Technology, Lausanne, Switzerland

ABSTRACT : A numerical model has been developed for the study of the spatial movement of blocks striking a three-dimensional topography. The generalized movement of such blocks is determined by fundamental equations of the dynamics of rigid solids. When a block touches the topographical surface, a condition of either instantaneous contact (impact) or permanent contact (rolling or sliding) is imposed upon it.

The geometry of the blocks may be defined as prisms with polygonal section or ellipsoids. The block is taken to be absolutely rigid and the topography has either an inelastic or plastic behaviour coupled with a surface roughness.

The model has been fitted to the results of a case-study of a real rockfall, in which a million cubic meters are involved. The study permits a delimitation of risk zones in cultivated areas and underlines practical protection steps which may be implemented when rockfalls are at a minimum due to winter freezing.

RESUME : Un modèle numérique a été developpé pour l'étude du mouvement spatial de blocs impactant une topographie tridimensionnelle. Le mouvement général d'un bloc est régi par les équations fondamentales de la dynamique du solide rigide. Quand le bloc touche la surface topographique, on impose des conditions de contact instantané (impact) ou permanent (roulement ou glissement).

La géométrie des blocs peut être choisie en forme de prisme à section polygonale ou d'ellipsoïde. Le bloc est considéré comme indéformable, la topographie présente un comportement au choc soit inélastique, soit plastique, tout en possédant une rugosité de surface.

Le modèle a été calé sur un éboulement réel de l'ordre de un million de mètres cubes, sur la base d'observations faites in situ. L'étude doit servir à délimiter les zones de risques dans des cultures en exploitation et préconiser des mesures de protection à mettre en oeuvre durant la période de gel hivernal où l'activité de chutes est pratiquement stoppée.

ZUSAMMENFASSUNG : Es wurde ein numerisches Modell zur Untersuchung der räumlichen Bewegung von Blöcken entwickelt, die auf eine dreidimensionale Topographie prallen. Die allgemeine Bewegung eines Blocks wird durch die Grundgleichungen der Dynamik starrer Körper beschrieben. Wenn der Block die Geländeoberfläche berührt, wird augenblicklicher (Stoss) oder Dauernder (Rollen oder Gleiten) Kontakt vorgegeben.

Die Blöcke können prismatische Form mit Polygon-Querschnitt oder ellipsoïdische Form aufweisen. Der Block wird starr angenommen, und die Geländeoberfläche verhält sich entweder inelastisch oder plastisch mit Berücksichtigung einer Oberflächenrauheit.

Das Modell wurde mittels der Ergebnisse eines reellen Felssturzes von 1 Mio m^3 Ausmass geeicht. Die Untersuchung erlaubt die Festlegung von Risikozonen in landwirtschaftlich genutztem Gelände und von Schutzmassnahmen, die während der Winterfröste, die die Felsstürze praktisch zum Stillstand bringen, ausgeführt werden können.

1. INTRODUCTION

Near rock cliffs in mountainous regions, falling rocks and rockslides make dynamic studies of possible block trajectories necessary for the evaluation of the extent of the threatened zones as well as for the definition of adequate measures of protection.

Until now, existing models of computation are essentially two-dimensional (CUNDALL, 1971, AZIMI et al., 1982, BOZZOLO and PAMINI, 1982, HACAR et al., 1977, LOPEZ-CARRERAS, 1981, PITEAU and CLAYTON, 1976). They simply apply to a ground profile selected in a vertical plane or along the line of the deepest slope, without being able to take care of the lateral dispersion of the block trajectories. In many cases of complex topography, with corridors, funnels, conic slopes or any other shape, the modelling of a single profile can be very misleading.

Faced with a real rockfall of considerable extent, in which a million cubic meters are involved, the authors have developed a numerical model for the study of the spatial movement of blocks striking a three-dimensional topography.

2. DESCRIPTION OF THE NUMERICAL MODEL

The numerical model applies rigid body dynamics to simulate the motion of ellipsoidal or polygonal blocks. The driving force is gravity. During its motion the block encounters successive impacts on the topography. These impacts involve frictional contact and elastoplastic response.

SOLUTION ALGORITHM

The solution algorithm is split into the following steps. For each time step,

Motion :

Compute the acceleration, velocity, displacement and position. Euler's equations apply, the first in any convenient referential, the second in a body fixed frame :

$$m \frac{d\underset{\sim}{v}}{dt} = \underset{\sim}{F}$$

$$\underset{\sim}{J} \frac{d\underset{\sim}{\omega}}{dt} = \underset{\sim}{M} + \underset{\sim}{N}(\underset{\sim}{\omega}) \qquad (2.1)$$

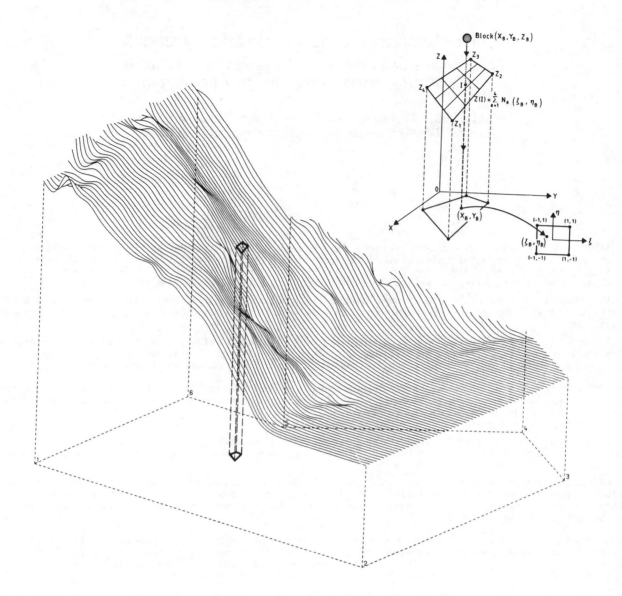

Figure 1. Discretization of the topography

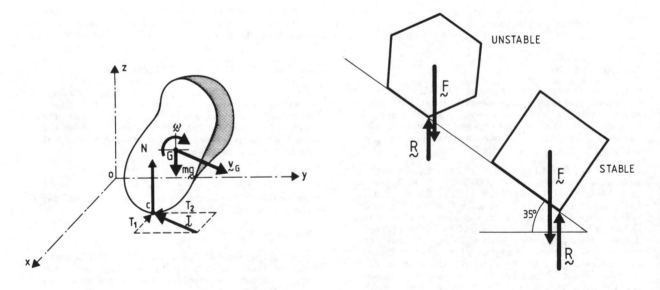

Figure 2. Equilibrium of block in contact   Figure 3. Influence of block shape

338

where m is the mass, $\underset{\sim}{v}$ the translation velocity of the center of gravity, F the applied force (- mg for gravity), $\underset{\sim}{J}$ the rotary inertia, $\underset{\sim}{\omega}$ the rotation velocity, $\underset{\sim}{M}$ the applied moment, $\underset{\sim}{N}\left(\underset{\sim}{\omega}\right)$ is a nonlinear expression in $\underset{\sim}{\omega}$.

The trapezoïdal algorithm is used for time discretization of Equations (2.1). This results in Equations (2.2).

$$\underset{\sim n+1}{a} = m^{-1} \underset{\sim n+1}{F} \quad ; \quad \underset{\sim n+1}{\dot{\omega}} = \underset{\sim}{J}^{-1} \left[\underset{\sim n+1}{M} + \underset{\sim}{N} \left(\underset{\sim n}{\omega} \right) \right]$$

$$\underset{\sim n+1}{v} = \underset{\sim n}{v} + \frac{\Delta t}{2} \left(\underset{\sim n}{a} + \underset{\sim n+1}{a} \right); \quad \underset{\sim n+1}{\omega} = \underset{\sim n}{\omega} + \frac{\Delta t}{2} \left(\underset{\sim n}{\dot{\omega}} + \underset{\sim n+1}{\dot{\omega}} \right)$$

$$\underset{\sim n+1}{d} - \underset{\sim n}{d} = \Delta t \underset{\sim n}{v} + \frac{\Delta t^2}{4} \left(\underset{\sim n}{a} + \underset{\sim n+1}{a} \right) = \underset{\sim}{\Delta d} \quad (2.2)$$

$$\underset{\sim n+1}{\dot{\theta}} = \underset{\sim n+1}{\dot{\theta}} \left(\underset{\sim n+1}{\omega} \right); \quad \underset{\sim n+1}{\theta} = \underset{\sim n}{\theta} + \frac{\Delta t}{2} \left(\underset{\sim n}{\dot{\theta}} + \underset{\sim n+1}{\dot{\theta}} \right) \quad ;$$

$$\underset{\sim n+1}{x} = \underset{\sim n}{x} + \underset{\sim}{\Delta d}$$

where $\underset{\sim n+1}{a}$ and $\underset{\sim n+1}{\dot{\omega}}$ are the accelerations at time $t = (n+1) \cdot \Delta t$, $\underset{\sim n+1}{d}$ the displacement, $\underset{\sim n+1}{\theta}$ the total rotation in Euler angles, and $\underset{\sim n+1}{x}$ the new position.

Localization :

Localize the block with respect to the discretized topography.

The topography has to be discretized. This is done as shown in Figure 1. The horizontal map is discretized into quadrilateral elements in a way which approximates the topography as closely as possible. Bilinear interpolation is performed to localize points on the surface within each element (Eqns. 2.3, 2.4) with respect to the coordinates of nodal points $\underset{\sim a}{X}$.

$$\underset{\sim}{X}(\xi, \eta) = \sum_{a=1}^{4} N_a (\xi, \eta) \underset{\sim a}{X} \quad (2.3)$$

$$N_a = 0.25 \left(1 + \xi_a \xi\right) \left(1 + \eta_a \eta\right) \quad (2.4)$$

$$-1 \leq \xi, \eta \leq 1 \; ; \; \xi_a, \eta_a = \pm 1$$

Possible impacts are identified by comparison of $\underset{\sim n+1}{x}$, the location of the block to $\underset{\sim}{X}$, the corresponding surface point.

Contact-impact :

If impact occurs, balance laws must be applied to determine the reflected motion (essentially $\underset{\sim}{v}^+$) from the incident motion (essentially $\underset{\sim}{v}^-$).

• Incident motion :

Define the incident motion as the one which occurs between t^- (limit of $t - \Delta t$ when Δt tends to zero) and $\bar{t}$ (paroxysm of the shock).

Assuming that the velocity of the contact point is zero at $\bar{t}$ (Eqn. 2.5.3), the following coupled equations apply, in a body fixed referential :

$$m \left(\underset{\sim}{\bar{v}} - \underset{\sim}{v}^- \right) = \underset{\sim}{I}^-$$

$$\underset{\sim}{J} \left(\underset{\sim}{\bar{\omega}} - \underset{\sim}{\omega}^- \right) = \underset{\sim}{r} \times \underset{\sim}{I}^- \quad (2.5)$$

$$\underset{\sim}{\bar{v}} + \underset{\sim}{\bar{\omega}} \times \underset{\sim}{r} = 0$$

The incident impulse, I^-, results from (Eqn. 2.5).

• Impact conditions :

A coefficient of restitution, e_n and a coefficient of friction are used to describe nonconservative impact conditions.

Let $\underset{\sim}{I} = \underset{\sim}{I}^- + \underset{\sim}{I}^+$ and split $\underset{\sim}{I}^-$ into normal and tangential components. The reflected impulse, $\underset{\sim}{I}^+$, is defined by :

$$I_n^+ = e_n I_n^-$$

$$I_t = 2 I_t^- \leq \mu I_n \quad \text{(see note)} \quad (2.6)$$

where :

$$\underset{\sim}{I}^+ = m \left(\underset{\sim}{v}^+ - \underset{\sim}{\bar{v}} \right) \quad (2.7)$$

by analogy with $\underset{\sim}{I}^-$.

The quantities e_n and μ can sometimes by assumed to be constants. Here, only μ is assumed constant and

$$e_n = \min \left(e_{ref}, \; m^{0.5} \cdot c/I_n^- \right) \quad (2.8)$$

where e_{ref} is the maximum value that e_n can take and is a function of the materials involved. c is a site-dependent constant equal to the maximum elastic energy of deformation of the ground considered to be elastic-perfectly plastic. c can be derived from theory or tuned until the maximum rebound duration corresponds to the one observed at the site. Obviously alternative definitions are possible.

Note : If Equation (2.6.2) is not satisfied by the initial solution then System (2.5) must be solved iteratively with Equation (2.5.3) replaced by $(I_t^-)^i = 0.5 \cdot \mu \cdot (I_n)^i$.

• Reflected motion :

Once the total impulse is known, the reflected motion results from :

$$m \left(\underset{\sim}{v}^+ - \underset{\sim}{v}^- \right) = \underset{\sim}{I}$$

$$\underset{\sim}{J} \left(\underset{\sim}{\omega}^+ - \underset{\sim}{\omega}^- \right) = \underset{\sim}{r} \times \underset{\sim}{I} \quad (2.9)$$

• Rolling and sliding (Fig. 2)

If v_n^+ is zero the block remains on the ground and Equations (2.10) apply.

$$m \underset{\sim}{a} = \underset{\sim}{F} + \underset{\sim}{R}$$

$$\underset{\sim}{J} \underset{\sim}{\dot{\omega}} = \underset{\sim}{r} \times \underset{\sim}{R} + \underset{\sim}{N} \left(\underset{\sim}{\omega} \right) \quad (2.10)$$

$$\underset{\sim}{a} + \underset{\sim}{\dot{\omega}} \times \underset{\sim}{r} = 0$$

The first two repeat the equations of motion with the ground reaction, $\underset{\sim}{R}$, added. The third maintains rolling contact. Again, this last equation is only valid if the normal and tangential components of the reactions, R_t and R_n, satisfy :

$$R_T \leq \mu R_n \quad (2.11)$$

If this is not true, Equation (2.10.3) must be replaced by $R_t = \mu R_n$ and System (2.10) will be solved iteratively.

• Arrest :

Friction progressively diminishes the velocity of the block. However, the dominant contribution to arrest is given by the shape of the blocks. Since the assumption is made that contact is always located at a single point, this point will always be at an edge if the block is polygonal. Therefore, the reaction, $\underset{\sim}{R}$, alternatively opposes and activates the motion and the necessary condition for a block to be able to stop is that its shape allows a stable static equilibrium (Fig.3).

3. APPLICATION TO THE ROCKFALL OF LES CRETAUX IN VALLIS

During the late spring of 1985, signs of instability of an important rock mass of carboniferous black schists were observed. The mass was situated some 1'000 meters above the Rhone Valley, near Riddes, with a mean dip of the general slope of about 40°. A site survey showed velocities of displacement on the order of one centimeter per day, with a large scattering. There was, however, no significant correlation with the weather conditions.

A rapid acceleration occurred in August, 1985 (Table 1), the rock mass loosened itself and many blocks of 1 to 10 m³ fell down on the woods and on the grapevines situated below.

Table 1. Upper rock mass velocities

Period	Velocity ahead [cm/day]	Velocity behind [cm/day]
28-31.08.85	400 to 100	300 to 80
01-15.09.85	100 to 50	80 to 30
16-30.09.85	50 to 15	30 to 13
01-10.10.85	15 to 10	13 to 8

Since then, the movements of the upper rock mass continue, continually checked by extensometers, with rare additional block falls (Fig. 4). The displacements are clearly affected by changes in season, especially the frost-thaw period. Carried away by underground waters and rains, the material deposited in the corridor runs out periodically in the form of a mud flow, now embanked and evacuated, in order to protect the cultivated areas.

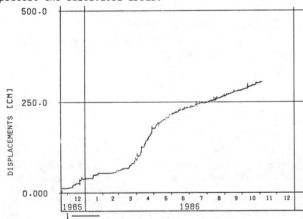

Figure 4. Les Crétaux - Upper rock mass displacements

Severe damages to the site resulting from the critical period of August, 1985, can still be seen in the woods; activities in the vineyard remain very dangerous and the main road passing through the site is still closed to traffic. Risks of further rockfalls are not yet under control.

Apart from the usual studies involving geological investigations, interpretation airphotos of and a detailed site survey, a quantitative analysis using the numerical model described above has been performed to delimit different risk zones and to design remedial measures which may be implemented when rockfall probability is at a minimum during winter freezing.

A view of the finite element mesh of the site topography is shown in Figure 5. After some preliminary tests for tuning the model with the observed block rebounds and velocities (estimated from cine-camera takes), a series of 8 runs was simulated.

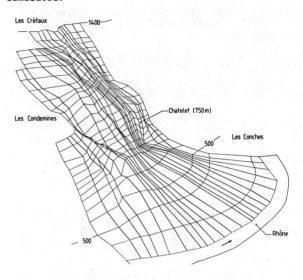

Figure 5. Les Crétaux - Perspective view of the discretized topography

The basic data is composed of the following information :
- Starting zone : 10 different block positions, spaced every 20 m, in front of the rock mass at the top of the corridor.
- Geometry of the block : hexagonal prisms (Fig. 6), best fitting the shape of the fallen blocks.
- Coefficient of restitution : $e_n = 0.85$ (corridor) $e_n = 0.40$ (vineyard).
- Size of the blocks : 4 m³ (mean observed size) or 10 m³ (max.).
- Initial falling height: 2 or 20 m above topography
- Coefficient of friction: $\mu = 0.50$ (probable value) or $\mu = 0.30$ (unfavourable value).
- Coefficient of plastic shock (site tuning) : $c = 331$ kg$^{1/2} \cdot$m$\cdot$s^{-1} or 2 c or c/2.

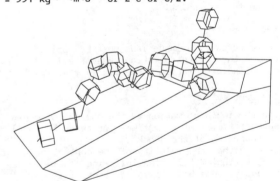

Figure 6. Perspective view of a block striking a simplified 3D-surface

From the large quantity of graphical output obtained for one single blockfall, some interesting results have been selected for presentation :
- The block trajectory with points of contact (Fig.7)
- The variation of the coefficient of restitution (Fig. 8).
- The variation of the incident normal velocity (Fig. 9).
- The variation of the tangential velocity after impact (Fig. 10).

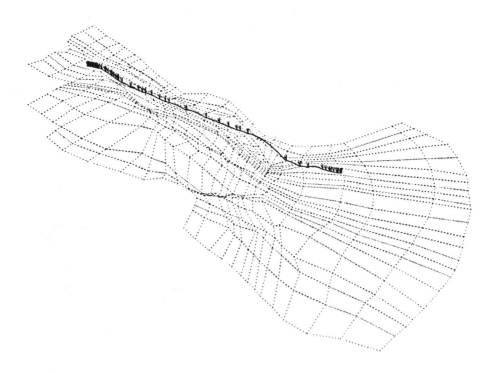

Figure 7. Les Crétaux - Block trajectory for the case : Hexagonal Prisma
0.99 m × 0.73 m × 0.86 m, c = 331 kg$^{1/2}$·m·s^{-1}, e$_n$ = 0.85 - 0.40, μ = 0.50. Initial position 5 - h = 2 m

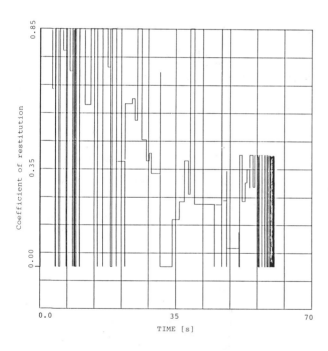

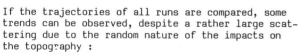

Figure 8. Les Crétaux - Coefficient of restitution, for the same block

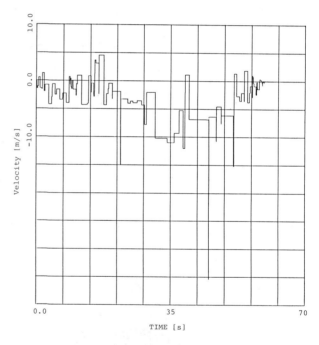

Figure 9. Les Crétaux - Incident normal velocity, for the same block

If the trajectories of all runs are compared, some trends can be observed, despite a rather large scattering due to the random nature of the impacts on the topography :
- Bigger blocks have a straighter trajectory, due to an inertial effect and their arrest point lies somewhat higher on the site, according to the plastic response of the ground to impact (Figs. 11 and 12).
- The initial falling height has little or no influence, the potential energy being rapidly dissipated in the first plastic shocks.
- A decrease in the coefficient of friction causes an increase in the velocities and the length of the rebounds.
- A decrease in the coefficient of plastic shock causes a decrease in the velocities and an earlier arrest, sometimes even in the upper part of the corridor.
- The zone of arrest is in good agreement with the observed real one, as are the velocities (mean value is on the order of 25 m/s), the total falling time (50 to 100 seconds), the height of the flights above surface (only a few meters in general), etc.

(Figs. 11 and 12 show the case of high c values, thus the trajectories are somewhat longer than the actual ones)

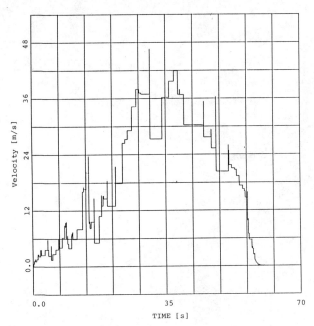

Figure 10. Les Crétaux - Tangential velocity after impact, for the same block

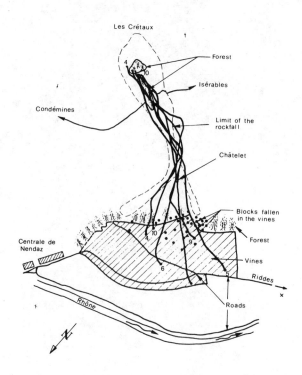

Figure 12. Les Crétaux - Trajectories of 8 4 m³ - blocks (Horizontal projection)

6. CONCLUSION

The authors' three-dimensional numerical model is still under development in order to improve its predictive power in cases where actual data of rock falls on a specific site are lacking; for example, additional laboratory tests are needed to check the theoretical derivation of the ground response under elastic-plastic impact. Nevertheless, its current stage of development is sufficient to allow for its use as a valuable tool for quantitative analyses and design considerations for protective measures.

ACKNOWLEDGEMENTS

The contribution of E. Davalle in the computer implementation of the model and the financial support of the canton of Vallis are gratefully acknowledged.

BIBLIOGRAPHY

Azimi C., Desvarreaux P., Giraud A. 1982. Méthode de calcul de la dynamique des chutes de blocs. Application à l'étude du versant de la montagne de La Pale (Vercors). Bull. Liaison LCPC 122.
Bozzolo D., Pamini R. 1982. Modello matematico per 10 studio della caduta dei messi. Laboratorio di Fisica Terrestre ICTS Lugano-Trvano.
Cundall P.A. 1971. A computer model for simulating progressive, large-scale movements in blocky rock systems. ISRM Symp. Rock Fracture Nancy : II-8.
Hacar Bemitez M.A., Bollo M.F., Hacar Rodriguez M.P. 1977. Bodies Falling Down on Different Slopes - Dymanic Study IX ICSMFE Tokyo II, vol. 2. 91-95.
Lopez Carreras C. 1981. Dynamique des déplacements de matériaux sur les pentes naturelles. Inst. polyt. de Lorraine (S.p.).
Piteau D.R., Clayton R. 1976. Rock fall characteristics affecting the design of protection measures. Landslides Analysis and Control, Special Report 176 : 218-228.

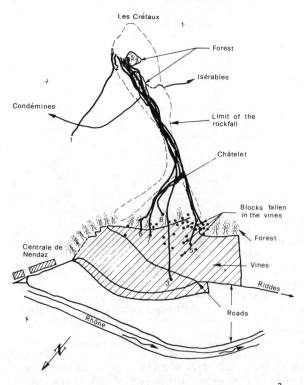

Figure 11. Les Crétaux - Trajectories of 8 10 m³ - blocks (Horizontal projection)

Baugrunduntersuchungen für tiefe Einschnitte in Gebirge geringer Festigkeit

Site investigations for deep cuts in rock masses of low strength
Etudes du sous-sol pour des tranchées profondes en terrain de résistance faible

K.EFFENBERGER, Ingenieur-Geologisches Institut, Dipl.-Ing. S.Niedermeyer, Bundesrepublik Deutschland
S.NIEDERMEYER, Ingenieur-Geologisches Institut, Dipl.-Ing. S.Niedermeyer, Bundesrepublik Deutschland
W.RAHN, Ingenieur-Geologisches Institut, Dipl.-Ing. S.Niedermeyer, Bundesrepublik Deutschland

ABSTRACT: In the submitted paper, recommendations on the performance and on the report of site investigations are presented by means of diagramms and tables, valuable especially for cuts in the course of large scale projects in railway and highway engineering. Furthermore the procedure of geotechnical site investigation during construction is demonstrated. A field example completes the explanations.

RESUME: Recommandations pour l'exécution et pour le rapport des études du sous-sol en liaison avec tranchées encors des grands projets dans la construction des routes et des chemins de fer sont présentées dans cet article en moyen des diagrammes et des tableaux. En plus le procédé en rapport avec les recherches pendant la réalisation d'une construction est presenté. Un exemple complète les presentations précédentes.

ZUSAMMENFASSUNG: Es werden Empfehlungen für die Vorgehensweise des Baugrundsachverständigen bei der Erstellung des Baugrundgutachtens sowie für die Baugrunduntersuchung von Einschnitten im Zuge von Großprojekten des Verkehrswegebaues anhand von Diagrammen und Tabellen vorgestellt. Des weiteren wird die Vorgehensweise bei baubegleitenden Untersuchungen aufgezeigt. Die Ausführungen werden durch ein Beispiel ergänzt.

1. EINLEITUNG

Es wird allgemein anerkannt, daß Einschnitte ebenso Ingenieurbauwerke sind wie z. B. Tunnel und Brükken. Dennoch wird häufig für die Planung von Einschnitten ein im Verhältnis zu Brücken und Tunneln geringer Erkundungsaufwand für erforderlich gehalten. Dies hat zur Folge, daß wichtige planerische Entscheidungen in die Bauphase verlegt werden. Diese Vorgehensweise birgt die Gefahr in sich, daß in der Bauphase umfangreiche Modifikationen und fallweise auch Änderungen der Planungskonzeption erforderlich werden. Dies kann neben Kostenmehrungen und Bauzeitenverzögerungen auch zu vertragsrechtlichen Auseinandersetzungen führen, was insbesondere bei tiefen Einschnitten gravierend sein kann. Erfahrungen zeigen, daß derartige "Überraschungen" sich häufig durch intensivere Baugrunderkundungen vermeiden lassen.

Aufgrund der Erkenntnisse aus den weitgehend abgeschlossenen Baumaßnahmen für den rund 100 km langen Südabschnitt der Neubaustrecke Hannover-Würzburg der Deutschen Bundesbahn werden Empfehlungen für die Vorgehensweise des Baugrundsachverständigen bei der Erstellung des Baugrundgutachtens sowie für die Baugrunduntersuchung von Einschnitten im Zuge von Großprojekten des Verkehrswegebaues vorgestellt. Diese Empfehlungen gelten insbesondere für tiefe Einschnitte mit Einschnittstiefen ≥ 20 m unter humiden klimatischen Verhältnissen. Sie wurden für Gebirge geringer Festigkeit, bestehend aus söhlig bis flach ($0 \ldots \leq 30°$) gelagerten Sedimentgesteinen entwikkelt. Das Gebirge geringer Festigkeit ist durch eines oder mehrere der nachstehend aufgeführten Merkmale gekennzeichnet:

(1) Wechsellagerungen, bestehend aus Peliten und Sand- und/oder Kalksteinen. Die einachsigen Druckfestigkeiten der Festgesteine sind wie folgt nach ISRM zu klassifizieren:
  - Pelite
    sehr gering bis gering fest
  - Sandsteine
    geringe bis hohe Festigkeit, je nach Kornbindung und Verwitterungsgrad
  - Kalksteine
    mäßige bis hohe Festigkeit, je nach Tonanteil und Verwitterungsgrad

(2) Einheitlich aufgebaute Gebirge aus Festgesteinen sehr geringer bis geringer Festigkeit

(3) Auflockerung und Entfestigung des Gebirges infolge gravitativer Massenbewegungen (Hangkriechen, Auslaugungsstrukturen, Rutschungen)

(4) Tiefreichende Paläoverwitterung verbunden mit einer starken Gebirgsentfestigung

(5) Verwitterungsempfindliche Festgesteine

(6) Quellfähiges Gebirge

(7) Kriechfähiges Gebirge

Es wird bei den nachstehenden Betrachtungen davon ausgegangen, daß die Trassenführung und die Trassenelemente feststehen.

2. EMPFEHLUNGEN FÜR DIE VORGEHENSWEISE DES BAUGRUNDSACHVERSTÄNDIGEN BEI DER ERSTELLUNG DES BAUGRUNDGUTACHTENS

Die zu empfehlende Vorgehensweise des Baugrundsachverständigen bei der Erstellung des Baugrundgutachtens wird in Abbildung 1 gezeigt. Zu dieser Abbildung sind im wesentlichen folgende Anmerkungen zu machen:

(1) Wie die Erfahrungen zeigen, müssen zur Optimierung der Baugrunderkundung und der Aussagen im Baugrundgutachten erschöpfende Datenerhebungen vor der eigentlichen Baugrunderkundung ausgeführt werden. Dies bedeutet, daß diese nicht auf den trassennahen Bereich beschränkt werden dürfen. Das heißt z. B., daß unter Umständen selbst Aufschlüsse, die in einer Entfernung von 10er-km von der Trasse liegen, geotechnisch beurteilt werden müssen. Diese Vorgehensweise ist eigentlich selbstverständlich und auch einleuchtend. In der Praxis wird dies jedoch oftmals aus zeitlichen und finanziellen Gründen nicht beachtet. Diese "Sparsamkeit" kann dem Bauherren unter Umständen teuer zu stehen kommen.

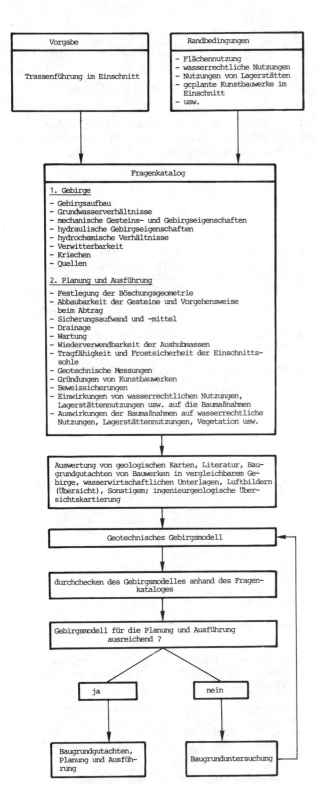

Abb. 1: Vorgehensweise der Baugrundsachverständigen bei der Erstellung des Baugrundgutachtens

(2) Obwohl ein Einschnitt oftmals als Linienbau-
werk betrachtet wird, ist zur geotechnischen Beurteilung ein dreidimensionales Gebirgsmodell erforderlich. Dies bedeutet, daß die Bohrungen und
Schürfe räumlich angeordnet werden müssen; jedoch
sind starre, geometrische Bohrraster nicht unbedingt sinnvoll. Nach unseren Erfahrungen sollte
der Baugrundsachverständige durch die Datenerhebungen vor der Baugrunderkundung und die Ergebnisse
der Baugrunduntersuchungen in der Lage sein, alle
ca. 25 - 50 m ein geotechnisches Querprofil zu erstellen, ohne die "Geophantasie" zu bemühen.

(3) Obwohl die Baugrunduntersuchung zur Erstellung des Baugrundgutachtens erschöpfend sein sollte, ist es fallweise zweckmäßig, verschiedene Untersuchungen in die Bauphase zu verlegen.
   Die Datenerhebungen vor der Baugrunduntersuchung
und die Baugrunduntersuchung selbst müssen die in
Abbildung 2 aufgeführten Randbedingungen und Eingabedaten für das Baugrundgutachten liefern. Diese
sind Grundlage für die ebenfalls in Abbildung 2
aufgezeigte Entwicklung eines geotechnischen Gebirgsmodells sowie für die gutachterlichen Aussagen
und Empfehlungen.

3. EMPFEHLUNGEN FÜR DIE BAUGRUNDUNTERSUCHUNG

3.1 Umfang

Die Baugrunderkundung dient dazu, die Lücken, die
nach der Auswertung der vorangegangenen Datener-
hebung noch vorhanden sind, zu schließen. Dies bedeutet, daß der Untersuchungsumfang von dem vorliegenden Datenmaterial sowie von den jeweiligen Randbedingungen abhängt. Des weiteren ist der Untersuchungsumfang häufig auch durch zeitliche und finan-
zielle Restriktionen begrenzt. Eine "Normierung"
des Untersuchungsumfanges - wie verschiedentlich
vorgeschlagen wurde - ist im allgemeinen nicht
sinnvoll.

3.2 Untersuchungsmethoden

Empfehlungen zu den Untersuchungsmethoden sind in
Tabelle 1 enthalten. Hierzu sind i. w. folgende
Anmerkungen zu machen:
   (1) Auf als "generell erforderlich bzw. empfeh-
lenswert" eingestufte Untersuchungen kann natürlich verzichtet werden, wenn diese bereits durchgeführt wurden (z. B. Untersuchungen im Zuge wis-
senschaftlicher Arbeiten).
   (2) Unter Ausnahmefälle werden solche Fälle verstanden, die im Zusammenhang mit geologischen Besonderheiten (z. B. Hangkriechen, Karst, Erdfälle
usw.), besonderen Randbedingungen (z. B. künstliche
Hohlraumbauten in unmittelbarer Trassennähe) oder
mit anderen nur selten auftretenden Fragestellungen
und Phänomenen stehen.
   (3) Der Einsatz geophysikalischer Verfahren in der
Baugrunderkundung ist unangemessen wenig verbreitet.
Nach unseren Erfahrungen hat sich z. B. die Refraktionsseismik als hervorragendes Hilfsmittel für die
räumliche Erkundung der Felsauflockerungszone bewährt.
   (4) Obwohl die Verwitterbarkeit in Gebirge geringer Festigkeit häufig eine wichtige Rolle spielt,
ist die Prognose der Verwitterbarkeit nur auf der
Grundlage gängiger Versuche nicht selten proble-
matisch. Man wird also häufig nicht darum herum-
kommen, die Verwitterbarkeit des Gebirges in natürlichen und künstlichen Aufschlüssen intensiver zu
studieren.
   (5) Für die Beurteilung der Scherfestigkeit des
Gebirges wurde in den letzten Jahren eine Versuchs-
technologie zur Prüfung von Großbohrkernen im Labor
entwickelt. Über diese Technologie haben Natau et
al (1983) anläßlich des 5. Internationalen Kongresses der ISRM berichtet. Für dieses Verfahren wird
zur Zeit eine ISRM-Empfehlung ausgearbeitet.

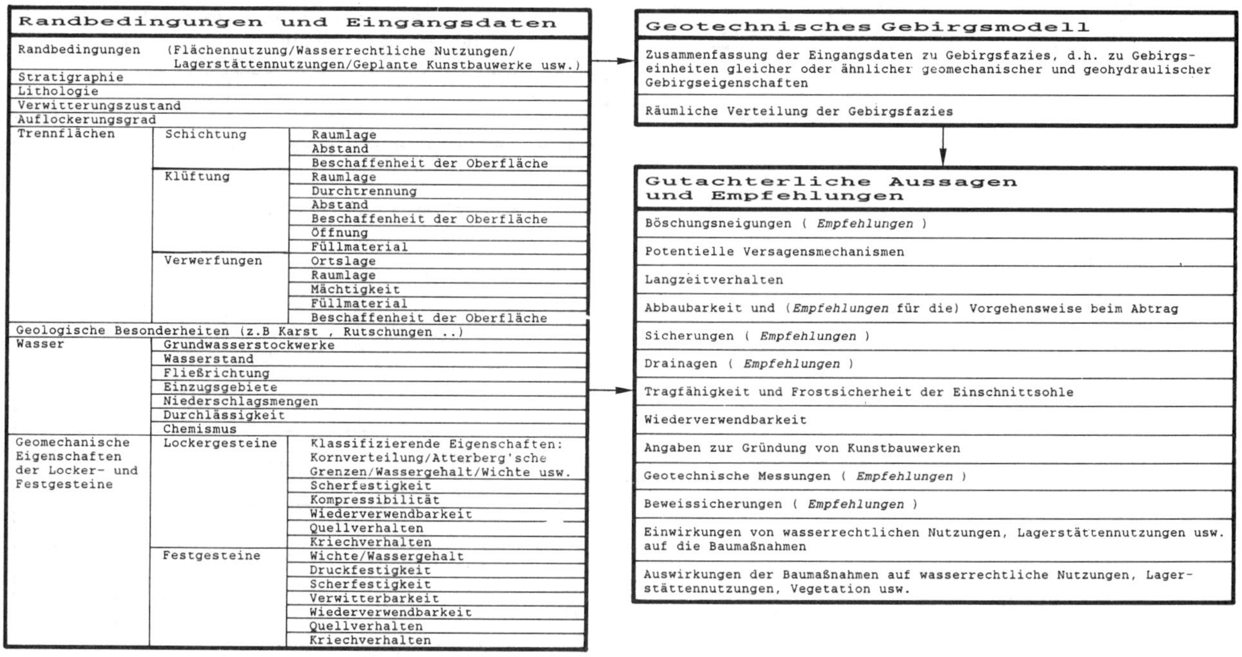

Randbedingungen und Eingangsdaten

Randbedingungen	(Flächennutzung/Wasserrechtliche Nutzungen/Lagerstättennutzungen/Geplante Kunstbauwerke usw.)	
Stratigraphie		
Lithologie		
Verwitterungszustand		
Auflockerungsgrad		
Trennflächen	Schichtung	Raumlage
		Abstand
		Beschaffenheit der Oberfläche
	Klüftung	Raumlage
		Durchtrennung
		Abstand
		Beschaffenheit der Oberfläche
		Öffnung
		Füllmaterial
	Verwerfungen	Ortslage
		Raumlage
		Mächtigkeit
		Füllmaterial
		Beschaffenheit der Oberfläche
Geologische Besonderheiten (z.B Karst , Rutschungen ..)		
Wasser	Grundwasserstockwerke	
	Wasserstand	
	Fließrichtung	
	Einzugsgebiete	
	Niederschlagsmengen	
	Durchlässigkeit	
	Chemismus	
Geomechanische Eigenschaften der Locker- und Festgesteine	Lockergesteine	Klassifizierende Eigenschaften: Kornverteilung/Atterberg'sche Grenzen/Wassergehalt/Wichte usw.
		Scherfestigkeit
		Kompressibilität
		Wiederverwendbarkeit
		Quellverhalten
		Kriechverhalten
	Festgesteine	Wichte/Wassergehalt
		Druckfestigkeit
		Scherfestigkeit
		Verwitterbarkeit
		Wiederverwendbarkeit
		Quellverhalten
		Kriechverhalten

Geotechnisches Gebirgsmodell

Zusammenfassung der Eingangsdaten zu Gebirgsfazies, d.h. zu Gebirgseinheiten gleicher oder ähnlicher geomechanischer und geohydraulischer Gebirgseigenschaften

Räumliche Verteilung der Gebirgsfazies

Gutachterliche Aussagen und Empfehlungen

Böschungsneigungen (*Empfehlungen*)

Potentielle Versagensmechanismen

Langzeitverhalten

Abbaubarkeit und (*Empfehlungen* für die) Vorgehensweise beim Abtrag

Sicherungen (*Empfehlungen*)

Drainagen (*Empfehlungen*)

Tragfähigkeit und Frostsicherheit der Einschnittsohle

Wiederverwendbarkeit

Angaben zur Gründung von Kunstbauwerken

Geotechnische Messungen (*Empfehlungen*)

Beweissicherungen (*Empfehlungen*)

Einwirkungen von wasserrechtlichen Nutzungen, Lagerstättennutzungen usw. auf die Baumaßnahmen

Auswirkungen der Baumaßnahmen auf wasserrechtliche Nutzungen, Lagerstättennutzungen, Vegetation usw.

Abb. 2: Inhalt des Baugrundgutachtens

Tab. 1: Empfehlungen für die Untersuchungsmethoden

Untersuchungsmethoden		generell erforderlich	generell empfehlenswert	fallweise erforderlich/empfehlenswert	in Ausnahmefällen erforderlich/empfehlenswert
ingenieurgeologische Detailkartierung		x	-	-	-
Detail-Luftbildauswertung		-	-	x	-
Kernbohrungen	vertikal	x	-	-	-
	schräg	-	-	x	-
Schürfe		x	-	-	-
Sondierungen		-	-	x	-
Oberflächengeophysik	Refraktionsseismik	-	x	-	-
	Reflexionsseismik	-	-	x	-
	Geoelektrik	-	-	x	-
	andere Verfahren	-	-	-	x
Untersuchungen im Bohrloch	geophysikalische Bohrlogs	-	-	x	-
	Bohrlochseismik (z.B. cross-hole-Messungen)	-	-	-	x
	optische Sondierungen	-	-	x	-
	Lugeon-Tests	-	-	x	-
	Pumpversuche	-	-	x	-
	Auffüllversuche	-	-	x	-
Instrumentierung von Bohrlöchern	Pegel	-	-	x	-
	Piezometer	-	-	-	x
	Verformungs- und Neigungsmeßeinrichtungen	-	-	-	x
Laborversuche	Lockergestein Klassifizierung	x	-	-	-
	Scherfestigkeit	x	-	-	-
	Wiederverwendbarkeit	-	x	-	-
	Kompressibilität	-	-	x	-
	Quellverhalten	-	-	-	x
	Kriechverhalten	-	-	-	x
	Festgesteine Klassifizierung (Punkt-Last-Tests, einachsige Druckversuche)	x	-	-	-
	Scherfestigkeit (Gestein)	-	-	x	-
	Scherfestigkeit (Trennflächen)	-	x	-	-
	petrographische Zusammensetzung	-	x	-	-
	Verwitterbarkeit	-	x	-	-
	Wiederverwendbarkeit	-	x	-	-
	Quellverhalten	-	-	-	x
	Kriechverhalten	-	-	-	x
	hydrochemische Eigenschaften	-	-	x	-
Großversuche	Scherfestigkeit (Gebirge)	-	-	-	x
	Scherfestigkeit (Trennflächen)	-	-	-	x

4. BAUBEGLEITENDE UNTERSUCHUNGEN

Das Ablaufschema baubegleitender Untersuchungen ist in Abbildung 3 enthalten. In diesem Diagramm sind Umplanungen, die nicht durch den Baugrund bedingt sind, nicht erfaßt.

Baubegleitend sollte auf jeden Fall eine ingenieurgeologische Kartierung des durch die Baumaßnahmen aufgeschlossenen Gebirges durchgeführt werden. In den Empfehlungen der ISRM für Baugrunderkundungsmethoden wird die baubegleitende ingenieurgeologische Kartierung von Einschnitten als erforderlich eingestuft. Sie dient der Überprüfung der Aussagen im Baugrundgutachten auf der Grundlage der tatsächlichen Gebirgsverhältnisse und ermöglicht somit dem Planer und Bauherrn bei gravierenden Abweichungen von der Prognose schnell zu reagieren. Des weiteren wird hiermit eine Grundlage für nachträgliche Planungen von ergänzenden Baumaßnahmen im Einschnitt und etwaiger späteren Sanierungsmaßnahmen geschaffen. Leider wird die baubegleitende ingenieurgeologische Kartierung nicht überall standardmäßig durchgeführt. Wichtig, häufig, aber unbeachtet, ist, daß die Kartierergebnisse in Planunterlagen - ähnlich wie es für Hohlraumbauten üblich ist - niedergelegt werden. Wem nützt schon eine Notizblattsammlung?

5. BEISPIEL

Die Baugrunduntersuchung für einen tiefen, ca. 530 m langen Einschnitt zwischen 2 Eisenbahntunneln bestand aus folgenden Untersuchungen (vgl. auch Abbildung 4):
(1) ingenieurgeologische Detailkartierung
(2) Detail-Luftbildauswertung
(3) 12 Kernbohrungen
(4) 2 Schürfe
(5) Refraktionsseismik
(6) Laborversuche zur Ermittlung von geomechanischen Eigenschaften der Locker- und Festgesteine

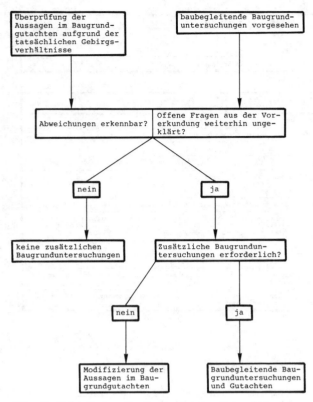

| Überprüfung der Aussagen im Baugrundgutachten aufgrund der tatsächlichen Gebirgsverhältnisse | baubegleitende Baugrunduntersuchungen vorgesehen |

| Abweichungen erkennbar? | Offene Fragen aus der Vorerkundung weiterhin ungeklärt? |

nein — ja

| keine zusätzlichen Baugrunduntersuchungen | Zusätzliche Baugrunduntersuchungen erforderlich? |

nein — ja

| Modifizierung der Aussagen im Baugrundgutachten | Baubegleitende Baugrunduntersuchungen und Gutachten |

Abb. 3: Ablaufschema baubegleitender Untersuchungen

    (6.1) Lockergesteine
       - Klassifizierung
       - Scherfestigkeit
    (6.2) Festgesteine
       - einachsige Druckfestigkeit
       - Scherfestigkeit (Gestein und Trennflächen)
       - petrographische Zusammensetzung
       - Verwitterbarkeit
       - Quellverhalten
    (7) dreiachsige Druckversuche an Großbohrkernen (Scherfestigkeit des Gebirges).

Das Gebirge setzt sich aus Sand- und Ton-/ Schluffsteinfolgen des Oberen Buntsandsteins, der von zum Teil mächtigen Fließerden überlagert wird, zusammen. Die Fließerdenüberlagerung erschwerte die ingenieurgeologische Detailkartierung erheblich. Besonders effektiv waren die refraktionsseismischen Messungen zur räumlichen Erkundung der Felsauflokkerungszone. Die Ergebnisse dieser Untersuchungen zeigten, daß der Einschnitt im wesentlichen innerhalb der hier sehr mächtigen Felsauflockerungszone liegt (Abbildung 4). Es wurde ein geotechnisches Gebirgsmodell entwickelt und die Kennwerte für die Standsicherheitsberechnungen entsprechend dem geomechanischen Gebirgsverhalten gewählt (Tabelle 2). Das Langzeitverhalten der hier sehr gering festen Ton-/Schluffsteine wurde aufgrund von Beobachtungen an bestehenden Böschungen sowie von Verwitterungsversuchen durch die Annahme einer 3 - 4 m mächtigen, ± parallel zur Oberfläche der Entwurfsböschung verlaufenden Verwitterungsschwarte (Lockergestein) berücksichtigt. Vorgabe des Bauherren waren freie, wartungsarme Böschungen. Die Standsicherheitsberechnungen ergaben erforderliche Böschungsneigungen von 1 : 2 und 1 : 1,7 (bergseitige Böschung) bzw. 1 : 1,8 (talseitige Böschung). Zur Vermeidung der damit verbundenen starken Eingriffe in die Landschaft wurde umgeplant. Die ehemals höchsten Böschungsbereiche werden nun auf einer Länge von ca. 250 m mittels Tunnel, der zum Teil in Deckelbauweise erstellt wird, durchörtert.

Tab. 2: Kennwerte für die Standsicherheitsuntersuchungen der Böschungen

Gebirgsfazies	γ (kN/m³)	φ' (°)	c' (MN/m²)
Fließerden (QL); Konsistenz ≥ halbfest; Teufen: bis ca. 1,5 m u. Geländeoberfläche (GOK)	20,0	29	0,006
Fließerden (QL); Konsistenzen halbfestfest; Teufen: ca. 1,5 m u. GOK bis zum 1. Refraktionshorizont	21,5	25	0,020
Rötquarzit (so4Q)	22,0	30...35*	0,005.. 0,010*
Untere Röttonsteine (so3T) oberhalb des 1. Refraktionshorizontes	21,5	25	0,020
Untere Röttonsteine (so3T) unterhalb des 1. Refraktionshorizontes	22,0	25	0,035
Grenzquarzit (so3Q)	23,5	30...35*	0,010*
Plattensandstein (so2)	23,5	30...35*	0,010*
Parallel zur Oberfläche der Entwurfsböschung sich ausbildende Verwitterungsschwarte in den Unteren Röttonsteinen (so3T)	20,0	29	0,006

* "ideelle" Kennwerte für böschungsstatische Berechnungen

6. DANKSAGUNGEN

Wir möchten der Deutschen Bundesbahn, Bundesbahndirektion Nürnberg, Herrn Abteilungspräsidenten Dipl.-Ing. Nußberger, für die Erlaubnis danken, Informationen zu verwenden, die im Auftrag der Deutschen Bundesbahn erarbeitet wurden.

Quellenverzeichnis

Natau, O., Fröhlich, B., Mutschler, T. (1983). Recent developments of the large triaxial test. Proc. 5th Int. Cong. ISRM, Vol 1, pp A65-74, Melbourne, Australia

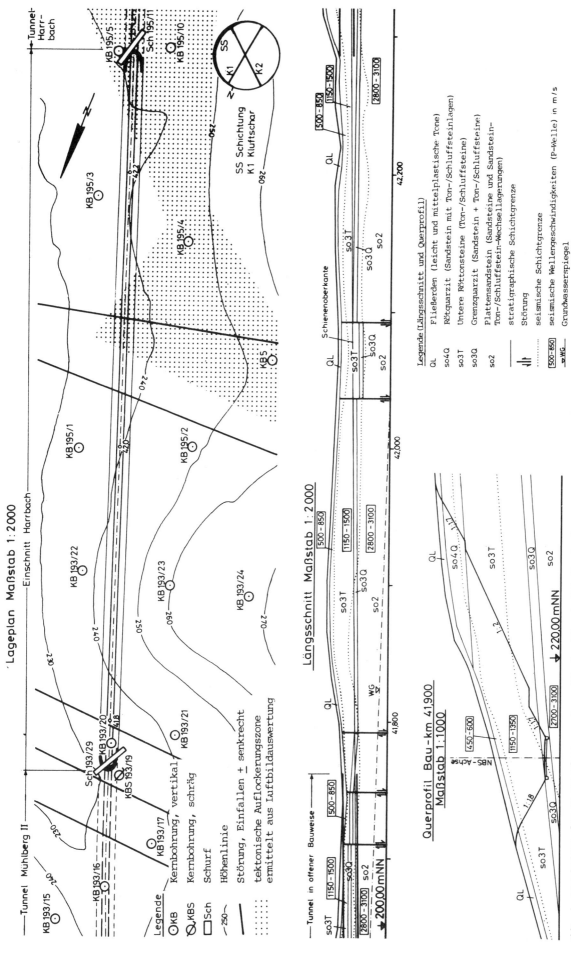

Abb. 4: Ergebnisse der Baugrunduntersuchung für den Einschnitt Harrbach

Stability investigations for ground improvement by rock bolts at a large dam
Etudes de stabilité d'une fondation de barrage améliorée par ancrages passifs
Standsicherheitsuntersuchungen für eine mittels Felsbolzen verbesserte Talsperrengründung

P.EGGER, Rock Mechanics Laboratory, Swiss Federal Institute of Technology, Lausanne
K.SPANG, Rock Mechanics Laboratory, Swiss Federal Institute of Technology, Lausanne

ABSTRACT : On the island of Sardinia, Italy, the construction of a 100 m high concrete gravity dam has been planned. The rock encountered after the excavation of the overburden exhibited poorer material properties than anticipated. As a result neither the global nor the local dam stability could be assured. Investigations showed the extent of the necessary rock improvement. Various possibilities have been investigated and a solution with passive anchors has been selected. But the traditional calculation methods were unable to show the effect of passive anchors in a sufficiently correct way. With the aid of a 3-dimensional finite element calculation the attempt has been made to determine the 3-dimensional effect of an anchor in the surrounding rock. In addition large scale in-situ shear tests have been carried out as a calculation check.

These investigations in conjunction with several security mesures permitted the design of a pattern of rock bolts able to assure the necessary improvement of the foundation rock of the dam.

RESUME : Sur l'île de Sardaigne en Italie, la construction d'un barrage-poids a été prévue. Puisque les propriétés géotechniques trouvées pendant l'excavation de la roche étaient moins bonnes que prévues, ni la stabilité globale, ni la stabilité locale n'étaient assurées. Des calculs à l'aide de cercles de glissement et aux éléments finis montraient les dimensions d'une amélioration de la roche nécessaire. Différentes possibilités d'amélioration ont été examinées et une solution par ancrages passifs a été choisie. Mais les méthodes traditionnelles de la prise en compte d'ancrages passifs dans les calculs ne donnaient pas satisfaction. A l'aide de calculs aux éléments finis, la tentative a été faite de déterminer l'effet 3-dimensionnel des ancrages dans la roche avoisinante. En plus, des essais de cisaillement in-situ à grande échelle ont été effectués pour vérifier les résultats des calculs.

Les études effectuées complétées par certaines mesures de sécurité permettaient de dimensionner les ancrages passifs pour assurer l'augmentation nécessaire de la stabilité du barrage.

ZUSAMMENFASSUNG : Auf der Insel Sardinien, Italien, ist der Bau einer 100 m hohen Gewichtsstaumauer geplant. Da sich beim Ausbruch der Felsgründung ungünstigere Kennwerte für den Fels ergaben als ursprünglich angenommen, war weder die globale noch die lokale Standsicherheit ausreichend gewährleistet. Gleitkreis- und ebene Finite-Element-Berechnungen zeigten das Ausmass einer erforderlichen Felsverbesserung auf. Verschiedene Möglichkeiten wurden untersucht und eine Lösung mit passiven Ankern gewählt. Ihre rechnerische Erfassung erwies sich jedoch als schwierig. Mit Hilfe einer 3-dimensionalen Rechnung wurde versucht, den räumlichen Effekt eines Ankers im umgebenden Fels zu ermitteln.

Zusätzlich wurden zur Ueberprüfung der Rechnung grossmassstäbliche In-Situ-Scherversuche durchgeführt. Mit Hilfe der verschiedenen Untersuchungen war es schliesslich möglich, in Verbindung mit einer Reihe von Sicherheitsmassnahmen eine Ankerung für die geforderte Erhöhung des Standsicherheit der Staumauer zu entwerfen.

1. INTRODUCTION

In the southwest of the island of Sardinia, Italy, the construction of a concrete gravity dam 100 m high and 600 m long has been planned in order to confine the TIRSO river for irrigation purposes. The horizontal forces resulting from the confinement, have to be transferred into the rock mass by means of the foundation footing. This produces high stresses near the dam rock interface. The rock encountered after the excavation of the overburden exhibited poorer material properties than anticipated. The global and the local stability of the affected rock mass have been examined, the latter with the aid of the finite element method. The section of Rock Mechanics of the Institute for Soil Mechanics, Rock Mechanics and Foundation Engineering at the Swiss Federal Institute of Technology in Lausanne, Switzerland, had been charged with the execution of these calculations.

2. DIMENSIONS AND GEOLOGY

2.1 Dimensions of the dam

Figure 1 shows a plan and Figure 2 a cross section of the construction. The dam consists of 15 m large blocks with 4 m wide cavities at the block borders.

Fig. 1. Plan of the TIRSO dam

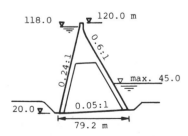

Fig. 2. Cross section at the highest part of the dam

349

2.2 Geological conditions of the foundation rock

The rock encountered in the planned foundation area consists mainly of strongly laminated metamorphic gneisses and granites. The various joint systems have very different orientations and the joint distances are in the small to medium range. The joints are either closed or filled with cohesive gouge material. The rock mass under consideration is moderately to highly weathered.

Stability investigations of large rock sections, therefore, have to take a high degree of separation and random jointing into account.

2.2.1 Originally assumed geotechnical properties

Based on the preliminary geotechnical investigations the following material properties had been proposed :

angle of friction	$\phi = 40^{\circ}$
cohesion	c = 0.2 MPa
modulus of elasticity (depth 0 to 20m)	E = 4000 MPa
modulus of elasticity (depth 20 to 40m)	E = 6000 MPa
modulus of elasticity (depth over 40m)	E =10000 MPa

2.2.2 Revised geotechnical properties

In order to check the material properties of the rock, six large scale in-situ-shear tests have been carried out. Various places in the planned foundation area with different degrees of weathering had been selected for these tests by the geological adviser. As a result of the tests the following properties of the foundation rock were found :

angle of friction	$\phi = 30^{\circ}$
cohesion	c = 0.32 MPa
modulus of elasticity (depth 0 to 20m)	E = 1000 MPa
modulus of elasticity (depth 20 to 40m)	E = 2000 MPa
modulus of elasticity (depth over 40m)	E = 4000 MPa

3. REVIEW OF THE DAM STABILITY BASED ON THE NEW MATERIAL PROPERTIES

3.1 Global stability

The review of the global stability has been carried out by the project's author, Prof. ing. F. Arredi, Rome, by conventional methods such as slip circles and slip surfaces in critical parts of the foundation rock.

These investigations resulted in some cases showing a factor of global safety inferior to the smallest allowed value of 2.0.

3.2 Local stability of the rock mass

3.2.1 General information

Concentrated forces and moments caused by the dam's dead weight and the water pressure, have to be transmitted to the rock. As a consequence, high compressive stresses are likely to occur in the downstream part of the dam foundation and small compressive or even tensile stresses in the upstream portion. Therefore, the impact of the changed rock conditions had to be examined particularly for these critical zones, and the extent of the latter with respect to length and depth had to be defined.

3.2.2 Description of the selected procedure

The requested investigations have been done using a computer program for geotechnical purposes based on the finite element method. Plane strains were assumed for the calculations. In the case of elastoplastic material behaviour the Mohr-Coulomb failure criterion was used.

First, the stability of the dam was examined for an ideally elastic rock material and then in each element the local safety factor was determined.

Following this a calculation was carried out with the assumption of an elastic-ideally plastic material behaviour.

The finite element mesh was composed of 186 nodes and 120 elements (Fig. 3).

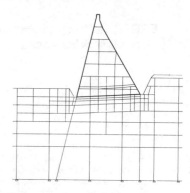

Fig. 3. Finite element mesh for the evaluation of the local stability

3.2.3 Loads

In a first calculation step only the dead weight of the rock mass was considered. The following steps simulated the excavation of the dam, the last of which representing the final state without water. In the case of service load forces resulting from impounding the 100 m high dam were taken into account, the uplift pressures corresponding to the Italian Standards.

3.2.4 Results

The finite element calculations described above resulted in node displacements and element stresses. The stresses are presented in Figure 4 for a typical case. Figure 5 shows the extent of the plastified area of the rock mass.

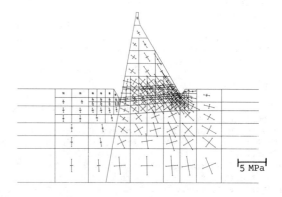

Fig. 4. Principal element stresses in the final state

Fig. 5. Plastified zones of the rock mass

For a better quantitative interpretation of the results a safety factor (M.S.) was defined. This factor is given by the ratio between stresses at failure and the existing ones (Fig. 6).

350

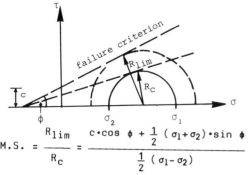

$$M.S. = \frac{R_{lim}}{R_C} = \frac{c \cdot \cos\phi + \frac{1}{2}(\sigma_1 + \sigma_2) \cdot \sin\phi}{\frac{1}{2}(\sigma_1 - \sigma_2)}$$

Fig. 6. Definition of the safety factor M.S.

The calculation of the safety factor for all elements resulted in a large zone where the required value M.S. = 1.5 was not attained (Fig. 7). This zone extended over the whole foundation area and reached considerable depth on the downstream toe. Some elements were even in a state of tension.

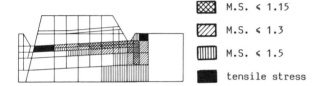

▨ M.S. < 1.15

▨ M.S. < 1.3

▥ M.S. < 1.5

■ tensile stress

Fig. 7. Zone of low safety factor M.S.

4. METHODS FOR IMPROVING THE DAM STABILITY

The stability calculations based on the revised rock properties proved that neither a local safety of 1.5 nor a global safety of 2.0 could be ensured for the dam foundation. Therefore various methods for improving the dam stability were studied.

4.1 Comparison of possible methods

An evaluation of the possible improvement methods included an assessment of their efficiency in increasing the safety as well as economical and practical aspects. The fact that almost all of the site installations had already been set in place had to be taken into consideration. Different possibilities for increasing the dam stability are discussed in the following :

4.1.1 Widening of the dam

This simple method of increasing the foundation safety could be achieved only to a limited extent because of practical constraints. For example, the concrete mixing plant would have to be displaced, if the dam were enlarged considerably.

4.1.2 Prestressed anchors

To increase the dam stability to the required degree, the dam itself and the stilling basin would have to be anchored. Finite element calculations with various arrangements of tendons resulted in anchor forces of at least 20'000 kN per meter of dam. The following points discourage the use of prestressed anchors :
1. The necessity of extremely heavy anchors. Development work would have to be done for the anchors as well as for ancillary equipment such as jacks. This additional work increases costs.
2. Longtime experience with this kind of heavy anchors is extremely limited.
3. Continuous and regular checking of the anchors by

experienced staff would be necessary.

4.1.3 Rock improvement by grouting

Water pressure tests showed a very low permeability of the foundation rock; this was confirmed by frequently observed clay filled joints. Consolidation grouting was therefore excluded.

4.1.4 Rockfill dam

This alternative for obtaining an increased dam stability was excluded for cost reasons.

4.1.5 Passive anchoring by rock bolts

The following arguments are in favour of this improvement method :
1. Rock bolt steel is much less sensitive to corrosion than prestressing steel.
2. Passive anchoring demands no extra monitoring in addition to usual dam monitoring.
3. It is less expensive than the other solutions.
4. If necessary, additional safety is provided by the possibility of subsequent anchoring in the cavities between the blocks.
5. Existing technology and experience in rock bolting can be used.
The sum of these positive arguments finally led to the acceptance of the above method for increasing the dam stability, in spite of the limited rock bolting experience for structures of this size.

4.2 Consideration of passive anchors in calculations

Current knowledge and research experience do not supply satisfactory calculation methods for the consideration of systematic patterns of passive rock bolts nor for the determination of the bearing capacity of a single bolt. Several investigators are working on this problem but have not yet arrived at generally accepted solutions. However, a sufficiently exact estimation of the anchoring effect in the present case could be accomplished by the following combined approach :
1. Interpretation of existing publications on the problem such as Londe, Bonazzi (1974), Dight (1982), Egger, Fernandez (1983), Schubert (1984) and Spang (1986/1).
2. Use of experiences and results of research work currently in progress at the Swiss Federal Institute of Technology, Lausanne (Spang (1986/2)).
3. Innovative investigations and calculations applied to this problem, presented in this paper.
The effect of a single bolt has been evaluated by a finite element study of a 3-dimensional anchor-rock-model as well as by site tests. Then, these results had to be extrapolated to the large scale ground improvement of the dam foundation.
Based on test results and publications (Wullschläger, Natau (1983), Egger (1978) and Bjurström (1974)), the influence of fully-bonded passive anchors on the rock mass has been assumed to increase the cohesion of the rock. However, this effect is not valid along the bolt axes. Therefore, an isotropic increased cohesion was assumed for the rock mass with weakness directions parallel to the anchors. In these directions, no improvement was considered. The same assumption was used in the determination of the safety factor.

4.3 Determination of the local dam stability on the base of an increased cohesion

To assure a global safety factor of 2.0 for the most unfavourable block, Prof. Arredi determined a required increase in cohesion of 0.6 MPa. The investigations described in chapter 3 were repeated with consideration of the ground improvement by passive rock bolts. Figure 8 shows zones with local safety factors inferior to 1.15 and 1.5 respectively.

Two elements are influenced by the notch effect at the downstream toe of the dam and have a M.S. of less than 1.15. However, this effect can be diminished by a thick concrete sheet. Two other elements remain slightly below the required safety factor of 1.5, but the overall effect of the envisaged rock improvement was deemed satisfactory.

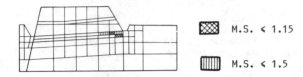

M.S. < 1.15

M.S. < 1.5

Fig. 8. Safety factor M.S. in the rock elements after bolting (increase in cohesion of 0.6 MPa)

The investigations described above supplied the rock mass improvement necessary to guarantee the required global and local stability of the dam. As a second part of the investigations, design criteria for the passive anchors had to be defined. For this purpose numerical studies were performed using a 3-dimensional finite element model, and a series of large scale site tests were carried out.

5. NUMERICAL STUDIES ON A THREE-DIMENSIONAL ANCHOR-ROCK MODEL

5.1 General information

The aim of these investigations was to determine the local effect of a single anchor bolt, loaded by an imposed joint displacement. In particular, the 3-dimensional fields of deformations and stresses in the surrounding rock were to be determined, which would not have been possible with 2-dimensional calculations. In contrast to the former investigations only a small section of the foundation rock was considered in the FE model.

5.2 Description of the code

The finite element programme used for this study allows for consideration of elastic or elasto-plastic material behaviour and for calculation of 2- and 3-dimensional models. In the present case, the rock mass was considered as a quasi-continuum, the high degree of jointing being taken into account by a general reduction of the material properties.

The programme calculates node displacements, element deformations and element stresses. The material properties of the ideally elastic elements are defined by the modulus of elasticity and Poisson's ratio. To describe the behaviour of the elasto-plastic materials, the Drucker-Prager failure criterion has been used.

5.3 Description of the model

The model had been designed in order to simulate direct shear tests. There are two rock blocks which are separated by a joint, represented by a layer of elements with material properties such as determined in the in-situ shear tests. The anchor (ø = 40 mm) and the mortar (ø = 100 mm) are represented by an annular array of elements. Because of model and loading symmetry, only one half of the system is considered. The horizontal dimensions (1.25 m × 1.00 m) correspond to the planned bolting pattern and a steel density of 1 $^0/_{00}$.

Two models, one with anchors normal to the shear surface and the other with inclined anchors (25^0 with respect to vertical) have been investigated. The following figures show the 3-dimensional model which consists of 450 nodes and 180 elements, each of them defined by 20 nodes.

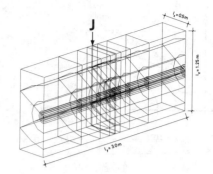

Fig. 9. 3-dimensional representation of the model J ... joint = shear surface

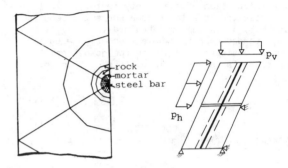

Fig.10. Plan of the 3-D-model   Fig.11. Loading system

5.4 Material properties and loads

Four different materials have been used in the model. The steel and the ciment mortar were considered to behave as elastic, the rock and the joint as elasto-plastic materials. The material properties are as follows :

steel : E = 210'000 MPa ν = 0.30
mortar : E = 20'000 MPa ν = 0.20
rock : E = 1'000 MPa ν = 0.25
         c = 1.60 MPa ϕ = 30^0
joint : E_j = 100 MPa ν = 0.25
         c_j = 0.32 MPa ϕ_j = 30^0.

For an appropriate consideration of the stresses in the upstream (us) and the downstream (ds) part of the dam foundation, loading cases with a high vertical stress of 2.2 MPa (ds) and without any vertical stress (us) have been calculated. Based on the assumption that the cohesion will increase by c_A = 0.6 MPa because of the passive rock bolt, a maximal shear stress of $p_v \cdot \tan \phi_j + c_j + c_A$ was applied. Hence for $p_v = 0$, τ_{max} is 0.92 MPa and for $p_v = 2.2$ MPa, it is 2.2 MPa.

5.5 Results of the calculations

The results of the calculations (Table 1) can be interpreted as follows :

	Load case	$\alpha = 0^0$/$\sigma = 2.0$	$\alpha = 25^0$/$\sigma = 2.2$	$\alpha = 25^0$/$\sigma = 0$	$\alpha = 0^0$/$\sigma = 0.2$
Mortar	max σ_m	+7.2	+25.5	+39.5	+39.6
	min σ_m	−65.0	−28.2	−10.1	−14.5
Steel	max σ_a	−199.6	+129.0	+332.6	+238.6
	min σ_a	−489.2	−118.1	+12.1	−21.2

Table 1. Maximal and minimal calculated stresses in the bolt and in the borehole mortar [MPa]

1) For the same vertical stress, the inclined bolt shows higher tensile stresses than the vertical one.

As a consequence, the inclined bolt has a higher blocking effect than the vertical one even at smaller displacements. Also the rock mass improved by inclined bolts is stiffer and has smaller displacements at failure.
2) The inclination of passive bolts reduces the compressive stresses in the mortar but also increases the calculated tensile stresses behind the bar. The latter effect breaks the bond between the steel and the mortar what is usually inevitable for large shear displacements as confirmed by numerous test results.
3) High vertical stresses reduce the maximum tensile stress in the anchor bar and in the mortar but also increase the maximum compressive stresses.

The value p_v = 2.2 MPa applied in the calculations is higher than the real maximum rock stresses and constitutes a very unfavourable loading case.

As an illustration of the obtained results, Figure 12 shows the deformed mesh and Figure 13 the displacements of the upper block, in the case of the inclined bolt.

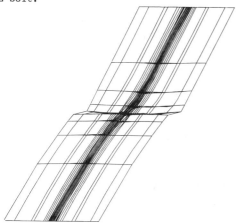

Fig. 12. Deformed inclined system at maximum load

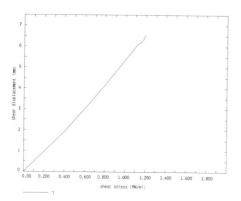

Fig. 13. Displacements of the inclined model vs. applied shear stresses

6. FIELD TESTS ON FULLY-BONDED ROCK BOLTS

As a complement and for a possible adjustment of the finite element calculations, a series of large scale in-situ shear tests have been carried out on the foundation rock of the dam. Figure 14 shows the test installations.

The concrete block acted as a stiff shear box and the jointed rock mass was simulated by concrete cylinders of the same strength. The rock bolts were either normal to the shearing plane or inclined (10° and 25° with respect to the vertical direction). Fully-bonded DYWIDAG steel bars of the GEWI type with 40 mm diameter were used. The boreholes had 101 m diameter and were filled with high-strength cement mortar.

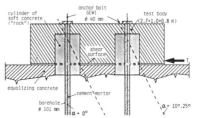

Fig. 14. Installation of the in-situ shear tests

The tests were particularly meaningful because they were carried out at full size. In addition, they enabled comparisons with test results obtained with very different steel diameters (8 mm for laboratory tests and 40 mm for site tests).

The Figure 15 shows the load-displacement curves of the site tests. The test II.1 with vertical bolts could note be continued until failure because of limited possible shear displacements.

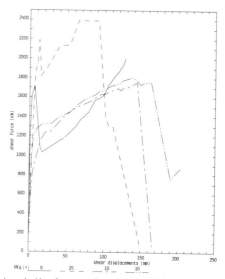

Fig. 15. Load-displacement curves of the site shear tests

Based on the test results, the increase of the shear strength due to one rock bolt (ΔT_A) has been recommended to be expressed as follows :
$$\Delta T_A = P_y \cdot f_{st}$$
with P_y = yield load of the steel bar
and f_{st} = parameter depending on the bolt inclination : f_{st} = 0.7 for α_A = 0°
f_{st} = 0.9 for α_A = 10°
f_{st} = 1.1 for α_A = 25°

7. CONFRONTATION OF THE RESULTS OF THE FINITE ELEMENT CALCULATIONS AND THE SITE INVESTIGATIONS

For a better understanding of the results, the main differences between the numerical model and the in-situ shear tests are summarized as follows :
1) Mortar and steel were considered as elastic materials in the calculations, whereas they obviously behaved in a brittle or elasto-plastic way, respectively, in the field tests.
2) In the calculations, the cohesion along the shear surface was assumed to remain constant for any shear displacement (elastic-ideally plastic behaviour); the field tests, however, showed a marked strain-softening behaviour of the concrete cylinders.
3) Contrarily to the observations in the site tests, the finite element model showed a significant dilatance of the joint filling. This resulted in stiffening the model and, therefore, in reducing the shear displacements.

For an easier comparison, the results of the two investigation methods are presented in a dimensionless form ($f_T = T/P_y$). In the case of the test results, the shear strength of the cylinders has been deducted from the total shear load as has the cohesion of the joint filling material in the calculations. Figure 16 shows the modified load-displacement curves obtained in the numerical model (a) and in the field tests (b), for the case with inclined (25°) bolts.

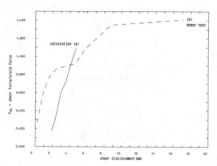

Fig. 16. Modified load-displacement curves from the numerical model (a) and field test (b) ($\alpha_A = 25°$)

The slopes of the two curves show a good agreement for small displacements. As anticipated, the finite element model is stiffer than the field test block, at higher loads.

Because the allowable displacements of the planned dam are very limited, we remain in the domain of good agreement between calculation and field tests.

8. PROPOSED PATTERN OF ROCK BOLTS

From the foregoing investigations the following pattern of passive rock bolts was proposed for the rock improvement at the TIRSO dam site :
steel bars : DYWIDAG GEWI 500/600, ø = 40 mm
boreholes : ø = 101 mm, filled with high-strength ciment mortar
Inclination of the anchors : alternating 10° and 25° versus upstream. Bolting density : 1 anchor bar ϕ_2 = 40 mm for 1.0 m × 1.25 m, i.e. 10 cm^2 steel per m^2 of foundation surface. The bolting density is gradually reduced with the depth below the foundation surface of the dam, depending on the required ground improvement.

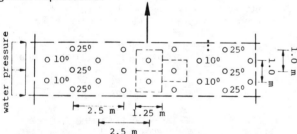

Fig. 17. Schematic outline of the proposed bolting system

9. CONCLUSIONS

The stability investigations presented in this paper were to show the local behaviour of the foundation rock of a gravity dam. They are based on the investigations concerning the global stability of the dam, carried out by Prof. ing. F. Arredi, Rome.

The first model examined yielded information about the deformations and stresses in the rock situated below the dam. With the help of plane-strain numerical calculations, critical zones in the foundation rock were detected. There the stresses were near the failure criterion in shear or tension. As a result, the need to modify the dam or to improve the foundation rock became obvious. Based on a technical and economical comparison of various possible methods, the decision was made in favour of a ground improvement by fully-bonded passive bolts. This method had to furnish an increase in cohesion in the range of 0.6 MPa to ensure the required dam stability.

In order to enable the design of the bolting system, numerical calculations with a 3-dimensional model and large-scale in-situ tests were carried out. The results of these investigations yielded conclusions concerning stresses in the steel bars, in the mortar and the surrounding rock. Therefore, they enabled a safe design of the rock bolting system.

Thus a large-scale ground improvement for the foundation of a gravity dam, using fully bonded rock bolts, could be proposed for the first time. To ensure a sufficient safety of the dam, the following conditions have to be respected :
1) Carefully limiting the permitted displacements and stresses in the steel.
2) Sufficient protection against corrosion.
3) Careful observation of the displacements and deformations of the dam, during the first impounding and later.
4) Maintain the possibility of installing additional rockbolts in the cavities between the dam blocks, if required.

These investigations have also shown that generally accepted and sufficiently tested design methods for passive rock bolts do not exist yet. The results obtained for the present case cannot be generalized without appropriate modifications. More detailed research work is needed for finding general rules.

The investigations presented are being continued at the Swiss Federal Institute of Technology, Lausanne, and their results will be published in due time.

10. ACKNOWLEDGMENTS

The writers wish to thank the Project Owner and the Designer and his advisers for their excellent collaboration. They also want to express their particular gratitude to the Contractor GRASSETTO S.A., Padua, for the possibility to carry out the extensive calculations and field tests as well as for numerous fruitful discussions.

BIBLIOGRAPHY

Bjurström, S. 1974. Shear Strength of Hard Rock Joints reinforced by Grouted Untensioned Bolts. Proc. III Conf. ISRM Denver 1974, 2B 1194-1199.
Dight, P.M. 1982. Improvement to the Stability of Rock Walls in Open Pit Mines. Ph. D. Thesis, Monash University Australia.
Egger, P. 1978. Neuere Gesichtspunkte bei Tunnelankerung. Trans. Techn. Publ. Clausthal 1978, p. 263-276.
Egger, P., Fernandez H. 1983. Nouvelle presse triaxiale - Etude de modèles discontinus boulonnés. Proc. V. Conf. ISRM Melbourne 1983, A 171-175.
Londe, P., Bonazzi D. 1974. La roche armée. Proc. III. Conf. ISRM Denver 1974, 2B 1208-1211.
Schubert, P. 1984. Das Tragvermögen des mörtelversetzten Ankers unter aufgezwungener Kluftverschiebung. PH. D. Thesis Montanuniversität Leoben.
Spang, K. 1986/1. Felsbolzen im Tunnelbau. Geotechnik, 1/86, p. 9-19.
Spang, K. 1986/2. Bemessungskriterien für passive Anker. XXXV. Geomechanik Kolloquium, Salzburg 1986, to be published.
Wullschläger D., Natau O. 1983. Studies of the Composite System of Rock Mass and Non-prestressed Grouted Rock Bolts. Proc. Int. Symp. Rock Bolting, Abisko 1983, p. 75-85.

Fondation des grands barrages – Les écrans d'étanchéité en paroi
Large dams foundation – The positive cut off
Dammfundament – Dichtungswand

G.Y.FENOUX, Directeur Technique Soletanche, France

ABSTRACT : The watertightness of a dam should be technically perfect, at a minimum cost. Numerous bully sandstones, amoungst others, pose continual problems to engineers (in North America and Southern, in Europe also). Into such a rock, the traditional grouting method is rarely effective, and always onerous. Up until the last few years there were no machines available to produce concrete diaphragm walls (anglo-saxon positive cut off). The Hydrofraise, recently produced, contains a solution to these problems.

The communication quickly shows the basis of its procedure amnd the progress which it engenders. Solid examples are given : the Igelsbach, Brombach and Kleine Roth dams (Bavaria, West Germany), the Fontenelle dam (Wyoming, U.S.A.), and the St. Stephen dam (Southern Caroline, U.S.A.).

RESUME : L'étanchéité des barrages doit être assurée de façon techniquement sûre, et au moindre coût. Les massifs gréseux (entre autres), nombreux (Amériques du Nord et du Sud, Europe) posent continuellement un problème aux ingénieurs. L'injection traditionnelle est rarement efficace, toujours onéreuse. Jusqu'à ces dernières années, aucun outillage ne permettait de réaliser des parois en béton (positive cut off anglo-saxon). L'Hydrofraise, procédé nouveau, apporte une solution à ces problèmes.

La communication expose rapidement les bases du procédé et les progrès qu'il engendre. Des exemples concrets sont présentés : les barrages d'Igelsbach, de Brombach et de Kleine Roth (Bavière, R.F.A.), le barrage de Fontenelle (Wyoming, U.S.A.), et le barrage de St. Stephen (Caroline du Sud, U.S.A.).

ZUSAMMENFASSUNG : Die Dichtheit von Talsperren muss in Fehlerloser und eindentiger Art auf dem technischen Plan garantiert sein; dies mit dem geringst möglichen Kostenafwand. Die zahlreichen Sandsteinmassive (Süd und Nordamerika, Europa) stellen ein standiges Problem für den Ingenieur dar. Die kostspieligen herkömmlichen Injektionen sind selten wirkungsvoll. Bis in die letzton Jahre gestatette es keine Maschine Betonschlitzwände in diese Art von Boden zu realisieren. "Hydrofraise", ein neues Verfahren lost dieses Problem.

Der Vortrag erläutert auf welchen Grundlagen dieses Verfahren basiert, und die Fortschritte die damit erzielt werden. Konkrete Projekte sind ausgeführt worden : Die Talsperren von Igelsbach, Brombach und Kleine Roth (Bavern, Deutschland), die Talsperre von Fontenelle (Wyoming, U.S.A.) und die Talsperre von St. Stephen (Sud-Caroline, U.S.A.).

1 AVANT-PROPOS

L'étanchéité des fondations des grands barrages pose continuellement des problèmes difficiles à résoudre.

La question 58 du dernier Congrès International des grands barrages fut consacrée à ce thème, et particulièrement aux techniques d'injection (Lausanne, juin 1985). Jusqu'à ces dernières années, en effet, seule l'injection permettait d'assurer l'étanchéité des barrages construits sur des massifs rocheux.

L'injection est généralement bien connue. Les progrès sont continuels, en particulier dans le domaine des coulis à haut pouvoir de pénétration.

Toutefois, l'expérience montre que l'injection ne permet pas (ou mal) de résoudre le problème de l'étanchéité lorsqu'on se trouve confronté au cas des massifs gréseux, ou à celui des massifs granitiques altérés.

Le grès présente deux perméabilités : une perméabilité de fissures, pour laquelle on fait appel à des coulis de ciment dilué, et une perméabilité de masse, comparable à une porosité, pour laquelle il faudrait employer des coulis chimiques à très faible viscosité.

En pratique, on se limite à l'emploi des coulis de ciment, pour d'évidentes raisons d'économie. Le coulis de ciment s'essore très vite. Le rayon d'action des forages est très faible (1 à 1.5 m). Il faut multiplier les lignes d'injection. Les résultats sont rarement bons. La pérennité des écrans ainsi réalisés est souvent mauvaise, et les exemples sont nombreux de barrages à réparer pour cette raison.

Le granite présente généralement des degrés d'altération allant jusqu'au sable, de façon aléatoire, de sorte que le projeteur ne sait pas comment faire les écrans. En effet, il faudrait pouvoir marier des injections de rocher (fissures) à des injections de terrain meuble fin (tube à manchettes). Les compromis utilisés laissent place aux imperfections.

Pour résoudre de façon radicale les problèmes ainsi posés par le grès, le granite, ou d'autres types de roches à comportement analogue vis-à-vis de l'injection, la solution idéale consiste à faire des écrans d'étanchéité en paroi.

La technique des parois, sous des formes diverses, s'est imposée pour la réalisation des écrans d'étanchéité dans les terrains meubles. Pour les roches, le problème de la perforation était jusqu'à ces dernières années un obstacle technique et économique.

Le procédé Hydrofraise est venu à point pour combler cette lacune.

2 LE PROCEDE HYDROFRAISE

Le procédé est caractérisé par un outillage et une méthode de travail.

2.1. L'outillage se compose de 3 éléments : porteur lourd, bâti hydrofraise, station de boue.

Le porteur lourd manipule le bâti et fournit l'énergie sous forme hydraulique.

Le bâti hydrofraise, métallique, a 15 m de hauteur et pèse entre 16 et 20 t. Il contient à sa base 3 moteurs hydrauliques compacts de 100 kW.

Deux moteurs, à axes horizontaux parallèles, entraînent directement des tambours munis de pics. Tournant lentement (10 à 20 tours/minute), avec un couple élévé (4.000 mkg), les pics désagrègent le terrain.

Le troisième moteur anime une pompe à boue dont l'orifice d'aspiration est situé juste au-dessus des tambours. La boue, débitée à plus de 300 m3/h, entraîne à la surface les déblais produits par les pics. La boue est refoulée jusqu'à la station de traitement d'où elle ressort criblée et dessablée. Elle est alors renvoyée dans la saignée. Il s'agit d'une circulation inverse de la boue de forage.

La station de boue se compose d'unités de stockage de bentonite et de boue, d'unités de fabrication et d'unités de criblage et dessablage.

2.2. La méthode de travail est simple : mise en station, perforation d'un élément de paroi vertical sans bouger, sans manoeuvrer, jusqu'à la profondeur désirée.

L'élément de paroi vertical a 2.40 m d'ouverture, et une largeur comprise entre 0.65 et 1.50 m.

Les panneaux primaires, constitués d'un certain nombre d'éléments, sont séparés par des merlons de 2.2. m. Ces merlons sont les panneaux secondaires que l'outillage perfore en mordant de 0.10 m dans le béton des deux panneaux primaires adjacents. Aucun dispositif spécial n'est utile pour faire le joint.

Le bâti est suspendu au porteur par un vérin à longue course. Ce vérin amortit les à-coups de manoeuvre. Il peut être asservi soit pour un poids constant sur les pics (terrain dur), soit pour une vitesse d'avance constante (terrain meuble).

Les pics qui désagrègent le terrain constituent les outils d'usure. Ils sont changés périodiquement. La consommation varie selon la dureté du sol. Position, nombre, disposition, type, forme, nuance d'acier, angle de coupe, sont les nombreux paramètres qui conditionnentl'efficacité.

L'efficacité maximale est obtenue lorsque les déblais produits par les pics sont de la plus grosse dimension évacuable par la pompe à boue.

2.3. Les points forts, les progrès par rapport aux autres procédés sont :
- la perforation en continu (pas de mini éboulements par laminage des parois à chaque passage de l'outil),
- la séparation entre la perforation et l'évacuation des déblais (pas de contrainte due aux camions de déblais),
- le dessablage permanent (cake mince et boue légère),
- l'absence de chocs, de vibrations, d'ébranlements,
- le passage des horizons durs (jusqu'à près de 100 MPa de résistance),
- la vitesse de perforation,
- la possibilité d'atteindre de grandes profondeurs, sans réduction de rendement,
- la qualité des joints (interpénétration due à la forme des tambours porte-pics),
- la possibilité, lorsqu'il y a des armatures, de mettre deux cages très proches l'une de l'autre dans deux panneaux différents,
- la précision de perforation (conséquence de la méthode de travail),
- le contrôle permanent de verticalité (clinomètres de précision, enregistrement des paramètres de forage),
- la possibilité, dans les cas extrêmes, de rectifier des déviations.

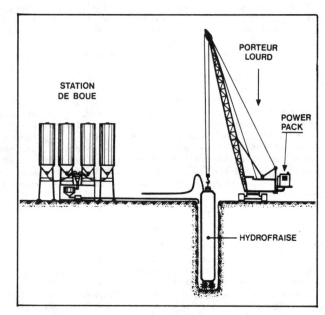

Figure 2. Composition schématique d'un atelier.

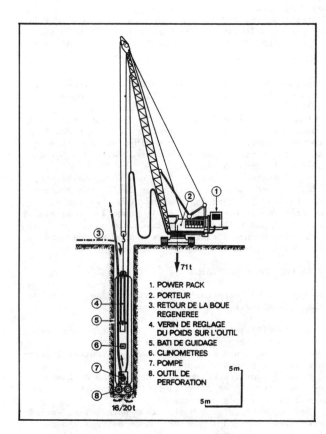

Figure 1. Schéma du bâti Hydrofraise.

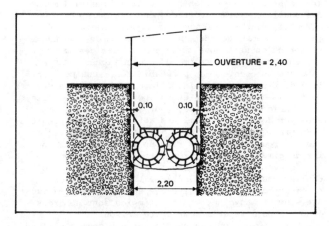

Figure 3. Réalisation des joints entre panneaux.

356

3.1. L'aménagement Main-Danube en République Fédérale d'Allemagne

Situé en Bavière, le canal Main-Danube, long de 130 km environ, fait partie de la liaison fluviale Mer du Nord-Mer Noire par le Rhin, le Main et le Danube.

Une succession de biefs et d'écluses permet le franchissement de la ligne de crête séparant le bassin versant du Danube, au Sud, de celui du Main, au Nord.

L'eau alimentant le canal est pompée dans le Danube, puis relevée par étapes jusqu'au point haut d'où elle peut redescendre alimenter le Main.

Ainsi, le canal assure un transfert des eaux du Sud de la Bavière, bien irriguée, vers le Nord où la pluviométrie est trois fois moindre.

Lors des basses eaux du Danube, le transfert devient insuffisant. Pour assurer un fonctionnement régulier de cet ensemble assez complexe, divers aménagements ont été prévus au voisinage du point haut du canal, ayant pour fonction de créer des réserves d'eau. Il s'agit des aménagements de BROMBACH et de KLEINE ROTH.

L'aménagement de BROMBACH comprend trois digues en terre. BROMBACH primaire et IGELSBACH créent des plans d'eau fixes. BROMBACH principal crée une retenue à plan d'eau variable, sa hauteur est de 40 m.

Les terrains de fondation sont essentiellement composés d'un socle de grès dur (30 à 50 MPa de résistance à l'écrasement en simple compression) avec passages de Marnes et argiles raides. Le grès est siliceux, fracturé et perméable.

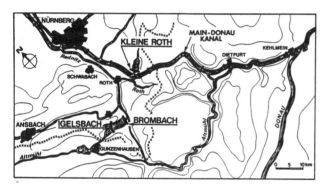

Figure 4. Plan schématique de l'aménagement.

Les spécialistes en injection savent combien le grès est un terrain difficile à étancher. A l'heure actuelle, un grand nombre de barrages, de notre connaissance, ont des problèmes de fuites dues aux roches gréseuses. En effet, le grès est poreux; les coulis de ciment perdent leur eau et s'épaississent très vite; les rayons d'action sont faibles; les refus en pression sont obtenus prématurément.

Un tel scénario s'est déroulé pour le barrage d'I-GELSBACH. En dehors du *thalweg* constitué de matériaux sablo-graveleux pour lesquels il était facile de faire un écran d'étanchéité en paroi moulée, des injections étaient prévues. Sur la rive droite, le traitement s'est révélé très coûteux en regard d'une efficacité médiocre. Trois lignes de forages espacés de 1.50 m et des coulis classiques n'ont pas permis de descendre la perméabilité en dessous de 10 U.L.

Sur la rive gauche, sur une surface de 5 200 m2, un essai de coupure en paroi hydrofraise a donc été décidé. Il fut jugé très positif aussi bien sur le plan technique que du point de vue économique (1983).

En conséquence, le Maître d'Ouvrage a modifié son projet initial de coupure étanche par injection et retenu pour tout l'écran de BROMBACH une paroi forée à l'hydrofraise. L'épaisseur est 0.65 m, la profondeur maximale 40 m, il s'agit de béton plastique. La surface totale est de l'ordre de 40 000 m2 (1984).

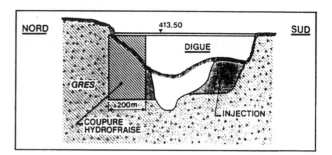

Figure 5. Coupe type du barrage d'IGELSBACH.

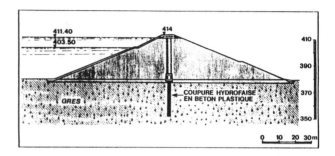

Figure 6. Coupe type de l'écran de BROMBACH.

La digue de KLEINE ROTH, dont les travaux ont débuté cette année, pour le même Maître d'Ouvrage, avec des problèmes analogues, est traitée de la même façon. La digue a 25 m de hauteur maximale, sa longueur en crête est 1.700 m. L'écran en paroi descend à 20 m environ, la surface forée à l'hydrofraise est de 27 000 m2 (1985 et 1986).

3.2. Le barrage de Saint-Stephen (U.S.A.)

Le barrage de Saint-Stephen est situé dans l'Etat de CAROLINE DU SUD, sur la rivière Cooper. Il s'agit d'une digue en terre coupée en deux par une usine hydroélectrique. La hauteur de la retenue est de 35 m.

Assise sur un substratum de Marne sableuse raide, cette digue présentait des fuites, aussi le U.S. ARMY CORPS OF ENGINEERS a-t-il décidé de la réparer.

Le projet comportait une paroi en béton rigide, avec joints réalisés au tube-joint. De plus, étant donné qu'il s'agit d'un site sismique, chaque joint était doublé à l'amont par un panneau en sol-bentonite de 3 m d'ouverture.

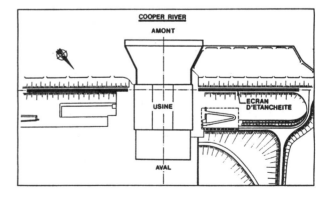

Figure 7. Vue en plan du barrage de Saint-Stephen.

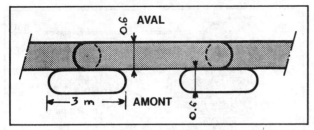

Figure 8. Représentation schématique de l'écran prévu à l'appel d'offres (Saint-Stephen).

Les difficultés prévisibles de perforation au voisinage du contact digue-substratum (avec le risque de déviations importantes lors du passage du corps de la digue relativement meuble au rocher relativement dur) et celles correspondant à la réalisation des panneaux supplémentaires amont (avec le risque de dévier sur les excroissances de béton de la paroi principale), ont conduit à retenir l'outillage hydrofraise.

De ce fait, le U.S. ARMY CORPS a accepté de modifier son projet.

Les panneaux primaires ont une ouverture de 9 m. Les panneaux secondaires mesurent 2.20 m. Les joints sont faits selon le procédé hydrofraise. Sur la partie verticale d'écran située dans la digue, un panneau complémentaire de 5.50 m d'ouverture est foré à l'amont, avec un ancrage de quelques mètres dans le substratum. Un tel anneau couvre ainsi deux joints. La paroi est en béton rigide. Les panneaux complémentaires sont en sol-bentonite. L'épaisseur est égale à 0.65 m. La profondeur est de 36 m. La surface totale forée est de 12 000 m2.

Les travaux se sont déroulés en 1984, il s'agissait du premier chantier de l'hydrofraise aux Etats-Unis.

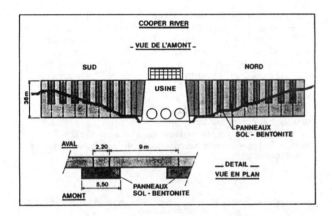

Figure 9. Elévation de la paroi réalisée à l'hydrofraise (Saint-Stephen).

3.3. Le barrage de Fontenelle (U.S.A.)
Le barrage de Fontenelle est situé dans l'Etat du WYOMING. Il dépend du Bureau of Reclamation.

Il s'agit d'une digue en terre, dont la longueur en crête est de 2 000 m environ, créant une retenue de 40 m de hauteur. Il est situé à 2 000 m d'altitude sur un socle de grès dur fracturé et fissuré.

L'étanchéité initiale comporte un écran injecté, avec des forages espacés de 3 m (10 pieds) sur une seule ligne.

Dès sa mise en eau, en 1965, des fuites avec érosion sont apparues. La retenue a été vidée et des réparations faites. La mise en service a eu lieu en 1967.

En 1982, des fuites sont de nouveau apparues. Petit à petit le niveau de la retenue a été abaissé. Actuellement le niveau est environ au 1/3 inférieur. Ce niveau est nettement inférieur à celui nécessaire pour l'exploitation et la production d'énergie.

Depuis 1982, un important programme de reconnaissance par puits et piézomètres a été réalisé. Un projet de réparation a été fait. L'appel d'offres lancé en début d'année concernait simplement un essai pour une paroi en béton rigide forée depuis la crête et descendant jusqu'à une profondeur de 55 m.

Le Maître d'Ouvrage a reçu 6 offres et retenu la solution hydrofraise.

Deux tronçons d'essai sont en cours, un à chaque extrémité. Ils couvrent environ 200 m et représentent 10 000 m2, dont une grande partie dans du grès très dur. La paroi a 0.65 m d'épaisseur.

D'après le Bureau of Reclamation, l'essai ainsi que les investigations coûtent 7.2 millions de dollars. Le coût de l'ensemble des réparations, si l'essai est considéré comme positif, sera compris dans la fourchette 25 à 50 M de dollars.

Compte tenu de la latitude et de l'altitude, on ne peut travailler que 6 mois par an. La fin des réparations est programmée pour 1990.

4 CONCLUSION

Le procédé de réalisation des parois est encore relativement jeune, surtout si on fait la comparaison avec les techniques traditionnelles de génie civil.

Toutefois, en 25 années, les progrès ont été considérables. Ainsi, lors des premières applications, en 1960, on ne perforait que des terrains meubles, avec des rendements industriels de perforation de l'ordre de 1 m2/heure. On peut désormais l'appliquer aux roches, avec des rendements économiquement réalistes : soit plus de 5 m2/heure dans une roche de 50 MPa de résistance à l'écrasement. On peut même traverser des couches ou des bancs rocheux de résistance voisine de 100 MPa.

L'expérience acquise sur quelques grands chantiers, et tout particulièrement des barrages, permet maintenant aux ingénieurs de projeter des écrans d'étanchéité sûrs et économiques, à des profondeurs de 100 m, de façon courante, dans la plupart des types de massifs rocheux (grès, schistes, calcaires, granites moyennement durs ...).

L'emploi des parois pour des réparations doit également être mis en évidence, car il élimine tous les aléas de coût prévisible, contrairement à ce qui souvent se produit lorsqu'on doit avoir recours à l'injection (ceci est un reproche courant).

Le remplissage de la paroi, en béton rigide ou plastique, en coulis, ou même en sol-bentonite, est effectué de façon parfaitement contrôlée : des garanties précises de résultat (un coefficient de perméabilité, ou un débit résiduel) peuvent être données.

Ces progrès sont à porter au crédit de l'hydrofraise, dont la longue mise au point a demandé plusieurs années. Initialement, "hydrofraise" était une marque déposée par l'entreprise SOLETANCHE, inventeur, qui a assuré le développement. Maintenant, le nom est passé dans le langage courant, pour désigner des outillages de paroi d'une nouvelle génération, aux grandes possibilités d'emploi. Des copies sont apparues, qui rendent ainsi indirectement hommage à l'hydrofraise originale.

358

General regularities of shear failure in rocks
Lois générales de la rupture des roches au cours du cisaillement
Allgemeine Gesetzmässigkeiten des Scherbruchs der Felsgesteine

YU.A.FISHMAN, 'Hydroproject' Institute, Moscow, USSR

ABSTRACT: The paper shows that all brittle materials and rocks, irrespective of the body shape and its structure, feature a common intrinsic regularity of shear failure - formation of the tension cracks at the initial phase and the zones of compression and crushing in the principal stress surfaces at the final phase when attaining the ultimate bearing capacity. In the long run, the peak bearing capacity of the rock mass is conditioned by the resistance to compression, the magnitude and the pattern of the stressed state on the specified plane of shear.

RESUME: Le rapport fait voir que tous les matériaux fragiles et les roches, indépendamment de la forme du corps et de sa structure, sont soumis à la même loi interne de la rupture au cisaillement: formation des fissures d'extension au stade initial et des zones de compression et de fragmentation des matériaux, suivant les aires principales, au stade final quand la limite de la capacité portante est atteinte. Finallement, la capacité portante limite du massif rocheux est déterminée par la résistance des roches à la compression, par la valeur et la nature de l'état de contrainte dans le plan de cisaillement envisagé.

ZUSAMMENFASSUNG: Es ist im Vortrag aufgezeigt worden, daB allen spröden Materialien und Felsgesteinen unabhängig von der Form und dem Gefüge des Körpers die gleiche innere GesetzmäBigkeit des Scherbruchs eigen ist und zwar die Bildung der Zugrisse am Anfangsstadium sowie die der Druck- und Ruschelzonen den Hauptflächen entlang am AbschluBstadium zum Zeitpunkt; wo die Bruchfestigkeit erreicht worden ist. Letzlich wird die Bruchfestigkeit des Felsgesteins durch die Druckfestigkeit sowie den Spannungswert und das Spannungsbild in der jeweiligen Scherebene bestimmt.

Papers presented at the IV and V ISRM Congresses (Fishman 1979; Fishman, Ukhov & Fadeev 1983) and some other publications indicate that failure of the rock foundations under concrete structures differs in some cases from the conventional shear mechanism which is usually presented as the sliding of the structure on the foundation. If there are no shear-prone surfaces of weakening in the foundation (large fractures, interbeds), the typical pattern of failure consists in the formation of a main tension crack on the upstream side of the structure and a compression zone in which crushing of the rock mass coincides with attainment by the system of the peak bearing capacity on the downstream side. And as indicated (Fishman, Panfilov & Sarabeev 1976) principal minimum and maximum stresses are respectively responsible for the failure.

The subsequent investigations have established that the described mechanism of shear failure is an intrinsic regularity of the brittle materials and rocks. But its outward manifestation could vary depending on the scale of the unit in question and extent of the material continuity.

It is known that a typical feature of the brittle material failure is the formation of tension cracks which are most distinct in uniaxial compression and tension tests. At tangential stresses causing shear strain one or several echelon-like tension cracks are formed which are oriented at an angle to the shear direction (Brace & Byerle 1961). Under the "pure shear" conditions, i.e. when only tangential stresses act, increase in the latter causes the growth of tension cracks and subsequent splitting of the body or the sample. In case of a complex stressed state when in addition to the tangential stresses, the normal stresses are involved, the formation of the tension cracks is not necessarily to lead to the ultimate failure of the material and it constitutes only an initial phase of the failure process. Experiments conducted by many authors show that the complete shear failure of brittle bodies takes place through the joining of the tension cracks and formation of the main shear surface (Obert 1972; Lajtai 1967; etc.).

But the mechanics of material failure between tension cracks has not been well studied. While its identification is of fundamental importance because it determines the culmination of the failure process, i.e. attainment by the system of the limit bearing capacity. We have experimentally and theoretically found that the failure of undisturbed material between the tension cracks results from their compression in the surfaces of principal stresses. Hence the pattern of brittle material failure at the microlevel (on the scale of one tension crack and one compression zone) resembles the failure of the rock foundation of the concrete structures at the macrolevel. The described pattern of failure is observed in the various forms of systems being sheared and contact configuration or in the given shear plane.

Fig.1 shows failure phases on 10 various models made of brittle materials (in this

particular case they were fabricated using
gypsum-bound material).

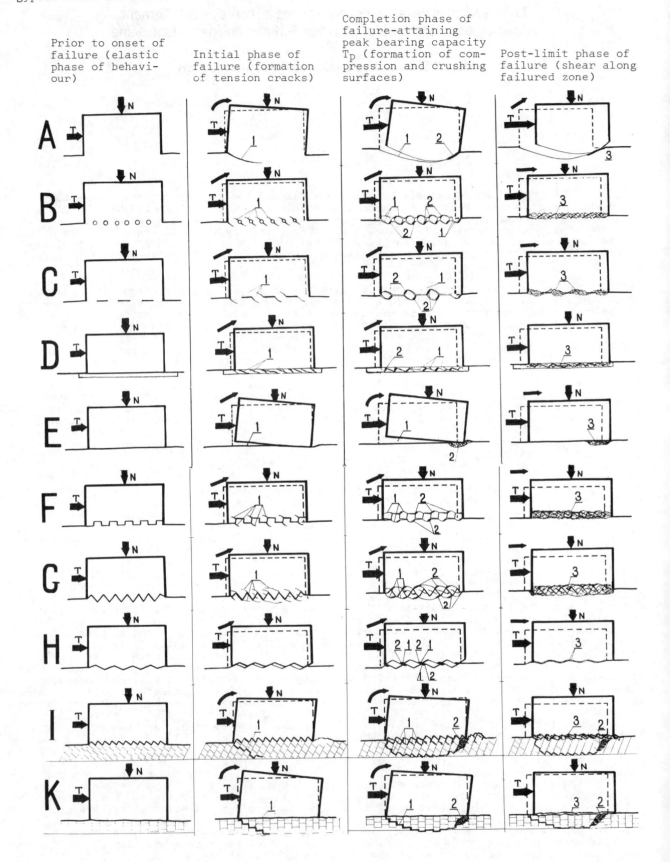

Figure 1. Pattern of model failure in shear: 1 - tension cracks; 2 - crushing surfaces;
3 - failured zone.

Model A (monolithic contact between the structure and the foundation) illustrates the failure pattern described in work (Fishman 1979) (see also Fig. 2).

Models B and C (the contact is weakened by holes, intermittent fissures or slots) illustrate the failure pattern described above, i.e. development of a serious inclined tension cracks propagating along the trajectory of principal stresses and subsequent failure of undisturbed material due to their compression by the maximum principal stresses (see also Fig. 3.4).

Figure 2. Model A at instant of completing phase of failure.

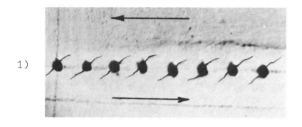
1)

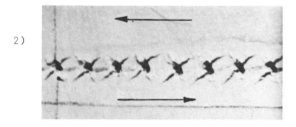
2)

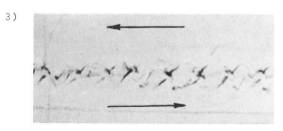
3)

Figure 3. Contact zone fragment of model B in three phases of failure: 1 - formation of tension cracks; 2 - formation of compression surfaces normal to tension cracks; 3 - shear along failured zone.

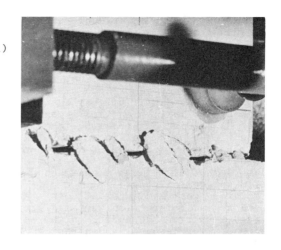
1)

2)

Figure 4. Failure of model C: 1 - after formation of crushing surfaces; 2 - post-limit phase of failure.

Attention should be drawn to the localization of failure by compression in the narrow zone of surfaces linking up the ends of the tension cracks and holes or slot ends. It resulted in the formation of a complex failure surface, and further shear displacement along this surface in the post-limit phase lead to splitting, crushing and grinding of the fragments and asperities that had been formed before.

In models F, G, H (dental contact) the pattern of failure does not principally change. When the dents are of a rectangular shape (Model F), each dent tends to fail like a small shear block according to type A (see also Fig. 5) which was confirmed by the tests of E. Lajtai (Lajtai 1967).

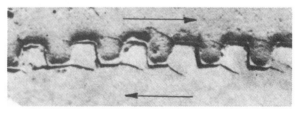

Figure 5. Failure of model F contact zone (E. Lajtai 1967).

The failure pattern of triangle-shaped dents depends on the angle of inclination and the magnitude of normal load N. At steep angles and large magnitudes N, triangle dents tend to fail similar to the rectangular ones (Model G) with the difference that the tension crack sometimes jumps at a time, over two or even three dents. At gentle angles of the dent inclination and small magnitudes N (Model H) sliding occurs along the ascending face of the dent, but in any case there occurs tension and crushing of the material at the dent crest when the stresses on the contact reach the ultimate magnitudes due to mutual displacement.

In Model D (weakened interbed), the tension cracks and compression surfaces are of an irregular pattern depending on the random position of microdefects in contrast to the preceding cases when a regular pattern of the tension and compression cracks is predetermined by an artificial contact configuration.

Model E (completely separated contact) is described in (Fishman, Ukhov & Fadeev 1983). It should be only noted that here break away occurs not along the surfaces of the principal stresses but on the contact whose tensile strength is zero. The general pattern of failure does not differ from the failure of Model A. We'd like to remind that such a pattern of failure takes place only at a high value of force N and a low compression strength of the foundation, while in the remaining cases, a classical picture of the test plate sliding over the base is observed. But here also there occurred microprocess of failure in the form of tension - compression of asperities and irregularities on the contact plane.

And finally in case of a discrete structure of the foundation (model I, K) the surface of rupture tends to develop through the opening of natural joints between the rock blocks. The main direction of the rupture surface coincides with the trajectory of tension cracks in the continuous media. Similarly the zone of compression are formed and, like in continuous media, these zones can take the form of a single element of area of crushing or a series of them. But in case of a "smooth" foundation surface or the presence of extensive joint, shear may occur in the form of sliding on the contact or along the joint.

The given above examples show that the mechanism of rock shear failure in the form of tension-compression features a common pattern in spite of diversity in its outward manifestation.

For example on models B,C,D,F,G etc. the shear block displacement with respect to the base in terms of macrolevel could be interpreted as sliding, while at the scale of separate dents or joint blocks (the microlevel) it is a failure with break-compression and rotation of the failing elements.

The similar shear along a natural joint in a rock mass regarded from the classic viewpoint is the Coulomb shear while this shear actually takes place (neglecting the filler material) as breaking and crushing of irregularities of the joint walls. In the long run, the resistance to shear, at the initial phase of failure, is controlled by the tensile strength of the rock, while at the terminal phase, it depends on the compression strength.

Identification of the above regularity is of principle importance because it provides a clue to the insight into many processes taking place at failure of brittle materials and rock, helps to explain the findings of some experimental studies which do not fit into the existing strength theories and provides a tool for solving some practical problems. For example, the nonlinear pattern of relationship $T_p = f(N)$ or average stresses: $\tau_p = f(\sigma)$ can be attributed to the fact that compression and crushing of the material is caused not only by the normal stresses but also by the tangential ones and therefore resistance to shear T_p tends to increase unproportionally to normal load N - the higher N, the smaller increase in T is needed to crush the rock.

In this connection shear resistance parameters $\mathrm{tg}\,\varphi$ and C are not and cannot be constants of a rock mass. They depend on the compressive strength of the material R_{cr} which is a more stable value. The established regularities of rock failure have been used to work out a new method of stability analysis for concrete dams included in the State Building Code of the USSR (Construction Code... 1986).

REFERENCES

Brace, W.F. & J.D. Byerle 1966. Proceedings of the 8th Rock Mechanics Symposium, Univ. of Minnesota, AIME, New York.

Construction Code (СНиП) 2.02.02-86. Foundations of hydraulic structures. Moscow: 1986 (in Russian).

Fishman, Yu.A. 1979. Investigation into the mechanism of the failure of concrete dam rock foundations and their stability analysis. IV Congress of the ISRM, Montre.

Fishman, Yu.A., Ukhov, S.B. & A.B. Fadeev 1983. Main principles and methods of investigations of in-situ rock masses. V Congress of ISRM, Melbourne.

Fishman, Yu.A., V.C. Panfilov & V.F. Sarabeev 1976. Photoelastic model studies of rock foundation failure in shear. Proceedings of VNIIG, vol. III: Publishing House "Energia" (in Russian).

Obert, L. 1972. Brittle failure of rocks. Fracture, vol. 7.

Lajtai, E.Z. 1967. The influence of interlocking rock discontinuities on compressive strength (model experiments). Rock Mechanics and Engineering Geology, vol. V/4.

Research on the slope stability of the left abutment at Liujiaxia Hydro-power Station

Recherches sur la stabilité de l'appui gauche de la centrale hydroélectrique de Liujiaxia
Die Böschungsstandfestigkeituntersuchung des linken Widerlagers des Wasserkraftwerkes Liujiaxia

FU BING-JUN, Institute of Water Conservancy and Hydroelectric Power Research, Beijing, People's Republic of China
ZHU ZHI-JIE, Institute of Geophysics, Academia Sinica, Beijing, People's Republic of China
XIA WAN-REN, Institute of Water Conservancy and Hydroelectric Power Research, Beijing, People's Republic of China

ABSTRACT: In order to justify the rock slope stability of the left abutment at Liujiaxia Hydropower Station, a series of rock mechanics tests was completed on the basis of detailed geological investigation. According to the established geomechanical model, stability analysis has been accomplished by using various methods including FEM etc.. Comparing these results led to a conclusion that the slope is considered to be stable but should be monitored in-situ carefully.

RESUME: Une série de tests mécaniques sur des roches se font basés sur des engnêtes geologique détaillées afin de justifier la stabilité des roches a la rive gauche de la centrale Hydraulique de Liujiaxia. L'analyse de la stabilité a été réalisée en adaptant des méthodes variees selon le modèle de géomecanique y compris la méthode FEM. La justification est faite: stable est la rive gauche, pourtant le contrôle sur les lieux est nécessaire.

ZUSAMMENFASSUNG: Für die Beweisführung der stabilität des Abhangs über den Felsen des rechten Ufers des Liujiaxia Wasserkrafwerkes hat man auf dem Ground der eingchenden geologischen untersuchungen die vershiedenen Versuchstudien der Felsmechanik durchgeführt. Nach dem bestimmten Muster der Felsmechanik hat man mit den verschiedenen Rechnungsmethoden und FEM Methode die stabilitäts analyse durchgeführt. Durch die Beweisführung meint man, deB der Abhang Stabit ist, aber er muB aufmerksam an Ort und stelle kontrolliert werden.

1. INTRODUCTION

Liujiaxia Hydro-power Station is a huge project situated on the upper reaches of the Yellow River. The main dam is of concrete-gravity type with a maximum height of 147 m. The total generating capacity is 1,225 million KW. Owing to tectonic movements, although the individual rock blocks are hard enough with high strength, the rock masses were weakened by various sets of joints and faults. After the completion of this project, not long after the impounding of the reservoir, serious seepage at the left abutment was observed which would lead to rock sliding and thus endanger the normal operation of power house situated just under these potential unstable rock blocks and behind the dam. This unfavourable phenomenon aroused the attention of engineers. So a comprehensive research was conducted which will be illustrated as follows.

2. GEOLOGY

2.1 Geological condition of the dam site

The dam is situated in a narrow V-shaped gorge with steep slopes. The bed rock is predominately composed of micaceous schist and hornblende schist which were partly intruded by a small amount of granite veins and lamprophyre dykes. the geological structure of the dam site is governed by a slightly plunged Hongliugou anticline with its axial plane striking NNW (Fig. 1). According to detailed geological survey, there are five sets of discontinuities distributed in this area (Fig. 2).

From the regional geological point of view, the dam site of Liujiaxia dam is situated in a comparatively stable region. No any active fault has been found to occur since Q_3. This has also been proved by the long-term monitoring data of earthquakes since the first impounding of the reservoir in 1969.

1. loess, 2. micaceous quartz schist, 3. granite vein, 4. fold axis and dips, 5. tectonic and bedding fissure, 6. tectonic crushing belt, 7. normal fault, 8. reverse fault, 9. thrust, 10. boundary of strata, 11. dip and strike of beds, 12. fault zone, 13. outline of the dam

2.2 Geological features and boundary condition of the stability in the left abutment

The rock slope of left abutment stretching N 75°E is 110 m high (with the elevation from 1610 to 1720 m) and 300 m wide. The micaceous schist and hornblende-quartz schist strike N 10°-30°E and dip NW 15°-45° towards the downstream river bed. there is a series of tectonic crushing zones developed nearly parallel to the strata (named as R_p in this paper) with thickness

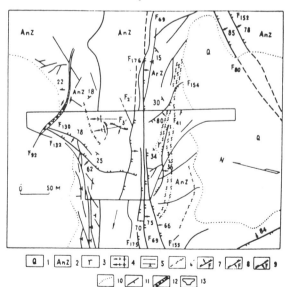

Fig. 1   Geological map of the dam area

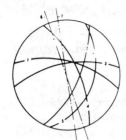

1. shearing fractures
2. shearing fractures
3. tensile fractures
4. compressive shearing fractures
5. bedding compressive shear zone (tectonic crushing belt)
6. direction of the axial plane of Hongliugou anticline
7. direction of the dam axis

Fig. 2 Stereographic projection of the discontinuities in the dam site (by means of the projection of upper hemisphere)

of 1 - 40 cm. Among them the longer ones behind the dam in this area are Rp30, Rp39, R_p5, R_p4, Rp19 etc.. The materials contained in the crushing zones are breccias, mica flakes, chlorite, calcite and clay. In addition to R_p, the main fractures which dominate the rock mass stability are T_{510} (nearly parallel to the slope) and T_{147}, T_{148} (nearly perpendicular to the slope as shown in Table 1.

Table 1   Occurrence of fractures and slope

No.	Occurrence	No.	Occurrence
T_{510}	N60°E, NW, ∠80°	R_p39	N30°E, NW, ∠30°
T_{147}	N40°W, SW, ∠80°	R_p5	N20°E, NW, ∠30°
T_{148}	N30°W, SW, ∠80°	R_p4	N20°E, NW, ∠32°
R_p30	N8°E, NW, ∠15°	R_p25	N25°E, NW, ∠25°
rock slope	N75°E, NW, ∠75°		

According to the occurrence of the fractures and crushing zones in Table 1 and considering their combination and the free boundary of the natural rock slope, the following potential unstable rockmass blocks can be determined by means of stereographic method as shown in Table 2 and Fig. 3.

Table 2   Potential unstable rockmass blocks

Block	Cutting boundaries	Restricted boundary	Sliding boundary	Type of sliding	Volume of block (m³)
A	T_{510}, T_{147}	T_{148}	R_p30	double plane sliding	6,700
B	T_{510}, T_{147}	T_{148}	R_p39	ditto	15,000
C	T_{510}, T_{147}	T_{148}	R_p5	ditto	17,000
D	T_{510}, T_{147}	T_{148}	R_p4	ditto	21,000

3. SEEPAGE ANALYSIS OF THE LEFT ABUTMENT

In view of the above-mentioned geological structure, it seems that the main seepage paths would be along the tensile fractures, such as T_{510}, etc.. So comprehensive measures of grouting and drainage with drainage predominant was adopted (Fu, 1983). The drainage system consists of a curtain of drainage wells, a drainage gallery and a diversion tunnel. In addition, three exploration adits formerly driven for geological investigation at elevations of 1,631 m, 1,660 m and 1,690 m, were reconstructed also as horizontal drainage galleries. On the left bank, one row of grouting boreholes with depths varying from 80 to 120 m were arranged to minimize the influence of seepage flow around the dam. Owing to the existence of a very good drainage system, the long term observation

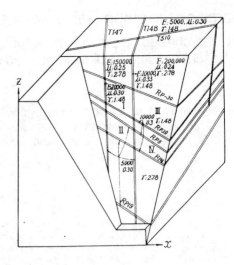

Fig. 3   Sketch of the potential sliding blocks and their properties expressed by E and μ

Note:
Block no.	E kg/cm²	μ
1	150,000	0.25
II	100,000	0.30
III	200,000	0.24
IV	150,000	0.25

in borehole No. 169 located in the region under study shows that the ground water table has always been lower than ele. 1,680 m since the impounding of the reservoir in 1967, although it has raised rapidly and the disalination of water has been observed (Fig. 4). By comparing the above mentioned elevation 1,680 m of water table with the elevation 1,683 m of the lowest sliding boundary R_p4, it is obvious that the seepage flow would not influence the natural stability of the analysed rockmass blocks.

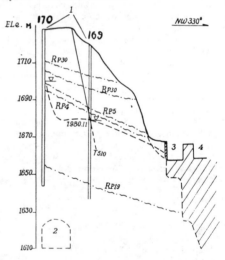

Fig. 4   Geological profile of left abutment 1:1000
1. Borehole,   2. Diversion tunnel,
3. Spillway channel, 4. Power house

4. MECHANICAL PROPERTIES OF WEAK SEAMS AND ROCKMASS

4.1 Microscopic analysis on weak seams

The mechanical genesis of the weak seams was examined under polarized microscope. Besides, the clay mineral contents and the characteristics of the weak seams have been analysed by using: a. The X-ray diffraction, thermo-spectral analysis and infrared spectral analysis; b. Scanning and transmission electron-microscopic observation; c. Chemical analysis; d. Granulometric analysis of the grains sized < 0.1 mm. Test results show that the mineral and chemical properties of the weak seams are favourable for the stability.

4.2 Mechanical test of the weak seams

4.2.1 Medium sized shear tests

23 undisturbed rock samples sized about 20 X 20 X 20 cm³ containing weak seams were taken from the drainage adits and near the surface of left abutment by using pneumatic drill lapping method. And then, these samples were sent to laboratory for shear test. The weak layers generally 1 cm in thickness were controlled as the shearing plane during testing. From the test results, it can be noticed that:
  a. Failure mostly occurred along the contact surface of the hard rock and weak seams, except a few samples failed through the thick clay intercalations.
  b. The shear stress-deformation curves appeared to be elasto-plastic approximately. The shear strengths got from the initial testing with multiple rock samples are nearly the same as that from the repeatative testing with one sample.
  c. The roughness and the characteristics of the filling materials should be considered as the main factors influencing the shear strength. Owing to the wide variety of the weak seam properties in natural condition, the test results were found to be scattered.
  d. The shear strength for crushing zones R_p can be expressed as: $f = 0.42 - 0.52$, $c = 0.9 - 1.9$ kg/cm²; and for other fracture groups T, $f = 0.47$, $c = 1.0 - 1.2$ kg/cm².

4.2.2 Laboratory test for disturbed clayey weak seams

Two kinds of shear tests, including 18 groups of conventional direct shear tests and 4 groups of triaxial tests have been accomplished. The test results are as follows: $f = 0.57 - 0.84$, $c = 0.15 - 0.80$ kg/cm² for group R_p in conventional test; $f = 0.62 - 0.67$, $c = 0.14 - 0.64$ kg/cm² for group R_p, and $f = 0.86$, $c = 0.2$ kg/cm² for group T in triaxial tests. Note that the breccias content in all samples is rather high. Grains with size greater than 2 mm were between 36 - 61%.

4.3 Laboratory test for the hard rock blocks

The conventional mechanical tests as well as the triaxial tests with the servo-controlled testing machine MTS.815 - 03 series have been carried out in laboratory. The test results show that the rock quality of the cryctallized schists are quite well as it is expressed by these figures: unit weight $\gamma = 2.78 - 2.80$ g/cm³, compressive strength $\sigma_c = 1000 - 2500$ kg/cm², modulus of elasticity $E = 300 - 500 \times 10^3$ kg/cm², and the average tensile strength $\sigma_t = 140$ kg/cm². From all these results, it can be seen that the rocks under study can be treated as isotropic linear elastic materials, possessing enough load bearing capacity.

5. ROCK SLOPE STABILITY ANALYSIS

5.1 Three dimensional finite element analysis

5.1.1 Model establishment and its boundary conditions

According to the established geomechanical model, the rock masses were classified into: a. linear elastic element; b. weak seam element. And then, they were divided into thirty kinds of materials. During calculation, spacial tetrahedral elements were adopted. The calculation model consisted of 3,021 nodes and 14,506 elements, including 684 elements of weak layers (Fig. 5). A three dimensional linear FEM programme named KDXK revised after G.H. Shi (Fortran language) was used for calculation.

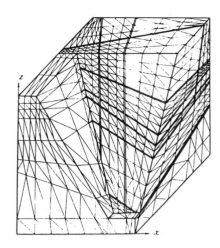

Fig. 5   Calculation scheme for three-dimensional FEM

The scope of the calculation boundary is 100 m long, 90 m wide and 110 m high, which covered the range doubled the analysed potential sliding rock masses approximately.

The geometrical boundaries adopted were proposed as: a. no displacement along the X - direction both on the upstream and downstream sides; b. it is constrained along Y - direction from the inner side; c. no displacement along Z - direction on the bottom.

The mechanical boundaries are as follows: a. stress field due to the overburden of rockmass itself, b. hydro-static pressure γh from the upstream side which is transformed into loads applied on nodal points; c. seepage pressure applied to the rock masses below R_p4, considered as a linear distribution; d. seismic load in the Y - direction, taking $a=0.12g$. Design parameters:
Basically $f = 0.45 - 0.47$ were adopted. moreover, seven different parameters ranging from 0.30 to 0.47 have been taken into consideration for sensitivity study. Cohesion was always neglected during calculation. Other parameters adopted can be seen in Fig. 3.

5.1.2 Calculation results

a. Combination of calculated loads:
  Scheme   I: loads from the weight of rockmasses;
  Scheme   II: loads from rockmasses and seepage pressure
  Scheme III: loads from rockmasses, hydro-static pressure and effect of earthquake.

b. Calculation results
In the calculation of three dimensional FEM, the stresses on the nodal points within the whole rockmass blocks are expressed in Cartesian coordinate system. For the convenience of calculation, the coordinates have been transformed, so as to express the stresses on a local coordinate system with X'Y' plane coinciding with the surface of discontinuities and Z'-axis normal to it. Having the stress data on the local corrdinate system, the normal stress σ'_N and shearing stress τ_N of each nodal point acting on the surface of discontinuity can be obtained. After integrating the stresses on all the nodal points, the normal and shear forces applied on the surface of discontinuity may be solved. And then, projecting them on the whole system, force F_i (F_{ix}, F_{iy}, F_{iz}) applied on the discontinuity i can be known. After summing up the forces applied on all the discontinuities, we can get the resultant force F (F_x, F_y, F_z) applied on the potential sliding rock blocks. Having the values of F, the equilibrium equations can be established by which the safety factor for sliding resistance K may be calculated quantitatively. We can take the case of double plane sliding as an example:

$$S l_o + R_1 l_1 + R_2 l_2 = F_X$$
$$S m_o + R_1 m_1 + R_2 m_2 = F_Y$$
$$S n_o + R_1 n_1 + R_2 n_2 = F_Z$$
$$k = \frac{R_1 f_1 + A_1 C_1 + R_2 f_2 + A_2 C_2}{S}$$

where S ———— sliding force;

R_1, R_2 ——— normal pressures perpendicular to the discontinuities 1 and 2 respectively;

l_o, m_o, n_o; l_1, m_1, n_1; m_2, n_2, l_2 ——— direction cosines of the vectors;

A_1, A_2 ——— area of the discontinuities 1 and 2 respectively;

f_1, f_2; C_1, C_2 ——— frictional coefficients and cohesions respectively.

The calculation results are as follows (Table 3).

Table 3

No of blocks	cases	0.30	0.35	0.37	0.40	0.43	0.45	0.47
A	I	2.365	3.132	3.311	3.580	3.848	4.027	4.206
	II	2.234	2.606	2.755	2.979	3.202	3.351	3.500
	III	1.331	1.552	1.641	1.774	1.907	1.996	2.084
B	I	1.018	1.188	1.256	1.358	1.460	1.527	1.595
	II	0.956	1.115	1.172	1.274	1.370	1.433	1.497
	III	0.682	0.796	0.842	0.910	0.978	1.024	1.069
C	I	1.069	1.247	1.318	1.425	1.532	1.603	1.674
	II	0.992	1.157	1.223	1.322	1.422	1.488	1.554
	III	0.751	0.876	0.926	1.001	1.076	1.126	1.176
D	I	0.851	0.993	1.050	1.135	1.220	1.277	1.334
	II	0.814	0.950	1.004	1.088	1.167	1.222	1.274
	III	0.630	0.736	0.777	0.841	0.904	0.946	0.983

Stability safety factor K

5.2 Stereographic projection method and limit equilibrium analysis

By means of the stereographic projection the spacial combination of the discontinuities and the key blocks tending to sliding can be found conveniently (Goodman and Shi, 1985). And then, the safety factors for sliding were calculated by limit equilibrium method. For this purpose, a SSARP micro-computer programme has been compiled.

5.2.1 The foundamental data are the same as those mentioned above. The design parameters adopted are as follows: a. $f = 0.4$, for group R_p and $f = 0.47$ for group T under normal condition; b. $f = 0.34$ for group R_p and $f = 0.40$ for group T; considering the long term strength due to theological effect. In all cases the cohesion c has not been considered.

5.2.2 The calculation schemes are considered as follows:

I. load from the weight of rock masses under natural condition.
II. considering the decreased f due to time effect.
III. horizontal earthquake forces adding to the normal condition.
IV. earthquake forces directing along the line of intersection of two planes.

The results are shown in Table 4.

6. CONCLUSION

All the results show that the left abutment rock slope is stable under natural condition except that under more unfavourable conditions. But if a macroscopic study is taken into consideration, the following factors would be of importance:

Table 4

No. of blocks	K for cases			
	I	II	III	IV
A	3.06	2.60	1.76	1.79
B	1.02	0.87	0.77	0.81
C	1.27	1.08	0.95	0.99
D	1.19	1.01	0.90	0.94

a. The dam site is situated in a comparatively stable region, no abnormal phenomenon of earthquake has been observed.

b. After impounding of the reservoir, although the underground water seepage has changed much, the seepage would not influence the stability of the rockmasses under study.

c. Certain safety factors must have been contained in all the calculations:

1) All the surfaces of the discontinuities are considered as smooth and continuous planes, but in reality they are rough and discontinuous and some times dislocated by other fractures.

2) Cohesion C has always been neglected, although it is about of the order of 1 kg/cm^2.

3) There is about more than 40% of gravel and sand grains contained in the weak seams. The actual shear strength must be higher.

4) In the calculation for double plane sliding type, the restraining forces are only considered from two planes. Actually, other fractures also possess certain restraining effect.

Based on the above mentioned conditions, it is suggested that:

a. Strengthen the seepage control and in-situ slope monitoring especially for the region around the potential sliding blocks. The surface drainage is also important.

b. Strengthen the earthquake prediction and observation.

c. For the time being, engineering treatment would not be considered except when abnormal instable phenomenon occurs according to in-situ monitoring data.

ACKNOWLEGEMENT:

Sincere thanks are due to Mr. Yang Yingqun, Zeng Liqing and Ms. Zhang Mingyao for their kind help.

REFERENCES:

Fu Bingjun, Zhu Zhi-jie et al. 1983. Analytical experience on the stability of the high dam foundation of the Liujiaxia Hydropower Station. Pro. of 5th Congress ISRM, Melbourne, Australia.

Goodman R.E., Gen-hua Shi, 1985. Block theory and its application to rock engineering, Prentice-Hill Inc.

Hsu Tseng-yan, Ho Sui-hsing, Fu Ping-chun, 1976. Control of seepage through the dam foundation at the Liuchiahsia Hydropower Station. Transaction of the 12th ICOID. Mexico.

Tan Tjong-kie, 1982. The mechanical problems for the long-term stability of underground galleries, Chinese Journal of Rock Mechanics and Engineering, Vol. 1, no. 1.

Testing the axial capacity of steel piles grouted to rock
Essais in situ pour déterminer la capacité axiale de pieux d'acier injectés dans le rocher
Bestimmung der axialen Belastungsfähigkeit von im Gestein einbetonierten Stahlpfählen

A.GARCÍA-FRAGÍO, Hispanoil, Madrid, Spain
E.JAMES, Fugro Ltd, London, UK
M.ROMANA, Intecsa, Madrid, Spain
D.SIMIC, Intecsa, Madrid, Spain

ABSTRACT: The behaviour of steel piles inserted and grouted into pre-drilled holes in a rock mass in studied, reviewing models available in the literature and checking their results by means of in situ tests. The influence of a wide set of variables such as rock mass properties, pile size, installations, procedures and type of loading has been considered. Only side resistance effects have been analyzed, in relation to load-deflection behaviour and load capacity of the pile. Test results are interpreted on the basis of empirical criteria as well as on the understanding of the measured stress state around the pile shaft, and emphasis in places on a good geomechanical assessment of the rock mass to gain a proper insight of the phenomena anayzed.

RESUME: Le comportement des pieux en acier injectés dans forages executés au prealable dans un massif rocheux est etudié. On analyse les modeles qui sont disponibles dans la bibliographie à la lumière des resultats d'essais in situ. L'influence d'un grand nombre de variables tel que les propietés de la roche et du pieu ainsi que le modele de chargement est consideré du point de vue de la resistance au frottement lateral.

ZUSAMMENFASSUNG: Der Einfluss von Gesteinseigenschaften, Pfahlgrösse, Installation und Belastungsprozess auf die Belastbarkeit von Pfählen in bezug auf das Belastungs/Verformungsverhalten wurde bestimmt. Ansätze aus der Literatur wurden mit in situ Messungen verglichen.

1. INTRODUCTION

The Spanish public-owned firm ENIEPSA is the operator of a group who installed a production platform 10 km offshore in the Gulf of Biscay, in a site 100 m of water depth where sea bottom is formed by a thin layer of detritic sediments overlying a calcareous claystone. The proposed template type platform rests on four legs, each of them transferring the foundation loads to a group of five piles drilled and grouted to the rock mass designed only to mobilize skin friction. This kind of foundation can be termed as singular given that, although its closest typology is the rock socket used on-shore, specific differences make it difficult directly apply the design criteria of socketed piles to the platform foundation:

.   the piles will have higher slenderness ratio (length ofer diameter) than that of the rock sockets studied in the literature.

.   the offshore construction of the piles (drilling, inserting and grouting) in a severe environment could result in the rock degradation, lower quality of the grout and, in general, in the need of a good construction control.

These and other aspects lead ENIEPSA to undertake the study of the behaviour of this type of piles by means of field tests. Obvious feasibility reasons advised to set an onshore test site in similar rock and model piles of smaller size, carefully simulating the off-shore construction procedures.

2. EXPERIMENTAL STUDY OF THE PILE BEHAVIOUR

The above considerations led to the design of a test programme aimed at:

-   Checking the validity of the empirical rules to obtain the average side resistance at the grout/rock interface for large diameter piles (475 mm diameter). The Williams and Pells (1980) criterium, which states that the ultimate bond strength is a percentage of the rock compressive strength, is the most widely used model.

-   Studying the influence of installation procedures on the overall behaviour of the piles.

Measuring the stress distribution along the grouted pile section in order to gain a deeper insight in the local mobilisation of strength and the corresponding strain path, which - should be useful for the study of the mechanism of shearing and the pile/rock interaction.

These ends require a careful control of the variables that could have influence in the test results. They are comprehensively shown in table 1 in the form of a flow chart to point out their influence in the test's strategy.

3. ROCK MASS PROPERTIES AT THE TEST SITE

Geologically described as a reddish-brown calcareous claystone pertaining to the Upper Cretacic - Eocene flysch, this soft rock is closely fractured with joints exhibiting slickensided surfaces and calcite infill. Geomechanical properties are shown in table 1.

Table 1
Geomechanical properties of the rock

Rock Matrix

- Bulk density (KN/m^3)	25.4	
- Porosity (%)	7.8	
- Mineralogy:		
. Calcite (%)	60	
. Caly minerals (%)	30	(mainly Illite, traces of smectite)
. Quartz (%)	10	
- Unconfined compressive strength (MPa)	21.5	
- Tension strength (MPa)	2.4	

Discontinuities

- Number of families	3	(bedding + 2 sets)
- Bedding dip	41º	
- Spacing (cm)	20-30	
- RQD (%)	73	

Geomechanical classifications

- RMR (Bieniawski, 1979)	50
- Q (Barton, 1975)	1.8

This rock mass can be described as a fair rock (Class II according to Bieniawski) or a poor rock (according to Barton et al; 1974). As the bedding is the main discontinuity system, the other two sets being scattered and randomly orientated, the hole walls after drilling remain quite smooth as measured by the caliper logs. Asperities somewhat smaller than 3 mm were detected. Neverthelss, the mineralogy of the rock suggests a not very high slaking resistance which could result in hole wall weathering.

4. TEST PILES

The programme included two series of axial load tests:

(i) Nine tension tests on small diameter piles (158 mm dia) performed to check the effect of various drilling fluids and additives on the rock/grout side resistance. Three sets of three piles each were constructed, respectively using seawater only, an inhibitor and a polymer. This piles were 7 m long, having grouted their bottom 1 m.

(ii) Five axial loading tests of large diameter piles (460 mm dia). Three of the tests were in tension loading and the ofher two in compression. This piles were 9 m long, having grouted their bottom 3 m.

All piles were provided with circunferential shear connectors on their outside surface down the grouted length. A poliuretane soft-toe was appended in their bottom in order to eliminate end bearing effects. Figure 1 shows the grouting arrangements of a pile.

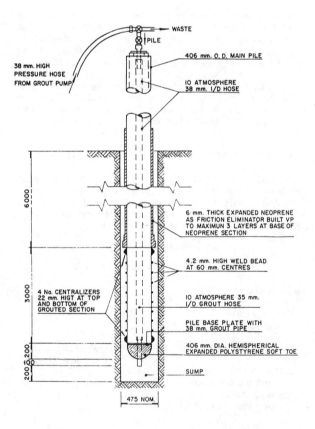

Figure 1. Grouting arrangements of a test pile.

5. INSTALLATION PROCEDURES

All piles were drilled with tricone rock roller bits. Direct circulation methods were used for the Minipiles and reverse circulation for the larger ones.

Different drilling fluids were used as follows:

- Minipiles 1, 2 and 3; Main Piles 1, 2, 3 and 4: sea water.

- Minipiles 4, 5 and 6; Main Pile 5 sea water with 30 lb/bbl of KCL.

- Minipiles 7, 8 and 9: Sea water with 30 lb/bbl of KCL and 2 lb/bbl of a polyanionic cellulosic polymer.

All shafts were flushed with a pyrophosphate scouring agent prior to grouting. The grout mix was based on a class G oilwell cement mixed with seawater. It was prepared in a colloidal mixer and the specific gravity was checked with a mud balance. A high pressure grout pump sent the mix through a stinger inside the pile down the hole bottom, from where it filled the pile/rock annulus. Cube samples were taken and stored in a curing bath and their strength was obtained at different time intervals to check the resistance gain. Minimum strengths of 40 MPa were specified at the pile load test, which were attained within 2-6 days after grouting.

6. PILE INSTRUMENTATION

Pile head movement was measured using Linear Variable Displacement Transducers (LVDT'S). Two LVDT's were fixed at diametrically opposite locations on each pile, so that by averaging the displacement reading the effect of bending of the pile was eliminated. The LVDT's were mounted on ring clamps on the mini-piles, and bolted to the triangular pile supports on the main piles. The LVDT's were bearing on to glass plates fixes to reference beams which passed each side of the pile.

An independent check of pile displacements was made using a Wild T-3 one second micrometer level, sighting on a target attached to the pile.

The primary method of measuring the load transferred to the pile was by the use of extensometric load cell of 1.000 tonne range, giving 0,5 per cent accuracy. The load cells were placed between the jack and a hemispherical seating which bore onto the pile head. The load cells were calibrated in an official Spanish laboratory prior to and during the testing programme.

The input voltage to the load cell strain gauge circuits was supplied from a Hewlett Packard stabilized voltage supply, and the signal output was monitored from each load cell by the data logger. The output from the load cells was therefore monitored and recorded each time the strain gauge reading and jack oil pressure readings were taken.

The oil pressure at the inlet to the jacks was monitored using pressure transducers. The oil pressure can be used with the cross-sectional area of the jack cylinders to give an estimate of the load applied to the pile. The actual load applied to the pile head will vary from this estimate by the amount of friction in the jacks.

The load distribution along the pile was monitored with strain gauges. Bonded foil gauges, consisting of four resistor elements per gauge and were mounted such that two elements were aligned axially and two circumferencially. The four resistors were connected in series and all four corners of the gauge were then connected to the instrument cabin using 4 core shielded cable.

Protection of gauges, stress relief and cable connections was formed using an initial plastic rubber coating, a second coating of harder rubber a thin metal shim and a final sealing adhesive layer (final thickness approximately 10 mm and area approximately 50 mm x 50 mm). A metal protection plate was also fixed just below each gauge.

Each instrummented level of the pile had four gauges at 90º spacing. Each pile had up to 14 intrumented levels, the first 12 located within the grouted section.

Lead connections were made in the instrument cabin so that some gauges were monitored as axial quarter bridges, some gauges as circumferential quarter bridges and some gauges were monitored ad full bridges. For those gauges monitored as axial of circumferential quarter bridges, the bridge circuit was completed using a "dummy" circuit of three balancing resistors.

Specially fabricated units were fixed to the main piles to detect cracking within the grouting annulus near the pile surface, in the body of the grout or near the rock surface.

The analogue outputs from all strain gauge transducer circuits and detectors were monitored using a 100 channel data logging unit under the control of a desk top computer. Sequencing of signals and the time interval of recording pulses were set at the computer. Analogue to digital conversion was performed and the digital outputs were stored on magnetic tape cassettes in the computer.

7. TESTS SET-UP AND LOADING SEQUENCE

All tension tests were carried out transmitting the reaction loading to two I beams resting on the foundation RC beams located at each side of the piles, as shown in figure 2.

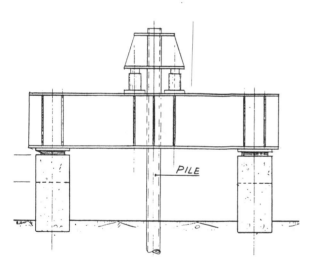

Figure 2. Tension tests.

300 tonne capacity jacks were used to apply the load to the pile head, using two units for the Minipiles and four units for the larger tests. For the compression tests, the arrangement was somewhat different, using as a reaction two adjacent piles which had been previously tested in tension and re-grouted (see figure 3). The same I beams were used as a load frame in this case.

Loading sequences were devised to reach a maximum average grout/rock - shear stress of 2,50 MPa in the case of the Main Piles and 4,70 MPa in the Minipiles. Loads were applied by increments maintained for a specified minimum duration. Rates of deformation at the pile head were less than 0,25 mm/hour before subsequent loading. After the performance of each static tests, a constant rate of uplift (or penetration) test was carried out at a rate of 0,5 mm per minute.

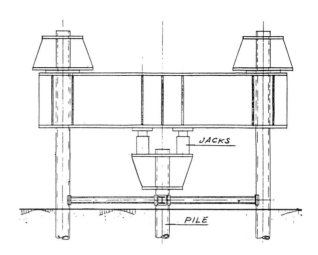

Figure 3. Compression tests.

8. TEST RESULTS

Some significant load/deflection curves obtained from static test are shown in figures 4 and 5.

The distributions of radial and shear stresses at the pile/grout interface are shown in figures 6 and 7 against depth down the grouted section for a typical test pile during the loading steps. It can be seen that the shear stress distribution generally follows the expected form given by elastic solutions (Poulos and Davis, 1.980) with higher values both at the top and bottom of the grouted section. On the contrary, the radial stresses reach values significantly higher than the predictions made by the elastic models (Frank, 1.975), even at low loading stages. It can be also seen that significant radial stresses are locked into the pile during unloading, suggesting an interface behaviour governed by a dilatancy mechanism.

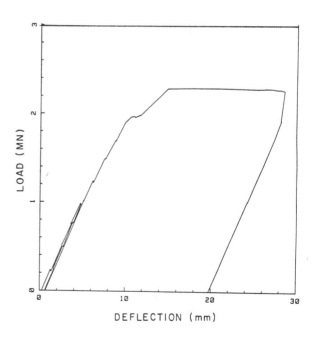

Figure 4. Load-deflection curve for a Minipile test.

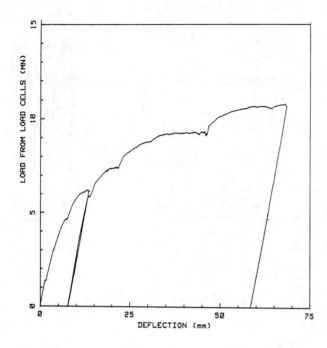

Figure 5. Load-deflection curve for a Main Pile test.

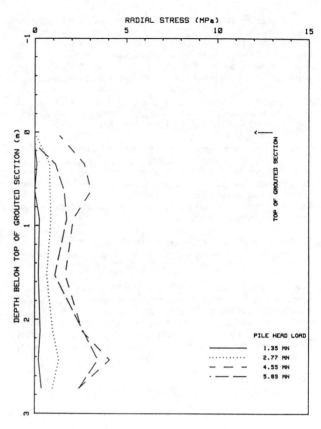

Figure 7. Pile-grout radial stress.

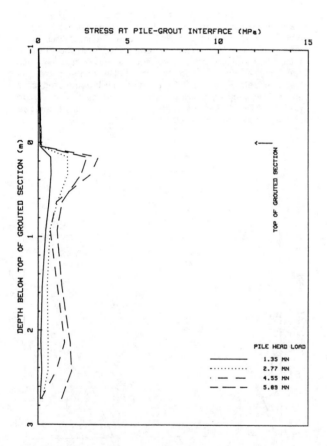

Figure 6. Pile-grout shear stress distribution.

CONCLUSSIONS

It is confirmed from the pile tests and from considerations of other data that the Williams and Pells approach is an appropiate method for assessing the ultimate bond shear strength for piles grouted to rock. In our case, the use of their formula gives a value of about 2,00 MPa for the ultimate bond shear strength of the grout/rock interface, which is a conservative estimate of the ultimate strengths measured on the 475 mm main piles. However, some uncertainties still exist in the pile design of which the prime ones are:

a) the inability of the William and Pells formulation to accomodate the varying stress along the length of the pile.

b) the limited applicability of the formula to a range of length to diameter values.

c) rock degradation may occur, particularly if shafts remain open for prolonged periods.

ACKNOWLEDGEMENTS

The authors wish to thank to ENIEPSA for their kind permission to publish the tests result. They are also grateful to the consulting firms INTECSA (Madrid) and FUGRO (London and Leidschendam) who carried out the tests.

370

REFERENCES

ASTM, 1981. Standard Method of Testing. Piles under static axial compressive load. Annual Book of ASTM standards, Part 19.

ASTM, 1981. Standard Method of Testing. Piles under static tensile load. Annual Book of ASTM standards, Part 19.

API, 1979. Planning designing and constructing fixed offshore platforms.

BARTON, 1975. Classification of rock masses. Preprint nº 75-AM-337. Soc. Mining Engnrs.

BIENIAWSKI, 1979. The geomechanics classification in rock engineering applications. Int. Symp. Rock Mech. Montreal.

POULOS, H.G., DAVIS, E.H., 1974. Pile foundations analysis and design. John Wiley and Sons.

WILLIAMS, A.F., PELLS, P.J.N., 1980. Side resistance of rock sockets in sandstone, mudstone and shale. Can. Geotech. J. Vol. 18.

TABLE 1

PILE SIDE RESISTANCE UNDER AXIAL LOADING
DEFINITION OF TEST PARAMETERS

Geomechanics of the surrounding rock

- **Rock matrix propererties**
  - Unconfined compressive strength
  - Tension strength
  - Modulus of elasticity
  - Mineralogy
- **Discontinuities**
  - Geological definition of the sets
  - RQD
  - Spacing
  - Direction
  - Persistence
  - Aperture and filling

Rock mass behaviour
- Strength
- Deformability
- Weathering
- Seawater saturation

SELECTION OF THE TEST SITE IN A REPRESENTATIVE GEOLOGIC FORMATION

Failure mechanism under axial load

- **Steel / Grout failure**
  - Radial stiffness of steel pile and grout
  - Radial stiffness of the rock mass
  - Connectors of shear stress for steel / grout interface
- **Grout / rock failure**
  - Rock strength
  - Hole wall roughness and cleanness
- **Uplift of a rock mass**
  - Discontinuities direction
  - Shear strength along discontinuityg plane

Pile characteristics
- Scaling of pile length, diameter and wall thickness
- Hole diameter
- Steel strength
- Grout strength
- Shear connectors: height and spacing

DESIGN OF SMALL DIAMETER TEST PILES FOR PARAMETRIC PURPOSES

DESIGN OF LARGE DIAMETER TEST PILES FOR QUANTITATIVE ASSESSMENT

Pile construction

- **Drilling**
  - Hole wall roughness and cleanness
  - Rock weathering at the wall
  - Rotation vs. percussion drilling
- **Grouting**
  - Density of the mix
  - Workability
  - Setting time
  - Shrinkage

Installation procedures
- Direct vs. reverse circulation methods of cutting's extraction
- Drilling muds for wall stability
- Use of inhibitors of rock weathering
- Use of polymers to increase drilling rates
- Type of cement. Use of seawater
- Water cement ratios
- Setting time controls

TEST OF DIFFERENT DRILLING FLUIDS

SELECTION OF CONSTRUCTION PROCEDURES SIMILAR TO OFFSHORE PRACTICE

Load pattern and test monitoring

- **Ultimate load**
  - Static vs. constant rate of increment
  - Compression vs. tension loading
  - Faligue effects
- **Deformability**
  - Recoverable and unrecoverable
  - Rate of deflection
  - Strain distribution along pile shaft

Stress - Strain parameters
- Radial stress distribution
- Shear stress distribution
- Pile deflections
- Pile head loads
- Location of failure interface
- Temperature

DESIGN OF PILE INSTRUMENTATION AND DATA ACQUISITION SYSTEM

About the mean area of a joint set

Sur l'aire moyenne d'une famille de diaclases
Über die mittlere Fläche einer Kluftschar

NUNO FEODOR GROSSMANN, Laboratório Nacional de Engenharia Civil, Lisboa, Portugal
JOSÉ J.R.D.MURALHA, Laboratório Nacional de Engenharia Civil, Lisboa, Portugal

ABSTRACT: The general expression, giving the mean area of a joint set as a function of the mean trace length of that set on an observation surface, as well as the formulas for the usual types of observation surfaces — general closed convex surface (ISM borehole core), general plane convex surface (rectangle, square, circle), and limited stretch of an adit with a rectangular cross-section — are deduced, assuming that the joint areas follow a Bessel distribution. A case history, concerning a location where three different joint samplings — on limited adit stretches, on ISM boreholes, and on circular surfaces — were successively carried out, is presented, showing a good agreement between the three independent mean area values obtained for each joint set. Finally, some remarks about an attempt to deduce those mean areas, based on the sampling performed in three neighbour parallel boreholes at the same location, are made.

RÉSUMÉ: L'expression générale, donnant l'aire moyenne d'une famille de diaclases en fonction de la longueur moyenne de la trace de cette famille sur une surface d'observation, ainsi que les formules pour les types usuels de surfaces d'observation — surface fermée convexe générale (carotte de sondage ISM), surface plane convexe générale (rectangle, carré, cercle), et tronçon limité d'une galerie avec une section droite rectangulaire — — sont déduites, admettant que les aires des diaclases suivent une distribution de Bessel. On présente une étude de cas, concernant un endroit où trois échantillonnages différents des diaclases ont été réalisés successivement — ces échantillonnages portant sur des tronçons limités de galeries, sur des sondages ISM, et sur des surfaces circulaires —, laquelle montre une bonne concordance entre les trois valeurs indépendantes de l'aire moyenne obtenues pour chaque famille de diaclases. Finalement, on fait quelques remarques sur une tentative pour déduire ces aires moyennes, ayant pour base l'échantillonnage effectué au même endroit dans trois sondages voisins parallèles.

ZUSAMMENFASSUNG: Die allgemeine Gleichung, die die mittlere Fläche einer Kluftschar als eine Funktion ihrer mittleren Ausbißlänge auf einer Beobachtungsfläche darstellt, sowie die Formeln für die gewöhnlichen Beobachtungsflächentypen — allgemeine geschlossene konvexe Fläche (ISM Bohrlochkern), allgemeine ebene konvexe Fläche (Rechteck, Quadrat, Kreis), und begrenzte Strecke eines Stollens mit rechteckigem Querschnitt — werden abgeleitet, unter der Annahme, daß die Kluftflächen einer Besselverteilung folgen. Eine Fallstudie, die eine Stelle betrifft, an der drei verschiedene Klufterfassungen — an begrenzten Stollenstrecken, an ISM Bohrlöchern, und an kreisförmigen Flächen — nacheinander durchgeführt wurden, wird vorgestellt, wobei sich eine gute Übereinstimmung der drei unabhängigen Werte der mittleren Fläche zeigt, die für jede Kluftschar erhalten wurden. Schließlich, werden einige Bemerkungen über einen Versuch gemacht, jene mittleren Flächen mittels der an der gleichen Stelle, in drei benachbarten parallelen Bohrlöchern durchgeführten Klufterfassung abzuleiten.

1 INTRODUCTION

Of the four basic geometric parameters characterizing a joint set — attitude, spacing, aperture, and area— — the area is the one less spoken of, because in most cases no direct sampling of the joint areas can be performed. However, the areal extent of the joints being a factor of great influence in many engineering problems, an easier to sample parameter — the intersection, trace length, or persistence — is usually studied, in order to get an idea on the relative size of the joints.

In the following, a relation between the mean area of the joints of a given joint set, and the mean intersection of those joints with the chosen observation surface, will be deduced, and a practical example given, showing the good agreement between the results obtained for several independent samplings on different observation surfaces of the same rock mass.

2 MEAN INTERSECTION

Let us consider a joint j, having an attitude (σ, δ) (σ being the strike, and δ the dip) and an area A. Let us further suppose that, when the centre of the joint j is located at the volume element dV, an intersection i with the observation surface S is obtained

ed. The mean intersection $\bar{i}$ of the joint j with the surface S, for all possible centre locations, will then be given by

$$\bar{i} = \frac{\iiint i\, dV}{\iiint dV} \tag{1}$$

provided that the integration domain of both integrals is limited to those volume elements for which

$$i > 0 \tag{2}$$

The integral of the numerator in expression (1) can be easily calculated (Grossmann 1967), and yields

$$\iiint i\, dV = A \iint_S \sin \alpha \, dS \tag{3}$$

α denoting the angle between the normal to the joint j, and the normal to the surface element dS of the observation surface S. The integral of the denominator in expression (1) corresponds to the volume $V_A(\sigma, \delta)$ in the neighbourhood of the surface S which is the locus of all possible centres for a joint, having an attitude (σ, δ) and an area A, and intersecting that surface.

Although it is not possible to write the analytical expression for the volume $V_A(\sigma, \delta)$ in the general case, several upper limits (and some exact solutions for special cases) could be deduced with the help of the notion of equivalent radius R (Grossmann 1984), which is defined by

$$R = \sqrt{\frac{A}{\pi}} \qquad (4)$$

For a single closed convex observation surface S, delimitating the volume V, and with a distance d between its two tangent planes having the attitude (σ, δ), for instance, we may state

$$V_A(\sigma, \delta) \leqslant V + R \iint_S \sin \alpha \ dS + \pi R^2 \ d \qquad (5)$$

So far, we have examined the mean intersection $\bar{I}$ of a joint j (having an attitude (σ, δ) and an area A) with the observation surface S. The results obtained are, however, equally valid for the mean intersection $\bar{I}$ of the joints of a joint set J (having all that same attitude (σ, δ) and that same area A) with that same observation surface S.

3 JOINT AREA DISTRIBUTION

If we want to eliminate the condition that the joints of the joint set J present a constant area distribution, we shall have to replace both terms of the fraction in expression (1) by the corresponding expectation for the area distribution.

The new numerator will be

$$\int_0^\infty (\iiint i \, dV) f(A) \, dA = \bar{A} \iint_S \sin \alpha \ dS \qquad (6)$$

where $\bar{A}$ stands for the expected value (mean) of the area distribution of the joint set J. The new denominator shall correspond to the volume $V(\sigma, \delta)$, given by

$$V(\sigma, \delta) = \int_0^\infty V_A(\sigma, \delta) f(A) \, dA \qquad (7)$$

All upper limits or exact solutions for $V_A(\sigma, \delta)$ known so far, obey to the general expression

$$V_A(\sigma, \delta) \leqslant a + bR + c\pi R^2 \qquad (8)$$

in which a, b, and c represent three different values, depending only on the observation surface geometry and the attitude (σ, δ). Using expressions (4), (7), and (8), we may, therefore, write

$$V(\sigma, \delta) \leqslant a + b \bar{R} + c \bar{A} \qquad (9)$$

where $\bar{R}$ stands for the expectation of the equivalent radius, for the area distribution.

4 BESSEL FUNCTION DISTRIBUTION

Up to now, we have assumed a general area distribution. There are, however, reasons to believe that the areas of the joints of a joint set follow a Bessel function distribution.

To begin with, experience has shown that most joints end at other joints. Kikuchi, Kobayashi, Inoue, and Izumiya (1985), for instance, present a case study in which, from 616 joint ends, only 11,5% where considered as "disappearing", this number still being an upper limit, as it is possible that some of those joints continued as "non-visible" microfissures. As a first approach, we will, therefore, suppose that all joints end against other joints.

This supposition implies that the joints are polygons, their area, consequently, depending basically on the product of two distances between "opposite" sides of the polygon. These distances, however, correspond to distances between sucessive joints along straight lines, for which already Priest and Hudson (1976) have shown that they usually follow a negative exponential distribution. The joint area distribution can, therefore, be obtained by multiplying two negative exponential distributions, an exercise which

yields a Bessel function distribution (Grossmann 1985).

On the other hand, Hudson and Priest (1979) already showed a good agreement between area data from 10 different rock masses, and Bessel function distributions.

In terms of the mean joint area $\bar{A}$, the probability density function f(A) of the joint area A of a joint set J, presenting a Bessel function area distribution, is

$$f(A) = \frac{2}{\bar{A}} K_o (2 \sqrt{\frac{A}{\bar{A}}}) \qquad (10)$$

where $K_o(2 \sqrt{A/\bar{A}})$ denotes the modified Bessel function of the 2nd kind and zero order, with the argument $(2 \sqrt{A/\bar{A}})$.

The expectation $\bar{R}$ of the equivalent radius, for the Bessel function area distribution, will be, using expression (4)

$$\bar{R} = \int_0^\infty \sqrt{\frac{A}{\pi}} \cdot \frac{2}{\bar{A}} K_o (2 \sqrt{\frac{A}{\bar{A}}}) \, dA = \frac{\sqrt{\pi}}{4} \sqrt{\bar{A}} \qquad (11)$$

5 GENERAL EXPRESSION

Using expressions (6), (9), and (11), we may now write

$$\bar{I} = \frac{\bar{A} \iint_S \sin \alpha \ dS}{a + \frac{\sqrt{\pi}}{4} b \sqrt{\bar{A}} + c \bar{A} - \Delta} \qquad (12)$$

where Δ stands for the non-negative quantity which must be added to the right-hand side of expression (9), in order to transform this inequality into an equality.

From expression (12), we may finally obtain

$$\bar{A} = \pi \left[\frac{1 + \sqrt{1 + \frac{1}{\pi}(a - \Delta)(\frac{8}{b})^2 (\frac{1}{\bar{I}} \iint_S \sin \alpha \ dS - c)}}{\frac{8}{b}(\frac{1}{\bar{I}} \iint_S \sin \alpha \ dS - c)} \right]^2 \qquad (13)$$

6 SPECIAL CASES

For the single closed convex observation surface S, pertaining to expression (5), expression (13) changes into

$$\bar{A} = \pi \left[\frac{1 + \sqrt{1 + \frac{64(V - \Delta)}{\iint_S \sin \alpha \, dS} \cdot (\frac{1}{\bar{I}} - \frac{d}{\iint_S \sin \alpha \, dS})}}{8(\frac{1}{\bar{I}} - \frac{d}{\iint_S \sin \alpha \, ds})} \right]^2 \qquad (14)$$

with

$$\Delta < V \qquad (15)$$

and for an ISM sample, with a diameter Ø and a length L, and for which ε is the angle between the corresponding borehole axis and any normal to the joints of the joint set J (Grossmann 1977), we may obtain

$$A = \pi \left[\frac{1 + \sqrt{1 + \frac{16 \, Ø \, L}{2E(\varepsilon)L + \frac{\pi}{2}Ø \sin \varepsilon} \cdot (\frac{1}{\bar{I}} - \frac{L \cos \varepsilon + Ø \sin \varepsilon}{Ø [2E(\varepsilon)L + \frac{\pi}{2}Ø \sin \varepsilon]})}}{8 (\frac{1}{\bar{I}} - \frac{L \cos \varepsilon + Ø \sin \varepsilon}{Ø [2E(\varepsilon)L + \frac{\pi}{2}Ø \sin \varepsilon]})} \right]^2 \qquad (15)$$

where $E(\varepsilon)$ denotes the complete (between 0 and $\pi/2$) 2nd kind elliptic integral, with the modular angle ε.

The deduction of this last expression assumes

$$\Delta \approx 0 \qquad (17)$$

and, as for most practical cases

$$L \gg \emptyset \ tg \ \varepsilon \qquad (18)$$

expression (16) may still be simplified, yielding

$$\overline{A} = \pi \left\{ \frac{\left[1 + \sqrt{1 + \dfrac{8\emptyset}{E(\varepsilon)} \cdot \left[\dfrac{1}{\overline{i}} - \dfrac{\cos \varepsilon}{2\emptyset E(\varepsilon)} \right]} \right]^2}{8 \left[\dfrac{1}{\overline{i}} - \dfrac{\cos \varepsilon}{2\emptyset E(\varepsilon)} \right]} \right\} \qquad (19)$$

It should be noted that the quantity $2\emptyset E(\varepsilon)/\cos \varepsilon$ corresponds to the length of a full intersection in the central zone of the integral sample, by a joint of the joint set J.

For a single plane convex observation surface S, expression (13) reduces to (Grossmann 1984)

$$\overline{A} = \frac{\pi}{4 \left(\dfrac{1}{\overline{i}} - \dfrac{d}{S \sin \alpha} \right)^2} \qquad (20)$$

yielding for a rectangular observation surface, with sides m and n, and for which ω is the angle between the projections on the observation surface plane of the normals to the joints of the joint set J, and a side m of the rectangle (Grossmann 1977)

$$\overline{A} = \frac{\pi}{4 \left(\dfrac{1}{\overline{i}} - \dfrac{m|\cos \omega| + n|\sin \omega|}{m \ n} \right)^2} \qquad (21)$$

for a square observation surface, with sides m

$$\overline{A} = \frac{\pi}{4 \left(\dfrac{1}{\overline{i}} - \dfrac{|\cos \omega| + |\sin \omega|}{m} \right)^2} \qquad (22)$$

and for a circular observation surface, with a radius R_o (Grossmann 1977)

$$\overline{A} = \frac{\pi}{4 \left(\dfrac{1}{\overline{i}} - \dfrac{2}{\pi R_o} \right)^2} \qquad (23)$$

The analysis of expressions (20) to (23) shows that for any single plane convex observation surface

$$\overline{A} \gtrsim \pi \left(\frac{\overline{i}}{2} \right)^2 \qquad (24)$$

The difference between the exact solution and this lower limit diminishes with the increase of the observation surface.

For a limited stretch of an adit, with a height h, a length l, and a width w, and which allows the study of the joints only on both walls and the roof, we may write (Grossmann 1977)

$$\overline{A} = \frac{\pi}{16 \left[\dfrac{1}{\overline{i}(1+\Delta)} - \dfrac{h|\cos \psi_h| + l|\cos \psi_l| + w|\cos \psi_w|}{(2h|\sin \psi_h| + w|\sin \psi_w|)l} \right]^2} \qquad (25)$$

with

$$0 \lesssim \Delta \lesssim 1 \qquad (26)$$

if the angle defined by the normals to the joints of the joint set J with the normals to the roof of the stretch, with the adit axis, and with the normals to the walls of the stretch, is denoted by ψ_h, ψ_l, and ψ_w, respectively.

7 CASE HISTORY

In a granitic rock mass in northern Portugal, studies for the best emplacement of an arch dam have been carried out in successive stages, some of them having included jointing characterizations.

The jointing of the lower zone of the right embankment (Fig. 1) was studied in three of those stages,

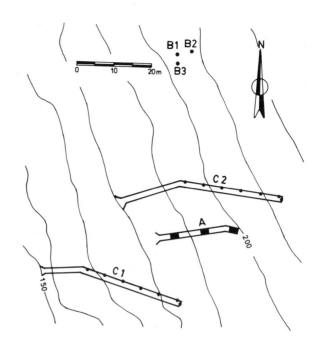

Figure 1. Location of the different joint samplings
A - limited stretches of the adit
($h \approx 1,85$ m; $l \approx 1,80$; $w \approx 1,35$ m)
B - ISM boreholes
(vertical; $\emptyset = 71,5$ mm; $L \approx 35$ m)
C - circular observation surfaces
($R_o = 0,75$ m)

using each time a different sampling technique. Some of the results of the three jointing characterizations of the considered zone, each of which was carried out independently, are summarized in table 1.

Table 1. Joint set parameters

Set	Sampling	Strike (^{o})	Dip (^{o})	Spacing (m)	Mean Area (m^2)
H^{13}_{160}	Stretch	167	25	1,47	3,16
	Borehole	144	11	0,62	1,48
	Circle	164	12	0,27	0,50
	Mean	160	13	0,50	1,07
H^{38}_{313}	Stretch	325	32	0,88	1,01
	Borehole	307	45	1,18	4,21
	Circle	307	40	1,09	0,32
	Mean	313	38	1,03	1,73
V^{75}_{336}	Stretch	334	79	2,36	7,02
	Borehole	332	67	1,01	1,02
	Cicle	338	78	0,51	0,40
	Mean	336	75	0,89	1,41
V^{79}_{213}	Stretch	209	78	2,05	∞
	Borehole	215	78	0,87	4,79
	Cicle	212	81	1,32	15,61
	Mean	213	79	1,25	>10,84
V^{82}_{86}	Stretch	83	86	1,39	0,69
	Borehole	91	79	0,72	1,75
	Circle	85	83	0,34	0,86
	Mean	86	82	0,59	1,08

The analysis of table 1 shows that the joints occurring at this site belong mainly to 5 different joint sets, which do not differ significantly as concerns their intensity (the spacings vary between 0,50 m and 1,25 m). Their mean area, however, presents a clear difference between the joint set V_{213}^{79} (a value of more than 10,84 m^2) and the 4 other joint sets (values in the range 1,07-1,73 m^2). The analysis of the data referring to the 3 independent jointing studies, further shows that for all of them the joint set V_{213}^{79} posesses the largest mean area.

Although a first glance at table 1 gives the impression that the 3 values, calculated from the 3 joint samplings for a same joint set parameter, differ very much from each other, a variability analysis showed that the dispersion of those 3 values is of the same magnitude as e.g. the dispersion for the 3 corresponding values, calculated using only the data of one ISM borehole at each time.

A further interesting aspect of the jointing at this site is that all 5 joint sets make an angle of about 55° with the attitude P (strike - 36°; dip - 48°), and that for each of those joint sets, the direction of minimum dispersion is approximately parallel (deviations of less than 20°) to the great circle containing the poles of the attitude P and the poles of the mean attitude of the considered joint set.

8 THREE NEIGHBOUR PARALLEL BOREHOLES

Due to the fact that the three ISM boreholes of the above-mentioned case history were parallel, and laid very close to each other (distances of about 5 m), an attempt was made to deduce the mean area of the joint sets present through a different approach. It consisted in determining, for each joint set present, the proportion of joints which cut only one, any two, or all three boreholes, and comparing these figures with the theoretical ones, calculated from known area distributions.

The assessment of whether two joint traces, appearing on two different boreholes, belonged to a same joint, was established by guaranteeing that, in both ways, the intersection with the second borehole of the prolongation of the plane, defined by one joint trace, fell in a certain vicinity of the other joint trace. The vicinity corresponded to the locus of all intersections with the second borehole, obtained allowing for a 5° inaccuracy in the definition of the attitude of the afore-said plane.

In order to consider that three joint traces, appearing in three different boreholes, belonged to a same joint, the above-mentioned criteria had to be satisfied by all three joint trace pairs.

The theoretical proportions of joints of a given joint set, cutting only one, any two, or all three boreholes, were evaluated, in the first place, under the assumption of a constant area distribution for the joint set, by establishing the volumes around the three boreholes which are the loci of all possible centres of the joints with a given area, and belonging to the considered joint set, cutting only one, any two, or all three boreholes. With the help of those proportions, the figures for the different proportions of joints of that set, cutting only one, any two, or all three boreholes, under the assumption of a Bessel function area distribution for the joint set, were obtained by

$$p_{Bi}(\bar{A}) = \int_0^\infty p_{ci}(A) \cdot \frac{2}{\bar{A}} K_o\left(2\sqrt{\frac{A}{\bar{A}}}\right) dA \qquad (27)$$

where $p_{ci}(A)$ stands for the proportion of i borehole cuts in the case of a constant area distribution with an area A, and $p_{Bi}(\bar{A})$ for the proportion of i borehole cuts in the case of a Bessel function area distribution with a mean area $\bar{A}$.

The comparison between these theoretical proportions and those obtained in the field, yielded, for the joint set H_{160}^{13}, a mean area of less than 7,8 m^2, and, for the joint set V_{336}^{75}, a mean area of less than 6,7 m^2.

These figures are much higher than those presented in table 1, but it must be borne in mind that, when the spacing of a joint set is small in comparison with the chosen borehole length, in which we look for the intersection of a joint, detected in another borehole, the simple existance of a trace of a joint with a similar attitude is not sufficient evidence that we are dealing with the same joint. So, the afore-said figures must be considered only as upper limits for the mean area, because any decrease in the proportions of joints cutting two or three boreholes would lead to a smaller mean area value.

The only way to avoid this problem in a future similar study, is to decrease the distance between the boreholes enough, so that the existence of fortuitous traces of joints of the same set in the inspected borehole lengths, becomes improbable. The decrease of the borehole distances will, however, imply also a decrease in the accuracy of the mean area determinations.

9 REFERENCES

Grossmann, N.F. 1967. Intervention in the Discussion of Theme 2 (Description of Rocks and Rock Masses with a View to Their Physical and Mechanical Behaviour). In Proceedings of the 1st Congress of the ISRM (Lisboa PORTUGAL, 1966 September 25-30), Vol. 3, p. 232-234. Lisboa: Laboratório Nacional de Engenharia Civil.

Grossmann N.F. 1977. Contribuição para o Estudo da Compartimentação dos Maciços Rochosos (Contribution to the Study of Jointing in Rock Masses), a research officer thesis. Lisboa: Laboratório Nacional de Engenharia Civil.

Grossmann, N.F. 1984. The Sampling Quality, a Basic Notion of the Discontinuity Sampling Theory. In E.T. Brown & J.A.Hudson (eds.), ISRM Symposium Design and Performance of Underground Excavations (Cambridge UK, 1984 September 03-06), p. 207-212. London: British Geotechnical Society.

Grossmann N.F. 1985. About the Volume of the Blocks in a Rock Mass. In ISRM Int. Symp. on the Role of Rock Mechanics in Excavations for Mining and Civil Works (Zacatecas MEXICO, 1985 September 02-04), Paper I.20. Mexico: Sociedad Mexicana de Mecánica de Rocas.

Hudson J.A. & S.D. Priest 1979. Discontinuities and Rock Mass Geometry. Int J. Rock Mech. Min. Sci. 16: 339-362.

Kikuchi K., T.Kobayashi, M.Inoue, & Y. Izumiya 1985. A Study on the Quantitative Estimation of the Joint Distribution and the Modelling of Jointed Rock Masses. Tokyo: Tokyo Electric Power Services Co., Ltd.

Priest S.D. & J.A. Hudson 1976. Discontinuity Spacings in Rock. Int. J.Rock Mech. Min. Sci. 13: 135-148.

Design methods for structures in swelling rock

Méthodes d'analyse et de conception de structures dans des roches gonflantes
Die Verfahren für die Analyse und den Entwurf für Projekte im quellfähigen Fels

M.GYSEL, Motor-Columbus Consulting Engineers Inc., Baden, Switzerland

ABSTRACT: The ISRM-Commission on Swelling Rock is currently working on analysis and design procedures for structures in or on swelling rocks. The following interim report evaluates this work. In conclusion, it is found that quite advanced design calculation methods are available. However, fundamental swell testing should definitely be improved. Furthermore, practical considerations concerning favorable tunnel cross sections, construction procedures or materials may be as important as the swell calculations themselves.

RESUME: La Commission ISRM sur les roches gonflantes est en train d'évaluer les méthodes d'analyse de structures dans les roches gonflantes. Le rapport suivant donne une synthèse interimistique sur ce travail. Tandis que les méthodes elles-mêmes sont dejá assez développées, il faut encore beaucoup améliorer les essais de base concernant les mécanismes du gonflement. Les calculs de gonflement ne forment qu'une partie de l'ensemble des travaux de projet pour ce genre de structures. Sans doute les considérations générales sur la forme de la section transversale du tunnel, sur les méthodes et phases de construction ou sur les matériaux à employer sont d'une importance identique.

ZUSAMMENFASSUNG: Die ISRM-Kommission "Quellfähiger Fels" arbeitet gegenwärtig an der Analyse der bis heute bekannten Berechnungsmethoden für Strukturen in quellfähigem Fels. Aus der Analyse kann gefolgert werden, dass bereits recht gut entwickelte Methoden zur Verfügung stehen. Der springende Punkt der weiteren Entwicklung scheint in der versuchstechnischen Untersuchung der grundlegenden Quellmechanismen zu liegen. Es wird auch darauf hingewiesen, dass die Berechnungen nur einen Teilaspekt der Projektierungsarbeiten darstellen. Gute Formgebung des Tunnelquerschnittes, geeignete Bauverfahren und Bauabläufe) sowie Baumaterialen spielen eine ebenso wichtige, oder vielleicht wichtigere Rolle.

INTRODUCTION

The Commission on Swelling Rock of the International Society for Rock Mechanics (ISRM) has compiled and analyzed the relevant and accessible work on how tunnels, caverns, shafts, as well as structures above the rock surface, may be treated (designed) under various conditions of swelling rock.

The purpose of this paper is to provide an interim report on the work of the Commission on Swelling Rock, regarding its Task D: analysis and design procedures for structures in or on swelling rocks.

This interim report shall highlight and make accessible some information and conclusions which have been compiled to date. A formal commission report on the above-mentioned Task D is to be published at a later date.

Furthermore, reference is made to other tasks of the Commission concerning the characterization of swelling rock, identification of problem areas (underground excavations, foundations, slopes, and surface excavations), as well as test procedures. It is noted that together the test procedures and the analysis/design procedures should provide the essential tools for engineering practice in swelling rock.

The following considerations primarily treat underground structures. The term "tunnel" is used in a general sense throughout the paper, and may also be used in the sense to describe cavern structures or inclined tunnels (shafts), etc.

Surface excavations or structures may be analyzed on the basis of the same swell property tests and swelling laws. However, in most surface situations, the swell deformation is more important than developing swell pressures. To prevent excessive heave, the use of prestressed anchors still calls for combined consideration of heave and pressure. Later discussed design methods given by Grob (1972), Einstein et al. 1972), or finite element solutions are directly applicable to analzye surface excavations as well as structures.

BASIC CONSIDERATIONS

The definition or characterization of swelling rock was given by the Commission on Swelling Rock (ISRM 1983). Recommendations on the test procedures to identify swelling materials and to predict swelling pressures or movements will be published by the same ISRM Commission.

Laboratory tests, from which swelling parameters and mathematical swelling laws are derived, are the basis of current analytical or numerical design methods for structures on, or more frequently in, swelling rock. For later design purposes, laboratory tests should provide information on the expected swelling pressures, swelling strains and ideally on the swelling stress as a function of axial permitted strain change (swelling law). Accordingly, there are three groups of laboratory swelling tests:

1. The "swelling pressure" test is intended to measure the maximum axial pressure necessary to constrain an undisturbed radially confined rock specimen at constant height when it is immersed in water.
2. The "swelling strain" test is intended to measure the maximum axial and radial free swelling strain developed when an unconfined, undisturbed rock specimen is immersed in water.
3. The "swelling stress as a function of permitted axial strain" test results in the axial strain necessary to reduce the axial swelling stress of a radially confined rock specimen immersed in water from a maximum value to a, for design purposes, sufficiently low value.

The "swelling pressure" and the "swelling strain" tests thus result in maximum values of the confined and unconfined specimen, whereas the "swelling stress as a function of permitted axial strain" test results in the connection between these values. This is called a swelling law.

Although the first two test types may be used for any swelling rock material, the third type is limited to rock with a reversible swelling behavior. In construction practice, the frequently found clay mineral type swelling falls under this category.

Huder and Amberg (1970) are believed to be the first to describe the test results for marls. Grob (1972) uses these results to define a one-dimensional swelling law. Three-dimensional hypotheses or laws formulated by Einstein et al. (1972) or Wittke and Rissler (1976) are also based on the tests and interpretations by Huder, Amberg and Grob. The design methods which are discussed below are based mainly on this line of swelling laws. However, some design methods provide the freedom for optional other swelling laws.

The Huder-Amberg-Grob type of swelling law, which gives a logarithmic function between the relieved stress and the obtained swelling heave, also was used successfully for gypsum Keuper which contains swelling clay minerals as well as anhydrite. The full reversibility of the swell strains and stresses is not valid in this case. However, the action of the "physical" swelling of the clay minerals is predominant.

The anhydrite swelling, while changing to gypsum, is not reversible and requires design considerations which cannot be treated directly by the design approaches which are discussed later. The swelling law should then be adjusted. Furthermore, it has to be checked that the considered rock element does not switch from decompressing to compressing for a specific calculation step, or vice versa.

It is understood that some authors, e.g., Wittke 1978, 1984, deal with irreversible swelling by means of considering "positive" swelling only, and by excluding automatically "negative" swelling as induced by a compression or recompression.

However, this procedure may not always be successful; the excavation of a tunnel in marl could lead to swell heaves which are partially reversed through later installed prestressed rock anchors. In this case, the calculation method should be able to handle "negative" swelling values. On the other hand, for the same case, the swell deformations would naturally be nonreversible when assuming anhydrite.

Presently, the swelling law in most cases is derived from oedometer tests. Owing to a number of factors, it is quite difficult to extrapolate the in-situ swelling behavior from oedometer tests only. Therefore, it is advisable to check or correct the laboratory-based swelling parameters by means of deformation or pressure measurements on the tunnel construction itself, or even by means of and investigation tunnel or test structure. The design calculation should be regarded as one step of a more comprehensive design procedure for structures in swelling rock.

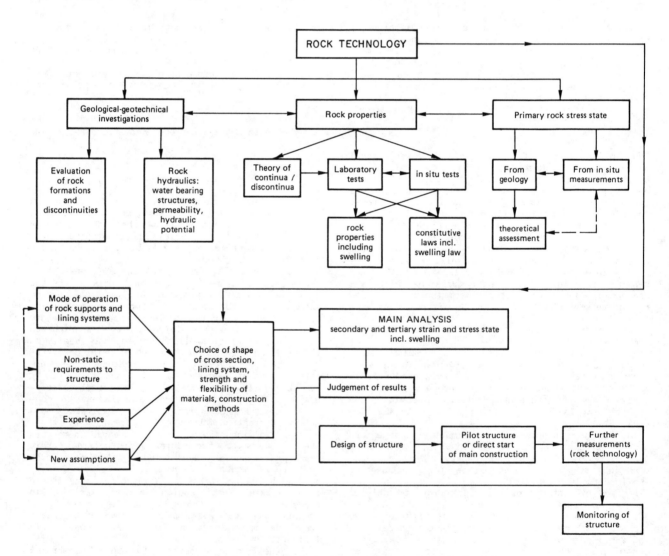

Figure 1. Flow chart for the design of underground structures, based on Gysel (1970)

378

Design procedure for structures in swelling rock:
1. Swelling tests in the lab → swelling parameters, possibly determination of other rock parameters and possibly determination of in-situ stress state.
2. Analysis and performance prediction for a test tunnel or test chamber (location and dimensions comparable to actual structure).
3. Construction of test structure and comparison of predicted and observed performance (mainly displacements and, if possible, stresses. This is used to modify the original swelling parameters.
4. Analysis and design of structure.
5. Construction control measurements to:
   - check displacements and stresses
   - adapt dimensions, if necessary
   - back-calculate actual swell parameters
Note: Steps 2 - 3 can be eliminated, i.e., for small structures.

This procedure may be used regardless of the type of basic swell test in the laboratory. Depending on improved knowledge concerning the real in-situ behavior and on the improved basic test techniques, steps 2 and 3 of the procedure may be deleted. Furthermore, in many cases a project-related test tunnel or chamber may not provide results in due time because of the slow in-situ development of the swell process.

In a personal note to the author, Lombardi (1986) points out that water from the rock or from the water tunnel, which is necessary to induce swelling of a potentially swelling rock under rock stress change, has to travel through the rock. Hydraulic parameters such as rock permeability, or changing rock permeability govern the development of swelling in time. To date, design procedures mainly analyze limit states before and at the end of the swelling only. The limit state and the end of the swelling may be considered to be an upper boundary of the real behavior, since the real in-situ swelling may not fully take place depending on the actual rock hydraulic situation.

Before discussing analysis methods, it must be considered that the analysis of the structure (step 4 of procedure) may be seen in a wide, general scheme of the design approach for underground structures. Figure 1 (see above) shows that practical design considerations concerning the tunnel cross-sectional form, the construction methods, sequences, or the choice of special materials are as important as the calculation itself.

DESIGN ANALYSIS METHODS FOR UNDERGROUND WORKS

The development of static calculation methods (for step 4 of the above-mentioned design procedure) somehow was triggered by the failure of the concrete invert of the Belchen highway tunnel in Switzerland. Grob (1972) analyzed the underlying swelling mechanisms. On the basis of earlier work by Golta 1967, Grob identified the decompressed areas near the tunnel as potential swelling zones. He connected the oedometric swell heave (Huder, Amberg 1970) which results from unloading, with the swell heave of a decompressed rock element in situ. In this manner, Grob defined a one-dimensional swelling law which neglects stress/strain changes in other (lateral) directions. His law is conservative in the sense that any radial stress decrease was considered to induce a swell heave. However, secondary stresses which remain above the maximum oedometric swell stress prevent any swelling of the rock element. When the horizontal rock stresses are smaller than the vertical stresses, the decompressed zones are found below the tunnel invert and, theoretically, above the tunnel crown. Grob, therefore, calculated the total heave of the tunnel invert by integrating the contributions of the rock elements along a radial path. The difference between the primary and secondary radial rock stress was previously calculated for the case of the unlined (or lined) tunnel for the rock elements. Grob's approach was meant primarily to provide information on

the swell heave. However, it was used also to calculate approximately the stresses in the tunnel lining by considering the swell heave as a loading of the tunnel invert. The heave is partly prevented by the elastically imbedded lining. Consequently, normal forces and bending moments result in the invert part.

Prior to a further discussion on generalizations of Grob's approach, it should be mentioned that the so-called methods of "characteristic lines" may also be used to analyze the swelling heaves and pressures around a tunnel. The axi-symmetric characteristic line method in elastic-plastic formulation by Lombardi (1970, 1973, 1977) may be used as follows. The calculated plastic zone around the tunnel is identified to be the main area of potential swelling The swelling law is introduced numerically into the main calculation which comprises elements such as time-dependent supporting of the tunnel or rock properties which change with time. It is evident that specific non-axisymmetric problems (swelling below the invert only) are not accessible to axi-symmetric characteristic line methods. In fact, a contradiction exists between axi-symmetric methods and the three-dimensional swell theory. A tunnel bored into a rock formation, which is under a hydrostatic stress state, should induce no swelling. Admittedly, the swelling does not striclty follow the continuum theory. The plastic or loosened zone around a tunnel may be susceptible to swelling regardless of the nature of the primary rock stress state (Gysel 1985).

Einstein and his co-workers (Einstein et al. 1972) introduced the assumption that the swelling is induced by the reduction of the 3-dimensional stress state (first invariant) in the rock. The swelling pressures try to bring the first invariant to its original value provided that the statical system of the structures allows it. Later formulations of 3-dimensional swelling laws are all based on this theory of the first invariant, regardless of some different interpretations. Einstein et al. (1972) at that time used their own model, e.g., to analyze tunnel inverts. Here, they applied characteristic lines for the invert and the swelling ground. The characteristic line of the ground included the marl swelling in function of the counterpressure of the invert lining. Grob's approach (1972) was refined. While observing and integrating the heave on a radial (vertical) line below the tunnel, the lateral directions were also considered. Furthermore, discrepancies between the oedometer stress state and the stress state below the tunnel invert were corrected. It is worthwhile to mention that Einstein et al. (1972) checked or calibrated the results of their calculations by field measurements. Additionally, Einstein et al. (1972) pointed out that the main uncertainties concern the knowledge of the primary stress state and the laboratory tests.

The author's approach (Gysel 1977) was to analyze a circular tunnel lining in swelling rock. Here, an attempt was made to overcome the former limitations of the characteristic lines which were restrained to axi-symmetry of stress, or to the analysis of special points (inverts). The analytical concept begins with the three-dimensional stress state. The secondary elastic stress state around the tunnel is modified where swelling occurs. The tertiary stress state is dependent on the yet-unknown pressure distribution on the tunnel lining. Einstein's hypothesis (1972) was used to calculate the swelling heave. However, an unpdated version of the method uses the direct implementation of the three-dimensional swelling law according to Wittke and Rissler (1976), Gysel's method makes further use of Morgan's tunnel analysis (1961) which was later modified by Muir Wood (1975). Instead of giving a direct solution, Gysel uses simultaneous characteristic lines for the tunnel invert (crown) and for the horizontal lateral direction. The solution has to fulfill equilibrium at both points simultaneously.

Normal force and bending moment of the tunnelling result from this analytical approach which is rapidly run on a personal computer without any need of intermediate graphical work. Morgan's assumption of an elliptical load/deformation distribution on the tunnel lining somehow restricts the application with regard to swelling. Extreme point-wise loads (narrow angle swell zone) are overestimated by the method. It was seen from parallel calculations with the finite element program Rheostaub (Kovari et al. 1983) that Gysel's adapted method provides conservative results which may be used for designs. The method is also useful to determine whether the expensive finite element tools should be used for selected cases, and for which cases.

It was a logical development to implement swelling behavior into finite element programs for tunnel analysis. Wittke and Rissler (1976) did so by using Einstein's theory of the change of the first invariant of the stress tensor which is linked to the volumetric swell term (Einstein et al. 1972). Wittke and Rissler (1976) introduced a second hypothesis in order to split the volumetric swell deformation to the chosen three axis directions. The same concept was adopted by Kovari et al. (1983) in their finite element program Rheostaub. However, the underlying swelling law may optionally be chosen as linear or semi-logarithmic. Both types may further be modified by using higher modes of polynomial functions (n>1) in order to adapt the swelling law to test results. Theoretically, Rheostaub is capable of combining rheological effects with swelling behavior. However, no investigations of this kind are known to the author. In any case, Rheostaub may be used for stage-wise excavation problems where each stage may trigger a new wave of swelling. Schwesig and Duddeck (1985) extensively discuss Wittke's approach; they find it beneficial to modify the formulation and use a differential instead of a finite implementation of the swelling law. Furthermore, similar to Kovari, they give room to adapt the swelling law to test results. The finite element solutions of Wittke, Kovari or Schwesig-Duddeck may be discussed yet from another point of view. It is known from inhibited oedometer swell tests that a well defined axial stress σ_0 inhibits any swell deformation. Therefore, it is necessary to define the criterion for which stress states no swell deformation of the rock elements or finite elements takes place (criterion of exclusion). Otherwise, the swelling influence on the tunnel structure will be overestimated. Mention is made that Richards (1984) or Sun Jun et al. (1984) are extending swelling analysis by studying the stress-strain problems of the swell induced stress-strain problems in combination with changing permeability and water content of the rock. It is noted that swelling rock formations such as marls may usually have a water content of 2 to 5 %. This is enough to feed the potential later swelling process. Effective permeability through pores, cracks and fissures play an important role. It is assumed that the loosened (decompressed) zones tend to have an increased permeability.

CONCLUSIONS

The discussed design methods for structures in swelling rock mark some points of development over the past 15 years. By discussing the work and the design approaches of the relevant authors, an attempt has been made to draw some light on the historical development of this discipline of rock mechanics, and more importantly, to illustrate the capabilities and limitations of design methods which are available to date. This status report must limit authors and methods; it is recognized that much more work has been undertaken which, unfortunately, cannot be treated here. The author's intention is to outline several main and common features of how to approach the design of structures in swelling rock.

In summary, the tunnel design engineer to date may work with a number of design methods which, in combination with the proposed comprehensive design procedure, provides a reasonable tool for treating the statics of a structure in swelling rock.

It is desirable, above all, to improve the calculation methods in reference to the underlying swelling law. Therefore, important work is to be undertaken in the testing field before the methods may be substantially and not only mathematically improved.

It is pointed out that in the past the swelling parameters were mainly determined from oedometer tests by using the theory of elasticity, the validity of which was seldom checked or discussed. Expressly, plasticity has an influence on the swelling parameters both during a test and in situ. Therefore, the entire swelling test should be improved:
- The basic swell test should be triaxial (see Pregl et al. 1980).
- The swelling law should be improved in the elastic domain and newly formulated for elastic-plastic conditions (role of deviatoric stress change)

Improved swelling laws may later be incorporated into the design methods. For purely analytical methods, this may lead to important mathematical changes. The same may be valid for numerical solutions. However, some of the known finite element programs (Kovari et al. 1983), or Schwesig and Duddeck, 1985) already contain the option of a free choice of the swelling law.

Anisotropic swelling and non-reversible swelling (anhydrite) or partly non-reversible swelling (mixed swelling of clay minerals and anhydrite) still represent a large and open field for future development for both swelling tests and design methods.

BIBLIOGRAPHY

Commission on swelling rock 1983. Task A, Characterization of swelling rock. International Society for Rock Mechanics.

Einstein, H.H. et al. 1972. Verhalten von Stollensohlen in quellendem Mergel. Proceedings International Symposium on Underground Works. Lucerne.

Golta, A. 1967. Sohlenhebungen in Tunneln und Stollen. Report: Wehrli, Weimer, Golta, Consulting Engineers. Zurich.

Grob, H. 1972. Schwelldruck im Belchentunnel. Proceedings International Symposium on Underground Works. Lucerne.

Gysel, M. 1970. Felsmechanik - Theorie und Anwendung. Strasse und Verkehr 56: No. 1.

Gysel, M. 1977. A contribution to the design of a tunnel lining in swelling rock. Rock Mechanics 10: 55-71.

Gysel, M. 1985. Auflockerungszonen um Stollen und Kavernen im Valanginienmergel des Oberbauenstocks. Publication NTB 85-30 of the Nationale Genossenschaft für die Lagerung radioaktiver Abfälle, NAGRA. Baden.

Huder, J. & G. Amberg 1970. Quellung in Mergel, Opalinuston und Anhydrit. Schweiz. Bauzeitung 88: 975-980.

Kovari, K. et al. 1983. RHEO-STAUB, User's Manual, Version 30.04.1983. Federal Institute of Technology, Zurich.

Lombardi, G. 1970. The influence of rock characteristics on the stability of rock cavities. Tunnels and Tunnelling: January - March.

Lombardi, G. 1973. Dimensioning of tunnel linings with regard to constructional procedure. Tunnels & Tunnelling: July.

Lombardi, G. 1977. Long-term measurements in underground openings and their interpretation with special consideration to the rheological behavior of the rock. International Symposium on Field Measurements in Rock Mechanics. Zurich.

Lombardi, G. 1986. General considerations concerning Task D, Commission on Swelling Rock, ISRM. Report 102.2-R-58, D34 to the author. Locarno.

Morgan, H.D. 1961. A contribution to the analysis of stress in a circular tunnel. The Institution of Civil Engineers. London.

Muir Wood, A.M. 1975. The circular tunnel in elastic ground. Géotechnique 25, No. 1: 115-127.

Pregl, O. et al. 1980. Dreiaxiale Schwellversuche an Tongestein. Geotechnik 3: Heft 1.

Richards, B.G. 1984. Finite element analysis of volume change in expansive clays. 5th International Conference on Expansive Soils. Adelaide, Australia.

Schwesig, M. & H. Duddeck 1985. Beanspruchung des Tunnelausbaus infolge Quellverhaltens von Tonsteingebirge. Bericht Nr. 85-47, Institut für Statik. Technische Universität Braunschweig.

Sun Jun et al. 1984. The coupled-creep effect of pressure tunnels interacted with its water-osmotic swelling viscous elasto-plastic surrounding rocks. Proceedings. Peking.

Wittke, W. & P. Rissler 1976. Bemessung der Auskleidung von Hohlräumen in quellendem Gebirge nach der Finite-Element-Methode. Veröffentlichungen des Instituts für Grundbau, Bodenmechanik, Felsmechanik und Verkehrswasserbau: Heft 2. RWTH Aachen.

Wittke, W. 1978. Grundlagen für die Bemessung und Ausführung von Tunneln in quellendem Gebirge und ihre Anwendung beim Bau der Wendeschleife der S-Bahn Stuttgart. Veröffentlichungen des Instituts für Grundbau, Bodenmechanik, Felsmechanik und Verkehrswasserbau: Heft 6. RWTH Aachen.

Wittke, W. 1984. Felsmechanik - Grundlagen für wirtschaftliches Bauen im Fels. Berlin, Heidelberg, New York: Springer.

A computer analyzing system used for the study of landslide hazards

Etudes des risques de glissements de terrains à l'aide de l'ordinateur
Die Anwendung der Computertechnik für die Erforschung der Erdrutschgefahr

HUANG SHABAI, Institute of Seismology, State Seismological Bureau, People's Republic of China

ABSTRACT: The major concerns in this paper are the applications of remote sensing technique and computer science in the study of landslide. The following studies have been done: (i) proposing and setting up a landslide computer analyzing system (LSCAS) and a more complete landslide data base (LSDB), (ii) introducing the theory and method of digital terrain model (DTM) into the study of landslide and (iii) proposing an unified inversion method of shear strength and slip plane function in which the multi-variable statistic theory is used.

In addition, one set of calculation methods about slidebody's volume, surface area and slope as well as the extraction of landslide border has been proposed on the basis of DTM.

The proposed system and methods have been applied to the case of Yunyang landslide happened in July of 1982 along the Yangtze River and encoured results have been obtained.

RESUME: Cet article consiste principalement à l'application de la télédétection et la technique d'ordinateur sur l'étude de la pente glissante pour résoudre les problémes ci-dessous: (i) poser et fonder un systéme complète d'analyse d'ordinateur sur la pente glissante ainsi que une banque des donnéés suffisantes de pente glissante; (ii) introduire la théorie du modèle digital du terrain (DTM) et son méthode dans le domaine d'étude de la pente glissante; (iii) employer la théorie d'analyse multiélementaire statistique dans l'étude de pente glissante, et suggérer une méthode d'inversion unitaire de la tensité de contre-cisaillement de terrain glissante et la fonction de distribution de la surface glissante.

En outre, on a fondé une mode de calculation de la surface, du volume du degré d'inclinaison et de la frontière de la pente glissante sur la base de DTM.

On a obtenu de succès satisfaissant dans l'essai de l'application sur la pente glissante de Yunyang dans la Province de Sichuan avec système et la méthode posés par cet article.

ZUSAMMENFASSUNG: DieserArtikel handelt hauptsachlich von der Anwendung der Fernerkundung und der Computertechnik in der Forschung des Erdrutschens, und lost besonders die folgenden Probleme: (i) Vorlegung und Grundung eines funktionell vollstandigen analytischen Systems des Erdrutschens mit Hilfe des Computer, und einer verhaltnismassig vollkommenen Erdrutschdatenbank; (ii) Einfuhrung der Theorie und Methode des Modells des Digitalterrains (DTM) in den Bereich der Erdrutschenforschung; (iii) Einfuhrung der Theorie der multivariablen statistischen Analyse in die Erdruschen forschung und vorlegung einer vereinigten Inversionsmethode der Gegenscherfestigkeit der Gleitflachenerde und der Verteilungsfunktion der Gleitflache.

Ausserdem hat man auf dem Grunde des DTM eine Methode fur die Berechnung der Flache, des Vilumens, der Neigung und der Grenze des Erdrutschens errichtet.

INTRODUCTION

Landslide is one of the unexpected geological hazards which often leads to serious losses in economy and property, especially for large engineering works. Therefore the harmfulness of landslide has increasingly been brought to the experts attention in various countries.

There sre a lot of factors which affect the generation and development of landslide, for example, geology, hydrology, weather and etc.. Earthquake is also one of the affecting factor The occurrence of earthquake is usually accompanied by the generation of a series of landslides. In order to study the process of generation and development of landslide, then to predict landslide and to provide the consultation of landslide prevention correctly, it is necessary to study the method of dynamic monitoring and stability analysis of landslide and to make a further study on the correlation among the developing ways, forming condition, affecting factors and patterns of landslide hazards simultaneously. Since the phenomena of landslide and deformation may be a precursor of earthquake, it would be possible to predict landslide by studying deformation, and to forcast earthquake further.

Landslide research, as a synthetic subject, was pushed forward by many other subjects during its development. With the rapid development of remote sensing technique and computer science, they have gradually been used for the study of landslide. But through analyzing present information on the applications of remote sensing technique and computer science in the field of landslide research, the following conclusions could be reached:

(i) The application of remote sensing technique in this field is basically at the level of qualitative description yet; the application of computer science only concerns the analysis of slope stability.

(ii) With regard to the landslide research, to set up a landslide computer analyzing system with stronger analyzing functions and a more complete landslide data base will further its development in the following aspects:

(a) the study of the generation and mode of landslide.

(b) the dynamic monitoring and medium or long term prediction of landslide.

(c) various analyses and evaluations during the design stage of large projects and after slope falling.

(d) thematic mapping of landslide hazards.

According the above information, the major concerns of the paper are:
1. proposing and setting up a landslide computer analyzing system (LSCAS) with more than 30 analyzing functions.
2. proposing and setting up a more complete landslide data base (LSDB).

1 LANDSLIDE DATA BASE (LSDB)

1.1 Structure of LSDB

Data base is the set which describes the files with certain completeness and satisfies some requirements in structure, access, redundancy and safety. For a special data base, the factors and items to be stored in it must be determined in accordance with its purpose and the user's requirements. As a special landslide data base, the basic factors about landslide area are: (a) geology; (b) hydrology; (c) terrain; (d) deformation and (e) slipplane distribution.

1.1. A Structure of LSDB

LSDB consists of three stages, that is, external, conceptual and internal schema. Their relation is shown in fig.1.

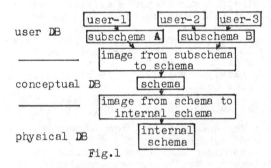
Fig.1

User level DB corresponds to external schema which is the logical structure of data visual to the user. Its function is to provide an int terface between the user and system. The user can operate the data stored in LSDB by inquiring language or user program.

Conceptual level DB corresponds to conceptual schema, whose function is to describe the whole logical structure of DB which is seen by DB's administrator (DBA). The main content of schema is a data model.

1.1. B LSDB's data model

Considering the relation between the factors stored in LSDB, a hierachical model is adopted as the·data model of LSDB. It is a sequential tree which takes every record type as its node. The character of the data model is that every node except the root one only has a father. The LSDB's data model is presented in fig.2.

1.1. C LSDB's file organization

File organization means to organize the concerned records as data file following certain logical structure, to store them in some storage devices to compose a physical file organization with a physical access which reflects the above logical structure. According to the character of landslide research, LSDB provides

a two-stage hierachical-indexed-sequential access method (HISAM) which is shown in fig.3.

In this way, the index file forms a tree with index blocks and record blocks as its nodes. When setting up LSDB,i.e., when stored raw data into LSDB, every record must be stored in storage units in hierachical sequential. The fashion of main file's access structure is sequential,i.e., in the storage space of computer, every record is stored in the order of absolute address.

1.1. D LSDB's file searching and maintenance

The file searching of LSDB is a random searching process, that is, giving the value of key word and demanding to search its corresponding record. The operating process of LSDB is to search the highest level index block first, then to search the second level index from left to right, and finally to search the value of special data item using sequential searching method when the main files have been reached. Its process is shown in fig.4.

In LSDB, the index block is stored in the internal storage permanently, hence it is not necessary to read in index block from external storage.

The maintenance includes file modification, deletion and reorganization. When it is demanded to delete or modify a record, the above searching process must be repeated first in order to find the required record, then a deleting sign is put in the record prefix. Every record in the main file is stored in a block whose size is integral times larger than that of the physical block. Therefore, it is only necessary to make a room that large for inserting a new record.

In order to improve LSDB's running status after operating for a period of time, the files in LSDB must be reorganized. It contains two parts, one is the reorganization of index list and another is the reorganization of main file They are carried out by reading out the original files,i.e., the file which have been deleted, modified and inserted for many times, classifying them in the order of key word's value and then restoring them into internal or external storage.

1.1. E LSDB's data format

In LSDB, the grid data format is adopted in order to satisfy the special requirements of landslide research. The greatest advantage of the format is that the data management and operation will be very simple and highly efficient and the data can be easy to be used to registrate with other information, so that it can be applied very efficiently in stability analysis, correlation analysis of various and consultation.

1.2. Process of setting up a LSDB

1.2. A Determination of space coordinate system (SCS)

Before setting up a practical LSDB, the SCS must be determined first, with which LSDB can provide a reliable controllong network for landslide monitoring and manage all data concerned efficiently.

For a special LSDB, the rule of SCS design is that the SCS should not be destroyed by the occurrence of landslide. Based on such a rule, the relative rectangular SCS is adopted

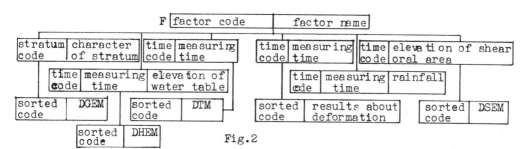

Fig.2

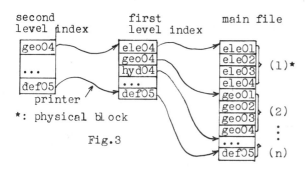

* : physical block

Fig.3

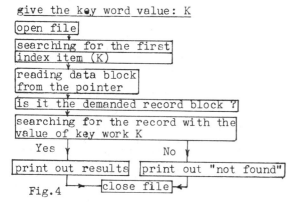

Fig.4

2 STRUCTURE AND FUNCTION OF LSCAS

2.1 The structure of the LSCAS

LSCAS is a synthetic analysis system used for the study of landslide, so it must include the following parts simultaneously: data classification, data acquisition, processing and analysis. The LSCAS consists of three subsystems.

(1) information acquisition subsystem: its function is to collect information obtained mainly by aerial and ground remote sensing with supplementary factors of geology, hydrology, weather and etc., and then to digitalize, registrate and classify them before inputting all data into LSDB.

(2) information processing subsystem: the purpose of which is to set up a practic LSDB.

(3) information application subsystem: it also consists of three subsystems: (a) stability analysis and dynamic monitoring subsystem (ST) which provides ten analysis methods for user to analyze the situation of slidebody's stability, to inverse shear strength of slip plane's rock and slipplane function and to set up the relation between deformation data and other factors relevant to landslide; (b) information consultation system (CON)which provides thirteen quantitatively or semi-quantitatively consulted results about landslide which will help the user to find out the situation of landslide very efficiently and (c) thematic mapping subsystem (TM), in which there are four available drawing fashions for user to select.

The LSCAS is a man-machine interface software system. Under the allowable condition of hardware, the users can select anyone of the functions shown in menu in the fashion of man-machine interface according to their needs.

The subsystem's name, subroutine used and their functions of LSCAS are shown in table-1.

3 EXPERIMENT AND CONCLUSION

In the experiment of LSCAS, the Yunyang landslide which lies along the Yangtze River (happened in July of 1982) was selected as an experiment area. The experiment included the following contents:

(a) Stability analysis. Table-2 is the results of safety factor given by three geological departments and table-3 the results got by LSCAS which are obtained by calculating seven sections on slidebody by five methods. From the above two tables, it can be seen that they differ from each other not much. However by adopting LSCAS, only several seconds were needed to get the above results.

(b) Multi-section shearstrength inversion. Table-4 and table-5 are the results of shear strength given by geological department using two-section method and by LSCAS using multi-section method. The results in table are basically consistent and only a little rising tendency exists in the results of C-value, which can be explained in practice and theory.

(c) Slipplane function's inversion. In order

which must be set on the stable ground surrounding the slidebody. Usually, the origin of SCS is put on a distinct ground point and Y-axis (parallel to landslide direction) is taken to pass through another distinct point as far as possible. Fig.5 shows the structure of SCS.

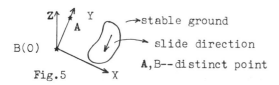

A,B--distinct point

Fig.5

Before the digitalization of the above factors, all information must be registrated in SCS first, then the digitalization is made in the grid data form with the size of 10x10 m. The whole process consists of three parts: data compilation, digitalization and input. The flowing chart is shown in fig.6.

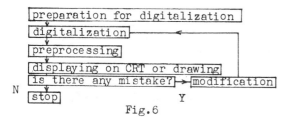

Fig.6

to get detailed information of slipplane, about 40 boreholes were drilled on the slidebody. In contrast, at most only 14 boreholes are required to get the analytic function of slipplane more accurately by using LSCAS.

From the above results, we can see that using LSCAS could save one half of costs at least in in the case of Yunyang landslide. Particularly it is obvious that three-dimensional information about slipplane obtained by LSCAS can express the status of slipplane better than point-by-point information only out of borehole. The root-mean-square errors of four inversions are:

$$M(14)*=2.80 \ (m) \quad M(9)=2.58 \ (m)$$
$$M(8)=2.40 \ (m) \quad M(7)=2.43 \ (m)$$

They are all about 6% of the average slidebody depth and can satisfy the requirements of other analysis.

In addition, the calculations of the push inside slidebody, slidebody's volume, surface area and border length have also been done in

table-1

name of subsystem		subroutine	function
TM	STEREO	output: stereo-drawing	
	GREY	:two-dimensional drawing expressed by grey level	
	ALPH	:two-dimensional drawing expressed by alphabet	
	DBXS	:section drawing in any direction	
	ELE	:slidebody's surface information	
	GEO	:geological information	
	HYDRO	:hydrological information	
	SLIDE	:slipplane information	
ST	SWE	Swedish's method	
	JAN	Janbu's method	
	BISH	Bishop's method	
	MP	Morgerstern-Price's method	
	PC	Push conveying method	
	CA	synthetic correlative analysis of multi-factor	
	TWOSTA	unified inversion method of shear strength and slipplane function	
	THRSTA	three-dimensional stability analysis method	
	LSSA	direct inversion method of slip plane function	
	DSSR	direct inversion method of shear strength	
CON	WDF	output:all information about drill hole	
	GLC	:characters of every stratum	
	BS	:deformation information of slide body	
	ZJ	:border information	
	PP	:the optimum scheme of landslide prevention	
	SLOPE	:slope of slidebody's and slipplane	
	AREA	: surface area of slidebody	
	VOL	:volume of slidebody	
	WAVE	:wave height caused by landslide	
	VELO	:slide velocity of slidebody	
	PUSH	:push distribution inside slide body	
	LBL	:border length of landslide	
	OPDATA	:output the above information in digital form	

Table-2

factor of safety	dept A	dept B	dept C
K	1.1496	1.2395	(1) 1.2117
			(2) 1.1507
			(3) 1.1490
			(4) 1.1460

Table-3

section No.	BISH's	PC's	MP's	SWE's	JAN's [#]
46	1.1570	1.0160	1.0454	1.0883	1.1571
48	1.1560	1.0170	1.2900	1.0801	1.1562
50	1.1570	1.0783	1.0766	1.0872	1.1570
52	1.1440	1.0450	1.0806	1.0820	1.1441
54	1.1930	1.0442	1.0887	1.1329	1.1930
56	1.2690	1.0354	1.1689	1.2061	1.2691
58	1.2770	1.1352		1.2154	1.2773

Table-4

location	C	F
eastern part	0.343	0.2053
middle part	0.319	0.1674

Table-5

section NO.	C	F
32,33,34,35,36,37,38,39,40, 41,42,43,44,45,46,47,48,4950	0.4984	0.1617
37,38,39,40,41,42, 43,44,45,46,47	0.4654	0.1663
54,56,58,60	0.4287	0.1863
38,40,42	0.4201	0.1681
57,60	0.3774	0.1870

the experiment. From the above experiment and analysis, the following conclusions can be drawn:

(1) By setting up a LSCAS and LSDB with strong function, the efficiency and reliability of various analyses and calculation concerned can be effectively increased, and the purpose of data share and unified management can be attained. Besides, it also makes the landslide research computerized and provides a new method of multi-factor analysis.

(2) The introduction of digital terrain model (DTM) into landslide research has made a great progress in realizing the quantitative andlocalized research of landslide which is advantageous as compared with qualitative ones.It also enables us to get the three-dimensional information about slipplane and push distribution easily and accurately.

(3) For the dynamic monitoring and forcast of landslide hazards, the LSCAS can provide the real-time factor of safety and push distribution in slidebody according to measured information of underground water or river water table.

(4) The LSCAS has provided an effective method of management for the user to search and consult rapidly any information concerned and also provided several methods to output the results of analysis in optional ways.

REFERENCES

Fredlund, D.G. 1978 1) Usage,requirements and feature of slope solfware. Canadian Geotechnical Journal. Vol 14, No.4 pp.83-95 2) Comparison of slope stability methods of analysis. Vol 14, NO3 pp.429-439 3) Use of computer for slope stability analysis. Journal of Soil Mechanics and Foundation Eng.. Vol 93, pp.519-541

* J.H. Duncan 1980. The accuracy of equilibrium method of slope stability analysis. pp.247-254

* M.Saito 1980. Reverse calculation method to obtain C & F on a slip surface. pp.285-289

* A.N. Patal & H.S. Mehta 1980. Application of digital technique in indentifying landslide prone area. pp.25-27

*: Int. Symp. on Landslide. India 1980

#: Note: For the meaning of BISH,PC,MP,SWE andJAN, see table-1.

Résistance au cisaillement de joints dans des roches stratifiées

Shear resistance of joints in stratified rocks
Die Scherfestigkeit von Klüften im geschichteten Gestein

P.J.HUERGO, Université Libre de Bruxelles, Belgique
M.K.K.MUGENGANO, Université de Lubumbashi, Zaïre

ABSTRACT: This paper discusses the shear tests results performed on strata joints of samples of phyllite and quartzphyllite. The joint behaviour is evaluated from the roughness geometry, assessed by the Moiré method, and from the compression and shear parameters assessed by mechanical tests.

RESUME: Cette communication analyse les résultats d'essais de cisaillement effectués sur des joints de stratification dans des échantillons de phyllades et de quartzophyllades. Le comportement des joints est évalué d'après la rugosité caractérisée par la méthode de Moiré et d'après les paramètres de compressibilité et de cisaillement déterminés par des essais mécaniques.

ZUSAMMENFASSUNG: Dieser Artikel ist eine analyse vor Scherversuchsresultate, die auf Schichtungsfugen in Phylliten und Quarzphylliten vorgenommen wurden. Das Verhalten der Fugen ist durch die mit der Moiré Methode bestimmte Rauhigkeit und durch die mechanisch festgestellten Zusammendrückbarkeit- und Scherenparameter gekennzeichnet.

1. INTRODUCTION

Les résultats expérimentaux décrits dans cette communication s'inscrivent dans le cadre d'une recherche fondamentale sur les mécanismes de déformation et de rupture des tunnels et des cavernes entreprise à l'Université Libre de Bruxelles. En effet, le comportement de ce type d'ouvrage dépend essentiellement des caractéristiques physiques et mécaniques du massif encaissant comme assemblage d'éléments rocheux de forme, de volume et/ou de nature différents. Or, sous l'action des sollicitations imposées, la réponse mécanique du massif varie non seulement en fonction des propriétés du matériau rocheux mais surtout en en fonction de la nature et de la géométrie des discontinuités y présentes. Par conséquent, le développement d'un modèle de calcul comporte nécessairement des éléments théoriques ou expérimentaux permettant d'évaluer le comportement des discontinuités - notamment leur résistance au cisaillement et leur compressibilité - et de l'intégrer dans le calcul global des performances de l'ouvrage.

2. NATURE ET PROPRIETES DES ROCHES ESSAYEES.

La recherche susmentionnée concernant particulièrement les ouvrages creusés en milieu anisotropique, l'ensemble des travaux expérimentaux a porté principalement sur des roches métamorphiques stratifiées. Les essais de cisaillement dont on fait état ci-après ont été effectués sur des éprouvettes de phyllades (Ph) du Siegenien supérieur et des quartzophyllades (Qph) du Siegenien moyen prélevés dans un dôme anticlinal découpé par de nombreux joints et diaclases. Ces roches microgrenues, finement stratifiées, de schistosité parallèle à la stratification, à teinte grise sombre, se distinguent par les difficultés particulières que l'on rencontre lors des travaux souterrains.

Du point de vue pétrographique, les phyllades et quartzophyllades se différencient par leur teneur en quartz. Ces dernières sont constituées à 70% de quartz tandis que les premières sont essentiellement des chlorites avec secondairement du quartz, du feldspath, de la pyrite et de la muscovite.

Les éprouvettes ont été choisies parmi les fragments des carottes présentant un joint naturel - en particulier, les joints de schistosité - dont le comportement est considéré comme le plus significatif du point de vue expérimental. Le diamètre des éprouvettes varie de 42mm (BX) à 72mm, les longueurs maximales des grands axes des joints pouvant atteindre 144mm ($\beta_{max} \simeq 60°$). Les essais ont toujours été effectués sur des joints descellés, le descellement étant matérialisé, si nécessaire, par percussion ou sciage des ponts de matière. La détection des joints étant parfois difficile - la schistosité se confondant souvent avec la stratification - les plans de faiblesse ont été identifiés à l'aide de mesures combinées de vitesse d'impulsion ultrasonore et de fréquence de résonance longitudinale et transversale.

Les caractéristiques physiques et mécaniques de la matrice intacte des roches étudiées sont résumées dans le tableau I ci-après.

Tableau I.

	γ (kN/m³)	E x10⁶(kPa)	ν (0)	σ_c x10²(kPa)	σ_t x10²(kPa)
Ph	24,5-27,4	7,3-10,2	0,13	681-954	103-135
Qph		8,8-10,8	0,09	764-920	143-150

où γ : poids volumique;
  E : module d'élasticité (pendage variable de 20 à 60°);
  ν : module de Poisson;
  σ_c : résistance à la compression simple (pendage variable de 20 à 60°);
  σ_t : résistance à la traction (compression diamétrale).

3. DISPOSITIF D'ESSAI

3.1. Essais de cisaillement.

Les essais ont été effectués à l'aide d'un appareil de cisaillement plan (fig.1) similaire à celui mis au point par Krsmanovic et al (1964).

L'éprouvette à essayer (10) est placée dans la boîte métallique fendue en son milieu de façon à ce que le plan du joint coïncide avec le plan théorique de cisaillement $\overline{AA}$. L'éprouvette est fixée dans cette position en noyant les extrémités, après la pose d'une armature, dans des blocs en béton de haute résistance (3 et 4). Le montage de l'ensemble boîte-blocs-éprouvette est maintenu calé dans les glissières verticales (8) pendant la manutention de façon à conserver intacte la configuration origi-

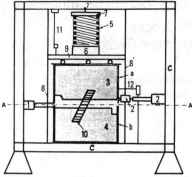

Fig.1. Schéma de l'appareil de cisaillement plan utilisé.

nale du joint. La partie supérieure (a) de la boîte peut glisser sous l'action du vérin (2), la partie inférieure (b) restant immobile par fixation au bâtis (c). La contrainte normale σ_n est appliquée en exerçant une force verticale à l'aide du vérin (6) maintenu en place sur le plateau coulissant (9). Les déplacements verticaux de celui-ci sont absorbés au moyen du ressort étalonné (5) qui prend appui dans sa partie supérieure sur un plateau à rotule (7). Les efforts et les déplacements horizontaux et verticaux sont mesurés respectivement par les capteurs de force (2') et (7') et les capteurs compensés de déplacement (12) et (11). L'essai de cisaillement plan est réalisé pour des paliers croissants de contrainte normale σ_n allant jusqu'à 4 MPa. A chaque palier, l'effort horizontal appliqué à la partie mobile (a) du dispositif d'essai est augmenté progressivement jusqu'au cisaillement intégral du joint. Une fois en bout de course, la partie mobile est ramenée à sa position initiale et un nouveau cycle de cisaillement est effectué sous un autre palier de contrainte.

3.2. Quantification de la rugosité.

La géométrie du profil de rugosité des joints soumis à cisaillement a été déterminée à l'aide de l'effet de Moiré, technique optique qui permet d'établir un relevé topographique intégral des surfaces auscultées. Le schéma du montage utilisé est représenté en figure 2. Le réseau R formé de traits rectilignes et parallèles alternativement opaques et transparents, est placé devant la surface à analyser. La source lumineuse (SL) projette des rayons incidents qui forment un angle θ avec la normale au réseau.

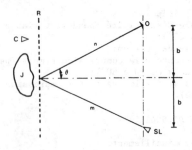

Fig.2. Montage optique pour la mesure du relief par l'effet de Moiré.

L'image de celui-ci et de son ombre sur l'objet (J), est créée, au niveau de l'objectif (O) par deux ensembles de rayons dont l'interférence produit des franges qui sont le lieu géométrique des points de même niveau. En effet, si la ligne joignant (SL) et (O) est parallèle au plan du réseau, les points d'intersection des lignes adjacentes de projection $\underline{m}$ et de visée $\underline{n}$ déterminent un plan qui coupe la surface de l'objet selon des lignes de contour dont l'image est la frange de Moiré correspondante (fig.3).

D'autres lignes de contour sont décrites par des plans déterminés par l'intersection des lignes $\underline{m}$ et $\underline{n}$ espacées de 2d, 3d, etc ... La distance Δz entre franges, autrement dit la différence de niveau entre lignes de contour, peut être calculée d'après l'expression suivante (Pirodda,1982) :

$$\Delta z = \frac{dL}{b} \left(1+\frac{z_i}{L}\right)^2 \left[1+\frac{d}{b}\left(1+\frac{z_i}{L}\right)\right]^{-1} \qquad (3.1)$$

où

L : distance selon la normale au réseau R entre celui-ci et la source lumineuse (SL);

2b : distance entre la source lumineuse (SL) et l'objectif O;

z_i : distance séparant le réseau de l'objet au niveau de la ligne de contour i, soit :

$$z_i = L \frac{id}{b-id} \qquad i = 1,2,3,... \qquad (3.2)$$

d : pas du réseau.

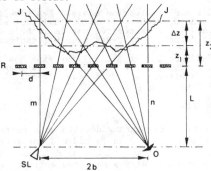

Fig.3. Création des lignes de contour par l'intersection des lignes de projection $\underline{m}$ et de visée $\underline{n}$.

Si $z_i/L \ll 1$ et $d/b \ll 1$, la distance Δz dite facteur de sensibilité prend la valeur quasi constante :

$$\Delta z \simeq \frac{d}{tg\theta} \qquad (3.3)$$

On constate que l'angle θ est inversément proportionnel au facteur de sensibilité Δz. Plus θ augmente, plus grand est le nombre de franges. Ceci peut conduire à un degré de précision élevé mais difficile à atteindre dans le cas des joints étudiés. En effet, au dépouillement des résultats, il s'est avéré que, pour des précisions élevées, l'identification des différentes franges n'est matériellement pas possible. Par conséquent, un choix judicieux des paramètres d et θ basé sur des expériences préalables doit être effectué avant la réalisation des essais sur les joints à étudier. Dans le schéma utilisé pour cette analyse (fig.2), on a choisi $\theta = 28°$ pour un pas $\underline{d}$ constant de 0,5 mm. En effet, il a été déterminé que pour $\theta > 28°$, il y avait confusion des franges malgré l'agrandissement de l'image et que pour $\theta < 28°$, le facteur de sensibilité est supérieur à la valeur minimale prescrite (< 0,1 mm).

La validité des valeurs Δz déterminées par l'équation (3.3) a été vérifiée expérimentalement en les comparant aux valeurs obtenues d'après les franges observées sur un objet repère de forme et dimensions bien déterminées - en l'occurence un cône en métal (C) d'une hauteur égale à 1 cm - placé immédiatement à côté de la surface du joint ausculté. La mise en place est jugée correcte lorsque le Δz de l'objet repère, c'est-à-dire, le rapport entre la hauteur du cône et le nombre de franges - image du cône est égal au Δz calculé.

La représentation graphique des franges sur un calque support est matérialisée par des courbes tracées, à l'aide d'un projecteur de profil, au milieu de chaque frange (fig.4). On choisit, par comparaison

de l'image avec l'objet, un point de cote zéro afin
de coter le relief en positif (sommets) ou en négatif
(creux). Les franges ouvertes ou discontinues représentent les crêtes.

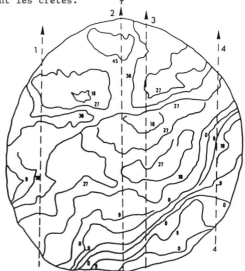

Fig.4. Relief du joint (le nombre de lignes de contour a été réduit pour la clarté de la ré-
présentation). Amplitude: 5×10^{-5} (m).

Une fois l'image cotée, on déterminée,
en se servant d'un analyseur d'images, l'amplitude et
l'intervalle des épontes ainsi que leur positionnement
par rapport à un système de coordonnées x, y. Une
carte intégrale du relief de la surface est ainsi
obtenue. Lorsque l'on veut reconstituer un profil de
rugosité selon un axe de cisaillement déterminé, on
parcourt l'image cotée à l'aide d'un crayon magnétique et on obtient un relevé des épontes sur toute la
longueur balayée. On peut donc, en comparant les profils pour un même axe relevé sur les images des surfaces supérieures et inférieures du joint, apprécier
l'imbrication et la complémentarité des aspérités des
deux surfaces.

4. ANALYSE DES RESULTATS EXPERIMENTAUX.

4.1. Compressibilité des joints.

La détermination de la variation statique de l'ouverture des joints en fonction de la contrainte normale
σ_n appliquée a été mesurée sur un certain nombre de
joints intacts en provenance de la même série de
carottes.

La figure 5 donne des exemples caractéristiques des
courbes σ_n-dh obtenues lors du cisaillement des phyllades et des quartzophyllades pour une gamme des contraintes normales allant jusqu'à 3,8 MPa.

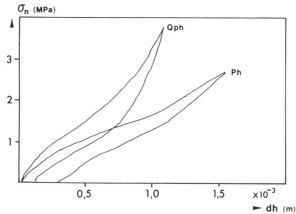

Fig.5. Courbes caractéristiques de compressibilité
des joints.

Les diagrammes en chargement et en déchargement ne
présentent pas la même allure, tous deux sont non
linéaires mais d'une façon plus marquée dans le cas
du déchargement. Après un cycle de charge, on constate une déformation permanente relativement importante du fait de la déformation plastique et de la
destruction des épontes. Pour des niveaux de contrainte normale élevés, on observe, en général, une
phase de relations contraintes-déformation approxi-
mativement linéaire en rapport avec un comportement
quasi-élastique du joint vers la fin du cycle de
chargement. Dans certains cas cependant, des phénomènes non linéaires localisés peuvent se produire par
un effet de bielle dû à l'intercalation de débris
entre les surfaces en contact. Avec la répétition des
cycles de chargement et déchargement, les diagrammes
tendent à se confondre en rapport avec les cassures
et l'imbrication progressive des aspérités.

Il y a lieu de remarquer que, pour l'ensemble des essais effectués, le module de raideur K_n varie de 1 à
4×10^3 MPa/m, les phyllades tendant vers les valeurs
inférieures et les quartzophyllades vers les valeurs
supérieures de l'intervalle. Ceci peut être attribué
à la présence dans ces dernières, des concrétions
dures de cristaux de quartz, ce qui donne des amplitudes d'épontes plus grandes, c'est-à-dire des
rugosités à l'origine plus importantes et peu cassan-
tes sous l'effet de charges et décharges normales
répétées.

4.2. Résistance au cisaillement.

La figure 6 montre quelques diagrammes typiques des
relations contrainte de cisaillement τ - déplacement
tangentiel dℓ obtenus dans les deux roches étudiées
pour différentes valeurs de la contrainte normale σ_n.

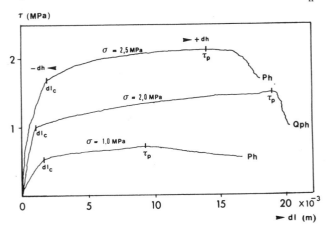

Fig.6. Courbes déplacement tangentiel dℓ - résistance au cisaillement τ.

On peut observer que les courbes présentent, dans
la phase initiale de l'essai, une allure quasi linéaire caractérisée par des déplacements tangentiels
faibles pour des augmentations de la résistance au
cisaillement importantes. Ce comportement correspond à une phase de contraction (dilatance négative)
dont la valeur maximale ($-dh_{max}$) marque approximati-
vement un point d'infléchissement dℓ_c de la courbe
dℓ- τ. Au-delà de ce point dℓ_c de courbure maximale,
les déplacements tangentiels deviennent très importants pour de faibles accroissements de la résistance au cisaillement, la courbe dℓ-τ prenant une
allure plus étalée en rapport avec la fin de la
phase de la dilatance. La valeur pic τ_p est atteinte lorsque la dilatance parvient à un maximum. Il y
a lieu de noter que les points d'infléchissement dℓ_c
et les contraintes de cisaillement pic τ_p tendent à
se produire en rapport avec des déplacements de
grandeur croissante avec l'augmentation de la contrainte normale appliquée. Dans les limites de l'in-

tervalle expérimental, le comportement est en accord avec celui du modèle à raideur constant proposé par Goodman (1974).

Le déplacement tangentiel $d\ell$ nécessaire pour atteindre la résistance au cisaillement pic τ_p est de l'ordre de 5 à 8% de la longueur maximale de l'axe de cisaillement du joint alors que la littérature rapporte des déplacements moyens d'environ 1% (Barton et al,1977) ou encore moins dans le cas de joints artificiels (Herdocia,1985). Toutefois, on doit considérer que les valeurs τ_p obtenues sont affectées, du fait des dimensions relativement réduites des joints carottés, d'un effet d'échelle dont il y a lieu de tenir compte lors de l'exploitation de ces résultats en vraie grandeur. La gamme de variation des modules de cisaillement $K_s(\tau_p)$ obtenus est donnée dans le tableau II ci-après.

La figure 6 représente les droites de régression enveloppes des faisceaux de relations contraintes normales σ_n résistance au cisaillement τ obtenues pour l'ensemble d'essais réalisés sur les deux roches étudiées.

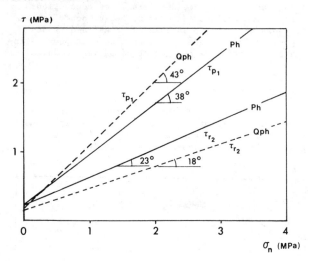

Fig.7. Relations contraintes normales σ_n - résistance au cisaillement τ.

Les valeurs limites des angles de frottement et de rugosité i= $\varphi_p - \varphi_r$ (Patton,1966) sont données au tableau II. La comparaison des résultats met en évidence que la résistance à la rupture et au polissage des épontes des quartzophyllades sont supérieures à celles des phyllades et ce vraisemblablement en rapport avec leur plus grande teneur en quartz.

Dans le cas des phyllades, le fait que $i_1 < i_2$ peut être attribué soit à ce que la rupture des indentations lors du premier cycle de chargement n'implique que les sommets des épontes de grande amplitude, ce qui conduit, pour les cycles subséquents de chargement et de déchargement, à une mobilisation d'un nombre croissant d'aspérités de faible amplitude, soit à une augmentation systématique de l'imbrication, suite à l'aplanissement des épontes de plus grande amplitude.

La caractérisation du relief des surfaces en contact peut être effectué en évaluant le coefficient de rugosité des joints JRC à l'aide de l'expression de Barton (1974) liant la résistance au cisaillement à la contrainte normale σ_n:

$$JRC = \frac{arctg(\tau_p/\sigma_n) - \varphi_r}{log(JCS/\sigma_n)}$$

où
τ_p : résistance maximale de cisaillement;

σ_n : contrainte normale appliquée au niveau du joint;
JCS : résistance à la compression simple des épontes obtenue d'après la compressibilité moyenne des joints;
φ_r : angle de frottement résiduel.

Si l'on considère la loi de Patton qui lie τ_p et σ_n à l'angle de frottement pic τ_p:

$$\tau_p = \sigma_n tg\, \varphi_p = \sigma_n tg(\varphi_r + i)$$

on peut exprimer le coefficient JRC en termes de l'angle de rugosité i :

$$i = \varphi_p - \varphi_r = JRC\, log(\frac{JCS}{\sigma_n})$$

Les coefficients JRC moyens obtenus pour les phyllades et quartzophyllades sont donnés au tableau II ci-après.

Tableau II.

	φ_p (°)	φ_r (°)	K_s x10^2MPa/m	$i_{1,2}$ (0)	$\overline{JRC}$ (0)
Ph	33-38	23-31	0,6-1,5	7 10	4,5 7,0
Qph	30-43	18-30	1,0-5,0	13 12	9,0 8,6

4.3. Caractérisation statistique de la rugosité.

Le relief de la surface du joint, déterminé à partir des mesures par la méthode de Moiré, est caractérisé à l'aide d'une série de paramètres statistiques calculés pour des axes caractéristiques de frottement décelés par examen au microscope minéralogique (concentration des zones de cassure, stries de frottement). Ainsi, on peut définir la rugosité d'après l'amplitude moyenne.

$$Z_o = \frac{1}{L} \int_0^L |y|\, dx \qquad (4.1)$$

où
L : longueur de l'axe de frottement considéré;
y : amplitude des épontes par rapport à un plan de référence.

En général, le paramètre Z_o est calculé d'après le relief de l'axe central de la surface du joint, c'est-à-dire le diamètre ou le grand axe du plan du joint dans la direction du cisaillement, pour les carottes de section circulaire. Pour cette étude, la rugosité a été mesurée sur plusieurs axes de frottement (par exemple les axes 1 à 4 dans le cas de la figure (4) de façon à mieux caractériser le profil du joint. Le nombre d'axes considérés peut être augmenté jusqu'à former un réseau très fin couvrant toute la surface auscultée, ce qui permet une définition intégrale de la rugosité.

Les autres paramètres statistiques utilisés sont la déviation moyenne de l'amplitude (Reeves,1985)

$$Z_1 = Z_o \sqrt{\pi/2} , \qquad (4.2)$$

le gradient d'amplitude,

$$Z_2 = \frac{1}{L} \int_0^L (\frac{dy}{dx})^2\, dx$$

et la déviation moyenne de la dérivée du gradient définis par Myers (Tse et al,1979)

$$Z_3 = \frac{1}{L} \int_0^L (\frac{d^2y}{dx^2})\, dx$$

Le tableau III résume les valeurs moyennes des paramètres statistiques obtenus pour les phyllades et les quartzophyllades.

Tableau III.

	$\bar{Z}_0$ $\times 10^{-3}$(m)	$\bar{Z}_1$ $\times 10^{-3}$(m)	$\bar{Z}_2$ (0)	$\bar{Z}_3$ $\times 10^{-3}$(m^{-1})	$\bar{i}$ (0)	$\overline{JRC}$ (0)
Ph	1,35	1,70	0,233	0,400	8,5	5,8
Qph	1,46	1,83	0,267	0,414	12,5	8,8

5. CONCLUSIONS

Le relevé topographique des surfaces des joints en vue de l'analyse statistique de la rugosité a été effectué à l'aide de la méthode de Moiré. Cette technique optique rapide et aisée permet de mettre en évidence la rugosité et l'imbrication avant et après essai des éprouvettes avec une très bonne précision (10^{-5} à 10^{-4} m). Les paramètres statistiques de rugosité ont été déterminés pour des profils parallèles à la direction de cisaillement d'après les données digitales fournies par balayage électronique de l'image du relief.

Les essais mécaniques de compressibilité montrent que les joints des phyllades sont plus compressibles que ceux des quartzophyllades et ce en rapport avec la présence, dans ces dernières, d'une teneur en quartz plus élevée, ce qui donne des épontes plus dures et résistantes à l'écrasement. De même, lors des premiers cycles de chargement et déchargement, la déformation permanente est plus importante dans les phyllades.

Les courbes de fermeture montrent une non linéarité plus marquée au déchargement, leur concavité changeant de sens selon le niveau de contrainte normale σ_n appliquée. Pour les paliers les plus élevés de σ_n, on constate, lorsqu'on atteint la fermeture maximale du joint, un comportement quasi-linéaire. Ce même type de phénomène a été observé par Herdocia (1985) sur des joints artificiels. Par contre, l'allure des courbes obtenues n'est pas conforme au comportement hyperbolique observé par Goodman (1974) et par Bandis et al (1983). Ceci est vraisemblablement dû au fait que la déformation des épontes - et ce notamment dans les quartzophyllades - suit essentiellement une loi élasto-plastique, perturbée par des ruptures par écrasement localisées et par des effets de bielle dus à la présence de débris. Ce comportement se traduit par une variation relativement importante des valeurs du module de raideur K_n lesquelles sont en valeur absolue plus élevées pour les quartzophyllades (2.5 à 4 MPa/m) que pour les phyllades (1 à 3 MPa/m). Par ailleurs, ces valeurs tendent à augmenter avec la répétition des cycles de chargement et déchargement, ce qui est en accord avec l'effet de précompression décrit par Hungr et al (1978) et par Herdocia (1985).

Les résultats des essais de cisaillement montrent que la variation de l'effort de cisaillement τ en fonction du déplacement tangentiel du joint dℓ pour différents paliers de contrainte normale σ_n croît de façon quasi linéaire, d'abord jusqu'à un seuil dℓ_c caractérisé par une croisance importante des τ pour des faibles déplacements dℓ, et puis selon un chemin caractérisé par des faibles augmentations des τ pour de grands déplacements dℓ, jusqu'à atteindre la résistance au cisaillement pic τ_p. Ce comportement est en rapport avec le développement d'une première phase de contraction dont la valeur maximale correspond au déplacement seuil dℓ_c, et par la suite, d'une deuxième phase de dilatance dont le maximum est atteint pour une valeur égale ou proche de τ_p. Lorsque la contrainte σ_n augmente la valeur du déplacement nécessaire pour atteindre τ_p augmente aussi, donc, la résistance maximale au cisaillement est dépendante du niveau de contrainte normale appliquée. Ces résultats ne concordent pas avec ceux rapportés par Goodman (1974).

Les relations contrainte normale σ_n - résistance au cisaillement τ obéissent au critère de Coulomb. La variation de l'angle de frottement entre les droites enveloppes correspondant aux τ pic et aux τ résiduels est plus importante pour les quartzophyllades caractérisées par des épontes de grande amplitude, de faibles intervalles, dures mais fragiles et aisément destructibles au cisaillement, que dans les phyllades de rugosité moins marquée. Les valeurs des paramètres statistiques d'amplitude Z_1 et de texture Z_2 et Z_3 caractérisant les deux roches étudiées, tombent dans la gamme des relations JRC-Z_2 trouvées par Barton et al (1977) et Z_1-Z_2-Z_3-JRC compilées par Reeves (1985).

6. REMERCIEMENTS

Le premier auteur remercie la Division d'Affaires Scientifique de l'OTAN pour le subside qui lui a été accordé pour la recherche coopérative n° 434/84 dont les résultats présentés ci-dessus en font partiellement partie. Les auteurs remercient le Professeur J.C.Verbrugge de la Faculté des Sciences Agronomiques de l'Etat à Gembloux et le Professeur J.Ebbeni de l'Université Libre de Bruxelles pour leurs avis sur le le sujet.

7. BIBLIOGRAPHIE

Bandis, S., Lumsden, A.C. et Barton, N.R. 1983. Fundamentals of rock joint deformation. Int.J.Rock Mech.Min.Sci. 20: 249-268.

Barton, N. et Choubey, V. 1977. The shear strength of rock joints in theory and practice. Rock Mechanics, 10:1-65.

Goodman, R. 1974. The mechanical properties of joints, Comptes rendus du 4e Congrès ISRM, Denver, 1:127-140.

Herdocia, A. 1985. Shear tests of artificial joints, Licenciate Thesis, Lulea University, Suède.

Hungr, O. et Coates, D.F. 1978. Deformability of joints and its relation to rock foundation settlements. Revue Canadienne de Géotechnique, 15:239-249.

Krsmanovic, D. et Langof, Z. 1964. Large scale laboratory tests of the shear strength of rocky material, Feldsmechanik und Ingenieurgeologie , Supplementum I : 20-30.

Patton, F.D. 1966. Multiple modes of shear failure in rock. Comptes-rendus du 1er Congrès S.I.M.R. Lisbonne, 1:509-514.

Pirodda, L. 1982. Optical methods of non destructive testing in Italy. A short selection, Industrial applications of holographic non destructive testing, SPIE Proceedings 349:1-16.

Reeves, M.J. 1985. Rock surfaces roughness and frictional strength, Int.J.Rock Mech.Min.Sci., 22:429-442.

Tse, R. et Cruden, D.M. 1979. Estimating joint roughness coefficients. Int.J.Rock Mech.Min.Sci., 16: 303-307.

New system for borehole TV logging and evaluation of weathered granites according to their colour indices

Nouveau système de diagraphie des sondages et évaluation des granits décomposés selon leur indice de couleur

Neue Methode zur TV-Bohrlochmessung und Bewertung von verwitterten Graniten nach dem Farbindex

YUNOSUKO IIZUKA, Shimizu Construction Co. Ltd, Tokyo, Japan
TAKASHI ISHII, Shimizu Construction Co. Ltd, Tokyo, Japan
KOJI NAGATA, Shimizu Construction Co. Ltd, Tokyo, Japan
YOSHITAKA MATSUMOTO, CORE Inc., Tokyo, Japan
OSAMU MURAKAMI, CORE Inc., Tokyo, Japan
KOJI NOGUCHI, Department of Mineral Industry, Waseda University, Tokyo, Japan

ABSTRACT: A new system for borehole TV logging, "BOREHOLE SCANNER", has been developed so that the scrolled colour images of the borehole-wall can be seen speedily. The video images of nine core samples ranging from well-weathered to fresh granites were processed and the colour indices were obtained. The colour indices are closely related to the physical properties of the core samples.

RESUME: "BOREHOLE SCANNER", un nouveau système de diagraphie des sondages, a été développé pour permettre une visualization immédiate en continu sur un écran de la couleur de la paroi du trou de sondage sur la surface. Les images vidéo de neuf carottes s'échelonnant due granite se trouvant dans un état avancé de décomposition au granite frais furent traitées afin d'en obtenir les indices de couleurs. Ces indices sont associés de très pres aux propriétés physiques des carottes.

ZUSAMMENFASSUNG: Eine neues TV-Bohrlochmessungssystem, der "BOHRLOCH-SCANNER", wurde entwickelt, um an der Oberflache eine sofortige Begutachtung der sich an der Bohrlochwand abrollenden Farbe zu ermöglichen. Die Bilder von neun Bodenproben, von stark verwittertem bis zu frischem Granit, wurden verarbeitet und Farbindexe wurden erstellt. Die Farbindexe stehen in enger Beziehung zu den Festigkeitseigenschaften der Bodenproben.

1 INTRODUCTION

TV logging is the most effective method for identifying fractures and beddings in boreholes. But speedy logging is difficult because only a very limited area, i.e. about 2 cm x 3 cm, on the inner wall can be seen in a borehole of a small diameter. The new TV system, i.e. "BOREHOLE SCANNER" has been developed so that the surrounding inner wall of the borehole can be seen immediately as continuously and longitudinally scrolled images. With this system, investigation of the fracture system and the colour variations of deeper bedrocks is easy.

In order to investigate whether the numerical information on the colour of rocks can be used for evaluation of weathered rocks, colour indices and properties of nine core samples ranging from well weathered to fresh granite were measured.

2 OUTLINE OF BOREHOLE SCANNER

The borehole scanner comprises a probe which receives optical signals from the inner borehole wall and converts them into electrical signals, a cable which transmits the electrical signals to the surface, a winch for manipulating the probe vertically, means which displays the transmitted signals as colour video images of the inner borehole wall and a video tape recorder which records the image signals. The system is similar in appearance (see Figure 1) with the conventional borehole TV system.

The scanning head and the mirror for lighting rotate at the speed of 3000 rpm. Optical signals from the borehole wall are received at the scanning head, converted into digital signals, transmitted to the surface and displayed on the TV monitor as the colour images of the borehole wall synchronously

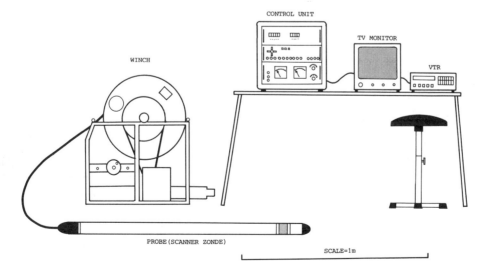

Figure 1. General view of the equipments of BOREHOLE SCANNER

with the speed with which the probe is lowered in the borehole (see Figure 2). The borehole scanner provides substantially full-scale video images of the inner borehole wall for the angle of 360°. Images of the inner borehole wall can be obtained in a scrolling manner as the probe is lowered along the borehole wall, and a geologist can study and analyze these images for beddings and fractures on the spot at the surface and at a speed similar to that of the geological survey conducted on the surface. The specification of the system is shown in Table 1.

Table 1. Specification of the borehole scanner

Borehole diameter	66 – 86 mmø
Water proof pressure	1200 m
Cable length	200 m (can be extended up to 1200 m)
Ambient temperature	1 – 60°C
Resolution	0.1 mm
Method of recording	NTSC signal recording (video tape recorder) and digital signal recording
Velocity of probe	> 30 m/h

The borehole scanner enables faster TV logging with reduced cost, and will be useful in investigating rock foundation and slope as well as underground facilities in the future. The image signals can be stored in a video tape recorder by converting them into NTSC signals. Moreover they can be outputted as digital image data, so that when inexpensive recording means become available in the future, the system can be adapted for automatic measurements such as shown in Table 2 by processing the images.

3 DETERIORATION INDICES OF WEATHERED GRANITE

The colour logging of the borehole wall, i.e. acquisition of numerical data on the colour variations in borehole wall, will add greatly to the objective geological data on the borehole. The data on the colour variations is, as a general rule, interpreted by a geologist. In order to adapt such data in the characterization of weathered rocks and the quality control of rock grouting and soil-cement, it is necessary to study the possible correlation of the colour variation data and the properties. The present paper discusses the preliminary investigation conducted on the correlation of the properties and the colour data of the weathered granites.

Various indices have been proposed for determining the degree of weathering in granites. Core samples of weathered RYOKE GRANITE taken at different depths (ranging from 0 m to 116 m below surface) in one location were measured for their chemical and physical properties. The samples were assessed for the correlation of the depth with the depth-dependent factors and the correlation between each different properties. The study indicates that weathering potential index (WPI) (Reiche 1943), porosity, tensile strength, and elastic wave velocity are suitable indices in evaluating the degree of deterioration of granites by weathering. Figures 3, 4 and 5 show the relations between the properties, i.e. deterioration indices of weathered granites.

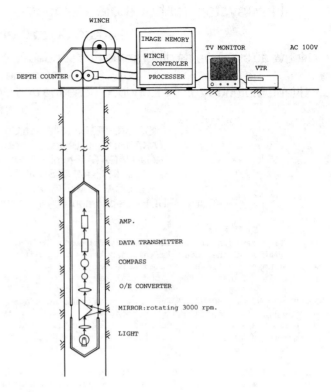

Figure 2. Conceptual mechanism of BOREHOLE SCANNER

Table 2. Items of automatic measurements by processing video images of the borehole wall

- Automatic measurement of strike, dip and average width of observed fractures
- Evaluation of deterioration indices of weathered rocks on the basis of rock colours
- Quality control of the rock grouting and soil-cement through measurement of colour distribution

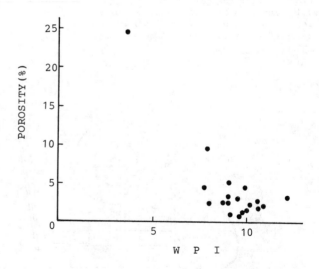

Figure 3. Relation between porosity and WPI of weathered granites

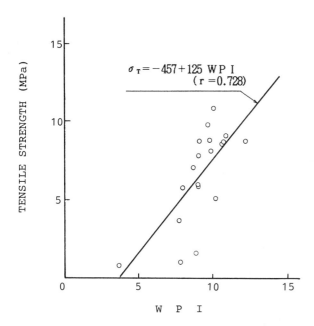

Figure 4. Relation between tensile strength
and WPI of weathered granites

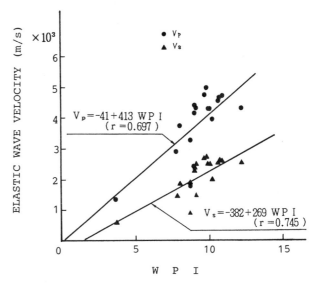

Figure 5. Relation between elastic wave velocity and
WPI of weathered granites

4 MEASURING METHOD OF COLOUR INDICES

Preliminary measurement on the colour of the weathered granites was conducted prior to completion of
a borehole scanner. Nine typical core samples representative of well weathered to fresh granites
were selected. Video images of the core samples
were formed in a manner shown in Figure 6 and recorded in a floppy disc as digital image data.
Approximately 17,000 pixels were sampled from the
data. Using an image processing unit (NEXUS 5500,
manufactured by Kashiwagi Giken Co.) capable of
discriminating R, G and B into 16 gradations each,
the colour indices were obtained.

Among those minerals constituting granites such
as quartz, potassium feldspar, plagioclase, biotite
and holnblend, most of particles which show colour
variations due to weathering are plagioclase.
Photometry of average colour variations of 17,000
(131 x 131 dots) pixels is not sufficient to enable
characterization of the colour variations reflecting
the degree of weathering. The following processings
were therefore necessary.

. Minerals such as quartz which are substantially
  white in colour are excluded from the measurement: Filters are optimally adjusted to block
  the white gradations.
. The number of pixels (Nb) in black gradations
  is counted.
. The number of pixels (Nr) in brown gradations
  is counted.

. Colour index = $\dfrac{Nr}{Nr + Nb}$ is

  calculated.

5 RELATION BETWEEN COLOUR INDICES AND ROCK PROPERTIES

The relations between the colour indices and the
measured properties of the nine core sample are
shown in Figures 7, 8, 9 and 10.

The colour indices and the deterioration indices
of the rock samples show a positive correlation.
The colour indices will be useful as geological information on the weathered granites.

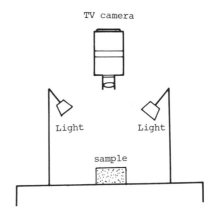

Figure 6. Conceptual Drawing of taking
colour image of a weathered granite

6 CONCLUSION

In case of the conventional borehole investigation,
identification of rock colour used to depend mainly
on the judgement of an individual geologist and
since no reference colour names were available, the
judgement tended to be subjective. The development
of a borehole scanner indicates the possibility of
continuous measurement on the colour variations in
the direction of the depth. A preliminary study on
the colour variations of the weathered granites and
the chemical and physical properties thereof will
enable estimation of rock deterioration in weathered granites by adequately processing the colour
image data on the borehole wall.

The development of techniques for the image processing of the borehole wall image data in colour
was initiated only recently. Time and efforts
should be exerted in developing methods and devices
for items listed in Table 2.

REFERENCE

Reiche, P 1943. Graphic representation of chemical
weathering, J. Sed. Petrology, vol. 76

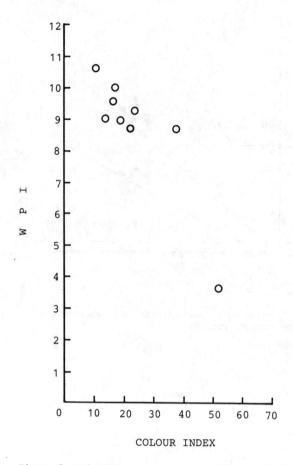

Figure 7. Relation between colour index and WPI of weathered granites

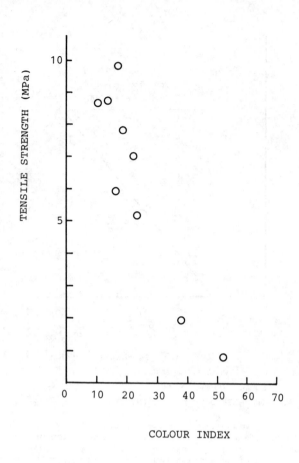

Figure 9. Relation between colour index and tensile strength of weathered granites

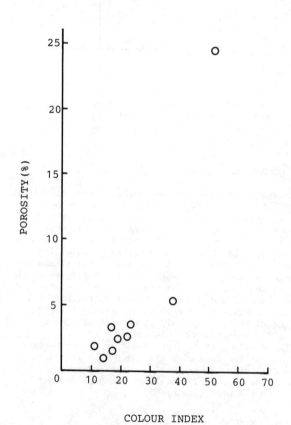

Figure 8. Relation between colour index and porosity of weathered granites

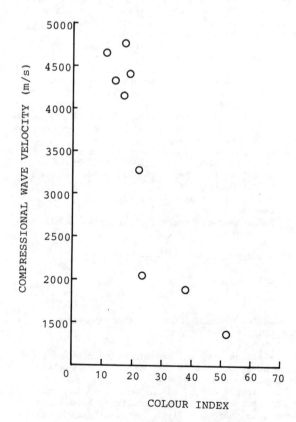

Figure 10. Relation between colour index and compressional wave velocity of weathred granites

396

A complex slope failure in a highly weathered rock mass
Défaillance complexe d'une pente dans un massif rocheux très altéré
Progressiver Böschungsbruch im einem stark verwitterten Felsmassiv

T.Y.IRFAN, Geotechnical Control Office, Hong Kong
N.KOIRALA, Geotechnical Control Office, Hong Kong
K.Y.TANG, Geotechnical Control Office, Hong Kong

ABSTRACT: A 37 m high cut slope excavated in a highly weathered, poor quality rock mass on the flanks of a large hillside in Hong Kong has been showing signs of progressive failure since 1983. This paper describes the results of the extensive investigations and discusses the possible mode and mechanism of this progressive failure as related to the complex geology and groundwater conditions.

RESUME: Une pente taillée haute de 37 mètres et creusée dans une masse rocheuse de mauvaise qualité et extrêmement désagrégée sur les côtes d'un grand coteau à Hong Kong faillit progressivement depuis 1983 à la suite d'une précipitation importante. Cet exposé décrit les résultats des études approfondies enterprises et discute le mode et le mécanisme possibles de cette faillite progressive en ce qui se rapporte à la géologie complexe et les conditions d'eaux de fond.

ZUSAMMENFASSUNG: Ein 37 M. hoher, abgegrabener Hang, ausgegraben in einer stark verwitterten, schlechter Qualitaet, Gesteinmasse, auf den Seiten eines grossen Huegels in Hong Kong, hat seit dem Jahre 1983 Symptome eines progressiven Verfalls gezeight. Dieser Aufsatz beschreibt die Ergebenisse ausfuehrlicher Untersuchungen und bespricht die moegliche Art und Weise und den Mechanismus dieses progressiven Verfalls mit Beziehung auf die komplizierte Geologie und Grundwasser-Verhaeltnisse.

1 INTRODUCTION

A large number of steep cut and fill slopes formed in the urban areas of Hong Kong directly overlook major roads or multistorey residential, commercial or community developments. Tropical cyclones and low pressure troughs bring short intense rainstorms and longer periods of heavy rainfall to Hong Kong. The effect of such rainfall is to cause a large number of landslides in the steep hilly terrain, which is extensively covered with a thick mantle of colluvium and deeply weathered rock.

The failures are generally sudden in both the weathered mantle and the underlying rock and these are usually related to rapid changes in pore pressure and the consequent reduction in shear strength. Some slow moving and progressive failures have been reported in weathered rocks from Hong Kong and elsewhere (Wolle 1985, Sassa 1984) but these are not common in Hong Kong. Only a few of the failures in weathered rocks have been monitored closely thus precluding detailed studies of failure mechanisms and critical examination of the methods of analysis used which traditionally treated them either as conventional soil mechanics or rock mechanics problems (Powell & Irfan 1986). Brand (1984) stressed that because of the nature of weathered rock profiles and colluvium, stability assessments in these materials must necessarily combine the classical analysis of soil mechanics and rock mechanics with a knowledge of the engineering geology of each situation, and with sound engineering judgement and experience. However, there are siutations where the complexity of the site geology confounds conventional analysis (Hencher et al 1984).

A 37 m high cut slope in a highly weathered, poor quality rock mass underlying a colluvial mantle at Tin Wan Hill, Hong Kong, has been showing signs of progressive failure since 1983. Sudden failures in the colluvial mantle have also occurred. The slope was excavated in 1963 on the southwestern flank of a large hill (Mt Kellett) in connection with the development of a housing estate (Figure 1). The cut slope originally had three berms with interberm angles of 45° to 50°, giving an overall slope angle

of 34°. The detailed investigation and monitoring undertaken by the Geotechnical Control Office (GCO) to design remedial works for the failed slope, which is threatening an urban road, has provided insights into the mode and mechanism of such failures.

This paper describes the history of the failure and the extensive investigation undertaken, and discusses the possible mode and mechanism of this progressive failure as related to the complex geology and groundwater conditions in a poor quality rock mass.

2 HISTORY OF SLOPE MOVEMENTS

An examination of aerial photographs and the few available records indicate that the slope has been subject to continuous instability since it was cut in 1963, and major remedial works in the form of regrading, surface protection (sealing) and subsurface drains were carried out between 1968 and 1982. Detailed study by the GCO commenced in early October 1983 when localised instability in the form of some minor rock falls and displacement of a surface drainage channel by 50 to 100 mm at the toe of the slope were observed. These movements indicated that the toe of the slope was being overstressed, which subsequently led to a larger scale shearing in mid-October 1983 after a heavy rainfall. Following a total of 175 mm of rainfall over three days related to Typhoon Joe, a slip scar 1.5 to 2.5 m high with a tension crack at least 2 m deep was formed at the crest of the cut slope along the central sector (Figure 1).

Remedial works were designed following site investigation, limited slope monitoring, and back analysis in late 1983 and 1984. These consisted of the construction of a reinforced concrete retaining structure founded on competent rock at the toe, trimming of the slope, installation of up to 20 m long raking drains, and provision of adequate surface cover in the form of shotcrete and surface drainage channels. Very little movement was recorded until 13 April 1985 when a relatively minor slip occurred in the weak rock mass on the south (Figure 1) following five

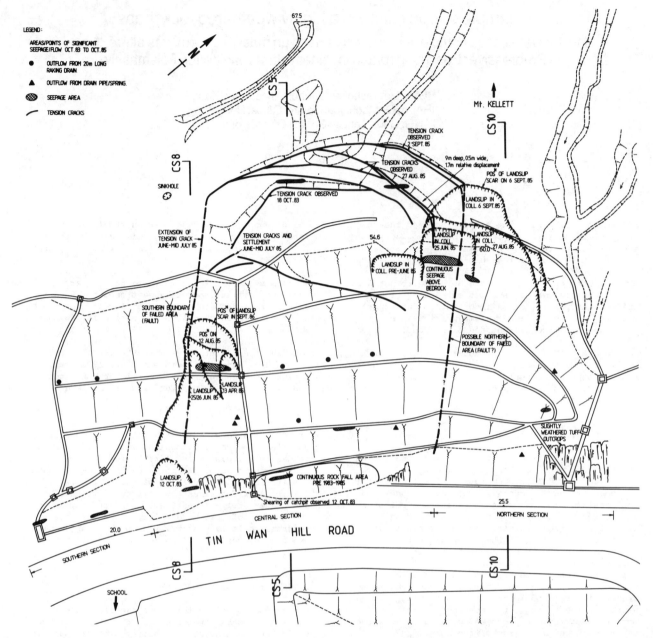

Figure 1. Site plan showing the original slope configuration and failure history at Tin Wan Hill Road

days of continuous rainfall (220 mm). Slope movements were re-activated again along the central
section, resulting in the extension and opening up
of tension cracks following 135 mm of rainfall on
25th June 1985; the steep colluvial face suddenly
failed on the same day, producing approximately 200
m³ of slide debris. The movement accelerated on 12th
July during the excavation at the toe for one of the
central panels of the retaining wall, which was
subsequently stabilized by emplacement of crushed
rock fill.

The most recent phase of movements started at the
end of August 1985 following intense rainfall of 119
mm on 26th and 79 mm on the morning of 27th August.
The failure progressed upslope with the opening up of
new tension cracks at least 9 m deep, rapid movements of the surface zone and landslips in colluvium
on the central and northern sectors of the slope
(Figure 1).

A detailed study, including a new programme of
drilling, monitoring and engineering geological
mapping, was undertaken after the progression of the
failure upslope. Investigations and detailed mapping
continued throughout the remedial works.

3 ROCK MASS CONDITIONS

The cut slope consists of pyroclastic volcanic rocks,
mainly acidic fine ash tuffs of Jurassic age, overlain by a variable thickness (0 to 9 m) of old very
dense weathered insitu colluvium, which is in turn
overlain by relatively recent and loose bouldery
colluvium near the crest region and beyond (Figure 2).
The rocks are deeply weathered, down to 30 m, to weak
rock and soil-like materials, particularly along
several shear/fault zones crossing the slope.

3.1 Weathering pattern

The weathering profile of the rock mass in the slope
is complex. The boundary between the competent and
strong rock mass (fresh to moderately weathered) and
the incompetent weak rock mass (highly to completely
weathered), loosely termed 'rockhead' for convenience, is at about the road level and almost
horizontal in the southern section of the slope.
Along the central section of the slope, which is the
main subject of this paper, rockhead rises up both

398

upslope and towards the north (Figure 2).

The approximate boundaries between the various grades of rock and the colluvium interface are all approximately parallel to each other and dip at 20° to 30° towards the cut slope face. This general pattern is broken along the shear zones, where 'valleys' occur in rockhead due to deeper weathering within these very closely fractured zones.

3.2 Discontinuity pattern

The rock mass is heavily broken into a blocky to tabular fabric by three major and two minor joint sets. The intensity of jointing increases upwards in the weathering profile and also along the shear zones,

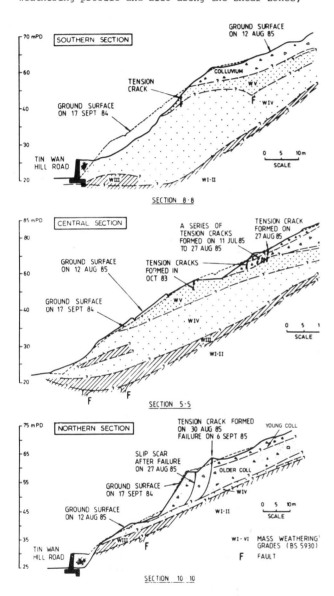

Figure 2. Typical engineering geological cross-sections (central sector)

where the weak rock mass contains closely to very closely spaced joints.

The critical discontinuity sets are the two subvertical to vertical joint sets striking almost parallel and perpendicular to the slope face, hence forming toppling/back release and side release faces respectively, and the subhorizontal joints which daylight on the cut slope face and have varying dip angles from about 8° to 40°. As a result of

faulting, the joint pattern is slightly different along the central section of the slope, where many joints of the subhorizontal set dip more steeply (20° to 40° in the direction of 150° to 170°) than the joints of the same set on the stable northern section of the slope, where the dips are gentler (5° to 20°) and in the direction of 135° (Figure 3). The majority of the joints along the central section contain white clayey infills of a plastic nature, which are considered to be kaolin veins.

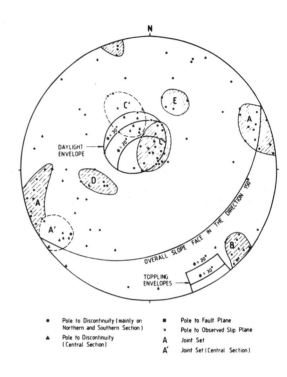

Figure 3. Stereoplot of discontinuities and kinematic stability analysis

In addition to simple sliding failures (Figure 3), a number of other sliding/shearing mechanisms are possible (e.g. stepped sliding) as well as shearing through the mass in this highly weathered, heavily jointed, poor quality rock mass. Failure mechanisms are considered in detail later in the paper.

3.3 Hydrogeological conditions

The catchment area above the slope is very large, and colluvium on the surface is highly permeable. The surface water descending from the steep side slopes of Mt Kellett is channelled into the landslip area by a number of ephemeral stream courses (Figure 1).

The limited piezometric data, the borehole evidence, the surface observations of seepage and the slope geology all indicate that a complex groundwater regime exists within the slope. In addition to a permanent water table in the closely jointed rock mass, instantaneous and/or semi-permanent perched water tables form along weathering grade boundaries, along kaolin infilled joints, along the colluvium/insitu weathered rock interface, and also within the colluvium of different ages due to differences in layer permeabilities. Long horizontal drains installed in the slope showed that there was a possibility of groundwater being channelled within the slope by the fault zones striking almost perpendicular to the slope face. A number of pipes (preferential flow conduits) were observed in the weathered rock as well as within the colluvium. It is highly likely that a network of pipes exist in the slope. The flow in such a jointed rock mass is

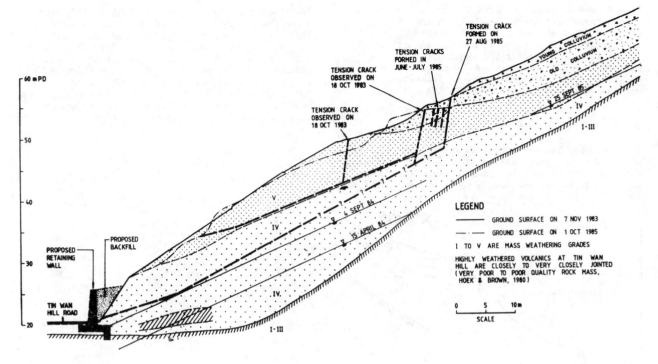

Figure 4. Postulated failure model (central section)

irregular and of widely differing velocities (Sharp et al 1972). The effect of all of these features on the groundwater regime and the stability of the slope cannot be estimated with any accuracy.

3.4 Main groundwater table

The depth of the groundwater table at the time of the 1983 and 1985 movements is not well known. However, the minimum and maximum groundwater levels reached in 1984 (Figure 4) were established from more extensive piezometric information. 1984 was a rather dry year by Hong Kong standards, with a total of 1550 mm of rain recorded near the failure area, compared to an annual average of 2240 mm recorded at the Royal Observatory. A seasonal (general) rise of 4 m of the main water table above the dry season (winter) level was recorded. In comparison, both 1982 and 1983 were exceptionally wet years with a total of 3010 mm and 2705 mm of rainfall respectively.

It is possible that the water table in October 1983 stood about 7 to 8 m above the lowest recorded level in 1984. Based on piezometres installed in new boreholes following the movements in June to September 1985, the maximum 1985 groundwater table was estimated to be about 2 to 2.5 m above the maximum level of 1984. The total amount of rainfall until 25th September 1985 was about 1920 mm.

4 MODE AND MECHANISM OF THE FAILURE

This slope failure can be considered 'progressive', as small movements of the order of tens of mm have been occurring since at least the summer of 1983, and there has not yet been any dislodgement of large volumes of rock from the central portion of the slope.

After the most recent phase of movements in the autumn of 1985, which resulted in the progression of the failure upslope, a more well-defined surface expression of the landslip has emerged.

4.1 Description of the failed mass

On the south, the moving mass is bounded by an al-

most straight and vertical shear surface trending at about 150°, i.e. slightly oblique to the cut slope direction of 135°. The northern release surface is not so obvious as it is masked by the slip debris from the colluvial face and the shallow rockhead. The upslope limit is the almost vertical tension crack showing 1 to 2 m relative displacement that was formed in August 1985 (Figure 1). It is up to 9 m deep and slightly curved in plan.

The photographic records of the slope show that the southern and possibly the northern release surfaces of the moving mass may coincide with the location of ancient shear/fault zones trending in a NW-SE direction. It was thus evident that the main direction of movement is oblique, by approximately 15°, to the dip direction of the cut slope. This direction is also consistent with the dip of the kaolin vein-bearing subhorizontal joints at the central section, pointing to a strong structural control on the mode of failure.

The exact depth and nature of the failure surface (or surfaces) are still open to discussion, as the failed mass has not been dislodged, and direct observations of the failure surface have not been possible. At least two levels of movement have been postulated from the combined evidence, however, and further evidence supporting this view has been obtained during the remedial trimming of the slope. A number of what were considered to be possible failure surfaces or zones (e.g. minor offset of joints and slicken-siding, plastically deformed layers, polished kaolin surfaces and other features such as infilled pipes) were observed from inspection of undisturbed samples and cores, and by surface inspection during trimming works.

4.2 Shallow failure surface

Evidence from the boreholes in the failed area indicate that a number of shallow near-planar failure surfaces may exist from 2 to 5 m below ground surface, mainly controlled by kaolin-bearing relict joints, but also occurring as shears through the material within the extremely weak, completely weathered rock mass. The deepest of these shear surfaces coincides with the approximate interface between highly weathered and completely weathered rock (Figure 4). The location where bulging of the slope was first observed in October 1983, half way up the cut slope face, would coincide with the daylighting of such a shearing surface. This failure surface has a general dip of approximately 20° to 22°.

During the trimming works, parts of the shallower failure surface were observed at a number of locations. At each locality, a completely weathered soil-like mass of a rather disturbed nature was sliding over a closely jointed, highly weathered rock mass composed of slightly stronger rock, with parts of the slip plane formed of a number of gently to moderately dipping joints of the subhorizontal set.

It is considered that the near surface failure of the slope is a combination of progressive surface creep in the soil-like mass and sliding along various kaolin-bearing, relatively shallow dipping relict joints. Additional evidence for this type of mechanism comes from the 'stable' area on the south, where a 2 to 3 m thick weathered rock section of the slope has moved by a couple of centimetres along a relatively extensive kaolin vein-bearing joint. The boundary of the movement upslope is a subvertical joint which has opened up. The jointed, weak rock mass within the slightly moved block shows disturbance of mass fabric in the form of outward curved joints near the ground surface and minor offset along the joints near the toe.

4.3 Deep failure surface

The postulated deeper seated failure surface, or more correctly the lower bound surface of the moving/disturbed rock mass shown on Figure 4 is drawn based on the borehole evidence, movement records and site observations. The latest tension crack formed was observed to be at least 9 to 10 m deep. A slip indicator installed in a borehole within the failed area was blocked at about 8.2 m depth sometime before June 1985. Shearing of the toe channel (100 mm displacement) first noticed after the October 1983 movement, heaving of the road pavement, said to have occurred sometime before the October 1983 failure, and rockfalls from the toe slope at various times, all point towards a failure surface daylighting at or just below the toe of the slope. This deep failure surface may also explain the slow moving creep-like nature of this landslide.

Because of the limited movements that have occurred and the closely jointed nature of the rock mass, it was not possible to observe any distinct shear plane within the poor quality rock mass during the recent phase of remedial works. The rock mass fabric was, however, observed to be disturbed above the road level with slippages and dilation both along the subhorizontal and vertical joint sets.

The discontinuity characteristics and the weathering state of the closely jointed rock mass suggest that two failure modes may be possible along this almost planar deep shear surface :
(a) The failure is joint controlled and a stepped surface defined by the subhorizontal joint set and one of the vertical joint sets exists, or
(b) the failure is a narrow shear band through the rock mass.
In the latter case, failure of the rock mass will involve the interaction of sliding along the joints, dilatancy, separation and rotation of the blocks and possible fracture of the intact rock (Ladanyi & Archambault 1972). For either mechanism, the greater part of the sliding surface is likely to be controlled by the subhorizontal joint set due to its unfavourable orientation and the low shear strength along its kaolin infillings. Some field evidence, and experiments carried out by Kanji (1976), suggest that along the rock-kaolin contacts the residual effective angle of internal friction could be as low as 12°.

5 DISCUSSION ON FAILURE MECHANISM

It is not known when the original failure commenced; but from the records it is clear that there was some movement of the slope even before the October 1983 failure. It is highly probable that the excavation of the 37 m high slope in 1963 and the subsequent stress release led to some, perhaps progressive opening of the vertical joints striking parallel to the slope. Minor movements on unfavourably oriented joints due to localised excess pore water pressures may have caused further opening of joints leading to increased infiltration of water and the general deterioration of the slope. Finally, general sliding/shearing occurred in October 1983 resulting in appreciable movement of several hundred mm, most probably as a result of a critically high water table. Time-dependent and strain-dependent weakening of the highly to completely weathered rock may also have contributed to the delayed failure.

After the 1985 movements, it was realized that a simplified model would not be appropriate for the complex and progressive nature of the failure. However, limit equilibrium analysis using the Janbu's method were still carried out to determine the general factor of safety of the slope to uniform groundwater rise so that further movements could be stabilised by remedial works. It was assumed in these analyses that failure was through the poor quality rock mass as defined by Hoek & Brown's (1980) empirical shear strength criteria. The analyses indicated that the overall factor of safety along the deep shear surface was reduced to below unity when the groundwater level reached 1.5 to 4.5 m above the 1984 maximum level, depending on the intact rock strength (Figure 5).

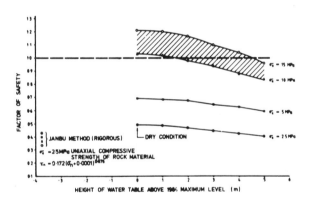

Figure 5. Plot of sensitivity analyses

After the initial general shearing of the rock mass, the shear strength along the failure surface reduced to less than the peak value. Therefore, movement could have initiated again in 1985 along the same shear surface at lower groundwater levels than the one recorded in 1983. Loss of support due to earlier movements and limited excavations for retaining wall construction, possible reduction of shear strength due to saturation of kaolin-bearing joints and localised pore pressures during moderate rainfall, are sufficient factors to explain the initiation of further movements in June to September 1985.

The observed shallow movements in the slope and the landslips in colluvium may have resulted from the movements occurring deeper within the slope causing or aggravating the formation and opening up of tension cracks in the overlying completely weathered material and colluvium. The 'sudden' landslips recorded from 1983 to 1985 which tended to occur during or just after moderate amounts of rainfall would then be caused by perched water tables or the filling up of open joints or tension cracks.

6 CONCLUSIONS

The earliest phases of movements on this slope are considered to be creep movements due to re-distribution of stress, with movements within part of the rock mass occurring along some of the kaolin vein-bearing joints. The movements may have been accommodated in part by internal deformation of the mass, especially the parts composed of much weaker plastic soil-like material. Stresses would then be transferred to stronger sections which acted as 'buttresses' until they also failed. This type of mechanism would require that zones of overstress or failure existed within the slope prior to overall failure of the slope. These overstressed zones may have contributed significantly to the progressive nature of the failure.

Conventional limit equilibrium methods give little or no information concerning localised failure zones (Chowdhury 1978). In conventional stability analyses based on simplified mechanisms of slope failure, as commonly applied in Hong Kong and elsewhere to rocks weathered to great extent, only a static factor of safety is computed. As shown by this failure, it is dangerous to accept this factor of safety without full regard to the dynamic nature of the slope's equilibrium.

Failures have occurred in this slope, both in the weathered rock mass and in colluvium, as a result of unfavourable and complex geological and groundwater conditions. It is therefore important that the engineering geology of the slope be clearly addressed in all future assessments of slope stability, and that realistic geotechnical models of the slope conditions be developed.

This case history also demonstrates that the design of slopes in weathered poor quality rock masses should account for stress history and weathering effects. Indeed, a slope may be stable when first excavated, but because of a gradual deterioration due to the physical and chemical effects of weathering and stress relief, it may become unstable with the passage of time.

ACKNOWLEDGEMENTS

This paper is published with the permission of the Director of Civil Engineering Services of the Hong Kong Government.

REFERENCES

Brand, E.W. 1984. Landslides in Southeast Asia: a state-of-the-art report. Proceedings of the Fourth International Symposium on Landslides, Toronto, 1 : 17-59.

Chowdhury, R.N. 1978. Slope Analysis. Elsevier, Amsterdam: 423 p.

Hencher, S.R., Massey, J.B. & Brand, E.W. 1984. Application of back analysis to some Hong Kong landslides. Proceedings of the Fourth International Symposium on Landslides, Toronto, 1 : 631-638.

Hoek, E. & Brown, E.T. 1980. Empirical strength criterion for rock masses. Journal of Geotechnical Engineering Division, American Society of Civil Engineers, 106, GT9 : 1013-1035.

Kanji, M.A. 1976. Shear Strength of Soil-Rock Interfaces. M.S. thesis, Department of Geology, University of Illinois, Urbana, USA.

Ladanyi, B. & Archambault, G. 1970. Simulation of the shear behaviour of a jointed rock mass. Proceedings of the 11th Symposium on Rock Mechanics, Berkeley, California : 105-125.

Powell, G.E. & Irfan, T.Y. 1986. Slope remedial works for differing risks. Proceedings of the First International Conference on Rock Engineering and Excavation in an Urban Environment, Hong Kong : 347-355.

Sassa, K. 1984. Monitoring of a crystalline schist landslide - compressive creep affected by "underground erosion". Proceedings of the Fourth International Symposium on Landslides, Toronto, 2 : 179-184.

Sharp, J.C., Maini, Y.N. & Harper, T.R. 1972. Influence of groundwater on the stability of rock masses. Transactions of the Institution of Mining and Metallurgy, London, 81, Bulletin No. 782 : A13-20.

Wolle, C.M. 1985. Slope stability. Pecularities of Geotechnical Behaviour of Tropical Lateritic and Saprolitic Soils, Progress Report, Theme 3.2 : 164-214. Brazilian Society for Soil Mechanics, Sao Paulo.

Evaluation of bearing capacity of foundation ground and stability of long slope for weathered granite

Evaluation de la capacité portante d'une fondation et stabilité d'un haut talus dans un granit altéré
Die Tragfähigkeit eines Fundamentes und die Stabilität einer hohen Böschung im verwitterten Granit

KOJI ISHIKAWA, Chuo Kaihatsu Corporation, Japan
KEIJI MIYAJIMA, Chuo Kaihatsu Corporation, Japan
KATSUJIKO YAMADA, Honshu Shikoku Bridge Authority, Japan
MAMORU YAMAGATA, Honshu Shikoku Bridge Authority, Japan

ABSTRACT : Bearing capacity of a foundation and stability of a high cut slope, which consist of weathered granite, are investigated in this study. Primary stress conditions and changes in stresses in the ground caused by external forces are determined by measurements or estimation. Mechanical properties of the ground are shown to be greatly affected by rock class and stress conditions. The finite element method is used for analysis and characteristics specific to weathered granite rock masses are explained.

RÉSUME : La capacité portante de la fondation et la stabilité du talus haut, qui consistent en granit altéré, sont examinés dans cette étude. Les conditions de la contrainte naturelle et le change ment des veceuees contraintes à cause de la force exterieure dans la terre sont décides par le mesure ou l'estimation. La propriété mécanique de la terre sont motrées à subir assez d'influence de la désignation des roches et des conditéons du vecteur contrainte, La méthod d'élément tini sert à résoudre les problems susdits et les Caractéres mécaniques de la massif rocheux granitique sont expliqués.

ZUSAMMENFASSUNG : Die Tragfähigkeit eines Fundaments und Stabilität einer hoher Böschung aus verwittertem Granit wird in dieser Studie untersucht. Die Anfangsspannunges und die Anderungen in der Spannungen des Bodens, verursacht durch äuBere Kräfte werden durch Messungen oder Schätzung bestimmt. ⌐Dabei zeigt sich, daB die mechanischen Eigenschaften des Bodens stark von Bezeichung des Fels und Spannungsbedingungen beeinfluBt wird. Die Finite Elemente Methode wird zur Analyse verwendet, und typische Eigenschajten der verwitterten Granitsteingebirgen werden erläutert.

1. Introduction

A precise investigation and evaluation method for bearing capacity of the large foundation of long-span bridge and stability of high cut slope, which consist of weathered granite, is proposed in this paper. Investigation method of the foundation rock is mainly carried out by several geophysical loggings, such as velocity, resistivity and density logging. Rockmass porosity is calculated based on these logging results and this porosity is proved to relate to deformation charachter of the rockmass. This deformation character is affcted according to the material properties by the primary stress in the ground and overburden pressure. Further, the effect on the deformation character by loosening and stress relief and compaction effect on this character by incremental load are studied. An evaluation method of bearing capacity of the foundation is proposed considering material properties and confining pressure dependency of the deformation character. Evaluation of rockmass properties by the investigation and slope stability analysis by FEM are performed considering material properties, stress anisotropy, stress relief and loosening of the slope forming rock.

2. Rockmass properties

Relationship between the rockmass porosity calculated from several geoplysical logging results and deformation modulus by pressuremeter is shown in Fig.1. Further, deformation modulus vs. measuring depth relationship is also exhibited in this figure. From the relationship in Fig.1, Eq (1) can be derived as follows :

$$Esp = \alpha \cdot n_R{}^{\beta_1} \cdot Z^{\beta_2} \qquad (1)$$

where, Esp : deformation modulus by pressuremeter
n_R : rockmass porocity calculated by geophysical logging
Z : measuring depth
α, β : material constants

Fig.2 shows deformation character vs. several confining pressure of laboratory tests relationship. The same relationship of in-situ tests are also shown in Fig.2, assuming $\sigma_c' = \gamma t'$. Z. From the relationship in Fig.2, Eq (2) can be derived as follows :

$$Ec = \alpha_1 n_R{}^{\beta_1} \cdot \alpha_2 \sigma_c'{}^{\beta_2} \qquad (2)$$

where, Ec : deformation modulus of laboratory test
σ_c' : consolidation pressure ($= \sigma_3'$)

Fig.3 exhibits the relationship between horizontal and vertical initial stresses Po' by pressuremeter and overburden depth Z. From this figure, it is concluded for the ground to be isotropic and overconsolidated stress conditions.

Subsurface strain under the loading plate is measured by embeded strain gauges utilizing the plate loading test as shown in Fig.4. Based on the analysis

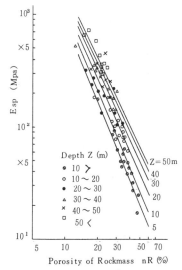

Fig. 1 Relation of porosity of rock mass for physical logging vs. deformation modulus for bore hole test at each depth

result for measured strain values, it is recognized that the deformation modulus in the ground has a tencency to increase toward depth direction by the overburden pressure. Deformation character in the ground surface is affected by the loosening and stress relief and the deformation modulus has a tendency to increase by compaction of the ground with incremental loads.

Stress ratio γ $(= \sigma_1'/\sigma_{1y}')$ vs. deformation modulus ratio K $(= E_1/E_{1y})$ relationship based on the laboratory test results is shown in Fig.5. Non-linearity of stress-strain relationship is formulated in this figure. From the relationship in Fig.5, Eq (3)~(6) can be derived as follows :

When σ_1'/σ_{1y}' > 0.5 $(\sigma_c' \leq 1000KN/m^2$, Esp $\leq 100MPa)$

$$K = -4.0 \times 10^4 \, Esp^{-1.77} (\sigma_1'/\sigma_{1y}' - 1.0) + 1.0 \quad (3)$$

However, when K > 0, and Esp $\leq 100MPa$

$$\sigma_{1y}' = 0.27 \, Esp^{0.66} \cdot (0.026 \, Esp^{0.66} + 1.0) \quad (4)$$

When $\sigma_1'/\sigma_{1y}' \leq 0.5$

$$K = 2.0 \times 10^4 \, Esp^{-1.77} + 1.00 \quad (5)$$

$$E_1 = K \cdot Ey \quad (6)$$

3. Evaluation of bearing capacity of the foundation

3.1 Simple analysis

Change of the stress conditions and deformation character of the ground under the loading test plate are estimated at the loading test point in the test pit. An analysis for the deformation of the ground is carried out by means of a cyndrical tube model accumulating slice plates as shown in Fig.6.

Analysis resalt simulated the load-displacement curve coincide well with the measured result as shown in Fig.8 Relationship between yielding bearing capacity qy by the in-situ test and material property representing by deformation modulus Es is shown in Fig.7. In order

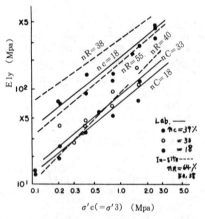

Fig. 2 Relation of deformation modulus at yield strain of samples and bore hole test vs. effective confining pressure

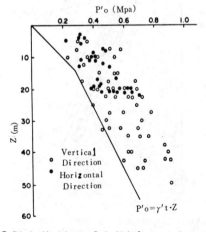

Fig. 3 Distribution of initial ground pressure in depth direction

to research the relationship between in-situ test and laboratory test, vertical yielding stress in loboratory test Vs. deformation modulus Es (=2.0Esp) relationship is also plotted in Fig.7. supposing stress conditions in the ground. In this calculation, the lateral yielding stress by laboratory test is assumed to be equivalent to yielding stress Pf by pressuremeter in the ground. In Fig.7, both bearing capacity vs. deformation modulus relationship shows a similar tendency.

3.2 Analysis of deformation characteristics

FE analysis is carried out by means of elastic theory for the axial symmetry problem and the analysis results of subsurface strains and surface settlement are compared with measured values. In this analysis, the elestic modulus E of initial condition is determined

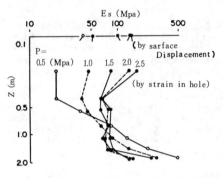

Fig. 4 (a) Variation of deformation modulus in depth direction at each load step

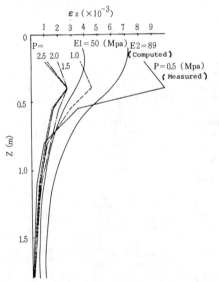

(b) Distribustion of subsurface strain obtained by testing and caluculation

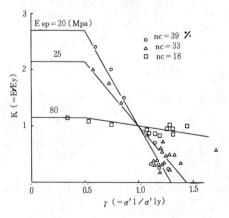

Fig. 5 Relation of the ratio of deformation modulus vs. stress ratio at each standard index

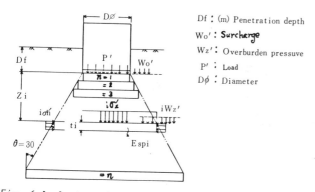

Df : (m) Penetration depth
Wo' : **Surcharge**
Wz' : Overburden pressure
P' : Load
Dφ : Diameter

Fig. 6 Analysis model obtained by thin cylindrical tube for plate loading test results

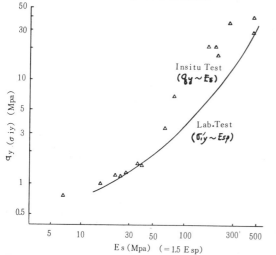

Fig. 7 Relation of bearing capacity and strength at yied point vs. deformation modulus

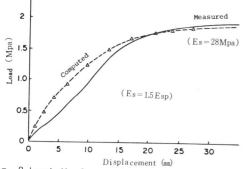

Fig 8 Load-displacement curve at loading test and its curve by simulation analysis

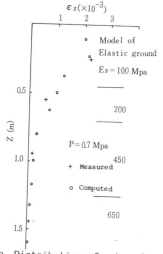

Fig. 9 Distribution of subsurface strain obtained by testing and calculation

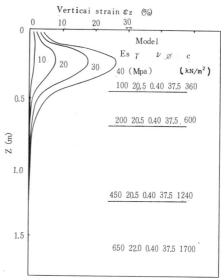

Fig.10 (a) Distribution of subsurface strain in depth direction at each load step

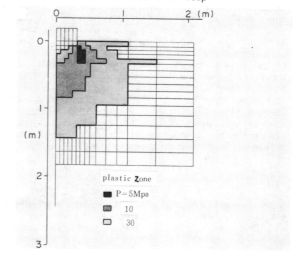

(b) Disribution of extent in plastic area at each load step

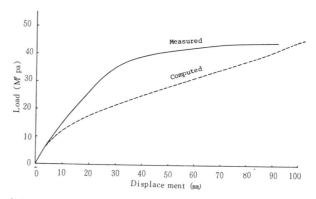

(c) Load-displacement curve at loading test and its curve by simulation

Tab. 1 Mechanical properties of granite

Rock class	DL	DH	CL	CM	CH	B
γ (KN/m³)	18	20	22	24	25	26
E (Mpa)	120	250	500	1,000	2,000	4,000
C (KN/m²)	100	250	400	600	900	1,500
ϕ (deg.)	30.0	32.5	35.0	37.5	40.0	42.5
ν	0.3	0.3	0.3	0.3	0.3	0.3

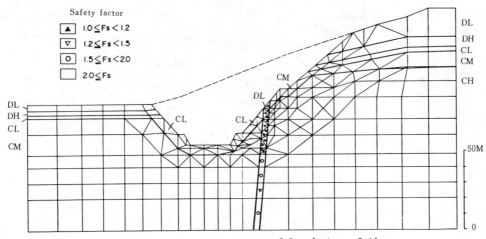

Safety factor

▲ $1.0 \leq Fs < 1.2$

▽ $1.2 \leq Fs < 1.5$

○ $1.5 \leq Fs < 2.0$

☐ $2.0 \leq Fs$

Fig.11 Geomechanical model and safety factor of local shear failure

based on measured strain $\varepsilon_{\bar{z}}$ vs. subsurface stress σ relationship assuming the ground to be an isotropic and homegeneous elastic body. The analysis result simulated subsurface strains and surface settlement agree well with these measured values as shown in Fig.9.

3.3 Analysis of bearing capacity

FE aualysis is carried out by means of elasto-plastic theory based on the geomechanical model which is prepared by the bearing test and rock shear test results. In Fig.10, development of the plastic zones by incremental load and load-displacement relationship are shown on the basis of analysis results. The analysis results simulated the ultimate bearing capacity and total surface settlement coincide with measured values at the field.

4. Stability analysis of high cut slope

Stress conditions in the rock slope are changed by overburden stess relief accompanying the excavation. In the cut slope surface, loosened zone is formed and mechanical properties of this zone is affected by the excavation. Therefore, it is necessary for the slope stability analysis to consider the change of stress conditions and mechanical properties in the slope forming rock.

Stability of a 70-M high cut slope composed of granite is studied by elastic FEM before the excavation. Heavily weathered class D granite is distributed in the slope surface, gradually changing the weathering degree to slightly weathered class $CM \sim CH$ rocks toward the depth. There is a fissure zone of high dip in the lower portion of the slope forming class DLrock. Geomechanical properties of each rock class for the granite are shown in Table 1.

The depth of the loosned zone caused by excavation is investigated by seismic prospecting. P-wave velocity (V_1) of 1st velocity layer in cut slope surface, which corresponds to the loosened zone, exhibits smaller seismic velocity than those (V_2) prior to excavation. The average P-mave reduction ratio (V_1/V_2) of granite slope around the study site was nearly 0.4.

It was found that the thickness of loosened zone was in the range of $2 \sim 5m$ below cut slope surface. Hori and Kawashima (1982) charified that the deformation modulus of the loosened zone due to excavation was reduced to about $40 \sim 60\%$ of the inherent defoumation modulus of the rocks.

Geomechanical model of FE analysis for the cut slope stability is shown in Fig.11. This model is prepared based on the results of geological recnnaissance and seismic prospecting. The loosened zone of 5m thickness is set up in the cut slope surface of class $CM \sim CH$ rocks. Mechanical characters of the lossened zone is evaluated to inferior rock class than that of inherent slope foming rocks. The effect of the

confining pressure for rockmass properties is not considered, because the great portion of the slope consists of hard rocks of class $CM \sim CH$.

Primary stress relief method by the excavation is adopted for FE analysis. This stress is estimated by the overburden load and Poisson's ratio of the slope forming rock, because of the lack of measuring data of the stress. Fig.12 shows an analysis result of safety factor of local shear failure to the shear stress in the slope. In this analysis, the safety factor is determined as a ratio of shear strength to shear stress in each elment of the FE mesh of the slope. As shown in this figure, the ratio of the elements except for the fissure zone shows more than 2.0 but that of the fissure zone is $1.0 \sim 1.5$. Analysis results of stress distribution indicates that a little tensile stress takes place in the fissure zone and some portions of cut slope surface.

From this analysis, it is judged that the slope secures its stability as a whole, thought there is a possibility of local instability in the fissure zone and loosened zone in the slope surface. Toppling failures caused by steep fissure zones and surface slope failures took place actually in the vicinity of the studied slope after excavation. Thus, FE analysis of this study explains considerably the condition of slope stability.

5. Conclusions

Rockmass properties of weathered granite aus relatively affected by the lossening and stress relief caused by excavation. Thus, the ordinary rock tests and measurements cause a problem for the estimation of engineering proporties of weathered granite. The following conclusions are obtained by this study :

(1) The results of FE analysis simulated both the kaboratory test for the sample and the bearing capacity test taken place at the field coinside well with the results of measured values.

(2) Precise evaluation of rockmass properties and numerical analysis considering stress dependency for the deformation character may produce more reasonable stability analysis for the foundation and cut slope.

REFERENCES

Ishikawa, K., Miyajima, K., Yamada, K. & Yamagata, M. Rockmass classification for weathered granite and evaluation of bearing capacity of foundation ground of long span bridge. Proc. 6th Japan symposium on Rock Mechanics.1984.

Ishkawa, K. & Yamada, K. 1986. Evaluation of bearing capacity of foundation ground for weathered granite. Proc. Engineering in Complex Rock Formations.

Hori, M. & Kawashima, Y. 1982. Excavation of inclined tunnel by tunnel boring machine and design of steel lining. Proc. 14th Symposium on Rock Mechanics.

The influence of the dispersion of the mechanical properties on the stability of rock foundation

Effet de dispersion des propriétés mécaniques sur la stabilité des fondations en roche
Die Wirkungen von Streuung der mechanischen Eigenschaften auf die Stabilität des Gründungsfelses

H.ITO, Central Research Institute of Electric Power Industry, Chiba, Japan
Y.KITAHARA, Central Research Institute of Electric Power Industry, Chiba, Japan

ABSTRACT: In this paper, the analytical method and its computer program to consider the influence of disper-sion of the mechanical properties were developed, and in order to clarify the influence of its dispersion, the numerical simulation of in-situ rock shear test and the analysis of rock foundation of a large structure were performed. The results are as follows: (1) The applicability of the developed program was confirmed by the comparison between the calculated values and the measured values. (2) The influence of the dispersion of the mechanical properties of rock mass used to the stability analysis does not appear notably to the settlement and sliding safety factor of the structure foundation. (3) The design considered the dispersion of mechanical properties can carry out sufficiently by the method using the deterministic values from the results of ana-lytical parameter survey. In this case, the values of mechanical properties should be evaluated as $[X$ (mean value) $\pm K$ (coefficient) x σ_x (standard deviation)$]$.

RESUME: Pour éclaircir les influences de la dispersion des valeurs physiques du sol sur la stabilité du sol, nous examinons les méthodes d'évaluation sur la dispersion des valeurs avec le développement d'une méthode analytique et son logiciel, l'essai in situ tenant compte de la dispersion des valeurs et l'analyse de la stabilité du rocher de fondation supportant des ouvrages importants. En conséquence: (1) nous avons pu vérifier l'aptitude du logiciel mis au point "RASRM" par la simulation numérique de l'essai in situ de cisaillement des roches; (2) les influences de la dispersion des valeurs physiques du sol sur les paramètres d'évaluation tels que le glissement et le tassement ne sont pas importants par rapport à la largeur de la dispersion et à la forme de distribution des valeurs physiques utilisées pour l'évaluation; (3) nous avons qu'une des méthodes de théorie de certitude peut être appliquée comme méthode de conception tenant compte de la dispersion des valeurs physiques qui présentent seulement des caractéristiques de dispersion; {(valeur moyenne) $\pm$ (coefficient) x (déviation standard)}.

ZUSAMMENFASSUNG: Es wurden die Entwicklung einer analytischen Methode and derer Programms, einen Baustellen-versuch mit Rücksicht auf die Dispersion sowie eine Stabilitätsanalyse des Grundfelsbodens für die wichtigen Bauwerke u.a. durchgefürt, um die Beeinflussung der Dispersion von den Bodenmaterialwerten auf die Boden-stabilitätabzuklären. Daraus ergaben sich die folgenden Punkte: (1) Die Gültigkeit des entwickelten Programms "RASRM" wurde durch eine nummerishe Verstellung des Baustel enfelsschubversuchs festgestelit. (2) Die Beeinflussung der Dispersion von den Bodenmaterialwerten auf die Schätzungsstücke z.B. Gleiten und Setzung des Bodens ist klein im Vergleich zu der großen Dispersion und Abweichung der zur Schätzung verwendeten Materialwerten und Vertellungsformen. (3) Auf den Entwurf kann die Wahrscheinlichkeitsrechnung gut angewendet werden. Dabei sind die einzusetzenden Materialwerte in der folgenden Form auszudrücken, damit das Verhalten der Dispersion berücksichtigt wird: Materialwert = (Mittelwert) $\pm$ (Koeffizient) x (Normalabwelchung).

1 INTRODUCTION

The values representing the physical and mechanical properties of rock mass are usually obtained with some dispersion even on the same type and quality of a rock at the same point. Accordingly, it is neces-sary to naturally consider the influence of the dis-persion of the values of mechanical properties in investigating the stability of the rock foundation in important structures or large slope. To consider the influence of the dispersion of mechanical prop-erties, there appears the following two methods; 1) the method treating the individual value of mechan-ical properties as a random variable (referred to hereafter as the probabilistic method). 2) the method using the representative value of the data as mechanical properties (referred to hereafter as the deterministic method). That is, the former method considers the distribution properties as being di-rectly incorporated into the estimating method, while treating the individual value of mechanical properties as the random variable following the probability distribution. The latter is the method analyzing the representative value (mean value or the most frequent value) or the allowable constant value used, while considering the fluctuation lati-tude of the dispersion of the individual value of mechanical properties. However, the relation between both methods concerning the influences exerted on the results of the stability analysis by the dis-persion of the mechanical properties of rock mass has not been made clear. Therefore, we developed in this study a finite element method program describing the non-linear mechanical properties while treating the mechanical properties as a random variable, to quantitatively clarify the influence of the dis-persion of mechanical properties exerted on the stability analysis of rock mass. Moreover, a case study is performed to examine the influence which is exerted on the settlement and sliding safety factor of an important structures, according to both the probabilistic and deterministic methods along with parameters of the distribution properties of the spatial dispersion of mechanical properties by using the developed program.

The outline is mentioned below (Ito 1984).

2 NUMERICAL ANALYSIS METHOD AND ITS SUITABILITY

2.1 Outline of numerical analysis method

The analysis program "RASRM" shown in Figure 1 has been developed so as to study the influence of the dispersion of mechanical property exerted on the deformation behaviour and the sliding fracture of the rock mass of the structural foundation. The developed program is mainly composed of three parts:

(1) Random sampling of individual value of mechanical properties by means of a Monte Carlo analysis.

(2) Nonlinear elastic analysis by means of a finite element method.

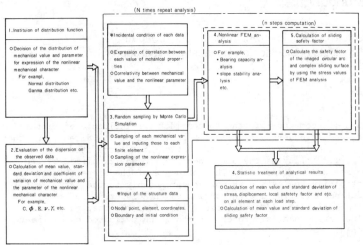

Figure 1. Flow of calculation program for probabilistic method.

(3) Statistical treatment of analyzed results.

The random sampling method used here is a method to generate pseudo-random numbers following an individual probability distribution based on uniform random number generation by the congruent method. The number of individual mechanical properties for sampling is the element number of finite elements, which are given to individual elements in order as input data for nonlinear elastic analysis. The FEM analysis method applied here is a nonlinear elastic analysis based on the two dimensional plane strain theory which introduced nonlinear strength and deformation properties of rock material, behaviour difference after shear or tensile rupture as the effect of the anisotropic deformation characteristics. As for the treatment of mechanical properties after rupture or the nonlinear properties of rock mass in this method, refer to reference (ITO 1981).

In (3), the mean value and the standard deviation of the stress, displacement and safety factor in the element at each step is to be calculated as many times as set in the analysis of (1) and (2).

2.2 Suitability of analysis method

The primary suitability of the nonlinear analysis

method in "RASRM" is examined comparing the analysed results and the measured value after a numerical simulation of the in-situ rock shear test of mudstone.

Mechanical properties used in numerical analysis are shown in Table 1. These values are the mean mechanical properties obtained by laboratory tests of undisturbed samples taken from the neighbourhood of the shear test point in adit by the block sample method so as to simulate the rock shear test as faithfully as possible.

Figure 2 is an example in which the measured values and calculated values of shear loading vs. displacement of rock shear test are comparatively shown. This figure is the case for vertical loading where $(\sigma N)=5kgf/cm^2$. Judging from the figure, analytical process to the rupture in which the horizontal displacement, settlement and rebound of rock block are included, lead to results which can sufficiently explain the behaviours of the measured results.

Hence the adequacy of the analysis method, such as the treatment of material constants and the introduction method of nonlinear mechanical properties in "RASRM", is assured.

3 DISPERSION OF MECHANICAL PROPERTIES OF ROCK MASS AND STABILITY ESTIMATION

3.1 Case study of rock shear test considering the dispersion

The influence of the dispersion of mechanical properties on deformation behaviour and rupture phenomenon is analytically examined. That is, having executed 30 cases of numerical analysis of a rock shear test of mudstone according to the proba-

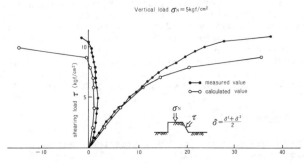

Figure 2. Shear loading ~ displacement of numerical analysis.

Table 1. Material constants for numerical analysis.

section of analytical domain	γ_t (gf/cm³)	low pressure parts		high pressure parts			
		τ_R (kgf/cm²)	$\sigma y / \tau_R$	C (kgf/cm²)	ϕ (degree)	E_0 (kgf/cm²)	ν_0
zone 1	1.60	6.55	0.65	13.4	2.0	5830	0.32
zone 2	1.60	7.10	0.65	14.1	2.0	6540	0.32
zone 3	1.60	7.64	0.65	14.8	2.0	7250	0.32

(Remaks)

(1) Strength character
- low pressure parts ($\sigma < 15kgf/cm^2$)

$$\tau = \tau_R \sqrt{1 + \frac{\sigma}{\sigma_t}}$$

- high pressure parts ($\sigma \geq 15kgf/cm^2$)

$$\tau = C + \sigma \cdot \tan\phi$$

(2) Nonlinear expression of the modulos of deformation and poisson's ratio

$$E/E_0 = 1.29R^{0.35} \quad (R \leq 0.483)$$
$$E/E_0 = 1.00 \quad (R > 0.483)$$

$$\frac{0.45 - \nu}{0.45 - \nu_0} = 1.43R^{0.25} \quad (R \leq 0.239)$$
$$\frac{0.45 - \nu}{0.45 - \nu_0} = 1.0 \quad (R > 0.239)$$

in which, $R = Min (d_1/D_1, d_2/D_2, d_3/D_3)$

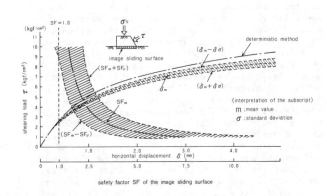

Figure 3. Relationship between shear loading and displacement, sliding safety factor.

408

bilistic method under the conditions of Table 1 mentioned above, the dispersion degree of that result is examined. In this analysis, the modulos of deformation and shear strength which have a large influence on the rupture process of rock mass are also treated as random variables.

Figure 3 shows the relationship between shear loading of the rock block and horizontal displacement. In this figure, the results by the deterministic method used a mean value, not taking the dispersion of the mechanical properties into consideration. In consequence, the shear strength of the rock obtained by the probabilistic method is lower than the results by the former deterministic method using mean values. Compared with the results of the deterministic analysis for the strength when the rock block is ruptured in the practical test (shear loading τ = 7kgf/cm^2), the mean curve of the probabilistic analysis results is lower by 12%, superior and inferior curve by taking the standard deviation of the analyzed results into consideration, become lower by 7% and 15% respectively.

3.2 Case study on the stability of rock foundation of major structures

On settlement of rock foundation of major structures dealing with weathering granite and mudstone and the sliding stability in case of earthquakes, case studies considering the dispersion of mechanical properties have been done for 50 cases and the influence of dispersion is examined. Correlation between results by the probabilistic method thus obtained and results by the deterministic method when the {mean value(μ)},{mean value(μ) ± standard deviation($\sigma\mu$)},{(μ) ± 0.5($\sigma\mu$)} etc. are used is examined. The outline will be mentioned below.

The mechanical properties used in the case study are as shown in Table 2. Each value of material propertie and the nonlinear mechanical properties shown in the table is obtained by an in-situ test and laboratory test, respectively. In Table 3, the distribution characteristics of the dispersion of mechanical properties used in the case study are shown. The settlement and sliding safety factor of the foundation of structure obtained under the above analysis condition are shown in Table 4. Figure 4 shows the settlement of the foundation under the ordinary loading conditions of structures in the histogram. Figure 5 shows the sliding safety factor of rock below the concrete mat foundation under the seismic load condition in the histogram. In these figures, in addition to results by the deterministic method, the distribution of results by the probabilistic method for 50 cases is also shown. Moreover, in order to compare the results by the probabilistic method and deterministic method, the rela-

tion between the coefficient K and the probability R that the former results is less than the latter results obtained by using the property value = {mean value(μ)} ± {influence coefficient (K)} x {standard deviation ($\sigma\mu$)} is arranged. In figure 6, these results concerning the settlement and sliding safety factor are shown. Here the probability R is the value defined by the following equation when the distribution of the analyzed results of the probabilistic method is assumed to have a normal distribution.

$$ R = \int_{-\infty}^{x} \frac{1}{\sqrt{2\pi} \cdot \sigma_x} e^{-\frac{1}{2}(\frac{x-\bar{x}}{\sigma_x})} dx $$

Here;
$\bar{x}$: mean value of sliding safety factor or settlement for all cases obtained by the probabilistic method.
σ_x: standard deviation of results by the probabilistic method.
x : sliding safety factor or settlement obtained by the deterministic method using the property value estimated by the coefficient K value.
R : coefficient to show the variation width by magnification of the standard deviation.

From these results, the following results are obtained. (1) Teh mean value of the settlement by the probabilistic method is slightly larger than that obtained by the deterministic method using the mechanical properties (μ). The mean value of the sliding safety factor by probabilistic method is comparatively smaller than the deterministic results, though their difference is small. (2) Although the coefficient of variation of the mechanical properties

Table 3. Dispersion condition of mechanical properties for probabilistic method

Cases of analysis		Dispersion condition of mechanical property	Remarks
Weathered granite rock mass	CASE-01	• (C.V.)$_C$=0.4, (C.V.)$_E$=0.4 • Normal distribution • Sampling section···±2σ	○ No consider the diupersion of mechanical property on pertial rock mass thatis, just under the structure
	CASE-02	• (C.V.)$_C$=0.2, (C.V.)$_E$=0.2 • Normal distribution • Sampling section···±2σ	
	CASE-03	• (C.V.)$_C$=0.4, (C.V.)$_E$=0.4 • Normal distribution • Sampling section···±1σ	
	CASE-04	• (C.V.)$_C$=0.4, (C.V.)$_E$=0.4 • Lognormal distribution • Sampling section···±2σ	
Mudstone rock mass	CASE-05	• (C.V.)$_C$=0.4, (C.V.)$_E$=0.4 • Normal distribution • Sampling section···±2σ	○ Consider the dispersion of mechanical property on all rock mass
	CASE-06	• The same as above	
	CASE-07	• (C.V.)τ_R=0.214, (C.V.)$_E$=0.248 • Normal distribution • Sampling section···±2σ	
	CASE-08	• The same as above	○ The same as CASE-05

(note) (C.V.)$_{C or \tau_R}$;Coefficient of variation of cohesion
        (C.V.)$_E$;Coefficient of variation of modulos of deformation

Table 4. Results of stability analysis by probabilistic method

CASE		Weathered granite rock mass				Mudstone rock mass			
		01	02	03	04	05	06	07	08
Settlement (cm)	$\bar{x}$	4.89	4.67	4.72	4.89	5.37	5.29	5.60	5.65
	$\bar{\sigma}_x$	0.17	0.07	0.12	0.28	0.20	0.22	0.09	0.10
	(C.V.)	0.035	0.015	0.025	0.057	0.036	0.041	0.016	0.018
sliding safety factor	$\bar{x}$	6.08	6.18	6.16	6.02	3.97	3.92	3.94	4.02
	$\bar{\sigma}_x$	0.37	0.16	0.19	0.45	0.25	0.25	0.09	0.17
	(C.V.)	0.060	0.026	0.031	0.074	0.062	0.064	0.023	0.042

(note)
x ;mean value by probabilistic analysis $\bar{\sigma}_x$;standard deviation of probabilistic analysis
(C.V.) ;coefficient of variation by probabilistic analysis

Table 2. Mechanical properties for stability analysis of the weathered granite rock mass

Section of analytical domain	unit weight γ_i(gf/cm^3)	cohesion C(kgf/cm^2)	internal friction angle ϕ(degree)	intial modulos of defrmation E$_s$(kgf/cm^2)	poisson's ratio ν_0
zone 1	2.54	9.92	45.0	7200	0.25
zone 2	2.54	10.47	45.0	7906	0.25
zone 3	2.54	11.03	45.0	8632	0.25

(Note)
Nonlinear expressin of the modulos of deformation

$E/E_0 = 0.411 \log R_0 + 0.997$
  $(0.0372 < R_0 < 1.0)$
$E/E_0 = 0.001$ $(R_0 \leq 0.00372)$
$E/E_0 = 1.0$ $(R_0 \geq 1.0)$
in which
$R_0 = \frac{R}{0.46} = \frac{d/D}{0.46} = \frac{1}{4} \cdot \frac{1}{0.46} \left(\frac{C \cdot \cos\phi + \frac{\sigma_1 - \sigma_1}{2} \cdot \sin - \frac{\sigma_1 - \sigma_1}{2}}{C \cdot \cot + \frac{\sigma_1 + \sigma_1}{2}} \right)$

(rupture envelop line)
$\tau = C + \sigma \cdot \tan\phi$

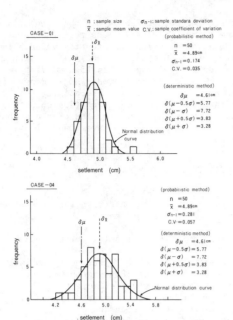

Figure 4. Results of stability analysis by probabilistic method

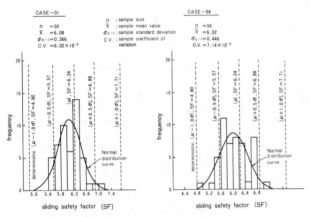

Figure 5. Histogram of sliding safety factor just
under the structure

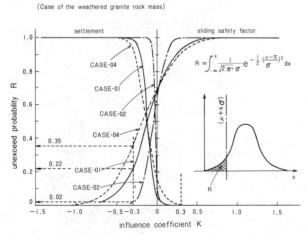

Figure 6. The relationship between unexceed probability and influence coefficient

are large (0.2 to 0.4) according to the probabilistic method, the coefficient of variation of the settlement and sliding safety factor obtained as the result of the stability analysis are very small. That is, the coefficient of variation of the settlement is 0.014 to 0.057 and the sliding safety factor is 0.023 to 0.075. (3) The results of the probabilistic method, set to 0.2 - 0.4 of the coefficient of variation of the mechanical properties for analysis, are distributed within the range of the deterministic result obtained by using $\{(\mu) \pm 0.3 \ (\sigma\mu)\}$ of the mechanical property for the settlement. For the sliding safety factor, almost all except the results of a small percentage are distributed within the range of the probabilistic results using $\{(\mu) \pm 0.5 \ (\sigma\mu)\}$ of the mechanical property. That is, the sensitivity for the variation of the values for analysis is a little higher in the sliding safety factor than that in the settlement. (4) As a result of the examination of the difference of the distribution shapes of the dispersion of the mechanical property by the normal distribution and the logarithmic normal distribution, the latter result is larger in the mean value and standard deviation than the former result, while the coefficient of variation, which indicates the degree of dispersion, shows a small value, 0.057. (5) Based on the relation between R and K by this example if K = 0.3, the approximately 22% of results by probabilistic method, that is set the coefficient of variation to 0.4, is lowered deterministic method and the case of setting the coefficient of variation to 0.2, the results of approximately 2% is less than.

Judging from the above results, we knew that the design evaluation considering the dispersion of the property values had been fully quantitatively possible by deterministic method.

4 CONCLUSION

The results in this study are as follows: (1) as a result of the numerical simulation of the in-situ test using the developed analysis program, the suitability of the analysis method and the program is ensured by the coincidence of the analysis value with the measured results. (2) According to the probabilistic method, although the coefficient of variation of the mechanical properties of rock mass used as an analysis condition, is as large as 0.2 - 0.4, the coefficient of variation of the settlement and sliding safety factor obtained by the analysis shows small value less than 0.08. The explains why there does not appear to be a great difference of the influence of the dispersion of the mechanical properties exerted on the evaluation items including the sliding and settlement, in comparison with the great difference of the range and distribution shapes of the dispersion of the mechanical properties used for analysis. (3) It is fully possible to apply the deterministic method as a design method considering the dispersion of the mechanical property. The property value, used in this case, may be considered the expressing the properties of dispersion such as $\{(\text{mean value}) \pm (\text{coefficient}) \times (\text{standard deviation})\}$. According to the example in this study, it is made clear to evaluate the safety side with allowance if the value of the above coefficient is set to 0.5.

REFERENCES

Ito, H. & Y. Kitahara 1981, Proc. Int. Sym. on Weak Rock, Tokyo, pp.653-658.
Ito, H. & Y. Kitahara 1984, Tech. Report of CRIEPI, No.384026 (In Japanese)

Stability of the Flysch coastal slopes of the Adriatic Sea in the static and seismic conditions

Stabilité des versants côtiers de l'Adriatique en Flysch dans des conditions statiques et dynamiques
Die Stabilität der Flyschküstenabhänge der Adria unter statischen und dynamischen Bedingungen

IBRAHIM JAŠAREVIĆ, Professor, Dr.Sci., Civ. Eng., Civil Engineering Institute, Faculty of Civil Engineering, University of Zagreb, Yugoslavia
VLADIMIR JURAK, Lec. Dr.Sci., Geol.Eng., Faculty of Mining, Geology and Petroleum, University of Zagreb, Yugoslavia

ABSTRACT:

The stability and integral functionality of the flysch costal slopes of the Adriatic in the static and seismic conditions have been analysed, based on the extensive engineering geological, geophysical, geotechnical laboratory and "in situ" investigations.
The design model of the endless long slope is made on basis of the geotechnical models which had to be done before. The design model was used for the stability analyses. The method of the slope mass balance analyses of the method which uses seismic coefficient. During the observations, in the area of flysch slopes a lot of cases of slow (longlasting) progressive failure have been found out. This is not the consequence of the change of loading stresses but the change of the shear strength parameters.

RÉSUMÉ:

Les études approfondies et les essais géologiques, géophysiques et géotechniques en laboratoire et en place ont permis l'analyse de la stabilité et de la fonctionnalité intégrale des versants cotiers de l'Adriatique dans des conditions statiques et sismiques.
L'analyse de la stabilité a nécessité un modele gólogique et un mod ele géotechnique, a partir desquels a été formé le modele de calcul d'un versant de longueur infinie. La méthode de l'équilibre limite du bloc glissant est plus fiable dans les conditions sismique que la méthode de coefficient sismique. Dans la région des versants en flysch ont été observés un grand nombre de cas de ruptures lentes (de longue durée) et progressives, entraînées par la variation des parametres de résistance lors du cisaillement et non pas par la variation des charges extérieures.

ZUSAMMENFASSUNG:

Auf Grund ausführlichen ingenieurgeologischen, geophysischen, geotechnischen, laboratorischen und "in situ" Untersuchungen, sind die Stabilität und Integralfunkcionalität der Flyschküstenabhänge der Adria in statischen und Erdbebenbedingungen analysiert worden.
Um die Stabilitätanalyse durchzuführen, ist es notwendig ein ingenieurgeologisches, bzw. geotechnisches Modell zu machen, auf dessen Grund ein "Design Modell" des endlosen Abhangs hergestellt ist.
Die Methode des Grenzengleichgewichtes des schiebenden Blocks in den Erdbebenbedingungen ist zuverlässiger als die Methode des Erdbebenkoeffizienten. Während der Betrachtungen der Flyschabhänge sind viele Fälle langsamen, langzeitigen, progressiven Bruches festgestellt. Das ist die Folge einer Scherparameteränderung und nicht der Änderung der äusseren Belastung.

1. INTRODUCTION

Taking part in the work of a bigger investigation team on the project: "The Flysch Investigations in the costal Area with the Aim of More Rational, Safer and More Economic Building", financed partly by Self - managing Community of interest for the scientific work in the Socialist Republic of Croatia, two areas have been chosen. The complete investigations have been performed in the last 30 months. It enabled the systematization of the up to now experiences in the process investigation - planning - building - instrumentation and observation at various objects.
The following areas are separated:
- clastic deposits in the tectonically disturbed area of the town of Rijeka
- the disturbed clastic deposits in the zone of the main tectonic contact with the carbonate rocks (the greater area of the town of Dubrovnik).
The up to now experiences in investigation, planning, building and instrumentation of

objects built in flysch, are not systematized. The main reason is probably the fact that there is the difference in Eocene clastic sediments in composition, tectonic, state and characteristics, which makes investigation, planning, building and instrumentation of the objects in these areas more complicated.

With regard to great damages that appear while building and exploitition of the objects in Eocene clastic sediments of the costal zone of the Adriatic Sea as well as to the fact that over 50% of the building area of the coast is placed in flysch and having in mind the future building of the important turistic and industrial objects, the realisation of the investigation programme began in the year 1984.

The basic investigation aim has been formulated: systematization of the knowledge concerning flysch and changing the obtained results in recomendations (directives) for

411

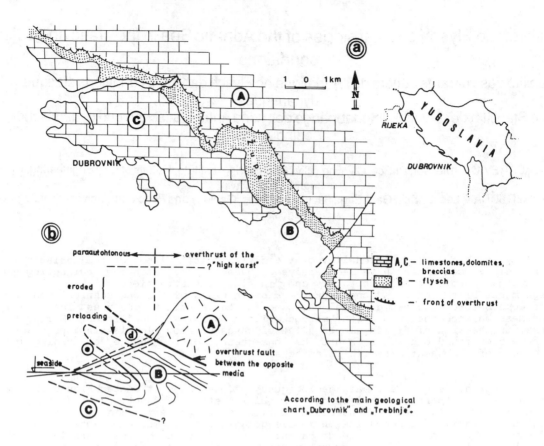

Fig. 1. The area of eocene flysch (B) in the costal belt region of the town of Dubrovnik (a) and schematic layout of the structural complex (b).

more rational processing: investigation - planning - building - instrumentation and the observation of the system object - flisch rock mass.
Important problems that appear in building on flysch costal slopes are:
- slope stability in conditions determined by their natural surroundings,
- the artificial slope stability caused by some human activity in the changed conditions in comparison to the natural ones.
To analyse the stabilities it is necessary to make the engineering geological, geotechnical model respectively, with the quantified parameters essential for the calculation in static and seismic conditions.
In town planning it is necessary to make geotechnical evaluation of the area which spread over the flysch slopes. For geotechnical evaluation of the area, using the system of the dimensionless coefficients which is usual in geotechnical practice, it is necessary to establish, by preliminary analysis the stability factor in static and seismic conditions and permanent displacements on the potential sliding plane.

2. ENGINEERING GEOLOGICAL CHARACTERISTICS

2.1. GEOLOGY

The regional characteristics on the longer part of the Yugoslav Adriatic coast is the presence of two lithogenetic complexes: the Mesozoic carbonate complex (M) and Eocene clastic complex - flysch ($E_{2,3}$). The prevailing characteristics of the tectonic texture of the costal slope present they carbonate and flysch inverse relation. The schematic

layout of the texture of the surroundings of the town of Dubrovnik illustrates this statement (Fig. 1 Jašarević and al., 1986).
The contact surface between two lithogenetic complexes, with opposite physical - mechanical properties, is the overthrust in some parts (the area of the town of Dubrovnik) and reverse fault in the other ones (the area of the town of Rijeka). In both cases the two rock masses are differentiated in relief and height (A and B, Fig. 1 and 2).

Placed in the area of the tectonic compressibility of the masses, both rock masses are intensively deformed. It refers particularly to Eocene flysch under overthrust. Isoclinal and recumbent folds of the southwest overturn can appear there. The solid and plastic components have deformed differently because of their exchange inside flisch in the stress field caused by the compression forces. It is possible therefore to talk about the internal tectonics of the flysch comples.

With the described structural complex there appear the transversal rupture deformations - younger faults - that intersect both rock masses and reach the sea.
The morphogenesis of the costal slope is caused by the presented relation of the rock masses and the intensity of the physical - geological processes in the costal area through the youngest age (Quaternary).
The consequences of the recent texture and slope morphogenesis are the hydrological and engineering geological circum - stances and the field stress of the rock mass. During the formation of the costal slope the quality of the original rock and the accumulation of the wear products changed. Two genetic cathegories have been differentiated on the

Fig. 2. The typical costal flysch slope under the carbonate rock complex

slope: d - overlaying products most often delluvial - colluvial deposits and e - eluvial deposit weathered flysch (Fig. 1). Autochthonous products of the weathered original rock mass (eluvial deposit; Q_e) and the allochthonous overlaying products (delluvial - colluvial deposits; Q_d) are of close or opposite physical - mechanical properties, depending on the genetics of the overlaying products, its grain size distribution and particular cohesion, i.e. the origin of the components from the carbonate or clastic rock mass.

TABLE 1. GRAIN SIZE DISTRIBUTION

	GRAVEL (%)	SAND (%)	SILT (%)	CLAY (<4 μm) (%)
* RIJEKA	3 – 25	9 – 32	34 – 44	15 – 53
** DUBROVNIK	0 – 2	2 – 7	41 – 80	15 – 51

2.2. LITHOLOGIC AND MINERALOGIC COMPOSITION

Eocene clastic complex has lithologic heterogeneity vertically and lateraly with the exchange of the solid and soft rock deposits. A lot of various siltstones and sandstones are present as well as not many coarse grained clastic deposits (breccias and conglomerate rocks). As the siltstones are rocks of low range of diagenesis and lithification and they are dominant, they give the main characteristics to the flysch complex behaviour. The are supple to atmospheric wearing and expressively inconstant in water. The enclosed tables show the particle participations concerning dimensions (table 1) and mineral composition of fine clastic components (table 2). It is visible that in the soft deposit siltstone, the part of the phyllosilicate is about 50% or more with the presence of the minerals inclined to swelling.

2.3. INVESTIGATION PROCESSES

The including engineering geological mapping, the analyses of aerial photographs, drilling investigation with sampling and standard penetration test, installation of piezometers, excavation of pits which showed themselves

TABLE 2. MINERALOGY OF FLYSCH SAMPLE (FINE CLASTIC COMPONENTS) IN APPROXIMATE PERCENTAGE MASS PASSING [%] (Slovenec, Faculty of Mining, Geology and Petroleum, Zagreb)

SAMPLE FROM THE LOCALITIES		Q	P	K	Ca	D	Σ (%)	T	Mo	C	C-V	Kl	ANF	ΣFA (%)
THE AREA OF THE TOWN OF RIJEKA	BAKAR - RETAINING WALL	20 – 30	6 – 15	+	18 – 25	0	45 – 70	20 – 30	0	SIGNIFFI CANTLY		SIGNIFFI CANTLY		30 – 55
	BAKAR - THE HARBOUR OF BULK CARGO *	24 – 34	10 – 13	0 – 8	3 – 7	0	34 – 52	22 – 27	< 5	< 5	SIGNIFFI CANTLY	0	0	ABOUT AND >50
	OREHOVICA	13 – 16	3 – 6	0	15 – 18	+	30 – 40	22 – 25	0	10 – 15	0	0 – 5	25 – 35	60 – 70
THE AREA OF THE TOWN OF DUBROVNIK	SREBRENO **	12 – 14	5 – 9	0	25 – 38	1 – 2	50 – 56	22 – 26	< 5	ABOUT 5	SIGNIFFI CANTLY	0	SIGNIFFI CANTLY	44 – 50
	CAVTAT	15	7	0	30	1	53	20	0	SIGNIFFI CANT	SIGNIFFI CANT	0	SIGNIFFI CANT	≤ 50

Q - QUARTZ
P - PLAGIOCLASE
K - Ca FELDSPAR
Ca - CALCITE
D - DOLOMITE
T - MICA
Mo - MONTMORILLONITE
C - CHLORITE
C-V - MIXTURE OF CHLORITE - VERMICULITE
Kl - KAOLINITE
ANF - AMORPHOUS MATERIAL (ALLOPHANE) + UNSTABLE PHYLLOSILICATE

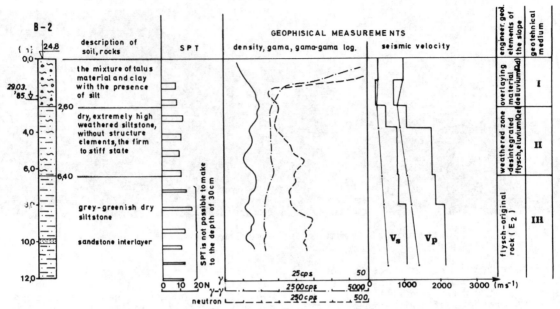

Fig. 3. Borehole log ("Geofizika", Zagreb)

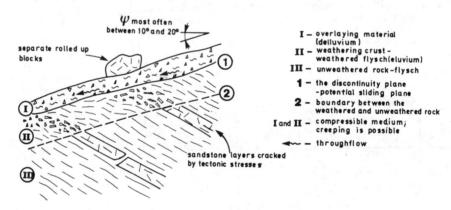

I — overlaying material (deiluvium)
II — weathering crust - weathered flysch(eluvium)
III — unweathered rock - flysch
1 — the discontinuity plane - potential sliding plane
2 — boundary between the weathered and unweathered rock
I and II — compressible medium; creeping is possible
⟵⟵ — throughflow

Fig. 4. Engineering geological elements in the geology of the costal flysch slope - the base for the geotechnical model on several microlocalities

very useful for getting information about the composition and state of the overlaying material and hypodermic water, geophysical investigations (resisting sounding and seismic surveying and logging by down - hole method. Fig. 3 shows the sample of the typical borehole log.

2.4. ENGINEERING GEOLOGICAL ELEMENTS AND THE GROUNDWATER CONDITIONS

Fig. 4 shows engineering geological elements of the flysch costal slope in case when overlaying material consists of clay and has equal thickness from place to place. The separated engineering geological elements of the slope are at the same time special geotechnic median with definite characteristics.

If the flysch complex is under the outer influence or in case of cutting its behaviour is dictated by siltstone. Its nonresistance towards atmospheric wearing results in decreasing its geomechanical parameters. The unweathered as well as the weathered siltstone in the weathering zone (eluvium) is geomechanically treated as the coherent materials.

The flysch complex as the whole represents impermeable mediu, weathered flysch from the weathering zone (eluvium) too, or its permeability is very low while the overlaying materials have great veriety of conductivity.

The pit excavations and drilling boreholes on the slope show that seepage of the hypodermic groundwater occurs on the discontinuity plane, i.e. the border of the weathered flysch and overlaying material which represent paleorelief of the former slope. There exists in the vertical direction the prefered seepage zone, the same zone exists in the lateral direction because in the same hydrologic season some boreholes are filled with water and the others are dry. It means that the groundwater seepage down the slope is a distinguished process in space and time.

The groundwater level on the slope varies because of the microrelief of the boundary plane of the overlaying material and weathering plane, and because of the variety of the hydraulic conductivity of the overlaying material.
The overlaying material becomes saturated to the surface on the particular slope parts which is important for the analyses of the global and local stability of the slope (Anderson and Burt 1978).

414

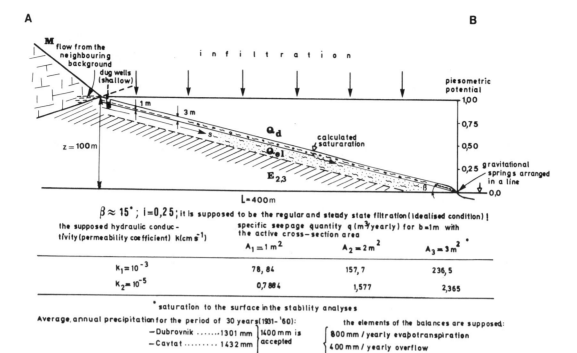

A

B

M flow from the neighbouring background

i n f i l t r a t i o n

dug wells (shallow)

1 m

3 m

s

Q_d

calculated saturaration

Q_{el}

$E_{2,3}$

z = 100 m

piesometric potential

1,00

0,75

0,50

0,25 gravitational springs arranged in a line

0,0

β

L = 400 m

$\beta \approx 15°$; $i = 0,25$; it is supposed to be the regular and steady state filtration (idealised condition)!

the supposed hydraulic conductivity (permeability coefficient) k(cm s^{-1})

specific seepage quantity q(m³ yearly) for b=1m with the active cross-section area

	$A_1 = 1\,m^2$	$A_2 = 2\,m^2$	$A_3 = 3\,m^2$ *
$k_1 = 10^{-3}$	78,84	157,7	236,5
$k_2 = 10^{-5}$	0,7884	1,577	2,365

* saturation to the surface in the stability analyses

Average, annual precipitation for the period of 30 years (1931-'60):
— Dubrovnik .... 1301 mm
— Cavtat .... 1432 mm
1400 mm is accepted

the elements of the balances are supposed:
800 mm / yearly evapotranspiration
400 mm / yearly overflow
200 mm / yearly infiltration

Fig. 5. Seepage of the hypodermic water through the delluvial overlaying material with the elements of balance

Fig. 5. shows the seepage of the hypodermic water on particular profiles of the costal slope with the elements of the hydrologic balance for the area of the town of Dubrovnik (profile in fig. 2).
With the suposstion of the relative modest yearly infiltration of 200 mm one gets the quantity of 0,2 m³/ yearly/m² of the slope which gives specific quantity of seepage q=80 m³/yearly in the distance of 400 m to the seaside. With the active unit area of corss section A=1m² and relative high con-ductivity the quantities are the same (Fig. 5). For the greater active unit area of cross section with the same hydraulic conductivity the hidden flow from the neighbouring rocky background (cracked solid rock or talus) or greater infiltration are necessary. Gravitational springs of low capacities arranged in a line paralell or near the coast and wells placed immediately under the contact with the folded carbonate complex indicate that there is a hidden flow in the overlaying material on some parts.

TABLE 3. RESULTS OF GEOTECHNICAL LABORATORY INVESTIGATIONS

AREA		ROCK KINDS	GRANULOMETRIC COMPOSITION			CONTENT OF CaCO₃ (%)	MOISTURE CONTENT (%)	LIQUID LIMIT (%)	PLASTIC LIMIT (%)	PLASTICITY INDEX	CONSISTENCY INDEX	NATURAL VOLUME WEIGHT (Mg / m³)	SPECIFIC GRAVITY (Mg / m³)	SOIL CLASSIFICATION AC
			CLAY (< 0.002 mm)	SILT (0.002 - 0.06 mm)	SAND (0.06 - 2 mm)									
THE AREA OF THE TOWN OF RIJEKA	HARBOUR OF BULK CARGO BAKAR	SANDY - CLAYEY SILTSTONE	7 - 12	30 - 70	15 - 65	11 - 36	6 - 10	29 - 34	14 - 21	16 - 19	0.7 - 1.3	2.32 - 2.34	2.72 - 2.74	CL
	BAKAR-RETAIN. WALL		4 - 8	50 - 70	20 - 45	11 - 20	6 - 10	31 - 38	18 - 22	12 - 18	1.40 - 1.80	2.36 - 2.55	2.74 - 2.77	CI/CL
	OREHO-VICA		12 - 18	40 - 70	15 - 50	15 - 16	4 - 10	33 - 38	18 - 19	15 - 20	1.50 - 1.90		2.70 - 2.72	CI/CL
THE AREA OF THE TOWN OF DUBROVNIK	SRE-BRENO	CLAYEY - SANDY SILTSTONE	20 - 45	30 - 70	10 - 30	8 - 20	10.60-15.42	40 - 44	7 - 20	23 - 33	0.90 - 1.15	2.11 - 2.22	2.64 - 2.72	CI/CH
	CAVTAT		15	80	5	27.2	7.8	36.5	13.3	23.2	1.24	2.45	2.69	CI/CL

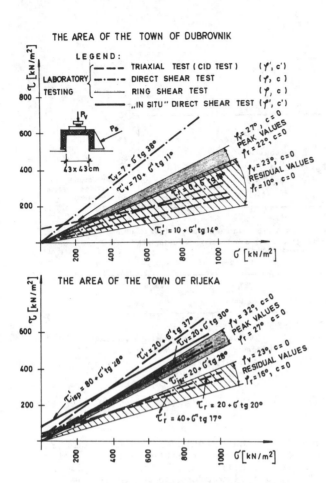

THE AREA OF THE TOWN OF DUBROVNIK

LEGEND:
LABORATORY TESTING
- — — — TRIAXIAL TEST (CID TEST) (φ', c')
- —·—·— DIRECT SHEAR TEST (φ, c)
- ‖‖‖‖ RING SHEAR TEST (φ, c)
- ———— "IN SITU" DIRECT SHEAR TEST (φ', c')

THE AREA OF THE TOWN OF RIJEKA

Fig. 6. The compared diagrams of the laboratory shear strength parameters and the "in situ" ones

The considerations show the real base for adopting the extreme hydrological condition - saturation to the surface, in the analyses of the global slope stability.
The plane of discontinuity is considered to be the potential sliding plane - the contact between the two engineering geological elements of the slope (1 in fig. 4). The type of sliding - shallow with translational dis-

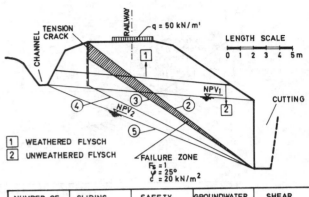

① WEATHERED FLYSCH
② UNWEATHERED FLYSCH

FAILURE ZONE
$F_s = 1$
$\varphi = 25°$
$c = 20\ kN/m^2$

NUMBER OF ANALYSES	SLIDING PLANE KP	SAFETY FACTOR F_s	GROUNDWATER LEVEL	SHEAR STRENGTH PARAMETERS
10	②	0.899	NPV_2	$\varphi = 25°$ $c = 20\ kN/m^2$
11	③	0.965	NPV_2	$\varphi = 25°$ $c = 20\ kN/m^2$
12	④	1.307	NPV_2	$\varphi = 25°$ $c = 25\ kN/m^2$
13	⑤	1.583	NPV_2	$\varphi = 25°$ $c = 20\ kN/m^2$

Fig. 7. The back analyses of the stability for the soil failure caused by digging the foundation of retaining wall (the harbour of Bakar - Rijeka)

placement of the overlaying material in the weathering zone - is defined in this way. Creeping displacements as the consequence of longlasting weathering of the physically mechanical parameters parallel to the weathering of the original rock (flysch) which often results by progressive failure, are possible in eluvial horizon.
The elements for fig. 4 are used for the analyses of the global slope stability. The slope is completely saturated. The gradients of hypodermic water are the consequence of the inclination of the contact plane of the overlaying material and the eluvial horizon.

3. LABORATORY AND "IN SITU" GEOTECHNICAL INVESTIGATIONS

Laboratory investigations are performed on the samples taken from the investigation bo-

COMPARISSON OF GEOTECHNICAL CHARACTERISTICS OF SOME LOCALITIES   TABLE 4.

DESCRIPTION OF THE CHARACTERISTICS		LOCALITIES		
		THE AREA OF THE TOWN OF RIJEKA	THE AREA OF THE TOWN OF DUBROVNIK	ALGEIR (HARBIL)
(Q+Ca+P+K)		34-70	30-70	33-35
PHYLLOSILICATE		30-70	44-50	65-67
MONTMORILLONITE		5	5	30
MICA		20-30	20-26	25
CONTENT OF $CaCO_3$ (%)		11-36	8-27	4-10
GRAIN SIZE DISTRIBUTION (%)	SAND 0.06-2 mm	15-65	5-30	2-4
	SILT 0.002-0.06 mm	30-70	30-80	29-35
	CLAY 0.002 mm	4-18	15-45	59-61
LIQUID LIMIT/PLASTICITY INDEX		29-38 / 12-20 CI/CL	37-44 / 23-33 CL/CI/CH	73-79 / 51-61 CH
MOISTURE CONTENT (%)		4-10	8-15	5-10
SHEAR STRENGHT PARAMETRES $\varphi(°)$ $c(kN/m^2)$	RING SHEAR TEST	$\varphi r = 16-23°, Cr = 0$ $\varphi v = 27-32°, Cv = 0$	$\varphi r = 10-23°, Cr = 0$ $\varphi v = 22-27°, Cv = 0$	$\varphi r = 6°$ $Cr = 0$
	DIRECT SHEAR TEST "IN SITU"	$\varphi r = 28°$ $Cr = 20$ $\varphi v = 28°$ $Cv = 80$	—	$\varphi r = 19°$ $Cr = 150$ $\varphi v = 19°$ $Cv = 245$
	BACK ANALYSIS	$\varphi r = 25°$ $Cr = 20 (kN/m^2)$	—	

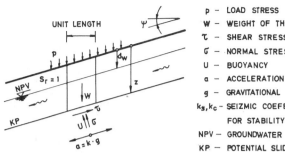

p - LOAD STRESS
W - WEIGHT OF THE SLIDING BLOCK
τ - SHEAR STRESS AT KP
σ - NORMAL STRESS AT KP
U - BUOYANCY
a - ACCELERATION
g - GRAVITATIONAL ACCELERATION
k_s, k_c - SEIZMIC COEFFICIENT (OR CRITICAL ACCELERATION)
FOR STABILITY (OR PERMANENT DISPLACEMENTS) ANALYSES
NPV - GROUNDWATER LEVEL
KP - POTENTIAL SLIDING PLANE
∿ - SEEPAGE FORCE

Fig. 8. The design model for the static and seismic analyses of the stability

RESULTS OF THE PARAMETRIC STABILITY ANALYSES   TABLE 5

SLIDING PLANE	GEOMETRICAL DATA	SHEAR STRENGHT PARAMETERS c(kN/m²)	K_s (K_c)	SLOPE STATE	p (kN/m')	$F_{s \ stat}$	$F_{s \ pst}$	U_m (cm)
DEEP	Ψ = 16° z = 9 m	c = 30 φ = 25°	0,08 (0,1094)	UNDRAINED dw=0	0	1,432	1,091	23
			0,08 (0,1695)	DRAINED dw=2,5m	0	1,670	1,277	6
SHALLOW	Ψ = 16° z = 2,5 m	c = 6 φ = 20°	0,10 (0,0175)	UNDRAINED dw=0	0	1,067	0,764	
			0,10 (—)		10	0,643	0,477	
			0,10 (0.2006)	DRAINED dw=2,5m	0	1,773	1,287	5
			0,10 (0,0696)		10	1,219	0,926	68
DEEP	Ψ = 14,3° z = 8 m	c = 0 φ = 22°	0,12 (0,0044)	DRAINED dw=2,0m	0	1,019	0,641	600
		c = 6 φ = 20°	0,12 (0,0389)			1,167	0,764	115
		c = 10 φ = 22°	0,12 (0,0619)			1,268	0,829	68
		c = 50 φ = 20°	0,12 (0,2709)			2,161	1,440	0,43

reholes in the area of the towns of Rijeka and Dubrovnik in order to clasify and get knowledge about the necessary geotechnical properties presented partly in the table 3. By classification tests the samples of the original rock-flysch (part III fig. 4) belong to CI/CL, respectively CI/CH materials with realtive consistency $C_r > 1$. The grain size distribution curves of the samples from both areas Rijeka and Dubrovnik are very similar. Petrographic analyses show that they are clayey siltstone of low cemented calcite binder, with more than 50% phyllosilicate, the typical clay mineral. The shear strength parameters are determined on the undisturbed samples by triaxial (CID) test and direct shear test. They are shown in fig. 6 for the area of the towns of Dubrovnik and Rijeka.

"In situ" investigations included geophysical loggings in the boreholes, SPT tests and "in situ" shear strength test of large blocks. For the original rock (III - Fig. 3) the number of blows of SPT is in a way unreliable indicator because of the sandstone interlayers which appear in flysch (Fig. 4) and which doesn't show the real state of the rock mass during SPT.

"In situ" direct shear stresses on large blocks (43 x 43 cm) are performed in the area of the town of Rijeka in the zone of normal stresses 300-500 kN/m². The obtained values are shown in Fig. 6.

The determination of the representative shear strength parameters is a delicate task. The best way to determine them is by use of the back analyses of the eventual existing landslides. In the area of Rijeka (in Bakar, the harbour for bulk cargo) the failure zone has been found while digging the foundation for the retaining wall (Fig. 7). By the back analyses it is found out for F=1 that the shear strength parameters are $\varphi=25°$ and c=20 kN/m² (Fig. 7).

The residual values of the parameters established for "in situ" shear test of great blocks are $\varphi=28°$ and c=20 kN/m² (Fig. 6). The table 4 has been made based on the geotechnical investigations of flysch in the areas of Rijeka, Dubrovnik and Algiers (Hudec and Jašarević 1986). The dependence of the shear strength parameters on mineralogic composition, grain size distribution, liquid limit and plasticity index is obvious, Comparing the results of the geotechnical investigations in flysch at the three mentioned places, one can see that the greater percentage of sand gives greater angle of friction and greater quantity of clayey silt particles enlarge cohesion.

4. THE ANALYSES OF THE SLOPE STABILITY IN STATIC AND SEISMIC CONDITIONS

Based on extended sliding planes of the glysch slopes (Fig. 2 and 4) the model of the "endless" long slope (Fig. 8) is suggested as the design model.

The calculations of the static stability are performed by the method of slope mass balance analysis with additional hypothesis:
- the slip surface is a plane with the in-

417

clination equal to the slope inclination
- interlamellar block forces are parallel to the slope inclination
- steady state flow is parallel to the slope inclination
- the material above the water level is completely saturated 100% (S_r = 1).

For preliminary analyses, which must be done during geotechnical investigations of the glysch slopes, it is possible to use the expressions or normalized diagrams to determine the safety factor (Had - Hamou and al. 1985).
For seismic analyses by use of pseudostatic method the following additional hypothesis are necessary:
- acceleration is constant along the sliding body
- the direction of the acceleration is parallel to the slope
- the sliding mass is effected by the pseudostatic inertial force proportional to the weight of the potential sliding body.

For the practical analysis we use the modified expressions Had - Hamou and al.
The permanent displacements of the sliding mass are calculated by use of the modified Newmark's method (Newmark 1985) which implies the above mentioned hypothesis (Chang and al. 1984).
The stability and the integral functioning of the slope at seismic loading can be disturbed in two ways:
a) the slope failure because of the liquefaction or because of the loss of material strength at changeable loading,
b) slope failure because of sliding of the potential sliding masses under the influence of the inertial forces caused by earthquake.
The analyses of the slope stability are done by:
- the Newmark's method of the sliding block
- using the seismic coefficient K_s.
The analysis by using the seismic coefficient is not considered reliable in seismic analysis of the stability of the natural slopes. The table 5 shows partial results of the performed parametric analysis in static as well as in seismic condition by use of the method of the sliding block.
The seismicity of the area of the town of Dubrovnik has been treated in order to estimate the seismic stability.
- the strongest earhquake with epicenter in Dubrovnik happened on 6th April 1667. (M=6,5-7,0 I_O= X degrees MCS, h = 15 km).
- the eathgquake in the southern part of Montenegro happened on 15th April 1979 (M = 7,0, I_O = IX-X degrees MCS, h=17 km).
The following data for flysch slopes in the surrounding of the town of Dubrovnik are accepted:
- maximal acceleration 0,36 g
- magnitude M = 7,0
- seismic coefficient K_s = ($\frac{1}{2}$ to $\frac{1}{4}$) of

a_{max}, value 0,08-0,12.

According to the analysis of the performed calculations, it is obvious that special care must be given to estimate and choose the shear strength parameters. Based on series of tests, performed on flysch samples which are taken from the areas of the towns of Rijeka and Dubrovnik (Fig. 6), relative great differences of internal friction angle (φ) and cohesion (c), depending on the type of the performed test, have been found out. The result analysis from the area of the town of Rijeka shows good agreement of the shear

strength parameters, defined by "in situ" test (the maximal shear corss-section area of 1500 cm²) with the result of the back analysis for one failure of the glysch slope (Fig. 7).
The differences in the volume of cohesion which greatly influences the stability analysis, with the excess of thangential stresses which response to the values above the level of the elastic deformations, can be very big, which is probably the consequence or rheologic phenomenon important in flysch. During the observation of many slopes in the area of the towns of Rijeka and Dubrovnik it is visible that they are stable in static and seismic conditions. Artificial slopes, as the results of human activities can be unstable because of irregular interventions (cutting, filling up earth, drainage and channeling of the surface and underground water, incorrect object load stress, etc). During the observations in the area of flysch slopes, a lot of cases of slow (longlasting) progressive failure have been found. This is not the consequence of the change of loading stresses but the change of the shear strength parameters.

5. CONCLUSION

1. By programming geotechnical investigation in flysch, the care should be taken about:
   - large heterogeneity and varieties in composition and texture of the flysch slopes
   - spacious variabilities of geotechnical
   - parameters
   - influence of the macroporosity on the
   - permeability of the flysch slopes
   - slow (longlasting) influence of the progressive failure on the stability of the flysch slopes
   - influence of the sample size by shear strength testings
   - rheologic parameters dependence on the hystory of stress-strain behaviour.

2. The coreelation between dinamic and static cohesion can be used in estimating the undriained parameters which are necessary for seismic analyses of stability and the calculation of permanent displacements (Ishihara, 1985). According to the performed investigations on the cleys, Ishihara established that the static and dynamic failure envelope has the same friction angle and the relation of C_d/C_s cohesion is 1,6 to 2,4. It goes primarily out of viscous nature of flysch. Because we do not posses the data of dynamic failure envelope for the areas of the towns of Dubrovnik and Rijeka, the stabile slopes in the seismic conditions can be explained by the enlarged values of cohesion.

REFERENCES

1. Chang, C-J., Chen, W. F., Yao, J.T.P. 1848: Displacements in Slopes by Limit Analysis. Journal of Geotechnical Engineering, Vol. 110, No. 7, pp 860-874.
2. Had-Hamou, T., Kavazanjian, E., 1985: Seismic Stability of Gentle Infinite Slopes. Journal of Geotechnical Engineering, Vol. 111, No. 6, pp. 681-697.
3. Huđec, M., Jašarević, I., Simić, R., 1986: Some experiences in underground constructions in flysch, International Congress on Large Underground Openings, vl. 2, pp. 205-212, Firenze.
4. Ishihara, K., 1985: Stability of natural deposits during Earthquakes, Proc. of the Eleventh ICSMFE, San Francisco, Vol. 1, pp. 321-376. A.A. Balkeme, Rotterdam.

Foundation of high valley bridges in triassic sediments

Fondations de ponts au-dessus de vallées profondes dans des sédiments triasiques
Gründung hoher Talbrücken im Buntsandstein

R.KATZENBACH, Dr.-Ing., Consulting Engineer, Darmstadt, FRG
W.ROMBERG, Dipl.-Ing., Consulting Engineer, Darmstadt, FRG

ABSTRACT: By example of an approx. 750 m long and 45 m high valley bridge it is shown that bridge foundations within the sand- and mudstones of triassic rock formations are especially influenced by the high tectonic strengthening of the rock, by the inclination of the layers and by the leaching of the deep salinar karst. In the report the foundation exploration and the results of pile- and foundation-load-bearing tests and seismic measurements are described. Based on the rock mechanical investigations the foundation design had been developped. The measurements at the ready built bridge show that the chosen pile foundations and the hillside protection system stand the test; thus, the described foundation concept could be applied to several other bridges.

RESUME: A l'aide de l'exemple d'un pont de vallée qui est long de 750 m et haut de 45 m nous montrons que des fondations de ponts dans les grès et les argiles compactes du triassic sédiment sont surtout déterminées par la forte tension tectonique et l'inclination des couches ainsi que par le lessivage du salinar karst profond. Dans ce rapport nous décrivons les études du sous-sol et le résultat des essais de chargement des pieux et des fondations ainsi que des mesures sismologiques. Le projet de fondation se base sur ces essais mécaniques de roches. La fondation sur pieux qu'on a choisie s'est révélée comme avantageuse comment les mesures du pont achevé montrent et elle est pour cette raison appliquée aussi lors de la construction d'autres ponts.

ZUSAMMENFASSUNG: Am Beispiel einer etwa 750 m langen und 45 m hohen Talbrücke wird gezeigt, daß Brückengründungen in den Sand- und Tonsteinen des Buntsandsteins insbesondere von der hohen tektonischen Beanspruchung des Gebirges sowie von den Auswirkungen der Salzauslaugungen in dem tiefen Salinarkarst geprägt werden. In dem Bericht werden die Baugrunderkundung und das Ergebnis von Pfahl- und Fundamentprobebelastungen sowie die seismischen Messungen beschrieben. Der Gründungsentwurf, der aufbauend auf den detaillierten felsmechanischen Untersuchungen erstellt worden ist, hat sich nach den Messungen am fertigen Bauwerk bewährt und kommt daher auch an zahlreichen anderen Brücken zur Anwendung.

1. INTRODUCTION

The foundation of high valley bridges in the triassic sedimentary rock formations (Bunter Sequences) are connected with several severe rock engineering problems which are arising due to the high tectonic strengthening of the sand- and mudstones, the inclination of the layers and the leaching of the deep salinar karst. At the hillsides the critical inclination of the layers towards the valley and the low shear strength of the strata joints lead to a low stability of the natural rock slopes, so that those bridge piers which are situated at unstable hillsides, need special sheetings for their skid proof foundation. As an effect of the deep salinar karst frequent disturbances at the solution edge, solution depressions and often narrow fossil collapse pipes occur. The collapse pipes are filled with collapsed tertiary clay and silty sandy collapsed loose triassic sediments. The filling of the collapse pipes is much more deformable than the surrounding sand- and mudstones. Therefore, bridge piers which are situated within or at the edge of such collapse pipes need special foundation measures in order to avoid an inadmissible deviation of the piers.

At five of altogether ten, 300-1600 m long and up to 70 m high railway valley bridges (total length: 6.900 m) in the middle part (PA 14 and 15) of the new rapid rail transit line Hannover-Wuerzburg (Engels 1985 and 1986) the bridge foundations are influenced by those special manifestations of Bunter sequences in eastern Hesse (West Germany) between the cities of Fulda and Kassel. By example of one of these bridges the rock engineering problems and their solutions are described in detail.

The described railway valley bridge is 748 m long and 45 m high and consists of 17 fields, each 44 m long and designed as single-span beams. The foundations of the abutments are loaded with vertical loads of altogether V=30 MN and horizontal loads of H= 13 MN; the foundations of the piers are loaded with altogether V=40-50 MN and H=1 MN. About 60 % of the loads are caused by the dead weight of the construction system, the other 40 % belong to the live load which will be caused by the train service after the year 1991.

2. FOUNDATION EXPLORATION AND FIELD TESTS

The subsoil was investigated by 30 soundings and by 61, 20-70 m deep core drillings with a total length of 1.900 m. Within the borings SPT and dilatometer tests (Goodman et al. 1968; Smoltczyk & Seeger 1980) had been carried out. In addition to the borings the inclined rock formations were investigated by 7 test pits at the hillsides. In the middle part of the bridge (axis 10-11; Fig. 1), were 3 collapse pipes had been investigated, a 20x30 m big test pit had been carried out in order to prove the deformability and the bearing capacity of the collapse masses. There four 15 cm thick and 4-10 m long injection piles, two 70 cm thick 6 m long bored piles, two 1,50 m thick bored piles with a cement injection of the skin surface and four 1,5x1,5 m big test foundations had been proved. In addition to the borings and the field tests it was proved by use of seismic methods (downhole- and crosshole-measurements) if there would be hollow blocks or loose zones within the collapse pipe. At last between the piers of axis 6 and 7 a large scale pier-tensile-test had been carried out to get detailed data about the stiffness of the ready-built foundation.

419

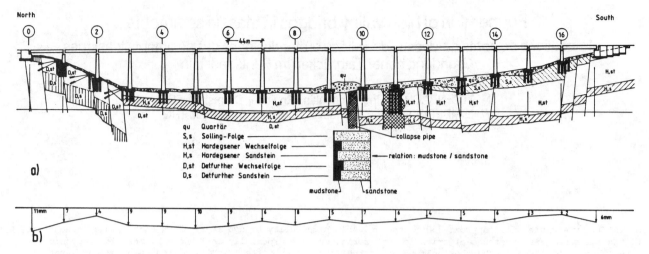

Fig. 1. Longitudinal section in the axis of
the bridge
a) geological survey
b) measuered settlements (Sept. 1986),
caused by dead weight

3. GEOLOGICAL AND HYDROLOGICAL SITUATION

The described bridge if founded in the middle Bunter
which consists of laminated and fissured sand- and
mudstones. The mudstone content is 1-30 % within the
different sequences of the middle Bunter (Fig. 1 and
2). The strata joints and the fissures have a
distance of 5-50 cm. The strata joints are wavy,
rough and sandy as well as plain, smooth and clayey.
In some cases the strata joints are covered with
1-2 cm thick micaceous material. The fissures are
rough and sandy; about 50 % are opened 1-5 cm wide
and are filled with sandy silt, the other 50 % have
no filling.

From the northern abutment to bridge pier 3 the
layers are dipping ± parallel to the natural slope
surface with an angle of inclination of $ß_\varsigma$=20-30°
(Fig. 1). There, the stability of the layers which
are near to the ground surface is only $f ≈ 1,0$
because of the low shear strength of the strata
joints. The stability of the northern hillside,
which is covered by 2 m thick quarternary silty
stoney sand, increases with depth, as the deeper
parts of the inclined rock formations find an
abutment beneath the valley surface.

In the valley (axis 3-10) the sand- and mudstones
are layered ± horizontal and are covered with 4-7 m
thick quarternary soils which consist of soft loam
and loose sand and gravel. At axis 10 and 11 (Fig.1)
the sand- and mudstones of the middle Bunter are
roughly interrupted by altogether 3 collapse pipes
which are filled in the upper 20-30 m with clay and
disturbed mudstones. Beneath the clay follow sandy

		Solling - Sandstein	35 m	
Mittlerer Buntsandstein	Solling-Folge			
	Hardegsen-Folge	Hardegsener Wechselfolge	±35m	
		Hardegsener Sandstein	15 m	
	Detfurth-Folge	Detfurther Wechselfolge	30 m	
		Detfurther Sandstein	±25 m	
	Volpriehausen-Folge	Volpriehausener Wechselfolge	45-60 m	
		Volpriehausener Sandstein	25 m	

Fig. 2 Stratigraphic classification of the middle
Bunter in eastern and northern Hesse
(after Prinz 1980)

and stoney masses of collapsed sandstones. The
pipes are built very irregular (Fig. 3), follow the
geological faults and have diameters from 10 to 30 m,
sometimes even more. At the bridge, described here,
the pipes are 600 m deep and lead down to the
salinar karst where leaching had caused the collapse
of the Bunter. At the southern hillside (axis 12-16)
the quarternary soils are up to 16 m thick and
consist of loose silty sandy talus material. Beneath
the quarternary soils the strong Solling-sandstone
occurs (Fig. 1). At axis 16 and at the southern
abutment a 7 m thick quarternary sliding mass had
been explored covering the sandstone.

In the valley the ground-water table is situated
near to the ground surface. At the hillsides the
height of the ground-water table increases less than
the ground surface so that the water is there 10-20 m
deep below the natural slope surface. The groundwater is aggressive against concrete because of the
contents of 40-70 mg CO_2 per litre.

4. MECHANICAL DATA OF THE SUBSOIL

4.1 Sand- and mudstones

The sand- and mudstones have a weight of 23-25 kN/m³.
The uniaxial compressive strength of the sandstones
(strong core samples ∅ 10 cm) amounts to 25-60 MN/m².
The young's modulus differs from 5.000 to
20.000 MN/m² for the strong sandstone. The shear
strength of the rock formation was investigated by
triaxial compression tests at 120 cm high rock specimen with a diameter of 60 cm; the testing procedure
is described by Gudehus & Wichter (1980). The triaxial shear tests delivered a shear strength of
ϕ_{RF}^{ι} = 30-50° and c_{RF}^{ι} = 0,1-0,5 MN/m². The shear
strength of the strata joints is considerably lower
than the shear strength of the rock formation and
was investigated by laboratory test series and by
back analysis of slidings (Romberg & Katzenbach 1986)
to ϕ_S^{ι} = 15-20° and c_S^{ι} = 0. Höwing & Kutter (1985)
report about similar values of the shear strength of
filled rock joints with a clay content of 45-65 %.

4.2 Clayey collapse mass of the pipe

The disturbed and collapsed clays have a weight of
20-22 kN/m³. The water content is about 20 %, the
plasticity index varies from 15-30 %. Because of the
disturbance and the inhomogenities of the collapse
mass only a few specimen could be proved in the
compressive and shear test; the uniaxial compressive
strength was measured in the laboratory to
0,06-0,12 MN/m². The drained shear strength amounts
to ϕ' = 20-25° and c'=0,005 MN/m².

5. RESULTS OF THE FIELD TESTS IN THE CLAY OF THE COLLAPSE PIPE

In the dilatometer tests in the 70 m deep core drilling deformation modulus of E_{v1}=15-20 MN/m² were measured in the depth of 0-35 m and E_{v1}=40 MN/m² in the depth of 35-70 m with a scattering of $\pm$ 200 % caused by the inhomogenities of the collapse mass. The static penetration testing delivered a skin friction of nearly constant q_S=0,025 MN/m² and a point pressure of q_C=5 MN/m² in the uppest 5 m and below this depth q_C=10-20 MN/m².

The geophysical crosshole- and downhole-measurements which were carried out up to a depth of 80 m below the the ground surface showed values for the velocity of the shear waves in the clayey material of c_t=400 m/s up to a depth of 15 m. The deeper parts brought c_t=600 m/s. The velocity of the compression waves increased from c_1=1.900 m/s in the upper parts of the clayey collapse masses to c_1=2.500 m/s in the deeper regions. With the equations (1) - (3)

Eq. (1): $\quad G_{dyn} = \varrho \cdot c_t^2 \qquad (\varrho=2,1-2,2 \text{ t/m}^3)$

Eq. (2): $\quad \nu_{dyn} = (c_1^2 - 2c_t^2)/(2c_1^2 - 2c_t^2)$

Eq. (3): $\quad E_{dyn} = 2 \cdot G_{dyn} \cdot (1 + \nu_{dyn})$

the dynamic young's modulus can be computed to E_{dyn}=1.000-2.300 MN/m² for the clayey collapse mass; that is 100times greater than the measured static deformation modulus. The shear waves within the surrounding sand- and mudstones had a velocity of c_t=1.000-1.200 m/s; the velocity of the compression waves is c_1=3.000-3.400 m/s in the rock formation. This leads with eq.(1)-(3) to dynamic young's modulus of E_{dyn}=7.000-10.000 MN/m² for the sand- and mudstones.

The failure load of the 1,5x1,5 m big test foundations was 0,35 MN/m² (found. 1) resp. 0,2 MN/m² (found.2). In the initial loading the deformation modulus was measured to E_{s1}=10 MN/m²; the recharge-modulus obtained a value of E_{s2}=30-40 MN/m².

Three of the four 15 cm thick injection piles had at failure a skin friction of 0,20 MN/m²; one pile carried a skin friction of 0,08 MN/m² at failure. The failure load of the three 7,4 and 10,0 m long injection piles was 1,1 MN. The total settlements were measured to 11 mm from which 7 mm are caused by elastic settlements.

The failure load of the 6 m long bored pile (ϕ 70cm) was 0,6 MN. From the strain gauges within the piles and from the pressure cell at the toe of the pile the skin friction at failure was backanalised to q_S=0,04 MN/m² and the point pressure to q_C=0,25 MN/m².

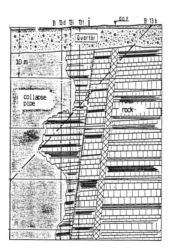

Fig. 3. Edge of the collapse pipe (result of 4 core drillings).

This rather low and unsufficient bearing capacity of the bored piles on the one hand and the rather high load bearing capacity of the injection piles on the other hand led to the idea to found the bridge pier at axis 11 on big bored piles with a cement injection of the skin. In a large scale test a 1,5 m thick and 30 m long bored bile with an injection of the skin was proved. The pile was loaded with a vertical load of 20 MN which was fully carried by the skin friction. The maximum settlement of the pile head was 11,5 mm, that of the pile toe 4 mm. The intermediate skin friction at the maximum load of 20 MN was 0,14 MN/m². The pile behaved fully elastic during all phases of the tests and could not be loaded to its failure load because of the bearing capacity of the 24 anchors which held the press crown about the pile.

6. DESIGN OF THE BRIDGE FOUNDATION

6.1 General

The permissible soil-pressure was limited to 0,8 MN/m² for shallow foundations with respect to the allowable settlements of the abutments and the bridge piers. According to several load bearing tests on test-piles in sand- and mudstones (Sommer et al. 1985) the foundations on bored piles were designed with a permissible skin friction of 0,15 MN/m² and a permissible point pressure of 1,0 MN/m² within the layers of the Bunter. These permissible pile load data belong to an estimated pile settlement of 2 cm.

6.2 Bridge abutments

Both bridge abutments are founded shallow and so deep that the foundation horizont is situated below the unstable layers of the hillside. The minimum safety-factor against sliding was defined to f=1,3 referring to a sliding mechanism from the top of the hillside down to the valley; the computation models are described by Katzenbach & Romberg (1986) and Laemmlen & Katzenbach (1986).

6.3 Bridge piers at the hillside

On the bridge pier 1 acts a horizontal slope thrust of S=25 MN. At the piers 2 and 3 the slope thrust has a value of S=40 MN per pier. The slope thrust is caused by the inclined unstable layers of the triassic sediments. The computation model for evaluating the slope thrust includes several geological and rock mechanical assumptions which are described in detail by Moll & Katzenbach (1985). The bridge pier is not able to carry such a big horizontal load without additional measures. Thus, a hillside protection system consisting of anchored bored diaphragm walls was built above the piers 1, 2 and 3 (Fig. 1). The diaphragm walls consist of 15-20 m deep bored piles with 1,5 m diameter and a horizontal distance of 3-4 m. The head of the wall is anchored by permanent anchors with a permissible load of 0,6 MN; each anchor was proved by tension tests up to 0,9 MN. Each bridge pier is founded deep on 10 bored piles with 1,5 m diameter. The length of the piles is given by the fact that
1. the vertical load must be carried with the permissible pile parameters given in chapter 6.1
2. and that the toe of the pile must reach the depth for which a safety against sliding of f=1,30 can be computed.

6.4 Bridge piers in the valley

Each bridge pier is founded on 10 bored piles with a length of 12-15 m and a diameter of 1,5 m. Each pile had got a cement injection of the skin surface of

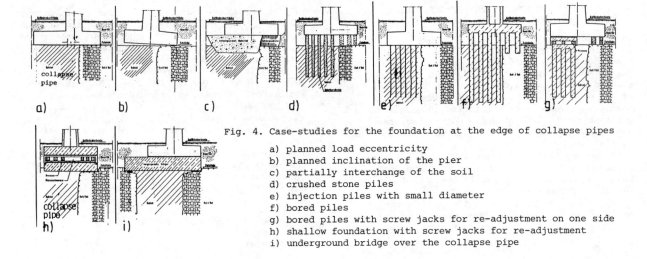

Fig. 4. Case-studies for the foundation at the edge of collapse pipes

a) planned load eccentricity
b) planned inclination of the pier
c) partially interchange of the soil
d) crushed stone piles
e) injection piles with small diameter
f) bored piles
g) bored piles with screw jacks for re-adjustment on one side
h) shallow foundation with screw jacks for re-adjustment
i) underground bridge over the collapse pipe

about 1,5 t cement per pile. The permissible load of a pile is 7-8 MN. The minimum distance between the piles is 3,75 m (=2,5 diameter). All piles are inclined 8:1 orthogonal to the bridge axis.

6.5 Bridge piers in the collapse pipes

As it was not possible to change the general bridge design concept, i.e. to enlarge the distance between the bridge piers, it was a fact that one or two foundations would be situated within the soft clay of the collapse pipes. At axis 10, which is surrounded by 2 collapse pipes, it was possible to avoid a foundation in the pipe by partially rotation of the foundation system. However, at axis 11 the geological conditions were more unfavourable: The eastern part of the foundation is fully situated in the pipe an the western part on the strong sand- and mudstones of the Bunter. In several case studies the fundamental possibilities for the foundation were discussed (Fig. 4). The solutions, shown in Fig. 4a-4e are possible for subordinate buildings or in those cases where the subsoil is more homogeneous, but they seemed to be not sufficient in the described case and are eventually too riskful for the foundation of a 45 m high bridge pier. To build an underground bridge over the collapse pipe seemed to be too expansive (Fig. 4i); however, such a solution is just under construction at a nerby valley bridge which will be described elsewhere. It was decided to build a system as shown in Fig.5, which is a mixture of Fig. 4f and 4e. The chosen foundation consists of fifteen 30 m long 1,5 m thick bored piles, which are founded in the collapse pipe, and of 9,5 m long and 1,5 m thick bored piles in the sand- and mudstones. The piles in the collapse pipe had got a cement injection of the skin with a volume of altogether 61 m³ (= 55 t cement). Between the foundation in the pipe and the foundation in the rock a 4 m thick concrete-plate is spanned which carries the load of the bridge pier. The plate can be re-adjusted by screw jacks on the side of the pipe if necessary.

7. MEASUREMENTS

Before the superstructure had reached the piers 6 and 7, there a large scale pier-tensile-test had been carried out. The aim of the test was to get true data about the stiffness and the rigidity of the ready-built pile foundations. Indeed, the results of this test could not be used to optimize the foundations of the discribed valley bridge but they delivered the parameters for a special design

concept of a similar railway-bridge. The heads of the two piers were stretched together by use of pre-stressing cable with a maximum load of H=700 kN. The moment, acting on the foundation, was M=34 MNm; it caused an inclination of the foundation of 0,15 to o,20 °/$_{oo}$. The horizontal displacement of the head of the pier was measured to 3 cm; about 23 % of this displacement were caused by the inclination of the foundation, the other 77 % were caused by bending of the concrete pier shaft. The stiffness of the foundation is

Eq. (4) $K_\varphi = M / \varphi$ = 200.000 MNm.

Horizontal displacements did not occur at the foundations during the tensile test respectively they had been so small, that they had not been measured (i.e. less than o,1 mm).

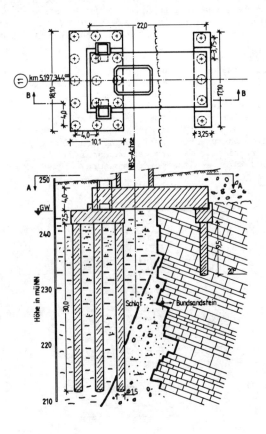

Fig. 5. Chosen foundation at the edge of the collapse pipe.

The measured settlements, caused by the dead weight of the bridge are shown in Fig. 1b. The shallow founded abutments have settled 12 mm (north) resp. 6 mm (south). The settlements of the pier foundations are scattering between 2 and 10 mm. The maximum settlements are measured at the piers 3-8 which are founded in the Hardegsen series. The lowest settlements occur at the foundations in the strong Solling-sandstone. The measured settlements are at time not greater than the predicted settlements.

8. FINAL REMARKS

On the bases of detailled soil- and rock-investigations a safe und economical foundation system was designed for the 45 m high valley bridge under difficult geological conditions. Solutions, which had been developped at the described bridge, are meanwhile applied to other bridges. All rock engineering problems, which are known from foundations in triassic rock, come together at the described valley bridge; they had been solved by a narrow cooperation between the employer (German Federal Railway), the constuction contractor and the consulting engineers.

REFERENCES

Engels, W. 1985. Die Neubaustrecke Hannover-Würzburg im Mittelabschnitt, eine 111 km lange Baustelle. Die Bundesbahn: 541-552.

Engels, W. 1986: Neubaustrecke Hannover-Würzburg im Mittelabschnitt, Halbzeit in der Bauausführung. Die Bundesbahn: 767-774.

Höwing, K.-D. & Kutter, H.K. 1985. Effect of filler composition on the mechanical behaviour of filled rock joints. Proc. 6th Nat. Rock Mechanics Symp., p.21-26. Essen: DGEG.

Goodman, R.E., Van, T.K. & Heuzé, F.E. 1968. Measurement of rock deformability in boreholes. Proc. 10th US Symp. Rock Mech., Austin.

Gudehus, G. & Wichter, L. 1980. Verformungs- und Festigkeitseigenschaften zweier Keupermergel. Berichte 4. Nat. Tagung über Felsmechanik, p. 199-205. Essen: DGEG.

Katzenbach, R. & Romberg, W.. Construction of a tunnel and an adjoining railway bridge in rockmasses with low stability. Proc. Int. Symp. on Engineering in complex rock formations. p.522-527. Beijing: Science Press.

Laemmlen, M. & Katzenbach, R. 1986. Foundation and stability investigations on natural rock slopes in Bunter Sandstone as a basis for the planning of slope-stabilizing measures - a case study from the Hattenberg, Eastern Hesse. Geol. Jb. Reihe C, 44. Hannover: BGR.

Moll, G. & Katzenbach, R. 1986. Design of a tunnel and a bridge foundation in a rock slope endangered by sliding. Proc. 7th Nat. Rock Mechanics Symp. Essen: DGEG.

Prinz, H. 1980. Manifestation of the Deep Saline Karst at the Route of the Developing Section Hannover-Würzburg. Rock Mechanics, Suppl. 10. p. 23-33. Wien New York: Springer.

Romberg, W. & Katzenbach, R. 1986. Back analysis of the shear strength of a folded rock. Proc. Int. Symp. on Eng. in complex rock formations. p. 798-801. Beijing: Science Press.

Sommer, H., Wittmann, P. & Ripper, P. 1985. Bearing capacity of large-diameter piles in the Detfurther alternating sequence of north Hessian Buntsandstein. Proc. 6th Nat. Rock Mech. Symp., p.55-62. Essen: DGEG.

Smoltczyk, U. & Seeger, H. 1980. Soil investigations by using the Stuttgart lateral pressure in bore-hole device. Geotechnik 3: 165-173.

Stochastic estimation and modelling of rock joint distribution based on statistical sampling

Calcul probable quantitatif et l'élaboration d'une maquette de la répartition des fentes rocheuses selon l'échantillonnage statistique
Stochastische Schätzung und Modellkonstruktion der Kluftverteilung, auf Grund statistischer Probenahmen

KOKICHI KIKUCHI, Tokyo Electric Power Services Co. Ltd, Japan
HIDETAKA KURODA, Shimizu Construction Co. Ltd, Tokyo, Japan
YOSHITADA MITO, Graduate School of Science and Engineering, Waseda University, Tokyo, Japan

ABSTRACT: In order to clarify the fact of geometrical distributional characters of joint systems, we examined distributional character using massive joint data in granite area, and have suggested a sampling, estimation and modelling method of joint distribution.

RESUME: Afin de pouvoir éclaircir les caracteristiques géometriques et la répartition des fentes rocheuses, nous en avons analysé les characteristiques grace a de fort nombreuses données receuillis dans la zone graniteuse. Ceci nous a ainsi permis de préconiser une méthode pour l'échantillonnage, le calcul quantitatif et l'élaboration d'une maquette de la répartition des fentes.

ZUSAMMENFASSUNG: Um die charakteristiken geometrischer Verteilung in Kluftsystemen aufzeigen zu können, haben wir die Verteilungsbedingungen mit Hilfe von umfangreichen Daten über Kluften in Granit untersucht. Das Ergebnis war die Entwicklung einer Methode zur Probennahme, Schätzung und Modellkonstruktion von Kluftverteilung.

1 INTRODUCTION

It is considered many joints, which exist in rock mass, function an important role for mechanical and hydraulic behaviors of whole rock mass. Since, however, it is said that joint system has a complex distribution in three-dimensional space, form, and character, effective survey method or estimaion method are not yet established and only few examples have been reported. Therefore, there are several unclear points even on the most basic geometrical character of joint system.

In order to clarify the fact of distributional characters, we suggested, at first, statistical sampling and estimation method. This sampling method includes a survey for area division and a survey for joint elements, and the estimation method highlights on (1) orientation, (2) persistence, (3) density (spacing), and (4) connectivity as the influence elements of geometrical character of joint systems. As for the above case (2) and (3), stochastic estimation is also included. This sampling method and estimation method were applied to a joint survey of outcrop at a small island (I-Island) consists of granite, and the distributional character of those (1) to (4) shall be clarified using massive actual field data. Moreover, based on the knowledge obtained through this survey, we suggested a method how to make a geometrical model of joint system on the two-dimensional plane, then tried to apply it to the I-Island.

2 SUGGESTION FOR JOINT SAMPLING

2.1 Survey for area division

It is natural distributional character of joint systems is changed depends on the location in a wide region. In case of considering a model in a wide region, unless we consider this kind of change of joint distribution character, a model expressing averaged distribution character will be simulated and rock mass character, which should be expressed as the model, will not enough expressed. Therefore, we suggested a method that to devide the survey region into some equivalent joint distribution character region, then to simulate a model for each devided region. When deviding a region, orientation of the joints in the survey region shall be paid attention, then devide according to a condition whether a prevail orientation is existing or not. The steps are as follows:

1. The region to be surveyed shall be fractionalized by grid system, and measuring points shall be set in each small region.
2. Scanline shall be set at each measuring point, then the orientation of a joint which crossing this line shall be measured.
3. Analyzing method based on polar coordinate shall be carried out on the obtained orientation data (detailes are written in below), and obtain a peak of each small region.
4. Analyzing method based on polar coordinate shall be carried out on the peak of each small region, and obtain a peak-set of section standard.
5. Devide the survey region according to the condition whether the peak of each small region is included into the peak-set obtained above 4.

2.2 SURVEY FOR JOINT ELEMENTS

In order to simulate a model of each devided small region based on survey result for area division, detailed survey of joint element shall be required. To carry out a sure sampling of joint elements, which are explained below, we suggested a plane sampling method by unit square of 1.5 x 1.5 ^{m} written below. Since the size of this unit square can be applied to the exploration edit, it is able to obtain a conformable data between the ground surface and rock mass inside.

The steps are as follows:
1. Set the grid system in the survey region and the square of 1.5 x 1.5 ^{m} shall be set in three-dimensionally as possible at the crossing points. Then the following steps of 2. to 4. shall be carried out for each square.
2. Joint trace shall be sketched on only joints which have trace length of more than 30^{cm} appearing on the square surface as the objects. If the trace

is extending to the outside of the square, the edge condition shall be remarked.

3. Based on each sketch of the joints as objects, strike, dip, trace length outside the square, aperture, sort of fillings, and width shall be recorded.

4. Some joint plane shall be selected by each prevail joint-set as a roughness sampling, and their two-dimensional profiles shall be drawn.

3 APPLICATION OF SURVEY FOR AREA DIVISION

The surveyed area is a small island of 3 km in North-South and 1 km in East-West consists of Cretaceous granite (I-Island). Fig. 1 shows an example of two-dimensioanl area division based on the actual field data. The grid interval is 200 m and the total number of small regions is 82. Fig. 1(a) shows an example of devided regions which were divided by a condition whether the peak of each small region belongs to the peak-set 1 (N 88 E 83 SE) or not, and Fig. 1(b) shows an example whether the peak of each small region belongs to the peak-set 2 (N 40 E 85 SE). Fig. 1(c) shows an example of three kinds divisions which peak of each small region is of only peak-set 1, of only peak-set 2, and of peak-set 1 and 2.

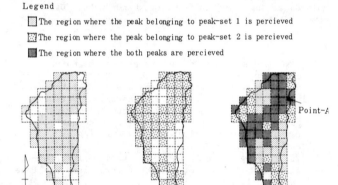

Legend

- ▨ The region where the peak belonging to peak-set 1 is percieved
- ▦ The region where the peak belonging to peak-set 2 is percieved
- ■ The region where the both peaks are percieved

Point-A

0   1 km

(a) based on peak-set 1   (b) based on peak-set 2   (c) based on peak-set 1 & 2

Fig. 1. The application of the area division method (I-island)

4 ESTIMATION OF JOINT ELEMENTS

As the elements which provide distributional character of joints, (1) orientation, (2) persistence, (3) spacing and density, (4) connectivity, (5) aperture, (6) roughness, and (7) fillings are considered. With regard to these elements of (1) to (4), we examined on the data of 1,097 joints obtained through the survey for joint elements at the Point-A (see Fig. 1(c)) in the area of 6 m x 18 m, and clarified the distributional character of each joint element.

4.1 Orientation

As the method to obtain a prevail orientation distribution of joints ("Analyzing Method Based On Polar Coordinate"), spherical net can be used to calculate the orientational density distribution. The density peak shall be obtained, and the joints group of more than 4% of total number, which were included within the area of 15° at its centeral angle from the peak shall be defined as a joint-set (see Fig. 2). Further, the ratio of number of joint of a joint group corresponding to the number of whole joint shall be defined as the concentrative ratio (CR). At the Point-A, the conter diagram

shown in Fig. 3 was obtained, and two joint-sets were perceived. The peak orientation of set-1 was N 75 E 85 NW and the concentrative ratio was 22 %, and set-2 was N 5 W 75 NE and the concentrative ratio was 21 %.

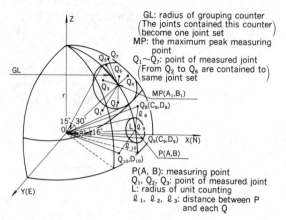

GL: radius of grouping counter (The joints contained this counter become one joint set)
MP: the maximum peak measuring point
$Q_1 \sim Q_7$: point of measured joint (From Q_2 to Q_6 are contained to same joint set)

$MP(A_1,B_1)$

P(A, B): measuring point
Q_1, Q_2, Q_3: point of measured joint
L: radius of unit counting
ℓ_1, ℓ_2, ℓ_3: distance between P and each Q

Fig. 2. Analysis method of joint orientation

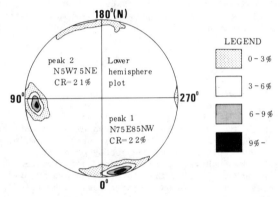

180°(N)

peak 2
N5W7 5NE
CR=2 1%

Lower hemisphere plot

90°   270°

peak 1
N75E85NW
CR=2 2%

0°

LEGEND
▨ 0 - 3%
□ 3 - 6%
▨ 6 - 9%
■ 9% -

Fig. 3. Result of the orientation analysis (Point-A)

4.2 Persistence

Fig.-4 shows trace length distribution of set-1 at the Point-A, which corresponds to the probability density function of exponential distribution very well. It is said sometimes that trace length distribution is similar to log-normal distribution, though, as shown in Fig. 4, according to the result of further survey on short joints, we obtained the result that shorter joint has larger number of joint. Further, chi-square test result shows that even if distribution is regarded as exponential distribution at the 5 % confidence level, there is no significant difference. Therefore, it shall be consider that trace length distribution is similar to exponential distribution.

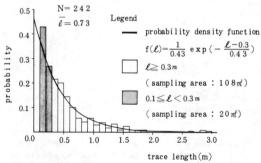

N= 24 2
$\bar{\ell}$ = 0.73

Legend

— probability density function

$f(\ell) = \frac{1}{0.43} \exp\left(-\frac{\ell - 0.3}{0.43}\right)$

□ $\ell \geqq 0.3\,m$ (sampling area : 108 ㎡)

▨ $0.1 \leqq \ell < 0.3\,m$ (sampling area : 20 ㎡)

trace length(m)

Fig. 4. Joint trace length distribution (Point-A: Set 1)

426

4.3 Spacing and density

Fig. 5 shows spacing distribution of set-1 at the Point-A, which is also corresponding well to the probability density function of exponential distribution as well as trace length distribution. Moreover, since the same result was obtained for set-2, it can be assumed that the spacing distribution is corresponding to exponential distribution. Further, chi-square test result shows that even if distribution is regarded as exponential distribuion at the 5 % confidence level, there is no significant difference. If the reciprocal of averaged spacing means number of joints appearing in the unit length interval is ρ_1, and the number of middle point of the joint trace per unit square is ρ_2 (which is defined as density), the ρ_1 of set-1 and set-2 become 1.67 (1/m), 2.56 (1/m) and ρ_2 value become 2.01 (1/m^2) and 2.00 (1/m^2).

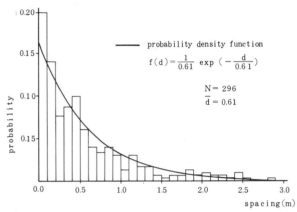

Fig. 5. Joint spacing distribution (Point-A: Set 1)

4.4 Connectivity

In case of considering the connectivity of each joint in two dimensional plane, connection type can be classified into 5 types as shown in Fig. 6. Table 1. shows the compositional ratio of connection types at the Point-A.

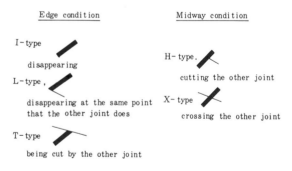

Fig. 6. Classification of connection type of joint

Table 1. Composition (%) of connection type(Ponit-A)

Type	Set-1	Set-2	Non-set	System
I	15.4	15.6	17.3	16.4
L	5.1	3.8	5.1	4.7
T	24.3	10.8	25.8	21.1
H	16.2	27.5	19.4	21.1
X	39.1	42.3	32.4	36.7

5 SAMPLING DENSITY FOR JOINT ELEMENTS

It is natural to be impossible to survey whole region because of several restrictions when surveys

for joint elements are carrying out. Therefore, it is necessary to estimate the whole distribution through the obtained data of unit square sampling. The estimation method through $\bar{\ell}$, ρ_1, ρ_2 by unit square sampling of 1.5 m, which we have considered[1] is as follows:

$$\bar{\ell} = \frac{\sum_{i=1}^{N} \left(\frac{L\ell i}{L+\ell i} \right)}{\sum_{i=1}^{N} \left(\frac{L}{L+\ell i} \right)} \qquad (1)$$

$$\rho_1 = \frac{\sum_{i=1}^{N} \ell_{Ii}}{L^2 N_A} \qquad (2)$$

$$\rho_2 = \frac{\sum_{j=1}^{N_A} N_{Ej}}{2L^2 N_A} \qquad (3)$$

where:

N: Total number of joints
L: Side length of unit square (=1.5 m)
li: Actual trace length
lIi: Trace length inside unit square
NA: Number of unit square
N_{Ej}: Number of trace edge observed in a unit square j

When the relation of $\rho_1 = \bar{\ell}\rho_2$ is confirmed[1], and if we obtain two kinds of value of $\bar{\ell}$, ρ_1, ρ_2, it is possible to estimate the rest parameter. Fig. 7 shows the relationship of Pearson's coefficient of variation among the value of $\bar{\ell}$, ρ_1, ρ_2 obtained through sampling density at the Point-A. This figure shows the value movement obtained through each parameter estimation using the above formula (1), (2), and (3) with a certain sampling density.

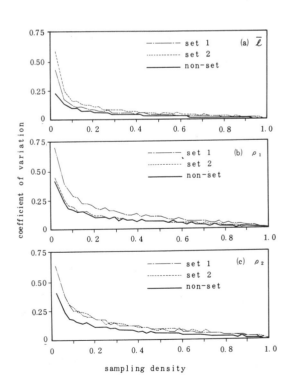

Fig. 7. Relationship between sampling density and coefficient of variation of the obtained parameters

6 MODELLING METHOD OF JOINT SYSTEM IN TWO-DIMENSIONAL PLANE

In order to create a typical model of the objective region based on the joint survey results, it is necessary to find out a character of the joint system and to reflect the survey results to the model. As above-mentioned, we obtained the result that joint length has a statistical regularity when we devide the joint system by their orientation into joint-sets, and if we survey characteristics of joint elements by each joint-set. Therefore, here we suggest stochastic modelling method of joint systems in two-dimensional plane, which is reflected the statistical regularity of these joint elements.

The steps are as follows:
1. Devide joint system by its orientation into joint-sets.
2. Decide number of joint to be created in the model by each joint-set based on the orientation of survey surface and joint systems, and on the number of surveyed joints.
3. Decide the middle point of joints by each joint-set. Since joint spacing is corresponding to exponential distribution, use the uniform random numbers for this work.
4. Decide the orientation using the uniform random numbers within a range of strikes or dips of each joint-set.
5. Decide the length according to exponential distribution obtained by each joint-set.
6. Simulate different models depend on the used sequence of random numbers in the above steps 3. to 5. Adopt one of these models as the real model, of which composition of the connecting type is the most similar to the actual one.

In these steps, orientation as one of joint elements is considered in step 4., persistence is in step 5., density is in step 2. and 3., and connectivity is considered in step 6. in each model. Fig. 8 shows an example of applying this modelling method to the joint survey result at the Point-A, with a sketch of outcrop.

Moreover, Fig. 9 shows a flow-chart of modelling procedure.

CONCLUSIONS

Conclusions of this paper can be summarized as follows:
1. A method to decide the modelling region was suggested and such method was applied to an actual site.

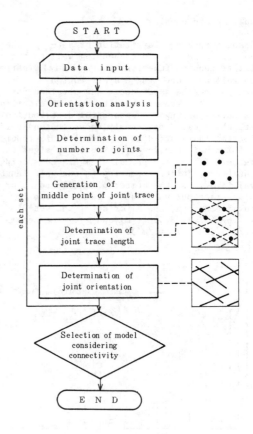

Fig. 9. Flow-chart of modelling procedure

2. Trace length distribution and spacing distribution of joints are corresponding to exponential distribution very well.
3. The relationship between the estimation result of the survey data using unit square of 1.5 m on the actual site and sampling density was shown.
4. Modelling method of joint system based on the estimation results was suggested, and its applying case was shown.

REFERENCE

1) Kikuchi, K.; T. Mimuro; Y. Izumiya and Y. Mito. A joint survey and determination of joint distribution. Procedings of 2nd International Symposium on Field Measurements in Geomechanics. Kobe, Japan. 1987.

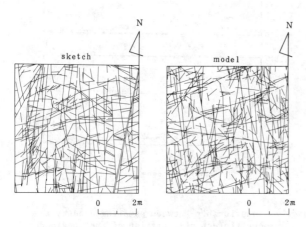

Fig. 8. Joint trace map and model (Point-A: 6x6 m^2)

Relation between physical anisotropy and microstructures of granitic rock in Japan
Les propriétés physiques et les microstructures du granit de la région sud-ouest du Japon
Verhältnis zwischen physischer Anisotropie und Mikrostrukturen der Granitfelsen in Japan

YOZO KUDO, Tokuyama Technical College, Japan
KEN-ICHI HASHIMOTO, Tokuyama Technical College, Japan
OSAM SANO, Yamaguchi University, Ube, Japan
KOJI NAKAGAWA, Yamaguchi University, Ube, Japan

ABSTRACT: Microstructures, physical properties, and the in situ orientations of the planes of anisotropy were investigated for the specimens from 40 granite quarries in the Inland Sea area in Southwestern Japan. Thin section analysis showed that the marked microstructures were open cracks and healed cracks in quartz and healed cracks in feldspar. Sound velocity, tensile strength and compressive strength exhibited obvious anisotropy, which can be explained by the preferred orientation of such cracks. Microcracks in granitic rocks were similar in preferred orientations at almost all the quarries in this area.

RESUME: Cet article a pour objet de montrer la correspondance entre les microstructures et les propriétés physiques, et de plus de montrer l'orientation in situ du plan d'anisotropie du granit dans 40 carrières de la région sud-ouest du Japon. Les examens au microscope optique ont montré que les microstructures se caractérisent par microfissures ouvertes et microfissures cicatrisés dans le quartz et par microfissures cicatrisés dans le feldspath. La vitesse de propagation des ultra-sons, la résistance à la tension et à la compression observées ont mis en évidence une anisotropie des microfissurations. L'orientation des fissures dans le granit et les massifs granitiques est apparue presque constante dans la région sud-ouest du Japon.

ZUSAMMENFASSUNG: Mit den Exemplaren, die an den 40 Steinbrüchen im Seto Binnenmeer Gefiet von Südwest Japan gesammelt worden waren, wurden die Mikrostrukturen, die physische Eigenschaften und die in-situ-Richtungen der Anisotropiesflächen in der Granitfelsen untergesucht. Die Dünn-Schnitt-Analyse zeigt, daß die auffallenden Mikrostrukturen in Quarz die gräffnete Risse und die geheilte und in Feldspat die geöffnete Risse sind. Die Geschwindigkeit der elastischen Welle, die Zug-und Druckfestigkeit der Examplare stellen eine äugenfallige Anisotropie dar, die von den hervorragenden Richtungen der Microrisse erklärt werden können. Meiste Richtungen der Mikrorisse in der Granitfelsen in diesem Gefiet sind miteinander ähnlich.

INTRODUCTION

Granitic rocks are typical of rock types usually assumed to be isotropic for the engineering field. But they contain numerous small defects which are preferentially oriented along three mutually perpendicular or nearly perpendicular planes. Several authors have asserted that these defects significantly affect the physical properties of the granitic rocks, e.g. sound velocity [Birch, 1960], tensile strength [McWilliams, 1966; Peng and Johnson, 1972; Kudo et al., 1986], uniaxial compressive strength [Dale, 1923; Osborne, 1935; Douglass and Voigt, 1969; Peng and Johnson, 1972]. Some of these physical properties indicate strong anisotropy, while others denote only slight anisotropy.

Estimation based on the isotropic theory can bring about extremely misleading conclusions when a physical property indicates strong anisotropy. The analysis of the planes of anisotropy up to now, however, is very extensive and time-consuming. It should be desirable for the engineering practice that the full description of the microstructure is not always needed. Because the anisotropic symmetry is controlled mainly by preferred orientation of cracks [Douglass and Voigt, 1969; Scholz and Koczynski, 1979; Plumb et al., 1984], it would be possible to establish the compliant directions only by the measurements of cracks.

It is naturally recognized that the characteristics of the rocks varies with distance. Quantitative discussions in geostatistics [Clark, 1979; Hudson, 1986] are beyond the scope of this study. Is there anything in common qualitatively within the information concerning the preferred orientation of the cracks in all of the granitic rocks?

In this paper, first, we investigate the microstructures, physical properties and in situ orientations of the microcracks for the specimens of

granitic rock quarries. Second, we discuss the similarities among the specimens from the quarries within the extent of a few kilometers. Finally, we search the similarities in the Inland Sea area in Southwestern Japan over hundreds of kilometers.

FABRIC OF GRANITIC ROCKS IN THE INLAND SEA AREA

All granitic rocks posses three sets of nearly orthogonal planes. These planes are called, in quarryman's terminology, rift plane (mé, in Japanese), grain plane (niban) and hardway plane (shiwa) in order of ease of splitting.

With a polarizing microscope, three thin sections oriented parallel to the three quarry planes were analyzed. Except for intergranular cracks, following types of cracks were distinguished: intragranular open cracks and intragranular healed cracks (bubble tracks) in quartz grains, and intragranular healed cracks and some open cracks in feldspar grains.

In order to estimate the similarities within a region of a few kilometers in extent, test specimens from four quarry sites on Oshima island (region 5, 47 km² in area) and eight quarry sites on Kitagishima island (region 7, 7.4 km² in area) were sampled.

Fig. 1(a) and Fig. 1(b) show photomicrographs of Oshima granodiorite. In the quartz, there can be seen a number of microcracks. Their features are; 1) Open cracks and healed cracks showing strong preferred orientation parallel to the rift plane. 2) Open cracks showing secondary preferred orientation parallel to the grain plane. 3) The other preferred orientation of healed cracks parallel to the hardway plane. Also in feldspar, there can be seen many healed cracks, especially the ones parallel to the hardway plane (Fig. 2). The microcrack fabric observed on thin sections from five quarries on Oshima island were similar to each other. Fig. 3(a) and

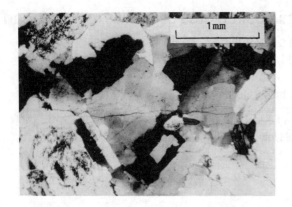

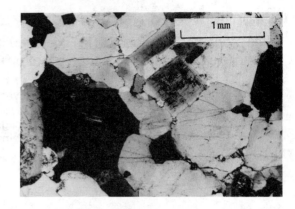

(a)   (b)

Fig. 1. Photomicrographs showing typical rift cracks within quartz grains found in a thin section of Oshima granodiorite (region 5). Thin sections were cut parallel to grain plane.

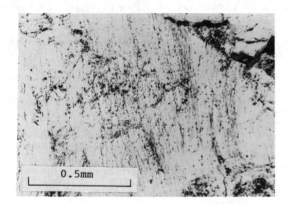

Fig. 2. Photomicrograph of Oshima granodiorite (region 5) showing typical healed cracks within plagioclase (which are approximately parallel to the hardway plane). Thin section was cut parallel to rift plane.

Fig. 3(b) show photomicrographs of Kitagishima granite. By the thin sections from eight quarries on Kitagishima island, it can be concluded that the characteristic of microcrack distribution is similar to that on Oshima island. However, there are fewer healed cracks within feldspar parallel to the hardway plane for Kitagishima granite. These observations suggest that the microstructures only change slightly within the range of a few kilometers in distance.

Microstructures characterizing three quarry planes of eight regions in the Inland Sea area are classified from thin section analyses and listed in Table 1.

PHYSICAL PROPERTY OF GRANITIC ROCKS

Thin section analyses showed that the microcracks in the granitic rocks have strong preferred orientation. Accordingly, we can expect that the physical properties will indicate strong anisotropy. Diametral compression test was performed on Oshima granodiorite. The size of the specimen was 43 mm in diameter and 21 mm long. The tensile strength for diametral directions was measured at 15° intervals, from H-axis (normal to the hardway plane) to G-axis (normal to the grain plane), from R-axis (normal to the rift plane) to H-axis and from G-axis to R-axis. Sound velocity normal to the splitting plane was also measured for each specimen preceding the fracture tests.   Fig. 4 and Fig.5 show the variations of the sound velocity and the tensile strength, respectively, for disks cored parallel to R-, G- and H-axis. It shows that the tensile strength and the sound velocity varies strongly with direction. There are some valleys corresponding to the three quarry planes. Low tensile strength in these directions can be attributed to the preferred orientations of cracks.

A block of Kurokamijima granodiorite(region 3) was sawed into a disk (200 mm thick and 500 mm in diameter) whose axis was parallel to G-axis, and cored into radial direction at 15 interval. The size of the specimen for uniaxial compression test was 43mm in diameter and 112 mm in length. Each test was carried out at a given strain rate. Results are shown in Fig. 6. Curves of the uniaxial compressive strength versus inclination to the plane of anisotropy are concave upward and specimens loaded at an angle of 30° to the rift plane exhibit the lowest strength.

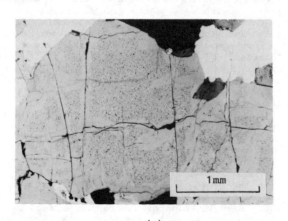

(a)   (b)

Fig. 3. Photomicrographs showing typical microcracks of Kitagishima granite. Thin section is cut: (a)parallel to rift plane (longitudinal direction; grain plane). (b)parallel to grain plane(longitudinal direction; rift plane).

TABLE 1. Description of microcracks

Name	Region Number	Dominant cracks		
		Rift plane	Grain plane	Hardway plane
Kurokamijima granodiorite	3	OCQ	HCQ OCQ	OCQ
Kurahashijima granite	4	OCQ	OCQ HCQ	HCF HCQ
Oshima granodiorite	5	OCQ HCQ	HCQ OCQ	HCF
Akasaka granite	6	OCQ	OCQ HCQ	HCF OCQ
Kitagishima granite	7	OCQ	OCQ HCQ	OCQ HCQ
Aokishima granodiorite	8	OCQ HCQ	OCQ	ND
Aji granite	9	OCQ HCQ	HCQ OCQ	HCQ OCQ
Mannari granite	10	OCQ	OCQ HCQ	ND

OCQ, open cracks in quartz; HCQ, healed cracks in quartz; HCF, healed cracks in feldspar; ND, not well developed.

ORIENTATION OF CRACKS

In situ orientations of the planes of anisotropy, i.e. the preferred orientation of the microcracks, were measured at several quarries on Oshima and Kitagishima islands based on thin section analyses with the aid of the empirical knowledge of quarrymen. Fig. 7(a) and Fig. 7(b) show the preferred orientations of microcracks in both islands. In all quarries studied, the rift plane is horizontal and is characterized by open cracks in quartz grains. The grain plane is vertical and is also characterized by open cracks and healed cracks in quartz. At almost all the quarries on either islands, the orientation of the grain plane strikes nearly parallel with each other. However, at two quarries on Kitagishima island, the grain plane shifted from the dominant orientation (N55°E). But the secondary preferred orientation was found to be parallel to the dominant orientation. We could, therefore, conclude that the crack distribution is similar within the region of a few kilometers in extent.

In Fig. 8, the average preferred orientations of microcracks at several quarry sites are plotted on the map showing the Inland Sea area in Southwestern Japan. At almost all the quarries in this wide region of hundreds of kilometers in distance, microcracks in quartz grains parallel to the rift plane have strong preferred orientation along the horizontal plane. The vertical microcracks parallel to the grain plane also seem to have a similar orientation. In this region, there is one more preferred orientation of microcracks nearly perpendicular to both the planes, which may reflect the history of tectonic stresses.

CONCLUSION

Thin section analyses on microcracks and experimental results strongly support the concept that the anisotropy in the physical properties of granitic rocks is attributed to the preferred orientation of microcracks. It was found that the planes of anisotropy had a close correlation with quarry planes and their in situ orientations were more or less constant over hundreds of kilometers in Southwestern Japan.

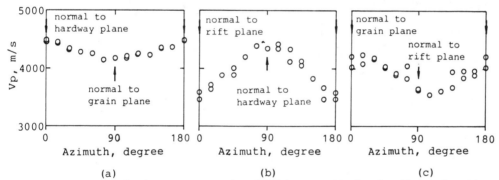

Fig. 4. Sound velocity measurements of Oshima granite (region 5) as a function of azimuthal variations. (a) Disks with flat surface parallel to rift plane. (b)Disks with flat surface parallel to grain plane. (c)Disks with flat surface parallel to hardway plane.

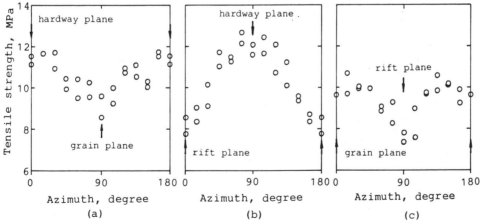

Fig. 5. Tensile strength measurements of Oshima granite as a function of diametral loading directions. (a) Disks with flat surface parallel to rift plane. (b)Disks with flat surface parallel to grain plane. (c)Disks with flat surface parallel to hardway plane.

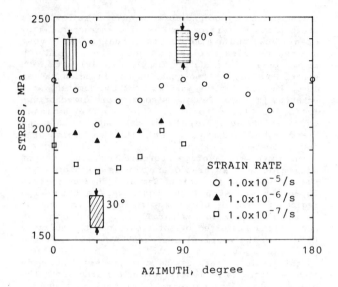

Fig. 6. Compressive strength measurements of Kurokamijima granodiorite (region 3). The axes of all specimens are parallel to grain plane.

The strike of the grain plane was roughly parallel to the longer axis of the Cretaceous Hiroshima granite-belt and to the Median Tectonic Line. This indicates that the microstructure of the Hiroshima granites strongly reflects the history of tectonic stresses.

Within granites in Southwestern Japan, we often found oriented bubble tracks nearly parallel to the hardway plane which is normal to both the rift plane and the grain plane. These cracks may suggest the possible alteration of the maximum (or intermediate) principal stress.

The region surveyed in this study is restricted within the Inner Zone of Southwestern Japan. It should be noted that the crack distribution has different characteristics within the Outer Zone of Southwestern Japan and the Northeastern Japan (unpublished data).

LITERATURE
Birch, F., J. Geophys. Res., 65, 1083-1102, 1960.
Clark I., Practical geostatistics, Appl.Sci., London, 1979.
Douglass, P.M. and Voigt, B., Geotechnique, 19, 1969.
Hudson J. A., Proc. of the 7th West Japan Symposium of Rock Engineering, 4-20, 1986.
Kudo Y. et al., Proc of JSCE, 370, 3-5, 189-198, 1986(with English abstract).
Peng S.S. and Johnson, A. M., Int. J. Rock. Mech. Min. Sci., 9, 37-86, 1972.
Plumb et al., J. Geophys. Res., 89, B11, 1984.
Scholz, C. H. and Koczynski, T. A., J. Geophys. Res., 84, B10, 5525-5534, 1979.

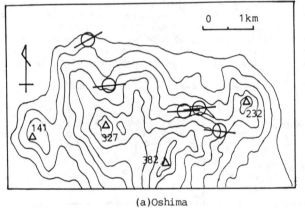
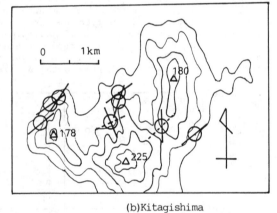

(a)Oshima   (b)Kitagishima

Fig. 7. Map of preferred orientations of microcracks. Circles indicate that the rift plane (plane of highest crack concentration) is quasi-horizontal. Solid lines represent the orientations of grain plane (plane of preferred orientation of vertical microcracks). Dotted lines represent the secondary preferred orientations of vertical microcracks.

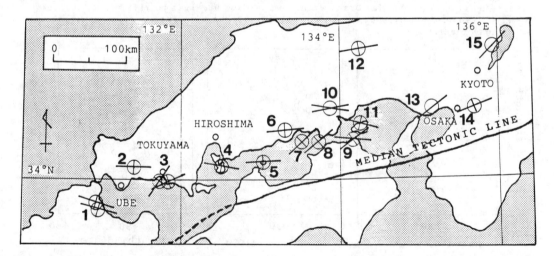

Fig. 8. Map of preferred orientations of microcracks. Circles indicate that there is a strong preferred orientations of microcracks within quasi-horizontal plane. Longer bars represent the average preferred orientatins of vertical microcracks. Shorter bars represent secondary preferred orientations of vertical microcracks.

Contact effects at the interface between rock foundations and concrete dams with power plants at their toes
Effets de contact à l'interface fondation-barrage en béton avec une centrale située au pied du barrage
Das Verhalten der Fuge zwischen Beton und Felsgründung von hohen Talsperren mit einem am Fuss liegenden Kraftwerk

A.N.MARCHUK, The B.E.Vedeneev All-Union Research Institute of Hydraulic Engineering (VNIIG), Leningrad, USSR
A.A.KHRAPKOV, The B.E.Vedeneev All-Union Research Institute of Hydraulic Engineering (VNIIG), Leningrad, USSR
YA.N.ZUKERMAN, The B.E.Vedeneev All-Union Research Institute of Hydraulic Engineering (VNIIG), Leningrad, USSR
M.A.MARCHUK, The V.V.Kuibyshev Moscow Institute of Civil Engineering, USSR

ABSTRACT: Analysis of field observation data revealed an opening of concrete-rock contact beneath the upstream face of some large Siberian dams. The calculations performed enable one to establish possible causes of this phenomenon.

RESUME: Les études in situ ont mis en évidence l'ouverture du joint de contact beton-rocher sous les parements amont de grands barrages en Sibérie. Les calculs ont permis de définir la cause possible de ce phénomène.

ZUSAMMENFASSUNG: Die Naturbeobachtungen haben das Offnen der Kontaktfuge zwischen dem Beton und der Felsgründung unter den Wasserseiten von hohen Talsperren in Sibirien gezeigt. Die Berechnungen erlaubten eine mögliche Ursache dieser Erscheinung zu ermitteln.

In the USSR layouts with power plants located at dam toes are most commonly used and believed to be highly economical. However, analysis of field observation data on stress-strain behaviour of the largest Siberian dams revealed certain limitations of this structure arrangement. In particular, it was found that beneath the upstream face of some dams an opening of concrete-rock contact occurs which results in subsequent increase of uplift pressure over an extended area of the dam toe. This phenomenon is more distinct under the power house sections.

Apart from the well-established causes of disruption of contact between rock and concrete (Eidelman, Durcheva 1981; Marchuk 1983) one must mention the effect which is specific for power plant sections: the increase in deformability of foundation below the downstream wedge of the dam.

Since the downstream wedge transfers maximum compressive stresses to the rock foundation, its pliability contributes to the total displacements of the dam profile. The stress-taking rock step cut by the foundation pit for the power house is always more deformable than the rest of the rock mass underlying the dam toe. Decrease of the step rock deformation modulus can be explained by stress-relieving of the rock mass during excavation as well as by rock loosening due to blasting and frost weathering at the construction stage. The relevant investigations (Marchuk, Khrapkov, Zukerman, Marchuk 1985; Rubtsov 1965) confirm that rock excavation is responsible for stress-relieving of the rock mass overlying zone and cracking due to reduction of the vertical component of the natural stress tensor.

Analysis of the geophysical data on rock foundations of the Ust-Ilymskaya, Boguchanskaya and some other hydro power plants showed that zones of technogeneous stress-relief extend to a depth of 15-20m and the deformation modulus E of the overlying subzone of the intensive stress-relief is 6-12 times less than that of the underlying subzone of weak stress-relief. This stress-relief of the power house section is further facilitated by blasting and frost weathering effects.

Investigations of frost weathering action at the Zeiskaya dam revealed that in 4.5 years cracking of unweathered diorite showed 38% increase for a depth of up to 4 m. Cement consumption in the consolidation grouting beneath the downstream monoliths ap-

pears to be 1.3-1.7 times higher than that under the first upsream monolith, which is indicative of intensive foundation cracking under the downstream dam wedge due to excavation of the pit for the power house foundation.

It can be stated that by the time of the reservoir filling the deformation modulus of the rock foundation along the dam toe has an irregular distribution. This irregularity results from varied degree of stress-relief throughout the rock mass and different pattern of concrete distribution in the structure. The maximum value of the deformation modulus is observed under the upstream toe.

The subsequent concrete placement in the downstream dam wedge and power house section permits neither to restore the initial rigidity characteristics of rock within the stress-relieved zone of the foundation, nor to obtain the values of deformation modulus equal to those under the other sections of the structure.

During reservoir filling the stress and settlement curves of the contact cross-section undergo transformation due to decrease in vertical compressive stresses under the upstream toe and (in certain conditions) due to development of corresponding tensile stresses. The analysis of this problem from the viewpoint of the elasticity theory shows that the necessary condition of the tensile stress development is the increased rigidity of the rock foundation under the upstream face at rather economical cross-section of the dam.

This condition is met in the construction of all sections of the Bratskaya and Ust-Ilymskaya dams; however, field observations reveal opening of contact under the power house sections. At that much lesser opening is detected below the commissioning sections which take the water load when their construction is nearly completed (Fig. 1). The situation is worse for the sections which are put in service prior to placing concrete in the power house pit. These sections exhibit maximum opening of the contact joint which is twice as much as the design value.

The analysis confirms, that the poor performance of the power house sections compared to overflow and non-overflow ones is conditioned by such factors as weakening of the dam cross-section due to penstock laying and freezing of concrete around the pipes in cold season.

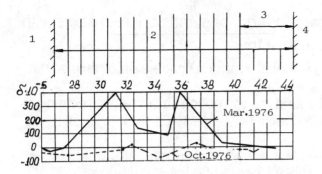

Figure 1. Opening of the concrete-rock contact beneath the upstream toe of the power house section of the Ust-Ilymskaya dam:
1 – dividing abutment; 2 – power house sections;
3 – commissioning sections; 4 – erection site.

Consider the influence of one more factor, that is the rock unconsolidation in the vicinity of the downstream wedge. The fact of this influence is confirmed by the field observations conducted at the Bratskaya and Ust-Ilymskaya (Fig. 2) hydro power plants. To study the effects of rock foundation weak-

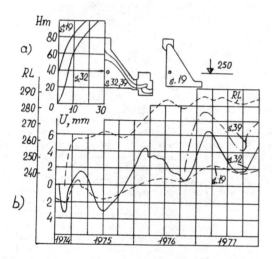

Figure 2. Horizontal displacements of power house and overflow sections of the Ust-Ilymskaya dam:
a) flexure of sections along the upstream face;
b) curves of seasonal displacement of points along the upstream face of overflow and power house sections.

ening in the overlaying zone of the power house pit and below the downstream toe the stress–strain state of the 32nd (power house) section of the Ust-Ilymskaya dam ∂ .. was estimated by the finite-element method (Solovieva 1979) . A number of alternatives of the unconsolidated rock in the dam foundation beneath the downstream face were considered at four different values of the deformation modulus $E: \frac{E}{E_O} = 0.1 + 1.0$ ($E_O = 2.25 \cdot 10^4$ MPa is the deformation modulus of undisturbed rock) . A plane problem of the elasticity theory is solved. The deformation modulus in the penstock zone is assumed to be equal to the product of the initial modulus by the ratio of the concrete layer thickness to that of the entire section B.

As illustrated in Fig. 3 the finite element mesh has in all 968 elements and 535 nodes. The stresses in the dam and foundation caused by the major static loads and seasonal variations of temperature are studied.

Fig. 4 shows the curves of the normal contact

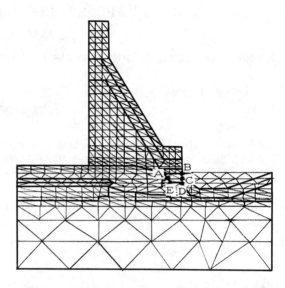

Figure 3. Finite-element mesh for stress-strain calculation of the 32nd section of the Ust-Ilymskaya dam.

stresses σ_y along the dam toe near the upstream face caused by static loads. The curves are given for the following alternative of the weakened zone location (Fig. 3) : the zone of low deformation modulus spreads as a belt of 7.5 m thickness along the the power house pit profile and joins some ABCDEF outlined area immediately below the dam body. The analysis of the curves obtained reveals that the foundation rock weakening results in higher

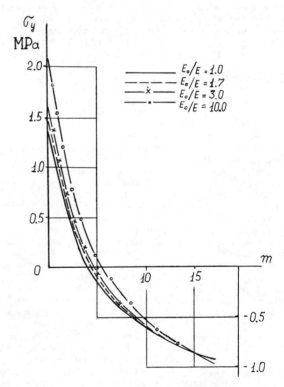

Figure 4. Curves of the contact stresses σ_y along the upstream dam toe caused by static loads.

depth of the tension zone near the upstream toe and in higher tensile stress values. Fig. 5 presents the winter curves of normal stresses caused by combined action of static loads and seasonal variations of

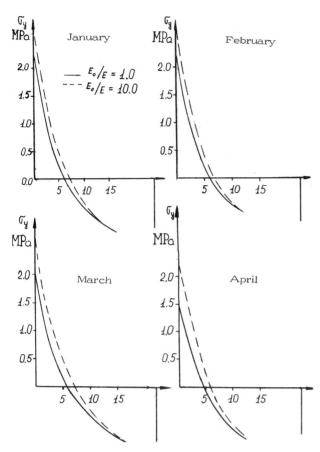

Figure 5. Curves of the contact stresses σ_y along the upstream dam toe caused by static loads and seasonal variations of temperature.

temperature.

Results of the calculations performed can be considered as the upper bound of unfavourable effects of the rock unconsolidation zone beneath the downstream toe on the dam behaviour in operation period.

With the aim to overcome the above drawback of the power house arrangement some measures shall be taken for increasing foundation rigidity under the downstream wedge.

With this aim in view it is necessary to increase the distance between the downstream toe and the turbogenerator axis, to avoid notable difference in elevations of the dam and power house foundations, to provide grouting and reduce the interval between rock excavation and concrete placement in the power house pits.

REFERENCES

Eidelman, S.Ya., V.N.Durcheva 1981. The Ust-Ilymskaya concrete dam. Moscow: Energia.

Marchuk, A.N. 1983. Static behaviour of concrete dams. Moscow: Energoatomizdat.

Marchuk, A.N., A.A.Khrapkov, Ya.N.Zukerman, M.A.Marchuk 1985. Characteristics of the static performance of concrete pressure bearing structures at hydroelectric stations adjoining the dam. Gidrotekhnicheskoe stroitelstvo. 12: 12-16.

Rubtsov, V.K. 1965. Study of massive rock disruption after blasting by filling bore holes with water. Gidrotekhnicheskoe stroitelstvo. 2:33-34.

Solovieva, Z.I. 1979. Deformation of the Bratsk dam rock foundation. Gidrotekhnicheskoe stroitelstvo. 8:43-46.

Application of rock reinforcement and artificial support in surface mines
Utilisation du renforcement de roches et du soutènement artificiel dans les mines à ciel ouvert
Die Anwendung von Felsverankerung und künstlicher Abstützung im Tagebau

DENNIS C.MARTIN, President Piteau Associates Engineering Ltd, Geotechnical and Hydrogeological Consultants, West Vancouver, British Columbia, Canada

ABSTRACT   Rock reinforcement and artificial support using rock anchors, rockbolts or dowels are becoming increasingly attractive as a viable technology for development of open pit mines. Both localized support and systematic pattern support can be used for significant reductions in mining costs through improved safety and increased slope angles. This paper discusses some of the main considerations in installing artificial support in surface mines and presents selected case examples of recent experience with artificial support.

RESUME   L'armature par le roc et l'appui artificiel à l'aide d'ancres de rocs, de boulons de rocs ou de goujons deviennent de plus en plus attrayants comme technologie viable pour le développement des mines à ciel ouvert. On peut réduire les coûts miniers de façon significative en se servant d'appui local et d'appui type systématique parceque la sécurité est améliorée et l'inclinaison des angles est augmentée. Cet exposé décrit quelques-unes des considérations principales lorsqu'on installe un appui artificiel à la surface des mines et donne des exemples de cas choisis d'expériences récentes avec les appuis artificiels.

ZUSAMMENFASSUNG   Felsverstarkung and künstliche Abstützung unter Benutzung von Felsverankerungen, Felsbolzen oder Dübeln werden als machbare Technologie im Tagebau in steigendem Masze attraktiv. Sowohl die örtlich begrenzte, als auch die systematisch gegliederte Abstützung kann zur beachtlichen Kostensenkung durch verbesserte Sicherheit und steilere Abhangsneigung benutzt werden. Dieser Beitrag setzt sich mit den Hauptüberlegungen bei der Installierung künstlicher Abstützung im Tagebau auseinander, und zeigt neuere Erfahrungen mit künstlicher Abstützung anhand von ausgewählten Beispielsfällen auf.

INTRODUCTION

Rock reinforcement refers to the various remedial measures used to mobilize and conserve the inherent strength of a rock mass so that it becomes self supporting (Brady and Brown, 1985). Artificial support refers to those cases where the rock mass is truly supported by structural elements which carry the weight of individual rock blocks. Although a range of innovative and effective methods of rock reinforcement and artificial support have been widely used in civil engineering and underground mining applications for some time, rock reinforcement and artificial support systems have only recently been evaluated and applied in open pit mines.

Rock anchors and cable anchors have been extensively used in civil engineering for such projects as tieback retaining walls, stabilization of dam abutments, slope stabilization, etc. Until recently, most stabilization systems were cumbersome, expensive and difficult to install.

Recent technological advances and experience in underground mining have shown that pre-reinforcement using cable dowels, rock anchors and shotcrete, is economically feasible to improve stability and enable safer and more efficient mining of an orebody.

Recent experience in open pit mines has demonstrated the feasibility of using support and reinforcement systems in surface mining projects.

DEVELOPMENT HISTORY

A summary of a number of documented applications of artificial support in open pit mines is given in Table I and briefly summarized in the following:

Early Experience

Research into the use of rock anchors in open pit mines was first reported in the early 1970's. Trial installations of cable anchors and welded wire mesh were sponsored by CANMET at the Hilton Mine in Quebec in 1969 (Barron et al, 1971). In 1972, an approximately 60m long by 60m high section of slope at the Twin Buttes Mine in Arizona was stabilized using 40 cable anchors (Seegmiller, 1976). The performance of this small trial section indicated that pattern support could be instrumental in stabilizing pit slopes.

An extensive rock anchoring program using 360 cable anchors at Nacimiento Mine in New Mexico was undertaken in 1974 and 1975 to stabilize the east wall of the open pit. Seegmiller (1982) documented several problems that were encountered during installation and reported that subsequent failure of a large number of anchors, which had not been fully grouted after tensioning, resulted in instability of about one third of the slope.

In 1975, a large program of remedial support was undertaken at Inco's Pipe Mine in Northern Manitoba. This project was originally envisaged to consist of 40 cable anchors; however, difficult access, scheduling and other problems only enabled 22 anchors to be installed (Sage, 1977). The costs of this project were high.

The early support systems were based on experience and technology gained from civil engineering practice. Materials were costly and installation was time consuming and interfered with mining production. Most of these projects were undertaken for research or were installed on an emergency basis with limited long range planning and execution. The limited success and high cost of some of these projects did not generate enthusiasm for artificial means of rock reinforcement in the mining community.

Recent Advances

In the mid to late 1970's, increased costs of waste stripping and deeper development of some mines made the installation of artificial support more attractive for open pits. Innovative use of artificial

437

TABLE I

SUMMARY OF SIGNIFICANT APPLICATIONS OF ROCK ANCHORING IN OPEN PITS

DATE	MINE/LOCATION	REASON FOR SUPPORT	EXTENT OF SUPPORT	DESCRIPTION OF SUPPORT	EFFECTIVENESS	REFERENCE
1969	HILTON, QUEBEC	Research project	Local	Post-tensioned cable anchors, reinforced concrete stringers and welded wire mesh		Barron et al, 1971
1972	TWIN BUTTES, ARIZONA	Research	Pattern	40 tensioned cable anchors in 60mx60m section of slope	Successful but expensive	Seegmiller, 1976
1973	HILTON, QUEBEC	Potential wedge failure on ramp	Local	Two rows vertical grouted dowels installed using blasthole rig and old drill steel	Highly successful	
1974	NACIMIENTO, NEW MEXICO	Possible failure of overall slope in plane shear	Pattern	360 tensioned cable anchors 20m to 80m long installed	One-third of slope became unstable as a result of stess corrosion and failure of anchors	Seegmiller, 1976
1974/1975	PIPE, MANITOBA	Rockfall and stability problem below haulroad	Pattern	22 cable anchors installed on high, inaccessible slope	Extremely costly and not possible to complete the work in specified time	Sage, 1977
1977/1978	PIPE, MANITOBA	Stabilize benches and also enable steep slopes in lower benches at end of mine life	Local/ Pattern	Vertical grouted dowels installed to stabilize benches and haulraod	Moderately successful	Janeson, 1979
1978	MARY KATHLEEN, AUSTRALIA	Stabilize bench crests for rockfall control and steeper slope	Pattern	Predowelling using inclined cable dowels for 240m high slope	Highly successful	Rosengren, 1986
1979	PALABORA, SOUTH AFRICA	Stabilize haulroad in a zone of weak rock	Local	Vertical grouted dowels installed using blasthole drill and scrap rail	Enabled completion of mining phase	
1980	CARDINAL RIVER, ALBERTA	Toppling of beds at base of footwall slope	Local/ Pattern	10m long grouted rebar dowels installed at -20°	Enabled extraction of coal at pit base	Piteau et al, 1982
1981/1982	ISLAND COPPER, BRITISH COLUMBIA	Large wedge/plane failure in weak altered rock	Pattern	Deep cable anchors and threadbar anchors installed	Enabled mining to continue below major fault zone	
1982	MT. CARBINE, AUSTRALIA	Instability of residual soil and weathered rock near pit crest	Pattern	Cable dowels, constructed from old hoist cable, installed as the slope was mined	Stable slope was developed in residual soils	Brachmanski, 1984
1984/1986	PALABORA, SOUTH AFRICA	Control ravelling and stabilize fault zones	Local	Grouted rails, grouted shovel cable dowels, shotcrete and mesh	Effective where zone to be supported is of limited extent	Martin et al, 1986
1985/1986	PALABORA, SOUTH AFRICA	Stabilize bench crests in hard rock to control rockfalls	Local/ Pattern	Inclined grouted cable dowels installed prior to blasting and after excavation	Effective where good bond established between dowels and rock	Martin et al, 1986
1984/1985	GREGG RIVER, ALBERTA	Large slab on a footwall slope due to unfavourable fault offset	Local	Threadbar anchors installed along edge of slab	Some anchors failed due to high stresses, but mining was successfully completed	Karst, 1985
1984/1985	LINE CREEK, BRITISH COLUMBIA	Potential instability in roll on unbenched footwall slope	Local	Fully grouted dowels installed to secure roll area	Effective when dowels installed before mining	Hannah, 1986
1984 to 1986	DOYON, QUEBEC	Toppling failure on hanging wall of ore zone	Pattern	10m long Swellex rock bolts installed as mining proceeded	In progress; very successful	Lachance, 1986
1984/1986	MT. WHALEBACK, AUSTRALIA	Stabilize lobes of ore in footwall and enable steeper footwall	Local/ Pattern	Untensioned cable bolts	In progress	Rosengren, 1986
1985/1986	WOODCUTTERS, AUSTRALIA	Support berms and overall slope in weak rock	Pattern	Untensioned cable bolts and local grouted drill rods	In progress	Rosengren, 1986
1985/1986	SMOKY RIVER, ALBERTA	Potential slab failure on 65° footwall slope	Pattern	6m to 10m long post tensioned grouted threadbar rock anchors installed on 3.1m x 4.6m pattern for 500m long by 100m high slope	In progress; considerable cost savings over waste stripping	Martin et al, 1985

438

Figure 1 Installation of inclined cable dowels using an airtrack drill at Palabora Mine. Vertical rails were installed using a blasthole drill.

Figure 2 Predowelling using used 100mm diameter shovel cable in conjunction with shotcrete and mesh to stabilize a fault zone at Palabora mine.

support in underground mining and the development of cheaper, and simpler support systems which could be installed more rapidly and efficiently within the mining cycle contributed to the renewed interest in artificial support. Use of scrap rail, old shovel cable, old hoist cable, etc. became popular because of the low cost and ease of installation of these materials. In addition, the newer generation of rock anchors, such as continuous cold rolled threadbar, split sets and swellex bolts, made installation faster and simpler. Today, there is extensive experience with a variety of support systems, all of which are simple and easy to install.

CRITERIA FOR SUPPORT

Artificial support may be installed using two basic criteria.
   1. Support is installed in local areas in response to actual instability to ensure the integrity of the rock mass and enable the design slope angles to be achieved on a local basis.
   2. Possible rock reinforcement is assessed and pattern support designed based on cost benefit analyses. In these cases, slope designs are scientifically evaluated and slopes steepened so that installation of support results in a defined cost saving in the mining plan.
   The importance of these criteria and their impact on use of artificial support are discussed below:

LOCALIZED SUPPORT

Support of localized or "problem" areas in open pits has been used in many mines in response to immediate requirements such as safety, possible loss of access or equipment damage. Such systems are often installed rapidly with little planning. However, it is essential that the basic engineering geology and rock mechanics controls on stability are identified to enable design of adequate support.
   The requirements and design for localized support systems often cannot be predicted until the slopes are exposed. Hence, materials and equipment must be available and versatile enough that a range of installations is possible. If a small airtrack drill (Figure 1) or blasthole drill can be made available, use of inexpensive or discarded materials such as old rail, shovel cable or hoist rope can be very effective for localized support.

Hilton Mine

In the mid 1970's Hilton Mine in Quebec installed dowels to stabilize a potential wedge failure which

threatened to sever the ramp. This work consisted of drilling two rows of vertical holes along the edge of the ramp using an available blasthole drill. Old drill steel was placed in the holes and grouted. Although a part of the wedge subsequently failed, the dowels limited the extent of the failure and the ramp was not lost. This simple technique was inexpensive and effective for maintaining access to the open pit.

Palabora Mine

At Palabora the rock is relatively competent except in the vicinity of major fault zones. Procedures have been adopted to stabilize the pit benches in the vicinity of fault zones using a combination of shotcrete, wire mesh, steel rail and discarded shovel cable. The steel rail and shovel cable are grouted into drillholes up to 15m deep at a variety of angles (Figure 2). Initial installation in 1979/1980, using the in-pit blasthole drill, proved successful. Subsequently, an airtrack drill capable of drilling holes at any angle was purchased for installation of inclined dowels and groundwater depressurization holes (Figure 1). Mine personnel have concluded that the use of localized support in site specific areas is essential to maintain steep interramp slopes in the open pit (Martin et al, 1986).
   At the present time, support systems are selected and designed as specific problem areas arise. The main applications for support are:
   1. Stabilization of bench crests above the haulroad for rockfall protection, using cable dowels.
   2. Stabilization of fault zones, using grouted cable dowels, shotcrete and mesh.
   3. Stabilization of the edge of the haulroad, using grouted rails.
   4. Use of draped mesh to protect the haulroad, transformers or other installations from rockfalls.
   Figure 3 summarizes the engineering geology, rock mechanics and stability analyses approach used to design a system of cable dowels to support selected bench crests at Palabora. Even though a localized support system is being used, detailed investigation and design were required. Results of geological mapping (Figure 3a) were used to define the main failure mechanisms of bench crests (Figure 3b). This information was used to prepare a failure model and conduct the necessary stability analyses (Figure 3c). Design of the cable dowel system is shown in (Figure 3d) and completed installations are shown in (Figure 3e).

Cardinal River Coal

Another application of localized support is the use of grouted rebar dowels or threadbar rock anchors.

a) Lower Hemisphere Equal Area Projection of All Joints and Faults in Area Requiring Support

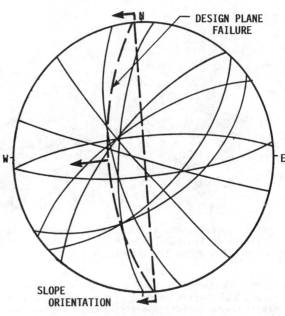

b) Lower Hemisphere Equal Area Projection of Planes Representing Peak Discontinuity Set Orientations

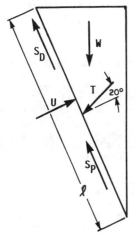

W = Weight Force

T = Tensile Force Due To Dowels

S_D = Shear Force Due To Dowels

U = Uplift Force Due To Groundwater

S_P = Shear Force Mobilized Along Failure Plane

c, ϕ = Cohesion, Friction Mobilized Along Failure Plane

FREEBODY DIAGRAM OF POTENTIAL FAILURE BLOCK

At Limiting Equilibrium:

$$S_P = W \cos65° \tan\phi + c\ell$$

Disturbing Forces, D

$$D = W \sin65°$$

Resisting Forces, R

$$R = (W \cos65° + T \cos20° - U) \tan\phi + c\ell + S_D$$

Factor of Safety, $F = \dfrac{R}{D}$

LIMITING EQUILIBRIUM ANALYSIS METHOD

c) Failure Model and Analysis Method for Design of Dowel Support System

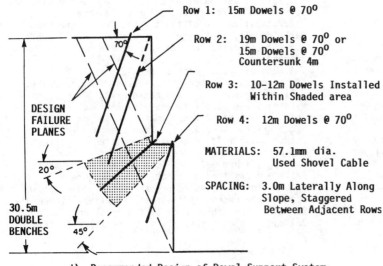

Row 1: 15m Dowels @ 70°

Row 2: 19m Dowels @ 70° or 15m Dowels @ 70° Countersunk 4m

Row 3: 10-12m Dowels Installed Within Shaded area

Row 4: 12m Dowels @ 70°

MATERIALS: 57.1mm dia. Used Shovel Cable

SPACING: 3.0m Laterally Along Slope, Staggered Between Adjacent Rows

d) Recommended Design of Dowel Support System

e) Completed Dowel Installations on 30.5m High Double Benches

Figure 3   Analysis, design and installation of cable dowels to stabilize berms and provide better rockfall control at Palabora Mine, South Africa

440

SLOPE HEIGHT: 30m
SLOPE ANGLE: 89°
DIP OF BEDDING JOINTS: 80°
SPACING OF BEDDING JOINTS: 0.6m

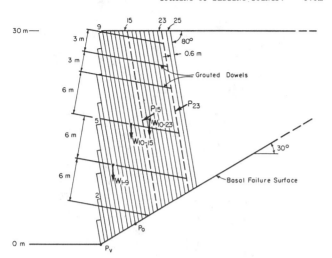

a) Stability analysis model and recommended rock reinforcement.

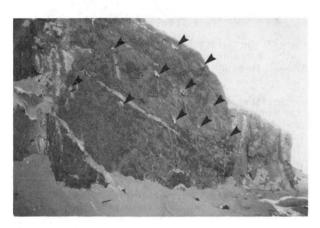

b) Completed dowel installations.

Figure 4 Use of fully grouted rebar dowels to control toppling in the lower section of a footwall slope at Cardinal River Coal Limited, Alberta (after Piteau et al, 1982).

Figure 4 shows the analysis results for the use of grouted rebar dowels which were installed to control toppling failure where the bedding was overturned near the base of a high footwall slope at Cardinal River Coal Mine in Alberta.

PREPLANNED PATTERN SUPPORT

The author is aware of several projects where the mine planning and geotechnical studies indicated that it was feasible to use artificial support to steepen overall slopes and reduce waste stripping costs. These projects required detailed geotechnical investigations to determine the possible failure mechanism and support requirements. These investigations were followed by detailed mine planning and economic studies to evaluate the cost benefit of artificial support.

Mary Kathleen Mine

At Mary Kathleen Uranium Mine in Australia, a program of predowelling, using inclined cable dowels manufactured from discarded hoist rope, was implemented to control wedge failures on benches. By controlling the wedge failures, higher benches and narrower catch

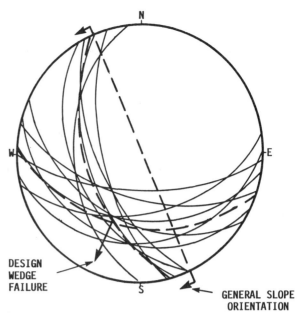

a) Lower hemisphere equal area projection of joint sets showing wedge failures which required cable dowelling.

b) View of bench crest after cable dowelling.

Figure 5 Installation of cable dowels at Mary Kathleen Uranium Mine, Queensland.

berms were feasible, and the slope could in effect be steepened to an overall slope angle in excess of 50° (Rosengren, 1986). Figure 5 shows the geotechnical aspects, and details of the cable dowelling program for the 240m high slope at Mary Kathleen Mine.

Smoky River Coal

One of the most extensive rock reinforcement projects to date is currently in progress at Smoky River Coal Limited in northern Alberta. The footwall of the Upper East Limb Pit at No. 9 mine consists of interbedded sandstones, siltstones, shale and coal, which have an average bedding dip of about 60° to 65°.
Detailed geotechnical assessments indicated that the main mechanism of possible instability was by simple slab failure involving planar sliding along bedding planes (Martin et al, 1985). This failure mode was particularly important because a number of reverse faults, which strike subparallel to and dip opposite to bedding, were mapped in the footwall rocks. These faults result in daylighting of 1m to 2m thick slabs in the footwall. The analyses indicated that if rock anchors could be installed, the slab would be stabilized, and an unbenched slope would be feasible.

In order to conduct a rational design of the support system, a cost benefit analysis was conducted by mine planning personnel. In terms of assessing the optimum slope geometry, it was considered most practical to optimize the cost of excavation and the cost of anchoring.

Coal at the base of the pit has a much higher overall stripping ratio than coal at the crest. Hence, if the lower sections of the slope were steepened, a significantly reduced amount of waste stripping would be required. A combination of benched and unbenched slopes with remedial measures provided the most economic benefit.

A typical example which compares a slope excavated entirely by benching with slopes which are developed by a combination of benching and anchoring is shown in Figure 6.

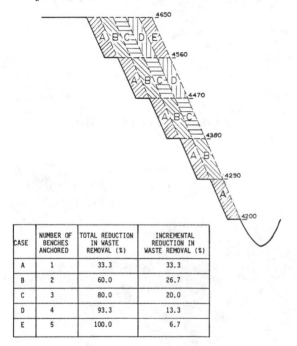

CASE	NUMBER OF BENCHES ANCHORED	TOTAL REDUCTION IN WASTE REMOVAL (%)	INCREMENTAL REDUCTION IN WASTE REMOVAL (%)
A	1	33.3	33.3
B	2	60.0	26.7
C	3	80.0	20.0
D	4	93.3	13.3
E	5	100.0	6.7

Figure 6 Optimization of rock anchoring program at Smoky River Coal based on the incremental cost of anchoring.

The waste material to be excavated is divided into a number of blocks of equal area defined by the berm width and bench height. If rock anchors are used to stabilize the lowest bench (A) a total of five blocks or 33% of the waste excavation is eliminated. For each additional bench where rock anchors are used, the volume of waste excavation which is eliminated is reduced by a successively smaller amount.

It is clear that there is an optimum elevation to which rock anchors should be placed. This optimum elevation depends on the cost of anchor installation, the number of anchors required and the cost of waste removal. At Smoky River, the lower 100m was excavated without benches using rock anchors and the upper 60m was excavated by benching.

A total of about 2000 threadbar bar anchors are being installed on an approximate pattern of 3.1mx4.6m (Figure 7). Cost savings as a result of the anchoring program are expected to be approximately one half to two thirds of the cost of waste removal.

CONCLUSIONS

Installation of artificial support on a local basis for specific problem areas or in a systematic pattern for general rock reinforcement are viable technology for open pit mines. Evaluation of the feasibility of rock reinforcement or artificial support requires not

Figure 7 Installation of fully tensioned resin grouted threadbar anchors on a 500m long by 100m high footwall slope at Smoky River Coal Limited, Alberta.

only detailed geotechnical and rock mechanics assessments to determine support requirements, but also a detailed economic assessment to determine the costs of the support or reinforcement system compared with alternative solutions such as slope flattening, buttressing or crest unloading. Optimization of the support requirements, mine planning requirements and costs often results in a combination of waste stripping and artificial support.

A well organized and well executed program of installation is mandatory to achieve the cost savings indicated from the design study. Detailed monitoring and documentation is required to evaluate the success of the program for subsequent application to other mining situations.

As documented case histories become available concerning the succcessful application of artificial support in open pits, the use of these measures will become more widespread and will be found to be extremely beneficial. Development of technology specifically suited to open pit mining situations will increase the economic advantages of using artificial support.

REFERENCES

Brachmanski, T., 1984. Personal Communication, Mt. Carbine Mine Cairns, North Queensland.

Barron, K., Coates, D.F. and Gyenge, M., 1971. Support for Pit Slopes. CIM Bulletin, March, 1971, pp. 113-120.

Brady, B.H.G. and Brown, E.T., 1985. Rock Mechanics for Underground Mining. George Allen & Unwin, London, 527 p.

Hannah, T., 1986. Personal Communication. Crows Nest Resources Ltd., Sparwood, B.C.

Janeson, J., 1979. Wall Control Techniques at Inco's Pipe Open Pit. CIM Bulletin Volume 72, Number 805, May, pp. 76 to 80.

Karst, R., 1985. Personal Communication, Chief Geologist, Gregg River Resources Ltd., Hinton, Alberta.

Lachance, M., 1986. Personal Communication. Chief Engineer, Doyon Mine, Quebec.

Martin, D.C., Sheehan, P. and Fawcett, D.A., 1985.
Geotechnical Assessments and Design of Optimum
Method of Excavation of a Footwall Slope at Smoky
River Coal Ltd., CIM District, 5 Meeting, Hinton,
Alberta, September, 9 p.

Martin, D.C., Steenkamp, N.S.L. and Lill, J.W., 1986.
Application of a Statistical Analysis Technique to
Design of High Rock Slopes at Palabora Mine, South
Africa. Mining Latin America, IMM Conference,
Santiago, Chile, November, pp. 241-255.

Piteau, D.R., Stewart, A.F. and Martin, D.C., 1982.
Design Example of Open Pits Susceptible to
Toppling. Chapter 29 in Third International
Conference on Stability in Surface Mining, Brawner,
C.O. ed AIME Vancouver, June 1981, pp., 679-712.

Rosengren, K.J., 1986. Wall Reinforcement in Open
Pit Mining. Proceedings of Large Open Pit Mining
Conference, Mount Newman Australia, October 9 p.

Sage, R., 1977. Mechanical Support. Chapter 6 in
Pit Slope Manual CANMET Report No. 77-3, 111 p.

Seegmiller, B.L., 1976. A Case History of Support at
Nacimiento Mine. CANMET Report 76-27, December.

Seegmiller, B.L., 1982. Artificial Support of Rock
Slopes. Chapter 10 in Third International
Conference on Stability in Surface Mining, Brawner,
C.O. ed. AIME Vancouver, June, 1981, pp. 249-288.

Block nonlinear stability analysis of jointed rock masses
Analyse non linéaire de la stabilité des massifs rocheux jointés
Nichtlineare Block-Stabilitätsanalyse für geklüftetes Gestein

A.C.MATOS, Faculty of Engineering, Oporto, Portugal
J.B.MARTINS, University of Minho, Braga, Portugal

ABSTRACT: A nonlinear model ffor the calculation of the deformation and of the safety factor for jointed rock masses subjected to selfweight and external foces (hydraulic, nayling, prestressing, etc.) is presented. The constitutive laws for the joints allows for soil or weathered rock filling and also for surface of contact reduction due to opening of the joints during the loading process.

RESUMÉ: Un modèle non linéaire pour le calcul des déformations et du coefficient de sécurité des masses rocheu-ses avec des joints soumises a des charges extérieures (hydrauliques, de précontrainte, etc) est présenté. Les lois constitutives pour les joints considèrent la ramplissage avec du sol et aussi la réduction de l'aire de contact à cause de l'ouverture des joints pendant le procès de chargément.

ZUSAMMENFASSUNG: Ein nichtlinearer Ansatz für die Berechnung der Deformation und des Sicherheitsfaktors für geklüftetes Gestein wird beschrieben. Berücksichtigt werden Schwere, Zusatzlasten, gefüllte und verwitterte Klüfte, und die Kontaktveränderung während der Verschiebung von Küften.

1 INTRODUCTION

The deformation and stability of rock masses depends mostly on joints behaviour and little on the pro-perties of the sound rock between joints.
Therefore, the analysis of the stability of jointed rock masses can be done considering the blocks between joints as rigid.
The very large difference between the stiffness of the joints and that of the sound rock of the blocks creates numerical difficulties in the use of the common finite elements in the phase of collapse. Also, the several mechanisms of collapse cannot be well represented. The main mechanisms of collapse are a combination of the following three: sliding between joints, rotation and tension(or compression). In the sliding, a yield relationship of the "bi--linear" type is considered together with a non-linear shear-distortion function, with a "peak" and a "residual" shear stress. Therefore, the model considers a "peak" cohesion c, a "peak" friction angle $\emptyset$ and a "peak" distortion . It considers also a residual cohesion c_R, a residual friction angle $\emptyset_R$ and a residual distortion .
The model contemplates further the reduction of shear strength due to the rotation between blocks with the reduction of contact area, and, eventwally the collapse by overturning of the blocks.

2 CONSTITUTIVE LAWS FOR ROCK JOINTS

The modelling of rock joints is not an easy task. Even in the case wether the joints are filled with soil, their deformation behaviour is very different from that of large masses of the same soil. This is due to the fact that the thicknesses of the joints are usually small when compared with the dimensions of their surfaces.
In the case of a joint without soil inside, the behaviour is essentially controlled by the friction angle of the two walls of the joint, $\emptyset_u$, and the "imbrication" of their asperities, i.e., the dimensions a of the asperities and the average inclination i of the slopes of the asperities. Imbrication implies dilation of the joint during the deformation process.
On the other hand, the inclination i is a decreasing function of the average normal stress (Fig. 2.1).We may assume:

$$i = i_o (1 - \sigma / \sigma_T)^k , \qquad 2.0$$

where k is a parameter to be obtained experimentally

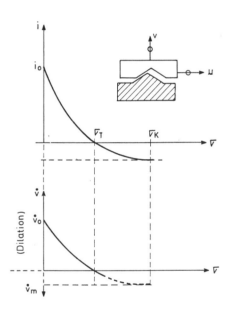

Fig. 2.1 Imbrication behaviour

The overall friction angle for the joint, $\emptyset$, is such that

$$\tan \emptyset = \tan (\emptyset_u + i)$$

Therefore, imbrication and the corresponding dila-tancy are favourable to stability. However, at rupture i is very small and, therefore, $\emptyset = \emptyset_u$. In (2.0) is the average normal stress at no volume change.

2.1 Characterization of the rock joint of the model

A number of types of rock joints can be defined taking into account the thickness of the joint, the sizes of the asperities, the existence or not of filling, the nature of the filling material, etc.

445

(C.Matos, 1985/6).
In what concerns the model we propose here, we
assume that in the stress-strain relationship, the
joint is characterized by:
i) the distortion   at peak stress (γ_P) (Fig.2.1.1)
ii) a linear relationship between   and average
normal stress .

$$\gamma_P = m\,\sigma\ ; \qquad\qquad (2.1)$$

iii) the distortion γ_R at the beginning of the step
of residual stress (τ_R) (or by the difference

$$\gamma_S = \gamma_R - \gamma_P \quad \text{(Fig. 2.1.1)}$$

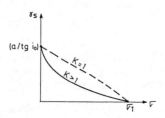

Fig.2.1.1 Behaviour:fragile $\gamma_S = 0$; ductile $\gamma_S > 0$

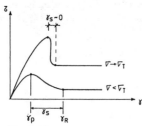

Fig.2.1.2 γ_S Variation

Experimental results seem to suggest that γ_S is a
decreasing function of the average normal stress
(Fig. 2.1.1)

$$\gamma_S = (a/\tan i_o)\ (1 - \sigma/\sigma_T)^k \qquad (2.1.1)$$

In regard to the shear strength a bilinear
relationship τ, σ is assumed (Fig. 2.1.3) and
defined by the initial friction angle   and initial
cohesion c and the residual friction angle $\emptyset_R$ and
residual cohesion c_R.

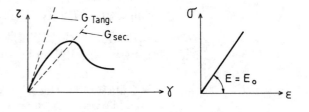

Fig. 2.2.1 Behaviour in tension – compression

2.2 Constitutive law for compression – tension ($N = f_3 - (\varepsilon)$)

The relationship of Fig.2.2.1 between the normal
force on the joint and the normal displacement   d
has been suggested by Goodman (1972, 1977).
However, in our case we assume the linear relationship of Fig. 2.1.4, where $\varepsilon = d/H$, d is
the closing displacement and H the thickness of the joint.
The value of E_o must be chosen according to the
type of joint. For open joints without filling
($d < d_v$, max), a value of E_o near zero should be
taken. For a closed joint without filling ($d > d_{vmax}$)
   we should chose a value near the Young's
modulus of the rock.
For a joint filled with soil the value of E should
approach the average value of E for the filling
material.
Meanwhile it must be said that an "exact" value
for E_o is not needed since it operates only as a
starting parameter to obtain the stifness matrix,
in the non linear iterative calculations.

2.3 Constitutive law for shear ($\tau = f_1(\sigma, \gamma)$)

To obtain the relationship $\tau = f_1(\sigma, \gamma)$ matching
the experimental curves of Fig. 2.1.1 we have
resorted to one "analog elastic bar".(C.Matos,1986)

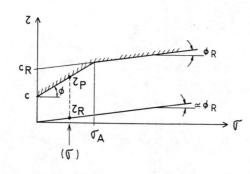

Fig.2.1.3 Limit behaviour under residual and peak conditions.

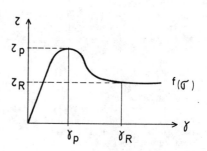

Other parameters the required to operate the model
are the thickness of the joint H, and the "elastic"
parameters E_o and G_o (Fig. 2.1.4).

It can be shown (Matos,1986) that this analog gives
the following expression for $\tau = f_1(\sigma, \gamma)$:

$$\tau = \frac{\sigma_e \, tg\phi + c}{(1 - \frac{k}{\sqrt{1+k^2}})} \left(1 - \frac{R}{m\sigma_e \sqrt{1+k^2}}\right) \quad ;$$

$$R = \sqrt{H^2 + (m\sigma_e - \gamma)^2}$$
$$0 < \gamma \leq \gamma_p \qquad\qquad\qquad (2.3.1)$$

In (2.3.1) $\sigma_e = N/S_e$,2.3.2), where N is the normal force at the joint and S_e the actual area of contact $m = \gamma_p/\sigma_e$ is obtained from Fig.2.1.1 .
The expression (2.3.1) is valid for $\gamma < \gamma_p$ only. A more general expression can be set up (C.Matos, 1986).

2.4 Constitutive law for relative rotation:
$$M = f_2(N,\theta)$$

The distribution of normal stress in the joints due to increasing external loads is not, in general, uniform. This fact gives rise to moments M and relative rotations θ (Fig.2.4.1).
In our model we consider a linear distribution of normal stress.
However, when in some part of the joint tensions exceed the unconfined strength of the filling material q_u , there will be a reduction of contact area. Therefore two cases are considered:

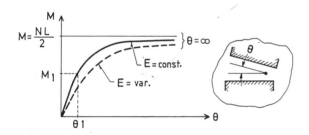

Fig.2.4.1 Behaviour $M = f_2(\theta)$ for cohesion c=o

a) $M \leq M_1$ (There is compressions, only).
In this case the relationship between M and θ is:
$$M = (EL^3/12)(\theta/H), \qquad\qquad (2.3.2)$$
where H is the thickness of the joint and L its length.
The limit value M_1 of M is:
$$M_1 = (N/L - q_u)(L^2/6) \qquad\qquad (2.3.3)$$
and the corresponding relative rotation is
$$\theta_1 = M_1/EL^3/12H \qquad\qquad (2.3.4)$$
b) $M > M_1$ (partial contact area)
If $q_u = 0$, after some algebra we have the following result:
$$M = 1/2 \, (LN - \sqrt{8N^3H/9E\theta}) \qquad (2.3.5)$$

If $q_u > 0$, there will be local plastification in tension.
By a development similar to that leading to (2.3.5) we obtain
$$M = 1/2 \, [L(N+q_u.L) - [8H(N+q_u)^3 / (9E\theta)]^{1/2}]$$
$$\qquad\qquad\qquad\qquad (2.3.6)$$
As it can be seen from (2.3.6), the overturning moment ($\theta = \infty$) is
$$M = LN/2 + q_u L^2/2 \qquad\qquad (2.3.7)$$

3. THE SET UP OF THE MODEL
The rock mass is supposed to be composed of rigid blocks with deformable joints between them. The model considers as external displacements the horizontal and vertical displacements, $d_{H,G}$ and $d_{V,G}$ and the rotation of each block about its mass center. The deformations are defined at the joints and are the normal extension ε , at the centers, the distortion γ and the relative rotation θ.
The corresponding forces are $F_G (M_G, V_G, H_G)$ at the center of the block and $F_i(M_i,N_i,T_i)$ at the center of joint i (Fig. 3.1).

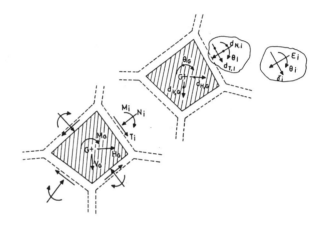

Fig.3.1 Forces and displacements of a block at G and deformations at a joint i

The set of blocks behaves as a structure with rigid nodes of finite dimensions, the joints being the deformable members.
The working stifness matrix for each joint is given by
$$k_j = \begin{bmatrix} E_o \ L^3/(12H) & 0 & 0 \\ 0 & E_o \times L \times 1 & 0 \\ 0 & 0 & G_o \times L \times 1 \end{bmatrix} \quad (3.1.1)$$
where E_o is the normal elasticity modulus, and G_o is the shear modulus. This matrix is such that
$$[M,N,T]_j^T = k_j [\theta, \varepsilon , \gamma]_j^T, \qquad (3.1.2)$$
where
$$\theta_j = d_{M,m} - d_{M,n} ; \ \varepsilon_{Nj} = (d_{N,m} - d_{N,n})/H; \ \text{and}$$
$$\gamma_j = (d_{T,m} - d_{T,n})/H, \qquad\qquad (3.1.3)$$
m and n being the blocks in contact at joint j.

To construct the stiffness matrix of the structure the following strategy will be stated.
First we set the stiffness matrix of a block connected to the exterior by elastic joints in the general axes, starting from the stiffness matrix in the joint (local) axes.
Then we see the incidence of forces on the neighbour blocks.
Let us consider the block m connected to block n by

joint j.

If we take all the blocks fixed except block m the following relationships can be writen.

$$[M,N,T]_j^T = k_j \cdot B_j \cdot \tilde{d} \qquad (3.1.4)$$

where

$$B_j = \begin{bmatrix} 1 & 0 & 0 \\ 0 & 1/H & 0 \\ 0 & 0 & 1/H \end{bmatrix} \qquad (3.1.5)$$

and

$$\tilde{d} = [d_{M,m}, \ d_{N,m}, \ d_{T,m}]^T$$

The local displacement $\tilde{d}_j$ can be transformed to the general sistem of axes and the same can be done to general force $\tilde{F}_j$

we have

$$\tilde{d}_{m,j} = T_{m,j} \cdot \tilde{D}_{m,j} \quad \text{which imply} \quad \tilde{F}_{m,j} = [T]_{m,j}^T \cdot \tilde{Q}_{m,j} \qquad (3.1.6)$$

Putings (3.1.4) into the second of (3.1.6)

$$\tilde{F}_j = [T]^T \cdot k_j \cdot B_j \cdot T_j \cdot \tilde{D}_j \qquad (3.1.7)$$

where

$$T = \begin{bmatrix} 1 & 0 & 0 \\ -y & 1 & 0 \\ x & 0 & 1 \end{bmatrix} \times \begin{bmatrix} 1 & 0 & 0 \\ 0 & \cos\alpha & \operatorname{sen}\alpha \\ 0 & -\operatorname{sen}\alpha & \cos\alpha \end{bmatrix} \qquad (3.1.8)$$

and $x = x_{m,j} - x_{m,G}$ and $y = y_{m,j} - y_{m,G}$ (3.1.9)

(3.1.7) means that the stiffness of block m for joint j, in the general sistem of coordinates, is

$$K_{m,j} = [T_{mj}] \ k_j \cdot B_j \cdot T_{m,j} \qquad (3.2.0)$$

The stiffness of the block will be

$$K_{m,m} = \sum_{j=1}^{N \, J \, m} K_{m,j} \qquad (3.2.1)$$

Forces on block m imply simetric forces on neighbour block n, through joint m . Therefore

$$K_{m,n} = - [T_{n,j}]^T \ k_j \ B_j \ T_{n,j} \quad , \ n \neq m \qquad (3.2.2)$$

In this case there is no place for any summation, since there is only one joint in contact of two blocks.

Element matrices $K_{m,m}$ and $K_{n,m}$ will be spread in the general stiffness matrix in the usual way.

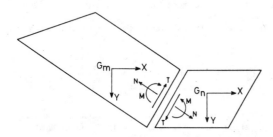

Fig.3.2

The equilibrium equations will be given by

$$K \cdot \tilde{D} = \tilde{F} \qquad (3.2.3)$$

The force vector may include other than the weight of the blocks: the resultant of pore pressures on the joints, percolation forces, anchoring or presstressing forces, seismic forces, and so on.

3.1 Introducing the nonlinearity

We have applied the method of initial stresses [Zienckiewicz,1974] with a constant stiffness matrix K.

To start with, we calculate $\tilde{D}_1$ from (3.2.3) and from the first of (3.1.6) we calculate $\tilde{d}_{m,j}$ and, in a similar way, $\tilde{d}_{n,j}$ and from the difference we obtain Θ_j, $\varepsilon_{N,j}$ and γ and from (3.1.2) the values of

$$[M,N,T]_{j,1}^T$$

Further with N and Θ_j, we obtain M_2 from (2.3.2) or (2.3.7) . Also $T_2 = \tau \times L$ is obtained from $\tau = f_1 (\sigma_e, \gamma)$, $\sigma_e = N/S_e$, S_e the reduced contact area, obtained from M_2 when $M_2 > M_1$ (2.3.5).

The corrections to be made to M,N,T are

$\Delta M = M_2 - M_1'$, $N = N_2 - N_1 = 0$, $\Delta T = T_2 - T_1 \rightarrow$

$\rightarrow \Delta \tilde{Q}_2 = \tilde{Q}_2 - \tilde{Q}_1$ with $\Delta \tilde{Q}_2$ a correction value for for $\tilde{F}$ will be obtained from the second of (3.1.6)

$$(\Delta F)_m = \sum_j \Delta \tilde{F}_{m,j} = (T_{m,j}^T) \ \Delta Q_{m,j} \qquad (3.2.4)$$

To obtain the factor of safety we first run the program with a factor of safety F_s equal to 1 (more or less) in $tg\phi/F_s$ and C/F_s and then iterate with greater (or smaller) values of F_s up to a step of no stability in the calculations, which reflects the actual lost of stability of the rock mass.

3.2 The computer program

The computer program has 9 subroutines and can be run in any microcomputer.

4 EXAMPLE

The program has been applied to the case refered by Gen-Hua Shi and R.E. Goodman(1985) with the followings parameters: $\gamma=23$ kN/m³, $\phi=45°$, $\phi_R=30°$, c= 10kN/m², $c_R = 20$ kN/m², $E_0= 10^8$ kN/m², $G_0=10^5$kN/m² m=10^{-4}, s = 10^{-2} and H = 10^{-2} m.

A safety factor F_s=0.86 has been found has the first failure by overturning of block 5/6/7. It follows F_s 0.95 for the sliding of the group of blocks 2/3 and 1/4 along the joint of the base of block 2/3.It should be noticed that some joints fail with ϕ_R and others do not reach the residual strength situation as happened for the joint at the base of block 1/4.

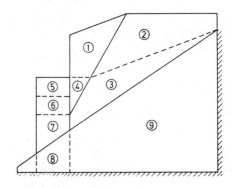

Fig.4.1 Failure of a jointed rock mass (a case study)

REFERENCES

Gen-Hua Shi and R.E. Goodman 1985 "Two Dimensional Discontinuous Deformation Analysis" Int J.Num.Anal. Meths.in Geomech.9:541-556.

Goodman,R.E. 1972 "Duplication of Dilatancy in Anal. of Jointed Rocks", J.of S.M.F.Div.ASCE,V.98,SM4: 400-422.

Goodman, R.E. 1977 "Analysis in Jointed Rocks" in "Finite Element in Geomechanics",G.Gudehus (ed),N.Y., John Wiley.

Matos, A.C. 1986 Ph.D. Thesis (in Portuguese), U. OPorto.

Matos, A.C. & J.B.Martins 1985 "The Stability of Blocks with Soil filling Joints",Proc.XI Int.Conf. Soil Mech.F.E., S.Francisco, 9/A/9:2335-2337.

Zienckiewicz, O.C. 1974 "Viscoplasticity as a Computational Model for Deformation and Collapse Studies" Symposium on Non-linear Techniques, Crowthorne.

Cortes-La Muela hydroelectric project – The foundations of Cortes and El Naranjero Dams

Le projet hydroélectrique de Cortes-La Muela – Les fondations des barrages Cortes et El Naranjero

Das Cortes-La Muela Wasserkraftwerk – Fundamente der Cortes und El Naranjero Dämme

N.NAVALÓN, Hidroeléctrica Española, S.A., Madrid, Spain
J.M.GAZTAÑAGA, Hidroeléctrica Española, S.A., Madrid, Spain
J.M.LÓPEZ MARINAS, Hidroeléctrica Española, S.A., Madrid, Spain

ABSTRACT: The construction of two dams on the river Júcar in Spain has provided the opportunity for studying the foundation rock at both sites. This has included detailed geological mapping, microseismic tests and consolidation grouting. An analysis of the relationships between the geological features and the results of the geophysical testing is performed. A comparison is made with the data collected during the investigation stage.

RESUME: La construction de deux barrages sur le fleuve Júcar en Espagne a permis l'étude de la roche de fondation dans les deux cas. On a réalisé la surveillance géologique, les essais microséismiques et les injections de consolidation. On analyse la relation entre les caractéristiques géologiques et les resultats des essais géophysiques, et on les compare avec les données relevées lors des investigations préliminaires.

ZUSAMMENFASSUNG: Der Bau der beiden Talsperren am Jucar in Spanien hat Gelegenheit geboten, die Gründungsgesteine beider Standorte zu untersuchen. Die Studien umfassten die Erstellung ausführlicher geologischer Landkarten, mikroseismische Versuche und die Verfestigung der Vergussmassen. Es wurden die Beziehungen zwischen den physikalischen Merkmalen, den Ergebnissen der geophysikalischen Versuche und dem Abbindevorgang der Vergussmasse untersucht. Die einzelnen Daten werden mit denen, die während der Untersuchungsphase erhalten wurden, verglichen.

1. INTRODUCTION

Cortes and El Naranjero dams, on the river Júcar (Valencia, Spain), form part of the Cortes-La Muela pumped storage scheme. They are located in middle and upper Cretaceous formations on the edge of a diapiric structure created in favour of tension faults with an average N 70° E direction, through which outcrops the Keuper.

The existence of this diapir means that the layers have a sharp dip at its edge that tends to smooth out further away. In the Cortes site, the stratification dips upstreamwards at some 18°. In the Naranjero site, the dip is downstreamwards with an average of 23°.

The occurrence of the same levels throughout the Júcar valley means that the foundation of both dams is based on the same materials-crystalline limestones and marlish limestones with innumerable clayey interlayers of millimetre thickness. This simplifies work to a certain extent and allows experience to be extrapolated from one site to the other.

2. INVESTIGATION PRIOR TO THE CORTES DAM DESIGN

The Cortes dam is a 116 m. high arch-gravity structure with a crown length of 312 m. The geological survey at the site was supported by a wide ranging investigation that is described hereinafter.

Three trenches were dug, two on the left bank and one the right, allowing the stratigraphy to be studied in areas covered by debris.

There were 15 boreholes, carried out in two different campaigns, totalling 139.5 and 565 m., respectively. Other campaigns covered a long section of the river downstream and analysed quarries and special problems.

Eight galleries were drilled, four in each bank, located at different levels and totalling 994 m. This allowed the internal state of the rock to be known, as well as a mapping of the fractures and definition of the stratigraphy with precision.

Finally a seismic geophysical test was carried out comprising 12 profiles, 3 running across the valley and nine longitudinal. This investigation allowed the geotechnical characteristics of the rock and the debris thickness to be determined.

This work demonstrated the existence of a monoclinal structure dipping upstream by 18°. Four large stratigraphic units were distinguished in it, that affected the dam and are are summarised in table 1.

The lower part of the dam is supported on the C4 unit, that is rigid but permeable. This requires a waterproofing treatment down to the C3 package, which is impermeable.

The abutments are supported on the C5 formation that is considered to be impermeable normally to the strata, given the innumerable clayey layers and the somewhat marlish nature overall. In principle, no treatment will be given in the abutment zone and future criteria will be defined as the reservoir is filled up.

The C6 unit is only affected by the upper part of the dam and is considered to be permeable. It will have to withstand very low water pressure, so no serious problems are foresighted.

The fracturing systems detected by the measures taken both at the surface and in the galleries are displayed in table 2 and in figure 1.

As it can be seen, a certain dispersion exists although a system slightly predominates.

The geophysical prospection results are shown in table 3. The celerities obtained are somewhat lower than was to be expected from this type of formation. There is a great similarity between both banks, with equal coefficients of deformation and dynamic elastic moduli, with the thickness of the decompressed zone being somewhat deeper ot the left bank. In the river bed, as it is logical, the figures are higher and the decompressed zone considerably less.

3. INVESTIGATIONS DURING THE CONSTRUCTION OF THE CORTES DAM

In order to complete and refine the data obtained from the preliminary investigations, data continued to be gathered during the construction stage and a

Table 1. Stratigraphic entities defined through the investigation

Investigation for the design		Investigation during construction
Turonian C 6	Crystalline limestone. Some 50 m. Sandy-clayey marls. 5 m. Crystalline limestone. 5 m.	
Cenomanian C 5	Monotonous series of crystalline, marlish and detritic limestone with interlayers. 65 m.	18. Tabular marlish limestone 17. Breccoid, porous, powdery, white-grey limestone 16. White-yellow panelled, marlish limestone 15. Breccoid limestone and marlish limestone 14. Tabular marlish limestone 13. Powdery, breccoid, crystalline limestone 12. Tabular marlish limestone 11. Green clavey marls 10. Yellowish tabular marlish limestone
Cenomanian C4	Calcareous bed of massive appearance with 30 m. deep void levels	9. Tabular, breccoid, cryptocrystalline limestone 8. Yellow tabular crystalline limestone 7. Porous tabular dolimitic limestone 6. Greenish-yellowish tabular limestone 5. Tabular dolomitic limestone 4. Powdery, breccoid dolomitic limestone with voids 3. Yellowish-grey dolomitic limestone with voids 2. Breccoid limestone 1. Powdery, porous dolomitic limestone
Albian C3	Predominately marlish formation with some calcareous interlayering 14 m.	

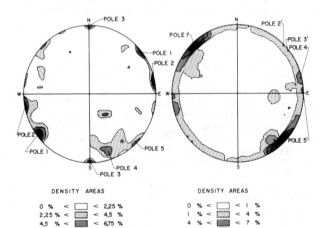

DENSITY AREAS
0 %	<		<	2,25 %
2,25 %	<		<	4,5 %
4,5 %	<		<	6,75 %
6,75 %	<		<	9 %
9 %	<		<	11 %
9 %	<			

DENSITY AREAS
0 %	<		<	1 %
1 %	<		<	4 %
4 %	<		<	7 %
7 %	<		<	11 %
11 %	<		<	15 %
15 %	<			

Fig. 1. Cortes Dam. Fracturing density graphs, a) Investigation for the design (obtained manually), b) Investigation during construction (obtained by computer).

Table 2. Fractures systems

Measured at the design stage

Pole	Direction	Dip	Density
1	N 36 W	V	11,2%
2	N 10 W	V	10,1%
3	E-W	V	6,7%
4	N 72 E	75° N	5,6%
5	N 45 E	V	5,6%

Measured during construction

Pole	Direction	Dip	Density
1'	N 47 E	V	15%
2'	N 39 W	V	4%
3'	N 45 W	V	4%
4'	N 13 W	V	4%
5'	N 3 W	V	4%

geological mapping of the foundation was performed.

The methodology employed was as follows. Once the foundation rock was perfectly clean, before the pouring of concrete, a detailed geological mapping was carried out with the help of surveying, basically comprising:
a) Determination of the significant beds and its ceiling and wall.
b) Preparation of a detailed stratigraphic column of the strata affected by each block foundation.
c) Detailed investigation of the major techtonic discontinuities where representation on the selected cartography (scale 1:100) was feasible.
d) Gathering of data on discontinuities for a statistical analysis of fracturing.
e) Topographical definition of water surgences found and analysis thereof.
f) Topographical definition of potholes or signs of karsticity.

g) Seismic prospection of the excavation for each dam block.

Once these data have been collected, many emphasised by signals, a photograph was taken of the whole block, a complement to the foundation drawing and for consultation purposes in the case of doubt, together with detailed photographs of the most interesting zones.

This survey allowed certain zones or points to be determined for improval or changing prior to concrete placement.

All data obtained allowed the following documentation to be prepared for each dam element:
1. A foundation drawing at a scale of 1/100.
2. A stratigraphic column at a scale of 1/50.
3. A fracturing density graph that allowed the following to be defined:

I) Depth to which consolidation treatment should be taken, obtained from the geophysical data.

II) Impermeabilisation cutoff, taking into account the situation of the impermeable layer and the major fracture families.

Apart from these two immediate objectives, in the future any event occurring in the dam relating to the foundation can be analysed thanks to a complete set of documentation.

The new investigations have led to more detailed knowledge on the site. From assuming a monoclinal structure of 18°, it would seem that the dip falls off gradually as it moves upstream, to around 13° behind the dam. Another very interesting aspect was the detection of a nearly total absence of signs of karsticity as those observed, linked to fractures, were of no importance whatsoever. This is noteworthy in a mainly calcareous rock.

A minor fault was detected in the right bank, with displacement in its own plane of some 3 m. It had gone completely undetected previously as, in the steep canyon walls, covered with a calcareous crust, the different levels could not be distinguished and the feature was confused with a fracture. This discontinuity does not imply any problems for the structure. Its location can be seen in figure 2.

Improved stratigraphic knowledge has allowed the system shown in table 1 to be marked out.

Systems 1 (partially) and 4 are full of voids and therefore have greater permeability. System 11 is the only different one throughout the foundation, on consisting of decimetre levels of modular calcareous marls and clayey marls, its total thickness being 1.33 m.

The fracturing measured demonstrates the existence of five poles that were indicated from 1' to 5' in table 2 and can be seen in figure 1. There is one predominant family, the 1' with a N 47 E direction and a vertical dip, having a density of over 15% that will correspond with pole 5 of the investigation for the design.

Poles 2' and 3' are really the same family and would correspond to the 1; Finally, the 4' and 5', also a single system, coincide with 2. The E-W and the N 72 E, 3 and 4 are not individualised, being disseminated along the ample 1% line.

As demonstrated by figure 2, the river bed zone shows little sign of fractures and the observed ones belong to the main system. A denser fracturing band is detected, also with the direction of the main family, affecting elements 8, 9 and 10.

The results obtained by the new seismic tests are shown in table 3. As can be seen in figure 4, the foundation excavation was generous, seeking to find support on areas with the greatest guarantee, with rates of at least 3.5 km/s, and even higher figures in the river bed. The preliminary design was modified so that the structure was not directly supported on any of the marl layers of a certain thickness that exist, which leads to two cuts on the right bank and one on the other.

The test carried out during construction detected a slight deterioration of the rock due to the action of the explosives, leading to a narrow strip of decompressed rock, of the same thickness on both banks.

The intermediate rate zone of over 2 km/s and less than 3.5 Km/s is small on the right bank and much more notable on the left bank, also having lower rates.

By virtue of these results, consolidation treatment was carried out in the worst zones up to a depth of 10 m.

4. PRELIMINARY INVESTIGATIONS AND DESIGN INVESTIGATIONS IN THE NARANJERO SITE

The Naranjero dam is an arch gravity dam curved in plan, with a height of 84 m. and a crown lenght of 191 m.

The lithological similarity of the site with that of Cortes meant that investigations were not so extensive in this case. Two investigation galleries were drilled, one per bank, totalling 492 m., six boreholes distributed over a transverse profile totalling 498 m. and finally a seismic prospection was carried out with the implementation of 10 profiles: six longitudinal and four transverse.

The dam is supported on the same formation as Cortes, so that the previous paragraphs are valid in this case. Strata direction is nearly normal to the river and the dip varies, growing from upstream towards downstream, reaching 29° 30' at the site.

Nevertheless, it should be emphasised that a sizeable debris thickness exists on the right bank, due to an old slide, with a high degree of cementation.

On trying to define precisely the structure, a level difference was found between the strata on both banks, caused by the presence of a fault on the right bank, the exact location of which it was not possible to find in the dam zone due to the alluvial cover.

The fracturing measured, as shown in table 4 and figure 5, demonstrated a single clearly predominant system with a N 74° 45' E direction and a dip of 62° S.

Other fractures were very disperse and of little density.

Geophysical test results are shown in table 5 and in figure 8. Three zones are essentially defined (another more superficial zone is of no importance and corresponds to the debris and alluvian). The

Table 3 Cortes Dam. Geophysical Investigation results.

		Investigation for the design				Investigation during construction			
		h (m)	α (Km/s)	Eo (10^3MPa)	D (10^3MPa)	h (m)	α (Km/s)	Eo (10^3MPa)	D (10^3MPa)
Left Bank	Decompression zone	18	2,5-2,7	12,74	7,74	0,7-3,2	0,7-2,0	0,98-8,82	0,39-3,92
	Intermediate zone					2,8-10	1,75-2,2	6,86-10,78	3,43-4,9
	Deep zone		3,5-3,6	26,46	17,15		3,3-3,6	24,99-29,4	19,7-17,64
River Bed	Decompression zone	8,0-10,0	3,0-3,1	22,54	12,74	0,3-4,6		1,47-12,74	0,49-6,37
	Intermediate zone					1,4-6,60		15,19-24,99	7,35-14,21
	Deep zone		4,0	37,24	27,44			37,24-38,71	24-25,97
Right Bank	Decompression zone	15,0	2,5-2,8	12,74	7,84	0,6-3,8		1,96-11,76	0,78-5,39
	Intermediate zone					1,4-6,5		7,84-23,03	3,4-12,74
	Deep zone		3,5-3,6	24,46	17,15			97,93-36,26	16,17-23,52

h = thickness; α = average celerity; Eo = dynamic modulus of elasticity; D = Coefficient of deformation.

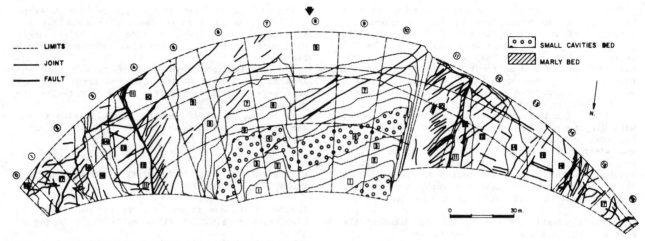

Fi. 2. Cortes Dam. Geological plan of the foundation.

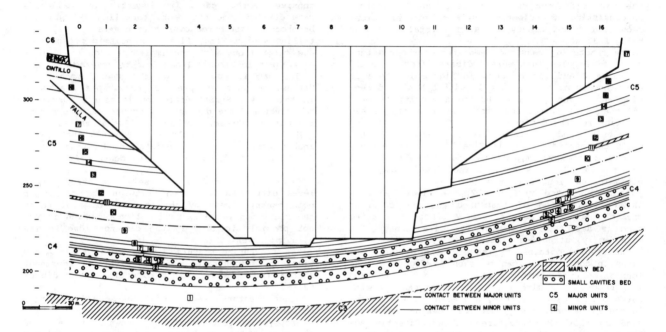

Fig. 3. Cortes Dam. Cross section through the reference surface, showing the geology. The different layers drawn are those indicated in table 1.

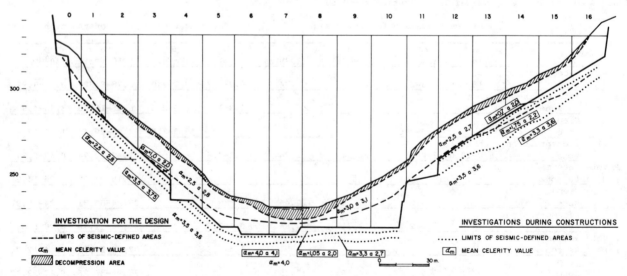

Fig. 4. Cortes Dam. Cross sections through the reference surface, showing the results of the geophysical tests.

decompressed zone thickness is somewhat greater in the river bed and right bank than in the very steep left bank, where the occurrence of debris was minimum. It should be noted that the presence of the fault was detected, even though somewhat displaced from its real location.

Table 4. Fractures systems

Measured in the design studies

Pole	Direction	Dip	Density
1	N 74°45'E	62°5	14%

Measured in the construction phase

Pole	Direction	Dip	Density
1	N 66° E	65 S	20%

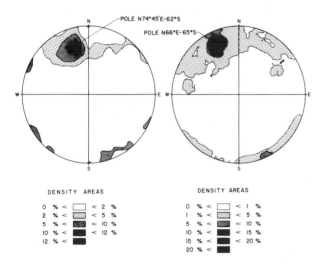

DENSITY AREAS

0 % < ☐ < 2 %
2 % < ▦ < 5 %
5 % < ▤ < 10 %
10 % < ■ < 12 %
12 % < ■

DENSITY AREAS

0 % < ☐ < 1 %
1 % < ▦ < 5 %
5 % < ▤ < 10 %
10 % < ■ < 15 %
15 % < ■ < 20 %
20 % < ■

Fig. 5. El Naranjero Dam. Fracturing density graphs. a) Investigation for the design (obtained manually) b) Investigation during construction (obtained by computer).

5. INVESTIGATION DURING CONSTRUCTION OF EL NARANJERO

Geological surveys have allowed the stratification, structure and fracturing to be defined in great detail. Essentially, stratification is similar to that of Cortes, but some minor differences exist in determining the different systems as, in this case, lithology is more homogeneous. The thickness of the C5 formation is greater than in Cortes, whereas the thickness of the marl level at its base is reduced to one half. This may be due to a slide along it that produced lamination.

The structure defined in the studies for the design is practically confirmed, with slight modifications with regard to dip.

The fracture analysis, as can be seen in figure 5, confirms the predominance of a very slightly different system to that defined in the design, with a direction of N 66° E, dip 65° S and a concentration of 20%. Other fractures indicated are very disperse.

The fault could be determined with great detail. It is considerably twisted and in some sections adapts itself to the strata. The average direction in the river bed is N 42 E, with vertical dip, an opening of some 30-40 cm and displacement according to the fault plane is some 7 m. Other features of lesser importance have been detected from the same family as the above that, together with it, limit a zone with dense fracturing. See figures 6 and 7.

The existence of this strip explains the displacement of the weakest rock zones, indicated by the geophysics, with respect to the fault situation. Nevertheless, as can be appreciated in figure 6, the fractures are not abundant and clearly indicate the predominance of a system.

The total absence of solution phenomena, despite the calcareous nature of the site, is a remarkable fact, in an even more accentuated way than for Cortes site.

The foundation excavation, as can be seen in figure 8, has practically eliminated the whole decompressed zone and in some points, i.e. at the foot of the left abutment, reaches the higher quality zone.

New geophysical tests demonstrate that the rock altered by the excavation represents a minimum strip. Also the denser fractured zone is insinuated in them, as the thickness of lower quality zones increases.

As a result of this, the consolidation treatment was limited to 7 m.

Table 5. El Naranjero Dam. Geophysical investigation results.

		Investigation for the design				Investigation during construction			
		h (m)	α (Km/s)	Eo (10^3MPa)	D (10^3MPa)	h (m)	α (Km/s)	Eo (10^3MPa)	D (10^3MPa)
Left Bank	Decompression zone	1-2 / 1-7	1,7 / 2,7	13,72	6,80	0,4-2,3	0,9-2,3	1,47-12,25	0,68-5,88
	Intermediate zone	8-19	3,5	22,54	14,21	2-5,8	2,6-2,7	15,19-16,66	7,35-8,33
	Deep zone		4,5	37,73	25,0		3,5	24,5-27,93	14,21-16,17
River Bed	Decompression zone	2,0-7,0 / 8,0-15	1,0-1,4 / 2,6	10,78	5,88	0,2-1,8	1-2,4	1,96-12,74	0,88-6,37
	Intermediate zone	9-15	3,5	22,54	14,21	1,5-4,6	2,6	15,19	7,35
	Deep zone		4,6	39,20	25,97		3,5	27,93	16,17
Right Bank	Decompression zone	5,7 / 7-14	1,0-1,2 / 2,6	10,78	5,88	0,3-2,4	1,5-2,5	4,9-12,74	1,96-6,37
	Intermediate zone	4-12	3,5	22,54	17,21	1,4	2,6	15,19	7,35
	Deep zone		4,6	39,20	25,97		3,5	27,93	16,17

h = thickness; α = average celerity; Eo = dynamic modulus of elasticity; D = Coefficient of Deformation.

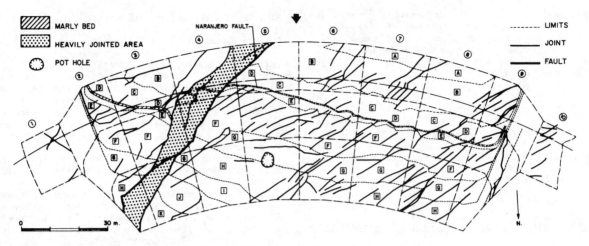

Fig. 6. El Naranjero Dam. Geological plan of the foundation.

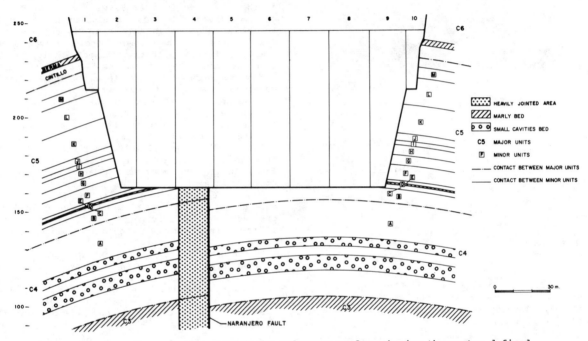

Fig. 7. El Naranjero Dam. Cross section through the reference surface showing the system defined.

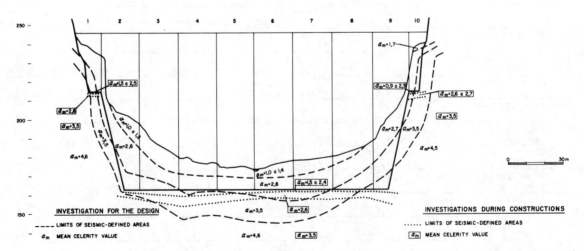

Fig. 8. El Naranjero Dam. Cross section through the reference surface, showing the results of the geophysical tests.

454

6. CONCLUSIONS

The Cortes and El Naranjero dams foundation studies demonstrate the need to continue geological and geotechnical investigations throughout construction, as they allow the data obtained in the exploration stage for the design to be considerably improved upon.

This leads to a better quality of the foundation treatment and, therefore, of its connection with the structure, allowing the consolidation grout depth and its spatial differentiation to be defined. Similarly, definition of the impermeable cutoff is considerably improved.

Independently of these direct actions, the documentation obtained allows for future treatment if necessary, without the need for new investigations or reducing them to a minimum.

7. REFERENCES

Navalon, N. 1986 "The construction of Spain's Cortes La Muela scheme" Water Power & Dam Construction. February pp. 37-41.

Sáenz Ridruejo, C. 1973. "Estudio Geológico del emplazamiento de la presa de Cortes de Pallás (Río Júcar)".

Sociedad de Reconocimientos Geofísicos, S.A. 1972. "Estudio geotécnico de la cerrada de Cortes de Pallás en el río Júcar".

Sociedad de Reconocimientos Geofísicos, S.A. 1986. "Definición de las características elásticas de la cimentación de la Presa de Cortes II".

Hidroeléctrica Española, S.A. 1984 "Presa de El Naranjero" Proyecto de Construcción. Anejo no. 2. Estudio Geológico.

Sociedad de Reconocimientos Geofísicos, S.A. 1986. "Definición de las características elásticas de la cimentación de la presa de El Naranjero".

Flexural buckling of hard rock – a potential failure mode in high rock slopes?

Flambement flexural de roche dure – Un genre de rupture en pentes abruptes?
Knickung von hartem Fels – Ein potentieller Bruchmechanismus in hohen Böschungen?

B.NILSEN, Norwegian Institute of Technology, Trondheim

ABSTRACT: A method is presented for analysing flexural buckling of rock slabs subjected to axial load in two perpendicular directions. The presentation is primarily concentrated on analyses carried out for the Ørt-fjell open pit, which is planned with maximum slope heights of more than 300 m. It is concluded that in general a very special combination of geology, stress condition and rock character has to exist for flexural buckling of hard rock to occur.

RESUMÉ: Une méthode pour analyser flambement flexural des roches sousmises a une force axiale dans deux directions perpendiculaires est ici présentée. Premièrement cette présentation est concentrée sur des ana-lyses faites sur le grisement de fèr a ciel ouvert de Ørtfjell. Ce grisement est projeté avec des hauteurs des pentes plus que trois cents meters. En general il faut une combination particulière de géologie, con-dition de contraintes et caractère de roche pour briser des roches dures par flexure.

ZUSAMMENFASSUNG: Eine Methode zur Analyse der Knickung von Gesteinsplatten unter zweiachsigem Druck wird presentiert. Sie ist in Zusammenhang mit dem geplanten Ørtfjell Tagbau entwickelt worden, wo maximale Böschungen von 300 m vorgesehen sind. Die Analyse zeigt, dass ganz spezielle Kombinationen von Geologie, Spannungszustand und Gesteinscharacter existieren müssen, um Knickung von hartem Fels hervorzurufen.

1 INTRODUCTION

Two main criteria have to be fulfilled for buckling failure of slabs or plates of rock to occur in the walls of high slopes or large rock caverns. Firstly, the degree of jointing of the rock mass has to be small, and with continuous joints parallel to the walls representing the predominant joint set. In cases with extreme high stresses in the rock mass, such joints can theoretically be initiated by the high stress level. Secondly, for buckling to occur, the slabs which are separated by joints have to be subjected to axial stresses which are high com-pared to the mechanical strength of the respective rock.

Buckling failure is most commonly reported from layered coal deposits, where very low friction materials between the layers are quite common. In such cases buckling is often initiated solely by the weight of the slab itself (Kutter 1974, Cavers 1981). Because of the plastic nature of soft rock material as coal, buckling failure normally developes slowly and gradually in such cases.

In the sidewalls of highly stressed excavations buckling failure occasionally plays a significant role in the failure of more competent and elastic rocks (Hoek & Brown 1980). At the toe of high slopes in hard rock buckling may theoretically cause failure and sliding due to undercutting.

In most cases, methods for analysing buckling are apparently taking into consideration axial stress only in one direction. This may represent a gross oversimplification. High tectonic stresses, for instance, may often cause a situation where the axial stress in horizontal direction is larger than the axial stress in the dip direction of the slope face.

At the deepest part of the Ørtfjell open pit mine in Norway, high tectonic horizontal stresses are expected. As part of the stability analysis pro-gramme for this mine, the probability of flexural buckling of plates of rock subjected to axial stress in two perpendicular directions has been studied. This paper will present the main principles of the method which has been applied together with some results for the Ørtfjell case.

This presentation is restricted to the flexural type of buckling, and is not dealing with hinge types of buckling as for instance discussed by Cavers.

2 ACTUAL CASE

Figure 1 shows a cross section through the planned Ørtfjell open pit mine for the deepest alternative of final floor level. All results for the Ørtfjell case which are presented in this paper, are based on this maximum depth alternative.

Mica schist of Cambro-Silurian age ("Dunderland schist") is the predominant rock in the pit area. This schist has a well defined jointing along the foliation and a uniaxial compressive strength of approx. 65 MPa (measured on Ø 32 mm water satu-rated drillcores).

The strike of the foliation is roughly parallel to the long axis of the pit (perpendicular to the cross section in fig. 1), while the dip is steep to the North (approx. 70° as an average). The South wall is therefore the foot wall of the pit, and the wall which may theoretically be subjected to buckling.

The pit is planned to be mined down 300 - 400 metres as indicated in fig. 1. As a starting point, and based on experience from the operations of nearby pits, the pit wall design is based on benches of 13 metres width, 30 metres height and bench sur-face slopes of 70°.

Based on three dimensional stress measurements and two dimensional finite element analyses, it has been concluded that the major principal stress (σ_1) is horizontal and orientated parallel to the foliation. For a final depth of the pit as shown in fig. 1, the analyses indicate that σ_1 will be of a considerable magnitude (σ_1 = 48 MPa). The inter-mediate principal stress in the foot wall will run parallel to the dip direction of the wall, and have a considerably lower value (σ_2 = 17 MPa). The

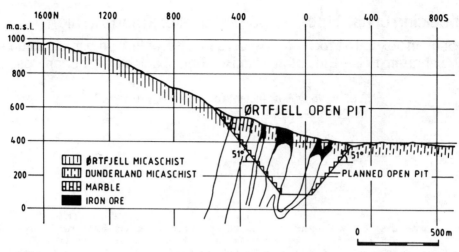

Figure 1. Cross section through the planned Ørtfjell open pit, Northern Norway for the deepest alternative of final floor level.

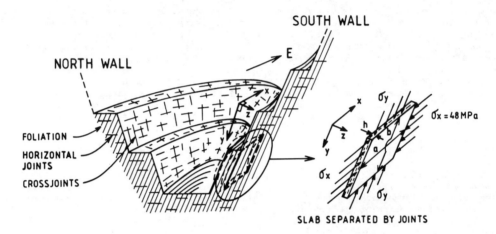

Figure 2. Sketch indicating the potential buckling situation at the toe of the foot wall.

principal stress perpendicular to the slope face (σ_3) will be approx. equal to zero close to the face. More details concerning geology and rock mechanic measurements at the Ørtfjell open pit mine are given by Broch & Nilsen (1979) and Nilsen (1979).

The sketch in fig. 2 is an attempt to visualize the potential buckling situation at the Ørtfjell open pit mine. As can be seen, the analysis is based on the potential flexural buckling of slabs separated by three different joint sets, and on the existence of axial stresses in two perpendicular directions.

3 METHOD FOR ANALYSES

The study of potential buckling is based on a general method for analysing elastic, flexural buckling of rectangular plates (Timoshenko & Gere 1961). This method makes it possible to analyse slabs compressed in two perpendicular directions as indicated in fig. 2.

As is the case for all methods for analysing buckling of rock slabs, also the method presented by Timoshenko and Gere is based on several idealizations and simplifications. The most important are:
- Homogeneous, isotropic and elastic material.
- Planar slab or plate (no curvatur).
- Constant thickness of plate.

- Fixed corners of plate.
- No load perpendicular to the plate.
- No shear stress on the surface of the plate.

For the Ørtfjell case the assumption about isotropy is probably representing the major restriction. The assumption about homogenity and elasticity are believed to represent only minor restrictions, and also the assumption of planar slab and fixed corners. Potential buckling slabs in the pit walls are located very close to the wall (see fig. 2), and hence the front load is negliable compared to the axial loads. According to finite element analyses, the magnitude of the shear stresses are negliable as compared with the axial stresses on the slab. Hence, for the Ørtfjell case this method is believed to be relevant for an evaluation of the probability of buckling failure to occur at the foot wall.

Also more generally the majority of uncertainties and idealizations are believed in many cases to represent only minor restrictions. This method for analysing flexural buckling is therefore believed to be of a wider interest.

The general principle of the method may be illustrated by the simple model in fig. 3 of a bar subjected to axial load. P_k is the critical load, or the buckling load. If the load P is increased above P_k, the situation is labile, and buckling failure will occur.

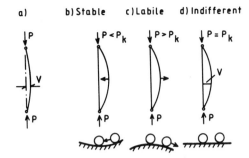

a) b) Stable c) Labile d) Indifferent

$P < P_k$ $P > P_k$ $P = P_k$

Figure 3. Conditions of equilibrium for centrically loaded bar.

As is the case for the simple bar model, critical values of axial loads are also connected to the model presented by Timoshenko & Gere, and buckling failure will be the result if these critical values are exceeded. The plate model, however, is loaded in two perpendicular directions. For any load or stress in x-direction (σ_x), there is accordingly a critical load or stress in y-direction ($\sigma_{y(cr)}$), and vica versa, see fig. 2.

For the actual case, the analyses of potential buckling have been carried out by keeping the stress in x-direction (σ_x) constant, while the critical stress in y-direction has been calculated for various combinations of the dimensions of the slab (a, b and h). In the pit wall the parameters a, b and h are the spacings for crossjoints, horizontal joints and foliation joints, respectively. Based on the plate model, and according to the energy method, the critical stress in y-direction ($\sigma_{y(cr)}$ may be calculated as (Timoshenko & Gere 1961):

$$\sigma_{y(cr)} = \frac{\sigma_e (m^2 + n^2 \cdot \frac{a^2}{b^2})^2 - \sigma_x \cdot m}{n^2 \cdot \frac{a^2}{b^2}} \text{ , Where:}$$

$$\sigma_e = \frac{\pi^2 \cdot D}{a^2 \cdot h} = \text{"reference stress" (MPa)}$$

$$D = \frac{E \cdot h^3}{12(1 - \nu^2)} = \text{elasticity of flexure (MPa)}$$

E = Youngs modulus (MPa)

ν = Poissons ratio

a, b and h are dimensions of plate

m and n are constants of iteration

According to this model buckling will have the character of bending in two perpendicular directions. The bending will have a maximum value at the center of the place. Statically, however, the system is indetermined. The critical deformation for failure, hence, can not be determined. For a more detailed description of the method, reference is made to the interesting discussions presented by Timoshenko & Gere.

4 RESULTS

For the Ørtfjell case the analyses have been concentrated on the toe of the foot wall, and on a floor level corresponding to the final depth of the pit (see fig. 1 and 2). This situation is representing the case when buckling most likely will occur.

Analyses have been carried out for alternative spacings between foliation joints (h) within the range 0,25 - 1,0 m, which according to joint mapping is the normal variation in the pit area. Primarily, the analyses are based on a Youngs modulus of E = 18 · 10³ MPa and a Poisson ratio of ν = 0,15, respectively. These are the mean values from laboratory tests.

Figure 4 shows the resulting critical stress in y-direction ($\sigma_{y(cr)}$) for h-values of 0,25 m, 0,5 m and 1,0 m, respectively. Five values of a and a range of b-values are included in each diagram. Hence, the stippled lines in the diagrams represent values of $\sigma_{y(cr)}$ for a wide variety of combinations of joint spacings. Also, the diagrams may be interpreted as giving information of "critical joint spacings" for a given stress level. In this example the stippled lines will be used for defining critical b-values for given values of h, a and $\sigma_{y(cr)}$.

The horizontal, solid lines in the diagram show the stress in y-direction resulting from stress measurements and finite element analyses ($\sigma_{y(FEM)}$ = 17 MPa), see chapter 2. As previously discussed, buckling will occur when the stress in y-direction is exceeding the critical value. Hence, for each h-value the diagram is defining a set of critical joint spacings a and b. For a defined a-value, the critical value of b is defined as the abscissa of the point of intersection between stippled and solid lines.

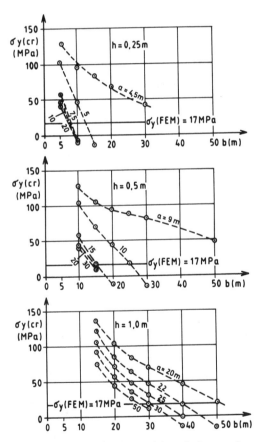

Figure 4. Critical stress ($\sigma_{y(cr)}$) for various combinations of h, a and b (spacing for foliation joints, crossjoints and horizontal joints, respectively).

Based on fig. 4 the critical joint spacings a and b have been evaluated for the respective h-values and for $\sigma_{y(cr)} = \sigma_{y(FEM)}$. The resulting values are presented in fig. 5. The curves should be interpreted as shown in the upper right corner of the figure, i.e. buckling will occur if the plot of the joint spacings a and b is above the curve representing the actual h-value. For instance, for squarical slabs buckling will occur for slab dimensions larger than approx. 9 · 9 m for h = 0,25 m, 13 · 13 m for h = 0,5 m and 30 · 30 m for h = 1,0 m.

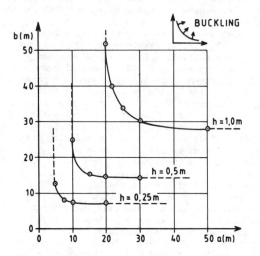

Figure 5. Critical joint spacings a and b for alternative values of h.

As part of a comprehensive parameter study for the Ørtfjell case, analyses have been carried out also for an alternative and considerable lower value of σ_x (σ_x = 8 MPa) while the other input parameters are left the same. Figure 6 shows the buckling curves for h = 0,25 m corresponding to the original σ_x-value of 48 MPa and the alternative σ_x-value of 8 MPa, respectively. As should be expected, the situation is very clearly in favour of buckling for the case of the highest σ_x-value. Due to opposite directions of the major axial stress for these two cases, the locations of the curves in fig. 6 are oppositely displaced relatively to the coordinate axes.

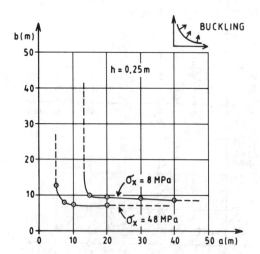

Figure 6. Critical joint spacings a and b for h = 0,25 m corresponding to original and alternative value of σ_x.

Results for h = 0,5 m and alternative values of Youngs modulus are presented in fig. 7. As clearly indicated by these results, even for a moderate increase of Youngs modulus a considerable increase of the critical joint spacings will be the result.

Minor variations of the value of Poisson's ratio (0,15 ± 0,05) have according to the analyses a

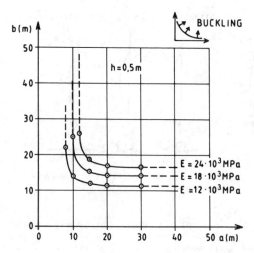

Figure 7. Critical joint spacings a and b for h = 0,5 m and alternative values of Youngs modulus (E).

neglibile influence on the final result (less than ± 2% influence on the calculated $\sigma_{y(cr)}$).

5 DISCUSSION

For buckling failure to occur, each one of the following conditions have to be fulfilled:

- The spacing between foliation joints (h) has to be smaller than a certain critical value.
- The joint spacings for crossjoints and horizontal joints (a and b, respectively) have to be larger than the critical values corresponding to the actual stress-level and the actual h-value.
- The joints have to be very continuous, so that slabs having the actual dimensions are separated by joints.

In order to evaluate the probability of buckling failure it is necessary to evaluate the probability of each of the joint conditions. If these conditions are mutually independent, the probability of buckling failure may be calculated by multiplying the probabilities of each one of them to be fulfilled.

Mapping carried out in the Ørtfjell area indicate that in this case, the actual joint parameters are apparently mutually independent. Hence, it is assumed that no major mistake is done by calculating the probability of buckling failure as described above. It should be emphasized, however, that in other cases a distinct interdependence may exist between the major joint sets. In such situations, which for instance is believed to be the normal case for sedimentary rocks, the product referred to will not correctly express the probability of buckling.

Ideally, the probability of the joint spacing h of being smaller than a certain limit l_1, i.e. P (h < l_1), may be expressed by the relative number of cases when joint mapping has indicated a value lower than l_1. Correspondingly, the probabilities of joint spacings a and b to be larger than certain limits l_2 and l_3, i.e. P (a > l_2) and P (b > l_3), respectively, may ideally be expressed by the relative number of cases when joint mapping has indicated values higher than l_2 and l_3.

For the Ørtfjell case, the probability of buckling has been analysed for various combinations of joint spacings. Here, the discussions will be restricted to a spacing between foliation joints of h ⩽ 0,5 m. According to joint mapping this is fairly common in the actual area, with an apparent probability of:

P(h ≤ 0,5 m) ∿ 0,35

As discussed in the previous chapter, for buckling of a squarical slab to occur a critical spacing a · b > 13 m · 13 m will be needed for a h-value of 0,5 m (see also fig. 5). Based on joint mapping, the probability of the respective spacings to be larger than 13 m is:

P(a > 13 m) ∿ 0,15
P(b > 13 m) ∿ 0,10

According to this evaluation the probability of buckling failure will be:

P(buckling) = P(h ≤ 0,5 m) · P(a > 13 m) ·

P(b > 13 m) = 0,005

This result may be interpreted as a probability of buckling failure to occur of 0,5%, or that buckling failure has to be expected in 0,5% of the area of the deepest part of the foot wall.

In general a potential major restriction for an analysis like this is connected to the continuity of the joint sets. Based on mapping in the actual area and in nearby pits, however, this is believed to represent no major limitation for the Ørtfjell case. The situation in fig. 8, which is from a nearby pit, is demonstrating this. In this case a rock slab with dimensions approx. 30 m · 30 m · 1 m is separated by continuous joints, and the pit wall is defined by the foliation. According to fig. 6 buckling failure would be the result for a situation comparable to this at the toe of the Ørtfjell pit.

Figure 8. Pit wall following distinct jointing along foliation. Slope height: 30 - 40 m.

For a situation as illustrated in fig. 2, an extra safety against buckling will normally be represented by front load on the potential buckling slab and potential gaps of intact rock along the joints. In other situations, water pressure may increase the danger of buckling. Normally, however, this water pressure can be reduced by drainage.

Taking all the uncertainties into consideration, the conclusion for the Ørtfjell case has been that for the deepest alternative of floor level, local buckling of the foot wall has to be expected (Nilsen 1979). At a potential buckling situation the most efficient methods for stabilization will undoubtly be drainage and the installation of large capacity rock anchors. The installation of rock anchors, if properly done, may have the very favourable effect of increasing the thickness (h-value) of the potential buckling slab.

6 CONCLUSION

The results which are presented in this paper are restricted to one single open pit mine case. The basic principles of this simple method, however, are also believed to be of a wider interest for analysing flexural buckling of rock slabs under bi-axial stress conditions.

Based on the analyses carried out for the Ørtfjell case, it can be concluded that a very special combination of geology, stress conditions and rock character has to exist for flexural buckling of hard rock to occur. Briefly characterized, this combination has to include a very low degree of jointing with joints parallel to the wall representing the major joint set, tangential stresses in the wall which are very high compared to the mechanical strength of the rock, and preferably also a rock type having a relatively low Youngs modulus.

REFERENCES

Broch, E. & Nilsen, B. 1979. Comparison of calculated, measured and observed stresses at the Ørtfjell open pit. Proc. 4th ISRM-congr., p. 49 - 56. Rotterdam: Balkema.
Cavers, D.S. 1981. Simple methods to analyse buckling of rock slopes. Rock Mechanics 14: 87 - 104.
Hoek, E. & Brown, E.T. 1980. Underground excavations in rock, p. 234 - 235. London: IMM.
Kutter, H.K. 1974. Mechanisms of slope failure other than pure sliding. In L. Müller (ed.), Rock Mechanics, p. 213 - 220. Udine: Springer.
Nilsen, B. 1979. Stability of high rock slopes (in Norwegian). Norw. Inst. of Techn., Dept. of Geol., Rep. 11, p. 218 - 233. Trondheim: Dept. of Geol.
Timoshenko, S.P. & Gere, J.M. 1961. Buckling of thin plates. In Theory of elastic stability, p. 348 - 360. New York: McGraw-Hill.

A numerical analysis of earth anchor retaining wall constructed on soft rock
Analyse numérique sur un mur de retenue d'une ancre de terre construit sur de la roche meuble
Numerische Analyse einer auf weichem Fels errichteten Erdanker-Stützmauer

TOSHIMICHI NISHIOKA, Tokyo Electric Power Co. Inc., Japan
AKIO NARIHIRO, Tokyo Electric Power Co. Inc., Japan
OSAMU KYOYA, Tokyo Electric Power Co. Inc., Japan

ABSTRACT: In this paper, the defining method of the input property was examined using FEM analysis based on the result of field measurement of a large scale earth anchor retaining wall constructed on soft rock. Furthermore, applicability of a beam-spring model that allowed easy handling were examined.

RESUME: Cet article traite de la méthode de définition de la propriété de charge par l'analyse FEM sur la base des résultats de mesures effectuées sur un mur de retenue d'une ancre de terre de grandes dimensions construit sur de la roche meuble. L'article examine également l'applicabilité d'un modèle à poutre élastique permettant une manutention simple et les méthodes concrètes de mise en application.

ZUSAMMENFASSUNG: In diesem Papier wurde unter Verwendung der FEM-Analyse, die auf dem Ergebnis der Feldmessung einer großangelegten, auf weichem Fels errichteten Erdanker-Stützmauer beruhte, die Definierungsmethode der Eingabeeigenschaft untersucht. Darüber hinaus wurden die Anwendbarkeit eines Balkenfeder-Modells, das leichte Handhabung erlaubte seiner Anwendung untersucht.

1 Preface

In Japan regulations on the site condition of nuclear power plants are strict and require Tertiary deposit as a bed rock of those structures.

Furthermore, considering the conditions on the distance from active faults and the standy supply of cooling water for generators, our company have chosen soft rock as a bed rock of nuclear power reactor buildings.

As a large scale excavation is necessary to obtain this bed rock, the retaining wall excavation has been carried out in order to use spaces efficiently and to economize costs.

Conventionally, the retaining wall is used for relatively soft ground such as the alluvial deposit.

And it is designed using the simplified beam model where the active earth pressure or the apparent earth pressure based on a number of measured result on axial force of struts is applied as an external load. That is, it examines only the equilibrium of forces. Further, for the design of the retaining wall which excavation depth is 15m or more, a modified method considering the deformation as well as the equilibrium of forces is used. However, even in the modified method applied external loads are same as that of simpler one.

On the other hand, the stability analysis of the excavation of hard rock is done by a method considering the forces released by excavation.

In case of a large scale excavation on soft rock, which is classified in the middle of above two cases, it is necessary to consider the characteristics of initial earth pressure and the interaction between the retaining wall and the ground. In this paper, according to these considerations, the behavior analysis and the defining method of input property used on the beam-spring model (BSM) were examined by means of FEM analysis based on the field measurement result of large scale earth anchor retaining wall constructed on soft rock. Furthermore, applicability of BSM that could simulate the actual behavior was examined.

2 Outline of excavation work

The excavation work was carried out by the vertical cutting with soldier pile type earth anchor retaining wall from GL±0.0m down to GL-29.15m. Excavation was done in 9 stages which was the same stages as the anchors, and to prevent the surface exfoliation a lagging work for the clayey soil portion and a rock bolting and mortar spraying for the mudstone portion were executed as a protection work. In addition, anchors were prestressed. Fig.1,2 show the plan and section of the excavation work. Excavation site is consisted of sandy and silty soil alternate layers down to GL-13.6m and the mudstone further down below as shown in Fig.2.

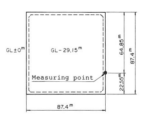

Fig. 1. Plan of excavation

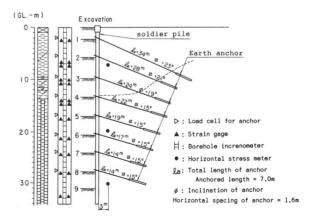

Fig. 2. Section of excavation

3 FEM analysis

In order to interpret the ground behavior during the excavation on soft rock and examine the defining method of the input property used in BMS, a numerical simulation was carried out using FEM analysis based on the measured result.

3.1 Conditions for FEM analysis

FEM analysis was carried out under the two dimensional plane strain and elastic condition. Retaining wall and anchors were modeled by beam and truss elements respectively, and prestressed force introduced into the anchor was evaluated by a nodal force. Analyzed model is shown in Fig.3. The validity of the boundary condition and element size of this model were examined in advance.

In addition, in order to simulate the interaction between ground and retaining wall, spring elements with vertical spring value $kv \fallingdotseq 0$ and horizontal spring value $kh \fallingdotseq \infty$ were introduced.

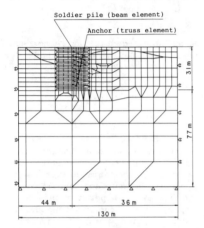

Fig. 3. Analyzed model

3.2 Input property

It was considered that the deformation modulus of ground should be determined based on the test result that could simulate the unloading process of excavation. Therefore, the deformation modulus during unloading process of cyclic loading of tri-axial compression test with $\overline{CU}$ condition using a sample core obtained by the boring near the retaining wall before excavation was employed as input property.

Comparing the deformation modulus of borehole holizontal loading test carried out before and after the excavation, its change due to excavation was considered to be negligible because there was not any noticeable difference.

In order to define an initial earth pressure, a high pressure consolidation test was carried out since the site was in the over consolidation condition. Initial earth pressure and Poisson's ratio before excavation were determined according to the unloading process after the consolidation yield stress applied in tri-axial Ko consolidation test.

Table 1 gives the soil properties used in the analysis.

Additionally, it is clarified that the deformation modulus during unloading process of cyclic loading and the initial earth pressure obtained by borehole horizontal loading test were almost equal to that obtained by above-mentioned laboratory tests.

3.3 Outline of measurement

Fig.2 shows the arrangement of the equipment.

Table 1. Input properties

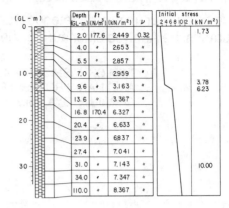

Depth (GL-m)	γt (N/m³)	E (kN/m²)	ν	Initial stress (kN/m²)
2.0	177.6	2449	0.32	1.73
4.0	"	2653	"	
5.5	"	2857	"	
7.0	"	2959	"	
9.6	"	3.163	"	3.78 / 6.23
13.6	"	3.367	"	
16.8	170.4	6.327	"	
20.4	"	6.633	"	
23.9	"	6837	"	
27.4	"	7.041	"	
31.0	"	7.143	"	10.00
34.0	"	7.347	"	
110.0	"	8.367	"	

3.4 Comparison with measured result

Fig.4 and 5 show the displacement and bending moment diagram of retaining wall at the excavation stage 2nd, 4th, 6th and 8th. As shown in the figures, the analyzed result and the measured one corresponded each other well. However, near the excavation level, the analyzed bending moment was greater than the measured one.

This difference could be caused by the plastic or yielding behavior of the excavated ground that alleviated the curvature of retaining wall displacement and decreased the bending moment by the elastic analysis.

In order to evaluate this plastic or yield behavior in the elastic analysis, the deformation modulus was

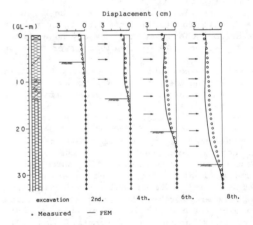

Fig. 4. Distribution of wall deflection

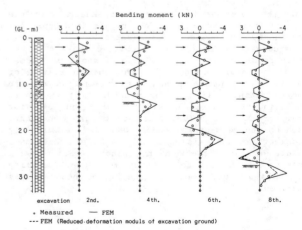

Fig. 5. Distribution of bending moment

reduced tentativity to 1/6 of the initial value for the area where the holizontal stress at excavation side was greater than that Rankin-Lezzale's passive earth pressure.

Re-analyzed result is also shown in Fig.5 and agrees with measured result very well. The displacement diagram of re-analyzed result not shown here because it was hardly effected.

Fig.6 shows the holizontal ground stress distribution at excavation stage 2nd, 4th, 6th and 8th.

Fig.7 shows the change of anchor axial force at stage 1st, 3rd, 5th and 7th. As shown in these figures, the analyzed result and the measured one were corresponding to each other very well.

According to these, it was clarified that the behavior of retaining wall excavation on soft rock was dominated macro-scopically by the elastic behavior of ground due to the forces released by the excavation and this behavior could be interpreted by FEM elastic analysis.

Moreover, the defining method of the initial earth pressure and the deformation modulus was considered to be appropriate.

4 Examination of BSM model

FEM analysis made it possible to simulate the behavior of the retaining wall excavation on soft rock. Though, the applicability of a BSM which might allow for easy handling was examined here.

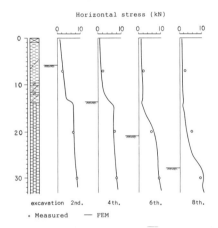

Fig. 6. Distribution of horizontal stress behind wall

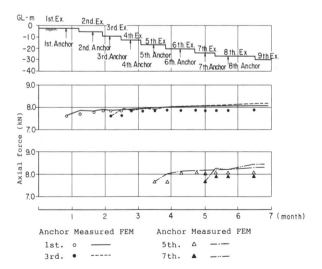

Fig. 7. Axial force of anchor

The earth pressure change was great because the initial earth pressure was great. Therefore, in the BSM it is necessary to evaluate this change properly.

Also the effect of the retaining wall to the ground during anchor prestressing was necessary to be evaluated properly because the anchor prestress was great. For this purpose, a model introducing a ground spring at the both side of the retaining wall was applied. This model was considered to be similar to FEM model.

4.1 Outline of BSM

In this BSM, forces f_1, f_2 released by the excavation and prestressed force f_3 of anchor were applied to the retaining wall supported by ground springs as enternal loads. According to the actual excavation procedures, this incremental analysis for each excavation stage was repeated.

Here the lower limit value of earth pressure at the back of retaining wall was assumed to be the active earth pressure and the upper limit value of reaction force of ground spring at the excavation side was supposed to be the passive earth pressure. General idea of analyzed model and analyzing flow are shown in Fig.8 and 9.

4.2 Initial earth pressure and anchor spring value

The initial earth pressure before excavation was defined in the same way as that in FEM analysis.

The stiffness of free length of anchor was used for the anchor spring value.

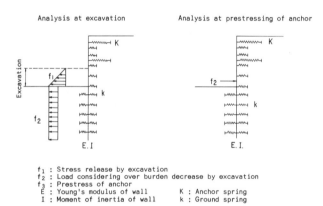

f_1 : Stress release by excavation
f_2 : Load considering over burden decrease by excavation
f_3 : Prestress of anchor
E : Young's modulus of wall   K : Anchor spring
I : Moment of inertia of wall   k : Ground spring

Fig. 8. Analysis model

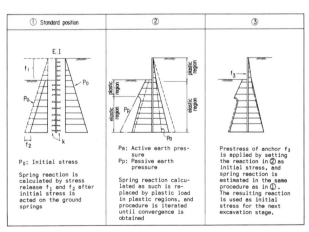

Fig. 9. Analysis flow

4.3 Ground spring value

The ground spring in the BSM could be calculated theoretically if the stiffness matrix of the ground in FEM analysis and the displacement of the nodal points consisting the retaining wall modeled by beam elements were given. However, the process was not easy because non-linear spring elements were introduced between the ground and the retaining wall.

Furthermore, the ground spring value k was considered to be a function of strength and deformation modulus of ground, depth of excatation, a shape of excavation area and soil formation. However, the quantitative evaluation of these effects into ground spring value k was very difficult. For the estimation of the retaining wall behavior at coming excavation stage during construction, the ground spring could be evaluated by the back analysis based on the measured result.

In order to examine the defining method of ground spring for design, the ground spring value was supposed to be determined by k = α × E (deformation modulus).

Using α as a parameter, analysis was conducted and its results was compared with the measured displacement and bending moment of retaining wall.

Here for the judgement of plastisity, Rankin-Lezzale's active and passive earth pressure were used.

As the result, namely, α was set to almost 1/2000 for the active side ground spring and 1/1000 for the passive side one. Furthermore, FEM analyzed result showed that the change of horizontal ground stress at each excavation stage occurred within about 5m upward from the excavation level and beyond that distance there was little change. Therefore, in this area no ground spring at the back of retaining wall was assumed.

4.4 Comparison with measured result

Fig.10 and 11 show the displacement and bending moment of retaining wall at excavation stage 2nd, 4th, 6th and 8th. Fig.12 shows the change of the axial force of anchor at stage 1st, 3rd, 5th and 7th. As shown in these figures, while the displacement did not show a good agreement, the bending moment of analyzed and measured result corresponded each other very well.

Since the ground spring was evaluated as elastic-plastic material, the bending moment near the exca-vation level was alleviated and corresponded with measured result.

The analysis axial force of anchor is slightly greater than that of measured result. This could be caused by the limitation of the modelling because it dealt the end of anchor as fixed, and that could not represent the elastic displacement of ground in-

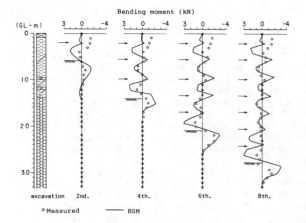

Fig. 11. Distribution of bending moment

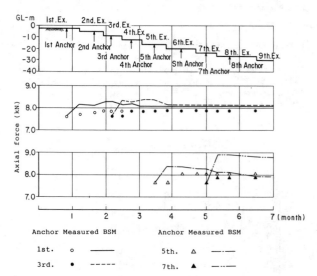

Fig. 12. Axial force of anchor

cluding the anchorage part. Furthermore, the greater difference between measured result and analyzed result could be seem as going down to lower anchors. This reason was considered that the effect of the fixed condition was reflected greater because the length of anchor was shorter in lower anchors and the spring value was relatively stiff compared with upper ones.

5 Conclusion

According to the result, the design of retaining wall excavated on soft rock can be done by FEM analy-sis.

On the other hand, in order to evaluate the ground spring of the BSM, FEM analysis has to be carried out before the excavation. However, the model is confirmed to be effective as far as it is used for the prediction of the sectional force of retaining wall during excavation.

Here after we are trying to establish a designing method with further study on a simpler method of defining the ground spring value.

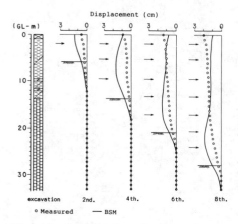

Fig. 10. distribution of wall deflection

466

Direct shear behaviour of concrete-sandstone interfaces

Comportement au cisaillement direct des contacts béton-grès
Verhalten von Beton-Sandstein Zwischenflächen bei direkter Scherung

L.H.OOI, University of Sydney, Australia
J.P.CARTER, University of Sydney, Australia

ABSTRACT: A laboratory study has been made of the direct shearing of concrete-sandstone interfaces under conditions of constant normal stiffness. The results demonstrate that the peak and residual strength envelopes (stress ratio versus normal stress) are independent of the test path.

RESUME: Une étude de laboratoire a été faître sur le cisaillement direct de la surface séparanté béton-grès sous la condition de rigidité normale constante. Les résultats montrent que les enveloppes de la résistance maximale et de la résistance résiduelle (rapport des contraintes vs la contrainte normale) sont indépendentes du chemin d'essai.

ZUSAMMENFASSUNG: Eine Laboruntersuchung der direkten Scherung von Beton-Sandstein Zwischenflächen unter konstant normalen Steifheitbedingungen wurde durchgeführt. Die Resultate zeigen, dass die maximalen und die residuellen Widerstände (Spannungsverhaltnis verglichen mit der Normalspannung) unabhängig sind von der Spannungstestrichtung.

1 INTRODUCTION

The mechanical behaviour of interfaces between concrete and rock can have an important influence on the response of certain foundations systems to applied loading. A particular example occurs when a concrete pile, formed in a rock socket, is used to support structural loads. Resistance to the loading will be provided by the development of shear stresses at the cylindrical interface along the length of the shaft. Some resistance will also be provided in end bearing at the base of the pile, but, unless the pile is extremely short or the magnitude of the loading is extremely high, most of the applied loading will be carried in side shear.

Laboratory studies of the shear behaviour of concrete-rock interfaces are important, not only because they provide basic data essential in the design of rock-socketed piles, but also because they provide insight into the fundamental behaviour. Conventionally, the testing of interfaces has been carried out in the direct shear apparatus with constant normal stress applied across the shear interface. This type of testing provides a reasonable model for cases in practice where no constraint is placed upon the normal displacements accompanying the shearing, e.g. when rock blocks slide freely under gravity. However, in cases such as a concrete pile contained within a rock socket, the stress normal to the plane of shearing can be far from constant. When this type of pile is loaded axially, the pile shaft will displace vertically and at large enough loads (often within the working load range for the pile) relative displacements (slip) will occur between the shaft and the surrounding rock. If the socket containing the pile has a rough surface, then the relative displacement will be accompanied by some dilation at the interface. Because the surrounding rock mass tends to restrain this dilation, the normal compressive stresses acting on the side of the pile will not remain constant but will increase. This phenomenon has been measured and described previously, e.g. by Johnston (1977) and Williams (1980), and it has been shown that, to sufficient accuracy, the normal stiffness of the surrounding rock mass is approximately constant. Hence, it is reasonable to expect that direct shearing under conditions of constant normal stiffness will provide a more realistic laboratory model of the shaft behaviour of concrete piles socketed into rock, than would direct shear testing under conditions of constant normal stress.

Several direct shear devices, capable of applying a variety of constraints on the normal mode of deformation, have already been described in the literature, e.g. Lam and Johnston (1982), Desai et al (1985), Natau et al (1979), and some data are available for interfaces of concrete and artificial mudstone (Johnston and Lam, 1984). This paper describes a new constant normal stiffness, direct shear device capable of applying static and cyclic shear loading to a sample containing one of a variety of possible interfaces. The sample may contain a specially prepared interface such as a concrete-sandstone interface of arbitrary roughness and bonding, it may be formed of two blocks from either side of an artificially prepared or a natural discontinuity (e.g. a joint or a bedding plane), or it may be an intact specimen of rock or cohesive soil which does not contain a pre-existing discontinuity plane. When used in a servo-controlled testing machine the device is capable of applying either load or displacement controlled static or cyclic shear loading. The cyclic shear loading may be either one-way or two-way in nature. The device has been used to study the behaviour of concrete-sandstone interfaces with reference to pile sockets, and a comprehensive set of results is presented in this paper.

2 CONSTANT NORMAL STIFFNESS CONDITION

As suggested above, if dilation accompanies shearing at the interface between a pile and the surrounding rock formation, then to a first approximation the stiffness of the formation with respect to the normal displacement can be regarded as constant. If the pile has a circular cross-section then the formation stiffness can be estimated from the solution for the expansion of a long cylindrical cavity in an elastic continuum. It can be shown that the radial displacement at the cavity boundary is given by:

$$u = (p/2G)a \qquad (1)$$

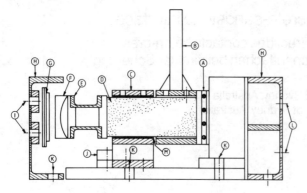

Figure 1. Schematic layout of CNS device.

in which u = the radial displacement at the cavity
wall,
p = the radial normal stress at the cavity
wall,
a = the radius of the cavity, and
G = the shear modulus of the elastic rock
mass.

Equation (1) can be rearranged to give the radial
stiffness as:

$$k = 2G/a \qquad (2)$$

Thus the units of k are stress per unit length.

3 APPARATUS

The constant normal stiffness direct shear device
has been described in detail already by the authors
(Ooi and Carter, 1986) and a sketch of the major
components is given in Fig. 1. The device consists
of two collars which grip a core specimen up to 80
mm in diameter. One of the collars (Item B in Fig.
1) is connected directly to the cross-head of a
servocontrolled testing machine (an INSTRON Model
TT-KM). The cross-head applies the force necessary
to cause direct shear on a circular plane between
the two collars and normal to the axis of the
cylindrical specimen. The second collar (C) is
restrained so that movement can occur only in the
direction along the axis of the specimen. Movement
in this direction will be due to dilation (or
contraction) of the specimen in the immediate
vicinity of the shear plane. Reaction to the normal
displacement at the shear plane is achieved through
first a load cell (E), then a spherical seat (F),
and finally a reaction beam (G). The reaction beam
is used to provide the constant normal stiffness
condition.

4 SAMPLE PREPARATION

The materials used for preparing the interface
samples described here were Hawkesbury sandstone and
concrete. The sandstone, which came from the Sydney
Basin in New South Wales, Australia, typically has a
porosity of about 16%. The unconfined compressive
strengths of the samples used in this testing
program were typically between 15 to 20 MPa. The
strength of the concrete used was generally in
excess of 40 MPa at the time of shear testing.

Samples were manufactured from 76 mm diameter
sandstone cores taken from large rock blocks.
Samples with a smooth interface were prepared by
cutting flat ends of the core with a diamond saw
perpendicular to the core axis. Concrete was then
cast onto one of the smooth ends in a mould to
complete the specimen. Samples with a "rough"
interface were prepared by creating triangular
asperities at one end of the sandstone core and then

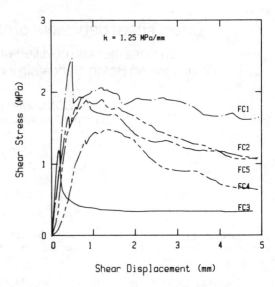

Figure 2. Typical shear stress-displacement curves.

casting concrete onto this end in a manner similar
to that for the smooth interface samples. The
triangular asperities were manufactured by filing
the end of the sandstone core with a carborundum
block until it conformed to the contours of a steel
template. The cross-section of each asperity was an
isosceles triangle with an overall height of 2.5 mm
and a base width of either 9.5 mm or 15.2 mm, giving
the interface a "saw-tooth" appearance in
cross-section. An asperity angle i was defined as
the height of each asperity divided by one-half of
its base width. For the "smooth" interface this
angle was 0°, and for the rough interfaces this
angle was either 27.8° or 18.2°.

For some of the samples, bonding was permitted at
the interface by casting the concrete directly onto
the sandstone. For the remainder, an unbonded
interface was prepared by separating the sandstone
from the concrete by a thin plastic film. All
samples were cured in a humid environment for 28
days after preparation.

5 TEST PROGRAMME

A number of variables will influence the development
of side shear resistance in socketed piles. In order
to study the effects of some of these parameters a
large number of direct shear tests were carried out
in the laboratory under conditions of constant
normal stiffness. The parameters studied, together
with the ranges investigated, are listed below:
(a) initial normal stress - 0.05 to 1.20 MPa,
(b) normal stiffness   - 0.39 to 4.13 MPa/mm,
(c) interface roughness   - i = 0°,18.2°,27.8°,
(d) bonding.

6 TEST RESULTS

A typical set of results is presented in Fig. 2 and
the relevant values of the parameters for these
tests are listed in Table 1. Fig. 2 shows the
average shear stress applied to the interface
plotted against the relative (or shear) displacement
of each half of the specimen. In all cases a peak
shear strength was observed, followed by a reduction
of the strength with further relative displacement.
In most cases the peak occurred at a shear
displacement of about 1 mm or less and a residual
response occurred at a displacement of 5 mm or more.
The tests presented in Fig. 2 were all carried out
at the same normal stiffness and approximately the
same value of initial normal stress. Fig. 2 thus

468

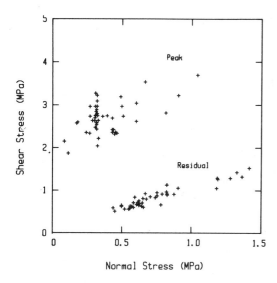

Figure 3. Strength envelopes at peak and residual.

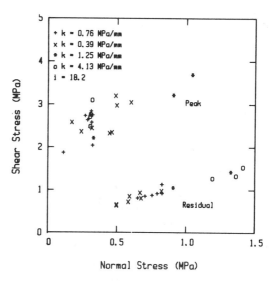

Figure 4. Effect of normal stiffness on strength.

reveals the influence of the interface roughness and bonding on the shear behaviour. A comparison of the curves for samples FC1 and FC5 indicates that the bonded interface was stronger in both the peak and post peak response. The same trend can be observed by comparing FC2 with FC4. Furthermore, it can be observed that the rate of decrease in shear strength after the peak is higher in the case of an unbonded interface, i.e. the behaviour of an unbonded interface is more "brittle" than that of a bonded sample. Fig. 2 also indicates that interface samples with an asperity angle of 18.2° appear to be slightly stronger in the peak and post peak response than samples with an asperity angle of 27.8° (compare FC1 with FC2 and FC5 with FC4). In all cases a rough interface ($i > 0°$) was stronger than a corresponding sample with a smooth interface ($i = 0°$).

Fig. 3, in which shear stress is plotted against normal stress, presents a summary of all the measured peak and residual strengths for rough interfaces ($i > 0°$). It is evident that the residual shear strength is only a function of normal stress and is independent of the angle of roughness, the bonding condition and the normal stiffness during shearing. A residual friction angle of approximately 48° is indicated. The results for the peak shear strength show a large variation, but a trend is fairly well defined. There appears to be no marked difference between the peak strength envelopes for the two types of rough interface tested here ($i = 18.2°$ and 27,8°). This is not to say that the peak strength envelope will be completely independent of the interface roughness when a wider range of asperity angles (including $i = 0°$) is considered. Furthermore, it appears that the peak strength envelope is fairly insensitive to the normal stiffness, as indicated by the selected data plotted in Fig. 4.

In Fig. 5 the data of Fig. 3 have been represented as a plot of the ratio of shear stress to normal stress against the normal stress, and it is interesting to note that the spread in the peak strength data has reduced, and a unique curve can now be fitted to the failure envelope. This demonstrates that both the peak and residual failure envelopes are independent of at least the initial normal stress and the normal stiffness during shearing, i.e. the strength envelopes are path-independent. Leichnitz (1985) has also demonstrated such path-independent strength behaviour for artificially created sandstone interfaces.

Although the peak and residual strength envelopes (stress ratio versus normal stress) are independent of the path taken during testing, the actual values of shear and normal stress at any stage of shear testing, of course, will depend on parameters such as the initial normal stress, the normal stiffness, the interface roughness and the bonding condition. The higher the initial normal stress and the higher the normal stiffness, the more difficult it will be for the interface to dilate. The stress paths plotted in Figs 6, 7 and 8 clearly demonstrate this. Each plot shows the initial test condition prior to shearing and then the stress path from the peak, through softening, to the residual condition. For clarity, the paths from the initial condition up to the peak have not been drawn. In Figs 6 and 7 the normal stiffness is constant at 1.25 MPa/mm and it may be observed that for both asperity angles the amount of dilation, as indicated by the increase in normal stress, is greater at lower initial normal stress levels. Fig. 8 presents stress paths for a relatively high normal stiffness of 4.13 MPa/mm. By comparing this data with Figs 6 and 7 it can be seen that, when all other parameters are constant, a greater increase in normal stress results from a higher normal stiffness during shearing. Although the dilational behaviour tends to be suppressed at higher normal stiffness, the product of the high normal stiffness with a small dilation still results in a large change in normal stress.

Figs 6 and 7 show data from tests at the same normal stiffness on samples with different asperity angles. However, it is difficult to tell from the data what influence, if any, this change in asperity angle has upon the post peak dilation.

7 CONCLUSION

A study has been made of the direct shear behaviour of concrete-sandstone interfaces under conditions of constant normal stiffness. The results of a large number of tests have been presented with most of the data coming from tests on rough interfaces. It was found that the peak and residual strength envelopes are independent of the initial normal stress and the normal stiffness during shearing, i.e. they are independent of the test path. However, the actual stress paths during testing were highly dependent on the initial normal stress and the normal stiffness and to some degree dependent on the roughness of the interface and the condition of bonding. Bonding of the concrete to the sandstone was found to improve the peak and residual strength and also to decrease

469

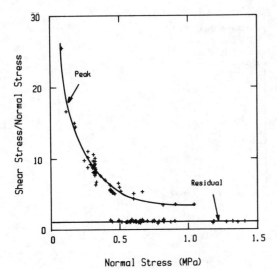

Figure 5. Stress ratio versus normal stress.

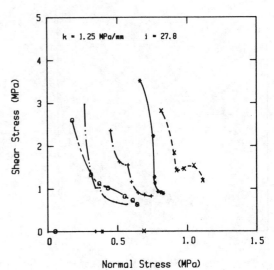

Figure 6. Typical post-peak stress paths.

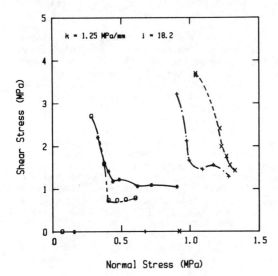

Figure 7. Typical post-peak stress paths.

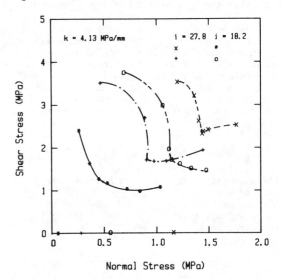

Figure 8. Typical post-peak stress paths.

the rate of post peak softening. Smooth interfaces were found to be extremely brittle in behaviour, particularly when initially bonded.

It appears that the CNS test is useful for modelling the path dependent behaviour of a concrete-rock interface in socketed pile foundations and for determining the path independent strength envelopes.

ACKNOWLEDGEMENT

Support for this work from a University of Sydney Research Grant is gratefully acknowledged.

REFERENCES

Desai, C.S., E.C. Drumm, & M.M. Zaman 1985. Cyclic testing and modeling of interfaces. J. Geotech. Engg, ASCE, 111(6):793-815.
Johnston, I.W. 1977. Rock socketing downunder. Contract J. 279(5155):49-67.
Johnston, I.W. & T.S.K. Lam 1984. Frictional characteristics of planar concrete-rock interfaces under constant normal stiffness conditions. Proc. 4th Australia-New Zealand Conference on Geomechanics, Perth. Inst. Engrs, Australia. 2:397-401.
Lam, T.S.K. & I.W. Johnston 1982. A constant normal stiffness direct shear machine. Proc. 7th South

East Asian Geotechnical Conference, Hong Kong. pp908-820.
Leichnitz, W. 1985. Mechanical properties of rock joints. Int. J. Rock Mechanics and Mining Sciences. 22(5):313-321.
Natau, O., W. Leichnitz & K. Balthasav 1979. Construction of a computer-controlled direct shear testing machine for investigations of rock discontinuities. Proc. 4th Int. Congress on Rock Mechanics, Montreux, Switzerland. 3:241-243.
Ooi, L.H. and J.P. Carter 1986. A constant normal stiffness, direct shear device for static and cyclic loading. Submitted to Geotech. Testing J., ASTM; also Research Report No. 520, School of Civil and Mining Engineering, Univ. Sydney, Australia.
Williams, A.F. 1980. The design and performance of piles socketed into weak rock. Ph.D. Thesis, Dept. of Civil Engg, Monash University, Australia.

Table 1. Test details.

Sample No.	Normal Stiffness (MPa/mm)	Initial Normal Stress (MPa)	Asperity Angle (°)	Bonding
FC1	1.25	0.320	18.2	Yes
FC2	1.25	0.314	27.8	Yes
FC3	1.25	0.312	0.0	Yes
FC4	1.25	0.311	27.8	No
FC5	1.25	0.276	18.2	No

Comparison of rock mass deformabilities determined using large flat jack technique and plate jacking test at two dam sites

Comparaison de déformabilités des massifs rocheux determinées par des essais de 'Flat jack' large (vérin) et d'essai de plaques portantes à deux chantiers de barrage
Vergleich der Gebirgsverformbarkeiten, die durch Belastungversuche und Grossdruckkissenversuche in zwei Talsperren festgestellt wurden

A.ÖZGENOĞLU, Middle East Technical University, Ankara, Turkey
A.G.PAŞAMEHMETOĞLU, Middle East Technical University, Ankara, Turkey
C.KARPUZ, Middle East Technical University, Ankara, Turkey

ABSTRACT : This paper gives the results of the several large flat jack (LFJ) and plate jacking tests carried out at two dam sites to determine the rock mass deformability. The moduli of deformability and elasticity obtained with these methods are compared, and the difficulties associated with the application of LFJ technique in hard rock masses are touched on.

RESUME : Cette article donne les résultats de plusieurs d'essais de "Flat jack" large (LFJ) faites à deux chantiers de barrages pour déterminer la déformabilité des massifs rocheux. Les module de déformation et d'élasticité obtenues avec ces methodes sont comparées, et les difficultés associées avec l'application de LFJ technique à des rocheux dures sont discutees.

ZUSAMMENFASSUNG : Der vorliegende Aufsatz veröffenlicht die Ergebnisse der mehreren Belastungsversuchen und Grossdruckkissenversuchen die um Bestimmung der Gebirgsverformbarkeiten in Zwei Talsperren durchgefürt wurden. Die, durh diese methode erhaltene Verformbarkeitsmodul und Elastizitätsmodul sind vergleicht und die mit Grossdruckkissentechnikverwendungen im hart Felsmassiv verbundenen Schwierigkeiten sind erwähnt.

1. INTRODUCTION

Several dams are planned to be constructed on the river Ceyhan which flows into the Mediterranean, on the south of Turkey. Sır and Düzkesme dams are two of these dams. This paper deals with part of the investigations carried out to evaluate the deformability of the rock mass for the final design of aforementioned dams.

Moduli of deformability and elasticity of the rock mass have been determined with two different methods, namely with large flat jack (LFJ) technique, and with plate jacking (PJ) tests. A total of nine LFJ and nine PJ tests have been carried out at both dam sites. Three of the LFJ tests have been repeated after consolidation grouting to assess the improvement in rock mass characteristics. Six of the pre-grouting and all of the post-grouting LFJ tests were associated with seismoacoustic measurements to determine the wave-front velocities.

In this paper, the rock mass deformability obtained with LFJ and PJ methods are compared and the problems associated with the LFJ technique in the hard,brittle and abbrasive rock formation (metaquartzitic - sandstone) are briefly given. Since the LFJ test results relevant to post-grouting tests and seismo-acoustic measurements have been given elsewhere (Paşamehmetoğlu et al. 1984) they are not included in this paper.

2. GEOLOGY

The dominant rock type at both dam sites has been identified to be metaquartzitic-sandstone (Özgenoğlu et al. 1982). At Düzkesme dam axis, however, there occurs same phyillite intercalations within this formation. The metaquartzitic-sandstone series have acquired a weak schistosity as a result of the regional dynamothermal metamorphism.

The major discontinuity at both dam sites is the bedding plane along which some sliding and shearing have apparently taken place. The evidence for this movement are the well-polished faces and slicken-sides on these planes. Less frequently, however, shear planes developed at an angle to the bedding. Two or three joint sets have been observed as the other discontinuity planes besides bedding in the area. Apertures between the shear-slide surfaces are generally filled with silt-clay filling, and in places with quartz veins.

3. LARGE FLAT JACK TESTS

The large flat jack technique developed by the LNEC-Portugal (Rocha and Silva 1970) have been used. The tests were conducted using six loading-unloading cycles, whenever possible. The cyles consisted of sufficient number of loading and unloading increments. Stabilized deformations have been measured and used to evaluate the moduli values for each cycle and for each deformeter of the jack.

Seven LFJ tests have been performed at Sır dam site, and two at Düzkesme dam site. Locations of the testing points and positions of the jacks are given in Table 1.

Table 1. LFJ test locations and jack positions.

Dam site	Test no.	Location	Chainage	Position
Sır	FJ1	LA-106	25.50 m	N26E 90 Bottom
"	FJ2	LA-106	23.00 m	N38E 28NW Right wall
"	FJ3	RA-105	30.50 m	N07W 29NE Left wall
"	FJ4	RA-103	41.00 m	N15W 65NE Bottom
"	FJ5	LA-104	29.50 m	N38E 60NW Bottom
"	FJ6	LA-106	47.00 m	N30E 27NW Right wall
"	FJ7	LA-102	38.50 m	N10E 27NW Face wall
Düzkesme	FJ8	RA-202	30.50 m	N45E 90 Bottom
"	FJ9	LA-201	77.00 m	N10E 75NW Bottom

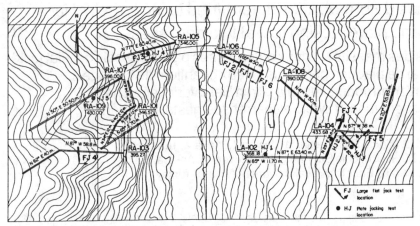

Figure 1. Layout of the large flat jack and plate jacking test locations in Sır dam adits.

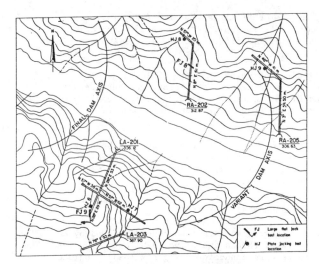

Figure 2. Layout of the large flat jack and plate jacking test locations in Düzkesme dam adits.

Figures 1 and 2 show the layout of the testing places in Sır and Düzkesme dam adits respectively.

In calculating the deformation modulus (D) and modulus of elasticity (E), the LFJ constant, K, has been taken as 196.7 cm. LNEC documents and the work of Loureiro-Pinto (1981) as well as the field observations have been considered in assessing the value of K.

3.1. Test results

LFJ tests yielded deformation and elasticity moduli values showing quite a variation for both dam sites. The variation in test results depended on:
  a. test location
  b. pressure interval, and
  c. deformeter position.
Deformation moduli varied between 1 260 MPa and 23 700 MPa at Sır dam site, and between 5 950 MPa and 27 350 MPa at Düzkesme dam site. Modulus of elasticity values, as expected, have been higher than the deformation moduli. The lowest and the highest E were 2 905 MPa and 25 398 MPa for Sır dam site, and 11 580 MPa and 52 282 MPa for Düzkesme dam site.

Almost in all tests, the bottom deformeters yielded higher modulus values as compared to the top ones. However, pressure level had an inconsistent effect on modulus values. While the modulus value in some tests and for some deformeters has been dropping with increasing pressure, it increased in the others. The increase in modulus value with increasing pressure

have been observed in tests at Düzkesme dam site mostly. Test results, for the minimum and maximum pressure intervals applied at each location, as the average modulus value of the top deformeters and the bottom deformeters separately, are given in Table 2.

Table 2. Large flat jack test results.

Test no.	Pressure interval (MPa)	Deformation Modulus (MPa)		Modulus of Elasticity (MPa)	
		Mean of top deformeters	Mean of bottom deformeters	Mean of top deformeters	Mean of bottom deformeters
FJ1	0.2-1.0	5 282	22 208	13 553	24 566
	0.2-5.0	4 154	12 225	6 315 (x)	13 155
FJ2	0.2-1.0	5 569	9 392	8 423	14 006
	0.2-3.5 (a)	2 417	3 165	–	–
FJ3	0.2-1.0	2 058	5 864	7 404	16 727
	0.2-7.0	2 924	5 263	9 666	7 912
FJ4	0.2-1.0	5 525	8 567 (x)	10 729	8 994
	0.2-5.0	3 658	28 292 (b)	6 315	15 243
FJ5	0.2-1.0	6 102	9 919	13 661	18 813
	0.2-5.0	2 122	3 830	3 994	6 590
FJ6	0.2-1.0	1 742 (x)	8 663	2 905 (x)	10 844
	0.2-3.5	3 935	7 537	13 771	10 270
FJ7	0.2-1.0	1 911	5 309	8 339	20 925
	0.2-5.0 (a)	3 670	7 812	4 666	8 567
FJ8	0.2-1.0	9 154	26 325 (x)	15 640	34 470 (x)
	0.2-5.0	9 831	15 079	13 849	24 728
FJ9	0.2-1.0	9 669	11 138 (x)	24 041	28 955 (x)
	0.2-6.0	12 797	15 662 (x)	20 756	29 355

(x) The value obtained for one of the deformeters.

(a) The value found for zero-time displacement.

(b) Anomalies in strain readings.

3.2. Difficulties in the practice

There have been some difficulties associated with the LFJ tests. These difficulties can be divided into two groups as the troubles pertaining to slot preparation and the difficulties in attaining the foreseen maximum pressure of 7.5 MPa.

It has been very difficult both to drill the 17 cm diameter central hole and to cut the 100 cm wide slot due to the hard, abrasive and fractured character of the formation (quartzitic sandstone). Especially, the drilling and cutting of the bottom 15-20 cm necessitated great efforts because of the squeeze of the drill and the saw. Moreover, the rapid wearing of the diamond drill bit and diamond disc saw, and the damage on these equipment resulting from the small rock fragments that have fallen into the hole or slot have caused significant delays in the program.

The difficulty in attaining the foreseen maximum pressure (7.5 MPa) has arisen mainly from the central hole which was produced inevitably. The filling of this hole with half cylinders of hardwood did not serve the purpose and it had to be filled with mortar.

Table 3. Plate jacking test results.

Dam site	Test no.	Horizontal test					Vertical test				
		Maximum pressure (MPa)	Deformation Mod. (MPa)		Mod. of Elasticity (MPa)		Maximum pressure (MPa)	Deformation Mod. (MPa)		Mod. of Elasticity (MPa)	
			Left plate	Right plate	Left plate	Right plate		Top plate	Base plate	Top plate	Base plate
Sır	HJ1	6.0	6 129	12 440	16 794	55 979	6.0	5 183	6 151	47 982	6 586
"	HJ2	6.0(x)	3 076	6 114	12 348	62 976	6.0	9 707	4 262	62 198	9 542
"	HJ3	7.5	2 548	5 495	87 467	20 580	7.5	1 066	6 323	6 560	18 743
"	HJ4	6.0	7 056	5 872	27 989	33 587	6.0	4 396	11 196	59 977	59 977
"	HJ5	6.0	6 129	14 731	38 167	167 936	6.0	3 400	5 089	43 061	47 982
Düzkesme	HJ6	7.5	38 874	22 572	149 943	209 920	6.0	7 302	8 035	37 319	83 968
"	HJ7	7.5	10 496	15 435	52 480	51 200	6.0	15 129	10 765	152 669	36 508
"	HJ8	6.0	7 565	6 940	38 167	18 254	6.0	5 383	4 798	52 480	15 695
"	HJ9	6.0	7 668	7 959	119 954	79 970	7.5	6 560	7 497	26 240	26 240

(x) It has been 4.5 MPa for the right plate.

Therefore, none of the jacks could be recovered at all. Furthermore, some tests had to be stopped before the maximum pressure was reached due to the causes such as the premature formation of the crack in the slot plane resulting from the holes drilled on both ends of the jack for seismoacoustic measurements, and the oil leakage from the wiring inlet point of the jack.

4. PLATE JACKING TESTS

The moduli of deformability and elasticity of the rock masses at Sır and Düzkesme dam sites have also been determined by plate jacking test which is the most common in-situ test used in rock mechanics. These tests have been carried out by the state owned establishment E.İ.E. (General Directorate of Electrical Survey Administration) at nine locations of which five were at Sır dam site and four at Düzkesme dam site. Both the horizontal and vertical tests were performed at each location. Test locations are given in Figures 1 and 2.

In majority of the tests maximum pressure of 6.0 MPa were attained with six cycles. The maximum pressure was 7.5 MPa in few tests. Test results for both plates are given in Table 3.

Plate jacking test results for both moduli have shown a high scatter similar to LFJ test results. This scatter has been less for Düzkesme dam site as was the case in LFJ tests. Plate jacking tests yielded values for modulus of deformation varying between 1066 MPa and 14 731 MPa, and for modulus of elasticity varying between 6560 MPa and 167 936 MPa at Sır dam site. In other words, the maximum modulus/minimum modulus ratio has been 13.82 for deformation modulus and 25.60 for elasticity modulus. This ratio has been 38 874/4 798= 8.10 for deformation modulus and 209 920/15 695= 13.38 for elasticity modulus at Düzkesme dam

site. A similar situation was seen when the horizontal and vertical test results were evaluated separately. The average of the ratio of the modulus values for the opposite plates (i.e. modulus value for right plate / modulus ratio for left plate or modulus value for top plate/ modulus value for base plate), both for horizontal tests and for vertical tests, and both for modulus of deformability and modulus of elasticity, has been higher at Sır dam site when compared with Düzkesme dam site. When the horizontal and vertical tests were compared, the average of the ratio of the modulus values for opposite plates has been higher for vertical tests.

5. COMPARISON OF LFJ AND PJ TEST RESULTS

Large flat jack tests and plate jacking tests, in general, yielded modulus of deformability results which are in harmony for both dam sites. A similar harmony exists for the minimum values of modulus of elasticity but not for the maximum values which have been found higher in plate jacking test. In other words, while the variation in modulus of elasticity has been between 2 905 MPa and 25 398 MPa at Sır dam site, and between 11 580 MPa and 52 282 MPa at Düzkesme site for the large flat jack tests, it was between 6 560 MPa and 167936 MPa at Sır dam site, and between 15 695 and 209 920 MPa at Düzkesme dam site for plate jacking tests.

To compare the LFJ and PJ tests on location basis, five of these tests, which were located closely, have been considered. The results are pretty consistent as seen in Table 4. Especially when the mean values are taken into consideration, it is evident that the results for the tests FJ7 and HJ2, and for FJ5 and HJ3 are very close. Test results, excluding the FJ8 and HJ8, further indicate that the mean deformability moduli obtained from the LFJ tests are smaller than

Table 4. Comparison of large flat jack and plate jacking tests on location basis.

Test no. (Maximum pressure in MPa)	Deformation modulus (MPa)					
	Large flat jack test		Plate jacking test			
	Top deformeters	Bottom deformeters	Horizontal test		Vertical test	
			Left plate	Right plate	Top plate	Base plate
FJ3(7.0)-HJ4(6.0)	2 924	5 263	7 056	5 872	4 396	11 196
	Mean: 4 094			Mean: 7 130		
FJ7(5.0)-HJ2(6.0)	3 670	7 812	3 076	6 114	9 707	4 262
	Mean: 5 741			Mean: 5 790		
FJ5(5.0)-HJ3(7.5)	2 122	3 830	2 548	5 495	1 066	6 323
	Mean: 2 976			Mean: 3 858		
FJ9(6.0)-HJ6(7.5)	12 797	15 662	38 874	22 572	7 302	8 035
	Mean: 14 230			Mean: 19 196		
FJ8(5.0)-HJ8(6.0)	9 831	15 079	7 565	6 940	5 383	4 798
	Mean: 12 455			Mean: 6 171		

473

the values obtained from the PJ tests. This is not
surprising because for LFJ tests, the volume of rock
subjected to load is greater than for plate jacking
tests.

6. CONCLUSIONS

The large flat jack and plate jacking tests carried
out at Sır and Düzkesme dam sites yielded consistent
modulus values. This consistency was found to be
better for deformation modulus as compared to modulus
of elasticity.
  Large flat jack technique, although comprises
difficulties in hard, brittle rock formations can be
recommended as a second method for the assessment of
the rock mass deformability if more than one method
is needed.

ACKNOWLEDGEMENT

Acknowledgement is made to the authorities of E.İ.E.
for providing the data relevant to plate jacking
tests.

REFERENCES

Loureiro-Pinto, J. 1981. Determination of the
  deformability modulus of weak rock masses by means of
  large flat jacks (LFJ). Proc.Int.Symp. on Weak
  rock (Tokyo), p. 447-452.
Özgenoğlu, A., T. Ataman and others 1982. Sır and
  Düzkesme dam sites rock mechanics investigations
  (Final Report). Middle East Technical University,
  Ankara.
Paşamehmetoğlu, A.G., A. Özgenoğlu, C. Karpuz and
  A.Bilgin 1984. Response of rock mass to grouting:
  a case history. Proc.Int.Conf. on In situ soil and
  rock reinforcement (Paris), p. 173-177.
Rocha, M. and J.N. Silva 1970. A new method for the
  determination of deformability in rock masses.Proc.
  2nd Congr. ISRM (Beograd), Vol.1, p.423-437.

Beitrag von Bauwerksmessungen zur Beurteilung der Gründung einer Staumauer

The role of measurements for the judgement of a dam foundation
Contribution des mesurements pour le jugement de la fondation d'un barrage

P.RISSLER, Dr.-Ing., Ruhrtalsperrenverein Essen, Bundesrepublik Deutschland

SUMMARY: The paper describes devices for and findings of long-term observations at a gravity dam operated since 80 years and the conclusions concerning the stability of the foundation.

RESUME: La contribution décrit les installations et les résultats des observations à long terme d'un barrage-poids de 80 ans et les conclusions concernant la stabilité de la fondation.

ZUSAMMENFASSUNG: Der Beitrag beschreibt Einrichtungen und Erkenntnisse jahrzehntelanger Bauwerksmessungen an einer seit 80 Jahren in Betrieb befindlichen Gewichtsstaumauer und die Schlußfolgerungen hinsichtlich der Standsicherheit der Gründung.

1. EINLEITUNG

Wir sind gewohnt, uns die Beanspruchung eines Bauwerks als Spannungen oder als Kräfte vorzustellen. Diese setzen wir in Beziehung zu sogenannten zulässigen und beschreiben das Verhältnis zwischen zulässigen und aufnehmbaren durch Sicherheitsbeiwerte. Dieses Konzept versagt, wenn bei natürlichen Baukörpern, wie z.B. dem Felsuntergrund, die Größe der aufnehmbaren Spannungen bzw. Kräfte nur in weiten Grenzen abgeschätzt werden kann. Eine Sicherheitsbetrachtung nach dem Schema

$$\frac{\text{zulässige Spannung}}{} = \frac{\text{aufnehmbare Spannung}}{\text{Sicherheitsbeiwert}} \quad (1)$$

ist dann fragwürdig. Sie kann sogar zu falschen Schlußfolgerungen führen, je nachdem, ob der Sicherheitsbeiwert den Unsicherheiten zufällig gerecht wird oder nicht. Es ist hier daran zu erinnern, daß Sicherheitsbeiwerte in der Vergangenheit zumeist nicht rational abgeleitet wurden, sondern oftmals der Tradition entspringen.

Sicherheitsbeiwerte dürfen daher nicht als tatsächlicher Abstand zwischen Beanspruchung und Bruch interpretiert werden. Ebenso wenig dürfen sie als Maßzahl für die Sicherheit eines Bauwerks angesehen werden. Sie sind vielmehr ausschließlich ein Maß für unsere Einschätzung über die Größe diesbezüglicher Kenntnislücken zum Zeitpunkt der Berechnung.

Diese Kenntnislücken sind zur Zeit der Planung größer als zur Zeit der Bauausführung und da wiederum größer als nach einer 80-jährigen Betriebszeit. Von daher könnte man im Prinzip, mit zunehmendem Lebensalter des Bauwerks bzw. mit dem angesammelten Wissen über sein Verhalten, bei Überprüfungen der Standsicherheit die rein rechnerische, formale Sicherheitsspanne verringern. In der Praxis wird darauf gewöhnlich verzichtet, weil

- die ursprünglichen Annahmen auf der sicheren Seite liegen,

- es schwierig wäre, dem jeweiligen Kenntnisstand entsprechende Sicherheitsspannen festzulegen.

Man kann diesem Umstand jedoch dadurch Rechnung tragen, daß eine Beurteilung der Standsicherheit nicht nur auf Berechnungsergebnisse, sondern auch auf andere zuverlässige Informationen, wie z.B. Bauwerksmessungen, abgestützt wird.

2. DER STELLENWERT VON VERFORMUNGSMESSUNGEN

Berechnungen und Bauwerksmessungen ergänzen sich gegenseitig.

Berechnungen allein fehlt der Bezug zur Wirklichkeit. Ihre Ergebnisse sind Hypothesen über das Verhalten des Bauwerks, mehr nicht. Der Nachweis der Übereinstimmung mit der Realität bleibt offen.

Bauwerksmessungen (gemeint sind hier im wesentlichen Verformungsmessungen) allein vermitteln zwar Hinweise auf das tatsächliche Verhalten, jedoch nur punktuell und für bestimmte Zeitpunkte. Sie sind immer lückenhaft. Es fehlt ein Bindeglied, um aus diskreten Meßwerten ein Gesamtbild des Verhaltens zu formen und Zwischenwerte sinnvoll zu interpolieren.

Das eigentliche Aufgabenfeld der Berechnung ist die Prognose, also die Vorausschau auf einen noch nicht existierenden Zustand. Die gängigen Berechnungsverfahren sind durchwegs für den Entwurf entwickelt worden.

Die Verformungsmessung am Bauwerk stellt hingegen fest, was gegenwärtig ist. Dementsprechend ist ein Standsicherheitsnachweis vor der Ausführung nur durch Berechnung möglich. Für ein im Betrieb befindliches Bauwerk sollte dagegen die Bauwerksmessung - vorausgesetzt, es liegen ausreichend umfangreiche und sorgfältig erstellte Ergebnisse vor - die eigentliche Beurteilungsgrundlage sein. Der Berechnung kommt dabei die Aufgabe der Interpretationshilfe zu.

Am Beispiel einer 80 Jahre alten Gewichtsstaumauer wird in der Folge der Beitrag von Verformungsmessungen zur Beurteilung der Gründung einer Staumauer aufgezeigt.

3. BEISPIEL FÜRWIGGEMAUER

3.1 BAUWERK, MESSEINRICHTUNGEN, EINPRESSUNG

Die Gewichtsstaumauer der Fürwiggetalsperre wurde 1902 - 1904 erbaut. Sie ist 29 m hoch und steht auf groß- und engräumig verfalteten unterdevonischen

Abb. 1 Fürwiggemauer

Schichten graublauer und grünlichgrauer Schiefer mit
Einlagerungen von Grauwackengesteinen. Erkennbar sind
in zwei kürzlich aufgeschlossenen Großschurfen eine
etwa talparallel streichende Schichtung mit unterschiedlichem, doch im wesentlichen flachem Einfall-
winkel sowie vier steil einfallende Kluftscharen, von
denen jedoch nur zwei (K1 und K2) so weit durchtrennt
sind, daß sie als mechanisch relevant angenommen
werden müssen. Die Schichtung tritt als Trennfläche
kaum in Erscheinung. Über die Gründungsgeometrie
liegen ausführliche Unterlagen vor.

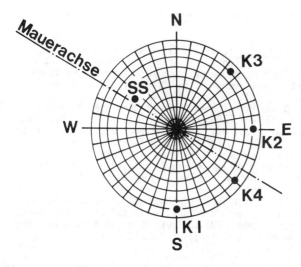

Abb. 2 Trennflächensystem (linker Hang)

Bereits vor dem ersten Einstau wurde an der Mauerkrone eine Alignementmeßeinrichtung geschaffen. Diese
bestand aus einer Visierstrecke A, D und zwei Zielmarken B, C, die auf Brüstungspfeilern nahe der Hoch-
wasserentlastungsanlage montiert waren. Das System
ist irgendwann zerstört worden. Derzeit sind noch die
Basisplatten der Zielmarken auf den Brüstungspfeilern
zu erkennen. Es liegen jedoch lückenlose Meßergebnisse aus der Zeit von 1904 bis 1928 vor.

1961 wurde Mauer und Untergrund zum wiederholten
Male verpreßt. Dazu wurde der Speicher entleert. Dies
wurde auch zum Anlaß genommen, ein geodätisches Meßsystem einzurichten. Es besteht aus zwei nach rück-
wärts versicherten, massiven Beobachtungspfeilern und
14 Meßmarken an der luftseitigen Maueroberfläche
(Abb. 3).

Diese Marken werden seitdem zweimal jährlich, im
Frühjahr und im Herbst (bei annähernd gleichen Temperaturbedingungen), eingemessen. Daraus werden radiale
und tangentiale Verschiebungsbeträge gerechnet. Anzu-

merken ist, daß die Nullmessung und eine Wiederho-
lungsmessung 1961 vor der Entleerung der Sperre
stattfanden.

Aus den geodätischen Messungen ist zu entnehmen,
daß die Mauer seinerzeit beim Einpressen geringfügig
in Richtung Wasserseite, in Talmitte und nahe der
Krone etwas stärker, an den Seiten und unten etwas
schwächer, bewegt wurde (Abb. 4). Im Mittelschnitt
durch die Mauer sind dies translatorisch 2,8 mm und
eine Verdrehung zur Wasserseite von etwa 1,7'.

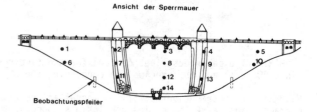

Abb. 3 Geodätisches Meßsystem

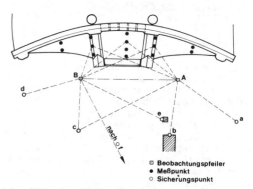

Abb. 4 Irreversible Verschiebungen während der
         Sanierung

Anhand der Bauberichte läßt sich aus dem aufgenom-
menen Einpreßgut im Untergrund eine Gesamtstärke des
Zementsteins von 4,3 cm herleiten. Natürlich ist dieser über eine große Zahl von Trennflächen verteilt.
Aus der Tatsache, daß das Bauwerk sich beim Einpressen verdreht hat, dürfte zudem zu schließen sein, daß
das Einpreßgut ziemlich weit bis unter die, zu diesem
Zeitpunkt entlastete, luftseitige Gründung vorgedrungen ist, vermutlich im wesentlichen über die Klüfte
K1 und K2.

3.2 BEURTEILUNG DER VERFORMUNGSMESSUNGEN SEIT 1961

Fürwiggemauer und Gründung reagieren zunächst im
wesentlichen elastisch auf die Wasserdruckbeanspru-

chung. Es ist außerdem zu erkennen, daß sich das nahezu symmetrische Bauwerk auch nahezu symmetrisch verformt. Die Auslenkung ist in Talmitte und an der Mauerkrone etwas größer als an den Seiten und unten (Abb. 5).

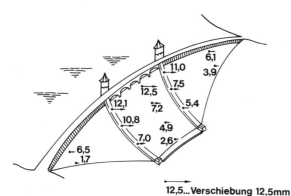

Abb. 5   Verschiebungen durch Wasserdruck

Es sei an dieser Stelle angemerkt, daß die derzeitigen elastischen Verformungen des Bauwerks infolge Wasserdruck gut mit denen der alten Alignementmessungen korrespondieren. Daraus darf - im wesentlichen für die Mauer selbst - geschlossen werden, daß sich die Steifigkeit seit damals nicht verändert hat.

Für die Beurteilung der Standsicherheit der Gründung ist insbesondere die zeitliche Entwicklung der Verformungen interessant. Wir konzentrieren uns dazu hier auf den Meßquerschnitt in Talmitte, welcher den Ort der größten Beanspruchungen des Bauwerks repräsentiert. Der Meßpunkt 14 liegt dort nahe an der Gründungssohle und spiegelt daher weniger das Mauerverhalten, als vielmehr das Verhalten der Gründung wider. Daneben soll auch der oberste Meßpunkt, 3, betrachtet werden.

Wir haben zu beiden Meßpunkten die zeitliche Entwicklung der Meßwerte (nur für Zustände annähernd oder gleich Vollstau) und den zeitlichen Verlauf der Stauhöhe dargestellt. Die Stauhöhe ist zudem auf die resultierende Wasserdruckkraft W umgerechnet (Abb. 6 und 7).

Es sind deutlich zwei Bereiche zu unterscheiden:

1961 bis 1969:
Die Talsperre war praktisch immer voll. Der Untergrund stand unter mehr oder weniger konstanter Dauerlast. Die Meßwerte waren - im Rahmen der üblichen Streubreite - konstant.

1970 bis zur Gegenwart:
Der Stauspiegel schwankte in höherem Maße. Der Untergrund wurde zyklisch beansprucht. Beide Meßpunkte zeigen Bewegungen zur Luftseite an.

Ähnliche Tendenzen sind auch bei den übrigen Meßpunkten erkennbar. Die seit 1970 gemessenen Bewegungen sind zwar durchwegs sehr klein. Dennoch ist eine sorgfältige Analyse angezeigt.

Erste Hinweise liefert ein Vergleich mit den Verschiebungen während des Einpressens 1961 (Abb. 8 und 9). Dazu sind für die Meßpunkte 3 und 14 alle Meßwerte seit 1961 der jeweils zum Meßzeitpunkt vorhandenen resultierenden Wasserdruckkraft W gegenübergestellt.

Es ist zu erkennen, daß die 1961 vor der Maßnahme vorhandene Position der Mauer noch nicht wieder erreicht ist. Die Translation ist zum Teil zurückgegangen, die Rotation kaum.

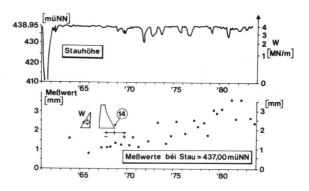

Abb. 6   Meßpunkt 14 - Zeitliche Entwicklung der Meßwerte

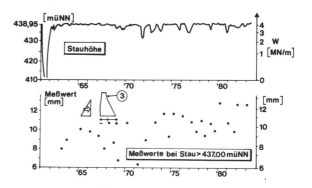

Abb. 7   Meßpunkt 3 - Zeitliche Entwicklung der Meßwerte

Dies legt den Schluß nahe, daß beim Einpressen eine Spannungsumlagerung stattgefunden hat, die nunmehr durch Kriechvorgänge allmählich rückgängig gemacht wird. Die Richtigkeit dieser These würde die Prognose erlauben, daß die irreversiblen Verschiebungen spätestens dann zum Stillstand kommen, wenn das Bauwerk seine Lage von 1961 (vor der Einpressung) wieder erreicht hat.

Die Messungen belegen, daß die Kriechvorgänge nicht bei Vollstau, sondern erst bei Wasserspiegelschwankungen einsetzen und offenbar sich auch dadurch weiter entwickeln. Warum dies so ist, kann bisher theoretisch nicht im einzelnen begründet werden. Es ist jedoch festzustellen, daß die Bewegung nicht zeitabhängig, sondern abhängig von Häufigkeit und Größe der Wasserspiegelschwankungen bzw. von Häufigkeit und Größe der Schwankungen aus der Wasserdruckbeanspruchung ΔW verläuft. Die zu einem beliebigen Zeitpunkt vorhandene irreversible Verschiebung Δl_{irr} ist daher eine Funktion der Summe aller bis dahin zu verzeichnenden Schwankungen der Wasserdruckbeanspruchung

$$\Delta l_{irr} = f(\Sigma \Delta W). \qquad (2)$$

Damit wird eine Einschätzung der vergangenen, gegenwärtigen und auch zukünftigen irreversiblen Verschiebungen möglich, nämlich aus der jeweiligen, auf $\Sigma_\Delta W$ bezogenen Veränderung Δs, dem Differenzenquotienten

$$\Delta s = \frac{\Delta(\Delta l_{irr})}{\Delta(\Sigma \Delta W).} \qquad (3)$$

Dieser ist ein Beurteilungskriterium dafür, ob eine Tendenz zur Progression oder zur Beruhigung besteht.

477

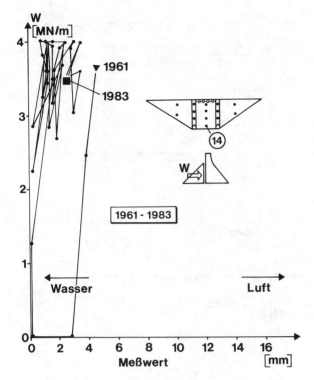

Abb. 8 Radialverschiebungen Meßpunkt 14

Es spricht nach dieser Auswertung alles dafür, daß die Bewegungen in absehbarer Zeit zum Abschluß kommen werden. Die weiter oben skizzierte These, daß die Bewegungen als Anzeichen für den allmählichen Abbau einer beim Einpressen im Gründungsbereich erzeugten Spannungsumlagerung anzusehen sind, wird dadurch gestützt.

Die Gründung der Fürwiggemauer ist daher gegenwärtig als sicher zu beurteilen. Das Bewegungsverhalten der Mauer auf dem Untergrund wird auch in Zukunft laufend überwacht. Zusätzlich werden wir die Meßeinrichtungen ergänzen. Damit ist diese Beurteilung, gestützt auf die bisherigen Informationen, auch in Zukunft jederzeit möglich. Wir glauben, daß das Verhalten des Bauwerks gegen Abgleiten durch diese Messungen besser beurteilt werden kann, als durch jeden rechnerischen Nachweis.

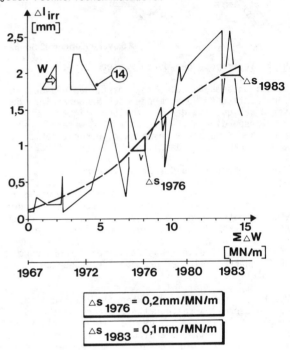

Abb. 10 Kriechneigung Meßpunkt 14

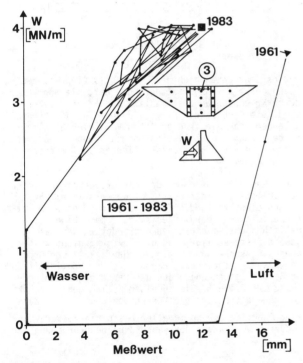

Abb. 9 Radialverschiebungen Meßpunkt 3

Gleichung 2 ist für die Meßpunkte 3 und 14 anhand der Meßergebnisse ausgewertet worden (Abb. 10 und 11). Jeder Knickpunkt entspricht einem Wertepaar ΔW und Δl_{irr}, der verbindende Polygonzug der chronologischen Reihenfolge. Diese wird zusätzlich durch eine horizontale, nicht lineare Zeitachse veranschaulicht. Eine ausgleichende Kurve (gestrichelt) schaltet Sekundäreinflüsse (Temperatur, Meßfehler usw.) aus. Die Steigung, identisch mit Δs gemäß Gleichung 3, spiegelt für jeden Zeitpunkt die augenblickliche Kriechneigung wider. Für beide Meßpunkte ist übereinstimmend festzustellen, daß diese etwa 1976 ihr Maximum erreichte und bis 1983 deutlich abgeflacht ist.

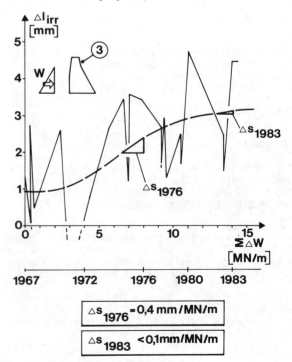

Abb. 11 Kriechneigung Meßpunkt 3

478

Développement des modèles numériques dans l'analyse de la propagation des éboulements rocheux

Development of numerical models for the analysis of propagation of rock-falls
Entwicklung numerischer Modelle zur Analyse der Ausbreitung von Felsrutschen

L.ROCHET, Laboratoire des Ponts et Chaussées, France

ABSTRACT:

The study of the propagation phenomena of rock-falls takes advantage of the development of numerical modelization and computer-simulation methods. This paper presents the main results obtained from the investigations performed in this field by the Laboratoire des Ponts et Chaussées (France).

In the field of rock-falls which are the most frequent it has been defined a process of propagation with small interaction (independent process). Two types of modelization have been developed :
- models with envelope trajectories, aiming at the analysis of the limit conditions of propagation
- models with random variables, aiming especially at analyzing dispersion

Concerning the rock-falls implying large masses, the interaction mechanisms of moving materials are predominent. A unit approach is proposed via the modelization of large propagation phenomena taking account of the variation in functions which characterize the internal energy dissipation as a function of the moving mass.

The various models presented here are illustrated by the application to the analysis of several typical sites.

RÉSUMÉ:

L'étude des phénomènes de propagation des éboulements rocheux bénéficie du développement des méthodes de modélisation numérique et de simulation sur ordinateur. La communication présente les principaux résultats des recherches développées dans ce domaine par le Laboratoire des Ponts et Chaussées (France).

Dans le domaine des éboulements les plus fréquents on définit un mode de propagation à faible interaction (mode indépendant). Deux types de modélisation ont été développés :
- modèles à trajectoires enveloppes, orientés vers l'analyse des conditions limites de propagation.
- modèles à variables aléatoires orientés plus particulièrement vers l'analyse de la dispersion

Dans le domaine des éboulements en très grande masse les mécanismes d'interaction des matériaux en mouvement deviennent prépondérants. Une approche unitaire est proposée à travers une modélisation prenant en compte une variation des fonctions caractérisant la dissipation interne d'énergie en fonction de la masse en mouvement.

Les différents modèles présentés sont illustrés par des applications à l'analyse de plusieurs sites caractéristiques.

ZUSAMMENFASSUNG:

Die Untersuchung der Ausbreitungsvorgänge von Felsrutschen ist durch die Entwicklung numerischer Modellmethoden sowie Rechner-Simulations-methoden begünstigt. Dieser Text beschreibt die Hauptergebnisse von Untersuchungen, die durch den Laboratoire des Ponts et Chaussées (Frankreich) durchgeführt werdeN.

Im Gebiet der häufigsten Rutschen wird ein Ausbreitungsprozess mit schwacher gegenseitiger Beeinflussung (unabhängiger Prozess) definiert :
- modelle mit Umhüllen-Bahne, zur Analyse der Ausbreitungs-Randbedingungen
- modelle mit regellosen Variablen, besonders zur Analyse der Verbreitung

Im Gebiet der umfangreichen Rutschen sind die Wechselwirkungsvorgänge überwiegend. Eine einheitliche Annäherung wird mittels einem Modellmachen der umfangreichen Ausbreitungsvorgänge vorgeschlagt, die eine Annäherung der die innere Energiedissipation gemäss der bewegenden Masse charakterisierenden Funktionen.

Die verschiedene beschriebene Modelle sind durch Anwendungen auf die Analyse mehrerer kennzeichnender Baustellen.

1 - INTRODUCTION

Les mécanismes de rupture et d'instabilité qui affectent les massifs rocheux superficiels (versants rocheux, falaises) participent à l'évolution naturelle des pentes. Les phénomènes d'éboulements rocheux constituent une phase ultime d'évolution spectaculaire et brutale. Pour les éboulements importants, l'énergie mise en jeu peut être considérable et le pouvoir destructeur des masses en mouvement très élevé.

La prise en compte des risques naturels liés aux instabilités rocheuses repose sur l'analyse de trois problèmes principaux :
- la localisation des zones d'instabilité potentielle, zones origine du risque, à l'échelle de temps considérée
- l'évaluation de l'aléa de rupture générateur de l'éboulement. Le développement de la rupture jusqu'à sa phase ultime dynamique apparait comme un phénomène aléatoire dont l'incertitude croît inversement à l'échelle de temps considérée

- l'estimation de la propagation des masses éboulées résultant de l'interaction de multiples paramètres. La complexité des mécanismes, l'incertitude sur les paramètres ou l'inaccessibilité de la plupart d'entre eux déterminent un caractère fondamentalement aléatoire des phénomènes de propagation.

Le développement des méthodes de modélisation numérique permet, comme dans d'autres domaines de la mécanique des roches, la mise au point d'outils de simulation et d'analyse sur ordinateur des phénomènes de propagation des éboulements rocheux. Les principaux résultats des travaux effectués au Laboratoire des Ponts et Chaussées (France) sont présentés dans cette communication. Ces modèles de simulation font l'objet de nombreuses applications à l'étude des risques et des mesures de protection pour les sites exposés.

2 - PRINCIPAUX TYPES D'EBOULEMENTS

D'une manière générale la définition du domaine des éboulements rocheux peut être étendue à l'ensemble des phénomènes de déplacement rapide de masses rocheuses mises en mouvement à l'issue d'un processus de rupture. Contrairement aux phases de rupture et de déformation en place, l'énergie cinétique du système constitué par la masse en mouvement devient prépondérante. Celui-ci entre dans une phase dynamique que l'on peut définir par le terme général d'"écoulement rocheux".

Les écoulements rocheux traduisent une évolution naturelle avec diminution rapide de l'énergie potentielle du système. La dissipation d'énergie met en jeu différents phénomènes (fragmentation, déformation, déplacement, vibration, dissipation thermique...) dont l'importance relative peut être très différente suivant les cas. Le volume total de la masse en mouvement constitue un paramètre déterminant. On peut définir une classification des éboulements rocheux en quatre types principaux, liés au volume des masses en mouvement, relevant de mécanismes de propagation différents : (tableau I)
- chutes de pierres et de blocs isolés
- éboulements en masse
- éboulements en très grande masse
- déplacements en masse

Du point de vue des mécanismes mis en jeu, cette classification repose sur l'importance et la nature des interactions qui se développent au sein de la masse en mouvement et au contact du substratum.

Pour les deux premiers types d'éboulement on définit un mécanisme de propagation en mode indépendant caractérisé par un niveau d'interaction nul ou faible entre les éléments en mouvement. Le niveau d'interaction devient très important pour les éboulements du troisième type pour lesquels on définit un mode de propagation en grande masse

TABLEAU 1 - DIFFERENTS TYPES D'EBOULEMENTS ROCHEUX

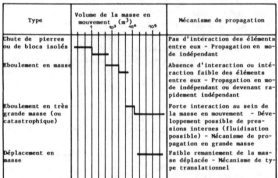

3 - NATURE DES PHENOMENES DE DISSIPATION ET DE TRANSFERT D'ENERGIE

Le développement d'un écoulement rocheux se fait par réduction du niveau d'énergie potentielle et dissipation de l'énergie ainsi libérée suivant deux types de phénomènes :
- des phénomènes internes :
  - transformation de la diminution d'énergie potentielle en énergie cinétique
  - transfert d'énergie cinétique entre les éléments en mouvement
  - dissipation d'énergie par frottement interne dans la masse
  - dissipation d'énergie cinétique par fragmentation interne
- des phénomènes externes traduisant les interactions entre les matériaux en mouvement et le substratum :
  - déformation plastique des zones de contact et dissipation par frottement
  - dissipation d'énergie liée à la fragmentation lors des chocs
  - transfert d'énergie cinétique par la mise en mouvement d'éléments du versant
  - dissipation sous forme d'énergie vibratoire transmise au substratum

- conversion éventuelle d'énergie cinétique résiduelle en énergie potentielle.

4 - FACTEURS PRINCIPAUX DES MECANISMES DE PROPAGATION

Les mécanismes de propagation sont complexes et résultent de l'influence de nombreux facteurs. On peut distinguer les principaux facteurs suivants :
- le volume de la masse en mouvement. De celui-ci dépend principalement l'importance des interactions qui se développent au sein de la masse en mouvement et au contact du substratum
- la topographie dont dépend l'énergie potentielle mise en jeu au cours de la propagation. La morphologie du versant joue un rôle déterminant dans la cinématique du phénomène :
- la nature des terrains sur lesquels se développe l'éboulement, qui détermine l'importance des phénomènes de dissipation d'énergie lors des impacts
- la nature et la fissuration naturelle préexistante du rocher dont dépend en partie la fragmentation des éléments en mouvement
- la dimension et la forme des blocs qui jouent un rôle essentiel dans la cinématique de la propagation
- le type du mécanisme de rupture à l'origine de l'éboulement. La cinématique de la rupture influe sur les conditions initiales de la propagation à travers les phénomènes de dissipation d'énergie par fragmentation et frottement.
- l'existence éventuelle d'obstacles naturels ou d'ouvrages de protection sur le trajet de l'éboulement.

5 - PROPAGATION DES EBOULEMENTS EN MODE INDEPENDANT

5 - 1 - Types de modèles

La propagation d'un éboulement rocheux apparait comme un phénomène complexe traduisant l'influence sur l'ensemble des éléments en mouvement des paramètres multiples mis en jeu le long des trajectoires. La complexité des mécanismes d'échange d'énergie lors des impacts, l'incertitude sur la détermination des paramètres, l'hétérogénéité et variabilité locale des conditions d'impact conduisent à considérer les trajectoires comme des fonctions de variables aléatoires. Ceci conduit à développer deux types de modèles numériques de simulation :
- modèles à trajectoires enveloppes. Ceux-ci sont orientés vers l'analyse des conditions limites de propagation : limites d'extension probables, caractéristiques limites probables des trajectoires (hauteur, vitesse, énergie cinétique). Les valeurs affectées aux paramètres de calcul correspondent à une estimation de leurs limites probables. Le seuil de probabilité attaché aux modèles à trajectoires enveloppes est associé aux limites d'extension effectivement observées sur un ensemble de sites caractéristiques différents.
- modèles à variables aléatoires. Ce type de modèle est orienté vers l'étude de la distribution aléatoire des trajectoires à l'intérieur de la limite d'extension de l'éboulement (dispersion des points d'arrêt, probabilité de dépassement d'un seuil donné). Les paramètres de calcul pris en compte à chaque impact sont des variables aléatoires auxquelles sont affectées des lois de probabilité.
Les modèles utilisés sont généralement bidimensionnels, ou tridimensionnels lorsque la morphologie du versant l'impose.

5 - 2 - Paramètres pris en compte

Les études théoriques et l'analyse des observations effectuées sur différents sites d'éboulements récents conduisent à définir un ensemble de paramètres principaux à prendre en compte dans les modèles :
- les données topographiques. Celles-ci doivent permettre d'établir une description détaillée de la morphologie à une échelle convenable, à deux ou trois dimensions. Les données topographiques sont généralement fournies par un plan photogrammétrique comportant le tracé des courbes de niveau.
- la caractérisation des terrains du point de vue des échanges d'énergie au cours des contacts entre le versant et les blocs en mouvement. On définit cinq classes types de terrains auxquels correspondent des va-

leurs caractéristiques prises en compte dans les fonctions de transfert traduisant les échanges d'énergie lors des impacts. La classification des terrains est indiquée dans le tableau II.
- la forme et la dimension des blocs. La forme d'un bloc est un paramètre essentiel du fait de l'importance de la rotation dans les conditions de propagation. La forme d'un bloc est caractérisée par son élancement cinématique E qui est voisin de son élancement géométrique défini par le rapport entre les diamètres maximum et minimum du bloc, mesurés dans un plan normal à son axe d'inertie principal.
- un critère de fragmentation. La prise en compte de la fragmentation en cours de propagation est plus particulièrement intéressante pour les modèles à variables aléatoires. Pour les modèles à trajectoires enveloppes visant à définir les limites probables de propagation, celle-ci peut être prise en compte globalement à partir d'une analyse du site, des éboulements antérieurs et d'une estimation de la blocométrie maximum probable.
- l'existence éventuelle d'obstacles naturels ou artificiels (versants boisés, ouvrages de protection). On remarquera que les effets destructeurs liés à la propagation du front de l'éboulement peuvent supprimer ou réduire considérablement l'efficacité des éléments de protection.

TABLEAU II - CLASSIFICATION DES TERRAINS APPLIQUEE AUX MODELES DE PROPAGATION D'EBOULEMENTS ROCHEUX

Classe	Type	Désignation
A	Rocher sain	Rocher compact, parties d'ouvrage massives en béton ou en maçonnerie
B	Rocher altéré	Rocher fissuré, zones d'altération de surface, roche tendre
C	Eboulis compact	Eboulis consolidé, remblai compacté plateforme ou terrain compacté
D	Eboulis meuble	Eboulis vifs, terrain naturel non compacté
E	Terrain meuble	Terrain très déformable, terrain de couverture, matériaux lâches

5 - 3 - Mécanisme de propagation

L'analyse des observations effectuées sur des sites d'éboulements naturels ou provoqués met en évidence le rôle déterminant de la rotation des blocs dans la propagation. Leur forme très généralement anisotrope (caractérisée par un élancement cinématique supérieur à 1) induit sous l'influence de la rotation une déviation de la vitesse réfléchie, entraînant un relèvement et un allongement de la trajectoire jusqu'à l'impact suivant. Ce mécanisme constitue un facteur essentiel de la propagation en accentuant les phases aériennes des trajectoires (avec accroissement de l'énergie cinétique) et en espaçant les zones de contact avec le versant (avec dissipation d'énergie cinétique). Le phénomène s'inverse dans les zones de vitesse faible où le roulement devient prépondérant et plus favorable aux blocs de plus faible élancement cinématique. C'est le cas en particulier dans la phase terminale de la propagation.

Entre deux impacts successifs la trajectoire d'un bloc est définie essentiellement par l'influence de la gravité, la résistance de l'air pouvant être négligée.

Les variations d'énergie cinétique lors des impacts sont définies par un modèle d'impact.

5 - 4 - Modèle d'impact

Les relations entre l'énergie cinétique incidente juste avant l'impact et l'énergie cinétique résiduelle du bloc à la sortie de l'impact sont définies par des fonctions de transfert :

Dans le modèle proposé les relations entre E_i et E_r sont définies par quatre fonctions caractéristiques :
- Ψ_n fonction de transfert normale caractérisant l'énergie cinétique de translation transmise dans la direction normale. Ψ_n traduit la partie élastique dûe au rebond proprement dit. Dans le cas général celle-ci est très faible.
- Ψ_g fonction de transfert de glissement caractérisant l'énergie cinétique de translation transmise dans la direction tangentielle. Ψ_g traduit l'influence du mécanisme de glissement associé à un déplacement tangentiel au cours de l'impact.
- Ψ_ρ fonction de transfert de rotation caractérisant l'énergie cinétique transmise, liée à la rotation du bloc. Celle-ci est prépondérante.
- K_ρ facteur de roulement. Cette fonction dont la valeur est comprise entre 0 et 1 est liée à la variation de l'énergie cinétique entre deux impacts successifs. Elle traduit l'influence relative des mécanismes de rotation et de glissement au cours de l'impact
(roulement sans glissement à la fin de l'impact pour $K_\rho = 1$).

Les valeurs des fonctions de transfert dépendent de plusieurs paramètres principaux : la nature du terrain dans la zone d'impact (type de terrain), le volume des blocs, l'énergie incidente.

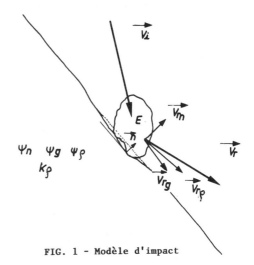

FIG. 1 - Modèle d'impact

Le vecteur vitesse $\vec{V_r}$ définissant la vitesse réfléchie à la sortie de l'impact peut être exprimé sous la forme d'une somme vectorielle

$$\vec{V_r} = \vec{V_{rn}} + \vec{V_{rg}} + \vec{V_{r\rho}}$$

les composantes $\vec{V_{rn}}$, $\vec{V_{rg}}$, et $\vec{V_{r\rho}}$ ne dépendant que des mécanismes de transfert d'énergie cinétique correspondants :

$$\vec{V_{rn}} = \vec{F_n} (\vec{V_i}, \vec{n}, E, \Psi_n)$$

$$\vec{V_{rg}} = \vec{F_g} (\vec{V_i}, \vec{n}, E, \Psi_g, K_\rho)$$

$$\vec{V_{r\rho}} = \vec{F_\rho} (\vec{V_i}, \vec{n}, E, \Psi_\rho, K_\rho)$$

481

Les composantes $\vec{V}_{rn}$, $\vec{V}_{rg}$ sont orientées respectivement suivant les directions normale et tangentielle du versant dans le plan de la trajectoire émergeante. L'angle de déviation de la composante $\vec{V}_{r\rho}$ dépend de l'élancement cinématique E du bloc.

Le rapport entre l'énergie cinétique de rotation et l'énergie cinétique de translation dépend de la valeur du facteur de roulement K_ρ, de la forme et de l'élancement cinématique E du bloc.

5 - 5 - Calage du modèle

La complexité des mécanismes mis en jeu dans la propagation des éboulements rocheux ne permet pas une détermination directe des paramètres de calcul. L'estimation des valeurs caractéristiques correspondantes repose essentiellement sur l'analyse de données expérimentales et d'observations effectuées sur des sites différents d'éboulements récents pour lesquels on dispose de données précises.

Dans les applications il est parfois possible d'effectuer un contrôle du calage du modèle à partir des données d'éboulements antérieurs observés sur le site étudié.

5 - 6 - Etude tridimensionnelle

Dans le cas fréquent où la morphologie du versant est régulière ou sans variation latérale brusque susceptible de déterminer une déviation importante des trajectoires, l'étude des conditions de propagation peut être dissociée en deux phases :

- la détermination du tracé en plan des axes de propagation à partir de l'analyse des lignes d'écoulement sur le versant, permettant de définir des profils cinématiques caractéristiques (fig. 2).

- l'analyse des trajectoires suivant les profils cinématiques définis, au moyen d'un modèle bidimensionnel. Celui-ci correspond au développement d'une surface réglée à génératrice verticale ayant pour directrice la trace en plan du profil cinématique considéré.

Lorsque la morphologie du versant présente des effets de déviation latérale importants l'étude de propagation nécessite l'utilisation d'un modèle tridimensionnel permettant de déterminer le tracé des trajectoires dans l'espace (Fig. 3).

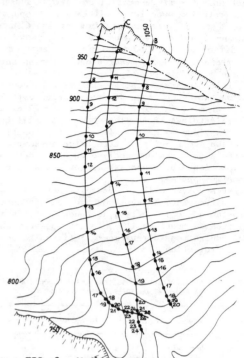

FIG. 3 - Modèle tridimensionnel
Tracé en plan des trajectoires

5 - 7 - Exemple d'application - Eboulement d'AIGUEBLANCHE

Un éboulement rocheux d'un volume de 15.000m^3 environ provenant de la rupture d'un pan de falaise dominant la ville d'AIGUEBLANCHE (Savoie, France) a provoqué la destruction de plusieurs maisons d'habitation et la coupure de la route nationale RN 90 sur une longueur de 500m. La présence au niveau de la falaise de niveaux de brèche compacte a favorisé la formation dans l'éboulis de nombreux gros blocs de 5m^3 jusqu'à 200m^3, dont la plupart se sont propagés jusqu'à la base du versant, après un parcours de l'ordre de 1 km et une dénivelée de 600m.

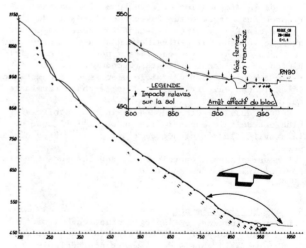

FIG. 4 - Aigueblanche - Exemple de trajectoire enveloppe (Profil cinématique Co)

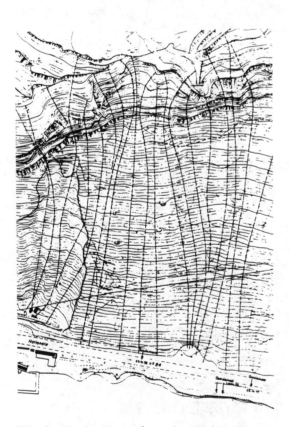

FIG. 2 - Etude du tracé en plan des axes de propagation - Détermination des profils cinématiques caractéristiques

La simulation de cet éboulement au moyen d'un modèle à trajectoires enveloppes permet d'analyser les conditions de propagation sur le versant. Celles-ci apparaissent assez nettement différentes suivant les zones, en liaison avec des variations latérales de la topographie. Les observations de terrain et les photographies aériennes effectuées après l'éboulement permettent une reconstitution des trajectoires relatives à un certain nombre de blocs marquant la limite d'extension de l'éboulement. La comparaison de plusieurs trajectoires limites caractéristiques relatives à des zones différentes du site montre un bon accord entre les trajectoires limites calculées et les trajectoires limites observées.

6 - EBOULEMENTS CATASTROPHIQUES - PROPAGATION EN GRANDE MASSE

6 - 1 - Influence du volume sur les conditions de propagation

L'analyse des données connues relatives aux éboulements en très grande masse étudiés par divers auteurs permet de mettre en évidence une relation entre le volume de la masse éboulée et la distance relative de transport (caractérisée par l'angle de site s du sommet de la falaise initiale par rapport au point extrême atteint par l'éboulis). Les corrélations proposées montrent une variation générale de type logarithmique entre tg (s) et le volume V de l'éboulement (Scheidegger, 1973). Parallèlement, l'analyse des données relatives à la longueur d'étalement L_e des matériaux met en évidence une variation générale en fonction du volume V de la forme :

$$L_e = K \, V^a \quad \text{avec} \quad a \simeq 1/3$$

traduisant, d'une manière générale l'existence d'une certaine relation de similitude dans l'étalement des matériaux éboulés (Davies, 1981).

Les corrélations proposées fournissent une indication intéressante sur l'évolution d'ensemble du phénomène. Toutefois l'analyse plus détaillée des données montre une dispersion souvent importante autour des relations moyennes calculées, traduisant l'influence probable d'autres paramètres tels que la topographie, les conditions de confinement liées à la morphologie du versant, la présence éventuelle d'eau en quantité importante au moment de la rupture.

6 - 2 - Modélisation

6-2-1 - Eboulement en masse de type régressif

Le développement d'un phénomène d'éboulement en masse de type régressif résultant de la succession dans le temps d'éboulements partiels indépendants, peut être modélisée à partir d'un modèle itératif déduit des modèles de propagation à trajectoire enveloppe en mode indépendant. La topographie du versant ne peut plus être considérée comme constante. Celle-ci évolue du fait des éboulements successifs et de l'étalement des matériaux éboulés.

Les observations effectuées sur les sites d'éboulements, l'analyse de la morphologie des pentes d'éboulis vifs et les résultats expérimentaux d'études sur modèles réduits en laboratoire, permettent de proposer un modèle de génération de la courbe d'étalement des matériaux éboulés définie par sa pente locale $\mathfrak{p}(x)$ sous la forme :

$$\mathfrak{p}(x) = f(p_o, p_t, \propto_1, \Delta V, \quad x)$$

où p_t, $\propto_1$, ΔV, x sont respectivement la pente topographique locale, la pente limite de l'éboulis, l'incrément du volume de l'éboulement, la distance de la limite d'extension.

- p_o représente la pente de la courbe limite d'étalement sur un versant de pente inférieure nulle. Les observations de terrain et les essais de laboratoire conduisent à proposer une expression de $\mathfrak{p}_o$ sous la forme :

$$\mathfrak{p}_o(x) = A \, x^b \quad \text{avec} \quad |\mathfrak{p}_o| \leqslant \propto_1$$

A dépend du volume ΔV, b est une constante comprise entre 0,5 et 1.

6-2-2 - Eboulement en très grande masse

Le mode de propagation en grande masse est caractérisé par une forte interaction des matériaux en mouvement. Il apparait intéressant d'étendre à ce domaine la notion de fonction de transfert développée dans le cas de propagation en mode indépendant. On peut considérer qu'à chaque instant les échanges d'énergie ne s'effectuent plus entre un élément en mouvement et un substratum fixe, mais entre un élément mobile à la surface de l'écoulement et une zone plus interne elle même en mouvement. Les transferts d'énergie entre ces différentes zones s'expriment alors en fonction des vitesses relatives. Celles-ci sont inférieures aux vitesses absolues des éléments en mouvement. Il en résulte une plus faible dissipation d'énergie cinétique dans la zone externe de l'éboulement.

Exprimées dans un repère fixe, en fonction des vitesses absolues, les fonctions du transfert apparaissent comme des fonctions croissantes de la masse en mouvement. Elles peuvent être exprimées sous la forme :

$$\Psi_i = \mathfrak{f}_i \left(\Psi_{o_i}, \quad \emptyset(V) \right) \quad (i = n, g \text{ ou } \rho)$$

avec Ψ_{o_i} les valeurs correspondantes des fonctions de transfert en mode indépendant ,

$\emptyset(V)$ une fonction de masse (variant entre 0 et 1), traduisant l'influence du volume total de la masse en mouvement sur la variation des fonctions de transfert.

L'existence de corrélations générales, de type logarithmique entre l'extension des éboulements en très grande masse et leur volume conduit à proposer pour la fonction $\emptyset(V)$ une expression de la forme :

$$\emptyset(V) = 1 - e^{-aX^c} \quad \text{avec} \quad X = \log \frac{V}{V_o}$$

où a, c, Vo sont des constantes.

Le calcul est conduit de manière itérative par incréments de volume successifs. Le modèle calcule pour chaque incrément, un état transitoire fictif de la morphologie du versant détermine les conditions du calcul de l'incrément suivant.

6-2-3 - Exemple d'application - Site de SECHILIENNE

Le versant rive droite de la vallée de la Romanche (Alpes, France) témoigne, dans le secteur de SECHI-LIENNE, de l'existence de mécanismes d'instabilité importants affectant le massif rocheux à l'échelle du versant. Une évolution importante s'est manifestée dès l'automne 1984 sur une partie du secteur et s'est intensifiée en 1985 et 1986 en donnant naissance à de nombreuses chutes de blocs et à des éboulements superficiels fréquents. L'étude et l'auscultation du site ont permis de conclure à la déstabilisation d'une zone de 2 à 3,5 millions de mètres cubes et à l'existence d'un risque de rupture en masse d'un secteur particulièrement disloqué.

Une étude de simulation d'un éboulement en grande masse d'un volume de 2 à 3,5 millions de mètres cubes a été effectuée afin d'analyser les conséquences de la rupture sur l'occupation du fond de la vallée (route nationale, RN 91, lit de la Romanche, bâtiments), et définir les travaux de protection à réaliser.

Le calage du modèle numérique de propagation en grande masse a été effectué en l'appliquant à l'analyse de l'écroulement du Mont Granier (Savoie, France) (volume 500Mm³) pour les valeurs du volume supérieures à $10^8 m^3$, et, sur les éboulements classiques connus pour lesquels on dispose de nombreuses données pour les valeurs du volume inférieur à $10^5 m^3$.

Les résultats des calculs du modèle sont représentés par le tracé des courbes d'étalement des matériaux et la restitution de la topographie finale du versant après l'éboulement. La figure 5 présente un résultat de la simulation suivant un profil de calcul pour V = 2Mm³. Ces résultats montrent une obstruction du lit de la Romanche et un comblement partiel du fond de la vallée permettant d'envisager la réalisation des travaux préventifs de déviation de la route RN 91 et du lit de la rivière sur la rive gauche de la vallée.

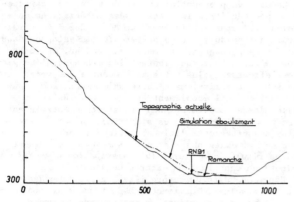

FIG. 5 - Séchilienne - Simulation d'un éboulement en grande masse (volume 2Mm^3)

7 - CONCLUSIONS

La propagation des éboulements rocheux met en jeu des mécanismes complexes caractérisés par des transferts d'énergie au sein de la masse en mouvement et avec le substratum.

Des modèles numériques de simulation sur ordinateur des phénomènes de propagation ont été développés dans le domaine des éboulements classiques (volume inférieur à 0,1Mm^3) : modèles à trajectoires enveloppes, modèles à variables aléatoires, (modèles bidimensionnels ou tridimensionnels).

L'étude des phénomènes de propagation en grande masse conduit à proposer une approche unitaire de la modélisation de la propagation des écoulements rocheux, à travers la généralisation de la notion de fonction de transfert, établie pour le mode de propagation indépendant, et la définition d'un modèle de propagation permettant de prendre en compte l'influence du volume de la masse en mouvement sur les conditions de propagation.

BIBLIOGRAPHIE :

AZIMI C, DESVAREUX P., GIRAUD A., MARTIN-COCHET J. (1982) - Méthodes de calcul de la dynamique des chutes de blocs - Application à l'étude du versant de la Pale (Vercors) - Bulletin de Liaison des L.P.C. n° 122, 93-102

BOZZOLO D., PAMINI R. (1982) - Modello matematico per lo studio della caduta dei massi - Départimento della publica educazione - Lugano - Trevano

BROILI L. (1977) - Relations between scree slope morphometry and dynamics of accumulation proces ses - Istituto Sperimentale Modelli e Strutture - n° 90 Sept. 1977 - Bergamo

CAMPONOUOVO G.F.(1977) - ISMES' experience on the model of S. Martino - Istituto Sperimentale Modelli e Strutture - n° 90 Sept. 1977 - Bergamo

DAVIES T.R.H. (1981) - Spreading of rock avalanche debris by mechanical fluidization - Rock Mechanics 15, 9-24

FALCETTA J.L. (1985) - Un nouveau modèle de calcul de trajectoires de blocs rocheux - Revue française de Géotechnique 30, 11-17

GOGUEL J. (1978) - Scale dependent Rockslide Mechanisms - In : Rockslides and avalanches, 1 (Voight B, ed) Dév. Geotech. Eng - 14A, 693-705 - Amsterdam, Oxford, New York : Elsevier

KORNER HJ. (1976) - Reichweite und Geschwindigkeit von Bergstürzen und FleiBscheelawinen - Rock Mechanics 8, 225-256

PITEAU D.R., CLAYTON R. (1978) - Methods of protection - Rock-falls characteristics effecting the design of protection measures - Landslides Analysis and Control, 176 - Washington

SCHEIDEGGER A.E. (1973) - On the prediction of the reach and velocity of catastrophic landslides - Rock Mechanics 5 - 231 - 236

Reinforcement of slopes under Denia Castle (Spain)
Consolidation d'une pente rocheuse sous le château de Denia, Espagne
Stützung von Böschungen unter der Denia Burg (Spanien)

M.ROMANA, Professor of Geotechnical Engineering, Polytechnical University of Valencia, Spain
F.A.IZQUIERDO, Assistant Professor, Polytechnical University of Valencia, Spain

ABSTRACT: The Northern slope below the Denia Castle (Spain) was excavated as a limestone quarry in the last years of XIX Century. Several rock falls happened during quarry explotation and afterwards. As a consequence, the North Tower was undermined and has suffered severe fissuring but without collapse. In 1984 a study was done and the slope support necessary was designed including bolts, shotcrete and seven 1000 kN cable anchors. In 1985 the designed works were built and the tower was succesfully underpinned. Movements in the slope were controlled during the anchoring. The paper describes design and construction of slope correction.

RESUME: Le talus Nord sous le Château de Denia (Espagne) fut excavé comme une carrière de calcaire á les dernières années du Siècle XIX. Conséquenment, la Tour Nord fut minée et elle souffrit une sévére fissuration mais sans collapser. A 1984, une étude fut réalisée et la stabilisation nécessaire fut projetée au moyen des boulons, du béton projeté et de 7 ancrages de 1000 kN. A 1985, les travails projetés furent construis et la Tour fut stabilisée avec de succés. Les mouvements ont été contrólés pendant les operations d'ancrage. Cet article décrit le projete et les ouvrages pour la correction du talus.

ZUSAMMENFASSUNG: Der nördliche Schütt unter der Burg von Denia (Spanien) wurde als Kalksteinbruch in den letzten Jahren des vorherigen Jahrhundert ausgegraben. Verschiedene Felsfälle geschahen während des Tagebaues und nachher. Die Nachfolge war, daß der Turm geschrämt wurde und ernste Risse auftauchten, jedoch ohne einen Zusammenbruch. In 1984, wurde eine Untersuchung durchgenommen und der notwendige Ausbau durch Felsnagel, Spritzbeton und Felsanker von 1000 kN. entworfen. In 1985 wurden die entworfenen Bauten konstruiert und der Turm mit Erfolg stabilisiert. Während der Felsankerung wurden die Verschiebungen im Schütt gemessen. Die Mitteilung beschreibt den Entwurf und die Ausführung der Schüttstabilisierung.

1 INTRODUCTION

The North slope of the Denia Castle is the cut of an old cretaceus limestone quarry used for Denia breakwater. Both rock joints and slope dip have originated instabilities with fallen rocks at the slope base. A plane slide scar just below North Tower undermined it. There were flexion and shear fissures in the fabric (figure 1).

The North Tower is at the top of a circular slope with dip direction changing from 270º to 360º, and dip between 60º (in the base) and 90º (at the top). The undermining of the North Tower was due to plane slides along a joint of a system almost vertical, parallel to North slope (dip direction 330º), and sometimes open. Below the Tower a second joint of the same system could be seen, deeper in the slope, and could cause a bigger slide with danger of fall for the entire Tower.

In 1984, the regional Valencian Government, through the Fine Arts General Directorate, asked the geotechnical Engineering Department of the Polytechnical University of Valencia, to study the slope stability and the remedial measures to assure the Tower safety.

To underpine the North Tower, the selected solution was filling with shotcrete of the slide car below it and the installation of 7 presstressed 1000 kN anchors. The solution was built in the summer of 1985 and to the date the behaviour of slope and tower has been normal.

2 GEOLOGICAL AND GEOTECHNICAL DESCRIPTIONS

The Denia Castle is on a rocky hill from Upper Cretaceus formed by sandy and/or marly microcristalline limestones. There are karstic features along vertical joints systems with some decimetric solution tubes. Some joints are calcite-cemented and others are fi-

Figure 1. View of the undermined North Tower. Vertical fissures can be observed in the fabric.

lled with red clay ("terra rosa").

Regional tectonics is extremely complex. Fracture directions are (Alonso, 1982):
- F1 (0º-15º/170º-180º) - Normal or siniestral shear faults.
- F2 (20º-35º) - Associated with coastal catalan fractures, very common, almost vertical. Normal faults with fall ot the east block giving mountain ranges along the coast.
- F3-4 (35º-45º/60º-70º) - Main local tectonic di-

rection. Generally siniestral shear faults almost vertical.
- F5-6 (85º-95º/105º-120º) - Shear faults or thrust folds.
- F7-8 (135º-145º/150º-155º) - Active shear or inverse faults (Estrella, 1977).
      Range of fracture directions
At the studied slope joints range mainly in the F2, F3 and F4 systems.

Rock is limestone, in beds more than 1 m thick, with joints dipping 70º and often vertical. Due to heterogeneities the rock properties change from point to point. Representative values are:
- Unconfined compression strength (C_o): 47-62 MPa.
- Weathering degree: W_2 (sligthly weathered)
- R.Q.D.: 75-90%
- Water content: Usually dry. Dipping with rain.
- Joints: Four families (table 1).
- Bedding: Almost always horizontal. Sometimes dipping 40º.
- Classification: R.M.R.=79-84 (Bieniawski, 1978) Without adjustement ratings (Romana, 1985).

3 DESIGNED SOLUTION

Figure 2 summarises plane slide and toppling analysis for a 70º slope with three dip directions (270º, 330º, 360º). Friction angle is estimated between 30º (residual) and 40º (medium). Analysis shows some risk of <u>toppling</u> at family D_1, but in the field there is no evidence of it. In all cases <u>plane slides</u> can occur with families D_1 and D_3. Evidence can be seen at the slope with falls through family D_1, joints almost parallel to slope dipping more than 60º, sometimes open. Figure 3 presents <u>wedge analysis</u> for all cases. Only with dip direction of 270º families D_1-D_3 and D_2-D_3 could originate some dangerous wedges. There is no field evidence.

Falls happen along joints of D_1 family, and are only possible with 40% of the joints, which dip less than slope.

It was possible the fall of an important rock mass below the Northern Tower along a D_1 joint. Therefore it was necessary an estimation of its shear strength to asses the actual instability risk and to design

Table 1. Joints characteristics. Denia Castle Hill.

Family	Representative dip direction	Representative dip	Spacing	Persistence	Roughness	Opening	Filling	Weathering
Bedding B	-----	Horizontal	0'6-2 m	Continuous	Slightly rough	Closed	None	None
D_1	152º(332º)	85º	0'6-2 m	Almost continuous	Slightly rough	Open 1-5 mm	None or calcite	Slightly
D_2	190º	85º	2 m	Continuous	Rough	Open >5 mm	Calcite clay	Karstified
D_3	295º	70º	2 m	Continuous	Rough	Closed	None	None

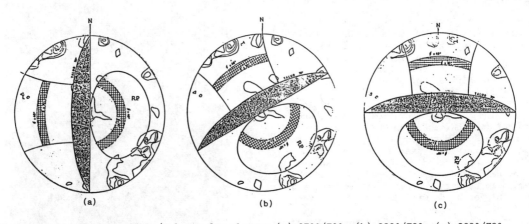

Figure 2. Plane slide and toppling analysis for slopes: (a) 270º/70º, (b) 330º/70º, (c) 360º/70º

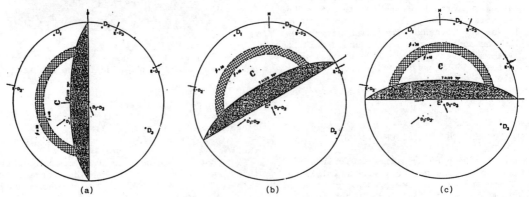

Figure 3. Wedge failure analysis for slopes: (a) 270º/70º, (b) 330º/70º, (c) 360º/70º

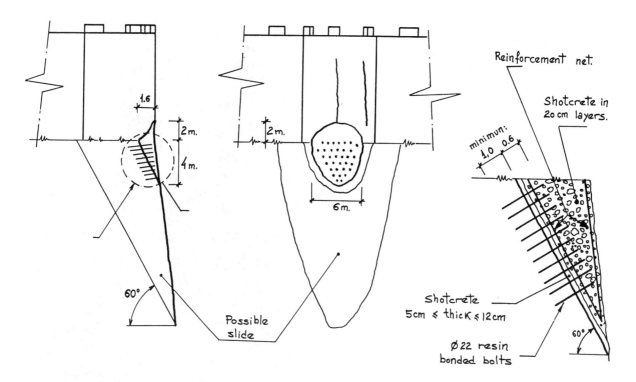

Figure 4. Small slide. Recommended remedial measures.

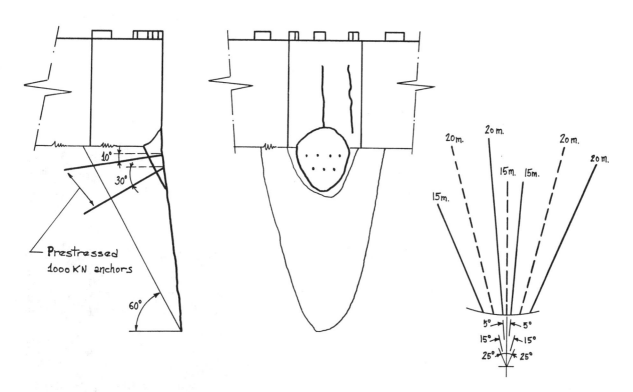

Figure 5. Possible big slide. Scheme of the prestressed anchors system.

appropiate remedial measures.

This joint dips 60º, is almost continuous and seems closed. The parameters of the Barton (1977) formula are: $\emptyset_b$ = 33º; JRC = 10; JCS = 50 MPa.

Back-analysis of the small slide allows an estimation of cohesion due to cementation and/or rock bridges.

The slide happened along a D_1 joint dipping 65º. Weigth of slide mass, and part of the tower, is 1025,6 kN and the normal stress at the joint is 0'0048 MPa.

Friction peak angle according Barton is:

$$\emptyset_p = 33º + 10 \log \frac{50 \text{ MPa}}{0,048 \text{ MPa}} = 63,16º$$

Without cohesion safety factor is:

$$F = \frac{\text{tg } 63,16º}{\text{tg } 65} = 0,92$$

With F = 1, cohesion force (F_{coh}) is given by:

$$F = 1 = \frac{1025,6 \cos 65º \text{ tg } 63,16º + F_{coh}}{1025,6 \sin 65º}$$

F_{coh} = 72,9 kN and c = 0,003 MPa

Figure 4 shows a scheme of the recommended remedial measures.

In the case of the possible big slide (figure 5) weigth of rock mass and tower is 18,500 kN, normal stress at the joint is 0,14 MPa and peak friction angle is:

$$\emptyset = 33º + 10 \log \frac{50 \text{ MPa}}{0,14 \text{ MPa}} = 58,53º$$

Without cohesion safety factor is:

$$F = \frac{\text{tg } 58,53º}{\text{tg } 60º} = 0,94$$

sligthly bigger than for the actual slide.

With the assumed cohesion value (c = 0,003 MPa) the Tower is at the limit of stability.

To reach a safety factor of 1,3 a prestressed anchors system has been designed to be built after the Tower underpinning (figure 5).

Back-analysis using Barton formula to estimate the peak friction angle, has given satisfactory values. Safety factor for the actual fallen rock mass is only sligthly smaller than for potential bigger slide.

4 CONSTRUCTION

Construction was done according the following sequence of operations:

- Placing of gypsum "witnesses" on the fissured rock and tower fabric (No cracking was observed during the works).
- Scaling by hand the instable rocks.
- Surface shotcreting (5-6 cm) below the Tower.
- Installation of 18 resin bolts ($\emptyset$ = 25 mm, l = 4 m) to support shotcrete and rock during the drilling of anchors.
- Shotcreting, by layers, the void below the Tower.
- Installation of seven prestressed anchors of a nominal tension of 1000 kN each.

Figures 6 and 7 show different construction stages and the big automobile crane (weight 40 tons, height 70 m) which was used for all operations.

Figure 6. View of a first stage of underpinning. Surface shotcreting, bolting and installation of hollow tubes for the anchors.

Figure 7. Installation of anchors from a platform suspended by a 70 m high crane.

488

5 MONITORING

To control the displacements of the Tower and rock
slope during the stressing of anchors, three movements were placed (figure 8) supporting topographical marks.

Movements were determined with a electronic distan
ciometer and they are given in table 2.

Table 2. Absolute displacements

Measurement	Location	Displacement (mm)
1	Tower	3.0
2	Below anchors on shotcrete mass	2.2
3	On original slope	2.2

The measurements made possible to check the defor
mability of the joint. For the normal stress vs
closure curve of a joint, Goodman (1976) proposed
the empirical hyperbolic function:

$$\sigma_n = \left(\frac{\Delta V_j}{V_m - \Delta V_j} \right) \sigma_i + \sigma_i$$

in the low stress region, where, ΔV_j is the joint
closure under a given σ_n, V_m is the maximum closure
and σ_i is the initial stress level.

The deformation modulus of the rock can be estima
ted by (Bieniawski, 1978):

$$E_m = 2\ RMR - 100$$

For $RMR \simeq 80$ $E_m \simeq 60$ GPa.

The forces due to the anchoring (7000 kN) are dis
tributed about 144 m². So the incremental normal
stress in the deep open joint was:

$$\Delta\sigma = \frac{7000\ kN}{144\ m^2} = 0{,}05\ MPa$$

and rock deformations are negligibles.

If $\sigma_i = 0{,}14$ MPa and $\Delta V_j \simeq 2$ mm, Goodman formula
gives:

$$V_m = 7{,}6\ mm$$

value which is according with field observations
about the opening of this family (see Table 1).

6 CONCLUSIONS

At design stage, back-analysis using Barton-Choubey
(1977) formula to estimate the peak friction angle
has given satisfactory values.

Rock mass displacements were monitored during the
stressing of anchors. All deformation is due to the
almost regular closing of the deep joint (which gave origin to the danger of the big slide). Measurement results were in agreement with estimations obtained through the Goodman (1976) empirical function
for normal stress-closure of open joints at low
stresses.

Figure 8. Monitoring of the anchoring operations by
three   monuments (situated at 1, 2 and 3).

7 ACKNOWLEDGMENTS

The Authors take this oppotunity to thank the Polytechnical University of Valencia and the Fine Arts Ge
neral Directorate of the Valencian Regional Government for granting permission to submit this paper and
for the facilities made available.

Design details of this case history were presented
previously (Romana and Izquierdo, 1985).

REFERENCES

Alonso-Matilla, L.A. 1982. Estudio de las rocas íg-
neas de Castellón, Valencia y Alicante. Tesis Doctoral. Universidad de Salamanca. 980 pp.
Barton, N.; Choubey, V. 1977. The shear strength of
rock joints in theory and practice. Rock Mechanics
10: 1-54
Bieniawski, Z.T. 1978. Determination of rock mass de
formability: experience from case histories. Int.J.
Rock Mech. and Min. Sci. Vol.15: 237-248
Bieniawski, Z.T. 1979. The geomechanics classifica-
tions in rock engineering. 4th. Cong. Rock Mech.
Montreux. Tomo 2: 41-48
Goodman, R.E. 1976. Methods of Geological Enginee-
ring in discontinrcus rock. West. New Yorker
Rodriguez-Estrella, T. 1977. Síntesis geológica del
Prebético de la provincia de alicante.II.Tectónica.
Bol. Geol. Min. Tomo LXXXVIII: 273-299
Romana, M. 1985. New adjustment ratings for aplica-
tion of Bieniawski classification to slopes. Int.
Symp. Role Rock Mech. in Excav. for Min. and Civil
Works. Zacatecas. México: 59-63
Romana, M.; Izquierdo, F.A. 1985. Slope stability of
near vertical slopes under Denia Castle. Int. Symp.
Role Rock Mech. in Excav. for Min. and Civil Works.
Zacatecas. México: 107-112.

Preplaced cable bolts for slope reinforcement in open cut mines
Le pré-placement de câbles d'ancrage pour le renfort des talus dans les mines à ciel ouvert
Vor dem Abbau installierte Kabelbolzen zur Stützung von Böschungen in Tagebaubergwerken

K.J.ROSENGREN, Principal, Golder Associates Pty Ltd, Brisbane, Australia
R.G.FRIDAY, Associate, Golder Associates Pty Ltd, Melbourne, Australia
R.J.PARKER, Director, Golder Associates Pty Ltd, Melbourne, Australia

ABSTRACT: The use of fully grouted cable bolts is rapidly becoming an important method of slope stabilisation in open pit mines throughout Australia. This paper outlines a simple analytical technique which enables assessment of different methods of cable bolting by comparing displacements at which maximum shear capacity is developed.

RESUME: L'emploi de cables de tension entièrement scellés au ciment se révèle rapidement comme une méthode importante pour la stabilisation des talus dans les mines à ceil ouvert d'Australie. Cet article présente sommairement une technique d'analyse simple qui permet d'assesser les différentes méthodes d'ancrage et de tensionnement des cables par la comparaison des déplacements pour lesquels la capacité de cisaillement maximale est engendrée.

ZUSAMMENFASSUNG: Sehr schnell verbreitet sich überall in Australien die Anwendung von völlig mörtelverpreß ten Kabelbolzen als eine wichtige Methode zur Böschungsfestigung in Tagebaubergwerken. Dieser Beitrag beschreibt ein einfaches analytisches Verfahren, welches die Begutachtung verschiedener Kabelbolzenmethoden durch den Vergleich der Verschiebung ermöglicht, an welchem die maximale Scherfähigkeit auftritt.

1. INTRODUCTION

The use of untensioned and tensioned cable bolts is rapidly becoming an important method of slope stabilisation in many open cut mines throughout Australia. The use of anchors for slope stabilisation is not a new concept and there are many examples reported in the literature for civil engineering projects. Rock anchors have been used in a number of open cut mines (Sage 1977, Seegmiller, 1974 and 1975) although in most cases this has involved high capacity tensioned bolts which have proved to be too expensive for routine use.

Although untensioned fully-grouted cable bolts are used extensively in underground mining there has not been significant use made of them in open cut mining until very recently. In 1976 untensioned fully-grouted cable bolts were used in conjunction with smooth-blasting techniques to improve final wall stability in the reopened Mary Kathleen Uranium Mine. The authors are not aware of any further use of systematic cable bolting for slope reinforcement in Australia until late 1984. Since then, cable bolting has been instigated on a routine basis at a number of mines including the Woodcutters Project near Darwin, Mt. Newman Mining's Mt. Whaleback Mine, Griffin Coal's Muja Open Cut and numerous small open cut gold mines throughout Western Australia.

The current techniques available for cable bolting were outlined by Rosengren (1986). A brief description of cable bolting programmes at Woodcutters, Mt. Whaleback and Muja were provided in Rosengren's paper. A more detailed account of the reinforcement programm being introduced at Mt.Whaleback was included in a paper by MacFarlane, et al (1986).

In recent practice, cable bolting has been used to minimise rock mass dilation during excavation, thus relying on the inherent strength of the rock to maintain stability. Both untensioned and tensioned fully grouted cables have been used. Wherever possible, cables are preplaced, so that they are installed in down-holes from current working levels, as illustrated in Figure 1. Pretensioning of cables can also be used to increase the stiffness of the system where the sliding surface is relatively smooth or there is a high risk of failure.

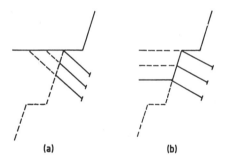

Figure 1 Preplaced Cable Bolts

At present there are no simple analytical techniques available for design of cable bolts which rely on the prevention of rock mass dilation to develop strength. To date, design has been largely based on engineering judgement, although design experience is rapidly being gained.

This paper provides a simplified analysis of the contribution to shear force along a sliding surface, from an individual cable bolt. The analysis develops a relationship between increase in shear force, shear displacement and dilation angle. The intent of the analysis is to provide a basis for comparison of various factors which influence cable bolt behaviour.

2. BACKGROUND

An analytical approach for the design of cable bolts was provided by Dight (1983). In simple terms, Dight's approach was similar to that adopted in this paper, the major difference being that he used his analysis to develop a serviceability design criteria. He concluded that an acceptable displacement during excavation was 0.05% of joint length and thus prepared design curves which related increase in shear strength along a sliding surface to angle of installation of the cable bolt.

Heuze (1981) demonstrated the importance of dilation in cable bolt performance. His model used a system of springs to represent a reinforcing member. Thus, as a rough joint attempted to slide, there was an increase in normal force from the springs and a corresponding increase in shear strength.

3. ANALYSIS

3.1 Approach

In developing the model, a dilation angle i, was used to express the ratio of normal displacement to shear displacement, $\tan i = D_n/D_s$.

Any relative movement along the sliding surface will cause local resistance to be developed involving shear and tension. The mechanism considered in this analysis is illustrated in Figure 2. As movement occurs the cable remains intact but causes local crushing of the grout. Movement is resisted by frictional drag of the curving cable as it slides on the crushing grout. Beyond the curve, sliding of the cable through the surrounding grout is resisted by bond stress.

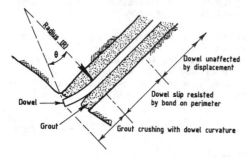

Figure 2   Mechanism of Sliding

Assuming the cable to be acting in its elastic range with respect to tension, and that is has negligible bending strength and shear strength, the shear force contribution along the sliding plane, from each cable can be estimated from:

$$F_s = P[\cos(\alpha+\theta)\tan(\phi+i)+\sin(\alpha+\theta)] \qquad (1)$$

where   P   is tension in the cable
          α is the cable installation angle
          θ is rotation of the cable
          ϕ is the basic friction angle
          i   is the effective dilation angle

as defined on Figure 3.

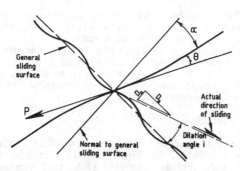

Figure 3   Definitions

Two major assumptions have been made in the analysis, as follows:

(a)   That lateral resistance to the cable crushing the grout is constant at wf, where

w   is the effective width of the cable
f   is the crushing strength of the grout.

(b)   Sliding of the straight cable through grout is resisted by a constant force up, where

u   is a constant bond stress
p   is the perimeter length on which sliding occurs.

3.2 Theory

Curved Portion

If it is assumed, for simplicity, that the curved portion of the cable has a constant radius R, then the relationship between radius and tension is given by the hoop tension formula:

$$P = Rwf \qquad (2)$$

Since the tension reduces from P to Q over the curved portion, due to frictional drag, the above relation is modified to use the average tension.

$$(P+Q)/2 = Rwf \qquad (3)$$

The frictional drag is taken into account with

$$P-Q = \mu wf\, R\theta \qquad (4)$$

where μ is the coefficient of friction of the dowel sliding on the grout.

Combining the previous two equations to eliminate Rwf allows Q to be calculated from P

$$Q = P(1-\mu\theta/2)/(1+\mu\theta/2) \qquad (5)$$

Cable extension over the curved portion is

$$\delta_c = [(P+Q)/2-T]R\theta/(aE) \qquad (6)$$

where   a   is cable cross-sectional area
          E   is cable elastic modulus
          T   is the initial cable tension

Substituting for R From (4) gives

$$\delta_c = (P+Q-2T)(P-Q)/(2aE\mu wf) \qquad (7)$$

Straight Portion

The cable tension is assumed to vary linearly from Q to T over the bond length (Q-T)/(up). Cable extension in this portion, is

$$\delta_s = (Q-T)^2/(2upaE) \qquad (8)$$

Displacement

Displacement can be considered in two parts:

(i)   Cable extension of $2(\delta_c + \delta_s)$ at an angle of $\theta + \alpha$ to the normal to the general sliding surface and

(ii)   Curvature of the cable resulting in:
  (a)   displacement transverse to the cable, $2R(1-\cos\theta)$
  (b)   shortening parallel to cable, $2R(\theta-\sin\theta)$

Normal and shear displacement are therefore given by:

$$D_n = B\cos(\theta+\alpha) - 2R[(1-\cos\theta)\sin\alpha + (\theta-\sin\theta)\cos\alpha] \quad (9)$$

$$D_s = B\sin(\theta+\alpha) + 2R[(1-\cos\theta)\cos\alpha - (\theta-\sin\theta)\sin\alpha] \quad (10)$$

where $B = [(P+Q-2T)(P-Q)/(\mu wf) + (Q-T)^2/(up)]/(aE)$

The two displacement components are related by the dilation angle of the rock surfaces

$$\tan i = D_n/D_s \quad (11)$$

Solution

A computer program was developed to find the relationship between shear force and shear displacement for a number of cable types. A solution for F_s and D_s for a particular P and i was found by first calculating R from (2). A value of θ was then assumed so that Q could be calculated from (5). D_n and D_s were then calculated and if D_n/D_s was not close to tan i, then θ was adjusted and Q recalculated. Finally F_s was calculated from (1).

4. RESULTS

4.1 Assumed Values

In deriving the relationship between increase in shear strength on a sliding plane and displacement, the following values were assumed:

- lateral crushing strength of grout in hole, f = 40,000 kPa

- friction coefficient of curved cable sliding on grout, $\mu = 1.0$

- elastic modulus of steel, $E = 2 \times 10^8$ kPa

- bond stress of cable in grout u = 2000 kPa

- rock on rock friction angles of $\phi = 25^o$ and 30° were considered.

Cable bolts are typically made up from 15.2mm diameter prestressing strands which have an ultimate tensile strength of 250 kN. The most common cable capacities used are 500 kN (two strands) and 1000 kN (four strands). Cable strands can also be unravelled (birdcaged) to improve bond strength. Estimated values of wf, aE and up for typical cable bolts are listed in Table 1.

TABLE 1

Cable Bolt	wf kN/m	aE kN	up kN/m
500kN	900	57,200	194
500kN, birdcaged	2840	57,200	446
1000kN	1200	114,000	294

4.2 Shear Force Versus Displacement

The relationship between calculated shear force and displacement is presented graphically for 500 kN cables with $\phi = 25^o$ and $\phi = 30^o$ in Figures 4 and 5, respectively. Similar results are also presented as Figure 6 for a birdcaged cable and as Figure 7 for a cable pre-tensioned to 250kN. Figures 4 to 7 are for cables installed normal to the sliding surface. The influence of cable inclination is shown as Figures 8 and 9 for cables installed at angles of -10° and $+15^o$ from normal (negative implies a steeper angle and positive a flatter angle than normal). The relationship between shear force and displacement for 1,000kN untensioned cables with $\phi=25^o$ is also presented graphically as Figure 10. Some care is necessary in interpreting the results, particularly at large displacements where interference could occur between individual strands or strands and rock.

4.3 Dilation Angle

To interpret Figures 4 to 10 account needs to be taken of the nature of the sliding surface since dilation angle is included as a variable. In general terms, dilation angle will be high at very small displacement, progressively reducing as displacement increases. A simple representation of dilation angle versus displacement is presented as Figure 11. This considers maximum and minimum dilation angles for different scale lengths.

These dilation angles are assumed to be cumulative to produce the combined values shown as Figure 11(d). As a further simplification, a smooth curve has been used to represent average dilation angles, as shown. The assumed relationship between dilation angle and displacement has been applied to produce a relationship between shear force and displacement presented on Figures 4 to 10.

Maximum shear loads for each type of cable bolt and the displacement required to achieve it are summarised as Table 2.

TABLE 2
CABLE BOLT CAPACITIES

Cable*	Max.Shear Load (kN)	Shear Displ. (mm)
500kN	500	48
500kN, $\phi = 30^o$	570	48
500kN, birdcaged	555	16
500kN, pretensioned	935	0
500kN, $\alpha=-10^o$	455	90
500kN, $\alpha= 15^o$	580	30
1000kN	975	70

* $\phi = 25^o$ unless otherwise noted

5. IMPLICATIONS FOR CABLE BOLT DESIGN

The results on Table 2 indicate a wide range of maximum shear capacities depending on how the bolt is installed, but more importantly a wide range of shear displacements at which maximum shear capacity is reached.

The effect of increasing friction angle from $\phi = 25^o$ to $\phi = 30^o$ is to increase shear capacity by about 14%. There is no change in shear displacement required to achieve maximum capacity since friction angle has no effect on cable deformation geometry.

The birdcaged cables show an increase in capacity at a substantially reduced displacement. The increase in bond strength and greater bearing area make these cables stiffer than the intact strands. Some care needs to be taken with this interpretation since other factors may influence the performance of these cables. For example, after the onset of grout cracking, the curves in the individual wires will straighten, so that in

493

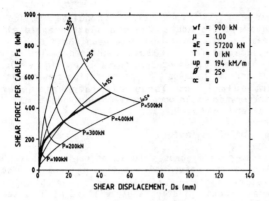

Figure 4 500kN Cable

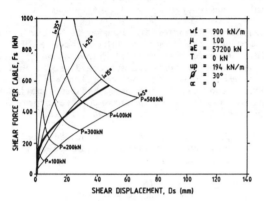

Figure 5 500kN Cable, φ=30°

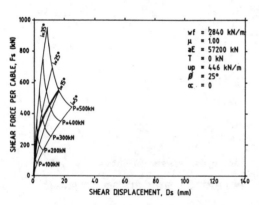

Figure 6 500kN Cable, Birdcaged

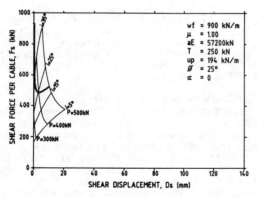

Figure 7 500kN Cable, Pretensioned

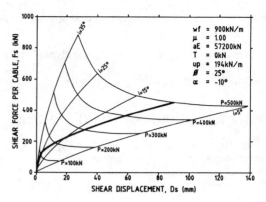

Figure 8 500kN Cable, α=-10°

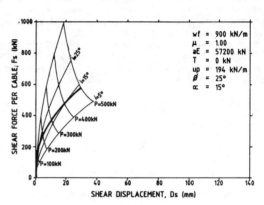

Figure 9 500kN Cable, α=15°

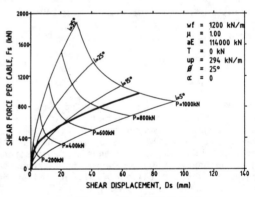

Figure 10 1000kN Cable

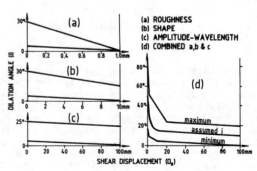

Figure 11 Assumed Dilation Angle

494

fact, over part of the displacement range, these cables may be less stiff than the intact strands. The stretching of birdcaged strands was observed in one case of a slope failure involving cable bolts.

Pretensioning of a 500kN cable bolt to 50% of its ultimate tensile strength (250kN) has a significant influence on shear capacity. Because of the high i values assumed at very small displacement, the pretensioning results in a shear capacity of around 935kN at zero displacement. This decreases with displacement as i rapidly decreases, but nevertheless an ultimate shear capacity of 520 kN is developed at a displacement of 10mm; very much lower than the 48mm displacement for an untensioned cable.

The combined effects of dilation angle and pretensioning clearly have a significant influence on the behaviour of cable bolts. Pre-placed cable bolts which are not actively tensioned during construction develop tension during excavation and thus derive significant strength in a dilatant rock mass. In recent applications active tensioning of cable bolts has only been used where dilation angle is relatively small, i.e. the sliding surface is relatively smooth.

The flatter cable installation angle of $\alpha = 15^o$, resulted in a 16% increase in shear force, at a significantly reduced displacement of 30mm, compared to a bolt installed normal to the sliding surface. However, for a 500kN bolt installed 10° steeper than normal, the ultimate shear capacity of 455kN is not achieved until a displacement of 90mm.

For the 1000kN cable bolt, a maximum shear capacity of 975kN was achieved at a displacement of 70mm. The displacement at which maximum capacity is achieved is significantly higher than for a 500kN bolt. Therefore, it is not necessarily acceptable, to replace two 500kN cable bolts with one 1000kN cable bolt since stiffness of the system will be reduced.

Examination of Figures 4 to 10, for 500kN cable bolts indicates that significant shear capacity, of the order of 200kN, is mobilised at displacements of less than 5mm. The effect of pretensioning is to develop high shear capacity prior to any displacement, and furthermore develop ultimate capacity at significantly reduced displacement.

Most of the cable bolting performed to date in Australia has been with untensioned cables. However, an important feature of their use has been pre-placement, so that in effect, even the so-called untensioned bolts have some pre-tensioning resulting from excavation.

The results of this study suggest that any deviation in installation angle, steeper than normal to the sliding surface, will significantly reduce the performance of the cable bolt and installation angles should be kept as flat as possible because of uncertainties in sliding surface inclination and drilling alignment. However, drilling cost and installation difficulties become more significant for pre-placed cable bolts in very flat drill-holes.

6.   CONCLUSIONS

A simple analytic technique has been provided to enable comparison of different cable bolting techniques. The cases studied have demonstrated various methods by which the stiffness of the cable bolt-rock mass system can be increased to improve performance. The major increase in stiffness is provided by pretensioning which can result from excavation for pre-placed cables or from active tensioning of cables. Stiffness can also be increased by birdcaging of the cable strands or by flattening the angle of installation.

7.   ACKNOWLEDGEMENT

Part of the work included in this paper was performed during geotechnical studies for the South Wall of the Mt. Whaleback Mine which is operated by the Mt. Newman Mining Co. Pty. Limited. The authors wish to acknowledge the contribution of Mr. G.A. MacFarlane in formulation of wall reinforcement programmes at Mt. Whaleback.

8.   REFERENCES

Dight, P.M., 1983. A case study of the behaviour of a rock slope reinforced with fully grouted bolts, in Rock Bolting (ed. O. Stephansson), pp 523-538, (A.A. Balkema: Rotterdam).

Heuze, F.E., 1981. Analysis of bolt reinforcement in rock slopes, Proc. of 3rd Int. Conf. on Stability in Surface Mining, Vancouver, Soc. of Mining Engineers of AIME.

MacFarlane, G.A., Parker, R.J. and Swindells, C.F., 1986. Slope stability investigations for the South Wall of the Mt. Whaleback pit - how geotechnology can pay its way, AusIMM/IEAust. Newman Combined Group, Conference on Large Open Pit Mining.

Rosengren, K.J., 1986. Wall reinforcement in open pit mining, AusIMM/IEAust., Newman Combined Group, Conference on Large Open Pit Mining.

Sage, R., 1977. Pit slope manual, Chapter 6-Mechanical support, CANMET report 77-3.

Seegmiller, B.L., 1974. How cable bolt stabilisation may benefit open pit operations, Mining Engineering, 26(12): 29-34.

Seegmiller, B.L., 1975. Cable bolts stabilise pit slopes, steepen walls to strip less waste, World Mining, July, pp 37-41.

A new design method for drilled piers in soft rock: Implications relating to three published case histories

Méthode de conception pour l'emplacement de piliers forés dans une masse rocheuse molle évaluée pour trois cas documentés
Methode für den Entwurf von Bohrpfählen in weichem Gestein (an 3 Fällen erprobt)

R.K.ROWE, Faculty of Engineering Science, University of Western Ontario, London, Canada
H.H.ARMITAGE, Golder Associates, London, Canada

ABSTRACT: It is shown that a new design procedure for piers in soft rock would result in a reasonable, safe pier design which satisfies both settlement limitations and endbearing requirements for three well documented cases. The implication of the field test data with respect to design is then discussed.

RÉSUMÉ: Une nouvelle procédure pour la mise en pratique de l'emplacement de piliers dans une roche molle donne des résultats raisonables et sécuritaires en satisfaisant les éxigences de tassement ainsi que ceux de capacité.

AUSZUG: Es wird gezeigt, dass ein neuer Entwurfsprozess für Pfähle in weichem Stein zu einem praktischen und sicheren Pfahl führt, der in drei gut dokumentierten Fällen die Bedingungen der Senkungsbegrenzungen und Spitzentragfähigkeit erfüllt.

1 INTRODUCTION

The behaviour and design of piers socketed into soft rock has been the subject of considerable research in recent years (eg. see Osterberg & Gill, 1973; Ladanyi, 1977; Pells & Turner, 1979; numerous papers in the Structural Foundation on Rock Conference, 1980; and many others). As part of this research, a number of well documented field tests have been conducted (eg. Williams, 1980; Horvath, 1980; Horvath et al., 1983; Glos and Briggs, 1983 and others). These tests have provided a wealth of field data which indicates that socketed piers may be safely designed to carry loads well in excess of those commonly adopted.

Rowe and Armitage (1984) have made a detailed study of the published field cases and have developed an empirical correlation for estimating the available side shear resistance and rock mass modulus for piers socketed into soft rock. This study also provided statistical data which allows an estimate to be made of the probability of the mass modulus and side shear resistance being less than a presumed design value. Based on this information and a series of theoretical solutions (Rowe & Armitage, 1987a), Rowe and Armitage (1987b) have proposed a new limit state design method for drilled piers in soft rock. The procedure is relatively simple and a socket can be designed in a few minutes using a set of design charts. The design method is based on
(1) satisfying a specified design settlement criterion (allowing for possible slip at the pier/ rock interface); and
(2) checking to ensure that there is an adequate "factor of safety" against collapse.

The objective of this present paper is to illustrate the application of the proposed procedure with reference to three well documented cases (Glos & Briggs, 1983; Williams, 1980; Horvath, 1980; Horvath et al., 1983).

2 GENERAL FEATURES OF THE ROWE & ARMITAGE (1987) APPROACH

The design method involves firstly selecting a maximum groundline settlement which is consistent with good performance of the structure. The maximum pier head settlement, ρ_m, is then calculated by subtracting the compression of the pier or column above the actual socket from the maximum groundline settlement. The design settlement for the socketed pier is then taken to be this maximum pier head settlement divided by two, i.e., $\rho_d = 0.5 \rho_m$.

The design parameters for the rock socket are obtained by applying partial factors to the expected values of the deformation and strength parameters. The probability of exceeding the design settlement can then be related to the choice of partial factors. Thus the design procedure consists of determining a pier geometry such that:
(1) the probability of exceeding the design settlement ρ_d is acceptably small;
(2) the bearing pressure at the base of the pier under design conditions is less than a specified allowable value q_{ba}; and
(3) the maximum pressure that would develop at the base of the pier if only 30% of the expected side shear resistance were mobilized q_{bu} will not exceed one third of the estimated base bearing capacity q_{bm} (i.e., $q_{bu} < q_{bm}$).

Condition (1) ensures that the serviceability limit state criterion is satisfied. Condition (2) is intended to limit yielding of the rock beneath the base of the pier under design conditions.

Normally, one would want the probability of exceeding the design settlement ρ_d to be less than 30%; this will give a probability of exceeding the maximum allowable settlement ρ_m (which is twice the design settlement) of less than 3%. Consequently, the expected observed settlement of a pier designed to satisfy conditions (1) to (3) above will be less than the design settlement (i.e., the design settlement is not the settlement that would be predicted for piers with the selected geometry assuming the most probable rock and pier properties).

Considerable research has been conducted into available side shear resistance of concrete sockets under short term conditions (see Rowe & Armitage, 1984). However, very little research has been conducted into the long term load-transfer behaviour of socketed piers. Based on the very limited available data (eg. Ladanyi, 1977; Horvath, 1980), it would appear that some additional load transfer to the base of the socket will occur with time due to the time dependent characteristics of the rock and concrete. Because of this uncertainty as to exactly how much load will eventually be carried in side shear, condition (3) above is intended to ensure that even if only 30% of the expected side shear resistance is mobilized, there will still be a "factor of safety" of at least 3 against bearing capacity failure of the pier.

Full details concerning the step by step procedure are given by Rowe & Armitage (1987b).

3 CASE I - GLOS & BRIGGS, 1983

Full scale field load tests on two complete rock socketed piers (i.e., with endbearing and side shear) have been reported by Glos and Briggs (1983). These tests, which were performed for a project near Farmington, New Mexico, USA, were used to modify the design of socketed piers which were originally designed on the basis of presumptive endbearing and sideshear values.

The sockets were constructed in the Picture Cliffs Formation, which generally consists of interbedded sandstones and shales. From borehole logs (Glos, personal communication, 1983), the rock grades from a highly weathered state near the ground surface to a slightly weathered condition at 13.5 m below the ground surface. The test sockets (denoted East and West respectively) were located in this slightly weathered zone approximately 15.5 and 14.2 m below the ground surface. The discontinuities at the level of the socketed piers were described as being close to medium close (i.e., 50 mm - 1 m) and clean (i.e., no fracture filling material). Additional details concerning the measured properties of the rock mass are reported by Glos and Briggs (1983).

The test piers had a nominal diameter of 610 mm and socket lengths of 1.4 m and 1.47 m at the east and west pier respectively. Both sockets were artificially roughened to produce two sets of grooves 100mm high and 76 mm deep.

To illustrate the design procedure, consider the redesign of the West socket as follows:
 (1) assumed maximum permissible pier head settlement of ρ_m = 10.6 mm and hence a design settlement ρ_d = 0.5 ρ_m = 5.3 mm
 (2) pier diameter D = 0.61 m
 (3) design load Q_t = 4.5 MN (the design load for the production pier)
 (4) modulus of the piers E_p = 39 000 MPa
 (5) representative uniaxial compressive strength of the rock, σ_c = 8.9 MPa (from Glos and Briggs, 1983).

Rowe and Armitage (1984) have correlated the socket side shear resistance and rock mass modulus deduced from a large number of field tests, with the average uniaxial compressive strength of soft rock deposits in which the socket is founded. These correlations which were developed for rock where there were no open discontinuities within the zone of influence of the socket may provide an initial estimate of the expected side shear resistance and mass modulus as follows:

$$(6a) \quad \tau = 0.45 \sqrt{\sigma_c} \quad \text{for regular clean sockets}$$

$$(6b) \quad \tau = 0.6 \sqrt{\sigma_c} \quad \text{for clean rough, R4, sockets}$$

$$(7) \quad E_r = 215 \sqrt{\sigma_c}$$

where τ = expected side shear resistance in MPa

   σ_c = average uniaxial compressive strength of rock core (ASTM D2938), in MPa

   E_r = expected rock mass modulus of the rock in MPa.

Thus for a rough socket,

$$(6b) \quad \tau = 0.6 \sqrt{8.9} = 1.79 \text{ MPa}$$

$$(7) \quad E_r = 215 \sqrt{8.9} = 640 \text{ MPa}$$

The partial factors are chosen to meet a serviceability limit state such that the probability of exceeding the design settlement is less than 30% (i.e., $f_\tau = f_E = 0.7$) so that the design parameters are given by:

$$(10) \quad \tau_d = f_\tau \cdot \tau = 0.7 \times 1.79 = 1.25 \text{ MPa}$$

$$(11) \quad E_d = f_E \cdot E = 0.7 \times 640 = 450 \text{ MPa}$$

$$(12) \quad E_p/E_d = 39\,000/450 = 87$$

Based on the bore log data, it is assumed that the modulus of the rock beneath the socket (E_b) is the same as that along the sides of the socket (E_r) and hence

$$(13) \quad E_b/E_r = 1$$

The design procedure makes use of two key dimensionless parameters $(L/D)_{max}$, I_d (for full details see Rowe & Armitage, 1987b) where: $(L/D)_{max}$ is the dimensionless socket length that would be required if all the load was carried in side shear, viz.

$$(14) \quad (L/D)_{max} = Q_t/(\pi D^2 \tau_d) = 3.1$$
and I_d = is the dimensionless design settlement factor, viz.

$$(15) \quad I_d = \rho_d E_d D/Q_t = 0.32.$$

From a knowledge of $(L/D)_{max}$, I_d, E_p/E_d, E_b/E_r, the dimensionless socket length $(L/D)_d$ corresponding to a design settlement of ρ_d can be readily determined using design charts published by Rowe & Armitage (1987a,b). These charts also indicate the proportion of the total load (Q_t) which is carried in endbearing (i.e., Q_b/Q_t). In this particular case, Figs. 2a, 3a from Rowe and Armitage (1987b) give respectively,

$$(16) \quad (L/D)_d = 2.4 \quad (\text{i.e., } L_d = 1.46 \text{ m})$$

$$(17) \quad (Q_b/Q_t) = 23\%$$

It then remains to check that the load carried in endbearing is acceptable. The bearing pressure q_b is given by:

$$(18) \quad q_b = (Q_b/Q_t)Q_t/(\pi D^2/4) = 3.55 \text{ MPa}$$

According to the recommendations given by Rowe and Armitage, for q_b to be acceptable we require

$$(19) \quad q_b \leq q_{ba}$$

(where $q_{ba} = \sigma_c$ is recommended)

which is satisfied in this case.

It is also desirable to check that even if there is additional load transfer from side shear to the base of the socket, there will still be an adequate "factor of safety" against collapse. Assuming that only 30% of the possible side shear resistance τ is mobilized, the endbearing pressure q_{bu} would be

$$(20) \quad q_{bu} = 4\left(\frac{Q_t}{\pi D^2} - 0.3(\frac{L}{D})_d \overline{\tau}\right) = 10.2 \text{ MPa}$$

For this to be acceptable, we require that

$$(21) \quad q_{bu} \leq q_{bm}$$

where $q_{bm} = 2.5\ \sigma_c$ is recommended

and since 10.2 < 22.2 MPa, this is satisfactory.

This design indicates that a pier of length 1.46 m satisfies all the design requirements with a design settlement of 5.3 mm. Thus we would expect the actual settlement to be less than 5.3 mm. This length of 1.46 m is close to the actual length of Glos and Briggs's West test pier (i.e., 1.47 m) which experienced a settlement of 2.5 mm at the load of 4.5 MN. Clearly, the design settlement of 5.3 mm is satisfied and there is also a substantial factor of safety against collapse of the pier which was eventually loaded to 8.9 MN without any sign of collapse.

Based on the results of Glos and Briggs's load tests on these two piers with artificially roughened rock sockets, it was noted that:

1) The observed endbearing loads (Q_b/Q_t) were consistent with theoretical solutions;

2) The maximum side shear resistance mobilized by the test piers was approximately 8 times larger than empirical values initially used to design the production piles, thereby resulting in a savings of \$2 million (US) in subsequent construction cost (Glos and Briggs, 1983, 1984).

4 CASE II - WILLIAMS (1980)

The work reported by Williams (1980) represents one of the most extensive socketed pier testing programmes undertaken to date. These tests have covered a wide range of socketed pier geometries, socket roughness and pier types (eg. endbearing only, complete piers etc.).

The load tests were conducted at various sites in a rock formation known locally as Melbourne mudstone (Melbourne, Australia). Melbourne mudstone consists of interbedded claystone (rare), siltstone and sandstone, with siltstone predominating (Williams, 1980). Both the uniaxial compressive strength of the rock and the frequency of discontinuities vary from each test site considered. The two sites of particular interest are referred to as the 1) Stanley Ave. site and 2) Middleborough Rd. site. At the Stanley Ave. site, the piers were constructed in highly weathered mudstone, where the uniaxial compressive strength of the rock ranges from approximately 0.44-0.83 MPa, and joint frequency is generally less than 5 joints/meter (see Williams, 1980). The piers at Middleborough Road were constructed in either medium or highly weathered mudstone. Compressive strengths ranged from 2.0-3.4 MPa with joint frequency less than 1 joint/m.

In the previous section, the piers at Glos and Briggs's site were redesigned using the procedure proposed by Rowe and Armitage (1987b). This redesign involved the use of an empirical correlation to obtain the expected rock mass modulus as well as the use of partial factors to obtain a lower mass modulus for use in design. The justification for applying partial factors to the modulus arises from a statistical study (Rowe & Armitage, 1984) of the variation in rock mass modulus which can occur for rock which has apparently similar characteristics (including uniaxial compressive strength). Williams's data provides a good means of illustrating this variability.

The rock mass modulus values backfigured by Williams from the load-displacement curves for the 17 piers tested at the Stanley Ave. site vary from a low of 74 MPa to a high of 562 MPa. The reason for this variability is not immediately obvious since the piers are not located at great distances from one another (i.e., the test site was enclosed by an approximately 20 m by 6 m rectangular area) and the two available core logs do not indicate any significant changes in the stratigraphy of the rock mass.

The modulus values can be categorized into three groups, depending on their magnitude as follows:

ZONE	RANGE OF BACKFIGURED MODULUS (E_r)
I	$E_r < 100$ MPa
II	$100 < E_r < 250$
III	$E_r > 250$

It happens that these zones correspond to the central, north and south sections of the test site respectively.

Several factors which may contribute to the variability in mass modulus at this site are discussed below.

  i) Since only two core logs are available, it is difficult to accurately assess what type of rock surrounds and/or lies beneath the test piers. The north-core log showed beds of highly-extremely weathered zones of mudstone between highly weathered (pink and yellow) mudstone. Also, a bed of sandstone (approximately 200-300 mm thick) was located about 2.5 m below the rock surface. Consequently, it is thought that some variation in the backfigured modulus could be expected depending on the position of the test pier in the rock mass.

  ii) Possibly of greater significance is the testing arrangement and methods used to perform the pier load tests.
  (a) Piers in Zones I and II were all tested using a hydraulic jack reacting against a fixed reaction beam. Also, the diameters of the sockets are all less than .4 m. Endbearing was prevented in the sideshear sockets (for Zones I and II) by dissolving a sytrofoam disc placed between the rock and pier base.
  (b) In Zone III, the piers all had diameters of 0.6 m or larger. The method of developing reaction for the applied load on Zone III piers was considerably different from the other pier tests at the Stanley Ave. site. A collapsible steel base was used to prevent endbearing for these sideshear sockets, however, calculated vertical stresses (from strain gauge measurements) near the base of one Zone III pier, indicated that significant endbearing loads were in fact developed (eg. $Q_b/Q_t = 48\%$ at $Q_t = 1.76$ MN).

These observations suggest that the difference in backfigured rock mass modulus at the Stanley Ave. site may be influenced by,
(1) Interaction effects due to the selected testing arrangement;
(2) Variations in the rock mass at each socket location;
(3) Differences in the diameter and length of the piers; and
(4) The method used to prevent endbearing on rock for sideshear resistance piers.

These results serve to illustrate that caution is required in estimating rock mass modulus. Careful consideration should be given to the factors discussed above and it should be clearly recognized that the deduced modulus may depend on the nature of the test performed to determine that modulus (see also Rowe & Armitage, 1987a). Furthermore, even with the same test method, significant variation in modulus can occur over a relatively small distance at a given site. It is for this reason that it is recommended that the design modulus be obtained by applying partial factors to the expected mass modulus, even if the modulus has been determined from a load test at a particular site for which the piers are being designed.

4.1 Design calculations for pier M8

The results of design calculations for a complete socketed pier are described in this section. Complete piers (i.e., involving endbearing and sideshear) were tested only at the Middleborough Road site, therefore, the redesign of pier M8 (see Williams, 1980) will be considered.

The design calculations for pier M8 were based upon Rowe and Armitage's (1987b) empirical correlations using data given by Williams (1980). From core logs, the socket is to be located in medium weathered mudstone with uniaxial compressive strength of 2 MPa and a joint frequency less than 1 joint/m with no evidence of clay seams. The test pier was recessed .6 m to avoid a zone of highly weathered mudstone.

The parameters adopted in the redesign of the pier were as follows:
(1) maximum permissible pier head settlement ρ_m=20 mm and a design settlement ρ_d=0.5ρ_m= 10 mm (assumed)
(2) pier diameter D = 0.66 m
(3) design load Q_t = 3.5 MN
(4) modulus of pier E_p = 35 000 MPa
(5) uniaxial compressive strength σ_c = 2 MPa

Using expected sideshear resistance (τ = 0.64 MPa) and mass modulus (E_r = 304 MPa) values deduced from empirical correlations (Eqs. 6a, 7), and partial factors f_τ = f_E = 0.7, a pier length L_d = 1.85m (Q_b/Q_t = 48%) was determined by following the design procedure. Since this socket geometry is similar to that of pier M8, the design settlement of 10 mm may be compared to the observed settlement of 2.6 mm at a load of 3.5 MN.

Clearly, from consideration of settlement, the design is satisfactory (conservative). However, a check on endbearing pressures (eg. similar to calculations [18]-[21] for the Glos and Briggs case) indicates that the allowable values of q_{ba} and q_{bm} are exceeded (i.e., q_b = 4.9 MPa > q_{ba} = 2 MPa; q_{bu} = 8 MPa > q_{bm} = 5 MPa) for this combination of L, D and Q_t. Thus, following the design procedure proposed by Rowe and Armitage (1987b), it is necessary to increase the pier length (or diameter) to reduce the endbearing pressures. However, it should also be noted that the actual test pier with the same geometry was loaded to Q_t = 5 MN before the design settlement of 10 mm was reached and was eventually loaded to 8 MN without collapse (the load displacement curve was still rising). Thus, even though the pier is judged unacceptable in the current design procedure (due to excessive endbearing load), the "factor of safety" against exceeding the design settlement was 1.4 (i.e., a load 40% greater than the design load would be required to give a settlement of 10 mm) and the factor of safety against collapse was greater than 2.25.

Because the pier geometry selected for a design settlement of 10 mm gave excessive endbearing pressure (compared with the recommended values), the pier length was increased to ensure that both endbearing criteria were satisfied. Adopting a more stringent design settlement of 4.5 mm results in a design socket geometry (i.e., $(L/D)_d$) which is nearly 2.5 times larger than the length of the test pier. While the observed settlement is not known for an $(L/D)_d$ = 7, it is likely to be less than the 2.6 mm exhibited by the actual pier M8 with (L/D) = 2.7. The redesigned pier now satisfies both settlement and allowable endbearing requirements. (When endbearing governs it may prove more economical to increase the socket diameter rather than the length; both options can readily be evaluated for a given case.)

5 CASE III - HORVATH (1980); HORVATH ET AL. (1983)

Six full scale socketed pier load tests were performed with the objective of examining and comparing the behaviour of socketed piers which were, (1) normally constructed with a standard single flight auger, (2) using a special grooving device, (3) preloaded at the socket base in order to "stiffen" the load-displacement response. The tests were conducted at a quarry site near Burlington, Ontario, where the rock mass is exposed at the ground surface. The Queenston formation in which the sockets were constructed is a predominantly red shale (with inclusions of harder green shale) of Upper Ordovician age. The heads of the socketed piers were located below the highly weathered surface rock at a depth of approximately 0.6 m. The rock below this level is described as being increasingly massive with depth, with predominantly horizontal joints typically 0.5 to 1 m apart and occasionally filled by up to 37 mm of clayey soil (see Horvath, 1980). The average uniaxial compressive strength was 6.75 MPa. Further details of rock properties and test results are available from Horvath (1980) or Horvath et al. (1983).

For the purposes of the present discussion, attention will be focused on the behaviour of the two complete piers (i.e., both endbearing and sideshear resistance) P2 and P4.

Both piers were initially formed using conventional drilling techniques however socket P4 was then artificially roughened using a special pneumatic roughening tool which produced 25 mm deep x 40 mm high grooves at a spacing of 150 mm.

The load distribution curves for these two piers have been reproduced in Fig. 1. Data used by Horvath to construct these figures were obtained from:
(1) Strain gauges located at the mid-depth of the socketed pier tests section; these gauges were used to measure the radial and vertical strains from which the stresses were calculated.
(2) Load cells at the top and bottom of the pier (to measure the applied load and the load transferred to endbearing).
It is noted that these two different techniques may be expected to have different accuracies. In particular, the stresses (and hence load) determined from the strain gauges at mid-depth should be viewed with particular caution since
(1) they are based on the assumption of a homogeneous elastic pier behaviour. Thus the calculation of stress from strains may be affected both (a) by changes in the modulus and Poisson's ratio of the rock as the load increases; and (b) by local stress concentration and microcracking of the concrete
(2) they are based on the assumption that the stress distribution across the pier is uniform
(3) they are based on the use of the stress-strain relationship

$$\sigma_z = \frac{E_p(1-\nu_p)}{(1-2\nu_p)(1+\nu_p)} \left[\varepsilon_z + \frac{\nu_p \varepsilon_r}{1-\nu_p} + \frac{\nu_p \varepsilon_\theta}{1-\nu_p}\right]$$

and it has been assumed that $\varepsilon_r = \varepsilon_\theta$ near the edge of the pier where the vertical and radial strains ($\varepsilon_z, \varepsilon_r$) were measured. Since the tangential strain ε_θ was not measured, one cannot be certain as to the validity of this assumption. However, based on theoretical considerations, one would not expect $\varepsilon_r = \varepsilon_\theta$ at the location of the instruments. At this location, the relationship between ε_r and ε_θ will depend on (a) the ratio of pier to rock modulus E_p/E_r and (b) the amount of slip which has occurred along the socket. [For example, if there is no slip, and $E_p/E_r \approx 1$ then $\varepsilon_\theta \approx 4\varepsilon_r$ for $\nu_p \approx 0.27$.]

As shown in Fig. 1, the load distribution curves have been normalized with respect to length of the socketed portion of the pier. The distribution of load for pier P2 suggests that the shear along the pier-rock interface is non-uniform at low load levels (i.e., in the elastic range of behaviour). This nonlinearity of shear stress along the pier shaft has been previously recognized by Williams et al. (1980) from instrumented piers socketed into Melbourne mudstone.

Also from Fig. 1a, it can be seen that as the applied load increases, the shape of the load distribution curve changes. The results imply that as the load increases, less and less load is carried in sideshear in the upper half of the socket while more and more of the load is supported by sideshear in the lower portion of the pier. At an applied load of 8 MN, the load distribution curve indicates that none of the applied load is carried between the head and mid-section of the pier. Therefore, the sideshear resistance along the periphery of the pier is zero. This implies a complete loss of sideshear resistance along the top half of the socket and is contrary to the test results given by Williams (1980) and Williams et al. (1980). For example, Fig. 2 shows the results of a load test performed by Williams on a complete socketed pier. Even after significant yielding at the pier-rock interface (Fig. 2a), the shear stress near the top of the pier is non-zero and is uniformly distributed along the pier shaft (Fig. 2c). Thus the present writers would be rather hesitant to infer a complete loss of sideshear resistance along the top half of Horvath's pier P2 particularly in view of (a) the large sideshear which must otherwise have been mobilized along the lower half of the pier and (b) the potential of erroneous measurement of strains at the pier midpoint combined with the question of interpretation as previously discussed.

The distribution of vertical load for the roughened socket P4 deduced by Horvath is shown in Fig. 1b. The interpretation of the midpoint load is even more difficult in this case than for the smoother socket P2 because of (a) erratic strain gauge readings and (b) the assumptions used in deducing the stress from strain certainly do not seem appropriate for an artificially roughened socket.

Thus the present writers are hesitant to draw conclusions about the real distribution of load along the socket based on Horvath's data. However, considering the more reliable data at the top and bottom of the socket it does appear that

(a) only approximately 18% of the total load is carried to the base of the pier at low load levels;
(b) the proportion of load carried to the base increases with increasing load;
(c) at low load levels (i.e., in the elastic range), the proportion of load at the base of the pier is essentially the same for both the conventional (P2) and artificially roughened socket (P4). As the load increases, slip appears to occur more readily for the conventional socket and the proportion of load carried to the base increases relative to that for the artificially roughened sockets.

All of the foregoing findings are in agreement with theoretical expectations (Rowe & Armitage, 1987a).

An additional point of note is the increase in load carried to the base of the pier which was observed when load was sustained for a period of 36 to 40 hours. At socket P2, the ratio Q_b/Q_t increased from 23 to 28% for a load Q_t = 4.45 MN. At socket P4, the corresponding change appears to be from 27 to 30%. There is insufficient data to draw any firm conclusions however there does appear to be a time dependent load redistribution. Considerably more test data on full scale sockets is required before the effect of time dependent load transfer can be directly considered in design. Nevertheless, recognizing this possibility, the design procedure proposed by Rowe and Armitage (1987b) allows for an adequate factor of safety against endbearing failure even if only 30% of the expected sideshear capacity is eventually mobilized.

5.1 Design calculations for piers P2 and P4

Rowe and Armitage (1987b) have given complete "redesign" calculations for these two sockets using empirical correlations to obtain τ_r and E_r as well as backfigured values of these parameters. The results from these calculations are given in Table 1. It is evident from these results that there is substantial

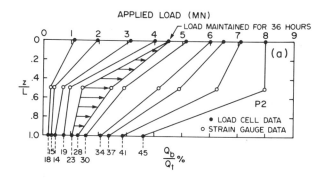

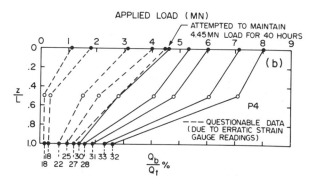

Fig. 1. Vertical load distribution curves for field tested complete socketed piles performed by Horvath et al. (1983) (a) conventional socket (b) roughened socket

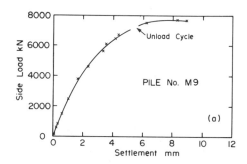

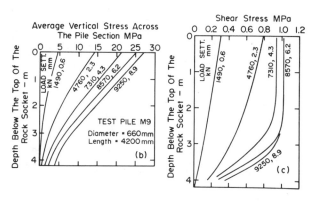

Fig. 2. Load test results for Williams (1980) pier M9

Table 1. Redesign of Horvath's piers

Pier	P2		P4	
Method of Determining $\bar{\tau}_r, \bar{E}_r$	Empirical	Backfigured	Empirical	Backfigured
Design Settlement ρ_d (mm)	8.5	11.5	6.0	8.3
Pier Diameter D (m)	0.71	0.71	0.71	0.71
Applied Load Q_t (MN)	4.5	4.5	4.5	4.5
Pier Modulus E_p (MPa)	37000	37000	37000	37000
Uniaxial Compressive Strength σ_c (MPa)	6.75	6.75	6.75	6.75
τ (MPa)	1.17	1.45	1.56	1.8
E_r (MPa)	560	283	560	377
f_τ	0.7	0.7	0.7	0.7
L_d (m)	1.35	1.35	1.35	1.35
Q_b/Q_t (%)	45	32	29	27
Observed Settlement (mm)	8.4		5.9	

variation in backfigured modulus over a relatively short distance between piers P2 and P4. In this case, the empirical correlation provides a value of E_r which exceeds the value backfigured from the tests, although the reverse trend is true for the value of τ. However, despite these variations, when the appropriate partial factors are applied in the redesign of these piers, it is found that in all cases the observed settlement is less than the design settlement. Thus these calculations show that piers designed to have the same geometry as Horvath's test piers P2 and P4 would have satisfied the design settlement criteria while having a proven "factor of safety" against collapse of at least 2. (The test piers were not brought to collapse and so the actual "factor of safety" is unknown.) Since the maximum permissible settlement is twice the design settlement, it is evident that there is also at least a "factor of safety" of two against the settlement exceeding the maximum permissible settlement (i.e., the settlement at which damage would begin to occur).

6 CONCLUSIONS

An examination of three well documented cases has shown that:

(1) at low load levels, most of the load supported by a socket is carried in sideshear
(2) as the load level increases, there is slip along the socket and an increasing proportion of load is carried in endbearing
(3) the deduced rock mass moduli will vary depending on the details of the test used
(4) on a given site, the rock mass modulus may vary significantly over relatively small distance and it is recommended that design modulus values be obtained by applying partial factors to the expected mass modulus.

The test piers for each were redesigned using the procedure proposed by Rowe and Armitage (1987b) and it was shown that piers designed on this basis would have proved satisfactory in that they had satisfied the design settlement criterion and had a proven "factor of safety" of more than two against exceeding the maximum permissible setlement and against collapse. (The actual factor of safety against collapse is unknown since none of the piers could be loaded to collapse.)

7 ACKNOWLEDGEMENTS

The research described in this paper was supported by the National Research Council of Canada under contract No. ISU83-00082. Additional funding was provided by the Natural Sciences and Engineering Research Council of Canada under grant A1007. Many thanks are due to Dr. M. Bozozuk, of the National Research Council, who acted as Scientific Advisor on this project.

8 REFERENCES

Glos III, G.H. & O.H. Briggs, Jr. 1983. Rock sockets in soft rock. J. Geotech. Engrg., ASCE, 109(4): 525-535.
Horvath, R.G. 1980. Research project report on load transfer system for rock-socketed drilled pier foundations. For the National Research Council of Canada, DSS File No. 10X5.31155-9-4420, Contract Serial No. 1SX79.000531.
Horvath, R.G., T.C. Kenney & P. Kozicki 1983. Methods of improving the performance of drilled piers in weak rock. Can. Geot. J., 20(4): 758-772.
Ladanyi, B. 1977. Friction and endbearing tests on bedrock for high capacity socket design: Discussion. Can. Geot. J., 14(1): 153-155.
Osterberg, J.O. & S.A. Gill 1973. Load transfer mechanisms for piers socketed into hard soils or rock. Proc. 9th Can. Symp. on Rock Mech., Montreal, Quebec, Inf. Canada, 235-262.
Pells, P.J.N. & R.M. Turner 1979. Elastic solutions for the design and analysis of rock-socketed piles. Can. Geot. J., 16(3): 481-487.
Rowe, R.K. & H.H. Armitage 1984. The design of piles socketed into weak rock. Fac. of Engrg. Sc., UWO Research Report GEOT-11-84.
Rowe, R.K. & H.H. Armitage 1987a. Theoretical solutions for axial deformation of drilled shafts in rock. Can. Geot. J., 24(1).
Rowe, R.K. & H.H. Armitage 1987b. A design method for drilled piers in soft rock. Can. Geot. J., 24(1).
Williams, A.F. 1980. The design and performance of piles socketed into weak rock. Ph.D. Thesis, Monash University, Melbourne, Australia.
Williams, A.F., I.W. Johnston & I.B. Donald 1980. The design of sockets in weak rock. Proc. Inter. Conf. on Struct. Found. on Rock, Sydney, Vol. 1, 327-347.

Assessment of rock slope stability by Fuzzy Set Theory
Détermination de la stabilité des versants rocheux par la théorie des ensembles flous
Stabilitätsanalyse von Böschungen mit Hilfe der Fuzzy Set Theorie

S.SAKURAI, Professor of Civil Engineering, Kobe University, Japan
N.SHIMIZU, Research Associate, Kobe University, Japan

ABSTRACT: This paper presents a new approach to assessing the stability of rock slope. The approach is developed by mathematically incorporating judgement and experience of practising engineers, with conventional limit equilibrium analysis. Fuzzy Set Theory is used for evaluating design parameters, and for assessing slope stability. Some examples are presented to illustrate the procedure for this proposed method.

RESUME: La détermination de la stabilité des versants rocheux est abordée de façon nouvelle. La méthode combine le traitement mathématique du jugement et de l'expérience de l'ingénieur praticien avec l'analyse traditionnelle par équilibre limite. La théorie des ensembles flous est utilisée pour l'évaluation des paramètres de dimensionnement et pour le calcul de la stabilité de la pente. Quelques exemples sont présentés pour illustrer la démarche de la méthode proposée.

ZUSAMMENFASSUNG: Die vorliegende Arbeit beschreibt ein neues Verfahren zur Bestimmung der Stabilität von Felsboeschungen. Das Verfahren beruht auf der Kombination von Ingenieurerfahrung und -verständnis mit der Grenzgleichgewichtsmethode. Die Fuzzy Set Theorie dient zur Bestimmung der Entwurfsparameter und der Stabilitaet von Boeschungen. Einige Beispiele illustrieren die Anwendung des vorgeschlagenen Verfahrens.

1 INTRODUCTION

Mechanical characteristics of rock masses contain various types of uncertainties. These uncertainties cause great difficulty in the determination of mechanical constants as definite values, even for the same grade classification of rock masses. In order to overcome this difficulty, the Fuzzy Set Theory (introduced by Zadeh 1965) is a useful tool, which allows us to consider the mechanical constants to be uncertain values. Thus, when analyzing rock engineering problems, the Fuzzy Set Theory must be adopted. It is particularly important for the assessment of rock slope stability, because in rock slope problems, the results of analyses are greatly influenced by the uncertainties of mechanical characteristics. Of course, one can use an ordinary probabilistic approach to solve rock slope problems (Marek & Savely 1978, Chowdhury 1986). However, compared with materials such as steel and concrete, the determination of a probability density function for the mechanical constants of rock masses is extremely difficult. In other words, there is no reliable way to determine the input data for the probabilistic approach. This means that the probabilistic approach may be less applicable to practical engineering problems.

On the other hand, the Fuzzy Set Theory can easily provide all the input data necessary for the analyses on the basis of engineers' subjective judgements. Therefore, the Fuzzy Set Theory is recommended. Recently the Fuzzy Set Theory has begun to win recognition as a potential tool for solving rock mechanics problems (Fairhurst & Lin 1985, Nguyen & Ashworth 1985a, Nguyen 1985b, c, Shimizu & Sakurai 1986).

In this paper, a method based on the Fuzzy Set Theory for determining the strength of rock masses with uncertainties, is described. This method determines the strength parameters by considering their applicability to actual engineering practices. An evaluation method of the stability of rock slopes, using the strength parameters obtained on the basis of the Fuzzy Set Theory, is therefore proposed. According to the Fuzzy Set Theory, the uncertainties in mechanical constants of rock masses are considered as "fuzziness" instead of "randomness" defined in probability theory. The fuzziness is defined as the uncertainties appearing due to either the complexity of the characteristics, or to the lack of knowledge in understanding the characteristics (Yao & Furuta 1986).

2 ESTIMATION METHOD FOR THE STRENGTH PARAMETERS OF ROCK MASSES BASED ON ROCK CLASSIFICATIONS BY THE FUZZY SET THEORY

2.1 Fuzzified rock mass classification

The authors have proposed a rock classification method based on the Fuzzy Set Theory (Shimizu & Sakurai 1986). This classification consists of two major steps;
1. The ambiguous description for the judgement of parameters of rock classifications, for instance, "strength is high", "spacing of discontinuties is very close", are firstly represented by a fuzzy set.
2. The process of giving an evaluation of the parameters of rock classification is mathematically formulated by fuzzy integral and other fuzzy operations.

The proposed rock classification gives the distribution of "fuzzy expected value". This distribution assigns a grade signifying to what degree the rock mass belongs to each rock mass class. The classification parameters and the results of the evaluation are shown in Table 1 and Fig. 1.

2.2 Estimation of the internal friction angle and the cohesion of rock masses

For the strength parameters of rock masses, internal friction angle and cohesion, values are given for each class of rock classification, although they have uncertainties (Bieniawski 1979). In this study, we express these ambiguous values by fuzzy numbers.

A fuzzy number can model an ill-known quantity whose value is "approximately M" (denoted by $\underline{M}$) (Dubois & Prade 1980). For instance, "approximately 2" is

Table 1. Descriptions of the judgements for each parameter of the fuzzified rock mass classification

Parameter	Descriptions for the judgements				
Strength of intact rock material*	Very high	High	Medium	Low	Very low
Drill core quality RQD*	Excellent	Good	Fair	Poor	Very poor
Spacing of discontinuities*	Very wide	Wide	Moderate	Close	Very close
Condition of discontinuities*	Very good	Good	Fair	Poor	Very poor
Ground water*	Completely dry*	Damp*	Wet*	Dripping*	Flowing*

----: example
* Taken in part from Bieniawski's RMR system (1979)

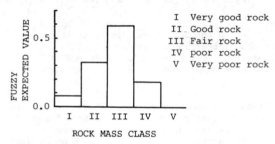

Figure 1. Results of the fuzzified rock mass classification

defined by the membership function shown in Fig. 2. The membership function assigns a grade between 0 and 1 to the objects that belong to a fuzzy set. Namely, 2 belongs to fuzzy number "approximately 2" with a membership grade of 1, and 1.7 with a grade of 0.7, etc. In this study, a membership function denoted by the following four-parameter representation is used (Dubois & Prade 1980).

$$\underline{M} = [\ m_1,\ m_2,\ m_3,\ m_4\ ]$$

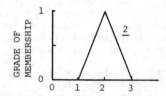

Figure 2. Membership function of fuzzy number "approximately 2"

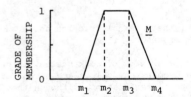

Figure 3. Membership function of fuzzy number $\underline{M}$ represented by the 4-parameter $[m_1,\ m_2,\ m_3,\ m_4]$

Internal friction angle and cohesion, expressed by fuzzy numbers (denoted by $\underline{\phi}$ and $\underline{c}$, respectively), are given in Fig. 4. For example, for poor rock,

$$\underline{c} = [\ 25,\ 40,\ 60,\ 75\ ]\ (kPa)$$

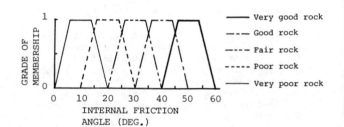

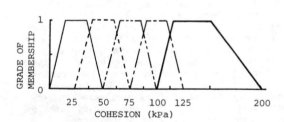

Figure 4. Membership function of internal friction angle and cohesion expressed by fuzzy numbers

Now, taking into account the fuzzy expected value obtained by the rock classification explained in 2.1 as a weight, the fuzzy numbers $\underline{\phi}$ and $\underline{c}$ of each class given in Fig. 4 are superimposed. A method proposed by Nishimura et al (1986) has been modified and used in this superimposition. For an example of Fig. 1, $\underline{\phi}$ and $\underline{c}$ are given in Fig. 5.

3 EVALUATION OF THE STABILITY OF ROCK SLOPES

3.1 Fuzzified factor of safety

For the plane sliding surface of the rock slope shown in Fig. 6, the factor of safety is expressed as follows by the conventional limit equilibrium analysis.

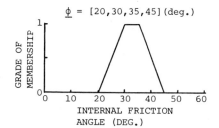

$\underline{\phi}$ = [20,30,35,45] (deg.)

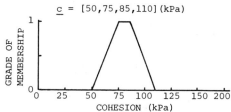

$\underline{c}$ = [50,75,85,110] (kPa)

Figure 5. Membership function of $\underline{\phi}$ and $\underline{c}$ of rock mass obtained by the proposed method

$$F = \frac{K_1 c + K_2 K_3 \gamma \tan\phi}{K_2 K_4 \gamma} \qquad (1)$$

where, $K_1 = (H-z)/\sin\psi_p$, $K_2 = [(H^2-z^2)\cot\psi_p - H^2\cot\psi_f]/2$, $K_3 = \cos\psi_p$, $K_4 = \sin\psi_p$, ϕ = internal friction angle, c= cohesion, γ = the unit weight of rock mass, H= height of slope, z= depth of tension crack, ψ_f = slope angle, ψ_p = inclination of failure plane.

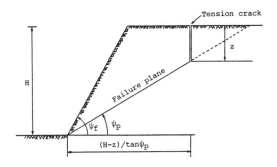

Figure 6. Geometry of slope

Let ϕ , c and γ be fuzzy numbers, $\underline{\phi}$ =[ϕ_1, ϕ_2, ϕ_3, ϕ_4], $\underline{c}$=[c_1, c_2, c_3, c_4] and $\underline{\gamma}$=[γ_1, γ_2, γ_3, γ_4], respectively. The fuzzified factor of safety is obtained from Eq. (1) by using the fast computation formulas (Dubois & Prade 1980).

$$\underline{F} = [\frac{K_1 c_1 + K_2 K_3 \gamma_1 \tan\phi_1}{K_2 K_4 \gamma_4} , $$
$$\frac{K_1 c_2 + K_2 K_3 \gamma_2 \tan\phi_2}{K_2 K_4 \gamma_3} , \frac{K_1 c_3 + K_2 K_3 \gamma_3 \tan\phi_3}{K_2 K_4 \gamma_2} , $$
$$\frac{K_1 c_4 + K_2 K_3 \gamma_4 \tan\phi_4}{K_2 K_4 \gamma_1}] \qquad (2)$$

Let us now consider an example in which a slope (H=60m, z=20m, ψ_p=35 , $\underline{\gamma}$ =[23.5,24.5,25.5,26.5] (kN/m^3))is taken. The fuzzified factor of safety is thus obtained in Eq. (2) as shown in Fig. 7, for ψ_p= 40°, 50°, 60°, 70° and 80°, respectively.

3.2 Evaluation of slope stability

We define a classification for evaluating the stability of slopes on the basis of the fuzzified factor of safety $\underline{F}$ = [f_1, f_2, f_3, f_4] as shown in Fig. 8, that is, "unstable ($f_3 < 1.0$)", "poor ($f_2 < 1.0 \leq f_3$)", "fair ($f_1 < 1.0 \leq f_2$)" and "stable ($1.0 \leq f_1$)".

It is thought that in general many slopes fall into the "fair" class. Therefore, in order to classify them in further detail, we introduce a "stability index (S.I.)" and define it as follows:

$$S.I. = (f_2 - 1) / (f_2 - f_1) \qquad (3)$$

Equation (3) is derived from the concept of "necessary measure" (Dubois & Prade 1980) and it expresses the degree of certainty for the judgement, "This slope is stable".

The above procedure is used in the evaluation of slope stability for the example problem shown in 3.1, and the results are shown in Table 2.

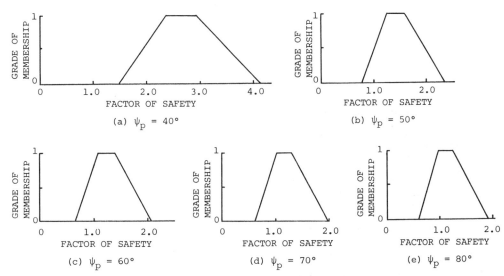

(a) ψ_p = 40°

(b) ψ_p = 50°

(c) ψ_p = 60°

(d) ψ_p = 70°

(e) ψ_p = 80°

Figure 7. Membership function of the fuzzified factor of safety (H = 60m, z = 20m, ψ_p = 35°)

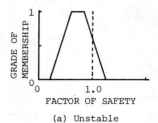 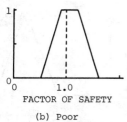 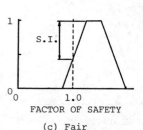 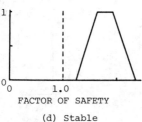

| (a) Unstable | (b) Poor | (c) Fair | (d) Stable |

Figure 8. Class of slope stability

Table 2. Results of evaluation of slope stability

Slope angle	Class	Stability Index
40°	Stable	/
50°	Fair	0.51
60°	Fair	0.19
70°	Fair	0.02
80°	Poor	/

4 CONCLUSION

In this paper, the authors have proposed a method for determining internal friction angle and cohesion as fuzzy numbers by using the fuzzified rock mass classification. According to this method, it is relatively easy to determine the distribution of internal friction angle and cohesion of rock mass, which contain uncertainties. Rock slope stability problems are taken into account as an example in geotechnical engineering. A method of evaluating the stability of slope has been proposed through the use of internal friction angle and cohesion as fuzzy numbers. The special feature of the proposed method of evaluating the slope stability is that the strength parameters of rock masses are dealt with as fuzziness instead of randomness in probability theory.

ACKNOWLEDGEMENTS

This study has been financially supported, in part, by the Japanese Ministry of Education.
 The authors thank Prof. F. Descoeudres and Dr. P. Fritz for their assistance in writing the abstract in French and German. Thanks are also due to Ms. H. Griswold for proofreading and typing the manuscript.

REFERENCES

Bieniawski, Z.T. 1979. The geomechanics classification in rock engineering applications. Proc. 4th Int. Cong. on Rock Mech., ISRM, Montreux, Vol. 2, pp.41-48. Rotterdam, Balkema.

Chowdhury, R.N. 1986. Geomechanics risk model for multiple failures along rock discontinuities. Int. J. Rock Mech. Min. Sci. & Geomech. Abstr. Vol.23, No.5, pp.337-346.

Dubois, D. & Prade, H. 1980. Fuzzy sets and systems, Theory and applications. London, Academic press.

Fairhurst, C. & Lin, D. 1985. Fuzzy methodology in tunnel support design. Proc. 26th US Sympo. on Rock Mech., Rapid city, Vol.1, pp.269-278. Rotterdam, Balkema.

Marek, J.M. & Savely, J.P. 1978. Probabilistic analysis of the plane shear failure mode. Proc. 19th US Sympo. on Rock Mech., Nevada, Vol.2, pp.40-44. University of Nevada-Reno

Nishimura, A., Fujii, M., Miyamoto, A. & Tomita, T. 1986. Engineering dealing of subjective uncertainty in bridge rating. Report of the construction engineering research institute foundation, Kobe, No. 28. (in japanese)

Nguyen, V.U. & Ashworth, E. 1985a. Rock mass classification by fuzzy sets. Proc. 26th US Sympo. on Rock Mech., Rapid city, Vol.2, pp.937-945. Rotterdam, Balkema.

Nguyen, V.U. 1985b. Overall evaluation of geotechnical hazard based on fuzzy set theory. Soils & Foundations, Vol. 25, No.4, pp.8-18.

Nguyen, V.U. 1985c. Some fuzzy set applications in mining geomechanics. Int. J. Rock. Mech. Min. Sci. & Geomech. Abstr. Vol.22, No.6, pp.369-379.

Shimizu, N. & Sakurai, S. 1986. A study on rock mass classification by fuzzy set theory. Proc. JSCE, No.370/III-5, pp.225-232. (in japanese).

Yao, J. & Furuta, H. 1986. Probabilistic treatment of fuzzy events in civil engineering. J. Probabilistic Eng. Mech., Vol.1, No.1, pp.58-64.

Zadeh, L.A. 1965. Fuzzy sets, Information and Control, Vol. 8, pp.338-353.

506

Strength and bearing capacity of rock foundations of concrete dams
Résistance et capacité portante des fondations rocheuses des barrages en béton
Festigkeit und Tragfähigkeit der Felsgründungen von Betonstaumauern

D.D.SAPEGIN, The B.E.Vedeneev All-Union Research Institute of Hydraulic Engineering (VNIIG), Leningrad, USSR
R.A.SHIRYAEV, The B.E.Vedeneev All-Union Research Institute of Hydraulic Engineering (VNIIG), Leningrad, USSR
A.L.GOLDIN, The B.E.Vedeneev All-Union Research Institute of Hydraulic Engineering (VNIIG), Leningrad, USSR

ABSTRACT: An approach to the estimation of safety of concrete dams on rock foundations is discussed from the viewpoint of strength and bearing capacity of the latter. Design and experimental aspects of the problem in question are considered.

RESUME: L'approche à l'évaluation de la sécurité des barrages en béton sur fondations rocheuses est discutée. On examine les aspects analytiques et expérimentaux du problème.

ZUSAMMENFASSUNG: Es wird einen Ansatz zur Sicherheitsbewertung von Betonstaumauern auf Felsuntergründen unter Gesichtspunkt der Tragfähigkeit und Festigkeit der letzteren beschrieben. Es wird berechnete und experimentelle Aspekte dieses Problems betrachtet.

This report discusses the approach being employed at the B.E.Vedeneev VNIIG for estimating safety of concrete dams on rock foundations from the viewpoint of strength and bearing capacity of the latter.

In accordance with the Design Codes the said estimation should be performed by carrying out two kinds of calculations: the calculations of general stability (for all structures) and those of local strength of the foundation (for Class I magnitude structures only).

The use of conventional methods is necessary and sufficient for calculating the general stability of a structure provided that the design shear force does not exceed the ultimate shear strength (with safety factors taken into account). This condition is to be checked for all potentially hazardous design shear surfaces (both flat and broken ones). At that, two kinds of shear are possible for the broken surfaces: the longitudinal shear (along the arrises) and the lateral shear (across the arrises).

When defining ultimate shear forces it is recommended to take into account the shear strength of the downstream toe (undisturbed rock or backfill). The value of this strength depends on the relationship between the deformation modulus of the toe material and that of the foundation and varies from $0.7 E_p$ to E_r, where E_p and E_r are the passive Coulomb pressure and the pressure at rest, respectively.

The local strength of the foundation is calculated (1) to decide on the necessity of measures preventing possible damage to the watertight elements, (2) to be taken into account when elaborating measures meant to increase the strength and stability of the structure and (3) to check whether the ultimate value of the strain is reached when estimating the stress-strain state of the structure and foundation.

Local strengths of a foundation are calculated proceeding from the condition that their values do not exceed those of ultimate tensile strengths at different sections and maximum tangential stresses according to the Coulomb hypothesis at differently oriented design shear surfaces.

It must be pointed out that the local strength calculations are finding ever increasing use for estimating general stability of structures as well. For some reasons, however, these methods are not widely used in to-day practice and so far are not covered by the Norms and Codes for designing water retaining structures.

To calculate local strengths of the foundation one needs to know the stress state of the structure-foundation system. Up to now the basic method of estimating this state remains to be a solution of an elasticity theory plane problem for the domain including the dam profile and a fragment of the foundation large enough to obviate the effect of its boundary conditions on the stresses near the structure.

At present the most popular method of solving the elasticity theory problems is the finite element method which, simply enough, takes account of the complex geometry of the design area, heterogeneity of the dam and foundation materials and different types of boundary conditions.

An important issue in the structural analysis of a dam-foundation system is the adoption of the deformation model of the rock. Most widely used is the model of the linearly elastic isotropic or anisotropic medium.

The predicted results can be brought closer to the real characteristics of the material by using different elastoplastic models also realized within the framework of the finite element method.

When estimating the rock mass stress state one shall take into consideration the presence of joints. In the finite element grid large (primary) joints can be allowed for explicitly as cuts with a double numbering of nodes. Boundary conditions at the surfaces of these cuts - the possibility of joints opening, the non-admission of interpenetration of joint surfaces, the presence of friction - are realized within the FEM framework by using either unilateral constraints or special finite elements.

When the rock mass contains many discontinuites (e.g. has a distinct block-jointed structure) the explicit assignment of the joint geometry leads to the formulation of problems which are very large in volume. There are programs designed specially for their solution. Besides, the methods are being developed which enable the problem on equilibrium of the jointed meduim to be reduced to that of the continuous medium but having special properties which, from the mathematical point of view, are equivalent to plasticity.

Apart from the aforesaid design methods for estimating stability of concrete dams and strength of their foundations the experimental geomechanical simulation methods are also widely used at the B.E.Vedeneev VNIIG. Their employment makes it possible to allow for many important features of rock masses which cannot be properly assessed by the design methods. These are primarily the block-jointed structure of the rock mass and consequently its non-linear deformability, plasticity, anisotropy, complex fracture mechanism, etc. Use is made of both large-scale (geometrical scale $\alpha \gtrsim 0.02$) and small-scale ($\alpha \leqslant 0.005$) geomechanical models. The former are made of high-strength (R_{comp} = 4 - 10 MPa), relatively low-deformable (E = (4 - 10) 10^3 MPa) and heavy (Δ = 2.5 3.5 t/m^3) equivalent materials prepared on the base of gypsum and/or cement binder with aggregates of sand, lime, diatomite, cast iron shot, latex, etc. The latter are made of low-strength (R_{comp} = 0.07-0.6 MPa), highly-deformable (E = 20-100 MPa) and heavy enough (Δ = 2-5 t/m^3) pressable materials prepared on the base of mineral oil or rosin alcoholic solution with fillers of blue powder, zinc oxide, barium oxide, chalk, iron minium, etc.

To meet similarity conditions the following parameters of block-jointed rock foundations are generally simulated in the geomechanical models: the uniaxial compressive and tensile strengths of the rock masses, the shear strength along the joints, the deformability and density of rocks in the mass.

In the model investigations the local strength of rock foundations is mostly estimated proceeding from the results of measurements and observations of strains, opening of existing joints and formation of new ones, appearance of shattered rock zones. The stability estimation is based on the analysis of loads corresponding to failure (limit equilibrium) condition of the models.

For making the above design and experimental (model) estimations of dam stability and local strength of dam foundations one needs to know the values of strength and deformation parameters of rock foundations. For their determination a wide range of field and laboratory investigation methods and broad choice of the associated equipment are available at the B.E.Vedeneev VNIIG.

To define the deformation characteristics of a rock mass assumed to be isotropic or orthotropic, flexible or rigid loading plates are employed for the compression of rock surfaces (in the workings or pits).In so doing the characteristics are defined by the measurements of displacements of different external and internal points of the plate foundations.

The strength characteristics of rock blocks are determined by the method of uniaxial compression and tension of generally cylindrical core specimens.

The shear strength characteristics along isolated and systematic joints in the rock foundation are defined by the shear of rock pillars and those along contact surfaces - by the shear of concrete plates concreted to the rock. At that, in the last few years not only field but also laboratory methods become more and more popular. They are used to advantage due to elaboration of special techniques and equipment which put the roughness of the joint walls of the specimen in correspondence with that of the prototype foundation. Besides, for determining the characteristics of shear strength along major joints having complex non-uniform structures recourse is sometimes made to simulation or experimental prediction methods. These methods are based on the schematization of the joint and adjoining rock, determination of its physico-mechanical properties for the portions selected during the schematization, and the reproduction of these portion-wise in the physical or computational models.

To define the characteristics of shear strength along the shear surfaces not confined to the joints the geomechanical simulation methods are chiefly used. In so doing the characteristics of the joints and the rock blocks which are essential for modelling are determined by the aforementioned methods. Model tests are performed by shearing the jointed rock pillars or concrete plates along the jointed rock foundations. In certain conditions the characteristics in question are obtained by the design methods which are based on their dependence on the shear strength along the joints and on the roughness of weakness contours formed by different joint systems.

It should be emphasized that the shear strength characteristics of the rock foundations are determined similarly in the calculations of both the general stability of structures and the local strength of their foundations. The feasibility of the same and suitability of the local strength conditions are substantiated (so far for triaxial compression tests only) by the special model investigations performed by shearing the jointed rock pillars and by applying the uniform stress field to the block-jointed models with different variation of all stress components.

Finally, it should be noted that the employment of the complete combination of the above methods permits of making reliable and economically effective assessments of concrete dams stability and local strength of rock foundations. The approach discussed in this report has been used for substantiating the safety of some large USSR dams (the Ust-Ilymskaya, Boguchanskaya, Bureiskaya, Toktogulskaya hydro power plants, etc.) .

Probabilistic analysis of intensely fractured rock masses
Analyse de probabilité de masses rocheuses soumises à de fortes fractures
Probabilitätsanalyse von intensiv frakturierten Felsmassen

JAMES P.SAVELY, Inspiration Consolidated Copper Company, Claypool, Ariz., USA

ABSTRACT: Slope stability in rock masses that have been intensely fractured in-place should be analyzed differently than soil or blocky ground. Interlocking of discrete blocks adds strength, but under certain loading conditions blocks can become mobile and cause instability by sliding, rolling, crushing and shearing. This interaction of small, discrete blocks in the base of a slope seems to explain the release of failing masses when no visible structure is present and when there is high rock substance strength. A criteria for recognizing failure modes in these masses is presented followed by a stability analysis using a capacity-demand approach to calculate probability of failure.

RESUME: La stabilité de pente dans masses rocheuses qui ont été soumises à de fortes fractures sur place doit etre analysée différemment que des sols ou les terrains en quartiers de roche. L'emboitement de blocks discontinus aide à renforcer, mais dans certaines conditions de charge, les blocks peuvent devenir mobiles et causer une certaine instabilité suite aux effets de glissement, de roulement, de concassement et de cisaillement. L'interaction de petits blocks discontinus à la base d'une pente semble expliquer le relachement de masses défaillantes en l'absence de toute structure visible et en présence d'une puissante force de substance rocheuse. Nous présentons ici des critères de reconnaissance pour les cas de chutes dans ces masses, suivis d'une analyse de stabilité faisant appel à la demande en capacité pour calculer la probabilité de chutes.

ZUSAMMENFASSUNG: Abhangsstabilität in Felsmassen, die in situ intensiv frakturiert wurden, sollte anders analysiert werden als Boden odr klotzige Erde. Das Ineinandergreifen von separaten Blöcke sich bewegen und durch Rutschen, Rollen, Zerbröckeln und Abscheren Unbeständigkeit verursachen. Diese gegenseitige Beeinflussung von kleinen, separaten Blöcken am Fuss eines Abhangs scheint das Loslassen von geschwächten Massen zu erklären, wenn keine sichtbare Struktur vorhanden ist und wenn grosse Steinsubtanzfestigkeit besteht. Ein Kriterium für das Erkennen von zum Versagen neigenden Kompositionen in diesen Massen wird vorgelegt, und darauf folgt eine Stabilitätsanalyse, die eine Kapazitäts-Anforderungs-Methode zur Errechnung der Wahrscheinlichkeit eines Versagens benutzt.

1 INTRODUCTION

Intensely fractured rock masses are important middle ground between competent rock and soil. Stability analysis is commonly done by either assuming the failure process resembles that of a soil, or that small rock blocks will behave the same as large blocks and will slide with a minimum of block interaction.

When joints, faults, and other weak geologic features form a continuous failure path and they are adversely oriented to a slope, they will cause failure. Even a rock mass composed of small rock blocks can be satisfactorily modeled as a rigid block when the failing mass moves as a single unit, as though the blocks were hinged together.

If an intensely fractured mass is completely loosened or if rock substance is so weak that it is better characterized as a soil, behavior will tend to be plastic. Rotational shear methods, such as Bishop's Method of Slices, will be adequate to assess stability. The assumptions in using these methods are that shear strength is mobilized along the entire failure surface simultaneously, and stresses are completely distributed between slices.

A rock mass that is intensely fractured, but in-place is not loose aggregate. Instead, it is an assemblage of small, interlocked, often discrete rock blocks in a maximum packing arrangement. This tight packing inhibits mobility of the rock blocks and adds apparent strength to the rock mass. Blocks cannot move relative to adjacent blocks without crushing or the mass dilating sufficiently to overcome interlocking.

Line load and point load stresses develop at the contacts of the rock blocks as a result of internal adjustments caused by blasting and excavation. Failure is progressive rather than simultaneous. At low stress levels, which are common to slopes, constituent rock blocks slip relative to each other, corners shear off, blocks rotate, or they crush in induced tension. Stresses redistribute, which overstresses proximate blocks and failure progresses. It may take an extended period of time before failure fully develops, or portions of the slope may fail leaving hanging segments.

A method to assess stability of slopes in intensely fractured rock is needed. The analysis presented considers the mass to be comprised of small, initially interlocked rock blocks with substance strength greater than the shear strength along the sides of the blocks.

Statistical and probabilistic methods are used for three reasons. First, exact values for stress and strength are difficult or impossible to obtain. It is easier to estimate a range of values than it is to estimate one value. Statistics describes this variability by estimating distributions for the variables. Probabilistic analysis accounts for the variability.

Second, for rock masses involving complex geology, variability in parameters is often high, and it is necessary to analyze for simultaneous occurrence of different failure modes. Multiple failure modes can be thought of as series-connected components of the rock mass system. Definite boundaries between material behavior do not have to be uniquely specified if reliability analysis is used. Calculated probabilities of failure of each component determine the composite reliability of the rock mass, and quantify the contribution of each failure mode to the composite reliability. This information can be used to focus design and remedial efforts.

Third, probability can be used as a mathematical expectation in optimization of designs. Failure costs can be weighted by probability of failure to determine expected costs. These can be compared with construction or excavation costs for optimization.

2 FAILURE MODES

In a slope analysis all failure modes are considered viable until it is determined that some do not apply or that their probability of occurrence is so low that further consideration is meaningless. Implied is simultaneous occurrence of different failure modes, each with a probability associated with stability. Thus, a rock mass that is considered to have failure modes associated with intensely fractured rock may also have components of soil-like shear failure or rigid block sliding.

Scale is important. An intensely fractured mass at one scale is an assemblage of large blocks at another. By changing dimensions of the problem, inhomogeneity, discontinuity, and anisotropy of the rock mass also changes. Therefore, mechanics and analytical technique change.

When constituent block size is the same order of magnitude as the slope, failure will most likely involve movement of large blocks along continuous discontinuities. If block sizes are small, say an order of magnitude or more smaller than the slope, interaction of numerous discrete blocks can influence deformation. Orientation of blocks as represented by discontinuity set orientations is a criterion for block interaction.

Interaction of rock blocks in the base of the slope explains release of failing masses when no visible or interpreted structure is undercut and when rock substance strength is sufficiently high to prevent failure through intact material. Rotation of blocks can be significant.

Mechanics of rolling blocks enters the area of post-deformation behavior. Usually, stability is assessed at the onset of failure, but for a rolling mechanism to develop some deformation must occur. Then, the mass must remain dilated for rolling to continue.

Stability may be predicted from expected sliding resistance. However, disturbance of the mass by excavation or blasting, may loosen the mass, which is generally not considered in a stability analysis. Loosening allows block rotation and stability can partially depend on rolling shear strength. The problem is similar to residual strength versus peak strength analysis.

Development of rolling blocks is probably rare even in intensely fractured masses. First, it occurs when joints or joint paths are continuous, and rock blocks are discrete. Block shape must be conducive to rolling, so equidimensional blocks are most susceptible, and orientation of continuous joint sets or continuous paths must be in the same direction as the field stress. Second, other modes are also possible so rolling is only one component of several that may develop.

A synthesis of results from field observations and published model studies suggest four components to failure in intensely fractured rock masses. These are:

1) <u>Sliding</u> on continuous joints or continuous joint paths and slippage between blocks.

2) <u>Shear failure through intact blocks</u> and shearing of corners of blocks.

3) <u>Rotation of discrete blocks</u> in a defined zone (kink band deformation, also called buckling, roller bearing, or internal block rotation).

4) <u>Crushing of intact blocks</u> by induced tensile failure.

Some of these occur independently, but most occur in combinations with each contributing a strength component.

When a continuous joint system or paths dip in the same direction as shear stress, a negative sense as defined by Hayashi (1966), high dilatancy is expected. The slope moves as a large mass away from thin surfaces of separation. This is where kink band and block rotation occur in a basal zone under low normal stress. When continuous joints or paths dip in a direction opposed to shear stress, a positive sense, low dilatancy is expected and deformation will be distributed throughout the mass.

Observations on failure and anisotropy in rock mass dilation suggest some recognition criteria for failure analysis in intensely fractured rock masses (Figure 1). Although only two discontinuity orientations are shown for clarity, the intent is that orientations of continuous features must be identified. When criteria call for continuity with a dip steeper than the slope angle, this means all identified continuous paths must be steeper. If criteria call for discontinuity, all paths must be discontinuous, regardless of the actual number of fracture sets.

Nature is more complex than simple models. With this realization a number of failures will have complex geometries not suited to simple models. It is common to have influence from major structure with failure involving some aspect of soil-like or rock mass failure. Analyses of these complex modes are generally done using ingenuity and judgment to combine analytical techniques. The recognition criteria is to aid in completing these complex models as well.

3 STRENGTH CRITERIA

Strength criteria is required for sliding, rolling, shearing of rock substance, and crushing of rock substance. Probability density functions (PDFs) of these strengths are compared with those of stress in the slope to determine probabilities of failure associated with specific strength properties and deformation modes. Probability calculation provides an accounting of various strength and failure mode combinations.

3.1 Sliding Strength

Criteria for sliding along discontinuities and rock block surfaces is the power failure curve proposed by Jaeger (1971). The general equation takes the form:

$$\tau_s = \tau_o + K \sigma_N^M$$

where:
τ_s = sliding shear strength
τ_o = intrinsic shear strength (shear strength at zero normal stress)
σ_N = normal stress on the shear surface
K, M = power curve parameters from nonlinear regression

Variance about the mean shear strength relationship should be developed by the method proposed by Call (1981) which was developed further by Miller and Borgman (1984). Normal load is calculated in stability analysis and used to obtain the appropriate PDF from the power curve relationship and the standard deviation lines (Figure 2).

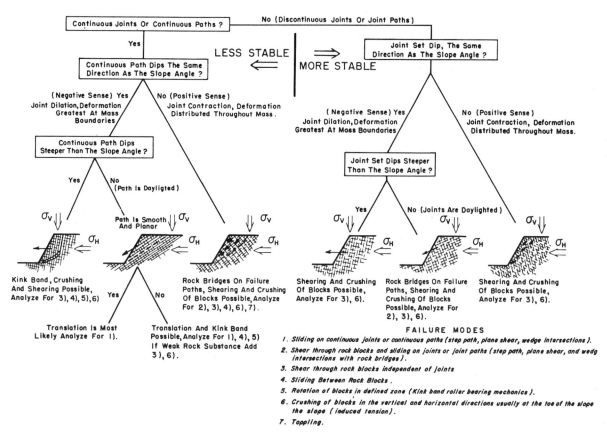

Figure 1. Recognition criteria for analysis of failure modes in intensely fractured rock.

FAILURE MODES
1. *Sliding on continuous joints or continuous paths (step path, plane shear, wedge intersections).*
2. *Shear through rock blocks and sliding on joints or joint paths (step path, plane shear, and wedge intersections with rock bridges).*
3. *Shear through rock blocks independent of joints*
4. *Sliding Between Rock Blocks.*
5. *Rotation of blocks in defined zone (Kink band roller bearing mechanics).*
6. *Crushing of blocks in the vertical and horizontal directions usually at the toe of the slope the slope (induced tension).*
7. *Toppling.*

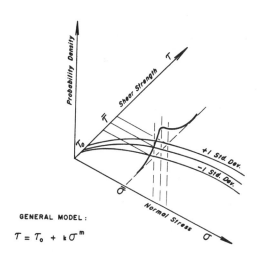

GENERAL MODEL:

$$\mathcal{T} = \mathcal{T}_o + k \sigma^m$$

Figure 2. Power shear strength criteria showing variability for a specific normal load.

3.2 Rolling Strength

Ladanyi and Archambault (1980), allow for kink band development, which implies rotation of blocks, in their shear strength equation. Parameters that allow this condition are derived empirically from their model studies. The general equation is:

$$\mathcal{T} = \sigma'(1 - a_\mathbf{B})\left(\mathring{V} + \tan \phi_\mathrm{J} + \frac{c_\mathrm{J}}{\sigma'}\right) + a_\mathbf{B}\eta\sigma_\mathbf{c}\frac{\sqrt{1 + n} - 1}{n}$$

$$\left(1 + \frac{n\sigma'}{\eta\sigma_\mathbf{c}}\right)^{1/2} \bigg/ 1 - (1 - a_\mathbf{B})\,\mathring{V}\tan\phi_\mathrm{J}$$

and

$$\mathring{V} = \left(1 - \frac{\sigma'}{\eta\sigma_\mathrm{T}}\right)^\mathrm{K}\tan i$$

$$a_\mathbf{B} = 1 - \left(1 - \frac{\sigma'}{\eta\sigma_\mathrm{T}}\right)^\mathrm{L}$$

$\mathcal{T}$ = shear strength of rolling blocks
σ' = effective normal stress
$a_\mathbf{B}$ = shear area ratio
$\mathring{V}$ = dilation rate due to shear
ϕ_J = sliding friction angle
c_J = cohesion for sliding
$\sigma_\mathbf{c}$ = uniaxial compressive strength of rock
n = ratio of uniaxial compressive strength to uniaxial tensile strength
σ_T = transition normal stress
η = degree of interlocking on the failure surface

For kink band development and rolling blocks:

$K = 5$
$L = (2 / n_\mathrm{r})^3 \tan i$
$\eta = 0.5$

where: n_r = number of rows of blocks in the kink band (usually 2 to 5)

An estimate of variability is needed for the probabilistic evaluation. This can be determined by Monte Carlo simulation of the PDFs of the input variables. Alternatively, second moment methods or point estimates might be used to approximate the variability for a given normal load (Rosenblueth, 1975).

The most serious difficulty is obtaining representative distributions of input parameters, so variabilty is high. This makes calculated probabilties high due to uncertainty. It is believed that this strength is probably a peak strength with regard to rolling.

3.3 Shearing Strength of Rock

The most applicable equation for intact rock shear strength is a power curve or linear relationship. The equations are:

$$T_R = C_R + K \sigma_N^M$$

$$T_R = C_R + \sigma_N \tan \phi_R$$

where: T_R = intact rock shear strength
  C_R = intact rock cohesion
  σ_N = normal stress
  K,M = power curve parameters for intact rock
  ϕ_R = internal friction angle of rock

Variance about the mean strength relationship is determined as in the sliding strength calculation, or from regression analysis of failure stress versus confining stress of triaxial test data.

3.4 Crushing Strength of Rock

Best estimates for crushing strength come from point load tests or Brazilian disk tension tests. However, it is likely that Brazilian strength is a lower bound and uniaxial compression strength an upper bound for crushing.

The process of assigning a strength is one of weighting crushing strength by block size gradation, but using soil strength to represent the strength of the fraction of very small sizes, of say 10mm and less. The 10mm size is arbitrary, but in a practical sense the objective is not to assign a rock crushing strength to soft gouge and crushed materials that would be better represented by soil tests.

Size gradation of rock blocks, which is determined from structural analysis, can be expressed as a histogram of sizes (Figure 3). Strength is assigned to each interval based on mid-point size and strength values from testing. It is possible at this point to include a size-strength relationship if it is known or can be assumed. For large blocks that are impractical to test, it may be necessary to assume a size-strength relationship, or assume no reduction in strength.

A weighted strength is calculated by multiplying the percent of material in the size class by the strength value. This is done incrementally for each size and summed over the entire range of sizes for the rock mass crushing strength estimate. Variability in strength is determined by similarly weighting the dispersion found from testing various size ranges. Confinement is not considered because for crushing to develop, blocks must adjust to produce line and point contacts. In this configuration the blocks are unconfined.

4 STRESS DETERMINATION

A clastic mechanics approach has potential for predicting load-deformation behavior of fractured rock masses based on measureable geologic properties. It is predictive at least to the extent that rigid block models represent slope behavior in masses composed of large blocks or slip surface models represent soil shear behavior.

In a clastic model the discontinuum is represented by an assemblage of discrete rock blocks, described

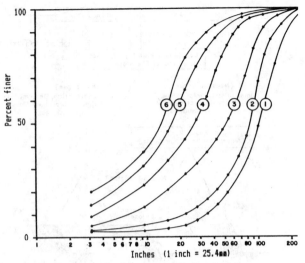

Figure 3. Rock block size gradation curves.

geometrically by distribution angles. These angles are different for various block shapes and arrangements. They determine the directions of stress transmission through the rock mass. The result is a stress estimate based on discrete block characteristics and slope geometry.

Trollope (1968), presents equations to determine the state of stress for any point in a slope. He separates the slope into zones bounded by slope geometry and stress distribution lines that are determined from block shape and block arrangement. The outer zone near the toe is where shear failure would occur. Maximum compressive stresses occur at a point defined by the base of the slope and a stress distribution line near the face. Trollope's equations are lengthy and can be obtained from the reference.

Full arching conditions are assumed in this analysis, which produces maximum horizontal stress. The arching condition could be treated as a variable that adds dispersion to the stress calculation. The orientation of discontinuity sets will determine block shape, block arrangement, thus, directions of stress transmission. These stress distribution angles are true variables in real rock masses.

5 CASE EXAMPLE

The example presented is stability analysis for an open pit mine slope. The planned slope will have a dip direction of 098. It will be 340m long in plan and have heights to 150m. The rock is an intensely deformed metamorphic unit derived from volcanics and sediments. Foliation is obscure or absent. Major structure is favorably oriented to the wall, so structural analysis concentrated on defining joint sets and their characteristics. Six sets were defined:

SET	DDR	DIP	LENGTH (m)	SPACING (m)	WAVINESS (deg)
	mean/sd	mean/sd	mean/sd	mean/sd	mean/sd
1	071/8	52/9	7,8/5.5	.47/.33	1.5/0.6
2	259/8	74/20	11.4/6.9	.30/.30	11.5/0.1
3	129/18	40/14	8.7/9.1	.30/.14	3.5/4.5
4	203/20	53/18	7.4/7.0	.37/.11	2.4/1.7
5	309/15	45/15	1.5/1.5	.21/.16	1.8/0.5
6	021/11	47/12	4.0/5.6	3.04/1.24	1.8/0.2

These set properties represent better portions of the rock mass because the most intensely fractured areas are difficult to map. A step-path geometry is possible from Sets 1 and 2 that will form continuous paths with step-path angles in the same direction as the slope dip. Step-path angles were determined by a numerical approximation to have a mean value of 67 degrees with coefficient of variation of 23 percent. Step-path angles can also be determined by simulation (Call and Nicholas, 1978).

From Figure 1, the path on the left side applies. The difference between the mean step-path angle and the dip of the joint set where sliding occurs (Set 1) is 15 degrees. This is considered an effective roughness and the failure path cannot be considered smooth and planar. Therefore, failure modes 1), 4), 5), and possibly 3) and 6) should be analyzed.

Strength criteria for joint surfaces, gouge and crushed rock are shown in Figure 4. Intact rock properties are:

	mean (MPa)	sd
Uniaxial Compression	57	25
Brazilian Disk Tension	7	2
Coefficient of Internal Friction	.831	.166
Intact Rock Cohesion	13	6

Joint set relationships indicate that steep stress path angles are likely because of the rock block geometry. It is believed that an angle of 15 degrees from the vertical axis is representative, which would be roughly parallel to Joint Set 2. Since the dip of Joint Set 2 has a coefficient of variation of 30 percent, the same dispersion was used for the stress path orientation.

Stresses were calculated at the boundary between Zone 1 and Zone 2 as defined by Trollope (1968). This is where maximum shear and vertical stresses are expected to develop. By considering the stress

path to have a variable orientation, it is possible to develop an expected dispersion in the stresses for various slope heights and angles.

The slope profile shows the spatial relationship of rock classes in the slope that relate to size gradation curves (Figures 3 and 5). Intact rock crushing and shearing were not viable because the rock strength is sufficiently high for slopes under about 300m. Viable modes of rock mass failure involve interblock slippage and perhaps rolling of blocks.

The distribution of shear stress was compared with rolling strength as calculated from Ladanyi and Archambault's equation, and with sliding strength to represent interblock slippage as represented by the criteria in Figure 4. The strength was weighted according to the size gradation for each rock class that occurred in the base of the slope for each slope height and slope angle.

The probability calculation was formulated as a capacity-demand problem. Calculation was done using the standard normal PDF. The following probability values show the contribution of rolling and interblock slippage to possible instabilities. The erratic values are due to the different strength properties that are present in the slopes as represented by the rock class encountered on each level.

The composite probability of failure is calculated as a series-connected reliabilty expressed as unreliability where:

$$Pf = 1 - \prod_{i=1}^{n} (1 - P_i)$$

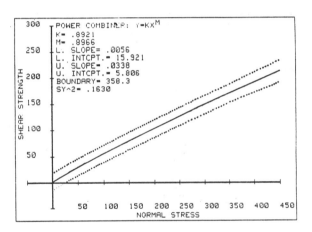

Figure 4a. Strength criteria for joint surfaces (psi units, 145 psi = 1 MPa).

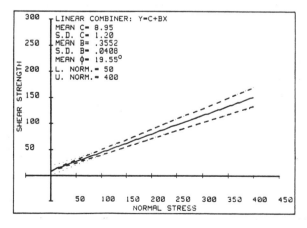

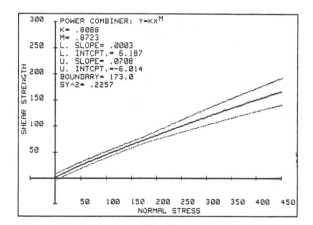

Figure 4b. Strength criteria for gouge and crushed rock (psi units, 145 psi = 1 MPa).

Pf = composite probability of failure
P_i = probability of failure of the i-th failure mode
n = number of failure modes

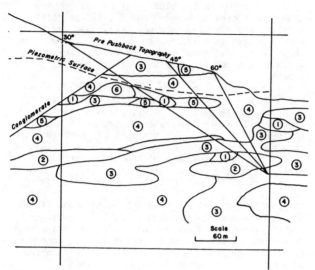

Figure 5. Slope profile showing spatial occurrence of rock classes.

ROLLING BLOCK PROBABILITIES

SLOPE HEIGHT	SLOPE ANGLE		
(m)	30	45	60
30	.079	.015	.015
60	.057	.009	.010
90	.058	.011	.008
120	.073	.014	.006
150	.054	.015	.006

INTERBLOCK SLIDING PROBABILITIES

SLOPE HEIGHT	SLOPE ANGLE		
(m)	30	45	60
30	.183	.043	.043
60	.111	.008	.006
90	.139	.003	.002
120	.086	.004	.001
150	.069	.005	.001

STEP-PATH PROBABILITIES

SLOPE HEIGHT	SLOPE ANGLE		
(m)	30	45	60
30	.001	.004	.116
60	.001	.013	.236
90	.001	.015	.298
120	.001	.016	.350
150	.001	.018	.351

COMPOSITE PROBABILITIES

SLOPE HEIGHT	SLOPE ANGLE		
(m)	30	45	60
30	.248	.061	.167
60	.163	.030	.248
90	.190	.029	.305
120	.154	.034	.355
150	.120	.038	.356

6 CONCLUSIONS

Probabilistic analysis can be extended to intensely fractured rock masses. If the rolling block failure mode develops in the rock mass, stability can be much less than that predicted by the rotational shear analysis. If the mass does not dilate sufficiently to allow the blocks to roll, interlocking

of the blocks will add strength and a rotational shear analysis will underestimate stability. Similar stability inaccuracies result at the other end of the spectrum where only rigid block sliding analysis is used to evaluate the intensely fractured mass.

Stresses were calculated using clastic mechanics but other methods of stress analysis are applicable. The principle is to make a best estimate of stress in the slope. Of significance is the effect of continuous joint paths in relation to the stress orientation. This determines if low shear strength failure modes will develop, such as the rolling mechanism in a dilated rock mass.

REFERENCES

Call, R.D., 1981. "Evaluation of Material Properties", Proc. of Soc. Min. Eng. Workshop on Nonwater-Impounding Mine Waste Structures, AIME, Denver, CO, 25pp.

Call. R.D. and Nicholas, D.E., 1978. "Prediction of Step-path Failure Geometry for Slope Stability", preprint, 19th U.S. Symp. on Rock Mechanics, Stateline, Nev., 8pp.

Hayashi,M., 1966. "Strength and Dilatancy of Brittle Jointed Mass - The Extreme Value Stochastics and Anisotropic Failure Mechanism". Proc. 1st ISRM Cong., Lisbon, v.1, pp.295-302.

Jaeger, J.C., 1971. "Friction of Rocks and Stability of Rock Slopes", Geotechnique, v.21, no.2, pp.97-134.

Ladanyi, B. and Archambault, G., 1980. "Direct and Indirect Determination of Shear Strength of Rock Mass", Preprint No. 80-25, paper presented at Las Vegas Annual Meeting AIME., 16pp.

Miller, S.M. and Borgman, L.E., 1984. "Probabilistic Characterization of Shear Strength Using Results of Direct Shear Tests", Geotechnique, pp. 273-276.

Rosenblueth, E., 1975. "Point Estimates for Probability Moment", Proc. Nat. Acad. Sci. USA, v.72, no.10, pp. 3812-3814.

Trollope, D.H., 1968. "The Mechanics of Discontinua or Clastic Mechanics in Rock Problems", in Rock Mechanics in Engineering Practice, K.G. Stagg and O.C. Zienkiewicz, ed., John Wiley and Sons, New York.

Regularities of alteration of elastic waves velocities in footing massif of Inguri Arch Dam during its operation

Régularités de la variation des vitesses d'ondes élastiques dans le massif de fondation du barrage-voûte d'Ingouri pendant son exploitation

Regelmässigkeiten der Geschwindigkeitsänderung elastischer Wellen im Felsuntergrund der Bogenstaumauer Inguri während der Nutzung

A.I.SAVICH, 'Hydroproject' Institute, Moscow, USSR
M.M.ILYIN, 'Hydroproject' Institute, Moscow, USSR
V.A.YAKOUBOV, 'Hydroproject' Institute, Moscow, USSR

ABSTRACT: For a long time "Hydroproject" Institute (Moscow) is carrying on observations over deformation processes at the foundation of the Inguri dam (Western Georgia), and the geophysical methods are being used. It has been found that each stage of the project construction and operation corresponds to a certain type of alteration in velocities of elastic waves in supersonic and seismic frequencies range. The reasons of the observed alterations of the velocities, as well as possibilities of usage of geophysical data for quantity evaluation of deformation processes and their forecast are discussed in this paper.

RESUME: L'institut "Hydroproject" (Moscou) fait depuis longtemps des observations sur les processus de déformation dans la fondation du barrage-voûte d'Ingouri (Géorgie Ouest) en utilisant les méthodes géophysiques. On a constaté que chaque phase de construction et d'exploitation de l'aménagement correspond à la variation certaine des vitesses d'ondes élastiques dans les gammes de fréquences ultra-sonores et sismiques. Les raisons des variations observées des vitesses, ainsi que la possibilité de l'utilisation des données géophysiques pour la détermination quantitative des processus de déformation et pour la prévision des ces derniers, sont discutées.

ZUSAMMENFASSUNG: Seit längerer Zeit werden vom Institut Hydroprojekt (Moskau) die Deformationvergänge im Felsuntergrund der Bogenstaumauer an der Wasserkraftanlage Inguri (Westgrusien) mit geophysikalischen Verfahren beobachtet. Es ist festgestellt worden, daB jeder Bau- und Betriebsstufe der Wasserkraftanlage ein bestimmter Charukter der Geschwindigkeitsänderung elastischer Wellen im seismischen und Ultraschall-Frequenzbereich entspricht. Die Ursachen der beobachteten Geschwindigkeitsänderungen sowie die Möglichkeiten der Anwendung der geophysikalischen Daten zur qualitativer Einschätzung der Deformationsvorgänge und deren Vorhersage werden erörtert.

1 INTRODUCTION

The Inguri hydropower scheme with its arch dam, 271.5 m high, which has been built under complicated engineering and geological conditions, operates successfully during several years, and is a unique project. For the said period of time the water reservoir level reached, and not once, its maximum design elevation, while an amplitude of its fluctuations achieved 100 m. During this period significant techno-genetic loads transferred upon the structure footing and those loads have changed the original strain-deformed state of the massif and have caused development of various deformation processes there (Savich et al. 1979; Savich et al. 1983).

To keep under control the mentioned processes a system of comprehensive field investigations has been created at the dam section and a component part of the system are the regime seismo-acoustic observations (Savich et al. 1979). Such investigations with usage of elastic waves in seismic (f≈ 100 Hz) and ultrasonic (f≈ 30-50 KHz) range of frequencies were being carried on at the Inguri HPS already more than 10 years, and thanks to them some certain regularities in alteration of the massif properties have been found in a zone of its interaction with the dam at different stages of construction and operation of the structure. The received conclusions in many aspects are based upon the results of a quantity geomechanical interpretation of observed space-time variations of

velocities of the elastic waves (longitudinal Vp and lateral Vs) of the seismic (V^c) and ultrasonic (V^{y^3}) ranges of frequencies. Transition to the level of quantity estimation of processes which go inside the massif demands further development of the geomechanical models. The possible reasons for velocities alteration and methods of quantitative interpretation of the data found at the regime seismoacoustic observations are discussed in the present paper on the basis of an analysis of the measured variations in time of the velocities V^c and V^{y^3} for one stretch of the left-bank abutment of the arch dam.

2 CONDITIONS OF EXPERIMENT

The massif of the left-bank abutment of the arch dam is represented with a strata of thick-layer Low-Cretaceous limestone within the limits of which several differently orientated systems of tectonic and lithogenetic fractures are developed (Fig. 1). The mostly spread are the fracture of system I having modulus of jointing M =2.5-10, of system II with M = 2-5, of system III with M = 0.2-0.3, of system IV with M =0.2-0.25, and as well of systems V and VI with modulus M< 0.1. The filler of the joints is a various one, mainly it is fault gouge and carbonate landwaste; a part of the joints has no filler at all (Dzhigauri 1980). The massif is subjected to significant tectonic stresses, and the maximum compressive stress σ_3 is ≈ 20-30 MPa, it

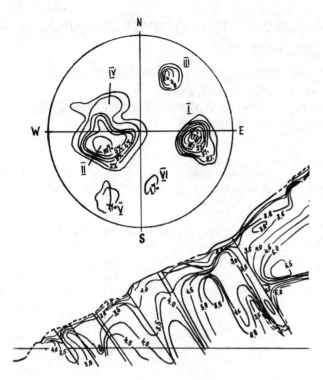

Figure 1. Spatial distribution of longitudinal waves velocity (in isolines of velocity,
km/s) and outline diagram of fissures system singled out in the left bank Inguri dam
massif.

is horizontal, and oriented along the Inguri
river valley. At the subsurface part, down
to some 50-60 m, the massif is unloaded due
to unexpected decrease of the vertical com-
ponent $\sigma_z = \sigma_I = \gamma h$, where: γ - volume
weight, h - thickness of the above laying
rock (Savich et al. 1981).

The area of the massif which is located
between the grouting and drainage galleries
is investigated with seismoacoustic methods;
the area is located at a distance of
20-200 m from the footing of the dam (Savich
et al. 1979; Savich et al. 1979a). Here, in
the open-cut workings, at different lifts
some stationary centers of excitation and
receiving of seismic vibrations are equipped
and mounted, and they provide for estimation
of elastic waves running time by radioscopy
methods; it is carried on in the seismic
range of frequencies, at different directions, with an error of not more than 1%. A
system of control boreholes have been
drilled to carry on the regime ultrasonic
observation here. The used investigation
methodics provides for accuracy of the velocities $V_p^{y_3}$ and $V_s^{y_3}$ measuring on the basis
of 5-10 m not worse than 0.3-0.5% (Savich
et al. 1979). Space division into districts
according to values of V_p^c has been done,
and a space model in a form of isolines of
longitudinal waves was drawn on the basis
of close study of the mentioned massif in
the pre-construction period of time. Some
quasi-homogeneous blocks have been singled
out within the limits of this model; the vo-
lume indicatrixes of V_p^c, $V_p^{y_3}$ and $V_s^{y_3}$ velocities at different directions have been found
for those blocks; those indicatrixes characterize quasi-anisotropism of the massif

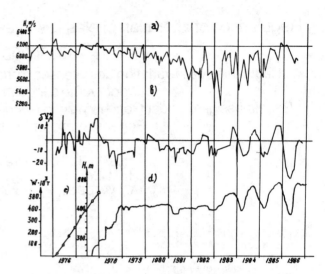

Figure 2. Example of correlation of ultrasonic velocity (a) and relative seismic velocity (b) with weight of concrete (c) and
with reservoir level fluctuations (d).

elastic properties at two different scale levels: at the scale of structural elements
with linear sizes about some 1.0 m, and at
the scale of blocks with linear sizes of some
dozens of meters (Savich et al. 1983; Savich
et al. 1983a).

During construction and in the period of
the reservoir filling, practically speaking
monthly, the repeated estimations of the elas-
tic waves velocities V_p^c, $V_p^{y_3}$ and $V_s^{y_3}$ have
been measured, it as well has been done ac-
cording to the standard schemes of observations. As a result, for each fixed direction
for the basic rock under investigation the
temporary rows of values $V = f(t)$ have been
found, and they reflect the "life" of the
massif under influence of various techno-
genetic and natural factors.

3 ALTERATION OF ELASTIC WAVES VELOCITIES
   IN THE PERIOD OF THE DAM CONSTRUCTION

On the basis of graphs $V^c = f(t)$ and $V^{y_3} = f(t)$ one can form a notion of the elastic
waves velocities alteration in the massif
under observations (see Fig. 2). Judging by
these data, the deformation processes which
take place in the massif affect significantly upon the elastic properties of rock and
cause relative alterations of V_p^c velocities
by 30-40%, and as to alterations of $V_p^{y_3}$ and
$V_s^{y_3}$ values - by 10-15% (Fig. 2). And, by the
way, due to the velocities irregularities at
different directions the transformation of
corresponding volumetric indicatrixes is observed (Fig. 3). It is essential that at
some individual periods of time the graphs
of $V = f(t)$ well correspond to the alteration of this or that exterior factor which
at this present stage causes the deformation
processes in the massif. Thus, for the period
before the reservoir filling the correlation
of the graphs $V = f(t)$ to the season changes
of temperature and the volume of concrete
which had been places in the dam body is observed (Savich et al. 1983a). At the initial
period of the reservoir filling a drastic
reduction in velocities is observed at

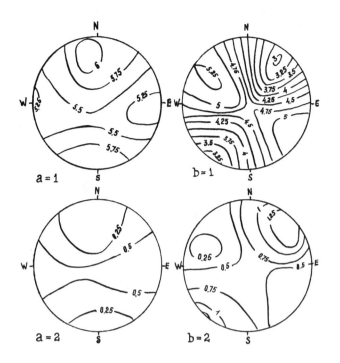

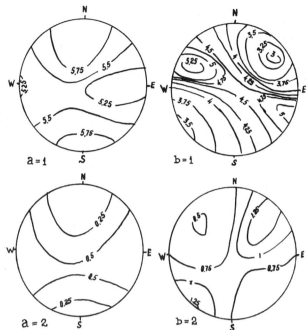

Figure 3. Stereographic projections of experimental volumetric indicatrixes of elastic waves velocities (a) and their alteration in time (b) under weight of concrete placed in left-bank part of the Inguri dam for ultrasonic (1) and seismic (2) ranges of frequencies.

Figure 4. Stereographic projections of theoretical volumetric indicatrixes of elastic waves velocities (a) and their alterations in time (b) under weight of concrete for ultrasonic (1) and seismic (2) ranges of frequencies.

the rise of water level, while after that - a practically synchronous alteration of the velocities and the reservoir level fluctuations takes place (Fig. 2).

The alterations of the elastic waves velocities which have been observed prior to the reservoir filling well corresponded to those which had been expected and forecast on the basis of ratio $V = f(\sigma)$, which had been found at the large-scale field tests (Savich & Yashchenko 1979). At this period of time an alteration of the effective elasticity of the rocks under investigation took place mainly due to the squeezing of the various-scale joints existing in the massif, i.e. elastic packing processes prevailed in the massif then. A model of media having a great number of non-interacting joints can serve the theoretical model (Yakubov 1985), and which takes in account the dependence of jointing parameters (for instance, their concentration) against the stresses which act in the media. On the basis of such model an analysis of volumetric indicatrixes for V_D^c and $V_D^{y^3}$ velocities for the area was effected, and those velocities correspond to the natural massif and to the surcharged with the dam body one. The said indicatrixes (Fig. 4) show good conformity of the experimental and analythical data and as well confirm that the mentioned theoretical model can be used for description of the deformation processes at the discussed stage of the massif "behaviour". However, the description of the further alterations of V and finding of relationship between V and fluctuations of the reservoir, within the framework of the mentioned model, encounter serious difficulties.

4 PECULIARITIES OF VELOCITIES ALTERATION IN THE PERIOD OF THE RESERVOIR FILLING AND OPERATION

Judging by the found experimental data, the main factor which controls the alteration of velocities of the elastic waves in the zone of interaction between the massif and the arch dam is the depth of the reservoir H (Fig. 5), or the accompanying this H values of storage water total weight, the loads P from the structures transferred upon the massif, etc. (Savich et al. 1983a). The connection between values of V and H shows itself the most distinctly when smoothing over the temporary rows of $V = f(t)$ and $H = f(t)$ with the help of the spline-function when you are lucky to reduce the distorting effect of the other factors (Fig. 5).

An analysis of the smoothed over temporary rows shows that since some moment after beginning of reservoir filling the elastic waves velocities in the massif under investigation alter themselves proportionally to H; however, the extreme values at the graphs of $V = f(t)$ are shifted relatively to the corresponding alterations $H = f(t)$ by a certain value Δt-alteration of velocities "delays itself" a little bit. The value of this delay changes itself for different sections of the massif and for different cycles of loading, and it is a kind of indicator of "inertia" of the massif, a reaction for the reservoir fluctuations. Depending on local engineering-geological conditions and some other factors the value of $\delta V = \dfrac{\Delta V}{V_{min}}$ alters itself (as well as the value Δt) where: V_{min} - the value of velocity at minimum level of reservoir. As the total pressure P_Σ upon the massif surface arises, and this

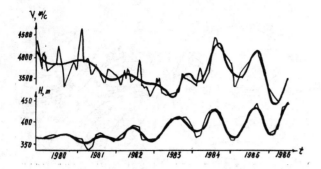

Figure 5. Example of alteration in time of
elastic waves velocities in seismic range of
frequencies and reservoir level fluctuations
(1 - experimental data; 2 - smoothing curve).

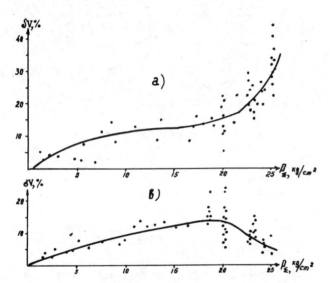

Figure 6. Relative alteration in time of velocity of longitudinal seismic (b) and ultrasonic (a) waves depending on pressure caused
by dead weight of dam (0÷15 kg/cm²) and by
reservoir levels fluctuations (15÷25 kg/cm²).

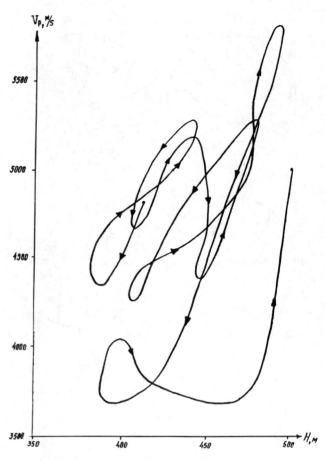

Fig. 7. Alteration of longitudinal seismic
waves velocity depending on reservoir level
fluctuations.

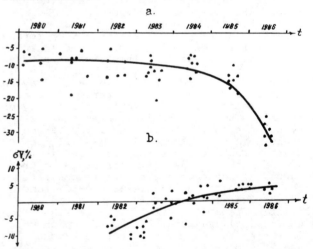

Fig. 8. Relative alteration in time of minimum velocities of longitudinal seismic (a)
and ultrasonic (b) waves.

value depends on the dead weight of the dam
and the weight of water in the reservoir, a
significant growth of δV^c values for the
elastic waves of the seismic range is observed, while the alterations of δV^y for
waves of ultrasonic range attain more complicated nature (Fig. 6). An interesting peculi-
arity of the graph $\delta V^c = f(P_\Sigma)$ is a bend in
the sphere of $P_\Sigma \approx 2$ MPa and a sharp change
of δV^c at $P_\Sigma > 2$ MPa; while the value of P_Σ
arises, the value of δV^y_3 starts to attenu-
ate, and it takes place approximately at the
same value of $P_\Sigma \approx 2$ MPa. It is evident that
the alike behaviour of the graphs $\delta V = f(P_\Sigma)$
cannot be forecast with the well known theoretical solutions, and therefore, for description of the observed phenomenon it is necessary to complicate the initial geomechanical (physical) model.

To create such a model it is important to
find the reasons which bring with themselves
the abnormal alteration of the elastic waves
velocities. For this purpose the nature of
V alteration at various cycles of filling of
the reservoir, and as well the behaviour of
V_{min} in time were analysed. It was found for
the sections which have approximately synchronous changes of V and H the graphs
V=f(H) for each loading cycle have a quasi-

convertible nature with distinctly shown
loops of hysteresis when the velocity value
during the reservoir filling ($\underline{V}$) in the majority of cases is less than at drawdown ($\overline{V}$)
of it (Fig. 7).

Such a behaviour of the graphs V=f(H) is
typical mainly for the values of $P_\Sigma < 2$ MPa.
At the values of $P_\Sigma > 2$ MPa for a part of the
graphs $\delta V^c = f(P_\Sigma)$ correlation between the
values $\underline{V}$ and $\overline{V}$ changes itself for the reverse one (i.e. $\underline{V} > \overline{V}$), and a significant

518

decrease of V_{min}^c takes place. This is distinctly seen in the graph $V_{min}^c = f(t)$ (Fig.8), where in 1984 year the values of P_Σ exceeded 2 MPa and there was abrupt change of δV_{min}. It is significant that no alike phenomenon for the velocities of ultrasonic waves, i.e. for small blocks, are observed (Fig. 8).

5 GEOMECHANICAL INTERPRETATION OF SEISMO-ACOUSTIC DATA

The following interpretation of deformation phenomena which take place in the massif is given on the basis of generalized material and peculiarities of alteration of the elastic waves velocities. At insignificant values of P_Σ the massif "works" quasi-elastically, i.e. when water level in the reservoir rise the existing joints and voids at its bottom partially close themselves, while at drawdown of water they open again. At $P_\Sigma > 2$ MPa a widening of some large cracks takes place, and as a result the jointing of large blocks increases, and the corresponding values of V_{min}^c decrease. This process is caused with surplus pressure in the relatively well penetrated large cracks. When the surface of such fissures is a large one, even insignificant surplus joint pressure can cause concentration of significant stresses at the apex of fissures and at their intergrowth. In small blocks the intensity of such concentration of stresses is significantly weaker, and therefor no intergrowth of fine fissures takes place. Moreover, as the widening of large fissres taking on the squeezing of "inter-joint space" increases, i.e. of small blocks, and this brings with itself the observed growth of V_{min}^y values. It should be mentioned that the correctness of the said interpretation is confirmed with dependence of the V_{min}^c decrease intensity on the rate of reservoir drawdown: at the rate of drawdown which is less than water permeability velocity from the massif, the values of V_{min}^c are not changed practically; while at the rate of drawdown which exceeds the permeability velocity - a sharp decrease of the V_{min}^c takes place (Fig.5). Thus, a nature of V alteration in the massif is determined not only with peculiarities of static stresses distribution which depend on the dead weight of the structures and pressure of water, but as well on dynamics of the processes pertaining to permeability at the footing of the structures.

Taking into account all said above, it is possible to explain the abnormal "behaviour" of the $\delta V^c = f(P_\Sigma)$ graph (Fig.6) as follows: it is known that as pressure grows the intensity of V alteration significantly depends on the initial value of V_0, which corresponds to the massif at P=0. At the same values of P in more weak rocks which are characteristic of small values of V_0, the more significant alterations of velocities δV have to be observed, and on the contrary, the more V_0, the less the relative changes of δV velocities (Savich & Yashchenko 1979). Consequently, the abnormal growth of δV^c at $P_\Sigma > 2$ MPa can be explained with the developing jointing in the massif at the last cycles of reservoir filling, and this is confirmed with the corresponding decrease of V_{min}^c. In such a case the final part of the $\delta V^c = f(P_\Sigma)$ graph can be identified with the sphere of standard curves transition $\delta V^c = f(P_I V_0)$, while P_Σ grows, to the equations which corresponds to the decreased values of V_0. Development of the said process can bring to dangerous deformations at the footing of the structures, and this predetermines the necessity of constant observa-

tions over variations of the velocities and development of methods for reliable forecast of possible consequences.

CONCLUSION

The above-discussed regularities of the elastic waves alteration in the rocks of the Inguri arch dam footing show intricacy of deformation processes which take place in the massif during construction and at various stages of the project operation.

It was found that each stage has its own specific nature of alteration in velocities of elastic waves in supersonic and seismic frequencies ranges, and that it depends on the peculiarities of deformation processes development. Therefore, the observed graphs of alteration of V^c and V^y in time can serve good indicators of the processes which develop inside the massif. However, diversity of factors which affect significantly the variations of V values makes it quite impossible to use the existing simple theoretical models for description of the observed phenomena.

Therefore, the nearest task to develop the seismo-acoustic methods of control over deformation processes, which take place at the footings of high dams is the development of a physical model which will reflect the above-described regularities of alteration of elastic waves velocities at different stages of a project "life".

REFERENCES

Savich,A.I. et al. 1979. Geophysical methods for study of deformation processes in foundations of large hydraulic structures and storage reservoirs. Bull.Int.Ass.Engng Geology. 20: 58-61.
Savich,A.I. et al. 1979a. Distinguishing features of deformation development in the foundation of the Inguri arch dam. Proc. IV Congr. Int.Soc.Rock Mech. Montreux, Switzerland: 585-588.
Savich,A.I. et al. 1983. Geophysical studies of rock masses. Proc. V Congr.Int.Soc.Rock Mech. Melbourne, Australia, 1983, V.A.:19-30.
Savich,A.I. et al. 1983a. Long-term geophysical observations on the Inguri dam rock foundation. Bull.Int.As..Engng Geology, Paris: 28: 315-319.
Dzhigauri,G.M. et al.1980. Ingurskaya plotina na reke Inguri. In: Geologiya i plotiny. M: Energia, v. 8: 7-30.
Savich,A.I., V.I.Koptev & A.M.Zamakhaev 1981. Osnovnye zakonomernosti raspredeleniya polya estestvennykh napryazhenij v massive porod na uchastke plotiny Inguri GES po dannym seismoakusticheskikh izmerenij. In: Geologo-geofizicheskie issledovaniya v rajone stroi-tel'stva Ingurskoj GES. Tbilisi: Metsniereba: 74-93.
Savich,A.I. & Z.G.Yashchenko 1979. Issledovaniya uprugikh i deformatsionnykh svoistv gornykh porod seismoakusticheskimi metodami. M.: Nedra.
Salganik 1973. Mekhanika tel s bol'shim chislom treshchin. Izv. AN SSSR. Mekhanika tverdogo tela. 4: 149-156.
Yakubov,V.A. 1985. Lineino-uprugaya azimutal'no-anizotropnaya model'verkhnei mantii. Izv. AN SSSR, Fizika Zemli, 3:86-90.

Felsverformungen während der Errichtung der Bogenmauer Zillergründl
Rock deformations during the construction of Zillergründl Arch Dam
Déformations de la roche au cours de la construction du barrage-vôute de Zillergründl

P.SCHÖBERL, Tauernkraftwerke Aktiengesellschaft, Salzburg, Österreich

ABSTRACT: In the course of the construction of the 186 m high Zillergründl arch dam the underground deformations were measured already from the beginning of the rock excavation works. During dam concreting the correlation noticed between settlements and concrete load was a distinctly linear one. The check computation of the modulus of deformation confirmed the values obtained in the design stage.

RESUME: Lors de la construction du barrage-vôute de Zillergründl, 186 m de haut, on a commencé à mesurer les déformations du sous-sol dès le début des travaux de fouille. Au cours du bétonnage du barrage on a constaté une corrélation entre tassements et charge du béton nettement linéaire. Le calcul de contrôle du module de déformation confirme les valeurs obtenues déjà au stade de l'étude du projet.

ZUSAMMENFASSUNG: Bei der Errichtung der 186 m hohen Bogenmauer Zillergründl wurden die Untergrundverformungen bereits ab Beginn der Aushubarbeiten erfaßt. Während der Sperrenbetonierung ergab sich ein ausgeprägt linearer Zusammenhang zwischen Setzung und Betonauflast. Die Rückrechnung der Verformungsmoduls bestätigt die im Projektierungsstadium ermittelten Werte.

EINLEITUNG

Als vorläufig letzte Ausbaustufe der Kraftwerksgruppe Zemm-Ziller der TKW AG wurde von 1979 bis 1985 die 186 m hohe Bogenmauer Zillergründl mit einer Kronenlänge von 525 m errichtet. Die Sperrenstelle liegt im Zentralgneis der westlichen Hohen Tauern auf etwa 1.700 m Seehöhe. Die rechte Talflanke besteht aus mehr massigen Gneisen, die linke Flanke aus einem straff geregelten Schiefergneis, der steil gegen den Berg einfällt. In Talmitte erreicht die Überlagerungshöhe etwa 40 m (sh. Abb. 1,2).
Bei der Planung und Ausführung der Überwachungseinrichtungen wurde besonderer Wert auf die Möglichkeit frühzeitiger Installation und Inbetriebnahme jener Meßeinrichtungen gelegt, mit denen die Verformungen des Untergrundes während des Überlagerungsaushubes und der Sperrenbetonierung erfaßt werden können. Zusätzliche Bedeutung erhielten diese Messungen durch eine nahe der luftseitigen Begrenzung der Sperre liegende, steil zur Luftseite hin einfallende Störungszone. Ein kontinuierlicher Verlauf der Setzungsmulde über diese Störungszone hinweg wäre als Hinweis zu deuten, daß zumindest unter diesen Beanspruchungen das Spannungs- und Verformungsbild im Untergrund durch die Störungszone nicht wesentlich beeinflußt wird.
Die Auswertung dieser Messungen während der Bauzeit sollte eine Verbesserung der im Projektierungsstadium ermittelten felsmechanischen Kennwerte und damit eine verbesserte Vorhersage des Untergrundverhaltens während der ersten Stauperioden ermöglichen.

MESSEINRICHTUNGEN

Die Untergrundverformungen werden in 7 Meßquerschnitten erfaßt. In jedem Meßquerschnitt können die Verschiebungen in horizontal-tangentialer und vertikaler Richtung nahe der Luft- und Wasserseite der Mauer, in radialer Richtung und normal zur Aufstandsfläche mit Extensometern erfaßt werden. Durch diese Extensometeranordnung ist erstmals auch die Ermittlung der Verdrehungen um die vertikale Achse möglich. Die Länge dieser Extensometer entspricht der doppelten Breite der Aufstandsfläche. Klinometer nahe der

Abb. 1 Bauzustand Betonierbeginn

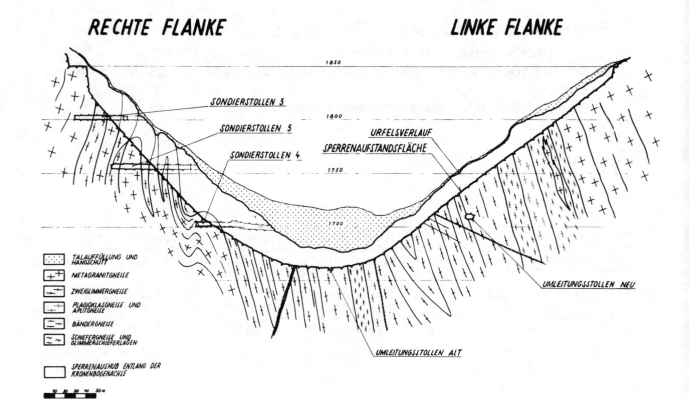

RECHTE FLANKE   LINKE FLANKE

SONDIERSTOLLEN 3
SONDIERSTOLLEN 5
SONDIERSTOLLEN 4

URFELSVERLAUF
SPERRENAUFSTANDSFLÄCHE

UMLEITUNGSSTOLLEN NEU

UMLEITUNGSSTOLLEN ALT

TALAUFFÜLLUNG UND HANGSCHUTT
METAGRANITGNEISE
ZWEIGLIMMERGNEISE
PLAGIOKLASGNEISE UND APLITGNEISE
BÄNDERGNEISE
SCHIEFERGNEISE UND GLIMMERSCHIEFERLAGEN
SPERRENAUSHUB ENTLANG DER KRONENBOGENACHSE

Abb. 2   Übersichtslängenschnitt in der Kronenbogenachse

Wasserseite, in der Mitte und nahe der Luftseite der Mauer ermöglichen einen Vergleich der direkt gemessenen Verdrehungen um die tangentiale Achse mit den aus den vertikalen Extensometern gewonnenen Daten. (sh. Abb. 3)

Um zumindest im Mittelschnitt, also in der Talfurche, die zu erwartenden Hebungen und Setzungen von Beginn der Bauarbeiten an verfolgen zu können, wurde im Entwässerungsstollen, der vor Aushubbeginn fertiggestellt war, ein Nivellement eingerichtet, das von einem Festpunkt etwa 300 m luftseits der Sperre ausgeht. Am wasserseitigen Ende dieses Entwässerungsstollens wurden zwei vertikale Extensometer mit 15 m und 50 m Meßlänge eingerichtet.

Um die Verformungen zufolge Eigengewicht der Mauer ab Beginn der Betonierung verfolgen zu können, wurde in den unteren 5 Meßquerschnitten vor Betonierbeginn an der Wasser- und Luftseite der Mauer je ein elektrisches Extensometer in vertikaler und horizontal-tangentialer Richtung mit 27 m bis 20 m Meßlänge eingebaut, also jeweils etwa 1/3 der einige Monate später eingebauten oben erwähnten mechanisch gemessenen Extensometern.

Dazu kommen noch in jedem Meßquerschnitt je 5 auf den Radialschnitt verteilte Telepreßmeter zur Messung der Druckspannungen normal zur Aufstandsfläche.

Sobald der Bauzustand die Messungen in den unteren Kontrollgängen der Sperre ermöglichte, wurden die Nivellementschleifen auf den Sohlgang der Sperre, den Injektionsgang im wasserseitigen Vorboden, auf den Gang am luftseitigen Mauerfuß und die radialen Verbindungsgänge in der Meßblöcken ausgedehnt.

Bei den Messungen wurde besonderer Wert auf eine möglichst frühzeitige, abgesicherte Nullmessung gelegt; zeitweilige Verlängerungen der Meßintervalle wegen des Baubetriebes beeinträchtigen die Aussagekraft dieser Messungen nicht.

BAUABLAUF

Die Bauarbeiten im engeren Sperrenbereich begannen 1979 mit dem Abtrag der Überlagerung in den Talflanken, mit der Errichtung von Umlauf- und Entwässerungsstollen und einiger Sondierstollen. 1981 begannen die Felsaushubarbeiten in den Talflanken. Im Frühjahr 1982 wurde der Überlagerungsaushub im Talboden begonnen. Die Felsaushubarbeiten wurden im Frühjahr 1983 abgeschlossen. Insgesamt mußten 1,1 Mio. m³ Überlagerungsmaterial und rund 600 000 m³ Fels ausgehoben werden. Bezogen auf den Meterstreifen im Talboden bedeutet dies eine Entlastung um etwa 55 MN, verteilt auf eine Länge von etwa 55 m.

Nach dem Einbringen von rund 20 000 m³ Probebeton an der linken Flanke noch im Herbst 1982 wurde im Sommer 1983 die eigentliche Sperrenbetonierung begonnen. Bis Mitte November 1983 waren rund 284 000 m³ Beton eingebracht. 1984 wurden dann von Anfang Mai bis Mitte November 731 000 m³ Beton eingebracht und auch die Anschüttung des luftseitigen Sperrenvorplatzes durchgeführt. Im Betonierjahr 1985 wurde mit der Einbringung von 346 000 m³ Beton die Betonierung der Sperre im September abgeschlossen. Zusammen mit den Nebenbauwerken wurden rund 1,4 Mio. m³ eingebracht.

Der Mittelblock 12 hat an der Basis eine Stärke von 40 m, die bis zur Krone hin auf 6 m abnimmt. Die Breite jedes Blockes beträgt 20 m an der Krone und nimmt gegen die Aufstandsfläche bedingt durch die Sperrenform etwas zu. Der Mittelblock hat eine Kubatur von rund 90 000 m³, das ergibt bei einem spezifischen Gewicht von 25 kN/m³ für den Sperrenbeton ein Blockgewicht von 2 252 MN, das sind etwa 110 MN je m Aufstandsfläche, verteilt auf 40 m Blocklänge. Nach Fertigstellung der Mauer entspricht dies im Mittelschnitt Spannungen von 5,4 N/mm² Druck an der Wasserseite und rund 0,5 N/mm² Zug an der Luftseite.

522

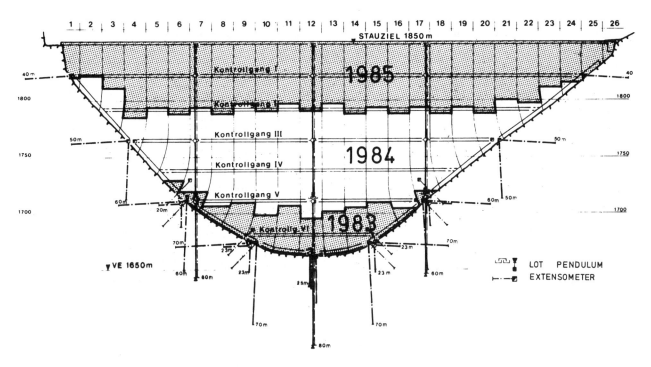

Abb. 3 Längenschnitt mit Meßeinrichtungen und Jahresfugen

MESSERGEBNISSE WÄHREND DES AUSHUBES

Zur Erfassung der Felsverformungen während dieser Bauphase standen das Nivellement im Entwässerungsstollen sowie die beiden lotrechten Extensometer am wasserseitigen Ende dieses Stollens zur Verfügung. In Abbildung 4, 5 sind der zeitliche Verlauf des Aushubes im Talboden und die Ergebnisse der Extensometermessungen dargestellt. Der Aushub bis Oktober 1982 bewirkte eine Hebung von rund 1 mm, die während der Arbeitspause bis Feber 1983 etwa gleich blieb. Der forcierte Überlagerungs- und Felsaushub ab Frühjahr 1983 ergab zunächst eine Hebung des Meßpunktes um weitere 7,5 mm. Die Hebungen des Untergrundes nahmen dann bis September 1983, dem eigentlichen Beginn der Sperrenbetonierung, im Talboden noch geringfügig zu und erreichten mit 8,27 mm, bezogen auf die 50 m lange Meßstrecke, ihren höchsten Wert.

SPERRE ZILLERGRÜNDL

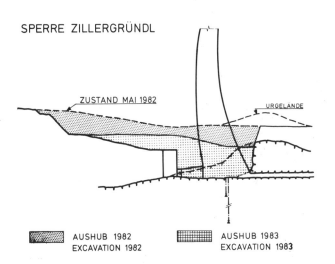

ZUSTAND MAI 1982

URGELÄNDE

AUSHUB 1982
EXCAVATION 1982

AUSHUB 1983
EXCAVATION 1983

Abb. 4 Überlagerungs- und Felsaushub im Talboden

Das 15 m lange Extensometer zeigt eine Verlängerung der Meßstrecke von 7,55 mm an. Erwatungsgemäß sind daher die Hebungen auf die oberste Felsschwarte beschränkt, im Bereich von 15 bis 50 m sind die Auswirkungen des Felsaushubes nur mehr gering.

Die Nivellementmessungen im Entwässerungsstollen zeigen nur geringfügig größere Werte und bestätigen daher die Ergebnisse der Extensometermessungen.

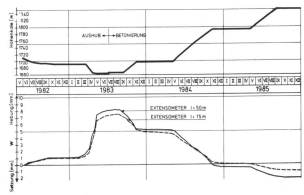

Abb. 5 Zusammenhang Baugeschehen-Extensometermeßwerte

MESSERGEBNISSE WÄHREND DER SPERRENBETONIERUNG

Mit dem Beginn der Sperrenbetonierung und der damit verbundenen Belastung begannen die Setzungen des Untergrundes. Der zeitliche Verlauf der Extensometermeßwerte und der Nivellementmessungen spiegeln deutlich den Bauablauf wider. Während der Betonierphase nahmen die Stauchungen des Untergrundes zu, während der Winterpausen werden nur äußerst geringfügige Verformungszuwächse gemessen.

Auf Grund des unterschiedlichen Zeitpunktes der Nullmessungen konnten die Meßergebnisse der kurzen und langen Extensometer, aber auch jene der Klinometer nahe der Aufstandsfläche zunächst nicht direkt miteinander verglichen werden. Daher wurden die Meßwerte in Relation zur Betonauflast dargestellt. Ab-

bildung 6 zeigt die nahezu lineare Abhängigkeit der
Setzungen von der Betonauflast und der Differenz der
Setzungen zwischen Wasser- und Luftseite vom Moment
zufolge der Betonauflast. Die Winterpausen sind durch
geringfügige Sprünge im Meßwertverlauf zu erkennen.
Auch die Ergebnisse der Nivellementmessungen zeigen
eine nahezu lineare Korrelation zur Betonauflast.

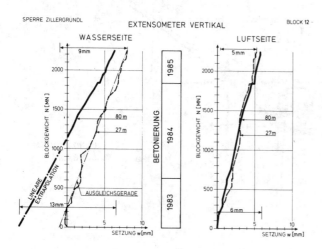

Abb. 6   Zusammenhang Blockgewicht-Extensometermeß-
         werte

Somit konnte eine lineare Ausgleichsrechnung durchgeführt werden und die Ausgleichsgerade - sofern erforderlich - über den Meßbeginn hinaus bis zum Betonierbeginn verlängert und damit die Gesamtsetzung
der Sperre während der Betonierung ermittelt werden.
Zusammen mit den Ergebnissen der Nivellements konnte nach diesen Auswertungen die Setzungsmulde zufolge der Sperrenbetonierung konstruiert werden. Die
größte Setzung ergab sich für die Wasserseite des
Mittelblockes mit rund 14 mm. Wie auch aus dem Mittelschnitt ersichtlich ist, klingen diese Setzungen
wasserseits der Sperre relativ rasch ab; unter dem
luftseitigen Sperrenfuß sind die Setzungen der
Beanspruchung entsprechend erheblich geringer (im
Mittelblock rund 5 mm), erfassen jedoch zufolge der
bis zu 40 m hohen Anschüttung des Sperrenvorlandes
einen weiten Bereich. Zu den Flanken hin verläuft

die Setzungsmulde weitgehend symmetrisch, sodaß im
Lastfall "Eigengewicht" kein Unterschied im Ver-
formungsverhalten der beiden Talflanken erkennbar ist.
Auch bei der Überquerung der Störungszone zeigt die
Setzungsmulde keine Unstetigkeit.
Aus dem Vergleich mit den Nivellements ist erkenn-
bar, daß mit den kurzen Extensometern rund 2/3, mit
den langen Extensometern mehr als 90 % der Gesamtsetzung erfaßt werden. Die Verankerungspunkte der
Extensometer in 80 m Tiefe setzen sich zufolge der
Errichtung der Sperre somit um weniger als 1 mm.
Die Ergebnisse der Klinometermessungen stimmen mit
jenen der Extensometermessungen gut überein. Die
größten Verdrehungen im Mittelblock betragen etwa
0,3 mm je m in Richtung Wasserseite.
Auch die Meßwerte der wasserseitig angeordneten
Telepreßmeter zeigen eine lineare Abhängigkeit von
der Betonauflast. An der Wasserseite des Blockes 12
wurden mit 5,0 N/mm² am Ende der Sperrenbetonierung
fast genau die rechnerischen Spannungen gemessen.
Für die seitlichen Meßquerschnitte wurden die gleichen Verfahren angewendet, die Ergebnisse sind in
der Darstellung der Setzungsmulde enthalten.

RÜCKRECHNUNG DES VERFORMUNGSMODULS

In Abbildung 8 wird ein Vergleich der gerechneten
und gemessenen Setzungen im Mittelschnitt darge-
stellt. Die gerechneten Setzungen wurden für den
Lastfall "Eigengewicht" unter Annahme unabhängiger
Blöcke ermittelt, wobei die Felsverformungen nach
Vogt bzw. Mladyenovicz mit einem Verformungsmodul
für den Untergrund von 20 kN/mm² berechnet wurden.
Wie dieser Vergleich zeigt, stimmen Messung und Rechnung hinsichtlich der Setzung und der Neigung der
Aufstandsfläche gut überein, wenn mit einem globalen
Verformungsmodul von 22 kN/mm² für den Untergrund
(entgegen der ursprünglichen Rechnungsannahme von
20 kN/mm²) gerechnet wird.
Weiters wurde für den Mittelschnitt der Sperre eine
ebene FEM-Vergleichsrechnung des Lastfalles "Eigengewicht" durchgeführt und eine Übereinstimmung mit
den in den verschiedenen Tiefenstufen gemessenen
Untergrundverformungen unter Annahme folgender Verformungsmoduln für den Untergrund erreicht:
- für den obersten, durch den Felsaushub entspannten
  und nachträglich wieder konsolidierten Bereich bis
  etwa 20 m Tiefe .... 10 kN/mm²
- für den daran anschließenden Bereich bis etwa 60 m
  Tiefe .... 20 kN/mm²
  und
- für den tiefer liegenden Bereich .... 40 kN/mm².

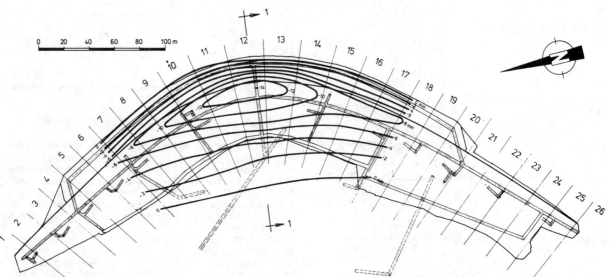

Abb. 7   Schichtenlinien der Setzungsmulde zufolge Betonierung der Sperre

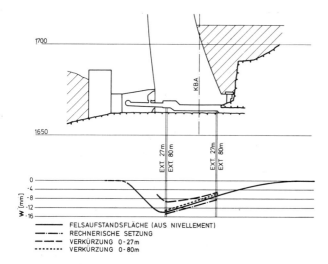

Abb. 8 Schnitt durch die Setzungsmulde in der Tal-
        achse

Damit ergibt sich für die rund 100 m mächtige Felsschwarte, die durch die Sperrenbetonierung beeinflußt
wird, ein mittlerer Verformungsmodul von rund 22 kN/
mm².

ZUSAMMENFASSUNG

Bei der Errichtung der 186 m hohen Bogenmauer Zillergründl wurden die Untergrundverformungen möglichst
frühzeitig erfaßt.
Der Aushub von rund 40 m Überlagerung im Talboden
bewirkte eine Hebung von 8 mm. Während der Sperrenbetonierung wurden durchwegs lineare Zusammenhänge
zwischen Betonauflast und Setzungen der Aufstandsfläche festgestellt, die eine Extrapolation bis zum
Betonierbeginn und somit eine Abschätzung des Einflusses der Errichtung der gesamten Sperre auf den
Untergrund zuließen. Durch die Sperrenbetonierung
wird eine rund 100 m mächtige Felspartie beeinflußt.
Die maximale Setzung in Talmitte beträgt rund 14 mm,
die Setzungsmulde verläuft zu den Flanken hin weitgehend symmetrisch, zur Wasserseite hin ist sie relativ scharf begrenzt.
Die Rückrechnung des globalen Verformungsmoduls
ergibt einen Wert von 22 kN/mm² und bestätigt damit
die aus den geologischen und felsmechanischen Erkundungen angenommenen Werte.

LITERATURHINWEISE

WIDMANN, R.
   Gründungsprobleme bei der Bogenmauer Zillergründl
   Zeitschrift Felsbau, Jahrgang 1 Nr. 3/4 1983
NOWY, W.
   Sperre Zillergründl - Beeinflussung des Sperren-
   aushubes durch tiefgehende Hangauflockerung
   Zeitschrift Felsbau, Jahrgang 2 Nr. 4 1984
SCHLEGEL, G.
   Geodätische Probleme beim Bau des Zillerkraftwerkes
   Vortrag beim 7. nationalen Felsmechanik Symposium
   Aachen 1985
WIDMANN, R.
   Grundlagen für den Entwurf der Bogenstaumauer
   Zillergründl
   Zeitschrift Wasserwirtschaft, Jahrgang 74, Nr. 3
   1984
VOGT, G.
   Über die Berechnung der Fundamentdeformation
   Det Norske Videnshaps Akademi, Oslo 1925
MLADYENOVICZ, V.
   Calcul des Barrages-Voûtes par resolution d'
   equations lineares
   Travoux, 1958

Strata buckling in foot wall slopes in coal mining
Flambage des strates dans l'éponte inférieure des pentes dans une mine de charbon
Schichtenwölbung im Liegenden eines Kohletagebaus

T.SERRA DE RENOBALES, Empresa Nacional Hulleras del Norte, S.A., HUNOSA, Asturias, Spain

ABSTRACT: The matter of study is the colapse of a foot wall slope in an Open cast bitominous coal mining, in Central Asturian Carbonferous Basin (Spain), because it is considered that may it contribute to the establishment of definition methods for slopes. Through the geomechanical analysis of the rocky massif and the geometry of the failure, a hypothesis of the possible mechanism of failure is established. The quasi-plane failure, parallel to the stratification and with a concentration of blocks at the foot make us think that the mechanism of failure has been the one denominated "Buckling of strata". This shows the need of considering, in the definition of foot wall slopes, the failure by buckling of strata and we describe a methodology to prevent this failure mechanism, remarking the importance of doing a detailed study of the strata thickness of our rocky massif in relation to the depth.

RÉSUMÉ: Nous allons étudier l'éboulement d'un talus de mur à une Explotation à ciel ouvert, dans le Bassin Houillier Central Asturien, parceque nous pensons que ça peut contribuer au établissement de méthodes pour la définition des talus. Parmi l'analyse géomecanique du massif rocailleux et de la géométrie du éboulement on peut établir une hypothèse sur le mécanisme de rupture. La rupture presque-plaine, parallèle à la stratification et avec une concentration de blocs à la base nous fait considérer que le mécanisme de rupture est cel qu'on conaît comme "gauchissement des strates". Nous pouvons démontrer que quand-on parle de la définition des talus de mur, on est obligé à parler de la rupture à cause du gauchissement des strates et on développe une méthode pour prévenir ce mécanisme de rupture en détachant comme il est important l'étude en détail des épaisseurs des couches de notre massif rocailleux par rapport avec la profondeur.

ZUSAMMENFASSUNG: Es wird das Zusammenbrechen Ciner Mauerhalde beim überirdischen Steinkohleabbau im zentralasturianischen Kohlebecken (Spanien) erforscht, in der Annahme, dadurch zur Erstellung von Methoden für die Definition von Halden beitragen zu können. Durch die geomechanische Analyse des Felsmassivs und der Fall geometrie wird eine Hypothese des möglichen Bruchmechanismus aufgestellt. Der gleichsam ebene Bruch, parallel zur Lagerung und mit einer Anhäufung von Steinblöcken an der Base, Lässt annehmen, dass es sich bei dem Bruchmechanismus, um die sogenannte "Schichtenwölbung" handelt. Es zeigt sich die Notwendigkeit, bei der Definition von Mauerhalden den Bruch durch Schichtenwölbung in Betracht zu ziehen Jund die Wichtigkeit der Durchführung eines detaillierten Studiums der grössten Schichtmächtigkeiten unseres Felsmassivs in Bezug zur Tiefe hervorzuheben.

1 INTRODUCTION

The optimizing of operation costs in open cast mining, generally leads to design mining slopes up to the maximun angles which permit their stability calculation being the determination of the sloppe stability one of the most important aspects in mine design and planning.

The problem is, therefore, the perfect definition of the geotechnical parameters of the material as it will be found under the excavation conditions, taking into account that there are several factors which have some influence in the strenght characteristics and materials strain; the most important are the scale factor and the decompression and alteration of the rocky massif.

Therefore, the characteristics of intrinsic strenght must be estimated taking into account not only the results of the geotechnical investigations, but also the analysis of failures in the existing slopes, in order to compare the intrinsic strenght values to interpret these cases and the resulting values of laboratory tests. Moreover, the analysis of failures in the existing slopes permits the observation and definition of the kinematically possible failure mechanisms.

The causes of a talus colapse in Coto Bello, an open cast explotation of coal in the central Asturian Coal Basin (north of Spain), are analysed in this article with the purpose of showing a practical experience which may contribute to the establishment of methods for the talus definition.

2 GEOMETRY OF THE TALUS AND FAILURE IDENTIFICATION

Since the maximum inclinations that can be applied to wall slopes of coal beds, are necessarily in accordance with the stratification in order to minimize the volume of mine spoils to be extracted, the talus had been designed parallel to the main discontinuity, that is, going next to the stratification in a single plane, with a mean inclination of 45°.

The types of possible failures with this geometry are the ones indicated in figure 1.

The possible plastification of the foot requires a great height of the talus and a very reduced thickness of the rock (Manera and Ramírez, 1.986). In the other types, the failure is originated along the preexisting mechanical discontinuities in the rocky massif, and the way of failing depends on the orientation values of these discontinuities.

The talus colapse consited of a slide of a plaque (1,5 m. to 3 m. thickness, 125 m. length and 55 m maximun height) which dragged down 92 bolts tantered to 25 t., and laid in two rows at 30 m. and 35 m from the top. After the failure took place it could be seen that it was quasi-plane, parallel to the stratification, with a concentration of rock blocks of all sizes piled up at the talus foot.

3 GEOMECHANIC CHARACTERIZATION OF THE ROCKY MASSIF

The decisive influence of the discontinuities on the stability conditions of the wall-talus on

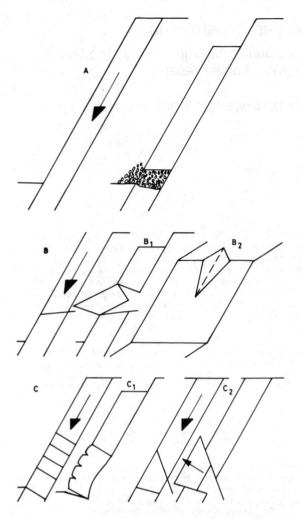

Figure 1. Type A: Failure due to foot plastification. Type B: Failures due to discontinuities cropping out on the excavation hole: B1 polygonal failure; B2 wedge failure. Type C: Failures due to discontinnuities not cropping out to the excavation hole: C1 buckling; C2 failure in two blocks.

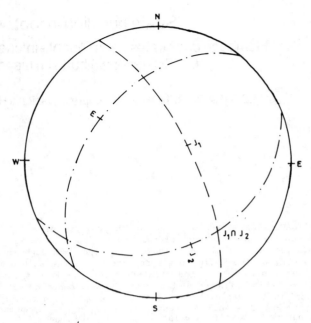

E 306,44/ 44,82
J_1 63,15/ 69,48
J_2 156,54/33,03

Figure 2.

rocks requires special attention therefore, the analysis of the orientation distribution of the discontinuity planes was made with the help of statistic criteria and the Fractan computer programme developed by Shanley and Mahtab, which encloses the adjustment of a distribution, the realization of statistic tests and the technique of clustering.

It was determined the mean orientation of the stratification 306.44°/44.82° and two sets of joints were obtained, J1 and J2, of which the mean orientation are 63.15°/69.48° and 156.54°/33.03°. These sets of joints have been checked with new measurements of the fallen down blocks once they had been placed in their original position.

The stratification and the sets J1 and J2 are represented in figure 2 by maximum circles; the tool employed was Schmidt grid.

The spatial distribution of the sets and their intersection with the stratification makes impossible the faiture caused by the discontinuities which crop out on the surface of the excavation; (type B, figure 1). On the other hand, subvertical and subparallel discontinuities to the layers which should act as a desining and limiting element of the two blocks that determine the failure in two blocks have not been detected (Type C2, figure 1). This type of failure is largely described by Manera and Ramirez (1.986).

Therefore it is considered that the failure me-

chanism has been the so called "buckling of strata" (type C1, figure 1).

Besides the directional analysis of the discontinuity planes, the characteristics of the rock strenght by simple compression tests, Brazilian tests and the measurement of the longitudinal and transverse strains in order to obtain the modulus of elasticity, E, determination of the density, friction tests in closed joints and open joints, all of them coincident with the stratification, either natural or obtained by banging the test piece, in order to determine the cohesion values, So, and the interior friction angle of the rock and of these discontinuities were obtained.

These laboratory tests were done upon drill cores of the three holes that had been drilled in the talus before the failure took place, in the area which afterwards colapsed when the excavation height was of 20 m. The holes were drilled perpendicular to the stratification, with a depth of 15 m. each.

The results obtained are the following:
- Mean density = 2710 kg/m3, after having done 10 tests.
- Mean strenght at simple compression: 50 MN/m2 after having done 10 tests. The obtained values in function of the depth are shown in figure 3.
- Strenght to the shearing strain of the rock: Interior friction angle, $\emptyset > 45°$ both in samples tested at natural wetness and in submerged samples after having done 4 tests.
- Strenght to the shearing strain in discontinuities:Three tests were made with samples tested natural wetness and two tests were made with submerged samples; the values of the interior friction angle vary from 20° to 37°. The cohesion was almost zero; the reason may be that discontinuities had smooth surfaces. The results of the friction tests are shown in figure 4.
- Modulus of elasticity. In the tests at simple compression it was determined with this formula:
$E = \Delta\sigma y / \Delta\varepsilon y$ where $\Delta\sigma y$ is the vertical tension increment, $\Delta\varepsilon y$ is the vertical strain increment.
The formula $E = 16p/\pi \cdot dl(\Delta\varepsilon_1 + 3\Delta\varepsilon_2)$ was employed in the Brazilian tests, where p is the total charge applied to the sample, d is the sample diameter, l is the length, $\Delta\varepsilon_1$ is the perpendicular strain to the charge, $\Delta\varepsilon_2$ is the parallel strain to the charge.
The obtained values were:

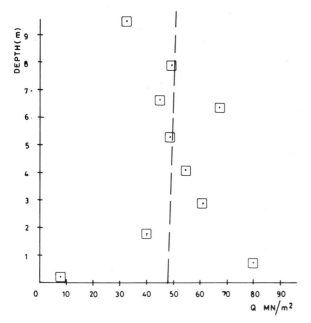

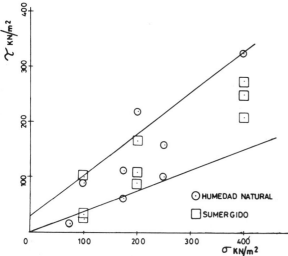

Figure 3 and figure 4.

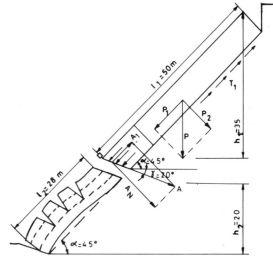

Figure 5.

- Perpendicular strain to the stratification:700 MN/m2.
- Parallel strain to the stratification:8,5 GN/m2.

4 ANALYSIS FAILURE

The failure caused by the buckling of the strata together with the acting forces has been sketched in figure 5. The lower strata bulge and the upper part slides parallel to the stratification. The water pressures on the discontinuities have not been represented because a perimetral water ditch at the crest of the talus was made along with a grid of horizontal drillings of drainage which eliminated the interstitial pressures. On the other hand, both, the anchoring force and its components have been represented, given that, as we have previously said, two rows of bolts at 30 m. and 35 m. from the top had been placed.

The security factor is defined taking into account the forces that favour the slide, and the ones opposing it. The only force which favours movement is the upper block weight. As resistant forces we have the resistant tangential force, T_1, of the upper part, the anchoring force, A, and the buckling critic charge, Pcrit, of the lower part.

The resistant tangential force, T_1, resolves into two terms, being each of them associated to the cohe-

sion and friction forces respectively. Following Coulomb theory, we obtain:$T_T=Sol_1+Pntg\emptyset$ where, So is the cohesion, l_1 is the upper block length; Pn is the normal weight component, and $\emptyset$ is the interior friction angle.

The anchoring force, A, has a component which is directly opposed to the sliding forces and another component that increases the normal forces thus increasing the sliding resistance; that is: $A=A[\cos(\alpha+\gamma)+sen(\alpha+\gamma)tg\emptyset]$ where, α is the angle formed by the discontinuity planes, stratification, with the horizontal; γ is the anchoring angle in relation to the horizontal and $\emptyset$ is the interior friction angle.

The buckling critic charge will be measured with the Euler formulae for a column, submitted to a centred compression, embedded at the bottom (Oteo and González, 1.982). It will be equal to:Pcrit= $\pi^2 EI/4l^2$ where, E is the modulus of elasticity of the material, I is the inertia moment of the buckling column and l is the lenght of the bulged area.

Therefore, to take place the buckle of the strata it is necessary that: $P_1-T_1-A \gtrless$ Pcrit. And the development is $\gamma dl_1 sen\alpha -Sol_1-\gamma dl_1 cos\alpha tg\emptyset -A[\cos(\alpha+\gamma)+sen(\alpha+\gamma)tg\emptyset] \gtrless \pi^2 Ed_1^2/48l_1^2$ where, E is the modulus of elasticity of the "undamaged rock", and d_1 is the thicknees of an undamaged rock column equivalent to a column with a thickness d forme by several strata. We may employ the columns buckle theory for a single part, taking into account that there is not slide among the strata. Since the elasticity modulus of the "undamaged rock", E, is logically greater than the elasticity modulus, E', of the set of strata, the "undamaged rock column" and the thickness $d_{\not\equiv}$ will be equivalent to the column formed by the set of the strata and the thickness d (figure 6).

Applying the remarked values in figure 5 and the results of the laboratory tests, we get a value $d_1 \leq 1.89$ m. for the mean thickness of the slided plaque d=2.25 m.

It has been taken the minimun value of $\emptyset$, obtained in the laboratory tests, $\emptyset=20°$, and the value of the modulus of elasticity determined after those test pieces which had been tested with the charge applied parallel to the stratification, E=700 MN/m²

After the results of the strain tests we can say that the elasticity characteristics of our buckling material depend on the petrographic anisotropies.

The slide was former by several strata of a maximun thickness, 1 m. As this thickness is largely lower than 1.89 m. we can conclude by seying that the failure of the talus was caused by the mechanism so call "buckcling of strata". Moreover, $d/d_1 = \sqrt{E/E'}$ from which it is deduced that $E \leq 414890$ KN/m², that is, the 60% of the value obtained in the laboratory.

529

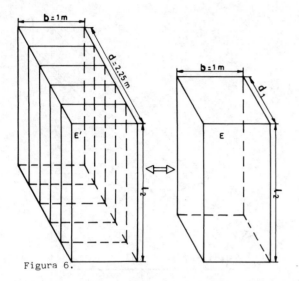

Figura 6.

5 METHODOLOGY TO PREVENT THE BUCKLING IN ROCK SLOPES

After Laboratory tests we can determine the intrinsic characteristics of our materials γ, So, $\emptyset$ and E in our rocky massif. With a conventional geological investigation: investigation ditches and boring samples we can know the stratigraphic column of our rocky massif, being necessary to make a detailed study of the strata thickness in relation to the depth. This information will permit us calculate -for a certain talus length with an inclination α- the thickness of the plaque able to buckle. And also, reckoning different values of d for different lengths of the talus, we will know the limit height that can reach a certain talus, in order to avoid the buckling mechanism.

Remember that the condition for the buckling to be produced is: $\gamma dl_1 \operatorname{sen}\alpha - Sol_1 - \gamma dl_1 \cos\alpha tg\emptyset \geqslant \pi^2 Ed_1^3/48l_2^2$ $\geqslant 0$. It must be checked that: $d \geqslant So/\gamma \operatorname{sen}\alpha - \gamma\cos\alpha tg\emptyset$ For the real value nearer to the value that verifies this condition we will calculate, reckoning for different lengths of the talus, the thickness value, d_1, of the "undamaged rock" column equivalent to the column formed by the whole of the strata, with a thickness d, which complies with the condition "being lesser than" the thickness of the greater stratum measured in the interval d, (reckoning for different talus lengths).

We must make a hypothesis about the length of the area which slides and the length of the area that buckles. The most conservative hypothesis is to consider $l_1 = l_2 = 0.5$ lt; where, l_t is the total length. There is an example presented in chart 1:

Chart 1.

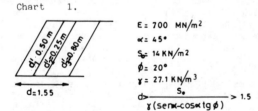

$E = 700$ MN/m^2
$\alpha = 45°$
$S_o = 14$ KN/m^2
$\emptyset = 20°$
$\gamma = 27.1$ KN/m^3
$d > \dfrac{S_o}{\gamma(\operatorname{sen}\alpha - \cos\alpha tg\emptyset)} > 1.5$

l_T (m)	$l_1=l_2$ (m)	d (m)	d_1 (m)		
10	5	1.55	$\leq$ 0.16	$d_1 < d'_3$	It doesn't buckle
20	10	1.55	$\leq$ 0.32	$d_1 < d'_3$	It doesn't buckle
30	15	1.55	$\leq$ 0.48	$d_1 < d'_3$	It doesn't buckle
50	25	1.55	$\leq$ 0.81	It $d_1=0.81>d'_3$	It buckles
49	24.5	1.55	$\leq$ 0.79	limit height for F=1 34.65m	
				$d_1 < d'_3$	It doesn't buckle

If our materials have a cohesion So=0, the limit height that a talus of a certain inclination α can reach, will be limited by the thickness of the first stratum. In this case d=d'=d_1 and the buckling mechanism will be produced if:
$l_1 \geqslant \sqrt{\pi^2 Ed'^2/48 \gamma} (\operatorname{sen}\alpha - \cos\alpha tg\emptyset)$ being d' the thickness of the first stratum. If the first stratum is not very thick, we will have very small limit heights. Its failure may not cause economical problems, but it may cause safety problems to persons. We must make the talus with its faces parallel to the stratification, with berms that avoid surpassing the limit critic height in each of the benches.

An example is presented in chart 2.

Chart 2.

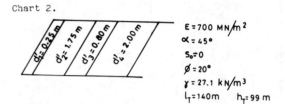

$E = 700$ MN/m^2
$\alpha = 45°$
$S_o = 0$
$\emptyset = 20°$
$\gamma = 27.1$ kN/m^3
$l_T = 140$m $h_T = 99$ m

d' (m)	$l=l=0.5 l_t$ (m)	limit h (m)	h for F=1.2 (m)
0.25	9.04	12.78	11.67
1.75	33.07	46.77	42.69
0.80	19.62	27.75	25.33
2.00	36.15	51.12	46.66

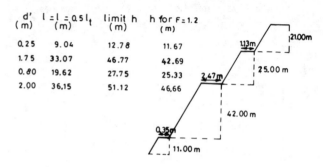

6 CONCLUSIONS

It has been proved the need of considerating the mechanism of failure by strata buckling when designing foot wall slopes of coal mining.

In order to avoid this type of failure, it is necessary to make a detailed study of the strata thickness in relation to the depth. This study will permit us define the limit height which a certain talus can reach, or for a given height of explotation, it will permit us calculate the design of the talus leaving it faces parallel to the stratification, with some berms which avoid surppassing the limit height in each bench.

If our materials have a cohesion $\geqslant$ 0, the limit height is conditioned by the value which complies that $d \geqslant So/\gamma \operatorname{sen}\alpha - \gamma\cos\alpha tg\emptyset$.

If our materials have a cohesion =0, the limit height is conditioned by the thickness of the first stratum. Nevertheless, if the first stratum is not very thick and next to it we have a much thicker stratum, the miner must value the convenience of removing the first stratum, (the weak one), the fall of which may be a risk for people, so that the limit height of the talus is conditioned by the thickness of the following stratum.

The methodology presented in this paper has permited the analysis of the causes of a foot wall colaps in the Spanish mine Coto Bello.

AKNOWLEDGEMENTS

The authoress is grateful to the mining enterprise HUNOSA for the given facilities to make this paper, and for the permission to public this information.

REFERENCES

Ayala, F.J.; Granda, J.R.; Starti, A. (1.984): Abacos de estabilidad para la minería de carbón y sedimen-

taria a cielo abierto con capas inclinadas. Madrid,
IGME.

González, L. y Oteo, C.S. (1.982): Diseño de
cortas mineras de carbón en condiciones geológicas
complejas. In 7º Simposio Nacional Obras de
Superficie de Mecánica de Rocas. Madrid.

Manera, C.; Ramírez, P. (1.986). Rotura en dos
bloques de los taludes de muro de explotaciones
de carbón. In Boletín Geológico y Minero
97-3:361-366.

Excavation, reinforcement and monitoring of a 300m high rock face in slate
Excavation, renforcement et surveillance d'un mur d'ardoise d'une hauteur de 300m
Aushub, Verbesserung und Messungen für eine 300m hohe Felsböschung im Schiefer

J.C.SHARP, PhD, Consultant in Rock Engineering, Geo-Engineering, Jersey, UK
M.BERGERON, P.Eng, General Foreman, JM Asbestos Inc., Asbestos, Quebec, Canada
R.ETHIER, P.Eng, Senior Geologist, JM Asbestos Inc., Asbestos, Quebec, Canada

ABSTRACT: A 300m high rock face, forming part of the Jeffrey Mine in Canada, has been excavated through a slate rock mass. The primary structural control is bedding which dips at 55 deg in the same sense as the face. The paper describes the investigation, design and excavation/rock support procedures together with the monitoring systems used to check stability conditions.

RESUME: L'un des murs de la mine Jeffrey au Canada consiste en une excavation d'une hauteur de 300 metres dans une masse d'ardoise. La structure predominante est une stratification ayant un pendage moyen de 55 degres dans la même direction que le mur. L'exposé decrit la recherche, la conception, l'excavation, le renforcement ainsi que la methode de surveillance utilisée pour verifier les conditions de stabilité.

ZUSAMMENFASSUNG: Ein 300m hohe boeschung, ein part der Jeffrey tagebau in Canada, wurde in einer schieferfelsmasse ausgegraben. Die primaere schichtflaechen haben eine neigung von 55 grad in der boeschungsrichtung. Der artikel beschreibt die untersuchungen den entwurf (ausgrabung, ausbau-systeme zusammen mit massungen um die standsicherheits-bedingungen zu kontrollieren)

1 INTRODUCTION

The north wall of the Jeffrey Mine at Asbestos, Quebec, is formed by a 300m high cut in slates which dip at approximately 55 degrees into the open pit. Plate 1 shows the completed face of a major expansion through the slate carried out between 1976 and 1986. The expansion has involved the excavation of relatively steep faces with individual 70 or 80ft (21,3m or 24,4m) high benches undercutting the general bedding dip. The location of the wall is shown in Ref 1, Fig 1.

Reinforcement of the entire face by means of tensioned bolting has been undertaken in order to ensure stability of individual benches and safety of the working area below. The overall design approach was based on the use of rock reinforcement to enable steeper slopes to be mined (typically of the order of 5 degrees overall) thereby minimising waste excavation.

The paper outlines the geological conditions which are largely based on mapping during excavation and assesses these in relation to the original assumptions based on surface mapping projections and drillhole core observations. The use of controlled excavation procedures is discussed as an essential element in forming sound final profiles by means of presplit blasting at an acute angle to the bedding. Stabilisation measures comprising both rock reinforcement and drainage are outlined.

The entire project was evaluated during execution using various monitoring systems. A series of deep boreholes was drilled into the slope face to check the assumed geological conditions at depth and permit the installation of seven precision extensometers up to 100m in length, designed to observe the response of the relatively stiff rock mass to ongoing excavation. Towards the toe of the slope, a precise hydraulic levelling system developed specifically for the site and capable of monitoring ± 1mm over a 300m length has been installed.

The paper concludes with a summary of predicted and observed performance and assesses the benefits of monitoring control in large scale mining environments. The mine planning is based on the imperial measurement system and hence levels, bench intervals etc. are generally quoted in these units. (1m = 3,281ft).

2 GEOLOGICAL CONDITIONS

2.1 General lithology

The north wall slopes are formed largely in a dipping, low grade slate, which is composed of black phyllite, impure quartzite and quartz-sericite-chlorite schist. The base of the slope lies in the ultrabasic orebody which is separated from the slate by a sheared contact zone. A section through the north wall showing the overall geology is given on Fig 1. The overall wall orientation is designed to be approximately parallel to the strike of the bedding.

The majority of the wall is composed of black phyllite banded with quartzite up to 7m thick. Toward the western end, the thickness of black phyllite decreases exposing metagreywacke and quartz-sericite-chlorite schists. Further to the west lies an intrusive ultrabasic zone termed the contact fibre zone (CFZ). The black phyllite is thin-bedded, crumpled and altered adjacent to the intrusive contact. Foliation which is almost parallel to bedding, is used synonymously for the bedding.

The intrusive series containing the ore material rest unconformably on the slate, the orientation of the footwall contact being steeper than the overall bedding dip as shown on Fig 1. Adjacent to the contact is a footwall zone composed of heavily sheared, incompetent periodotite.

The slate is overlain by some 30 to 50m of overburden composed of dense materials ranging from silty sands to sandy gravels.

2.2 Structure of the slate rock mass

A surface mapping programme of the previously exposed faces in slate carried out in 1971 recorded some 240 measurements of foliation or bedding planes, joints, and shears/faults, which were used in the original design assessment. The data were collected primarily from the black phyllite rock type. Exposures of several faults were mapped. These were located towards the western end of the wall adjacent to the contact fibre zone and at sections 3E and 7E. The faulting appeared to be of limited extent and in general was

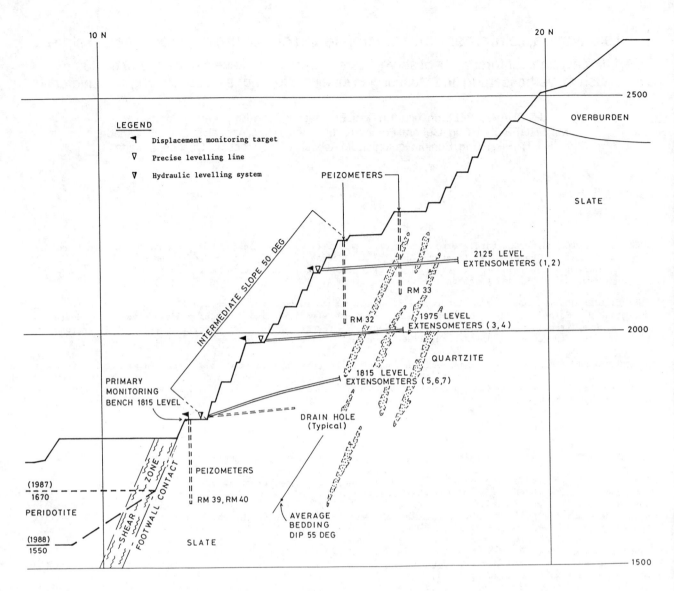

Fig 1   Typical Cross Section through North Wall showing general Geology and Monitoring, Drainage Arrangements

not adversely oriented so as to influence stability.

The geological conditions were mapped in detail during excavation of the current face in order to check the original design assumptions or allow modifications to be implemented as required.   A comprehensive structural survey of the final face was also undertaken in 1984-85 and was complemented by aerial sub orthogonal photography to provide an interpretation base for larger scale joints and faults as illustrated on Plate 2. The 1984-1985 survey, comprising some 1500 measurements along 30 separate traverses was further supplemented by borehole core observations from the extensometer boreholes.

In summary the bedding is characterised by large scale (~50°) dip variations of between 40 and 75 deg with an average bedding dip of 55 deg as illustrated on Fig 1.   Locally, minor flexures and rolls in the bedding lead to increased dip variability as indicated on Fig 2.   Considerable attention was given to such dip variability during excavation in order to provide a design basis for the rock reinforcement.   The bedding, as in many sedimentary rocks, formed both a 'fabric' to so-called 'intact' rock and discontinuous partings.   Disturbance, as a result of excavation, could also lead to the generation of discontinuities along the weak bedding fabric.

The jointing was generally subordinate to the bedding structures and no significant joints at unfavourable attitudes were observed.

3 STRENGTH CHARACTERISTICS OF THE SLATE

3.1 Uniaxial compressive strength

The uniaxial compressive strength of the slate was determined from 'intact' samples with bedding fabric at various angles relative to the core axis.   Typical strengths of 50 MPa were obtained for the samples with foliation approximately normal to the core axis. Lower strengths of the order of 15 - 20 MPa were determined for those samples where the fabric was at a critical angle to the core axis.

3.2 Shear strength

The shear strengths of bedding planes and joints were determined from a comprehensive testing programme on both small (50 x 50mm) and large (250 x 250mm) samples as part of the 1971 investigation programme.   Tests were conducted to determine the strength characteristics of bedding discontinuities as well as 'intact' bedding planes.   Field roughness measurements were also conducted.

The results of the testing programme are summarised on Table 1.

In addition field checks were conducted of the existing slate faces to assess strength conditions of the bedding in situ.

Plate 1   Jeffrey Mine North Wall - General View of Completed Excavation in Slate

Table 1. Summary of shear strength characteristic of Slates

	Angle of shearing resistance	Apparent cohesion
Smooth, bedding discontinuities	28 deg	0
Medium-rough, bedding discontinuities	34 deg	0
Rough, bedding discontinuities	40 deg	0
'Intact' bedding planes	60 deg	0,70 MPa

4 GROUNDWATER CONDITIONS

Observations on piezometric conditions within the slopes made since 1971 have indicated a groundwater table close to the slope face and groundwater pressures that increase with depth at a rate greater than hydrostatic.   This can be explained by the anisotropy of the slate rock mass.

Typical permeability values obtained from packer tests in 1971 gave values in the range 3×10^{-6} cm/sec to 1×10^{-4} cm/sec with a mean of 4×10^{-5} cm/sec.   These values indicate that the slate in general is fairly tightly jointed and of 'low' to 'medium' permeability. It was recognised at an early stage that the permeability parallel to bedding would probably be significantly greater than that normal to it which could lead to unfavourable development of relatively high pressures close to the slope face.

Drilling for drainage purposes (horizontal holes drilled into the bench faces) and extensometer installations in recent years has indicated that where groundwater is present in significant quantities it is concentrated within discrete fissures.   Some of these fissures in the rock have yielded continuous flows of the order of 1,5 - 4 litres/sec (20-50 gpm).   Typical groundwater intersection patterns are shown on Fig 3 (1815 level).

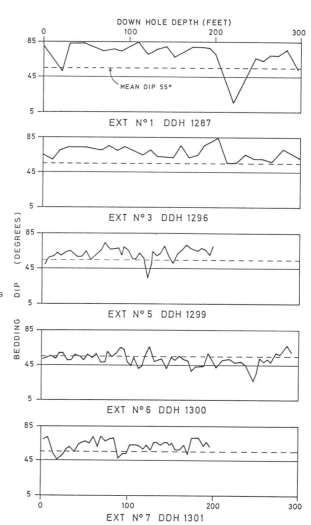

Fig 2   Bedding Dips recorded in Extensometer Drillholes

535

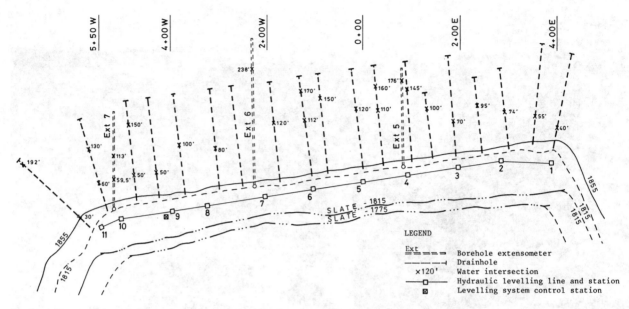

Fig 3   Plan of the 1815 Bench Level showing Monitoring and Drainage Arrangements

LEGEND

Ext	Borehole extensometer
	Drainhole
×120'	Water intersection
	Hydraulic levelling line and station
	Levelling system control station

5. SUMMARY OF SLOPE DESIGNS

The comprehensive investigation and design studies in 1971 produced a general design basis for the north wall slopes covering individual benches up to 105ft (32m) in height (3 x 35ft sub benches), inter-ramp (intermediate) slopes of the order of 120m in height, and overall slope profiles up to 300m in height.

Physical and theoretical models of slope failure were studied to determine general stability trends.

Based on field observations, where undercutting had taken place on the bench faces created prior to 1971, the faces often remained stable indicating that the bedding planes were in general stronger than the test results on bedding discontinuities would indicate. In other areas, peeling off to bedding had occurred, particularly where large scale bulk blasting techniques had been used.

By adopting controlled blasting methods for the final bench faces, it was considered that stable benches could be created at a steeper face angle than the bedding with a degree of additional support to ensure that safe working conditions would be created on a multiple bench (intermediate) slope.

The potential failure of intermediate slope sections was recognised as a serious hazard based on the relatively brittle nature of the material (sudden collapse) and the threat to both lower working areas and ramp stability. For this reason it was considered prudent to limit the overall slope of the intermediate sections to an angle less than the average bedding dip.   Large scale slope anchorage for intermediate slope sections was examined but was not cost effective.

Kinematic failure models for intermediate slope angles less than the mean bedding dip, require the presence of low angle joints (dipping out of the face), local flattening of the bedding or combinations of both.   Based on conservative assumptions, factors of safety against failure of the order of 1,5 were predicted using a multiple block-wedge model.

Based on the observed geological structures, the stability of the intermediate slope sections was considered to be satisfactory owing to the lack of major continuous low angle joints or bedding planes that could undercut significant areas of the slope face. (If more adverse conditions had been observed during excavation of the face, it was recognised that additional stabilisation could be provided particularly in the toe region by an extension of the bench anchorage system).

6 DESIGN UNCERTAINTIES

Owing to anticipated variabilities in the slate at depth and the limitations of borehole evaluations it was recognised that the proposed slope design would require on site checking and evaluation during execution.   This would be achieved by means of geological mapping of exposed bench faces, selected core drilling, drainage assessments and overall displacement monitoring of the completed faces.

A further uncertainty was the influence of the footwall contact on the overall stability which would only be evaluated once the true geometry of the contact at depth was determined.   The effect of the contact which undercuts the overall slope profile is partly offset by the limited base width of the slope at this depth and the buttressing effect of the east and west walls.   Studies on the stability of the lower slope zone are currently being carried out.

7 EXCAVATED PROFILE

Using the general design guidelines established from the 1971 study, the slope geometry was laid out on the basis of the following parameters (in order of priority) :
- intermediate slopes that would not undercut the bedding dip (resultant slope selected: 50 deg)
- benches 70ft (21m) in height (2 x 35ft sub benches) excavated with face angles of 65 degrees to give a 19ft wide (5,8m) bench and 7ft (2,1m) sub-bench
- ramps as required for overall wall access.

This profile was adopted for the final wall down to the 1975 level as shown on Fig 1.

Below the 1975 level, as part of a general revision to mining layouts, a change to a 40ft (12,2m) sub-bench height was made giving an overall bench height of 80ft (24,4m).   For the 40ft (12,2m) benches the face (drilling) angle was increased to 70 degrees to give an increased width of catch bench of 31ft (9,4m) again with a 7ft sub-bench offset.   The intermediate slope angle of 50 degrees remained unchanged.

8 EXCAVATION PROCEDURES

As an integral part of developing steeper slopes in the north wall slates it was recognised that it would be necessary to cut benches with faces some 10 to 15 degrees steeper than the bedding angle.   This would call for controlled excavation procedures to limit

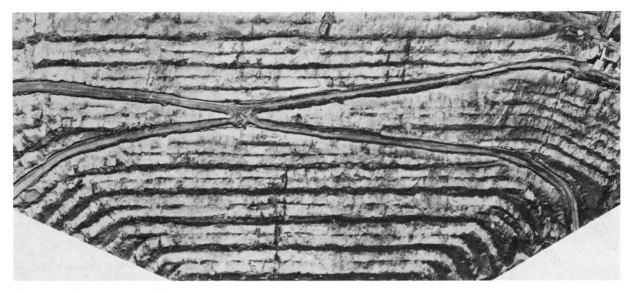

Plate 2   Sub orthogonal View of North Wall used for Structural Interpretation of major Geological Features

damage beyond the nominal bench profile and in particular eliminate disturbance of, and slippage along, bedding planes within the bench profile.

During the initial excavation stages in 1977, a trial programme of drilling and blasting with computer simulation studies was carried out in order to develop an optimum presplit/buffer blasting pattern for the particular conditions.

The primary feature of the controlled blasting pattern is a perimeter pre-split pattern comprising 100mm dia holes at a spacing of 1,5m, loaded with a small diameter continuous tube explosive.   The overall blasting layout for the 40ft bench interval is shown on Fig 4.

The resultant bench faces have corresponded closely to the design profiles and a high percentage of perimeter holes has been visible over the majority of the face as indicated on Table 2.

Table 2. Summary of Bedding Dips and Percentage exposure of perimeter drillholes

	AVERAGE	BEDDING	DIP			PERCENT DRILLHOLES
LEVEL	6W	3W	0	3E	6E	6W – 6E
2300/2265	61	60	55	49		68%
2230/2195	72	65	63	53		53%
2160/2125	68	64	57	55		52%
2090/2055	66	63	59	52		49%
2020/1975	63	60	51	53		36%
1935/1895	66	63	57	60		60%
1855/1815	60	62	60	53		60%

9   BENCH REINFORCEMENT

The main purpose of the benches is to retain any loosened material that may fall from the upper levels and thus prevent rockfalls onto the working area below. Stabilisation in the form of rock reinforcement has thus been provided to ensure the long term stability of the bench faces and hence the catch benches themselves.

Preliminary analyses showed that for the assumed strength parameters of bedding discontinuities given in Table 1, undercut bedding planes would always fail. (Factor of Safety = 0,6).   In practice it was found that provided controlled excavation procedures were used, the faces remained generally stable with only localised breaking back to bedding generally near the bench crest.

To provide reinforcement with normal factors of safety for such installations (1,3 to 1,5) for the entire bench face based on the worst case of slippage

along a 55 degree bedding plane with an angle of shearing resistance of 34 degrees was considered to be unnecessarily conservative based on the observed behaviour of bench faces excavated with controlled blasting.   (To maintain theoretical equilibrium for an angle of shearing resistance of 34 degrees requires a horizontal anchor force of 53 ton/m run.)

Based both on field observation of inferred bedding plane strengths with controlled blasting and the fact that bolting would be installed at an early stage and tensioned to inhibit separation of the bedding planes, it was decided to adopt an active anchorage force of some 30 ton/m run (43 ton/m run ultimate).   To achieve the maximum possible benefit from active effect of the reinforcement, a layout was chosen comprising 3 rows of anchors in the upper sub-bench as indicated on Fig 5.   The anchors, which are made up of 16mm dia. 7 strand tendons are used in conjunction with a 200 x 200 x 20mm faceplate, have an ultimate capacity of 26 ton and are tensioned to some 18 ton on installation.

The adoption of 40ft benches with a 70 degree bench face angle increased the mass of potentially unstable material based on a 55 degree bedding plane from some 80 ton/m run to some 190 ton/m run.   However, at the levels concerned, the bedding dip was closer to 60 degrees than 55, and this reduced the theoretical anchor requirement to some 65 ton/m run.   Based on the observed performance of the face down to the 1975 level, it was considered adequate to continue with the

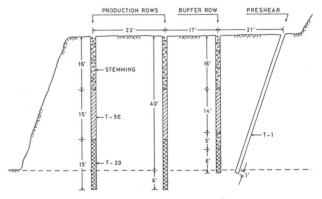

T – 1   Tovex water gel – small dia. cont. tube
T – 5E   Tovex water gel – 1,15 gm/cc   4300 m/sec
T – 20   Tovex water gel – 1,28 gm/cc   5500 m/sec

Fig 4   Blasting Layout for 40ft Bench Section

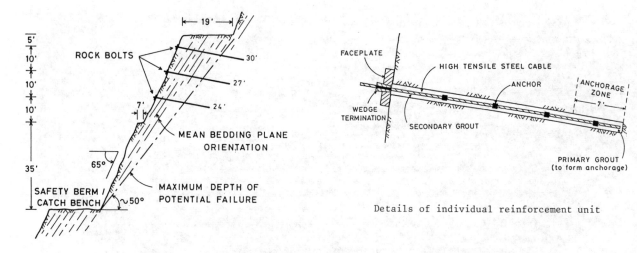

Details of individual reinforcement unit

Fig 5   North Wall Benches - Rock Reinforcement Arrangement for 70ft Bench Height (80ft benches similar)

same bolting layout even though this was considerably less than the theoretical requirement.   The decision again indicated the beneficial effect of the controlled blasting.

10 DRAINAGE

Drainage is not being relied upon as a fully effective stabilisation measure (owing to the difficulty of intercepting the relatively discrete water bearing fissures) but the beneficial effect of it is being realised through a routine drainage drilling programme.

Over the majority of the face, drainage drilling has been carried out on a vertical interval of 70ft, with drains of the order of 30m (100ft) long spaced at 15 - 30m (50 to 100ft).   Piezometric observations of drainage measures are of limited value owing to the discrete fissure related conductivity of the slate rock mass.

At the lower levels, (notably the 1815ft bench) higher groundwater pressures and flows were encountered during extensometer hole drilling (see Fig 3).   Based on these observations a more extensive drainage pattern (60m long holes at 15m spacing) was introduced.

Careful logging of groundwater interceptions and water pressures have allowed an assessment of fissure continuity to be made.   Typical interconnections are between the 1815 level (west) and the 1975 level (centre) indicating fissures (bedding discontinuities) subparallel to the face.   Such observations correlate with earlier predictions of permeability anisotropy related to bedding.   Further evidence for such effects comes from three deep piezometers below the 1815 level where artesian pressures have been observed.

11 DEFORMATION - DISPLACEMENT MONITORING

Lack of precedent in the excavation and reinforcement of high steep slopes led to the adoption of routine monitoring. The objectives were both to check the existence of stable conditions and to permit design modifications to take account of the geological and rock mass conditions actually encountered.

For the slate rock mass, which can be considered as relatively brittle, the deformation behaviour should be essentially elastic, except perhaps for localised surface zones.   In order to check this behaviour, high resolution monitoring would be required, capable of providing accurate and repeatable data over a long time period.   The requirements of the monitoring system were identified as follows : significant areal coverage; high resolution to detect small, essentially elastic, movements; resolution between near surface and at depth movements; ability, in part, to monitor continuously and provide instantaneous warning of early

changes; proven performance and reliability under all conditions (particularly prolonged winter period).

Systems that were evaluated in detail included borehole extensometers (tensioned rod or wire types capable of accommodating localised near surface shear movements), borehole deflectometers and automated precise levelling systems.

In addition it was planned to use the precise distance monitoring system already in use for other slopes at Jeffrey (Ref 1).

11.1 Extensometer installations

Deep precision extensometers were chosen as the most suitable method for monitoring the rock mass response and the overall excavation faces.   Installations with anchors at 6, 15, 30, 90m have been provided at the 2125, 1975 and 1815 levels as shown on Fig 1.   Considerable attention was given to the design of the units and improvements were made through time.   The latest installations (1815 level) incorporate individually sleeved aluminium alloy rods tensioned to 40kg and equipped with hydraulic anchors.   Large numbers of centerers were used along the borehole and each unit was subjected to a post installation pull test.

Figure 6 shows a typical extensometer record and the response of the rock mass to mining below the 1975 level.   The observed response is compared with deformation predictions from finite element analyses of the excavation stages to confirm that the deformations are

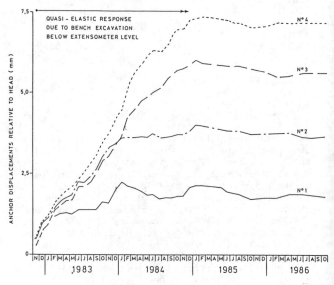

Fig 6   Typical Extensometer Response (EXT No.3)

Plate 3   Hydraulic Levelling Station - Details of
           Float Chamber (prior to burial)

essentially elastic.

The practical resolution of the instruments allowing
for thermal effects is of the order of 0,5mm for depths
up to 30m which is considered extremely good considering the harsh thermal environment and the long measurement lengths.

11.2 Hydraulic level monitoring system : 1815 level

In order to provide a greater degree of resolution of
possible movement zones along the slope face, thus
permitting lateral interpretation of any extensometer
observations, an automated precision levelling system
was installed on the 1815 level to observe the performance of the wall during future mining below this
level.   Prior to this installation, precise level
surveys using conventional survey equipment and a
series of closely spaced level stations had been
undertaken at the 2125 and 1975 levels.

A hydraulic levelling system was chosen following
the successful application of this system for precise
level measurements at Mundford (UK) in 1966 (Ref 3).
Following an appraisal of commercially available
systems a decision was taken to develop a static fluid
levelling system suited specifically to the requirements at Jeffrey.   An overall system accuracy under
significant temperature variations of ± 1mm over a
300m base length was specified.   Following successful
prototype trials, Geokon Inc (USA) were selected to
build the complete installation comprising 11 stations
at intervals of approximately 30m.

The layout of the system is shown on Fig 1,3 and a
typical level station arrangement on Fig 7.

Each station consists of a chamber partially filled
with fluid and connected to the hydraulic line.   The
upper part of the chamber is vented to a common vent
line.   The fluid level relative to the instrument is
monitored by means of a float and a DCDT transducer
with a range of ± 25mm.   Fluid temperature is measured at each station by thermistor.   The base of the
station comprises a tube grouted into rock at a depth
of 5m and protected by a greased sleeve through the
frost penetration zone.

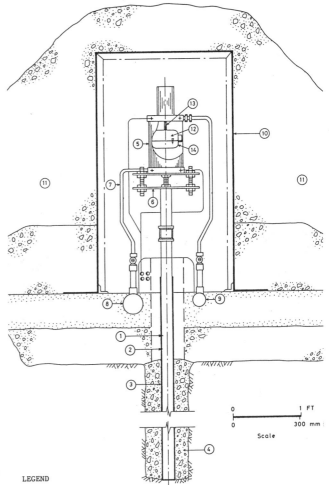

LEGEND

1   Reference tube
2   Plastic collar through frost zone
3   Grease coating through frost zone
4   Cement grout
5   Float housing
6   Levelling plate
7   Hydraulic line
8   Hydraulic manifold
9   Air manifold
10   Thermally insulated housing
11   Selected fill
12   Level sensing float
13   DCDT transducer
14   Glycol/ water mixture

Fig 7   Cross Section through Level Station showing
         prinicipal operating Components

The fluid is contained in a large diameter line laid
on a +1% grade to the west and connected to a 750
litre fluid reservoir at the east end (considered to
be the stable reference area).   The fluid consists
of a mixture of 60% glycol and 40% water to permit
year round operations in an ambient temperature range
of -35°C to +50°C.   To minimise differential temperatures within the system, particularly due to variable
sunlight exposure during the day, the entire system
is insulated and results in a maximum temperature
variation between stations of 2 deg.

The system has been installed for some 2 years, a
period of minimal mining activity in this area and
the system stability has therefore been evaluated in
detail.   After initial problems with entrained air
in the system had been overcome, a working resolution
of better than ± 1mm has been established after
allowing for temperature variations.   Automated data
processing and a remote readout/alarm system are used
for monitoring control.

Results of level change testing after full commissioning and operation of the system showed that for a
sudden level change of 10mm (effected by reservoir

lowering) system stabilisation was achieved to within 1mm after a period of 20 minutes. This degree of system response is obviously essential if potentially rapid movements are to be detected.

12 CONCLUSIONS

The north wall of the Jeffrey Mine has been mined to a steep overall profile through the use of stabilisation measures comprising rock reinforcement and drainage. Monitoring systems, designed to permit observation of the elastic response of the slate rock mass, have confirmed the overall stability of the wall although some localised zones of overstress have been observed. Mining is continuing down the slate contact as indicated on Plate 4 to form the final phase of the north wall expansion.

The overall steepening of the north wall by 5 degrees beyond a naturally stable profile has resulted in waste excavation savings of 10m tonnes. The cost of the measures required (controlled blasting Ca$ 0,60m; rock reinforcement Ca$ 0,98m; drainage Ca$ 0,015m; monitoring Ca$ 0,35m) have been far outweighed by the excavation savings achieved.

This case history illustrates the substantial benefits that can be achieved from investigation, stabilisation and performance monitoring in the excavation of major rock slopes. It should however be realised that such measures can usually only be justified in cases where the stability of the slope is governed by a dominant structural control such as bedding, cleavage, or regular, pervasive joint structures.

REFERENCES

Sharp, J.C., C. Lemay & B. Neville August 1986. Observed behaviour of major open pit slopes in weak ultrabasic rocks. Proc. 6th Int. Congress Rock Mech. Montreal.
Golder Associates, 1970 - 1986. Project technical review - slope stability.
Ward, W.H., J.B. Burland & R.W. Gallois 1980. Geotechnical assessment of a site at Mundford, Norfolk, for a large proton accelerator. Geotechnique, Vol.18(4).

ACKNOWLEDGEMENTS

The permission of JM Asbestos Inc to publish this paper is gratefully acknowledged. The paper reflects the contribution of many people and organisations including Dr. L. Oriard (blasting), Golder Associates and Geokon Inc. (instrumentation development).

Plate 4   Mining down the Slate Contact (1986) showing Meshing, Bolting and Drainage Measures

Plate 5   General View of the completed Slope down to the 1815 Monitoring Level (1986)

Observed behaviour of major open pit slopes in weak ultrabasic rocks
Observation du comportement des pentes majeures dans des roches ultrabasiques peu compétentes dans une mine à ciel ouvert
Verhalten einer hohen Tagebauböschung in sprödem ultrabasischem Fels

J.C.SHARP, PhD, Consultant in Rock Engineering, Geo-Engineering, Jersey, UK
C.LEMAY, P.Eng, Plant Engineer, JM Asbestos Inc., Quebec, Canada
B.NEVILLE, P.Eng, Senior Mining Engineer, JM Asbestos Inc., Quebec, Canada

ABSTRACT: Slopes in ultrabasic rocks at the Jeffrey Mine in Canada have a history of major failures over the period 1970 - 86.   Limiting stable angles are in the range 20 - 25 deg for 300m high slopes.   The paper describes the major failures that have occurred and from the results of performance monitoring establishes design criteria.   Details of instrumentation systems and drainage measures are also given.

RESUME: Les pentes dans les roches ultrabasiques de la mine Jeffrey au Canada presentent un historique de mouvements importants de 1970 à 1986.   Les angles pour des pentes stables sont de l'ordre de 20 - 25 degres pour des murs de 300 metres de hauteur.   L'exposé decrit les mouvements importants qui se sont produits et indique comment les resultats de surveillance des pentes ont permis de definir des criteres de conception. Une description detaillée d'instruments de mesure et de methodes de drainage est aussi presentée.

ZUSAMMENFASSUNG: Boeschungen in ultrabasichem fels in Jeffrey Tagebau, Canada, haben historische, grosse rutschungen waehrend der periode von 1970-86.   Die grenzwinkel fuer 300m hohe standfeste boeschungen liegen bei ca. 20-25 grad.   Der artikel beschreibt die groessten rutschungen die in der vergangenen jahren geschenen sind.   Die ergebnisse der messungen geben jetzt gute entwurfskriterien.   Messungseinrichtungen und draenage-bzw. entwaesserungsmassnahmen sind auch beschrieben.

1 INTRODUCTION

The south east area of the Jeffrey Mine at Asbestos, Quebec, Canada, has a history of major slope failures since 1970.   In 1970 when the overall pit depth was some 240m (800ft), studies were commenced to try and determine the behaviour of the rock mass in relation to slope stability.   These studies were prompted by the need for efficient ore and waste extraction and the occurrence of major instabilities in 1970-71, 1975 and in 1983, all below the adjacent town area as indicated on Fig 1.   An aerial view of the 1970-71 failure is given on Plate 1.

Excavation under marginal stability conditions is common to many mining operations.   However at Jeffrey the proximity of the town area to the mine has led to the particular development of comprehensive monitoring systems to provide an overall control on stability conditions.

The ultrabasic rocks are extremely complex structurally and in relation to strength characteristics. Groundwater also has a major influence on slope stability conditions at Jeffrey and is strongly influenced by seasonal variations.   Owing to the complexity of the material behaviour, slope stability predictions at Jeffrey are based largely on observed performance of past and existing slopes, coupled with generalised back analyses of major failures.

The paper summarises the geological and groundwater conditions at Jeffrey and gives an outline of past and current failures and their inferred mechanisms.   A description of monitoring systems suited to the large scale slopes is provided and typical results are given. Recent developments at the mine are discussed together with results of drainage control measures.

2 GEOLOGICAL CONDITIONS

The occurrence of asbestos is often associated with weak, ultra-basic rocks such as peridotite, dunite and serpentinite that contain significant percentages of weak minerals such as olivine that can be hydrothermally altered or weathered to chlorite.

A section through the most critical slopes with respect to ore recovery and town location is given on Fig 2 and shows a sequence of alternating peridotite and dunite strata dipping into the slope.   Major shear zones up to 50m in overall width occur within the rock and are accompanied by many lesser sub-parallel features.   These features consist of small platy rock fragments in a matrix of talcose and chloritic fines.

The more competent rock strata are key elements in the overall slopes and have been the subject of several detailed appraisals to identify structural patterns that could control potential failure mechanisms. Although some localised structural patterns were identified and these in isolated cases could be linked to small scale wedge failures over one or two benches, the

Plate 1   Aerial View of 1970-71 South East Failure

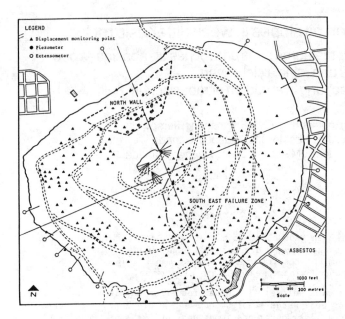

LEGEND
▲ Displacement monitoring point
● Piezometer
○ Extensometer

NORTH WALL

SOUTH EAST FAILURE ZONE

ASBESTOS

1000 feet

100 200 300 metres
Scale

N

Fig 1   General Plan of Jeffrey Mine and Failure Area

rock was generally devoid of any structural consistency or significant throughgoing weaknesses that could adversely influence stability conditions.

Attempts have been made to characterise the rock mass through extensive borehole drilling and surface mapping programmes.   From these studies a generalised descriptive system of very poor to good rock, mainly based on drill core RQD assessments has been derived.   Although such characterisations have been useful in providing a qualitative zoning for the rock mass, they have not correlated well with observed slope performance.

3   ROCK MASS STRENGTH

The assignment of representative strength parameters and the derivation of realistic behavioural stress-strain models present major difficulties in weak rock masses devoid of systematic jointing.   This situation is typical of the ultrabasic rocks at Jeffrey and in spite of considerable effort, representative modelling has to date proved an insoluble problem.   It is however evident from the deformation behaviour of the rock mass in situ that the presence of low strength discontinuities even in a quasi random pattern leads to a dramatic reduction in the strength of the mass, particularly in a slope configuration where confinement may be limited.

The strength of the sheared rock zones can be determined reliably from small scale samples.   For the more competent rock types, uniaxial and triaxial tests have yielded strength and deformability data as given on Table 1.   The applicability of these data to rock mass behaviour is however questionable as will be discussed later.

Table 1. Summary of deformation and strength properties

	Young's Modulus (GPa)	Poisson's Ratio	Effective Angle of Shearing Resistance (deg)
Overburden	0,05	0,4	25 (1)
Very Poor Rock	0,14	0,35	23
Poor Rock	–	–	30
Fair Rock (3)	4,6	0,25	40 (2)
Good Rock (3)	8,6	0,20	60
Transition Zone (3)	1,4	0,3	35

Notes :   (1) Cohesion c' = o for all materials at zero
              normal stress
          (2) Interpolated value
          (3) Discontinuity strength values

4   GROUNDWATER CONDITIONS

Before significant movements of the slope developed, packer tests indicated the rock mass to be of extremely low permeability.   To determine initial conditions, deep piezometers with a low volume response were therefore used.   With the onset of significant slope de-

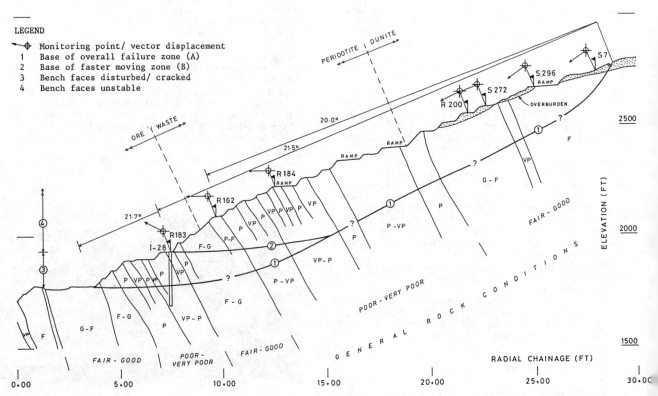

LEGEND
↦⊕ Monitoring point/ vector displacement
1   Base of overall failure zone (A)
2   Base of faster moving zone (B)
3   Bench faces disturbed/ cracked
4   Bench faces unstable

Fig 2   Cross Section through Current Slope Profile showing general Geology and Failure Zones

formations leading to a general increase in permeability, standpipe piezometers are currently used with 2 – 3 units per hole to typical depths of 65m (200ft).

The monitored groundwater conditions generally correspond to a phreatic surface at or close to the ground surface for the depths at which typical failure surfaces are inferred.

5 PREDICTED STABILITY CONDITIONS

The south east corner slopes can be generally subdivided into an upper zone of 'poor' or 'very poor' rock and a lower zone of 'fair' to 'good' material, the division corresponding to the ore/waste boundary. It could thus be expected that stability conditions would improve once the ore levels were reached and that increased slope angles could be adopted from this point down.

A simplified slope model as shown on Fig 3 was investigated in 1979 as part of a parametric study of governing stability parameters using a linked finite element – limit equilibrium approach. This model demonstrated above all the importance of the buttress concept whereby the lower more competent material becomes stressed by the deformed weaker zones forming the upper part of the slope. Results of limit equilibrium analyses showed however for the strength conditions given in Table 1, overall slopes would be stable even for relatively steep slopes in the lower 'buttress' zone. Such predictions unfortunately are not borne out by the observed behaviour of the slopes in practice and it is evident that the true response of the slope material is much more complex than that predicted by simple analyses, even accounting for progressive failure events.

6 SUMMARY OF OBSERVED SLOPE FAILURES 1970 – 1983

A number of significant failures of the south eastern slopes at Jeffrey have occurred as summarised on Fig 4. The initial failure in 1970-71 involved rock slopes with up to 100m (300ft) of superimposed overburden material comprising sands, silts and glacial tills. The 1974-75 failure was a reactivation of the 1970-71 event and occurred with only a limited expansion pushback. The 1983-86 failure occurred generally in

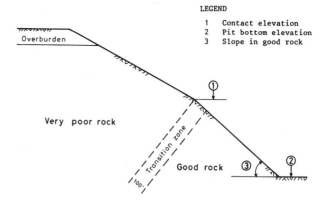

Fig 3   Simplified Cross Section through Slope used for Stress Analyses

undisturbed material after a significant expansion pushback.

The failure 'surfaces' illustrated on Fig 4 are a simplified interpretation of the maximum depth of significant shear movements. The actual failure mechanism is complex and involves secondary internal shear movements as well as overall progressive straining of the rock mass.

6.1 1970-71 Failure (See Fig 5)

During 1970, an ore skipway facility on the west side of the eventual slide area underwent an extension of the rails which was originally attributed to creep of the fill materials on which the rails had been laid. Deeper seated movements were later demonstrated following the installation of long, wire borehole extensometers. During December 1970 and January 1971, a major overburden slide developed in an adjoining area to the north involving part of the town of Asbestos. Ongoing movements were influenced by the leakage of town service water into existing cracks forming the rear scarp to the slide.

In January 1971, a localised wedge failure occurred below the main haul road in relatively competent rock (see Fig 6). The failure was controlled by localised jointing although some extension of fractures may have occurred due to small, overall movements of the slope.

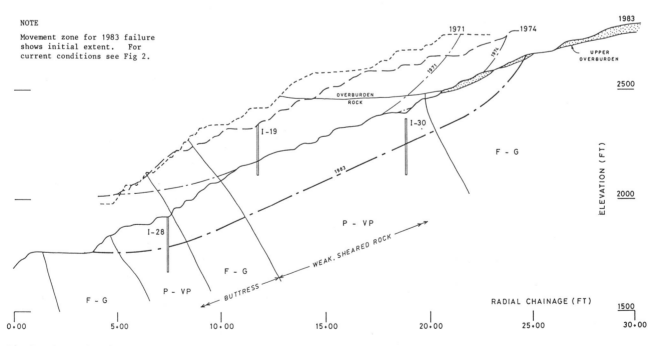

Fig 4   Cross Section through 1971, 1974 and 1983 Failure Zones showing relative Slope Locations

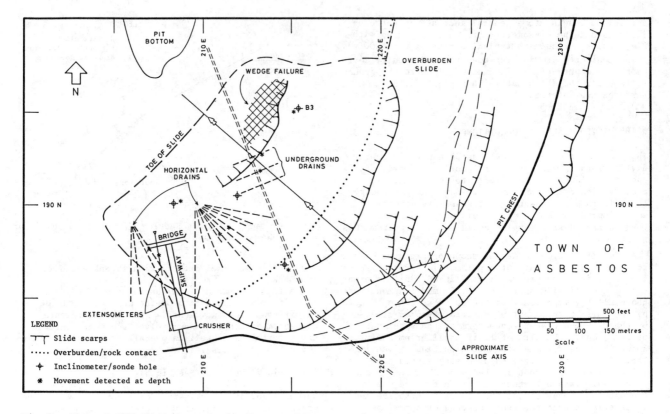

Fig 5   Plan of 1970-71 Failure showing Extent of Movements in relation to Plant and Town Areas

Mining was being carried out near the toe of the slope (approximate elevation 2,100ft) at the beginning of 1971, relatively rapid progress having been made during the latter part of 1970 with mining below the 2,300ft level.

In February 1971, a massive, overall movement of the south east corner was suspected.   A movement monitoring programme was initiated to determine both the overall extent of surface movement and the depth to which movements were occurring. Approximate piezometric conditions within the slope were also determined from the same boreholes.

Measurements on the surface and in boreholes (inclinometer measurements or casing deformation observations using sondes) indicated movement of the entire slope some 700ft in height, 2,000ft in width and at depths up to 250ft (see Fig 5,6).   Approximately 20m tons of material were contained within the slide zone.

Maximum movements were evident at the centre of the slide area with a gradual decrease in movement rates towards the lateral slide boundaries.   As movements continued, scarps at the slide crest became evident. Observations correlated directly with extensometer data from the skipway area.   During March and April 1971, the spring break-up period, major movements of the slope occurred with typical horizontal movements of the order of 1500mm/month.   The movements resulted in major disruption to the haul road system and ravelling of the more competent benches forming the lower part of the slope.

The primary cause of instability was related to mining of the more competent lower buttress zone in ore (see Fig 6) to a state where it was no longer able to support the overlying shear zones/very poor rock and overburden material.   The build-up in shear stress through the buttress was thought to have led to the extension of unfavourably oriented joints within the ore zone coupled with a general overall deformation of

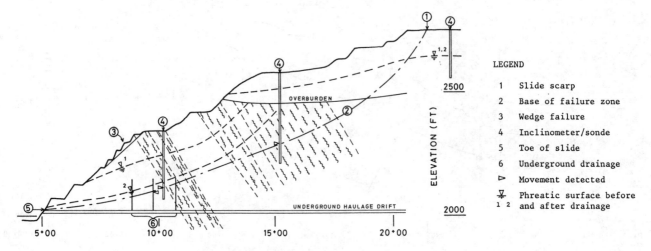

Fig 6   Cross Section through 1970-71 Failure showing Failure Zone and Groundwater Conditions

the buttress zone. A second but significant factor was the occurrence of adverse groundwater conditions, particularly during the spring break-up period.

Once the nature and extent of the failure had been confirmed, steps were taken to stabilise the area by a combination of drainage and unloading measures. Drainage was effected from an underground drift below the slide with holes being drilled up into the sheared and dilated rock mass (Ref 1).

Only limited surface and subsurface monitoring data were available from this early failure. However, it could be concluded both from inclinometer displacement profiles and from the nature of the rock in the base of the slide as encountered in the drainage holes, that shear displacements at the base of the failure were occurring over a zone of significant width and were accompanied by associated fracturing and dilation of the rock. Dilation effects were particularly evident due to increased permeability of the rock in this area resulting in excellent drainage response.

6.2 1974-75 Failure

The 1974-75 failure was essentially a reactivation of the 1971 failure with regression of the back scarp as shown on Fig 4. Although a new slope profile had been generated by mining a subsequent expansion, the slope materials themselves had been subjected to prior displacements. The instability was initiated by the creation of an unfavourable slope geometry due to phasing of ore/waste expansions.

Movements were detected in 1974 and were typically of the order of 300-500mm/month by mid 1974. By January 1975, as a result of ongoing mining of the lower slopes, they had increased to some 1000-1500mm/month. A major overburden failure in January 1985 halted mining in the entire area and an intensive programme of investigation was commenced.

The results from deep inclinometers in previously disturbed material indicated progressive straining with depth and a wide shear zone resulting from the slope displacements as shown on Fig 7. Although the installation was in predominantly sheared material, it provided a valuable insight into the degree of internal deformation within the movement zone and confirmed the lack of a well-defined basal failure plane with small scale strains below the inferred shear zone. Data on surface vector movements for this failure were relatively limited but the data showed for the mid and upper slope regions, displacement vectors subparallel to the slope profile.

By September 1975, conditions had effectively stabilised as a result of mining and water control measures. Taking into account overall movements up to this time, cumulative displacements of the order of 30m had occurred for both the 1970-71 and 74-75 events. For both the 1970-71 and 1975 failures, no significant tendency for increasing movement rates with cumulative displacement was observed. The lack of such an effect was perhaps contrary to expected strength loss of the material with ongoing deformation.

6.3 1982-1986 Failure (See Fig 2,8)

Prior to the development of an overall failure of the newly mined expansion a relatively shallow failure of a localised mid slope section had occurred in November 1981. This failure was attributed to possible deep disturbances from the prior slide that had weakened the near surface material.

In late 1982, a major zone of movement developed over the south east corner area as shown on Fig 8 as a result of mining the A expansion below elevation 2000 without the planned waste stripping of Expansion B. It was considered based on previous predictions that the zone of more competent rock in the vicinity of El 2000 and below might inhibit the downward progression of movements with ongoing mining. This was not the case however and movements continued to follow the expansion bench level down to mining cessation at El 1780.

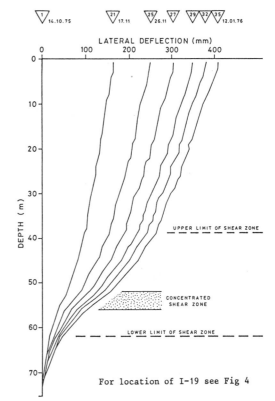

Fig 7   Inclinometer I-19 Displacement - Depth Response

Movement rates for the overall failure were initially small, typically 20mm/month in late 82/early 83, reducing to some 10mm/month during the summer period and increasing again to some 30mm/month after October 1983 until the year end in response to increased precipitation. During 1984 movement rates were typically 10 - 20mm/month and these rates persisted until March 1985, when increases in movements of up to 50mm/month were observed over a 2 month period corresponding to the spring break-up. After May, movements returned again to the typical rate of some 10 - 20mm/month until October 1985.

Although no mining had occurred directly within the slide area, an expansion had been proceeding immediately to the south as indicated on Fig 8. During 1985, a relatively rapid but shallow failure occurred over a localised area which later formed a faster moving zone of more significant extent, termed Zone B. (The extent of the Zone B area is shown on Fig 8 as of April 1986.) Further enlargement of this area occurred with time although movement rates during summer 1985 were generally favourable, a condition attributed to less than average precipitation.

During October/November 1985, a rapid increase in movement rates for Zone A and B was experienced with an extension of the Zone B area. Conditions for the two zones up to October 1986 are given on Table 2 and shown on Fig 8.

Table 2. Movement rates for zones A and B (mm/month)

Period	Zone A	Zone B	Precipitation
Summer 1985	10 - 20	50 - 75	25mm/month
Oct - Dec 85	100	300	50* "
Jan - mid Mar 86	20 - 30	40 - 50	25 "
Mid Mar-end Apr 86	200-500	1000-2000	30 "
End Apr-Oct 86	50 -100	400 - 750	50 "

* Excluding Dec. 85

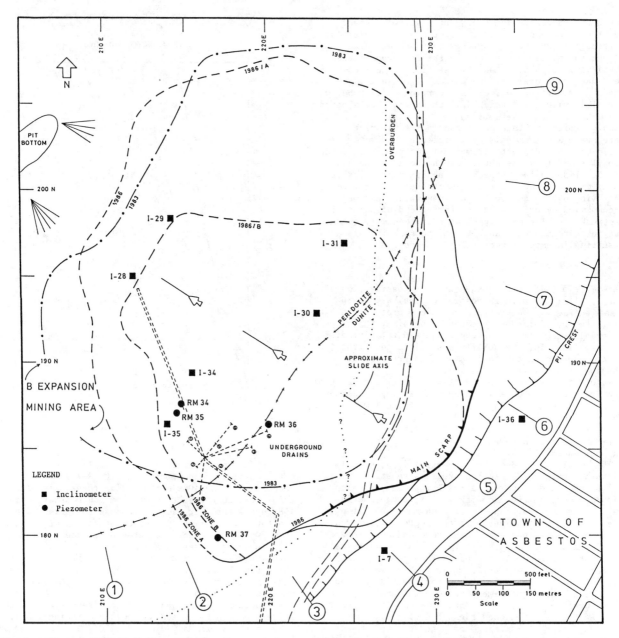

Fig 8   Plan of 1983-86 Failure showing Movement Zones, Monitoring Locations and Underground Drainage System

The increased movement rates are attributed directly to above average precipitation levels and enhanced infiltration as break-up of the slope occurred.   The spring break-up in 1986 was also extremely rapid and severe in terms of surface water infiltration into the slopes.

Overall cumulative movements in October 1986 are of the order of 2m for Zone A (the overall failure zone) and up to 10m for Zone B.   With time, an increase in the depth of movement has been inferred, leading to the conditions shown on Fig 2.   The total mass of the moving material is estimated to be in the range 40-50 million tons.

7   GROUND DISPLACEMENT PROFILES AND INFERRED FAILURE MECHANISMS

Movements of a large number of surface points and sub-surface data from inclinometers have allowed general inferences to be made on the failure state.   Particular attention has been focussed on the question of on-going movements and whether or not accelerating movement trends could be experienced for the prevailing geometrical and strength conditions.

7.1   Slope displacement vectors

The slope vector displacements are shown on Fig 2 for the current slope failure.   The vectors show a sub-parallel trend to the slope profile in the mid/upper regions with a more pronounced vertical component close to the upper slide scarp.   Towards the base of the failure the resultant surface movements become approximately horizontal, whilst in the toe zone, particularly during the initial displacement phase, the overall movement trend is above the horizontal.

Assuming an inferred base of significant movements as indicated on Fig 2, it is possible to infer that whilst the upper and mid slope sections are undergoing displacement under constant volume, the lower section is undergoing a volume increase with displacement. The dilation of the lower zone is directly attributed to complex shear mechanisms within the more competent strata and an overall break up of the zone rather than the generation of a single failure surface.

The limited data available from previous failures indicate that the degree of volume increase/dilation is greatest during the initial displacement stages and tends to zero with significant cumulative movements (> 5m).

7.2 Subsurface inclinometer data

Based on inclinometer data, it is recognised that slope movements have initiated as a progressive straining with enhanced movements at the surface and decreasing movements with depth. From the observations of I-19 (see Fig 7) it was estimated that some 250mm of collar deflection was required before a shear zone would develop at depth. The typical depth at which movement was inferred in this case was about 55m with a shear zone thickness of some 25m.

The data from the inclinometer installations in the south east corner installed since 1983 have been somewhat inconclusive. Summary data for all installations are given on Table 3. The response has typically shown an increasing strain gradient with time with little shear kinking prior to blockage. This is the case even where large displacements have occurred such as for I-34, I-35. It is also evident from comparisons of inclinometer and surface movement data that a significant component of the movement appears to occur beyond the bottom of the inclinometer casing.

Table 3. Summary of inclinometer observations

	Period of Operation	Cumulative Movement Axis A	Date/Depth Of Blockage
I-28	02.83 - 06.86	613 mm	26.06.86 / 4,3m
I-29	02.83 - 02.85	166 mm	12.05.85 / 38 m
I-30	02.83 - 10.84	160 mm	12.11.84 / 47 m
I-31	02.83 - 05.84	90 mm	17.08.84 / 38 m
I-34	04.85 - 03.86	563 mm	04.04.86 / 20 m
I-35	04.85 - 04.86	426 mm	01.05.86 / 27 m

In spite of the overall strain pattern within the slopes, it is considered likely that the depth at which shear failures eventually develop is within the casing interval. Deeper seated strains are obviously occurring due to shear stress transfer into the slope, but the effect of such strains on material behaviour is unclear. Overall it is expected that the slope would seek to move along a single weaker zone of material (shear zone as developed from the slope movements) rather than maintain a complex pattern of internal shear deformations.

7.3 Drillhole observations

Strong evidence that concentrated zones of shear displacement do develop at depth comes from the 1972 and 1986 underground drainage programmes. Both programmes showed the presence of a dilated, highly fractured zone containing significant quantities of water under pressure. In addition the 1986 programme inferred the presence of grey, silty material (gouge ?) at the base of the zone at locations very close to those inferred from overall movement observations and shown on the overall slide section (Fig 2).

7.4 Inferred failure mechanisms

The inferred mechanisms of slope failure involve the development of a basal shear zone through which shear displacements are concentrated. Above this zone the rock mass moves subparallel to the slope profile except at the upper and lower zones when an accommodation to the failure surface geometry must occur. The importance of understanding how the material may behave with ongoing deformation is linked to the potential for loss in shear strength as shear deformations concentrate into a single zone. Such a reduction in shear strength is a commonly observed trend in rock masses particularly where extension fractures are formed and can lead to a corresponding decrease in slope stability and accelerating movement trends.

The evidence from past failures is that the rock does become weakened as a result of slope deformations, as can be inferred by comparing maximum stable slope angles in undisturbed and disturbed material as shown on Fig 9. What is important in predicting future slope behaviour is the degree of strength loss that may still occur. The inclinometer data from the current failure would indicate that there is still potential for further development of concentrated zones of weaker, sheared material which could result in reduced stability.

From precedent experience of previous slope failures where large deformations have occurred, (primarily the 1970-71 and 1974-75 failures) it is evident that even for overall deformations in excess of 10m, an increase in movement rates compatible with an expected significant loss in strength with deformation has not occurred.

From the overall deformations that have occurred to date on several failures, it therefore appears that the rock mass may reach a residual strength after surface movements of some 5 - 10m have been experienced and a preferred zone of shearing has developed towards the

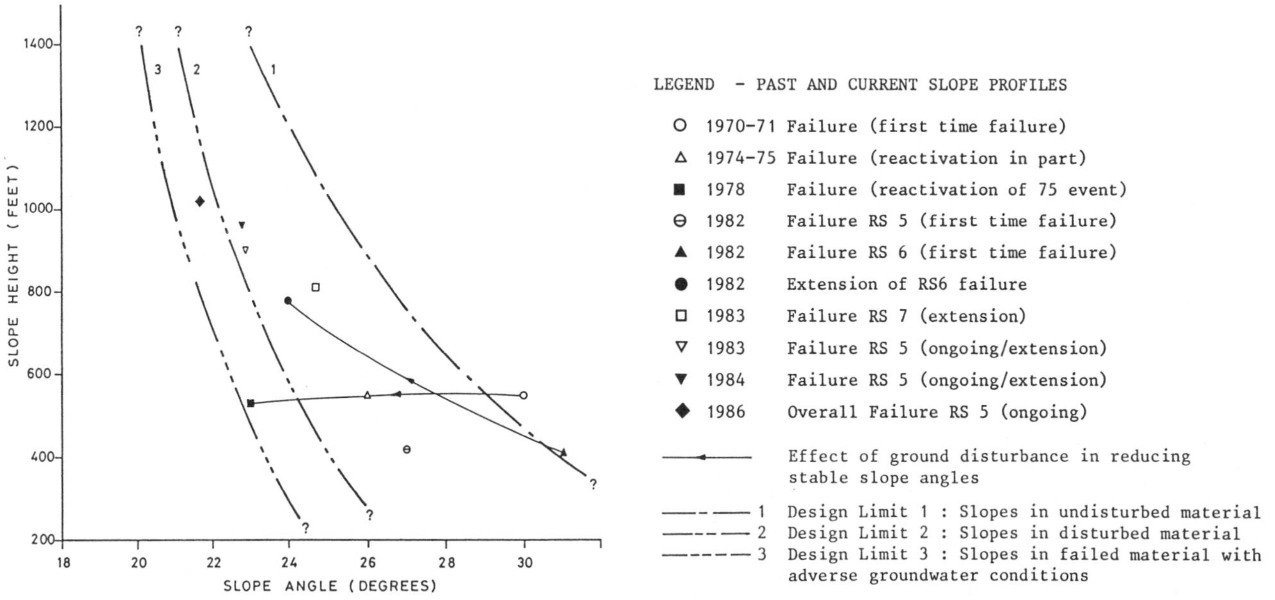

LEGEND - PAST AND CURRENT SLOPE PROFILES

O   1970-71 Failure (first time failure)

Δ   1974-75 Failure (reactivation in part)

■   1978   Failure (reactivation of 75 event)

⊖   1982   Failure RS 5 (first time failure)

▲   1982   Failure RS 6 (first time failure)

●   1982   Extension of RS6 failure

□   1983   Failure RS 7 (extension)

▽   1983   Failure RS 5 (ongoing/extension)

▼   1984   Failure RS 5 (ongoing/extension)

◆   1986   Overall Failure RS 5 (ongoing)

———◄———   Effect of ground disturbance in reducing stable slope angles

———·——1   Design Limit 1 : Slopes in undisturbed material

———··——2   Design Limit 2 : Slopes in disturbed material

——————3   Design Limit 3 : Slopes in failed material with adverse groundwater conditions

Fig 9   Summary of Slope Performance and Design Data for South East Corner Slopes in Ultrabasic Rock

base of the movement zones. Although it cannot be conclusively demonstrated, it would appear that for the very large slope failures that occur at Jeffrey, the long term behaviour of such failures will be characterised by an overall planar failure subparallel to the slope which moves at reasonably constant rate with time. Although some increase in velocity can be attributed to strength loss within the rock mass, the loss is such that accelerating movements that could lead to a sudden high velocity failure are considered unlikely. Significantly greater variations occur as a result of seasonal groundwater changes.

The depth to which failures occur does not appear to be controlled by particular geological weaknesses. It is apparently a function of the overall failure geometry and can be reliably deduced for new failures from precedent observations.

8 GROUNDWATER OBSERVATIONS 1985 - 86

Monitoring of groundwater conditions within the failure area has been carried out during 1985, 1986 using data from deep piezometers and water level measurements in inclinometers at the locations shown on Fig 8. The general groundwater conditions measured away from drainage facilities are shown on Table 4 and are plotted on Fig 10 (RM 34B).

Taking into account precipitation data for 1985 and 1986 and recognising that 1985 was a drier than average year up to October, an overall rise of groundwater levels of some 15 - 20m is inferred for the period mid 1985 - mid 1986. Such rises are obviously very significant in terms of stability changes. The spring break-up in 1986, a very sudden thaw, led to significant infiltration and overall groundwater rises of some 13m in one month. The resultant effect on movement rates is evident from Table 2.

Table 4. Changes in Groundwater Pressure (m head) with time

PERIOD	Lower Slopes		Mid Slopes	Inferred Change
	RM 34B	I-35	I-31	
15.10-07.11.85	+ 4,0m	0,0?	+ 5,8m	+ 3,6m
07.11-12.12.85	+ 6,5	+ 2,4m	-	+ 4,9
12.12-12.03.86	+ 5,5	+ 0,6	+ 6,1	+ 4,9
12.03-15.04.86	+11,6	+19,8	+15,2	+13,4
15.04-01.05.86	+ 1,5	+ 3,0?	0,0?	+ 1,8
01.05-31.07.86	- 3,0	-	- 4,6	- 3,4

9 S.E. CORNER DRAINAGE PROGRAMME

Underdrainage of large failures was first carried out at Jeffrey in 1971 using access from a disused haulage drift under the slide area. The findings of the 1971 programme which resulted in the final stabilisation of some 20m tons of material are described in Ref 1.

In 1985 it was decided to attempt a partial stabilisation of the south flank of the current failure using similar methods in an attempt to optimise mining of the south area of the pit.

9.1 Results of drilling programme

The drainage programme drilling was commenced in January 86 and completed during February. A total of 8 holes was drilled into the south flank of the slide from an underground location as indicated on Fig 8. As a general rule the holes encountered a zone of very fine material, possibly comprising the base of the sheared zone, immediately below a fractured water bearing shear zone corresponding to the base of the failure.

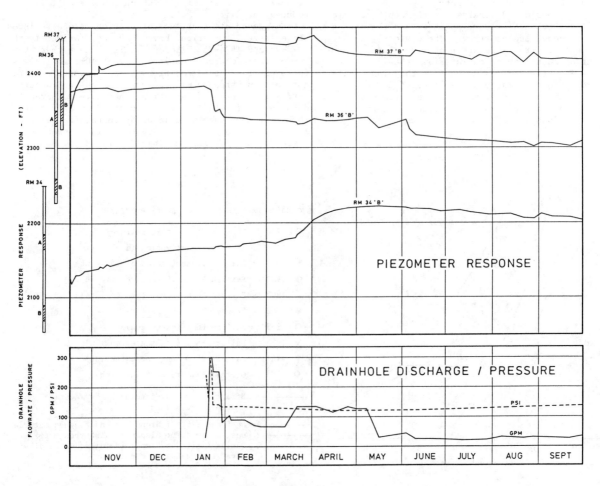

Fig 10   Piezometric Observations and Drainage Response for South East Corner Slopes

Hole numbers 1,2,4,5,6 encountered initial flows of 30 - 60 g.p.m. which reduced to some 2 - 20 g.p.m. after four days. Hole number 3 which was drilled on a plan orientation of approximately due east from the underground drift encountered large quantities of water at elevation 2235ft approximating to the base of the slide. Initial flows were of the order of 300 g.p.m. under a significant head. Flows were sustained at a high level, in excess of 200 g.p.m. for some 7 days, and then fell to about 100 g.p.m. at the end of January, declining to 70 g.p.m. in mid March. Flows from this hole then doubled with the spring break-up and were sustained at 130 - 140 g.p.m.

9.2 Groundwater response

The effect of the drainage programme has been to cause a localised but significant fall in piezometric levels as indicated directly by the response of Piezometer RM 36 (see Fig 10) and possibly by the less dramatic response of RM 37. The response of RM 36 to the drainage programme was a fall of some 15m for the lower unit between mid January and mid February 1986. A minimal rise was noted as a result of the spring break-up in spite of the enhanced flows from the drainage holes and by mid April the overall decline was re-established.

The benefits of the programme can be interpreted as not only a fall of groundwater pressures but a prevention of water pressure increases arising from the spring break-up, and much higher movement rates than those experienced. Taking initial conditions as for January 1986 a head difference of some 30m has been established between unit RM 34B (typical of the area outside the drainage zone) and unit RM 36B within the drainage influence area. Since mid April, an overall decline in piezometric levels has continued with a further drop of some 8m from mid April to August compared to an overall drop of less than 3m for piezometers at a similar depth outside the drainage zone.

10 OVERALL DESIGNS FOR SLOPES IN WEAK ULTRABASIC ROCK

The observed performance of both stable and unstable slope sections at Jeffrey has been used to derive design criteria for future mine planning. As is evident from the slope cross sections (Fig 2), the overall rock slopes are relatively flat and stability is adversely influenced by the weak nature of the rock and high groundwater levels.

The principal observations at Jeffrey can be summarised in terms of slope height and slope angle as shown on Fig 9 for the particular geological and geometrical conditions pertaining to the south east corner. The data indicate the summarised experience over 16 years for slopes up to 350m in height.

From the observed response, it has been possible to infer three design criteria for slopes in undisturbed, disturbed and failed ground with adverse groundwater conditions. Although apparently crude, these criteria provide the most reliable predictions for overall slope design. Particular features of slope geometry or groundwater conditions are further investigated by limit equilibrium based sensitivity analyses in order to prepare final slope layouts.

11 SLOPE MONITORING SYSTEMS

Monitoring of the rock slopes at Jeffrey requires both the gathering of a significant amount of overall long term performance data and the acquisition of continuous movement records to check the ongoing safety of the operation. Although data collection is common to many mines, few open pits operate with a town located on a large part of their periphery. The particular conditions at Jeffrey thus require accurate monitoring, high data availability and rapid reporting and interpretation.

The main control method used for the relatively flat,

ultra-basic slopes utilises electronic distance measurement (EDM) from base stations on the opposite side of the pit. The field application of laser measurement techniques was pioneered at Jeffrey in 1972 and some 1000 reflectors and 14 base stations have been installed since that time. The current layout of reflectors is illustrated on Fig 1.

Currently state-of-the-art infra red distance monitoring is utilised with some 300 single prismatic reflectors and sight distances up to 1800m. Using a protected, temperature stabilised base station tied in to known stable bench marks, an accuracy of better than ± 5mm is generally achieved except in very adverse conditions of rain, snow or mist. Prism icing is minimised by the use of hoods, developed at the mine.

The frequency of distance readings on the 300 targets currently installed varies from a few hours to a week and to determine the actual movement resultant of these targets precision theodolite readings of vertical and horizontal angles are made at regular time intervals.

In addition to direct distance monitoring, triangulation and levelling, other monitoring systems as described in this paper are utilised. These include precision slope indicators (typically 60m deep, total no 30+), piezometers (typically 60m deep, 2 tips, total no 30+) and in the vicinity of critical structures, deep extensometers.

12 CONCLUSIONS

In an open pit operation, on the scale of the Jeffrey Mine, with complex rock types and no overall structural control to the geology, monitoring is an extremely important element of the mining programme. The monitoring information is used for a variety of crucial functions including daily safety control, evaluation of current mining plans and future slope design.

With the extensive and precise nature of the system at Jeffrey, monitoring provides an 'active' input into short term mine planning. The very early identification of movement zones allows steps to be taken to minimise the impact of mining on stability by the implementation of corrective measures and at the same time provides for optimum ore extraction. The system contrasts strongly with more common 'passive' systems that frequently only record the occurrence of an event for subsequent post mortem examination.

If reliance is to be placed on active monitoring systems to optimise mining in areas of potential slope instability, guidelines on reading, evaluation and communication procedures must be clearly established. This permits early and confident decision making by management both for safety purposes and for optimum excavation sequencing.

Owing to the complex nature of the ultra-basic rock masses at Jeffrey, it has proved impracticable to derive meaningful constitutive models of the rock mass behaviour for use in slope design. Instead it has been necessary to use historic performance data as the primary basis for planning future slopes taking account of, as necessary, prior slope deformations. Major unresolved factors are the degree of apparent strength loss with large scale straining of the slope profile and the nature of the basal shear zones.

Groundwater has been a major influence on stability in the ultra-basic rocks and will continue to affect to a large extent the behaviour of ongoing movement zones. Control measures in the form of drainage can be successfully applied, but only on a limited scale and both their implementation and benefit are influenced by ongoing movement trends.

REFERENCES

Sharp, J.C. May, 1979. Drainage to control movements of a large rock slide in Canada. Proc. 1st Int. Mine Drainage Symposium, Denver.
Golder Associates. 1970-1986. Project technical reviews - slope stability.

Plate 2   Jeffrey Mine – General View of the South East Corner Rock Slide – 1971

ACKNOWLEDGEMENTS

The permission of JM Asbestos Inc to publish this
paper is gratefully acknowledged.   Among the great
number of people who have contributed to the project
over the past 20 years, the long term contribution
of Mr. Bernard Coulombe, the mine manager, should be
mentioned.   The authors are also grateful to Dr. Jack
Crooks of Golder Associates who provided a detailed
review of the paper.

550

Protection against rockfall – Stepchild in the design of rock slopes

Protection contre le chute de pierres – En deuxième plan dans la conception des pentes rocheuses
Schutz gegen Steinschlag – Ein Stiefkind beim Böschungsentwurf im Fels

R.M.SPANG, Geotechnical Consultants Dr.R.M.Spang and Dr.R.W.Rautenstrauch, Witten, FRG

ABSTRACT: In the following paper the causes behind and the geotechnical conditions for small scaled rockfall are first of all explained. This background is followed by a brief description of general protective measures against falling rock. Distinction is made between the so-called "active" and "passive" measures as to their respective functions - i.e., whether the measures prevent instable boulders from moving or measures to stop or divert them once they are in motion.

Subsequent to the review of measures, the kinetics of falling rock are discussed and a computer program is presented which permits calculating rockfall trajectories, velocities, and kinetic energies. Proposed passive measures like fencing with wire rope nets can be installed at crucial sites, thus permitting optimization of the locations and dimensions of those protective measures. The paper concludes with general design criteria for passive measures developed from the results of parametric studies. Practical examples are given for illustration.

RESUME: Une définition du terme chute de pierres est données auparavant. Les conditions et causes d'une formation de chutes de pierres sont discutées ensuite. Les différentes possibilités d'une protection côntre les effets des chutes sont decrites, en différenciant entre les mesures actives et passives. Les mesures actives empêchent la formation des chutes, les mesures passives s'opposent à la propagation des masses éboulées par protéction ou par déviation.

Le deroulement cinétique de la chute est étudié et un programme d'ordinateur est decrit, qui permet la calculation des trajectoires de pierres et de l'energie cinétique à chaque point du trajectoire. Les effets de mesures passives peuvent être introduit dans la calculation à n'importe quel point du trajectoire. Leur arrangement et leur géométrie peuvent être ameliorés ainsi. Plusieurs critères pour la conception de mesures passives provenant d'etudes paramètriques sont donnés à la fin, expliqués par des examples.

ZUSAMMENFASSUNG: Es wird zunächst eine Definition des Steinschlags gegeben. Anschließend werden Voraussetzungen und Ursachen für die Entstehung von Steinschlag erörtert. Die verschiedenen Möglichkeiten von Steinschlagschutzmaßnahmen werden beschrieben, wobei zwischen aktiven und passiven Maßnahmen unterschieden wird. Aktive Maßnahmen verhindern die Entstehung des Steinschlags, passive Maßnahmen fangen ihn ab oder leiten ihn um.

Im Anschluß daran wird die Kinetik des Steinschlags untersucht und es wird ein Computerprogramm beschrieben, das die Berechnung der Steinschlagbahnen und der kinetischen Energie in jedem Bahnpunkt gestattet. An beliebiger Stelle der Bahn können passive Maßnahmen in die Berechnung eingeführt werden. So können deren Anordnung und Geometrie optimiert werden. Der Aufsatz schließt mit einer Reihe von Entwurfskriterien, die sich aus Parameterstudien ergeben haben. Diese werden an Beispielen erläutert.

INTRODUCTION

In October 1985 the West German Federal Railway (Deutsche Bundesbahn) established an investigative commission to study the matters of supervision and maintenance of earth and rock structures along all of the Federal Railway's 28,000 km of track network. Marking 150 years of railway traffic in the Federal Republic of Germany, it was evident that there should be a systematic check of the rock and soil embankments and slopes. Shortly after commencement of the commission's work, its attention was focused upon the steep rock slopes along tracks in Germany's higher elevations.

The commission's investigative analysis of incidents occurring during the previous ten years showed that in most rail interruptions, accidents or serious derailments it was not a case of spectacular slides from major slopes or those of especially steep gradient, but rather from occurrences of small-scaled rockfalls. The commission's findings were in accord with those of a special conference on rockfall protection of railway lines which had been held in 1978 at Kandersteg/Switzerland under the auspices of the International Association of Railways (Union International des Chemins de Fer). Besides noting detailed description on a wide range of protective measures, the international conference demonstrated that the designs of such measures resulted largely from the whims of personal intuition and experience, but not as the outcome of rational or reproducible design criteria. To the author's knowledge this situation is still dangerously prevalent, and doubts must be raised as to the certainty of various so-called "safety factors" and ultimately about the subsequent economics of inappropriately selected solutions.

From the above described situation it could be recognized that there would be an increasing demand for protective measures, but also a critical lack of reliable design tools. It therefore seemed appropriate to improve the actual design procedures by a combination of field studies and mathematical rockfall models.

WHAT IS ROCKFALL?

Generally spoken rockfall is a mass movement which
- is correlated to steep slopes
- is located at the slope surface
- requires loose rock pebbles or boulders
- starts suddenly without any announcement
- is not pre-visible as a single event.
Typically it is characterized by
- zero velocity at the beginning

- high acceleration during its descent
- temporary loss of ground contact
- high kinetic energy when reaching the foot of the slope.

There is no generally accepted definition of the rock volume which characterizes a rockfall event. Within the literature, this volume ranges from gravel sized pieces up to some hundreds of cubic meters and more. Because this analysis is restricted to volumes, which can be intercepted by walls, fences, ditches and similar protective measures, the volume is restricted to some few cubic meters per event. In terms of kinetic energy the single rockfall is by practical reasons restricted to a maximum of about 500 kNm. Rock volumes, which by their position above an endangered object will lead to higher kinetic energies, in most cases may be better fixed in place or removed than to be allowed to get in motion.

WHAT IS ENDANGERED BY ROCKFALL?

Rockfall may endanger buildings, civil engineering or mining sites, roads, railway lines and so on, which are situated within or near the foot of steep slopes. It can result in serious damages, loss of lifes, circulation problems and so on.

UNDER WHICH CONDITIONS ROCKFALL CAN OCCUR?

Not every slope may be a source of rockfall. There exist some clear-cut conditions which have to be fullfilled.
- On the slope's surface loose rock pebbles or boulders must exist.
- The slope's inclination must be steep enough to make these boulders instable. (Creeping is not considered).
- The slope's inclination additionally must be steep enough to prevent boulders, once being in motion, to come to rest before they arrive at one of the above described objects.
- There must act a triggering force.

These conditions are valid for uniformly inclined slopes as well as for slopes with changing inclinations. Let us now consider these conditions in detail.

Loose rocks on a slope surface may develop by weathering, either caused by physical or chemical factors. Especially frost and the action of roots may weaken a formerly sound rock surface, separating the originally intact rock mass into an agglomeration of isolated pebbles and boulders. MÜLLER (1963) stressed that in hard rock slopes first detachments often occur 20 to 30 years after their excavation and that an age of 70 - 80 years often limits their service life.

Loose rocks naturally may result from erosion, where pebbles and boulders are washed out of a matrix of cohesive residual soil.

Man-made rockfall frequently comes from the excavation itself, if the excavated slope surface is affected by blasting or ripping. Also the dumping of debris from the top of a slope can result in rockfall problems.

According to RITCHIE (1963) rockfall is limited to slope inclinations greater than about 35°. It is worthwhile however, to check this critical value for each slope individually, because it is obvious that the geometry of the rock element and the quality of the slope surface can rise this angle up to 45° and more. So the critical inclination, above which protective measures have to be taken, will increase from spherical to cubic elements and may reach its maximum for slender slabs (Figure 1). In the same way a layer of loose soil mostly increases the critical angle, depending however on the mode of motion.

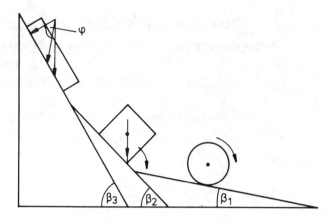

Figure 1: Effect of rockfall geometry on the critical inclination of a slope.

Analogous to the angle which limits the generation of rockfall, another critical value exists, which decides if rockfall from the steeper upper part of a slope, may come to rest before it reaches the interesting object. This angle depends on a number of factors, amongst which are
- the mode of motion
- the quality of the slope surface, characterized by the degree of attenuation, the angle of friction, and the roughness in respect to the volume of rockfall
- the existance, kind and density of vegetation, the effect of which, however, is mostly over-estimated. Rockfall can be triggered, for example, by
- great temperature chances, especially by changes between insolation and frost
- vibrations from blasting, earthquakes and traffic
- local fracturing of rock due to stress concentrations
- heavy rainfalls, causing cleftwater pressures and erosion
- animals and men, moving on a slope.

WHAT CAN BE DONE AGAINST ROCKFALL?

In principle two totally different kinds of protective works exist. The first one, which shall be called "active" prevents the generation of rockfall. The following variations exist.
- selection of slope inclinations smaller than the critical angle.
- removal of all instable pebbles and boulders.
- covering the slope surface by a dense vegetation, by wire mesh or shotcrete.

Slope angles lower than the critical one require no further control, maintenance or repair, but the method is costly and stands often in contradiction to landowners and ecologists.

Perfect cleaning of a slope surface will, in most cases, not be feasible, and weathering will go on anyway. Cleaning therefore has to be repeated within annual intervals.

Covering a slope surface with vegetation is an ecological must, but it requires slope inclinations below 45 to 50°. Already these values require additional measures to prevent the topsoil from sliding or from being eroded.

The second approach is the "passive" one, which catches the rockfall, dissipates its kinetic energy or diverts its paths. The following possibilities exist.
- relocation of endangered objects
- berms
- earth fills and gabions
- intercepting ditches
- catch nets, fences and walls
- rock sheds and tunnels.

Some of these possibilities are shown in figure 2. A more detailed description is given by SCHUSTER & KRIZEK (1978).

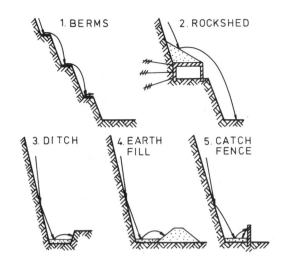

Figure 2: Some passive measures and their effect on rockfall trajectories.

Active measures have to cover the whole area, from which rockfall can start, whereas the passive measures can be limited to one or few linear arrays. Therefore active measures in general will be much more expensive than passive ones. On the other hand, active measures can easily be designed whereas the mostly unknown dynamic behaviour of passive measures causes severe design problems.

WHICH INPUT DATA ARE REQUIRED FOR THE DESIGN OF PASSIVE MEASURES?

The answer depends on the selected measure. As for the relocation of endangered objects the new location must be outside the endangered area. Berms have to be broad enough to be hit by the rockfall and must be able to stop it or at least essentially reduce its kinetic energy. Rockfall protection tunnels and rock sheds must sustain the dynamic loading without damage. Furthermore they shall not divert rockfall to sensitive areas, which may not have been endangered before. Catch structures shall catch the rockfall and dissipate its complete kinetic energy. Walls, fences, and nets should sustain the impact without being destroyed. All structures shall be optimized as for their location, kind and geometry.

These requirements can be fullfilled if the geometry of the "design rockfall", and its possible or probable path are known. From this path impact velocity, angle and kinetic energy can be derived.

HOW CAN THESE INPUT DATA BE ESTABLISHED?

There are three possible approaches.
- The historical approach. It applies observations from previous rockfalls to risk evaluation and design of protective measures. If the observations are thoroughly made and analysed and the results are applied to the slope, where the observations came from, the historical approach can lead to reliable results. However, for the design of new rock cuts, such observations are not available. Because of the many factors, which govern rockfall paths, the application of experiences from other slopes will not be possible. Therefore the historical approach can not be taken as a standard design tool.

- The experimental approach. It uses artificially triggered rockfall to establish design criteria and input data. By application of this approach RITCHIE (1963) established general rules for the design of ditches and fences within and below uniformly inclined slopes, whereas BROILI (1977) used the method to design a distinct rockfall protection structure in Italy.
As already stressed above the generalization of results may be delicate. However, the tests can be executed on each slope, if they do not endanger those structures for which protective measures shall be designed. Besides this limitation the method is costly. Therefore even the experimental approach is restricted to favorable situations.
- The analytical method. It applies laws of motion and the theory of particle collision to the calculation of rockfall paths. The method first was used by PITEAU (1977). It can be universally applied. Its results cover completely the requirements of structural design, it allows parametric studies to check safety factors and optimizes geometry and location of protective measures. Of course, the analytical approach has its limitations too, which will be discussed below.

WHAT DETERMINES THE PATH OF A ROCKFALL?

Not taking into account the possibility of artificial triggering, a rockfall starts by one of the following primary modes of failure. In the case of a free rock surface this primary mode of failure will be identical with the failure modes of rock slopes, i.e. sliding on planar or wedge shaped discontinuities, toppling or buckling. In the case of loose rocks on a soil covered or free rock surface the primary mechanisms are sliding and rolling. Vertical free falling is restricted to overhanging rock slopes. The primary mechanism is followed by free falling, sliding, rolling or bouncing.

In some cases the rockfall will keep sliding or rolling until it comes to rest. More often - and obligatorily in the case of vertical free falling - the initial mode of motion ends by impact at the slope surface, where not only the mode of motion, but also the translational and angular velocities, the direction and the amount of kinetic energy change. After the collision the rockfall can continue in the bouncing mode or slide or roll until another collision leads to a further change or the pebble or boulder comes to rest. Besides by collision, the mode of motion also can be altered by changing slope inclinations.

Not only the initial mode of motion and the geometry of the slope surface determine the paths of rockfall, but also the mechanical quality of the surface itself.

In the sliding mode, the coefficient of kinetic friction governs the energy loss of a stone for a given slope inclination. The energy consumption by friction increases with decreasing slope inclination. If loose soil covers a moderately steep slope, the rockfall frequently merges at its front, which may lead to a complete deceleration. In other cases the merging of the front will lead to subsequent toppling and to the transition from sliding to rolling.

In connection with the slope inclination, the coefficient of static friction determines whether sliding as an initial mode of motion is possible or not. In the case of rolling the same coefficient decides upon the question, if pure rolling or a combination of rolling and sliding takes place. Note, that in the case of pure rolling no energy is consumed by friction, but a certain amount of energy may be lost by the rolling resistance.

The influence of rough rock surfaces or debris layers on rolling rock particles was described by RITCHIE (1963). He observed that the deceleration of rockfall by these conditions increases, if the

asperities or the size of the debris are in the same
order of magnitude or even greater than the rockfall
itself.

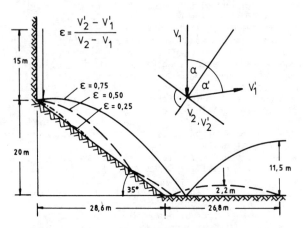

Figure 3: Influence of the coefficient of restitution on rockfall trajectories.

As shown by figure 3 the change of rockfall velocity
resp. the amount of energy loss at each impact is
governed by a factor ε , which is called the
coefficient of restitution. A soil covered surface
will have a low ε - value, which means a high
loss of kinetic energy, whereas a sound hard rock
surface will have a high ε -value. This factor, by
the way, has to be determined experimentally.

Free falling leads to the maximum kinetic energy.
In general the amount of kinetic energy decreases
from rolling to bouncing and sliding. Sliding mostly
is the less critical mode of motion. Free falling
also leads to the maximum rebound height. The maximum distance between two impacts will be archieved
by the bouncing mode.

Energy is also consumed by air resistance, which
is assumed to increase linearly with velocity. In
the case of large distances of freefall the effect
of air resistance may not be neglected.

In the rolling mode, and under certain conditions
in the free falling and bouncing mode too, the rockfall gets an angular velocity. This angular velocity
also is subject to sudden change during impact.
Angular acceleration leads to a reduction in translational kinetic energy. Thus a sphere, which rolls
down an inclined plane has a considerably lower
translational velocity than a sliding sphere on the
same, but frictionless plane. The practical effect
is that the trajectories of rotating rockfall are
narrower. On the other hand, the maximum reach often
results from rolling, because the energy consumption
by rolling resistance is smaller than that by sliding resistance or impact. Generally the reach is
small for freefall, but increases from sliding to
bouncing and rolling.

The path of a rockfall is also influenced by the
geometry of the pebble or boulder itself. Slabs, as
they result either from the original joint pattern
or from weathering or stress relief (GERBER and
SCHEIDEGGER, 1965) may come to rest on slopes,
where spherical pebbles or boulders will get additional acceleration, depending on the position of
the rotation axis.

Along a rockfall's path potential energy is transformed into kinetic energy and is partially consumed
by the above described resistances. Depending on the
slope geometry also the transformation of kinetic
into potential energy is possible. Rockfall comes to
rest, if its kinetic energy has become zero in a
stable position on a gently inclined or horizontal
part of a slope or on its foot, or if it hits an
obstacle, which takes its full impulse and leaves
it in a stable position.

WHICH ARE THE CHARACTERISTICS OF THE ROCKFALL COMPUTER PROGRAM?

For mathematical treatment of rockfall an algorithm
was established which was based on NEWTON'S laws of
motion, including his theory of particle collision
(SZABO, 1964, and MARION, 1970). The adequate computer program was written in basic and runs on an
IBM-AT 02 personal computer. The periphery consists
of a colour graphic screen and a Hewlett-Packard
7475 A-plotter.

The program requires two-dimensional slope profiles, which are divided into slices with vertical
boundaries. The slope inclination must be uniform
between two neighbouring boundaries. Slice boundaries are established where the slope inclination or
relevant surface properties change. The intersections
between boundaries and slope surface are related to
a global rectangular cartesian reference frame and
given in coordinates.

The other input data come from geotechnical mapping, which includes a joint survey and -if it is
feasible - some rockfall tests in situ. From these
investigations, the following parameters are
derived.
- failure mechanism
- geometry of rock element resp. potential rockfall
- specific gravity of rock
- rock strength (not used in the program)
- coefficients of restitution
- coefficients of static and kinetic friction
- rolling resistance
- start conditions, including start point in coordinates, start velocities - translational and
  rotational -, angle of throw, if rockfall starts
  in trajectory.

Additionally existant or planned passive structures
are given by the coordinates of two points each, and
their coefficient of restitution.

The actual program version is based on the kinetics of particles. The rolling mode is regarded as
sliding with low friction. The angular velocity is
neglected. The change of the mode of motion generally is restricted to the slice boundaries, except
in the free falling and bouncing mode.

No rock desintegration during impact is considered. Each run only deals with one rockfall,
interactions between two or more stones, descending
at the same time, are thus excluded. Sliding and
rolling at the same time is not regarded.

The output first comprises all input data; to
facilitate the control of the system geometry, the
slope and any passive structure is drawn on the
screen or plotted. At the end of each run the path
of the rockfall is presented either by coordinates,
by plotting or on the screen. In case of impact at a
passive structure, the angle of impact vs the horizontal axis, the impact velocities and the adequate
kinetic energy is given too. These data present the
design parameters for the subsequent structural
design of these structures.

WHICH GENERAL DESIGN CRITERIA CAN BE GIVEN?

From parametric studies and field experience the
following general design criteria can be derived.
- First it should be checked by a computer rockfall
  program or by field tests, if a rockfall risk
  exists. If it is affirmed and protective measures
  have to be taken, it has to be decided whether
  active or passive measures shall be applied. Geotechnical, economical and ecological aspects shall
  be taken into account.
- If passive measures shall be executed it should be
  proved, if the slope inclination allows for covering the surface by topsoil and vegetation. The
  execution of fascine work may reduce erosion and
  shallow slides as well as decelerate or even stop
  rockfall from upper parts of the slope.

- When designing a rock cut, the berms should be inclined towards the mountainside, because this will reduce the width of trajectories. Dewatering may be achieved by ditches at a distance between 10 and about 50 m, which are located perpendicular to the striking of the slope.
- Berms should be covered by rock blocks, the size of which shall be at least similar or even larger than the expected rockfall. If this is not feasible, because berms must be practicable, cohesive soil or sand also can be used.
- If the inclination or width of a berm is unfavorable, the additional use of fences at the valleyside of the berms should be considered.
- If the slope is very steep a broader horizontal foot area with a ditch and guardrails or fences may be better than berms. This consideration is in good accordance with the fact that berms within rock slopes often have an unfavorable effect on the global slope stability.
- Rolling and sliding rockfall mostly can be catched by simple ditches. The walls of the ditches should be as steep as possible, because well rounded and flat walls may increase the risk that a stone can overcome the ditch by rolling.
- Rigid walls are quite more sensitive against impact than flexible structures like wire mesh fences are. Reinforced concrete walls often show detachments or fissures after being hit by larger rockfalls. Thus the reinforcement steel is no longer protected and subject to corrosion. Besides rigid walls can not so easily be repaired as for example fences or catch walls made by rails and ties.
- Fences on slopes, which are not uniformly inclined, will be situated mostly on flatter portions of the slopes. If the inclination of such a flatter portion is flat enough to decelerate rockfall, the fence shall be erected at the valleyside of this portion. Otherwise it may be better to install the fence just on the mountainside.
- Fences behind ditches or at the foot of a slope should be located outside the fallout area resp. should be located so that rockfall first hits the ground and then hits the fence. Then the fence may be one order of magnitude lighter than otherwise. Additionally the place behind the fence shall be covered by soft fills, which show a low ε-value.
- Fences at the foot of a slope are more easily to be checked, maintained and repaired than fences within a slope. Therefore a heavy and high structure at the foot may be better than two light ones within the slope. By the way, good experiences were made by removing larger rockfalls from fences within a slope, when they were destroyed in situ to gravel sized pieces by blasting.
- Fences should be as flexible as possible and absorb the impulse by a high elastic and plastic deformations without being destroyed. This requirement is fullfilled by a combination of vertical or inclined steel girders with a lagging of steel wire rope nets which are anchored by steel wire ropes to the mountainside. Within these ropes special rope brakes are installed which are patenteed by the Swiss Kabelwerke Brugg AG. These brakes begin to slide if rockfall induced tension in the ropes exceeds a certain amount of shear resistance, thus limiting the loading of the structure after the use of the full elastic deformation of the nets and transforming an additional part of the kinetic energy into heat.

SOME PRACTICAL EXAMPLES

Case 1: When planning a playground for children, the designer noted some recently fallen rockblocks of about 2 - 3 m³ each within the proposed playground area. The subsequent investigation revealed that these rocks came down from an adjacent major slope which was completely covered by a forest. Within the upper part of this slope, an old quarry was found. Its face was about 16 m high and nearly vertical. At the foot of the face a gently inclined rock surface was exposed which was cut by a debris covered slope dipping 30 to 35° towards the planned playground. The situation is shown by figure 4.1. The unfavorable orientation of discontinuities favoured the occasional toppling of rock elements from a thickly bedded sandstone.

Obviously the coefficient of restitution was relatively high for the rock surface and considerably lower for the debris slope. The rockfall computer program proved the existence of a rockfall risk but showed that the situation at the foot of the slope was favorable for a relatively low catch structure. Because of the relatively high kinetic energy of the rockfall, a catch wall made out of a combination of gabions and earthfill was executed.

Case 2: When a steep mountain slope above a national road had been cleared of an old forest by its private owner, many weathered rock outcrops and loose rock blocks on its upper boundary became visible. The geotechnical mapping showed considerable rockfall risks. Because active measures were not economically feasible, catch fences were to be designed.

The rockfall computer program showed, as depicted by figure 4.2, that a small berm below the cliffs was quite useless. Because of the shown trajectories also a catch structure at the outer edge of the berm would not have been effective. The optimal location of catch structures obviously was near the foot of the slope. The high kinetic energies of the calculated rockfalls led to a heavy fence with inclined steel girdes, and a lagging out of a steel rope mesh, which were anchored to the mountainside.

Case 3: A natural slope was mainly covered by a thin layer of residual soil, but also showed extended rock outcrops of weathered shales and sandstones. These outcrops were the source of occasionally small scaled rockfalls up to a volume of 0.1 m³, which endangered a road. The situation is explained by figure 4.3. By use of the rockfall computer program a catch fence at the slope's foot was tested first, but the required height of this fence was not feasible. Therefore a second fence was located just above the steeply inclined foot, where the slope inclination was less steep. Because of a step above the proposed location the rockfall showed a relatively high kinetic energy at the impact, requiring a medium catch structure. When inducing a third fence just above the step, all three fences could be executed as light onces, which was preferred by the owner.

Case 4: Figure 4.4 presents a similar situation. A rock slope above an urban street showed a gentle lower and a steep upper profile. Whereas the lower part was completely covered by rock debris, the upper part showed extended outcrops of moderately weathered and jointed siltstones. The expected rockfall showed a maximum volume of about 0.5 m³. Because the rockfall risk was already known during the construction of the street, the planners had provided a ditch at the foot of the slope. By a systematic check of all rock slopes of that city the above described slope also was subject to a risk evaluation. The investigation proved that the ditch was able to catch the "design rockfall", but its storage volume was too small to permit the desired large control intervals. Therefore guardrails were installed at the outer edge of the ditch.

FURTHER INTENTIONS

On the base of the above described considerations an extended version of the rockfall computer program is just in progress. It will enable us to consider rigid bodies of general geometry. A good part of the

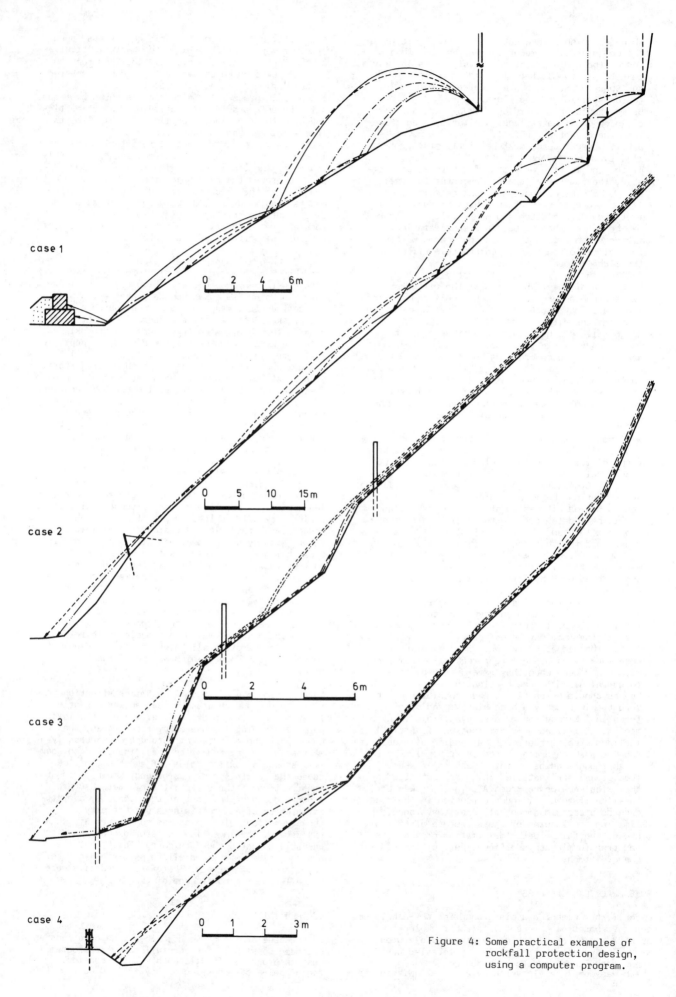

case 1

0   2   4   6m

case 2

0   5   10   15 m

case 3

0   2   4   6m

case 4

0   1   2   3 m

Figure 4: Some practical examples of
rockfall protection design,
using a computer program.

above described limitations thus will be overcome,
including the consideration of rotational velocities
and acceleration.
   The limited experiences regarding the coefficient
of restitution of walls and especially of fences of
different kind shall be extended by field tests as
well as by structural analyses. This work will be
done in cooperation with producers.
   By these investigations we hope to contribute to
a more rational, economic and safer design of rock
slopes.

BIBLIOGRAPHY

BROILI, L. (1977): Relations between scree slope
   morphometry and dynamics of accumulation pro-
   cesses; ISMES, 90, 11 - 23, Bergamo.
GERBER, E.K. & SCHEIDEGGER, A.E. (1965): Probleme
   der Wandrückwitterung, im besonderen die Ausbil-
   dung Mohr'scher Bruchflächen. - Felsmech. Ing.
   geol., Supl. II, 80 - 87.
JOHN, K.W. & SPANG, R.M. (1979): Steinschläge und
   Felsstürze, Voraussetzungen - Mechanismen -
   Sicherungen. - UIC-Tag. Schutz der Bahnanlagen
   gegen Steinschlag und Felssturz, Schlußbericht,
   Kandersteg.
PITEAU, D.R. (1977): Computer rockfall model. -
   ISMES, 90, 127, Bergamo.
MARION, J.B. (1970): Classical dynamics of particles
   and systems. - New York.
MÜLLER, L. (1963): Der Felsbau, I. - Stuttgart.
RITCHIE, A.M. (1963): Evaluation of rockfall and its
   control. - Highway research record, Vol. 17,
   13 - 28.
SCHUSTER, R.L. & KRIZEK, R.J. (1978): Landslides,
   analysis and control. - Nat. Acad. Sci, Transp.
   Research Board, special report 176, Washington
   D.C.
SZABO, J, (1964): Höhere Technische Mechanik. -
   Berlin.
SPANG, R.M. (1980): Geologische und geotechnische
   Aspekte des Steinschlagproblems. - Disputation
   Fakultät für Geowissenschaften, Ruhr-Universität
   Bochum (unpublished).

Economical design and construction of large scale mudstone cut slopes
Etude à coût réduit et construction d'un grand talus excavé dans la pélite
Wirtschaftliche Konstruktion grosser Hangeinschnitte aus Schlammstein

TORU SUEOKA, Technical Research Institute of Taisei Corp., Japan
MASASHIGE MURAMATSU, Technical Research Institute of Taisei Corp., Japan
YASUO TORII, Toyota Motor Co. Ltd, Japan
MICHIAKI SHINODA, Toyota Motor Co. Ltd, Japan
MASAKATSU TANIDA, Project Manager of Toyota Test Course, Taisei Corp., Japan
SHOJI NARITA, Vice Project Manager of Toyota Test Course, Taisei Corp., Japan

ABSTRACT: In the large scale slope excavation of mudstone ground with severe slaking, cut slope failures occurred at several tens of locations. By use of the various statistical characteristics of failed slopes, an economical slope design minimizing the total cost was obtained and a special construction method for this type of bedrock was prepared and executed.

RESUME: Lors de l'excavation sur une grande échelle d'un terrain pélitique sérieusement détrempé, des ruptures de talus se sont produites en plusieurs dizaines de points. L'utilisation des diverses caractéristiques statistiques des talus ayant subi une rupture a permis, d'une part, de réaliser économiquement l'étude du talus, minimisant ainsi le coût global, et d'autre part, de préparer et d'appliquer une méthode de construction spéciale pour ce type de roche de fond.

ZUSAMMENFASSUNG: Bei großen Ausgrabungen in Schlammsteinboden mit schwerem löschverhalten traten an –zig Plätzen Hangeinschnittbrüche auf. Unter Zuhilfenahme der verschiedenen statistischen Eigenschaften der Hangbrüche wurde ein wirtschaftliches Hangdesign zur Herabsetzung der Gesamtkosten erarbeitet sowie ein besonderes Bauverfahren für diese Art von Schichtgestein vorbereitet und durchgeführt.

1 INTRODUCTION

Although subsurface exploration is performed in advance, as the ground is a natural feature, the safety of a cut slope depends on many unknown factors, and there have been cases where slopes have failed after excavation has been started. In such cases, civil engineers are required to make important engineering judgements to design and construct the optimum cut slope within the given construction period and the budget.

This paper describes how civil engineers arrived at the optimum solution for the Toyota Motor Corporation in the excavation of large scale mudstone cut slope at the car speed test course, located in a cold northern district of Japan. In other words, despite the fact that the cut slope failed at many locations, the most economical method to stabilize the slope from the viewpoint of not only safety factor, but also the best gradient in the case of a cut slope, the best method of construction to prevent a failure from occurring on a slope with a high probability of failure, etc., were examined.

2 SITE DESCRIPTION AND TESTING

The test course of the Toyota Motor Corporation is located 12 km to the west of the suburbs of Shibetsu City in Hokkaido on the hilly country along the Inuibetsu River, a branch of the Teshio River, which forms a gentle slope of about 170 m above sea level. The geology near the site consists mainly of mudstone of the Central Ezo layer group belonging to the Cretaceus Period of the mesozoic era, the thickness of which is about 3,000 m. The strike is approximately in the north-south direction, sloping basically toward the eastern side, but it changes intricately depending on the location.

The mudstone contains montmorillonite and is actively slaking. At some location the mudstone contains tuff clay which has a large CEC and considerable swelling, and is an active clay.

The excavation commenced in May 1982 with a design slope of 1 : 1.5 with reference to the standard slope of the cut slope in Japan as shown in Table-1. After the excavation had started, slope failures occurred at more than 20 locations, as shown in Fig.-1, and the restoration work and a new design specification for the future excavation of a new slope became necessary.

3 INVESTIGATION OF FAILED SLOPE

To investigate the cause of the failure of the slope and to make the data useful for the future design in the site, the failure configuration was mainly examined on the slope shown in Fig.-1.

It was discovered that more dip slope structue of the ground had failed, and that 95% of the slope failed within 40 days after the cutting had been completed, indicating the slope failure occurred significantly soon after the excavation (Fig-2). More than 75% of the slope failures were $Tr - C_L$, $D - C_L$ in the geology (degree of weathering with reference to the classification for dam engineer), indicating that the C_L layer was considered to be the one that constituted the cause, assuming that the failure occurred in the lower part of the slope.

Table-1. Resistance of mudstone to swelling and weathering and suitable slope gradients (compiled by the Japanese Society of Soil Mechanics and Foundation, cut-slope gradient)

Classification by observation	Example	Increase ratio in water absorption by repetition of drying and wetting	Slope gradient	
			no ground water	ground water
1 Highly consolidated	Shale, consolidated tuff of before the polaengene period	lower than 1%/time	1:0.8	1:1.0
2 Relatively low in consolidation	Aphiolite of the Nesgene period	1.0 - 2.0	1:1.0	1:1.2
3 Extremely low in consolidation	Tuff mudstone clay of the Pliocene or the diluvial period	higher than 2.0	1:1.2	1:1.5

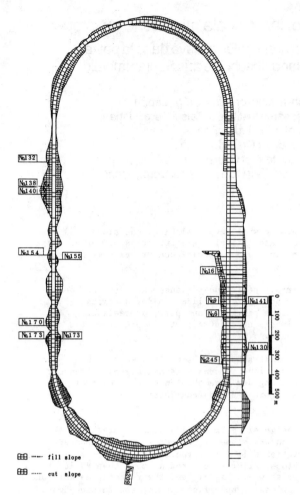

Fig.-1 Test Course and Failed Slopes

70% of the failures fall between 250 - 2,000 m² in size (Fig.-3). Examining the magnitude of the failures, most were 2 - 3 m in thickness and 10 - 20 m in length of the top from the bottom of the failed slopes as shown in Fig.-4. Most showed a degree of weathering of C_L. More than 60% of the failures occurred within 0 - 2 m from the top of the slope. As a result, more than 70% were 5 - 15 m in vertical height, as shown in Fig.-5.

4 ANALYSIS OF STABILITY OF SLOPE

From the investigation of the failures, the direct cause of failure may be considered to include: (1) a weak clay layer that exists in the mudstone, (2) slaking and swelling of the mudstone itself are considerable, (3) earth pressure in the horizontal direction of the natural ground is large and therefore the slope being excavated tended to collapse.

In any event, unlike earth slope, local surface discontinuities, heterogenious in quality, swelling nature, etc., are considered to affect the stability of the overall natural ground. As it is difficult to evaluate the strength of the overall natural ground even with the present levels of technology, there seems to be no established method of analysis. However, civil engineers are required to solve the problems presently faced as rationally as possible.

In this paper, the circular slip method is used as the analysis technique, even though not all the failures necessarily demonstrated a circular slip. In other words, the phenomena occurring in the slope during excavation are unknown and the average C and ϕ values used for the analysis do not necessarily represent the true phenomena. However, from an engineering viewpoint it is considered a rational "black box" method of establishing the true phenomena, examining the stability of the slope, obtaining the engineering parameters C and ϕ, and applying them to the new slopes being excavated (Fig.-6).

In the analysis, the configuration of the failed slopes was first put into order (Table-2). Next, a safety factor of 0.95 was applied to the failed slope. With regard to the soil parameters C and ϕ, the calculation was performed assuming that they were of a uniform soil layer, and the relation as shown in Fig.-7 was obtained. As the values of C and ϕ cannot be specified from Fig.-7 alone, the

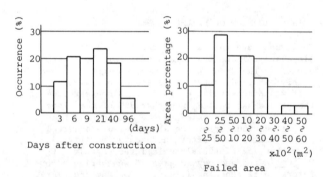

Fig.-2 Days before Collapse of Slope   Fig.-3 Size of Collapsed Slope

Table-2. Size of Collapsed Slope

	Lower end of slide a (m)	Upper end of slide b (m)	Vertial thickness of slide d (m)
Average	11.7	2.6	3.4
Standard deviation	3.5	2.8	1.9

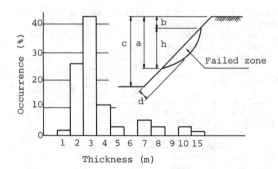

Fig.-4 Thickness d of Collapsed Slope

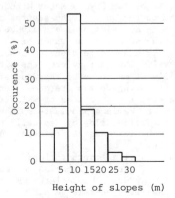

Fig.-5 Height of Collapsed Slope

560

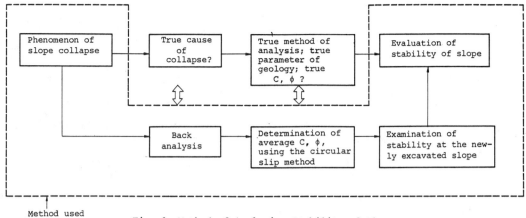

Fig.-6 Method of Analyzing Stability of Slope

upper end and the lower end of the failed slope and the value of C were used as parameters to determine ϕ, and the circular arc Rn.min. that gives the minimum safety factor was obtained as shown in the flow in Fig.-8. The correlation between Cn and Rn.min. was obtained as shown in Fig.-9 and the values of C and ϕ were determined. The values of C and ϕ thus obtained become Cm = 5.9 kpa, ϕm = 17°, respectively, with the scattering being as shown in Fig.-10.

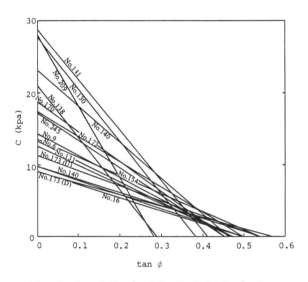

Fig.-7 C - ϕ Obtained by Back Analysis from Examples of Collapse

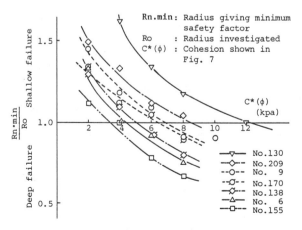

Fig.-9 Cohesion C Assumed from Depth of Circular Slip

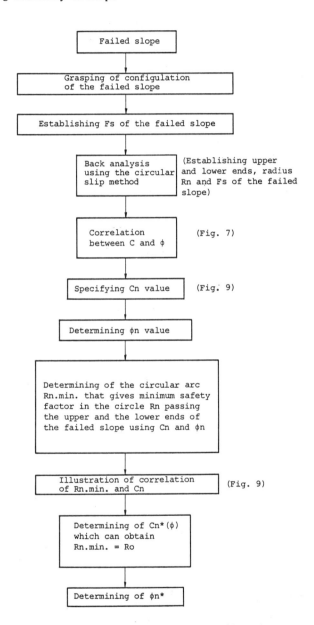

Fig.-8 Determining of C, ϕ of the Failed Slope

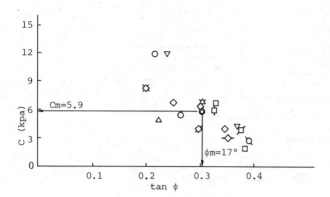

Fig.-10 C, φ Obtained by Back Analysis from
Failed Slope Examples

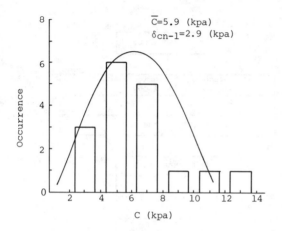

$\overline{C}$=5.9 (kpa)
δ_{cn-1}=2.9 (kpa)

Fig.-11(a) Distribution of Cohesion C of
Failed Slope

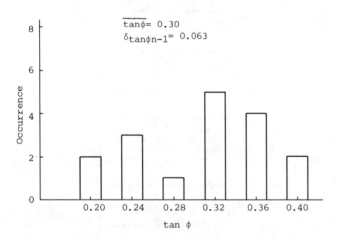

$\overline{\tan\phi}$= 0.30
$\delta_{\tan\phi n-1}$= 0.063

Fig.-11(b) Distribution of Angle of Friction φ
of Failed Slopes

5 ECONOMICAL DESIGN AND CONSTRUCTION OF SLOPE

In the maintenance of the failed slopes and the
design of the slope to be excavated in the
following year, an investigation of numerous
examples of failed slopes and site experiments on
construction methods for the stabilization of
slopes using a slope gradient of 1 : 1.5 were
performed, and comprehensive judgement was made
taking into account the reliability, cost,
workability, construction period, safety in
construction, etc., of each construction method
(Table-3). As a result, it was found that the most
suitable construction method for slopes lower than

10 m is the one of which the slope is reexcavated,
relieving the swelling energy of the natural
ground, and banking performed (Fig-12). For slopes
higher than 10 m the most economical design was
considered to be the one in which the slope is cut
on the basis of a gradient of 1 : 2.5 and
maintained by repairing the slight failure. The
excavation was performed based on this policy.

About 3 years have elapsed since the construction
of the slope, but no major failures have occurred,
verifying that the above engineering judgement was
not erroneous.

Table-3. Evaluation of Slope Protection Work
(for 5 to 15m height slope)

	① Reliability	② Economy	③ Work-ability	Construc-tion ④ period	⑤ Safety	Relative Composite evaluation
	100-failure ratio	relative value (%)	1 - 5	1 - 5	1 - 5	①×②×③ ④×⑤ ×1/100
Anchar pin (1.5m)	12.2	81.3	4	5	4	79
Anchar pin (2.0m)	25.0	74.0	3	5	3	83
Anchor pin + wire	0.5	76.1	4	4	5	8
Concrete grid frame	40.9	14.6	2	2	3	7
Quilt cage drain cage	64.3	51.9	3	2	4	80
Replacement with gravel	21.5	79.6	3	4	4	82
Re-banking	93.6	48.7	3	4	4	219
Restoration	0.5	97.9	5	5	5	6

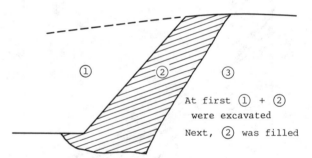

At first ① + ② were excavated

Next, ② was filled

Fig.-12 Banking stabilization method for slopes
lower than 10 m

6 CONCLUSION

In the design and construction of this large scale
mudstone cut slope, the statistical characteristics
of the configurations of the failures, based on the
numerous slope failures that occurred at the site
were obtained. From the correlation of C and φ and
the configuration of the failed slope, a method was
proposed in which C and φ can be specified, Cm and
φm values were obtained and applied to the slope to
be excavated, and the design and construction of
the slope were performed taking into account the
maintenance cost when the slopes fail, so as to
minimize the total cost. The results of a 3 year
observation verified the adequacy of the
construction method proposed here.

REFERENCE

D.A.Salcado and F.H.Tinoco; A rock slide in an
urban area: A case history, the International
Society of Rock Mechanics. 1983, C73-81.
H.George; Characteristics of varved clays of the
Elk Valley, British Columbia, CANADA, Engineering
Geology, 23, 1986 p.59-74.

562

Empirical failure criterion for a jointed anisotropic rock mass
Critère empirique de rupture d'un massif rocheux fissuré et anisotrope
Empirisches Bruchkriterium für ein klüftiges und anisotropisches Felsmassiv

K.THIEL, Institute of Hydroengineering of the Polish Academy of Sciences, Gdańsk
L.ZABUSKI, Institute of Hydroengineering of the Polish Academy of Sciences, Gdańsk

ABSTRACT: A criterion for failure of a jointed rock mass built up of rock that shows shear and tensile strength anisotropy is proposed. The criterion is based on the modified Coulomb condition, on the relationships describing the variation of cohesion and uniaxial tensile strength with direction, proposed by the authors, and on the shear failure criterion set out by Jaeger for a rock with a single joint. An example is given of how the criterion can be used for describing failure of rock mass built up of metamorphic mica schists and possessing two sets of joints. The rock mass considered was the foundation of a dam. The proposed criterion can be used in interpreting the results of strength tests carried out in connection with a stability analysis of the foundation of an engineering structure or of a slope.

RESUMÉ: On propose un critère de rupture d'un massif rocheux constitué des roches se caractérisant par und anisotropie de la résistance au cisaillement et à la traction. Ce critère est basé sur le critère de rupture modifié de Coulomb, la proposition de la variation de la cohésion et de la résistance à la traction uniaxiale en fonction de la direction, ainsi que sur le critère do rupture par cisaillement de la roche avec une fissure de Jaeger. On présente un exemple de l'application de ce critère pour décrire la rupture d'un massif rocheus composé des schistes métamorphiques à mica avec deux systèmes de fissures constituant la fondation d'un barrage. Le critère proposé peut être appliqué dans l'interprétation des résultats d'essais de résistance liés avec l'analyse de la stabilité de la fondation des barrages ou des talus rocheux.

ZUSAMMENFASSUNG: Das Bruchkriterium des klüftigen Felsmassives welches die anisotropische Scher und Zugfestigkeit hat vorgeschlagen wurde. Dieser Vorschlag auf dem modifizierten coulombischen Bruchkriterium, der eigenen Proposition der Beschreibung der Kohesion und der einachsialen Zugfestigkeit als Funktionen der Richtung, und das Jaeger-Scherkriterium von Fels mit einzelner Kluft, gegründet wurde. Ein Beispiel der Anwendung dieses Kriteriums für die Beschreibung des Bruches des metamorphischen Glimmerschiefers mit zwei Kluftsysteme, vorgestellt wurde. Das vorgeschlagene Bruchkriterium kann bei der Interpretation von Versuchsergebnissen der Festigkeit mit der Stabilitätanalyse von Dammboden und Böschungen gebundenen, benützt werden.

1. INTRODUCTION

The anisotropy of strength of a rock mass results from the lithological features of rock mass (lithogenic anisotropy - see Model I, Fig.1), and from the presence of sets of joints in it (tectonogenic anisotropy, Model II , Fig. 1).

When describing the lithogenic anisotropy we begin with empirical relationships developed on the basis of data obtained from studies on metamorphic mica shists carried out by the authors. This relationships describe how the cohesion and the uniaxial tensile strength vary with direction. To describe the tectonogenic anisotropy, on the other hand, the known Jaeger's criterion that defines the shear strength of a rock dissected by a single joint is used, in conjunction with the condition of limited tensile strength.

2. FAILURE CRITERIA FOR MODELS I AND II

The modified Coulomb criterion that takes account the anisotropy of the parameters of the shear and tensile strength has the form

a) Model I
- shear

$$\sigma_1 = [S_0(\alpha)(1+tg^2\beta) + \sigma_3 \cdot tg\beta(1+\mu(\alpha)\cdot tg\beta)] \cdot \frac{1}{tg\beta - \mu(\alpha)} \quad (1)$$

- tension

$$\sigma_1 + \sigma_3 \cdot tg^2\beta + T_0(\alpha)(1+tg^2\beta) = 0 \quad (2)$$

where: $S_0(\alpha), \mu(\alpha), T_0(\alpha)$ are the cohesion, the coefficient of internal friction, and the uniaxial tensile strength, respectively, of the rock; the three quantities are expressed as functions of the angle α (for α see Fig.1).

We can write the criteria (1) and (2) in explicit forms by substituting empirical formulae for $S_0(\alpha)$, $\mu(\alpha)$, and $T_0(\alpha)$. Jaeger (1960) gives such a formula for the cohesion

$$S_0(\alpha) = S_1 - S_2 \cos 2\alpha \quad (3)$$

where S_1 and S_2 are empirical constants.

McLamore and Gray (in Goodman, 1980) have also found a formula for the variation of the coefficient of internal friction, the

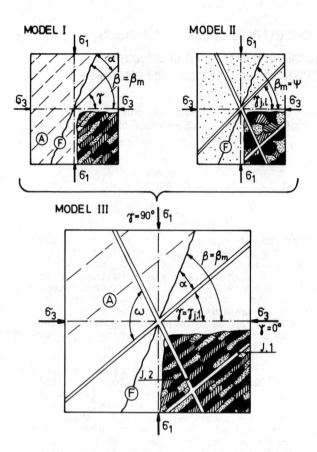

MODEL I MODEL II

MODEL III

Fig.1. Models of an anisotropic rock mass
Model I - lithogenic anisotropy, Model II -
tectonogenic anisotropy, Model III - combined
lithogenic and tectonogenic anisotropy
A - anisotropy surface in the rock, F -
failure surface, J.1, J.2 -sets of joints.

character of this variation being similar
to that of the variation of cohesion in Eq(3)

There are no however empirical relationships for $T_0(\alpha)$, and this restricts practically the range of use of these criteria
to compressive stresses.

b) Model II
- shear of rock

$$\sigma_1 = 2 \cdot S_0 \cdot tg\psi + \sigma_3(1 + 2 \cdot \mu \cdot tg\psi) \qquad (4)$$

where: $\psi = \pm (45° + \phi/2)$
S_0, μ - the cohesion and the coefficient of internal friction of
the rock, respectively, the quantities which are constant and
independent of direction.
- tension of rock

$$\sigma_3 = - T_{0,r} \qquad (5)$$

where: $T_{0,r}$ is the uniaxial tensile
strength of the rock, constant and
independent of direction,
- shear along the i-th set of joints

$$\sigma_1 = [S_{0,j,i}(1 + tg^2\gamma_{j,i}) + \sigma_3 \cdot tg\gamma_{j,i}(1 + \mu_{j,i} \cdot tg\gamma_{j,i})]\frac{1}{tg\gamma_{j,i} - \mu_{j,i}} \quad (6)$$

where: $S_{0,j,i}$ - the cohesion of a joint
in the i-th set,
$\mu_{j,i}$ - the friction coefficient
of a joint in the i-th set

$\gamma_{j,i}$ - see Fig.1,
- tension along the i-th set of joints:

$$\sigma_1 + \sigma_3 \cdot tg^2\gamma_{j,i} + T_{0,j,i}(1 + tg^2\gamma_{j,i}) = 0 \qquad (7)$$

where: $T_{0,j,i}$ - the tensile strength of
a joint in the i-th set.

It follows from equations (4) and (6),
which are the Jaeger criteria, that failure
may occur through shearing off the rock or
through a slide along the joints. Taking in
addition equations (5) and (7) into account,
we consider the possibility of extension
fracture through the body of the rock or
along the joints.

3. PROPOSED CRITERION OF FAILURE FOR MODEL
III.

The general form of the criterion has been
obtained by replacing equations (4) and (5)
in model II by equations (1) and (2) , which
is equivalent to replacing the rock with
isotropic strength represented by model I
by a rock whose strength is anisotropic.
The general form of the failure criterion
thus obtained for a rock mass showing both
lithogenic- and tectonogenic strength anisotropy comprises equations (1) , (2) , (6),
and (7), cf. Zabuski (1986).

The explicit expression for $S_0(\alpha)$ in equation (1) has been derived from the results
of direct shear tests made on mica schist
specimens of 40x60x40 cm. The specimens
were sheared in parallel, at an angle of 45°
and perpendicularly to the foliation (i.e.
$\alpha = 0°$, 45°, 90°). This expression is

$$S_0(\alpha) = A_i + B_i\sqrt{\alpha} \qquad (8a)$$

where: A_i and B_i - empirical constants,
$i = 1,...,n$, where n is the number of
the intervals into which the α - variation
range, from 0° to 90° , is divided in
order to make the curve fit better the
experimental results; in practice n=1 or 2.

For the schist examined equation (8a) has
the form (cf. Fig.2)

$$S_{0,1}(\alpha) = 0.083 + 0.032\sqrt{\alpha} \text{ , MPa , } \alpha \in \langle 0°, 45° \rangle$$
$$S_{0,2}(\alpha) = 0.146 + 0.0225\sqrt{\alpha} \text{ , MPa , } \alpha \in \langle 45°, 90° \rangle \qquad (8b)$$

The assumption that $S_0(\alpha)$ varies according to formula (8a) is supported by the
observed fact that, if the shear occurs
along the foliation ($\alpha = 0$), the individual
particles of micaceous minerals slide one
upon another, while in shear effected in a
direction inclined to the foliation, even at
a small angle, the particles themselves are
sheared. Therefore, since the mechanisms of
shear in two cases are different, one can
expect that the change from $\alpha = 0$ to an α
only slightly greater than zero will result
in a considerable increase of cohesion. At
greater values of α the shear mechanism
remains unchanged with changes of α and,
consequently, S_0 varies within a narrow
range and its variations are not violent. In
Fig.2, ΔS_0 for the change of from 0° to 45°
is about 0.215 MPa, whereas at greater

564

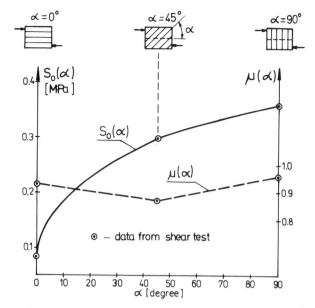

Fig.2. Shear strength parameters of the rock, plotted as a function of the angle α

values of α and identical increase in α, from 45° to 90°, gives ΔS_0 as small as about 0.062 MPa.

The results of the studies on shear also indicate that the value of the internal friction coefficient is approximately constant, irrespective of α (cf. Fig.2), and thus $\mu(\alpha) = const$ has been taken.

With the parameters $S_0(\alpha)$ and $\mu(\alpha)$ so defined, the criterion (1) can be put in the form

$$\sigma_1 = [(A_i + B_i\sqrt{|\beta_m - \gamma|})(1 + tg^2\beta_m) + \sigma_3 \cdot tg\beta_m(1 + \mu \cdot tg\beta_m)]\frac{1}{tg\beta_m - \mu} \quad (9)$$

where: $|\beta_m - \gamma| = \alpha$, cf. Fig.1.

The tensile strength of the rock has been determined using the method of transverse compression at an angle of 0°, 45°, 90° with respect to the foliation of the rock. $T_0 = T_0(\alpha)$ has been taken to vary in the way defined by the function (curve 1 in Fig.3):

$$T_0(\alpha) = T_{0,0} + K \cdot \alpha \quad (10a)$$

where: $T_{0,0}$ - uniaxial tensile strength in the plane of foliation,
K - empirical coefficient,
For the schists examined we have obtained

$$T_0(\alpha) = 0.0716 + 0.0283 \cdot \alpha, \text{ MPa} \quad (10b)$$

In the studies on tests mentioned above, it has been observed that when the load is applied at an angle of 45° with respect to the foliation, the rock was not fractured but sheared. This suggests that for $\alpha = 45°$ the tensile strength is greater than that found from the studies (see points for $(90° - \gamma) = 45°$ in Fig.3), and thus the theoretical line (10b) lies above these points. To verify the hypothetical function (10a), and its special case (10b), the straight line 1 has been compared with the curves obtained by Barla (1974) for rocks whose degree of the tensile strength anisotropy $T_{0,90°}/T_{0,0°}$ is close to that for mica schists. Curves 2 and 3 in Fig.3 represent the tensile strength plotted as a function of the angle $(90° - \gamma)$.
If the value of the angle $(90° - \gamma)$ ranges from 0° to about 50°, then extension frac-

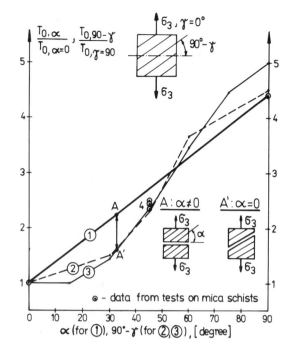

Fig.3. Uniaxial tensile strength as a function of the angle $\alpha(1)$, or of the angle $(90° - \gamma)$ ((2), (3)).
1 - relationship for $T_0(\alpha)$, 2 - tests on gneiss, 3 - tests on serpentinite schist.

ture occurs at $\alpha = 0°$, and not at $\alpha \neq 0°$ marked on the horizontal axis of the diagram. This indicates that the tensile strength at an angle $\alpha \neq 0°$ must be greater than that found experimentally and represented by curves 2 and 3. This is why the theoretical curve 1 lies above curves 2 and 3. If the angle $(90° - \gamma)$ approaches 90°, then $\alpha = (90° - \gamma)$, and curves 1, 2 and 3 may be positioned close one to another.

Taking $T_0 = T_0(\alpha)$ as defined by equation (10a), the criterion (2) has the form

$$\sigma_1 + \sigma_3 \cdot tg^2\beta_m + (T_{0,0} + K|\beta_m - \gamma|) \cdot (1 + tg^2\beta_m) = 0 \quad (11)$$

where: $|\beta_m - \gamma| = \alpha$, cf. Fig.1.

Equations (9) and (11) define the resultant limit surface for a rock with lithogenic anisotropy. However, when $\beta_m \neq \gamma$, then it is not possible to determine the failure direction and thus the strength analytically. The angle β_m should be determined numerically, by seeking a minimum (equation (9), equation (11)) of σ_1 for various values of β and constant values of σ_3 and γ, or by seeking numerically such $\beta = \beta_m$ that satisfies the condition $d\sigma_1/d\beta = 0$ and then substituting β_m thus found in equation (9) or (11).

Assuming in criterion (9) that $\beta_m = \gamma$ or $\beta_m = \psi = 45° + \phi/2$ gives a simplified form of this criterion. Fig.4. shows the sections of the limit surface (9) by the planes $\gamma = const$ and $\sigma_3 = const$. Curves I, Ia, Ib, Ic correspond to shear along the foliation and curves II, IIa, IIb to shear at the angle $\beta_m = \psi$. Curve III in Fig.4a represents the approximate section of the limit surface by the plane $\gamma = const$ for the condition $\beta_m \neq \psi$ and $\beta \neq \gamma$. It can be seen from the above that the surface determined for $\beta_m = \psi$ is the upper limit of the actual surface III. Numerical calculations have

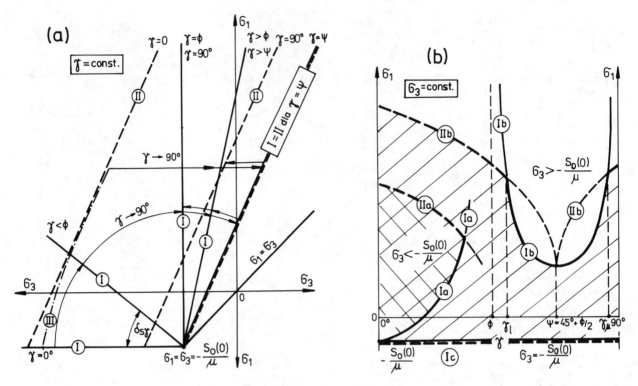

Fig.4. The limit surface for an anisotropic rock subjected to shear, referred to coordinates σ_1, σ_3, γ ; a) sections of the limit surface by the planes γ = const; b) sections of the limit surface by the planes σ_3 = const. I,Ia,Ib,Ic - shear along the surface of anisotropy ($\beta_m = \gamma$), II, IIa, IIb - shear transversly to the surface of anisotropy, at an angle $\beta_m = \psi$ III - shear along surfaces $\beta_m \neq \gamma$ and $\beta_m \neq \psi$. ▨ safe region for $\sigma_3 > -S_0(0)/\mu$ ▨ safe region for $\sigma_3 < -S_0(0)/\mu$

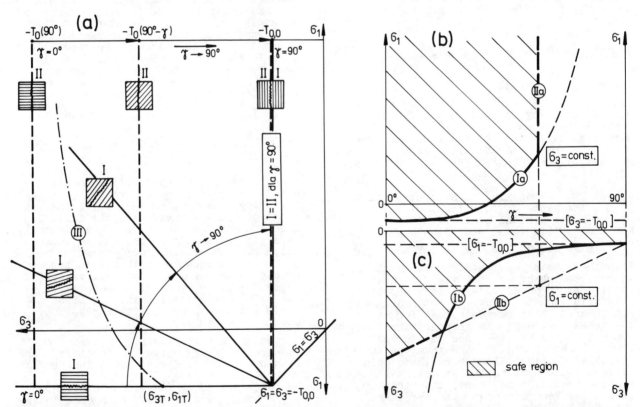

Fig.5. The limit surface for an anisotropic rock subjected to tension, referred to coordinates, σ_1, σ_3, γ ; a) sections of the limit surface by the planes γ = const. b) section of the limit surface by the plane σ_3 = const; c) section of the limit surface by the plane σ_1 = const. I, Ia, Ib - fracture along the surface of anisotropy $\beta_m = \gamma$, II,IIa,IIb - fracture transversely to the surface of anisotropy, at an angle $\beta_m = 90°$, III - fracture along surfaces $\beta_m \neq 90°$.

shown that the maximum difference between 6_1, found for $\beta_m = \psi$, and the actual 6_1 does not exceed ca. 5% so that in practical calculations we can use the simplified limit surface.

Fig.5 shows the sections that the planes $\gamma = $ const, $6_3 = $ const, $6_1 = $ const cut through the limit surface defined by (11). Also in this case curves I,Ia, Ib define the limit surface that corresponds to extension fracture along the foliation, whereas curves II, IIa, IIb define the limit surface corresponding to fracture at the angle $\beta_m = 90°$. Here the surface found for $\beta_m = 90°$ is the lower limit of the actual surface (i.e. for $\beta_m = 90°$) whose section by the plane $\gamma = $ const is shown schematically by curve III.

Numerical studies for various $T_{0,0}$ and K have shown that the value of the angle β_m in equation (11) can be found from the relationships

$$\beta_m(6_{3,\gamma}) = \frac{6_{3,\text{II}} - 6_3}{6_{3T} - 6_{3,\text{II}}} \cdot (23.2° + 0.482 \cdot \gamma) + 90°, \ \gamma \in \langle 0°, 45° \rangle, \ (12a)$$

$$\beta_m(6_{3,\gamma}) = \frac{6_{3,\text{II}} - 6_3}{6_{3T} - 6_{3,\text{II}}} \cdot (89.8° - 0.9978 \cdot \gamma) + 90°, \ \gamma \in \langle 45°, 90° \rangle, \ (12b)$$

where: $6_{3,\text{II}} = - \left[T_{0,0} + K(90° - \gamma) \right]$

$$6_{3T} = - T_{0,0} + K \cdot (\beta_{mT} - \gamma) \frac{1 + tg^2 \beta_{mT}}{tg^2 \gamma - tg^2 \beta_{mT}}$$

$$\beta_{mT} = 66.8° - 0.482 \cdot \gamma \ , \ \gamma \in \langle 0°, 45° \rangle \qquad (12c)$$

$$\beta_{mT} = 0.2° + 0.9978 \cdot \gamma \ , \ \gamma \in \langle 45°, 90° \rangle \qquad (12d)$$

Equations (12a) and (12b) have been set out based on two facts, namely, that the angle β_{mT} defining the direction in which the rock is fractured at the point 6_{3T} (cf. Fig. 5) is dependent only on γ (equations (12c) and (12d)) and that the angle β_m for a given γ is a linear function of 6_3, that is this angle increases linearly from the value β_{mT} to 90°, as 6_3 decreases from 6_{3T} to $6_{3,\text{II}}$.

The criteria (9) and (11) established above predict the following properties of a lithogenetically anisotropic rock:

a) for $6_3 \geqslant 0$:

- the compressive strength Q, in particular the uniaxial compressive strength Q_0, is always maximum when $\gamma = 0°$ and is minimum when $\gamma = \psi = 45° + \phi/2$,
- the index of anisotropy as defined by $Q_{\gamma=0}/Q_{\gamma=\psi}$ is maximum when $6_3 \simeq 0$ and decreases with increasing 6_3: if $6_3 \rightarrow +\infty$ then the index of anisotropy tends to unity,
- the direction in which the rock is sheared coincides with the direction of the foliation or is close to ψ ($\beta_m \approx \psi \pm 10$); the value of β_m tends to ψ as $6_3 \rightarrow +\infty$

b) for $6_3 < 0$:

- over a relatively wide range of γ, failure may be initiated by shear and not by fracture,
- the direction of failure through fracture coincides with the direction of the foliation or is inclined to it, with β_m not equal to 90°, except when $6_1 = +\infty$.

In considering the tectonogenic anisotropy it has been assumed that generally a rock is dissected by n sets of joints. The shear strength of the i-th set of joints is given by equation (6) where, with the reference system as in Fig.1 if the angle $\gamma_{j,i}$ is positioned in the 2nd quadrant, the system is represented in the 1st quadrant so that the angle defining its orientation is $\gamma^*_{j,i} = (180° - \gamma_{j,i})$, The criterion for fracture along the i-th set of joints is given by (7) where, if $\gamma_{j,i} \in (90°, 180°)$, then $\gamma^*_{j,i} = (180° - \gamma_{j,i})$

4. EXAMPLE OF USING THE CRITERION FOR A DESCRIPTION OF FAILURE OF MICA SCHISTS

We shall consider a rock mass built up of mica schists in which the parameter $S_0(\alpha)$ varies according to (8b), $\mu = 0.9227$, and the parameter $T_0(\alpha)$ varies according to (10b), and which is dissected by two sets of joints J.1 and J.2 whose strikes are approximately parallel to the strike of the surface of the rock foliation. The joints in the two sets have identical strength parameters $S_{0,j,i} = 0.0$, $\mu_{j,i} = 0.577$ and $T_{0,j,i} = 0.0$.

The angle between the J.1 and J.2 is $\omega = 67°$ (cf. Fig.1).

Individual segments of the limit surface have been determined from (6), (7), (9), and (11). By superposing these segments we obtain the resulting limit surface referred to the axes $6_1, 6_3, \gamma$.

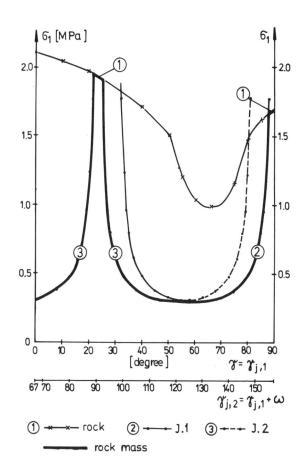

Fig.6. Section of the limit surface for the strength of jointed mica schists by the plane $6_3 = +0.1$ MPa.

As an example, Fig.6 shows the section of this surface by the plane $\sigma_3 = + 0.1$ MPa. It can be seen that the strength of the rock mass is determined first of all by its tectonogenic anisotropy and that in this case over about 90% of the γ range ($0°$ ÷ $90°$), failure occurs by shear along one of the two sets of joints. This percentage decreases with increasing σ_3 and, for $\sigma_3 = + \infty$, which in practice corresponds to $\sigma_3 >> S_0(\alpha)$, shear along the joints occurs over about 70% of the γ - range.

5. CONCLUSIONS

The failure criterion proposed in this paper for an anisotropic rock mass permits us to find the strength, the direction of possible failure, and also the mechnism of failure by shear or fracture along or across the surface of joints or along or across the surface of foliation, layering, schistosity, etc. in all possible states of stress. To use the criterion one needs to know the strength parameters of the rock and of the joints, but these parameters can be determined by relatively simple strength tests.

The proposed failure criterion for an anisotropic rock describes well the strength and failure conditions of a rock in which planes of weakness (e.g. foliation or schistosity planes) are clearly marked.

The criterion presented here has been verified by the results of investigation on mica schits. It would be useful to verify it by tests on other rock of this type. The authors suggest that this criterion can be applied to a wide group of lithogentically and tectogenetically anisotropic rock masses.

REFERENCES

Barla G. (1974) . Rock Anisotropy - Theory and Laboratory Testing, In: L. Müller (ed.), Rock Mechanics, Courses and Lectures, No. 165, Udine 1974, Springer-Verlag, Wien New York, pp. 35-69.
Goodman R.E. (1980). Introduction to Rock Mechanics, John Wiley and Sons, New York - Chichester-Brisbane-Toronto
Jaeger J.C. (1960). Shear Failure of Anisotropic Rocks, Geol. Mag. No. 97, pp.65-72.
Jaeger J.C., Cook N.G.W. (1969).Fundamentals of Rock Mechanics,Chapman and Hall Ltd., Science Paperbacks, London.
Zabuski L. (1986) Failure criterion for a jointed anisotropic rock mass, Inst. of Hydroengineering of the Polish Academy of Sciences Gdańsk (PhD Thesis) (Kryterium zniszczenia spękanego masywu skalnego o anizotropowych własnościach wytrzymałościowych).

Acoustic emission monitoring on foundation grouting for rock masses
Contrôle d'émission acoustique d'étanchement par injections de masses rocheuses
Akustische Emissionsüberwachung von Fundamentguss für Gesteinsmassen

TAKAO UEDA, Takenaka Technical Research Laboratory, Tokyo, Japan
HIDEHIKO NAKASAKI, Takenaka Technical Research Laboratory, Tokyo, Japan

ABSTRACT: As a monitoring technique of behavior of rock masses during grouting, an acoustic emission method was introduced, and field tests were conducted at various dam sites. As a result, it was found that the method was effective in detecting the behavior of cracks in the rock masses and particularly in detecting hydrofracture at the time of grouting, and newly a grouting control technique employing the AE method has been proposed.

RÉSUMÉ: En tant que technique de contrôle de perturbation des masses rocheuses lorsque'un coulis leur est injecté, une méthode d'émission acoustique a été introduite et des essais sur le terrain ont été effectués sur divers barrages. En tant que résultat, il s'est avéré que cette méthod est efficace car elle permet de comprendre le comportement des fissures dans les masses rocheuses et en particulier de détecter les hydrofractures au moment du coulage, ce qui fait que récemment, une technique de contrôle du coulage avec la méthod AE a été proposée.

ZUSAMMENFASSUNG: Als eine Technik zur Überwachung von Störung in Gesteinsmassen beim Abpressen wird eine akustische Emissionsmethode vorgestellt, mit der an verschiedenen Dämmen Versuche vor Ort angestellt wurden. Diese Methode hat sich für die Erfassung des Verhaltens von Rissen in gesteinsmassen als wirkungsvoll erwiesen, insbesondere bei der Erfassung des Hydrofrac während des Abpressens. Fernerhin wurde eine neue abpreß-Kontrolltechnik unter Anwendung der AE-Methode vorgeschlagen.

1. INTRODUCTION

In spite of the fact that needs for the dam construction in Japan are still very high in recent years, suitable ground for it is decreasing in number. As a result, it has become necessary to construct dams on foundation ground such as soft or fractured rock masses, which was considered unsuitable for previously for dam construction. At such rock-masses, grouting technique which controls the seepage flow becomes important, although geological conditions are frequently complicated, and setting-up of appropriate grouting conditions is difficult. If grout injection pressure is raised too high in order to enhance grouting efficiency, rock-masses may fracture and a resulting adverse effect to the structure is unavoidable. There is also a possibility of causing an unforeseen accident due to leaking-out of grout.

In view of the above, the authors directed their attention to the acoustic emission (AE) method as a means of monitoring the behavior of rock masses during grouting, and have been examining the proper applicability of the AE method through laboratory and field tests. In this paper, the authors paid attention to water pressure tests, examined the relation between injection pressure – injected quantity – AE occurrence rate, and attempted to classify grouting characteristics from the above relationship. Further a reference was made to the grouting control technique which applied the AE method to rock mass grouting.

2. PROPERTIES OF FOUNDATION ROCK AND INJECTION CHARACTERISTICS

2.1 Classification of Injection Characteristics through Water Pressure Test

Grout injection characteristics are seemed to be greatly dependent upon the crack characteristics of the foundation rock masses. In particular, it is necessary to clarify the effects exercised by the distribution, development condition and properties of cracks as well as the presence or absence of void-filling material. F.K. Ewert attempted to classify grouting characteristics, when grout injection is given in fractured rock masses (Ewert 1985). Here, authors substituted his concept with a simplified model and obtained classification of types such as those mentioned below.

Fig. 1 shows the schematical relationship between injection pressure and injected quantity during the water pressure test. Here denote maximum injection pressure applied and the maximum injected quantity per time which is obtained during the grouting pressure application by Pmax and Qmax respectively, and also denote each injection pressure and the injected quantities per time which have been obtained up to then by non-dimensional quantities of P/Pmax and Q/Qmax respectively. Then grouting characteristics can be broadly classified into types shown in Fig. 2. Features of these types are shown next.

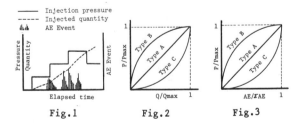

Fig. 1 Schematic P-Q-AE relationship during water pressure test.
Fig. 2 Idealized curves of non-dimensional P-Q relationships.
Fig. 3 Idealized curves of non-dimensional P-AE relationships.

Type A:

Injection pressure and the injected quantity per time are in a proportional relationship, which suggests that the flow is a streamline flow following Darcy's law. The surface of discontinuity is slightly open, but the rock masses have high elasticity, and their tensile strength is considered to be high.

Type B:

When injection pressure exceeds a certain pressure level, the injected quantity suddenly increases. The reasons for this are expansion and progress of cracks in the rock masses, or wash out, erosion, etc., of void-filling materials.

Type C:

Conversely to Type B, this type shows clogging of cracks when pressure becomes higher, and suggests the occurrences of clogging at the tips of cracks, blocking of the flow which is due to capillary force, etc.

Among the above types, Type B requires careful attention in grouting control, because this type is liable to induce hydrofracture and upheaval of rock masses which may result in harmful influence to the superstructure. However, such phenomena also liable to occur as a result of washing out of intrafracture void-filling materials, etc. In order to discriminate between these two types of discrepancies, careful examination on the P–Q curve is required.

2.2 Rock Mass Injection Characteristics viewed from Acoustic Emission

If AE measurement is used in combination with the above-mentioned water pressure test, it is possible to evaluate the rock mass behavior during grouting not only simply from the relation between injection pressure and the injected quantity, but also from the number of occurred acoustic emission, thereby obtaining a more effective means. In the same way as with the P–Q curve, therefore, grouting types can be classified by introducing non-dimensional quantity which is obtained by dividing the cumulative number of acoustic emissions (number of events), which are occurred in respective pressure stages, by the total number ΣAE which has been occurred up to the maximum pressure stage. In this way, the grouting types can be given by curves shown in Fig. 3 in the same way as before. Features of the respective types are given below.

Type A:

AE occurrence rate is constant with respect to grout injection pressure. Rock masses have high elasticity, and expansion of cracks is little. Thus the rock masses are not easily affected by injection pressure.

Type B:

The AE occurrence suddenly increases, when a certain pressure level is exceeded, thereby indicating the occurrence of the hydrofracture such as initiation and expansion of cracks.

Type C:

Comparatively greater AE occurrence is observed at the initial stage of pressure application. Although rock masses contain weak points at cracks, stabilization trends of rock masses appear, as injection pressure increases.

Now on the basis of Figs. 2 and 3, three parameters, P/Pmax, Q/Qmax and AE/ΣAE were combined and injection characteristics were classified. The result is shown in Table 1. Features of the respective types are shown in the Table. By the P–Q control method

currently adopted, it was difficult to judge Types 2 and 3. This was attributable to the fact that an increase in the flow rate accompanying the pressure rise includes an increase due to the progress of cracks and an increase due to wash out of void-filling materials. However, if the AE method is used in combination, it is possible to detect such features that in the case of the former flow rate increase, the number of AE conspicuously increases, whereas in the case of latter flow rate increase, no significant changes appear in the number of AE occurrence. Thus the behavior of intra-rock fracture with respect to grout injection has become clear.

2.3 Limit Injection Pressure

At present, there are two concepts about the limit injection pressure. In the first concept, the load only above the injection depth is taken as resisting force, whereas in the second concept, both of the overburden load, and the strength of the rock mass itself are considered as resisting force. If the above two concepts are sorted out while paying attention to the intra-rock fracture, the following discussion will become possible:

Now let's consider a most simplified model shown in Fig. 4, which is a rock mass where pre-existing fracture with a length of a exists at the periphery of the injection hole. In such a case, the fracture expanding pressure (P_b) will be given by the following equation, derived from fracture mechanics (Abe 1983).

$$P_b = \frac{K_1c}{f_2 \sqrt{\pi a}} + \sigma_3 \qquad (1)$$

here, K_1c : fracture toughness of rock
f_2 : coefficient given by function of a/b
b : radious of injection hole
σ_3 : minor principal stress

Otherwise, in the case of nonexistence of crack, limit pressure calculated by two-dimensional theory of elasticity is given as follows, for the permeable rock masses (Wong 1973).

$$P_b = \left(\frac{1-\nu}{1-N\nu}\right)(2\sigma_3+T) \qquad (2)$$

here, ν : poisson ratio of rock
N : proportion of fluid force used in expanding the hole
T : tensile stress of rock mass

Fig. 5 schematically shows the relation between the two equations on the basis of tests, etc., carried out by the authors (Ueda, et.al. 1986). At impermeable rock masses (when N = 1) where cracks are barely found, limit injection pressure to the degree of "minor principal stress plus tensile strength of rock" is demonstrated, but when "a→∞" is valid

Table 1 Classification of injection characteristics.

Types	Non-dimensional P-Q-AE relationships	Features	Illustrations	Rock types
1		**Stable injection** — Injection pressure, injected quantity and AE occurrence rate are in a proportional relationship, and rock mass keeps stable.		medium to hard rock mass
2		**Hydrofracture** — Both of injected quantity and AE occurrence rate increase at a certain pressure level.		soft rock, rock mass with joints or partly opend crack
3		**Wash-out or Erosion** — Injected quantity increases as pressure becomes higher, though AE occurrence rate doesn't change much.		jointed rock mass with void filling materials
4		**Clogging** — As injection pressure becomes higher, injected quantity and AE occurrence rate decrease.		

Legend: ----- injection pressure - injected quantity
——— injection pressure - acoustic emission (AE)

570

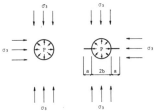

Fig.4 Schematic diagram of stresses acting around a injection hole.

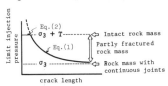

Fig.5 Limit injection pressure and pre-existing crack length.

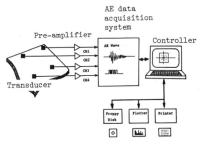

Fig.7 AE monitoring system.

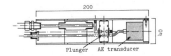

Fig.8 Detail of buried type AE transducer.

namely, when continuous cracks have developed, resisting force can be demonstrated only to the degree of the minor principal stress. In actuality, the limit injection pressure may assume an intermediate value depending upon the degree of development of the cracks. Therefore, if grouting efficiency is to be improved by raising injection pressure as high as possible, it will become necessary to obtain information about the degree of development of cracks in the rock mass, fracture toughness of the rock, etc. Further to prevent the occurrence of the hydrofracture which will lead to harmful deformation of rock masses in such pressure grouting, the AE method which is effective in detecting the above-mentioned phenomenon will increase its importance as an assistance to grouting control.

3. EXAMINATION BY LABORATORY TESTS AND FIELD MEASUREMENTS

3.1 AE Measuring Method

Fig. 6 shows a schematic diagram of the grout injection system and multi-channel AE measuring method. Figs. 7 and 8 show the configuration of the AE monitoring system and details of the newly-developed underground buried type sensor. For the sensor installation method in the field, the wave guide method and the underground burying method shown in Fig. 8 are used. In the type shown in Fig. 8, branches at the flanks of the sensor expand and stick closely to the hole wall. The authors use both types, but at the construction site, the underground buried type is more advantageous in eliminating the noise. Measurements are carried out mainly by the event (or ringdown count) measuring method, and sensors which are ordinarily used have frequencies ranging from about 10 to 30 kHz.

3.2 Laboratory Tests

To confirm whether or not the relation of injection pressure-injected quantity-AE occurrence rate shown in Table 1 represents the characteristics of grout injection to rock masses, laboratory tests were conducted prior to field tests. For the sample, a mudstone specimen of the rectangular pallalelepiped shape measuring 50 x 50 x 30 cm was used. After a grouting pipe of 3 cm in diameter was inserted into the center of the specimen and the packer was worked on it, pressure was applied in accordance with the procedure of the water pressure test. Fig. 9 (a) and (b) show the relationship of pressure-injected quantity-AE occurrence rate. Fig. 9 (a) shows the case in which hydrofracture occurred, and Fig. 9 (b) shows the result of the test on a specimen in which a mixture of gypsum and sand was injected into existing fracture so that the specimen would be in a state of reproducing the outflow of the so-called void-filling materials. In the case of the causing hydrofracture by injection, a sudden occurrence of AE and a simultaneously sudden increase in the injected quantity were observed, whereas in the case of the outflow phenomenon of the void-filling materials, no significant increase in AE was observed. Fig. 10 (a) and (b) show the relationship of non dimensional P-Q-AE. The difference between Types 2 and 3 shown in Table 1 can be clearly observed.

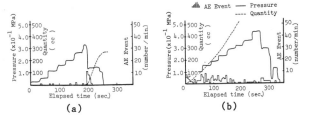

Fig.9 P-Q-AE relationships obtained by laboratory tests.

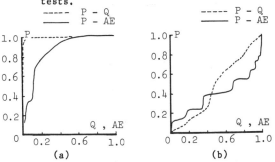

Fig.10 Non-dimensional P-Q-AE relationships.

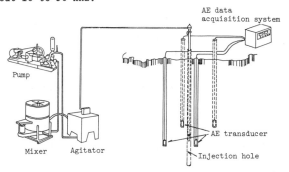

Fig.6 A conception of AE monitoring during grouting.

571

3.3 Field Measurements and Their Results

Table 2 shows measuring conditions at various dam sites which the authors so far conducted. Figs. 11 (a), (b), (c) and (d) show typical examples of the relation of pressure-injected quantity-AE occurrence rate for respective types. In particular, Fig. 11-(c) shows the results of the water pressure test conducted until the rock mass fracture occurred. In about 48 min. after commencing injection, significant AE occurrence was observed. The rock mass was fractured by water pressure and resultant leaking-out of water to the ground surface was recognized. Fig. 12 (a), (b) and (c) show the relation of non-dimensional P-Q-AE. At site N-13-1 shown in Fig. 11 (a), the rock was composed of comparatively harder tuffaceous breccia of the Tertiary Period and indicated a stabilized injection. This can be understood by the fact that the rock mass has sufficiently high limit injection pressure. At the site where O-E-11 was conducted, a great deal of sandy hornfels with developed cracks was observed and cover concrete was placed. Although no deformation of cover concrete was observed, a considerable quantity of injected water was found leaking from the peripheral part. Therefore, this rock was judged to belong to Type 3. These facts showed good correspondence to the type classification in Fig. 12, thereby suggesting that the classification in Table 1 was valid.

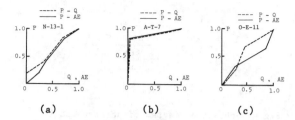

Fig.12 Non-dimensional P-Q-AE relationships.

4. CONCLUSION

The authors examined the AE method with the aim of introducing AE measurement into grout control and were able to confirm its effectiveness. How to adopt the AE measurement in the control of actual grouting process on the basis of the results of the abovementioned examination is exemplified in the flow chart shown in Fig. 13. First, grouting characteristics during the water pressure test at typical boreholes should be found out in order to obtain data regarding the water pressure property and limit injection pressure, etc., of the rock masses. Next, regarding the disturbance of the rock masses at the time of grout injection and injection condition, control reference values are set up on the basis of data obtained during the water pressure test, and then monitoring is started. To follow these steps is necessary, because the AE occurence condition greatly varies with the properties of rock masses, injection depth, etc., and without data gathered from the water pressure tests, it is difficult to obtain valid reference values to be used at the time of grouting.

Finally, the authors would like to express their deep appreciation to the staff concerned of Takenaka Doboku Co., Ltd. for their valuable cooperation given in the course of our obtaining the data mentioned in this paper.

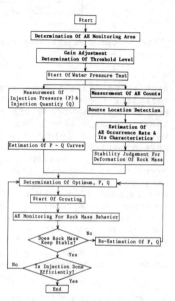

Fig.13 Flow chart of grout injection control technique by AE.

Table 2 Field monitoring sites and test conditions.

Trial No.	Location	Injection depth GL -m	Injection pressure x10⁻¹ MPa	Lugeon value	Rock type
N-13-1	N dam	5.4 - 11.2	1 → 2.5 → 1	4	Tuffaceous breccia
N-13-4		21.2 ∿ 27.5	1 → 6 → 1	14	
O-E-11	O dam	0.0 - 5.0	1 → 3 → 1	9	Sandy hornfels
A-T-7	A dam	3.15 - 8.15	0.5 → 3.5 → 1	1	Light tuff of the Quaternary Period

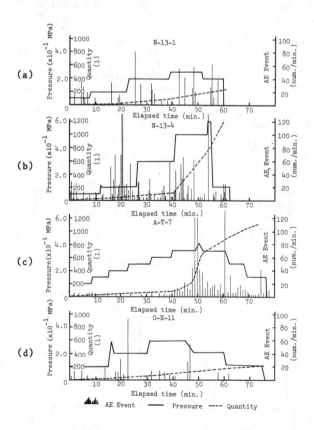

Fig.11 Typical test results of P-Q-AE relationships from field tests.

REFERENCES

Abe, H. & Takahashi, H., 1983. Crustal Fracture Mechanics for Geothermal Energy Extraction. Energy and Resources Vol. 4, No. 6, 515 - 522.

Ewert, F.K., 1985. Rock grouting. Springer Verlay, 276 - 299.

Ueda, T. et.al., 1986. Laboratory hydrofracture tests and AE characteristics for soft rocks. Proc. of the 41st Annual Conf. of JSCE. 3, 397 - 398.

Wong, H.Y. & Farmer, I.W., 1973. Hydrofracture mechanisms in rock during pressure grouting. Rock mechanics 5, 21 - 41.

Einfluss des Untergrundes auf die Spannungen von Bogenmauern nahe der Aufstandsfläche

Influence of the foundation on the stresses acting near the base of an arch dam
Les effets de la roche de fondation sur les contraintes agissant près de la base d'un barrage-voûte

R.WIDMANN, Tauernkraftwerke AG, Salzburg, Österreich
R.PROMPER, Tauernkraftwerke AG, Salzburg, Österreich

ABSTRACT: The stress field at the dam base of an arch dam is mainly determined through the deformations of the dam and its foundation. The studies deal with the problems involved in the selection of a representative mathematical model that allows a realistic calculation of the deformations at the dam base occurred as a consequence of the water load acting on the dam and the bottom of the reservoir. If the deformability of the foundation is lower than that of concrete, the relationship between the moduli of deformation of concrete and rock gains in importance. Areas exposed to tensile stresses should be excluded from load bearing. A strongly anisotropic behaviour of the foundation should also be taken into account, since through anisotropy forces and displacements may adopt different directions and thus considerably influence the stress pattern.

RESUME: Le champs des contraintes s'exercant sur la fondation d'un barrage-voûte est avant-tout déterminé par le comportement de déformation du barrage et du sous-sol. Les analyses se concentrent surtout sur les problèmes survenant lors du coix d'un modèle mathématique représentatif qui permette d'entreprendre un calcul réaliste des déformations dans la fondation qui sont provoquées par le poids de l'eau agissant sur le barrage de même que sur le fond du réservoir. Si le sous-sol se prête moins aux déformations que ne le fait le béton, la relation entre les moduli de déformation du béton et de la roche gagne en importance. Les zones exposées aux contraintes de traction ne devront généralement pas recevoir du poids. Une anisotropie très accentuée du sous-sol devrait également être prise en considération, étant donné que grâce à ce phénomène les forces et déplacements prennent des directions différentes et pourront donc jouer un rôle décisif dans la distribution des forces et contraintes.

ZUSAMMENFASSUNG: Das Spannungsfeld an der Aufstandsfläche von Bogenmauern wird wesentlich vom Verformungsverhalten von Mauer und Untergrund an der Kontaktfläche bestimmt. Die Untersuchungen behandeln die Probleme bei der Wahl eines repräsentativen mathematischen Modells für die wirklichkeitstreue Berechnung der Verformungen an der Aufstandsfläche zufolge der Wasserlast auf die Mauer und den Stauraumboden. Ist der Untergrund weniger verformbar als der Beton, gewinnt das Verhältnis der Verformungsmodeln von Beton und Fels an Bedeutung. Zugspannngsbereiche sollten grundsätzlich vom Mittragen ausgeschlossen werden. Eine stark ausgeprägte Anisotropie des Untergrundes sollte ebenfalls berücksichtigt werden, da die darauf zurückführende unterschiedliche Richtung von Kräften und Verschiebungen einen wesentlichen Einfluß auf die Spannungen haben kann.

1. EINFÜHRUNG

Die österreichische Gruppe der Felsmechaniker hat immer die Ansicht vertreten, daß große Felsbauten, seien es Hohlraumbauten oder Gründungen, als Einheit mit dem Fels, dem Gebirge betrachtet werden müssen. Selbstverständlich gilt dies auch für Bogenmauern, die relativ hohe Kräfte auf den Felsuntergrund übertragen und meist entlang ihrer Aufstandsfläche Felsbereiche unterschiedlicher mechanischer Eigenschaften überqueren müssen.

Bis in das vorige Jahrzehnt lag der Schwerpunkt der Bemühungen um wirklichkeitstreue Berechnungsverfahren für Bogenmauern auf der Berechnung des Mauerkörpers selbst. Trotzdem gibt es für die Berechnung der meist doppelt gekrümmten Gewölbemauerschale veränderlicher Stärke und entlang des Umfanges veränderlicher Einspannung nach wie vor kein theoretisch exaktes Berechnungsverfahren. Allerdings stehen insbesondere seit der Möglichkeit des Einsatzes von leistungsfähigen Computern ausreichend genaue Näherungsverfahren, die auf verschiedenen Vereinfachungen aufbauen, zur Verfügung.

Die beiden gebräuchlichsten Verfahrensgruppen, das Lastaufteilungsverfahren und die Finite-Element-Method, beruhen auf der Zerlegung des Flächentragwerkes in leichter berechenbare Einzelelemente. Beim Lastaufteilungsverfahren sind die horizontale und vertikale Traglamellen, in deren Kreuzungspunkten möglichst alle sechs Verformungskomponenten, die bekanntlich zur Festlegung der Bewegung eines Punktes im Raum erforderlich sind, übereinstimmen müssen. Bei der Finite-Element-Method werden in den Knotenpunkten die drei Verschiebungen aus dem stationären Wert der potentiellen Energie ermittelt.

Für die Spannungsverteilung im Element müssen bei beiden Verfahren Annahmen getroffen werden. Beim Lastaufteilungsverfahren wird meist eine lineare Verteilung der Normalspannungen und der Torsionsschubspannungen sowie eine parabolische Verteilung der Schubspannungen zufolge der Querkräfte vorausgesetzt. Eine Berechnung nach der Theorie gekrümmter Träger oder der Scheibentheorie wäre ohne weiteres möglich, der Mehraufwand dürfte jedoch nach bisherigen theoretischen Vergleichsuntersuchungen kaum sinnvoll sein. Bei der Anwendung der Finite-Element-Method gibt es eine Vielzahl verschiedener Elemente, die nach der Membrantheorie, der Theorie der dünnen oder dicken Schale, oder auch als Volumselemente unter der Voraussetzung bestimmter Normal- und Schubspannungsverteilungen berechnet werden können. Bei der Beurteilung der erzielten Genauigkeit wird man jedoch auf allfällige Diskontinuitäten im Spannungsverlauf an den Grenzen von Element zu Element achten müssen. Auch hat die bisherige Erfahrung gezeigt, daß im Gegensatz zum Lastaufteilungsverfahren die Ergebnisse nicht unwesentlich von der Art der gewählten Elemente und der Netzteilung abhängen.

Hier interessiert jedoch die Berücksichtigung des Felsuntergrundes bei diesen Berechnungen. Beim Lastaufteilungsverfahren werden Federkonstante oder auch Verformungen in den Auflagerpunkten eingeführt, mit deren Berechnung wir uns im folgenden noch befassen wollen. Bei der Finite-Element-Method können in den Knoten zum Untergrund ebenfalls Federkonstante oder Verschiebungen angebracht werden. Die Finite-Element-Method bietet aber erstmals die Möglichkeit, Mauer

und Untergrund gemeinsam durch ein allerdings sehr
umfangreiches mathematisches Modell darzustellen, das
auch die Struktur und verschiedene mechanische Eigenschaften einzelner Bereiche des Untergrundes beinhaltet.

Damit sind aber auch schon die beiden grundsätzlichen Schwierigkeiten angedeutet, die der wirklich-
keitstreuen Berechnung des Untergrundes entgegenste-
hen:
- Die Formulierung eines die Struktur des Untergrundes wirklichkeitstreu darstellenden, mathematischen
  Modells, das mit einem vertretbaren Zeitaufwand berechnet werden kann,
- die Einführung wirklichkeitstreuer Materialgesetze
  für das Verhalten der verschiedenen Bereiche, aus
  denen der Untergrund aufgebaut ist und
- die Einführung repräsentativer Kennwerte für die
  Sickerströmung im Fels.
Schon die geometrische Darstellung des Felskörpers in
einem mathematischen Modell erfordert wesentliche
Vereinfachungen, da weder der Geologe die wirkliche
Struktur bis in die erforderlichen Tiefen verläßlich
angeben kann, noch der Statiker komplizierte räumliche Strukturen auch wirklich mit einem einigermaßen
vertretbaren Aufwand berechnen kann. Die praktischen
Möglichkeiten beschränken sich daher meist auf die
Annahme eines geschichteten Untergrundes mit "ver-
schmierten Klüften", wie es Herr Professor Wittke
nennt, dessen Schichtflächen zumeist auf weite Bereiche eben sind.

Die Materialgesetze für die physikalische Darstellung der unterschiedlichen Felskörper können mit noch
so vielen und aufwendigen Versuchen nur sehr näherungs-
weise erfaßt werden, da auch die größten in situ-Versuche nur ein relativ kleines Felsvolumen erfassen
können im Vergleich zu jenem, das später durch die
Talsperre tatsächlich beansprucht wird. Auch entsprechen die aus den üblichen Kurzzeitversuchen abgeleiteten Materialgesetze kaum jenen unter der in Wirklichkeit zyklischen Langzeitbelastung.

Es ist daher immer erforderlich, im Rahmen von Parameterstudien durch extreme, aber im Bereich des Möglichen liegende Annahmen die Auswirkungen der unum-
gänglichen Vereinfachungen abzugrenzen.

2. UNTERGRUNDVERFORMUNGEN ZUFOLGE BELASTUNG DER BOGENMAUER

Das erste Problem, das hier behandelt werden soll, be-
zieht sich auf die Möglichkeit der Berücksichtigung
der Untergrundverformungen zufolge der Kämpferkräfte
der Bogenmauer bei der Berechnung des Mauerkörpers.
Beim Lastaufteilungsverfahren erfolgt dies bekanntlich mit elastischen Federn in den Auflagerpunkten
der einzelnen Traglamellen, wobei eine Koppelung dieser Federn zwar über den Mauerkörper, nicht aber über
den Untergrund gegeben ist. Wird bei der Finite-Element-Method der Untergrund in das mathematische Modell
unmittelbar einbezogen, so ist eine Koppelung dieser
Federn auch über den Untergrund erreicht.

2.1. Zur Wahl des mathematischen Modells

Das mathematische Modell, das zunächst den Untersuchungen zugrundegelegt wird, ist ein übliches Netz
aus finiten Volumselementen (Abb. 1). Mit diesem Modell sind die beiden Grenzfälle leicht zu simulieren:
- Mit dem Netz wird das räumliche Verhalten des Untergrundes erfaßt. In der Natur würde dies etwa
  einem isotropen Untergrund entsprechen.
- Löst man die Knoten zwischen den einzelnen Scheiben
  des Untergrundes, so werden damit die voneinander
  unabhängig gedachten elastischen Federn durch unabhängige Scheiben im Untergrund simuliert. In der
  Natur würde dies etwa einem radial geschieferten
  Untergrund entsprechen, in dessen Schieferungsflä-
  chen keine Schubspannungen übertragen werden können.
Mit diesen beiden mathematischen Modellen wurden zwei
Annahmen für die Untergrundeigenschaften untersucht.

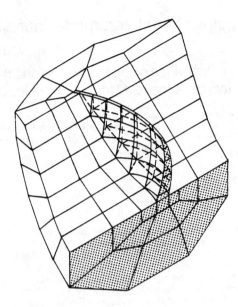

Abb. 1: Berechnungsnetz
        grid for analysis
        réseau de calcul

- Ein homogener Untergrund mit gleichen Verformungseigenschaften entlang der gesamten Aufstandsfläche
  und
- der Einfluß einer die Aufstandsfläche etwa radial
  querenden Schwächezone im Untergrund.
Die Ergebnisse derartiger Untersuchungen zeigen einen
zunächst überraschend geringen Einfluß der Vernach-
lässigung von Lastumlagerungen im Untergrund (Abb. 2).

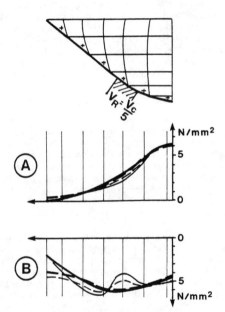

Abb. 2: Hauptspannungen zufolge Wasserlast
        A Wasserseite   B Luftseite
        ──── homogen   ── ── radial geschiefert
        ────   ── ── mit nachgiebiger Zone

        Principal stresses due to waterload
        A upstream   B downstream
        ──── homogeneous ── ── radially foliated
        ────   ── ── with yielding zone

        Contraintes principales causées par la
        pression de l'eau
        A amont   B aval
        ──── homogène   ── ── foliation radiale
        ────   ── ── avec zone élastique

574

Dieser Einfluß ist jedenfalls geringer als jener der Vereinfachungen in der geometrischen Darstellung der räumlichen Struktur des Untergrundes und in den Materialgesetzen. Dieser grundsätzliche Unterschied zur Berechnung der Auskleidung von Hohlraumbauten läßt sich einfach erklären.

Die Belastung der Auskleidung von Hohlraumbauten ergibt sich aus jenen Kräften, die nicht durch Lastumlagerungen im Fels von den verbleibenden Felsbereichen aufgenommen werden können. Die Kräfte werden daher umso kleiner, je mehr der Fels zum Mittragen herangezogen werden kann. Bei Gründungen hingegen ist die vom Untergrund aufzunehmende Gesamtbelastung gegeben. Durch unterschiedliche Nachgiebigkeit des Untergrundes entlang der Aufstandsfläche kann es Lastumlagerungen im Bauwerk, in der Mauerschale geben, die ja bei ausreichend feinem Berechnungsnetz erfaßt werden. Lastumlagerungen im Untergrund hingegen ändern nichts an der aufzunehmenden Gesamtbelastung und haben daher nur einen geringen Einfluß, der außerdem durch eine entsprechende Ermittlung der Federmatrizen weitgehend erfaßt werden kann.

Es dürfte daher im allgemeinen gerechtfertigt sein, den in Wirklichkeit räumlichen Untergrund durch voneinander unabhängige Federn in den Auflagerpunkten zu simulieren. Die Berechnung dieser Federmatrizen erfolgt meist unter Annahme eines elastisch isotropen Halbraumes, was wohl nur selten die tatsächlichen Verhältnisse auch nur halbwegs wirklichkeitstreu wiedergibt. Es bereitet jedoch keine Schwierigkeiten, diese Federmatrizen für die verschiedenen Belastungen unter Berücksichtigung einer vereinfachten Struktur, zum Beispiel für einen geschichteten Untergrund, zu ermitteln. Bei den Materialgesetzen wird auch zu berücksichtigen sein, daß die Aufnahme von Zugspannungen im Felskörper kaum gewährleistet werden kann.

2.2. Zur Berechnung der Federmatrizen

Bereits 1974 wurden Untersuchungen vorgelegt, in denen der Einfluß der Richtung von Kluftsystemen und der Topographie des luftseitigen Vorlandes ebenso wie jener der Ausschaltung von Zugspannungen behandelt wurde (Abb. 3).

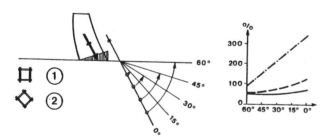

Abb. 3: Kämpferverformungen
Deformations of the abutment
Déformations sur la surface de fondation

	—·—·—	— — —
isotrop	Kluftsystem 1	Kluftsystem 2
isotrop	joint system 1	joint system 2
isotrop	système de joint 1	système de joint 2

Mit der Lage der Felsoberfläche im luftseitigen Vorland ändern sich auch die Spannungsverteilung im Untergrund und damit die Verformungen an der Aufstandsfläche. Dazu kommt noch der Einfluß der Struktur des Untergrundes, der in den damaligen Berechnungen homogen und isotrop oder mit zwei zueinander senkrechten Systemen von Kluftscharen, die entweder parallel und normal zur Aufstandsfläche oder jeweils unter 45 Grad zur Aufstandsfläche einfallend angenommen wurden. Obwohl der Verformungsmodul für diese Kluftscharen mit nur 1/50 von jenem des homogenen Felskörpers in

die Berechnung eingeführt wurde, hatten diese Kluftscharen nur einen sehr geringen Einfluß auf die Verschiebungen an der Aufstandsfläche. Wenn jedoch in diesen Kluftscharen, wie es ja auch der Wirklichkeit entsprechen dürfte, keine Zugspannungen aufgenommen werden können, dann ändern sich vor allem die Verschiebungen parallel zur Aufstandsfläche, insbesondere beim Kluftsystem parallel und normal zur Aufstandsfläche, ganz entscheidend.

Es dürfte also zweckmäßig sein, die Federkonstanten unter Ausschluß von Zugspannungen im Untergrund zu ermitteln. Als ungünstigster Grenzfall bietet sich hiefür der "Viertelraum" an, wobei eine Wasserlast auch auf die wasserseitige Begrenzung dieses Viertelraumes aufgebracht werden kann. Die Ergebnisse derartiger Untersuchungen zeigen im Vergleich zum Halbraum wieder einen geringen Einfluß auf die Vertikalverschiebungen (und damit auch auf die Verdrehungen von der Wasser- zur Luftseite), jedoch einen großen Einfluß auf die Horizontalverschiebungen. Die Annahme eines geschichteten Viertelraumes wird also in Übereinstimmung mit den früher erwähnten Untersuchungen zweifellos die größten Horizontalverschiebungen ergeben (Abb. 4).

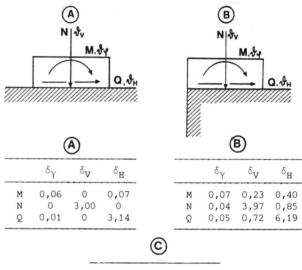

	δ_γ	δ_V	δ_H
M	0,06	0	0,07
N	0	3,00	0
Q	0,01	0	3,14

	δ_γ	δ_V	δ_H
M	0,07	0,23	0,40
N	0,04	3,97	0,85
Q	0,05	0,72	6,19

	δ_γ	δ_V	δ_H
M	1,17	–	–
N	–	1,32	–
Q	–	–	1,97

Abb. 4: Kämpferverformungen
    A   Halbraum
    B   Viertelraum
    C   Verhältnis B:A

Deformations of the abutment
    A   semi-space
    B   quarter-space
    C   relation   B:A

Déformations sur la surface de fondation
    A   demi-espace
    B   quart d'espace
    C   relation   B:A

Beim Vergleich gerechneter und gemessener Verschiebungen an der Aufstandsfläche können unterschiedliche Richtungen in einer ausgeprägten Orthotropie begründet sein. Schließt zum Beispiel die Schichtung mit der Kraftrichtung einen Winkel von 45° und erreicht das Verhältnis der Verformungsmoduln normal und quer zur Schichtung 5:1, so weicht die Verschiebungsrichtung um etwa 15° von der Kraftrichtung ab. Dieses Verformungsverhalten kann die Spannungen im aufstandsflächennahen Bereich beeinflussen und sollte daher berücksichtigt werden.

3. UNTERGRUNDVERFORMUNGEN ZUFOLGE STAURAUMBELASTUNG

Während bisher Belastungen behandelt wurden, die von
der Mauer auf den Untergrund übertragen werden, soll
nun die Wirkung einer Belastung des Talbodens durch
das gespeicherte Wasservolumen untersucht werden. Die
Sickerströmung im Fels oder, in grober Vereinfachung,
die Wasserauflast auf den Talboden, wird Untergrundverformungen hervorrufen, die zu zusätzlichen Zwängsspannungen im aufstandsflächennahen Bereich der Mauer
führen können. Dieser Einfluß kann in ebenen Schnitten
oder auch in einem räumlichen Modell untersucht werden.
Mit dem bereits früher verwendeten mathematischen Modell werden nun die Auswirkungen untersucht, die sich
aus einer Öffnung der vertikalen Blockfugen gegenüber
der monolithisch gedachten Mauer ergeben, was ja etwa
einem ebenen Schnitt entsprechen würde.

Die Ergebnisse zeigen ein grundsätzlich unterschiedliches Verformungsverhalten (Abb. 5). Während die unabhängigen Blöcke sich gegen die Wasserseite neigen,

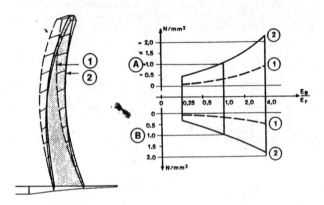

Abb. 5: Verschiebungen und Spannungen im Mittelschnitt
zufolge Stauraumbelastung
1   unabhängige Blöcke   A   Wasserseite
2   monolithische Mauer   B   Luftseite

Displacements and stresses of the central
section due to reservoir load
1   independent blocks   A   upstream
2   monolithic dam   B   downstream

Déplacements et contraintes vus en coupe
centrale causés la pression de l'eau retenue
1   blocs indépendants   A   amont
2   barrage monolithique   B   aval

überwiegt etwa in der oberen Hälfte der monolithisch
gedachten Mauer der Einfluß der durch den Wasserdruck
verursachten Talweitung und es kommt sogar zu einer
luftseitigen Verschiebung. Dieser wesentliche Unterschied in den Verformungen wirkt sich natürlich auch
in den Spannungen aus. Der Vergleich der Berechnungsergebnisse zeigt, daß die Zwängsspannungen zufolge
Stauraumbelastung bei Untersuchung in ebenen Schnitten
leicht unterschätzt werden können. Bei einer ebenen
Berechnung kann der Mittelschnitt den Verformungen des
Untergrundes weitgehend folgen, die zusätzlichen
Zwängsspannungen ergeben sich nur aus der Krümmung der
Aufstandsfläche und werden daher sehr rasch abklingen.
Im räumlichen Modell hingegen wird der Mittelschnitt
an diesen Verformungen durch die räumliche Tragwirkung
behindert, die zusätzlichen Zwängsspannungen werden
viel weiter in den Mauerkörper hineinreichen und etwa
den doppelten Wert wie bei den ebenen Berechnungen erreichen. Diese Studien können entweder am FE-Gesamtmo-
dell Sperre und Untergrund oder durch Vorgabe entsprechender Kämpferverformungen am Mauerkörper allein
durchgeführt werden.

Es ist selbstverständlich, daß die Stauraumbelastung
auf einem steifen, wenig nachgiebigen Untergrund geringere Zusatzspannungen auslösen wird als auf einem
weichen, sehr nachgiebigen Untergrund.

4. ZUGSPANNUNGEN AM WASSERSEITIGEN MAUERFUSS

4.1. Einfluß der Verformbarkeit des Untergrundes

Die Verformungen des Untergrundes hängen primär von
dessen Materialgesetzen, den Verformungsmoduln, ab,
die im allgemeinen in verschiedenen Richtungen unterschiedlich sein werden. In die Berechnung des Gesamt-
systems Sperre-Untergrund geht bekanntlich nicht der
absolute Verformungsmodul, sondern das Verhältnis der
Verformungsmoduln von Beton und Fels ein. Für die
folgenden Parameterstudien wurde angenommen, daß dieses Verhältnis über die ganze Länge der Aufstandsfläche konstant ist, jedoch parallel und normal zur Aufstandsfläche verschieden groß sein kann. Die errechneten Verschiebungen sind mehr für den Vergleich mit
den gemessenen von Interesse. Die Spannungen, deren
Messung wesentlich schwieriger ist, müssen gesondert
untersucht werden.

Maßgebend für die Beurteilung der Spannungen an der
Aufstandsfläche einer Bogenmauer sind meist die vertikalen Zugspannungen am wasserseitigen Mauerfuß, die
sich aus der Überlagerung der vom Verformungsmodul unabhängigen Druckspannungen zufolge Eigengewicht und
vom Verformungsmodul abhängigen Zugspannungen zufolge
Wasserlast ergeben. Relativ kleine Änderungen der Zugspannung zufolge Wasserlast können also große, absolute Änderungen der resultierenden Zugspannung zur
Folge haben (Abb. 6).

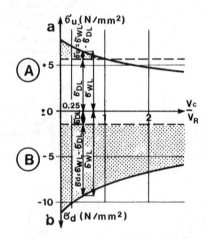

Abb. 6: Einfluß der Untergrundnachgiebigkeit auf die
Kämpferspannungen
A   Wasserseite   B   Luftseite
V   Verformungsmodul
c   Beton   r   Fels
σ   Vertikalspannungen
DL Eigengewicht   WL Wasserlast

Influence of the foundation yieldingness on
the abutment stresses
A   upstream   B   downstream
V   deformation modulus
c   concrete   r   rock
σ   vertical stresses
DL dead load   WL water load

Influence de l'élasticité du sous-sol sur les
contraintes dans les appuis
A   amont   B   aval
V   module de déformation
c   béton   r   roche
σ   contraintes verticales
DL poids mort   WL pression de l'eau

Bei dieser Untersuchung wurde davon ausgegangen, daß
bei gleichem Verformungsmodul von Beton und Fels die
Vertikalspannung am wasserseitigen Mauerfuß gerade
verschwindet, wie dies bei Gewichtsmauern üblich ist.

Wird nun der Untergrund normal zur Aufstandsfläche
weniger verformbar, also $V_B : V_F$ kleiner als 1, so steigen die vertikalen Zugspannungen rasch an und überschreiten bei einem Verhältnis von etwa 1:3 den zulässigen Bereich, der im internationalen Gewölbemauer-
bau etwa mit 1-2 N/mm² begrenzt wird. Wenn auch diese
Spannungen mit zunehmender Entfernung von der Aufstandsfläche wieder abklingen, so sind sie doch für
das Spannungsfeld im aufstandsflächennahen Bereich von
großem Interesse. Führen diese Zugspannungen zu Rissen am wasserseitigen Mauerfuß, so ändern sich übrigens
die Druckspannungen am luftseitigen Mauerfuß nur
wenig. Werden jedoch nur die Verschiebungen parallel
zur Aufstandsfläche größer, wie dies im wesentlichen
etwa der Annahme eines Viertel- anstelle eines Halbraumes entsprechen würde, ändern sich die vertikalen
Spannungen nur wenig.

4.2. Auswirkungen von Zugspannungen im Bereich des
     wasserseitigen Mauerfußes

Das Problem der Zugspannungen rund um den wasserseitigen Fuß von Gewölbemauern ist ja seit langem bekannt,
wobei vor allem über die allfälligen vertikalen Zugspannungen sehr viel gesprochen und geschrieben wird.
Selbst bei ebener Betrachtung des Problems erkennt
man jedoch sofort, daß es zumindest wasserseits der
Aufstandsfläche auch horizontale Zugspannungen geben
muß. Diese horizontalen Zugspannungen entstehen ja
letztlich dadurch, daß die horizontale Komponente der
Kämpferresultierenden Verschiebungen der Aufstandsfläche in Richtung Luftseite verursacht, an denen das
wasserseitige Felsvorland nur dann teilnehmen kann,
wenn horizontale Zugspannungen übertragen werden.
Während bekanntlich die vertikalen Zugspannungen am
wasserseitigen Mauerfuß durch entsprechende Formgebung und Mauerstärke beeinflußt oder auch, wie bei Gewichtsmauern, zum Verschwinden gebracht werden können,
ist dies bei den horizontalen Zugspannungen naturge-
mäß nicht möglich.
Hier sei noch darauf hingewiesen, daß in Wirklichkeit gedankliche Trennung von horizontalen und vertikalen Zugspannungen nicht gegeben sein kann, sondern
daß eben schräge Hauptzugspannungen vorhanden sein
werden, deren Richtung im wesentlichen durch das Verhältnis des horizontalen zu den vertikalen Zugspannungen bestimmt wird.
Mit diesem Problem hat sich Henny bereits 1929 aufbauend auf noch älteren Studien von Ottley und Bright-
more 1908 beschäftigt. Henny schreibt: "... Die Gefahr
einer solchen Zugspannung liegt in der Tatsache, daß
- sie mit der Höhe der Mauer zunimmt,
- mögliche Risse oder Erweiterungen vorhandener Fels-
  klüfte sich vom wasserseitigen Fuß aus unter die
  Mauer fortsetzen,
- die so gebildeten Wasserwege die angenommene vertikale Zone der natürlichen oder durch Injektionen
  erzielte Dichtheit der Gründung kreuzen können und
  als Folge eine direkte Verbindung zwischen dem Stau-
  becken und dem Dränagesystem der Gründung herbeige-
  führt werden kann, die zu erhöhtem Auftrieb und ero-
  dierender Durchsickerung führt. ..."
Casagrande hat in seiner ersten Rankine-Lecture 1963
dieses Problem für Gewichtsmauern untersucht, bei denen voraussetzungsgemäß keine vertikale Zugspannung
am wasserseitigen Mauerfuß vorhanden sein dürfte. Casagrande schreibt: "... zusätzlich sollte man berücksichtigen, daß der Wasserdruck auf die Mauer, der auf
den Untergrund übertragen wird, Klüfte entlang des
wasserseitigen Mauerfußes verursachen kann, die sich
öffnen, sodaß der volle hydrostatische Druck auf eine
vertikale Verlängerung der wasserseitigen Oberfläche
der Mauer wirken kann. Diese Klüfte müssen nicht parallel zur Mauerachse sein, jedes unregelmäßige Kluftsystem kann Verschwenkungen zur Folge haben."
Mit den Auswirkungen derartiger mehr oder weniger
steilstehender Klüfte, die vom wasserseitigen Mauerfuß ausgehen und sich unter die Mauer hin fortsetzen,
hat sich Fishman 1979 und 1983 sehr eingehend befaßt
(Abb. 7). Fishman hat in situ-Versuche an Betonblöcken

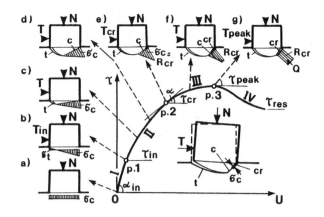

Abb. 7: Zusammenhang zwischen horizontalen Verschie-
        bungen und progressivem Versagen der Gründung
a - f   Stufen des Versagens
σ_c Druckspannung
σ_t Zugspannung
R_{cr} Grenztragfähigkeit der Gründung
Q   Bruchfestigkeit
t   Zugbruch
c   Druckbereich
cr   Bruchbereich

Connection between horizontal displacements
of a shear block and stages of foundation
failures
a - f   stages of failure
σ_c compressive stress
σ_t tensile stress
R_{cr} ulitmate crushing resistance of rock
        foundation
Q   crushing strength
t   tension fracture
c   compression zone
cr   crushing zone

Relation entre les déplacements horizontaux
d'un bloc de cisaittement et les étapes de
rupture progressive du sous-sol
a - f   étapes de rupture
σ_c contrainte de compression
σ_t contrainte de traction
R_{cr} résistance maximum à la rupture du
        sous-sol rocheux
Q   résistance à la rupture
t   rupture causée par contraintes de
        traction
c   zone de compression
cr   zone de rupture

auf Fels durchgeführt, bei denen als Belastung mit
einer etwas höher angreifenden Horizontallast Moment-
und Querkraft sowie durch eine Vertikallast auch eine
Normalkraft auf die Aufstandsfläche aufgebracht wurde.
Mit steigender horizontaler Belastung ergibt sich ein
charakteristischer, zur Luftseite hin fortschreiten-
der Rißverlauf. Zum Bruch kommt es nach Fishman erst
dann, wenn in der luftseitigen Mauerfuß verbleibenden Druckzone die Druckfestigkeit von Beton oder
Fels überschritten wird.

5. ZUSAMMENFASSUNG

Faßt man das Ergebnis der vorgestellten Untersuchungen für den Bereich der Aufstandsfläche zusammen, so
dürfte
- die Simulierung der Rückwirkungen des Untergrundes
  auf das Spannungs- und Verformungsverhalten des
  Mauerkörpers durch voneinander unabhängige Federmatrizen gerechtfertigt sein;
- die Ermittlung dieser Federmatrizen aber nicht unter Annahme eines elastisch isotropen Halbraumes,

sondern unter Berücksichtigung der tatsächlichen
luftseitigen Felsoberfläche, einer allfälligen Anisotropie und unter Ausschluß einer Zugspannungsüber-
tragung im Fels erfolgen müssen;
- eine Untersuchung des Einflusses der Stauraumbelastung bzw. der Sickerwasserströmung unter der Staumauer nur an der monolithischen Mauer, also am räumlichen Modell, sinnvoll sein;
- die Nachgiebigkeit des Untergrundes, insbesondere
normal zur Aufstandsfläche, einen wesentlichen Einfluß auf die Spannungen im Kämpferbereich haben, sodaß der mögliche Schwankungsbereich in Parameterstudien untersucht werden muß.

LITERATUR

Henny, D.C., 1929.
Classification, selection and adaption of high dams.
ASCE, Papers and Discussions, 11: 2327 - 2 336.

Casagrande, A., 1963.
Control of seepage through foundations and abutments
of dams. Geotechnique, 11: 161 - 183.

Magnet, E. und Widmann, R., 1974.
Gründungsprobleme der Gewölbemauer Kölnbrein.
ISRM, Vol. II B: 904 - 909.

Fishman, Yu.A., 1979.
Investigation into the mechanism of the failure of
concrete dams rock foundation and their stability
analysis. ISRM, Vol. II: 147 - 152.

Fishman, Yu.A., 1983.
Analysis of displacements of concrete shear blocks
and concrete dams on rock foundation by the field
measurement results.
Field measurements in geomechanics, Vol. 2: 865 - 874.

Baustädter, K. und Widmann, R., 1985.
The behaviour of the Kölnbrein arch dam.
ICOLD, Q. 57, R. 37: 633 - 651.

Softening behaviour of rock medium and instability in rock engineering

Affaiblissement du rocher et instabilité en mécanique des roches
Die Erweichung eines Felsmediums und die Instabilität im Felsbau

YIN YOUQUAN, Beijing University, People's Republic of China

ABSTRACT: The paper presents a constitutive relation of rock media and weak surfaces taking strain softening and permeation softening into consideration. It provides a general formulation of the instability theory of rock system due to softening of rock media and weak surfaces. It is applied specifically to the stability analysis in rock engineering.

RESUME: Get article présente la relation constitutive pour la masse rocheuse ainsique pour les surfaces de faiblesse qui est applicable au cas de l'adoucissement dû au deformations et au cas de l'adoucissement occassioné par l'ecoulement de l'eau. On propose une formulation générale de la théorie de la instabilité du systeme rocheux occassionée par l'adoucissement et s'applique au calculs de stabilité dans la génie géotechnique.

ZUSAMMENFASSUNG: In diesem Artikel werden die konstitutiven Gleichungs-systeme für Fels mit weichen Zwischenlagen unter Berücksichtigung des 'strain-softening' Verhaltens und der Sickerwasserströmung aufgestellt. Es wird eine allgemeine Instabilitätstheorie des Felssystems mit weichen Zwischenlagen vorgestellt und Beispiele für deren Anwendung bei der Analyse der stabilität von Projekten des Felshaüs gebracht.

1. INTRODUCTION

The strength (or yield limit) of a rock medium and a weak structural surface may decrease with the development of plastic deformation and water permeation. These phenomena are called strain softening and permeation softening respectively. Landslide, collapse and subsidence of rock engineering project due to excavation, water permeation or the action of the earthquake force can be regarded essentially as the loss of stability of rock system due to softening of rock media and weak surfaces. In order to discuss these instability phenomena by means of solid mechanics, this paper presents a constitutive relation for rock medium and weak surface taking strain softening and permeation softening into consideration. It also incorporates the incremental analysis of solid mechanics together with an energy criterion for determining the stability of a rock system. Thus it provides a general formulation of the theory for the instability of rock system. The theory can be used to calculate the critical load (water load or the excavation step at which the project would become unstable) of an engineering project and to study precursories of oncoming instability of the rock system, so that it provides a necessary basis for the design of an engineering project to have sufficient resistance against disaster. The paper also illustrates the implementation of the theory in the finite element program NOLM and its application to the stability analysis of rock slopes and underground openings.

2. A CONSTITUTIVE RELATION OF ROCK MEDIUM AND WEAK SURFACE IN A ROCK MASS

In uniaxial tests or triaxial tests under low confining pressure with an ordinary testing machine, the rock medium fails suddenly in a brittle manner, but with a stiff testing machine one can obtain a continuous stress-strain curve including both the hardening range and the softening range. In plasticity theory, when the yield stress of a medium increases or decreases with the development of a plastic internal variable κ, they are called strain hardening or softening respectively. Strain softening is an intrinsic property of the rock medium. On the other hand, most rocks have porosity to a certain degree

and pore water has a significant effect on the deformation character and the strength of a rock mass. Permeation of water will decrease the strength (or the yield limit) of rock media (e.g., the cohesion S_0 and the internal friction coefficient μ). This is called permeation softening. On the contrary, drainage will result in hardening.

The elasto-plastic constitutive relations accounting for strain softening has already been established (Wang and Yin, 1981). In order to extend these relations to the case of permeation softening, we have to make use of the concept of effective stress. It is the stress acting on the solid skeleton of the two phase fluid-solid model. We shall call the stress on the combined two phase model the total stress, it consists of the effetive stress σ' and the pore pressure p

$$\sigma = \sigma' + pe \qquad (1)$$

$$\sigma = [\begin{array}{cccccc} \sigma_x & \sigma_y & \sigma_z & \tau_{yz} & \tau_{zx} & \tau_{xy} \end{array}]^t \quad (2)$$

$$\sigma' = [\begin{array}{cccccc} \sigma'_x & \sigma'_y & \sigma'_z & \tau'_{yz} & \tau'_{zx} & \tau'_{xy} \end{array}]^t \quad (3)$$

$$e = [\begin{array}{cccccc} 1 & 1 & 1 & 0 & 0 & 0 \end{array}] \quad (4)$$

As the yield behaviour of the skeleton material relates not only to a plastic internal variable κ but also to a permeation parameter η (it can be taken as the degree of saturation or the moisture content), the yield condition can be written in a general form as

$$f(\sigma', \kappa, \eta) = 0 \qquad (5)$$

In case the pore pressure does not cause significant deformation of the solid skeleton, the total strain ε of the two phase medium can be assumed to consists of an elastic part and a plastic part

$$\varepsilon = \varepsilon^e + \varepsilon^p \qquad (6)$$

where

$$\varepsilon = [\begin{array}{cccccc} \varepsilon_x & \varepsilon_y & \varepsilon_z & \gamma_{yz} & \gamma_{zx} & \gamma_{xy} \end{array}]^t \quad (7)$$

$$\varepsilon^e = [\begin{array}{cccccc} \varepsilon^e_x & \varepsilon^e_y & \varepsilon^e_z & \gamma^e_{yz} & \gamma^e_{zx} & \gamma^e_{xy} \end{array}]^t \quad (8)$$

$$\varepsilon^p = [\begin{array}{cccccc} \varepsilon^p_x & \varepsilon^p_y & \varepsilon^p_z & \gamma^p_{yz} & \gamma^p_{zx} & \gamma^p_{xy} \end{array}]^t \quad (9)$$

The elastic strain relates to the effective stress by the Hooke's law ($\sigma = De^e$). The plastic strain increment obeys the normality rule, which means the plastic flow is so associated to the yield surface. In this manner, the obtained constitutive relation is

$$d\sigma = D_{ep}d\epsilon + D_\eta d\eta + edp \qquad (10)$$

$$D_{ep} = D - \frac{H(L)}{A} \frac{\partial f}{\partial \sigma'} (\frac{\partial f}{\partial \sigma'})^t D \qquad (11)$$

$$D_\eta = -\frac{H(L)}{A} D \frac{\partial f}{\partial \sigma'} \frac{\partial f}{\partial \eta} \qquad (12)$$

$$A = (\frac{\partial f}{\partial \sigma'})^t D \frac{\partial f}{\partial \sigma'} - \frac{\partial f}{\partial \kappa}h \qquad (13)$$

$$h = \begin{cases} \sigma'^t \frac{\partial f}{\partial \sigma'} & \text{when } \kappa = w^p \text{ (plastic work)} \\ e^t \frac{\partial f}{\partial \sigma'} & \text{when } \kappa = \theta^p \text{ (plastic dilation)} \\ [(\frac{\partial f}{\partial \sigma'})^t \frac{\partial f}{\partial \sigma'}]^{\frac{1}{2}} & \text{when } \kappa = \bar{\epsilon}^p \text{ (equivalent plastic strain)} \end{cases} \qquad (14)$$

$$H(L) = \begin{cases} 0 & \text{if } L \leq 0 \\ 1 & \text{if } L > 0 \end{cases} \qquad (15)$$

$$L = (\frac{\partial f}{\partial \sigma'})^t Dd\epsilon + \frac{\partial f}{\partial \eta} d\eta \qquad (16)$$

$L > 0$, $L = 0$ and $L < 0$ indicate plastic loading, neutral loading and unloading respectively. We note from eq. (16) that variations of both the strain and the permeation parameters can cause plastic loading or unloading. The matrix D_η in eq. (12) represents the coupling effect between permeation and plasticity.

In order to describe the discontinuous properties of joints, faults and weak structural surfaces in a rock mass, we introduce the concept of a displacement discontinuity surface. Due to the complexity of its behaviour (e.g., it has both dilation and softening) it is necessary not only to set up the discontinuity conditions for the stress and the displacement but also to establish a constitutive relation which relates the stress vector $\bar{\sigma}$ acting on the surface with the displacement discontinuity vector (u) across the surface. In doing so, we look upon the discontinuity surface as the limit of a material layer when its width approaches zero (Fig.1). This discontinuity surface is not a simple geometric surface but a surface consisting of material particles and possessing material properties. The displacement discontinuity vector

$$(u) = [\; u^+ - u^- \quad v^+ - v^- \quad w^+ - w^-]^t \qquad (17)$$

corresponds to the following stress vector acting on the surface

$$\bar{\sigma} = [\; \tau'_{ns} \quad \tau'_{nt} \quad \sigma'_n]^t \qquad (18)$$

The constitutive relation of the discontinuity surface is then (Yin and Zhang 1981; Yin and Zhang 1982)

$$d\bar{\sigma} = \bar{D}_{ep}d(u) + \bar{D}_\eta d\eta + \bar{e}dp \qquad (19)$$

$$\bar{D}_{ep} = \bar{D} - \frac{H(L)}{A} RR^t \qquad (20)$$

$$\bar{D}_\eta = -\frac{H(L)}{A} BR^t \qquad (21)$$

$$\bar{D} = \begin{bmatrix} k_t & 0 & 0 \\ 0 & k_t & 0 \\ 0 & 0 & k_n \end{bmatrix} \qquad (22)$$

$$\bar{e} = [0 \quad 0 \quad 1]^t \qquad (23)$$

$$\tau = (\tau^2_{ns} + \tau^2_{nt})^{\frac{1}{2}} \qquad (24)$$

$$R = k_t [\; \tau_{ns}/\tau \quad \tau_{nt}/\tau \quad \mu/m]^t \qquad (25)$$

$$m = k_t/k_n \qquad (26)$$

$$A = \mu^2 k_n + k_t + h(\frac{\partial S_0}{\partial \kappa} - \sigma'_n \frac{\partial \mu}{\partial \kappa}) \qquad (27)$$

$$h = \begin{cases} S_0 & \text{when } \kappa = w^p \\ \mu & \text{when } \kappa = \theta^p \\ (1+\mu^2)^{\frac{1}{2}} & \text{when } \kappa = \bar{\epsilon}^p \end{cases} \qquad (28)$$

$$B = -(\frac{\partial S_0}{\partial \eta} - \sigma'_n \frac{\partial \mu}{\partial \eta}) \qquad (29)$$

$$L = R^t d(u) + Bd\eta \qquad (30)$$

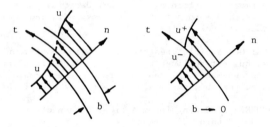

Figure 1. The discontinuity surface looked upon as the limit of a material layer.

Here k_t and k_n are the tangential and normal elastic stiffness of the material surface. They can be determined experimentally by Goodman's method (1976) of measuring the joint stiffness in a jointed rock sample. When $(\partial S_0/\partial \kappa - \sigma'_n \partial \mu/\partial \kappa)$ is positive, zero or negative, the material surface is strain hardening, perfectly plastic or strain softening, respectively. B 0 corresponds to permeation softening.

3 THE FORMULATION OF INSTABILITY IN ROCK ENGINEERING

We may regard the deformation process in rock engineering before the impending unstable state as a quasi-static one, and use the quasi-static method to define the critical load and to study the mechanical precursory phenomena of instability. We divide the real loading path or history (including permeation and drainage, excavation and construction) into many small loading increments and compute for each increment step by step. Let us consider a typical incremental step. At the beginning of this step, the stress σ and strain ϵ in the rock mass, the displacement discontinuity (u) on the discontinuity sruface together with the internal variables in the solid skeleton and the weak surface are all known. Denoting the corresponding pore pressure and permeation parameter by p and η, their increments by dp and dη, respectively, we can calculate them from the permeation problem. In this step, the increments of the traction dF acting on the boundary S_T, of the body load df in the rock mass and of the displacement du_0 at the boundary S_u are given, the displacement increment du, the strain and stress increments, dϵ and dσ, as well as the displacement discontinuity increment are to be determined. They should satisfy the following three sets of conditions:
  1. Equilibrium conditions

$$L \; d\sigma + df = 0 \qquad \text{in rock mass V} \qquad (31)$$

$$L_1 d\sigma = d\bar{\sigma} \qquad \text{on discontinuity surface A} \qquad (32)$$

$$L_2 d\sigma = dF \qquad \text{on boundary } S_T \qquad (33)$$

in which

$$
L = \begin{bmatrix} \dfrac{\partial}{\partial x} & 0 & 0 & 0 & \dfrac{\partial}{\partial z} & \dfrac{\partial}{\partial y} \\[2ex] 0 & \dfrac{\partial}{\partial y} & 0 & \dfrac{\partial}{\partial z} & 0 & \dfrac{\partial}{\partial x} \\[2ex] 0 & 0 & \dfrac{\partial}{\partial z} & \dfrac{\partial}{\partial y} & \dfrac{\partial}{\partial x} & 0 \end{bmatrix} \qquad (34)
$$

$$
L_i = \begin{bmatrix} l_i & 0 & 0 & 0 & n_i & m_i \\ 0 & m_i & 0 & n_i & 0 & l_i \\ 0 & 0 & n_i & m_i & l_i & 0 \end{bmatrix} \qquad (35)
$$

where l_1, m_1, n_1 and l_2, m_2, n_2 are the direction cosines of the outward normal on the discontinuity surface A and on the boundary S_T respectively.

2. Geometric conditions

$$L^t du = d\varepsilon \qquad \text{in rock mass V} \qquad (36)$$

$$d(u) = du^+ - du^- \qquad \begin{array}{l}\text{on discontinuity}\\ \text{surface A}\end{array} \qquad (37)$$

$$du = du_o \qquad \text{on boundary } S_u \qquad (38)$$

3. Constitutive relations

$$d\sigma = D_{ep} d\varepsilon + D_\eta d\eta + edp \qquad (39)$$
$$\text{for rock mass}$$

$$d\bar{\sigma} = \hat{D}_{ep} d(u) + \hat{D}_\eta d\eta + \hat{e}dp \qquad (40)$$
$$\text{for discontinuity surface}$$

in which

$$\hat{D}_{ep} = M^t \bar{D}_{ep} M \qquad (41)$$

$$\hat{D}_\eta = M^t \bar{D}_\eta \qquad (42)$$

$$\hat{e} = M^t \bar{e} \qquad (43)$$

M is the matrix of coordinate transformation from the local coordinate at the discontinuity surface to the general coordinate of the entire problem.

Eqs. (31)-(40) constitute a closed system of equations. Except for the simplest problems, one has to resort to numerical solution by finite element method. In that case, a weak equilibrium condition, namely the virtual work theorem, may be used in place of equations (31)-(33):

$$\int_V \delta(d\varepsilon)^t d\sigma dV + \int_A \delta[d(u)]^t d\bar{\sigma} dA \qquad (44)$$
$$= \int_V \delta(du)^t df\, dV + \int_{ST} \delta(du)^t dF\, dS$$

When the load is so small that the rock medium and the discontinuity surface are within the elastic and hardening ranges, the incremental solution obtained by the above formulation is uniquely determined and hence the corresponding stress and strain fields are stable equilibrium fields. With increasing load the medium and the surface will enter the softening range and the obtained field may be an unstable one. For such a case, an energy criterion which can be used to determine the stability of a rock system is given below.

Let us examine an equilibrium configuration in which the deformation and stress are expressed by u, (u), ε, σ, etc., and give this configuration a set of small virtual displacements, without violating the geometrical conditions (36)-(38), to obtain a new configuration. If the virtual work done by the external load does not exceed the increase of the internal energy (including the stored elastic strain energy and the plastic dissipation), the system is considered stable. If this condition is not met for some virtual displacements, then the excess energy will appear as kinetic energy. This indicates that the original configuration is unstable. The energy criterion for instability may be expressed as

$$\int_V \delta\varepsilon^t(\sigma + \delta\sigma)\, dV + \int_A \delta(u)^t(\bar{\sigma} + \delta\bar{\sigma})\, dA \qquad (45)$$
$$-\int_V \delta u^t f\, dV - \int_{S_T} \delta u^t F\, dS < 0$$

Since the considered configuration is in equilibrium, it satisfies the virtual work theorem and the above criterion can be rewritten into

$$\int_V \delta\varepsilon^t \delta\sigma dV + \int_A \delta(u)^t \delta\bar{\sigma} dA < 0 \qquad (46)$$

As the relations between $\delta\varepsilon$ and $\delta\sigma$, and between $\delta(u)$ and $\delta\bar{\sigma}$, are given by the constitutive Eqs.(39) and (40) respectively, the stability of a rock system is related directly to the characteristics of the constitutive matrix. For example, if there is no water permeation and the rock medium together with the weak surface are in the perfectly or the hardening plastic ranges, the matrices D_{ep} and $\hat{D}_{ep}$ are positive definite and the left-hand side of formula (46) is always positive, then the configuration under examination is definitely a stable one. Therefore the strain softening or the permeation softening behaviour of the rock medium and the weak structural surface is a necessary condition for the loss of stability of rock engineering.

To sum up, our method in studying instability of the rock engineering consists of dividing the real loading path into many small increments, solving stepwise the incremental boundary value problems, and using the energy criterion to judge the stability of the configuration. As soon as the instability criterion (46) is found to hold, it means the advent of unstable state and the critical load is obtained. By analyzing the deformation field at this moment, the failure pattern may also be obtained.

4 FINITE ELEMENT PROGRAM NOLM AND ITS APPLICATION IN STABILITY ANALYSIS

The above method of stability analysis is implemented in the program NOLM (Yin and Zhang, 1985). This program is mainly used for the stress-deformation and stability analysis of rock-soil engineering and for numerical simulation of some geological and seismological problems.

The program NOLM has been used to analyze a certain air shelter engineering. Its early stage of excavation was completed four years ago (Fig.2). The surrounding rock is granite and its cohesion $S_0 = 1.96 \times 10^6$ N/m^2 and the coefficient of internal friction $\mu = 1$. The soft layer is weathered rock, its $S_0 = 2.4 \times 10^3 N/m^2$ and $\mu = 0.59$. For the safety of subsequent stages of excavation, four excavation steps were taken in the computation according to the projected excavation process as shown in Fig.3. The computed result shows that during the excavation process the entire granitic region remains in the elastic stage while plastic deformation (shear failure) only develops locally in the soft layer and further shear failure of the soft layer is prevented by the dilation character of the rock medium. So the rock system will be stable in the excavation process and the anchorage measure is unnecessary in this project.

NOLM has also been used to study the mechanism of a landslide occurred in a certain phosphorus mine (Hu, Jia and Xu, 1983). The main slide section of the mountain is shown in Fig.4. Seven time steps were taken in the computation to simulate the permeation process of the rain water from the backland to the foreland of the mountain. Each step corresponds to a certain location where the rain water arrived. At the 5th time step, in the intermediate layer and the joint with large dip, a virtually interconnected plastic deformation region from the backland to the foreland of the sliding mass appears. At the 6th time step, a large scale stress drop appears and a ladder like fracture occurs, the mountain mass is now unstable.

581

The numerical result shows that the permeation softening character of the clay in the intermediate layer
controls the mountain landslide and that the seepage
of the rain water in the intermediate layer is the
main triggering factor of the landslide. An interesting fact which appeared in the computation is that
the back chasm at the top of the mountain showed an
accelerating tensile fracture process before the landslide. It is actually a precursor of the landslide.
The theoretical research and the in-situ monitoring
of such precursor of instability may have important
practical significance.

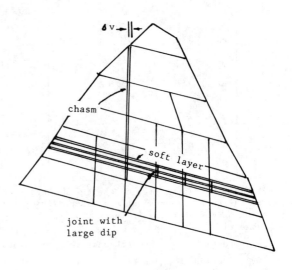

Figure 4. The main slide section of the mountain and
its finite element net.

Yin Youquan & Zhang Hong 1985. The finite element
program NOLM for elasto-plastic stress-strain analysis and stability analysis of the rock-soil
system. In "The Records of Geological Research",
48-56, Peking University Press.

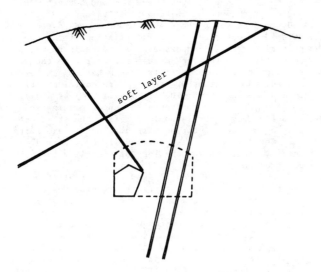

Figure 2. A typical cross-sectional view of the civil
air shelter analyzed.

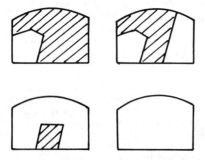

Figure 3. The projected excavation process of the
civil air shelter.

REFERENCES

Goodman, R.E. 1976. Methods of geological engineering in discontinuous rocks. West Publishing Co.,
St. Paul, Minn.
Hu Haitao, Jia Xuelang & Xu Kaixiang 1983. The mechanism and movement pattern of the mountain landslide in the Yanchi River Phosphorus Mine. 2nd National Conference on Engineering Geology.
Wang Ren & Yin Youquan 1981. On the elasto-plastic
constitutive equation of rock-like engineering
materials. Acta Mechanica Sinica. 13: 317-325
(in Chinese).
Yin Youquan & Zhang Hong 1981. The one-dimensional
joint element analysis of rock mass engineering.
Mechanics and Practice. 3:52-54 (in Chinese).
Yin Youquan & Zhang Hong 1982. A mathematical model
of strain softening for simulating earthquakes.
Acta Geophysica Sinica. 25: 414-425 (in Chinese).

An integral geomechanical model rupture test of Jintan Arch Dam with its abutment
Modèle géomécanique intégral d'un essai à la rupture du barrage-voûte de Jintan et de ses appuis
Ein integraler geomechanischer Bruchversuch des Widerlagers und der Bogenstaumauer Jintan

ZHOU WEIYUAN, Tsinghua University, Beijing, People's Republic of China
YANG RUOQIONG, Tsinghua University, Beijing, People's Republic of China
LUO GUANGFU, Tsinghua University, Beijing, People's Republic of China

ABSTRACT: In this paper, an integral geomechanical model test for Jintan arch dam is described. The dam is situated in complex rock formations. The failure mechanism, stability factors, and the interaction between arch dam and abutments are given.

RESUME: Ce document décrit le test de modèle géomecanique intégral pour le barrage-voûte de Jintan. L'ampleur du modèle en question constitue l'une de plus importantes en Chine. Le barrage se trouve dans des formations de roche complexes. Il est également abordé dans le document le processus d'échec, les facteurs de stabitité et l'interaction entre le barrage-voûte et son fondement. Ils correspondent parfaitement aux calculs des résultats.

ZUSAMMENFASSUNG: In der vorliegenden Arbeit wird ein integraler geomechanisch Modellversuch fuer Jinshuitan Bogenstaumauer vorgesstellt. Die Bogenstaumauer liegt auf der komplizierten felsformationen. Der Zerstoerungssmechanismus, die stabilitaetsfaktoren und die Interaktion zwischen der Bogenstaumauer und den seitlichen

1 Introduction

In China, a number of large arch dams which are more than 100m in height have been constructed. In the design of arch dams, the stability problem of abutments is acquiring greater importance since the geological conditions in dam projects are becoming more and more complicated.

Mathematical models and analytical methods have been extensively used in stability analysis. However, so far, they are inadequate in the case of geomechanical conditions concerning the stability of complex rock formations.

For nearly 30 years, a permanent research laboratory at Tsinghua University specialising in physical model tests of arch dams has been established. Geomechanical models can be used for a great variety of purposes, mainly to demonstrate the deformation processes of the dams from normal load to rupture, and to examine the stability factors of abutments. They can provide a basis of information for the validity of the analytical results.

The geomechanical model test for Jintan arch dam has been performed in our laboratory. It is a model test of the largest scale in China. Owing to the complex rock formations in Jintan arch dam project, the geomechanical model test has been utilized to check the results of numerical analysis. This paper will briefly examine this approach in analysing the stability of Jintan arch dam.

2. Geomechanical model of Jintan arch dam

Jintan dam, a dome-shaped arch dam of 110m in height, is situated on the upper reach of Ou River in Zhejiang Province in the southeastern part of China. It is under construction at present. The geology for the dam site mainly consists of granite and gneissic granite as shown in Fig.1. The rock masses are slightly weathered and relatively intact. However, the abutment rock masses are cut into blocks by several large faults and three sets of main joints. As shown in Fig. 1, there were found some remarkable faults, such as F17, F11, F8, F12, F14 in the right abutment, F23, F31, F32 in the left abutment, and some gently

dipped joints. Some infillings appear in these faults. Generally, these faults with their influenced zones are 0.1-0.3 meters wide. These discontinuities form the weak discontinuities in the rock abutments. The stability of abutments mainly involves in the behavior of these discontinuities. For the sake of assessment of Jintan arch dam stability and reliability, an integrated three dimensional model was utilized to investigate the dam behavior from under normal load to rupture. Considering the precision of observations and the size of testing table, the model scale was taken to be 1/100. The model has a length of 6m and a width of 5m.

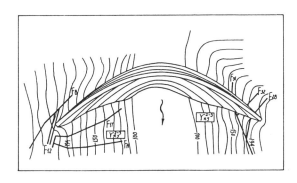

Figure 1. Geological map of Jintan arch dam

2.1 Reproduced Geomechanical conditions.

In the model, different geological conditions including deformability, mechanical characteristics of the foundation rocks and the main discontinuities were reproduced, as given in Table (1).

The average deformation modulus of intact rock in foundation and abutment of Jintan project is $2.2 \times 10^4 N/m^2$.

Table 1   Geomechanical parameters of discontinuities in abutments

No. of fault	geological description strike	dip angle	width of fracture zone (m)	width of effected zone (m)	modulus of deformation (N/m²)	shear strength of breccia f	c(N/m²)
F8	N50°E	80–85°SE	0.05	1.0	0.40	.25	0
F11	N58°W	55°NE	0.01	1.0	0.3	.42	0
F12	N20°E	75°SE	0.08	0.57	0.4	.42	0
F14	N15°E	53°SE	0.15	1.5	0.2	.35	0
F16	N10°E	55°SE	0.3	1.0	0.3	.25	0
F17	N75°W	57°NE	0.4	1.0	0.2	.25	0
F28	N25°E	75°NE	0.08	1.2	0.4	.35	0.02
F31	N45°W	45°NE	0.6	1.5	0.2	.25	0.05
F32	N35°W	80°SE	0.3	0.8	0.35	.42	0
F60	N75°W	50°NE	0.7	2.0	0.3	.25	0

Table 2.   Geological parameters of joints

| | left bank | | | | right bank | | |
No. of Joint	strike	dip	spacing(m)	No.of joint	strike	dip	spacing(m)
1	N10–20°E	70°SE	2–3	4	N40°E	80°NW	1.5
2	N45°E	50°NW	1–4	5	N40°W	40°NE	0.8
3	N70°E	75°NW	0.5	6	N75°W	60°NE	2.5

All the strength parameters for joint surface are assumed to be:
f=0.71   c=0.52 N/M²

2.2 Principle of similitude

The following similarity criteria should be satisfied.
Assuming $C_L = L_{pr}/L_m = 100$ $C_\varepsilon = 1$ $C_f = 1$, then:

$$C_L = C_c = C_E = C_\sigma = C_r/C_L \qquad (1)$$

C_r the ratio of density between prototype and model
C_ε the ratio of strain between prototype and model
C_f the ratio of coefficient of internal friction
C_c the ratio of cohesion
C_E the ratio of elasticity
C_σ the ratio of stress
C_L the geometric scale ratio

For rupture tests, it is important that the strains at geometrically similar points should be the same, i.e., $C_\varepsilon = 1$. It also means that the models are usable from elastic state to rupture.

Formula (1) also means that the strength ratio must be equal to geometric scale ratio. The gypsum briquettes assembled in blocks have the mechanical characteristics satisfying the conditions given above, therefore, they are widely used in geomechanical models to reproduce different geological features and various kinds of rock masses.

2.3 Technique of modelling

The Jintan model was made of heavy casting plaster, so that $C_r=1.0$. The material can be obtained by using baryte powder mixed with gypsum. The model was made of about 25000 small rectangular parallelepipedal briquettes capable of reproducing, when placed close to one another, the fracture discontinuities and the intact rock, as shown in Fig.2.

In each case, the dimensions of the briquettes and therefore the intervals between the joints are checked by testing the overall behavior of an assemblage of briquettes. In Fig.3, there are shown the stress-strain relations of briquettes under tridimensional tests which are similar to those of in-situ rock masses, and in Table (3) are given their mechanical parameters.

To reproduce the breccia in the faults, plastic sheets of different thicknesses were inserted between the contact surfaces of the model which give a fairly good imitation of the characteristics of the infilling material, as shown in Table (4).

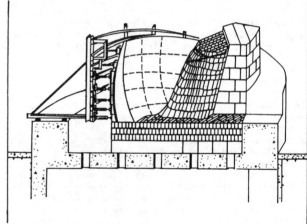

Figure 2.   Geomechanical model of Jintan arch dam

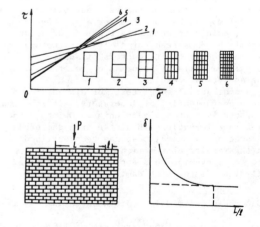

Figure 3.   Gypsum briquettes tests

2.4 Load applied to the model.

Hydraulic jacks were used to simulate the water load applied to the dam. In the rupture test, the increment load method was used, and $\lambda = p/p_o$ was assumed to indicate the load intensity, where p-applied water load; P_o -normal water load.

Table 3   Mechanical characteristics of briquettes

No.of material	type of material	r N/m^3	compressive strength N/m^2	modulus of single briquette	deformation N/m^2 assemblage of briquettes	tensile strength N/m^2
1	C43	0.24	0.382	314	15–70	0.053
2	C63	0.23	0.097	71	5–8	0.002
3	R82	0.26	0.05	32	2.5	0.001
4	R93	0.242	0.02	25	1.5	0.001

Table 4   Infilling materials and their mechanical characteristics

No.	Description of infilling material	coefficient of friction
1	polythene sheet	0.15
2	aluminium sheet	0.25
3	parchmyn paper	0.35
4	paraffin based paper	0.45
5	untreated surfaces in contact	0.65
6	powdered limestone	0.75

3 Analysis of experimental results

3.1 Deformation process of dam model

Under normal water load, the dam model behaved elastically and symmetrically. The maximum radial deflection, 5.4cm, was found at the crest of crown cantilever, whereas at the surface of foundation, 0.558cm, as shown in Fig. 4.

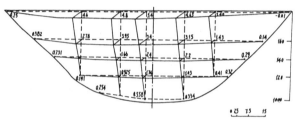

Figure 4. Deformations of Jintan dam model

When the over load factor λ exceeded 2.13, the dam behaved inelastically, as in Fig.4. The stiffness of cantilever elements became decreasing. After λ>5.0–6.0, the cantilevers cracked near the foundation, so the dam became a multi-arch element structure.

When λ > 1.87, the dam model behaved inelastically, and after overload factor λ =7-8, the deformation increased drasttically. Finally, the dam model collapsed at λ =9.0.

3.2 Propagation of cracks in model

Along the dam heel of the model, ten strain gages were stuck to the model surface to monitor the deformation of the model.

In Fig.5, as λ =1.87, the upstream dam heel of the model behaved in a plastic state; as λ>3.2, some fissures emerged near the foundation of crown cantilever beam. When λ approached 3.47, the cracks spreaded over the left part of the heel and gradually to the right part. When λ =5.33, the upstream heel became entirely cracked.

When λ =4.0, some fissures emerged in the downstream dam surface of the model, and near the foundation of the dam along the maximum principal stresses. As the load increased, these fissures propagated. They formed some open cracks horizontally as shown in Fig.6a. As λ>8.0, the dam model was cut visibly by cracks near the El. 160m. As λ =9.0, a large fractured zone appeared across the dam body, and then the dam collapsed. The cracks appeared along the trajectories of the maximum principal stresses and intercepted the abutments orthogonally.

On the dawnstream face, it was visible that a set of arch elements were formed at the final limit stage of the dam. In Fig.6b, the limit state of dam model observed by strain gages is given. It can be seen that the tensile cracks emerged earlier, however, they did not represent the final state of dam. The final state of dam model would come when the compressive cracks zone lost its load capacity. The compressive crack zone is a key zone for the load capacity of arch dam. It can be seen from Fig.6a that the deflection of cantilevers varied abruptly as λ>8.0. This can be understood as the appearance of compressive cracked zone.

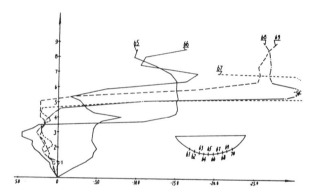

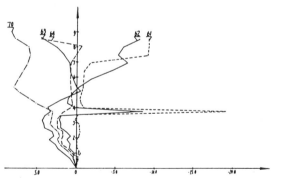

Figure. 5 Propagation of cracks along the dam heel.

3.3 Stress and strain distribution of dam model.

The stress distributions are shown in Fig.7. The maximum compressive stress occurred at the downstream surface is 6.6 N/m^2 whereas the maximum tensile stress is 1.76 N/m^2 at the crown cantilever. Both stresses occurred symmetrically near right and left abutments. However, owing to the assymmetric existence of discontinuities in the abutments,

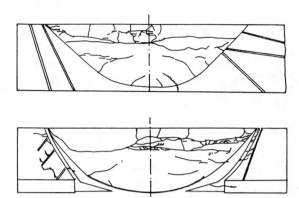

Figure 6a   The final rupfure feature of Jintan arch dam.

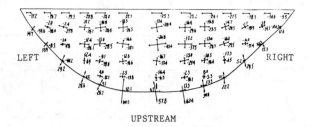

UPSTREAM

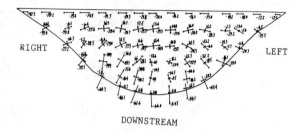

DOWNSTREAM

Figure 7.   Principal stresses in Jintan arech dam model

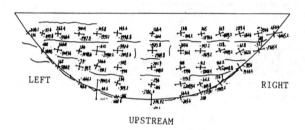

UPSTREAM

DOWNSTREAM .

Figure 6b.   Limit strain state of Jintan arch dam model

stress concentration was found in the stress distribution. As at the El. 180m, the compressive stress near the left bank is -1.5 N/m², whereas that near the right bank is -1.9. The second principal stresses in downstream surface are tensile.

From Fig.6b, the maximum compressive strains reached 2000-3000 , which is within the permissible limit of concrete. Along the abutmens of the dam model, is shown a maximum strain zone, which is a key one to the safety of dam model.

3.4 Deformation process of abutments

All deformations observed under normal load are within the limit of 0.1 cm. However, when λ=3.2, plastic deformation appeared at some measured points, e.g., the deformation at El. 170m reached 0.4 cm. After λ > 5.0, most points were subjected to plastic deformation, and finally, the maximum deformation reached 3.7 cm in the left abutment whereas 11.5 cm in the right abutment. It is obvious that the right abutment is not so stiff as the left abutment because of the existence of F_{17}, F_{11}, F_{12}, F_8.

The special feature of this experiment is the embedment of strain gages in the abutments of the model. The deformations on the geological faults in abutments may be attributed to the following points.

Fault F_{17} in right abutment gave an important effect to the deformation and stability of rock abutment. Its maximum deformation was 0.5cm under normal

load. After λ > 2.13, it behaved plastically and its compressive deformation reached 1.6cm and finally 11.5cm. F_{17} had large deformations when λ > 3.0. Its failure will lead to the rupture of dam, so that special attention must be paid to it henceforth.

Fault F_{11} located in the downstream of F_{17}, its maximum deformation reached 0.05cm under normal water load. This is fairly favorable for the safety of right abutment. Its maximum deformation is 2.0cm which corresponds to λ =8.0-9.0. From model test,all the measuring points on the right bank gave larger deformation values under normal load than those of the left bank.

From test, Fault F_{12} and F_8 in right abutment only effected the stability of abutment slightly.

As in Fig (8), after λ exceeded 3.2, the abutments behaved inelastically but the deformation in right abutment was larger than that of the left.The maximum deformation occurred in the lower middle part of the abutments.

In the right abutment, a slide of rock masses cut off by F17 anf F11 were found when λ > 8.0 as in Fig.9.

In the left abutment, large inelastic deformations were found when λ =3.2. However, no slide trend was found.

Some strain gages were embedded in the abutments to monitor the internal deformation of discontinuities. In the right abutment, F28, F31 show slight deformation even under high water pressure, whereas in right abutments F17 with F11 has a large deformation and fractured when the overload factor exceeded 3.2.

3.5 Stability and reliability of Jintan arch dam.

From the deformation given above, it can be seen that when λ > 1.87, the dam began behaving inelastically. So, $K1=P_1/P_0=1.87$, which referred to the elastic limit of dam. When λ > 9.0, the dam collapsed, then $K2=P_2/P_0$ > 9.0. K2 referred to the limit load capacity of dam. Similarly, the safety factor of dam abutments can be defined as follows:

K1=3.2   K2 ≥ 9.0. for left abutment,
K1=2.13 K2 ≥ 9.0 for right abutment.

For some discontinuities such as F17, F11 and F12 k1=2.13 and k2=9.0.

From the above observations, a conclusion is drawn that the integral safety factor is adequate.However, some local safety factors of discontinuities are inadequate some treatments of faults are needed.

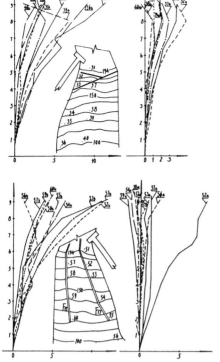

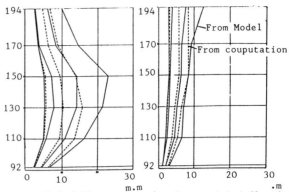

a. Radial deflection   b. Tangeatial deflection

Figure 10. A comparison of deformations between model tests and calculation by F.E.M..

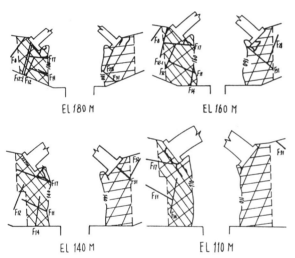

Figure 8. Deformation of model abutments

EL 180 M   EL 160 M

EL 140 M   EL 110 M

Figure 9. Cracked abutments of dam model

3.6 A comparison with the results of F.E.M.

The deformations observed in the model and evaluated by F.E.M. are shown in Fig. 10. The maximum deformations are 5.48 and 5.44 for model and numerical analysis, respectively. The nonlinear deformations emerged at 2.18 and 2.24 respectively. These results obtained by model and F.E.M. coincided very well.

3.7 Interaction between the dam and its foundation

In the limit state of dam shown in Fig.6b, the arch dam was supported by the "natural arches", i.e., the water load caused pure compression in the abutments, where as cracks caused tension lost. In the lower part of right abutment, some stresses were transferred to its adjacent parts owing to its inadequate stiffness. As in Fig.6b, an interaction was seen between them. The dam heel cracked and transferred its load to arch members,which gave a higher value of S.F.

4. Conclusions

1). Geomechanical model tests are suitable for demonstrating the deformation processes of dams up to failure, especially for arch dams.
2). Jintan arch dam, owing to its deficiencies in geological conditions, has an inadequate S.F. for local stability of abutment. Consequently, a treatment of these faults should be studied.
3). The techniques of reproduction and tests have undergone successive stages of refinement and evolution. The results of tests very well with that of F.E.M..
4). The interaction between arch dam and foundation reveals that arch dams have a higher reliability than other types of dams.

Acknowledgment

The authors are particularly indebted to the Research Group of Jintan Arch Dam Project, Tsinghua University. Ms.Shen Dali and Ms.Zhu Shouyi contributed much to the model tests. The tests were conducted under the auspices of Huadong Design Institute, China.

REFERENCES:

Fumagalli,E.1973. Statical and Geomechanical Models. Springer-Verlag/Wien
Oberti, G.and Fumagalli, E. 1963. Geomechanical models for testing the statical behavior of dams resting on highly deformable rock foundations. Rock Mechanics and Engineering Geology. 1/2 773-764
Zhou w.y., Yang. R.Q. 1984. The determination of stability factor of dam abutment using finite element method and geomechanical models. Proc. of 4th Newzealand and Australia Geomechanics Conf. 543-548.
Tokan0, M. 1960. Rupture studies on arch dam foundation by means of models. The Kansai Electric Power.
Paes de Barros, F., Colman, J.L. 1982. ITAPU project: The structural safety assessment through physical models. Proc. of 14th Inter. Commession on Large Dams. 1041-1058.

A three dimensional analysis of an arch dam abutment using a fracture damage model

Analyse tridimensionnelle des appuis d'un barrage-voûte à l'aide d'un modèle fissuré
Eine dreidimensionale Finite-Element-Analyse für das Widerlager einer Bogenstaumauer mit Hilfe eines Bruchschadenmodells

ZHOU WEIYUAN, Tsinghua University, Beijing, People's Republic of China
YANG RUOQIONG, Tsinghua University, Beijing, People's Republic of China
WU PENG, Tsinghua University, Beijing, People's Republic of China

ABSTRACT: In this paper, a three dimensional finite element analysis of Jintan arch dam abutments using fracture damage model is described. A damage mechanics model which involves in statistical parameters of rock strength is proposed. For jointed rock mass, the damage principles and tensors, and the damage propagation function are discussed. They are based on the triaxial test results of rockmass. In this paper, the numerical analysis results show that the damage model agrees very well with the phenomena of physical model.

RESUME: Dans ce document, a été faite une analyse d'élément limité de trois dimensions du fondement de construction de roche faisant appel au modèle de l'endommagement à fracture, analyse appliquée au fondement du barrage-voûte de Jinshuitan. Un modèle de la mécanique de l'endimmagement est proposé, modèle qui concerne les paramètres statistiques de la puissance de la roche. Le document aborde également la méthode répétitive moderne quant à l'èspace de l'usage excessif. Une analyse du fondement du barrage-voûte de Jinshuitan a été utilisée dans le programme mentionné ci-dessus.

Zusammenfassung: In der vorliegenden Arbeit wird eine dreidimensionale finite element Analyse fuer die seitlichen Gesteinsauflager mittels dem Bruchmodell Vorgestellt, das fuer Jinshitan Bogenstaumauer angewandet wird. Ein Bruchmechanismusmodell einschliesslich der statistischen Parameter der Gesteinfestigkeit wird entwickelt.

1. INTRODUCTION

Stability analysis of dam abutments is of paramount importance in dam design. In China, jointed rocks with deficient geolgical conditions are often encountered in dam sites.

The problems to be studied in the stability analysis of dam abutments are mainly involved in mechanical behavior and strength of jointed rock mass.

In this paper, a damage mechanics model is proposed. It treats the rock mass as a macroscopically non-homogeneous one and leads to the possibility of globally modelling the nucleation and the propagation of these discontinuities, including their effects on the stress-strain relations.

In this paper, presented there are the general theory and applications of damage model to the dam abutments. The generalization for tridimensional conditions which takes the anisotropic damage effects into consideration and some new proposals are made for evaluating the deterioration of rock masses.

2. BRITTLE FAILURE OF ROCK MASS AND PRINCIPLES OF DAMAGE MECHANICS

It is obvious that for rockmass, its geological formations form the main factors in mechanical characteristics. Generally, the rock mass behaves as mechanically anisotropic owing to the existence of a few sets of discontinuties which have their own specific orientations. From the results of rock specimen tests, the failure configurations of the specimens have shown that the failure surfaces are dependent on the pre-existing discontinuities in the rockmass. The fracture cracks would develop along some joints, bedding planes, fissures or faults under some loading conditions.

In Fig. 1, the fracture tests of multijoint gypsum specimens are shown. The cracks in these specimens, of which the fracture intensity factor ratios are greater than 5.0, initiated and propagated along the pre-existing joints apparently. Here the ratio is de fined between the intact rock and the weak joint surface.

In Jintan projects, three sets of joints are assumed in abutment rockmasses, of which the fracture toughness K_{Ic} and K_{IIc} are much smaller along the joint orientations than those which deviate from them. For example, from tests in laboratory, along joints, $K_{Ic}=15TM^{-3/2}$ $K_{IIc}=5.4TM^{-3/2}$, whereas in intact rock $K_{Ic}=160TM^{-3/2}$, $K_{IIc}=98Tm^{-3/2}$. This implies that the cracks are apt to extend along joint orientations.

From the above experiment results, some important conclusions could be drawn as follows:
- In rockmass, several pre-existing discontinuities could be found, which is refered to primary damage.
- These discontinuities form the weak links along which the fracture toughness is much smaller than those in intact rock. Hence, the damage effects of rock should be considered along these surfaces.

In rock masses, their damage propagations only involve in these weak links, i.e., the extension of weak joints.

If a damage tensor is specified for a rock mass, it implies that the damage configuration is prescribed. A damage model may be defined to describe this damage phenomenon.

Discontinuities in the rock mass such as joints, bedding planes flaws of certain-sized discontinuities in rock mass are assumed as a damage to intact rock. The damage effect may be considered as an deduction of area of rock mass or a deduction of modulus of elasticity.

3. DAMAGE MECHANICS MODEL

In case of anisotropic development of deterioration, the damage parameter is defined with a scalar D bearing upon the area of cross section. The cross sectional area of the rock mass is expressed as follows:
$$S = S_o \ (1-D)$$
Where D = 0 corresponds to intact rock
D = 1 corresponds to collapse
S and So represent the damaged and O riginal area respectively. As shown Fig.2, the tetrahydron has an surface area S.

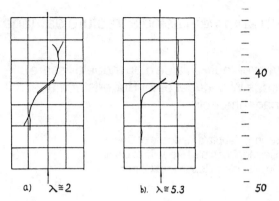

a) $\lambda \cong 2$ b). $\lambda \cong 5.3$

Figure (1)   Fracture tests of gypsum specimens

Let $\underset{\sim}{S}$ denote the surface vector, then
$$\underset{\sim}{S} = S.\underset{\sim}{n} = Si.ei \qquad (1)$$
After the rock damaged, the surface vector $\hat{S}$ is
expressed by: $\hat{\underset{\sim}{S}} = \hat{S}.\hat{n} = \hat{S}i.ei \qquad (2)$
Define $\hat{\underset{\sim}{S}} = (I-\underset{\sim}{D}).\underset{\sim}{S} \qquad (3)$
and substitute (1),(2) into (3), then
$$\hat{\underset{\sim}{S}} = (1-Di)si.ei \qquad (4)$$
In the above formula,"$\wedge$" denotes post-damage and
$\underset{\sim}{D}$ the damage tensor.
Its algebric form is
$$[I-D] = \begin{vmatrix} 1-D1 & & 0 \\ & 1-D2 & \\ 0 & & 1-D3 \end{vmatrix} \qquad (5)$$

Here, the tensor D is the so-called damage tensor. It is one of the second order. It is used to denote the deterioration of rock mass under a certain load.

Assume a force vector P is applied to the surface ABC as in Fig.2 then
$$\underset{\sim}{P} = \underset{\sim}{\sigma}.\underset{\sim}{s} = \hat{\underset{\sim}{\sigma}}.\hat{\underset{\sim}{s}} = \hat{\underset{\sim}{\sigma}}(I-\underset{\sim}{D}).\underset{\sim}{s} \qquad (6)$$
where $\hat{\underset{\sim}{\sigma}}$ is the net stress tensor.
From (6), the following expression is obvious.
$$\underset{\sim}{P} = \hat{\underset{\sim}{\sigma}}.\hat{\underset{\sim}{s}} \qquad (7)$$
Hence from (6) and (7),
and
$$\underset{\sim}{\sigma} = \hat{\underset{\sim}{\sigma}}(I-D)$$
$$\hat{\underset{\sim}{\sigma}} = \underset{\sim}{\sigma}(I-D)^{-1} \qquad (8)$$
Its algebric formula is
$$[\hat{\sigma}_{ij}] = \begin{vmatrix} \sigma_{11}\frac{1}{1-D_1} & \sigma_{12}\frac{1}{1-D_2} & \sigma_{13}\frac{1}{1-D_3} \\ \sigma_{21}\frac{1}{1-D_1} & \sigma_{22}\frac{1}{1-D_2} & \sigma_{23}\frac{1}{1-D_3} \\ \sigma_{31}\frac{1}{1-D_1} & \sigma_{32}\frac{1}{1-D_2} & \sigma_{33}\frac{1}{1-D_3} \end{vmatrix} \qquad (9)$$

4. ANISOTROPIC CONSTITUTIVE RELATIONS OF JOINTED ROCK

In (9), the tensor $[\hat{\underset{\sim}{\sigma}}]$ is used to express the damage effect on rock behavior under applied load.
Hence, $[\hat{\underset{\sim}{\sigma}}] = [E][I-\underset{\sim}{D}](\underset{\sim}{\varepsilon})$
and $[\underset{\sim}{\varepsilon}] = [I-\underset{\sim}{D}]^{-1}[E]^{-1}[\hat{\underset{\sim}{\sigma}}] = [C][\hat{\underset{\sim}{\sigma}}] \qquad (10)$
[C] is a fourth order tensor denoting the constitutive relations of rock masses. Generally, [C] can be obtained by superposition of deformations.
Let $[\underset{\sim}{\varepsilon}]_i$ denote the strain vector of rock mass which contains only the ith discontinuity in Fig.2. Then the total strain of the sample is:
$$[\underset{\sim}{\varepsilon}] = \sum_{i=1}^{n}[\underset{\sim}{\varepsilon}]_i = \{[C_1]+[C_2]+...[C_n]\}[\hat{\underset{\sim}{\sigma}}] \qquad (11)$$
where $([C_1]+[C_2]+...+[C_n])$ is a tensor denoting the constitutive relations of rock mass which contains n discontinuities.

In rock mass, the discontinuities are idealized as planes as shown in Fig.3. It is easy to express the damage tensor of rock mass with multiset discontinuities by superposition of damage tensors,
$$[D] = \sum_{i}^{n}[D]_i \qquad (12)$$

It should be noted that the comprehensive damage tensors for rock mass with multiset discontinuities

are easily expressed as (12) without evaluating fracture propagation of the comprehensive rockmass samples.

5. PROPAGATION OF DAMAGE

The damage propagation of rock mass can be expressed mainly by defining the propagation equations of damage for each set of discontinuities.

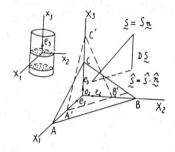

Figure (2)   Definition of damage tensor (after Murakami Ohno,1981

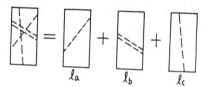

Figure (3)   Rock mass with multiset joints

For tridimensional problems, the damage criterion may be expressed at each stress state by a function of strains. It is obvious from experiments that tension ($\underset{\sim}{\varepsilon} > 0$) and shearing strains produce damage along some certain discontinuities, whereas the action of compressions prevent damage.

Generally, the damage phenomena accompany with strain softening behaviors. Thus, the damage criterion can be written as
$$Fi = \alpha J_1 + \beta J_2 - k = 0 \qquad (14)$$
where $J_1 = \varepsilon_1 + \varepsilon_2 + \varepsilon_3$
$J_2 = 2/3[(\varepsilon_1 - \varepsilon_2)^2 + (\varepsilon_2 - \varepsilon_3)^2 + (\varepsilon_1 - \varepsilon_3)^2]^{\frac{1}{2}}$
and K (ε_p) is a function of plastic strains.

Fi is defined in strain space. $[\varepsilon]$ lying on Fi will produce damage. Generally, Fi is a cone-shaped surface in strain space. K is a function of plastic strains. Therefore, Fi expresses set of damage surfaces in strain space. The criterion is proposed as follows.
If Fi = 0 and $\frac{\partial Fi}{\partial \varepsilon}.d\varepsilon \geq 0$,
$$d\hat{\underset{\sim}{\sigma}} = (I-\underset{\sim}{D}) E_p \cdot d\underset{\sim}{\varepsilon}$$
$$dD = \frac{\alpha K(1-A)}{(\alpha J_1 + \beta J_2)^2}dJ_2 + \frac{\beta A(\alpha+\beta)}{exp\,\beta(\alpha J_1 + \beta J_2 - k)}dJ_1 \qquad (15)$$

If $F_i = 0$ and $\frac{\partial Fi}{\partial \varepsilon}.d\varepsilon < 0$ $dD = 0$ (16)

Formula (15) expresses the damage effects caused by two factors:shearing strain and compressive strain which give dD < 0.

Fi is a damage criterion for a certain discontinuity in rock and is refered to as the ith set of dicontinuities.

6. DETERMINATION OF DAMAGE PARAMETERS BY FRACTURE TESTS

As proposed in the last section, Fi is defined by a function in terms of J_1, J_2 and K. Here, the parame

590

ters α, β and formula (15) can be obtained by experiments, i.e., fracture tests. The specimens are prepared by assuming specimens from rock and then making a prescribed crack in the specimens. For specimens as such, the area vector S is assumed as follows:

$$\hat{S} = S\ e_1 + S\ e_2 + S\ (1-D)\cdot e_3$$

Damage tensors $\underset{\sim}{D}$ may be defined as $\underset{\sim}{D} = \underset{\sim}{D}(\underset{\sim}{n} \otimes \underset{\sim}{n})$; where $\otimes$ denotes the tensor product, $\underset{\sim}{n}$ is the unit normal of the ith set of damage. In the avove case, $n = e3$ hence $\hat{\underset{\sim}{S}} = \underset{\sim}{S} - S(e3 \otimes e3)$

Triaxial tests of the specimens should be conducted, and α, β, K can be obtained. In other words, to conjecture the charateristics of rock mass, only the properties of rock specimens with discontinuities are needed.

7. A DISCRETIZATION PROCEDURE FOR F.E.M.

The F.E.M. discretization is carried out by the increment load method. Their anisotropic stiffness is determined by stiffness iterative method, i.e., initial stress method.

Suppose that [Ks] is the symmetric part, and [Kani] is the anisotropic part, then,

$$([Ks] + [Kani])\cdot(U) = (F)$$
$$[Ks]\ (U) = (F) - [Kani]\cdot(u)$$
$$(\Delta F) = [Ks]\ (\Delta U)$$

They can be solved by using iterative method.

In Jintan project, three sets of joints were assumed. Their geometric features and damage tensors are shown in Table 1.

Table 1. Geometric features and area of damage.

no.of joint	strike angle	dip angle	Area of damage D_i
1	N45°E	85°NW	0.5
2	N45°W	45°NE	0.5
3	N75°W	60°NE	0.5

Assume that the Cartesian coordinate system oxyz is a northward system, then the damage tensor

$$\underset{\sim}{D} = \underset{\sim}{D1} + \underset{\sim}{D2} + \underset{\sim}{D3}$$

$$\underset{\sim}{D1} = D2 \begin{vmatrix} 0.4665 & 0.4665 & 0.0595 \\ 0.4665 & 0.4665 & 0.0595 \\ 0.0595 & 0.0595 & 0.0076 \end{vmatrix}$$

$$\underset{\sim}{D2} = D3 \begin{vmatrix} 0.2066 & -0.2066 & 0.3482 \\ -0.2066 & 0.2066 & -0.3482 \\ 0.3482 & -0.3482 & 0.5868 \end{vmatrix}$$

$$\underset{\sim}{D3} = D3 \begin{vmatrix} 0.6998 & -0.1875 & 0.4183 \\ -0.1875 & 0.0502 & -0.1121 \\ 0.4182 & -0.6121 & 0.25 \end{vmatrix}$$

$$\underset{\sim}{D} = \begin{vmatrix} 0.686 & 0.036 & 0.413 \\ 0.036 & 0.362 & -0.401 \\ 0.413 & -0.401 & 0.422 \end{vmatrix}$$

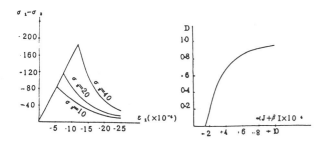

Figure (5)   Damage propagation parameters

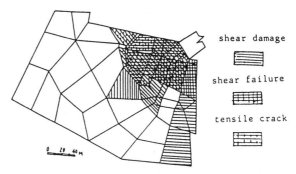

shear damage

shear failure

tensile crack

Figure (6) F.E.M. evaluation for Jintan arch dam abutment

From triaxial compression tests, α and β in formula (14) have been determined. From the stress-strain relations by uniaxial compression tests of Jintan rock samples, A and B are evaluated as below:

$$\varepsilon_a = -30\cdot10^{-4} \qquad \varepsilon_b = -60\cdot10^{-4}$$
$$\sigma_a = 14.0\ N/M^2 \qquad \sigma_b = -6.0\cdot10^{-4}N/M^2$$
$$J_a = 22.5\cdot10^{-4} \qquad J_b = 45\cdot10^{-4}$$
$$I_a = -18\cdot10^{-4} \qquad I_b = -36\cdot10^{-4}$$

Hence   A = 0.8048
         B = 0.4739 10^4

The damage propagation parameter D and predicted stress strain relations are shown in Fig. 5.

In Fig. 6, a three dimensional F.E.M. analysis of Jintan arch dam abutment is shown. Under increasing water load, the rock mass in abutment deteriorates from elastic deformation to shear damage, tensile cracks and shear failure. This analysis agrees well with the result from geomechanical model test (Zhou W.Y. et al., 1986).

CONCLUSION

To express the function of anisotropic rock mass behavior is difficult. However, by using the damage tensors and F.E.M, it is convenient to express the damage parameters.   It should be noted that by damage mechanics, the sample tests from discontinuities and from intact rock are sufficient to predict the behavior of jointed rock mass without the needs of considering the scale effect in them. It is effective to predict the stability of rock mass by means of damage mechanics.

REFERENCES:

T.Kyoya, Y.Ichikawa and T.Kawamoto. 1985. A damage mechanics theory for discontinuous rock mass. Fifth International Conference on Numerical Methods in Geomechanics.
Mazars, J.1985. Mechanical Damage and Fracture of Concrete Structures. 5th Int. Conf. on Numerical Methods in Geomechanics.
Jean Lemaitre.1984. How to use Damage mechanics. Nuclear Engineering and Design. 80: 233-245
D.Krajcinovic and G.U. Fonseka.1981. The continuous damage theory of brittle materials. Trans. ASME J.Appl. Mech. 48.
J.Lemaitre. 1984. A continuous damage mechanics model for ductile fracture. Trans. ASME J.Eng. Mater. Technol. 38: 223-230
Zhou W.Y., Yang R.Q and Luo G.F. 1986. An integral geomechanical model rupture test of Jinshuitan arch dam with its abutment. Proc. 6th Int. Congr. I.S.R.M.
Zaitsev, Y.and Wittmarn, F.H. 1973. Fracture of porous viscoelastic materials. 3rd Int. Conf. on Fracture. 1-323.

3

Rock blasting and excavation
Sautage et excavation
Sprengen und Ausbruch

Penetration rate prediction in percussive drilling
Prédiction du taux de pénétration dans le forage à percussion
Vorhersage der Bohrleistung für das Schlagbohrverfahren

J.BERNAOLA, School of Mines, Polytechnical University of Madrid, Spain
P.RAMIREZ OYANGUREN, School of Mines, Polytechnical University of Madrid, Spain

ABSTRACT   The drilling speed of a percussive drill may be predicted   from a drillability laboratory test as a function of the drill percussion power, the hole diameter and the bit cutting edge length.

RESUME   La vitesse réelle de forage d'un marteau peut être déterminée à partir de test de forage en laboratoire, comme une relation entre la puissance de percution, le diamètre de forage et la longueur du fil du taillant.

ZUSAMMENFASSUNG   Die Bohrgeschwindigkeit eines durchschlagshammers ist durch einen laborversuch vorawszusehen der auf die durchschlagskraft (-leistung), den durchnesser des bohrers und die Laenge der bohrerkante (-klinge) basiert ist.

INTRODUCTION

This report describes a method to estimate penetration rate in percussive drilling based on a simple laboratory test carried out on small rock samples. - The test is a reduced scale reproduction of the actual percussive drilling phenomenon. The small rock samples must be representative of the rocks existing at the site. Therefore it is recomended to do a geological study previously to the sample collection.

DRILLING EQUIPMENT FACTORS AFFECTING PENETRATION RATE

These factors can be classified in two groups:

- Factors depending on the drill rig
- Factors depending on the drilling rods and bits.

Among the first ones, the most important are: percussion power, rotation and feed force. From those depending on the drilling accessories, the diameter of the hole, the type of bit and the number of rods have to be mentioned.

Percussion power

The percussion is the main energy source in percussive drilling and hence it has the biggest influence on penetration rate. Percussion power is usually expressed as the product of the number of blows per minute by the kinetic energy of the piston at the - instant when it hits the drill string.

Rotation

The basic purpose of the rotation in percussive drilling is to provide indexed blows so that each - crater breaks into the preceding one, taking advantage of the additional free face created and increasing the drilling efficiency.

Theoretically, for each application, there is an ideal indexing angle and, consequently, and optimum rotary speed, which provides biggest cutting sizes and maximum efficiency.

Nevertheless, rotation in percussive drilling is just an auxiliary operation, which only means 10-15% of the power output.

Feed force

This force is necessary to mantain the bit in - contact with the rock in order to make the energy transfer between them possible.

The increase of this force beyond the value -- which assures a continuous interface contact, does not mean any substancial improvement in the penetration rate, but rather increases the wearing rate of the cutting tool.

Hole diameter

The diameter of the hole is one of the most important factors affecting the penetration rate, because the percussion energy is distributed over a -- larger area as the diameter increases.

Type of bit

The type of bit has also a certain influence on penetration rate. A button bit, for example, does not penetrate so much into the rock in each blow. However, the number and distribution of the buttons is - such, that fewer blows are necessary to cover the full surface of the hole. Therefore, the penetration rate achieved with this type of bits is a little higher that with cross bits.

For a similar reason, single chisel bits of integral steels provide lower penetration rates than cross bits.

Number of drillings rods

In drifter drilling the hole depth has also an influence on the penetration rate. In each meter of rod, 0'2-0'4% of the shock wave energy is converted into heat as a result of friction forces. In the same way, in each coupling, about 3% of the energy is reflected backwards and an additional 5'5% is lost due to friction. Therefore, it can be concluded that in each rod 9-10% of the available energy is dissipated and consequently the penetration rate will be reduced in the same proportion.

THEORETICAL FORMULATION OF PENETRATION RATE

The horizontal projection of the area of contact between the cutting blade and the rock in the case of a cross bit, is function of the penetration achieved and can be expressed (see Fig. 1) by the following formula:

$$a = 2hl \ tg \ \frac{\alpha}{2}$$

where:
- "a" is the projected area
- "h" is the penetration
- "l" is the length of the cutting edge
- "α" is the wedge angle of the cutting tool, which is approximately the same for all bits.

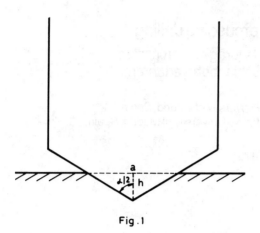

Fig.1

The "F" force necessary to achieve a penetration "h" can be obtained by using the expression:

$$F = \sigma.a = 2\sigma.1.h.tg\frac{\alpha}{2}$$

and:

$$\frac{F}{h1} = 2\sigma tg\frac{\alpha}{2} = k$$

where "σ" is the resistance of the rock to penetration.

From this expression it can be concluded that, assuming that   is a constant, the ratio F/h1 is only function of the mechanical properties of the rock - and will be referred as "k".

The energy which should be applied to the bit to get a penetration "h" can be calculated by:

$$T = \int_0^h 2\sigma.1.h\ tg\frac{\alpha}{2}.dh = k\ \frac{h^2 1}{2}$$

If "η" is the efficiency of the energy transfer to the rock and "E" the impact energy:

$$T = \eta.E = \frac{kh^2 1}{2}$$

$$h = \sqrt{\frac{2\eta E}{k1}}$$

The influence of the indexing angle on drilling efficiency was first studied by H.L. Hartman (1966) Nowadays the following proportion for the optimum indexing angle between two consecutive indentations, is generally accepted:

$$\theta_{opt} \sim \phi^{-1}.k^{-1}.E^{1/2}$$

Therefore, the minimum number of blows which are necessary for completion of a cylinder of "h" height with a cross bit, will be:

$$n \sim \phi.k.E^{-1/2}$$

And the drilling rate with a drill giving N -- blows/min:

$$V = \frac{N}{n}.h \sim N.1^{-1/2}\phi^{-1}.k^{-3/2}.E.\eta^{1/2}$$

The efficiency "η" of the energy transfer to the rock depends on several factors:

- The rock type
- The hole area
- The drill string mass
- The feed force
- The drill piston dimensions

Taking into account all these variables would require an specific study of each application and, on the other hand, it would be of limited practical value, due to the 1/2 power of the "η" factor.

For this reason, an average value of "η" may be assumed and the differences in the efficiency overlooked.

With this assumption, we can define a standard penetration rate:

$$Vs = \frac{V.1^{1/2}.\phi}{N.E}10^6$$

that, for each type of rock is only function of its physical properties, which will be reflected in the "drillability index".

The units of the variables in the "Vs" formule would be as follows:

V: Penetration rate (in/min)
E: Impact energy (lb-ft)
N: Blow frequency (b.p.m.)
ϕ: Hole diameter (in)
1: Length of the cutting edge (in)

DESCRIPTION OF THE LABORATORY TEST

A laboratory equipment, used by Joy Mfg. Co., was selected to carry out the drillability test, since it seemed to be one of the simplest ways to reproduce - the actual phenomenon to scale.

It consist on an electric percussive drill, which is free to move vertically along two guide shafts. - Under a certain constant feed, the machine drills some 10 mm holes in a rock sample which is inserted in a steel cup. The laboratory bit is a simple winged one, with a wedge angle of 120º. The drilling time is measured by means of an automatic timer controlling the electric drill motor. The cuttings are removed - by an air nozzle.

The samples should be first sized, unweathered - and representative of the rock to be drilled. They are cut and positioned showing a flat horizontal surface on top, that will serve as a reference plane to measure the hole depth.

At least three holes per sample should be drilled.

After having cleaned the holes thoroughly, the - depth from the surface to the deepest part of the hole is measured using a depth micrometer.

The penetration rate in the laboratory test is - calculated from the measured hole depth. and the timed drilling interval. The average penetration rate, expressed in inches per minute is taken as the "Ip drillability index" which is indicative of the rock susceptibility to be drilled.

PENETRATION RATE PREDICTION

In order to find the correlation between Ip and the standard drilling rate "Vs" a total of 20 different types of rock that ranged from the softest to - the hardest available ones, were tested.

Using several drills, which impact energy and - blow frequency were known, different diameter holes were drilled in the field and the drilling times for the first rod were measured. Afterwards, the corresponding standard penetration rates Vs were calculated.

From 340 field measurements, corresponding to - cross bits (see the attached table I), the following regression line, with a correlation coefficient close to 0.9 was obtained:

$$Vs = \frac{V.\phi.1^{1/2}.10^6}{E.N} = 51\ Ip + 90 \qquad (1)$$

Based on this formula, the drilling speed attainable for the first rod with a certain percussive - drill in a determinated rock, can be estimated as a function of:

- The rock drillability index Ip.
- The drill percussive power ExN
- The hole diameter ϕ

- The length of the bit cutting edge "1"

As mentioned before, all the above applied to -
percussive drilling with cross bits. Some data were
taken with button and chisel bits and the observed
discrepancies with the calculated Vs for cross bits
averaged a 17-18% difference, which was positive in
case of the button bits and negative for the chisel
bits.

However, in both cases a 15% difference would
possibly be closer to the reality.

Thereforem the V values calculated for the cross
bits should be corrected by either 1.15 or 0.85 fac
tor when button bits or chisel bits are respectively
used.

CONCLUSIONS

The penetration rate in percussive drilling with
cross bits seems to be directly proportional to the
drill percussion power ExN and inversely proportional to the hole diameter $\emptyset$ and to the square root -
of the bit cutting edge length.

On the other hand, the described laboratory test
is a good way of predicting actual penetration rates
in percussive drilling, provided that the tested --
samples are fully representative of the rock types
to be drilled.

LITERATURE

Hartman, H.L. 1966. "The effectiveness of indexing
    in percussion and rotary drilling" International
    Journal Rock Mechanics and Mining Sciences. pp.
    265-278.

Ip \ EN/$q^{1/2}$	22 000	35 000	43 000	51 000	53 000	54 000	56 000	61 000	62 000	66 000	68 000	74 000	77 000	84 000	85 000	87 000	88 000	90 000	93 000	95 000	103 000	108 000	113 000		129 000	130 000	143 000	169 000	182 000
2,81		271	267	275		259	250	279			250 257	277	247		259		261	244	247		252	241 259 296B				254	252	219	
4,35		329	326	333		324		336			309 316	324	305		306		324	306	290		316	306 306				315	301	266	
4,65		243	279	333		268	295	344			324 324	338	312		318		335	317	301	358	335	315 315 366B				327	311	278	
5		357	349	353		343	304	361			331 338	351	325		329	333B	347	328	312		350	329 329 380B		384B	477B	338	325	290	
5,15													331				352			253I		333				338		293	
5,23		357	360	363		352		377			353 346	365	338		341		352	339	323		359	338 343				346	336	296	
5,64													351				375			284I		352				362		308	
5,80		371	384	402		380		410			382 375	392	351		376		375	356	355		379	375 361 431B				369	371	314	
5,91		386	384	402		380		418			390 375	399	357		376		386	361	360		383	380 361 435B				373	374	317	
5,95		386	384	412		389	357	418			382 397	405	364		382		386	367	366		383	370 384 444B				377	378	320	
6,68		428	430	451		426		459			426 441	446	403		424		432	406	398		432	412 426 491B				419	420	355	
6,94		443	453	461		444		475			441 456	459	416		435		443	428	409		451	426 440 505B				431	430	367	
7,49		486	477	510		491		533 516			471 485	493	455		465		483	456	446		485	463 472 537B				469	462	396	
7,70	455	500	500	510		500	446	525			485 500	500	468		476		500	472	452		500	472 481 551B				477	476	408	
8,23		529	535	549		528		525		485	515 529	527	494	476	506		528	506	484		534	509 514 588B				508	507	432	
8,39		543	535	549		537		566			529 544	541	506	542	518		545	511	489		544	519 523 593B				523	514	444	
8,61		557	547	569		556		574			551 537	547	519		529		557	522	500		558	532 528 611B				531	521	456	475I
11,17		743	721	725		722		738			706 691	689	688		671		722	689	634		723	685 694				685		586	
11,78			860B	755		898B	670	770	984B		750 846B	716	727		706		761	833B	667			731 731 815B				723	699	615	
18,90						981																	1177B						

TABLE I: Standard penetrations rates calculated from field drilling speed.

NOTE: B: Button bit I: Chisel bit

Leistung und Grenzen des Fräsvortriebes beim Ausbruch eines 22 km langen Druckstollens

Advance and limits of TBM-driving during the excavation of a 22-km-long pressure tunnel
Effet et limites de l'avancement par tunnellier en suite de l'excavation d'une galerie en charge de 22 km

B.BONAPACE, Leiter der Projektierungsabteilung, Tiroler Wasserkraftwerk AG (TIWAG), Innsbruck, Österreich

ABSTRACT: The main part of the Strassen-Amlach Hydroelectric Power Scheme of TIWAG - currently under construction in Austria's East-Tirol - is the 22-km-long headrace pressure tunnel. The tunnel first cuts across an approximately 14-km-long section of massive dolomitic limestone in the "Lienzer Dolomiten", followed by the tectonically stressed layers of the Calcareous Alps and the crystalline basis of the Gail Valley.
In May 1985 two full-face tunnelling machines bore diameter 3.9o m started from two adits in upward direction. The TBM-driving for the extremely long Tunnel section in the "Norian" dolomite, 13,7 km long, was completed within 19 months at an average advance rate of 36 m per working day. In the second section, 8.2 km long, where the geological conditions often changed a geological accident accured. Water, mud and rock material rushing in after a gypsum-karst system of the "Raibler series" had unexpectedly been encountered, twice filled up the tunnel over a length of 65 m and also buried the tunnelling machine so that all tunnel driving work was stopped for lo months. The paper describes the geological conditions, the basic data of the tender, the experience gathered during tunnel driving as well as the measures taken in order to recover the buried tunnelling machine and to cope with the collapse zone.

RESUME: La galerie en charge de 22 km de long constitue l'élément principal de la Centrale Hydroélectrique Strassen-Amlach qui se trouve actuellement en voie de construction dans la province autrichienne du Tyrol de l'Est sur l'initiative de la Société Hydroélectrique du Tyrol. Dans le massif des Dolomites de Lienz le tracé de la galerie traverse sur une longueur de 14 km dolomite superieure, roche solide, ensuite une séries de structures tectoniques des Alpes calcaires et finalement la base cristalline de la vallée de Gail.
Depuis le mois de mai 1985 la galerie est excavée au moyen de deux tunneliers avec un diamètre de forage de 3,9o m. L'excavation est, en outre, réalisée en montant, depuis deux points d'attaque. Le tronçon situé dans la dolomite superieure a une longueur exceptionnelle de 13,7 km et a été excavé dans un délai de 19 mois avec un rendement moyen d'excavation de 36 m par jour ouvrable. Dans le deuxième tronçon, 8,2 km de long, faisant preuve des conditions géologiques fortement variables, eu lieu un accident géologique grave. Le tunnelier a roncontré imprévue un systéme karstogypseux de la série "Raibler" de lequel sorti des venues d'eau, d'argile et des matériaux pierreux qui out provoqué deux fois l'ensevelissement du tunnelier et le remblayage de la galerie sur une longueur de 65 m, ainsi les travaux d'excavation ont été interrompus pendant lo mois. Le rapport ci-après expos les conditions géologiques, les données de base contenues dans l'appel d'offres, les expériences acquises durant l'avancement du tunnel, les travaux de récupération du tunnelier ainsi que les méthodes adaptées pour franchir l'accident géologique.

ZUSAMMENFASSUNG: Der 22 km lange Druckstollen ist das Herzstück des Draukraftwerkes Strassen-Amlach, das derzeit von den Tiroler Wasserkraftwerke AG (TIWAG) in Osttirol, Österreich errichtet wird. Die Stollentrasse durchfährt im Gebirgsstock der Lienzer Dolomiten auf ca. 14 km standfesten Hauptdolomit, jedoch in der Folge die tektonisch beanspruchten Stockwerke der Kalkalpen und die Kristallinbasis des Gailtales. Der Stollen wird mit zwei Vollschnittfräsen, Ausbruchsdurchmesser 3,9o m, seit Mai 1985 von zwei Angriffsorten jeweils steigend vorgetrieben. Der Stollenabschnitt im Hauptdolomit, mit der außergewöhnlichen Länge von 13,7 km wurde in 19 Monaten mit Durchschnittsleistungen von 36 lfm/Arbeitstag aufgefahren. Im zweiten 8,2 km langen Stollenabschnitt mit stark wechselnden geologischen Verhältnissen erfolgten Einbrüche von Wasser, Lehm und Gesteinsmaterial aus einem unvermutet angefahrenen Gips-Karstsystem der Raibler Rauhwacken, die den Stollen samt Fräse zweimal auf 65 m Länge verschütteten und den Vortrieb lo Monate stoppte. Die geologischen Verhältnisse, die Grundlagen der Ausschreibung, die Erfahrung bei der Ausführung der Vortriebsarbeiten und die Maßnahmen zur Bergung der verschütteten Fräse und zur Bewältigung der Verbruchszone werden beschrieben.

1. GEOLOGIE

1.1 Geologische Verhältnisse

Der Druckstollen durchfährt in W-O-Richtung, parallel zum Drautal, die südlich davon liegenden Formationen des Gailtal-Kristallins und der Lienzer Dolomiten.

Die Schiefergneise, Amphibolite und Glimmerschiefer des Gailtal-Kristallins treten im westlichsten Abschnitt des Stollens zutage, sie bilden die Basis der Lienzer Dolomiten, die wiederum aus Schollen unterschiedlichen Alters vom oberen Perm bis zum unteren Jura aus Kalken, Dolomit, Mergel und Tonschiefer aufgebaut sind. Bei der Gebirgsbildung wurden die Schichtpakete verfaltet und verschuppt, sodaß die zeitliche Reihenfolge der Gesteinsserien oft nicht mehr erhalten geblieben ist. Der westliche Abschnitt der Lienzer Dolomiten mitsamt seiner Kristallinunter-

lage liegt im Einflußbereich der alpinen Großstörungen des Drautales und des spitzwinklig einmündenden Gailtales. Besonders in der Aufschiebungszone der Lienzer Dolomiten auf die Kristallinbasis, der sogenannten "Abfaltersbacher Schuppenzone", - dem 3 km langen Stollenabschnitt zwischen Hauptdolomit und Kristallin -, ist der Schollenbau durch tektonische Einwirkung in Unordnung geraten. Die baugeologische Prognose war hier äußerst erschwert. Der östlich anschließende Stollenabschnitt liegt in einem einheitlichen Dolomitstock. An seinem Ende tritt der Stollen auf einige Hundert Meter in die Kössener Schichten ein, die dem Hauptdolomit entlang des Drautales vorgelagert sind, Abb. 1.

Bei der Ausführung des Stollens fand man die prognostizierte Reihenfolge der Gesteine, - mit Ausnahme in der Abfaltersbacher Schuppenzone - gut bestätigt. Der Druckstollenabschnitt Amlach verlief nach den

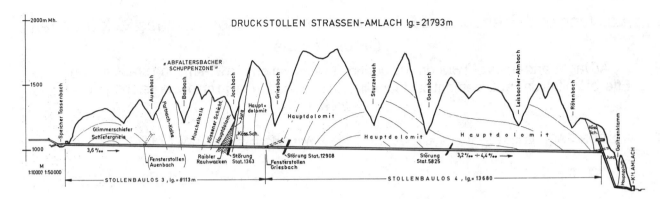

Abb. 1. Geologischer Längenschnitt

ersten 4oo m Kössener Schichten durchwegs im Hauptdolomit. Der Stollen des Bauloses Griesbach verläßt
nach 6oo m den Hauptdolomit und durchörtert in dich-
ter Folge Plattenkalke, Kössener Schichten, Jura-
kalk und ab der Störung bei Station 1363 Raibler
Schichten von ca. 45o m Mächtigkeit, die bei der obertägigen Kartierung nicht vorausgesehen werden konnten
Abb. 2. Die tektonisch angeschoppten Rauhwacken brach-
ten eine folgenschwere Behinderung des Vortriebes
durch das unvermittelte Eindringen der Fräse in ange-
füllte Lösungshohlräume der Gips- und Anhydriteinla-
gerungen, deren Inhalt aus Wasser, Schluff und bruch-
schotterartigen Lockermassen sich in den Stollen ergoß und die Vortriebseinrichtung zweimal begrub.

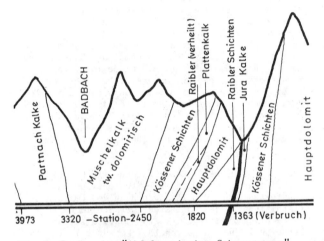

Abb. 2. Serien der "Abfaltersbacher Schuppenzone"

Der folgende 24o m lange Abschnitt der Raibler
Serien bestand aus einer kohäsionslosen Masse zer-
fallener Rauhwacken und Trümmerdolomit mit Einlagerungen von Gips, Lehm und Schwimmsanden, welche mit
dem Bergwasser auszufließen neigten. Diese Zone mußte
deshalb konventionell durchörtert werden, wobei
vorauseilend das Gebirge abschnittsweise durch Drainagebohrungen entwässert wurde. Ab dieser Schwächezone verhielten sich die Raibler Schichten standfest
und konnten gefräst werden. Anschließend wechseln
die Serien wieder in bunter Folge aber gut standfest,
zwischen Hauptdolomit, Gipsbrekzie, Plattenkalk und
die mächtigeren Kössener Schichten, Muschelkalk und
Partnachkalke bis Station 4ooo an der Grenze zum
Gailtal Kristallin.
Die Gesteine des Kristallin, vor allem Gneise, Glimmerschiefer und Einlagerungen von Amphibolit streichen
spitz zur Stollenrichtung. Sie neigen stark zu Nach-
brüchen aus stollenparallel verlaufenden, weichen
Glimmerschieferbändern im Verschnitt mit Kluft- und
Harnischflächen.

1.2 Hydrogeologische Verhältnisse

Die glimmerreichen Gesteine des Gailtaler Kristallins
und die tonreichen Wechsellagen in den Kössener
Schichten der Lienzer Dolomiten sind vorwiegend dicht,
während die Hartgesteine Amphibolit, die kalkreichen
Schichten und der Hauptdolomit, soweit sie bei der
Gebirgsbildung zerlegt wurden, wasserführend sind.
Durch eine sorgfältige Kartierung der Quellaustritte
im Projektsgebiet und Messung der Schüttung, Temperaturen, Leitfähigkeit und des Chemismus wurde versucht, die hydrogeologischen Verhältnisse klarzu-
legen.
Im Bereich des Gailtal-Kristallins treten die
meisten Quellen aus dem Hangschutt. In der Annahme,
daß die Amphibolite linsenförmig in den dichteren
Gesteinen der Gneis-Glimmerschiefer eingelagert sind,
werden in diesem Stollenabschnitt wenig Bergwasserzutritte erwartet. Dies ist beim Vortrieb bisher auch
bestätigt worden.
In der Abfaltersbacher Schuppenzone, wo dichte Ge-
steinspakete mit durchlässigen wechseln, war mit entsprechend unterschiedlichen Wassereintritten gerechnet worden. Bei der Aufnahme der Quellen unterschied sich die Heilquelle am Badbach deutlich von
den übrigen Bergwasseraustritten durch höhere Temperatur, Sulfatgehalt und Leitfähigkeit. Sie schien
aus tiefliegenden, obertägig nicht erkennbaren, gips-
führenden Gesteinsschichten aufzusteigen. Aufgrund
der Abnahme der Schüttung und ihrer Zusammensetzung
ist die Quelle jedoch mit dem kilometerweit entfern-
tem Wassereinbruch aus der Störung im Stollen Griesbach in Zusammenhang.
Dem Hauptdolomitstock sind zum Drautal hin die
Kössener Schichten vorgelagert. Deren dichte Ton-
lagen bilden einen Wasserstau, der sich durch Quellaustritte an der Grenze zum Dolomit verfolgen läßt.
Die Druckstollentrasse verläuft innerhalb dieses
Aquifers und wirkt als Drainage. Demnach sind im
Hauptdolomit starke Bergwasserzutritte und aus Klüften konzentrierte Wassereinbrüche von mehreren
loo l/s prognostiziert worden. Beim Vortrieb trat
das Bergwasser auf weiten Strecken flächig verteilt
in den Stollen, wobei die spitzwinklig zur Stollenrichtung verlaufenden Schichtfugen ausgezeichnete
Wasserwegigkeit aufwiesen und die Schüttung der Über-
laufquellen bis zu 15oo m vorauseilend beeinflußten.
Die Wassereinbrüche aus den Klüften waren nicht
so gewaltig wie vorausgesagt, verursachten jedoch an
zwei Störungen Einschwemmungen von bruchschotter-
artigem Material. Die Gesamtschüttung am Portal blieb
ab Stationsmitte immer über 6oo l/s, max. 81o l/s.

2. GRUNDLAGEN DER AUSSCHREIBUNG

2.1 Technischer Entwurf - Druckstollen

Der 22 km lange Stollen des Draukraftwerkes wird mit
maximal 11 bar Innendruck belastet und deshalb als
Druckstollen mit durchgehender Betonauskleidung ausgebildet. Die Auskleidung wird, soweit der Bergwas-
serspiegel nicht nachweisbar über dem Innendruck

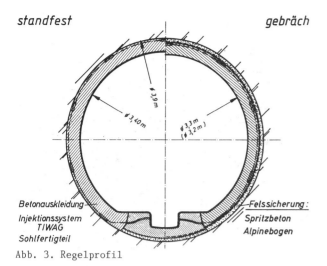

standfest   gebräch

Betonauskleidung ——————————— Felssicherung:
Injektionssystem   Spritzbeton
TIWAG   Alpinebogen
Sohlfertigteil

Abb. 3. Regelprofil

liegt durch Spaltinjektionen nach dem System TIWAG
vorgespannt.

Um genügend Überlagerungshöhe bei der Querung der
tief eingeschnittenen Seitentäler zu erhalten war es
notwendig, die Stollentrasse weit ins Berginnere zu
verlegen. Andererseits wurden dadurch hohe Überlage-
rungen bis zu 7oo m erreicht, bei denen Gebirgsdruck-
erscheinungen zu erwarten waren.

Das Regelprofil, Abb. 3 wurde so gewählt, daß je
nach erforderlicher Ausbaustärke für die Sicherung
noch ein mindestens 2o cm starker Betoninnenring eingebaut werden kann. Bei einem Ausbruchsdurchmesser
von 3,9o m muß die Betonschalung auf Innendurchmesser
3,2o bzw. 3,3o oder 3,4o einstellbar sein.

Aufgrund der baugeologischen Verhältnisse wurde der
Stollen in 2 Baulose unterschiedlicher Länge unter-
teilt, in das Baulos Amlach mit der außerordentlichen
Druckstollenlänge von 13,7 km und dem Fensterstollen
(Länge 24o m) sowie das Baulos Griesbach mit dem
8,2 km langen Druckstollenabschnitt und dem 69o m
langen Fensterstollen. Beide Abschnitte werden steigend mit Vollschnittfräsen, Ausbruchsdurchmesser
3,9o m aufgefahren. Im Hinblick auf den großen Berg-
wasserandrang aus dem Hauptdolomit wurde der Stollen
mit zunehmendem Gefälle von 3,2 bis 4,4 ‰ ausgelegt
und ein Sohltübbing mit integriertem Wassergraben vorgesehen, der mit dem Vortrieb mitlaufend verlegt wird.
Mit größtem Gefälle vermag er bis zu 25o l/s und bei
Überflutung bis auf Schienenoberkante ca. 1ooo l/s
Stollenwasser abzuführen.

2.2 Geologische und geotechnische Erkundung

Zur Beurteilung der Geologie und Hydrologie des Drau-
stollens wurde eine Detailkartierung des Gebietes
vorgenommen und geologische Lagepläne und Schnitte
ausgearbeitet. Durch ein geologisches Modell wurden
die komplexen Verhältnisse räumlich dargestellt.

Von allen Gesteinsarten wurden an der Oberfläche
Proben entnommen und im Triax-Labor-Versuch die geotechnischen Kennwerte, Scherfestigkeit, Druckfestigkeit und E-Modul sowie der Quarz- und Karbonatgehalt
ermittelt und samt Beschreibung der Gesteine und ihrer
Lagerung und Klüftung in einem Gesteinskataster dargelegt.

An insgesamt 42 repräsentativen Gesteinsaufschlüs-
sen wurden detaillierte Gefügemessungen durchgeführt
und statistisch erfaßt. Die Gefügedaten und die geotechnischen Kennwerte wurden nach den Bewertungssyste-
men von BIENIAWSKI, BARTON und WEBER ausgewertet und
auf dieser Grundlage für den Druckstollen ein baugeo-
logischer Prognoselängenschnitt ausgearbeitet.

In einem Sondierstollen im Hauptdolomit wurde ein
Radialpressenversuch durchgeführt und der Verformungsmodul ermittelt. An diesem Großversuch konnten seismische Meßwerte geeicht und Vergleiche mit den Werten

aus den Triax-Labor-Versuchen vorgenommen werden.

Auf Grundlage der Quellkartierung und Messung der
Schüttung, Temperatur und Chemismus aller Quellaustritte wurde eine Studie über die hydrogeologischen
Verhältnisse erstellt und die zu erwartenden Gebirgs-
wasserverhältnisse im Längenschnitt dargestellt.

2.3 Gebirgsklassifizierung

Die Vergütung des Ausbruches und der Vortriebssiche-
rung erfolgt nach Gebirgsklassen, soweit der Stollen
fräsbar ist. In Bereichen, die nicht mehr fräsbar
sind und Sondermaßnahmen erfordern, die nicht kalku-
lierbar sind, wird nach Aufwand vergütet, d.h. das
geologisch bedingte Risiko trägt der Bauherr.

Der Ausschreibung wurde eine Gebirgsklassifizierung
für jeden Gesteinsbereich, Hauptdolomit, Kössener
Schichten, Partnach Kalke, Gneisglimmerschiefer,
Amphibolit zugrundegelegt. Im Sinne einer Klassifi-
zierung wird jeder Gesteinsbereich hinsichtlich der
Beschaffenheit des Gebirges nach Klassen unterschieden, die gekennzeichnet sind durch:
- den Ort und den Zeitpunkt des Einbaues der erforderlichen Stützmaßnahmen
- das Ausmaß der erforderlichen Stützmaßnahmen.

Die TIWAG hat hiebei eine Gebirgsklassifizierung
auf Grundlage der österreichischen Norm vorgenommen,
die jedoch die Aufwendungen für die Ausbruchssicherung
und einer zumutbaren Wassererschwernis einschließt.
Die Unterteilung nach Klassen F1 bis F6 berücksichtigt
insbesondere die unterschiedliche Arbeitsbehinderung
der Vortriebsmaschine durch den Einbau der erforderlichen Stützmaßnahmen, die erschwerten Einbaubedingungen und die Kosten der Sicherungsmittel. Darüber
hinaus gibt es für jede Gebirgsklasse nach Menge ge-
staffelte Zuschläge für Wassererschwernisse über
5 l/s.

Die Stärke der erforderlichen Stützmaßnahmen wurde
durch die Berechnung des Ausbauwiderstandes nach dem
Kennlinienverfahren abgeschätzt. Die Berechnung stützt
sich auf Parameter für elastisch-plastisches Gebirgsverhalten und berücksichtigt unterschiedliche Über-
lagerungshöhen.

Bei der Ermittlung der felsmechanischen Kennziffern
wurden vor allem ausgewertet:
- die Meßergebnisse des Radialpressenversuches im
Hauptdolomit
- die geotechnischen Kennwerte der Gesteinsbohrkerne
aus dem Triax-Labor-Versuch
- Parameterstudien aufgrund der geologischen Kartierung und umfangreicher Gefügemessungen.

Als Beispiel sei der in Abb. 4 ermittelte Ausbauwiderstand für die Kössener Schichten bei 24o m Überlagerungshöhe angeführt. Zur Überprüfung der Rechen-
annahmen sind im Zuge des Vortriebes Durchmesserkon-
vergenzmessungen, Extensometermessungen und Labor-
Tests an Gesteinsproben vorgesehen.

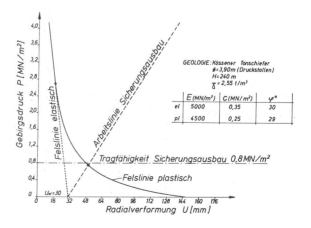

Abb. 4. Ausbauwiderstand - Kössener Schichten

2.4 Konzept der Vortriebseinrichtungen

Der Ausbruch des Druckstollens erfolgt mit Hartge-
steins-Vollschnittfräsen. An das Konzept der Vortriebsmaschine und der Nachlaufeinrichtung wurden Anforderungen gestellt, die nachstehende Arbeiten ermöglichen sollten:
- Sicherungseinbau und Ankerung gleich hinter dem
  Bohrkopf.
- Einbau der Sohltübbinge möglichst knapp hinter der
  Vortriebsmaschine und vor der Arbeitsbühne.
- Ausführung von Spritzbetonarbeiten und Sicherungseinbau mit Ringschluß von der Arbeitsbühne aus, die
  zwischen Vortriebsmaschine und Beladezone angeordnet ist.
- Bei fallweisem Rückziehen der Vortriebsmaschine muß
  die Verspanneinrichtung so eingerichtet sein, daß
  sie über den Sicherungseinbau schreiten kann.
- Eignung zur Steuerung und Verspannung auch bei
  nachgiebigem Gebirge.
- Erfassung und Fernübertragung von Leistungsaufnahme,
  Anpreßdruck und Arbeitshub.
- Ausführung von Erkundungsbohrungen mit einem leistungsfähigen Lafettenbohrgerät bis zu 15 m vor
  die Stollenbrust.

3. AUSFÜHRUNG DER VORTRIEBSARBEITEN

3.1 Baulos Amlach

Der Vortrieb im Baulos Amlach wurde mit einer Voll-
schnittmaschine ATLAS COPCO JARVA Mk 12 im Mai 1985
am Portal des 24o m langen Fensterstollens begonnen.
Der extrem lange Stollenabschnitt von 13.680 m konnte
in 19 Monaten aufgefahren werden. Dies ergibt eine
mittlere Vortriebsleistung von 36 m pro Tag, bei 2o
Arbeitstagen pro Monat.
Das baugeologische Verhalten des Hauptdolomit war
gut, es mußte relativ wenig gesichert werden. Die
flächig verteilten und in den Schichtfugen verstärkten
Wassereintritte verfolgten den Vortrieb auf weite
Strecken und belästigten die Mannschaft bei der Arbeit,
jedoch die Vortriebsleistung wurde dadurch nicht wesentlich beeinträchtigt. Die Spitzenleistungen betru-
gen 82 m pro Tag. Durch den Einbau eines Bahnhofes
hat sich die enorme Stollenlänge auf die Leistung ausgewirkt. Das Diagramm, Abb. 5, zeigt die
mittleren Tagesleistungen in den einzelnen Monaten und
die jeweiligen Monatsvortriebe im Baulos Amlach.
Die Ergänzung der Vortriebseinrichtung am Beginn,
der Einbau einer Californiaweiche als Bahnhof in Stol-
lenmitte und die Anlage eines 7o m langen Lüftungs-
schachtes zum Ansaugen von Frischluft bei Station
72oo drücken sich jeweils in einer geringeren Monats-
leistung aus. An zwei Störungen verursachten Wassereinbrüche bruchschotterartige Materialeinschwemmungen,
die vor dem Bohrkopf Kavernen entstehen ließen und den
Fräsvortrieb aufhielten.
Die erste Störung bei Station 5825 konnte, nachdem
der Stollen abgeschalt, die starken Bergwasserquellen,
anfangs ca. 1oo l/s, in Drainagerohren abgeleitet und

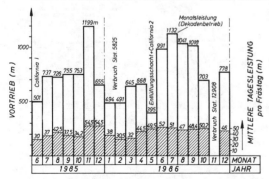

Baulos 4: Druckstollenabschnitt Amlach, 13.680m

Abb. 5. Monatsvortrieb und mittlere Tagesleistung

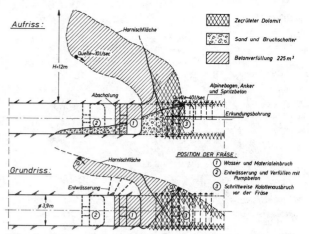

Abb. 6. Verbruch Station 129o8 Baulos Amlach

die Kaverne mit Pumpbeton verfüllt war, problemlos
nach einer Woche durchfahren werden.
In der zweiten Störung bei Station 129o8, Abb. 6,
mußte nach Wasserableitung und Verfüllen des Verbruchshohlraum mit 225 m³ Beton, die zerrüttete Zone
durch schrittweisen Kalottenvortrieb vor dem Bohrkopf
und Sicherung mit Alpineringen und bewehrtem Spritzbeton, durchörtert werden. Die Fräse konnte erst einen
Monat später, nach Überwindung dieser Schwierigkeiten
den Vortrieb wiederaufnehmen.
Der Stollenabschnitt Amlach ist unseres Erachtens
der längste, mit einer Fräse von einem Angriffsort aus,
aufgefahrene Stollen.

3.2 Baulos Griesbach

In diesem 8,2 km langen Stollenabschnitt wurde eine
Vollschnittfräse vom Typ ROBBINS Nr. 1212-228 eingesetzt. Nachdem die Fräse samt Nachlauf im Freien montiert worden war, begann der Vortrieb ebenfalls im
Mai 1985, 8o m innerhalb des Portals des 69o m langen
Fensterstollens. Anfänglich verliefen die Vortriebsarbeiten plangemäß, bis ein schwerwiegendes Ereignis
am 2o. September 1985 den mechanischen Vortrieb stoppte.
Bei Stollenstation 1363 wurde aus standfesten Jurakalken heraus eine Störung angefahren, aus der sich
zusammen mit starkem Wasserandrang ein murenartiger
Materialeinbruch in den Stollen ergoß und Fräse und
Nachlauf innerhalb von 1o Stunden auf 6o m Länge begrub. Noch bevor der Stollen bis zum Fräskopf wieder
freigelegt war, ergoß sich ein zweiter Materialein-
schub breiartig in den Stollen und verschüttete die
Fräse von neuem.
Die Maßnahmen zur Bewältigung dieser Schwierigkeiten
werden im nächsten Abschnitt 4 beschrieben. Wie sich
später herausstellte, quert die Stollentrasse eine
Zone Raibler Schichten die tektonisch eingeschuppt,
jedoch an der Oberfläche nicht erkennbar waren. In
deren Gips- und Anhydritlagen haben sich Lösungshohlräume ausgebildet, die mit Wasser, Lehm und schotter-
artigem Material angefüllt waren. Das Eindringen der
Fräse in dieses Karstsystem hat zu dem Schadensereig-
nis geführt. Der anschließende 24o m lange Abschnitt
der Rauhwacken mußte mit konventionellen Methoden vor
dem Bohrkopf der Fräse ausgebrochen und gesichert werden, weil das Gebirge zu einer kohäsionslosen Masse
zerrüttet und unter Wasserdruck zu schwimmsandartigen
Ausbrüchen bereit, den Fräsvortrieb ausgeschlossen
hätte. Für diese Strecke wurde ein eigenes Regelprofil,
Abb. 7 entworfen. Die verschweißten Alpinebögen und
die zusätzliche Bewehrung übernehmen den vollen Innendruck und eine PVC-Dichtfolie soll den Austritt des
Triebwassers in die Anhydrit- und Gipslagen verhindern.
Zur Sicherung des Fertigstellungstermines hat der
Bauherr mittlerweile beschlossen, ein zusätzliches
Stollenfenster anzulegen, um im Bedarfsfalle von die-
sem aus einen eigenen Stollenvortrieb für den restli-
chen 3,o km langen Stollenabschnitt bis zum Einlauf

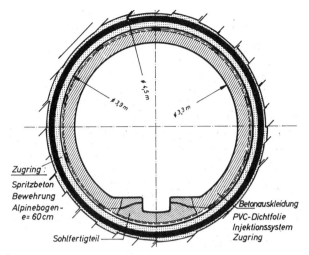

Abb. 7. Sonderprofil-Raibler Rauhwacken

einzurichten bzw. die Herstellung der Betonauskleidung in unabhängige Stollenstrecken zu unterteilen.

Erst nach lo Monaten Unterbrechung und nach Generalüberholung der Fräse im Stollen konnte der Fräsvortrieb am lo. August 1986 wieder aufgenommen werden. Im Durchlaufbetrieb wurden nunmehr in den standfesten Raibler- und Kössener Schichten, den Muschel- und Partnachkalken innerhalb von zwei Monaten eine unerwartet hohe Vortriebsleistung von 24oo m erzielt.

Der Eintritt der Druckstollentrasse in die Kristallinbasis brachte am Übergang nicht die erwarteten großen Schwierigkeiten, jedoch generell in dem bisher aufgefahrenen Abschnitt der kristallinen Gesteine sehr verminderte Vortriebsleistungen durch Ausbrüche in der Firste, stellenweise bereits über dem Bohrkopf. Dies erforderte einen Sicherungsausbau mit Ringschluss gleich hinter dem Bohrkopf.

Das Diagramm Abb. 8 zeigt die großen Unterschiede in den Vortriebsleistungen, die die stark wechselnden Verhältnisse in diesem Baulos widerspiegeln. Dementsprechend schwanken die erzielten Tagesleistungen von 2,1 lfm bis max. 83,4 lfm.

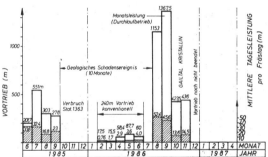

Abb. 8. Monatsvortrieb und mittlere Tagesleistung

3.3 Registrierung der Fräsdaten

Mit dem Ziel, mehr Klarheit über die Fräsbarkeit der einzelnen Gesteine und die Gebirgsklassifizierung nicht allein von einer rein individuellen Beurteilung abhängig zu machen, wurde die Registrierung des Bohrvorganges verlangt und von der Stollenaufsicht die Stillstandszeiten der Fräse, getrennt nach geologischer oder maschinenbedingter Ursache, festgehalten.

An der Tunnelbohrmaschine werden der Anpreßdruck, die Hubstellung und die Stromaufnahme der Antriebsmotoren mittels Meßwertgeber aufgenommen und per Telefonkabel aus dem Stollen in die Bauleitung übertragen. Dort erfolgt die Aufzeichnung durch Schreibgeräte und durch Datensammler. Die Schreiberaufzeichnung dient

zur Kontrolle des momentanen Baugeschehens im Stollen. Die drei Signale werden jede Sekunde gemessen und über eine Minute gemittelt. Dieser Mittelwert wird im Datensammler über einen Zeitraum von 24 Stunden gespeichert. In einer Rechenanlage, bestehend aus PC und Plotter, werden auf der Baustelle die Daten täglich ausgewertet und die zeitabhängig erfaßten Meßwerte in die stationsabhängigen Werte Anpreßdruck, Bohrgeschwindigkeit und Arbeit je m³ Ausbruch weiterverarbeitet und mit Hilfe des Plotters aufgezeichnet, Abb. 9.

Diese Werte unterscheiden sich je nach Zerlegungsgrad eines bestimmten Gesteines bzw. nach Gesteinsarten. Zusammen mit der Auswertung der geologisch bedingten Stillstände durch den Einbau der Stützmittel geben sie dem Bauleiter die Möglichkeit, auf Grundlage von Messungen zu entscheiden.

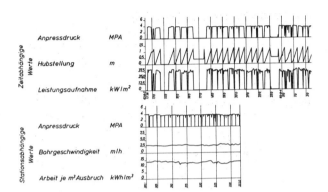

Abb. 9. Registrierung der Fräsdaten

4. BEWÄLTIGUNG DES VERBRUCHES

4.1 Hergang des Schadensereignisses

Am 2o. September 1985 ist die Fräse aus standfesten Jurakalken in eine weiche Zone eingedrungen, aus der anfänglich eine kleine Quelle von der Stirnseite in den Stollen eindrang, die rasch auf einen Zufluß von 3o-4o 1/s anstieg und Feinteile, Sand und Schotter mitführte. Mit zunehmender Fortdauer stieg die Wassermenge und der eingeschwemmte Materialkegel. Nach fünf Stunden erreichte der Zufluß mit 26o 1/s das Maximum, nach lo Stunden flossen noch 13o 1/s ab und nach einer weiteren Stunde stabilisierte sich die Zuflußmenge bei 12 1/s und beendete den Materialeintrag.

Das mitgerissene Material, das zunehmend grobstückiger geworden war, hat den Stollen bis auf 4o m Länge vollkommen angefüllt und der flach geneigte Schuttkegel reichte 65 m weit bis zu den Trafos der Vortriebseinrichtung. Insgesamt sind bei diesem Ereignis ca. 5oo m³ Material und rd. 7ooo m³ Wasser aus dem Gebirge eingeschwemmt worden. Die Zusammensetzung des Materials, das neben Jurakalk- und Dolomitbruchstücken auch reinen Gips, Anhydrit und schwarzen Mergel aufwies, deutete darauf hin, daß der Stollen unerwartet in Raibler Schichten eingedrungen war, die ober Tage nirgends aufgeschlossen waren.

Man ging nun daran den verschütteten Stollen auszuräumen, mit dem Ziel, die Vortriebseinrichtung freizulegen und den Verbruch durch Sondierbohrungen aus dem Raum hinter dem Bohrkopf abzutasten.

Während der Räumarbeiten, die bereits in den Bohrkopfbereich vorgedrungen waren und kurz den Blick in einen Hohlraum ober der Maschine freigaben, ereignete sich ein zweiter Materialeinstoß, der sich als Schlammasse in den Stollen ergoß und diesen auf 35 m total zuschoppte. Die flache Böschung reichte diesmal sogar 75 m in den Stollen. Ohne daß sich der Wasserzufluß von 12 1/s bei diesem Ereignis erhöht hat, sind innrhalb von 8 Stunden 6oo m³ Feststoffe als Mure eingeflossen. Das Material, Abb. lo, bestand aus wassergesättigtem schluffigen Feinsand bis Mittelkies, festen Lehmknollen, Kalk- und Dolomitbruchstücken.

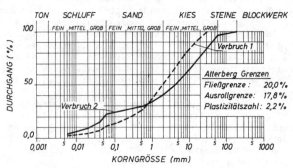

KORNVERTEILUNG

Abb. 1o. Kornverteilung der Verbruchsmassen

Im Sand war diesmal bis zu 2o % Quarz enthalten, aber
kein Gips. Es war offensichtlich, daß ein Karstsystem
angefahren worden war und der Inhalt aus ursprünglich
getrennten Kammern sich in den Stollen ergoß.

4.2 Erkundung, Entwässerung und Injektion

Durch Kernbohrungen sollte der Verbruchshohlraum abgetastet werden und Klarheit über die Beschaffenheit des
anstehenden Gebirges bringen. Abb. 11.
Aus sicherer Entfernung vom Verbruchsort wurden
39 m hinter der Brust die ersten Bohrungen in der Firste leicht steigend vorgetrieben. Beim Erreichen des
Verbruchsraumes verstopfte Schlamm und Sand die Bohrlöcher und verhinderte ein weiteres Vordringen. Der
anstehende Wasserdruck wurde mit 5 bar gemessen. Das
Wasser aus dem Verbruch unterschied sich gegenüber
dem übrigen Bergwasser in der erhöhten Temperatur von
11° statt 6° C, hohen Leitfähigkeitswerten bis 3ooo μs
und einem Sulfatgehalt bis zu 16oo mg.
Da nun erkannt wurde, daß der Verbruchsraum mit Material gefüllt war, das bereit war unter Wasserdruck
jederzeit wieder auszubrechen, war das nächste Ziel,
den Wasserdruck durch Entwässerungsbohrungen abzubauen
und das Material durch Injektionen zu verfestigen.

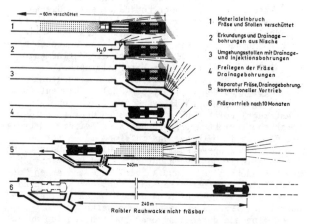

Abb. 11. Bewältigung der Störzone Station 1363

Der Bereich um die Maschine sollte mit PU-Schaum injiziert werden und der äußere Bereich mit reinem Zementgut. Aufgrund der Materialbeschaffenheit, Abb. 1o,
wurde auch die Anwendung anderer Injektionsmittel wie
Weichgel, Wasserglasgemische und Kunstharze erwogen.
Für dieses umfangreiche Bohrprogramm wurde, durch Auf-
weiten des Stollenprofiles 21 m hinter der Brust, ein
Arbeitsraum für 3 Bohrgeräte geschaffen. Die Fräse
verblieb noch eingepackt im Material als Sicherheits-
stoppel. Nun wurden Erkundungs- und Drainagebohrungen
fächerartig rund um den Stollen angelegt. Vom linken
Ulm aus gelang es schließlich Wasserwege anzubohren,
die anfangs bei 8 bar Druck bis zu 12o 1/s spendeten.
Durch Driftversuche mit Kochsalzlösung wurde eine

Verbindung mit dem 1,5 km entfernten Griesbach bestätigt. Die Schüttung ging langsam bis Mitte Dezember
auf 25 1/s zurück; der Wasserdruck im Verbruch konnte
auf 1,5 bar abgebaut werden. Gleichzeitig wurden im
Nahbereich des Bohrkopfes stollenparallele Bohrungen
bis in den Verbruch niedergebracht und mit PU-Schaum
injiziert, welcher die Wasserwege rund um den Bohrkopf gut verstopfte und das angrenzende Material zu-
sammenkittete. Um die zeitraubenden Bohrarbeiten ab-
zukürzen, wurde seitlich im standfesten Fels ein Um-
gehungsstollen angelegt und aus einer kleinen Bohr-
kaverne, Abb. 11, in 4 m Abstand zum Verbruchsrand 7
Injektionsfächer zu je 7 Bohrlöchern angelegt. Nahe
dem Bohrkopf wurde mit PU-Schaum mit geringem Druck
und im weiteren Bereich mit Zementgut mit hohem Druck
bis 3o bar injiziert. Darüber hinaus wurde nach der
Injektion durch Kontrollbohrungen die Konsolidierung
des Verbruchsraumes geprüft und die Umgebung durch
verrohrte Drainagebohrungen sorgfältig entwässert.
Insgesamt wurden über 5ooo lfm Bohrung hergestellt
sowie 1o.76o kg PU-Schaum und 4oo to Zement verpreßt.
All diese Arbeiten dauerten 4 Monate.

4.3 Bergung der TBM, Durchörterung der zerrütteten Zone

Das Material im Verbruchsraum war danach derart kon-
solidiert, daß von einem kurzen Stichstollen aus der
Bohrkopf freigelegt und ein Ringraum um ihn herum ohne
Schwierigkeiten, jedoch gesichert mit Alpinebogen und
bewehrtem Spritzbeton, hergestellt werden konnte.
Mittlerweile wurde der Stollen hinter dem Fräskopf
auch vollständig ausgeräumt und zusätzlich gesichert.
Obwohl die Umgebung der Fräse offensichtlich gut mit
PU-Schaum verpreßt worden war gelangte durch die, zwar
weiter abgerückten Hochdruckinjektionen Zementgut über
die Kabelanschlüsse ins Innere der Antriebsmotoren und
durch die Dichtungslamellen in das Hauptlager, sodaß
der Bohrkopf abgezogen und das Lager an Ort und Stelle
geöffnet und gereinigt werden mußte. Die Motoren wurden ausgebaut und im Werk gereinigt. Die übrigen Re-
paratur- und Überholungsarbeiten konnten alle im Stollen ausgeführt werden, sodaß die Fräse ab Anfang März
1986 wieder einsatzbereit war.
Weitere Aufschlußbohrungen zeigten auf, daß an den
8 m breiten Verbruchsbereich nur eine schmale kompakte
Dolomitlinse angrenzt, dahinter aber eine breite Zone
rauhwackiger Dolomite mit komplexer Wechsellagerung
von Schwimmsand, Lehm und zerrüttetem Dolomit ansteht
und zufolge des herrrschenden Bergwasserdruckes von
8 bar an einen Fräsvortrieb nicht zu denken war.
So wurde vor dem Bohrkopf eine provisorische Einrichtung für konventionellen Vortrieb installiert und
in Abschnitten von ca. 3o m Länge schrittweise mit
Kalotten- und Sohleausbruch vorgetrieben. Vorauseilend
wurde in Stollenachse eine verrohrte Kernbohrung und
über der Firste ein kegelförmiger Drainageschirm angeordnet. Letzterer hat den anfangs herrschenden Berg-
wasserandrang geordnet abgeleitet, ohne daß die angetroffenen Sandlinsen ausgewaschen wurden. Die Kernbohrung hat zu guter letzt noch Erdgas mit Methange-
halt angefahren, das aber rechtzeitig erkannt und ab-
gefackelt bzw. durch zusätzliche Ventilation so ver-
dünnt werden konnte, daß die zur Explosion neigende
Konzentration von 5-1o % Methan nie erreicht wurde.
Auf diese Art wurden 24o m Stollen vorgetrieben, im
Abstand von 6o cm Alpinebogen versetzt, Bereiche der
Kalotte mit Stahldielen verpfählt und der ganze Umfang
mit Baustahlgitter und Spritzbeton gesichert, Abb. 7.
Geschuttert wurde bis zum Bohrkopf mit einer kleinen
Laderaupe und von dort mit der mittlerweile einsatz-
bereiten Vortriebseinrichtung.

SCHLUSSBEMERKUNG

Wie dieses Beispiel zeigt, sind dem maschinellen Vortrieb im Karstgebirge Grenzen gesetzt. Die blockierte
Fräse bedeutet ein zusätzliches Hindernis zur Überwindung der Schwierigkeiten.

Some recent improvements in oil well drilling technology
Quelques améliorations récentes dans la technologie de forage de puits de pétrole
Neue Fortschritte in der Ölbohrtechnologie

G.BRIGHENTI, Mining Science Institute, University of Bologna, Italy
E.MESINI, Mining Science Institute, University of Bologna, Italy

ABSTRACT: This work briefly examines rock cutting mechanics in oil well drilling and leading criteria in the choice of bits to obtain the minimum drilling cost.
Two types of bits used since some time by industries are taken into account: "PDC" bits and "extended nozzle" bits. The conditions for their optimal use in the view of the results obtained in Italy and abroad are then discussed.

RESUME: On examine brièvement la mécanique de coupe de la roche dans le forage de puits de pétrole et les principes à suivre dans le choix des trépans afin de minimiser le côut de forage.
Deux types de trépans "PDC" et les trépans "à tuyère extendue", dont on discute les conditions pour leur utilisation optimale sur la base des résultats obtenus en Italie et à l'étranger.

ZUSAMMENFASSUNG: Es werden in Kürze die Mechanik des Felsschnittes bei Ölbohrungen und die Kriterien für die Wahl der Bohrmeissel zur kostengünstigsten Bohrung untersucht.
Es werden zwei Bohrmeisseltypen "PDC" und "Extended nozzle" in Betracht gezogen, die seit kurzer Zeit in der Industrie verwendet werden, sowie die Bedingungen für deren optimale Verwendung auf Grund der bei ihrem Einsatz in Italien und im Ausland erzielten Ergebnisse besprochen.

1. INTRODUCTION

In the last few years oil industry has been faced with several difficulties hindeing its development. Due to the fall of oil price, costs had to be reduced just when operative conditions were becoming harder both for the drilling of increasingly deeper wells and for the progressive expansion of off-shore drilling.

All aspects concerning oil well drilling technology should then be reexamined and further rationalized to make cost reduction possible. For this purpose there are several problems to be considered, e.g. choice of installations and drilling parameters.

This work examines the problems related to bits, paying special attention to two types of bits which have recently beeen used: "PDC" bits, a new type which is being proposed as an alternative to traditional roller ones, particularly when using downhole motors working at high speeds and low weight, and "extended nozzle" bits, which, even if devised in the Sixties, only recently become popular thanks to the use of new materials and new construction technologies.

2. OUTLINE OF ROCK CUTTING MECHANICS

In order to determine, at least from a qualitative point of view, the best application field of each bit, the mechanics of rock failure associated with drilling should be investigated.
If we consider rotation drilling (both rotary drilling and downhole motors drilling), bits mainly work by compression action, percussion action and scraping-cutting action; the choice of the mechanism to be used depends on type of tool, working conditions and type of rock.

As a general rule, cutting action is preferred in case of rocks with a plastic or pseudoplastic behaviour, while combined action of compression and percussion proved more suitable in case of brittle rocks.

On other hand, the rheological behaviour of a rock depends, besides on its mineralogical and petrographic conditions, also on temperature, stresses to which it is subjected, characteristics of the saturating fluid and how load is applied by the bit.

It often appens, therefore, that rocks which are usually elastic at the surface, show a plastic behaviour into the depths, particularly in case of high differences between hydrostatic drilling mud and formation pressure (differential pressure). Theoretical and experimental researches have been conducted on this subject also at the Mining Science Institute of the University of Bologna (Brighenti 1974, Brighenti 1978). Such researches have confirmed what mentioned above and have also pointed out the importance of the lithostatic stress horizontal component. In particular, they have shown (a fact which is little known), that when its value exceeds the differential pressure, surface elastic rocks do not show a plastic behaviour. Such researches, which are still under way, have also clearly confirmed that drillability depends on several factors and, therefore, it would not be correct to characterize it with just one parameter.

The drilling fluid also plays a very important role as regards both the stresses transmitted to the rock and the mechanical action of downhole cleaning and physico-chemical modifications induced in the rock.

The above considerations clearly point out the complexity of the rock drilling mechanism and show how difficult, if not impossible, it is to determine formulae to be used in tool designing and in performance

evaluation . Anyway, a good comprehension of the physical phenomenon is essential to orient bit designing
and the choice of optimal drilling parameters.

3. BIT CHOICE CRITERIA

Like all other installation components and drilling
modes, also the criteria to be followed in the choice
of the bits are of an economic nature.
The goal to achieve, in fact is the minimization of
drilling costs, which depend on cost of bits, auxiliary
equipment, drilling fluid, unit rig cost and drilling
time. The energy cost, and therefore the work necessary to drill the rock volume unit, does not play an
important role, and therefore the study on this aspect
of the problem even if interesting from a theoretical
point of view, is of little interest in practice.
As concerns the bit on the contrary, it is very important to consider its cost, rate of penetration and
life. The latter factor obviously becomes more and
more important as the depth of the hole increases,
since also trip times necessary to change the bits
increases parallely.
Special attention should be paid to some factors, often
neglected, which highly influence the drilling cost.
One of them is the increase in accidents (drillpipe
failure and sticking, well kick or even blowout, etc.)
due to the use of some drilling techniques or particular tools. Another factor is the necessity, imposed by
the use of some bits, to comply with special working
procedures (e.g. necessity of using particularly
stabilized batteries and completely eliminating vibrations). A third and last factor is the higher or lower
skill of the personnel and its familiarity with new
drilling techniques or new tools and equipment.

All these inconveniences and bonds, which are quite
difficult to be evaluated from a quantitative point of
view, in practice highly modify the drilling cost, thus
decisively influencing the bit choice.

As already said, such factors cannot pratically be
evaluated from a quantitative point of view, expecially
as their influence depends on the skill and experience
of the drilling personnel. However, it is unquestionable
that technologically simple bits easy to be used, such
as those which will be treated in the following sections,
should reduce the inconveniences mentioned above.

4. BITS RECENTLY USED IN OIL-WELL DRILLING

4.1 "PDC" bits

Polycrystalline Diamond Compact (PDC) bits represent
one of the most remarkable innovations in the Eighties
in terms of drilling technology (Varnado et Al. 1980,
Madigan et Al. 1981, Hoover et Al. 1981, Radtke et Al.
1984, Millheim 1986). They can be considered as an
extremely sophisticated elaboration of the far-off drag
bit. The PDC make use of cutting elements contituted
by synthetic diamond blanks. Synthetic diamonds are
cemented one to the other using high pressures and
temperatures, and in some cases also tungsten carbide.
The blanks so obtained, which are characterized by the
high hardness and wear resistance of diamond and by
the considerable impact resistance of tungsten carbide,
are subsequently bonded on suitable tungsten carbide
studs (fig. 1).
The elements so obtained, constituted by blanks and

POLYCRYSTALLINE DIAMOND
COMPACT

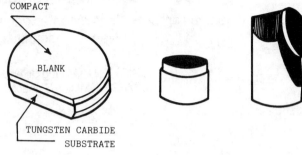

Figure 1. Polycrystalline diamond blank and some
diiferent cutting assembly on studs.

studs, are then pressed-matched either on highly-resi-
stant steel body or on tungsten-carbide body (matrix
body).

The result is pratically a monolithic tool with no
elements in relative movement (and therefore mechanically
simpler than roller bits) using the synthetic diamonds
blanks to remove rock chips mainly by a cutting action
(fig. 2), contrary to what happens with roller bits.

This action becomes particularly effective when a precise orientation is given to the studs (side rake angle)
and a suitable inclination to the blanks (back angle).

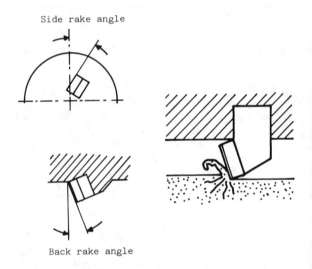

Side rake angle

Back rake angle

Figure 2. Rock cutting action of blank and orientation
on stud and on bit body.

The cutting mechanism demands a lower quantity of energy
as compared to compression breaking typical of roller
bits. However, PDC bits must be highly wear-resistant,
an essential requirement for a bit which is used in
oil-well drilling.

The main construction characteristics of the bit, which
make it suitable for certain type of formations, are
the following ones:
- material used for the body;
- number, exposure, inclination and orientation of
  studs and blanks;
- gage protection;
- thickness of synthetic diamond blanks;
- hydraulic configuration and type of bit body profile.
PDC were also devised both for traditional rotary drilling and for downhole mud motors. PDC drill bits, thanks
to their intrinsic characteristics, are particularly
suitable in the following cases:

- high rotation speed (turbodrilling);
- considerably reduced weight on the bit as compared to roller cone bits, and therefore improved control of deviation;
- use of oil muds or high specific gravity muds;
- drilling into high depths;
- drilling under overbalanced conditions;
- coring and slim-hole drilling.

PDC drill bits are very suitable to be used in soft and medium hard formations in which an oil-based mud is used (Madigan et Al. 1981, van Prooyen et Al. 1982, Turnbull 1982). Madigan et Al. 1981 report the results of a comparative analysis between PDC drill bits and traditional roller cone bits, in order to quantify the PDC technical performance. The analysis was performed on about a thousand field runs in the North Sea and in several USA basins, in heterogeneous formations constituted by evaporites, carbonates, marls, clays, gypsum and sandstones. Van Prooyen et Al. 1982 also report the results obtained using PDC in the North Sea, in Holland, West Germany and Middle East using both turbo-drilling and rotary systems.

However field application performed over the last few years have shown that PDC, though not suitable for hard formations (Hoover et Al. 1981), may be successfully used also with the more popular water-based muds in shaly formations. In fact, Radtke et Al. 1984, have reported the results obtained by drilling, with PDC, shaly formations of the Gulf Coast using water-based muds. The use of these more popular and economical muds was made possible adjusting: (1) downhole hydraulic parameters (flow rate, pressure drop, impact force); (2) weight and number of rotations of the bit.

Particularly advantageous economical conditions due to reduced drilling times as compared to traditional roller cone bit were found in Alaska (Kuparuk Field River), (Balkenbush 1985). In this field PDC were used at depths ranging from 1000 to 2000 m in sandy-shaly formations, in which the weight on the bit ranged from 133 to 178 kN, the rotation speed from 150 to 200 rev/min, the standpipe pressure was 27 MPa, the mud flow rate ranged from 0.025 and 0.0283 m^3/s, (always water-based mud with a weight equal to $1.22 \cdot 10^4$ N/m^3).

Recent results obtained using PDC in the Rocky Mountains (Pain et Al. 1985), (Green River and Powder River Basins) have shown that the choice of a particular bit design (bit profile, cutter density and shape, cutter exposure, side and back angles) is related to the type of formation from which both drilling speed and downhole bit life depend.

PDC bit were also used in Italy both in rotary and turbine drillings.

Table 1 shows some bit records related to field runs performed with both traditional tricone bits and PDC bits: they refer to wells Seregna 6, Villafortuna 1, S.Giovanni 1, Fiumetto 1 and Monteseggio 1.

More precisely it reports characteristics, size of the tools, depth out, actual drilling time (T), rotation speed (RPM), weight on the bit (WOB), rate of penetration (ROP) and cost per meter (C), obtained with the usual formula:

$$C \ (\$/m) = \frac{B + R(T + t)}{F} \quad \ldots\ldots\ldots\ldots (1)$$

where B is the bit cost, R the rig cost per hour, t the trip time (roughly evaluated in 1 hour each 300 meter of deepth).

Table 1 also reports, for PDC bits, the limited rate of penetration ROP* at which the bit must work in each drilling to give the same unit cost C_R as the offset tricone bits. Such a rate of penetration was calculated using the following formula (easily obtained from eq. (1)) :

$$ROP* = \frac{B^* + Rt}{C_R T} + \frac{R}{C_R} \quad \ldots\ldots\ldots\ldots (2)$$

where B* is the cost of the bit to be compared (in our case the PDC), possibly including the turbine cost.

On the basis of Table 1 it is possible to see that, in case of use of LX27 and R40 (run 1) in Seregna 6 and of S226 and S248 in Villafortuna 1, S.Giovanni 1 and Fiumetto 1 (run 1), PDC are more economical than traditional bits.

The failure in the case of some runs in Seregna 6

Table 1. Bit records of PDC bits and offset tricone bits related to field runs performed in Italy.

PDC bit type	Well	Lithology	Depth out (m)	Drlg time (h)	ROP (m/h)	WOB (Tonn[+])	RPM (rev/min)	Cost/m ($)	Offset bit type	ROP (m/h)	Cost/m ($)	ROP* (m/h)
LX27 DB 12 1/4"	Seregna 6	sh-marl	4067	240	2.28	6–11	140	283	1-3-4	1.45	454	1.45
R40 CH "	Seregna 6 st	shale	3314	71	0.65	3	325	1477	1-3-5	0.55	1500	0.59
" " "	" " "	"	3346	46	0.70	3	325	--	"	"	"	0.72
" " "	" " "	marl-lmst	4015	55	0.62	4–5	340	1288	"	0.90	830	1.22
S226 CH "	Villafortuna 1	conglom	4150	216	1.52	7–15	110	546	"	1.20	761	1.07
" " "	"	marl	5344	202	1.04	--	--	829	"	0.70	1308	0.68
" " "	"	"	5493	147	1.01	--	--	1237	"	"	"	0.78
S248 CH 8 3/8"	"	shale	5856	147	1.53	--	--	574	"	0.45	2086	0.43
" " "	"	"	5932	66	1.15	--	--	1163	"	"	"	0.64
" " "	"	"	6003	74	0.96	--	--	1175	"	"	"	0.60
S226 CH 12 1/4"	S.Giovanni 1	sh-sdstn	3014	105	1.95	12	100	496	1-1-6	1.50	598	1.37
D185 CH "	Monteseggio 1	shale	1938	33	1.61	12	320	767	2-1-7	1.00	643	2.08
" " 8 3/8"	"	sh-sdstn	3344	20	0.75	9–12	50–385	2015	1-3-5	1.00	869	2.18
S248 CH 8 1/2"	Fiumetto 1	shale	3272	214	1.24	9–10	45–370	609	5-1-7	1.20	1019	0.73
" " "	"	"	3317	31	1.13	6	45–380	1371	"	"	"	1.96

(+) 1 Tonn = 9.81 kN; DB = Diamant Boart; CH = Christensen.

(runs 2 and 3), in Monteseggio 1 and Fiumetto 1 (run 2) is probably due to incidental choices by the drilling personnel, who sometimes are obliged to extract the bit in advance, rather than to a wrong choice of the drilling tool. Anyway, acceptable penetration rate values were obtained also in these field run.

We may conclude therefore, that at present, in Italy, the PDC bits may be used in soft and medium hard formations (silt,clays, marls, sandstones), for depth above 2000 - 2500 m, obviously in optimal conditions of weight on the bit, rotation speed and hydraulic conditions.

Moreover, we believe that the improvement in construction technologies will reduce the gap between the cost of PDC bits and tricones, a fact which, together with a better use of the bit by the personnel, will make such tools even more competitive.

4.2 The "extended nozzle" roller cone bits

The extended nozzle (EN) bits are a slightly modified version of the roller bits (three and two cone bits).

As compared to similar traditional bits, in EN bits the nozzle from which the drilling fluid flows is extended. Moreover, a further central nozzle is added in the lower part of the bit.

The mechanical action regarding indentation of the bit into the rock remains, therefore, unchanged (breaking is mainly due to a compression-cutting combined action), while downhole hydraulic effect is considerably increased.

The idea of extending the nozzles had already been accepted at the beginning of this century in the case of drag bits, in order to improve their performances. However, when in 1933 the tricone bits started to become popular, together with drag bits the principle by which a minimum distance between the fluid-outlet section and bottomhole was also set aside.

It was only during the Fifties, (Eckel et Al. 1951), that experimental laboratory tests clearly pointed out that the impact force of the drilling fluid was inversely proportional to the nozzle/bottomhole distance.

The design and construction of EN bits trace back, however, to the Sixties (Feenstra et Al. 1964, Feenstra 1968), but it was only during the Seventies that a more reliable and competitive tool, able to provide good ROP's, was available (Pratt 1978).

This was made possible, thanks to both the contribution given by experimental studies (Sutko et Al. 1971, Sutko 1973) and to the elimination of mechanical defects, which were responsible for frequent failure of the nozzles.

An increased bottomhole hydraulic power is obtained using extended tubes, which facilitate cuttings removal keeping the hole clean and increase the impact force of fluid against the formation, thus facilitating the action of the bit.

Milled tooth EN bits and tungsten carbide insert bits (TCI) are currently available on the market, from 9 5/8" to 17 1/2" diameter. They are suitable to successfully drill soft and medium hard formations, characterized by a plastic or pseudolastic behaviour.

Table 2 reports the code numbers of EN bits available on the market, according to I.A.D.C. standards.

Figure 3 shows a milled tooth tricone EN bit, 12 1/4" diameter and a tungsten carbide tricone EN bit, 17 1/2" diameter.

As a general rule, nozzle extensions enable D/Ø =3-5 ratio, of the distance of the mud outlet section/bottomhole (D) to the nozzle diameter (Ø).

Figure 4 schematically illustrates the specific function of the central nozzle, which is necessary to sufficiently clean the cutters in the central part of the rollers. In fact, the approaching to the three remaining nozzles to bottomhole, decreases the fluid cleaning action on the cutters, which may lead to bit balling.

Table 2. I.A.D.C. code of EN rock bits available on the market.

	I.A.D.C.	1-1-4
	I.A.D.C.	1-1-6
	I.A.D.C.	1-2-4
milled teeth	I.A.D.C.	1-2-6
	I.A.D.C.	1-3-4
	I.A.D.C.	1-3-5
	I.A.D.C.	1-3-6
	I.A.D.C.	1-3-7
	I.A.D.C.	4-3-7
	I.A.D.C.	5-1-5
tungsten carbide	I.A.D.C.	5-1-7
inserts (TCI)	I.A.D.C.	5-1-9
	I.A.D.C.	5-3-5
	I.A.D.C.	5-3-7

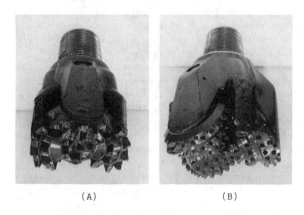

(A) (B)

Figure 3. (A) Milled teeth EN bit 12 1/4"; (B) Tungsten carbide inserts EN bit 17 1/2".

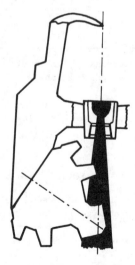

Figure 4. Cleaning action of the center jet mounted on an EN bit.

Such nozzle can be converging or diffusive. In case of a diffusive nozzle, the problem of caviation must be considered, due to which the nozzle may seriously be damaged, expecially in shallow holes. Generally speaking, cavitation hardly occurs below 1800 - 2000 m. Peschel et Al. 1985 suggest that the central jet should have an outlet section equal to 15 - 20% of the total one. Moreover, the central diffusive nozzle must be suitably dimensioned (Baker 1979) since it produces a lower pressure drop, its diameter being the same as the one of the converging nozzle. In particular, the diameter of a diffusive nozzle, calculated so as to ensure a certain mud flow rate, will have to be reduced as compared to a converging nozzle, due to a lower amount of energy used.

The first results obtained from the use of EN bits were published in an important work by Pratt in 1978. Results obtained in run field in Luisiana (on and off-shore) and in the Gulf of Mexico on more than 90 EN bits, have shown an increase in ROP by 15 to 40% as compared to traditional tricone bits, while a rise in ROP equal to 3 - 4% was enough to reach the same drilling cost obtained using traditional tricone bits. In economic terms, Pratt quantified a saving on the drilling total cost which, for off-shore uses ranged from 8 to 28% and from 1 to 26% for on-shore ones.

Sellers et Al. 1982 have reported the results obtained from the comparison of insert EN bits and offset similar conventional bits. The drilling concerned surface wells (550 m), in West Texas in soft and medium soft formations. The average increase in ROP was equal to 38%, a higher percentage as compared to offset bits, with a saving on the total drilling cost equal to 27%.

In a more recent work Perschel et Al. 1985, have reported the results of field experiments in the North Sea, where milled tooth and TCI EN bits were used, 12 1/4" and 17 1/2" diameters, in mixed formations characterized by clays, marls, limestones and anydrites. In particular, EN bits proved successful also in directional drillings, with angles of deviation above 58°. The Authors also report that EN bits were used to drill the cementing shoe. As concerns economic advantage, an average increase in ROP by 16% was calculated, as compared to standard bits. In the hard operating conditions in the North Sea it was calculated that already an increase in ROP by 4% gives rise to a breakeven cost, as compared to similar conventional tricone bits. In this way, a 24% saving on the drilling cost was obtained.

Other Authors have published results obtained from drilling experiments using EN bits, such as Robbibaro 1979 (drillings in Alaska and California), Zinger 1983 (drillings in the South-East Asia) and Selby et Al. 1985 (drillings in the Middle East).

EN bit performances have been tested also in Italy, even if slightly late. Between 1984 and 1986 EN bits

have mainly be used by AGIP for about 70 drilling field runs in directional wells in Cervia Mare, Cavone 14 (with side-track), Villafortuna 2-3, Concorezzo 1, Santalessandro 1, Suviana 1. Trecate 1, Manfria 1 bis, Prezioso 3 (Mesini et Al. 1986).

The use of such bits in directional wells in Cervia Mare was of particular interest since, in this case, a direct comparison with similar conventional bits was possible. The field runs performed using bits 12 1/4" diameter (I.A.D.C. code 1-1-4) were considered in four directional wells in Cervia Mare (19, 20, 21, 22), at depths ranging from 1500 to 3500 m. In two of them (19 and 22), four extended nozzle bits were used.

Using formulae (1) and (2) costs were analyzed as concerns these four directional holes. The corresponding results are reported in Table 3. Pliocene formations, constituted by sand and clay-layers, were drilled using EN bits. ROP's higher than 52% were recorded, as compared to conventional bits used in the offset bits 20 and 21, as well as a 25% breakeven ROP*. The resulting decrease by 24% in cost per meter, from 230 $/m to 174 $/m, has fully payed back the higher cost of the bit (diffeence of about 2000 $), as compared to standard bits.

Less successful results were obtained in the drilling of the 12 1/4" diameter in well Cavone 14, together with the surrounding tract drilled (side track) using EN bits. In fact, in drillings 1980 to 2638 m deep using the EN bits, the cost per meter was 326 $/m as against 278 $/m using conventional bits in the depth range 1993 - 2638 m in the vertical hole. It should be noted, however, that a reduced weight had to be applied on the bit, having to go back on the vertical. Therefore, EN performance can be considered satisfactory also in this case.

5. CONCLUSIONS

A full understanding of the modes by which the bit cuts the rock is essential to design a bit and to choose the best drilling techniques. In particular, some investigations carried out in the laboratory of Mining Science Institute of the University of Bologna, have pointed out the importance of lithostatic tension on the rheological behaviour of the rock and the impossibility of characterizing drillability with just one parameter.

However, many factors should be considered when choosing a bit, according to the classic principle of minimizing the drilling cost. Among such factors, besides rate of penetration and life of the bit, sturdiness and use simplicity of the tool should not be neglected: frequent accidents and the demand for unusual sophisticated techniques imply, especially in case of not particularly trained personnel, high rises in the average drilling time, and therefore a much in-

Table 3. Economical and technical data referring comparison between extended nozzle (EN) and conventional tricone bits performed in four off-shore wells (Cervia Mare - Italy).

Well	Bit runs	Total bits cost ($)	Interval drilled (m)	Drlg. time (h)	Trip time (h)	ROP (m/h)	Cost/m* ($/m)
Cervia M. 19D	7 (2EN)	21598	1936	264	64	7.3	180
Cervia M. 20D	8	19616	2028	373	74	5.4	230
Cervia M. 21D	11	26972	1906	318	93	6.0	230
Cervia M. 22D	7 (2EN)	20764	1915	233	67	8.2	168

(*) Cost per meter has been calculated using eq.(1) and considering an unit rig cost of 550 $/h.

creased expenditure due to the high incidence of unit rig cost on total drilling cost.

Extended nozzle bits (especially considering the improved construction technologies and a better skill in their use have highly reduced breaking ang clogging of the nozzle) and PDC bits (with no parts in relative movement and suitable for working at high rotation speeds and with low weights on bit), have proved to be plain tools easy to be used which, in soft or medium soils, are more economical as compared to traditional milled tooth or insert roller bits.

Further improvements in construction technique are to be expected (particularly as concerns techniques related to the construction of PDC bit cutters), which will involve a cost decrease as compared to traditional bit. We believe, therefore, that the bits examined in this work become the bits of the Nineties.

AKNOWLEDGEMENTS

This research has been supported by grants from Italian Consiglio Nazionale delle Ricerche (grant no.84.00280.05) and Ministero della Pubblica Istruzione.

Particular appreciation is expressed to AGIP S.p.A., Smith Tool Italy and Christensen Diamond Products Co Italy, that released data to us.

BIBLIOGRAPHY

Baker, W., 1979. Extended nozzle bits require precise nozzle sizing. Oil & gas Jou. March, 19 : 88- 97.

Balkenbush, R.J., J.E. Onisko, 1985. Application of Polycrystalline Diamond Compact bits in the Kuparuk River Fiel, Alaska. Jou. Petr. Tech. July: 1220-1224.

Brighenti,G., 1974. Influenza delle caratteristiche degli strati sulla perforazione rotary a grande profondità. Tech. report Mining Science Instite U. of Bologna: 1- 43.

Brighenti,G., 1979. Influenza della pressione di fondo pozzo e della tensione litostatica sulla perforazione profonda. Tech. report Mining Science Institute U. of Bologna: 1-14.

Eckel, J.R., W.J. Bielstein, 1951. Nozzle design and its effects on drilling rate and pump operation. Drill. and Prod. Prac. API.

Feenstra, R., J.J. van Leeuwen, 1964. Full scale experiments on jets in impermeable rock drilling. Jou. Petr. Tech. March: 329-336.

Feenstra, R., 1968. Bit with extended jet nozzles. U.S. Patent no.3,363,707 January, 16.

Hoover, E.R., J.N. Middleton, 1981. Laboratory evaluation of PDC drill bits under high-speed and high wear conditions. Jou. Petr. Tech. December:2316-2321.

Madigan, J.A., R.H. Caldwel, 1981. Applications for Polycrystalline Diamond Compact bits from analysis of carbide insert and steel tooth bit performance. Jou. Petr. Tech. July: 1171-1179.

Mesini, E., G. Serracchioli, 1986. Sull'impiego di scalpelli tricono ad ugelli prolungati nella perforazione rotary. Proc. of 'Giornata di studio su problemi di geoingegneria', Piacenza, October: 30-37.

Millheim, K.K., 1986. Advances in drilling technology (1980-1986) and where drilling technology is heading. Paper SPE no.14070, Beijing, March: 1-16.

Pain, D.D., B.E. Schieck, 1985. Evolution of Polycrystalline Diamond Compact bit designs for rocky mountain drilling. Jou. Petr. Tech. July: 12131-1219.

Peschel E., H.F. Abdullah & G. Walter, 1985. Economical drilling with rock bits equippeed with extended nozzle: field experiences in the North Sea. Paper SPE no.13461, New Orleans, March: 331-335.

Pratt, C.A., 1978. Increased penetration rates achieved with new extended nozzle bits. Jou. Petr. Tech. August: 1191-1198.

van Prooyen, G., R. Juergens & H.E. Gilberts, 1982. Recent field results with new bits. Jou. Petr. Tech. September: 1938-1946.

Radtke, R.P., D.D. Pain, 1984. Optimization of hydraulics for Polycrystalline Diamond Compact composite bits in Gulf Coast shales with water-based muds. Jou. Petr. Tech. October: 1697-1702.

Robbibaro, P.A., 1979. New bit improves ROP in sand shale sequences. World Oil, April: 191-194.

Selby, B.A., W.A. Sauvageot, 1985. Rock bits equipped with extended nozzle lower drilling cost in the Middle East region. Paper SPE no.13690, Baharain, March: 111-114.

Sellers, M., Vita P., 1982. Chevron succeeds with extended nozzle bits in West Texas. Petr. Eng. Int., January: 44-50.

Sutko, A.A., G.M. Myers, 1971. The effect of nozzle size, number and extensions on the pressure distribution under a tricone bit. Jou. Petr. Tech., November: 1299-1304.

Sutko, A.A., 1973. Drilling hydraulics - a study of chip removal force under a full-size jet bit. Soc. Petr. Eng. Jou., August: 233-238.

Turnbull, R.W., 1982. Turbodrill, PDC bits spell success in Mobil's Statfjord. Petr. Eng. Int., October: 40-56.

Varnado, S.G., C.F. Huff & P. Yarington, 1980. Studies aim at optimum design for PDC hard-formations bits. World Oil, March: 63-70.

Zinger, S., 1983. Drillers in South East Asia benefit using extended nozzle rock bits.Smith's internal report, March: 1-11.

Contribution à l'étude de destructibilité des roches en forage rotatif

Destructibility of rocks with rotation drilling bits
Die Zerstörbarkeit der Gesteine im Drehbohren

J.BRYCH, Professeur à la Faculté Polytechnique de Mons, Ir.,Ph.D., Ingénieur-Conseil, Belgique
NGOI NSENGA, Ingénieur Civil des Mines, Lubumbashi, Zaïre, M.Sc., et Candidat Docteur (F.P.Ms)
XIAO SHAN, Ingénieur Civil des Mines, Chine et candidat M.Sc. (F.P.Ms)†

ABSTRACT: To better define the problems between the mechanical properties of rocks and the geometric and technical properties of drilling tools, a series of experimental tests were done on the destructibility of rocks using different types of drilling tools and drilling systems. For this research, done in the Laboratory of Drilling and Rock Mechanics of the Faculté Polytechnique of Mons, Belgium, the following rotary drilling processes were used : Rotary Drilling System, Rotary Large-Diameter Drilling System and Turbo-Drilling Large-Diameter System. This results, both theoretical and experimental, have been presented in severals masters and Ph. D. Thesis of our Laboratory and appear in this paper in summary form. For this project a fully computorized portable punch test apparatus was developed in collaboration with the Diamant Boart Company of Bruxelles.

RESUME: Afin de pouvoir mieux cerner les problèmes de liaison entre les paramètrs mécaniques des roches d'une part et des paramètres géométriques et technologiques des outils de forage d'autre part, de nombreux essais de destructibilité des roches par différents types d'outils et de systèmes de forage rotatif ont été effectués au laboratoire de forages et de mécanique des roches de la Faculté Polytechnique de Mons. Pendant cette étude les procédés suivants du forage rotatif étaient visés : le forage Rotary classique, le forage Rotary à grand diamètre et le turboforage à grand diamètre. Quelques premières conclusions théoriques et expérimentales venant de différentes thèses de maîtrises et de doctorats (3), (4) et (8) sont présentées sous forme d'un résumé dans le cadre de cette communication. C'était aussi dans le cadre de ces travaux de recherche qu'une presse portative de poinçonnage, entièrement informatisée, a été récemment développée en collaboration avec la Société Diamant Boart.

ZUSAMMENFASSUNG: Um die Probleme des Zusammenhangs zwischen den Parametern der Gebirgsmechanik einerseits und den geometrischen und technologischen Parametern der Bohrwerkzeuge andererseits besser zu erfassen, wurden im Laboratorium für Bohrtechniek und Gebirgsmechanik der Faculté Polytechnique de Mons (Technische Universität von Mons) zahlreiche Versuche über die Zerstörbarkeit der Gesteine mittels verschiedener Werkzeugtypen und Drehbohrsystemen durchgeführt. Im Laufe dieser Arbeiten wurden folgende Drehbohrverfahren untersucht : das herkömmliche Rotary-Bohren, das sogenannte Big Diameter-Drehbohren und das Big Diameter-Turbobohren. Erste theoretische und experimentielle Schlußfolgerungen aus verschiedenen Doktor- und Magisterarbeiten werden im Rahmen dieses Berichts in Kurzfassung gegeben. Ebenfalls im Laufe dieser Forschungsarbeiten wurde kürzlich in Zusammenarbeit mit dem Unternehmen Diamant Boart eine vollständig komputergesteuerte tragbare Lochpresse entwickelt.

Pour étudier les problèmes de liaison entre paramètres mécaniques des roches d'une part et des paramètres géométriques et technologiques des outils de forage d'autre part, nous avons choisi le programme suivant pour nos travaux de recherche :

* LE MATERIAU ROCHEUX était simulé au départ de notre étude par différents types de matériaux équivalents permettant de présenter les différents types de résistance et de comportements mécaniques d'une part et d'abrasivité d'autre part. Actuellement, le passage du matériau équivalent vers des roches réelles est d'application.

* LE TYPE D'OUTILS A LAMES a été choisi pour notre travail expérimental de laboratoire du fait qu'il couvre à la fois les anciens types d'outils à lames et aussi les outils à lames de conception moderne, c'est-à-dire du type stratapax, diapax, etc...

* LES SYSTEMES DE FORAGE tels que rotatif classique ou rotatif à grand diamètre classique et le turboforage à grand diamètre ont été comparés expérimentalement au laboratoire en travaillant dans des matériaux équivalents de différents types, sous différents régimes de travail et en absence des fluides d'injection (les conditions expérimentales étaient telles que le fond du trou pouvait être considéré toujours comme propre et donc sans phénomènes de rebroyage).

Le tableau 1 résume les qualités des matériaux équivalents utilisés pour nos diverses expérimentations de laboratoire. Tous les types de résistance, déterminés habituellement dans des laboratoires de mécanique des roches, ont été pris en considération y compris la résistance au poinçonnage (σ_p) et celle de cisaillement (σ_{cp}) par poinçonnage. Cette dernière expérimentation a été effectuée à l'aide d'une nouvelle presse portative, entièrement informatisée (fig. 13).

Tableau 1. Caractéristiques mécaniques des matériaux équivalents.

Echant. n°	Composition	Propriétés mécaniques				
		σ_{cs} (MPa)	σ_t (MPa)	σ_{cis} (MPa)	σ_p (MPa)	σ_{cp} (MPa)
278/1	sable plâtre eau	0,984	0,102	0,221	42,660	12,259
196/9	plâtre craie eau	0,592	0,085	0,149	9,993	4,707
196/10	sable plâtre craie eau	2,616	0,343	0,628	56,420	19,614

En ce qui concerne les caractéristiques géométriques des outils à lames adoptés pour ce travail de recherche, le tableau 2 et la figure 1 donnent un aperçu à ce sujet.

Tableau 2. (Caractéristiques géométriques des outils à lames).

α(°)			e	ℓ (mm)	dm (mm)	γ(°)
-45°	0	+45				
	outil n°					
101	104	107	2			
102	105	108	4	12	94	
103	106	109	6			0
110	113	116	2			
111	114	117	4	36	70	
112	115	118	6			
119	122	125	2			
120	123	126	4	12	94	15
121	124	127	6			

où α – angle de coupe ; β – angle du tranchant ;
 γ – angle de dépouille ; ℓ – longueur des lames ;
e – nombre de lames ; d_m – diamètre moyen de l'outil

Figure 2. Banc d'essais de laboratoire pour étudier un seul cycle de destruction en matériaux équivalents

Figure 1. Quelques types d'outils utilisés dans le cadre de cette recherche.

Parmi les caractéristiques technologiques dont on a tenu compte pendant nos expérimentations, il s'agissait de :
n .... vitesse de rotation de l'outil
F .... poids axial sur l'outil
C, F_c ... couple de destruction qui au niveau de chaque lame se réduit en force de coupe F_c
v_t ... vitesse de pénétration (paramètre résultant de la combinaison des trois paramètres précédents).

Le but de nos travaux de recherche était entre autres aussi d'étudier en premier temps les différentes corrélations qui existent entre les divers paramètres intervenant dans le processus de destruction en forage rotatif et répondre à la question :
"QUELLE(S) EST (SONT) LA OU LES RESISTANCE(S) TYPE(S) QUI INTERVIENNENT REELLEMENT DANS LE PROCESSUS DE DESTRUCTION DES ROCHES EN FORAGE ROTATIF DES DIFFERENTS SYSTEMES MENTIONNES PLUS HAUT".

Tous les essais dans le cadre de cette étude ont été effectués dans des conditions ambiantes de laboratoire c'est-à-dire à la pression atmosphérique pour le moment. Le matériel expérimental qui a été utilisé dans le cadre de ces travaux de recherche est présenté aux figures 2, 3, 4 et 5.

En ce qui concerne les outils du forage rotatif classique (fig. 1), on peut résumer leur travail comme la recherche d'une fonction telle que :

$$F (G, T, R) = 0$$

Figure 3. Banc d'essais de laboratoire pour étudier la destructibilité des matériaux équivalents en forage rotatif.

Figure 4. Banc d'essais de turboforage à grand diamètre (fabrication F.P.Ms).

Figure 5. Banc d'essais de turboforage à grand diamètre (fabrication Diamant Boart).

où   G ... paramètres géométriques (α, β, γ,ℓ,e, d_m)
     T ... paramètres technologiques (F,Fc,c, n, V_t^m)
et   R ... paramètres mécaniques des roches (R, A)

ou bien $F(\alpha,\beta,\gamma,\ell,e,d_m-F,F_c,(c),n,V_t-R,A)=0$

ou en fonction de la vitesse de pénétration de l'outil :

$$v_t = f(\alpha,\beta,\gamma,\ell,e,d_m-F,F_c(c),n-R,A)$$

En vue de pouvoir réduire le nombre de paramètres
étudiés, parallèlement deux approches ont été adoptées :

a) l'approche du phénomène de destructibilité à partir d'une seule lame isolée en adoptant la notion de
la surface active qui englobe la plupart des paramètres géométriques sous un seul facteur SURFACE ACTIVE
D'UN OUTIL.

b) l'approche globale du phénomène à partir de la notion d'énergie spécifique de destruction. Cette approche au niveau de l'ensemble de l'outil fait appel
aux notions de puissance de destruction, liées à un
volume de roche abattue. Son intérêt principal est
de permettre une comparaison des performances des outils de géométries différentes travaillant tous dans
un même matériau avec des paramètres technologiques
différents.
   Les deux méthodes appliquées sont complémentaires
ce qui nous a permis, en combinant l'une avec l'autre,
d'arriver à une modélisation mathématique du phénomène.
En ce qui concerne le processus de destruction des
roches à l'aide des outils du type à deux lames (fig.
6) on a d'après (1) :

$$v_t = c.e.n\ \delta_t$$

où n ... vitesse de rotation de l'outil
   δ_t profondeur de passe (l'avancement de la lame
        par tour = v_t /c.e.n)
   c ... constante
   v_t . vitesse de pénétration

De nos travaux expérimentaux de laboratoire, effectués dans différents types de matériaux équivalents,
résulte que :
* la profondeur de passe est quasi indépendante de
  la vitesse de rotation
* la profondeur de passe dépend de la poussée sur
  l'outil
* la force de coupe est quasi indépendante de la vitesse de rotation
* la force de coupe dépend de la poussée sur l'outil

On peut donc calculer la pression latérale exercée
par chaque lame sur le matériau à forer en passant par
une surface active latérale, conventionnelle, égale à
la projection verticale de la face supérieure de la
lame en contact avec le matériau (fig. 7).

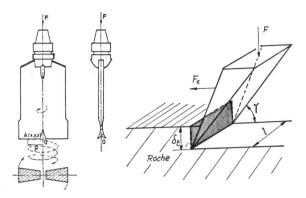

Fig.6. Outils à deux lames.   Fig.7. Surface active
                               latérale d'un outil à
                               lames.

D'après la figure 7, on peut écrire :

$$S.act.lat. = \delta_t \times \ell$$

d'où $$P_{lat} = \frac{F_c}{S_{act.lat}} = \frac{F_c}{\delta_t.\ell}$$

L'interprétation de nos résultats expérimentaux en matière de calcul des pressions latérales exercées sur
le même matériau équivalent, par différents types
d'outils fonctionnant avec différents régimes de travail (paramètres technologiques), permet de tirer les
conclusions suivantes :
* en ce qui concerne les pressions latérales de contact, ces dernières sont très élevées et de loin
  supérieures à la résistance à la compression simple
  des matériaux testés.
* les différentes valeurs de ces pressions oscillent
  autour d'une valeur moyenne constante très proche
  de la valeur de la résistance au cisaillement par
  poinçonnage (σ_{cp}), ce qui nous permet d'écrire
  après de nombreuses vérifications que :

$$P_{lat} = \text{résistance latérale} = \sigma_{cp} = \text{constante}$$

Cette pression latérale pourrait donc être considérée comme une caractéristique intrinsèque de chaque
matériau testé au laboratoire, ce qui était le cas
pendant toute notre expérimentation. On peut donc
écrire :

$$\sigma_{cp} = \frac{F_c}{\delta_t.\ell} \longrightarrow \delta_t = \frac{F_c}{\ell.\sigma_{cp}}$$

et $$v_t = c.e.n. \frac{F_c}{\ell.\sigma_{cp}}$$

Cette dernière équation montre l'importance de la sollicitation latérale (couple) et de la prise en compte
de la résistance du matériau à détruire (σ_{cp}) dans le
processus de destructibilité d'une roche par un outil
à lames (fig. 1, 6, 7 et 8).

En ce qui concerne le processus de destruction des
roches par outils de forage à lames en liaison avec
l'énergie spécifique, cette approche globale fait appel à la somme des efforts développés au niveau de
l'outil . L'énergie spécifique de destruction est
définie ici comme le travail que doit fournir un outil
à lames pour détruire un volume unitaire de roche (Loi
de KICK), c'est-à-dire :

$$W_{sp} = \frac{P}{Q} = \frac{P_c + P_a}{V_t . S}$$

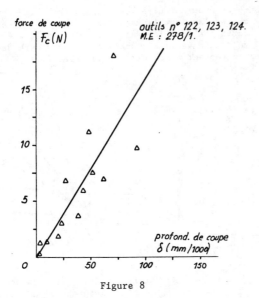

Figure 8

où P... puissance totale développée par l'outil
  Pc...puissance de coupe
  Pa...puissance axiale
  Q....débit de roche abattue
  S... surface horizontale détruite

On sait que :
$$d_m = (d_e + d_i)/2; P_c = C.\omega \quad ; \text{ et } P_a = F.v_t$$

d_m, d_e, d_i ... diamètres moyen, extérieur et intérieur
C ... couple de destruction
ω ... vitesse angulaire
F ... poussée axiale sur l'outil

d'où

$$W_{sp} = \frac{C.\omega}{V_t \cdot \pi \cdot d_m \cdot \ell} + \frac{F}{\pi \cdot d_m \cdot \ell} \qquad (a)$$

généralement

$$P_a \lll P_c \quad , \text{ ce qui entraîne :}$$

$$W_{sp} = \frac{C.\omega}{V_t \cdot \pi \cdot d_m \cdot \ell} \qquad (b)$$

D'une façon générale, la formule de l'énergie spécifique nous fait apparaître ici la prépondérance des paramètres de la rotation (couple et vitesse de rotation) par rapport à la poussée axiale sur l'outil. Ceci nous fait penser à devoir soulever une remarque fondamentale suivante : le problème de destructibilité des roches par outils de forage du type à lames devrait être abordé surtout à partir de la sollicitation latérale qui constitue la principale composante du processus de destruction et non à partir de la sollicitation axiale comme il en est très souvent le cas dans la littérature disponible.

La poussée axiale sur l'outil a sa signification importante, mais seulement dans le but de maintenir l'outil à la profondeur de passe voulue.

Les résultats de calcul de l'énergie spécifique pour chaque matériau équivalent testé nous ont montré que :

* la variation de l'énergie spécifique existe dans la même fourchette comme il en est dans le cas de la pression latérale
* l'énergie spécifique, exprimée en N/mm^2 ou en Pa a la dimension d'une pression d'une valeur élevée

* Quel que soit le type d'outil ou les paramètres technologiques adoptés pour assurer le fonctionnement de ce dernier, une constance des valeurs de cette énergie, pour ce qui concerne le travail dans un matériau donné, est observée autour d'une moyenne. Il en résulte que l'énergie spécifique de destruction pourrait être considérée comme une caractéristique importante du matériau qui s'identifierait à la RESISTANCE AU CISAILLEMENT PAR POINCONNAGE, c'est-à-dire :

$$W_{sp} = \sigma_{lat} = R_{cp}$$

* la pente des fonctions P-v_t pourrait caractériser la performance de chaque outil dans le matériau donné. Plus cette pente est faible, mieux le matériau est détruit par l'outil donné, d'où la performance de l'outil est inversément proportionnelle à la pente (fig. 9).

En se basant sur les travaux de Ngoi Nsenga (3,4,5, 6), on arrive à formuler l'équation suivante pour calculer la vitesse d'avancement d'un outil de forage du type à lames dans des matériaux équivalents de différentes qualités :

$$v_t = \frac{e.n}{60} \cdot \frac{F_c}{\ell(W_{sp} - \frac{F}{\pi.\ell.d_m})} \simeq \frac{e.n}{60} \cdot \frac{F_c}{\ell.W_{sp}}$$

où $W_{sp} \doteq \sigma_{cp}$, en considérant que $\frac{F}{\pi.\ell.d_m} \lll W_{sp}$

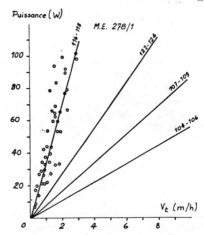

Figure 9. Puissance de coupe pour divers types d'outils.

En ce qui concerne l'influence de la poussée axiale sur l'outil (F) et la force de coupe (Fc), il faut signaler que ces deux forces ne peuvent pas être dissociées l'une de l'autre dans le processus de forage. Toute étude du processus de découpage qui tiendrait compte uniquement de l'un ou de l'autre facteur fausserait dès le départ les résultats obtenus. Ces deux forces doivent garder un certain équilibre afin que le processus de forage puisse se dérouler normalement.

La poussée sur l'outil est l'un des paramètres, qui influence considérablement le couple. Dans tous les matériaux équivalents testés dans le cadre de cette étude, la corrélation Fc-F était du type linéaire pouvant être exprimée par la formule mathématique suivante :

$$F_c = a. F + b$$

Tenant compte de la surface active axiale S.ax, on peut écrire :

$$P_{ax.min} = \frac{F_o}{S_{ax_t}} = -\frac{b}{a.S_{ax_t}}$$

614

où
Fo ... effort de poussée axiale minimum
Pax min ... pression axiale minimum
Sax_t... surface axiale totale de l'outil
S_{ax}... surface axiale par lame (Fig. 10)

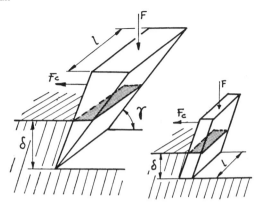

Figure 10

Les essais de laboratoire nous ont montré que la pente "a" (fig. 11) est quasi constante pour un même nombre de lames quelle que soit la géométrie de l'outil.

Il résulte des travaux de NGOI NSENGA (3, 4, 5 et 6) que :

$$F_c = \frac{K}{e} \left[F - (S_{ax_{ti}} + \theta.t) . \sigma_t \right]$$

où
$S_{ax_{ti}}$ ... surface axiale initiale de l'outil à lames
θ .... augmentation de la surface d'attaque initiale en fonction du temps de forage
t .... temps de forage
σ_t ...... résistance à la traction du matériau foré
K .... coefficient exprimant l'état du contact outil-matériau foré
e .... nombre de lames de l'outil

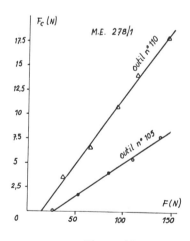

Figure 11

Sur la base des travaux expérimentaux de laboratoire, en utilisant des matériaux équivalents, la relation suivante peut être d'application pour exprimer le travail d'un outil de forage du type à lames :

$$V_t = \frac{n}{60} . K \left[\frac{F - (Sax_i .e + \theta.t) . \sigma_t}{\ell . \sigma_{cp}} \right]$$

Ce modèle mathématique n'est valable que dans le cas des poussées sur l'outil égales ou supérieures à une poussée axiale minimum. Celle-ci correspond en effet au début de la troisième zone du diagramme v_t-F (fig. 12). C'est en travaillant dans cette dernière qu'on peut obtenir des meilleurs résultats en matière de forage rotatif à l'aide des outils à lames. Ce modèle s'applique donc dans cette zone.

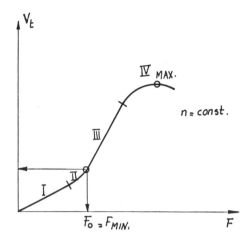

Figure 12. L'avancement en fonction de la poussée sur l'outil.

CONCLUSIONS GENERALES.

* Le couple de destruction joue un rôle prépondérant dans le forage rotatif à l'aide des outils à lames. Ce dernier est à la base de destruction de la roche par cisaillement au niveau de chaque lame. Une excellente corrélation est possible entre la valeur de la force de coupe, la surface latérale active de l'outil, et la résistance au cisaillement par poinçonnage, déterminée au laboratoire à l'aide d'une presse informatisée comme celle de la fig.13.

Figure 13

615

* La géométrie des outils et surtout de leurs lames joue un rôle très important dans le processus de destructibilité (S_{ax} et $S_{act.lat}$). La longueur des lames ainsi que le nombre de ces dernières influence beaucoup le processus de forage.

* Deux paramètres mécaniques de la roche à forer interviennent de façon très importante dans le processus de forage rotatif à l'aide des outils à lames, c'est-à-dire :

  σ_{cp} ... la résistance au cisaillement par poinçonnage et

  σ_t ... la résistance à la traction

  en utilisant l'outil dans la troisième zone du diagramme v_t - F (fig. 12).

* Une comparaison ayant été effectuée au niveau du laboratoire entre la méthode du forage classique Rotary et celle du turboforage à grand diamètre, il résulte d'après les travaux de Xiao Shan (8) qu'en ce qui concerne le contact "outil-roche", toutes autres choses égales, c'est-à-dire le diamètre du forage, les surfaces actives latérale et axiale, la géométrie des lames et la position de ces dernières sur l'outil, le principe de destruction de la roche correspondant à celui du turboforage à grand diamètre est de loin beaucoup plus intéressant que celui du forage rotatif classique.

(4) Ngoie Nsenga. 1985. Etude des relations fondamentales dans le forage rotatif. Etude bibliographique. Publication du laboratoire Mécaroches, F.P.Ms, n° 230, juin 1985.
(5) Ngoie Nsenga. 1985. Etude expérimentale au laboratoire sur le M.E278/1. Public. du laboratoire MECAROCHES F.P.Ms, n° 205, octobre 1985.
(6) Ngoie Nsenga. 1986. Etude expérimentale au laboratoire sur les M.E 196/9 et 196/10. Publication de laboratoire MECAROCHES F.P.Ms n° 229, décembre 86.
(7) Ziaja M.B., Miska S.,1982. Mathematical model of the diamond bit drilling process and its pratical application. Society of Petroleum Engineering Journ., déc. 1982.
(8) Xiao Shan. 1986. Etude comparative de deux systèmes de forage à grand diamètre par modélisation physique au laboratoire (Thèse M.Sc, F.P.Ms).
(9) Hunsdorfer E. Ermittlung optimaler bohrparameter und Konstruktion Nevartiger Bohrwerkzeuge für jeweilig Vorhandene Bohrtechnische bedingungen in der Schurfbohrtechnik (Nassovia Weilburg Forschungs Bericht 03 R 237).
(10) Hoffrichter et Broul. 1967. A comparison of four different methods of determining the shear strength of rocks (Geotechnical conference Oslo 1967, Proceedings).

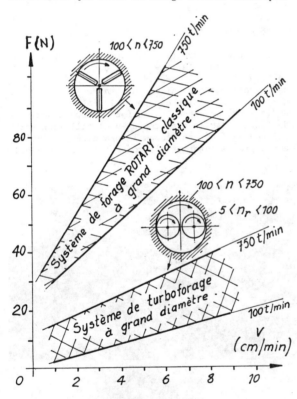

Figure 14. Comparaison de deux systèmes de forage à grand diamètre en matériaux équivalents au laboratoire (D'après Xiao Shan (8)).

REFERENCES BIBLIOGRAPHIQUES

(1) Brych J. 1976. Contribution à la recherche d'une méthode pour déterminer la forabilité des roches (Publication n° 30, juin 76, du laboratoire MECAROCHES, F.P.Ms).
(2) J. Brych. 1970. Turboforage à grand diamètre (Annales des Mines de Belgique, n° 1).
(3) Ngoi Nsenga. 1984. Considérations sur les problèmes de destructibilité des roches par les outils de forage à lames. Thèse M.Sc., F.P.Ms, 1984.

Seismic control of mine and quarry blasting in the USSR

Contrôle de l'action séismique des explosions industrielles
Überwachung der Erdbebenwirkung der industriellen Sprengungen in der UdSSR

A.B.FADEEV, Leningrad Civil Engineering Institute, USSR
L.M.GLOSMAN, VNIIGS, Leningrad, USSR
M.I.KARTUZOV, Mining Institute, Sverdlovsk, USSR
L.V.SAFONOV, NIIKMA, Leningrad, USSR

ABSTRACT: The evaluation of seismic effect of single and millisecond blasts is given. One shown the criteria which characterize seismic safety of buildings, structures and mine openings. The complex of engineering methods of seismic control of blasting is considered to reduce damage brought to mine engineering systems.

RESUME: On a donné une évaluation de l'effet sismique dû aux tins explosibles à une seule fois et pendant une milliseconde. On a présenté les critères caractérisant la résistance sismique des bâtiments, des ouvrages et des excavations. On examine l'ensemble des méthodes d'ingénieur visant à contrôler l'action sismique des explosions afin de diminuer les dommages aux systèmes de mine de génie civil.

ZUSAMMENFASSUNG: Die Einschätzung der Erdbebenwirkung der Sprengungen mit Moment- und Millisekundewzündung ist gegeben. Die Kriterien, die Erdbebenstandsicherheit von Gebäuden, Ingenieurbauwerken und bergmännischen Bauwerken charakterisieren, sind aufgezeigt. Es wird die Gesamtheit der Ingenieurverfahren für die Überwachung der Erdbebenwirkung der Sprengungen behandelt, die zur Reduzierung von Schäden der bergmännischen Ingenieursystemen beitragen.

The total mass of explosive charges blasted simultaneously in quarries and mines reaches hundreds and even thousands of tons. Seismic action of such explosions can cause considerable damage to mine openings, the surface industrial and civil structures.

Seismic action of an explosion results from strength effect of blast gases on rock mass.

To express seismic waves parameters according to explosive charge mass and the distance to the place of an explosion their interpretation by energetic and geometric similarity methods is generally adopted. The intensity of seismic effect is characterized by the velocity of soil vibrations when seismic blast waves propagate.

Beyond the vicinity zone (one of crushing and cracking) the dependence of soil vibrations velocity on the value of a deep-seated charge and the distance to a blast place is expressed by the equation:

$$V = K_1 \cdot K_2 \sqrt[3]{Q/r^3}, \quad cm/s \qquad (1)$$

where Q - the mass of charges blasted simultaneously; r - the distance to the place of a blast, m; K_1 - the coefficient depending on soil conditions in the foundation of structures; K_2 - the coefficient with regard for the number of free surfaces.

Assessing seismic action of a single or accidental explosion one should orient on average values of coefficient K_1, maximum values are taken for routine regular blasting (Table 1).

If charge blasting is carried out in the volume of the rock restricted from all the sides, the number of free surfaces is 8. When blasting a charge on the bench of a quarry there are 2 free surfaces: a surface and a bench slope. At underground ore breaking there is one free surface for one compensating slot.

Table 1.

Soil conditions	K_1	
	Average	Maximum
Rock, dense soils sans water content up to 5-10 m thick	200	300
Sand-clay soils sans or with weak water content more than 10 m thick	300	450
Loose and saturated soils	450	600

The dependence of coefficient K_2 on the number of free surfaces is given in Table 2.

Table 2.

Number of free surfaces	K_2
1	3.0
2	1.0
3	0.5
4	0.3

To reduce seismic effect millisecond blasting is used. In this case the next charge explosion occurs in the rock mass which is under stressed state because of the gas residual pressure of the previous group charges. Each charge explosion occurs at a new position of a free surface formed at the explosion of the previous group charges. Additional crushing occurs due to the collison of flying rock pieces.

As cine-metric observations showed, after the explosion of a cylindrical charge 100 mm in diameter the broken rock moved with acceleration during 20-40 ms, that is during 20-40 ms the residual pressure of explosive gases was higher than atmospheric one. The outflow of gases from the charging room occurs over a period of 60 ms at the explosion of a charge 250 mm in diameter.

It is known from blasting practice that optimum intervals of retardation for deep-holes 100 mm in diameter are within 10-15 ms, and for deep-holes 200-250 mm in diameter they are within 35-50 ms, that is the adopted intervals are shorter than the duration of the residual action of gases. It is obvious that the residual pressure of gases from the charges of previous retardation group may influence on the action of the next group charges.

One may consider the moment of first face soil advance as the formation start of a new exposed surface. According to a lot of information the first rock advances are atarted in 3-5 ms after the charge detonation. By the moment of the explosion of the next group charges the broken material is removed 0.5 m away.

Thus by the moment of the next group charges explosion when the retardation interval is 10-15 ms, a new exposed surface will be fully formed, and this factor may be well used to explain millisecond blasting mechanism.

The retardation interval sufficient for getting the minimum seismic effect may be determined by the equation:

$$T = 2W \sqrt{\gamma/q}, \quad ms \qquad (2)$$

where W - the shortest distance from a charge to a free surface, m; - the density of the rocks blasted, t/m³; q - the specific consumption of an explosive, kg per cubic meter of rock blasted.

In practice one should take the nearest millisecond interval created by retarding means.

A series of special experiments was carried out with blasting one, two and more charges simultaneously and with different millisecond intervals.

In Fig. 1 a number of seismograms from 6 blasts (Table 3) are represented. The blasts were carried out at one place under the same conditions and with seismographs being installed at the same places. Upper tracks conform to vibrations at point 1, the middle tracks - to vibrations at point 2 and the lower one - at point 3. Points 1 and 2 were on the rock surface of the bench blasted and at point 3 there were sand-clay overburdens about 3 m thick.

At explosion 1 from the oscillogram (Fig. 1a) it is seen that at point 1 which was 120 m far from the explosion, seismic vibrations had the character of one-fold pulse with the period duration of about 25 ms. It is impossible here to divide the fronts of longitudinal and transverse waves. The total duration of vibrations does not exceed 100 ms. At points 2 and 3 high frequency head part of vibrations is already distinctly isolated which conforms to longitudinal waves, and the succeeding vibrations with 10 Hz frequency which conform to surface waves. The total duration of vibrations extends to 0.5-1 s.

When blasting 7 charges consequently (II) with 20 ms retardation interval (Fig. 1b) at

Table 3.

Or-der Nos	Total charge mass, kg	Number of retardation groups	Maximum charge mass at a group kg	Retardation interval	Velocities, cm/s		
					T_1	T_2	T_3
I	320	1	320	-	4.1	0.6	0.37
II	2240	7	320	20	5.1	0.9	0.24
III	1280	4	320	50	7.5	0.75	0.14
IV	640	1	640	-	5	0.91	0.49
V	2100	4	640	40	9.1	1.0	0.49
VI	1500	1	1500	-	15	2.7	1.1

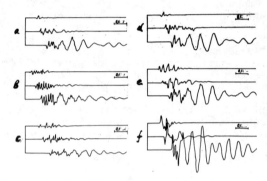

Figure 1. Oscillograms of vibrations.

point 1 vibration pulses from all 7 charges are distinctly and separately seen. At point 3 separate pulses from each charge are already not seen so distinctly; they are seen only against the background of low frequency vibrations. However, their maximum amplitude does not exceed the amplitude of vibrations from a single charge explosion (a).

When blasting 4 charges with 50 ms intervals (III) at point 1 (Fig. 1,c) pulses from individual charges come separately and pauses are seen between them. At points 2 and 3 the separation of pulses travel is seen worse, however the amplitude of vibrations does not exceed the one from a single explosion.

At simultaneous blasting of 2 charges located at the distance of 7 m from each other (IV), the interval between the moments of pulses coming from different charges (Fig. 1, b) does not exceed 2-3 ms. This spread is quite imperceptible when the duration of pulses is 25 ms, and the simultaneous explosion of 2 charges adjacently located is registered as one pulse (point 1). At more remote points 2 and 3 just as in the foregoing cases vibrations become of lower frequency character and extend with time.

When 7 charges are blasted at 4 groups (V) in the order of 1-2-2-2 with 40 ms interval at point 1 the pulses are seen as separate ones (Fig. 1, e), at points 2 and 3 individual pulses are almost not seen, however the amplitude of vibrations is a little lower than that of simultaneous blasting of 2 charges.

When 5 deephole charges with the total mass of 1500 kg (VI) are instantly blasted the amplitude of vibrations is much more than that of the previous millisecond blasts when the total weight of charges was as great as 2100 kg (Fig. 1,f).

Thus it is found that if an explosion of the next charge is carried out after the disintegration of the seismic focus from the previous charge, the seismic effect at any distance does not depend on the total number of charges blasted but is determined by the mass of charges of one retardation group.

Fig. 2 shows the oscillogram of the explosion carried out at 31 groups of retardation at a magnesite quarry. Maximum charge mass at a retardation group is 440 kg and the interval is 35 ms.

The observations of the seismic action of blasts showed that millisecond intervals determined by Eq. (2) fully provided the separate running of seismic pulses from each retardation group of charges in the near zone and the reduction of vibrations level to the effect of one group of charges regardless of the number of groups and total explosive mass in a blasted rock block.

The choice of the allowable velocity of vibrations is determined by the construction stability (by the strength of elements and units) and the purpose of a structure protected. The leading principle while specifying allowable velocities is (a) the preservation of load-carrying structures and people's safety at a single explosion and (b) lack of damages requiring repair according to technical or aesthetic considerations for routine regular blasts.

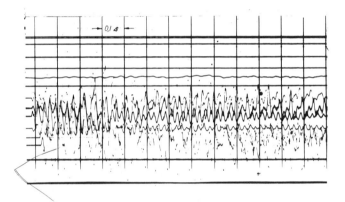

Figure 2. Oscillogram of 11 t explosive blast at 31 groups of retardation.

Table 4 shows allowable values of vibrations velocities for particular types of structures.

Table 4.

Structures	Allowable velocities of vibrations (cm/s) when blasting	
	Repeated	One-fold
Hospitals	0.8	3
Large-panel residential buildings and children's institutions	1.5	3
Residential and public buildings of all types except large-panel, office and industrial buildings having deformations, boiler rooms and high brick chimneys	3	6

Table 4 (continued)

Office and industrial buildings; high reinforced-concrete pipes, railway and water tunnels, traffic flyovers, saturated sandy slopes	6	12
Single-storey skeleton-type industrial buildings, metal and block reinforced-concrete structures, soil slopes which are part of primary structures, primary mine openings (service life up to 10 years); pit-bottoms, main-cross-entries and drifts	12	24
Secondary mine openings (service life up to 3 years), haulage breakthroughs and drifts	24	48

The definitions of antiseismic blasting parameters (of allowable mass of charges at a rated distance up to a structure or of a secure distance at rated mass of charges) can be performed by means of Eq. (1) or according to a nomogram (Fig. 3) constructed by Eq. (1).

When blasting on a surface the seismic action of an explosion is not taken into consideration.

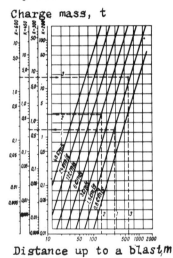

Figure 3. Nomogram for definition of safe blast parameters.

BIBLIOGRAPHY

Medvedev, S.V. 1964. Blasting seismics. Moscow: Nedra.
Mossinets, V.N. 1976. Crushing and seismic action of an explosion in rock formations. Moscow: Nedra.
Fadeev, A.B. 1972. Crushing and seismic action of explosions in quarries. Moscow: Nedra.
Safonov, L.V. & G.V. Kuznetsov 1964. Seismic effect of deephole charges blast. Moscow: Nauka.
Don Leet 1963. Seismic action of a blast. (In Russian. Translated from English): Gosgortechizdat.

Prediction of roadheader cutting performance from fracture toughness considerations

Prédiction de la performance d'une machine à attaque ponctuelle en fonction de la résistance à l'extension d'une fracture
Vorhersage der Schnittleistung von Streckenvortriebsmaschinen aus Bruchzähigkeitsbetrachtungen

I.W.FARMER, Ian Farmer Associates, Newcastle upon Tyne, UK
P.GARRITTY, Ian Farmer Associates, Newcastle upon Tyne, UK

ABSTRACT: Using the concepts of cutting energy, power transfer and pick wear together with rock descriptions and machine performance data a means of estimating rock cutting performance by roadheading machines is proposed.

RESUME: Utilisant ensemble les concepts d'energie coupante, de transfert de puissance et d'usure du pic avec la description de la roche et la performance de la machine meouree experimentalement un moyen d'estimer la performance d'une machine a attaque ponctuelle est propose.

ZUSAMMENFASSUNG: Um die schneidende Wirkung von Streckenvortricbmaschinen abzuschatzen di Autoren schlagen vor die folgenden charakteristischen Kennzeichnen in Betracht zu nehman wie Energie Aufnahme beim Schneiden, Kraftubertragung, Meiszel Abnutzung und Gebirgseigenschaften.

1 INTRODUCTION

In planning construction of a tunnel a major consideration is the choice of rock excavation method. This choice must be based to a large extent on the anticipated performance of available excavation systems in the predicted geotechnical environment.

The performance of machine excavation systems, such as full face tunnel boring machines (TBMs) and partial face machines (roadheaders), is related to the mechanical behaviour of the rock. However, there is at present no generally accepted criterion for prediction of instantaneous advance or cutting rates from any group of rock parameters. There is not even general agreement on which set, or sets, of parameters should be used.

One approach applicable to both roadheader and TBM performance is to consider the energy required to excavate the rock. Rock cutting specific energy is defined as the energy required to cut a unit volume of rock, and can be related to the concept of fracture toughness which is the strain energy stored in a unit volume of rock immediately prior to failure.

2 ENERGY CONSIDERATIONS OF TUNNELLING MACHINE PERFORMANCE

The energy, Wi, input to an excavation face from a tunnelling machine can be expressed as the cutting or specific energy per unit volume of rock. This is given by:

$$ Wi = \frac{Pk}{VER} \qquad \dots\dots\dots\dots\dots\dots (1) $$

Where P is the cutting head power, k is the energy transfer ratio or power efficiency, and VER is the volume excavation rate.

In order to devlop a satisfactory relation between available cutting energy and rock fracture, it is necessary to consider the simple mechanics of fracture of a brittle elastic material. Griffith's (1921, 1924) explanation of micro-crack extension in brittle materials is based on an energy balance for the solid volume of the material to be occupied by an extended crack. It equates the strain energy, Ws, stored in unit volume of rock as it fractures with the surface energy, Wa, required to satisfy unit area of the newly formed crack surfaces in the rock volume. This can be written simply as:

$$ Ws = \frac{\sigma_{cf}^2}{2E} \quad \alpha \quad WaSo \dots\dots\dots\dots (2) $$

Where σ_{cf} is the rock compressive strength, E is the deformation modulus, and So is the specific crack surface area per unit volume.

Thus by equating equations 1 and 2 a relation can be obtained from a simple energy balance in the form:

$$ \frac{Pk}{VER} = \frac{\sigma_{cf}^2}{2E} \quad \alpha \quad WaSo \dots\dots\dots\dots (3) $$

High energy input is therefore required to excavate rock when So is high and the cuttings are small, and when the term $\sigma^2 cf /2E$, which can be related to the concept of fracture toughness, is high, as for example occurs with high strength, medium modulus rocks such as strong sandstones, and medium strength, low modulus rocks such as some shales.

Since both uniaxial compressive strength and deformation modulus are readily determined in the laboratory it is informative to speculate on the degree to which such values can be applied to the effective fracture toughness of rock masses. The brittle nature of rock fracture means that strength reduces in proportion to the square root of specimen dimensions (Farmer, 1983), while the presence of natural discontinuities further reduces the effective mass strength of the rock. For many applications, therefore, laboratory indices have little direct correlation with rock mass response. However, in rock cutting only a small volume of rock at a free surface, and therefore under comparatively little confinement, is attacked at a particular time. Furthermore although natural discontinuities will tend to facilitate machine excavation (Johnson and Fowell, 1984), discontinuity frequency is also related to rock strength (Stimson, 1980). Thus for the particular circumstances of rock cutting, laboratory parameters may be considered to give an adequate representation of rock cutting behaviour.

Estimates of fracture toughness made from measured values of uniaxial compressive strength and deformation modulus should not be confused with apparent specific energies, calculated from measured values of machine head power consumption and volume excavation rate by ignoring the energy transfer ratio in equation 1. Specific energies measured in this way not surprisingly correlate closely with excavation rates since the former is calculated from measurements of the latter (see McFeat-Smith and Fowell, 1977). Apparent specific energy is not, therefore, a particularly useful concept when considering excavation performance.

Equation 3 indicates that excavation rate is theoretically proportional to the rate of energy input to the rock, which is directly proportional to machine head power for a constant specific energy. This implies that there is no upper limit to machine operation; only a gradual reduction in excavation rate with increasing rock toughness. Although cutting or product analysis is rarely carried out, equation 3 also shows that product size will reduce as cutting specific energy increases. An extension of this is that pick wear will increase. The degree to which the theoretical behaviour of machines operating in a range of rock types is modified in practice is discussed in the following section.

3 MACHINE EXCAVATION PERFORMANCE

In Figure 1 the plot of uniaxial compressive strength squared against twice the deformation modulus provides an energy balance requirement for excavation of various rock types. The geological data indicate that lithologically similar materials tend to fall into diagonally distributed groups in which large variations in modulus and strength fall into a narrow ratio range. This allows for a lithologically related index of brittleness (Deere and Miller 1966, and Hobbs 1974).

From Figure 1 it is clear that partial face machine operations which have experienced difficulties plot within the diagonally distributed group expected from theoretical considerations. Specific energies corresponding approximately to the upper and lower envelopes of the group are included. Along the lower energy input requirement line problems are experienced by the least powerful, hydraulically softest machines, especially when confronted by a full face of massive rock. Along the upper limit even the most powerful machines experience excessive tool wear and damage, and progress may be impossible in full face rock conditions. Between the limits specific surface area of product and cutter consumption increase and progress is reduced, and although favourable jointing patterns may facilitate excavation the presence of tougher bands reduces progress.

The theoretical ease with which such competent rocks as granites may be excavated warrants comment as these are recognised to be beyond the effective capabilities of partial face machines. The explanation lies in the necessity to employ shallow, less efficient depths of cut when excavating hard, high strength rocks to avoid excessive tool forces. As the specific surface area, So, of the product increases a proportionally larger amount of the machine power is used in comminution and grinding of the rock. Cutability and abrasivity tests therefore become increasingly important.

Figure 2 shows the relation between $\sigma c f^2/2E$ and volume excavation rate for machines operating in Coal Measures rocks (Barnett et al 1984, Bates et al 1981, 1982). The data are derived from the monitored performances of two powerful machines, a Thyssen-Paurat Titan and a DOSCO MK 3. These machines have nominal cutterhead motor powers when cutting at slow and fast speeds of 100kw and 200kw, and 93kw and 142kw respectively.

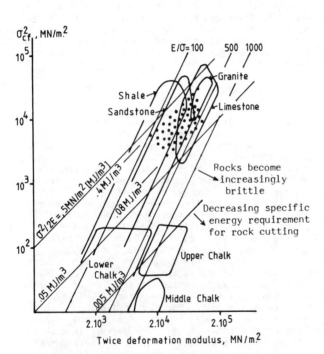

E/σ   Index of rock brittleness.

$\sigma_{cf}^2/2E$ Rock cutting specific energy requirement.

Region in which partial face machines experience difficulties in excavating rock.

Figure 1 Relation between rock cutting specific energy requirement and rock brittleness showing the region in which partial face machines have difficulty in excavating the rock.

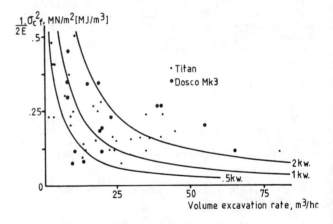

Figure 2 Relation between $\sigma^2 cf/2E$ and volume excavation rate.

Although there is appreciable scatter in the data it is clear that the theoretical head power requirement curves fit the general form of the data. Outlying points reflect assumptions made in the derivation of equation 3, errors in the estimation of deformation moduli, and the importance which discontinuities and rock structure may assume. There are a number of important observations which can be made from this plot.

1. Only a small fraction of the machine head power consumed, typically one or two per cent, is translated into actual rock breakage. This is particularly significant as both machines in the sample

are modern, powerful and hydraulically stiff.

2. The benefits of increasing machine power are most apparent in the excavation of rock with low specific energy requirements. High specific energy rocks show comparatively low excavation rates even with powerful machines.

3. The form of the cutting power requirement curves is such that they are increasingly parallel to the axes as the input power is reduced. A curve may therefore show a quite dramatic decrease in volume excavation rate for a small increase in $\sigma^2 cf/2E$. This may be an explanation for the acceptance of a maximum rock strength capable of being cut by early roadheading machines. Take for example a rock with a strength of 50MN/m^2 and a deformation modulus of 12GN/m^2. Such a rock plots at about the point of maximum curvature of the 0.5kW power curve. A further increase in strength for the same deformation modulus causes a significant fall in excavation rate. With increasing transfer ratio the power input curves become appreciably flatter and the effects of increasing specific energy requirements on performance are not so significant.

4. There is evidence that the energy transfer ratio, k, increases with decreasing rock specific energy requirements. Aspects of the tool geometry designed into the machine heads and lower energy usage in rock comminution and grinding with increasing average product size are important contributory factors.

5. For a constant value of $\sigma cf^2/2E$ the range of data is related to rock type, Figure 3. Typically sandstones are associated with the lower values of volume excavation rate and mudstones the higher, with siltstones located in an intermediate position. This can be explained in terms of the abrasivity and strength of the siliceous rocks which result in less efficient cutting due to the use of shallow depths of cut to avoid excessive tool forces, and increased energy usage in rock comminution and grinding.

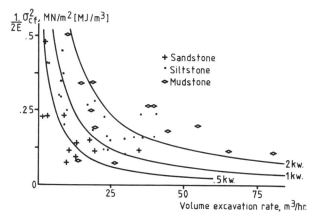

Figure 3 Influence of rock type on the relation between $\sigma^2 cf/2E$ and volume excavation rate.

4 TOOL CONSUMPTION

Tool consumption rates are an important consideration since tools can be expensive to replace in quantity. In British conditions tool consumption normally only becomes significant in the stronger, more abrasive sandstones, siltstones, limestones, ironstones, igneous formations and so on. Tool replacement rates are typically less than 2 picks per 100m^3 of mudstones excavated (Barnett et al 1984, Bates et al 1981, 1982, Burke 1981). This rate increases for abrasive siltstones and becomes really significant

in cutting sandstones when it may be necessary to replace more than 15 tools per cubic metre of rock excavated. It is, therefore, essential to consider that overall tool consumption rates during excavation of a face composed of more than one rock type are generally closely related to those in the strongest, most abrasive rock.

Since both abrasion and breakage contribute to tool consumption rates it is important to avoid transmission of excessive forces to tools. Table 1 compares tool forces generated by normal cutting operations.

Table 1 Comparison of tool forces during normal cutting operations (after Burke 1981 and Bates et al 1981).

		DOSCO Mk2A	DOSCO MH105
	Sumping	11.2	16.3
	Slew	3.4	3.0
Tool	Elevate	3.2	3.9
forces		Thyssen	
in		Titan	
kN			
	Cutting		
	Left	4.3	
	Right	3.9	
	Up	2.9	
	Down	5.3	

Tool forces generated during sumping are particularly significant and cautious sumping operations in strong rocks might therefore be expected to significantly improve tool consumption rates.

5 CONCLUSIONS

In Figure 4 an attempt is made to combine the data from Figures 1 and 2 with tool consumption information to provide a conceptual approach to rock cutability.

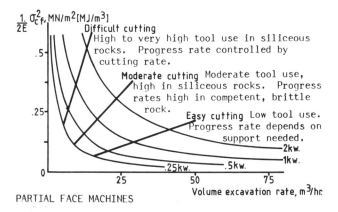

PARTIAL FACE MACHINES

0.5kw Hydraulically soft or worn machine. Poor floor conditions. Operating to the rise on gradients.

1kw Hydraulically stiff. Moderate floor conditions.

2kw Hydraulically stiff. High power, stable floor conditions. Water jet assisted. Operating to the dip.

Figure 4

This approach utilises the concepts of cutting energy, power transfer and pick wear together with rock descriptions and machine performance data to provide a method of estimating limits of cutability. The approach combines a fundamental statement of the

mechanics or rock breakdown under the boundary
conditions or rock cutting with extensive site data.
   The statements in Figure 4 may need refinement but
ultimately they provide a framework for preliminary
decision making

REFERENCES

Barnett, G.H., Beresford, C., Jarvis, E.E. & Mangham,
   B. 1984. Operational performance of the Thyssen-
   Titan E134C roadheader and the Robbins 193-214
   tunnel boring machine at Gascoigne Wood Mine,
   Selby, North Yorkshire Area. U. K. Mining Research
   and Devolpment Establishment Report No. 84/37.
Bates, J.J., Barnett, G.H., Jarvis, E.E. & Newman,
   P.R. 1981. Operational performance of the Thyssen-
   Paurat Titan roadway drivage machine at Silverdale
   and Warsop Collieries. U.K. Mining Research and
   Development Establishment Internal Report No. 81/26.
Bates, J.J., Jarvis, E.E., Barnett, G.H. & Beresford,
   C. 1982. Operational performance of the DOSCO Mk3
   roadway drivage machine at Silverdale and Daw Mill
   Collieries. U.K. Mining Research and Devolpment
   Establishment Internal Report No. 82/14.
Burke, K.J. 1981. DOSCO MH105 low height roadheader
   trials at Whitwick Colliery. U.K. Mining Research
   and Development Establishment Internal Report No.
   81/15.
Deere, D.U. & Miller, R.P. 1966. Engineering classif-
   ication and index properties for intact rocks.
   Air Force weapons Laboratory Report AFWL-TR-65-16,
   Kirkland A.F B. N.M.
Farmer, I.W. 1983. Engineering Behaviour of Rocks,
   Chapman & Hall, London.
Griffith, A.A. 1921. The phenomena of rupture and
   flow in solids. Phil. Trans. R. Soc., A221 : 163-
   198
Griffith, A.A. 1924. Theory of rupture. Proc. Ist
   Int. Congr. Appl. Mech. Delft, p55-63
Hobbs, N.B. 1973. The prediction of settlement
   structures on rock. Conf. Settlement of Structures.
   Cambridge, p579-610.
Johnson, S.T. & Fowell, R.J. 1984. A rational approach
   to practical performance assessment for rapid exca-
   vation using boom type tunnelling machines. Proc.
   25th U.S. Rock Mechs. Symp. Evanston, p759-765.
McFeat-Smith, I.M. & Fowell, R.J. 1977. Correlation
   or rock properties and the cutting performance of
   tunnelling machines. Proc. Conf. Rock Eng.,
   Newcastle upon Tyne, p581-602.
Stimson, B. 1980. Intact rock strength and fracture
   spacing relationships in a porphyry copper deposit
   Int. Jl. Rock Mechs. Min. Sci. & Geomech. Astr.
   17 : 67-68

ACKNOWLEDGEMENTS

This paper is based on work undertaken under contract
to CIRIA. The permission of the Director has been
given to publish this paper.

Tunnel excavation by combination of static demolisher with slit cut drilling

Excavation de tunnel par la combinaison de l'emploi d'un démolisseur statique et d'une haveuse/carotteuse à fente

Stollenausschachtung durch eine Kombination von Statikzerstör und Schlitzbohrung

K.FUKUDA, Shimizu Construction Co. Ltd, Tokyo, Japan
H.KUMASAKA, Shimizu Construction Co. Ltd, Tokyo, Japan
Y.OHARA, Shimizu Construction Co. Ltd, Tokyo, Japan
Y.ISHIJIMA, Hokkaido University, Sapporo, Japan

ABSTRACT: On the basis of linear fracture mechanics, a tunnel excavation method by combination of static demolisher with slit cut drilling was designed by giving due consideration to the effect of in-situ stresses. Several types of slit patterns were studied for their effectiveness on the strength of numerical analysis, laboratory experiments and a few experimental excavation in in-situ rock mass in order to propose the appropriate design plan for tunneling in underground formations.

RESUME: Une méthode d'excavation de tunnel utilisant un démolisseur statique et une haveuse/carotteuse à fente a été conçue sur la base de la mécanique des fractures linéaires, en tenant compte de l'effet des contraintes existant sur le site. Plusieurs types de fentes ayant des formes différentes furent étudies pour analyser leur efficacité sur la force d'une analyse numérique, d'expériences en laboratoire et de quelques excavations expérimentales dans la masse rocheuse du site afin de proposer un plan de conception appropriée pour percer des tunnels dans des formations souterraines.

ZUSAMMENFASSUNG: Auf der Basis der linearen bruchmechanik wurde eine Methode zur Stollenausschachtung Konzipiert, die eins Kombination von Statikzerstör und Schlitzbohrung verwendet und am Ort auftretende Druckauswirkungen in Betracht zieht. Verschiedene Arten von Schlitzmustern wurden durch numerische Analysen, Laborversuche und einige experimentelle Ausschachtungen in Gesteinsmassen vor Ort auf ihre Wirksamkeit untersucht, um einen angemessenen Konstruktionsplan zum Stollenbau in Untergrundformationen zu entwickeln.

1 INTRODUCTION

When adopting the conventional drill and blast method for tunnel excavation, it would be practically impossible to completely protect the surrounding rock from damage induced by the shock generated during the detonation of explosives even though the smooth blasting technique may be adopted. The excavation method described herein uses the so-called static demolisher which is a kind of expansive cement insted of explosives.

After the hole drilled in the rock mass is charged with the static demolisher in slurry state, the hole wall will be gradually pressurized while the slurry commences to harden and expand. In consequence, fractures initiate from the hole wall and they grow in a rather stable manner. Therefore this method incorporates a special characteristic of being able to induce fractures in the rock mass without vibration and to complete a tunnel having section of a prescribed dimension, free from any damage to the tunnel wall.

The attainable maximum pressure due to the expansion of this material is as high as 50 to 70 Mpa under favorable conditions. However, this value may be insufficient to effectively induce fractures into the rock mass which is loaded by the in-situ stresses and is located ahead of the tunnel with little available free surfaces area. Therefore, some suxiliary aids are required. In this paper a method of tunnel excavation by using static demolisher in combination with slit cut drilling is introduced with some theoretical considerations and both laboratory and field experimental results.

2 FRACTURING MECHANISM OF ROCK MASS BY USING STATIC DEMOLISHER

Pressure is gradually generating after injection of the static demolisher into the hole accompanying expansion of the material due to its recrystallization. It should be noticed that this pressure is completely retained within the borehole in the rock since the material of solid state can not intrude into the emanated cracks. This is disadvantageous from view-point of effective fracturing since the intense fracture propagation could only be realized under the condition that the pressure of substantial magnitude acts along the cracks. On the other hand, this brings easiness of precise prediction on the crack behaviors since the boundary condition of simplified nature is prescribed everywhere along the boundary.

A computing scheme which can numerically predict, based on the linear fracture mechanics, the crack behaviors developing from the charge holes of prescribed number and spatial distribution have been developed by using the body force method (Nishitani 1978). In this scheme, the effect of rock pressure can be also taken into consideration.

Some exemplified results obtained by using this scheme will be mentioned.

(1) Fracture behavior emanating from a single hole in an infinite plate

At the initial stage of fracturing, several small cracks initiate in the radial direction from the hole wall. Consequently, two symmetrical cracks grow from these during the course of pressurization. When these cracks extend to some length, then additional two cracks perpendicular to the first pair ones emerge and finally four predominant symmetrical cracks are formed.

(2) The fracture behavior emanating from holes of equal spacing arranged along a line pararrell to the free surface in a semi-infinite plate

In this case, growth of two equal length fractures toward the direction connection each of the borehole is most prominent, if the distance between the hole line and free surface is not so large. Example of fracture growth sequence is shown in Fig.1, which indicates that once the induced pressure attains the critical value of 26 Mpa then the fracture grow decreasing pressure due to the interaction on the neibouring cracks, forming finally a fracture surface penetrating between the holes. This type of fracturing can be apllied to bench-cut.

The compressive stresses existing in underground has a effect of retarding the growth of fractures. To evaluate this effect, calculation was made for the crack growth under a compressive stress of 0.5 Mpa acting in a vertical direction to the row of holes. As shown in Fig.1, maximum pressure of 30 Mpa is required in order for the crack to penetrate between the holes, whose magnitude is large by 4 Mpa than the one obtained without stress.

3 DESIGN OF TUNNEL EXCAVATION THROUGH THE COMBINATION OF THE STATIC DEMOLISHER AND SLITTING PROCESS IN DEEP UNDERGROUND FORMATIONS

To apply the static demolisher to the tunnel excavation under the influence of rock pressure, effectiveness of the slit-cut drilling method as an aid for fracturing has been investigated.

There are several numbers of function of the slit when combined with static demolisher.

1. The in-situ stresses in the region where is under the influence of the slits will be partially or totally released.

2. It will serve as the free surface for the static demolisher and while accelerating the fracture development, it will also control the direction of the fracture.

3. It will prevent the growth of the fracture across the excavation configulations when the slits are dispositioned along the boundary.

Generally speaking, increasing the number of slit is advantageous to accelerate the induced fractures under the action of the static demolisher. On the other hand, it would be advisable to limit the total slit length as small as possible from the view-point of excavation efficiency since the slitting operation consumes much time. Therefore, optimization of the number and the location of slits is essential for this method. Hence, several slit patterns as shown in Fig.3 are examined by conducting stress analysis for the following conditions:

1. Tunnel configulation and size; A rectangular shaped tunnel of 3.02m in hight by 6.1m in width advances 3.5m per round (See Fig.2).

2. In-situ stresses ; Of the three principle stresses, one stress is in vertical direction with a magnitude of 30 Mpa, with the remaining two stresses acting in a horizontal plane, one of which works in the axial direction of the tunnel and with a magnitude of 60.2 Mpa and the remaining stress with a magnitude of 42 Mpa.

3. Properties of the rock mass;
  Young's modulus E=70 Gpa
  Compressive strength σ_c =210 Mpa
  Tensile strength σ_t =18 Mpa
  Fracture toughness KIC=2.6 MN/m

4. Borehole and static demolisher's power ; The diameter of the hole is 52 mm and the maximum pressure attainable by the static demolisher is 30 Mpa.

In order to evaluate the effect of the slit precisely, it is necessary to carry the three dimensional stress analysis. However, in this paper, analysis was limited to the two dimensional one which was conducted on the vertical and parallel sections to the tunnel axis.

(1) Condition in the vertical section to the tunnel axis

The stress state existing ahead of the tunnel face will differ from the initial stress state. In this case, however, under the assumption that a remote stress state is equal to the initial stress, an anticipation was made on the disturbance generated when a slit was cut in a pattern as shown in Fig.3. In the event the charge holes for the excavation are arranged in a lattice as shown in Fig.2, the fractures induced by the static

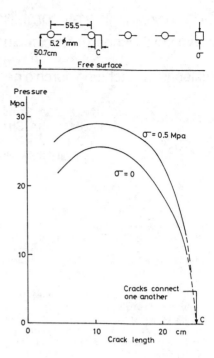

Figure 1. Pressure vs. growth of cracks emanating from holes arranged along a line parallel to the free surface.

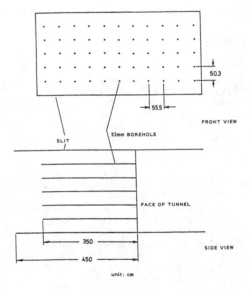

Figure 2. Arrangement of slits and holes for tunnel excavation.

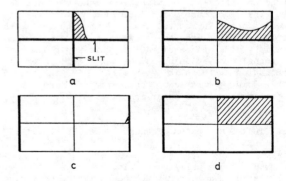

Figure 3. Stress relieved zone due to slits (Hatched area, only first-quadrant is shown).

626

demollisher will develop parallel to the slit. However, when the compressive stress is acting in the vertical direction to the fractures and its magnitude exceeds 0.5 Mpa, the fracture will not penetrate between the boreholes and effective demolition will not be accomplished (See Chapter 2). From this view-point, only the zone where the stress was under 0.5 Mpa shown as hatched area in Fig.3 could be demolished (This zone will be referred as the stress relieved zone).

From this figure, it is apparent that only the slit pattern shown in Fig.3-d would enable the demolition of the entire region of the proposed tunnel section. With a slit pattern as described in Fig.3-c, fracturing of the almost entire section of the tunnel may also be possible by adopting more condensed hole spacing. In the remaining slit patterns, only local stress relief is attained and even though the borehole spacing is reduced, it would be difficult to demolish the entire section of the tunnel.

(2) Condition in the parallel section to the tunnel axis

In relation to the two slit patterns as shown in Fig.3-c and 3-d, analysis was conducted for the stress state on the section parallel to the tunnel axis and results as shown in Fig.4 were obtained.

It is apparent that, for the slit pattern described in Fig.3-d, most of the inside region surrounded by the slit is stress relieved, as shown in Fig.4-d. Therefore, in this case, if the length of the charge hole is shorter by 80cm than the depth of slit, demolition along the entire length of the charge hole would occur.

On the other hand, in a slit pattern as described in Fig.3-c, the stresses within the slit would not be significantly released and therefore satisfactory demolition within this zone cannot be expected (See Fig.4-c). Based upon the foregoing analysis, it may be said that when the overburden is high only the slit pattern as described in Fig.3-d would be effective and reliable for tunnel excavation and an advance by a length of 3.5m per round could be possible. For the slit arrangement described in Fig.3-c, excavation under pressure may still be possible but the attainable advance per round will be a little shorter than the one by the former.

4 LABORATORY EXPERIMENTS

A laboratory test was conducted using moetar block to confirm whether or not the extent of demolition would coincide with predictions.

Slits and boreholes were drilled through one face of each cubic mortar block of 25cm x 25cm x 25cm in dimension. The slit pattern adopted was similar to the one shown in Fig.3-c. The slit was of 5mm width and 65mm in depth, while the hole was of 5mm diameter and 50mm in depth.

The block was set upon a bi-axial loading apparatus and stress of 9 Mpa was applied in both x and y directions and maintained during the test.

Fractures induced after injection of the static demolisher into the holes are as follows: The initial cracks developed connecting holes arranged on lines which are parallel to the central slit. And after the appearance of four dominant fractures running parallel to the slit, several fractures connecting these began to appear (See Fig.5). The upper and the lower faces of the central slit came into contact due to the deformation accompanying the breakage.

The fragments could be rather easily extracted from the specimen. The cavity created after extraction the fragments is of wedge shape as shown in Fig.6. At places remote from the slit, these were small regions that remained free from demolition (See Fig.6), although the formation of fractures along the line connection the bottom of the holes

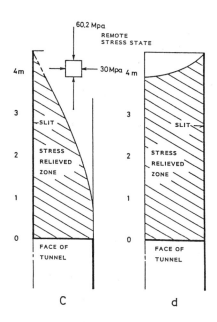

Figure 4. Stress relieved zone due to slits ahead of the face.

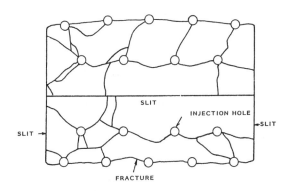

Figure 5. Sketch of fractures.

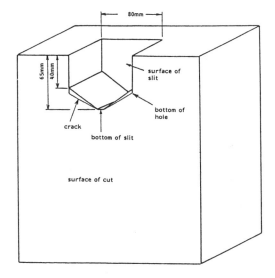

Figure 6. Sketch of fractured zone (After removal of fragments).

627

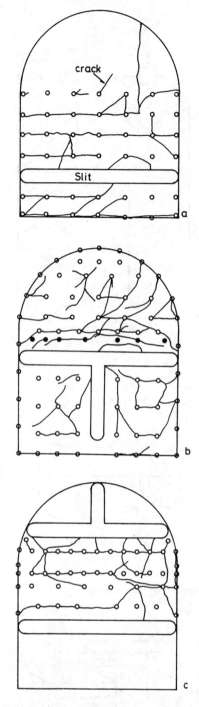

and is enough tough to be excavated without supports. The rock core taken from the tunnel wall yielded the following laboratory test results:

mean uniaxial strength = 88 Mpa
mean tensile strength = 10 Mpa
mean Young's modulud = 30 Gpa

Depth of the charge holes of 42mm in diameter, as well as of the slits of 92mm in thickness, is taken to be 50cm. The hole was charged with the static demolisher for its full length. The slitting operation was conducted as follows: Firstly the holes of 42mm were drilled along a prescribed line with narrow spacing. Then each hole was enlarged and connected one another.

Follwing three different fracturing design as shown in Fig.7 were investigated.

(1) Case of single horizontal slit (Fig.7-a)

Intense fractures were occurred along the four rows of hole parallel to the slit. Removal of the broken zone in the upper portion above the slit was easily conducted by using pick, the failured zone in the lower portion below the slit was restricted to the very vicinity of the tunnel face.

(2) Case of T-shaped slit (Fig.7-b)

Development of fractures was also predominant in the upper portion above the horizontal slit than in the lower one. Since the depth of the fractured zone was shallow, extra charge holes as shown in solid circles in the figure were added. As a result, rock breakage to a depth of about 40cm was attained. Mean size of the fragments was about 10cm in this portion.

(3) Case of double slits (Fig.7-c)

Central portion between the slits was well fractured and the broken rock was easily removed by using pick. In the lower portion below the slit, however, the fractured zone was limited to the vicinity of the face. On the whole, the result in this was best among the three and an advance of about 50 percent of the hole, that is of 25cm, per round (per shift) was attained.

6 REFERENCE

Nishitani, H. 1978. Solution of Notch Problems by Body Force Method. G.C.Shi (ed.) Stress Analysis of Notch Problems, Chap.1. Netherlands: Noordhoff.

Figure 7. Design of tunnel excavation and fracturing result.

and slits was noted. The existence of these regions may be said to be in harmony with the predictions noted in Chapter 3.

5 FIELD EXPERIMENTS

Field test on the prescribed excavation method was conducted in a horse-shoe shaped tunnel of 1.65m in cross sectional area. The test site was about 8m deep from the surface and it was near the portal where located in the urban area. Therefore elimination of the vibration accompanying the excavation is required.

The ground formation is constituted by green tuff and rhyolite which contains very little seams

A determination of vibration equation by empirical methods
Détermination empirique de l'équation de la vitesse de particules des vibrations des tirs à l'explosif
Bestimmung der Sprengerschütterungen auf empirische Weise

GINN HUH, Korea Professional Engineering Association, Seoul
KYUNG WON LEE, Korea Institute of Energy and Resources, Seoul
HAN UK LIM, Kangweon National University, Chuncheon, Korea

ABSTRACT: The effect of blasting pattern, rock strength and different explosives on the blast-induced ground vibration was studied to determine the maximum charge weight per delay within a given vibration level. The blasting vibrations were measured at 10 sites along the subway line in Seoul and the empirical particle velocity equation from over 100 test data was obtained. $V = K(D/W^{\frac{1}{3}})^{-n}$, where the values for n and K are estimated to be 1.7 to 1.5 and 48 to 138 respectively.

RESUME: L'effet, du type de tir, de la résistance des roches et de divers explosifs, sur la vibration de terrain provoquée par le tir à explosif, a été étudié à fin de déterminer le poids de charge maximum par delai dans une limite de vibration donnée.
Les vibrations de tir à explosif ont été mesurées sur les dix sites le long des deux lignes de métro séoulite et l'équation empirique de la vitesse de particules à partir des cent résultats d'essais à été obtenue. $V = K(D/W^{\frac{1}{3}})^{-n}$, où les valeurs de n et K sont supposées respéctivement de 1.7 à 1.5 et de 48 à 138.

ZUSAMMENFASSUNG: Der Einfluss von Sprengbild, Gesteinsfestigkeit und Sprengstoffsorten auf die Boden-erschütterung ist untersucht, um die maximale Sprengstoffmenge per Abschlag unterhalb der gegebenen Erschütterungsniveau zu bemessen. Die Erschütterungsmessungen waren in 10 verschiedenen Orten entlang der U-Bahn Linie in Seoul durchgeführt. Eine empirische Gleichung über die Durchlaufgeschwindigkeit der Partikel ist aus 100 Untersuchungsdaten so gewonnen, dass $V = K(D/W^{\frac{1}{3}})^{-n}$, wo die Werte von n und K als 1.7-1.5 und 48-138 entsprechend bestimmt worden sind.

1 INTRODUCTION

The blasting works for quarries and underground construction near urban areas has recently increased complaints of ground vibrations. The 3rd and 4th subway line in Seoul pass through the metropolitan area having tall buildings, large shopping centers and national cultural treasures. The distance from blasting sites to structures was generally within 40 meter and down to 14 m from cultural treasure site. In order to prevent the damage to structures, it was necessary to minimize the blasting vibrations.

Many investigators, Duvall et al(1961), Nicholls et al(1971), Siskind(1973), Langefors et al(1978) and Gusstafsson(1981) have studied on the effects of air and ground vibrations from blasting on residences and other types of structure.
Some reports by Woo et al (1967), Ryu et al(1979), Huh(1984) and others in Korea described the relationships between damage and ground vibrations from nearby blasting.
The excavating work for the construction of 3rd and 4th subway line includes open cuts of 33.7 Km and tunnel blasting of 15.0 Km. The geology of the sites is mainly composed of Precambrian gneiss which is intruded by Juriassic granite and overlain by alluvium with unconformity. So the rocks were granite and gneiss along the subway lines, the blasting method was adjusted corresponding to the rock type.
The objectives of this study are ; (1) to formulate the empirical vibration equation appropriate to each location around subway line, (2) to evaluate the effects of blasting pattern, rock strength and explosives on the blasting vibrations, (3) to determine the maximum charge weight per delay within allowable vibration values.

2 MEASUREMENTS OF GROUND VIBRATIONS

2.1 Characteristics of ground vibrations

The characteristics of blasting vibration have been well described and are summarized briefly here.

Particle velocity is more closely associated with damage to structure than either displacements or acceleration(Nicholls 1971). It should be observed in three mutually perpendicular directions : a vertical component, a horizontal component radial and a horizontal component transverse to the source.
Because the distance from shot to structures is too short, the maximum vibration level among three components was adopted.
The effect of distance and charge weight on the vibration level is basic to all blasting vibration studies. Many types of propagation equations have been proposed. The peak particle velocity of each component of ground motion (V, cm/sec) can be expressed by a equation as a function of distance (D,m) from the blast and maximum charge weight (W, kg) per delay, $V = K (D/W^b)^{-n}$: (1) where K, b, and n are constants associated with a given site. Empirical methods have been used to estimate values of b and n. Specially b may be about (square root scaled distance)or (cube root scaled distance). The measured data showed that cube root scaled distance was most applicable to this study.

2.2 Instrumentation

The shots were measured with three types of seismoraph. Sprengnether (VS-1200), VME (nitro-consult Co) and Rion (VM-12B). The former was mainly used and the latter two was auxiliary used. The natural frequency of Sprengnether is 5 to 200 Hz within the range of the observed frequencies(Nicholls 1971). Three portable seismographs were calibrated in accordance with the objectives of this research.

2.3 Experimental procedures and site

A total of 109 blasts were recorded at 10 sites. Blast-to-structure distances ranged from 8 to 84.2 meters, while charge weight varied from 0.1125 to 7.875 kgs per blast. The types of used explosives were gelatine dynamite, slurry explosives and ammonium dynamite which were produced in Korea. The characteristics of three explosives are represented in table 1.

Table 1. The characteristics of explosives

Explosives Classification	Gelatine dynamite	Ammonium dynamite	Slurry explosive
Diameter(mm)	25	28	25
Length(mm)	182-186	180-182	260-280
Weight (gr)	112.5	112.5	150
NG (%)	64-66	55-60	
Detonation velocity(m/sec)	5000 -5500	3500 -4000	3900

The blasting patterns have been divided into four kinds open cut by bottom blasting(O.B.B.), open cut by bench blasting(O.C.B), tunnel center cut (T.C.C), and tunnel cut by caving blasting (T.C.B).
In case of tunnel blasting, tunnel heading has only one free-face. After the center holes are blasted (T.C.C. or O.B.B), the rest works are implementations of bench cut (O.C.B., or T.C.B) against the opening to make the full sectional area.

3 RESULTS AND DISCUSSIONS

Geologic conditions, such as strength, the degree of weathering, and type of lineation and others have influences on wave propagation. Similar investigations were conducted in the same rock type over a certain area to determine whether amplitudes and attenuation rates were related. Table 2 and Fig. 1 show the results of measured ground vibration and scale distance for blasting vibration. The typical vibration constants are estimated to be 1.60 to 1.78 for n and 43 to 138 for k in the granite base while 1.5 for n and 17 to 87 for K in the gneiss base.

Table 2. Results of measuring blast-induced ground vibration

Site	Rock type	Sc (MPa)	Blasting type	Explosives	$V=K.(S.D.)^{-n}$ K	n
Miari	granite	88	TC	GD	90	1.67
		120	BH	GD	99	1.72
				SE	76	1.72
Sam-sungyo	granite	145	BT	GD	138	1.60
				SE	107	1.63
Kumho-dong	granite	123	BH	GD	93	1.70
Jang-chung-dong	granite	78	TC	GD	85	1.70
			BH	SE	58	1.70
				AD	47	1.70
Hongun-dong	granite	35	TC	SE	60	1.72
			BT	GD	94	1.64
Toege-ro 4	granite	39	TC	AD	48	1.54
Dongia-dong	gneiss	140	BT	GD	87	1.50
			BH	GD	49	1.50
Paksug-gogae	gneiss	85	TC	GD	56	1.50

Sc : Compressive strength
GD : Gelatine dynamite
AD : Ammonium dynamite
SE : Slurry explosives

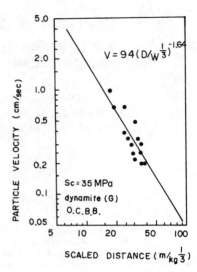

Figure 1. Scaled distance vs blasting vibration.

3.1 Laboratory test of rock samples

Some parameters were tested to find the relationships between hardness and strength. Rock samples were obtained from in-situ rock and prepared as cylindrical cores of 42 mm inner diameter. The shore hardness was determined as the average value of not less than 25 measurements at the same specimen. The uniaxial compressive tests were conducted in laboratory. The relations between Schmidt rebound hardness and uniaxial compressive strength could be represented by the following equation and Fig. 2

$$Sc = 0.0514 \times (SH)^{2.03} \quad : (2)$$

where Sc:uniaxial compressive strength (MPa)
SH:Schmidt rebound hardness.

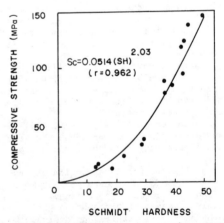

Figure 2. Relation between compressive strength and Schmidt rebound hardness.

If the value of Schmidt rebound hardness for in-situ rock was known, the compressive strength could be approximately estimated.

3.2 Effect of explosives type for vibration constants

The data from blasts were studied to determine the effect of explosives type on the vibration constants (n).

Because the value of n (exponent) at same site is
equal, the comparison of vibration constants among
different explosives (Table 3) could be represented
by an equation.

$$\frac{V \cdot {}_{S.E}}{V_{G.D}} = \frac{K_1 \cdot (SD)^{n1}}{K_2 \cdot (SD)^{n2}} = \frac{K_1}{K_2} \quad : (3)$$

Table 3. Comparision of blasting vibration constants
among explosives.

Site	Explo-sive	Blast-ing type	V=K.(S.D.)$^{-n}$		Ratio(r)	
			K	n		
Jangchu-ngdong	SE	BH	58	1.7	$\frac{K_{AD}}{K_{SE}}$	$= \frac{47}{58} = 0.81$
	AD	BH	47	1.7		
Umin-dong	SE	BH	24	1.5	$\frac{K_{AD}}{K_{SE}}$	$= \frac{17}{24} = 0.71$
	AD	BH	17	1.5		
Miari	GD	BH	99	1.7	$\frac{K_{SE}}{K_{GD}}$	$= \frac{76}{99} = 0.78$
	SE	BH	76	1.7		
Samsun-gyo	GD	BT	138	1.6	$\frac{K_{SE}}{K_{GD}}$	$= \frac{107}{138} = 0.76$
	SE	BT	107	1.6		
Dongj-adong	GD	BH	49	1.5	$\frac{K_{SE}}{K_{GD}}$	$= \frac{26}{49} = 0.56$
	SE	BH	26	1.5	$\frac{K_{AD}}{K_{SE}}$	$= \frac{20}{26} = 0.75$
	AD	BH	20	1.5		
	GD	BT	87	1.5	$\frac{K_{SE}}{K_{GD}}$	$= \frac{64}{87} = 0.74$
	SE	BT	64	1.5		

Based upon the data of table 3, the ratios (r) of
slurry explosives to gelatine dynamite appear to be
0.78, 0.76, 0.56, 0.74, and those of ammonium explosives to slurry explosives 0.81, 0.71, 0.75.
The recommended safe value of the ratio(r) is about
0.80 and 0.65(0.78 x 0.81 = 0.65).
These values are based on a consideration of the
gelatine dynamite. Vibration constant could also be
corrected for gelatine dynamite (Table 4).

Table 4. Corrected blasting vibration constant for
gelatine dynamite.

S_c (MPa)	Blasting vibration constant			Site	Rock type
	TC	BT	BH		
35	75*	94		Hongundong	
78	85		73** 78*	Jangchung -dong	
88 120	90		99 95*	Miari	gran-ite
145		138 134*		Samsungyo	
39	76*			Toegero 4	
85	56	(76)		Paksuggogae	gneiss
140	(67)	87	49	Dongjadong	

** : Corrected value of ammonium dynamite
 * : Corrected value of slurry explosive
() : Estimated value from blasting type

In this case, corrected values are 80% for slurry
explosives and 65% for ammonium dynamite.

3.3 Effect of rock strength and blasting patterns
for vibration constants.

Vibration constants were studied in terms of compressive strength of rock and blasting patterns. Fig. 3
and Fig. 4 show that vibration constant (K) generally
increases as compressive strength of rock increases

and represents the difference by blasting patterns.

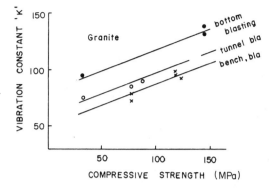

Figure 3. Compress strength vs blasting vibration
according to blasting types in granite

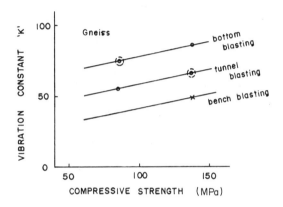

Figure 4. Compressive strength vs blasting vibration
according to blasting types in gneiss

Vibration constant could also be represented by
equation (4),(5).

$K_{(T.C.C)}=0.37$ Sc + 60(tunnel center cut)
$K_{(O.B.B)}=0.37$ Sc + 80(open cut by bottom blasting):(4)
$K_{(O.C.B)}=0.37$ Sc + 50(open cut by bench blasting)
Equation (4) was obtained in granite.
$K_{(T.C.C)}=0.21$Sc+40(tunnel center cut)
$K_{(O.B.B)}=0.21$Sc + 60(open cut by bottom blasting) :(5)
$K_{(O.C.B)}=0.21$Sc + 30(open cut by bench blasting)
Equation (5) was also obtained in gniess.

4. CASE STUDY

Many reports by Langefors(1978), Edwards(1960), 10-
year Bureau reports and others recommended a safe
vibration level for structures.
   To reduce the damage, West Germany standard (table
5) was adopted as the allowable vibration values.

Table 5. Allowable value of blasting vibration

Classification	Vibration value on ground(cm/sec)
Cultural treasure	0.2
Housing, Apt with partial crack	0.5
Shopping center	1.0
Factory & Rein-forced concrete Bld.	1.0 - 4.0

1. West-Germany Vornorm DIN 4150, Teil 3
2. Frequency up to 100Hz

To minimize the possibility of damage to nearby residential structure, some nomograms (Fig. 5) and tables (excluded in this report) for estimating safe charge weight and distance were given in this study.

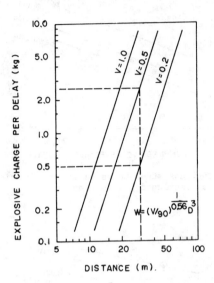

Figure 5. Explosive charge vs distance and allowable vibration value

Example:
Find the maximum charge weight per delay for gelatine dynamite of ammonium dynamite by empirical formula.
Conditions ;
  Kind of rock ; Seoul granite
  Rock strength; 85 MPa
  The nearest distance to structure ; 15 m
  Allowable vibration value ; 0.5 cm/sec
  Type of blasting ; tunnel center cut
Solution ;

$$V = K \ (D/W^{\frac{1}{3}})^{-n}$$

constant n equals to 1.7 for Seoul granite (Table 3)
constant K equals to 94 for gelatine dynamite and K=60 for ammonium dynamite (Table 6)
For gelatine dynamite

$$W = (V/K)^{3/n} \cdot D^3 = (0.5/94)^{3/1.7} \cdot 15^3 = 0.327 (kg)$$

For ammonium dynamite

$$W = (0.5/60)^{3/1.7} \cdot 15^3 = 0.723 (kg)$$

In addition to the vibration control, the control blasting method of pre-splitting and smooth blasting have been employed with the drilling patterns of burn cut type. The used explosives were mostly of low specific gravity and low velocity such as slurry, ANFO and CCR instead of gelatine dynamite. The firings were done by the use of multi-stage delay detonators and milli-second detonators instead of the electric detonators of common use.

5. CONCLUSIONS

To determine vibration equation by empirical methods, a total of 109 blasts were tested and analyzed.
The results can be summarized as follows
  (1) The vibration constant K is represented by an equation,

        K = Ei . (Ri.Sc + Qi)

where Ei : correction ratio according to explosives
      Ri : constant according to rock type
      Sc : compressive strength of rock
      Qi : correction value according to blasting type.

Table 6. Vibration constants K in accordance with explosives and blasting patterns

Explo-sives	Blasting pattern Compressive strength (MPa)	Open cut		Tunnel cut	
		O.B.B.	O.C.B.	T.C.C.	T.C.B.
Gela-tine dyna-mite	180 -150	147	117	127	97
	150 -120	136	106	116	86
	120 - 90	125	95	105	75
	90 - 60	114	84	94	64
	60	103	73	83	53
Slurry explo-sive	180 -150	117	94	102	78
	150 -120	109	85	93	69
	120 - 90	100	76	84	60
	90 - 60	91	67	75	51
	60	82	58	66	42
Ammo-nium dyna-mite	180 -150	96	76	66	63
	150 -120	87	68	74	55
	120 - 90	80	61	67	48
	90 - 60	73	54	60	41
	60	66	46	53	34

conditions : $V = KW^{0.57} D^{-1.7}$
             Seoul granite
             hole diameter : 38 m/m

(2) By an analysis of the measured data near Seoul subway line, cube root scaling might be more reasonable than square root scaling.
(3) To estimate the safe charge weight easily in the field, some nomograms and tables were given.

For further understanding about the effects of explosives, rock strength and blasting types on the vibration levels it is necessary to carry out more tests.

ACKNOWLEDGEMENT

The authors wish to express their thanks to Mr. Min-Kyu Kim and Dr. Hee-Soon Shin of KIER, who assisted in the field and laboratory tests.

REFERENCES

Duvall, W.I. & J.F. Devine. 1963. Vibrations from blasting at Iowa limestone quarries. U.S.B.M. RI. 5968. P.1-16.
Edwards, A.T. & T.D.Northwood. 1960. Experimental studies of the effects of blasting on structures. The Engineer. 210:538-546.
Gusstafsson,R. 1981. Blasting technique.Dynamite Nobel Wien.Vienna.P.217-230.
Hur,G. 1984. Blasting pattern of Seoul subway construction.Seoul.P.81-90.
Langefors,U. & B.Kihlstrom. 1978. The modern technique of rock blasting(3rd). John Wiley and Sons,Inc. NewYork.P.258-294.
Nicholls,H.R. C.F.Johnson & W.I.Duvall. 1971. Blasting vibrations and their effects on structures. U. S.B.M. Bull. 656:105.
Ryu,C.H., & C.I.Lee. 1979. A study on the effects of the ground vibration due to blasting on the structures.Jou.of the Korean Institute of Mineral and Mining Engineers. 16:14-15.
Stagg,M.s., & A.J.Engler. 1980. Measurement of blast induced ground vibrations and seismograph calibration. U.S.B.M. RI.8506.P.1-24.
Siskind,D.E. 1973. Ground air vibration from blasting SME Mining Engineering Handbook. V.I.P.11-111.
Woo,H.J., Y.H.Yom. B.K.Hyun. & K.W.Lee. 1967. Influence on blasting work to the plants and facilities. Jou.of the Korean Institute of Mineral and Mining Engineers. 4:81-93.

Dynamic behavior of tunnel lining due to adjacent blasting
Comportement dynamique du revêtement d'un tunnel lors d'un minage effectué à proximité
Dynamisches Verhalten von Tunnelauskleidungen bei benachbarten Sprengungen

Y.GOTO, Tokyu Construction Co. Ltd, Tokyo, Japan
A.KIKUCHI, Tokyu Construction Co. Ltd, Tokyo, Japan
T.NISHIOKA, Tokyu Construction Co. Ltd, Tokyo, Japan

ABSTRACT: On constructing twin parallel tunnels which are very closely located, we checked dynamic influences of blasting for lining. The tunnels were excavated by the rock bolting and shotcrete method, and the distance between lining surfaces of both tunnels are only 1.5-1.8m. The measurements of time series of strains and accelerations were carried out at two stages: The first measurements were executed on a test blasting of the first main tunnel, and sensors were set on a wall of a test pilot tunnel. Next, during excavation of the second main tunnel, we installed measurement devices on the wall of first existing tunnel. After the measurements, velocities and deformations were calculated by integrating the time series of acceleration. Then, results were evaluated in terms of deformations of the lining, relationships measured stains and calculated strains, the correlation of the velocities with distances between blasting holes and sensors, and so on.

RESUME: Lors de la construction de deux tunnels parallèles situés à proximité l'un de l'autre, nous avons contrôlé l'influence dynamique du minage sur le revêtement. Les tunnels ont été creusés par blutage de roches et projection de béton. La distance entre la surface des revêtements des deux tunnels n'était que de 1,5-1,8m. Des séries de mesures dans le temps des contraintes et accélérations ont été effectuées en deux étapes. La première série de mesures a été effectuée lors du minage d'essai du premier tunnel principal et des capteurs ont alors été placés sur la paroi d'un tunnel pilote d'essai. Nous avons ensuite installé, lors du perçage du deuxième tunnel principal, des instruments de mesure sur la paroi du premier tunnel. Une fois les mesures effectuées, on a calculé les vitesses et déformations en intégrant les valeurs d'accélération obtenues dans le temps. Les résultats ont été ensuite évalués en fonction des déformations du revêtement, du rapport entre les contraintes mesurée et calculées, de la corrélation des vitesses et distances entre les trous de minage et les capteurs, etc.

ZUSAMMENFASSUNG: Beim Bau von zwei parallelen Tunneln, die nahe beieinander liegen, haben wir den dynamischen Einfluss von Sprengungen auf die Auskleidung untersucht. Die Tunnel wurden im Deckenanker- und Spritzmörtelverfahren ausgeschachtet, wobei der Abstand zwischen den Auskleidungsoberflächen beider Tunnel nur 1,5-1,8m betrug. Die Zeitserienmessungen von Dehnungen und Beschleunigungen wurden in zwei Stufen durchgeführt: Die erste Messung wurde bei einer Testsprengung des ersten Tunnels durchgeführt, und Sensoren wurden an die Wand eines Testtunnels angebracht. Dann installierten wir während der Ausschachtung des zweiten Haupttunnels Messinstrumente an der Wand des zuerst vorhandenen Tunnels. Nach den Messungen wurden Geschwindigkeiten und Deformationen durch Integration der Beschleunigungszeitserien berechnet. Dann wurde das Ergebnis in Bezug auf Deformationen der Auskleidung, das Verhältnis zwischen den gemessenen und berechneten Dehnungen, die Korrelation der Geschwindigkeiten mit den Distanzen zwischen den Sprenglöchern und Sensoren usw. bewertet.

1 INTRODUCTION

The Ogitsu Tunnel reported here is actually two motorway tunnels located about 150 km northeast of Tokyo. Initially, the tunnels were to be constructed by the open cut method, but environmental considerations for the natural park above the tunnel forced construction to be changed to the tunneling method. Thus, the two parallel tunnels have their lining surfaces separated by only 1.5-1.8 meters, and were constructed by the rock bolting and shotcrete method, using blasting for excavation.

Finite element analysis and another method were used to evaluate the load bearing behavior of the rock and the support system, and the results indicated no great problems in statics. However, the dynamic influence of blasting upon the first existing tunnel, and the rock mass of the intermediate part upon excavating the second tunnel could not be estimated.

Therefore, after we carried out a blasting test with a test pilot tunnel (Fig.1-a) and examined the safety of the tunnels and the state of the crack development of the tunnel lining, we commenced construction of the main tunnels. This paper

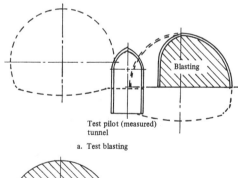

a. Test blasting

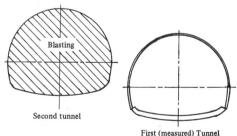

b. Main construction

Figure 1. Cross section of tunnel.

reports the results of the measurements for a test pilot tunnel and the main tunnel.

2 GENERAL DESCRIPTION OF THE MEASUREMENTS

Measurements were performed in two stages for the time series of the strain and the acceleration. The first measurements were arranged on a test blasting of the first main tunnel, and the pilot tunnel was regarded as the first existing tunnel. Sensors were set on the wall of the pilot tunnel (Fig.1-a).

Then, during the excavation of the second main tunnel, we installed measurement devices in the first existing tunnel (Fig.1-b). These tunnels in which the measurement devices were installed, is in the state where the concrete lining is not yet placed (the phase of forming a closed shell with a 20 cm layer shotcrete with steel fiber, 4 meter long rock bolts, steel liners and inverts). Slurry explosives and DS electric detonators were used for the blasting.

The rock in the area where the blasting test was performed belongs to the weathered mylonizing granodiorite group, and has many joints. Its material properties are shown in Table 1.

Table 1 Material properties.

Items	Measured value
Mass desity	2.4 – 2.5 g/cm^3
Uniaxial compression strength	10 – 20 MPa
Propagation velocity in core	
P wave velocity	15 – 3.0 km/s
S wave velocity	0.8 – 1.5 km/s
Propagation velocity in rock mass	0.9 – 1.2 km/s
Young's modulus	1 – 3 GPa

3 POSITION OF BLASTING HOLES AND SENSORS

Measurements were made using many different sensor arrangements. Here, we show the arrangement of the sensors and the position of the blasting holes that we studied in particular.

The case for the pilot tunnel:
Twelve piezoelectric accelerometers are arranged in the horizontal direction, and the positions for the blasting holes are the cut and the corner of the lifter holes (Fig.2).

The case for the main existing tunnel:
Twenty-four piezoelectric accelerometers are arranged perpendicular to the lining, and 30 mm strain gauges are put on the tunnel lining surface in the circumferential direction. The positions of the blasting hole are the cut and the corner of lifter holes in the upper section and the corner of lifter holes in the lower section (Fig.3).

These blastings in the corner of the lifter holes are carried out with one blasting hole per delay because of the need to avoid the influences of other detonations. After obtaining the acceleration, the particle velocities and deformations are calculated by integrating the time series for the acceleration data.

4 RESULTS OF MEASUREMENT

4.1 Correlation of peak particle velocities and distances between blasting holes and sensors

Generally, the control of blasting is executed by measuring the peak particle velocities. The

correlation between the peak particle velocities with the distances between the blasting holes and the accelerometers is shown in Fig.4. The charge weight is 300 g per blasting hole in both the case for the cut, and the lifter holes. In the case of cut, four holes are blasted at the same time.

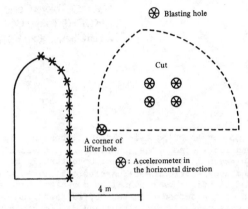

Figure 2. Position of blasting holes and measurement devices (case of a pilot tunnel).

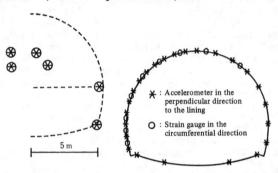

Figure 3. Position of blasting holes and measurement devices (case of a main existing tunnel).

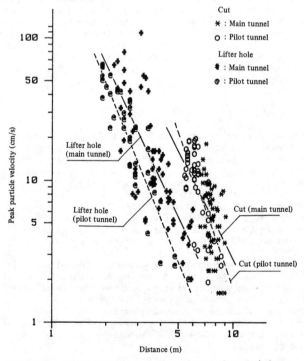

Figure 4. Relationship between the peak particle velocity and the distance.

Fig.4 shows that the velocities in the cuts are faster than for the lifter holes at the same distance. The reason for this is that there is almost no free surface and that four holes are blasted at the same time in the case of cuts. But, for example, using the results for the main tunnel, in the case of twin tunnels 10 m wide with a 5 m pillar of rock in between, the distance between a corner of the lifter hole and the tunnel lining is about 5 m. According to Fig.10, the velocity of the former is 2.5 cm/s, and the velocity of the latter is 7 cm/s, and in the case of a pillar of 2 m width,

the velocity is 7 cm/s in the cut and 60 cm/s in the lifter hole. Consequently, the shorter the distance, the more important the lifter hole is in the blast design.

4.2 Deformations and velocities

The deformation of the lining at every moment in the pilot tunnel are shown in Fig.5 and Fig.6, and those in the main tunnel are shown in Figs.7-9. In most cases, the deformation starts at the place adjacent to the blasting hole, and is greatest there, with

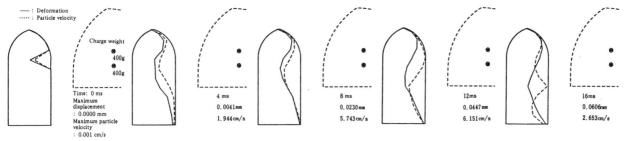

Figure 5. Deformations of lining at every moment (case of cut and a pilot tunnel).

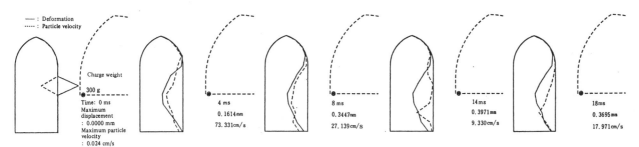

Figure 6. Deformations of lining at every moment (case of lifter hole and a pilot tunnel).

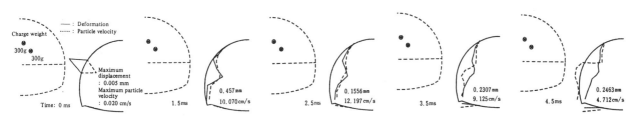

Figure 7. Deformations of lining at every moement (case of cut and a main tunnel).

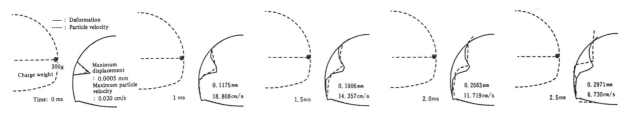

Figure 8. Deformations of lining at every moment (case of lifter hole and a main tunnel).

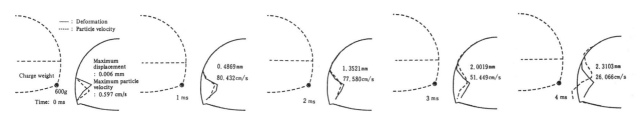

Figure 9. Deformations of lining at every moment (case of lifter hole and a main tunnel).

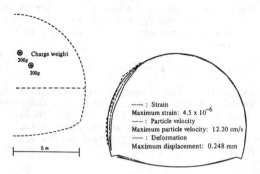

Figure 10. Distribution of strain, particle velocity
and deformation.

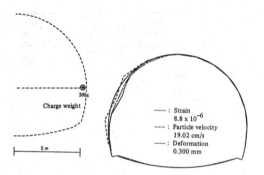

Figure 11. Distribution of strain, particle velocity
and deformation.

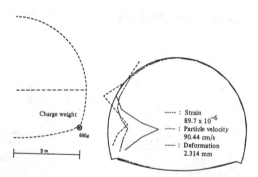

Figure 12. Distribution of strain, particle velocity
and deformation.

its shape almost symmetrical with respect to the top
of the deformation. However, in Fig.5, the
deformation starts at the place adjacent to the
blasting hole, and after that, the top of the
deformation shifts to the center of a side wall and
reaches its maximum value there. Accordingly, the
point of maximum displacement is not related to the
position of the blasting hole, but depends on the
shape of the tunnel and the rigidity of the support
system.

The distribution of the strain, the particle
velocities and the deformations are shown in
Figs.10-12. These figures represent the maximum
deformation. The lining of the pilot tunnel deforms
through the side wall, but that for the main tunnel
deforms locally at the place adjacent to the
blasting holes. In particular, in the case of the
main tunnel, particle velocities and deformations
almost do not occur at the crown and the opposite

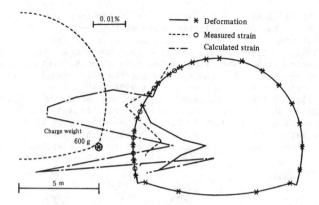

Figure 13. Comparison of measured strains and
calculated strains.

side wall. This is no different from the results of
blasting at a distance (Sakurai et al. 1977), where
velocities occur through the tunnel.

4.3 Measured strain and calculated strain

We can think of two factors that produce strain in
the tunnel lining. One is the strain produced by
propagations of plane waves, and the other is by
deformation of the tunnel lining. Here, we
attempted to study if the strain was caused by only
deformation of the tunnel.

The distribution of strain on the lining surfaces
measured using strain gauges and that calculated
from the static structural analysis by giving the
deformation calculated by integrating the time
series of acceleration are shown in Fig.13. The
distribution of the measured strain and the
calculated strain shows a similar tendency.
Particularly, at the point where tensile strain is
maximum, both strains almost agree. However, at the
point where compression strain occurs, both strains
are different. The reason why this occurs is that
the distance between the strain gauges was large.
We think that the measured strains would agree with
the calculated strains if we could lessen the
distance between the gauges. Therefore, the strain
on the lining due to blasting can be estimated from
the static structural analysis by giving
deformations which are calculated by integrating the
time series of the acceleration data.

5 CONCLUSION

We investigated the dynamic behavior of tunnel
linings due to very adjacent blasting.
Consequently, we obtained many kinds of results in
terms of deformation, particle velocity and strain
for tunnel linings.

REFERENCES

Sakurai, S. & Kitamura, Y. 1977. International
symposium on field measurements in rock mechanics.
Zurich.

An investigation of the influence of charge length upon blast vibrations
Investigation de l'influence de la longueur d'une colonne d'explosifs sur les vibrations de tirs
Eine Untersuchung über den Einfluss von Sprengladungslänge auf die Sprengschwingungen

J.R.GRANT, ICI Australia, Melbourne
A.T.SPATHIS, CSIRO Division of Geomechanics, Melbourne, Australia
D.P.BLAIR, CSIRO Division of Geomechanics, Melbourne, Australia

ABSTRACT: There are many factors which influence an observed blast vibration history at a monitoring location. One of the most dominant is the explosive charge length. When an explosive charge is initiated, a detonation front propagates at a characteristic detonation velocity so that the column of charge acts as an extended moving source of seismic energy. The amplitude and the rise time of the first arrival waveform, and some characteristic frequency of the blast vibration all appear to depend on both the charge length and this velocity of detonation.

A series of trial blasts were designed to investigate this phenomenon at a siltstone quarry at Mount Isa, Australia. The blast holes were drilled in a circular array of radius 20 m. with a suite of vibration detectors located at the centre of the circle. Nine PETN-TNT charges of 0.09 m. diameter were fired in lengths from 0.09 m. to 8.0m. These charges were spatially distributed and fired in a particular order to minimise the effect of previous blasts on the current blast. Prior to blasting, the volume of rock of interest was scanned using seismic pulse transmission methods to help quantify the initial ground condition and its anisotropy.

The results from the experiments suggest that increasing the charge length from 0.09-8.0 m. decreased the dominant frequency of vibration from 2 kHz down to 0.2kHz. Furthermore, the rise times and amplitudes of the first arrival waveform appear to increase with charge length although there is evidence for a plateau occurring for charge lengths greater than twenty blasthole diameters. The results are compared to those obtained from a simple blast model.

RESUME. Il y a beaucoup de facteurs qui influent sur l'histoire des vibrations résultantes d'une explosion observée à un point de contrôle. Un facteur des plus dominants est la longueur de la charge explosive. Quand une charge explosive est initiée, un front de détonation se propage à une vitesse caractéristique, de sorte que la colonne de charge se comporte comme une étendue source mobile d'énergie seismique. L'amplitude et la durée de montée de la première forme d'onde d'arrivée et quelque fréquence caractéristique de la vibration de l'explosion semblent toutes dépendre de la longueur de la charge et de cette vitesse de détonation.

Un programme expérimental a été préparé pour étudier ce phénomène dans une carrière à aleuronite à Mount Isa, en Australie. On a fait une série d'explosions expérimentales à une vingtaine de mètres d'un accéléromètre souterrain. On a fait éclater neuf charges PETN-TNT de diamètre 100 mm et ayant une longueur qui variait de 90 mm à 8,0 m. La distribution et la séquence des charges étaient telles que les explosions préalables de la série n'aient qu'un effet minimal sur la condition du sol, et donc sur les vibrations mesurées lors des explosions contiguës. On a balayé à l'aide d'une méthode de transmission d'impulsions seismiques le volume de roche relatif aux expériences pour contribuer à une quantification de l'état initial du sol et de son anisotropie.

Les résultats des expériences suggèrent qu'une augmentation de la longueur de la charge a déplacé les bandes d'énergie dominantes des explosions de 2 kHz à 200 Hz. De plus, les temps de montée et les amplitudes de la première forme d'onde d'arrivée semblent s'augmenter avec la longueur de la charge, bien qu'il y ait des indications qu'un palier se présente dans le cas des longueurs de charge supérieures à huit diamètres de trou de mine. On fait une comparaison entre les présents résultats et ceux obtenus au moyen de modèles simples d'explosions.

ZUSAMMENFASSUNG. Es gibt viele Faktoren, die ein an einer Kontrollstelle beobachtetes Sprengschwingungsmuster beeinflussen. Einer der wichtigsten ist die Länge der Sprengstoffladung. Beim Zünden einer Sprengstoffladung pflanzt sich eine Detonationsfront mit einer Eigengeschwindigkeit fort, so dass die Ladungssäule sich wie eine erweiterte, bewegte Quelle seismischer Energie verhält. Die Amplitude und die Aufstiegszeit der ersten Ankunftswellenform sowie irgendeine Eigenfrequenz der Sprengschwingung scheinen alle sowohl von der Sprengstoffladungslänge als auch von dieser Detonationsgeschwindigkeit abhängig zu sein.

Ein Versuchsprogramm wurde entwickelt, um dieses Phänomen in einer Siltsteingrube bei Mount Isa (Australien) zu untersuchen. Eine Reihe Versuchsdetonationen wurde mit einem Abstand von 20 m von einem festgelagerten, unterirdischen Beschleunigungsmessgerät durchgeführt. Neun PETN-TNT-Ladungen mit jeweils einem Durchmesser von 100 mm wurden bei Längen von 90 mm bis 8,0 m abgetan. Die Verteilung bzw. die Folge der Ladungen wurden dabei derart gewählt, dass frühere Schüsse der Versuchsreihe nur minimal auf den Bodenzustand, und daher auf die bei benachbarten Sprengarbeiten gemessenen Schwingungen auswirken würden. Das den Versuchen gehörige Gesteinvolumen wurde unter Verwendung seismischer Impulsgabe-Verfahren abgetastet, um zur Messung des Ausgangsbodenzustandes sowie seiner Anisotropie beizutragen.

Die Versuchsergebnisse deuten darauf hin, dass eine Zunahme der Sprengladungslänge die Hauptenergiebänder von 2 kHz bis 200 Hz versetzte. Ausserdem scheinen die Anstiegszeiten bzw. die Amplituden der ersten Ankunftswellenform mit der Ladungslänge zuzunehmen, obgleich es Hinweise darauf gibt, dass ein Plateau bei Ladungslängen von mehr als acht Sprenglochdurchmessern vorkommt. Die Ergebnisse werden mit denen, die aus einigen einfachen Sprengensmodellen erhalten wurden, verglichen.

INTRODUCTION

In underground mining, the blast process produces vibrations which may cause damage to the adjacent ground. Furthermore, when the vibrations reach the surface, they may create an environmental problem. Hence vibration levels are subject to strict statutory control. Both blast damage and environmental effects create a strong need for improvements to the blasting practices. Thus a joint research program was initiated between Mt Isa Mines Ltd., ICI Aust. and CSIRO in order to investigate the many factors that influence blast vibrations. One initial phase completed in 1982 involved the study of the influence of charge length upon the resulting vibration. This study was based upon the postulate by Harries (1981), that once a stable detonation was achieved in a cylindrical charge of commercial explosive, the resulting oscillation of the borehole wall and emanating stress wave would be dominated by the charge's length and the relative velocity of detonation (VOD) to the host material's stress wave velocities.

The problem was studied by using a full scale experiment as well as mathematical modelling. The mathematical modelling commenced with a simple superposition model following the work of Plewman (1965), Starfield (1968) and Larson (1982). The explosive charge was modelled as a series of stacked spheres, adapting Favreau's equilibrium solutions to a spherical explosion (1969). Initial results are demonstrated by Harries (1983). Analytical solutions for a detonating cylindrical charge have been partially attempted by many workers (Selberg 1952, Heelan 1953, Jordan 1962 and Abo Zena 1977), but the problem has still been found to be intractable, necessitating the stacked sphere approach, which ignores shear loading of the borehole. Recently, the modelling has been extended to include dynamic finite element techniques using the well known codes of Adina and Dyna 2D. Initial results, which are to be reported elsewhere, show the significant influence of the charge length upon both the dominant vibration frequency and peak p-wave amplitude. However, in the present study, only a simple blast model is used which sums phase delayed versions of the measured vibration from a short charge. The method is readily implemented in the frequency domain.

EXPERIMENTAL SET-UP

The experimental programme involved the measurement of particle vibrations for a sequence of cylindrical explosive lengths ranging from a length (L) to diameter (D) ratio of 1:1 to 88:1. At the same time every attempt was made to hold all other variables as constant as possible. A circular array of source holes around a central detector installation was used to provide intact travel paths for each explosive charge. A total of 9 charge lengths was fired at a travel distance of 20 m to the vibration detector (figure 1).

The site selected was a lower bench in a siltstone open cut mine. The rock is a medium bedded to massive siltstone, which is often strongly jointed. It has an unconfined compressive strength of 250-500 MPa, a density of 2700 kg/m^3 and a p-wave velocity of 5500-6500 m/s. The particular location was known to be in a very competent siltstone and had a high water table (4 m below the surface). A crosshole ultrasonic seismic study indicated that the rock was homogeneous in its seismic properties, and hence it was expected that any joints would have a uniform effect on the measured blast vibrations.

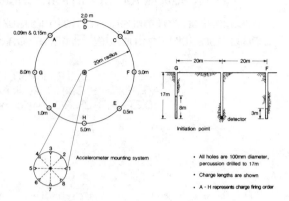

Figure 1. Experimental array showing location and firing sequence of the explosive charges.

Recording equipment used for both the crosshole seismic scans and the blast vibration measurements included a Racal 7D tape recorder and a digital Nicolet 1170 waveform analysis system with data subsequently transferred for more complete analysis onto a HP 1000 mini computer. This system has been developed along with ultrasonic crosshole seismic techniques for determining the attenuation characteristics of rock (Blair 1982a,1984, Spathis 1983,1985).

The hole depths were selected to be 17 m to avoid surface blast damage and the effect of free surface reflections upon the pulse onset. The drill hole size was 0.1 m and the explosives were specially prepared cylinders of cast ANZOMEX (a PETN-TNT mixture) of 0.09 m diameter. This explosive has a high VOD (7 km/s) which is uniform from sample to sample. All explosive charges were water coupled and stemmed. The charge was initiated by detonation at the base of the hole.

The central detector mount was constructed of aluminium having an acoustic impedance matched to that of the rock. The mount was drilled to house 8 radial Endevco 217E accelerometers. A gap of 5 mm between the housing and the hole perimeter was filled with a low-shrinkage grout, of matched impedance in an effort to avoid some of the transducer mount problems that can severely affect such experiments, (Blair 1982b, Grant 1983). Both the detector and the explosive initiation point were placed in the same horizontal plane. Each source hole had two accelerometers facing the charge, one at the front of the detector hole and one at the rear of the hole.

RESULTS

Figure 2 shows the acceleration amplitude spectra for the nine charge lengths. Similar results were obtained from the accelerometers which were aligned with opposite polarity.

It is clear from the amplitude spectra that as the charge length and thus L/D ratio increases, the dominant energy shifts to lower frequencies. When L/D equals one, the dominant frequency occurs near 2 kHz and as L/D increases to 88, the dominant frequency decreases to about 0.2 kHz. The apparent anomaly in the spectrum of the four metre charge length is attributed to the mount resonance of the accelerometer as neither the three nor the five metre charges indicate that significant frequencies occur above about 2 kHz.

In order to eliminate this unwanted resonance, all the records have been low pass filtered with the cutoff frequency set so that the peak amplitude and risetime of the pulse onset were altered by less than 10%.

An alternative measure of the 'frequency shifting' phenomenon is given by the mean frequency, $\bar{f}$, defined as,

$$\bar{f} = \frac{\int f\ B(f)\ df}{\int B(f)\ df}$$

where $B(f)$ is the acceleration amplitude spectrum. The results are shown in Figure 3 and they demonstrate that an increase in charge length causes a decrease in the dominant frequency of vibration.

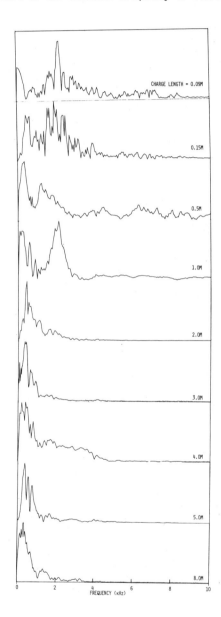

Figure 2. Front facing accelerometer frequency responses for each charge length.

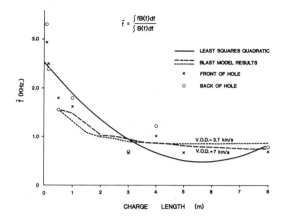

Figure 3. Mean radial acceleration frequency response versus charge length.

Figure 4 shows that in the time domain the peak particle acceleration occurs at the front of the record for all charge lengths and, also, the pulse broadening of the vibration appears to be proportional to the charge length. These two properties suggest that a time domain pulse width related to the pulse onset may be a useful diagnostic. Figure 5 shows a plot of pulse risetime (defined as the time taken for the vibration to rise from 10% to 90% of the pulse onset amplitude) against charge length. An increase in vibration risetime with charge length is evident. A similar trend occurs with the peak acceleration amplitudes although the data are more scattered than either the gain independent measures of f or risetime (Figure 6). All three parameters appear to reach limiting values for L/D greater than about 20.

A SIMPLE BLAST MODEL

A burning column of explosive acts as a moving source of seismic energy and can be simulated by using Fourier techniques to sum phase delayed versions of the measured vibration from a short (elemental) charge. If $G(f)$ is the spectrum of this elemental charge of length ΔL and V_0 the VOD of the explosive, then the spectrum $G_T(f)$, for the total vibration produced by a column of any length L can be shown to be

$$G_T(f)=G(f)\left\{COS[\pi f(L-\Delta L)/V_0]+Sin\frac{\pi f(L-\Delta L)/V_0}{\pi f\Delta L/V_0}\right.$$
$$\left.+ iSin[(\pi f(L-\Delta L)/V_0]\right\}$$

where f is the frequency and i the complex number, $\sqrt{-1}$.

This model does not account for geometric spreading effects apart from those implicity included by a measurement of the elemental charge. Although it is a simpler model than the stacked sphere model previously mentioned, it is more rapidly implemented using Fast Fourier Transform techniques. The inverse transform of $G_T(f)$ then yields the total vibration waveform required.

Results from the present simple model are shown in Figures 3, 5 and 6. Although there is generally poor agreement between the modelling and experimental results (due partly to the experimental scatter) it should be noted that both sets of results predict that an increase in charge length produces an increase in both amplitutude and rise time and a corresponding decrease in the average frequency.

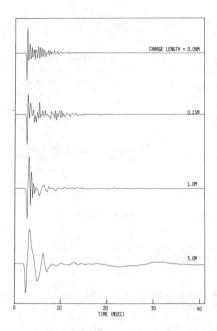

Figure 4. Examples of measured radial accelerations in the time domain.

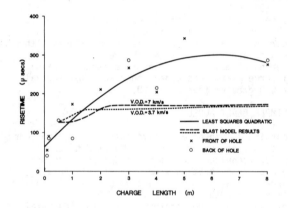

Figure 5. Radial acceleration risetime versus charge length.

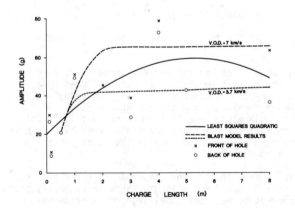

Figure 6. Radial acceleration amplitude versus charge length.

CONCLUSIONS

Both the experimental and modelling results show that the ground vibration frequency spectrum is strongly dependent upon charge length. This fact is of practical significance since the blast output vibration spectrum, in turn, has a strong influence on the excitation of structural resonances.Thus the present work implies that it should be possible to alter the charge length in order to reduce the amplitude of blast vibration as measured on resonant structures such as residential buildings, bridges and (underground) mine installations.Equally important is the present indication that vibration amplitude does not necessarily increase with increased charge weight as is universally claimed. Some of the success of decking as a mechanism for reducing vibrations may be attributed to the change in spectral content rather than the mere reduction in charge weight per delay.

However it should be appreciated that the present study involves only the near-field effects of the blast source for which the maximum charge length (8m) is a significant fraction of the wave path length (20m). For far-field measurements in which the charge length is a lot less than the wave path length, then the charge length would be expected to have a different influence on the vibration parameters.For example, it would be expected that the far field peak amplitude would increase with charge length L, for all values of L and not show the plateau as depicted by the present near-field results of Figure 6.

Furthermore, in the present study the peak particle motion occurs in the p-wave, however in the far-field the various wave types (p,s, Rayleigh) separate out and the peak particle motion is not necessarily confined to the p-wave. In particular, for surface measurements of surface (open cut) blasts, the far-field peak particle motion generally occurs in the Rayleigh wave whereas for surface measurements of deep underground blasts, the peak particle motion is generally observed in the p-wave. Thus the present study is, perhaps, most applicable to deep underground blasts.

A dynamic Finite Element study showing the wave type separation in going from near-field to far-field for both underground and surface blasts has been completed and is to be reported in the near future. Further experiments, coupled with more realistic dynamic numerical modelling procedures (such as Dynamic Finite Element) are planned in order to firmly establish the influence of both site dependent and blast design parameters upon the resulting vibration.

ACKNOWLEDGEMENTS

The authors are indebted to Gwyn Harries for his encouragement and assistance. The support given by Mount Isa Mines Limited, ICI Australia and CSIRO is gratefully acknowledged.Keith Mercer (formerly of ICI Australia) and Felix Leahy of Mount Isa Mines Limited have also given personal support for the work and their enthusiasm throughout the entire project is also gratefully acknowledged.

REFERENCES

Abo Zena, A.M. 1977. Radiation from a finite cylindrical explosive source. Geophys. 42, 1384-1393.
Blair, D.P. 1982a. Measurement of rise times of seismic pulses in rock. Geophys. 47, 1047-1058.
Blair, D.P. 1982b. Dynamic modelling of in-hole mounts for seismic detectors. Geophys. J.R. Astr. Soc. 69, 803-818.
Blair, D.P. & Spathis, A.T. 1984. Seismic source influence in pulse attenuation studies. J. Geophys. Res. 89, 9253-9258.

Favreau, R.F. 1969. Generation of strain waves in
  rock by an explosion in a spherical cavity. J.
  Geophys. Res. 74, 4267-4280.
Grant, J.R. 1983. An investigation into transducer -
  ground coupling techniques for surface vibration
  measurement. MSC. Thesis, University of Nth Qld,
  Townsville. Harries, G. 1981. Rock properties and
their effect
  upon blasting vibrations. Drilling and Blasting in
  Open Pits. AMF, Adelaide.
Harries, G. 1983. The modelling of long cylindrical
  charges of explosive. Rock Fragmentation by
  Blasting, 1st Internation al Symposium, Lulea.
Heelan, P.A. 1953. Radiation from a cylindrical
  source of finite length. Geophys. 18, 685-96.
Jordan, D.W. 1962. The stress wave from a finite
  cylindrical explosive source. Jnl. Math. Mech. 11,
  503-551.
Larson, D.B. 1982. Explosive energy coupling in
  geologic materials. Int. J. Rock Mech. & Geomech.
  Abstr. 19, 157-166.
Plewman, R.P. & Starfield, A.M. 1965. The effects of
  finite velocities of denotation and propagation on
  the strain pulses induced in rock by linear
  charges. Jnl. S. Afr. Inst. Min. Met.
Selberg, H.L. 1952. Transient compression waves
  produced from spherical and cylindri cal cavities.
  Ark. Fys. 5, 97-108.
Spathis, A.T., Blair, D.P. & Grant, J.R. 1983.
  Seismic pulse assessment of tunnel walls in rock.
  Proc. Int. Symp. on Field Measurements in Geomech.,
  Zurich.
Spathis, A.T., Blair, D.P. & Grant, J.R. 1985.
  Seismic pulse assessment of the changing rock mass
  conditions induced by mining. Int. J. Rock Mech.
  Min. Sci. Vol 22, No 5. pp 303-312.
Starfield, A.M. & Pugliese, J.M. 1968. Compressional
  waves generated in rock by cylindrical explosive
  charges: A comparison between a computer model and
  field measurements. Int. Jnl. Rock Mech. Min.
  Science, Vol.5 pp 65-77.

Essais de prédécoupage et mesures des arrière-bris dans les roches granitiques du Massif Central français

Presplitting tests and measurements of the induced back-breaks in granitic rocks of the French Massif Central

Versuchvorspalten und Auswirkungsmessung in Granitfels im Massif Central – Frankreich

HUBERT HERAUD, Chef du groupe Sols Roches, Laboratoire Régional des Ponts et Chaussées de Clermont-Ferrand, France
ANNE REBEYROTTE, Elève de l'Ecole Nationale Supérieure des Mines de Paris, France

ABSTRACT : This paper deals with presplit blastings realized on granitic rocks from two different techniques : standard presplitting with use of a detonating fuse of 70 g, and - presplitting with use of a continuous pipe small diameter water gel (that appears like a string of sausages) named Cisalite. Results are analysed from geometrical measures realized on the surface of the solope, and the importance of the rear-effects induced by blasting is studied by means of the pseudo-spectral measure of seismic wave and microseismic recordings. The conclusions bring out a smaller blast damage, when presplitting is carried out with sausagelike gelatinous explosive.

RESUME : La communication présentée traite de prédécoupage à l'explosif, réalisé dans des roches granitiques à partir de deux techniques différentes : prédécoupage classique au cordeau détonant de 70 g et prédécoupage au boudin continu de gel (cisalite). Les résultats sont analysés à partir de mesures géométriques effectuées en surface du talus et l'importance des effets arrière étudiée par mesure des pseudofréquences et enregistrements microsismiques. Les conclusions font ressortir une moindre dégradation du talus dans le cas d'un prédécoupage au boudin de gel.

ZUSAMMENFASSUNG : Dieser Beitrag behandelt die Vorspalten mit Sprengstoff, die in Granitfels mit zwei verschiedene Techniken ausgeführt wurden: klassische Vorspalten mit Knallzündschnur (70 g) - Vorspalten mit Sprenggelatine in durchgegenhendem Schlauch. Die Auswertung der Ergebnisse erfolgt mittels geometrischer Messungen auf der Böschung, und die Wichtigkeit des Rückwirkungseffekts wird aus der Messung der Scheinfrequenzen und der Microseismik geprüft. Zusammenfassen : wenn das Vorspalten mit Sprenggelatine durchgeführt wird, so ist der Schaden geringer.

INTRODUCTION

La réalisation d'ouvrages rocheux subverticaux et de grande hauteur conduit souvent à la mise en oeuvre de prédécoupage afin d'obtenir une paroi stable et aussi peu désorganisée que possible.

On utilise alors généralement un cordeau détonant de 40 g ou 70 g dans des trous de faible diamètre et de faible écartement.

L'apparition sur le marché de nouveaux explosifs en boudin continu du type gel, peut constituer une variante des techniques habituelles et plusieurs expérimentations ont été effectuées par le Laboratoire Régional des Ponts et Chaussées de Clermont.Ferrand sur les granites d'Auvergne en essayant de comparer les résultats d'un prédécoupage au cordeau de 70 g, à ceux d'un prédécoupage effectué au boudin de gel.

L'exemple présenté ci-après intéresse ces deux types de prédécoupage et décrit quelques méthodes de mesure permettant d'apprécier l'état de surface des talus, ainsi que les effets arrière dus aux tirs.

La roche prédécoupée est un granite à deux micas dont les caractéristiques géotechniques sont les suivantes :

- densité 2,6 T/m3

- vitesse microsismique 2100 à 3500 m/s

- R Q D moyen 64 %

- Intervalle entre discontinuités ID4
  (6 à 20 cm)

1 - CARACTERISTIQUES DES PREDECOUPAGES MIS EN OEUVRE.

Deux types de prédécoupage aux caractéristiques différentes ont été réalisés sur le même massif et sur un même talus.

Nature de l'explosif.

Cordeau détonant : un tube flexible de 11 mm de diamètre contient une âme de pentrite. Le grammage au mètre linéaire est de 70 g. La vitesse de détonation moyenne est de 6500m/s. Cisalite : Il s'agit d'un chapelet continu de gelsurite 2000 de 25 mm de diamètre, dont la vitesse de détonation est de 2850 m/s pour une charge linéaire de 400 g/ml. La cisalite est amorcée au cordeau détonant de 12 g accolé au boudin.

Foration.

La foration est effectuée en diamètre Ø89mm avec un écartement de 80 cm pour le cordeau détonant.

Pour les tirs à la cisalite, la foration se fait en diamètre 102 mm avec un écartement de 1 m.

Chargement - bourrage.

Sur le chantier le cordeau détonant est placé dans les trous de prédécoupage sans charge de pied et amené au fond du trou à l'aide d'un bourroir. Le remplissage se fait sur toute la hauteur au gravillon 4/6 mm. La cisalite est repliée en deux ou trois à l'extrêmité inférieure, formant ainsi charge de pied et l'ensemble est descendu dans le trou au bourroir. Le remplissage se fait avec un gravillon 4/6 sur le mètre supérieur.

Le tableau I ci-dessous résume les caractéristiques des deux types de prédécoupage mis en oeuvre ainsi que leurs variantes possibles.

	Cordeau détonant de 70 g.	Cisalite
- espacement des trous recommandé en m.	0,70 à 80	0,90 à 1 m
- diamètre de foration recommandé en mm.	(76 à) 89	(89 à) 102
- diamètre de l'explosif en mm.	11	25
- nature de l'explosif	généralement Pentrite	gelsurite 2000
- vitesse de détonation en m/s	6 500	2 850
- charge linéaire au m	70 g	400 g

Tableau I - Comparaison des caractéristiques et variantes possibles des prédécoupages au cordeau détonant de 70 g et à la cisalite.

Les tirs de prédécoupage sont systématiquement réalisés avant les tirs de masse et l'analyse de l'état du talus ainsi obtenu se fait après dégagement des matériaux en appliquant un certain nombre de paramètres qui permettent de définir la géométrie du talus et les caractéristiques physiques du talus réalisé.

Les tirs de masse réalisés à l'avant du prédécoupage ont pour caractéristiques principales : foration Ø 115 mm - maille moyenne 3 m x 3 m - explosif gelsurite 2000 en cartouche de 90 mm - grammage spécifique : 400 à 410 g/m3 - amorçage par détonateur fond de trou/Nonel.

2 - ANALYSE DE L'ETAT DE SURFACE DU TALUS

Certaines mesures simples effectuées directement sur la surface du talus donnent une bonne idée de la qualité géométrique du talus.

2.1. Relevé de profils horizontaux

En partant de la définition même du prédécoupage, il en découle que le front de taille du "talus idéal" doit avoir une surface très régulière dont les profils horizontaux seraient des lignes où les seuls accidents seraient des trous semi-circulaires des cannes de foration régulièrement espacées (Fig. 1).

On peut ainsi définir un paramètre Profil P =

$$P = \frac{\text{longueur Profil réel}}{\text{longueur "Profil idéal"}}.$$

Le lever de ces profils peut se faire manuellement à hauteur d'homme, ou par photorestitution à différentes hauteurs. Un lever manuel donne une précision suffisante.

FIG.1 - Comparaison profil idéal - Profil réel d'un talus relevé sur chantier.

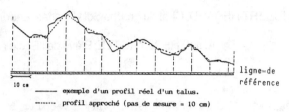

——— exemple d'un profil réel d'un talus.
------- profil approché (pas de mesure = 10 cm)

FIG.2 - Exemple de relevé manuel d'un profil horizontal de talus avec une mesure tous les 10 cm.

Dans la pratique le paramètre P suffit souvent pour donner une image du talus (Fig. 2). Dans les cas où l'on veut préciser la régularité de la géométrie, il est intéressant de donner la moyenne et l'écart-type de l'ensemble des valeurs distance ligne de référence.

2.2. Cannes de foration

Dans la théorie, le talus idéal devrait présenter des traces de cannes de foration parfaitement continues. On peut donc définir un paramètre C.

$$C = \frac{\text{longueur visible de cannes de foration}}{\text{longueur théorique des cannes}} \times 100$$

On peut également définir un paramètre N relatif au nombre de cannes visibles :

N = Nombre de cannes recoupées par une horizontale à une cote donnée.

Les résultats correspondant à l'analyse de la géométrie de l'état de surface du talus sont donnés dans le tableau II.

On remarque, pour les trois paramètres C. N et P les meilleurs résultats géométriques obtenus dans le cas du prédécoupage à la cisalite.

PREDECOUPAGE DE 11 m DE HAUTEUR - PENTE 200/100				
EXPLOSIF UTILISE	CORDEAU DETONANT 70 g		CISALITE	
C = cannes visibles / cannes théoriques %	31,5 %		50,5 %	
[Nombre de cannes théoriques]	[39]		[39]	
N = Nombre de cannes visibles				
N - à 5,5 m (à mi hauteur)	28		29	
N - à 11 m (en pied de talus)	11		24	
P = longueur profil réel / longueur profil "idéal"	P	ECART-TYPE	P	ECART-TYPE
P-manuel à 1 m du sol	1,12	23,15	1,06	9,16
P-photorestitution				
- à 1 m du sol	1,14	0,31	1,04	0,11
- à 4 m du sol	1,12	0,24	1,03	0,07
- à 9 m du sol	1,15	0,27	1,05	0,07

Tableau II - Prédécoupage au cordeau détonant et à la cisalite = Résultats comparatifs des paramètres C - N et P.

3 - ANALYSE DES EFFETS ARRIERE

Deux méthodes simples, directement applicables sur les chantiers de terrassement, permettent de préciser la qualité du talus réalisé par l'étude des caractéristiques physiques du matériau constituant le talus.

Mesures de pseudofréquences

La méthode consiste à quantifier l'incidence des tirs sur le massif, à partir de l'étude de la fracturation au niveau du talus. Les mesures en pseudofréquence approximent le signal sismique relevé sur le talus à partir de mesures faites à l'appareil Bison monotrace, par une fonction sinusoïdale équivalente. On calcule à partir de la première crête une première pseudopériode dont on déduira une première pseudofréquence et on relève la durée du signal (Fig. 3). Il existe une relation entre pseudofréquence/durée et l'état de fracturation du rocher. Le rocher est d'autant plus sain que "la fréquence" est forte et la durée du signal courte.

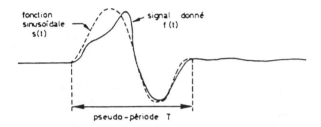

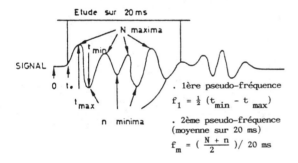

.Fig. 3 - Grandeurs pointées sur le terrain en vue de l'étude pseudofréquentielle (d'après Rasolofosaon).

L'étude pseudofréquentielle est effectuée sur chaque partie de talus à étudier sur une base de 1 m de longueur avec un géophone fixe scellé au plâtre et en déplaçant l'émetteur selon une circonférence centrée sur le géophone.

Les résultats sont donnés sous forme de diagramme pseudofréquence (f1) en fonction de la durée du signal (Fig. 4).

Malgré une assez grande dispersion des points de mesure on remarque que l'ensemble des points relevés sur le talus prédécoupé à la cisalite est très nettement décalé vers les valeurs de fréquences élevées et de courte durée du signal, par rapport à ceux relevés sur le talus prédécoupé au cordeau de 70 g/ml. Ces résultats traduisent donc une meilleure qualité du matériau après découpage à la cisalite.

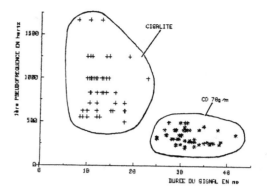

FIG.4 - Résultats des mesures de pseudofréquences pour 1 prédécoupage à la cisalite et 1 prédécoupage au cordeau détonant de 70 g.

Auscultation microsismique

L'étude microsismique est effectuée à partir d'un forage subhorizontal réalisé perpendiculairement à chacun des talus, à une hauteur d'environ 2,50 m et sur une longueur de l'ordre de 8 m. Le forage est préalablement rempli d'eau et la sonde est guidée par un tubage PVC dont l'extrêmité aménagée en demi-coquille laisse les émetteurs et récepteurs libres vers la partie supérieure. L'auscultation faite à partir de la sonde LCPC(Fig. 5) donne un enregistrement continu des vitesses microsismiques du rocher en fonction de la profondeur (Fig. 6).

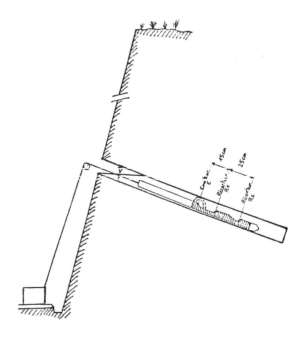

Fig. 5 - Sonde microsismique continue LCPC. L'eau assure la transmission des ondes entre la sonde et la paroi du forage.

645

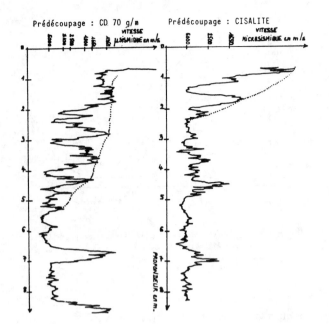

Prédécoupage : CD 70 g/m   Prédécoupage : CISALITE

FIG.6 - Enregistrements microsismiques obtenus à la sonde continue

Les vitesses les plus faibles, correspondant au matériau désorganisé de surface, intéressent une tranche d'environ 2,50 m dans le cas du talus prédécoupé à la cisalite, alors qu'elles atteignent 6 m d'épaisseur dans le cas du prédécoupage au cordeau détonant.

Une loi de décroissance ou d'atténuation des effets arrière avec la distance du tir peut être établie (Fig. 7). On suppose que les vitesses élevées de l'ordre de 5000 m/s correspondent au matériau intact non fissuré. Elles représentent un pourcentage de perte de 0 % ; une vitesse de 2500 m/s correspond à une perte de 50 %. On met ainsi en évidence la désorganisation des matériaux sur des épaisseurs importantes, en particulier dans le cas de l'abattage classique au cordeau détonant.

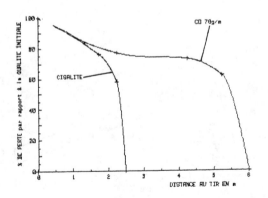

FIG. 7 - Courbes d'atténuation des effets arrière.
On retrouve la vitesse initiale V_0 à 2,50 m du tir pour la cisalite, à 6 m pour le cordeau détonant.

4 - COMPARAISON ECONOMIQUE.

A titre tout à fait indicatif, un calcul sommaire effectué à partir de prix moyens pratiqués pour les formations granitiques d'Auvergne en 1985-1986, fournit une comparaison entre les deux techniques de prédé-

coupage. Pour un prédécoupage au cordeau détonant de 70 g en diamètre 89 avec espacement de 0,70 m on obtient un prix moyen de 47 F. au mètre linéaire. Ramené au m2 de surface prédécoupée on a alors un prix moyen de 64 F./m2. Pour un prédécoupage à la cisalite en diamètre 102 avec espacement de 1 m on obtient un prix moyen de 58 F./ml ou 58F./m2.

Il apparaît donc que le prix de revient au mètre linéaire de forage est en faveur du cordeau détonant :

$$\frac{\text{prix ml cordeau détonant}}{\text{prix ml cisalite}} = 0,8$$

Les prix de revient calculés au m2 de parement réalisé ont des rapports qui s'inversent :

$$\frac{\text{prix m2 cordeau détonant}}{\text{prix m2 cisalite}} = 1,10$$

La photo ci-jointe donne un aperçu de deux types de prédécoupage.

prédécoupage à la   zone de prédécoupage
cisalite   faille   au CD 70 g.

Prédécoupage dans un massif granitique très fracturé.

A gauche : prédécoupage cisalite : la paroi est stable.
A droite : prédécoupage cordeau détonant 70 g : décrochement de dièdres - hétérogénéité de la surface - éboulements et accumulation en pied de talus.

5 - CONCLUSION.

Jusqu'à ces dernières années, les possibilités de réalisation de prédécoupage étaient limitées en raison, essentiellement, du manque de diversité des explosifs linéaires et continus offerts sur le marché. L'apparition du gel en boudin constitue une variante possible.

La cisalite, explosif à faible vitesse de détonation et à production de gaz importante, est bien adaptée au découpage des roches granitiques du Massif Central. Le cordeau détonant de 70 g à vitesse de détonation élevée et accompagnée d'une onde de choc importante, est beaucoup plus traumatisant pour le matériau restant en place.

Différentes méthodes directement utilisables sur les chantiers de terrassement permettent de caractériser l'état du talus, tant en surface qu'en profondeur : analyse de profils horizontaux et cannes de foration - mesures de pseudofréquences et enregistrements microsismiques continus.

L'application de ces différents paramètres aux cas étudiés, donnent des résultats plus favorables dans le cas de prédécoupage à la cisalite : la géométrie du talus est meilleure et les caractéristiques physiques mesurées traduisent la présence d'un matériau moins fracturé au niveau du talus. Néanmoins les mesures microsismiques montrent que même dans le cas le plus favorable (cisalite), le rocher est désorganisé sur une épaisseur de l'ordre de 2 m. Ces résultats méritent évidemment d'être confirmés sur d'autres types de roches.

Les considérations économiques ne doivent pas constituer un obstacle au développement de nouveaux essais ; il ressort en effet que même si l'explosif cisalite et la foration coûtent plus cher, le prix de revient au m2 de paroi prédécoupée reste du même ordre de grandeur, voire légèrement inférieur au prédécoupage classique.

D'une manière générale, il en ressort un gain indirect apporté par une meilleure tenue du talus et des coûts d'exploitation diminués en conséquence.

BIBLIOGRAPHIE.

FOURMAINTRAUX, D. SIFRE, Y. BEDAUX, R. 1983. Terrassement du rocher : Le sautage à l'explosif. Revue Générale des Routes et Aérodromes. 593 : 24-50.
PANET, M. WEBER, P. 1971. Découpage et prédécoupage à l'explosif. Aspects théoriques. Aspects pratiques. Revue de l'Industrie Minérale. Mines. n° spécial du 15.7.71 : 89-98.
RASOLOFOSAON, P. LAGABRIELLE, R. RAT, M. du MOUZA, J. 1983. Reconnaissance de formations par étude fréquentielle de signaux sismiques. Bulletin of the International Association of ENGINEERING GEOLOGY. 26-27: 285-293.
RAT, M. and all. 1982. Reconnaissance géologique et géotechnique des tracés de routes et autoroutes. Note d'information technique LCPC : 78-86.

The influence of water jets on the cutting behavior of drag bits
L'influence de jets d'eau sur le comportement de taillage des pics de taille
Der Einfluss von Wasserjets auf das Schneidverhalten von Schrämmeissel

M.HOOD, University of California, Berkeley, USA
J.E.GEIER, University of California, Berkeley, USA
J.XU, University of California, Berkeley, USA

ABSTRACT: An heuristic argument is used to predict the effects of jet pressure, jet flow rate, and cutting speed on the pick force reductions achieved by using water jets to assist in rock cutting. The primary influence of water jets is assumed to be the removal of crushed debris which forms ahead of the pick. It is predicted that the rate of debris removal is a function of the jet power dW/dt. From this it is shown that the reduction in mean cutting force is an increasing function of the characterizing parameter dW/dx, the ratio of dW/dt to cutting speed. This model is supported by the results of cutting experiments in Indiana limestone. Cutting force reductions are found to be a log-linear function of dW/dx, up to the point at which the jet energy is sufficient to slot the rock. Normal force reductions are also found to be a function of dW/dx.

RESUME: Un argument heuristique est utilisé pour prédire les effets de la pression d'un jet d'eau, de son débit, et de la vitesse de taille sur les forces requises au pic de taille quand un jet d'eau sous haute pression est utilisé pour aider l'excavation. La fonction primaire des jets d'eau sous pression est déblayer les débris qui se forment à l'avant de l'outil de taille. Le débit de déblaiement des débris est estimé être une fonction de la puissance du jet dW/dt. Par suite, on montre que la diminution de la force moyenne requise pour la taille augmente avec le paramètre dW/dx, qui est le taux de dW/dt par la vitesse de taille. Ce modèle est soutenu par les résultats d'expériences de taille dans un grès d'Indiana (Indiana limestone). La diminution de la force de taille requise est une fonction log-linéaire de dW/dx, jusqu'à ce que l'énergie du jet d'eau soit suffisante pour fendre la roche. La diminution de la force normale requise pour la taille est aussi une fonction de dW/dx.

ZUSAMMENFASSUNG: Um die Auswirkung des Düsenstrahldrucks, der Düsenstrahlgeschwindigkeit und Schnittgeschwindigkeit von Wasserdüsen beim schrämen auf die reduktion der Schrämmeisselandruckskraft vorauszusagen, wird ein heuristisches Argument herangezogen. Es wird angenommen, dass der Effekt des Wasserstrahls in erster Linie der Beseitigung des Bergekleins dient, das sich vor dem Schrämmeissel bildet. Es wird vorausgesagt, dass die Besietigung des Bergekleins eine Funktion der Wasserstrahlenergie dW/dt ist. Hierdurch wird gezeigt, dass die Reduktion der durchschnittlichen Schneidkraft eine ansteigende Funktion des Parameters dW/dx, das Verhältnis dW/dt zur Schnittgeschwindigkeit ist. Dieses Modell wird durch die Ergebnisse von Schnittversuchen im Indianakalkstein (Indiana limestone) unterstützt. Die Schnittkraftreduktion ist eine logaritmisch, lineare Funktion von dW/dx bis zu dem Punkt, an dem die Strahlenergie gross genug ist, das Gestein zu unterschrämen. Die Reduktion der Normalkraft ist ebenfalls eine Funktion von dW/dx.

1 INTRODUCTION

The discovery that moderate pressure (< 70 MPa) water jets can dramatically reduce the forces acting upon tools used in mechanical rock excavation was first reported over a decade ago (Hood, 1976). More recent studies have shown that these force reductions can be obtained in both hard and soft rock, using either drag picks or roller cutters (Ropchan et al, 1980; Dubugnon, 1981; Fenn et al, 1985).

These results represent a breakthrough in excavation technology, since they show that water jets provide an auxiliary means of introducing power into the rock cutting process, and thus they circumvent a fundamental limitation on the rate at which rock can be broken with mechanical tools. At present, mechanical excavation rates are limited by the power which can be applied via the pick, since the wear rate of the pick accelerates as the applied power increases; this acceleration in wear can be attributed to thermal degradation of the tungsten carbide tool face, caused by high wearflat temperatures which are a function of the applied power (Cook, 1984). Because water jets reduce the pick forces for a given depth of cut, deeper cuts can be achieved without increasing pick forces, and hence greater excavation rates are made possible without increasing the applied power and consequently the tool wear rate.

In addition to the reduction of wear, the effec-tiveness of water jets in reducing pick forces offers other benefits. Machinery which presently is only capable of excavating soft rock could be used in stronger ground. Alternatively, force reductions in softer rock would allow the use of less massive cutting head drive-train components, resulting in smaller, more maneuverable excavation machinery with production capacities equal to larger machines without water jets.

Although the dramatic benefits of this hybrid cutting system have been widely documented, a conclusive explanation of the fundamental process by which the jets act to reduce the pick forces is still not available. This lack of understanding of the breakage mechanism hinders commercialization of this technology because, in the absence of a mechanistic theory, the existing body of laboratory data cannot be extrapolated beyond the restricted range of cases studied. That is, the effects of the jet parameters (pressure, flowrate, and position with respect to the bit) as determined for a particular combination of machine parameters (cutting speed, pick geometry, etc.) and rock type cannot be extended to other combinations of these parameters. Instead, efforts to incorporate water jets into rock cutting systems must proceed by trial and error. Given the large number of variables which must be considered, and given the high cost of full scale testing, the need for a mechanistic model of water jet assisted cutting is urgent.

2 THEORETICAL DISCUSSION

It was hypothesized at the outset of this study that the principal mechanism by which water jets act to reduce forces on the pick is erosion of crushed material from in front of the pick. This mechanism was proposed by Tutluoglu (1984) after cutting tests in a lower speed range (50-250 mm/s) in Indiana limestone using the same apparatus as in the present study.

To determine the implications of this hypothesis, it was first necessary to study the rock cutting process in the absence of water jets, and from this to determine how and under what conditions the crushed material is produced. This study led to a conceptual model of how water jets might act to remove this material from in front of the pick, and how removal of this material might reduce the mean cutting force. This model was developed in a qualitative manner to give some indication as to how the degree of cutting force reduction might depend upon several of the parameters.

2.1 Dry cutting process

Based upon observations of the cutting process and a careful study of chip geometries, the following description of the cutting process seems plausible: A cutting cycle begins after the formation of a large chip, as depicted in Figure 1a. A certain amount of broken or crushed debris remains in the cut as a by-product of the chip formation process. This material is presumed to have formed before formation of the large chip. Hence it has no kinetic energy resulting from the release of strain energy by sudden fracture. As proposed by Tutluoglu (1984), this crushed material will behave as a semi-consolidated soil, and in order to remove this material the pick must work against Coulomb friction.

As the pick tip advances into the intact rock beneath this crushed material, more crushed material is formed ahead of the tip, as indicated in Figure 1b. The mass rate at which this material forms is roughly

$$dm_f/dt = \rho_r \cdot v \cdot w \cdot h(x), \qquad (1)$$

where ρ_r is the density of the intact rock, $h(x)$ is the height of the intact rock surface ahead of the pick (measured as the height above the pick tip) as a function of the distance x in the direction of pick travel, and w is the width of the pick. Until $h(x)$ reaches a critical height h_c, the contact stresses between the rock and pick produce only minor crushing and spalling, as in Figure 1c. When h_c is reached, an extension fracture forms which propagates to the free surface (Figure 1d) to form a second large chip. In this study, for 15 mm deep cuts h_c was observed to be about 5 mm.

An interesting feature of the large chips produced without water jets is that the fractures forming their bottom surfaces only rarely developed deeper than 10 mm below the surface, and the slope of this surface very near the pick was often horizontal or even downward, as in Figure 1d. This is contrary to detailed observations of chip formation using similar picks in Carthage Limestone (Friedman, 1983), in which the chips were seen to form by extension fractures propagating from the pick tip; however, in those tests the depth of cut was only 4 mm, which suggests that proximity to the free surface governed the extension of these fractures.

2.2 Mechanism of water jet assistance

It is proposed that water jets act to reduce cutting forces by eroding away a portion of the crushed

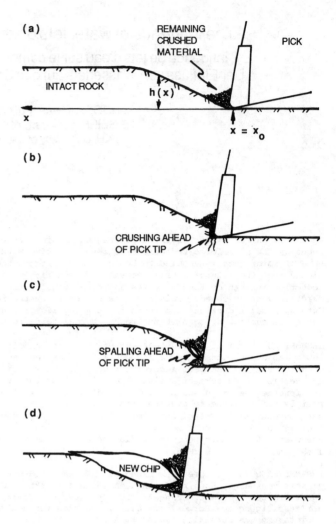

Figure 1. A typical chip formation cycle observed for 15 mm deep cuts, showing (a) crushed material remaining in path of pick after formation of a large chip, (b) crushing ahead of pick as it advances into the intact rock, (c) formation of small chips by spalling, and (d) formation of a second large chip.

material during the earlier stages of the chip formation cycle, that is, those stages represented by Figures 1a-c. In the last stage, represented by Figure 1d, the pick bears directly against the forming chip. At low to moderate jet energies the jet will not penetrate through this chip, and thus will not contribute further to cutting force reduction. Chips recovered from cuts made with jet energy per unit length on the order of 50 J/mm or more show that, at these energy levels, the jet does indeed penetrate the chip. The present analysis will consider only the effects of erosion during the early stages of the chip formation cycle.

In order to determine how water jets might reduce cutting forces by erosion of the crushed material during the earlier stages (Figures 1a-c) of the chip formation cycle, a relation must be obtained expressive of how the various parameters control the erosion process. Given such a relation, the second step is to determine how the removal of this crushed material will affect cutting forces. It is recognized that the physical phenomena which may play a role in these processes are complicated, and some of them, such as erosion and sediment transport by turbulent flow are in themselves challenging topics of current research. On the other hand, if research on water jet assisted cutting is to move beyond the descriptive stage, predictive models which can be tested by experiment are essential.

With this in mind, a qualitative approach was used to identify the parameters controlling the erosion process and the force reduction which is presumed to result from erosion of the crushed material. Dimensional analysis was used to show how these parameters can be grouped into a single parameter which is predicted to characterize the process of jet assisted cutting. The argument leading to this is predicated upon a number of heuristic assumptions, and in places is somewhat tenuous. However, this argument delivers a result which can be tested directly, and thus represents a first step toward a more rigorous theory.

To begin with, two fundamental assumptions are made. Firstly, the flow outward from the impinging jet is assumed to be fully developed turbulent. This is expected due to the high Reynolds' number (on the order of Re = 10^6) and due to the roughness of the surface over which the fluid flows, which is comparable to the thickness of the flow. Entrainment of particles further ensures rapid development of turbulence.

Secondly, it is assumed that the particles at the eroding surface are essentially unconfined and cohesionless. As noted by Leach and Walker (1966), the static pressure measured under a coherent jet impinging on a flat plate becomes negligible 2.6 jet radii from the point of impingement. Although the bulk of the crushed material is confined between the pick and the rock, the uppermost layers of particles cannot carry any appreciable stress because of the lack of confining stress normal to the free surface. Based upon this assumption, the mass rate of erosion is limited only by the rate at which the particles exposed to the flow can be carried away so that the particles underneath can be attacked by the water.

Now consider a portion q of the total flow Q, travelling away from the jet impingement point along some path parallel to the mean flow streamlines in steady turbulent flow. This flow channel is bounded on both sides by other flow channels whose geometry may be different if the mean flow field from the jet is not radially symmetric about the jet. The top of the channel is the free surface, and the bottom is the erosion interface, which in this simple model is assumed to be maintained at a constant level by the production of new crushed material.

When a particle of mass dm_r is eroded from the bottom of the flow channel, the fluid must perform a quantity of work dW_e on this particle to entrain it into the flow, resulting in an expenditure of power

$$dW_e/dt = dW_e/dm_r \cdot (dm_r/dt)_q \qquad (2)$$

if the rate of mass removal by q is $(dm_r/dt)_q$. Assuming that each particle must be moved through a certain distance before the particle beneath it can be removed, it is expected that the rate of mass removal will be proportional to the rate at which that work can be done, that is, the power supplied to the particle, which must be some function of the available fluid power. Since the flow is fully turbulent, this power transfer will be distributed more or less randomly among the fluid pathlines which comprise q, and it follows that the rate of erosion must be some function of the total available fluid power in the flow portion q, that is,

$$(dm_r/dt)_q = f_1[(dW/dt)_q] = f_1(p \cdot q), \qquad (3)$$

where p is the stagnation pressure of the jet. Thus for erosion by steady, turbulent flow it is expected that the total mass rate of erosion dm_r/dt will be controlled by the total jet power, $dW/dt = p \cdot Q$, according to some function $f_2(p \cdot Q)$ which is characteristic of the geometry of the mean flow paths. In the above, it was assumed that the entire bottom

surfaces of the flow channels consist of erosible, crushed material. In reality the availability of crushed material is limited by the rate at which it is formed, given by (1). The net rate of mass removal is the difference

$$dm/dt = dm_r/dt - dm_f/dt$$
$$= f_2(p \cdot Q) - \rho_r \cdot v \cdot w \cdot h(x). \qquad (4)$$

In this qualitative analysis the forms of the functions $f_2(p \cdot Q)$ and $h(x)$ are unknown. However, since the mass $\Delta m(t)$ of crushed material missing from in front of the bit due to action of the water jets is the integral of the above equation, we may write

$$\Delta m(t) = f_3(p \cdot Q, v, h). \qquad (5)$$

If it is assumed that the reduction ΔF_C of the cutting force due to the removal of this crushed material is a function of $\Delta m(t)$, then

$$\Delta F_C(t) = f_4(p \cdot Q, v, h). \qquad (6)$$

From size analyses of the rock chips formed while cutting (Tutluoglu et al, 1982), it is inferred that the geometry of the chip formation process is statistically repeatable along a length of cut, and hence the mean reduction in cutting force is a function only of $p \cdot Q$ and v:

$$\Delta F_C = f_5(p \cdot Q, v). \qquad (7)$$

The percentage mean cutting force reduction is given by:

$$R_C = (\Delta F_C / F_{CO}) \cdot 100\%, \qquad (8)$$

where F_{CO} is the mean cutting force measured without water jets. Dimensional analysis making use of the Buckingham Pi theorem (see McCormack & Crane, 1973: 93-101) yields the prediction that

$$R_C = R_C[p \cdot Q/(v \cdot F_{CO})]$$
$$= R_C[(dW/dx)/F_{CO}], \qquad (9)$$

and so for a given depth of cut (for which F_{CO} is constant) it is expected that R_C will depend only upon the ratio of jet power to cutting speed, which is the jet energy spent per unit length of rock, dW/dx.

This result implies that it is not important whether this jet power is supplied in the form of (i) jet pressure or (ii) flow rate. Also, it implies that the effect of increasing cutting speed can be offset by increasing jet power proportionally. However, examination of the assumptions leading to this simple result reveals that two restrictions on the form of the dependence (9) must hold:

Firstly, as the cutting speed v becomes so small that the rate of formation of crushed material becomes less than the potential rate of erosion, as given by (1), then the function R_C will asymptotically approach some upper limit corresponding to complete, instantaneous removal of the crushed material.

Secondly, the region in which crushed material forms is limited, and hence as flowrate Q is increased without increasing the pressure p, eventually a portion of the flow will never come into contact with the crushed zone, and will be wasted. Thus beyond some point, increasing Q will give diminishing returns in terms of cutting force reduction.

651

3. EXPERIMENTAL APPROACH

This study sought to test the hypothesis that erosion of crushed material from in front of the pick is the principal mechanism governing force reductions due to water jets. As was argued in the last section, this hypothesis requires that force reductions for a given depth of cut should depend mainly upon the ratio of jet power dW/dt to cutting speed v, that is, the jet energy used per unit length of cut, dW/dx, since

$$dW/dx = (dW/dt) \cdot (dt/dx). \qquad (10)$$

In order to test this hypothesis, a series of 15 mm deep linear cuts in Indiana Limestone was performed in which pick force reductions were measured over a range of dW/dx, while jet pressure, nozzle orifice diameter, and cutting speed were varied independently. These cuts were organized as a factorial experiment in order to facilitate statistical analysis. The dependence of force reductions upon jet energy levels was determined for each level of cutting speed and orifice diameter, and these results were compared to test the strength of the hypothesis.

3.1 Equipment

A modified industrial planer was used to perform the linear cutting tests. A block of the limestone was mounted on the planer bed and traversed beneath a stationary drag pick, which was mounted in a dynamometer used to measure forces on the pick. High velocity water jets were directed just in front of and parallel to the pick face. This equipment is described by Tutluoglu (1984).

3.2 Experimental plan

For statistical purposes these tests were designed as a factorial experiment in three variables: jet pressure (p), nozzle orifice diameter (a), and cutting speed (v). The nominal levels of these variables which were tested were:

```
p  =   10, 20, 35, 50, and 70 MPa,
a  =   0.65, 0.83, and 1.05 mm,
v  =   160 and 420 mm/s.
```

These values provide a range of two orders of magnitude in dW/dx of approximately 1 to 100 J/mm. Each of the possible $5 \cdot 3 \cdot 2 = 30$ combinations of these values was tested three times. In addition, in each layer a cut was made without jets. The potential benefits of using a factorial plan such as this have been described by Box et al (1978). For this study the principal advantage of this approach was that it allowed a more thorough testing of models in the variables p, a, and v.

The sequence of the cutting tests was randomized with respect to position on the block of rock using a balanced, incomplete Latin square arrangement (Box et al, 1978). The purpose of this arrangement was to eliminate effects of variations in rock properties across the breadth and depth of the block. Variations along the length of the block, of course, were accounted for by averaging the forces along the length of each cut.

4 EXPERIMENTAL RESULTS

The experimental part of this study gave data for four separate responses; the mean cutting and mean normal forces F_c and F_n respectively; and the mean peak cutting and mean peak normal forces F_{cp} and F_{np} respectively. These forces, defined in Figure 2,

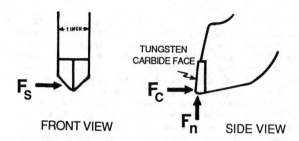

Figure 2. Drag pick used in cutting tests. Force components are defined as shown.

were measured as functions of the variables p, Q, and v.

The data presented in the following sections are shown in terms of force reductions, calculated from the experimental results according to the formula

$$R = [1 - F/F_0] \cdot 100\%, \qquad (11)$$

where F_0 = the force measured without water jets The magnitudes of the pick forces actually measured during these tests are illustrated by the representative data given in Table 1.

Table 1. Representative cutting force data

	F_c (kN)	F_{cp} (kN)	F_n (kN)	F_{np} (kN)
Cut without jet	3.382	7.860	1.534	2.898
Cut with jet*	2.409	7.065	0.880	1.979

* v = 420 mm/s, p = 35 MPa, a = 0.83 mm

From Table 1 it is seen that the magnitude of the mean and mean peak normal force measurements, were both much smaller than their cutting force counterparts. Also, for the dynamometer used the signal-to-noise ratio in the normal force direction was as small as 4:1 for the peak forces, as compared with a signal-to-noise ratio of more than 100:1 in the cutting force direction. Despite this, because F_n was calculated from more than 1000 data points per cutting test these results were presumably insensitive to the noise level. The values of F_{np}, on the other hand, were found from only 40 or so peaks per test. Consequently the F_{np} results were judged to be unreliable and are not presented here.

4.1 Mean cutting force reductions

The predicted dependence of R_c on dW/dx can be tested by comparing cutting tests for which dW/dx was similar, but for which the parameters p, Q, and v took on different values.

Figure 3a-c shows the 90 percent confidence bands for the percentage reduction of the mean pick cutting force, plotted as functions of dW/dx at both cutting speeds and for each of the 3 nozzle sizes. The confidence bands represent the 90 percent confidence intervals which were calculated for each set of replicated cutting tests, using Student's "t" distribution.

The high degree of overlap of the error bands in each of these plots is a strong indication that, at each jet diameter, the percentage force reduction is a function only of dW/dx when p and v are varied independently. The analysis shows that R_c increases in roughly log-linear fashion with increasing dW/dx, until dW/dx reaches approximately 20 J/mm. A log-linear regression fit for this region gave

$$\log R_c = 1.25 + 0.266 \log(dW/dx). \qquad (12)$$

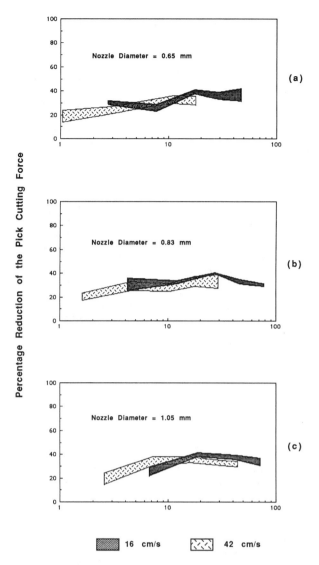

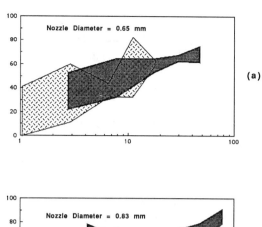

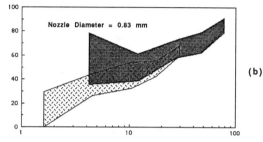

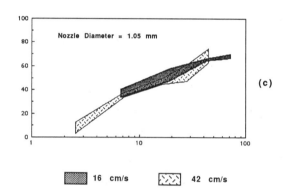

Percentage Reduction of the Pick Cutting Force

Percentage Reduction of the Pick Normal Force

Nozzle Diameter = 0.65 mm (a)

Nozzle Diameter = 0.83 mm (b)

Nozzle Diameter = 1.05 mm (c)

Nozzle Diameter = 0.65 mm (a)

Nozzle Diameter = 0.83 mm (b)

Nozzle Diameter = 1.05 mm (c)

16 cm/s   42 cm/s

16 cm/s   42 cm/s

Jet Energy per Unit Length of Cut (J/mm)

Jet Energy per Unit Length of Cut (J/mm)

Figure 3. Comparisons between 90% confidence bands
for R_c obtained for v = 160 and 420 mm/s, for orifice diameters (a) a = 0.65 mm, (b) a = 0.83 mm, and
(c) a = 1.05 mm.

Figure 4. Comparisons between 90% confidence bands
for R_n obtained for v = 160 and 420 mm/s, for orifice diameters (a) a = 0.65 mm, (b) a = 0.83 mm, and
(c) a = 1.05 mm.

Beyond dW/dx = 20 J/mm, the function R_c(dW/dx)
levels off, and for the 0.83 and 1.05 mm nozzles it
appears that R_c becomes a decreasing function of
dW/dx beyond dW/dx = 30 J/mm or so. This behavior
is consistent with that observed by Tutluoglu
(1984). In these experiments it was observed that
20 J/mm was approximately the threshold level beyond
which the jet began to cause noticeable erosion of
the intact limestone in the bottom of the cut. Thus
the point at which the jet begins to cut the rock
coincides with the point at which the benefits of
the jet assist begin to decrease. This suggests
that the decrease in jet effectiveness is somehow
due to the onset of jet cutting.

4.2 Mean normal force reductions

Figures 4a-c show the 90 percent confidence bands
for mean normal force reductions R_n achieved with
the three different nozzle sizes, comparing the results for the faster cutting speed with those for
the slower speed. It is apparent that the magnitude
of the mean normal force reductions (up to almost 90
percent) is much greater than the reductions seen in

the mean cutting force. The fact that the error
bars in these plots are generally greater than in
Fig. 3 is attributed to the higher signal-to-noise
ratio for the normal force.

Except for the lower energy end of the plot for
the 0.83 mm nozzle, there is obviously a very strong
agreement between the two speeds. Superposing all
of the plots in Figure 4 shows that there is no discernible difference between the different nozzles
sizes (i.e., different flowrates) and different
speeds. Hence the mean normal force reduction is
also a function of dW/dx. A best linear-fit line
over the entire range of dW/dx gives

$$R_n = 4.1 + 42.1 \log(dW/dx). \qquad (13)$$

This very interesting result is not directly
related to the earlier theoretical discussion, which
was primarily concerned with cutting force reductions. However, it can be argued that this behavior
is also due to erosion from in front of the bit.
Figure 5 suggests a possible erosion pattern as seen
from the front end of the bit. Presuming that the
crushed material immediately under the jet is plastically extruded outward beyond the region in which

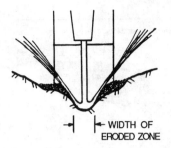

Figure 5. Possible erosion pattern ahead of bit.

confining pressure due to jet impingement prevents erosion (cf Rehbinder, 1976), the crushed material would be eroded completely away at the center of the bit, and the width of this zone of complete erosion would increase with increasing dW/dx. Thus the portion of the bit width which will be sliding on a cushion of crushed material will decrease with increasing dW/dx, with a resulting decrease in the normal force.

As the jet begins to cut the rock at dW/dx = 20 J/mm or so, a portion of the bit width will be relieved, causing even greater normal force reductions. From the mean normal force data there is no evidence that this produces a change in the slope of R_n(dW/dx). This is evidence that the apparent downturn in the cutting force reduction curves above dW/dx = 20 J/mm cannot be explained by the proposed erosion mechanism. Rather, a second mechanism must act at these high levels of dW/dx to offset a portion of the force reduction produced by the erosion of the crushed material from in front of the pick.

4.3 Mean peak cutting force reductions

In general, the mean peak cutting forces F_{cp} were reduced by a lesser degree than were the mean cutting forces. The maximum expected value for mean peak cutting force reduction was only about 25 percent, in comparison to the mean cutting force reductions, which exceeded 40 percent. The peak forces correspond to the formation of large chips. Since the duration of the peak forces and near-peak forces is observed to be only a small fraction of the total cutting time, this small reduction in peak forces is not sufficient to account for the major part of the observed reductions in mean cutting force. Thus the effectiveness of the jets in reducing the force required to form a large chip is small relative to the effectiveness of the jets in reducing average forces. This is further evidence in support of the proposed mechanism, which is hypothesized to be effective mainly during the early stages of the chip formation cycle, as depicted in Figures 1a-c.

5 CONCLUSIONS

A qualitative model was proposed which described the influence of pick speed, jet pressure, and jet flow rate on the cutting force reductions obtained when water jets are used to assist in mechanical rock cutting operations. This model was developed from the postulate that the primary influence of the jets is by the erosion of crushed rock which forms ahead of the pick. An important intermediate result obtained in the development of this model was the prediction that the rate of erosion of this crushed rock is a function only of the jet power dW/dt and the geometry of the flow field.

This led to the prediction that pick cutting force reduction R_c is a function only of the jet energy per unit length dW/dx, for a given jet/pick config-

uration and depth of cut. Experiments confirmed this prediction over the ranges of the parameters studied, up to the point at which dW/dx was sufficient for the jet to begin slotting the rock. Up to this point the relationship between R_c and dW/dx was found to be approximately log-linear. In the neighborhood of the optimal level of dW/dx, mean cutting force reductions were roughly 40 percent, while peak cutting force reductions only rarely exceeded 25 percent or so. This finding accords with a concept proposed in the formulation of the model, namely that the jets act mainly to reduce the force acting upon the pick during the relatively long intervals between the formation of large chips, since the peak forces are associated with these instants of large chip formation.

Experimental evidence also showed that the percentage reductions in mean normal force R_n were much greater than the achieved levels of R_c, a result noted by other workers (Hood, 1976; Ropchan et al, 1980; and Dubugnon 1981). A new finding was that R_n behaves as a function of dW/dx, according to a log-linear relationship. A mechanism for this force reduction was shown to follow directly from the proposed model of the cutting force reduction mechanism. The high normal force reductions obtained at high jet energies gave evidence that there is a second, distinct mechanism which diminishes cutting force reductions when the jet becomes powerful enough to slot the rock.

References

Box, G.E.P., W.G.Hunter & J.S.Hunter. 1978. Statistics for experimenters: An introduction to design, data analysis, and model building. New York: Wiley.
Cook, N.G.W. 1984. Wear on drag bits in hard rock. In: Canadian Mining & Metallurgical Bulletin, Special volume: Rock mechanics.
Dubugnon, O. 1981. An experimental study of water assisted drag bit cutting of rocks. In Proc of 1st US Water Jet Symposium, p. II-4.1 to II-4.11. Golden, Colorado: Colorado School of Mines Press.
Fenn, O., B.E.Protheroe & N.C.Joughin. 1985. Enhancement of roller cutting by means of water jets. In Proc of Rapid Excavation & Tunneling Conf, V.1, p. 341-357. New York: AIME.
Friedman, M. 1983. Analysis of rock deformation and fracture induced by rock cutting tools used in coal mining. Contractor report SAND83-7007. Albuquerque: Sandia Natl Laboratories.
Hood, M. 1976. Cutting strong rock with drag bits assisted by high pressure water jets. J.S.Afr. Inst.Min.&Metall. 77(4): 79-90.
Leach, S.J. & G.L.Walker. 1966. The application of high speed liquid jets to cutting -- Some aspects of rock cutting by high speed water jets. Proc of the Royal Society of London A, 260: 295-308.
McCormack, P.D. & L.Crane. 1973. Physical Fluid Dynamics. New York: Academic Press.
Rehbinder, G. 1976. Some aspects of the erosion of rock with a high speed water jet. Proc of the 3rd Intl Symp on Jet Cutting Technology, p. E1-1 to E1-20. Bedford, England: BHRA.
Ropchan, D., F.D.Wang & J.Wolgamott. 1980. Application of water jet assisted drag bit and pick cutter for the cutting of coal measure rocks (US DOE contract ET-77-A-01-9082, Colorado School of Mines). NTIS DOE/ET/12463-1.
Tutluoglu, L., M.Hood & C.Barton. 1982. An investigation of the mechanisms of water-jet assistance on the rock cutting process. Earth Sciences Division Annual Report LBL-15500, p. 40-43. Berkeley, California: Lawrence Berkeley Laboratory.
Tutluoglu, L. 1984. Mechanical rockcutting with and without high pressure water jets. PhD dissertation, University of California, Berkeley.

The effect of rock texture on drillability and mechanical rock properties

L'effet de texture des roches sur la vitesse de forage et les propriétés mécaniques des roches
Die Wirkung von Gefüge im Gestein auf die Bohrfähigkeit und andere mechanische Eigenschaften

D.F.HOWARTH, Department of Mining & Metallurgical Engineering, University of Queensland, St. Lucia, Australia
J.C.ROWLANDS, Department of Mining & Metallurgical Engineering, University of Queensland, St. Lucia, Australia

ABSTRACT: A quantitative measure of rock texture has been developed. Data is obtained from photographs of thin sections using an image analysis system. The 'texture coefficient' returns highly statistically significant correlations with mechanical property and drillability data in ten rock types.

RESUME: Le mesure quantitatif de texture de rocher a ete develope. Les donnees sont obtenuees des photographs des sections tres fines en utilisant une systeme d'analysis d'images. La "coefficient de texture" rendre les significatifs correlations statistiquement tres importants avec les proprietes mechaniques et donnees de vitesse de forage dans dix types de rochers.

ZUSAMMENFASSUNG: Das quantitative mass von felsengewebe ist entwickelt geworden. Daten wurden erhalten durch photographien von dunnen abschnitten mit the verwendung eines analysen systems. Die 'gewebe leistungs-fahigkeit' gibt hochwertige statistische bedeutung auf der mechanischen eigenschaft und die datan der bohrenfahigkeit in zehn felsentypen.

INTRODUCTION

It has been suggested (Olsson and Peng, 1976) that rock fracture is associated with four sequential events-crack nucleation, initiation, propagation and coalescence. Clearly, such textural features as grain size, shape, interlocking and orientation will influence the propagation and networking of cracks and, thus, the mechanical performance of rock. Experimental studies of microfracturing in a sandstone (Sangha et al., 1974) showed that failure occurred in the cement matrix rather than the quartz grains. Indentation fracture development that was induced by a sharp and truncated wedge (Lindqvist et al., 1984) and observed in a scanning electron microscope (SEM) indicated the influence of texture on fracture patterns. Fracture patterns in a fine-grained, dense limestone were straight with very few crack interactions, whereas cracks in a medium-grained, weakly grain-bonded marble showed many interactions and forking, which, presumably, followed weak grain boundaries.

The developed 'texture coefficient' can be used as a predictive tool for the assessment of drillability and rock strength properties. The coefficient also has the potential to be used for other applications; one example is a diggability index for large excavating equipment. In this application a muckpile would be photographed, the texture coefficient assessed and correlated with excavating machine loading forces or power consumed. The texture coefficient could provide a quantitative measure of rock fragmentation, and a measure of diggability providing that the correlations with machine operating characteristics were statistically significant.

DEVELOPMENT OF THE TEXTURE COEFFICIENT

The method of quantitative assessment of rock texture can be broken down into four parts:

i) Measurement and analysis of grain circularity
ii) Measurement and analysis of grain elongation
iii) Measurement and quantification of grain orientation
iv) Weighting of results based upon degree of grain packing.

The analysis is summarised in the following formula:

$$TC = AW \left[\left(\frac{N_O}{N_O + N_1} \times \frac{1}{FF_O} \right) + \left(\frac{N_1}{N_O + N_1} \times AR_1 \times AF_1 \right) \right] \dots (1)$$

where, TC = texture coefficient
AW = grain packing weighting
N_O = number of grains whose aspect ratio is below a pre-set discrimination level
N_1 = Number of grains whose aspect ratio is above a pre-set discrimination level
FF_O = Arithmetic mean of discriminated form-factors
AR_1 = Arithmetic mean of discriminated aspect ratios
AF_1 = Angle factor, quantifying grain orientation

Rock textures were assessed using microscopic image analysis of thin sections. Directional bias was investigated in anisotropic rocks by obtaining thin sections from three representative orientations, normal and parallel to the direction of mechanical performance assessment, and a random orientation. Image processing was performed using a DAPPLE SYSTEMS, IMAGE PLUS TM, automatic image analysis, with video camera input of thin section photographic prints.

Individual analysis consisted of selecting a reference area or "observation window", containing twenty to thirty rock grains, then processing this image to obtain the geometrical parameters; area, perimeter, length, breadth and angle, for each grain. Area and perimeter were calculated directly, however length and breadth were defined as being the maximum and minimum Feret;s diameters (Herdan and Smith, 1953) respectively, calculated every five degrees around the grain image. Feret's diameter is defined as being the perpendicular distance between two parallel, outer tangents to an object.

The final parameter 'angle' was defined as being the angle between the maximum Feret's diameter (length) and the horizontal direction. The maximum value of 'angle' is 180°.

The initial step of textural quantification was to analyse grain shape using two secondary geometrical parameters, aspect ratio and form factor. The first method of shape deviation (elongation), is best measured using the grain's aspect ratio. This is defined simply as the ratio of the grain's length to breadth. Thus for increasing elongation, the aspect ratio increases. Form factor is a measure of a grain's deviation from circularity. This deviation may occur in two ways; elongation of the shape or increased 'roughness' of the grain's perimeter. Roughness is best measured directly using the grain's form factor. Form factor is defined thus:

$$\text{FORM FACTOR} = 4 \times \pi \times (\text{AREA})/(\text{PERIMETER})^2 \quad \ldots \ldots (2)$$

For a perfect circle, the form factor is 1.0. As a shape deviates from circularity, (by elongation or increased roughness), the form factor decreases.

To differentiate between which mode of shape deviation measurement was to be used in the analysis, an aspect ratio discrimination level of 2.0 was introduced. Thus the arithmetic mean of the form factors was calculated for all grains falling below this level, whereas the arithmetic mean of the aspect ratios was calculated for all particles above this level. To maintain continuity in equation 1, the form-factor term was inverted to ensure that as a grain deviates from circularity, the result from either method of measurement, increases.

Angular orientation of grains was quantified by the development of an ANGLE FACTOR. This factor was only calculated for grains regarded as being elongated, i.e. their aspect ratio was greater than 2.0. The ANGLE FACTOR was calculated by a class weighted system applied to the absolute, acute angular differences, (i.e. $0^\circ - 90^\circ$), between each and every elongated grain. A full description of the development of the ANGLE FACTOR has been presented elsewhere (Howarth & Rowlands, 1987).

The final term in equation 1, AW, represents an area weighting, based upon the grain packing density in any observation window. The texture coefficient is scaled down according to the percentage area of grains in the total reference area. This factor is only apparent when dealing with sandstones.

The texture coefficients of the experimental rocks investigated are reported in Table 1.

DRILLING EXPERIMENTS AND ROCK TYPES

Diamond and percussion drilling tests were undertaken Operating conditions were maintained at constant levels throughout the tests. The type of drilling tool used was a thin walled impregnated bit with water flushing. In operation the drill was mounted vertically in a Nider N1/52 workshop drilling machine specially adapted for rock drilling work. The percussion drilling tool was a simple wedge indenter (tungsten carbide insert) located on the end of a drill steel which was driven by an Atlas Copco RH571 compressed air powered percussion drill with water flushing. In operation the drill and steel were mounted horizontally in a specially constructed drilling frame. All rock samples used in the drilling experiments were tested at ambient levels of moisture content.

Test conditions are reported in Table 2.

Details of the rocks tested and drill penetration rates have been reported elsewhere (Howarth & Rowlands, 1987).

DRILLABILITY - TEXTURE COEFFICIENT RELATIONSHIPS

In order to observe the effect of texture on drilling rate, the data was analysed on the basis of two groups of rocks, namely the sandstones and marbles on the one hand, and the igneous rocks and marbles on the other hand. The inclusion of the marbles in both groups of rocks was considered necessary as their textural and strength properties, and drillability fall between the properties of the

TABLE 1: Summary of texture coefficients

ROCK TYPE		AW	TEXTURE COEFFICIENT UNWEIGHTED*	WEIGHTED*
HELIDON	N[#]	0.37	1.62	0.61
SANDSTONE	P	0.32	1.63	0.52
	R	0.51	1.50	0.76
GOSFORD	N	0.65	1.63	1.06
SANDSTONE	P	0.63	1.79	1.12
	R	0.65	1.59	1.04
IPSWICH	N	0.49	1.51	0.74
SANDSTONE	P	0.49	1.56	0.76
	R	0.63	1.72	1.08
MT.CROSBY	N	0.57	1.47	0.83
SANDSTONE	P	0.62	1.44	0.89
	R	0.59	1.66	0.98
CARARRA	N	1.00	1.61	1.61
MARBLE	P	1.00	1.78	1.78
	R	1.00	1.21	1.21
ULAN	N	1.00	1.44	1.44
MARBLE	P	1.00	1.61	1.61
	R	1.00	1.13	1.13
MT.MORROW BASALT	R	1.00	2.80	2.80
ASHGROVE GRANITE	R	1.00	2.35	2.35
BEENLEIGH HORNFELS	R	1.00	1.60	1.60
MOOGERAH MICROSYENITE	R	1.00	2.01	2.01

* the weighted texture coefficient is the product of the unweighted texture coefficient and the area weighting (AW)
\# orientation to test direction N-normal, P-parallel, R-random

TABLE 2: Drilling test conditions

	THRUST	RPM	CONSTANTS AIR PRESSURE	WATER PRESSURE	WATER FLOW RATE
	N	-	kPa	kPa	l/min
DD A[1]	378	750	-	275*	5*
DD B[2]	378	750	-	275*	5*
PD C[3]	196	-	450	-	5#
PD D[4]	441	-	450	-	5#
DD E[5]	770	750	-	552*	2*

1. Bit (A). 31.8mm OD 27.9mm ID.
2. Bit (B). 31.9mm OD 28.1mm ID
3. Tool(C). 37.7mm wide
4. Tool(D). 29.0mm wide
5. Bit (E). 31.9mm OD 28.1mm ID

*average over the duration of the experiment
#not drilling
DD-diamond drilling
PD-percussion drilling

sandstones and igneous rocks. A graph of the random random texture coefficient against percussion drill penetration rates is shown in Fig.1

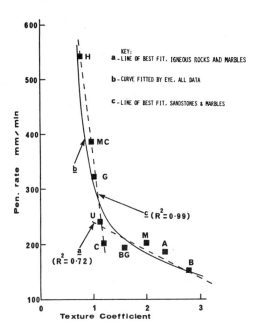

MECHANICAL ROCK PROPERTY - TEXTURE COEFFICIENT
RELATIONSHIPS

·Random texture coefficients are graphed against Uniaxial compressive strength, Brazilian disc tensile strength, static and dynamic Young's moduli and P wave velocity, and presented in Figs.2(a)-(e).

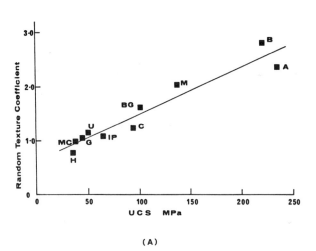

(A)

FIG.1: Random texture coefficient against percussion drill penetration rates - Exp.D.
(Rock types:
    H - Helidon sandstone sst.
    MC - Mount Crosby sst.,
    G - Gosford sst.,U- Ulan marble,
    C - Carrara marble,
    BG - Beenleigh hornfels,
    M - Moogerah microsyenite,
    A - Ashgrove granite,
    B - Mt.Morrow basalt
    IP - Ipswich sandstone)

Two lines of best fit can be drawn through the data which approximate the curve (b) presented. The first line is one of best fit through the igneous and marble data (R^2 = 0.72). The second is a line of best fit through the sandstone and marble data (R^2 = 0.99). The correlation is clearly significant. Penetration rate decreases with an increasing texture coefficient. Igneous rocks and sandstones behave differently. A small change in texture coefficient in the sandstones - which is basically caused by the quantity of phyllosilicate matrix produces a large change in drilling rate. Whereas in the igneous rocks a large change in texture coefficient - which is basically caused by the shape and degree of grain interlocking - produces a small change in drilling rate. Lindqvist et al (1984) showed that stress-induced cracking (caused by wedge indentation) was extensional in nature, with cracks propagating in the direction of maximum compression. Since percussion drilling is a wedge indentation process and recalling that the texture coefficient models grain-shape, orientation and degree of interlocking, in a qualitative sense it is apparent how texture influences drilling performance. An interlocked texture simply presents a physical barrier to crack propagation. It can be concluded that drilling processes and fracture mechanisms in three types of rocks, igneous, marbles and sandstones are related to, and dependent on, aspects of rock texture. Marbles represent the division between the igneous rocks and sandstones, since they display features of both groups. The calcite grains are crystalline, however, they are generally weakly bonded, as is the case with the sandstones.

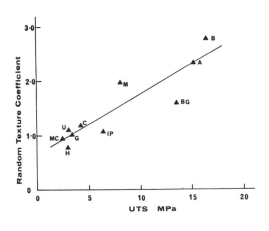

(B)

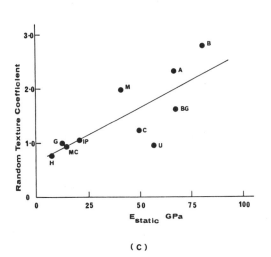

(C)

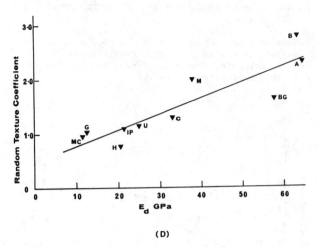

(D)

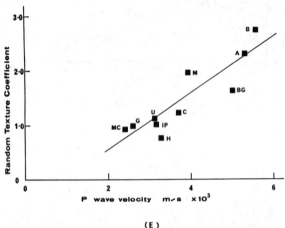

(E)

FIG.2: Random texture coefficients against rock
properties.
  a) Uniaxial compressive strength - dry
  b) Brazilian disc tensile strength - dry
  c) Static tangent Young's modulus - dry
  d) Dynamic Young's modulus - dry
  e) P wave velocity - dry
     (Rock types: See Fig.1)

These relationships indicate that the mechanical
properties of the experimental rocks are strongly
influenced by their texture. Coefficient of determination (R^2) values for these curves are presented
in Table 3. A similar analysis was undertaken using
the Schmidt hammer rebound number as the predictive
measure of mechanical rock property values. Coefficient of determination (R^2) values for these data
are also presented in Table 3. In all cases the
texture coefficient correlation was superior to that
obtained from the Schmidt Hammer test.

TABLE 3: Correlation of mechanical rock properties
         with the texture coefficient and Schmidt
         hammer rebound number.

ROCK PROPERTY	R^2 VALUES*	
	TC (1)	Sch.No. (2)
UCS (DRY	0.92	0.59
UCS (SATURATED)	0.91	0.63
BRAZIL TEST (DRY)	0.81	0.42
BRAZIL TEST (SATURATED)	0.89	0.38
STATIC YOUNG'S MODULUS (DRY)	0.64	0.56
DYNAMIC YOUNG'S MODULUS (DRY)	0.77	0.58
P WAVE VELOCITY (DRY)	0.76	0.50

*Linear Curves   (1) Random Texture Coefficient
                   (2) Schmidt hammer rebound No.

CONCLUSIONS

1. A dimensionless quantitative measure of rock
   texture - the texture coefficient - has been
   developed and applied to quantify the textures
   of ten igneous rocks, sandstones and marbles.
2. The texture coefficient describes, grain-shape,
   orientation, degree of grain interlocking and
   relative proportions of grains and matrix
   (packing density).
3. Measures of mechanical rock performance e.g.rock
   strength and percussion drilling rates return
   statistically highly significant correlations
   with the texture coefficients.
4. Observational and correlated data are strongly
   supportive of the suggestion that the texture
   coefficient is a measure of the resistance of
   the microstructure of a rock to crack
   propagation.
5. Igneous rocks had high texture coefficients,
   high strength and low drillability, whereas
   sandstones had low texture coefficients, low
   strength and high drillability. The properties of the marbles lie between those of the
   igneous rocks and sandstones.
6. In terms of the prediction of rock strength
   the texture coefficient is superior to the
   Schmidt Hammer Rebound test number.
7. The texture coefficient can be used as a
   predictive tool for the assessment of drillability and rock strength properties. The
   technique offers a useful approach in understanding fracture initiation and growth as
   controlled by the texture of intact rock
   samples.

REFERENCES

Herdan, G. & Smith, M.L. 1953. Small particle
statistics. Elsevier Publishing Company,
Houston, p.66.

Howarth, D.F. and Rowlands J.C. 1987. Quantitative assessment of rock texture and correlation with drillability and strength properties.
Rock mechanics & Rock Engineering (In Press).

Lindqvist P.A., Lai H.J. and Alm O. 1984.
Indentation fracture development in rock
continuously observed with a scanning electron
microscope. Int.J. Rock Mech. Min. Sci.,V21,
No.4, 165-82.

Olsson W.A. and Peng S.S. 1976. Microcrack
nucleation in marble. Int. J. Rock Mech. Min.
Sci., 13, 53-9.

Sangha C.M., Talbot C.J. and Dhir R.K. 1974.
Microfracturing of a sandstone in uniaxial
compression. Int.J.Rock Mech. Min. Sci., 11,
107-13.

Experimental study on the abrasive waterjet assisted roadheader
Etude expérimentale d'une machine de traçage auxiliaire équipée d'un jet d'eau abrasif
Untersuchung über die Leistung einer Vortriebsmaschine mit einem scheuernden Wasserstrahl

SHIGERU IIHOSHI, Technical Research Institute, Taisei Corporation, Japan
KENJI NAKAO, Dr., Technical Research Institute, Taisei Corporation, Japan
KYOZO TORII, Technical Division, Nihon Koki Co. Ltd, Japan
TADAO ISHII, Technical Division, Nihon Koki Co. Ltd, Japan

ABSTRACT: Abrasive water jet shows much better cutting performance on rocks, concrete and steel plates than the conventional water jet. This paper describes the results of experimental study in the development of an abrasive waterjet assisted roadheader aimed at cutting hard rocks, in order to seek after the possibility of its performance.
The paper describes the following contents:
(1) Calculation formula of rock cutting depth by abrasive jet
(2) Reduction in cutting power against mortar and rock samples which have been set with slots
(3) Concept of cutting method

RESUME: Le jet d'eau abrasif affirme de meilleures performances de coupe sur les roches, le béton et les plaques d'acier que le jet d'eau ordinaire. Cet exposé décrit les résultats d'une étude fondamentale sur la réalisation d'un jet d'eau abrasif et d'une machine de traçage auxiliaire appliqués à la coupe de roches dures de façon à vérifier ses performances.
Le texte décrit notamment les points suivantes:
(1) Calcul d'une formule de profondeur de coupe de roche par jet abrasif
(2) Réduction de la consommation de coupe sur du mortier et des échantillons de roches munies fente.
(3) Conception d'un procédé de coupe

ZUSAMMENFASSUNG: Mit dem untersuchten scheuernden Wasserstrahl läßt sich eine sehr viel bessere Leistung beim Schneiden von Gestein, Beton und Stahlblech erzielen als mit einem konventionellen Wasserstrahl. Diese Arbeit beschreibt die Ergebnisse der Grundlagenforschung zur Entwicklung eines scheuernden Wasserstrahls mit Hilfsvortrieb, der zum Schneiden von hartem Gestein bestimmt ist, und untersucht die damit erzielbare Leistung. Es werden folgende Themen behandelt:
(1) Formel zur Berechnung der Schneidtiefe des scheuernden Wasserstrahls in Stein
(2) Reduzierung der Schneidkraft durch Anbringen von Kerben in Mörtel- und Steinproben
(3) Konzept des Schneidverfahrens.

1 INTRODUCTION

Needs for economic excavation in hard rock tunnelling with small diameter have been increasing in Japan recently. Currently this is related mostly to small tunnel construction, i.e., local hydro-electric power stations, irrigations, waterwork systems, sewage, gas supplied systems and telecommunications. At present, there are several difficulties in cutting down the actual excavation cost of tunnelling, in conjection with the reduction of the excavation volume per unit length.

The waterjet assisted roadheaders that have been developed in recent years seems to improve such the situation in tunnelling. The results from various field tests (Plumpton et al., 1982) showed that this new type of machine has many advantages in reduction of cutting force, enhancement of pick life and in extending their range of use for hard rock excavation. Fairhurst et al., (1986) discussed the effect of waterjet assistance basing on the results of laboratory test and concluded that these benefits on this type of machine was not in terms of force reduction but rather in; tool-rock contact lubrication, thus reducing temperature build-up and consequently wear; tool tip cooling; rock debris flushing; and machine vibration reduction.

In the field of jet cutting technology, some recent studies (Hashish 1982) has shown that the use of abrasive waterjets have a more excellent cutting efficient when campared with waterjet by themselves.

Authors have been studying the abrasive waterjet assisted roadheader system in excavation of hard rock in order to construct the tunnelling techniques with small diameters, economically. This paper describes the results of basic study carried out before the production of trail machine.

2 ROCK CUTTING TESTS USING ABRASIVE WATERJETS

The detail of test method is shown in the reference (S.Iihosii et al., 1986). This chapter shows the additional data and the essence. The three types of rock specimens, granite, andesite and rhyolite were used. The physical and mechanical properties of the specimens are shown in Table 1. Fig. 1 shows the test apparatus for rock cutting. Range of test parameter is as follows;

Waterjet pressure	P ;	$180 - 340$ Mpa
Water flow rate	Qw;	$0.043 - 0.161$ kg/s
Traverse rate	T ;	$1.67 \times 10^{-3} -$
		2.50×10^{-2} m/s
Stand-off distance	S ;	$0.12 - 0.31$ m
Flow rate of abrasive	Qa;	$0 - 0.05$ kg/s
Type of abrasive	;	garnet, silica sand

The depth of cut (slot) in relation to the parameters has been expressed as the following formula.

$$H = K_1 \cdot \frac{W_{e,j}}{T \cdot S} + K_2 \cdot \left(\frac{Qa}{Qw}\right)^n \cdot \frac{W_{e,j}}{T \cdot S} \quad \cdots\cdots\cdots (1)$$

where H : Depth of slot (mm)
$W_{e,j}$: Effective power considering pressure loss (kW)

T : Traverse rate (mm/s)
S : Stand-off distance between rock surface and orifice for waterjet (mm)
Qa : Flow rate of abrasive, provided Qa ≤ Qa,opt (g/s)
Qw : Flow rate of waterjet (g/s)
n : Constant coefficient determined by the performance of abrasive jet nozzle
K_1, K_2: Constant coefficients determined by the type of rock and type of abrasive

In regard to flow rate of abrasive, there is the optimum flow rate of abrasive Qa,opt in relation to flow rate of waterjet flow rate Qw.
The relationship is as the following:

$$Qa,opt = \alpha + \beta \, Qw$$

where α, β; constant coefficients determined by the type of abrasive

Table 2 shows the constants obtained from multiple regression analysis.

Table 1. Rock properties

		Bulk density kg/cm^3	Sonic velocity		Uniaxial compressive strength Mpa	Tensile strength Mpa
			Vp m/s	Vs m/s		
Excavation test	Mortar	2,133	4,119	1,918	53.0	2.16
	Granite	2,654	4,526	2,114	166.8	7.89
	Andesite	2,640	3,902	1,831	209.8	6.83
Abrasivejet cutting test	Granite	2,641	4,300	2,230	150.3	7.80
	Andesite	2,664	4,250	2,290	128.5	8.50
	Rhyolite	2,431	3,820	1,920	96.2	7.00

Table 2. Constants in abrasivejet cutting

Type of rock	Garnet			Silica sand		
	K_1	K_2	n	K_1	K_2	n
Granite	440	5,500	0.764	440	2,930	0.762
Andesite	1,430	4,470	0.764	-	-	-
Rhyolite	728	3,602	0.764	-	-	-
	α = 17.5 β = 0.100			α = 8.92 β = 0.134		

3. ROCK CUTTING TESTS USING DRAG TOOL

Fig. 2 shows the configuration of the test apparatus. The physical and mechanical properties of the specimens are shown in Table 1. The dimension of specimens is width 140 mm x length 65 mm x height 120 mm. Heavy Duty tungsten - carbide - tipped drag tool was employed for cutting tests. The test parameters are as follows;

Depth of slot   Hs; 0 - 12 mm
Spacing of slot   2L; 0 - 37 mm
Depth of cutting t ; 0 - 24 mm

A mode of failure is roughly classified into four types as shown in Fig. 3. A crushed zone, resulting from initial tool penetration, is formed before the tool tip as shown by some reporters (Fowell et al., 1986).
Reduction of normal force and cutting force was observed in cases of II and III mode of failure. Fig. 4 shows the forces, in relation to depth of cut, measured on the specimens without slots. Fig. 5 shows the rate of reduction of normald force on motar specimens with dual-slots.
The rate of reduction is calculated on the basis of the values shown in Fig. 4. As for the cases on granite and andesite, the same patterns of reduction were obtained but the values of reduction themselves were rather small.
It is noticeable that enhancement of both normal and cutting force were observed in the area illustrated in Fig. 5 described as "90 - 150%". Fig. 6 shows the test results on the cases of small H/t using mortar specimens with single slot. This also indicates the same tendency. In most such cases, enlargement of crushed zone beneath the tip of drag tool was observed.

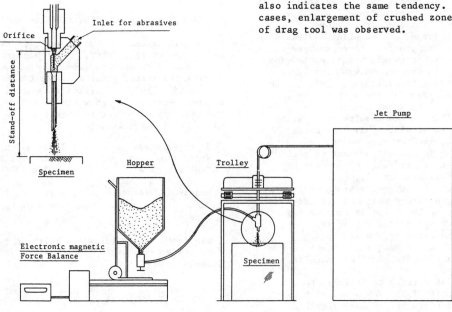

Fig. 1   Outline of Test Equipment

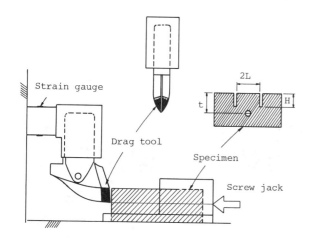

Fig. 2   Configuration of test aparatus

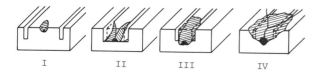

Fig. 3   Failure mode

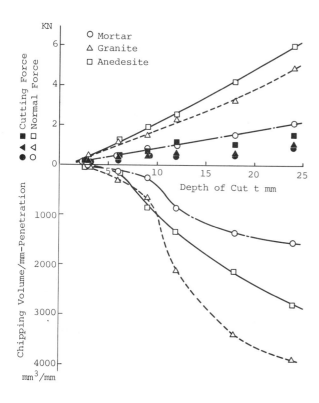

Fig. 4   The forces and chipping volume

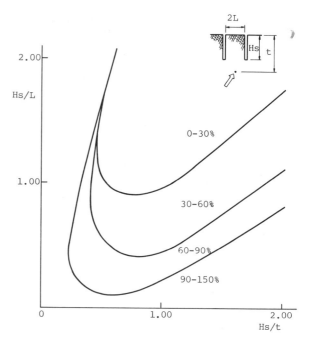

Fig. 5   Reduction Rate of Normal Force Derived
from Dual Slots (Specimen; Mortar)

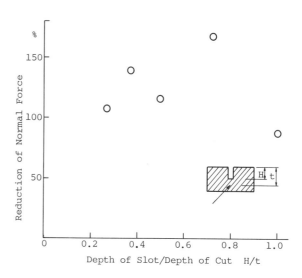

Fig. 6   Reduction of Normal Force in Relation
to H/t (Mortar, H = 6mm)

4. CONCEPTION OF CUTTING METHOD

Based on the test results and simple model as shown
in Fig. 7, a proper rate of energies has to be
supplied for waterjet and mechanical system is
discussed. Assume that the effective mechanical
energy "We,m" required for rock chipping by single
drag tool is the following formula.

$$We,m = 0.00981 \frac{1}{\eta} \cdot N \cdot Tc = 0.00981 \cdot \frac{1}{\eta} \cdot N \cdot r \ Pc \quad \ldots \quad (2)$$

where We,m: Effective mechanical energy for single
              drag tool (kW)
       η : Efficiency of machine (normally η = 0.9)
       N   : Number of rotations (S^{-1})
       Tc   : Torque (Nm)
       r   : Radius of cutter head
       Pc   : Normal force (N)

Assume that the traverse rate described in the formula (1) is due to rotation of cutter head,

$$T = 2\pi r \cdot N \quad \dots\dots\dots\dots \quad (3)$$

Let the formula (1) rearrange by substituting the formula (3) into formula (1).

$$We,j = \frac{2\pi rNHS}{K_1 + K_2\left(\dfrac{Qa}{Qw}\right)^n} \quad \dots\dots\dots \quad (4)$$

Based on the formula (2) and the formula (3), the both relationships between traverse rate and depth of slot using by abrasive waterjet, traverse rate and depth of cut using by drag tool are shown in Fig. 8 respectively. When H/t = 0.5 within proper limitation on H/L which causes the reduction of normal force is considered to give a beneficial reduction of normal force, the fitting points of energy are shown as the intersections of We,j and We,m line.

These points indicate that vast energy should be input into waterjet system as compared with energy supplied for drag tool-system and that cutting speed should be adjusted lower than the one that is employed in normal roadheader.

Fig. 9 shows the conceptional design in which the abrasive jet units are arranged, inside the cutter head and cutter shaft. The ideal cutter head of abrasive jet assisted roadheader in which many jets assist for each drag tool seems to be difficult to produce because of the cost and the technical problems.

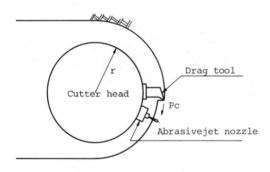

Fig. 7   Model for the calculation

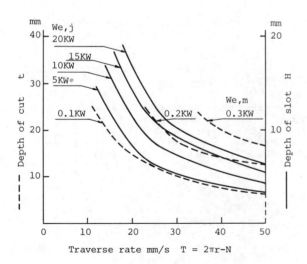

Fig. 8   Comparsion between waterjet energy and mechanical energy

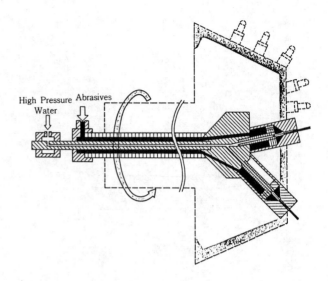

Fig. 9   Schematic Configuration of Cutter Shaft and Head

5. CONCLUSION

The following results have been obtained from the two types of laboratory tests for rock cutting using abrasive waterjet and drag tool.

(1) Emprical formula of rock cutting depth basing on the parameters such as jet pressure, waterflow rate, abrasive flow rate, traverse rate and stand-off distance
(2) Reuction of cutting forces on drag tool in relation to spacing and depth of slot and cutting depth.

Based on the above-mentioned items, the proper rate of energies supplied for waterjet system and mechanical system to reduce mechanical cutting force have been considered.

REFERENCES

Plumpton et al., 1982. The development of a Water Jet System to Improve the Performance of a Boom Type Roadheader. BHRA 6th Int. Sym. on Jet Cutting Technology, p.267-273.
C.E.Fairhurst et al., 1986. WATER-JET ASSISTED ROCK CUTTING - THE EFFECT OF PICK TRAVERSE SPEED. BHRA 8th Int. Sym. on Jet Cutting Technology, p.43-55.
M.Hashish 1982. Steel Cutting Abrasive Waterjets BHRA 6th Int. Sym. on Jet Cutting Technology, p.465-487.
S.Iihoshi et al., 1986. PRELIMINARY STUDY ON ABRASIVE WATERJET ASSIST ROADHEADER. BHRA 8th Int. Sym. on Jet Cutting Technology, p.71-77.
Fowell et al., 1986. WATER JET ASSISTED DRAG TOOL CUTTING; PARAMETERS FOR SUCCESS. BHRA 8th Int. Sym. on Jet Cutting Technology, p.21-32.

Evaluation of rock mass quality utilizing seismic tomography
Evaluation de la qualité de la masse rocheuse au moyen de la tomographie séismique
Qualitätsbewertung von Gesteinsmassen durch seismische Tomographie

T.INAZAKI, Public Works Research Institute, Tsukuba, Japan
Y.TAKAHASHI, Nippon Geophysical Prospecting Co. Ltd, Tokyo, Japan

ABSTRACT: This paper describes a tomographic method for estimating the spatial distribution of seismic velocity in rock mass, and presents the results of its application to cross-adit seismic data. With this method, a two- or three-dimensional velocity image can be reconstructed from the traveltime data of transmitted waves through the surveyed area. The performance of the reconstruction algorithm was examined using artificial model data. It was then applied to the seismic field data observed in adits at the dam site under survey, and compared with the inferred geotechnical data. These comparisons suggest that seismic tomography is useful in mapping the *in situ* mechanical properties of rock mass and in locating fractured zones and cracks.

RESUME: Ce document décrit une méthode tomographique d'estimation de la répartition de la vitesse séismique dans une masse rocheuse et présente les résultats de son application à des données séismiques simultanées de galeries à flanc de coteau. Avec cette méthode, une image à deux ou trois dimensions de la vitesse peut être reconstituée, à partir des données relatives au temps de déplacement des ondes transmises à travers la zone étudiée. Les performances de l'algorithme de reconstitution ont été examinées à l'aide des données d'un modèle artificiel. On l'a ensuite appliqué aux données séismiques réeles observées dans des galeries à flanc de coteux, sur le site du barrage faisant l'objet de l'étude et comparé aux données géotechniques déduites. Ces comparaisons suggèrent que la tomographie séismique est utile pour fixer en plan sur le terrain des propriétés mécaniques d'une masse rocheuse et pour situer les zones fracturées et les fissures.

ZUSAMMENFASSUNG: In diesem Papier wird ein tomographisches Verfahren zur Schätzung der räumlichen Verteilung seismischer Wellengeschwindigkeit in einer Felsmasse beschrieben und die Ergebnisse seiner Anwendung auf seismische Daten von Querstollen dargestellt. Mit diesem Verfahren kann ein zwei- oder dreidimensionales Geschwindigkeitsbild aufgrund der Wegzeitdaten der übertragenen Wellen durch das untersuchte Gebiet rekonstruiert werden. Die Leistung des Rekonstruktionsalgorithmus wurde unter Verwendung von Daten künstlicher Modelle untersucht. Es wurde dann auf seismische Felddaten, die während der Untersuchung in Tunneln an der Dammseite ermittelt worden waren, angewendet und mit den abgeleiteten geotechnischen Daten verglichen. Diese Vergleiche zeigen, daß seismische Tomographie für die Erfassung der mechanischen Eigenschaften von Felsmassen und der Anordnung von Bruchzonen und Rissen vor Ort nützlich ist.

1 INTRODUCTION

The mapping of weak zones, such as crack, fracture and fault, is one of the most important problem in many geotechnical projects. Although, it may be possible to detect such weak zones by drilling or tunneling, it is difficult to estimate the spatial distribution of weak zones through such means alone, because of its one-dimensional character.

Conventional geophysical techniques, which have been widely applied to geotechnical investigation, are also inadequate to determine the pattern of weak zones because they give information on the limited region close to the surface.

New methods to detect remotely weak zones covering effective depth and of sufficient resolution have been expected for long times. Nowadays, several methods have been developed as remote sensing techniques, and among them, tomographic method has been successfully applied to many geophysical applications: for example, Bois et al. (1972), Cosma (1983) and Gustavsson et al. (1986) used cross-hole tomography to reconstruct velocity structure between boreholes. Some field experiments of fracture mapping were presented by Mason (1981) and Ramirez (1986). Laine et al. (1980) applied it to geotechnical determination of grouted zone.

Tomography is useful for fracture-mapping and also for evaluation of rock mass quality, because there is a close relation between seismic parameters and mechanical properties.

In this paper, we describe the concept of seismic tomography used in our experiments, and present the result of model simulation and its application to corss-adit seismic data.

2. RECONSTRUCTION TECHNIQUE OF SEISMIC TOMOGRAPHY

Seismic tomography is a method for reconstructing unknown velocity distribution of the region of interest from traveltime data.

The techniques of seismic tomography are similar to those used in medical X-ray CT (Computed Tomography), but differ in restricted pattern of measuring geometry and ray bending. Namely, in medical X-ray CT, the object can be scanned from all direction by rotating the transmitter and detectors. On the contrary, the geometry of the sources and receivers in seismic tomography is fixed in space, restricted by the position of boreholes or adits, and provides the limited projection data. This makes the reconstruction techniques of seismic tomography sensitive to data noise. The second and much more important difference is in propagation property of wave. While the assumption of ray in simple straight line is adequate for X-ray CT, it is not appropriate for seismic tomography because spatial variations in the velocity cause ray to curve.

Consequently, the algorithm of reconstruction for X-ray CT may not be suitable for seismic tomography, and new algorithm taking account of ray bending must be employed. The resulting problem or algorithm of reconstruction may be non linear in case of ray-bending condition.

Iterative techniques are adaptable to the cases where the seismic rays are bent by refraction, because they can suitably deal with the above two problems.

Therefore, we adopted the algorithm of reconstruction which involves an iterative process of numerical ray tracing and linear inversion.

The spatially varing function, unknown slowness (reciprocal of velocity) n, along a raypath r, is related to the observed travel time t, of a ray by the line integral:

$$t = \int_r n \, ds \quad \dots\dots\dots\dots\dots\dots \quad (1)$$

To solve Eq. 1 by digital computer, it must be converted to the linear equation in discrete form:

$$t = R \, n \quad \dots\dots\dots\dots\dots\dots\dots \quad (2)$$

where, R is the two-dimensional matrix which contains the path-length data. Moreover, the region of interest is divided into a grid of square cells in each of which the slowness is assumed to be constant (Figure 1). Then, the observed traveltimes t_j of a ray j can be represented:

$$t_j = \sum_i \delta_{ji} \, n_i \quad \dots\dots\dots\dots\dots \quad (3)$$

where,

δ_{ji} = length of ray j intercepting cell i
n_i = slowness for cell i

Finally, we get the matrix R:

$$R = \sum_j \sum_i \delta_{ji} \quad \dots\dots\dots\dots\dots \quad (4)$$

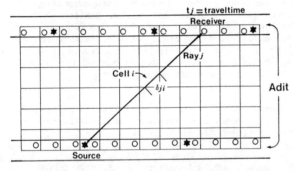

Figure 1. Seismic tomography model between adits showing a cell partition and a typical raypath.

Owing to the restriction in the obtained data on traveltimes, raypath must have been assumed in order to solve Eq.2 or Eq.3.

We first start the reconstruction from the assumption of a straight-line raypath, then, we obtain the slowness by using Back Projection Technique (BPT) which is most fundamental in conventional X-ray CT (Herman, 1980).

The slowness of each cell is given by:

$$n_i = 1/\left\{ (1/\sum_j \delta_{ji}) \cdot \sum_j (\delta_{ji} \cdot \sum_i \delta_{ji}/t_j) \right\} \dots (5)$$

where,

t_j = observed traveltimes along a raypath j

The following procedure of the reconstruction is ray tracing to calculate the traveltimes and raypaths corresponding to a given velocity distribution condition. It is treated as a two-dimensional case, i.e. seismic rays are supposed to transmit on a plane containing sources and receivers. In general, this algorithm has been based on the wave or ray equation as reduced to the differential equation (Bois et al., 1971, Lytle and Dines, 1980). The velocity is given discretely to each cell. To solve the differential equation, an interpolation process must be required to ensure the continuity condition of velocity distribution and its partial derivaties at arbitrary points. Accordingly, this algorithm becomes time-consuming computationally. Hence, we developed and employed a non-analytical simulation algorithm as ray tracing. With the algorithm we employed here, the raypath is traced as a line connecting nodes or central points of cells which distribute discretely, and the shortest time path is determined with simulative comparisons of each raypath ruled by Fermat's Principle. We call the algorithm of ray tracing "Explosion Method" after the similarity of its simulation procedure to the movement of wave front by explosion. The process of the algorithm is similar in concept to the shortest path problem in Network Theory.

The velocity distribution was obtained using the iterative reconstruction technique, which is identical in concept to those described by Dines and Lytle (1979) and Bois et al. (1971). The solution of Eq. 2 requires some precautions, since it contains noise in measured data as well as error due to the approximation of velocity distribution in the iteration process.

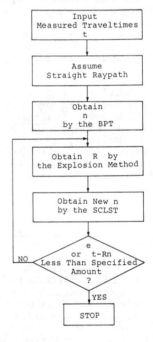

Figure 2. Flowchart showing the iterative reconstruction procedure of ray tracing and inversion

This type of error can be accounted for by rearranging Eq. 2:

$$t = Rn + e \quad \dots\dots\dots\dots\dots \quad (6)$$

where, e is a data error vector.

We adopted smoothness-constrained least squares technique (SCLST) to solve Eq. 6.

From a estimated n, new raypaths and traveltimes are computed and compared to the observed data, then revised n is computed basing on the difference between computed and observed traveltimes. This process is repeated until successful convergence is obtainesd (Figure 2).

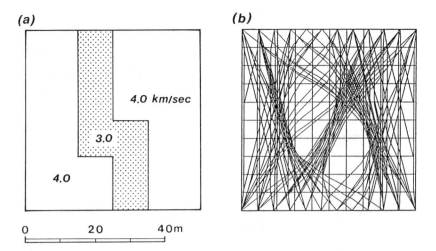

Figure 3. Example of synthetic models used in simulation. (a) The velocity distribution pattern representing crank-type low velocity zone. The region is 50 by 50 m in size, and divided into 10 x 10 discrete cells. (b) Synthesized raypaths using the Explosion Method as the ray tracing.

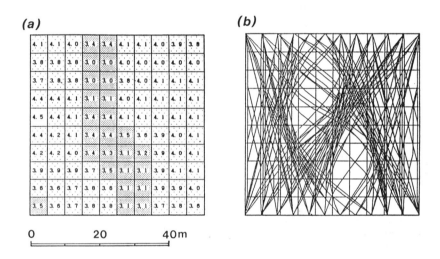

Figure 4. Results after 9 iterations corresponding to synthetic model shown in Figure 3. (a) Velocity solution. The number in each cell shows the computed velocity in km/sec. (b) Computed raypaths.

3. RECONSTRUCTION OF SYNTHETIC MODEL

Prior to field experiment, we examined the performance of the reconstruction algorithm using simulational model data. The major difficulty of the reconstruction is the non-uniqueness caused by limited projection and data noise. Therefore, we focused the effect of velocity structure (e.g. shape, contrast and orientation), scanning geometry, and data noise on the resolution of the tomography.

Figure 3 (a) shows one of the synthetic models where low velocity zone is embedded in higher velocity medium. 12 sources are located with equal spacing at the bottom and 12 receivers at the top side.

This model has the same geometry as that discussed by Dines and Lytle (1979), while they did not refer to the velocity and scale nuit. For synthesized 144 raypaths as shown in Figure 3 (b), traveltimes were calculated by using of the Explosion Method.

Figure 4 shows the results after 9 iterations. Both shape and velocity of low velocity zone are well defined (Figure 4 a), and Figure 4 (b) shows that the reconstructed raypaths are satisfactory in comparison with those as shown in Figure 3 (b). The error in the reconstruction of traveltimes is reduced to 0.08 msec in RMS after 9 iterations; the traveltimes of synthesized rays range from 12 to 18 msec.

The following conclusions can be drawn from these model simulations: First, the region of interest should be probed from all direction, if possible, to achieve high-grade resolution. Secondly, the resolution deteriorates abruptly with increasing data noise; successful correction of observed data and the smoothness constraint at inversion may diminish the influence of noise.

4. FIELD EXPERIMENT

Field experiment of seismic tomography method was carried out at a dam site under survey in the southwestern Japan. Granitic rocks of late Cretaceous age, which are overlain by weathered layer, crop out throughout the site. The thickness of this overburden increases with the height from river bed. The rock mass contains an extensive network of fractures which has orientation of NE-SW dominantly and conjugate orientation of NW-SE. Predominant dips of the fractures range from 70° N to vertical.

Sources and receivers were located along a line in the adits as shown in Figure 5. The receivers were spaced at 2 m intervals in each adit, and the sources at approx. 25 m intervals.

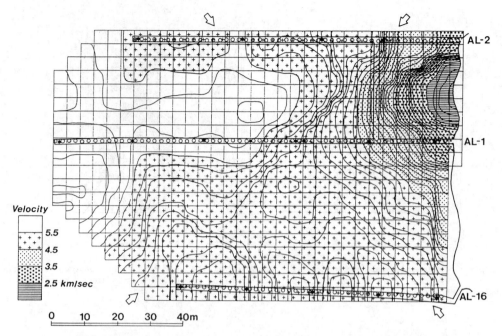

Figure 5. Reconstructed velocity image of the region between adits near to river bed. The adits can be treated as approximately on the same elevation. Sources are indicated by solid stars, and receivers by open circles. the velocity contour intervals is 100m/sec.

The number of traveltimes obtained for inversion was 549 in total.

Static correction on the observed traveltimes were performed to reduce the effect of the time delay due to relaxation adjacent to sources and receivers, and shift of time breaks.

The reconstructed region is about 80 by 120 m in size, and divided into 586 cells of dimension of 4 by 4m.

The result of inversion is shown in Figure 5. The corresponding residual for the traveltimes, which is 0.05 msec in RMS, coinsides generally with the error in the observed traveltimes. The principal feature on the image is a spot of low velocity near the surface and a increment of velocity with the depth. Note that the velocity image change apparently at the boundaries marked by open arrows.

The velocity image is compared with the results of geotechnical works for mapping fractured zones and for evaluating rock mass quality. Consequently, orientations and locations of velocity boundaries accord roughly with those of fractured zones observed in adits. Also, velocity structure is concordant with inferred classification map of rock mass quality: e.g. the region above 5.5 km/sec in velocity coincides roughly with the region of rock classified as B rank.

5. CONCLUSION

A tomographic method was examined by synthetic models where rock mass was assumed to include low velocity zone. The result shows that we can estimate the velocity distribution by using seismic tomography method.

Field experiment showed that the reconstructed velocity image is concordant with the data inferred from geotechnical works. These tests proves the usefulness of the present method for mapping the *in situ* mechanical properties of rocks and for locating fractures in them.

REFERENCES

Bois, P., La Porte, M., Lavergne, M. and Thomas, G., 1971, Essai de determination automatique des vitesses sismiques par measures entre puits, Geophys. Prosp., vol. 19, pp. 42-83.

Bois, P., La Porte, M., Lavergne, M., and Thomas, G., 1972, Well-to-well seismic measurements, Geophys. vol. 37, pp. 471-480.

Cosma, C., 1983, Determination of rockmass quality by the crosshole seismic method, Bull. of IAEG, no 26-27, pp. 219-225.

Dines, K. A. and Lytle, R.J., 1979, Computerized geophysical tomography, Proc. IEEE, vol. 67, pp. 1065-1073.

Gustavsson, M., Ivansson, S., Moren, P. and Pihl, J., 1986, Seismic borehole tomography - Measurement system and field studies, Proc. IEEE, vol. 74, pp. 339-346.

Herman, G., 1980, lmage Reconstruction from Projections, New York; Academic Press, 316p.

Laine, E. F., Lytle, R. J. and Okada, J. T., 1980, Cross-borehole observation of soil grouting, J. Geotech. Eng. ASCE, vol. 106, pp. 871-875.

Lytle, R. J. and Dines, K A., 1980, Iterative ray tracing between boreholes for underground image reconstruction, IEEE Trans. Geosci. Remote Sensing, GE-18, pp. 234-240.

Mason, I. M., 1981, Algebraic reconstruction of a two-dimensional velocity inhomogeneity in the High Hazles seam of Thoresby colliery, Geophys., vol. 46, pp. 298-308.

Ramirez, A. L., 1986, Recent experiments using geophysical tomography in fractured granite, Proc. IEEE, vol. 74, pp. 347-352.

Mechanisms of water jet assisted drag tool cutting of rock

Mécanismes d'assistance avec jets d'eau et pics dans des roches
Der Mechanismus der Gewinnung von Gestein mit Spatenmeissel und mitwirkendem Wasserstrahl

C.K.IP, Ian Farmer Associates, UK, formerly University of Newcastle Upon Tyne, UK
ROBERT J. FOWELL, Reader, Department of Mining Engineering, University of Newcastle upon Tyne, UK

ABSTRACT: High pressure water jets to assist the cutting action of drag tool equipped boom tunnelling machines are now available and recording improved cutting performance. The mechanisms of water jet assistance in reducing the forces experienced by the drag tools are explained in terms of phenomenological models based on the penetration characteristics of the jet. The importance of traverse speed, tool bluntness and spacing between tools on a cutting head are emphasised.

RÉSUMÉ: Les jet d'eau à haute pression utilisés pour assister l'action de perçage des machines équipées de pics sont maintenant disponibles et enregistrent de meilleures performances de perçage. Les mécanismes d'assistance avec jets d'eau dans le but de réduire les forces éprouvees par les pics sont expliquées en termes de models phénoménologiques basés sur les caractéristiques de pénétration du jet. L'importance de la vitesse de l'outil, de son (manque) de tranchant, et l'espacement entre les pics sur la tête perceuse y sont souliqués.

ZUSAMMENFASSUNG: Hochdruck-wasserstrahl unterstützung der schneidwerkzeug von Schneidkopfmachine mit spatenmeissel sind jetzt verfügbar, und registrieren abbauverbesserung. Der mechanismus der wasserstrahl unterstützung, und der verkleinerung die kräfte der spatenmeissel sind erklart, durch phännomenologisch modelle die als basis der bohrleistung der wasserstrahl haben. Der wichtigkeit von die werte, marschgeschwindigkeit, werkzeugabastumpfung und werkzeugzwischenraum auf einem schneidkopf sind unterstreichen.

1 INTRODUCTION

High pressure water jet assistance is now available for boom type tunnelling machines, which increases their range of applicability in terms of strength and abrasivity of the rock that they can economically excavate. Doubling of excavation rates in some cases has been reported (Barham et al. 1986) and dramatic increases in tool life have been established as a major advantage in employing water jet assistance (Morris et al. 1986).

A number of research projects have been conducted, in the USA (Ropchan 1980), France (Fairhurst 1985), and Germany (Baumann 1982), in hybrid drag tool studies following work undertaken by Hood (1976) in South Africa, and the promising field results obtained by Plumpton (1982) in the United Kingdom.

At the last Rock Mechanics Congress, held in Melbourne, the results from pilot studies on water jet assisted drag tool rock excavation studies were reported (Fowell et al. 1983). Since that time a major programme of research has been conducted within the University of Newcastle upon Tyne simulating more closely the operational variables to be found in practical situations.

The most important parameter for hybrid cutting has been found to be tool traverse speed. The encouraging results obtained from laboratory tests at low speeds, well below those found for drag tool machines, were not always obtained when higher traverse speeds were adopted for laboratory investigations. A phenomenological model has been proposed which explains the mechanisms of water jet assisted drag tool cutting of rocks and this paper demonstrates its application to practical situations where worn tools are employed rather than limited to new tool conditions (reported previously by Fowell et al. 1986). The influence of relieved cutting conditions where cutting arrays have been simulated are also described.

2 MODEL FOR HIGH PRESSURE WATER JET ASSISTED DRAG TOOL CUTTING

2.1 Unassisted Cutting

Normal drag tool cutting without water jet assistance may be considered to be composed of four components: large primary tensile chipping ahead of the tool; profiling of the irregular crater left by large chipping events to allow the tool to progress through the rock; crushing events associated with the wearflat on the bottom of the tool and initial penetration of the tool tip into the rock; secondary chipping prior to large primary chip formation provides the fourth component. Tensile chipping is preferred to crushing where possible, in order to promote more efficient excavation.

2.2 Water jet assisted cutting

Based on detailed observation of a representative chipping cycle by a drag tool and the research programme undertaken with water jet assistance, a model of the jet's action has been developed (Ip 1986) which explains the results obtained. The existence of an optimum jet pressure is predicted which provides maximum reduction in tool forces for the cutting conditions employed. The jet penetration of the rock surface was identified as the most important characteristic as it takes into account rock properties, the influence of traverse speed, jet nozzle diameter and position.

Three conditions have been identified dependent on the penetration depth of the jet acting alone on the rock surface expressed as a percentage of the mechanical depth of cut taken by the tool:

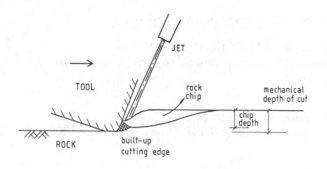

Figure 1 Hybrid Cutting Optimum Conditions

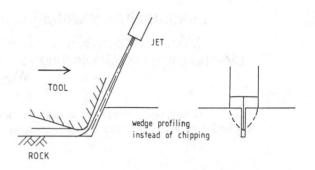

Figure 2 Hybrid Cutting Model Deep Penetration

Figure 3 Linear Cutting Rig

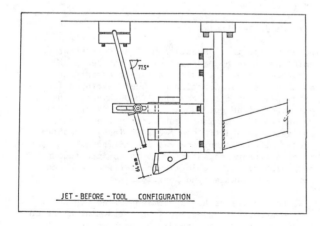

Figure 4 Jet-before-tool Configuration.

Figure 5 Sharp and Blunt Tool

668

Table 1. A Summary of Experimental Designs

Variable		Slot Depth	Tool Bluntness	Cutting Mode
Jet Pressure	(MPa)	N/A	0,18,44,70	0,18,44,70
Depth of Cut	(mm)	10	10	10
Traverse Speed	(m/s)	1.10	1.10	0.27 (PS) 1.10 (GS)
Nozzle Diameter	(mm)	0.9	0.9	0.9 (PS) 0.6,0.9,1.2 (GS)
Tool Bluntness		Sharp	Sharp,Blunt	Sharp
Cutting Mode		Unrelieved	Unrelieved	Unrelieved Relieved
Slot Depth	(mm)	0,3.7,5.9,7.4,10.9	N/A	N/A
Rock Type		GS	GS ML	GS PS

GS = Grindleford Sandstone; ML = Middleton Limestone; PS = Pennant Sandstone

Table 2. Mechanical and Physical Properties of Experimental Rocks.

Rock Properties	Dumfries Sandstone	Grindleford Sandstone	Pennant Sandstone	Middleton Limestone
1. Uniaxial Compressive Strength (MPa)	22.70	58.90	172.80	113.20
2. Tensile Strength (MPa)	1.99	8.65	23.39	12.58
3. Dry Density (kg/m^3)	1947	2362	2652	2562
4. Porosity (%)	23.47	8.75	1.58	3.79
5. Standard NCB Cone Indenter Number	0.90	2.83	4.30	3.45
6. Rebound Hardness	11.70	36.20	53.10	44.50
7. Plasticity (%)	60	19	9	8
8. Permeability (m/s)	2.76×10^{-6}	2.38×10^{-8}	$**** \times 10^{-10}$	2.26×10^{-11}
9. Dynamic Young's Modulus (GPa)	7.75	16.53	38.93	77.34

*** Impermeable.

(1) Jet Penetration Insigificant (strong rocks or high traverse speeds)

In this case, the water cannot reach the tool tip due to its inability to penetrate the rock surface. The only functions of the jet are cooling the tool/rock interface, flushing out fine debris and assisting chip removal once formed. Force reductions are marginal.

(2) Optimum Jet Penetration

At optimum jet penetration the water jet is able to reduce energy wasting crushing, whilst still retaining the energy-efficient chipping events. The object is to keep the jet penetration sufficiently deep to relieve the tool tip but not too deep to remove the crushed zone necessary to induce crack initiation (Figure 1).

(3) Excessive Jet Penetration (weak rocks)

The jet can penetrate to the tool tip and flush away the 'cutting edge' of the crushed material before the tool tip. As a result, reduced chipping is produced. The tool then operates through a profiling action. The net gain is reduced from the optimum condition. The tool track is left clear of crushed compacted debris, a common feature of dry rock cutting and there may be a slot left in the bottom of the track where excessive penetration is possible (Figure 2).

Previous applications of the model to explain the results obtained from nozzle diameter, jet position, mechanical depth of cut, rock material properties and traverse speed have been reported elsewhere (Ip et al. 1986, Fowell et al. 1985, 1986). The proposed model is now used to explain the action of the jet when applied to a worn tool and relieved tool cutting.

3 EXPERIMENTAL FACILITIES

An overall view of the instrumented planing machine used during the programme is shown in Figure 3..

The cutting rig has been extensively modified to enable an increase of the traverse speed from 0.25 to 1.10 m/s and cutting at depths up to 30mm in moderately strong sandstones. The rig can cut a rock sample of dimensions 1m x 1m x 0.7m and a more detailed description of the rig can be found elsewhere (Fauvel 1981). The high pressure water supply was provided by a 75-kW double-acting intensifier type pump with a capacity of 45 litre/min at 70 MPa.

The jet used for this series of experiments was of a continuous, steady type using a brass nozzle with a contraction angle of 45°. This was positioned so that the jet impinged 1-2mm before the tool tip. The impingment angle was $77\frac{1}{2}°$ towards the tool and the distance between the jet and the tool tip was 64mm (Figure 4).

A steel triaxial dynamometer was used to monitor the force components experienced by the tool during rock cutting. Amplified signals from the dynamometer, as well as from a pressure transducer, were recorded on an instrument tape recorder. The signals were later filtered, digitised, logged and analysed using a microcomputer.

Only one geometry of heavy duty radial tool was used during this experimental programme. When a blunt tool was required, an artificially induced blunt tool was used. The tool tip of a used tool was ground to give a wearflat of $21mm^2$ at an angle of about -4°, which is common for wearflats developed during rock cutting (Kenny and Johnson 1976). The blunt tool forces are contributed mainly by the wearflat. After grinding, the tool cut for 30m in Grindleford Sandstone to round off any sharp edges. The blunt tool used represents a moderately blunt tool by mining standards. Examples of the sharp and the blunt tools used are shown in Figure 5.

4 EXPERIMENTAL PROGRAMME

Since every type of rock exhibits heterogeneity to some extent, several replications of each test were required in order to provide a statistically significant value. The minimum number of tests required was related to the coefficient of variance and four replications of each test were carried out though only mean values are presented in the paper.

Typically a depth of cut of 10mm and a traverse speed of 1.10 m/s were adopted, though a slower speed was also included in some programmes. These represent a realistic cutting speed and a desirable depth of cut for a boom tunnelling machine cutting hard rocks. The experimental designs for the investigation of the effects of slot depth, tool bluntness, and cutting mode are summarised in Table 1.

The essential physical and mechanical properties of experimental rocks are tabulated in Table 2. Of these rocks, Middleton Limestone and Pennant Sandstone cannot be penetrated by the water jet, even at the highest pressure (70 MPa) and slow cutting speed (0.27 m/s). Grindleford Sandstone was chosen to represent the rock type with significant jet penetration due to its homogeneity and repeatable cutting characteristics. The jet penetration characteristics of Grindleford Sandstone are presented in Figure 6.

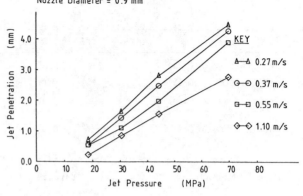

GRINDLEFORD SANDSTONE
Nozzle Diameter = 0.9 mm

Figure 6 Jet Penetration Characteristics of Grindleford Sandstone.

5 SIMULTANEOUS JETTING VERSUS PRESLOTTING

Slots of various depths were cut in a flat Grindleford Sandstone surface, using a water jet of 70 MPa at different traverse speeds. A 0.9mm diameter nozzle was used and the depths of jet penetration were measured. The sharp tool was then positioned in line with the slot and the unrelieved independent cuts were then taken. The traverse speed was 1.10 m/s and mechanical depth of cut was set to 10mm. The results are plotted in Figure 7, together with the results when the tool and jet acted together.

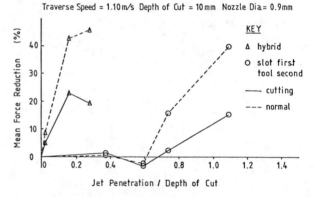

GRINDLEFORD SANDSTONE
Traverse Speed = 1.10 m/s Depth of Cut = 10 mm Nozzle Dia.= 0.9mm

Figure 7 Mean Force Reduction versus Slot Depth for jet acting together or separately with the tool : Grindleford Sandstone.

It is clear from the figure that the jet must be superimposed onto the tool, with the jet as near to the tip as possible in order to obtain maximum benefit. In hybrid cutting the water jet makes use of the post-chipping curvilinear surface to relieve the tool tip. The nearer the jet impingement to the tool tip (lead-on distance) the less jet penetration required for the same relieving effect. If the lead-on distance is larger than the chip length, hybrid cutting will lose most of the advantages of dynamic penetration and the cutting becomes one of cutting rock with a preformed slot in the tool path. Findings showing the benefits of reduced lead-on distance were reported elsewhere (Ozdemir and Evans, 1983; Tecen, 1982; Plumpton and Tomlin, 1982; Dubugnon, 1981; Hood, 1976), although no explanation was given. Figure 8 illustrates the difference schematically.

670

a) excessive lead-on distance

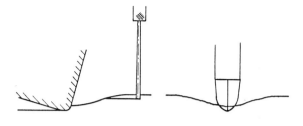

b) lead-on distance = 0-2 mm

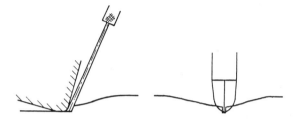

Figure 8 Schematic illustration of the effect of lead-on distance.

From the figure it can be seen that the beneficial effect of the slot in the tool-after-slot cuts is possible only when the slot is over 7mm, i.e. 70% mechanical depth of cut. It clearly shows that all the tool actions are concentrated at the bottom few millimetres of cut where the centre of the crack system is situated. The value of 70% corresponds to the chip depth which is also around 70% of the 10mm mechanical depth of cut (Table 3). Hence any slot depth less than the chip depth has no relieving

Table 3. Chip depth in unassisted cutting.

Mech. Depth of Cut (mm)	Chip Depth (mm)			Mean (mm)
	DS	GS	PS	
5	–	–	3.2	3.2 (64%)
10	7.2	7.1	7.3	7.2 (72%)
15	11.1	11.8	11.9	11.6 (77%)

DS = Dumfries Sandstone; GS = Grindleford Sandstone; PS = Pennant Sandstone.

effect on the crushing events at the tool tip. Obviously, the above statement is only applicable to the tool-after-slot configuration, and the slot depth requirement for the dynamic hybrid cutting is much reduced. However, the results do show the importance of the bottom few millimetres of cut and the slot must be deep enough to provide any benefic- ial relief effect to reduce the tool force.

6 INFLUENCE OF WEARFLAT ON CUTTING PERFORMANCE

The effect of tool bluntness on the tool forces is to increase the proportion of cutting and normal forces required to crush the rock underneath the wearflat. The results of cutting Grindleford Sandstone and Middleton Limestone with sharp and blunt tools at a fast traverse speed are plotted in Figure 9.

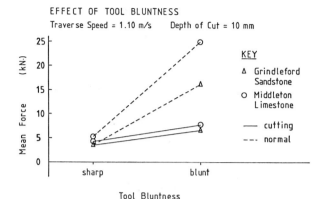

Figure 9 Mean Force versus Tool Bluntness for Grindle- ford Sandstone and Middleton Limestone; traverse speed = 1.10 m/s.

The introduction of this moderate wearflat caused a drastic increase in tool forces. Since the wearflat angle is small, most of the crushing force is provided by the normal force. The increase in the normal force component is slightly less than threefold at a cutting speed of 0.27 m/s and is more than fourfold when the cutting speed is 1.10 m/s. The increase in cutting force component is about 75%-100% and is similar for both speeds.

6.2 Jet Penetration Significant (Grindleford Sandstone)

As water jets are extremely effective in reducing crushing force components, when jet penetration is significant, hybrid cutting using a blunt tool produces a higher force reduction than when using a sharp tool (Figure 10). At a jet pressure of 70 MPa, the mean

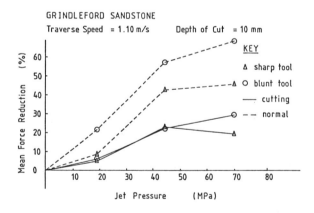

Figure 10 Mean Force reduction versus Jet Pressure for sharp and blunt tool = Grindleford Sandstone; traverse speed = 1.10 m/s.

normal force reductions were 70% for a blunt tool and 45% for a sharp tool for the jet-before-tool configu- ration cutting at 1.10 m/s. The higher force reduc- tion, together with the higher tool force for a blunt tool meant that the absolute force reduction was even more substantial (Figures 9 and 11). The results have significant practical implication as most tools used on a boom tunnelling machine are essentially blunt and the tool speeds employed are high. The machine is usually arcing force limited. When the water jet is applied and the rock can be penetrated by the jet, a

671

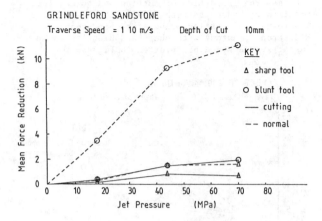

GRINDLEFORD SANDSTONE

Traverse Speed = 1 10 m/s   Depth of Cut   10mm

KEY

Δ sharp tool

○ blunt tool

— cutting

-- normal

Figure 11 Absolute Mean Force reduction versus Jet
           Pressure for sharp and blunt tool = Grindle-
           ford Sandstone; traverse speed = 1.10 m/s.

much improved machine performance can be expected.

Not only is blunt tool hybrid cutting producing a
force reduction but the optimum pressure is higher
when compared with that obtained for sharp tool hybrid
cutting. This highlights the important concept that
the force acting on a tool should be considered to be
composed of crushing and chipping components. The
water jet is useful in reducing the rock crushing
under the wearflat but, if excessive jet penetration
occurs, this tends to reduce the efficient chipping
events as well. The optimum jet penetration, and thus
optimum jet pressure, is a compromise between the
relative magnitude between chipping and crushing
force components. The greater the crushing events,
the higher the optimum jet penetration. Optimum jet
penetration hence depends on the degree of tool
bluntness, and may approach the mechanical depth of
cut in the case of very blunt tools.

A generalised figure showing the effect of bluntness
is given in Figure 12.

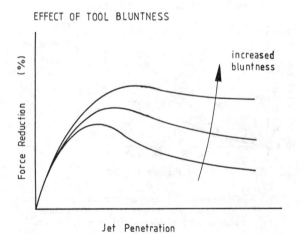

EFFECT OF TOOL BLUNTNESS

increased
bluntness

Figure 12 Generalised effect of tool bluntness on
           Force Reduction = Rocks with Jet
           Penetration.

6.3 Jet Penetration Insignificant (Middleton Lime-
     stone and Pennant Sandstone)

In general, the force reduction is small compared
with that of Grindleford Sandstone, where jet penetration was significant. For the jet-before-tool
configuration, the normal force reduction is around

25% for sharp tool and less than 15% for the blunt
tool (Figure 13). An interesting finding was that
there was a lower force reduction with blunt tool

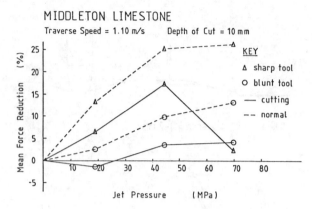

MIDDLETON LIMESTONE

Traverse Speed = 1.10 m/s   Depth of Cut = 10 mm

KEY

Δ sharp tool

○ blunt tool

— cutting

-- normal

Figure 13 Mean Force Reduction versus Jet Pressure
           for sharp and blunt tool = Middleton Lime
           stone; traverse speed = 1.10 m/s.

cutting than that for sharp tool cutting. This is
contradictory to the findings for the rocks with
significant jet penetration. The inferior performance of water jets with blunt tool cutting can be
explained by the inability of the jet to penetrate
and relieve the tool tip. The main actions of the
jet were debris clearance and aiding the chip
removal once formed, which contributes much less in
blunt tool cutting than in sharp tool cutting.

7 EFFECT OF CUTTING MODE

The cutting pattern adopted for this experiment
simulated a roadheader cutting head with two tool
spirals and one tool per line. The advance rate was
assumed to be 20mm per revolution and the spacing
between adjacent tool lines was 20mm (Figure 14).

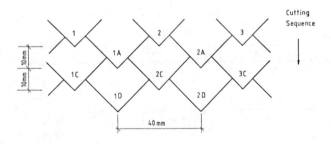

Figure 14 The cutting pattern for relieved cutting
           mode.

Although the depth of cut is changing on a production
rotary machine, this study nevertheless provides a
comparison between the results of unrelieved cutting
and the relieved cutting which is typical of the many
boom tunnelling machine cutting head designs used in
the UK.

7.1 Jet Penetration Significant (Grindleford
     Sandstone)

In this study, three nozzles of different diameter
were used and the traverse speed was 1.10 m/s, which
is comparable with the tool speed for a tunnelling
machine cutting hard rock. The tool normal force
reductions are plotted in Figure 15, together with
results obtained previously in unrelieved cutting,
using the same nozzle.

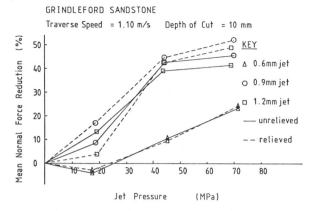

GRINDLEFORD SANDSTONE

Traverse Speed = 1.10 m/s   Depth of Cut = 10 mm

KEY

△ 0.6mm jet
○ 0.9mm jet
□ 1.2mm jet
— unrelieved
-- relieved

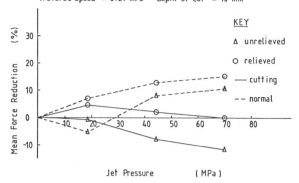

PENNENT SANDSTONE

Traverse Speed = 0.27 m/s   Depth of Cut = 10 mm

KEY

△ unrelieved
○ relieved
— cutting
-- normal

Figure 15   Mean Normal Force reductions versus jet
pressure for various nozzle diameters
in relieved and unrelieved cutting mode;
Grindleford Sandstone.

Figure 17   Mean Force reduction versus Jet Pressure
for unrelieved and relieved cutting mode;
Pennant Sandstone.

The results show that the force reduction trends
are very similar between the unrelieved and relieved
cuts. The 0.6mm nozzle still maintains its inferior
performance compared with the other nozzles, showing
the importance of jet penetration for force reduction.
For the 0.9 and 1.2mm jets, the force reductions for
the relieved cuts seem to be higher than those for
the unrelieved cuts, especially as the jet pressure
approaches 70 MPa. The reason is that there were
longer chips produced for the relieved cuts (Figure
16). Hence the tool forces are composed of higher

Figure 16   Major rock chips for relieved (left) and
unrelieved (right) cutting; depth of cut
= 10mm.

crushing force components and, therefore, more benefit
was derived from the tip relief provided by the water
jet. The optimum jet penetration, which is also
dependent on the proportion of crushing, is expected
to be higher. This seems to be the case as estimated
from the trends shown in the force reduction curves.

7.2 Jet Penetration Insignificant (Pennant Sandstone)

Although both force reductions were insignificant, as
expected in this category of rocks, there was greater
force reduction for relieved cutting compared with
unrelieved cutting (Figure 17). The results are
attributed, again, to the longer chips for relieved
cuts. Longer chips resulted in less chipping events
and more crushing events and hence the more work the
water jet can do in the debris clearance to minimise
grinding underneath the wearflat.

8   CONCLUSIONS AND FUTURE WORK

Through dedicated experimental designs the variables
likely to influence the water jet assistance have
been studied. In this paper the effects of water
jet assistance on tool bluntness, slot depth, and
the cutting mode are reported.

The most important finding was to recognise the
importance of jet penetration on tool force reduction.
The difference between the rocks with significant jet
penetration, which was not very clear when a pristine
tool was used, becomes clear when a blunt tool is
used. As all tunnelling machines cut rocks with
'blunt' tools, the practical implication is important.
It shows that unless the water jet is powerful enough
to penetrate the rock, only marginal force reductions
are to be expected.

When the rocks are too strong to be penetrated,
the benefits derived from the water jet are the
cooling of the tool tip and hence increased
resistance to wear. The reduced wear rate leads
to the preservation of an efficient cutting form for
the tool tip and longer tool life with maintained
cutting performance.

Most of the cuts taken during the main research
programme were unrelieved and carried out on a
trimmed rock surface. The relieved cutting results
demonstrated that the trends established for
unrelieved cutting are applicable to the practical
rock cutting situation, though there may be further
minor improvements to be gained with relieved
cutting.

The proposed model of water jet assisted drag tool
rock cutting was found to be successful in predicting
the trends for the effect of tool bluntness and
cutting mode. The optimum jet penetrations and
corresponding maximum force reductions were found
to depend on the tool bluntness and cutting mode as
well as traverse speed, depth of cut and rock type.

To validate the conclusions drawn from these linear
tests, a full sized instrumented boom tunnelling
machine research rig (Fowell et al. 1984) has been
modified to take high pressure water jets on the
cutting head. Pressures up to 140 MPa are to be
used in a variety of rocks.

Though this paper has only concentrated on the
reduced force components possible with the addition
of water jets to assist the action of a drag tool,
the other advantages of hybrid cutting; namely,

reduced respirable dust concentrations, improved tool life, and reduced methane ignition hazards, make hybrid cutting an attractive method of rock excavation on economic grounds, and also providing a safer working environment.

ACKNOWLEDGEMENTS

The authors wish to express their gratitude to Dr. S.T. Johnson, who made invaluable contributions to the success of this work. The main experimental programme was supported jointly by British Coal, through a contract with the European Coal and Steel Community, and the Science and Engineering Research Council, UK. The assistance and encouragement provided by Dr. C. Morris, Head of Rock Mechanics Branch, British Coal, Headquarters Technical Department, Burton on Trent, Staffordshire, is gratefully acknowledged.

REFERENCES

Barham, D.K. and Tomlin, M.G. High Pressure Water Assisted Rock and Coal Cutting with Boom-Type Roadheaders and Shearers. Proc. 8th Int. Symp. on Jet Cutting Technology, Durham, UK, 57-70. Sponsored by BHRA Fluid Engineering, Cranfield, UK, Sept., 1986.

Baumann, L. and Koppers, M. 'State of Investigations on High Pressure Water Jet-Assisted Road Profile Cutting Technology'. Proc. 6th Int. Symp. on Jet Cutting Technology, Surrey, UK, 283-300, 1982.

Dubugnon, O. 'An Experimental Study of Water Jet Assisted Drag Bit Cutting of Rocks' Proc. 1st US Water Jet Symp., Golden, Colorado, II4.1-4.11, 1981.

Fairhurst, C.E., Johnson, S.T. and Deliac, E.P. Report on the Franco-British Research Programme on Water-Jet Assisted Pick Cutting. Phase II: Effect of Traverse Speed'. Paris School of Mines, 1985.

Fauvel, O.R. 'Implications of Laboratory Rock Cutting for the Design of a Tunnel Boring Machine Cutter Head'. Unpublished Ph.D. Thesis, Dept. Mining Engineering, Univ. of Newcastle upon Tyne, 1981.

Fowell, R.J., Ip, C.K. and Johnson, S.T. 'Water Jet Assisted Drag Tool Cutting: Parameters for Success'. Proc. 8th Int. Symp. on Jet Cutting Technology, Durham, UK, 21-32, 1986.

Fowell, R.J., Johnson, S.T. and Ip, C.K. 'The Effect of High Pressure Water Jets on the Performance of Boom Type Tunnelling Machines'. Final Report to the National Coal Board. Dept. Mining Engineering, Univ. of Newcastle upon Tyne, UK, 81, 1985.

Fowell, R.J., Johnson, S.T. and Speight, H.E. 'Boom Tunnelling Machine Studies for Improved Excavation Performance'. ISRM Symp. on Design and Performance of Underground Excavations, Cambridge, UK. Brown, E.T. and Hudson, J.A. (Eds.) 305-312, 1984.

Fowell, R.J. and Tecen, O. 'Studies in Water Jet Assisted Drag Tool Rock Excavation'. Proc. 5th Int. Congress on Rock Mech., Melbourne, Australia, E207-213, 1983.

Hood, M. 'Cutting Strong Rock with a Drag Bit Assisted by High-Pressure Water-Jets'. J.South African Inst. Min. & Metall., Vol.77, No.4, Nov., 79-90, 1976.

Ip, C.K. 'Laboratory Water Jet Assisted Drag Tool Rock Cutting Studies at High Traverse Speed'. Unpublished Ph.D. Thesis, Dept. Mining Engineering, Univ. of Newcastle upon Tyne, 1986.

Ip, C.K., Johnson, S.T. and Fowell, R.J. 'Water Jet Assisted Rock Cutting and its Application to Tunnelling Machine Performance'. Proc. 27th US Symp. on Rock Mechs., Tuscaloosa, Alabama, 880-890, 1986.

Kenny, P. and Johnson, S.N. 'An Investigation of the Abrasive Wear of Mineral Cutting Tools'. Wear, 36, 337-361, 1976.

Morris, C.J. and MacAndrew, K.M. 'A Laboratory Study of High Pressure Water Jet Assisted Cutting'. Proc. 8th Int. Symp. on Jet Cutting Technology, Durham, UK, sponsored by BHRA Fluid Engineering, Cranfield, 1-7, UK., 1986.

Ozdemir, L. and Evans, R.J. 'Development of a Water Jet Assisted Drag Bit Cutting Head for Coal Measures Rocks'. Proc. Rapid Excavation and Tunnelling Conference, Vol.2, 701-719, 1983.

Plumpton, N.A. and Tomlin, M.G. 'The Development of a Water Jet System to Improve the Performance of a Boom-Type Roadheader'. 6th Int. Symp. on Jet Cutting Technology, Surrey, UK. Spnsored by BHRA Fluid Engineering, Cranfield, UK., 1982.

Ropchan, D., Wang, F.D. and Wolgamott, J. 'Application of Water Jet Assisted Drag Bit and Pick Cutter for the Cutting of Coal Measures Rocks'. Final Report to DoE, Contract No. ET-77-G-01-9082. Colorado School of Mines, USA, 1980.

Tecen, O. 'High Pressure Water Jet Assisted Drag Tool Cutting of Rock Materials'. Unpublished Ph.D. Thesis, Dept. Mining Engineering, Univ. of Newcastle upon Tyne, UK, 1982.

Factors influencing the stability of deep boreholes

Facteurs influençant la stabilité de forages profonds
Zur Stabilität von tiefen Bohrlöchern

P.K.KAISER, Professor of Civil Engineering, University of Alberta, Edmonton, Canada
S.MALONY, Research Associate, Civil Engineering, University of Alberta, Edmonton, Canada

ABSTRACT:Wellbore instabilities have been found to depend on not only the far-field horizontal stress regime and strength of the formation but also, to a large extent, on the rock structure near the borehole and the temperature gradients imposed by drilling fluid circulation. The significance of these factors on the prediction, prevention and back analysis of instability processes are discussed.

RÉSUMÉ: Il est montré que les instabilités dans la paroi des forages dépendent non seulement du régime à grande échelle des contraintes horizontales et de la résistance de la formation, mais aussi, en grande partie, de la structure du roc à proximité du forage et des gradients de température imposés par la circulation des boues de forage. L'importance de ces facteurs par rapport à la prédiction, la prévention et l'analyse à rebours des processus d'instabilité est discutée.

ZUSAMMENFASSUNG: Bohrlochinstabilitäten sind nicht nur vom horizontalen Spannungszustand und der Festigkeit des Gesteines abhängig, sondern auch von der Felsstruktur in der Bohrlochnähe und dem Temperaturgradienten, der durch die Zirkulation der Bohrflüssigkeit erzeugt wird. Die Bedeutung dieser Faktoren fuür die Vorhersage, Verhinderung und Rückrechnung von Instabilitätsvorgängen werden diskutiert.

1. INTRODUCTION

The high cost of drilling in resource exploitation, particularly where deep holes or holes in frontier areas are concerned, combined with current revenue reduction, has provided the impetus for increased research to improve wellbore stability. Factors contributing to borehole wall instabilities may be characterized as chemical or mechanical in origin. This paper reports on a current investigation, both numerical and experimental in scope, into the mechanical origins of wellbore instabilities, in particular, brittle rupture mechanisms.

The most relevant factors contributing to brittle rupture processes are:
1. Formation characteristics - strength of the intact rock and the spacing, persistence, orientation and strength of discontinuities;
2. Drilling fluid characteristics - density, temperature and temperature change due to circulation; and
3. Ambient stress field - magnitudes of the three principal stresses as well as their relative ratios and orientation with respect to the borehole axis.

The influence of pore pressure and radial flow to or from the borehole are also of great importance for the rupture process, but this aspect is not considered here.

2. NUMERICAL STUDIES

Numerical simulations of stress conditions near the bottom and around deep boreholes in rock were undertaken to provide input for scoping of the experimental program, to aid in data interpretation and to extend, by extrapolation, the range of applicability of the laboratory test results. In order to optimize the efficiency of this phase, it was decided to employ a simple numerical technique which will predict the extent of rupture initiation zones but not failure propagation mechanisms.

Sensitivity analyses by Kwong (1985) demonstrated that the spatial extent and failure mode around circular openings (i.e., boreholes) can be adequately predicted by comparison of the elastic stress distribution with the strength of rock and discontinuities. Fig.1 shows a comparison between results of this simplified analysis and those from a more rigorous finite element analysis (Kaiser et al., 1985). It may be seen that the simplified analysis yields a similar angular distribution of the two failure types (joint or intact rock mass failure). It underestimates the radial extent of the intact rock failure zone in the area where failure propagation due to shearing of joints occurs. Based on these findings, it was decided to employ this approach of comparing the elastic state of stress with the yield state of intact rock and discontinuities to asses the influence of various factors on borehole stability.

2.1 Influence of Drill Mud Density and Temperature

The finite element programs ADINA and ADINAT (Bathé, 1975) were employed to predict the elastic stress distribution near the bottom of a (2000 m) deep borehole in a typical rock during excavation and cooling (Kosar, 1985). The rock mass was assumed to be isotropic thermo-elastic with constant thermal conductivity and specific heat. Conditions of radial symmetry were assumed thereby allowing use of a two-dimensional axisymmetric formulation of the three dimensional problem.

Drilling fluid densities of 0.0 (air), 1.0 (water) and 2.1 (heavy mud) Mg/m³ and cooling periods of 1 day, 1 month and 25 years were investigated. To faciliate understanding of the results of these analyses it was decided to compare the propensity to failure under these conditions. This was accomplished by applying the Mohr-Coulomb failure criterion and calculating the angle of internal friction at which failure would initiate in a cohesionless material (Banas, 1986). This angle of internal friction, required to prevent yield initiation, is called 'critical friction angle', ϕ_c.

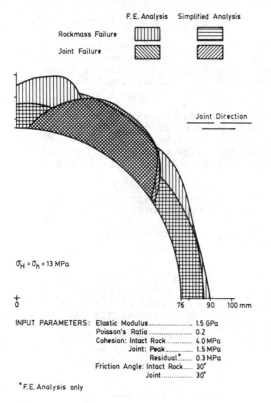

Fig.1 Comparison of type and extent of the failure initiation zones near a circular opening predicted by two methods of analysis (after Kwong, 1985).

As shown in Fig.2, for isothermal conditions ($\Delta T=0$), the critical friction angle increases rapidly as the borehole bottom approaches and reaches a steady maximum value within one diameter of its passing. Therefore, the most critical conditions are found far from the borehole bottom, where two dimensional solutions are applicable.

Cooling of the borehole wall by 100C° was found to negatively affect the stability and to change the probable mode of failure. (Note: in addition to tensile fracturing of the formation, three other modes of rock failure near a borehole are possible depending on the magnitude of the radial, tangential and vertical stresses (King, 1912)). Fig.2 shows a significant increase in the critical friction angle after cooling, particularly near the borehole bottom. Under a temperature reduction in the borehole, the critical region has shifted downward, necessitating three-dimensional analyses of the conditions at the borehole bottom. The zone with a critical friction angle of 90° indicates an area of tensile failure. These

results demonstrate that extremely critical conditions may be created at the borehole bottom due to extended cooling.

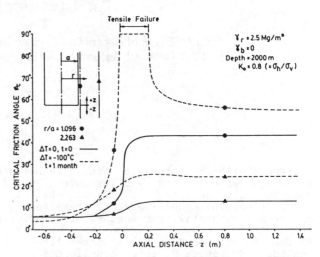

Fig.2 Stability requirements around an unsupported borehole excavated in a homogeneous, cohesionless medium under a uniform horizontal stress field.

Extended periods of cooling were also determined to be detrimental to the integrity of the borehole wall. There is a progressive increase, with time, in the critical friction angle radially away from the borehole. Therefore, it is possible to have a time-dependent propagation of the instability due to non steady-state temperature gradients.

In all cases, cooling by 100 C° promoted the development of the tensile failure mode in the immediate vicinity of the borehole wall. Further inside the rock, other modes due to either radial-tangential or vertical-tangential stresses exist. From parametric studies it was found that the extent of rupture initiation, for most homogeneous geologic materials, would be limited to the near wall region (within 1 borehole radius).

Under isothermal conditions, increasing the drilling fluid density was observed to improve the stability of the borehole as evidenced by a decrease in the critical friction angle (results not shown). However, at excessive fluid pressures hydraulic fracturing, as demonstrated by Bradley (1979), would occur. When the borehole wall is cooled, the benefit of borehole fluid pressure may be lost. For the conditions of our study, increasing the fluid density from 1.0 to 2.1 Mg/m³ resulted in a significant increase in the extent of the overstressed zone. For example, at a distance of 1.3 radii above the borehole bottom, the radial distance from the borehole wall to the point at which the critical friction angle has reached a value of 30° is 0.18 radii for water and 0.61 for heavy mud. Hence, the extent of the overstressed zone could be more than three times as great with the heavier fluid. On the other hand, with no drilling fluid the extent would also be large at 0.70 radii.

2.2 Effect of Weakness Planes

The role of weakness planes (bedding or

joints) in the development of wellbore
breakouts is currently in question. Babcock
(1978) suggested that breakouts are the
result of the drill encountering steeply
dipping fractures while Gough and Bell
(1982), amongst others, suggest that they are
caused exclusively by shearing in zones of
concentrated compressive stresses resulting
from non-uniform, far field horizontal
stresses. In reality, both factors are
probably contributing to the development and
propagation of breakouts. Stress
concentrations will dominate initiation and
weaknesses will influence propagation and
types of rupture mechanism.

In order to better assess the significance
of discontinuities for the borehole
instability problem, weaknesses were
introduced into the preceeding situation in
the following manner (Banas, 1986). The
elastic stresses were generated by finite
element analyses of a circular hole in an
elastic medium. The orientation of a single
set of weakness planes was selected and the
shear and normal stresses acting on those
planes around the borehole calculated. The
acting shear stress was then compared with
the allowable shear stress of the weakness
plane to determine whether failure could
occur. Prior to this it was checked whether
failure of the intact rock would occur.
Calculations were performed on six horizontal
slices, located above and below the borehole
bottom, to identify zones of possible
overstressing as shown in Fig.3.

The extent of these failure initiation
zones was found to depend on the orientation
and strength of the weakness plane, its
location along the borehole axis, and the
temperature gradient. Analyses were conducted
with weakness planes oriented such that their
strike was parallel to one of the horizontal
principal stress axes. For this condition the
extent of the overstressed zone increased
with an increase in dip angle to 60° and
decreased thereafter. Similar observations
were made by Desai and Johnson (1974).

The shape of this zone is non-symmetric,
particularly near the bottom of the borehole
as shown in Fig.3, even though the ambient
stress field is uniform. This suggests that
the influence of weakness planes could
mislead back analyses of the ambient stress
field.

The influence of cooling of the borehole
wall is dramatic as demonstrated by Fig.3.
Whereas, under isothermal conditions,
overstressing on weakness planes had been
limited to the near wall region, if cooled,
this zone can be of large extent (up to 7
borehole radii). This drastic sensitivity to
a temperature gradient may however not be of
great concern as the extent of the zone of
overstressing is highly dependent on the
strength parameters of the weakness plane (as
shown in Fig.3 at z/d = 4.0). Propagation of
failure will also be limited by the
constraint provided by the circular borehole
(i.e., motion cannot occur along
discontinuities unless failure of the intact
rock is created by stress redistribution).

These results indicate that, given suitable
conditions, rock structure may affect the
instability process. Laboratory verification
of these processes is currently in progress.

2.3 Shape of Rupture Zone

Zoback et al.(1985) and Haimson and Herrick
(1986) suggested that, for homogeneous
materials, the orientation and extent of the
breakout is a function of the strength of the
material and the in-situ stress field.
Therefore, knowledge of the breakout geometry
and rock strength properties should permit
determination of the acting field stress
ratio. A formula for the determination of
points of shear failure initiation around the
borehole, based on the elastic stress
distributions and the Mohr-Coulomb failure
criterion, was presented by the
aforementioned authors.

In order to better interpret our laboratory
test data, this relationship was employed to
prepare the breakout boundaries as a function
of the maximum horizontal stress (normalized
to the cohesion of the material) for various
values of the angle of internal friction and
horizontal stress ratio (σ_h/σ_H), as shown in
Fig.4. An important observation from these
plots is that, contrary to the common
assumptions, the maximum extent of the
breakout does not necessarily correspond to
the direction of the least horizontal stress.
Non-correspondence is found when the stresses
are high compared to the strength (cohesion)
of the rock (Fig.4.a: $\sigma_H/c = 10$) or when the
stress ratio is low (Fig.4.b: N = 0.3).
Fig.4.b further suggests that noncontinuous
breakouts that do not affect the entire
borehole are unlikely if the stress ratio is
between 0.75 and 1.3. In other words, a
significant difference in horizontal stresses
must exist before an unambiguous
determination of their relative magnitudes by
back analysis is possible. Furthermore, the
close proximity of the breakout boundaries
for N = 0.3 and 0.5 indicates that, even at
high stress ratios, the breakout shape is a
poor indicator of the magnitude of the field
stress ratio.

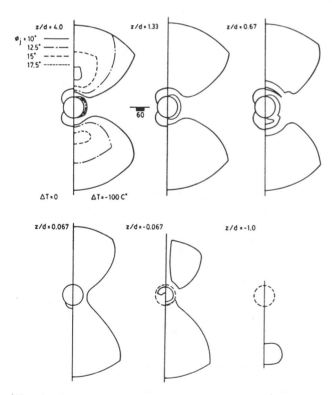

Fig.3 Plan views of the extent of possible
failure initiation zones at six elevations
and for ΔT=0° and 100° C (Intact rock:
c = 2 MPa, φ = 45°; Weaknesses: c = 0 MPa,
ϕ_j = 10°).

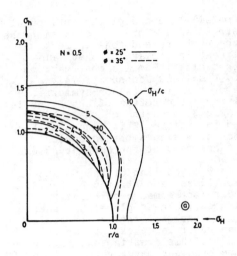

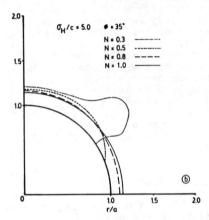

Fig.4 Extent of the rupture initiation zone near a borehole as a function of: a) the stress/strength ratio (σ_H/c); and b) the stress ratio (N=σ_h/σ_H).

3. LABORATORY STUDIES

In order to identify the effect of the drilling process and, in particular, to investigate the influence of non-uniform ambient stress fields on borehole stability, the large scale triaxial frame described by Kaiser and Morgenstern (1981) was modified to accomodate cubical specimens (300 mm a side) at field stresses up to 20 MPa. A partially penetrating borehole (44.5 mm in diameter) was drilled under stress in these specimens. Measurements of boundary stresses and displacements were made throughout each test. Extensometers were installed radially around the borehole in several specimens to monitor the radial strain during and after excavation. Prior to unloading, an epoxy resin was injected thereby permitting examination of the rupture surface after unloading and sample sectioning. The test apparatus can currently simulate isothermal conditions to a depth of 1100 m.

The model material selected for this study is an artificial sandstone constructed from sand and cement in a manner similar to that developed by Wygal (1963). Dry mix is rained through a particle distributor into a mould and water is then allowed to penetrate from the bottom by capillary action. The cube is allowed to hydrate for one week and air-dried prior to testing. The material has an elastic

modulus of about 1300 MPa and a Poisson's ratio of 0.25. It exhibits brittle behaviour at low confining pressures becoming ductile at higher levels. The strength of the intact material can be described, at low confining pressure, by a linear failure envelope with an intercept (cohesion) of 1.0 MPa and a slope (friction angle) of 35°. The density is 2.05 Mg/m^3.

3.1 Observed Breakout Shape

A series of six tests were conducted so far to investigate the breakout process. Horizontal stress ratios of 0.3 to 1.0 were applied. Borehole instabilities were succesfully generated and noted to orient with the direction of the minimum horizontal stress as shown in Fig.5 for two elevations in our test. There is reasonably good agreement between the observed breakout shape and that predicted using the relationship proposed by Zoback et al.(1985) The actual mechanism of breakout, shear failure or tensile spalling, has yet to be indentified.

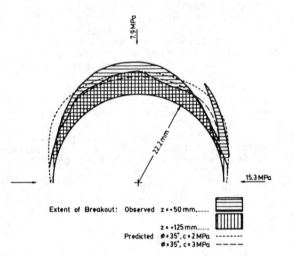

Fig.5 Comparison of predicted and observed borehole breakout - Test No.4-3.

Deviation from the predicted shape of the breakout can occur for a number of reasons. As was demonstrated in Fig.4.a, for a given field stress ratio the shape of the breakout is very much dependent on the cohesion particularly when the cohesion is low relative to the stress. This may be responsible for the deeper seated failure mechanism not aligned with the minor horizontal stress shown in Fig.5 (wedge on right side). The observed extent of the breakout may be less than expected due to material simply not falling away from the wall, or, in the case of the test specimen (Fig.5, z = +125), due to the proximity of the upper sample boundary. The three-dimensional conditions near the borehole bottom may also influence the shape and extent as suggested by the difference in geometry at the two levels shown. It was also observed in other specimens that the breakouts were larger near the borehole bottom. Both over and underestimation of the breakout geometry may, however, also reflect a time-dependent propagation of the breakout.

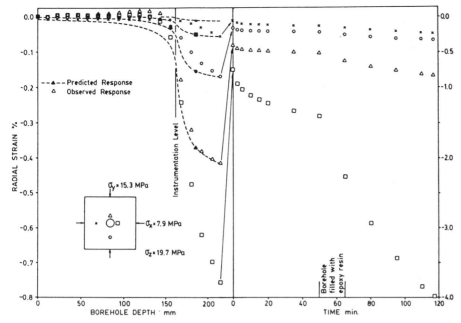

Fig.6 Development of radial strain near the borehole: Test No.4-3
(Note: change in ordinate scale for right side graph).

3.2 Breakout Propagation

The borehole breakout process is a progressive phenomenon. As suggested earlier, the development of a borehole instability reflects both the time-dependent stress redistribution resulting from the drilling process as well as the time-dependent response of the rock to the imposed stress. This is evidenced in Fig.6 by the response of the extensometer closest to the borehole in the direction of the least horizontal stress (square symbol). Yielding of the rock, as reflected by the excessive radial extension deviating from the response predicted for elastic rock by the method proposed by Barlow (1986), began ahead of the drill due to stress redistribution and continued later due to the time-dependent strength of the rock. It follows then that the extent and shape of the breakout changes with time. Back calculation of in situ stresses from breakout geometry will clearly be complicated by these time-dependent propogation processes.

4. CONCLUSIONS

Compressive stress concentrations arising from non-uniform horizontal stress fields constitute a major factor in the development of breakouts in homogeneous rocks. Our experimental studies confirm that the orientation of the major axis of the breakout is generally aligned with the direction of the minor principal field stress and breakouts initiate at the location of the maximum tangential compressive stress concentration. However, numerical analyses indicate that, under conditions of high stress/strength ratios or low horizontal stress ratios, N, the major axis of the breakout may be aligned differently or the depth of breakout may be nearly constant over a significant portion of the borehole circumference. This is supported by some observations in the model tests.

The occurence of weaknesses near the borehole wall may enhance the development of a breakout if the strength along the weakness planes is significantly less than that of the intact rock and if the discontinuities are suitably oriented. Breakouts similar in shape to those observed in homogeneous rock under non-uniform horizontal field stesses, may also occur in rock containing discontinuities even if the field stress is uniform as shown earlier by Kaiser et al.(1985).

Temperature effects were found to be very important for the breakout process. Cooling of the borehole wall assists in deepening the breakout, particularly where weak discontinuities are present. Furthermore, cooling may alter the mode of failure.

The development of wellbore instability is a time-dependent process due to time-dependent properties of most geologic materials. Drill advance and cooling add another time component.

Based on these observations, back calculation of the in-situ stress field from the borehole breakout geometry, as proposed by numerous authors, seems to be a highly speculative venture.

In terms of preventing the occurence of instabilities or of minimizing borehole damage, the driller has direct control over three of the critical factors that influence the onset and propagation of instabilities. These are: time, drilling fluid density and temperature. By carefully balancing the advantages of these three variables it should be possible to prevent or at least minimize borehole damage.

ACKNOWLEDGEMENTS

Financial support for this project by the Natural Sciences and Engineering Research Council of Canada through NSERC Strategic Grant No.G1336 is gratefully acknowledged.

REFERENCES

Babcock, E.A., 1978. Measurement of
subsurface fractures from dipmeter logs.
American Association of Petroleum
Geologists Bulletin, 62, pp.1111-1126.

Banas, L., 1986. Influence of weakness planes
on rupture mode near boreholes. Report 86-6
for NSERC Strategic Grant No.G1336, 37p.
plus 192 figures.

Barlow, J.P., 1986. Interpretation of tunnel
convergence measurements. M.Sc. Thesis,
Department of Civil Engineering, University
of Alberta, 235p.

Bathé, K.J., 1975 (rev. 1978). ADINA, a
finite element program for automatic
dynamic incremental analysis. AVL Report
82448-1, Department of Mechanical
Engineering, MIT.

Bradley, W.B., 1979. Failure of inclined
boreholes. Journal of Energy Resources
Technology, 101, pp.232-239.

Desai, C.S., and L.D. Johnson, 1974.
Influence of bedding planes on stability of
boreholes. Proceedings of the Third
International Symposium on Rock Mechanics.
Denver, 2B, pp.997-1002.

Gough, D.I. and J.S. Bell, 1982. Stress
orientation from borehole wall fractures
with examples from Colorado, east Texas and
northern Canada. Canadian Journal of Earth
Sciences, 19, pp.1358-1370.

Haimson, B.C. and C.G. Herrick, 1986.
Borehole breakout - a new tool for
estimating in situ stress? Proceedings of
the International Symposium on Rock Stress
and Rock Stress Measurements, Stockholm,
pp.271-280.

Kaiser, P.K., and N.R.Morgenstern, 1981.
Time-dependent deformation of small
tunnels- I. Experimental facilities.
International Journal of Rock Mechanics,
Mining Sciences & Geomechanics Abstracts,
18, pp. 129-140.

Kaiser, P.K., A. Guenot, and N.R.Morgenstern,
1985. Time-dependent deformation of small
tunnels- IV. Behaviour during failure.
International Journal of Rock Mechanics,
Mining Sciences & Geomechanics Abstracts,
22, pp. 141-152.

King, L.V., 1912. On the limiting strength of
rocks under conditions of stress existing
in the earth's interior. Journal of
Geology, 20, pp.119-138.

Kosar, K.M., 1985. Axisymmetric analysis of
borehole rupture problem. Report 85-1 for
NSERC Strategic Grant No.G1336, 22p. plus
181 figures.

Kwong, A., 1985. Review of numerical
techniques for borehole stability
assesment. Report 85-2 for NSERC Strategic
Grant No.G1336, 45p.

Wygal, R.J., 1963. Construction of models
that simulate oil reservoirs. Society of
Petroleum Engineers Journal, pp.281-286.

Zoback, M.O., D. Moos and L. Mastin, 1985.
Well bore breakouts and in situ stress.
Journal of Geophysical Research, 90,
pp.5523-5530.

Tensile strength of rock under biaxial point-loading
La résistance au charge ponctuel biaxial des roches
Zugfestigkeit von Gestein unter biaxialer Punktbelastung

H.K.KUTTER, Arbeitsgruppe Felsmechanik, Institut für Geologie, Ruhr-Universität Bochum, FRG
H.JÜTTE, Arbeitsgruppe Felsmechanik, Institut für Geologie, Ruhr-Universität Bochum, FRG

ABSTRACT: Tensile failure induced by multiaxial point loading frequently occurs in rock comminution and in jointed rock masses. In an experimental study with the simplest multiaxial point-load configuration, i. e. the biaxial one with the load axes normal to each other, the point-load strength of rock has been investigated. The ratio between biaxial point-load strength and uniaxial point-load strength turned out to be considerably larger than one, but constant for a given load ratio and independent of rock type, degree and orientation of anisotropy, and sample size and shape. An interpretation of these results is given on the basis of a critical energy criterion.

RESUME: Rupture tractive en conséquence de charge ponctuel multiple axial se passe souvent dans le concassage des roches et dans les roches crevassées. Dans une étude expérimentale avec l'essai ponctuel multiple axial plus simple, c'est à dire d'un essai ponctuel biaxial dont les axes de force font angle droit, la résistance au charge ponctuel des roches était déterminée. La relation de la résistance au charge ponctuel biaxial à uniaxial se trouvait beaucoup plus grande qu'un, mais restait constante pour une rélation donnée par les forces axiales et indépendante de la sorte des roches, du degré et de l'orientation de l'anisotropie et de la grandeur et forme des éprouvettes. A l'aide du critère d'énergie des résultats sont interpretés.

ZUSAMMENFASSUNG: Zugbruch infolge mehrachsiger Punktbelastung findet häufig bei der Gesteinszerkleinerung und im klüftigen Gebirge statt. In einem Versuchsprogramm mit der einfachsten mehrachsigen Punktbelastung, d. h. einer biaxialen mit zueinander unter einem rechten Winkel stehenden Lastachsen, wurde die Punktlastfestigkeit von Gestein ermittelt. Das Verhältnis von biaxialer zu einaxialer Punktlastfestigkeit stellte sich als beachtlich größer als eins heraus, jedoch auch als konstant für ein gegebenes Verhältnis der Achskräfte und unabhängig von der Gesteinsart, von Grad und Orientierung der Anisotropie und von der Probengröße und -form. Mit Hilfe des Energiekriteriums werden diese Ergebnisse gedeutet.

1 INTRODUCTION

The strength and deformational behaviour of rock and of rock masses are normally related to states of stress represented by distributed or uniformly distributed loads along the boundaries. Whereas this kind of loading may be applicable to most situations in rock engineering, there is nevertheless a large number of cases in practical rock mechanics where concentrated rather than distributed boundary forces are the cause of failure and collapse of the structure. Concentrated forces acting on rock blocks are for instance those generated by artificial support elements such as bolts, props, and steel ribs, or those which result from rotation and shear of adjoining rock blocks and subsequent point contacts at roughness peaks and block edges. Multiaxial point-loading also occurs in rock comminution, such as the breaking of rock lumps between the jaws of a crusher. Multiaxial point-loads are furthermore the prevailing type of loading of rock blocks in a rock fill mass or of large particles in a coarsely grained soil.

The critical aspect of these point-load systems is, that other than local zones of high compressive stress concentrations also regions of considerable tensile stress are produced. With the tensile strength of rock being much smaller than its compressive strength, systems under concentrated loads therefore are particularly prone to the formation of tensile cracks and subsequent tensile failure. This is well known and one practical application of this phenomenon is the familiar uniaxial point-load strength test (ISRM 1985) for rock classification.

Many experimental results exist for the uniaxial diametral point-load arrangement, and in addition several theoretical investigations of the stress distribution in splitting tests (Hiramatsu & Oka 1966, Davies & Bose 1968, Wijk 1978) have been conducted. But there is hardly any experimental or analytical information available on the strength of rock under multiaxial point-loading. The axial stresses in a cylinder compressed symmetrically by four equal line-loads have been analyzed by Jaeger (1967). Wijk (1979) explored a new multiple point-load test arrangement with eight concentrated loads evenly distributed around the cylindrical surface of the rock core. Assuming that for this test arrangement the axial tensile stress at the sample center is obtained by superposing four times the axial tensile stress evaluated for the biaxial test, he observed experimentally the multiple point-load strength to be at least 20 percent larger than the ordinary point-load strength index. He proposed that the difference may be due to a higher tensile strength of the rock when it is tensioned in only one direction and compressed in the two other directions (as in the multiple point-load test) rather than tensioned in two directions and compressed in the third direction (as in the conventional uniaxial point-load test). Another way of reasoning put forward by Wijk is that the increase in total contact area may be the cause of higher tensile strength when going from the ordinary two to eight load contact points.

It is, however, yet an open question whether and to what extent such an increase in tensile strength does also occur in systems with fewer than eight point-loads and if it is size-dependent. Therefore the simplest and practically most relevant multiaxial point-load system, the biaxial one, has been tested on various rock types, with sample size, orientation of the plane of weakness and direction of the plane of the two load axes as additional variables. It was not the purpose of this study to promote yet another type of index test, but to examine the variability of the tensile strength of rock with the loading system, particularly that of multiaxial point-loads.

2 THEORETICAL CONSIDERATIONS

In all splitting tests with diametrically applied point- or line-loads a zone of tension is produced, with the tensile stress oriented normal to the plane or axis of loading and distributed fairly uniformly over the central half diameter of the specimen. The principal stress in the direction of the loaded diameter is compressive and in the central region about three to five times the magnitude of the tensile stress. The third principal stress along this line is the same as the first one, i. e. tensile, for the point-load, but compressive for the line-load. Assuming linear elasticity and consequently the validity of superposition, the stress field in the specimen under biaxial point-loading can be determined by adding two stress fields of the type of the ordinary point-load test, with one rotated 90 degrees versus the other. The superposition of the stress fields allows then the determination of the location and magnitude of the maximum tensile stress in the biaxial test. If tensile failure is assumed to start when the maximum tensile stress component reaches a critical value, then the failure loads for a biaxial point-load test can be determined on this basis:

The stress field for a diametrically loaded cylinder or cube, as derived by Davies & Bose (1968), shows maximum tension to occur normal to the loaded diameter and fairly uniformly distributed over its central part. By superposition of the two stress fields the tensile stress there is doubled and consequently the critical biaxial load should be half the uniaxial one. Taking, however, the stress distribution, calculated more recently by Wijk (1978) for the point-load test on a spherical specimen, as presented in Figure 1, one notices that there the maximum tension does not occur in the central part but in the close vicinity of the point-load. Superposing two stress fields of this type which are oriented at right angles to each other results no longer in double the maximum tensile stress but only in an increase of less than 10 percent. This can be readily concluded from the distribution of $(\sigma_r)_{\theta=\frac{\pi}{2}}$ in Figure 1, which shows very little tension in the vicinity of the boundary. It follows from this analysis that the critical biaxial load should only by slightly smaller than the uniaxial one, based on the presumption, however, that failure is initiated in the close vicinity of the loading points. This result differs, of course, considerably from the previous one, leaving a wide range of uncertainty for the theoretically predicted biaxial failure point-load. The result might yet be different, if failure criteria other than that of maximum tensile stress are applied. An extensive experimental program has been initiated to clarify this matter.

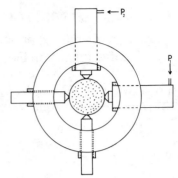

Figure 2. Biaxial loading frame

3 EXPERIMENTAL ARRANGEMENT

A special biaxial point-load test apparatus has been constructed to be used for ordinary (uniaxial) as well as biaxial point-load tests on cylindrical rock samples with diameters between 10 and 100 mm. The apparatus, a schematic drawing of which is shown in Figure 2, consists of a massive circular loading frame, two hydraulic pistons, two manometers, and four spherically-truncated ($r = 5$ mm), conical (60°) tungsten carbide load platens. The two passive load platens can be adjusted to the required position (one half specimen diameter from the frame center) by set screws. The two active load platens can be actuated either simultaneously by the same hydraulic pressure from a hand pump or separately by different pressures. Three different types of point-load tests could be conducted with this set-up: the ordinary uniaxial type, the standard biaxial one with equal forces on both axes, and a modified biaxial type, where the load on one axis was first raised and held constant at a value below the critical uniaxial point-load and then the load on the other axis was applied and increased up to failure of the specimen. The uniaxial tests were performed with the point-loads acting either in the diametral or the axial direction of the cylindrical samples, the biaxial tests with either both load axes in diametral direction or one in axial, the other in diametral direction, as shown in Figure 3.

The shape of all rock samples was that of a cylinder with a height-to-diameter ratio of 1:1. A potential size effect was taken into consideration by testing cores with diameters of 30, 41.5/44, and 61.5/63 mm. The effect of strength anisotropy was examined by orienting the load axis of the uniaxial test or respectively the load plane of the biaxial test (see Figure 3) either normal or parallel to the plane of weakness. Five rock materials were available for the tests: a medium-grained, weakly bedded Carboniferous Ruhr-Sandstone with a quartz content of 80 %, 10 % feldspar minerals, 5 % mica, 5 % additional minerals, a medium grain size of 0.3 mm, and practically no

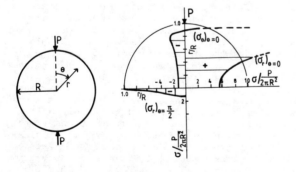

Figure 1. Stress distribution in spherical specimen under uniaxial point loading (after Wijk, 1978)

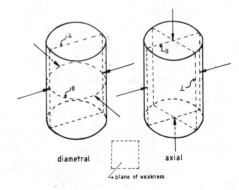

Figure 3. Types of load applications in biaxial point-load testing of anisotropic rock specimens

porosity; a Scandinavian granite with a quartz content of 40 %, 50 % feldspar minerals, 10 % mica, and a medium grain size of 1.5 mm; a diabase from the Rhenish massif with 40 % angite, 40 % albite, and 20 % chlorite; a Jurassic limestone from Southern Germany with 75 % cryptocrystalline carbonate and 25 % calcitic bioclasts (some of them dissolved and leaving pores up to a size of 0.1 mm); an Alpine gneiss commercially called Calanca Granite with 45 % quartz, 35 % feldspar minerals, 20 % mica, and a medium grain size of 1.5 mm. The sandstone and the gneiss are anisotropic.

The total number of tests was such that for every possible combination of rock type, sample size, type of loading, direction of loads and orientation of anisotropy at least five samples were tested.

4 TEST RESULTS AND DISCUSSION

4.1 Size effect in biaxial point-load testing and its correction

The test results from equally biaxially loaded cores show for both the axial as well as the diametral loading configuration clearly the existence of a size effect. This is evident from the plot of the test data in Figure 4, where the biaxial failure load P for a core of diameter D has been reduced to the point-load strength index $I_s(D) = P/D^2$. The increase of I_s with decreasing core size appears to follow a parabolic function very similar to that observed, e. g. by Greminger (1982) for the uniaxial point-load test on isotropic and anisotropic rock specimens. The same size correction procedure was therefore tried as has been suggested and successfully applied to ordinary point-load test data by Greminger (1982), i. e. for a reference standard size of 50 mm the corrected strength index $I_s(50)$ is for the diametral test with a circular fracture surface

$$I_s(50) = I_s(D) \cdot \left(\frac{D}{50}\right)^{0.5}$$

and for the axial test with a quadratic fracture surface

$$I_s(50) = I_s(D) \cdot \left(\frac{D}{50}\right)^{0.5} \cdot \left(\frac{\pi}{4}\right)^{0.75} .$$

No shape correction was necessary since the sample width was always equal to the sample height. In all tests the fracture surface was approximately planar and passed through the four loading points, even at those samples with the plane of weakness normal to the plane of the loads. The accordingly corrected strength data are also plotted in the diagrams of Figure 4. One can see that the quality of this correction is quite good since for each rock type tested a more or less constant strength index $I_s(50)$ is obtained for all sample sizes. The conclusion has therefore been drawn that the size correction procedure developed for the ordinary point-load test is equally valid for the biaxial test, including the modified type with different axial loads. Henceforth all point-load strength data discussed here are those corrected to reference size.

4.2 Comparison between uniaxial and biaxial point-load strength

All together three series of biaxial point-load tests have been performed: one with both axial loads raised simultaneously up to the fracture load, P_2; another series with one axial load held constant at a magnitude equal to half the uniaxial fracture point-load, P_1, and the other axial load, P, raised up to failure of the sample; and a third series with the same test procedure as the previous one, except that here the

first axial load is kept constant at a magnitude equal to half the ordinary biaxial fracture point-load, P_2. The first two series were run in the diametral and axial fashion (for definition see Figure 3), the last series only in the diametral one. The results - each data point represents the mean of at least five uniaxial and five biaxial tests respectively - are presented in Figure 5, with the uniaxial strength index plotted against the biaxial one. The biaxial strength index has always been evaluated for the failure load, P, as indicated in the insert drawings of Figure 5.

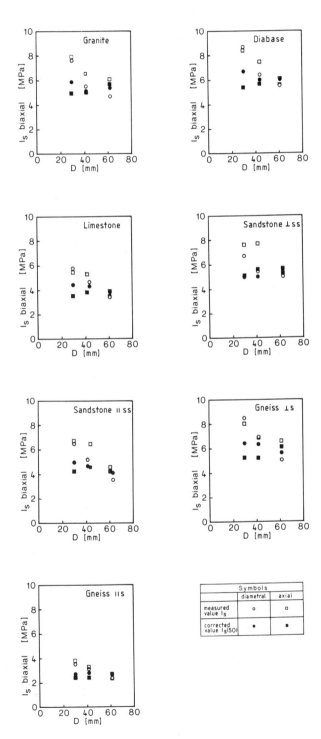

Figure 4. Size effect in biaxial point-load test results

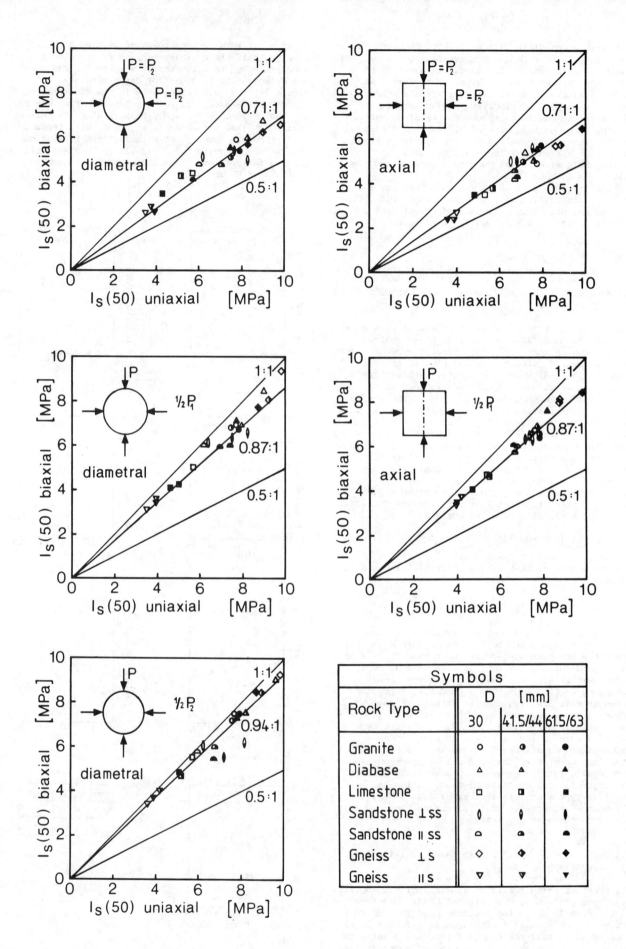

Figure 5. Comparison of biaxial with uniaxial point-load strength indices

It is quite obvious that the biaxial failure load, P_2, is not equal to half the uniaxial failure load, otherwise the data points would have to lie on the 0.5:1-line. In both, the diametral and the axial tests a fairly good fit is obtained by the 0.71:1-line, i. e. the biaxial fracture load is approximately equal to $1/\sqrt{2}$ times the uniaxial fracture load, P_1. With this finding it is not surprising that the failure loads in the modified biaxial tests turn out to be even higher, i. e. well beyond 0.8 times the uniaxial point-load in the second series and only slightly below the uniaxial point-load in the third series. A linear relationship between uniaxial and biaxial strength appears to exist for all three types of biaxial loading. Additional tests showed that the influence of the first load axis on the failure load of the second axis becomes negligible if the first load is not larger than a third of the uniaxial point load, P_1.

The conclusion that can be drawn from these experimental results is that evidently a simple critical tensile stress criterion of failure does not hold here. None of the two, on the basis of this criterion theoretically predicted biaxial failure loads, P_2, agree with the measured one. Consequently other failure criteria must be considered. A very simple argument for the apparent higher tensile strength in the biaxial point-load test can be deduced from the different states of stress in the interior of the sample. In the uniaxial test two principal stresses are tensile, i. e. tensile stresses exist in all directions normal to the load axis, and consequently the fracture plane will follow freely the direction with the weakest material elements. On the other hand, the orientation of the fracture plane in the biaxial test is predetermined since only one principle stress is tensile, i. e. it must coincide with the loading plane. The chance for this orientation to include the weakest elements is, of course, much smaller, and the strength is therefore higher.

Proof against this argument are, however, the results from uniaxial and biaxial tests on anisotropic rock samples with the load axis or load plane parallel to the plane of weakness. Although now the fracture plane is predetermined for both load situations, the ratio between uniaxial and biaxial point-loads still was 1:0.7. Another explanation must therefore be sought for the observed variation of tensile strength with the number of point-loads.

Rather than assuming a critical value of tensile stress required for failure, a general three-dimensional failure criterion might be more applicable here. The problem, however, is that practically all proposed three-dimensional failure criteria are formulated for compressive states of stress and do not properly extend into the tensile region. They are normally truncated by a simple constant tensile strength criterion, as for instance the Griffith criterion, or they suggest increasing tensile strength as the point of the bullet-shaped failure surface at the tensile side of the hydrostatic axis is approached (Kim & Lade 1984). The latter is in contradiction to the experimental findings.

A theoretical investigation by Brady (1970) on the effect of the intermediate principal stress on the fracture strength of brittle rock is the only study known to the authors which deals with the tensile stress state in more detail and finds the biaxial tensile strength to be lower than the uniaxial one. Assuming a brittle rock structure containing grain boundary cracks and microfractures Brady modelled mathematically an elastic isotropic continuum which contains a large number of statistically distributed and oriented narrow ellipsoidal cracks. Failure is postulated to occur when a sufficient number of microcracks develop and the probability of their joining to form a macroscopic fracture surface becomes large. The numerical example given by Brady indicates an influence of the intermediate principal stress similar to and in the same order of magnitude as the one

observed in the multiaxial point-load tests.

The most satisfactory and yet very simple explanation for the observed strength difference can be found on the basis of an energy criterion. This is strongly suggested by a striking feature of the test results: In all types of biaxial point-load tests the sum of the squares of the two axial loads, normalized to the uniaxial failure-point load, approach the value one. This is particularly striking in the ordinary biaxial tests with equal axial loads of a magnitude approximately 0.7 (i. e.$1/\sqrt{2}$) times the uniaxial failure load. In the modified biaxial point-load tests the normalized failure load is very close to 0.87 for a normalized first axial load of 0.5 and similarly close to 0.94 for a first load equal to half the biaxial load, P_2, i. e. in normalized form $\frac{1}{2}$ x 0.7. Since the work done by an external force acting on an elastic solid is proportional to the square of the force, the above results strongly suggest that the same amount of energy has to be available for failure to initiate in a rock sample under point-loading regardless if it is under ordinary uniaxial, biaxial, or modified biaxial loading. This corresponds very well with the hypothesis of Griffith (1922, 1924) which says that propagation of a crack is initiated when the rate of change of available energy, i. e. the difference between work done by the external forces and the change in elastic strain energy of the body, is equal to or larger than the surface energy of the new crack surface created. Since the change in strain energy due to crack growth is practically independent of the type of point-loading it follows that the external work required for failure initiation must be the same for all load arrangements. Hudson et al. (1972) suggested already earlier in a similar way the variation in tensile strength with different laboratory tests.

Moreover the critical energy concept accounts also for the size effect observed in point-load testing. The external work or the potential energy of the loading system is directly proportional to the square of the load, P, and to the diameter, D, of the sample. To obtain therefore the same required amount of critical potential energy in a sample of diameter D_2 as in one of diameter D_1, the failure load must be $\sqrt{D_1/D_2}$ times larger. This corresponds exactly to the size correction factor suggested by Greminger (1982) for uniaxial point-load testing and found valid for biaxial tests in this study.

Even the shape effect in point-load testing can be derived qualitatively from the critical energy concept. To obtain the same potential energy of the loading system in specimens with a circular, an elliptical and a rectangular fracture surface the critical load of the first one has to be smaller than that of the second and this again smaller than that of the third, since for the same point-load the displacement of the loading point is largest in the sample with a circular cross-section and smallest in one with a rectangular cross-section of greater width than height. This is again in full agreement with the experimental results and the shape correction suggested by Greminger (1982).

The earlier mentioned force limit below which the influence of the first axial load on the magnitude of the failure load on the second load axis becomes negligible can equally well be explained with the energy criterion. Assuming a contribution from the first force to the potential energy to be negligible if it is below 10 percent of the total required energy, the critical load limit follows as $\sqrt{0.10} \cdot P_1$, i. e. 30 percent of the uniaxial point-load at failure. The experimental results are in very good agreement with this theoretical finding.

5 CONCLUSIONS

(1) The maximum tensile stress reached in rock samples under biaxial point-loading is considerably higher than that in samples under conventional uni-axial point-loading. The concept that failure occurs when the maximum tensile stress component reaches a critical value therefore appears to be not relevant for point-loading.

(2) An energy criterion, however, turned out to be very relevant for failure in point-load testing. It has been shown experimentally that failure will occur when the work done by the loads reaches a critical limit.

(3) Size and shape effects observed in point-load testing can be satisfactorily accounted for by the energy criterion of failure.

(4) The influence of a second point-load, with its axis normal to the first one, becomes negligible if its magnitude is smaller than 10 percent of the uni-axial failure point-load. This is in full agreement with the energy criterion.

REFERENCES

Davies, J.D. & D.K. Bose 1968. Stress distribution in splitting tests. ACI Journal 65: 662-669.

Greminger, M. 1982. Experimental studies of the influence of rock anisotropy on size and shape effects in point-load testing. Int. J. Rock Mech. Min. Sci. & Geomech. Abstr. 19: 241-246.

Griffith, A.A. 1921. The phenomena of rupture and flow in solids. Phil. Trans. Roy. Soc. London A 221: 163-198.

Griffith, A.A. 1924. The theory of rupture. Proc. 1st Int. Congr. Appl. Mech., Delft: 55-63.

Hiramatsu, Y. & Y. Oka 1966. Determination of the tensile strength of rock by a compression of an irregular test piece. Int. J. Rock Mech. Min. Sci. 3: 89-99.

Hudson, J.A., E.T. Brown & F. Rummel 1972. The controlled failure of rock discs and rings loaded in diametral compression. Int. J. Rock Mech. Min. Sci. 9: 241-248.

ISRM Commission on Testing Methods 1985. Suggested method for determining point load strength (revised version). Int. J. Rock Mech. Min. Sci. & Geomech. Abstr. 22: 51-60.

Jaeger, J.C. 1967. Failure of rocks under tensile conditions. Int. J. Rock Mech. Min. Sci. 4: 219-227.

Kim, M.K. & P.V. Lade 1984. Modelling rock strength in three dimensions. Int. J. Rock Mech. Min. Sci. & Geomech. Abstr. 21: 21-33.

Wijk, G. 1978. Some new theoretical aspects of indirect measurement of the tensile strength of rocks. Int. J. Rock Mech. Min. Sci. & Geomech. Abstr. 15: 149-160.

Wijk, G. 1979. The multiple point load test for the tensile strength of rock. Swedish Detonic Research Foundation Report DS 1979: 23.

Measurement of rock fragmentation by digital photoanalysis
Analyse photographique par méthode digitale de la fragmentation induite par explosifs
Haufwerkverteilung durch digital photographishe Analyse

NORBERT H.MAERZ, Department of Earth Science, University of Waterloo, Canada
JOHN A. FRANKLIN, Department of Earth Science, University of Waterloo, Canada
LEE ROTHENBURG, Department of Earth Science, University of Waterloo, Canada
D.LINN COURSEN, E.I.du Pont de Nemours & Co. Inc., Wilmington, Del., USA

ABSTRACT: The authors describe a new method for measuring the size distribution of rock fragments produced by blasting, tested during recent experimental blasting in Virginia, USA. In this method, photographs of the broken rock are digitized, apparent size distributions are measured, and corrections are made for the effect of overlapping blocks. An unfolding (correction) function, has been derived empirically from measurements on small scale samples of crushed rock. Results of this research show further potential for the investigation of correlations between rock mass characteristics, types of explosives, blasting patterns, and fragmentation.

RESUME: Les auteurs décrivent une méthode nouvelle pour la détermination de la distribution granulométrique de fragments rocheux produits par sautage. Cette méthode a été appliquée sur un site experimental en Virginia, USA. Grâce à des photographies et à la technique digitale, la distribution des dimensions apparantes des fragments est obtenue en y intégrant une correction pour le chevauchement. C'est une correction empirique basée sur des measures effectuées a petite échelle sur des échantillons de rocs concassés. Les résultats de cette recherche sont encourageants et indiquent un potentiel pour l'étude de correlations entre les caractéristiques de la masse rocheuse, le type d'explosifs et de sautage et la fragmentation.

ZUSAMMENFASSUNG: Die Autoren beschreiben eine neue Methode zur Auswertung der Grössenverteilung von durch Felssprengung produzierter Bruchstücke. Die Methode wurde unlängst bei Versuchsprengungen in Virginia, USA, erprobt. Photographische Aufnamen der zersplierten Felsmasse von unterschiedlicher Grössenordnung wurden durch Digitalaufzeichnung transformiert. Die sich ergebende Grössenverteilung wurde gemessen und Korrekturen fur teilweise Überlagerung von Bruchstucken konnten vorgenommen werden. Eine Berichtigungsfunktion wurde von kleinen, mechanisch zerbrochenen Steinstücken empirisch abgeleitet. Die Auswertung der Ergebnisse ist Richtung weisend im Hinblick auf Möglichkeiten weiterer Erforschung der Korelationen von Felscharakteristik, Sprengmitteln, Sprengsatz Verteilung und Felszersplitterung.

1. INTRODUCTION

"Fragmentation" describes the size distribution of blocks produced by blasting. The ideal design of blast should produce a fragmentation closely matched to that required for a specific application such as for rockfill or armor stone, and reduce to a minimum the need for secondary blasting and crushing. Improved fragmentation in most applications means smaller blocks, and generally requires more drilling and more explosives. The costs, however, are offset by easier and cheaper loading, hauling, and crushing (MacKenzie 1966; Greenland and Knowles, 1969).

Because fragmentation is so closely related to the economics of the quarrying operation, it needs to be measured quickly and accurately in order to monitor blasting, and to optimise blast design, while reducing costs and environmental impact.

Currently, there are three methods for determining size distributions:

Sieving has been used extensively in scaled down blasting tests (Dick et al. 1973; Bhandari and Vutukuri 1974; Singh et al. 1980), but is prohibitively slow and expensive for full scale production blasts.

Predictions have been made from blasting parameters and rock mass properties, either using empirical formulae (Lovely 1973; Just and Henderson 1971), or from computer simulations (Gama 1984). These methods, however, do not measure the actual fragmention.

Photographic methods have been developed in which some parameter of block size, such as length or cross sectional area, is measured on the image either manually (Carter 1977; Aimone and Dowding 1983; Noren and Porter 1974) or using an image analysing computer (Gozon 1986). These methods measure only the wholly visible fragments, and not the ones overlapped by other fragments. This represents a serious sampling bias, as discussed below.

A new method of measuring fragmentation using digital photoanalysis has been developed at the University of Waterloo as part of a larger investigation to characterize rock fabric (Franklin and Maerz 1986). This method measures sizes of overlapping as well as non-overlapping fragments, and reconstructs the true size distribution. It was tested during the summer of 1986, by applying it to full scale blasting trials conducted in Virginia by E. I. du Pont de Nemours & Co. Inc.

2. MEASURING TECHNIQUES

2.1 Objectives

The process of deriving a size distribution from a photograph can be considered in four stages:

Photographic sampling: following a strategy designed to ensure that the size distributions in the photographs represent the muckpile as a whole.

Digitization of the photograph: either manually, or by an automatic process involving enhancement and edge detection.

Measurement of apparent block sizes on the photograph.

Conversion of apparent to real block size distributions.

2.2 Photographic Sampling

The muckpile is clearly heterogeneous with respect to fragment size. The largest sizes appear to have a tendency to be thrown to the forward fringes of the pile, and the smallest to cover the upper surface. Sizes appear to increase progressively from the back to the front of the pile, and lateral variations are also possible.

A photograph is a record only of a surface or section. The locations and directions of photography must be selected so that when the photographic data are extrapolated to three dimensions, they are representative of the whole muckpile. Three alternatives can be considered:

To photograph the complete muckpile from a balloon-mounted camera. Aside from the obvious practical difficulties, this method might give biased fragmentation measurements, because of the concentration of smaller fragments at the top of the muckpile.

To photograph a vertical cut through the length of the muckpile. Inevitable delays to the work of loading would hardly be welcomed by the quarry operator. Furthermore, any attempt to excavate a vertical face could easily introduce further errors because of sloughing and the plucking of larger blocks.

To photograph the broken rock product in the haulage trucks after loading from the muckpile. This method appears the least problematic, and was used in the Du Pont trials, mainly because it allowed sampling without delaying the loading and haulage operations.

Truckload samples were taken and photographed at regular intervals along the centerline of the muckpile, perpendicular to the quarry bench. This accounted for front-to-back variations. Vertical variations were averaged by getting the loader to lift a complete vertical section of the pile at each sampling point. A typical set of data, for Truck # 24 of the Du Pont blasting trials, is presented in Figs. 1 and 4.

While removing some biases, this method of sampling introduces others:

A perspective error: the closer blocks appear larger than the blocks further away.

A sorting error: the large blocks tend to slide to the bottom of the pile that forms in the

Fig. 1a: Broken rock in the haulage trucks.

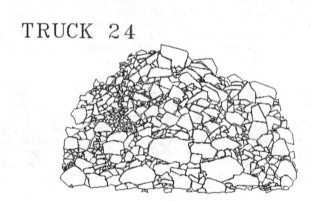

Fig. 1b: Digital image of the block profiles.

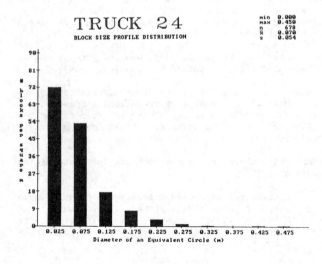

Fig. 1c: Distribution of the block diameters d_{ec}.

688

truck, whereas smaller ones tend to stay near the top.

An oversize error: very large blocks are not loaded into the truck.

To minimize the perspective error, photographs were taken using a telephoto lens, which flattens and compresses the depth of the image. The sorting error appears to be small and to affect the positions of blocks rather than the overall size distribution. Oversized blocks were counted separately. Research to date shows the total of these sampling errors appear to be quite small. Further studies of sampling methods and the associated errors are in progress.

2.3 Digitization

Two methods of digitization, manual tracing (vector), and automatic scanning (raster) methods are available (Franklin and Maerz, 1986).

For the Du Pont blasting trials, the manual method was used, in which photographs of the truckloads were digitized using an X-Y digitizing pad (Fig. 1b). "Profiles" of blocks, defined as the outlines of complete or partially overlapped blocks, were stored in digital form as the vertices of polygons.

In future studies, the authors expect to be making increasing use of the much faster automatic image analysis alternative. Using the manual method, each photograph took two to three hours to digitize. Techniques of image enhancement and edge detection are being developed to allow blocks to be "recognized" by the computer.

2.4 Measurement of Block Areas and Diameters

The area of each polygon (profile) was measured using the standard mensuration formula. Areas are difficult to visualize, so block sizes were expressed as the diameters "d_{ec}" of equivalent (equal-area) circles. These were then put into ten classes of equal class width, for display in the form of histograms (Fig. 1c). The frequency distributions were expressed as the number of blocks of a particular diameter class per square metre of surface on the muckpile (N_a).

2.5 Determination of True Block Size Distribution

Overview

This stage of analysis requires converting the measured distribution of diameters (d_{ec}) into a "true distribution"; the one that would be obtained if the particles were spread without overlaps. Block size must now be expressed three-dimensionally in terms of the diameter "d_{es}" of an equivalent sphere, one with a volume equal to that of the particle. This allows easy conversion to block weight or mass, as measured by sieving. We are much more concerned with weight than with numbers of fragments, particularly when considering small-sized particles.

Two methods were considered as described below, one analytical, and one empirical:

Analytical Approach

A somewhat similar problem has been studied and solved by stereologists in the fields of biology, metallography, and petrography: that of obtaining true particle size distributions from apparent ones observed in microscopic thin or polished sections (DeHoff and Rhines 1968; Underwood 1970; Weibel 1979, 1980). In these cases, the d_{ec} of a particle sliced at random is only some fraction of a diameter through its centroid. "Unfolding functions", derived on the basis of geometric probabilities, are used to convert from d_{ec} to d_{es} distributions.

At first an attempt was made to use the same type of function to correct for the overlap in the blast fragmentation. However, when one such function, given by Weibel (1979), was applied to some of the Du Pont data, it gave results that were obviously in error.

In a test to determine why, a box was filled with several hundred styrofoam balls, taking care to avoid regular packing. The spheres were photographed (Fig. 2a), the photographs digitized, and the diameters (d_{ec}) were measured. The results showed an abundance of very large and very small profiles (Fig. 2b) rather than the profile distribution derived using methods of statistical geometry for a slice through an assemblage of spheres (Fig. 2c). The abundance of large profiles corresponded to the many spheres that were fully exposed on the surface layer of the pile, whereas the very small profiles came from highly overlapped spheres in the second and third layers, seen through windows in the upper layer.

From this simple test, it was concluded that a different unfolding model would be needed from those of classical stereology. For present purposes, a semi-empirical unfolding model was the only practical alternative.

Semi-Empirical Approach

Simulations were carried out, using particles of crushed rock up to 50 mm diameter. A log-normal diameter (d_{es}) distribution was prepared by sieving and mixing, and was dumped into a scaled-down version of the box of a truck. The simulated pile was then photographed, digitized, and the diameters (d_{ec}) measured. The d_{ec} and d_{es} distributions were then compared.

Some mechanisms observed during this procedure are as follow:

Overlap of fragments: many observed diameters (d_{ec}) are smaller than their true diameters (d_{es}): i.e. the percentage of small sizes is greater than it should be.

Missing fines: smaller sizes are missing in the photograph, by virtue of falling into holes between and behind larger fragments, or because of insufficient photographic resolution: i.e. the measured percentage of small sizes is less than it should be. Note that this tends to offset the overlap effect.

Anisotropic stacking: anisotropic (platy) shape and stacking lead to a coarsening of the photographic measurements of block area: one tends to measure the largest dimensions of the block.

Observations also suggest that a significant portion of particles found on the surface of a pile are not overlapped, especially large ones. If the diameters of all particles adjacent to the surface are determined correctly (a hypothetical situation in view of overlap), the number of fragments of any class is proportional to the mean diameter of this class (d) and to the number of these particles per unit volume (N_v), i.e.

$$N_a(d) \approx d \, N_v(d) \ .$$

The relationship essentially follows from dimensionality considerations. The coefficient of proportionality in the above formula depends on the mean curvature of particles (Santalo, 1976) and is strictly unity for spheres.

To reflect the actual conditions of measurements the above formula must be modified to account for the fact that the area-equivalent diameters (d_{ec}) are determined instead of volume-equivalent diameters (d_{es}). An empirical unfolding function $f(d)$ is introduced as a coefficient of proportionality in the above formula that can be rewritten as follows:

$$N_a(d) = f(d) \, d \, N_v(d) \ .$$

The function $f(d)$ was determined from a model experiment that involved a log-normal fragment size distribution. The resulting "unfolding function" is illustrated in Fig. 3.

The fact that the value of the "unfolding function" approaches unity for large particle sizes confirms an intuitive conclusion that most large particles are visible on the surface without overlap. Deviation of $f(d)$ from unity is a measure of particle overlap for a given class.

Large values of the unfolding function for smaller diameters reflect the fact that the "visible" size of particles adjacent to the surface is necessarily smaller than the true particle dimension. The procedure of determining the area-equivalent diameters on the basis of the visible portion of particles unavoidably results in over-representation of small sizes. The "unfolding function" can be used to correct this situation and to convert the measured distribution of diameters (d_{ec}) into a true distribution (d_{es}):

$$N_v(d) = \frac{1}{d \, f(d)} \, N_a(d) \ .$$

The use of this unfolding function on samples of other sizes, shapes, or distributions essentially assumes that there are no scale effects when extrapolating from a crushed rock sample to the blast fragmentation sample; and that anisotropy, if present, of the blast fragmentation is assumed to be similar to that of the crushed rock sample.

As an example of the application of this unfolding function, Fig. 4a shows the results of converting the data of Fig. 1c to a volumetric size distribution. This may then be converted into a mass (M) distribution using the relationship:

$$M(d) = N_v(d) \, V(d) \, \rho \ ,$$

where V is the average volume of a block in the size class:

$$V(d) = \frac{\pi}{6} \, d^3 \ ,$$

and ρ is the rock density. Fig. 4b shows the results of converting Fig. 4a to a mass distribution.

Finally, one can present these data in more familiar format, by converting the mass distribution into a cumulative percentage mass distribution as shown in Fig. 4c. This is the same as the "cumulative percent passing" used in a standard sieve analysis, used when testing soils or aggregates.

Fig. 2a: Randomly packed assemblage of spheres.

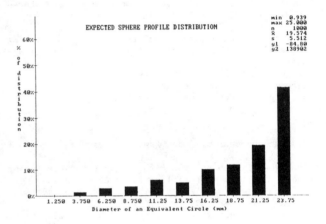

Fig. 2b: Expected d_{ec} distribution, of the spheres, based on geometric probabilities.

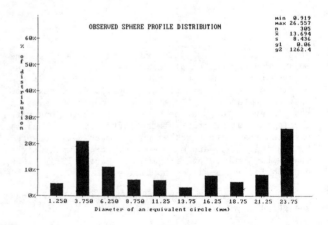

Fig. 2c: Observed d_{ec} distribution of the spheres, from the photograph.

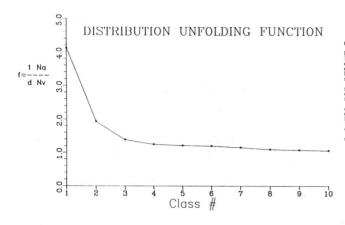

Fig. 3. Empirical unfolding function to relate d_{ec} to d_{es}.

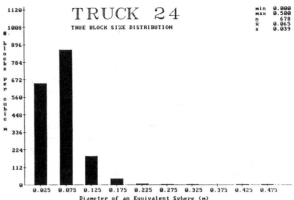

Fig. 4a: Unfolded distribution of block diameters (d_{es}).

Statistical Representation

Distributions are usually simplified by expressing them in terms of measures of central tendency (average or mean, median, mode) and of dispersion (variance or standard deviation, quartiles, coefficient of uniformity or sorting).

The two histograms of Figs. 4a and 4b show the relative abundance of blocks in each diameter class, according to their numbers and weights respectively. In this particular blast, the most common size (the "mode") was in the order of 7 cm, which was slightly greater or less than the "average" because of the non-symmetrical nature of the distribution. The mode has greater practical significance, being the size produced in the greatest quantity.

The cumulative percent distribution of Fig. 4c demonstrates the approximate log-normality of the size distribution, and is perhaps the most convenient way of visualising both central tendency and dispersion. Central tendency is measured as the "median", different from both the mode and mean, which is the size D_{50}: half of the sample weight is smaller and half larger than this size. Dispersion can readily be measured in terms of statistics such as D_{10}, D_{25}, D_{60}, D_{75} and D_{90}, defined the same way as D_{50}. Two alternative measures are the Coefficient of Uniformity C_u (D_{60}/D_{10}) and the Coefficient of Sorting C_s (D_{75}/D_{25}). The larger these coefficients, the greater the dispersion.

3. GEOLOGY AND EXPLOSIVES

Measurements were recently made on three full scale test blasts to evaluate fragmentation. The results are to be discussed more fully elsewhere (Maerz, Franklin & Coursen, 1987). They demonstrated, however, a link between block size before and after blasting. Rock that was initially more closely jointed, as measured by photoanalysis of the benches, tended to finish up in smaller fragments. An exception was that near-surface seamy rock still yielded a disproportionate percentage of oversized blocks, because of clay-filled joints and lack of confinement.

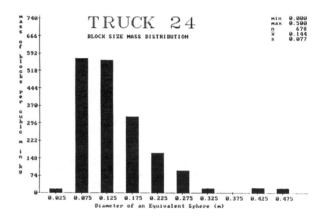

Fig. 4b: Distribution of mass in any given d_{es} class.

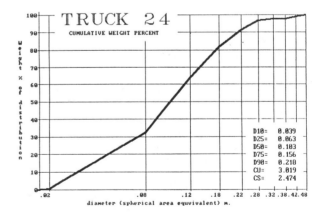

Fig. 4c: Cumulative mass distribution.

691

4. POTENTIAL OF THE TECHNIQUE

Despite being at an early stage in its development, the photoanalytical technique compares favorably with conventional methods of measuring fragmentation. Using the comprehensive photographic record, stored digitally, analysis can be carried out without disrupting production, and results can be re-analysed at a later date if necessary.

Methods of sampling and vector digitization have been developed and tested. Using scaled-down simulations, a semi-empirical method has been developed for correcting or "unfolding" the apparent size distribution to give a true volumetric or weight distribution of the broken rock. A start has been made in selecting and defining parameters of the size distribution, so as to adequately represent the efficiency of blasting and the quality and value of the product in its various applications.

The photoanalysis technique will, however, become much more efficient, and a really useful and practical tool, with replacement of the vector by the raster (automatic) method of digitization, and with further development of the unfolding formula, testing it for scale effects, different size distributions and block shapes.

5. ACKNOWLEDGEMENTS

The authors would like to thank CANMET and NSERC for their financial assistance and sponsorship of the development work, and Vulcan Materials Company for providing a site to test it.

6. REFERENCES

Aimone, C. T., and Dowding, C. H. 1983. Fragmentation Measurement Results for Fourteen Full-Scale Production Blasts: A Comparison with a Three Dimensional Wave Code. Proceedings of the 9th Conference on Explosives and Blasting Technique: 310-333.

Bhandari, S., Vutukuri, V. S. 1974. Rock Fragmentation With Longitudinal Explosive Charges. Proceedings of the 3rd Congress of the ISRM 2B: 1337-1342.

Carter, J. W. 1977. Analysis of a Simple Photographic Method Proposed for Determining Size Distribution of Ore Fragments. USBM Lab Report RBM 77-03: 1-19.

DeHoff, R. T., and Rhines, F. N. 1968. Quantitative Microscopy. New York: McGraw-Hill.

Dick, R. A., Fletcher, L. A., and D'Andrea, D. V. 1973. A Study of Fragmentation From Bench Blasting at a Reduced Scale. USBM Report 7704: 1-22.

Franklin, J. A., and Maerz, N. H. 1986. Digital Photo-Analysis of Rock Jointing. 39th Canadian Geotechnical Conference: 11-20.

Gama, C. D. 1984. Microcomputer Simulation of Rock Blasting to Predict Fragmentation. Proceedings of the 25st U.S. Symposium on Rock Mechanics: 1018-1030.

Gozon, J. S., Britton, R. R., and Fodo, J. D. 1986. Predetermining Average Fragment Size: A Case Study. International Symposium on Application of Rock Characterization Techniques in Mine Design: 190-195.

Greenland, B. J., and Knowles, J. D. 1969. Rock Breakage. Mining Magazine 120: 76-83.

Just, G. D., and Henderson, D. S. 1971. Model Studies of Fragmentation of Explosives. Proceedings of the 1st Australia-New Zealand Conference on Geomechanics 1: 238-245.

Lovely, B. G. 1973. A Study of the Sizing Analysis of Rock Particles Fragmented by a Small Explosives Blast. Australian Geomechanics Society National Symposium on Rock Fragmentation, Adelaide: 24-34.

MacKenzie, A. S. 1966. Cost of Explosives - Do You Evaluate it Properly? Mining Congress Journal, May 1966: 32-41.

Maerz, N. H., Franklin, J. A., and Coursen, D. L. 1987. Fragmentation Measurement for Experimental Blasting in Virginia. To be presented at the Society of Explosives Engineers 3rd Mini-Symposium on Explosives and Blasting Research, February, 1987.

Noren, C. H., and Porter, D. D. 1974. A comparison of Theoretical Explosive Energy and Energy Measured Underwater with Measured Rock Fragmentation. Proceedings of the 3rd Congress of the ISRM 2B: 1371-1375.

Santalo, L. A. 1976. Integral Geometry and Geometric Probability. London: Addison-Wesley.

Singh, D. P., Appa Rao, Y. V., and Saluja, S. S. 1980. A Laboratory Study of Effects of Joints on Rock Fragmentation. Proceedings of the 21st U.S. Symposium on Rock Mechanics: 400-410.

Underwood, E. E. 1970. Quantitative Stereology. Reading Massachusetts: Addison-Wesley.

Weibel, E. R. 1979. Stereological Methods. Vol. 1. Practical Methods for Biological Morphometry. London: Academic Press.

Weibel, E. R. 1980. Stereological Methods. Vol. 2. Theoretical Foundations. London: Academic Press.

Wirtschaftlichkeitsuntersuchungen bei Sprengvortrieben durch die 'Quantifizierung des Einflusses der abrasiven Minerale'

Rentability evaluation of underground excavation by blasting by means of 'quantification of the influence of abrasive minerals'

Evaluation de la rentabilité de travaux d'excavation souterrains à l'explosif moyennant la 'quantification de l'influence de minérais abrasifs'

ERIK MIKURA, Ing.Dr., Baugeologe, Korneuburg b.Wien, Österreich

SUMMARY The parameter "Abrasive minerals" is known to be a coefficient influencing rock excavation works; up to now, there has been no method of quantification. After analysing excavation and advance works through many years a method has been found.

The solution was based on advance work by full face milling machines and, over the evaluation of part face milling, proceeded to excavation by blasting. The advance is split up into cost brackets. The degree of abrasiveness is subsequently quantified according to the Rosiwal method, the value for quartz being supposed to be 100; the resistance to wear of other minerals is defined in relation to quartz. The weighting is effected according to the volume percent. Grain size is supposed to be 1 or cohesion equal to adhesion. The quantification of the change in cost with regard to the cost brackets (e.g. costs for bore holes, explosives consumption, presence of pellets in the debris, loading and transport, installation, ventilation) is made according to the different relations. These relations and the limits of the method are represented. The entire example applies only to cases of changing "abrasion" parameter in solid rock.

SOMMAIRE Le paramètre "minérals abrasifs" est connu comme un coefficient qui influence les travaux d'excavation, mais jusqu'ici il n'y avait pas de méthode de le quantifier. Après de analyses faites durant des années sur des travaux d'excavation et d'advancement, une méthode a été développée.

La solution commença par les travaux d'excavation moyennant des machines à pleine section, et passa par l'évaluation de machines à section partielle aux excavations à l'explosif. L'avancement est divisé en classes de coûts. Le degré d'abrasivité est ensuite qunatifié suivant la méthode de Rosiwal, le coefficient supposé du quartz étant 100; la résistance à l'abrasion d'autres minerals est définé en relation au quartz. La pondération se fait en fonction de la part de volume. La granulométrie supposée est 1 ou la cohésion égale à l'adhésion. La quantification de la modification des coûts relatifs aux classes de coûts (p.ex. frais pour le forage de trous, consommation en explosifs, présence de gros morceaux dans le débris, chargement et transport, installation, ventilation) est effectuée selon les relations différentes. Ces relations et les limites de la méthode sont montrées. L'exemple entier s'applique seulement aux changements du paramètre "abrasion" dans les rochers solides.

KURZFASSUNG Der Parameter "Abrasive Minerale" ist als Einflußgröße beim Felsausbruch bekannt, eine Methode seiner Quantifizierung fehlte bisher. Durch langjährige Analysen von Abbauen u. Vortrieben konnte jedoch ein Verfahren gefunden werden.

Den Ausgang nahm die Lösung bei den Vollschnitt-Fräsvortrieben und ging über die Auswertung bei Teilschnittmaschinen bis zu den Sprengvortrieben. Der Vortrieb wird in Kostengruppen aufgegliedert. Die Abrasivität wird dann nach der Methode von Rosiwal quantifiziert; dabei ist Quarz 100 und bei den anderen Mineralien wird ihre Schleißhärte zu Quarz in Relation gestellt. Die Gewichtung wird entsprechend dem Volumsanteil vorgenommen. Die Korngröße wird mit 1 angenommen oder die Kohäsion der Adhäsion gleichgesetzt. Entsprechend unterschiedlicher Beziehung wird dann die Quantifizierung der Kostenänderung zu den Kostenstellen (z.B. Bohrlochherstellung, Sprengmittelbedarf, Stückigkeit des Haufwerkes, Lade-u. Transportverschleiß, Einbauverschleiß u. Lüftungsaufwand) durchgeführt. Diese Beziehungen werden dargestellt. Die Grenzen des Systems werden angeführt. Das gesamte Beispiel betritt nur den Fall mit wechselndem Parameter "Abrasion" bei standfestem Gebirge.

1) AUSGANGSBASIS

Für die Anbotslegung ist eine verläßliche Vorbestimmung der Kosten notwendig. Diese Bestimmung der Kosten muß exakt quantifizierbar sein. Dazu müssen die Untergrundverhältnisse analysiert werden, und zu jedem Teilparameter muß der entsprechende Einfluß auf die Kosten festgestellt werden. Diese müssen in eindeutigen Abhängigkeiten dargelegt werden. Gleichzeitig mit der Aufnahme des Gebirges muß eine Analyse der Abbauvor-

gänge vorliegen. Auch hiebei sind die einzelnen technischen Schritte in ihren Kosten zu erfassen. Die Gegenüberstellung geologischer Parameter zu technischen Schritten muß dann zeigen, wie die gegenseitige Beeinflussung verläuft. Für Felslösungsvorhaben ist das Verschleißverhalten des Gebirges gegenüber den Materialien der Abbaugeräte (Stahl, Räder etc.) seit langem bekannt. Eine verläßliche Zuordnung war bisher jedoch kaum umfassend möglich.

2) BISHERIGER STAND DER BESTIMMUNG DES GEOLOGISCHEN PARAMETERS

a) Bestimmung des Quarzgehaltes (Modalanalysen)

Quarz ist das häufigste Mineral mit einer Härte größer Stahl. So wird der Volumsanteil von Quarz oft als Vergleichsbasis für den Verschleiß und andere damit zusammenhängende Kostenkriterien genannt.

ak)Kritik an der Heranziehung des Quarzgehaltes.

Es gibt andere Silikate ähnlicher Härte und andere Minerale, die dadurch vernachlässigt werden. Verbandseigenschaften und Kristallisation sind nicht berücksichtigt.

b) Bestimmung des Anteiles von SiO_2.

Hier werden chemisch alle SiO_2-Anteile ermittelt, auch solche in nicht Quarz-Form.

bk)Kritik an der Bestimmung des Anteiles von SiO_2.

Es gibt andere Minerale, die ähnliche Eigenschaften wie Quarz aufweisen, und ebenfalls wird die Verbandsfestigkeit u. Kristallisationsform vernachlässigt.

c) Versuche zur Bestimmung der Verschleißeigenschaften.

Hier ist das Feld sehr groß und viele Versuche sind auch für spezielle Zwecke geeignet, aber meist aufwendig und erfassen große Versuchsobjekte, z.B. Cerchar (Ritzhärte) Rosiwal 1896 (Schleifhärte) etc.

dk)Da die Beanspruchung auf die Abbaumaterialien meist "schleifender" Art ist, ist die Schleifhärte den tatsächlichen Gegebenheiten am ehesten entsprechend. Problematisch ist die direkte Versuchsdurchführung (Vergleichbarkeit des Schleifmittels u.-Vorganges, Zeitaufwand).

d) Kombination Schleifhärte der Einzelminerale u. Mineralaufbau. Schimazek-Knatz (1976) führen diese Kombination durch.

dk) Nicht von allen Mineralien bestehen geeignete Vergleichszahlen Rosiwals. Selbst durchgeführte Ergänzungen sind schwer in das System integrierbar.

3) BISHERIGE ANWENDUNG

Die Notwendigkeit für Verfahren zur Quantifizierung der Kosten der Abrasivität auf Untergrundlösung wurde bei den Fräsvortrieben erkannt. Die Kosten der Teilschnitt- u. Vollschnittmaschinen waren auch für weniger vertieft Arbeitende deutlich von den Verschleißeigenschaften abhängig. Bei Teilschnittmaschinen war auch die Grenze der technischen Einsatzmöglichkeit von den Verschleißdaten abhängig. So wurde auch in diesem Bereich der Weg von 2.a - 2.d gegangen. Die Bergbauforschung, die in ähnlichen Gesteinen starke Abweichungen der Leistungen feststellte, erkannte diesen Parameter und war auch das erste Beobachtungsgebiet für Sprengvortriebe. Die knappen Preise der Bauindustrie veranlaßten einige wenige Unternehmungen zu gezielten Forschungen auf diesem Gebiet. So ist der Verfasser von 1977 - 1987 in einem österr. Bauindustrieunternehmen mit dieser Problematik befaßt gewesen. Durch die Anwendung der Neuen österreichischen Tunnelbauweise gehörte diese Forschung zur Verfeinerung nicht der technisch-statischen Methode, wohl aber der Kalkulatorisch-Leistungsüberprüfenden und - bestimmenden Komponente. Damit wurde die Wirtschaftlichkeit durch Reduzierung des Preisrisikos weiter erhöht. Der laufende technische Fortschritt in der Maschinentechnik und Metallurgie verlangt laufende Adaptierungen der Forschungsergebnisse.

4) BESTIMMUNG DER TECHNISCHEN PARAMETER BEI SPRENGVORTRIEBEN.

a) Art der beeinflußten Parameter

Die Abrasivität beeinflußt folgende Leistungsteile primär:

Ausbruch:

- Bohrlochherstellung
- Sprengwirkung (Nachzerkleinerung)
- Schutterung (Ladeverschleiß, Fahrverschleiß, Einbauverschleiß)

Die Stützung - Bohrlochherstellung f. Ankerung, Auskleidung ev. Weiterverwendung - wird in diesem Artikel nicht weiterverfolgt. Auch der Verschleiß in der Gruppe Schutterung wird hier nicht näher untersucht.

Die Kostenfaktoren der Bohrlochherstellung

sind im wesentlichen der Bohrkronenverschleiß und die Bohrzeit. Auch die Bohrgeräte und besonders die Bohrstangen unterliegen Verschleißerscheinungen. Diese werden jedoch mit den direkten Verschleißkosten mitberücksichtigt. Bei diesem Verschleiß und diesen Reparaturkosten spielen viele subjektive Kriterien eine Rolle, wie z.B. das Alter der Geräte, die Sorgfalt der Bedienung und die Zuverlässigkeit der Wartung.

Für die Wirtschaftlichkeitsuntersuchung bleiben Bohrwerkzeugverschleiß und die Bohrzeit als objektiv den geologischen Verhältnissen zuordenbare Faktoren.

Die Kostenfaktoren der Sprengwirkung

sind ebenfalls zu untergliedern. Hier ist die Kleinstückigkeit bzw. Kornform des Haufwerkes und bei Bauvorhaben die Profilgenauigkeit zu nennen. Beides kann durch eine Änderung der Bohrlochabstände u. Sprengschemata beherrscht werden. Die Gesamtsumme u. Lage der Bohrlöcher ist aber wieder "geologiebedingt."

Für die Wirtschaftlichkeitsuntersuchung bleibt die Bohrlochanzahl ein objektiv den geologischen Verhältnissen zuordenbarer Faktor.

b) Die Abrasivität der Minerale beeinflußt folgende Parameter sekundär.

b)1 Kosten der Lüftung und Entstaubung

Die bei den Arbeiten "Bohren u. Sprengen", tw. auch Laden entstehenden Staubmengen u. Zusammensetzungen sind ebenfalls von den Mineralen der Abbaugesteine abhängig. Hiebei ist sowohl die Staubmenge als auch die Staubzusammensetzung von Bedeutung, wird doch in der Luftmenge wegen der Silikosevorsorge der Silikatanteil bestimmt.

Die Basisleistung der Lüftung muß auch bei längeren Bohrzeiten berücksichtigt werden.

b)2 Energiekosten

Sowohl erschwertes Bohren als auch länger laufende Lüftung und Beleuchtung verursachen sekundäre Kosten.

c) Die Abrasivität der Minerale beeinflußt folgende Kostenparameter tertiär.

Die folgenden Kostenparameter werden überwiegend durch die Arbeitszeitverlängerung, selten durch die Mehrmassen bestimmt.

c1) Bauinfrastruktur wie: Vorhalten der Einrichtung, Bereitstellung, Grundstücksmieten etc.

c2) Bauüberwachungspersonal (Bauregie) und Baustellenbezogene Kosten der Hauptverwaltung (Geschäftsführung u. Zentralregie).

d) Gewicht der beeinflußten Parameter

d1) Allgemein
Die Summe der von der Abrasivität beeinflußten Faktoren ist sehr hoch. Nur ihre große Bedeutung rechtfertigt den Aufwand für die Untersuchungen. Außerdem muß noch die Gruppe der sekundär beeinflußten Faktoren herangezogen werden. Dabei handelt es sich zum Beispiel um den Energieaufwand u. Lüftungskosten. Eine dritte "tertiäre" Einflußgruppe nicht unbedeutender Größe ist die Bauregie und dieser ähnlichen Kosten in den Zentralen (des Auftraggebers u. Auftragnehmers).

Abrasive Minerale können die Kosten des Ausbruchs um 50 % erhöhen.

Die durchgeführten Untersuchungen an Vortrieben nach der NÖT (NATM) zeigten diese Parametergröße deutlich. Dabei darf nicht vergessen werden, daß viele im Baujargon als "Härte" oder "Güte" eines Gesteins bezeichnete Eigenschaften tatsächlich den abrasiven Mineralien zuzuordnen sind. Die Prozentangabe von 50 % betrifft nur jene Ausbruchskosten, die direkt vom Vortrieb verursacht sind. Umgelegte Kosten aus der Stützung müssen vorher ausgeschieden werden. Auch manche Stützungsaufwände sind aber von der Abrasivität der Minerale beeinflußt, wie z.B. die Ankerbohrlochherstellung.

d)2 Aufgliederung
Bei Ausbrüchen wird in folgende Gruppen untergliedert:

Leistungsgruppe	Anteil an Gesamtaufwand
1) Bohren (der Sprenglöcher)	64 %
2) Laden-Schießen-Lüften	14 %
3) Schuttern	22 %
4) (Stützen/Sichern)	(nicht umgelegt)

Wenn man nicht nach dem Umlageprinzip arbeitet, so sind nur die Gruppen 1 - 3 zu berücksichtigen und somit 100 %. Im angeführten Beispiel sind die Zeitaufwände in % aufgegliedert. Dieses Beispiel betrifft einen Stollen mit geringem Querschnitt und relativ hartem, abrasivem Gestein. Die einzelnen Leistungsgruppen sind auch nicht scharf abgrenzbar, werden auch in den Polierberichten vieler Unternehmen nach dem Zeitaufwand erfaßt. Daher konnten vom Verfasser diese Übergruppen bei vielen Vortrieben analysiert und den Mineralanteilen gegenübergestellt werden. Etwaige nicht vom Mineralgehalt bestimmte Abweichungen konnten so auch erkannt werden (Geräteschaden, Schwankungen, Mannschaftsstärke etc.).

5) QUANTIFIZIERUNG DER ANTEILE VERSCHLEISSHOHER MINERALE

Gesucht wurde ein Verfahren, das mit kleinen Handstücken rasch zu einem objektiv, vergleichbaren Ergebnis führt.

Hiebei wurde vom Verfasser die Kombination

Rosiwalhärte

Druckfestigkeit

gewählt.

Die Bestimmung der Rosiwalhärte erfolgt dabei meist nicht im Direktversuch, sondern im Errechnen aus Modalanalysen vom Dünnschliff. Oft kann sogar die Mineralzusammensetzung aus Projektsbeschreibungen übernommen werden. So kann für die Genauigkeit der Angebotslegung ein ausreichendes Ergebnis erreicht werden.
Rosiwal setzt bekanntlich die Vergleichsgrenze Quarz = 100. (Mittel der Messungen parallel zu den verschiedenen Kristallachsen). Er mißt den Mineralverlust und nicht den Schleifmittelverlust.

Die Modalanalyse allein ermöglicht nicht die theoretische Errechnung der Schleifhärte von Gesteinen, die aus verschiedennen Mineralen aufgebaut für Teilschnittmaschinen eine Umrechnung unter Berücksichtigung von σ Druck- u. Scherfestigkeit sowie der Korngröße. Für die Anwendung im Sprengvortrieb sind diese Parameter anders zu gewichten, das Prinzip an sich bleibt aber gültig. So wird in den Fällen hoher Gesteinshärte (größer 600 kp/cm2) aus der Modalanalyse direkt ein theoretischer Verschleißbeiwert nach Rosiwal ermittelt. Die Korngröße kann hier in der Formel vom Schimazek-Knatz mit 1 festgesetzt werden.

6) QUANTIFIZIERUNG DER DRUCKFESTIGKEIT

Diese wird nach üblichen Versuchen, meist an Bohrkernen, die geometrisch genau bestimmt sind und deren Oberflächen ideal parallel und glatt hergestellt sind, einachsial festgestellt. Liegen Angaben des Projektes vor, werden nur an extremen Handstücken Eichversuche im eigenen Labor getätigt.

7) QUANTIFIZIERUNG DER ZUGFESTIGKEIT

Diese wird nur in Sonderfällen extrem zäher Gesteine berücksichtigt und nur in Relation zur Druckfestigkeit, dargestellt in ganzen Zahlen, in die Ermittlung aufgenommen.

8) SONSTIGE PARAMETER

8.1 Kohäsion-Adhäsion

Eine Unterscheidung der Bindung der Körner untereinander (Kohäsion) ist bei den meisten Kristallen nicht von der Internbindung (Adhäsion) trennbar. Daher erfahren diese beiden Größen keine differenzierende Bedeutung.

8.2 Korngröße

Wenn die Festigkeit hoch ist und wie in 8.1 erwähnt, Kohäsion und Adhäsion nicht getrennt feststellbar sind, ist die Korngröße auch eine nicht relevante Größe. In diesen Fällen wird sie "1" angenommen. Etwaige Sonderformen (z.B. auffallend große kompetente Kristalle in weicher Matrix) bleiben durch anders festgelegte Faktoren berücksichtigbar. Statistisch gesicherte Abhängigkeiten liegen mir dazu noch nicht vor.

8.3 Zerlegung

Dieser Parameter hat im allgemeinen beim konventionellen Vortrieb für den Bohrkronenverschleiß keine vortriebsbegünstigende Bedeutung. Eine starke Anisotropie oder Zerlegung des Gesteins im ungünstigen Winkel führt zu Vortriebserschwernissen. Bedingt durch die unterschiedlichen Ausbildungen der Trennflächen war mir bisher eine allgemein gültige Relation der Zerlegung zur Vortriebserschwernis nicht möglich. (Bei den Fräsen ist die Beeinflussung der Zerlegung nachweisbar und von Mikura 1978 und Wanner 1979 dargelegt).

Eindeutig führt der Wechsel unterschiedlich harter Gesteine oder offene Kluft zu hartem Gestein bei schleifendem Schnitt zur Bohrrichtung zu Bohrkostenerhöhung.

8.4 Bohrfortschritt und Bohrsystem

Grundsätzlich sind diese Relationen sowohl bei Drehschlag-, Drehbohrungen sowie unterschiedliche Anpreßstücke u. Kronengestaltungen gegeben. Für jede Gerätetype muß mit Eichversuchen oder aus den technischen Grunddaten die entsprechende Relation festgestellt werden. Sie kann in Grenzen des Feldes unterschiedlich verlaufen. Die rein mathematisch als gerade verlaufende (direkte) Abhängigkeit ist in Wirklichkeit leicht progressiv, doch ist auch diese Progression nur empirisch feststellbar.

9) BILDUNG DER VERSCHLEISSKOEFFIZIENTEN

Aus den Parametern Rosiwalhärte, ermittelt aus den Modalanalysen, wird bei Gesteinen großer, gleicher Härten ein Verschleißkoeffizient gebildet. Für diesen besteht eine lineare Abhängigkeit zur Bohrleistung in m/Zeiteinheit. [Abb. 1] Die unterschiedlichen Preisfaktoren ergeben in ihrer Summe auch eine lineare, direkte Abhängigkeit bei den Kosten [Abb.2a] Die Berücksichtigung der Zerlegung des Gebirges führt zu einer progressiven, nicht exakt bestimmbaren Abhängigkeit

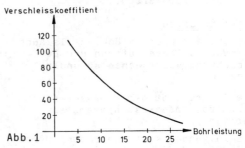

VERSCHLEISSKOEFFITIENT ZU BOHRLEISTUNG

Relation Verschleißkoeffizient zur Bohrleistung m/min.
Relation coefficient of abrasion - milling performance (m/min).
Relation coefficient d'abrasion - capacité de forage (m/min).

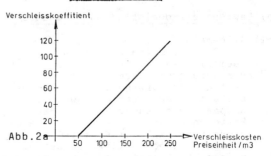

VERSCHLEISSKOEFFITIENT ZU BOHRKOSTEN

Relation Verschleißkoeffizient zu Verschleißkosten Preis/m3 . Direkte Abhängigkeit ohne Berücksichtigung der Zerlegung des Gebirges.
Relation coefficient of abrasion - wear costs (price/m3). Direct dependence without considering rock decomposition.

Relation coefficient d'abrasion - coûts d'abrasion (prix/m3). Dépendence direct sans considère la décomposition du rocher.

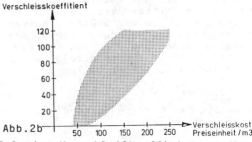

VERSCHLEISSKOEFFITIENT ZU BOHRKOSTEN

Relation Verschleißkoeffizient zu Verschleißkosten Preis/m3 Feld möglicher Lösungen mit Berücksichtigung der Zerlegung des Gebirges.
Relation coefficient of abrasion - wear costs (price/m3). Field of possible solutions under consideration of rock decomposition.
Relation coefficient d'abrasion - coûts d'abrasion (prix/m3). Solutions possible en considérant la décomposition du rocher.

[Abb. 2 b]. Ähnlich ensteht durch die
Überlagerung der Festigkeit mit der
Verschleißhärte eine progressive, un-
bestimmte Abhängigkeit.

LITERATUR

Jurecka, W. Kosten und Leistungen von
    Baumaschinen.
    Springer Wien-New York 1975.

Mikura, E.   Schnelle und verläßliche
    Verfahren zur Prognostizierung der
    Fräsleistung.
    Rock Mechanics 12, Springer Wien, 1980

Rosiwal, A. Neue Untersuchungsergebnisse
    über die Härte von Mineralien und Ge-
    steinen.
    Verh.k.k.geol.RA 1896, Nr.17 u. 18

Schimazek, J. und Knatz, H.   Die Be-
    urteilung der Bearbeitbarkeit von Ge-
    steinen durch Schneid- und Rollenbohr-
    werkzeuge.
    Erzmetall Bd. 29, H.3, 1976.

Wanner, H. und Aeberli, U.   Tunnelling
    machine performance in jointed rock,
    manuscript of International Congr.
    on Rock-Mechanics.
    Montreux 1979.

Weiss, E.H. und Mikura, E.   Geol.Kri-
    terien bei der Kostenermittlung von
    Abtragsarbeiten im kristallinen Unter-
    grund.
    Felsbau Jg.1,3/4 Glückauf Essen 1983

Rock weathering as a drillability parameter
L'altération des roches comme un paramètre de forabilité
Gesteinsverwitterung als ein Bohrbarkeitsparameter

A.MOURAZ MIRANDA, Professor, Instituto Superior Técnico, Technical University of Lisbon, Portugal
F.MELLO MENDES, Chairman Professor, Instituto Superior Técnico, Technical University of Lisbon, Portugal

SUMMARY: This paper deals with drilling results through a weathered sequence of rocks using two different drilling methods: down-the-hole hammer and diamond core drilling. Rate of penetration is log-related with rock physical parameters(density; Vickers microhardness; ultrasonic longitudinal velocity) and rock mass quality (R.Q.D.).
    Calculated specific energy also shows the same variations in the uppermost part of the drilling profile as the indicated physical parameters. It is this same portion of the hole the major difficult part for the drilling operation.

RESUMÉ: Dans le travail on présente des resultats de forabilité dans des sequences de roche alterées en utilisant deux méthodes de forages: marteaux en fond de trou et carottage rotatif avec outil diamantée. La vitesse de forage presente des relations diagraphiques avec des caracteristiques physiques des roches (densité; microdureté Vickers; vitesse longitudinal des ultrasons) et la qualité du massif de roche(R.Q.D.).
    L'energie specifique de forage calculée presente des variations importantes dans la partie supérieur du puit comme les paramètres physyques. Cette même partie du puit c'est normallement la plus difficile à traversée pour les foreurs.

ZUSAMMENFASSUNG: Dieses referat befabt sich mit bohrergebnissen durch eine verwitterbe gestensfolge, unter benutzung zweier verschiedener bohrmethodem: sinkbohrloch hammer und diamantkernbohrung.
Die bohrgeschwindigkeit ist mit den gesteinsphysikalischen parametern(dichte; vickers-mikroharte;ultraschall-longitudinalgeschwindigkeit) und der gebirgsqualitat(R.Q.D.) log-verbunden.
    Die berechnete spezifische energie zeigt im obersten teil des bohrprofils auch die gleiche variation wie die angegebenen physikalischen parameter. Es ist genau dieser abschnitt des bohrloches, der den schwierigsten teil fur den bohrablauf darstellt.

1 INTRODUCTION

Drilling through weathered rocks is always a concern for the drilling engineer. Nct only intact core samples are difficult to obtain from weathered materials but also when drilling through weathered rocks the sudden and sometimes unexpected changes in drillability requires special attention by the operators.
    Two types of drilling operations(waterwell drilling using down-hole hammers and diamond coring for geotechnical site investigation) supplied a large amount of data which can be considered of interest for drillability characterization of weathered materials. In a previous paper(MELLO MENDES & MOURAZ MIRANDA (1979)was shown how physical parameters varied along drilled cores on different grades of weathered greywacke.

2 FIELD WORK AND RESULTS

Down-hole drilling is a common practice by water well industry in granite and metassedimentary schist-greywacke areas in Portugal(MOURAZ MIRANDA, 1982) Drilling diameters are commonly 215 mm and 165 mm. The larger diameter is used for the first part of the hole to 10-15 m depth.
    Rotary diamond core drilling is normally conducted with 86 mm for the beginning of the hole and 76 mm for the rest of the drilling, for geotechnical site investigation.
    Using data from drilling recorders mounted on rigs it was possible to use as a field drilling parameter the instantaneous rate-of-penetration (R.O.P.).

Collected cuttings from down-the-hole drilling enabled same physical characterization of drilled rock .Vickers microhardness(VHN) and density(d). The diamond cored samples enabled the use of one more physical parameter:longitudinal ultrasonic velocity (VL).
    A typical section through weathered granite is shown in Fig.1 Diamond drilling was conducted close to a down-hole drilled hole.
    The intermediate top part of the weathered profile shows the most irregular variation of all the parameters (R.O.P.; large dispersion of VHN).
    Using data gathered from several drilled holes (± 50) in different rock types since top-soil to sand rock it was found that R.O.P. decreased from top to bottom of the hole with large variations in intermediate part of the profile.
    Calculating the specific energy of drilling, using BULLOCK(1976) formulae as in MOURAZ MIRANDA & MELLO MENDES(1983), it is possible to observe some correlation with the measured physical parameters.
Fig.2 indicates the calculated specific energy for the weathered profile of Fig.1 with reduction of the specific energy in the beginning of the hole followed by a steady increase with depth.
    Trying to understand the small variations on the production parameter(R.O.P.) in terms of geological variations the RQD was also noted(Fig.2). The upper part of the profile can be considered almost a soil and the intermediate upper portion a mixture of soil and rock.

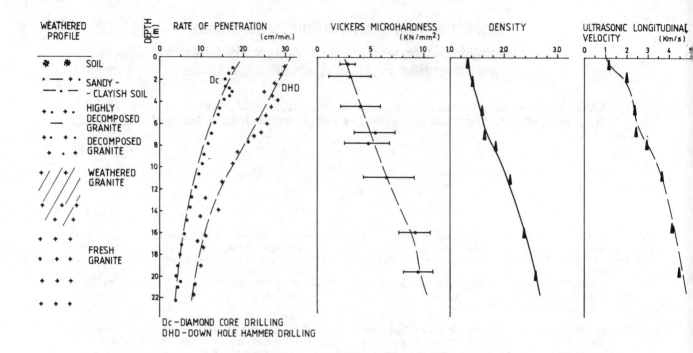

Fig.1. Variation of the rate of penetration(ROP) for down-hole-drilling and diamond core drilling, Vickers Microhardness(VHN), density(d) and ultrasonic longitudinal velocity(V_L) along a weathered granite profile.

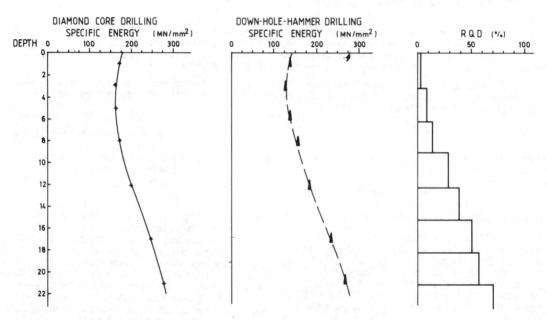

Fig.2. Calculated specific energy for a weathered profile

3 INDICATIONS FOR DRILLABILITY PREDICTION OF WEATHERED ROCKS

As per our knowledge it does not exist the universal optimum drilling method for a weathered sequence.

The upper part of the profile where drilling parameters show an irregular variation and to an unknown depth. Per drillers practice the selection of the drilling method is done empirically, with economic consequences, which could be optimised if the distribution of the various weathering degrees within a profile could be predicted in advance, particularly for the most weathered part.

Our aim of investigation was to predict the correction of drilling operation, optimizing it through the control of actual drilling parameters.

REFERENCES

MELLO MENDES, F.; MOURAZ MIRANDA, A.(1979)
Correlation of Rock Properties with a Bearing Workability
Proc.IVth. Cong.Int.Soc.Rock Mechanics,Montreaux,
v. III, p. 429-431

MOURAZ MIRANDA,A.(1982) Brief Survey of Portuguese Water-Well Drilling Industry
Proc. III Nat. Mining Eng. Meetings, Engineers Associations, Portugal

MOURAZ MIRANDA,A; MELLO MENDES,F. (1983)
Drillability and Drilling Methods
Proc. 5th Congr. Int. Soc. Rock Mechanics,Melbourne
v.5, E195-E200

Relations between conditions of rock mass and TBM's feasibility

Relations entre les conditions du massif rocheux et la faisabilité de TBM
Beziehungen zwischen den Gebirgsverhältnissen und der Ausführbarkeit der TBM

S.MITANI, Research Institute of Kumagai Gumi Co., Tokyo, Japan
T.IWAI, Research Institute of Kumagai Gumi Co., Tokyo, Japan
H.ISAHAI, Research Institute of Kumagai Gumi Co., Tokyo, Japan

ABSTRACT : In order to assess TBM's feasibility in various subterranean conditions, in-situ rock mass conditions were investigated following TBM excavation at two sites. The rate of penetration depends on durability of rock mass. The durability of rock mass is expressed clearly as energy consumption rate when the relation between the rate of penetration and consumed electricity is plotted, but is not substituted by proper geotechnical parameters as yet. The qualitatively classified rock mass conditions agree fairly well with correspondingly measured P-wave velocities and Schmidt Hammer reboundnesses distribution. The authors consider the rate of penetration, support requirement and the rate of utility of TBM by assessing these geotechnical parameters.

RESUME: Dans le but d'évaluer la faisabilité de TBM, dans des conditions rocheuses variées, les auteurs ont recherché des conditions du sol de deux sites où est exécutée la fouille par TBM. Le rapport de pénétration dépend de la résistance du massif rocheux. La résistance du massif rocheux est exprimée clairement quand est tracée la relation entre le rapport de pénétration et la consommation électrique, tandis que n'a pas été établie une relation entre ladite résistance et des facteurs géochimiques convenablement choisies. Les conditions des massifs rocheux classifiés suivant leurs qualités correspondent aux vitesses d'ondes P et a la distribution de rebondissement du marteau Schmidt. Le rapport de pénétration, les nécessités de soutenèment et le taux d'exploitation de TBM sont etudiés en évaluant ces facteurs géochimiques.

ZUSAMMENFASSUNG: Die Autoren haben die Geländeverhältnisse an zwei Orten, wo eine Aushöhlung durch TBM ausgeführt ist, untersucht, um die Ausführbarkeit von dieser Maschine in verschiedenen Felsverhältnissen abzuschätzen. Die Eindringgeschwindigkeit hängt von der Gebirgsbeständigkeit. Die Gebirgsbeständigkeit wird zwar klar wenn sich die Beziehung zwischen der Eindringgeschwindigkeit und dem Elektrizitätsverbrauch darstellt;die Gebirgsbeständigkeit wird jedoch auf geeinigten geotechnischen Faktoren noch nicht bezogen. Die qualitativ klassifizierten Gebirgsverhältnisse vertragen sich mit den diesbezüglich bemessenen Geschwindigkeiten der Druckwelle und mit der Verbreitung des Zurückprallens vom Schmidt-Hammer. Die Eindringgeschwindigkeit, die Notwendigkeit des Ausbaus und die Ausnutzungsquotient werden abgeschätzt.

Introduction

Tunnel Boring Machines have been used worldwidely in various geological conditions, and are to be used successively because of their following advantages: excavations can be done continuously without using any explosives, support requirement could be reduced largely because they don't damage the ground to be excavated so much, overbreak could be minimized, and so on. Moreover, there are great latent demand for small diameter TBMs in Japan because of on-going projects for development of small hydro powers.

However, it has not been cleared enough to assess TBM's performability in relation to variable ground conditions as are to be seen in Japan and other circum-Pacific regions.

The authors have carried out geological and geotechnical investigations following TBM driving at two sites for sewerage tunnel construction. The TBM used is made of KOMATSU, a ROBBINS type machine of 2.6 m in diameter, which is equipped with full shield intended to cope with difficult ground to be encountered. This paper presents the results of detailed studies on the assessment of TBM's performability, such as the rate of penetration, support requirement and the rate of utility of TBM in relation to rock mass conditions.

Geology of the sites

The two sites investigated locate in the central west part of Japan. Site A, length of the tunnel is about 1400 m, consists mainly of hard basement rocks which are of Palaeo-Mesozoic sedimentary rocks and dyke rocks intruded them at late Mesozoic Era, and at the upstream end soft Tertiary conglomerate overlies unconformingly, as are shown in fig. 1. The basement sedi-

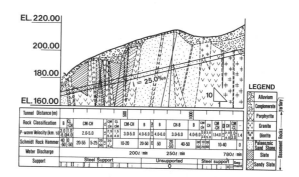

Fig.1 Geological Profile of Site A

mentary rocks are sandstone, slate and sandy slate, and the dyke rocks are porphylite and diorite. These basement rocks distribute like roof pendant being intruded by large granite body and thus they are jointed considerably. Along the plane of unconformity which dips gently to ahead of the excavation, the basement dioritic rocks are weathered considerably.

Site B, the length of the tunnel is almost 1800 m, consists mainly of granite which is weathered to varying extent, and at the downstream end soft Tertiary siltstone beds distribute in fault contact with granite as is shown in fig. 2. Granite is fairly massive and is classified its grade based mainly on the degree of weathering. There were two distinct fault zones and at two localities alluvial deposits make an appearance.

At both sites overburden is arround 10 to 30 meters and water inflow into the tunnel was very few.

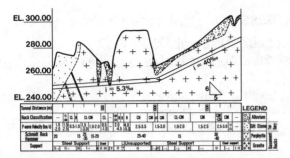

Fig.2 Geological Profile of Site B

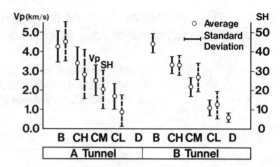

Fig.3 Relation between Vp, SH and Rock Classification

Conditions of rock mass

To make clear the relation between the performance of TBM and the conditions of rock mass, the authors have investigated with various survey methods following TBM driving.

First of all, the rock mass exposed at excavated surface was discriminated and mapped according to rock types and was classified qualitatively according to TANAKA's classification system which is employed widely in Japan, and joints and faults were also mapped and investigated their orientations and conditions. Then, measurement of P-wave velocity (Vp) was carried out at both side walls at each 1.0 to 1.5 m intervals and at the same time Schmidt Rock Hammer reboundness (SH) was measured there. Besides, coring and laboratory testing of their compressive strength, point load testing, etc. were also carried out.

As for TBM performance, thrust, torque and rate of penetration for each stroke of excavation (from 20 to 100 cms) were recorded as well as working cycletime and daily progress.

The qualitative classification system employed classifies rock mass into 6 categories, A, B, CH, CM, CL and D. The class A indicates that the rock mass is intact and the class D indicates that the rock mass is disintegrated and/or heavily weathered and is of very poor quality.

Table 1 Portion of each rock class at two sites

| Rock Class | Site A | | Site B | |
	Basement Rocks(1330m)	Tertiary(43m)	Basement Rocks(1600m)	Tertiary(115m)
B~CH	55.4%	—	19.6%	—
CM,CM'	35.8	100%	52.2	100%
CL~D	8.8	—	28.2	—

The portions of each rock class at two sites are shown in Table 1. Most portion of rock mass at site A consists of good to fair quality. Sandstones were massive and stable with few random joints. Slates were considerably unstable with a plenty of randomly distributed joints which cut the rock mass into small pieces, and were partly changed into clayey state. In porphylite bodies there distributed remarkably sets of joints which were parallel to subparallel with the axis of the tunnel and combined with randomly distributed joints made the rock mass at side walls of excavated surface very liable to collapse. The most poor quality rock mass was encountered at short intervals arround the unconformity between the basement rock and Tertiary conglomerate, where rock mass was highly weathered into soft material with high water content.

Most rock mass at site B consists of fair to poor quality. Granites were fairly massive with few remarkable joints but were weathered considerably. There were two distinct fault zones where granites were changed into fault clay which were 10 to 20

meters in width. Non-consolidated alluviums were treated by mortar injection from the surface before TBM driving and were changed into sufficiently stable ground.

Fig.3 shows the distribution of measured Vp and SH arranged for corresponding rock classes. Both Vp and SH distribution overlap at neighbouring rock classes but their average values represent each class of rock mass fairly well. Vp represents the average property of sufficiently large rock mass in which jointing and/or weathering develop. Instead, SH values are liable to be affected by the very local fractures or roughnesses of the surface of rock mass and is open to be scattered with the same rock class, and consequently it can not be solely used to represent the conditions of rock mass unless the rock mass is homogeneous. This is seen in fig. 3 where SH is scattered more at both sites than Vp, but at site B where granite is mainly affected by weathering and is fairly homogeneous with each rock class, the dispersion of SH is lesser.

Rate of penetration

The rate of penetration depends largely on the durability of rock mass and the force applied by TBM. The durability of rock mass is implied consequently in the relation between the rate of penetration and the consumed energy per unit volume of excavation as is seen in fig. 4. However, it has not been established yet to predict the rate of penetration in relation to in-situ rock mass condition.

Fig.5 shows the relation of the rate of penetration (Pe) vs P-wave velocity (Vp) and the penetration vs Schmidt Rock Hammer reboundness (SH) respectively. Penetration decreases as Vp and SH increase but Vp is more explanatory than SH, and when Vp and SH decrease extremely penetration seems to cease from increasing or rather to decrease.

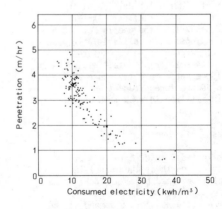

Fig. 4 Rate of penetration vs consumed electricity

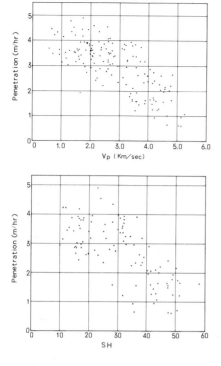

Fig.5 Rate of penetration vs Vp and SH

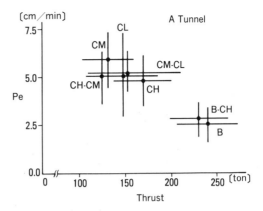

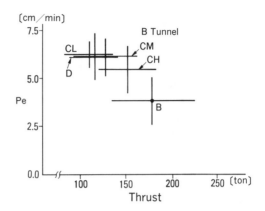

Fig.6 Relation between Penetration rate, Rock Class and Thrust

Fig. 6 shows the relation between penetration, thrust of TBM and rock classes. This figure is quite similar to fig. 5 when thrust is replaced by Vp or SH, but at site B thrust is smaller and penetration is higher than those of site A. From figures 5 and 6 we can conclude that the rate of penetration depends largely on the properties of rock mass, that there is a great gap between the fairly competent rock mass (B and B to CH) and the fractured and/or weathered ones (inferior than CH), and the latter require lesser energy to be excavated in spite of higher rate of penetration and thus it is deemed to be more economical from the viewpoint of excavation. The dispersion of points implies various factors' contribution such as crack distribution and their conditions, abrasion of cutters, tunnel alignment, and so on.

Support requirement

Table 2 shows the types of support installed. At both sites conventional support measures were adopted not only because considerable portion of poor quality ground was supposed to appear but also because shotcreting and rockbolting at small diameter tunnel is yet very unworkable when they are to be employed at excavating face area.

Fig. 7 shows the relation between the types of support installed and Vp corresponding to them. The distribution of Vp corresponds fairly well to types of support, however, there is a significant difference with the range of Vp values between the sites A and B. This is explained by the difference of characteristics of rock mass between the two sites as following; at site A, rocks are hard but are jointed to varing degree and there were repeated variations of rock types, so the characteristics of rock mass at both side walls often differ strongly. Thus, the excavated surfaces were rather unstable and required more support even though their Vp values are high. At site B, however, the rocks were mainly massive granite being weathered to varying degree with few jointing and were more homogeneous than site A, which introduced more stable ground despite of their lower Vp values.

Table 2 Type of Support Installed

Type	Support		Span (m)
O	Unsupported		
I₋₁	Steel Rib	⸢ 125×65	1.5
I₋₂	Steel Rib	H 100×100	1.5
II₋₁	Steel Rib	H 100×100	1.2
II₋₂	Steel Rib	H 100×100	1.0
III	Steel Rib	H 100×100	0.8
IV	Steel Segment		

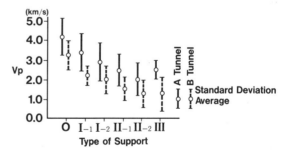

Fig. 7 Relation between Vp and each type of support

Thus, when we consider the stability of rock mass to be excavated, not one but several factors are to be assessed. The authors propose support criterion with conventional support measure for small diameter TBM as is shown in fig. 8. When performability of rockbolting and shotcreting for small diameter tunnel be im-

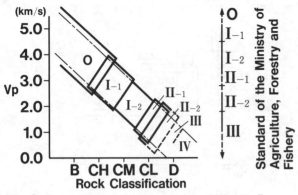

Fig.8 Proposed support criterion in relation to Vp and Rock Classification

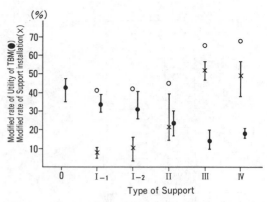

Fig.9 Relation between Type of Support and modified rate of utility of TBM and support installtion

proved they, too, should be employed as principal support measure. In this figure support types are distinguished with both Vp and rock class, and support criterion recommended by the Ministry of Agriculture, Forestry and Fishery for the same tunnel size to be excavated by blasting are shown with relation to Vp. Here, it is clearly noticed that the support requirement is reduced considerably by TBM excavation compared to conventional blasting method.

Rate of utility of TBM

The rate of utility of TBM depends largely on geologic conditions, and for small diameter tunnel where only single track is available for muck transportation, tunnel length also is a affecting factor due to the time required for waiting the muck cars to go and back.

Fig. 9 shows the modified rate of utility of TBM (●), modified rate of time required for support installation (x) and the sum of both (○) in relation to types of support installed, where effect of tunnel length is eliminated. The rate of utility of TBM is extremely low where heavy support is required (support type Ⅲ and Ⅳ) and the time required for the installation of support is particularly high there, and consequently TBM driving in such poor ground condition is far from economical one. On the contrary, when support requirement is fairly light (support type Ⅰ and Ⅱ) the drop of the rate of utility of TBM is slight and the time required for support installation is small in amount. The rate of penetration of TBM is larger with the rock mass which requires lighter support than with that which requires no support as aforementioned, thus the averge progress of tunnel driving is quasi-same with them, and TBM excavation here is economically acceptable (support type Ⅱ is regarded to be critical).

Discussion

The authors considered the rate of penetration, support requirement and the rate of utility of TBM to assess her performability in relation to rock mass conditions. The parameters employed to substitute the conditions of rock mass are selected supposing that they can represent the state of rock mass most adequately as well as that the investigation for them could be carried out at field easily. They are the qualitative rock mass classification system, P-wave velocity and Schmidt Rock Hammer reboundness.

The rock mass classification is valid practically, as it assesses integratedly the geotechnical properties of rock mass, density and conditions of jointing and degree of weathering, though, each rock class has some range of variation, which leads to some dispersion of corresponding geotechnical parameters essen-

tially. P-wave velocity is in good agreement with rock classes and is also one of the appropriate parameters to assess TBM's feasibility. However, the decrease in Vp value is caused by jointing as well as weathering and these effects can not be discriminated by Vp alone. In this sense Schmidt Hammer reboundness measurement or some other tests for measurement of hardness of rock mass make the necessary supplemental means.

The most practical way to assess the feasibility of TBM, the authors assume, is to examine these parameters synthetically, and the study for predicting the rate of penetration and of utility of TBM is undergoing with the said parameters inclusive of mechanical factors such as thrust and tunnel alignment. Neverthless, the orientation of joints as well as the abrasion of cutters are of much importance and these effects should be treated afterwards in a simplest way possible.

This report is based on the research carried out at two sites of small diameter TBM. Further investigations are necessary to cover wide varieties of geological and geotechnical conditions, and as for larger diameter machines support requirement is supposed to differ considerably from the results reported here, which also require further investigations.

Refferences

Bieniawski, Z. T. 1974. Estimating the strength of rock materials. J. of South African Ins. Min. Met., March

Tarcoy, P. J., et. al. 1975. Rock hardness index properties and geotechnical parameters for predicting tunnel boring. NTIS

Nord, G., et. al. 1979. European views on mechanical boring vs drill and blasting tunnelling. RETC Proc., vol. 1.

Farmer, I. W. & Glossop, N. H. 1980. Mechanics of disc cutter penetration. Tunnels and Tunnelling, July.

McFeat-Smith, I. & Tarcoy, P. J. 1980. Tunnel boring machines in difficult ground. Tunnels and Tunnelling, Jan./Feb.

Kikuchi, K. et. al. 1982. Geotechnically integrated evaluation on the stability of dam foundation rocks. Proc. 14 th ICOLD, Q .53, R. 4.

Tanimoto, C. & Ikeda, K. 1983. Acoustic and mechanical properties of jointed rock. Preprints of 5 th International Congress of ISRM, Melbourne.

Fujita, K. & Kasama, T. 1983. Geology of Kobe district Geological Survey of Japan.

Mitani, S. et. al. 1985. Geological condition and its impact on the feasibility of TBM. J. of the Japan Society of Eng. Geology, 26-1.

Mitani, S. et. al. 1985. On the prediction for the rate of penetration of TBM. J. of the Jap. Soc. of Eng. Geo., 26-4

On the performance and dynamic behaviour of a shield type blind raise boring machine

Sur la performance et le comportement dynamique d'une foreuse de cheminée en montant avec tubage
Der Bohrfortschritt und das dynamische Verhalten einer mit Schildmantel ausgerüsteten Blindschachtbohrmaschine

Y.NISHIMATSU, The University of Tokyo, Japan
S.OKUBO, The University of Tokyo, Japan
T.JINNO, The Dowa Mining Co. Ltd, Chiyoda-ku, Tokyo, Japan

ABSTRACT: This study was started to improve the boring capacity of the blind raise boring machine. The theoretical examination on the basis of the in-situ measurement preliminarily conducted indicates that an additional stabilizer is necessary to prevent the buckling of drill rod, otherwise boring capacity may be limited to 30m. After the comprehensive study including experimental work by a model rig to certify the safety of the new system, two test borings, where upper and middle stabilizers were used, were carried out to complete 39 and 43m raises. Though the depth of 43m is a new record for this machine, high increase rate of segment jacking force encountered at the depth more than 30m was found to be a new source of trouble remained for future research.

RÉSUMÉ: Nous avons commencé cette étude pour améliorer la capacité d'une foreuse montante de cheminée. L'examen théorique sur les données préalablement mesurées aux chantiers montre que l'équipement d'un stabilisateur supplémentaire est nécessaire pour prévenir la flexion des tiges, autrement la capacité de foration peut être limitée à trentaine mètres. Après l'étude complète y compris les essais dans un laboratoire par une foreuse modèle réduit pour attester la sûreté de ce nouveau système, nous avons exécuté deux test forages en utilisant des stabilisateurs supérieur et central, et pu aboutir à 39 et 43 mètres de foration. La force nécessaire pour fourrer des tubes dans un trou a augmenté rapidement au delà de trentaine mètres, bien que la foration de 43 mètres soit un nouveau record pour cette foreuse, ce qui est une difficulté nouvelle à résoudre par des recherches à venir.

ZUSAMMENFASSUNG: Die Verbesserung der Bohrfähigkeiten einer mit Schildmantel ausgerüstet Blindschacht-bohrmaschinen ist das Ziel dieser Forschung. Durch theoretischen Untersuchungen auf Basis der vorläufig durchgeführt Feldmessungen ist es gezeigt dass ist ein zusätzlich mittel Querträger notwendig vor dem Knick des Bohrgestänge zu beschützen und die Bohrfähigkeiten über 30m zu erhöhen.

Auf Basis der verschiedenen theoretischen Berechnungen und Modelversuchen wird eine verbesserte Bohrmaschine die ist mit dem Oberen und mittelen Querträger ausgerüstet entwicket und in einem Zeche eingesetzt. Der Erfolge dieser Verbesserungen ist mit dem leistungsfähigen Aufbrechungen der 39m und 43m hohe Blindschachte bestätigt.

Die Steigerung der Bohrfähigkeit von 30m auf 43m i.e. Erhöhung um rund 50% ist beachtbar, aber doch wird die bemerkenswürdigen Aufsteigung der Vorschubkraft des Schildmantel im über 30m hohe Aufbruch bemessen. Es ist ein nächste Problem das muss auf künftigen Untersuchungen gelösen sein.

1.INTRODUCTION

The development of the blind raise boring machine was a joint undertaking by the Dowa Mining Co.Ltd. and the Koken Boring Machine Co.Ltd.. This machine was built when it became apparent that the conventional shaft raising methods were not only uneconomical but also dangerous in difficult ground conditions frequently encountered in black ore mines such as Matsumine mine of Dowa Mining Co.Ltd.. The first unit was introduced into Matsumine mine in 1976 and more than 30 raises had been successfully completed at the end of 1981.

Basically the unit consists of a boring machine and a segment erection component. The segments of the shield frame are installed and jacked upward to keep the wall stable immediately after each boring cycle is over. Boring capacity of this machine was limited to 30m, and within this limit drill rods are strong enough with only one stabilizer located just beneath the cutter head.

This study was started to improve the boring capacity up to the depth of 60m. This improvement is necessary for extensive use of this machine in the boring of ventilation shaft, chute, pipe and cable leading pit in mines, tunnels and underground power stations.

First of all, the data was collected in-situ, such as thrust, torque and horizontal load to the stabilizer. The theoretical examination on the basis of the in-situ measurement indicates that an

additional stabilizer is necessary to prevent the buckling of the drill rod. After the comprehensive study including experimental work by a 1/40-scale model rig to certify the safety of the new system, test borings, where upper and middle stabilizer were used, were carried out.

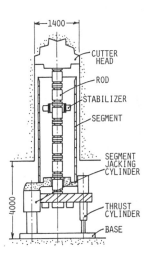

Figure 1. Schematic diagram of blind raise boring machine.

Table 1. Specifications of blind raise boring machine

Drilling Capacity	Dia. 1.4m
	Depth 30m
Rotation	by Hydraulic Motor
Speed	0 to 13rpm
Torque	max. 37kN•m
Thrust	by Hydraulic Cylinder
Load	max. 700kN
Stroke	1.2m
Shield Jack	by Hydraulic Cylinder
Load	max. 2000kN
Stroke	0.6m
Cutter Head	Dia. 1.4mx1.165m, 2390kg
Rod	Steel Pipe with Flanges
	Dia. 0.4mx1.0m, 265kg
Sield Segment	Steel Ring of Three Pieces
	Dia. 1.3mx0.5m, 110kg

2.PRELIMINARY DATA COLLECTION

In Figure 1, schematic drawing of the unit is shown and main feature are summarized in Table 1. The detail of the unit can be found in the previous work (Jinno et al. 1979).

Operational procedure is as follows: At the end of the boring stroke the chute and conveyer which discharges the cuttings are disconnected and removed. The drill rod holder is fastened and the drill drive head is lowered. A new drill rod is brought in by the drill rod positioner and connected. The drill rod holder is loosened and the segment jack-up saddle is lowered for the connection of another segment piece. After a complete ring of segments has been jacked up, the chute and conveyer are reconnected and a new boring cycle commences.

Field tests were carried out in the Matsumine mine in northern part of Akita prefecture, Japan. The difficult geotechnical conditions in this mine involves heavy rock pressure and clayey weak rock formation.

Test data of the following items were collected: (1)rotational speed of cutter head, (2)torque, (3)thrust applied to drill rod, (4)segment jack thrust, (5)force applied to stabilizer. The rotational speed is measured by a tachometer. The torque, the thrust of drill rod and the segment jack thrust are calculated from the oil pressure of actuator ram.

The stabilizer which consists of a steel ring supported by three arms as shown Figure 2(a) is located about 3m below the cutter head. The load cell was specially designed and fabricated to measure the three component of force applied to the steel ring. Basically the load cell is a grooved steel plate and strain gauges are bonded in the grooves as shown in Figure 2(b). The output of the load cell are recorded by multi-pen recorder through the drilling operation.

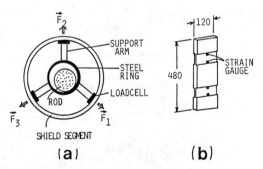

Figure 2. Plane view of stabilizer(a) and loadcell(b)

The results of the first and second field tests are shown in Figure 3 (a) and (b), respectively. In the figures, depth of raise is measured from the roof line of machine room and nearly equal to the total length of drill rods. The thrust force applied to

the cutter head is calculated by subtracting total weight of rods from the thrust jack force. The three components of force measured by load cells are periodically fluctuated corresponding to the rod rotation. An example of the test results is shown in Figure 4. The horizontal force F is calculated from the three components $\vec{F}_1, \vec{F}_2, \vec{F}_3$ by,

$$F = \left| \vec{F}_1 + \vec{F}_2 + \vec{F}_3 \right|_{max}.$$

In Figure 3, geological sections are also shown, which are determined from the cuttings sampled periodically. In Table 2, the approximate compressive strengths of rocks are shown together with the average penetration rate, classifying rock types into three categories (Nishimatsu et al. 1985).

For the first test result shown in Figure 3(a), the immediate roof strata was pyrite ore. From 2 to 8m of depth, boring was carried out through relatively soft layers and the horizontal force maintained in low level. At about 8m of depth, pyrite ore began to appear mixed with clay, and it was found that the horizontal force increased up to 100kN at the depth of 10m. From the depth of 10 to 22m, the strata of yellow ore was continued. The horizontal force decreased apparently from depth of 10 to 12m, but began to increase at the depth of

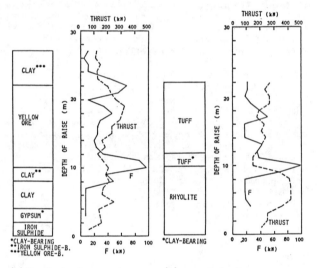

(a)first test   (b)second test
Figure 3. Result of the field tests on the prototype machine

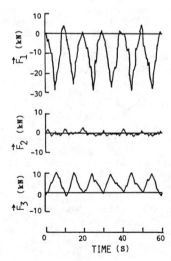

Figure 4. Force applied to stabilizer where tension is taken as positive. Measured in the first test at the depth of 8m and the rotational speed of bit is 6rpm.

706

Table 2. Typical rocks and penetration rate

Classification	Weak Rock	Medium Hard Rock	Hard Rock
Typical Rocks	Clay, Clay-Bearing Ore & Tuff, Mudstone	Tuff, Black Ore, Yellow Ore,	Siliceous Black & Yellow Ores, Rhyolite
Uniaxial Compressive Strength(MPa)	5 to 15	15 to 40	60 to 110
Penetration Rate(cm/min)			
Average	1.7	0.9	0.6
Maximum	4.3	1.0	0.8
Minimum	0.6	0.4	0.3

20m where yellow ore strata gradually changed to clayey strata.

Through the first test, thrust force was changed manually in the range of 100 to 300kN. The torque was roughly proportional to the thrust varying in the range of 4 to 9 kN•m. The rotation speed of cutter head was maintained at about 6rpm.

For the second test result shown in Fig.3(b), the layer of rhyolite was continued up to the depth of 10m, and thrust force was of high level in this region to keep a reasonable penetration rate. At the depth of 10m, the layer was changed from relatively hard rhyolite to soft clay bearing tuff. And the horizontal force took maximum value of 100kN at this depth where layer changed. After that, boring was conducted without any difficulty up to the scheduled depth of 22m. The rotation speed of cutter head was also maintained at about 6rpm, and the torque was changed proportional to the thrust .

3.ANALYSIS

The drill string(column) will be buckled if a force above critical value calculated by the Euler's column formula is applied;

$$P = n\pi^2 EI/l^2,$$

where P is a axial force, n is a numerical factor which depend on boundary condition, E is Young's modulus, I is geometrical moment of inertia and l is length of drill string i.e. total length of rods. Assuming both ends of drill string are free, the appropriate value for n is 1. Substituting the following values, P=600kN, E=210GPa and I=4.32•10^{-4}m^4 to the equation, a critical length is found to be 38.6m. Therefore, it is apparent that some modification to the unit is necessary to bore shaft to the depth of 60m.

Two possibilities were seriously examined; (1)Modification of the support system at the ends of drill string to the fixed end. (2)Settling an additional stabilizer. For the first case, n become 4 and the critical length is calculated to be 77m. However, more examination including a detail design of the support system and cost analysis showed that the second idea may be superior from the viewpoint of cost and reliability.

The second idea was, at first,carefully examined analytically including numerical simulation by a digital computer. The main results of the examination are as follows;
(1)One additional stabilizer which is located at the middle of the span may prevent the buckling if depth of raise is less than 60m.
(2)An upper stabilizer which is located just beneath the cutter head do not require modification,that is,the conventional stabilizer can be used for an upper stabilizer of the new unit.
(3)A horizontal force applied to the middle stabilizer which is located at the middle of span may be less than that for an upper stabilizer.
(4)A dynamic analysis shows that a critical rotational speed is more than 10rpm.

To certify the analytical results mentiond above, a

1/40-scale model of the unit was constructed.Changing the thrust, rotational speed and length of drill string, the experimental work was carried out. The result of this laboratory work showed that a boring capacity of the unit with two stabilizers were more than 60m as indicated also by the analysis.

4.IMPROVED BORING MACHINE

The analytical and experimental results encouraged the attempt to improve the boring capacity up to the depth of 60m. The results indicate that a conventional stabilizer is strong enough for upper and middle stabilizers. Then, stabilizers of the same design were used in the test boring, except a minor modification, to the middle stabilizer; the rubber of 2cm thickness was mounted on the three arms to protect the damage from the falling cuttings. The two test borings (the third and fourth) were conducted measuring forces applied to both upper and middle stabilizers.

The result of the third test boring is shown in Figure 5(a). In this case, a layer of tuff (middle hard rock) was extended up to the depth of 37m and boring was carried out in a satisfactory condition. The horizontal force to the upper stabilizer was less than 30kN when boring in the layer of tuff. It increased and took the maximum 52kN at the depth of 42m where boring was conducted through argillaceous tuff. However, the horizontal force to upper stabilizer was remained in low level and boring was able to be completed without difficulty. In this test boring, a middle stabilizer was located at 21.5m under the upper stabilizer. As expected on the basis of previous analytical and experimental work, the horizontal force applied to the middle stabilizer was also maintained in a low level, that is, less than 20kN.

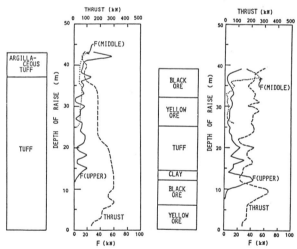

(a) the third test in which the middle stabilizer was set after boring 25m.

(b) the fourth test in which the middle stabilizer was set after boring 26m.

Figure 5. Result of the field tests on the improved machine.

The result of the fourth test boring is shown in Figure 5(b). As shown in this figure, the geological section in this case was more complex than that of the third case. Up to the depth of 12m, the layer of middle hard rock was extended and boring was conducted without difficulty. From the depth of 12m, the layer of clay (soft rock) appeared and horizontal force applied to the upper stabilizer increased up to 43kN. However, the horizontal force decreased gradually and maintained to be in low level until end of boring. The horizontal force applied to the middle stabilizer which was located also 21.5m under the upper one was kept in low level up to the depth of

30m, then increased to nearly 60kN. This result and the careful measurement of borehole after completion indicate that the center line of shaft was curved at the depth of 12-13m. The mis-alignment between the turn table(drill drive head), the middle and the upper stabilizers caused by a curved center line is considered to lead high horizontal force to the middle stabilizer.

5.DISCUSSION

Through the field test, it was found that the horizontal force to the stabilizer just under the cutter head, which may be approximately equal to the horizontal force to the cutter head, increases when boring the boundary of layers. It may be affected by the dip of boundary of layers. However, in this study, enough data to confirm this fact could not be obtained because of complex ground condition and shield segments which prevent us from direct observation of rock formation along shaft.

Concerning to the horizontal force to the middle stabilizer, it was found that the amplitude of that was usually smaller than that to upper stabilizer. The horizontal force to middle stabilizer may be mainly attributed to the mis-alignment which is caused by kink or dog-leg of the center line of the shaft. Before and during the field test, damage to the middle stabilizer by the falling cuttings was seriously examined. However, through the tests, no damage was happen to the middle stabilizer.

In this test, penetration rate was scheduled to be 1.0cm/min. For example, the measured penetration rate in case of the third boring test is show in Figure 6. It is found that the scheduled rate can be almost attained up to the depth of 40m. This result of penetration rate for the improved machine is better than the conventional average for medium hard rock shown in Table 2.

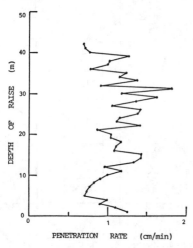

Figure 6. Penetration rate in case of the third test.

Through the test, it was found that the most critical factor remained for boring further depth was rapid increase of segment jacking force as shown in Figure 7. Improvement of jacking force capacity may be possible by some modifications, but, increase of segment strength may be not feasible from the viewpoint of cost analysis. The rapid increase of segment jacking force may be attributed to two reasons; (1)cuttings may drop and locked between the rock wall and outside face of segment. (2)Gradual time-dependent convergence of shaft cross section in case of soft layer under heavy rock pressure.

The study was started to improve the boring capacity. The analytical and experimental study indicated that boring of 60m-shaft will be possible by adding one more stabilizer. The test boring made

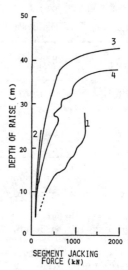

Figure 7. Segment jacking forces of the first to fourth tests.

a new record of depth for this raise boring machine. However, continuous increase of segment jacking force was remained as a problem which should be improved to bore the shaft up to 60m, economically.

REFERENCES

Jinno,T., Kotake,Y., Ohshika,H. & Yamatani,K. : Development of Blind Raise Boring Machine Used for Collapsible Formation, Proc. ICOMM held at Brisbane, Australia on July 26, 1979, p.339-344.

Nishimatsu,Y., Okubo,S. & Jinno,T. 1985. : A Trial to Improve the Performance of the Matsumine-type Boring Machine, J.MMIJ, vol.101, p.61-66.

Kalottenvortrieb in zerlegtem Gebirge mit geringer Überlagerung
Driving of the top heading in fractured rock with small overburden
Avancement de la calotte dans la roche fracturée avec petite couverture

R.PÖTTLER, Ingenieurgemeinschaft Lässer Feizlmayr (ILF), Innsbruck, Österreich
M.JOHN, Ingenieurgemeinschaft Lässer Feizlmayr (ILF), Innsbruck, Österreich

ABSTRACT: When driving the top heading for a railway tunnel it was possible, due to the small overburder, to study the behaviour of the fractured rock from above ground by means of deformation measurements and to compare them with the deformations of the shotcrete lining. Due to the low shear strength of the rock mass the deformations mainly occur above the excavation area. Although, on the whole, the settlements of the top heading extended up to the surface, only 43 % of the total surcharge act on the shotcrete lining. By means of the "Coupled Beam-Boundary Element Model (BE-BEM)" parameter studies were carried out, in which the deformation moduli of the rock mass and the shotcrete lining were recalculated considering the measurement results. The limit values of the interacting forces between the shotcrete lining and the rock mass as well as various coefficients of lateral stresses have been considered in the calculation. Finally it was shown by means of a FE-calculation that the characteristic values determined for describing the behaviour of the rock mass are in good agreement with the measurement results.

RESUME: Au cours de l'avancement de la calotte d'un tunnel de chemins de fer, il a été possible d'étudier le comportement de la roche fractureé au moyen des mesures des déformations. Les déformations de la roche surviennent surtout dans la zone du tunnel en raison de la transmission peu importante des efforts tranchants. Bien que les effets du tassement de la calotte se propagent jusqu'à la surface, ce ne sont que 43 % de la surcharge totale qui agissent sur la coque en béton projeté.
C'est au moyen d'éléments limitrophes du "modèle à barres couplés" que les –tudes paramètriques du module de déformation de la roche et de la coque en béton projeté ont pu etre détermin-es en concordance avec les résultats de mesure. Les valeurs limites pour l'interaction entre la coque en béton projeté et la roche ainsi que différents coefficients de pression latérales ont été considérés pour cette étude. En fin de compte on a prouvé par la méthode des élements finis que pour les caractéristiques ainsi déterminées, le comportement de la roche concorde avec assez de précision avec les résultats mesurés.

ZUSAMMENFASSUNG: Im Zuge des Kalottenvortriebes eines Eisenbahntunnels konnte aufgrund geringer Überlagerung das Verhalten des zerlegten Gebirges mittels Verformungsmessungen von Obertag erfaßt und mit den Verformungen der Spritzbeton- Außenschale verglichen werden. Die Gebirgsverformungen konzentrierten sich aufgrund geringer Schubübertragung auf den Tunnelbereich. Obwohl sich die Setzungen der Kalotte im wesentlichen bis zur Oberfläche fortpflanzten, wurde die Spritzbetonschale nur mit 43 % der vollen Auflast beansprucht.
Mit Hilfe des "Gekoppelten Stabwerk-Randelement-Modelles (BE-BEM)" wurde durch Parameterstudien in Abstimmung auf die Meßergebnisse der Verformungsmodul des Gebirges und der Außenschale rückgerechnet. Die Grenzwerte der Kraftübertragung zwischen Außenschale und Gebirge, sowie verschiedene Seitendruckbeiwerte wurden in die Untersuchung einbezogen. Abschließend wurde mit einer FE-Berechnung nachgewiesen, daß mit den so ermittelten Kennwerten das Gebirgsverhalten in ausreichender Übereinstimmung mit den Meßergebnissen wiedergegeben werden kann.

1. PROJEKTSBESCHREIBUNG

Der betrachtete Tunnel wurde im Oberen Muschelkalk aufgefahren. Dieser besteht aus einer Wechselfolge von Kalkstein-und dünnen Tonmergelzwischenlagen in söhliger Lagerung. Die Schichtmächtigkeit der überwiegenden Kalksteine beträgt bis 50 cm, jene der Mergelsteine bis 3 cm. Die tonig-mergelig belegten Schichtflächen sind stark verwittert. Die Klüfte verlaufen überwiegend quer zur Tunnelachse und fallen steil ein. Der Kluftabstand liegt im Meterbereich, die Klüfte sind ebenflächig, deutlich verwittert und teilweise mit einem dünnen Lehmfilm belegt. Bergwasser trat nur in Form von leichtem Tropfwasser an den Ankerköpfen auf.
Der Ausbruch des rd. 125 m2 großen Querschnittes erfolgte mit Unterteilung in Kalotte, Strosse und Sohle. Die 55 m2 große Kalotte wurde in Abschlägen von 1,0 m vorgetrieben und wie folgt ausgebaut:

Spritzbeton:	25 cm stark, Beton B 25
Betonstahlmatten:	1 Lage Q 188, BSt 500 M
Ausbaubogen:	Typ PS: 90/20/30, e=1,0 m
Vermörtelte Anker:	8 Stk l=4 m, 4 Stk l=6 m
Kalottenfuß:	Lastverteilerschienen Verbreiterung 0,5 m.

2. GEOTECHNISCHE MESSUNGEN

2.1 Verformungsmessungen

Bei einer Überlagerung von 13,8 m mit leicht geneigtem Geländeverlauf (siehe Abb. 1) wurden umfangreiche Messungen durchgeführt, um das Verhalten des Gebirges und die daraus resultierende Beanspruchung des Tunnelausbaues zu erfassen.
Die Oberflächensetzungen zeigen eine Setzungsmulde, welche im Querschnitt auf den Tunnelbereich beschränkt ist. 6 m links und rechts der Tunnelachse betragen die Setzungen rd. 40 % jener in Tunnelachse, im Abstand von 12 m wurden nur mehr 4 % - 10 % gemessen. Die Extensometer- und die Gleitmikrometermessungen ergaben, daß sich die Setzung der Kalotte bis zur Oberfläche fortpflanzen, die Auflockerung des Gebirges ist untergeordnet.
In Abb. 2 sind die Oberflächensetzungen als Funktion der Entfernung von der Ortsbrust aufgetragen. Die Setzungen eilen dem Vortrieb nur wenige Meter voraus. Die Vorverformung in Tunnelachse beträgt 13 % der gesamten Oberflächensetzung beim Kalottevortrieb, 6 m seitlich der Tunnelachse 8 % bis 12 %. In den Gleitmikrometer- und Extenso-

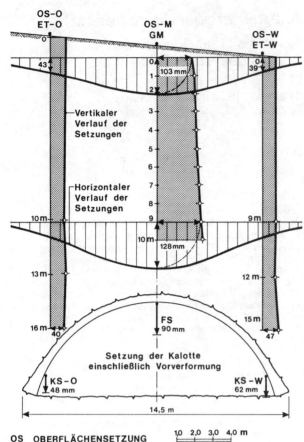

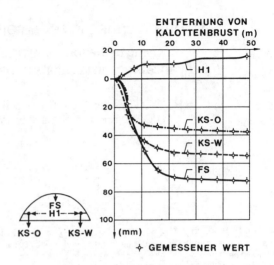

Abb. 3 Verformungen der Kalotte als Funktion der Entfernung von der Ortsbrust

OS OBERFLÄCHENSETZUNG
ET 3-fach EXTENSOMETER
GM GLEITMIKROMETER
-✛- GEMESSENER WERT

1,0 2,0 3,0 4,0 m
MASSTAB - BAUWERK
0 50 100 150 mm
MASSTAB - VERFORMUNGEN

Abb. 1 Setzungen des Gebirges und der Kalotte

metermessungen wurden die geringen Werte der Vorverformungen bestätigt. Die geringe Vorverformung ist auf die Klüftung und die damit verbundene geringe Schubübertragung in Längsrichtung zurückzuführen.

Die in der Kalotte gemessenen vertikalen Verformungen (siehe Abb. 3) zeigen einen zu den Oberflä-

chensetzungen analogen Verlauf. 15 m hinter der Ortsbrust, das entspricht dem Durchmesser des Ausbruchquerschnittes, werden 90 % der Gesamtverformungen gemessen, nach 25 m treten keine weiteren Veränderungen mehr ein. Wird zu der gemessenen Vertikalverformung die Vorverformung addiert, ergeben sich nahezu gleich große Gesamtsetzungen der Kalotte und Oberflächensetzungen.

2.2 Spannungsmessungen

Der Gebirgsdruck (Abb. 4) ist über den Umfang gleichmäßig verteilt (0,15 MN/m2 bis 0,3 MN/m2). Die tangentialen Druckdosen zeigen einen über den Querschnitt ungleichmäßigen Verlauf der Spritzbetonspannungen. Auf die tatsächliche Beanspruchung des Spritzbetonquerschnittes kann nicht rückgeschlossen werden, da der Meßwert neben örtlichen Einflüssen von der Lage der Meßdose abhängt. Außerdem wurden die Druckmeßdosen ohne Nachspanneinrichtung eingebaut, sodaß die Einflüsse aus Kriechen und Schwinden des Betons nicht kompensiert werden konnten. Die Spritzbetondruckdosen wurden daher nur zur Erfassung der Spannungsänderungen beim Abbau der Strosse und beim Sohlaushub herangezogen.

Die Ergebnisse der Ph. Holzmann-Meßträger (Baumann 1985) zeigen eine annähernd über den Querschnitt

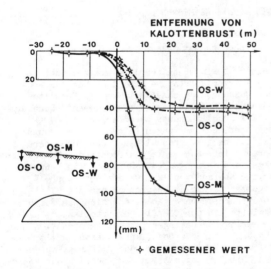

Abb. 2 Oberflächensetzungen als Funktion der Entfernung von der Ortsbrust

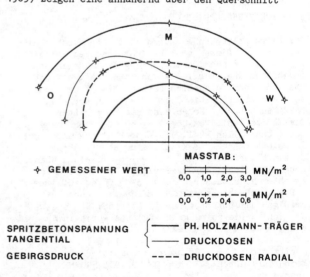

SPRITZBETONSPANNUNG ⎰ —— PH. HOLZMANN-TRÄGER
TANGENTIAL ⎱ —— DRUCKDOSEN
GEBIRGSDRUCK ---- DRUCKDOSEN RADIAL

Abb. 4 Beanspruchung der Spritzbetonschale der Kalotte

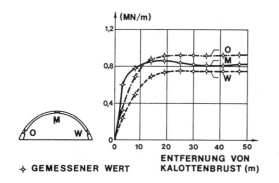

+ GEMESSENER WERT

Abb. 5   Normalkräfte als Funktion der Entfernung
         von der Ortsbrust (Ph. Holzmann Meßträger)

gleichmäßige Spannung in der Spritzbetonschale
von rd. 3,6 MN/m2. Aus der Normalkraft (N = 0,900 MN/m)
und dem Radius der Kalottenschale (7,0 m) errechnet
sich bei einem spezifischem Gewicht des Gebirges
von 0,021 MN/m3 eine Lasthöhe von rd. 6,0 m. Somit
wirken nur 43 % der vollen Auflast bis zur Geländeoberfläche auf die Spritzbetonschale.
    In Abb. 4 sind die in den Ph. Holzmann-Trägern
gemessenen Normalkräfte als Funktion der Entfernung
von der Ortsbrust aufgetragen. Die Beanspruchung
steigt in der Firste rascher an als in den Ulmen.
Einen Durchmesser hinter der Ortsbrust erfolgte
eine weitgehende Angleichung der Normalkräfte.
Die Beanspruchung der Schale im ostseitigen Ulm
ist größer als in der Firste und höher als im
westseitigen Ulm. Dies ist auf die geneigte Geländeoberfläche zurückzuführen.

3. RECHNERISCHE UNTERSUCHUNGEN

3.1 Rechenmodelle

Es wurden zwei Rechenmodelle angewandt. Das "Gekoppelte Stabwerk-Randelement-Modell" (BE-BEM), (Pöttler
1986/ Pöttler, Swoboda, Beer 1986/ Pöttler, Swoboda,
1986) ist ein zutreffendes Modell zur Erfassung
des Gebirgsverhaltens, ohne die bekannten Nachteile
der auf der Bettungsmodultheorie bzw. der auf
der Finite Element Methode beruhenden Modelle
aufzuweisen. Zur Beschreibung des Gebirges werden
Rand- oder Boundary-Elemente (BE) und zur Erfassung
der Spritzbetonschale Stabelemente verwendet.
Das Verhalten der Kontaktfläche Spritzbeton -
Gebirge wird durch Constraint-Elemente erfaßt.
Da im vorliegenden Fall durch die große Gesteinsfestigkeit und der im Vergleich zur Überlagerung
hohen Scherparameter mit keinen bzw. nur geringen
Bruchzonen im Gebirge zu rechnen ist, können die
BE das Gebirgsverhalten realitätsnah erfassen.
Besondere Bedeutung hat die Kraftübertragung an
der Kontaktfläche Spritzbeton/Gebirge (John, Pöttler,
Heissel 1985/ Pöttler 1985 a).
    Für die radiale Kraftübertragung sind zwei Zu-
stände möglich:
- gekoppelt   (volle Kraftübertragung)
- entkoppelt (keine Kraftübertragung)

Die tangentiale Kraftübertragung hängt im Unterschied zur radialen wesentlich von den Relativverschiebungen und den zwischen Spritzbetonschale
und Gebirge übertragbaren Kräften ab. Die Kraftübertragung wird durch das Mohr Coulomb'sche Bruchkriterium gesteuert. Es sind drei Zustände möglich:
- entkoppelt
- starr gekoppelt (Haftreibung)
- gleitend (Gleitreibung)

In der Regel ist die Beanspruchung des Constraint-

Elementes nicht von vorneherein bekannt. Die Berechnung muß daher iterativ erfolgen. Für die Entschei-
dung, welchen aktuellen Zustand (Iterationsschritt n)
das Constraint-Element besitzt, werden die im
Iterationsschritt (n-1) errechneten Verformungen
und Kräfte herangezogen.
    Als zweites Berechnungsmodell wird ein FE-Modell
angewandt. Das Gebirge wird durch ein multilaminares
Modell beschrieben (Pande, Xiong 1982). Die Kontaktfläche Spritzbeton/Gebirge wird durch "Thin-Layer-
Elemente" (Desai, Zaman, Lightner, Siriwardane
1984) erfaßt. Im Unterschied zu den Constraint-
Elementen handelt es sich dabei um Elemente, die
Spritzbeton und Gebirge elastisch koppeln und
damit bereits vor dem Erreichen der Grenztragfähigkeit eine Verschiebung, die durch die Materialpara-
meter (Elastizitätsmodul, Schubmodul und Querdeh-
nungszahl) bestimmt werden, zulassen. Als Grenz-
kriterium wird auch für die Thin-Layer-Elemente
die Mohr-Coulomb'sche Bruchhypothese herangezogen.

3.2 Parameterstudie mit BE-BEM

Für das Tragverhalten und Verformungsverhalten
des Gebirges und der Spritzbetonschale sind der
Elastizitätsmodul des Gebirges (E_G), der Seitendruckbeiwert k_o, der Elastizitätsmodul des Spritzbetons (E_S) und die Interaktion zwischen Spritzbeton
und Gebirge maßgebend. Zur Bestimmung dieser Werte
wurde eine Parameterstudie mit dem BE-BEM-Modell
gemäß Tabelle 1 durchgeführt.

Tabelle 1. Kennwerte für die einzelnen Berechnungen
mit dem BE-BEM-Modell

Kennwert	Ansatz 1	Ansatz 2	Ansatz 3	Dimension
E_G	60-250	75	75	MN/m2
E_S	8.500	1.750-8.500	2.500	MN/m2
k_o	0,5	0,5	0,0|0,5	-
c	∞|0	∞|0	0|0,25	MN/m2
φ	0|0	0|0	0|0|25	o
I	S|E	S|E	S|E|B	-
Abb.	6	7	8 9	

Legende:
E_G ... E-Modul des Gebirges
E_S ... Ideeller E-Modul des Spritzbetons
k_o ... Seitendruckbeiwert
c,φ ... Scherparameter der Constraint Elemente
         in tangentialer Richtung
I   ... Index in den Abbildungen

Zwischen Spritzbeton und Gebirge werden in radialer
Richtung nur Druckkräfte übertragen (tension cut
off).
Der Elastizitätsmodul des Spritzbetons stellt
einen ideellen Rechenwert dar, der die Vorverformung, die zeitabhängige Spannungsumlagerung, die
allmähliche Erhärtung des Spritzbetons, sowie
dessen Kriechen und Schwinden berücksichtigt (Pöttler
1985 b). Gegenüber dem 28-Tage E-Modul, welcher
im vorliegenden Fall rd. 21.000 MN/m2 betrug,
ist daher der ideelle E-Modul abzumindern.
Zur Bestimmung der gesuchten Kennwerte (Parameter)
wurden die Rechenergebnisse mit den Meßergebnissen
verglichen.
Die Firstsetzung betrug einschließlich Vorverformung
90 mm (Abb. 1). Bei den Firstsetzungen in den
Nachbarstationen wurden geringere Werte festgestellt,
dementsprechend ist eine Bandbreite von 75 mm
bis 95 mm zutreffend. Die Normalkraft in der Firste
lag bei 0,80 MN/m (Abb. 5). Die gemessenen Werte
ergaben einen Schwankungsbereich von 0,50 bis
1,1 MN/m. Bei Ausgleich der unterschiedlichen
Werte innerhalb eines Meßträgers war eine Bandbreite
von 0,7 bis 0,9 MN/m in die Beurteilung einzubeziehen.

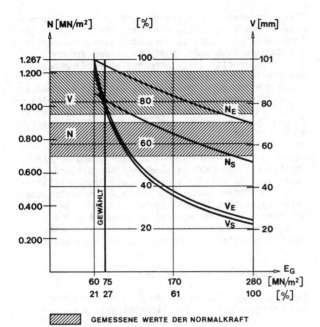

Abb. 6 Parameterstudie: Normalkräfte (N) und
        Firstsetzungen (v) für E_S = 8500 MN/m2
        und E_G = variabel, k_o = 0,50

In einer ersten Untersuchung (Berechnungsansatz 1
der Tabelle 1) wurde für einen E-Modul der Spritzbetonschale von 8500 MN/m2 der E-Modul des Gebirges
variiert. Die Normalkräfte (N) in der Firste sowie
die Firstsetzungen (v) sind in Abb. 6 dargestellt.
Es zeigt sich, daß die Firstsetzung in einem weit
stärkeren Maß vom E-Modul des Gebirges bestimmt
wird als die Normalkraft der Spritzbetonschale.
Dies gilt sowohl für die tangential starre Koppelung
zwischen Gebirge und Spritzbetonschale als auch
für die tangentiale Entkoppelung. Aus diesem Grund
wurden die Firstsetzungen zur Festlegung des E-Moduls
des Gebirges herangezogen. Der E-Modul des Gebirges

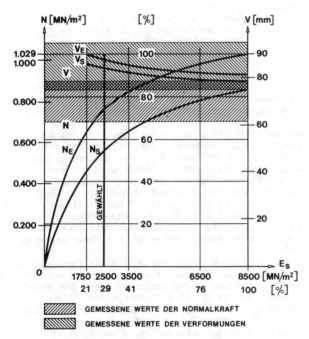

Abb. 7 Parameterstudie: Normalkräfte (N) und
        Firstsetzungen (v) für E_G = 75 MN/m2,
        E_S = variabel, k_o = 0,50

wird für die weiteren Berechnungen mit 75 MN/m2
festgelegt.
     In einer weiteren Untersuchung (Berechnungsan-
satz 2 der Tabelle 1) wurde der E-Modul des Spritzbetons variiert (Abb. 7). Der E-Modul des Spritzbeton hat nur einen geringen Einfluß auf die Firstsetzungen, die Normalkraft in der Spritzbetonschale
wird entscheidend von diesem bestimmt. Entsprechend
dieser Ergebnisse wurde den weiteren Berechnungen
ein E-Modul des Spritzbetons von 2.500 MN/m2 zugrunde
gelegt.
Abschließend wurde der Einfluß der tangentialen
Kraftabtragung und des Seitendruckbeiwertes (Berechnungsansatz 3 der Tabelle 1) untersucht. Zusätzlich
zur starren Koppelung und der Entkoppelung in
tangentialer Richtung wurde eine Koppelung unter
Ansatz der Scherparameter c = 0,250 MN/m2, ϕ = 25°
mit Abfall der Kohäsion auf c = 0 nach Erreichen
der Bruchgrenze für die Begrenzung der Kraftübertragung untersucht (Mohr-Coulomb/Tension cut off).
Der Verlauf der Normalkräfte und Verformungen
für diese 6 Berechnungen sind in Abb. 8 und 9
im Vergleich zu den Meßwerten gesetzt. Der Verlauf
der Normalkräfte ergab eine gute Übereinstimmung
mit den Meßergebnissen für den Seitendruckbeiwert
von k_o = 0,5. Der Einfluß des Seitendruckbeiwertes
auf die Verformung ist untergeordnet.
     Im vorliegenden Fall hat die tangentiale Kraft-
abtragung zwischen Spritzbeton und Gebirge sowohl
hinsichtich Normalkräfte wie auch Verformungen
keine erhebliche Bedeutung. Dies ist auf die geringen Relativbewegungen zwischen Spritzbeton
und Gebirge zurückzuführen (siehe Abb. 1).

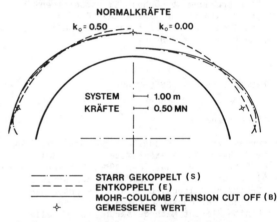

Abb. 8 Parameterstudie: Normalkräfte (N) für
        E_G = 75 MN/m2, E_S = 2500 MN/m2,
        unterschiedliche Interaktion zwischen
        Spritzbeton und Gebirge, sowie k_o = 0,0
        und k_o = 0,50

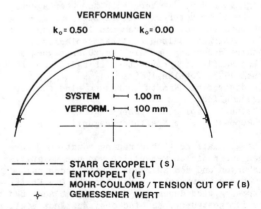

Abb. 9 Parameterstudie: Verformungen (v) für
        E_G = 75 MN/m2, E_S = 2500 MN/m2,
        unterschiedliche Interaktion zwischen
        Spritzbeton und Gebirge, sowie k_o = 0,0
        und k_o = 0,50

Die insgesamt beste Übereinstimmung ergaben
folgende Parameter:
E_G = 75 MN/m2 E_S = 2500 MN/m2 k_o = 0,5

Mit diesen Eingabeparametern wurde eine Berechnung
mit der FEM durchgeführt.

3.3 FE-Berechnungen

Die Berechnungen erfolgten mit dem Programmsystem
MISES 3, Rev. 6 entwickelt von TDV-Pircher, Graz.
Das zugrundegelegte Elementnetz ist in Abb. 10
dargestellt. Es umfaßt 456 isoparametrische Vier-
eckselemente mit 1215 Freiheitsgraden.

1215 KNOTEN / 456 ELEMENTE

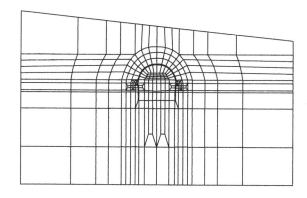

GEBIRGE :
  E = 75 MN/m²
  ν = 0.25
  k_o= 0.50
  γ = 0.021 MN/m³
  c = 0.25 MN/m²
  φ = 40°
  MAT.GESETZ : DRUCKER-PRAGER
  SCHICHTUNG: HORIZONTAL
  c = 0.10 MN/m²
  φ = 28°
  MAT.GESETZ : MOHR-COULOMB

SPRITZBETON :
  E = 2500 MN/m²
  ν = 0.16
  MAT.GESETZ: ELASTISCH

INTERAKTION SPRITZBETON-GEBIRGE :
  THIN LAYER ELEMENT :
  E = 35 MN/m²
  G = 30 MN/m²
  ν = 0.25

Abb. 10 Finite Element Netz

In Abb. 11 und Abb. 12 sind die Ergebnisse der
FE-Berechnungen jenen der BE-BEM Berechnung gegenübergestellt. Wie die Gegenüberstellung der Rechenergebnisse mit den Meßergebnissen zeigt, liegt
eine gute Übereinstimmung vor. Bruchzonen treten
keine auf. Das Gebirgsverhalten wird durch die
Elastizitätstheorie beschrieben. Darauf ist auch
die gute Übereinstimmung zwischen den Berechnungs-
ergebnissen des BE-BEM Modelles und den Meßwerten
zurückzuführen.

NORMALKRÄFTE

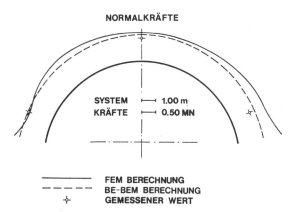

SYSTEM   ├── 1.00 m
KRÄFTE   ├── 0.50 MN

—————— FEM BERECHNUNG
— — — — BE-BEM BERECHNUNG
  ✛   GEMESSENER WERT

Abb. 11 Normalkräfte (N) in der Spritzbetonschale:
         Vergleich von FEM zu BE-BEM und Meßergeb-
         nissen

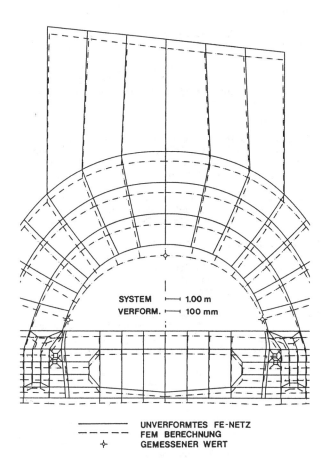

SYSTEM   ├── 1.00 m
VERFORM. ├── 100 mm

—————— UNVERFORMTES FE-NETZ
— — — — FEM BERECHNUNG
  ✛   GEMESSENER WERT

Abb. 12 Verformungen (v):
         Vergleich FEM zu Meßergebnissen

4. SCHLUSSBETRACHTUNG

Anhand der Rückrechnungen wird eine effektive
Methode zur rechnerischen Erfassung des Gebirgs-
tragverhaltens und der Beanspruchung der Spritzbetonschale aufgezeigt. Mit einem leistungsfähigen
wenig rechenzeitintensiven Modell (BE-BEM) wurden
umfangreiche Parameterstudien durchgeführt und
die maßgebenden Parameter ermittelt. Die mit diesen
Parametern anschließend durchgeführte FE-Berechnung
ergibt ein umfassendes Bild der Verformungen,
Spannungen und des Beanspruchungsgrades über den
gesamten Gebirgsbereich. Die Verwendung der BE-BEM
ist in diesem Fall zutreffend, da das Tragverhalten
des Gebirges aufgrund des guten Gebirges genügend
genau durch die Elastizitätstheorie beschrieben
wird.
Die Rückrechnung ergab, daß die tangentiale Kraftübertragung untergeordnet ist, was auch aus den
Meßergebnissen hervorgeht. Dies ist einerseits
auf die geringe Relativverformung zwischen Gebirge
und Ausbau, hervorgerufen durch das steif ausge-
bildete Kalottenfußauflager, bedingt. Andererseits
wird bei kleinen Relativbewegungen die tangentiale
Kraftabtragung noch nicht aktiviert. Die Aktivierung
der Verzahnung erfolgte im vorliegenden Fall erst
beim Strossenabbau.
  Die Gebirgstragwirkung wird aktiviert, es liegen
jedoch noch beträchtliche Reserven im Gebirgsge-
wölbe. Durch den relativ steifen Ausbau zieht
die Spritzbetonschale einen Großteil der Kräfte
an, im Gebirge treten keine Bruchzonen auf. Bei
Wegfall der Kalottenfußverbreiterung sind etwas
größere Verformungen, geringere Schnittkräfte
in der Spritzbetonschale und ev. Bruchzonen im
Gebirge zu erwarten.

Literatur

Baumann, Th. 1985. Messungen der Beanspruchung
in Tunnelschalen. Der Bauingenieur 60: 449-454.

Desai, C.S., Zaman M.M., Lighter, J.G. & Siriwardane,
H.J. 1984. Thin-layer element for interfaces
and joints. International Journal for Numerical
and Analytical Methods in Geomechanics, 8,
19-43.

John, M. & Pöttler, R. 1984. Application of the
Integrated Measuring Technique. In: Field Measurements in Geomechanics, Ed by K. Kovari; p 1063 -
1072. Rotterdam: Balkema.

John, M., Pöttler, R. & Heissel G. 1985. Vergleich
verschiedener Rechenmethoden untereinander
und mit den Meßergebnissen Felsbau, 3, 21-26.

Pande, G.N. & Xiong W. 1982. An improved multi-
laminate model of jointed rock masses. In:
Numerical Models in Geomechanics. Ed by R.
Dungar, G.N. Pande and J.A. Studer; p. 218-226.
Rotterdam: Balkema.

Pöttler, R. 1985 a. Analysis of tunnels in highly
jointed rock. In: Numerical Methods in Geomechanics Nagoya 1985; Ed. by T. Kawamoto and Y.
Ichikawa; p 1111-1118. Rotterdam: Balkema

Pöttler, R. 1985 b. Ideeller Elastizitätsmodul
zur Abschätzung der Spritzbetonbeanspruchung
bei Felshohlraumbauten. Felsbau, 3, 136-139.

Pöttler, R. 1986. Gekoppeltes Stabwerk-Boundary-
Element System zur Berechnung von Tunnelbauten
unter Berücksichtigung des nichtlinearen Mate-
rialverhaltens von Stahlbeton, Dissertation
TU Innsbruck.

Pöttler, R. & Swoboda G.A., 1986. Coupled Beam-
Boundary-Element Model (BE-BEM) for Analysis
of Underground Openings. Computers and Geotechnics, 2, 239-256.

Pöttler, R., Swoboda, G.A. & Beer, G. 1986. Non-
linear Coupled Beam-Boundary Element (BE-BEM)
Analysis for Tunnels on Microcomputers.
In Microcomputers in Engineering: Development
and Application of Software, Ed by B.A. Schrefler
and R.W. Lewis; 161-176 Pineridge Press Swansea
U.K.

Analyse des Zerkleinerungsprozesses beim Vortrieb mit Vollschnittmaschinen

Diagnosis of the disintegration process for full face tunnel boring machines
Diagnostique du processus de comminution lors du creusement des tunnels par des tunnelliers plein-profil

F.SEKULA, Bergbauinstitut der Slowakischen Akademie der Wissenschaften, Košice, Tschechoslowakei
V.KRÚPA, Bergbauinstitut der Slowakischen Akademie der Wissenschaften, Košice, Tschechoslowakei
M.KOČI, Bergbauinstitut der Slowakischen Akademie der Wissenschaften, Košice, Tschechoslowakei
F.KREPELKA, Bergbauinstitut der Slowakischen Akademie der Wissenschaften, Košice, Tschechoslowakei
A.OLOS, Bergbauinstitut der Slowakischen Akademie der Wissenschaften, Košice, Tschechoslowakei

ABSTRACT: The paper presents the results of the measurement of the instantaneous driving velocity, specific volume work of disintegration and of a new derived quantity on the tunnelling machine RS 24-27 of Czechoslovak production in mineralogical-petrographic formations of the mine in Rudňany.

RESUME: Dans ce travail sont présentés les résultats des mesures de la vitesse instantanée du creusement, du volume massique de la désintégration et d´une autre quantité dérivée sur la traceuse RS 24-27 de production tchécoslovaque dans les formations minéralogiques-pétrographiques de la mine a Rudňany.

ZUSAMMENFASSUNG: In der Arbeit sind veröffentlicht die Ergebnisse der Messungen der jeweiligen Vortriebsgeschwindigkeit, der spezifischen Volumenzerstörungsarbeit und einer weiteren abgeleiteten Grösse auf der Vollschnitmaschine RS 24-27 der tschechoslowakischen Produktion in mineralogishh - petrographischen Formationen des Grubenbaus in Rudňany.

Gegenwärtig besteht die allgemeine Tendenz für die Steuerung der technologischen Prozesse adaptive Systeme auszuarbeiten mit deren Hilfe der Prozess so zu steuern wäre, dass er aufgrund von bestimmten festgesetzten technologischen, bzw. ökonomischen Optimierungskriterien verlaufen würde. Eine Voraussetzung für die optimale adaptive Steuerung im Sinne von diesen allgemein erwähnten Kriterien ist die Kenntnis der Grundgesetzmässigkeiten von Prozessen, also insbesondere die exakte Kenntnis der funktionellen Beziehungen zwischen den Eingangs- und Ausgangsgrössen. In der Technologie des sprengstofflosen Grubenbau- und Tunnelbauvortriebes ist diese Frage höchst aktuell im Hinblick auf die relativ hohe Investitions- und Betriebskosten. Um den Einsatz dieser progressiven Technologie eindeutig ökonomisch effektiver zu machen, als die klassische Technologie des Vortriebes, besteht die Tendenz den eigentlichen Zerstörungsprozess durch systematische Entwicklung und Vervollkommnung der Technik, durch Vervollkommnung der Steuerungsarten bis zu den automatischen adaptiven Systemen zu verbessern.

Im Prozess des sprengstofflosen Vortriebes kann man die Eingangsgrössen in drei Kategorien einteilen. Steuerbare Grössen im Vortriebsprozess sind die Vorschubkraft F und die Drehzahl n in dem Falle, wenn deren Änderung die Konstruktion der Maschine ermöglicht. Die zweite Kategorie der Eingangsgrössen die nur vor dem Einsatz der Maschine beeinflussbar sind, bilden die Parameter wie die Anfangsgeometrie der Arbeitselemente G_1, derer Raumgestaltung auf dem Vortriebmaschinenkopf R_v, die Geometrie des Gesamtvortriebskopfes G_c und die technologischen Beschränkungen des Prozesses T_B gegeben durch die maximale Vorschubkraft F_{max}, maximale Drehzahl n_{max} und maximale Antriebsleistung P_{max}. Die dritte Kategorie der Eingangsgrössen bilden die Störungsgrössen, die durch die Naturbedingungen und die Qualität der metallurgischen Werkstoffe gegeben sind. Hierher gehören die mineralogisch-petrographischen und geologischen Faktoren der Lokalität G_1 in der die Vortriebsbmaschine eingesetz wird, und die Qualität der erzeugten Arbeitselemente G_2, aus dem materiallen Gesichtspunkt sowie aus dem Gesichtspunkt der Maschinenbaufertigung.

Im weiteren Text werden wir unter dem Begriff Eingagsgrössen nur die steuerbare Grössen die Vorschubkraft und die Drehzahl verstehen.

Für Ausgangsgrössen halten wir die jeweilige Vortriebsgeschwindigkeit und die Intensität der Abnutzung der Arbeitselemente des Vortriebsmaschinenkopfes. Positiver Effekt ist die jeweilige Vortriebsgeschwindigkeit und negativer Effekt des Prozesses ist die Abnutzungsintensität der Arbeitselemente. Für den optimalen technologischen, bzw. ökonomischen Effekt ist es notwendig eine geeignete Proportionalität von diesen zwei Ausgangsgrössen in verschiedenen Naturbedingungen zu wählen.

Die Erkenntnis des Vortriebsprozesses ist grundsätzlich durch zwei verschiedenen Verfahren möglich. Durch physikalische Modellierung oder durch empirisch-statistisches Verfahren. Gegenwärtig, auf dem Gebiet der physikalischen Modellierung gibt es kein exaktes Gesamtmodell, das alle komplizierten, den resultierenden Zerstörungseffekt beeinflussenden Faktoren berücksichtigen würde. Die Mehrkeit der veröffentlichten Arbeiten auf diesem Gebiet beschäftigt sich mit der Bestimmung von einigen Teilaspekten des verfolgten Prozesses (1),(2). Die veröffentlichten empirischen Erkenntnisse lösen die Beziehungen zwischen der jeweiligen Vortriebsgeschwindigkeit und den Eingangsgrössen des Prozesses (3), (4).

Die Autoren in ihren vorangehenden auf das drehende Bohren orientierten Arbeiten haben bewiesen, dass zwischen der Abnutzungsintensität der Arbeitselemente des Bohrwerkzeuges und der spezifischen Volumenzerstörungs-

arbeit w bei dem quasi homogenen Gestein die
lineare Abhängigkeit besteht. Die Analogie
zwischen dem Gesteinbohren mittels Rolldis-
ken und Vortrieb mit Vollschnittmaschinen
liegt darin, dass in beiden Fällen das Werk-
zeug, bzw. der Bohrkopf die Energie auf das
Gestein überträgt und so wenn wir ein Meter
Gesteins mit weniger Arbeitsaufwand in dem
gegebenen Art des Gesteins abbauen können,
dann besteht auch die Voraussetzung, dass
die Arbeitselemente des Bohrkopfes im Verlauf des Vortriebs energetisch weniger belastet werden und infolge dessen auch weniger abgenutzt.

Auf Grund dessen werden wir die Grösse w
für Ausgangsgrösse halten, die die Abnutzungsintensität der Arbeitselemente in der gegebenen mineralogisch-petrographischen Formation charakterisiert.Für eine komplexe Bewertung des Zerstörungsprozesses beim Vortrieb
haben wir den Quotienten der jeweiligen Vortriebsgeschwindigkeit v und der spezifischen
Volumenzerstörungsarbeit w eingeführt den
wir mit dem Symbol φ bezeichnen. Der Vortriebsprozess kann dann mit dem folgenden Schema
charakterisiert werden, wo alle Grössen dargestellt sind, die sich nach dieser neuen
Auffassung in der Beziehung zum Prozess be-
finden.

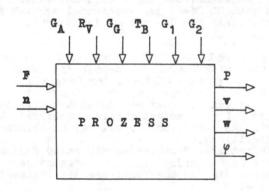

Abb. 1 Vortriebsprozessgrössen

Für die Überprüfung der Beziehungen zwischen den Eingangs- und Ausgangsgrössen des
Vortriebsprozesses haben wir an mehreren
Vortriebsmaschinen in der ČSSR Prüfmessungen
durchgeführt aus denen wir die Messergebnisse
auf der Vortriebsmaschine der tschechoslowakischen Produktion RS 24-27 anführen. Diese Vortriebsmaschine wurde auf der Lokalität
des volkseigenen Betriebes Eisenerzgruben,
Rudňany bei der Auffahrung einer Förderstrek-
ke auf dem Horizont XIX eingesetzt. Das
Schema des Grubenbaus mit der Kennzeichnung
des geologischen Profils des gemessenen Gru-
benreviers ist in Abb.2 dargestellt. Für die
Betriebsmessungen wurde am Bergbauinstitut
der SAW in Košice eine elektronische Indika-
tionseinrichtung entwickelt, die im Vortriebsprozess die Grössen v, w und φ kontinuierlich zu indizieren ermöglicht (5), (6). Die
Vorschubkraft F wurde von dem Druck des
hydraulischen Kreislaufes der Druckwalzen
des Bohrkopfes aufgenommen. Die Vortriebsmaschine wurde auf die konstante Drehzahl
n = 0,16 s^{-1} konstruiert, wodurch diese Ein-
gangsgrösse in unserem Falle den Charakter
eines Parameters erwarb und deshalb wurde
nicht aufgenommen.

Für die Bestimmung der Beziehung zwischen
der Vorschubkraft und den Ausgangsgrössen
v, w und φ wurden die Betriebsexperimente
so gesteuert, dass die Vorschubkraft von dem

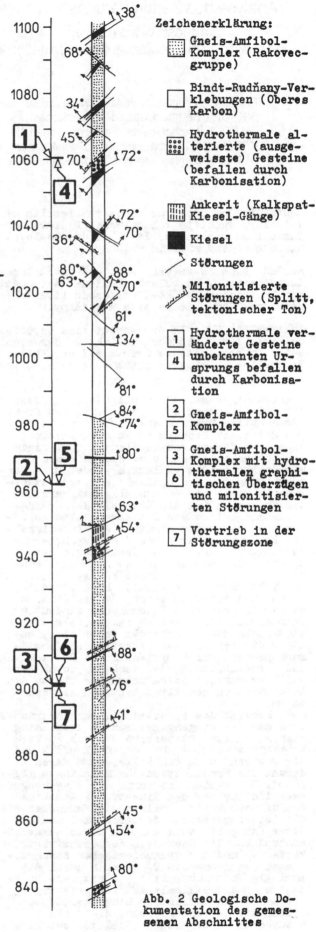

Zeichenerklärung:

Gneis-Amfibol-
Komplex (Rakovec-
gruppe)

Bindt-Rudňany-Ver-
klebungen (Oberes
Karbon)

Hydrothermale al-
terierte (ausge-
weisste) Gesteine
(befallen durch
Karbonisation)

Ankerit (Kalkspat-
Kiesel-Gänge)

Kiesel

Störungen

Milonitisierte
Störungen (Splitt,
tektonischer Ton)

1 Hydrothermale ver-
4 änderte Gesteine
unbekannten Ur-
sprungs befallen
durch Karbonisa-
tion

2 Gneis-Amfibol-
5 Komplex

3 Gneis-Amfibol-
6 Komplex mit hydro-
thermalen graphi-
tischen Überzügen
und milonitisier-
ten Störungen

7 Vortrieb in der
Störungszone

Abb. 2 Geologische Do-
kumentation des gemes-
senen Abschnittes

Nullwert bis zum Nominalwert allmählich ansteigen konnte. Die Änderungen der Ausgangsgrössen in der Abhängigkeit von der Vorschubkraft für verschiedene Stationierungen und
mineralogisch-petrographische Formationen
des aufgefahrenen Grubenbaues sind in Abb.3,
4 und 5 dargestellt.

Weitere Experimente fanden ohne unsere aktive Eingriffe in die Steuerung der Maschine statt. Die Verläufe der Vorschubkraft und
der Ausgangsgrössen in der Abhängigkeit von
dem Vortriebsfortgang sind für drei Typen
der mineralogisch-petrographischen Formationen in Abb.6 und 7 dargestellt. Die Stellen
der einzelnen illustrierten Messungen sind
auf dem Schema des Grubenbaues aufgetragen,
sich Abb.2.

Die Grössenverläufe v, w und φ in der Abhängigkeit von der Vorschubkraft dargestellt
in Abb.4 wurden während der Messungen im
Gneis-Amfibol-Komplex (das ältere Paleosoti-
kum) gewonnen. Die Messung wurde im Bereich
der Positionierung 962,09-962,12 m. Im Rahmen der aufgefahrenen Länge 0,03 m setzen wir
voraus keine markante Änderungen in dieser mineralogisch-petrographischen Formation aus
dem Gesichtspunkt der Zusammensetzung, aus
dem Gesichtspunkt der Grössen von Werten der
spezifischen Volumendislokation und nicht
einmal aus dem Gesichtspunkt der Beständigkeit gegen Vortriebszerstörung. Trotzdem
kann man bestimmte Unterschiede in den Ver-
laufwerten beobachten, besonders bei der
Grösse φ die am empfindlichsten auf die Änderung der Eigenschaften des zu zerstören-
den Massivs reagiert.

In der Abb.3 sind dargestellt die Verläufe
der Ausgangsgrössen, gemessen in den hydrothermal veränderten Gesteinen unbekannten
Ursprungs, befallen durch Karbonisation mit
milonitisierten Störungen. Die Messung wurde
im Intervall der Positionierung 1060,16 -
1060,20 m durchgeführt.

Die in Abb.5 dargestellten Verläufe wurden
gemessen im Gneis-Amfibol-Komplex mit Schlechten der hydrothermalen graphitischen Überzü-
ge und mit milonitisierten Störungen. Die
Messungen wurden im Intervall der Positionierung 900,56-900,60 m durchgeführt.

Die Verläufe der Ausgangsgrössen in der
Abhängigkeit von der Vorschubkraft haben
einen isomorphen Charakter mit den analogen
Abhängigkeiten, bestimmt während des drehenden Bohrens (7), (8), (9). Die jeweilige
Vortriebsgeschwindigkeit bei kleinen Vorschubkraftwerten steigt sehr langsam bis zum Wert
der sogenannten kritischen Vorschubkraft,
wenn es unter den Arbeitselementen (Disken)
zu einem Volumenausbrechen des Gesteins
kommt. Von diesem Vorschubkraftwert an, steigt
die jeweilige Vortriebsgeschwindigkeit stür-
misch an und in einigen Fällen kommt es bei
den höchsten Vorschubkraftwerten zur Dämpfung
der Geschwindigkeitssteigerung, bzw. zu ihrer
völligen Einstellung wegen der unzulänglichen
Wegschaffung der Zerstörungsprodukte von dem
Ort.

Die spezifische Volumenarbeit der Zerstörung w sinkt allmählich mit der ansteigenden
Vorschubkraft bis zum Minimum und in manchen
Fällen kommt es bei den grösseren Vorschubkräften zu der wiederholten Steigerung der
spezifischen Volumenarbeit wegen der sekun-
dären Zerstörung der unzulänglich beseitig-
ten Produkte von dem Ort.

Die Grösse φ hat einen ähnlichen Verlauf
wie die jeweilige Vortriebsgeschwindigkeit,
aber in manchen Fällen mit einem markanten
Maximum, verursacht dadurch, dass bei den
höheren Vorschubkräften diese empfindlicher

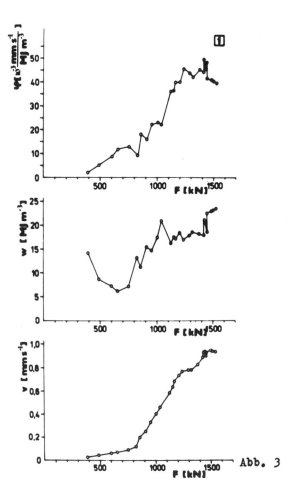

Abb. 3

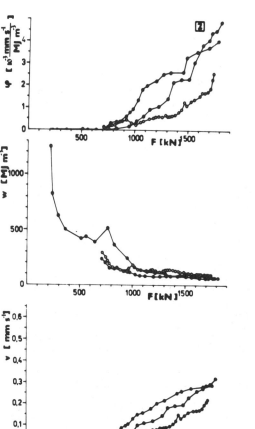

Abb. 4

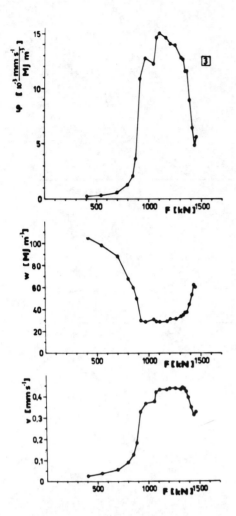

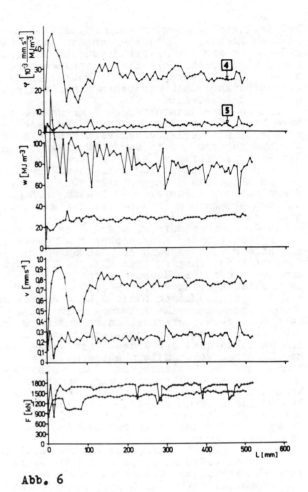

Abb. 6

Abb. 5

auf die sekundäre Zerstörung reagiert.
Wenn wir die in einzelnen mineralogisch-
petrographischen Formationen bestimmten Ausgangsgrössenverläufe vergleichen, so scheinen am leichtesten zerstörbar die hydrothermal veränderte Gesteine unbekannten Ursprungs
befallen durch Karbonisation mit milonitisierten Störungen, sieh Abb.3. Im Verlauf
der untersuchten Skalen der Vorschubkraft
sind die Werte der jeweiligen Vortriebsgeschwindigkeit die höchsten, die Werte der
spezifischen Volumenarbeit der Zerstörung
die kleinsten und die Werte der Grösse φ
die höchsten. Am schwierigsten zerstörbar
scheint der Gneis-Amfibol-Komplex, sieh
Abb.4, der in der gesamten untersuchten Vor-
schubkraftskala die kleisten Werte der jeweiligen Vortriebsgeschwindigkeit, die höchsten Werte der spezifischen Volumenarbeit
der Zerstörung und die kleinsten Werte der
Grösse φ. Die hydrothermale Veränderungen
im Gneis-Amfibol-Komplex verursachten eine
Senkung der Widerstandsfähigkeit des Gesteins
gegen Zerstörung wie es aus den Verläufen
der verfolgten Ausgangsgrössen in Abb.5 ersichtlich ist.
Die in Abb.6 und 7 dargestellten Ergebnisse, die durch keinen Eingriff in die Steuerung der Maschinen beeinflusst wurden, bestätigen die oben angeführten Schlussfolge-
rungen über die Zerstörbarkeit der einzelnen
mineralogisch-petrographischen Formationen.
Die Ausgangsgrössen dargestellt in Abhängigkeit von dem Vortriebsfortgang in Abb.6 und

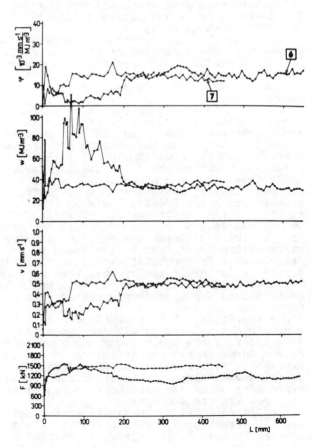

Abb. 7

718

7 ändern sich markant nur bei den Veränderungen der Vorschubkraft. Das bedeutet, dass in den angeführten Vortriebsintervallen die erwähnten mineralogisch-petrographischen Formationen sich als quasi homogene Systeme zeigen. Eine Ausnahme bilden die in Abb.7 [7] dargestellten Verläufe der Ausgangsgrössen von dem Vortriebsintervall 900,56-901,21 m, wo eine markante Störungszone vorkommt, die offensichtlich alle Ausgangsgrössenverläufe beeinflusste in dem Sinne, als ob die untersuchte mineralogisch-petrographische Formation schwieriger zu zerstören wäre. Wahrscheinlich wurde das durch die Tonfüllung in den tektonischen Schlechten verursacht, die hier als Bremsfaktor der Zerstörung unter den Arbeitselementen des Bohrkopfes wirkte. Von den vorangegangenen Messungen haben wir auch umgekehrte Erfahrungen , wenn in den tektonischen Zonen die Ausgangsgrössen solchen Charakter erhalten, als ob die gegebene mineralogisch-petrographische Formation leichter zu zerstören wäre. In solchen Fällen wurden aber die Störungsschlechten mit keinen anderen Komponenten ausgefüllt.

Die methodischen Aspekte sowie die in der Arbeit angeführten Ergebnisse zeigen, dass durch die angeführten Verfahren möglich ist komplexe und sofortige Informationen über den Zerstörungsprozess zu erhalten. Diese Informationen sind notwendig für weiteres Studium des Zerstörungsprozesses. Solch ein tieferes Studium des Vortriebsprozesses halten wir für sehr wichtig. Die erworbenen Kenntnisse über die Gesetzmässigkeiten des Vortriebsprozesses sollten den Übergang zur adaptiven Steuerung der Vortriebskomplexe aufgrund von exakten Kriterien ermöglichen.

LITERATURVERZEICHNIS

1. Roxborough, F.F. and H.R.Phillips 1975. Rock excavation by disc cutter. Int.J. Rock Mech.Min.Sci. and Geom.Abstr.12:361.
2. Sanio, H.P. 1983. Nettovortriebsprognose für Einsätze von Vollschnittmaschinen in anisotropen Gesteinen. Bochumer geol.u. geotechn.Arb.11:1-147 .
3. Weber, W. 1984. Entwicklung von Rollenbohrwerkzeugen für sehr harte und abrasive Gesteine und Einsatz von Hochdruckwasserstrahlen bei Tunnelbohrmaschinen. In IV. Symposium Theoretische und Technologische Aspekte der Zerkleinerung und Mechanischen Aktivierung Fester Stoffe (Ed.K.Tkáčová), Košice - unpublished res.
4. Wehrmann, W. and H.Otto 1982. Flacher Bohrkopf für Vollschnitt-Vortriebmaschinen. Bergbau-Forschung GmbH Essen.
5. Sekula, F., M.Merva and I.Hunsdörfer 1978. Anschluss des Indikators der Volumengesteinzerstörung beim Bohren. Amt für Erfindungen und Entdeckungen N.185857. Prag.
6. Krúpa, V., F.Sekula, J.Bejda and F.Krepelka 1986. Einrichtung für die Indikation der spezifischen jeweiligen Geschwindigkeit des Bohrens. Amt für Erfindungen und Entdeckungen N.246413. Prag.
7. Sekula, F., J.Bejda and G.Dunay 1975. Der Werkzeugverschleiss beim drehenden Gesteinsbohren in Abhängigkeit von den Bohrbedingungen. Glückauf-Forsch.-H. 5:195-198.
8. Sekula, F., V.Krúpa, J.Bejda 1980. Partiale Differentialgleichung der Bohrgeschwindigkeit. Sonderheft:43-47.
9. Krúpa, V., F.Sekula, J.Bejda, M.Koči 1984. "Über die weitere Entwicklung der mathematischen Modellierung des Diamantbohrens. Folia montana. Sonderheft:95-100.

An investigation into the effect of blast geometry on rock fragmentation

Investigations des effets de la géométrie d'un plan de tir sur la fragmentation du rocher
Der Einfluss der Sprengbohrlochanordnung auf die Gesteinszertrümmerung

D.P.SINGH, Banaras Hindu University, India
V.R.SASTRY, Banaras Hindu University, India

ABSTRACT: Rock fragmentation due to blasting is dependent on geomechanical, explosive and blast geometry parameters. A laboratory scale study was made to evaluate the effect of some blast geometry parameters (burden, spacing and bench height) on rock blasting. Experiments were divided into two groups - single hole tests to evaluate the effect of burden and bench height, and double hole tests to find out the effect of spacing and bench height. Optimum fragmentation burden was found to be 51 per cent of bench height. Significant increase in optimum breakage and fragmentation burdens was observed with increase in the bench height. The rate of fall in fragmentation with increase in burden was greater in smaller benches because of increased stiffness of burden rock. Spacing to burden ratio of 3.0 to 4.0 gave optimum fragmentation. Fragmentation at less than optimum fragmentation burden even at a spacing to burden ratio of 5.0 was finer compared to the values obtained at optimum and greater than optimum fragmentation burdens for all spacing to burden ratios, suggesting smaller burdens even with larger spacings for finer fragmentation. Considerable improvement in fragmentation was achieved both qualitatively and quantitatively with increase in bench height.

RESUMÉ: Une étude de laboratoire a été entreprise afin d'évaluer les effets des paramètres géométriques des sautages sur la granulométrie de la roche abattue. La fragmentation optimale a été réalisée pour un fardeau de 51% de la hauteur du banc. La fragmentation a été améliorée considérablement par une hautuer de banc supérieure, même à des rapports espacement-fardeau de 5,0.

ZUSAMMENFASSUNG: Eine wesentliche Verbesserung des Haufwerks wurde erreicht mit Hilfe von Laboruntersuchungen der Sprengparameter: Strossenhöhe, Bohrlochabstand und Vorgabe. Optimale Vorgabe war 51% der Strossenhöhe, und beste Erfolge wurden erzielt mit einem Bohrlochabstand, der das drei- bis vierfache der Vorgabe ist.

1 INTRODUCTION

It is a well known fact, that, presently only a meagre percentage of total explosive energy is being utilised in fragmenting and displacing rock mass. This may be due to the lack of knowledge of actual breakage mechanisms, as the fragmentation process involves extremely complex interactions among a large number of variables. A good blast requires judicious distribution of explosive energy at proper places in proper manner so that it produces a specific kind of fragmentation with significant looseness and displacement. Compared to the other two parameters, fragmentation is found to have a high degree of control on overall cost of mining.

A number of factors effect the rock blasting and they can be grouped into - explosive type (v.o.d., density etc.), blast geometry (burden, spacing, bench height, subgrade drilling, stemming etc.) and material type (physico-mechanical properties and structural discontinuities). An appreciable knowledge of the effect of all these parameters on rock blasting is essential for successful application of explosives for any given ground conditions. The topic of interest here, however, is blast geometry only.

The most critical of all geometric parameters of blasting is the burden. It is considered to have maximum influence on fragmentation as the stiffness of the rock is very sensitive to the beam depth (Mantell 1962; Allsman 1960; Langefors 1965). Singh & Sastry (1985) and Singh et al. (1985) concluded from their model scale investigations on the effect of discontinuities, that, compared to the influence of joints, burden has got more significant effect on fragmentation. Burden depends on a combination of variables including rock properties, nature of explosive, hole diameter and above all fragmentation desired. As burden increases the crater width narrows down and more of the hole bottom remains causing toes, with eventually at a certain point the radial cracked block freezes of stays locked in place which may be termed as critical burden (Ash & Smith 1976). The critical burden is an important parameter to be considered when describing the blastability of rocks (Rustan et al. 1983). For a given set of conditions, there exists a burden known as "Optimum fragmentation burden" which results in proper size distribution and throw alongwith minimum side effects like toe, air blast etc. Coarser fragments constitute a higher percentage for burdens greater than optimum fragmentation burden. When burden falls below optimum fragmentation burden strain wave fracturing increases. For very small burdens, strain wave fracturing occurs so rapidly that much of the gas energy is lost to the atmosphere as air blast and also results in excessive throw of rock.

Spacing controls mutual stress effects between blast holes and depends on burden, hole depth, charge length, initiation sequence and rock properties. Hagan (1983) reported that adequate results could be achieved at spacing to burden ratio (S/B) of 1.0. Kochanowsky (1964) recommended S/B values of 1.0 to 1.5 for field blasting. Ash & Smith (1976) found that S/B value of 2 gave better results with delay timing and when row of holes was located perpendicular to dominant joint planes. Langefors (1966) from his tests reported that S/B ratio of even upto 8 resulted in better fragmentation with alternate pattern multirow blasting. Singh et al. (1981) found that spacing to burden ratio ranging from 2.5 to 3.0 gave better results with instantaneous initiation. Spacing provided between the two holes in no case should be less than burden as it causes premature splitting of holes and early loosening of stemming in consequence causing sudden drop in blasthole pressure. According to flexural rupture theory, it was found that with the increasing bench heights and burdens, spacing can also be increased (Ash & Smith 1976). As spacing to burden ratio increases for a given set of bench height/burden (BH/B) conditions there will be an increase in larger size fractions.

Bench height plays a critical role in influencing the fragmentation results. Burden, spacing and bench height should not be treated as separate parameters, since a combination of these effects rock blasting. For each burden there is a maximum bench height to produce a full crater (Mason 1973). Bench height should never be less than burden which otherwise would leade to overbreak and cratering. The breakage angles for a given burden increase with increase in bench height upto a certain point beyond which no further significant change occurs (Atchison 1971). Smith (1976) concluded from his studies that with a constant burden and amount of stemming any increase in bench height causes greater flexure giving better fragmentation. Ash & Smith (1976) reported that increase in fragmentation was more significant when bench height increased from 1 to 2 times the burden than when the ratio was changed from 2 to 3 times the burden. Further increases in bench height may not improve fragmentation.

1.1 Statement of the problem

The aim of any blasting investigation must be directed to predict the fragmentation and displacement achieved by an explosive under given set of rock conditions and blast design. With an aim to achieve the above mentioned goal an attempt was made to investigate the influence of some of blast geometry parameters on fragmentation. The study was restricted to laboratory scale tests only. Selection of parameters for the study was, however, confined to burden, spacing and bench height. The experimental approach used recognizes the fact that it is not possible to fully simulate a field blast with a model blast. Hence, the conclusions drawn are of qualitative nature only.

2 EXPERIMENTAL

Small scale tests were conducted on models prepared from Chunar sandstone, which was found to be suitable for blasting studies by earlier research conducted in the department (Singh et al. 1980; Singh & Sarma 1983; Singh & Sastry 1985; Singh et al. 1985; Singh & Sastry 1986). It is a fine grained sedimentary rock and is well known for its homogenity and high blastability ratio. Experiments were conducted in two phases, single hole tests and double hole tests. Models for single hole tests were 500 mm x 400 mm x 150 mm and for double hole tests 1000 mm x 400 mm x 150 mm (see fig. 1). The bench

heights were 35 mm, 50 mm and 65 mm. Single hole tests were used to study the effect of burden and bench height whereas double hole tests were conducted to study the effect of spacing, burden and bench height on rock fragmentation. Optimum fragmentation burden, where mass surface area is maximum, was established for each bench height in single hole tests. Three burdens - optimum fragmentation burden, less than optimum fragmentation burden and greater than optimum fragmentation burden were used with spacing to burden ratios varying from 1.0 to 6.0 in double hole tests. In total 83 tests were conducted.

No subdrilling and no stemming were provided. Detonating fuse having a v.o.d. of 6500 m/s representing cylindrical charge was used to blast the models, keeping charge per unit length constant in all the cases. Charges were fired simultaneously in the case of double hole tests.

2.1 Fragmentation assessment

Fragments collected after blasting from each model were subjected to sieving. From sieve analysis results, various fragmentation measures such as, mass (MF), average fragment size (AFS), new surface area (NSA), mass surface area (MSA), (product of mass and new surface area created), fine fragmentation index (FFI), coarse fragmentation index (CFI), A1 (aperture through which 25% of total mass passes), A2 (aperture through which 50% of total mass passes) and A3 (aperture through which 75% of total mass passes) were calculated for analysing the fragmentation results. Further, analysis was done with the help of size distribution histograms.

3 DISCUSSION OF RESULTS

3.1 Single hole tests

3.1.1 Effect of burden on fragmentation

The mass of fragments (MF) increased with the increase in burden upto a certain value denoted as optimum breakage burden (OBB) and then started decreasing with further increase in burden (see fig. 2).

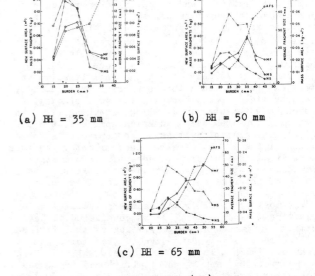

(a) BH = 35 mm   (b) BH = 50 mm

(c) BH = 65 mm

Figure 2. Mass of fragments (MF), average fragment size (AFS), new surface area (NSA) and mass surface area (MSA) vs burden.

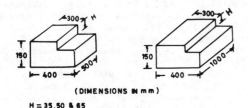

(DIMENSIONS IN mm)

H = 35, 50 & 65

(a) SINGLE HOLE TESTS   (b) DOUBLE HOLE TESTS

Figure 1. Dimensions of the models.

As bench height (BH) increased, the OBB also increased. On average, the OBB was around 76% of the BH. The average fragment size (AFS) also exhibited an increasing trend with increase in burden. Mass surface area (MSA), showed an increase with increase in burden upto a certain value, defined as "Optimum fragmentation burden"(OFB) and then started decreasing with further increase in burden values (see fig. 2). The OFB for 35 mm BH was found to be 20 mm, for 50 mm BH, 25 mm and for 65 mm BH, 30 mm. In general the OFB was around 51% of BH. Fine fragmentation index (FFI) decreased and coarse fragmentation index (CFI) increased with increase in the burden. As burden increases the effectiveness of strain wave fracturing decreases resulting in greater AFS and CFI values.

3.1.2 Effect of bench height on fragmentation

An interesting observation was that the OBB increased by 29.3% and OFB by 18.3% for every 15 mm increase in BH.

At the same burden higher benches become less stiff. As a result of this MF increased with increase in BH at same burdens. There was a decrease in AFS in 65 mm BH compared to 50 mm BH due to the lower stiffness in the former case (see fig. 3).

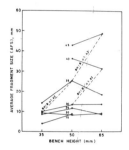

Figure 3. Average fragment size vs bench height.

The broken line in fig. 3 depicts the AFS for same BH/B ratio. There was an increase in AFS with increase in BH for equal BH/B ratios. In the present experimental set up the explosive used was taken on the basis of BH but not on burden volume basis. The diameter was kept constant (6.0 mm). This could be the reason for getting higher AFS for higher bench heights with same BH/B conditions. Therefore, to get lower AFS in higher BH the charge diameter should also be increased proportionately. At their respective OFB's the MSA obtained with 65 mm BH was 3.44 times that of 50 mm BH and 14.4 times that of 35 mm BH. The MSA obtained with 50 mm BH was around 4 times that of 35 mm BH. On average, the fragmentation improved by around 3.7 times for every 15 mm increase in BH at corresponding OFB values. It implies that, it is better to have one higher bench rather than two smaller benches. Also the number of holes could be minimised in the former case.

3.2 Double hole tests

3.2.1 Effect of spacing on fragmentation

The MF showed an increasing trend with increase in spacing to burden ratio (S/B) upto a certain value and thereafter decreased. The maximas varied from 3.0 to 4.0 (see fig. 4, 5 & 6). At larger spacings of 5.0 and 6.0, almost each charge behaved as an independent charge, producing separate craters

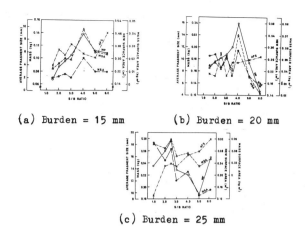

(a) Burden = 15 mm   (b) Burden = 20 mm

(c) Burden = 25 mm

Figure 4. Mass of fragments (MF), Average fragment size (AFS), new surface area (NSA), and mass surface area (MSA) for 35 mm bench height.

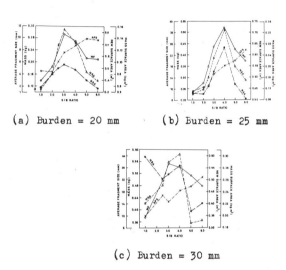

(a) Burden = 20 mm   (b) Burden = 25 mm

(c) Burden = 30 mm

Figure 5. Mass fragments (MF), average fragment size (AFS), new surface area (NSA) and mass surface area (MSA) for 50 mm bench height.

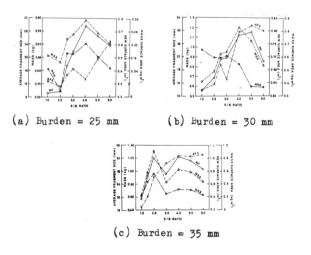

(a) Burden = 25 mm   (b) Burden = 30 mm

(c) Burden = 35 mm

Figure 6. Mass of fragments (MF), average fragment size (AFS), new surface area (NSA) and mass surface area (MSA) for 65 mm bench height.

having large humps inbetween two holes. It is expected that AFS should increase as S/B value increases and similar results were obtained in almost all the cases. At smaller spacings the strain waves from two holes interact in a better manner causing effective microstructural damage resulting in better utilization of gas energy for further fragmentation. This was evidenced by higher FFI and lower CFI values at smaller S/B ratios of 1.0 to 3.0 in all the cases.

For same BH and S/B, increase in burden produced coarser fragments, as was expected. This was evident by the incidence of higher percentage of fines (-1.68 mm size) at burdens smaller than optimum fragmentation burden for the same BH and S/B ratios. Histograms drawn between mass percentage and sieve sizes showed this phenomenon very clearly. Another interesting observation made was that in 35 mm BH models fragmentation produced at 15 mm burden was finer even at 5.0 S/B compared to any S/B at 20 mm or 25 mm burdens. This was confirmed by the lower AFS values obtained at smallest burden of 15 mm (see fig. 4). Similar was the case with 50 mm and 65 mm bench height models (see fig. 5 & 6). When the burden was greater than optimum fragmentation burden there was toe formation from S/B value of 1.5 onwards, which was true in all the cases, suggesting that in no case a burden more than OFB should be used, as the rock becomes more stiff and it is more difficult to move the bottom portion of bench.

Taking MSA and A1/M1 (aperture, A4, through which 25% of total mass, M25, passes/M25), A2/M2 (aperture A50, through which 50% of total mass, M50 passes/M50) and A3/M3 (aperture, A75, through which 75% of total mass, M75, passes/M75) factors into consideration it was observed that in 35 mm BH models with 15 mm and 20 mm burdens optimum fragmentation resulted at S/B values of 3.0 to 4.0 (see fig. 4). At 25 mm burden, which is greater than OFB, severe toe formation was there from S/B 1.5 onwards. In the case of 50 mm BH the optimum S/B ratio was 3.0 for 20 mm burden and it was 4.0 at 25 mm burden (see fig. 5). At 30 mm burden finer fragmentation resulted at S/B value of 1.0 and from S/B values of 2.5 onwards significant toe formation was there. In 65 mm BH models the optimum S/B value was found to be 4.0 in both the cases of optimum and less than optimum fragmentation burdens (see fig. 6).

3.2.2 Effect of bench height on fragmentation

For evaluating the effect of BH two burdens, one optimum and the other less than optimum fragmentation burdens were selected i.e. for 35 mm BH, 15 mm and 20 mm burdens, for 50 mm BH, 20 mm and 25 mm burdens and 25 mm and 30 mm burdens for 65 mm BH. Comparison was made in between these BH's at their corresponding optimum and less than optimum fragmentation burdens. Mass and mass surface area obtained at different BH-B values of 35-15, 50-20 and 65-25 are shown in fig. 7 and values obtained for BH-B values of 35-20, 50-25 and 65-30 are shown in fig. 8. From fig. 7 & 8 it can be observed that optimum and less than optimum fragmentation burdens, higher benches produced more MF as well as more MSA. The reason being that with increase in BH, burden rock becomes less stiff and breaks easily. There was a clear trend of increase in BH. The rate of increase in MF was more when BH increased from 35 mm to 50 mm than when BH was changed from 50 mm to 65 mm. At burdens less than optimum MF increased by 2.38 times on average, when BH was increased from 35 mm to 50 mm whereas, the increase was only 1.8 times when BH increased from 50 mm to 65 mm. The MF increased on average by 4.16 times when BH increased from 35 mm to 65 mm. Similarly, at OFB the MF increased by 4.6 times when BH increased from 35 mm to 65 mm.

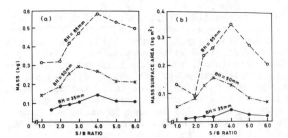

Figure 7. Mass and mass surface area vs S/B ratio for BH/B values of 35/15, 50/20 & 65/25.

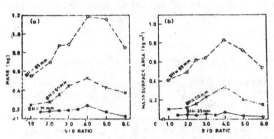

Figure 8. Mass and mass surface area vs S/B ratio for BH/B values of 35/20, 50/25 and 65/30.

Similar was the case with MSA. It increased by 5.72 times when BH was raised from 35 mm to 50 mm and by 1.81 times when BH changed from 50 mm to 65 mm at corresponding less than OFB values. An average increase of 9.85 times in MSA was observed when BH was raised from 35 mm to 65 mm. In the same manner at respective OFB values, 65 mm bench produced 12.35 times the MSA of 35 mm BH. This is a clear indication of better fragmentation in regard to size distribution as well as amount of yield in case of higher benches.

The aperture/mass (A/M) values of 25%, 50% and 75% passing lines for BH-B values of 35-15, 50-20 and 65-25 and 35-20, 50-25 and 65-30 are shown in fig. 9 & 10. From these it can be observed that fragmentation was better in 65 mm models compared to 35 mm and 50 mm BH models at corresponding OFB values (see fig. 10). At lower burdens than OFB also the fragmentation was better in higher benches (see fig. 9). The criterion taken was that, the lower the A/M value the better the fragmentation. These figures also indicate a greater improvement in fragmentation for higher benches at same S/B values due to reduced composite stiffness.

Overbreak was more prevalent at larger burdens. Toe formation was more in smaller bench of 35 mm compared to 50 mm and 65 mm benches with same burden and spacings. As the spacing distance increased toe formation was more, especially in cases, where, more than optimum fragmentation burdens were used.

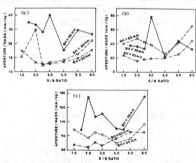

Figure 9. Aperture/mass vs S/B ratio for BH/B values of 35/15, 50/20 & 65/25 at passing lines (a) 25% (b) 50% & (c) 75%.

724

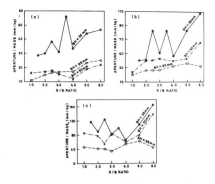

Figure 10. Aperture/mass vs S/B ratio for BH/B values of 35/20, 50/25 & 65/30 at passing times (a) 25%, (b) 50% & (c) 75%.

4 CONCLUSIONS

From the test results, the conclusions drawn were as follows :
- optimum breakage burden was around 72% of bench height and optimum fragmentation burden was around 51% of bench height.
- optimum fragmentation burden increased by 18.3% with every 15 mm increase in bench height.
- rate of fall in fragmentation with increase in burden was greater in smaller benches, due to increased stiffness of burden rock.
- spacing to burden ratio of 3.0 to 4.0 gave optimum fragmentation results.
- as spacing to burden ratio increased and reached the value of 5.0 each hole behaved as an independent hole leaving large hump between the holes.
- at burdens smaller than optimum fragmentation burden, the fragmentation was finer even at a spacing to burden ratio of 5.0 compared to results obtained at optimum and greater than optimum burdens with smaller spacing to burden ratios of even 1.0.
- at burdens less than optimum fragmentation burden mass surface area improved by 5.72 times in 50 mm bench height models and by 9.82 times in 65 mm bench height models compared to models with 35 mm bench height. Similarly, at respective optimum fragmentation burdens 65 mm bench produced 12.35 times the mass surface area of 35 mm bench, indicating the enhancement in fragmentation qualitatively as well as quantitatively.

5 ACKNOWLEDGEMENT

Authors express their gratitude to IDL Chemicals Ltd., Hyderabad, India, for providing the necessary funds for carrying out the experiments reported in this paper.

REFERENCES

Allsman, P.L. 1960. Analysis of explosive action in rock. Trans. SME/AMIE, New York: 217:468-478.
Ash, R.L. & N.S.Smith 1976. Changing borehole length improve breakage : a case history. Proc. 2nd conf. Explosives and Blasting Techniques, Louisville, Kentucky, 1-12.
Atichison, T.C. 1968. Fragmentation principles, surface mining, Ed. Pflieder, E.P., AIME, New York, 335-372.
Hagan, T.N. 1983. The influence of controllable blast parameters on fragmentation and mining costs. Proc. 1st Int. Symp. Rock Fragmentation by blasting, Lulea, Sweden, 31-52.

Kochanowsky, B.J. 1964. Developments in blasting techniques in opencast mining and quarrying. Symp. Mining, Quarrying and Alluvial Mining.
Langefors, U. 1965. Fragmentation in rock blasting. Proc. 7th Symp. Rock Mech. : 1-21.
Langefors, U. 1966. Fragmentation in rock blasting.Min. & Minerals Engng. 2 : 339-347.
Mantell, M.T. & J.C. Marron 1962. Structure Mechanics. Ronald Press Co. New York.
Mason, J.M. 1973. The effect of explosive charge length on cratering, M.S.Thesis, Univ. of Missouri, Rolla.
Rustan, A., V.S.Vutukuri & T Naarttijarvi 1983. The influence from specific charge geometric scale and physical properties of homogenous rock on fragmentation. 1st Int. Symp. Rock Fragmentation by blasting, Lulea, Sweden. 1:115-142.
Singh, D.P., Y.V.Appa Rao and S. S. Saluja 1980. A laboratory study of effects of joints on rock fragmentation. Proc. 21st U.S.Symp. Rock Mech., Univ. Missouri, Rolla, USA.
Singh, D.P., S.Ratan & S.Saran 1981. Effect of spacing and burden on rock fragmentation. Nat. Symp. Drilling and Blasting, Univ. of Jodhpur, Jodhpur, India, 82-91.
Singh, D.P. & K.S.Sarma 1983. Influence of joints on rock blasting, a model scale study. Proc. 1st Int. Symp. Rock Fragmentation by Blasting, Lulea, Sweden, 533-534.
Singh, D.P. & R.V.Sastry 1985. Rock fragmentation by blasting - influence of joint filling material. Nat. Sem. Mining : Present and Future, Banaras Hindu University, Varanasi, India.
Singh, D.P., R.V.Sastry & C.V.Suresh 1985. Fragmentation in jointed rock material - a model scale investigation. Nat. Sem. Indigenous Development of Mining Explosives and Accessories, Policies and Programme, New Delhi, India.
Singh, D.P. & R.V.Sastry 1986. Influence of structural discontinuities on rock fragmentation by blasting. Int. Symp. Intense Dynamic Loading and Its Effects, Beijing, China.
Smith, N.S. 1976. Burden rock stiffness and its effects on fragmentation in bench blasting, Ph. D. Thesis, Univ. of Missouri, Rolla.

A study for optimization of fragmentation by blasting in a highly cleated thick coal seam

Etude pour l'optimisation de la fragmentation par minage dans une couche de houille épaisse jointée
Optimierung von Sprengungen in einem geklüfteten mächtigen Kohleflöz

R.D.SINGH, Professor, Kothagudem School of Mines of Osmania University, India
VIRENDRA SINGH, Professor, Kothagudem School of Mines of Osmania University, India
B.P.KHARE, Reader, Kothagudem School of Mines of Osmania University, India

ABSTRACT: Geological discontinuities in the rock mass affect blasting results considerably. In an 18 m thick coal seam worked by opencast method at Prakasham khani, Manuguru of Godavari valley coalfields, India, the presence of cleats had resulted in production of large amount of oversize blocks of coal. This posed difficulties at the coal handling plant. A systematic study was taken up to optimize the blast geometry in order to achieve the required fragmentation. Field and laboratory trials with regard to the geological setting and blastability of coal helped in making suitable recommendations for selecting right type of explosive and blasting pattern.

RESUMÉ: Dans un lit de charbon de 18m exploité à ciel ouvert aux Indes, une étude a été initiée qui devait recommander les explosifs et procédures les plus susceptibles de réduire les problèmes des aux joints.

ZUSAMMENFASSUNG: In einem 18m mächtigen Kohleflöz, das im Tagebau gewonnen wird, ergaben Produktionssprengungen grosse Mengen an blockigem Haufwerk. Um ein besseres Haufwerk zu erreichen, wurden aufgrund von Geländearbeiten und Laborversuchen die Bohrlochanordnung und die Art des Sprengstoffs geändert.

1 INTRODUCTION

In India, surface mining is playing increasingly dominant role. Its contribution to coal production averages around 50% of the total which is expected to rise further to 60% by 2000 A.D. In other sectors too, opencast mining is dominating e.g. entire production of iron ore, limestone, gypsum comes from openpits.

Also, intense surface mining activity is projected to open new units and produce large tonnages in copper, lead and zinc sector. The mines are becoming larger and deeper and require handling of large volume of rock/ore or coal. Opencast mining is essentially a material handling process in which this activity shares about 40% of the total cost of production. In order to make it techno-economically successful, it is the degree of fragmentation that is of paramount importance. Good fragmentation not only increases the efficiency of shovel/loader/dragline but also of the transport system. Obtaining an optimum level of fragmentation is a difficult task, more so when the deposit is highly cleated and jointed. Presented here is the result of a study in a highly cleated thick coal seam for designing the blasting subsystems with a view to obtaining optimum fragmentation.

2 CHARACTERISTICS OF THE EXPERIMENTAL SITE

The study was carried at Prakasham khani in Manuguru Division of Godavari valley coalfield, India. The seam worked is about 18 m thick. It is banded, the bands being 0.30 - 2.74 m thick (Fig.1). The seam has two distinct sets of cleats crossing each other obliquely at an angle of about 112° and they are spaced generally 15 - 150 cm apart and are developed in both the directions (Fig.2). Also, the bedding plans are prominent along which different layers easily separate. There is a natural tendency for the coal to break in blocks limited by the cleats and bedding plans.

The average compressive strength of coal based on laboratory testing was found to be in the range of 10 - 15 MPa when loaded along the bedding planes and 8 - 20 MPa when loaded perpendicular to bedding planes.

3 THE PROBLEM

The coal had a tendency to break along cleats and bedding planes and thus fragments produced were in big lumps. The requirement was for the lumps to pass through a grizzly of 450 mm x 300 mm size. Larger size of blasted mass created problem of speedy clearance of output and also needed secondary breaking done manually, a slow process, and constrained the overall operational efficiency of the mine. This restricted the output that could be handled at the coal handling plant. Judging from the strength of coal it could not be characterized to be hard and therefore, ordinarily there should not have been any problem of fragmentation when blasted. But, contrary to the expectation the fragmentation was generally poor. It necessiated a review of the blasting system and a fresh look on other geological factors such as frequency and orientation of cleats. A study was, therefore, taken up to examine the blasting system including the design of blasthole geometry and the suitability of the explosives used vis-a-vis the cleat and bedding plane characteristics.

4 EXISTING BLASTING PRACTICES

The coal was won in two benches; the top bench was 9 - 12 m and the bottom bench was 5 - 6 m high. Vertical holes 150 mm dia were drilled in the top bench. 2.5 m apart in the same row and the burden was 4 m. In the bottom bench holes were spaced at 4 m in a row and the rows were 4 m apart. At a time one or two rows were blasted. The explosive used were ANFO and Dynax. Deck loading was practiced. Shots were blasted simultaneously in the same row with delays of 15 - 51 ms between the rows. The powder ratio used was 2.5 t/kg in the top bench and 4.5 t/kg in the bottom bench. Detonating cord was used for making connections.

5 EXPERIMENTAL STUDIES

The study at the mine site comprised the following:
(i)   Crater tests
(ii) Influence of cleats on explosion effect

727

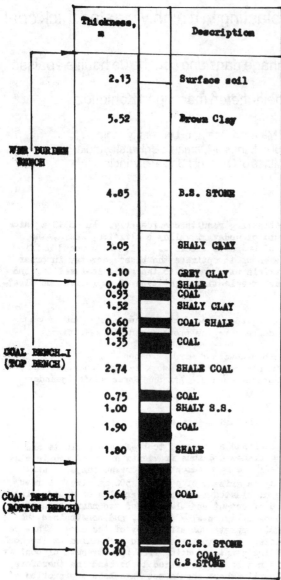

Figure 1. Borehole section, Prakasham khani, Ramagura (R.F. 1:200).

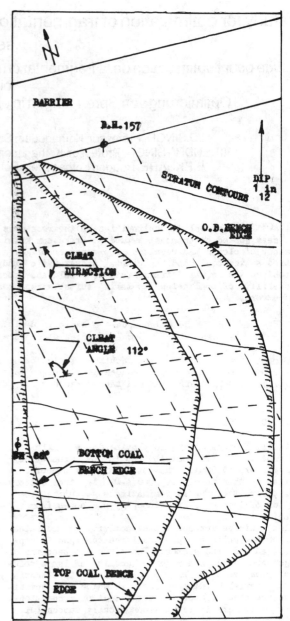

Figure 2. Orientation of cleats.

(iii) Evolving of a blast pattern based on above
(iv) Trial blast and comparison of results.

Crater tests were conducted in the top bench with both the explosives used in the mine. Explosive quantity of 2.3 kg was kept constant for each hole. Four holes of varying depth were drilled. Diameter of holes was 150 mm. Table 1 summarises the results of the cratering using Dynax. Calculation of maximum burden and maximum spacing for a 9 m bench using strain energy factor and optimum depth ratio gave figures of 3.85 m and 4.71 m respectively. For Dynax the following data were calculated.

Strain energy factor $E = 3.74$ m.kg$^{1/3}$

Optimum depth ratio, $\triangle = 0.48$

For the optimum breakage conditions the values of depth ratio for burden, spacing and charge depth were calculated using the following relationships.

$\triangle_B = 0.9$, $\triangle_S = 1.1$ and $\triangle_H = 1.4$

Keeping in view the frequency of cleats through which substantial energy of explosive escapes the pattern suggested was 3 m burden and 4 m spacing. When this pattern of holes was blasted it gave better fragmentation - oversize (+300 mm) was limited to 10 - 15% as against 25 - 30% with the earlier pattern used in the mine.

Table 1.

Depth of hole, m	Depth of crater, m	Volume of Crater, m³
0.5	0.59	1.17
1.0	1.19	3.29
1.5	0.2	0.04
2.0	0.0	0.00

6 RESULTS AND DISCUSSION

On the basis of cratering an attempt was made to calculate the maximum theoretical burden and spacing. These values needed modification to account for the presence of cleats and bedding planes. It was also observed that if the direction of the blast was designed to be out from the lesser angle of the rhombus of the cleat, it resulted in overbreak and coarse fragmentation. On the other hand, if blast was directed to be out from the open angle better fragmentation and straighter face was obtained. Recent studies (Fung 1981) (Ash 1963) and (Pugliese 1972) have shown that to achieve good fragmentation in massive or in-

728

frequently jointed rock a delay interval of 9 milli-
-seconds per metre of burden between holes in a row
gave good results. The delay interval between rows
should be 2 - 3 times between holes in a row.

Based on the above studies and author's own observations in the field, the blasting system was suitably modified which gave better results as far as fragmentation was concerned.

7 CONCLUSIONS

The study does point out that specific formula alone
can not give an adequate solution to a fragmentation
problem in a jointed and cleated coal seam. At every
site and for the type of explosive used trial study
have to be done to arrive at a workable blasting geometry which might also need modifications as and when
strata conditions show marked variation.

REFERENCES

Pugliese,J.M. 1972. Designing blast pattern using
   empirical formulae. USBM, IC, 8530. Washington D.C.
Ash, R.L. 1970. Cratering and its application in blasting. Unpublished report of mining and civil engineering, University of Minisotta, U S A.
Fung, R.(ed), 1981. Surface coal mining technology:
   Engineering & environmental aspects. Noyes data
   corporation, U S.A.

Low-frequency vibrations from surface mine blasting over abandoned underground mines

Vibrations à basse fréquence provenant de tirs à l'explosif de surface au-dessus de mines souterraines abandonnées

Niederfrequenzerschütterungen von Sprengarbeiten im Tagebau über verlassenen Gruben

DAVID E.SISKIND, Bureau of Mines, U.S. Department of the Interior, Minn., USA

ABSTRACT: Blasting for surface mining is a source of potentially damaging ground vibrations. Of particular concern are sites with low-frequency surface waves, those below 5 Hz. These vibrations can produce excessive structural response and strains in nearby buildings. A Bureau of Mines, U.S. Department of the Interior, study of one such site found the ground acting as the propagating medium primarily responsible for generating and sustaining these low-frequency waves. The observed waves were consistent with predictive models based on layer thicknesses equal to the depths to abandoned underground mines known to exist in the area. A thick zone of low-velocity glacial till also exists between the blasts and the affected homes. Because some influence was also found from the initiation delay timing, blast designs based on specific timing patterns could be used to reduce the generated vibrations at such sites.

RESUME: Le travail à l'explosif dans les exploitations minierès à ciel ouvert est une source d'ébranlement du sol pouvant avoir des conséquences néfastes. Les sites traversés par des ondes de surface de basse fréquence, au-dessous de 5 Hz, sont particulièrement préoccupants. Les vibrations peuvent produire une réponse structurale et des contraintes excessives dans les bâtiments avoisinants. Une étude d'un tel site par le Bureau of Mines, U.S. Department of the Interior, a indiqué que le sol jouait le rôle d'agent de propagation, et qu'il était le principal responsable de la génération et de la sustentation de ces ondes de basse fréquence. Les ondes observées étaient conformes aux modèles projetés, qui étaient basés sur des épaisseurs de couches égales aux profondeurs de mines souterraines abandonnées qui existent dans la région. Une zone profonde d'argile à moraines de faible vitesse existe également entre la zone dans laquelle ont lieu les explosions et les maisons concernées. Etant donné qu'une certaine influence du retard de la propagation lors de l'amorçage a également été notée, des séquences d'explosions basées sur l'application d'un chronogramme spécialement étudié pourraient être utilisées pour réduire les vibrations produites sur de tels sites.

ZUSAMMENFASSUNG: Im Zusammenhang mit dem Tagebau durchgeführte Sprengarbeiten verursachen potentiell schädliche Bodenerschütterungen. Gelände mit Niederfrequenz-Oberflächenwellen, d.h. unter 5 Hz, sind besonders besorgniserregend. Die Erschütterungen konnen übermäßige strukturelle Reaktionen und Beanspruchungen an Gebäuden hervorrufen. Eine von der Bergbaubehörde des US-Innenministeriums durchgeführte Untersuchung eines solchen Geländes stellte fest, daß der als Übertragungsmedium dienende Boden in erster Linie für die Erzeugung und Aufrechterhaltung der Niederfrequenzwellen verantwortlich ist. Die beobachteten Wellen stimmten mit den Voraussagemodellen überein, die sich auf Schichtenlagen gründen, die der Tiefe der bekannterweise in dieser Gegend liegenden, verlassenen Gruben entsprechen. Eine breite Zone langsamen Geschiebelehms besteht außerdem zwischen den Sprengbereichen und den in Mitleidenschaft gezogenen Gebäuden. Da auch festgestellt wurde, daß die Sprengverzögerungszeit einen Einfluß ausübt, könnten auf spezifischen Zeitmodellen basierende Sprengausführungen dazu benutzt werden, die an derartigen Stellen erzeugten Erschütterungen zu reduzieren.

1 INTRODUCTION

The Bureau of Mines studied a site in the western Indiana town of Blanford where surface coal mine blasting was producing unusual low-frequency, low-duration vibrations. At the request of two regulatory agencies, the Indiana Department of Natural Resources (DNR) and the U.S. Department of the Interior Office of Surface Mining (OSM), the Bureau investigated the influence of the ground structure, including extensive abandoned underground workings under both the active mining and Blanford. These workings were at several depths, between 30 and 110 m. The coalbeds and local sedimentary rocks are nearly horizontal.

The initial objective of the study was to determine if the site was unusual, as local homeowners claimed. After vibration data showed that the site was indeed unusual, researchers sought the causes including the structural conditions and factors in the blast designs used for the surface mining. By identifying relative influences, blast designs could be used to minimize the generated low frequencies at this and similar sites.

This paper briefly summarizes research, reported more fully in a recent Bureau of Mines Report of Investigation by the author (Siskind 1987).

2 ANALYTICAL PROCEDURES

Vibration records from 235 production blasts were available for analyses of amplitudes, frequencies, and total durations. These were originally collected by the Indiana DNR and the surface mine (Peabody Coal Company, operating the Universal Mine) at seven Blanford homes over a period of 11 months. Measurement distances ranged from about 300 to 3,000 m and as many as six homes were monitored at one time (fig. 1).

For purposes of identifying the generation and propagation influences, Bureau of Mines researchers instrumented seven blasts including two specially fired single charge shots. For these tests, measurements were made as close as 17 m from the blasts to determine the wave characteristics before they were modified by the propagation medium. The single charges were used to identify the specific vibration generation influences of the complex multideck, multidelayed blasts as contrasted to the simple vibration sources represented by the single-charge blasts.

All vibration records were plotted for propagation showing the generation and attenuation of particle velocity amplitudes. Researchers did this for each home, for various groups (by neighborhoods), and as an overall summary.

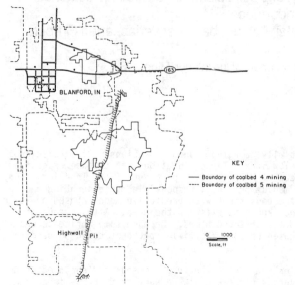

Figure 1. Town of Blanford, current surface mining highwall, abandoned underground workings and homes studied for vibrations and settlement.

In addition to vibration amplitudes, frequency characteristics were compared with records from various homes and various shot designs. Single charges and close-in monitoring revealed the ground's natural response frequency of 8 Hz. Shot design data were compared for the four blasting methods employed by Peabody during the study period. This allowed identification of delay sequencing influences on vibration frequency as opposed to ground influence.

Other site data were collected to identify areas around Blanford and the nearby surface mine undermined by previous mining. Depth data for these deep coalbeds were available from Peabody, which is conducting an extensive blasting vibration study of one Blanford home. Data were collected on subsurface composition and structure including the extent of the abandoned workings. Bureau researchers conducted level-loop surveys of eight homes in September 1985 and in April 1986 to identify possible blast vibration-induced subsidence, and long-time stability. Finally, a predictive model was used to determine if observed low frequencies were consistent with depths to the old workings.

3 RESULT OF ANALYSES

3.1 Propagation

Propagation plots revealed higher vibration amplitudes than those found in previous Bureau of Mines studies at surface coal mines (Siskind 1980). Figure 2 is a summary of vibration data collected by Peabody and DNR. The data are clustered, having been collected at seven Blanford homes rather than with widely spaced seismograph arrays. Most notable is that virtually all the measurements exceed the mean of the previous surface coal mine vibration summary, which is shown as maximum velocity in the Bureau of Mines report. Many even exceeded the envelope line of highest measurements, which was approximately two standard deviations (2σ) above the mean. For a given scaled distance (computed traditionally in $ft/lb^{1/2}$, Siskind 1980) and based on the traditional 8 ms

minimum time separation, vibration levels were 2 to 10 times higher than predicted from previous studies. Much of this difference is from interacting delayed charges, which the 8 ms minimum time separation is supposed to prevent. However, a significant

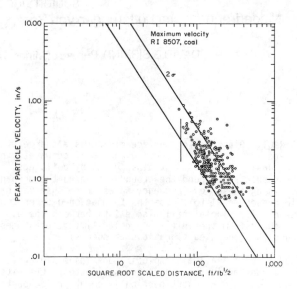

Figure 2. Propagation plot of Peabody and DNR data for all homes compared with coal mine blasting summary from previous studies published in RI 8507 (Siskind 1980).

contribution is caused by the generation of low frequency surface waves which have inherently lower attenuations and also contribute to high amplitudes by constructive wave interference.

Production and single charge data were obtained from the widely spaced Bureau of Mines propagation array (fig. 3). Vibration propagations from the

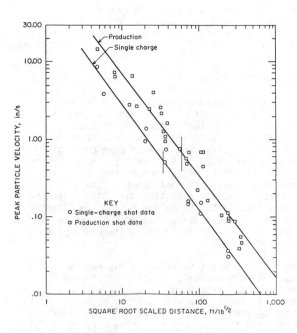

Figure 3. Propagation plot of Bureau of Mines tests at Blanford compared to coal mine blasting summary from previous studies published in RI 8507 (Siskind 1980).

732

single charges were close to those of previous studies. However, the production vibrations, for the same charge weight per delay, averaged three times higher. Evidently, vibrations from the delayed charges in the multihole, multideck production blasts are interacting. The somewhat shallower propagation mean line slopes, as compared with those from previous studies (summarized in RI 8507), are consistent with a low-vibration attenuation with distance. This was also previously suggested by the Peabody and DNR data.

3.2 Frequency character of vibrations

All vibration measurements available as time history records were analyzed for frequency content. Some had very low frequencies of about 4 Hz on all three vibration components, vertical, longitudinal, and transverse (fig. 4).

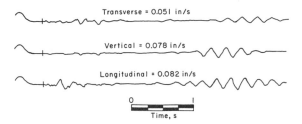

Figure 4. Vibration record from surface mine blast in Blanford, at a distance of 7,520 ft.

The source function for the blast shown in figure 4 lasted about 0.7 s, which is also the approximate duration of the high-frequency part of the vibration record. The low-frequency wave began at about 1.7 s and lasted at least 2 s for a total vibration duration of about 4 s.

Researchers compared vibration records for various blast designs and monitoring locations. Although the larger casting blasts frequently produced low frequencies, the three echelon designs that were tested also did so occasionally. Amounts of low frequencies and the numbers of vibration components involved varied around Blanford. However, they were generally consistent in character at any single site (Siskind 1987). Aside from site-to-site differences, only simple distance from the blast appeared to be a factor. At some homes, increasing distances correlated with more components showing low frequencies. At other sites, no such trend appeared.

The researchers attempted to identify the types of of waves observed based on expected components and their phases. Rayleigh waves are vertically polarized with retrograde elliptical particle motions. They have significant motion in the longitudinal and vertical directions, and little in the transverse. The generation of these waves requires only a single free surface, such as the ground-air interface, or any sharp acoustic contracting layer at depth.

Love waves are horizontally polarized shear waves. They are strong only in the transverse direction. Generation of Love waves requires a layer with top and bottom boundaries that have good reflecting properties. Extensive underground voids could provide such a reflecting surface, as could any low-velocity layer.

Six of the seven measurement sites had patterns of low frequencies resembling Rayleigh waves or Rayleigh and Love waves together. Two of these sites also occasionally had only higher frequencies (e.g., 20

Hz) at the same time and for the same shots when low frequencies were being measured elsewhere. One site (Jackson) had every kind of combination at different times. The seventh site (Verhonik) had only transverse low frequencies, suggesting Love waves. This seventh site was in a different location from the others, east of the blasting rather than north. For this site, there was a 30-m-thick layer of mine spoils in the propagation path.

3.3 Single charge blasts

The two special single charge shots were fired to identify the blast initiation sequencing influences on the wave form. They also show the complexity of the propagating medium. Figure 5 is the vertical vibration record from a 1-meter column of explosive, which took about 0.3 ms to fully detonate. After propagating 20 m (65 ft), the vibration duration is over 150 ms. At a distance of 355 m (1165 ft), the vibration is dominated by a 6.5-Hz wave lasting about about 2 s. This observation suggests that any blast at this site poses a potential low-frequency vibration problem.

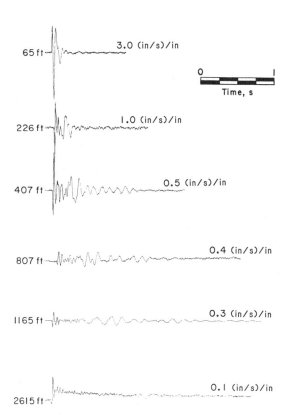

Figure 5. Vibration records from a single charge in Blanford, vertical component.

3.4 Theoretical model

The O'Brien model computes surface waves generated by multiple reflections of compressive body waves in a low-velocity layer. For a strong velocity contrast (strong reflector), the simplified relationship is

$$T = \frac{4h}{V_1},$$

where T is the surface wave period, h is the layer

thickness, and V_1 is the propagation velocity for the low velocity layer. This simple model is based on the fact that the velocity of the low-velocity layer, V_1, is much lower than that of the high velocity layer, V_2.

At Blanford, researchers measured a V_2 (high-velocity layer) of 3050 m/s and, using a 1.6 s arrival time difference, calculated a V_1 of 823 m/s. For a 6.5-Hz surface wave (T=0.15 s), a layer thickness (h) of 31 m is indicated. Similarly, for a 4 Hz surface wave, h is 51 m. Depths to the coalbeds as measured by Peabody at one site range from 26 to 120 m. The extensively mined No. 5 was at 69 m.

3.5 Level-loop surveys of homes

Researchers surveyed eight Blanford home foundations to identify surface changes from possible subsidence over abandoned workings. Most of the homes were out of level by significant amounts of up to 1 part in 65. However, a resurvey 7 months later showed no appreciable changes (Siskind 1987). With these data alone, it is not possible to tell if the homes are distorted or strained, or if these differences were originally built-in.

4 CONCLUSIONS

Research in Blanford, IN, found the propagating medium primarily responsible for the adverse vibration impacts through three mechanisms: (1) the medium favors generation of low-frequency surface waves of several types with frequencies between 4 and 10 Hz, (2) it has the appearance of reduced vibration attenuation (higher amplitudes) with distance as compared with other coal mine blasts, and (3) it produces interactions between delayed charges beyond what would be expected from the blasts as designed, because of constructive wave-interference for these long period waves.

Although further study of the subsurface conditions is needed in order to completely understand all of the factors, the observed surface waves are consistent with a strongly reflecting subsurface interface at a depth of about 51 m, or about the depth of the extensively mined No. 5 coalbed. This agrees with theoretical models that predict low-frequency waves from strongly reflecting near-surface horizontal layers.

The existing geologic structure between the main part of Blanford and the active pit is another possible or contributing cause of the low-frequency blast vibrations. This region has a coal cutout or zone where the coal and other sedimentary rock beds are missing, replaced by fill characterized as sandy, gravely drift. Vibrations propagating through such material often have abnormally low frequencies. Such a medium could also explain the rapid vibration amplitude attenuation observed between 407 and 807 ft (fig. 5).

Where there is a tendency to generate low-frequency waves, care must be used by the surface mine operator to avoid reinforcing periodicities of this order in the blast design.

BIBLIOGRAPHY

O'Brien, P.N.S., 1957. Multiply-Reflected Refractions in a Shallow Layer. Gphys. Prospecting. 5: 371-380.

Siskind, D. E., V. J. Stachura, and M. J. Nutting, 1987. Low-Frequency Vibrations Produced by Surface Mine Blasting Over Abandoned Underground Mines. Bureau of Mines Report of Investigations 9078.

Siskind, D. E., M. S. Stagg, J. W. Kopp, and C. H. Dowding 1980. Structure Response and Damage Produced by Ground Vibration From Surface Mine Blasting. Bureau of Mines Report of Investigations 8507.

Fragmentation studies in instrumented concrete models

Etudes de fragmentation sur des modèles en béton instrumentés
Sprengversuche in instrumentierten Betonmodellen

W.H.WILSON, Mechanical Engineering Department, University of Maryland, College Park, USA
D.C.HOLLOWAY, Mechanical Engineering Department, University of Maryland, College Park, USA

ABSTRACT: Laboratory scale bench blasting experiments were run in instrumented concrete models up to 1 m^3 in size to study the fracture and fragmentation mechanisms. Eight single-hole and two time-delayed two-hole tests were conducted. High speed photographs of the tests were analysed to determine the sequence and patterns of crack formation and growth, supplementing data from strain gages and crack detection gages. The photographs and accelerometer data were used to determine the period of viable gas pressure action. It was found that fractures important to the final fragmentation are initiated in the free surfaces nearest the charge by the reflecting P-wave, while later gas pressure acts mainly to extend existing fractures. Areas of the bench face to the sides of the borehole experience long periods of compression due to bending caused by the gas pressure.

RESUME: Des expériences de minage du front de taille de carrière out été simulées, à l'échelle du laboratoire, sur des modèles en béton équipés d'instrumentation de contrôle et de volume allant jusqu'à 1 m^3 afin d'étudier les mécanismes de fracture et de fragmentation. Huit essais avec simple orifice et deux essais avec orifices doubles et temps différé ont été entrepris. Des photographies rapides des essais ont été analysées afin de déterminer la séquence et les formes observées à l'origine et durant la croissance des fissures, complémentant ainsi l'information fournie par les jauges de contrainte et de détection de fissures. Les photographies ainsi que les données fournies par un acceleromètre ont été utilisées pour déterminer la période efficace des effets de pression des gaz. Il fut constaté que les fractures importantes pour la fragmentation finale sont engendrées dans les surfaces libres les plus proches de la charge par reflexion de l'onde de dilatation alors que la pression plus tardive des gaz agit principalement sur l'extension de fractures existantes. Les surfaces du front de taille qui sont de part et d'autre du trou de sondage subissent de longues périodes de compression dues au fléchissement causé par la pression des gaz.

ZUSAMMENFASSUNG: Sprengversuche wurden an einem Steinbruchmodell im Labor durchgeführt. Dazu wurden instrumentierte Zementmodelle in der Grössenordnung bis zu 1 m^3 benutzt, um die Riss- und Bruchmechanismen zu untersuchen. Acht Einfach- und zwei zeitverzögerte Doppel-Loch Versuche wurden durchgeführt. Fotografische Schnellverschlussaufnahmen wurden hinsichtlich des Rissauftretens und -ausbreitens untersucht. Zusätzlich zur Auswertung dieser Aufnahmen wurden Testdaten von Dehnmesstreifen und Rissdetektoren erhalten. Die fotografischen Aufnahmen sowie Beschleunigungsmessdaten wurden zur Bestimmung der Zeitspanne verwendet, in der rissfördernde Gasdrücke auftreten. Es wurde festgestellt, dass die Risse, die letztendlich den Bruch verursachen, zuerst an der freien Oberfläche direkt neben der Sprengladung auftreten, hervorgerufen durch reflektierende "P-waves". Später auftretende Gasdrücke dehen diese Risse dann aus. Die Bereiche des Steinbruchs an den Seiten der Bohrlöcher werden für eine längere Zeitspanne durch die vom Gasdruck verursachten Biegungen komprimiert.

1 INTRODUCTION

As a result of continued economic pressure, there is a need to make all forms of blasting more efficient, whether for mineral recovery or for construction. Improvements in blasting fragmentation would lead to direct savings in drilling and explosives costs, as well as in the processing costs of the blasted material. Full scale blasting and fragmentation experiments are difficult and expensive, due to the large number of variables and problems in observing and measuring the details of the processes. On the other hand, model experiments provide the ability to maintain better control over experimental parameters and to more easily monitor the fragmentation process, but they do involve simplifications of full scale conditions and materials. In some prior model investigations, the stress waves and developing fractures were observed by using transparent models. This includes the work of Kihlstrom (1969), Field and Ladegaard-Pedersen (1972) and Fourney, Holloway, and Dally (1975). Other investigations have employed model materials such as concrete, cement and sand mixtures, rock boulders, and rock outcroppings. The work by Bjarnholt and Skalare (1983), Bergman, Riggle, and Wu (1973), Bandari and Vutukuri (1974) and Winzer and Ritter (1985) are typical examples. As the number of such investigations has grown, the

experimental techniques for modeling and for instrumentation have also progressed.

This paper describes a program of ten fragmentation experiments run in concrete models as part of an effort funded by the U.S. Bureau of Mines to develop still better modeling and instrumentation methods, and to help clarify the mechanisms of stress waves and gas pressurization in bench blasting fracture development.

Seven tests employing single boreholes were run in blocks approximately ½ m^3 in size. One single-hole test and two tests with two time-delayed shots each were run in larger blocks measuring 1 m per side. The boreholes were typically 6.4 mm in diameter, from 10 to 20 cm deep, and were stemmed to about one-third depth. The explosive used was PETN, fully coupled and bottom detonated. Burden and spacing were chosen to simulate bench blasting conditions.

The generation of dynamic data during the model blasting was emphasized in these tests, through use of a number of instrumentation techniques and high speed photography. Strain gages were used to measure strain histories at various locations on the model bench faces during each test, and strain gage rosettes were used in one test to provide the principal strains and directions. Accelerometers were used on the model's top surface in the last two tests, and these clearly showed the ground motion pulses from each borehole. Several new instrumentation techniques were developed or modified for use in these tests. These included a fiber optic gage used to measure the detonation velocity of the borehole charges and graphite line crack-detection gages. A high speed camera was used to record the fracture development in the front face of all but one test, and also in the top face in four tests.

By considering all of the data generated by the instrumentation, analysis of the high speed photographs, and postmortem inspections, a more detailed picture of the mechanisms causing fracture and fragmentation was derived. This data showed that fractures playing an important role in fragmentation are formed very early in the blast at the bench face directly in front of the borehole and in the top surface above the borehole. These fractures are initiated at the free surfaces as the outgoing P-wave is reflected. While the early period of stress wave action was found to be important in initiating fractures, during the longer lasting period of gas pressure forces the predominant effect was growth of these fractures rather than creation of new fractures. In contrast to high tensions which occur in the bench face directly in front of loaded boreholes as the P-wave reflects, it was found that areas of the bench to the side of a borehole experience long periods of compression beginning shortly after passage of the stress waves. This compression is caused by the longer lasting gas pressure, which bends plate-like elements that are formed in the bench by radial fractures.

2 MODEL DESCRIPTION

The models were made from small aggregate, low water content concrete, having a compressive strength in excess of 35 MPa (5000 PSI) and a density of approximately 2200 kg/m^3 (136 lb/ft^3). For the first seven single-hole tests, the bench models were 46x46x50 cm in size (18x18x20 in.). For the last three tests, models 1 m^3 in size were made by cementing four large sections together. In the first of the two-hole tests in these larger models, it was learned that the model length did not provide the desired degree of end constraint, and so for the last test, additional blocks were cemented to each end of the model test section. These end-blocks acted as momentum traps and provided additional end constraint. This model configuration is illustrated in Figure 1.

All of the boreholes drilled into the models were 6.4 mm (1/4 in.) in diameter, except the 4.8 mm (3/16 in.) borehole used in Test 7. Burdens varied between 74 mm (2.9 in.) and 130 mm (5.13 in.), and the ratio of burden to diameter (B/D) used in the test program ranged from 11.5 to 20.5. The burdens and charge weights were varied in the first few tests to determine reasonable values for use with these models. A burden of about 89 mm (3.5 in.), giving a B/D ratio of 14, was found to give good model blasting results, and the later tests used a B/D close to this value. The B/D ratios used compared favorably to values ranging between 4 and 20 in model blasting tests by other investigators (Winzer and Ritter, 1985; Bjarnholt and Skalare, 1983; Bergman, Riggle, and Wu, 1973; Bandari and Vutukuri, 1974). Although a number of studies have recommended B/D values between 24 and 36 for optimum field blasting results (Ash, 1973; Pugliese, 1972), these higher values do not seem to scale well to the small boreholes which must be used in model studies.

PETN charges, ranging in mass from 1.5 g to 5 g, were in general bottom-detonated in the boreholes. Based on the early tests, a charge between 2 g and 3 g was determined to be appropriate for the model and burden dimensions being used. The charges were contained in plastic shrink-tubing, and the packing density averaged 1.4 g/cm^3. Borehole depths averaged 135 mm (5.3 in.), and generally the upper quarter to half of the borehole was stemmed using dry sand.

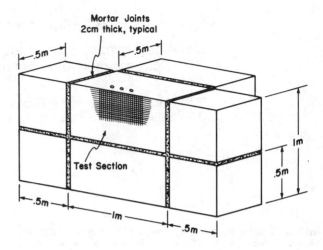

Figure 1. Diagram of the concrete model with end-constraint blocks, used in Test 10.

Figure 2 shows details typical of the borehole loading used in this test program. The two-hole tests both used roughly a square spacing pattern. Test 9 had a spacing of 86 mm between holes and a 89 mm burden, while Test 10 used a 103 mm spacing and had an average burden of 90 mm.

3 INSTRUMENTATION

Strain gages were used in at least two locations on the model bench face in each test. For the first nine tests, two gages were mounted directly in front of the borehole, oriented to measure horizontal strain. In Tests 8 and 9, additional gages were also mounted on the bench face one or two spacing dimensions to the side of the borehole, also measuring horizontal strains. For Test 10, two strain gage rosettes were mounted one spacing dimension to the side of the second borehole.

In Tests 9 and 10, B & K Type 8309 accelerometers, having a typical mounted resonant frequency of 180 kHz, were used to monitor the models' top-surface vertical motion. Two accelerometers were used in each case, mounted in threaded steel studs which were epoxied into the concrete at two and three burden distances directly behind the second boreholes. Three crack detection gages were also used in each of

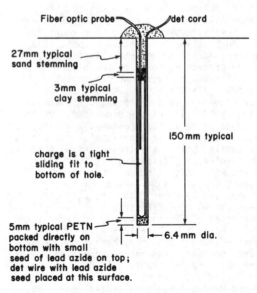

Figure 2. Typical charge loading used in the tests.

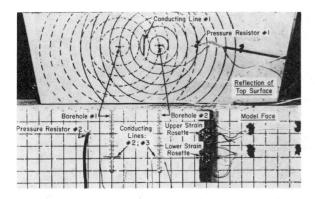

Figure 3. View of central part of model bench face used in Test 10, showing instrumentation. Grid and radial spacings were 25.4 mm.

these two tests. Adapting a technique used for crack detection in rock plates (Barnes, 1985), a soft pencil lead was used to draw lines on the model surfaces. Such lines are conductive and have little or no cohesion between the graphite particles along their length. Thus, the conductive path is broken almost instantaneously when a fracture opens in the material below. By monitoring conduction through lines on the model face in front of the boreholes and on the top surface between boreholes, the times of fracture in these areas were determined. Figure 3 shows a view of the model used in Test 10, in which the crack detection gages and strain gage rosettes can be seen.

In several tests, the velocity of detonation (VOD) of the borehole charge was measured using a newly developed fiber optic technique. Groups of individual fiber optic strands from a single optical cable were embedded to various depths in the PETN charges. As the detonation progressed along the charge, the light output was monitored. A sharp peak in intensity of the transmitted light occurred as the detonation front reached each group of fibers. The VOD was then calculated from a plot of the fiber group separation distances vs. time between light peaks. These VOD gages showed that the charge loading and initiation techniques used resulted in detonations having a typical velocity of 6 mm/μs.

A Beckman & Whitely Model 350 rotating-drum framing camera was used to photograph all of the tests except Test 9. In each case the bench surface was viewed; for the later tests the top surfaces were also viewed, either by tilting the model or by viewing the second surface through a mirror. The framing rate was nominally 35,000 frames per second, so that the interframe time was typically 28.6 μs.

4 TEST RESULTS AND OBSERVATIONS

In this section, some of the details of the instrumentation data and observations made from the high speed photographs and the postmortem inspections are given. How this information was interpreted to describe important fracture mechanisms and the resulting fragmentation in a bench blast is discussed in Section 5.

4.1 Crack development in the model top surfaces

The high speed photographs from each test were analyzed to determine the time, sequence, and pattern of fracture formation. This information was supplemented by data from the crack detection gages and the strain gages, and through postmortem inspections. As shown in a holographic study of explosively-induced fracture in rock plates (Holloway, 1982), a crack must form some time before it opens enough to become

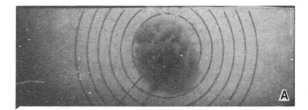

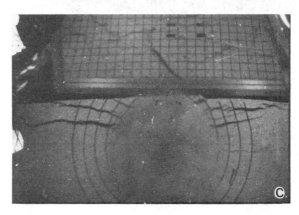

Figure 4. Three high speed frames from Test 8. Grid and radial spacings were 25.4 mm.
A: Top face 256 μs after detonation.
B: Top face 597 μs after detonation.
C: Top face and bench face 1421 μs after detonation.

visible, and the true tip of a propagating crack is some distance ahead of the apparent or visible tip. Thus, the high speed photographs can only establish an upper bound for the time of fracture. The sequence in which cracks become visible can also be misleading, since a slowly opening crack may form before a more rapidly opening one which becomes visible sooner. Nevertheless, it was useful to determine when and where cracks came into view at the model surfaces and how they appeared to grow.

The earliest fractures observed were radial cracks seen emanating from the boreholes in the top surfaces. In Test 8, two such cracks extending out approximately 25 mm and 38 mm from the borehole could be seen 60 μs after detonation. (These cracks can be seen in the first frame of Figure 4, which shows three of the high speed frames from Test 8.) In Test 10 a radial fracture extending 64 mm from the second borehole was visible 75 μs after its detonation. Crack detection gages used on the top surface near the boreholes confirmed that fractures occur there very soon after detonation. A gage 19 mm from the first borehole in Test 9 failed at about 34 μs after detonation, and a gage 64 mm from hole 1 in Test 10 failed 58 μs after detonation. The P-wave speed in the concrete models was determined from the strain gage and accelerometer records to be approximately 4.0 mm/μs. At this speed the time of arrival of the P-wave at the crack gages above would have been 26 μs after detonation for Test 9 and 33 μs after detonation in Test 10. Thus, radial cracks form near the boreholes within a very short time after the P-wave begins to arrive.

The radial cracks in the top surfaces opened and grew toward the front edge in subsequent frames, typically at angles between 10° and 25° with the

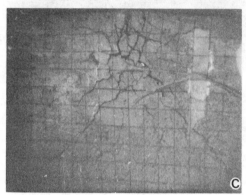

Figure 5. Three high speed frames from Test 3.
A: 277 μs after detonation. B: 420 μs after detonation. C: 764 μs after detonation.

front edge. Usually, explosive gas was seen venting
from these cracks, along their length from the borehole out to some intermediate point well behind the
visible crack tip. As the radial cracks approached a
boundary, they gradually turned to run at a shallow
angle or roughly parallel to the front edge before
finally breaking through into the perpendicular free
face. This sequence can be seen in the three frames
of Figure 4. The apparent propagation velocity of
these cracks, based upon the increase in crack
lengths measured from the high speed photographs, was
about 1 mm/μs near the boreholes, gradually slowing
to around 0.2 mm/μs as the crack tip moved far from
the borehole. These are minimum estimates for radial
crack velocities, since the true crack length always
exceeds the length visible in the photographs.

4.2 Crack development in the bench faces

The first fractures to appear in the bench faces were
invariably vertical cracks, in the area directly in
front of the boreholes. These fractures usually
appeared first near the upper edge, extending down
into the front face. Figure 5 shows three frames
from Test 3, and a vertical fracture is visible in
the first frame. For the nine tests photographed,
the times at which these first vertical fractures
appeared ranged from as soon as 120 μs to no later
than 225 μs after detonation; the average for all the
tests was 167 μs. Subsequent fracture development
observed in the bench faces followed a rather
repeatable pattern, except for Test 8. In all but
that case, after another 200 μs or so, a number of
fractures forming a network in a small area directly
in front of the borehole began to emerge. The fracture network which formed in Test 3 can be seen in
the later two frames of Figure 5. In these patterns,

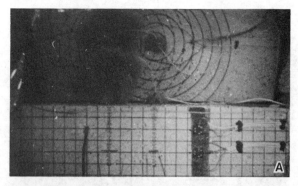

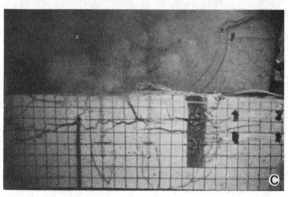

Figure 6. Three high speed frames from Test 10,
showing bench face and reflected view of top face.
A: 300 μs after detonation. B: 750 μs after detonation. C: 2100 μs after detonation.

typically from 6 to 12 cracks at various angles could
be seen, appearing to radiate from in front of the
midpoint of the borehole charge. Initially, these
crack networks covered an area of the bench face between 75 and 100 mm wide. Once the initial crack pattern and lines had formed in the central area, generally no new fractures were observed there throughout
the remaining photographic coverage, which lasted for
as long as 4 ms in the later tests. Rather, these
initial fractures were seen to progressively open and
to extend out into the bench face. The dominant
fractures tended to grow diagonally, so that an "X"
pattern of fracturing, with the crack network at its
center, was formed. None of these fractures in the
central area in front of the boreholes was observed
to vent explosive gas during any of the tests.

As had been the case in the top surfaces, data from
the crack detection gages in the bench face directly
in front of the boreholes indicated that fractures
formed there well before they became visible. In
Test 9, there were two crack detection gages in front
of the second borehole. The first was 35 mm below
the top edge; the other was 54 mm below the top edge.
Neither of these gages were affected by the first
detonation in the borehole 86 mm to their left, but
then both failed at about 55 μs after the second
detonation. The P-wave arrival time from the second

borehole at these two gages was approximately 30 μs after detonation. In Test 10, crack detection gages were located in front of each of the boreholes, 98 mm below the top edge. These gages can be seen in Figure 3. The gage in front of the first borehole broke 59 μs after detonation, while the gage in front of the second borehole did not fail until 256 μs, or 59 μs after the second detonation. The P-wave from each borehole would have arrived at the closest crack gage at about 24 μs after detonation. Thus, again it was found that fractures formed in front of the boreholes where the crack network patterns were first observed within microseconds after the P-wave arrived.

Assuming that occurrence of an open circuit signal on a strain gage channel was due to fracture at the strain gage, the strain records also indicated that fractures formed in this area of the bench face within this same early time frame. Over the course of the test program, for 14 different gages mounted on a bench face directly in front of a borehole, the time after detonation at which an open circuit signal occurred varied from 45 μs to 79 μs, averaging 53 μs. P-wave arrival times at these gage locations from the nearest boreholes varied from 22 μs to 35 μs after detonation. It is clear that cracks occurred in the region in front of the boreholes very shortly after first arrival of the P-wave.

Test 8 was a single-hole test shot in a large model block, and the fracture formation observed was somewhat different. After observation of a vertical fracture in front of the borehole, only two additional fractures could be seen in the bench face of this model for the next 700 μs, until radial fractures seen propagating in the top surface toward the front edge began to break through into the front face. As can be seen in Figure 4, these cracks then opened and grew rapidly down into the bench face, appearing to tear away large sheet-like fragments from either side of the borehole. Although a fracture network in front of the borehole was not seen in the high speed photographs from this test, postmortem inspection of the fragments revealed that a number of such fractures indeed had occurred. These fractures ran in directions characteristic of the crack network patterns observed in earlier tests, but extended only part way into the depth of the fragments formed. Thus, network fractures apparently initiated at the bench face and propagated inward but did not open enough in Test 8 to be seen in the high speed photographs.

The crack development observed in Test 10, a two-hole test in a large block with extended ends, showed characteristics both of the behavior seen in the small-block tests and that observed in Test 8. Figure 6 shows three of the high speed frames from this test. A vertical crack was first observed in front of the first borehole 188 μs after detonation, and a crack network was seen there in another 197 μs. However, the dominant crack first appearing after the second detonation in front of the second borehole was horizontal. At first, cracks from in front of both boreholes grew horizontally but later turned to run more or less diagonally. After about 700 μs, radial cracks from the top surface appeared to begin breaking through into the front face, forming large sheet-like fragments. Postmortem examination of larger fragments from the area in front of the boreholes again showed that a number of fractures had occurred at the bench surface and extended back into the model; these fractures appeared to radiate out in the bench surface from the charge center location.

At times much later in the explosive loading, between 1 ms and 1.3 ms after detonation, a third type of fracture pattern was observed forming in the bench faces. These were circumferential fractures emerging outside of the early crack network area. These cracks typically appeared first below the borehole; in later frames, they opened and extended, eventually forming a crater boundary around the borehole region of the bench face. Such circumferential cracks can

be seen in the last frame of Figure 6. The eventual volume of material removed in a test generally came from within this crater boundary (not counting the very large upper-corner fragments removed due to finite length effects).

4.3 Strain history in the bench face

The strain gage records obtained for the bench face area directly in front of a borehole in each of seven tests consistently showed rapidly increasing tension only, from time of stress-wave arrival until gage failure. Although it was assumed that gage failure occurred as fractures formed, it was difficult to judge the tensile strain magnitudes at time of fracture from these records, because they typically were increasing very rapidly just before gage failure. Tensile strains in excess of 2000 μm/m were commonly seen.

In Tests 8, 9 and 10, a total of eight strain gages were mounted in areas of the bench faces to the sides of the boreholes. As in the earlier tests, these gages were oriented to read horizontal strains and again very repeatable behavior was observed, but quite different from that observed directly in front of a borehole. For locations to the side, a very brief period of low magnitude tensile strain sometimes could be seen as the stress waves first arrived, but all records then showed very rapidly increasing compressive (negative) strain. Strains in these areas remained compressive for periods long in comparison to stress wave transit times, from 200 μs to as long as 500 μs. Magnitudes of -2500 μm/m were typically recorded. In Test 9, gages 1 and 2 were mounted directly in front of the second borehole detonated, and thus one spacing-distance, or 86 mm, to the right of the first borehole. The strain records from these two gages, Figure 7, show first the long-period compressive pattern typical for areas to the side of a borehole and then the sudden tension and gage failure typical for areas directly in front of a borehole. Both gages show compression starting about 40 μs after the first detonation, electrical noise from the second detonation at 195 μs, followed by a sudden tensile shift at around 240 μs, which is about 20 μs after arrival of the second P-wave. The records shown in Figure 8, from strain gages 1 and 4 of Test 10, illustrate further how strains depended upon position relative to the boreholes. These two gages were each horizontal elements of the strain gage rosettes which can be seen in Figure 3, mounted 207 mm, or two borehole-spacings, to the right of the first borehole, and 104 mm to the right of the second borehole. Two distinct periods of compressive stain can be seen in the records. After the first detonation, compression began at about 60 μs, reaching

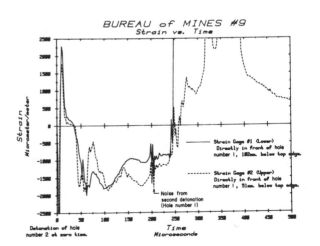

Figure 7. Strain records from two gages mounted to the side of first borehole in Test 9.

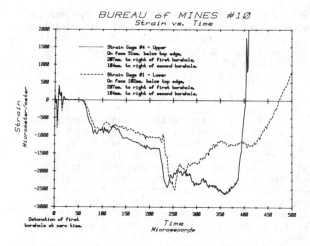

Figure 8. Strain records from the horizontal elements of the rosettes used in Test 10.

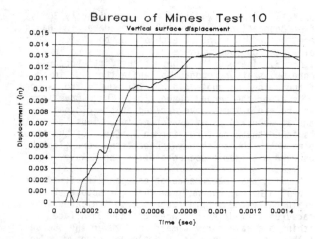

Figure 10. Vertical displacement history calculated for accelerometer 1 location on top face directly behind second borehole in Test 10.

levels around -1200 μm/m. Then, approximately 30 μs after the second detonation, a small perturbation due to the second P-wave can be seen, followed by an increase in compressive strain to around -2500 μm/m. Compression continued at gage 1 for around 400 μs. Gage 4, closer to the top edge, failed suddenly at 385 μs when a crack passed directly through it.

Figure 9 shows the principal strain calculated for the lower rosette. The plot of the larger principal strain is very similar to that of the horizontal gage at that location, also showing a long period of compression. This principal strain was in fact very nearly horizontal, acting consistently at about -10° from horizontal until the compressive period ended.

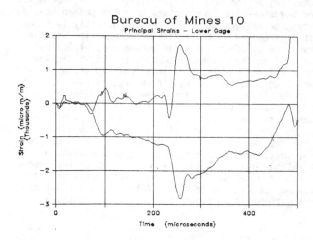

Figure 9. Principal strain history calculated for the lower rosette used in Test 10.

4.4 Gas pressure period of action

An estimate for the period of action of the gas pressure was obtained by further analysis of the high speed photographs and the accelerometer data. Gas was observed to vent from progressively greater lengths along the radial cracks in the top surfaces for long periods of time after detonation. This indicated that the gas pressure in the model interior remained sufficiently high to force its way progressively further into these cracks, and presumably to continue forcing them open. This period of continued expansion lasted for about 1.5 ms in Test 7, and 1 ms in Test 10. For the next 500 μs or so, gas continued to vent from the boreholes and along the cracks, but without any obvious further advance. After these times, gas venting appeared to diminish. The acce-

lerometer data from Tests 9 and 10 also provided some insight concerning the gas-pressure period of action. These records were integrated to produce vertical displacement plots, such as the one shown in Figure 10. This record is typical of those generated. The stress wave displacements from the two detonations can be seen, at the beginning of the response about 65 μs after detonation, and again as a small perturbation of the signal seen at about 250 μs, which was about 53 μs after the second detonation. These stress wave disturbances each lasted about 50 μs. In addition, the plots showed that after arrival of the first stress wave, the surface behind the boreholes was steadily driven up for about 1.2 ms to 1.4 ms after the first detonation. This long-term vertical displacement was attributed to the action of the gas in the boreholes and in the radial cracks. This data is consistent with observations of gas venting seen in the high speed photographs, suggesting that gas pressure continues to drive the material and to open the radial cracks for 1 ms to 1.5 ms after detonation. If this information is extrapolated to field conditions through a geometric scaling (assuming similar wave speeds, loading densities, fracture and gas propagation properties, and explosive performance) then a field blast is essentially complete after 13 to 16 ms per meter of burden (4 to 5 ms per foot).

4.5 Postmortem observations

In addition to the observations of fractures that had formed at the surface of the bench face, seen in the fragments from in front of the boreholes, postmortem inspections also revealed that most of the smaller fragments produced came from this same area, where the early crack network first formed. By comparing these fragments to the crack networks seen in the high speed photographs, it was observed that many of the fragments ultimately formed could already be identified early in the event. From Tests 6 and 7, in which the charge burial was deep for the particular burdens used, it was observed that no crater formed in the top surface, and that fragments which had come from in front of the unloaded part of the borehole were larger.

5 DISCUSSION

In determination of the fragmentation which will result in a bench blast, the test results described indicate that the fractures formed early in the event play a very important part. It is believed that the radial fractures which form around a borehole in the top surface and the crack network patterns which form in the bench face in front of the boreholes are caused by tangential stresses in the P-wave which become tensile and are intensified as the wave reflects at near-normal incidence from the free surfaces. These fractures open at the free surface and propagate back into the interior, and are initiated before any spall-type fracture surfaces are created. If the reflected wave is intense enough, then spall fractures will occur below the surface in planes roughly orthogonal to the radial fractures, creating fragments which form a crater around the borehole in the top surface. The initial fragmentation of the bench face is due to a similar combination of the early network fractures and spall fractures. In a holographic study of explosively-loaded rock plates, similar fractures, which were initiated at a free edge and propagated inward as the P-wave was reflected, have been observed (Holloway, 1982).

As the fractures in the bench face propagate inward, they provide the first relief for the burden. Meanwhile, radial fractures from the borehole will be selectively driven by the reflected stress waves as they pass back into the body, so that radial crack directions at the break-out angle toward the bench face will be favored (Field and Ladergaard-Pedersen, 1972). From this point on in the event, the energy used to create additional fracture surface is derived from work done by the gas remaining in the borehole and by gas penetrating into the radial cracks. These tests indicate that this continued fracturing is limited for the most part to extension of the cracks which have been created previously during the detonation and stress wave periods, rather than initiation of any more new fractures.

The radial fractures extending from the borehole create plate-like elements, loaded by a gas pressure distribution in the radial cracks and along their vertical centerlines by the borehole pressure. Each plate element is rigidly supported at its ends (until the radial fractures reach the bench face) and is weakened or broken at mid-span in front of the borehole by the crack network. Under this combination of loading and geometry, the plate elements bulge out in front of the borehole and are bent away from the body. The resulting stresses will tend to drive the radial cracks parallel to the bench, as they are opened by the pressure of gas continuing to penetrate into them from the borehole. The bending stresses in the bench face on either side of the borehole will be compressive as long as the plate elements are connected to the body and are being bent outward. Presumably, since the rear faces of these plate elements will be in tension, new fractures could be initiated there during this time. Such fractures would run forward into the thickness of the plates. It could not be determined in these tests whether such interior cracks occurred.

When the radial cracks reach the bench face and free the plate elements, a release wave propagates along the length of the plate back toward the center. Compression in the front face is relieved as this wave goes by, and oscillations of the element may even cause the front face to become tensile, as seen in the strain records of these tests.

The situation at the bottom of the borehole is somewhat different. Fractures generated there are also forced open by gas pressure, but as the plate elements above, in front of the borehole, curve out, they are constrained along their bottom edges while free at the top. This loading geometry would tend to cause the cracks propagating from the bottom of the borehole to angle down and toward the front, as was the case for the fracture surfaces which formed the lower part of the craters broken out of the bench faces in these tests.

Because the degree of constraint near the top is less, the velocity of cracks propagating there is greater than that of cracks coming from near the bottom of the borehole. Thus, the radial cracks reach the bench face first at the upper edge and then at progressively lower points. The circumferential

cracks seen forming in the bench faces around the lower part of the borehole region were caused as the more slowly advancing cracks from the bottom end of the boreholes reached the bench face. These fractures will continue to grow only as long as the gas pressure remains sufficiently high.

6 CONCLUSIONS

The experimental data obtained over the course of these tests from the model instrumentation, high speed photography, and postmortem inspections, has provided valuable information which supplements the findings of other investigations in describing the fracture development in bench blasting, and will help clarify the understanding of the mechanisms leading to fragmentation. It was found that very early in the event, radial cracks were formed around the borehole at the top surface, and a radial crack network pattern was formed at the bench face in front of the loaded borehole. These cracks formed as the P-wave reflected from these free surfaces, and the strain gage data showed that from the time of P-wave incidence at least until fractures occurred, increasing tensile strains acted in the bench area directly in front of the borehole. The early cracks were important in determining subsequent fracture growth and fragmentation. The smaller fragments from the bench were formed by the early network cracks and any spall fractures which occurred. Areas of the bench face to the sides of the boreholes were subjected to long periods of compression as radial fractures from the borehole formed plate-like elements which were bent out by the gas in the borehole and along the radial fractures. Larger fragments from the bench subsequently were formed as radial fractures were driven out from the boreholes. These fractures formed the final crater in the bench face.

For the model dimensions used, it was found that the period of viable gas pressure action was approximately 1 ms to 1.5 ms. This scales to about 13 ms to 16 ms per meter of burden.

These tests also provided improved instrumentation and modeling techniques for use in blasting studies, which will help make small scale blasting tests more advantageous in efficiently modeling field practice.

REFERENCES

Ash, R.L. 1973. The influence of goelogical discontinuities on rock blasting. PhD dissertation, Univ. Minnesota.
Bhandari, S. & V.S. Vutukuri 1974. Rock fragmentation with longitudinal explosive charges. Advances in rock mechanics, Washington, DC: Nat.Acad.Sci.
Barnes, C.R. 1985. A measurement technique for determining crack speeds in engineering-materials experimentation. Exp.Tech. (3):33-37.
Bergman, O.R., J.W. Riggle & F.C. Wu 1973. Model rock blasting - effect of explosives properties and other variables on blasting results. Int.J.Rock Mech.Min.Sci. 10:585-612.
Bjarnholt, G. & H.Skalare 1983. Instrumented model scale blasting in concrete. Proc.1st Int.Symp.Rock Frag.by Blasting, Lulea,Sweden.
Field, J.E. & A.Ladergaard-Pedersen 1972. Fragmentation processes in rock blasting. Dechema-Monographien 69.
Fourney, W.L., D.C. Holloway & J.W. Dally 1975. Fracture initiation and propagation from a center of dilatation. Int.J.Frac. 11:1011-1029.
Holloway, D.C. 1982. Application of holographic interferometry to stress wave and crack propagation problems. Optical Engineering 21:468-473.
Kihlstrom, B. 1969. Smooth blasting and presplitting. Estratto dal Bollettino Assoc.Min.Sub. 6:61-105.
Pugliese, J.M. 1972. Designing blast patterns using empirical formulas. Bur.Mines IC 8550, Washington, DC: US Dept. Int.

A numerical simulation method of the impact penetration system
Méthode de simulation numérique de pénétration d'un outil percuteur
Eine numerische Simulation des Stosseindringungssystems

XU XIAOHE, Northeast University of Technology, Shenyang, People's Republic of China
TANG CHUNAN, Northeast University of Technology, Shenyang, People's Republic of China
ZOU DINGXIANG, Northeast University of Technology, Shenyang, People's Republic of China

ABSTRACT:The method given in this paper can be used to obtain the force course at any position of an impacting penetration system and to evaluate the penetration efficiency of the system. By using the fundamental theory of longitudinal wave propagation, the computer programs become quite simple and clear.A model of the local deformation of the contact surfaces and temporary separation between the piston and rod which may appear in the impacting process is set up.Results of calculation are fairly consistent with those measured in the test.

RESUMÉ: La méthode que le présent papier donne pent obtenir le cours de la force en toute position dans le système de forage avec impact et l'efficacité de forage du système.Le programme de calcul est trés simple et très clair, parce qu'on utilise le principe fondamental de propagation d'onde longitudinale.On a établi le modèle apparaissint une déformation locale en surface à impact et une séparation temporaire au procédé d'impac.On compare les résultats de calcul aux résultats measurés, ils sont bien correspondance.

AUSZUG: Indieser Forschungsarbeit zeigender Methode könnten der Belaslungsvorgang und der Eindringwirkungs grad in jede Lage Stosseindringsystems ermittelt werden. Aufgrund der Benutzung der Grundlage von Querwellenübertragung ermöglichen das computer-programm sehr einfach und deutlich. Errichtete ein Modell von Stossflächen welche örtlich verformt und möglich momentan auseinander geht. Die Werte zwischen den Rechenergebissen und den speziellen Messungen sind sehr nah.

1.INTRODUCTION

It is a quick and precise way to get the numerical solution of the complicated impact penetration system by means of a microcomputer simulation. A computer simulation was used by P.K.Dutta (1980) to calculate the impact waves of different pistons. With the use of a recurrence formula, a program which can be used to calculate the incident waves and drilling efficiency of a two-section piston and rod system was given by Zhao Tongwu(1980).Another program for a three-section piston and rod system was given by Zou Dingxiang (1980). Recently, a microcomputer simulation method of stress wave energy transfer to rock in the percussive drilling was proposed by B.Lundberg (1982-1986). In his program, however, the local contact deformation on the impact surfaces is not taken into account. But thislocal deformation does exist in any case especially for the "short rod" condition and must be considered.

2.NUMERICAL CALCULATION METHOD

In order to calculate the transmission process of one-dimensional longitudinal wave,the piston together with the rod are divided into N small segments of equal lengths along the axis, as shown Fig.1,and the

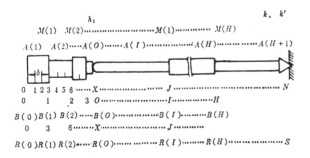

Figure 1. Model of the impact-penetration system employed in computer calculation

boundary sections through which the cross-sectional area changes are taken at the interface of two small segments. The values of the array B(I) (I=0,1,2,...0, ...H) denote the corresponding segment numbers showing the position of the interface respectively. The symbol O is used to denote the interface of the impacted section, its segment number is B(0). M(I) and R(I) express the transmission coefficients and forces of the corresponding interfaces,and A(I) express the cross-sectional areas on the left side of the corresponding interfaces. S is the penetration force of the bit into the rock, and we have

$$M(I)=2*A(I)/(A(I)+A(I+1)) \qquad (1)$$

2-1 Initial state

By using the impact equivalent theorem, the piston having an ensemble velocity V is equivalent to the piston having an initial forward wave P(J) and backward wave Q(J), which are

$$P(J)=SQR(E*RHO)*V*A(2)/2 \qquad (J \leqslant 0) \qquad (2)$$

$$Q(J)=-P(J) .$$

where E and RHO are the elastic modulus and density of the piston material. In the rod,

$$P(J)=0 \quad : \quad Q(J)=0 \qquad (J>0) \qquad (3)$$

By applying the properties of wave transmission, it is easy to give the micro-computer simulation program of the impact-penetration system. The main parts of the program are as follows:
```
190    T=T+1
200    FOR  J=0  TO  N-1
210    Q(J)=Q(J+1)  :  P(N-J)=P(N-J-1)
220    NEXT  J
230    P(0)=-Q(0)
240    FOR  I=1  TO  H
250    IF  I<>0  THEN  300
260    IF  R(0)=0  THEN  310
270    C=R(0)^(1/3)
280    R(0)=(2*P(R(0))+2*A(0)*Q(B(0))/A(0+1)
```

```
         +Z*C*)/(1+A(0)/A(0+1)+Z/C)
290   IF R(O) 0.1 THEN 310 ELSE R(0)=0 : GOTO 310
300   R(I)=M(I)*P(B(I))+(2-M(I))*Q(B(I))
310   G Q(B(I))
320   Q(B(I))=R(I)-P(B(I)) : P(B(I)
330   NEXT  I
340   Q(N)=(P(N)+W*(S-P(N)))/(1+W)
350   S=P(N)+Q(N)
360   IF F S THEN F=S
370   PRINT T, R(O), S
390   IF T O THEN 190
```

2-2 Normal propagation

Taking the length of every segment as DL and the transmission velocity of the longitudinal wave along the bar as a, the time for the wave to propagate a distance DL is DL/a, The forward and backward waves of every segment are calculated respectively by the microcomputer as shown in statements number 200-220 in the above program.

2-3 Propagation through the boundary

The force on the boundary between two sections can be got by the superposition of the forward and backward waves as shown in statement number 300 in the program. Then, by the superposition relation, the new forward and backward waves can be got as shown in statement 320.

2-4 Impact

Whem ignoring the local deformation on the impacted surfaces which are taken as common boundaries, a rough result can be easily obtained by applying the impact equivalent relation. But, in fact, the local deforma- tion exists in any case and must be taken into account. This problem can be treated as a Hartz contact problem. The result is much better. According to this theory, the two thirds power of the force acting on the impa- cted surfaces is proportional to the deformation of the impacted surfaces. The proportional factor is denoted by H1. The equation can be expressed by P(B (O) andQ(B(O)) as shown in statements number 270-280 in the program and Z is as follows

$$Z=2*E*A(O)/(3*H1*DL) \qquad (4)$$

2-5 Reflection in the bit

At the end of the rod, by supposing that the force on the loaded bit is proportional to the penetration (K is the proportional factor called penetration coeffi- cient), the reflected wave can be expressed as

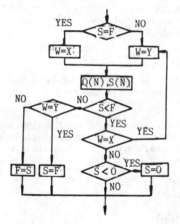

Figure 2. Flow chart for treating penetration and rebound in bit-rock interface

statement number 340 in the program, where W is as follow

$$W=X*E*A(H+1)/DL/K \qquad (5)$$

In view of the fact that the values of coefficient K for penetration and rebound are different, two cases are taken into account, namely, W=X for the penetration condition and W=Y for the rebound condition. The flow chart for treating this process is shown in Fig.2.

2-6 Temporary separation between piston and rod

In the impact of the piston with the rod, especially with the short rod,the temporary separation and contact again between piston and rod may appear. Suppose L is the gap size, then the piston and the rod separate as L o and treating this peoblem is shown in Fig.3.

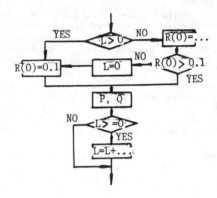

Figure 3. Flow chart for treating the separation between piston and rod

2-7 Penetration efficiency

The penetration efficiency is the ratio of work applied by bit penetrating into the rock to the kietic energy of the piston. The maximum force is F and the piston mass is M.So the penetration efficiency is as follows

$$EFFICIENCY=(F*F/K)/(V*V/M) \qquad (6)$$

3.COMPARISON OF EXPERIMENTS WITH CALCULATED RESULTS

The wave forms were measured to verify the reliability of the program.The mean value of local deformation coefficient H1 is 20x10 N/m.The values of penetration coefficient K for four kinds of rock are shown in table 1 (bit edge width, 35 mm).

The impact incident wave of type 7655 rock drill was measured which was shown in Fig.4.

Table 1. Penetration coefficient K

Rock type	Marble	Granit	Diorite	Gneiss
K (MN/M)	50	99	182	191

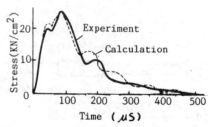

Figure 4. The incident wave of type 7655 rock drill

744

Also,the waves were measured for five kinds of
piston which are 1,2,4,5 and 8 kg in weight impacting
the short rod with a length of 0.4 m. The results are
shown in Fig.5.

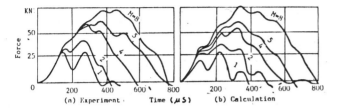

Figure 5. Wave measured in the middle of "short rod"

The correct simulation of impact-penetration system
depends of the precise bit-rock boundary condition,
that is, the load-penetration relation of the bit
penetrating the rock. The force acting on the bit is
measured from the output of the strain gages mounted
on the bit. The penetration history of the bit penetrating the rock is measured with two photoelectric
cells by receiving the light bem which passes through
a rectangular hole in the bit, and the cells are
circuited in a differential way as shown in Fig.6.

Photoelectric
cell

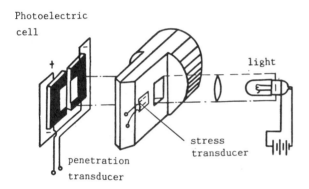

Figure 6. Schematic diagram of penetration measurement

The force and penetration are recorded by the transient
waveform storage through two channels to a 2x2048 byte
memory which is connected with a microcomputer so that
the curve of the load-penetration,the penetration
coefficient and the penetration efficiency can be got
immediately. Some of the load-penetration curves for

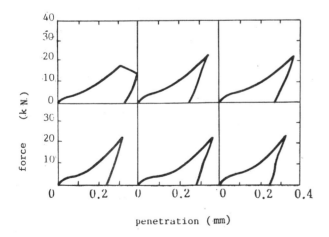

Figure 7.Load-penetration curves for marble

marble are illustrated in Fig.7.We know from the curves
that the slope of the curve in the initial part is
small because the contact between the bit and the rock
is not very close. The slope for the rebound of the
bit is much larger than that for the penetration,and
negative slope also appears some times.

Fig.8 gives a comparison between the dimensionless
maximum penetration force and the simulation results
(where m is the wave impedance). All of these show
that the simulation results agree quite well with the
experimental results.

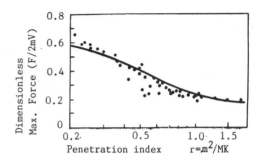

Figure 8.Experimental and theoretical maximum forces

4. CONCLUSIONS

It is of practical value to apply the numerical calculation method to simulate the impact-penetration system.
By applying the equivalent impact relation,the program
is very simple. As the local deformation and the temporary separation between the piston and rod are taken
into account, it is also very effective to apply this
method to the "short rod" case such as down-the-hole
dirlling and the calculation results agree quite well
with the experimental results.

REFERENCES

Dutta,P.K. 1980.The determination of stress waveforms
    produced by percussive drill piston of various
    geometrical designs.Int.J.Rock Mech.Min.Sci.5:501-
    518.
Zhao Tongwu.1980.A study of the efficiency of percu-
    ssive penetration by wave theory.Acta Metallurgical
    Sinica.16:109-120(in Chinese).
Zou Dingxiang.1980. Analysis and calulation of the
    stress state of impact penetration system by
    computer.Journalof Northeast Institute of Techno-
    logy. No1:109-120(in Chinese).
Lundberg,B.1982. Microcomputer simulation of stress
    wave energy transfer to rock in percussive drilling.
    Int.J.Rock Mech.Min.Sci.19:229-239.
Lundberg,B.1985. Microcomputer simulation of percussive
    drilling.Int.J.Rock Mech.Min.Sci.22;237-249.
Lundberg,B.Karlsson,L.G.1986. Influence of geometrical
    design on the efficiency of a simple down-the-hole
    percussive drill.Int.J.Rock Mech.Min.Sci.23:281-287.
Xu Xiaohe.1979.A study of longitudinal waves in drill
    rod exerted by stroke. Journal of Northeast Institute
    of Technology. NO.2:1-14(in Chinese).
Xu Xiaohe.1986.The fundamental theory of percussive
    drilling and its microcomputer calculation method.
    Shenyang:Northeast Institute of Technology (in
    Chinese).